ATOMISTICS OF FRACTURE

NATO CONFERENCE SERIES

I   Ecology
II   Systems Science
III   Human Factors
IV   Marine Sciences
V   Air-Sea Interactions
VI   Materials Science

VI MATERIALS SCIENCE

ATOMISTICS OF FRACTURE

Edited by

R. M. Latanision

Massachusetts Institute of Technology
Cambridge, Massachusetts

and

J. R. Pickens

Martin Marietta Laboratories
Baltimore, Maryland

Published in cooperation with NATO Scientific Affairs Division

PLENUM PRESS · NEW YORK AND LONDON

Library of Congress Cataloging in Publication Data

NATO Advanced Research Institute on Atomistics of Fracture (1981: Calcatoggio, Corsica)
  Atomistics of fracture.

  (NATO conference series. VI, Materials science; v. 5)
  Proceedings of a NATO Advanced Research Institute on Atomistics of Fracture, held May 22 – 31, 1981, in Calcatoggio, Corsica, France.
  Bibliography: p.
  Includes index.
  1. Fracture mechanics—Congresses. I. Latanision, R. M. II. Pickens, J. R. III. Title. IV. Series.
TA409.N39 1981   620.1′66   82-10141
ISBN-13: 978-1-4613-3502-3   e-ISBN-13: 978-1-4613-3500-9   AACR2
DOI: 10.1007/ 978-1-4613-3500-9

Proceedings of a NATO Advanced Research Institute on Atomistics of Fracture, held May 22 – 31, 1981, in Calcatoggio, Corsica, France

© 1983 Plenum Press, New York
Softcover reprint of the hardcover 1st edition 1983
A Division of Plenum Publishing Corporation
233 Spring Street, New York, N.Y. 10013

PREFACE

It is now more than 100 years since certain detrimental effects
on the ductility of iron were first associated with the presence of
hydrogen. Not only is hydrogen embrittlement still a major industrial problem, but it is safe to say that in a mechanistic sense we
still do not know what hydrogen (but not nitrogen or oxygen, for
example) does on an atomic scale to induce this degradation. The
same applies to other examples of environmentally-induced fracture:
what is it about the ubiquitous chloride ion that induces premature
catastrophic fracture (stress corrosion cracking) of ordinarily
ductile austenitic stainless steels? Why, moreover, are halide ions
troublesome but the nitrate or sulfate anions not deleterious to
such stainless steels? Likewise, why are some solid metals embrittled catastrophically by some liquid metals (liquid metal embrittlement) - copper and aluminum, for example, are embrittled by
liquid mercury. In short, despite all that we may know about the
materials science and mechanics of fracture on a macroscopic scale,
we know little about the atomistics of fracture in the absence of
environmental interactions and even less when embrittlement phenomena such as those described above are involved.

On the other hand, it is interesting to note that physical
chemists and surface chemists also have interests in the same kinds
of interactions that occur on an atomic scale when metals such as
nickel or platinum are used, for example, as catalysts for chemical
reactions. Such metals are very effective catalysts for the dissociation of molecular hydrogen. Indeed, considerable experimental
study has been directed toward the surface chemistry of hydrogen
adsorption, etc., on transition metal surfaces. But much of the
same uncertainity in terms of fundamental understanding pervades
this area of science. Why is it that nickel and platinum are
effective hydrogenation catalysts, but copper is not? And, why are
other metals effective catalysts for some chemical reactions, but
not for hydrogenation?

It seems likely that the development of fundamental understanding of the kind implicit in the above discussion of the
catalytic dissociation of hydrogen (embrittlement of the hydrogen

molecule!) would impact, as well, understanding of the fundamentals
of the embrittlement of metal surfaces (dissociation of metal atomic
bonds) in the presence of embrittling species such as hydrogen. It
seems quite clear now that much is known or may be developed by
organometallic chemists, quantum chemists, and others that bears
directly on problems of embrittlement and fracture of interest to
materials scientists and those involved with the mechanics of solids.
It is equally clear that the technological significance of the interactions of environments such as hydrogen, liquid metals, hot (perhaps
molten) salts, etc., with materials will become of even greater
consequence in the decades to come. It is important, therefore,
that direct and frequent communications be encouraged between the
kinds of disciplines described above, if we are to understand and
control environmentally-induced embrittlement.

The Advanced Research Institute on *Atomistics of Fracture* was
held on 22-31 May 1981 at the Hotel San Bastiano in Calcatoggio
(near Ajaccio), Corsica. Sixty-four delegates from thirteen nations
were in attendance. The conference format and organization followed
the precedent established in 1975 by the Advanced Study Institute on
Surface Effects in Crystal Plasticity and are described in detail in
the Proceedings of the latter. In short, the first four days of the
program were devoted to tutorial lectures - in this instance, on the
fracture of materials, surface reactivity and bonding, interfaces,
solution chemistry and mathematical modeling of fracture phenomena.
This was followed by a series of sessions organized in workshop style
on topics such as hydrogen embrittlement, intergranular embrittlement,
liquid metal embrittlement, and stress corrosion cracking. Each
session began with a survey lecture which was followed by short
contributed papers, and finally detailed discussions in the form of
four independent workshop groups. Edited summaries of the deliber-
ations of these groups were prepared by the session chairmen and
appear in these Proceedings as workshop summaries.

It is always something of an organizational gamble to bring
together groups of people from different disciplines. Without
general interests in common, it was not only necessary initially to
provide a tutorial program but also to stimulate discussion. While
progress was being made along these lines from the very beginning
of the conference, it was an unexpected event and a determined
session chairman which really excited the kind of dialogue that was
required. This occurred midway through the evening session on 25
May when, without warning, the hotel lost power and the conference
room became darkened. Without lights and electricity to operate
projectors, our first reaction was to delay the session until the
following morning. After lanterns were made available, however,
Session Chairman, Hans Bonzel, insisted that we continue rather
than lose the essence of discussions held earlier that evening.
The result was remarkable: in clarity and pertinence, the dialogue
that followed was superb and persisted in spirit throughout the

remainder of the conference.

We worked two full sessions each day and one day, 26 May, met in three sessions beginning at 9 a.m. and ending about 10 p.m.! Fortunately, even Nature cooperated as 26 May was the only totally rainy day during the conference! The photographs which appear at the end of this volume record both activities in the conference room and moments of relaxation in the beautiful surroundings of the hotel. Some photos will remind conferees of a visit to prehistoric Filitosa and of dinner one evening at an inland site high in the mountains of Corsica.

Thanks are due to the members of the Planning Committee for their help in developing the program and the workings of the conference in Corsica. We are grateful, as well, to the sponsors of the ARI without whose support and encouragement this conference could not have occurred. Finally, it is a pleasure to give recognition to the Conference Secretaries, Connie Martin and Gilberte Furstenberg, who handled a variety of logistical problems and mountains of typing with grace and good humor.

R.M. Latanision
February 1982   J.R. Pickens

ORGANIZATION

Chairman:

    R.M. Latanision
           Massachusetts Institute of Technology, U.S.A.

Planning Committee:

    A.S. Argon
           Massachusetts Institute of Technology, U.S.A.

    D.J. Duquette
           Rensselaer Polytechnic Institute, U.S.A.

    T.E. Fischer
           Exxon Research and Engineering Company, U.S.A.

    H. Gerischer
           Fritz-Haber-Institut, F.R.G.

    J.P. Hirth
           Ohio State University, U.S.A.

    K.H. Johnson
           Massachusetts Institute of Technology, U.S.A.

    N.H. Macmillan
           Pennsylvania State University, U.S.A.

    J. Oudar
           Ecole Nationale Superieure de Chimie, France

    P.L. Pratt
           Imperial College of Science and Technology, U.K.

    C.M. Preece
           Korrosionscentralen ATV, Denmark

    M.T. Thomas
           Battelle Northwest Laboratories, U.S.A.

    R. Thomson
           National Bureau of Standards, U.S.A.

    A.R.C. Westwood
           Martin Marietta Laboratories, U.S.A.

Sponsors:
    North Atlantic Treaty Organization
    U.S. Army Research Office - Durham
    U.S. Office of Naval Research

CONTENTS

CONTENTS

INTRODUCTORY LECTURES

Session Chairman: F.R.N. Nabarro

INTRODUCTORY LECTURES

GENERAL OVERVIEW: ATOMISTICS OF ENVIRONMENTALLY-INDUCED FRACTURE

R.M. Latanision

Department of Materials Science and Engineering
Massachusetts Institute of Technology
Cambridge, MA   02139, U.S.A.

INTRODUCTION

That the mechanical behavior of solids is affected by surface
and environmental conditions is now well-known and was, in fact, the
subject of an earlier NATO Advanced Study Institute on Surface
Effects in Crystal Plasticity.[1] While the plastics properties
(yield strength, work hardening rate, etc.) are sometines substantially affected by the presence of surface films, solvent
environments and the like, it is the remarkable effect of environments on the fracture of solids that is of most consequence in a
technological sense. Generally the latter interactions are considered to be adverse, and this is often a reputation that is well
deserved. Stress corrosion cracking[2,3], hydrogen embrittlement [4],
liquid metal embrittlement[5,6] and other such failure phenomena take
on catastrophic consequences. As engineers, we are of course typically concerned with the prevention and or control of such failures.
On the other hand, we should not forget that there are entire industries based upon the fragmentation of solids: materials removal
operations such as metal cutting and ceramic machining, grinding,
comminution, rapid excavation of hard rock, and others. Is it
possible that in these circumstances one might use controlled embrittlement to advantage in order to reduce the work of fracture
or fragmentation? While this approach is not typical of current
practice in industry, it seems clear that controlled embrittlement
is not only feasible but may well prove technologically attractive.
This matter will be discussed in detail by Westwood and Pickens [7],
and I will not pursue the issue further here.

My objective in this presentation is to develop a sense of
perspective regarding the kinds of embrittlement phenomena that are

3

of significance to us and some of the issues which this interdisciplinary gathering might treat. I will not attempt to review all of
the important aspects of these embrittlement phenomena. This will
be treated with more deliberation and detail by others at this
conference. What I will attempt to do, however, is to present a
cross section of the phenomenology, short descriptions of some
aspects of the current state of mechanistic understanding and,
finally, some of my views on how interdisciplinary interactions such
as these may prove useful in the future in developing an understanding of the atomistics of environmentally-induced fracture.

PHENOMENOLOGY OF ENVIRONMENTALLY-INDUCED FRACTURE

    Environmentally-induced fracture or embrittlement may be
considered from many points of view, all of which in our context
refer in essence to the premature failure of a metal or alloy in the
simultaneous presence of (tensile) stress and some usually specific
environment. For example, the stress corrosion cracking of aluminum
alloy 7075 in aqueous chloride environments may be thought of or
manifest in terms of a loss of load carrying capacity, a loss of
ductility, or an apparent loss of toughness (subcritical crack
growth) in a fracture mechanics sense. For the purpose of this
discussion, I wish to consider only certain aspects of the embrittlement of metals and alloys* in the presence of (a) (often only mildly
corrosive) electrolytes, (b) hydrogen, and (c) liquid metals with
the general purpose of then developing a case for studying the
atomistics of fracture from an interdisciplinary point of view. It
is not intended that this discussion will be comprehensive, and
much of the detail will appear in the presentations of others at
this conference. Brief, general treatments are found in the texts
by Tetelman and McEvily[8] and Hertzberg[9]. Many of my own detailed
views on these questions have been presented elsewhere[1,5,10-11].

    As is true of all of the phenomena mentioned above, suscepti-
bility to embrittlement is a complicated function of the chemical
environment which is present, the metallurgical history of the
materials, and mechanical conditions. For example, there is a
general tendency for susceptibility to environmentally-induced
embrittlement to increase as the strength of an alloy or family of
alloys is increased. Age hardenable aluminum alloy 7075 is most
susceptible to stress corrosion cracking in aqueous chloride
solutions in the peak hardened condition whereas the thermally
overaged alloy is more resistant. Likewise, high strength, quench
and tempered steels are particularly susceptible to hydrogen
embrittlement, whereas low strength steels are relatively immune.

---

* It should be appreciated that environmentally-induced failure
  occurs in virtually all classes of solids[1,3]. This treatment
  will consider only metals and alloys.

Similar temperature, stress level, grain size, strain rate, slip mode, etc., dependencies have been reported with reference to various embrittlement phenomena[6,10,12].

In what follows I have not particularly attempted to describe the specific and often exciting areas of controversy and overlap that have arisen in the multitude of studies of environmentally-induced embrittlement. These will surely be aired in the workshop sessions. What I have attempted to do, however, is to develop just enough of a perspective that in the spirit of this conference we might consider some relatively new, interdisciplinary approaches to understanding embrittlement phenomena.

A. Stress Corrosion Cracking

There are several characteristics of stress corrosion cracking - the premature failure of metallic alloys in the simultaneous presence of a tensile stress and an often only mildly corrosive electrolyte - that are intriguing:

a. It is often the alloys which are most resistant to general or uniform corrosion that are most susceptible to stress corrosion cracking. Examples are Fe-Cr-Ni (austenitic) stainless steels, some aluminum and titanium alloys, etc., which resist corrosion because of the presence of protective surface (passive) films.

b. While environmental specificity is. not now considered to be so restrictive as once thought, there are some alloy-environment couples that are particularly susceptible to stress corrosion cracking. Some of the latter are summarized in Table I. It is interesting, for example, that while transgranular stress corrosion cracking has been observed when austenitic stainless steels are exposed to hot Cl^-, Br^-, or F^- solutions, I^- inhibits stress corrosion cracking. Likewise, nitrates crack mild steels but inhibit stress corrosion cracking in austenitic stainless steels.

c. Pure, unalloyed metals are generally resistant to stress corrosion cracking.

d. Normally ductile alloys fail in what appears to be a brittle manner in the presence of certain environments.

Over the years much has been proposed in a mechanistic sense regarding models for the stress corrosion cracking of alloys[3,4]. Short descriptions of views which enjoy some current popularity are listed below, leaving details to literature cited.

Table 1: Environments Which Cause
Stress Corrosion Cracking of Certain Alloys

Aluminum Alloys	Seawater (Cl⁻ and other halides)
Copper Alloys	Ammoniated aqueous solutions
Nickel Alloys	OH^- (H_2S, pure water)
Mild Steels	OH^-, NO_3^-
High Strength Steels·	(water, moist air, H_2S)
Stainless Steels	
(austenitic)	OH^-, Cl^-, Br^-, F^-
Titanium Alloys	Seawater (Cl^-, and halides)

Mechanisms of Stress Corrosion Cracking. The various models of stress corrosion cracking can be divided into two basic classes: those which consider that crack propagation proceeds by anodic dissolution at the crack tip (dissolution models) and those which consider that crack propagation is essentially mechanical.

a. Crack propagation by the dissolution of film-free metal due to an increase in the number of active sites provided by plastic deformation at the crack tip. This mechano-chemical model was proposed by Hoar and Hines[13].

b. Crack propagation by the dissolution of metal at the crack tip as a consequence of the rupture of otherwise protective surface films by emergent dislocation. The film-rupture or slip step dissolution model has several variations, but is largely similar to that proposed initially by Champion[14] and Logan[15].

c. Propagation of cracks by the repeated formation and rupture of a brittle film growing into the metal at the crack tip as first described by Forty[16].

d. Adsorption (stress sorption) of surface active species, the consequence of which is a reduction in the surface energy required to form a crack and, therefore, reduced fracture stress. This adsorption model has been applied to various embrittlement phenomena since its proposal by Petch[17] and has been discussed with relation to stress corrosion cracking by Uhlig[18].

While I do not wish to debate here which model is correct - indeed, the consensus would probably agree that there is no single mechanism of stress corrosion cracking which applies to all circumstances - let's just look at the film rupture model as a demonstration of one possible series of events. This may be done sche-

matically with the aid of Figure 1[19]. In effect it is proposed that
the motion of dislocations, even at stresses below the macroscopic
yield point, ruptures the otherwise stable passive film on the alloy
surface thereby exposing clean, reactive metal which is subsequently
attacked. The rate of repassivation, or reformation of the surface
film, determines in this view the extent of dissolution. This event
is repeated and the crack which has initiated as described above
propagates discontinuously into the solid. While this model is
appealing in many respects, it is also deficient in some: for
example, slip step dissolution has been observed on alloys which are
susceptible to stress corrosion cracking as well as on pure metals
which are not. In effect, film rupture may be considered a necessary
though not sufficient condition for stress corrosion cracking to
occur. The same commentary could be developed with regard to all
the mechanisms that have been proposed - none is totally capable of
accounting for every aspect of stress corrosion cracking
phenomenology, though some elements of each model seem quite clearly
to be important. On this basis, a useful montage[20] showing chemical,

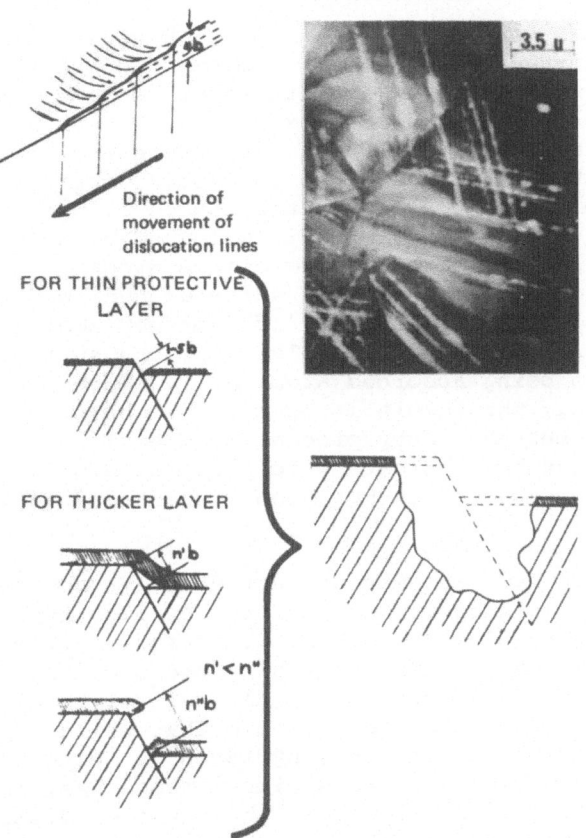

Figure 1:   Schematic drawing illustrating the geometrical aspects
            of slip step dissolution (Smith and Staehle[19]).

metallurgical and mechanical processes thought to be important in
stress corrosion cracking is given in Figure 2. It is difficult in
my view to avoid the involvement of passive or protective surface
films in the mechanism of stress corrosion cracking of austenitic
stainless steels, aluminum alloys, etc. We'll return to this point
later.

B.   Hydrogen Embrittlement

Hydrogen embrittlement is actually quite an old problem.
Reynolds[21] and Hughes[22] in 1874 and 1880, respectively, were perhaps
the first to associate certain detrimental affects on the ductility
of iron with the presence of hydrogen. Not only is hydrogen
embrittlement still a major industrial problem, but it is safe to
say that in a mechanistic sense we still do not know what hydrogen
(but not chlorine or nitrogen, etc.) does on an atomic scale to
induce this degradation. I will return to this issue later.

Let's first consider the sources of hydrogen. Molecular hydrogen will, of course, induce embrittlement of susceptible metals and
alloys (notice that hydrogen embrittlement is common to pure metals
such as nickel[23] as well as complex alloys such as Hastelloy C-276[24])
provided that dissociation of H_2 occurs thereby allowing for the
absorption of atomic hydrogen into the metal lattice. Hydrogen
embrittlement may also occur when the source of hydrogen is
electrolytic, i.e., when hydrogen is produced by the discharge of
protons in an electrolyte to form atomic hydrogen, provided in this
case that surface recombination of adsorbed hydrogen to form
molecular hydrogen is prohibited. If hydrogen adatoms are allowed
to combine, then hydrogen will evolve from the surface in molecular
form rather than being absorbed atomically. Once hydrogen enters a
susceptible metal, the result is expected to be the same - i.e.,
embrittlement - but the entry process is clearly dependent upon the
source of hydrogen and the catalytic properties of the surface
exposed to the source. This has been discussed in detail by
Berkowitz, et al.[25].

Electrolytic hydrogen may be introduced into materials, for
example, during cleaning (acid pickling) or electroplating operations
or as a consequence of electrochemical corrosion itself. In the
latter instance, the cathodic partial process which corresponds to
anodic dissolution in a corrosion cell is likely to be proton
reduction. Hence, it is quite possible that while a metal or alloy
surface is anodically dissolving, hydrogen is being absorbed into
that material. Awareness of this circumstance has, in fact, often
presented something of a dilemma in the sense of distinguishing
between stress corrosion cracking and hydrogen embrittlement in
electrolytes. Indeed, while there are some incidences of embrittlement which appear to be clearly associated with hydrogen (the

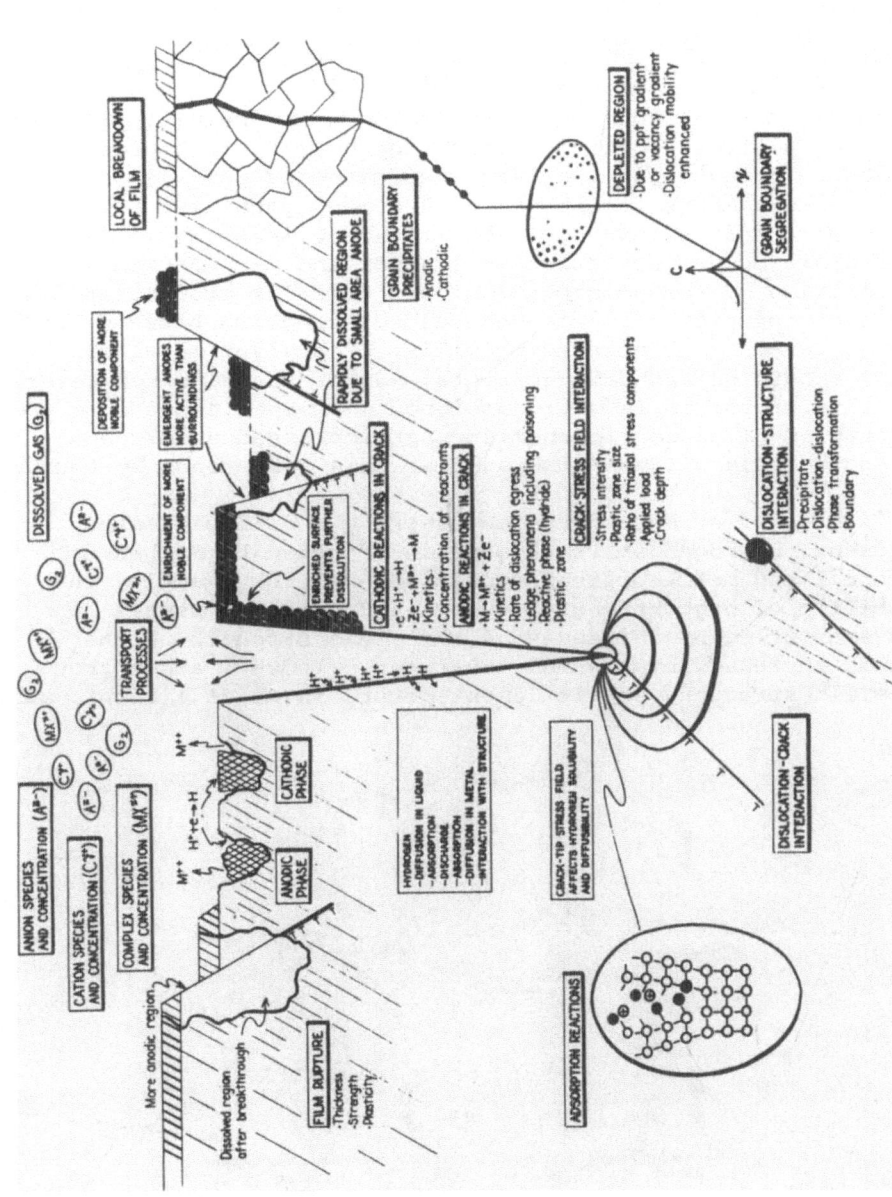

Figure 2. Montage showing important processes which affect stress corrosion cracking (after Staehle[3]).

embrittlement of high strength steels in moist air or water) and
others that appear to be anodically driven (the cracking of α-brass
in ammoniacal solutions) there are many other cases where the
distinction is less clear. For example, whether the embrittlement
of austenitic stainless steels and high strength aluminum alloys in
chlorides should be attributed to stress corrosion cracking or
hydrogen embrittlement is the subject of a considerable debate which
I have considered in detail elsewhere[10]. The essence of this problem
may be understood with the aid of Figure 3 which schematically shows
(A) active path stress corrosion cracking wherein crack propagation
is presumed to be anodically driven, the cathodic partial process
serving only as a means of consuming electrons generated by the
anodic process; and (B) hydrogen embrittlement in which case both
reactions again occur, but crack initiation and propagation is
driven this time by the absorption of cathodically produced hydrogen.
It might seem conspicuous that one could distinguish between these
two cases by examining the susceptibility of a given metal or alloy
at applied anodic or cathodic currents. Unfortunately, for several
reasons, this criterion is not unambiguous[10]. Some aspects of the
crucial problem of crack tip chemistry at impressed anodic and
cathodic potentials will be discussed at this conference by Pourbaix[27].

As is true of stress corrosion cracking and liquid metal
embrittlement, hydrogen embrittlement is affected by mechanical,
environmental and metallurgical conditions[12]. For example, the
susceptibility of high strength, quench and tempered steels to
hydrogen embrittlement increases as the yield strength of the
material is increased by thermal treatment. Indeed, above about
200 ksi yield strength, embrittlement occurs in moist air and water

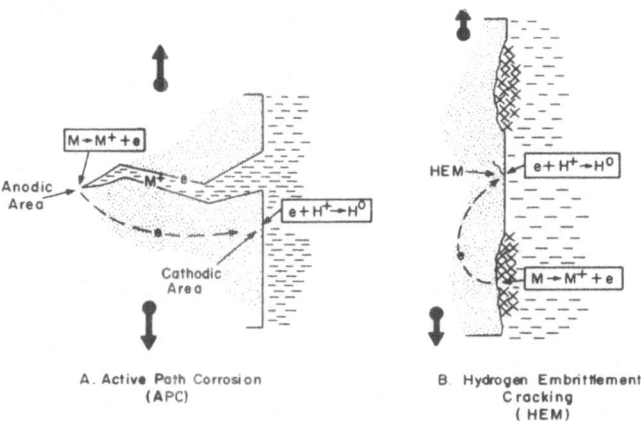

A. Active Path Corrosion
(APC)

B. Hydrogen Embrittlement
Cracking
(HEM)

Figure 3.   Schematic of cracking by active path corrosion (APC)
            and hydrogen embrittlement (HEM) mechanisms (after
            Wilde[26]).

vapor! On the other hand and with reference to environmental conditions, the presence of H_2S in aqueous environments limits the use of hardenable high strength steel in well drilling and oil and gas production equipment to a roughly 80-90 ksi yield strength. Sour gas (i.e., H_2S-containing) presents substantial concern in terms of the materials for use as tubulars in deep wells[24]. One of the most striking demonstrations of hydrogen embrittlement and its control is shown in Figure 4[28] which illustrates the effect of oxygen, argon, hydrogen and water on the cracking behavior of a high strength steel. Note that dry hydrogen and moisture accelerate cracking whereas oxygen arrests crack extention. It is concluded that molecular hydrogen chemisorbs dissociatively on iron[29] allowing atomic hydrogen to be absorbed into the matrix. In moisture, it is expected that hydrogen is produced as a consequence of the corrosion of the iron surface as explained above. In effect, whatever the source, absorbed hydrogen leads to crack growth. By contrast, it is considered that oxygen inhibits embrittlement by producing an oxide barrier which suppresses subsequent absorption.

Mechanisms of Hydrogen Embrittlement. Hydrogen induced losses in strength or ductility have been attributed to several mechanisms. These have been succinctly described by Louthan and McNitt[30] and the list below is essentially an expanded version of that which appeared in their publication.

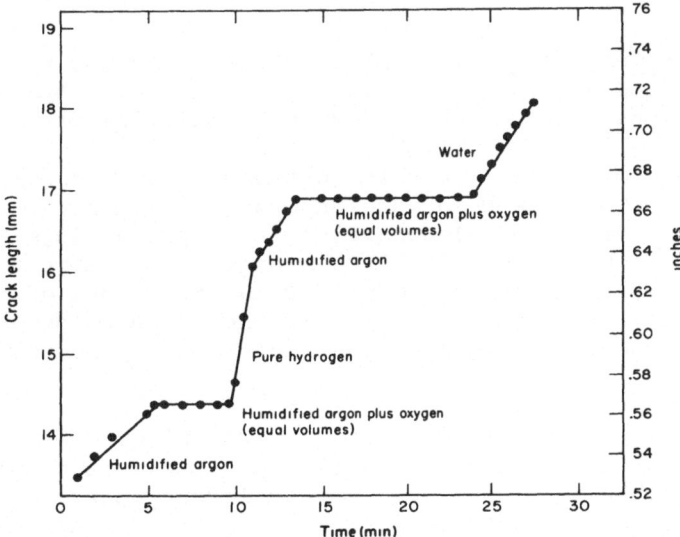

Figure 4. Fast crack growth of high-strength steel in water and hydrogen, but crack arrest in oxygen (after Hancock and Johnson[28]).

a. The accumulation of molecular hydrogen in internal voids
    and cracks exerts a pressure which lowers the apparent
    fracture stress. This pressure model was originally pro-
    posed by Zapffe[31].

b. A hydrogen-induced decohesion of the lattice proposed by
    Troiano[32] and modified by Oriani[33].

c. Adsorption of hydrogen to reduce the surface energy as pro-
    posed by Petch, i.e., the hydrogen equivalent of stress
    sorption in stress corrosion cracking.

d. Beachem's[34] suggestion that (absorbed) hydrogen-stimulated
    plastic deformation accelerates subsequent fracture. Though
    unspecific with regard to the means by which plasticity
    might be affected, recent field ion microscopy by Clum[35]
    suggests that hydrogen may reduce the work required to
    nucleate dislocations at the surface and, hence, induces
    plasticity. Lynch[36] has proposed similar behavior based
    on the view that chemisorption facilitates dislocation
    nucleation at crack tips, although, again, the mechanism
    by which this should occur is not well developed.

e. Formation of a hydrogen-rich phase (e.g., hydride) which
    has mechanical properties different than those of the
    matrix[37,38]. This seems quite clearly to be the case for
    Ti and Zr and their alloys.

f. Hydrogen-dislocation interactions which suppress glide and
    provide a means of producing locally large hydrogen accumu-
    lations that induce subsequent embrittlement[38-40].

It is not my purpose in this instance to critically assess the
above models. This has been done on many occasions by others and
will surely be discussed elsewhere in these proceedings as well.
It seems clear that there are circumstances where some models seem
to apply better than others. Considering the volume of literature
which has appeared on this subject, it should be no surprise that
considerable support as well as contradiction may be found for each
model.

As in the case of stress corrosion cracking, one may be led
to conclude on the basis of the above that there is probably no
one single mechanism of hydrogen embrittlement that applies
universally. Of the suggestions listed above, perhaps the one that
may most interest this interdisciplinary group is the electronic
model of decohesion suggested by Troiano[32] for the hydrogen
embrittlement of transition metals. I will return to this model

and its more modern successors which are just now developing, in a later section.

C. Liquid Metal Embrittlement

Liquid metal embrittlement has been less studied than the industrially more common problems of hydrogen embrittlement and stress corrosion cracking. The proposed use of liquid metal (Li or Na) coolants in various nuclear systems has stimulated renewed concern.

The prerequisites for the embrittlement of an otherwise ductile solid metal by an active liquid metal appear to be as follows[5,6]:

a. a tensile stress applied to the solid

b. a pre-existing crack, or some measure of plastic deformation and the presence of a stable obstacle to dislocation motion in the lattice

c. the presence of the embrittling species at this obstacle and, hence, at the propagating crack tip

d. a usually specific solid metal-liquid metal couple, the characteristic of which is limited mutual solubility and little tendency to form intermetallic compounds. Table 2 indicates the susceptibility of some common liquid metal-solid metal couples.

As is the case of stress corrosion cracking, the specificity indicated in Table 2 is not now considered as restrictive as once thought. For example, Figure 5 shows that polycrystalline pure aluminum is only slightly embrittled by mercury at room temperature, but additions of as little as 1-3% of a number of elements to the mercury produces marked effects on the severity of embrittlement[41]. Liquid metal embrittlement is usually more severe in polycrystals than in single crystal solids. Grain boundaries, of course, serve as stable obstacles to dislocation motion, so the observation is not unexpected. Likewise, failure ususally occurs in an intergranular manner although for highly anisotropic metals such as zinc, failure of polycrystalline specimens occurs predominantly by cleavage on basal planes.

In a mechanistic sense the most widely accepted view of liquid metal embrittlement may be understood with the aid of Figure 6[42]. This model, based on Kelly, et al.[43], suggests that an equilibrium crack in a solid subjected to an increasing force will propagate by cleavage or grow slowly by shear depending on whether the tensile

Table 2 - Susceptibility of Liquid
Metal-Solid Metal Couples

| | Liquid Metal | | | | |
Material	Hg	Li	Bi	Ga	Zn
Steel	NE	E	NE	NE	E
Copper Alloys	E	E	E	--	--
Aluminum Alloys	E	NE	NE	E	E
Magnesium Alloys	NE	NE	NE	NE	E
Titanium Alloys	E	NE	NE	NE	NE

NE - not embrittled   E - embrittled

fracture stress,σ , for the atom-atom bond or the shear stress, τ
to cause dislocation motion on a favorable slip system is achieved
first. Hence, as the ratio of σ/τ decreases, cleavage becomes
more likely and, conversely, shear failure becomes more probable
as the ratio increases. Embrittlement is associated in the case of
liquid metals with the reduction of the atom-atom bond strength due

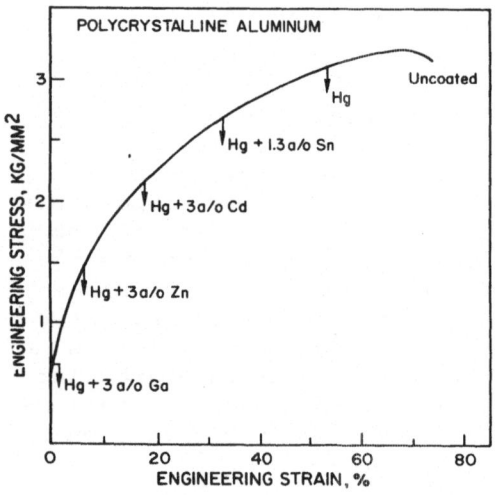

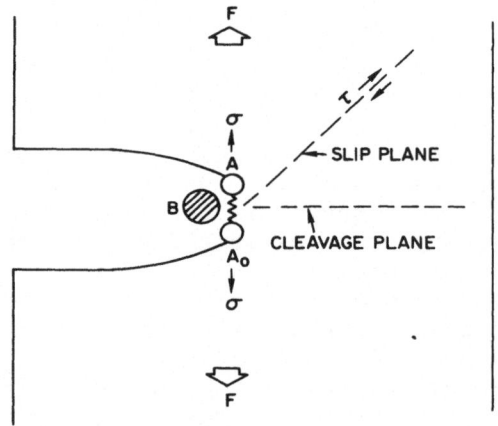

Figure 5. Embrittlement of polycrystalline pure aluminum by
various mercury solutions (after
Westwood et al.[41]).

Figure 6. Schematic of a crack
in a solid subjected to an increas-
force F. The bond A-A_O constitutes
the crack tip, and B is a surface-
active liquid metal atom (after
Westwood et al.[42]).

to the adsorption of the surface-active liquid metal atoms. Because
of conduction electron screening in metals, it is expected that the
yield stress or flow behavior of the solid should not be affected by
the adsorbed species, whereas the stress and strain at fracture may
well be influenced since the latter involves the consecutive
rupture of surface bonds as the crack propagates. While it is
in fact true that there are no measurable effects of liquid metals
on bulk elastic or plastic properties, which adds support to the
above model, Lynch[36,44] has recently argued that chemisorption may
facilitate the nucleation of dislocations at crack tips, although
the mechanism by which this might occur is unclear. The basis for
Lynch's suggestion is the high resolution fractographic evidence
for some crack tip plasticity (dimpling) even in the case of what
appears otherwise to be a most brittle fracture surface. While the
latter issue will surely be aired during the course of this workshop,
there is analogous support for the suggestion of Lynch in the work
of Clum[35], mentioned earlier, which suggests enhanced dislocation
injection in the presence of adsorbed hydrogen. It should be pointed
out that while such analogs are often informative, it is important
to develop such analogs with caution. For example, as pointed
out by Stoloff[6], while atomic hydrogen is quite mobile at ambient
temperatures and may therefore be easily transported particularly
into bcc solids, many embrittling liquid metal atoms are quite
large and relatively immobile. It is, therefore, somewhat more
difficult to account for the observed fractographic evidence of
enhanced plasticity during the fracture process in the presence of
liquid metals such as Hg than in the case of absorbed hydrogen.

SOME GENERAL OBSERVATIONS AND COMMENTS ON ATOMISTICS

     As mentioned at the outset, my intention in the preceding
section was not to attempt a comprehensive review of environmentally-
induced embrittlement. Rather, my hope was to develop just enough
of a perspective so that I might in this section describe areas
where in my view subsequent interdisciplinary interaction might
prove useful. As you will have recognized, there is considerable
opinion that phenomena such as stress corrosion cracking, liquid
metal embrittlement and hydrogen embrittlement have many similar
characteristics. In this section even more common ground will
appear, but I would caution the reader that common phenomenological
characteristics do not necessarily imply that a unique mechanism
(adsorption-induced decohesion, for example) is also common to
these embrittlement processes. We do not have the atomistic basis
yet to consider such a situation.

     In the following sections I would like to consider what, if
any, significance might be attached to (A) the solid/electrolyte
interface, (B) grain boundary segregation phenomena, and (C) the
experiences of organometallic and quantum chemists in understanding

environmentally-induced embrittlement.

A. The Solid-Electrolyte Interface

     In the sense that all of the embrittlement phenomena described
to this point may occur in electrolytes - i.e., stress corrosion
cracking occurs in aqueous environments, hydrogen may be generated
as the cathodic partial process in a corrosion cell, and mercury
may be deposited on zinc by a chemical replacement reaction
from an aqueous mercuric nitrate solution - it may be worthwhile to
consider the charge double layer which is characteristic of a
metal-electrolyte interface. The schematic in Figure 7 shows the
charge and potential distribution at a metal-dilute electrolyte
interface. This has been described in detail elsewhere[1] and will
not be repeated here. The point I wish to make at this stage is
that, in the sense of the electrical analog shown in Figure 7, we
have been typically more concerned with the Faradaic component -
i.e., anodic dissolution or hydrogen evolution - than with the cap-
acitive component. Should the charge double layer be of concern
otherwise? Consider that the electric field across the double layer
is likely to be of the order of 10^7 V/cm, typical of the field distribution present in a field ion microscope and capable of developing
stresses approaching the theoretical cleavage and shear strengths[45].In
addition, it is known that the surface energy, γ (or surface stress
in the case of a solid metal electrode[46]) is related to the surface

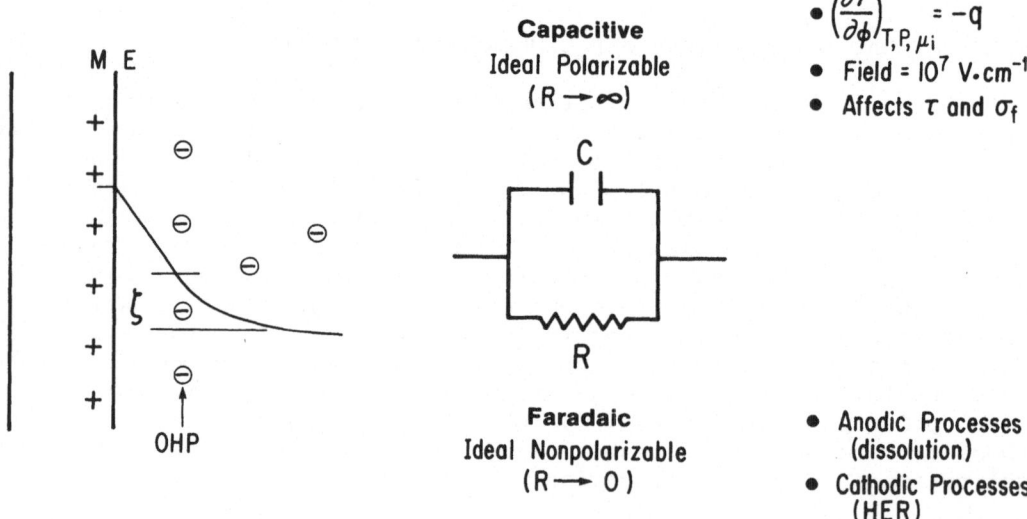

Figure 7. The charge and potential distribution at a metal-dilute
           electrolyte interface

charge density, q , and applied potential through the Lippmann equation[47]

$$\left(\frac{\partial \gamma}{\partial \phi}\right)_{T,P,\mu_i} = -q$$

At the potential of zero charge (pzc) the surface energy passes through a maximum. While there is considerable debate regarding the significance of changes in surface energy on the mechanical behavior of a variety of solids (see, for example, Macmillan[48] and Shchukin[49]), it does seem clear that changes in the surface charge density on metals can indeed affect the motion of dislocations which produce surface slip steps. In effect, the motion of such dislocations is resisted in part by the work required to produce the slip steps left in their trails and one expects that the pzc, where more work is required to create surfaces, the extent of glides of dislocations would be minimized. The net effect is that the creep rate of single crystals has been observed to pass through a minimum at the pzc[50,51] and the hardness passes through a maximum at the same potential[52].

One expects that changes in the charge density on a metal electrode may affect its fracture behavior as well, and there is evidence to support this view[52]. With reference to Figure 8, it appears that crystal plasticity, hence τ, may be affected by changes in the charge density, provided that glide dislocations produce surface steps. On this basis, one anticipates that the motion of such dislocations, Figure 8, away from sources near the crack tip may be inhibited (i.e., increasing τ) at the pzc encouraging cleavage. Likewise, because of the repulsion between like charges, the bond strength or cohesion between atoms in the surface layer of the crack tip, Figure 8, may also be affected by the charge density in the double layer. In this case one expects that at the pzc, where the surface charge is extinguished, cohesion of surface atoms would be maximized--i.e., cleavage would be discouraged.

It may be worth noting as well that the model proposed in Figure 8, would allow for thepossibility that adsorbates may significantly affect the fracture of metals since the propagation of a surface-initiated crack involves the consecutive rupture of surface bonds, and chemisorption, which is likely to involve some change in he distribution of charges in the double layer, may thus affect the bond strength or cohesion (perhaps as described above) between the atoms constituting the crack tip. Liquid metal embrittlement is a classic example of this. Note that in liquid metal embrittlement, the embrittling species is considered to be adsorbed directly on the solid surface and, hence, strongly affects the strength of bonds between surface atoms. Thus, the consecutive rupture of surface atomic bonds, induced by reduction in the cohesive strength of

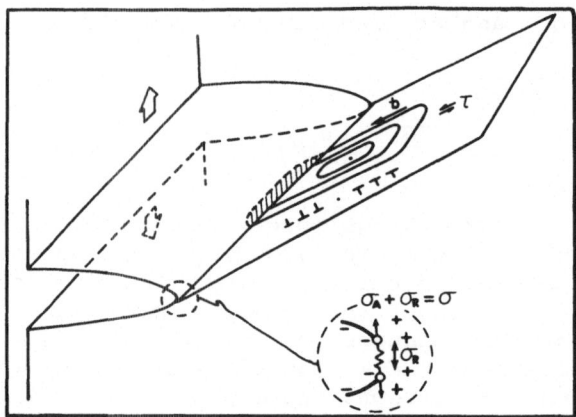

Figure 8.   Crack tip showing the possible influence of double
layer on σ and τ (Latanision[1]).

surface atoms, leads to catastrophic fracture. In contrast, in
aqueous electrolytes and in the absence of specific adsorption, ions
present in the outer Helmholtz plane, are effectively shielded from
the surface by water molecules, and are less likely to significantly
influence cohesive strength (i.e., fracture). If it occurs, specific
adsorption on the other hand (for example, of Cl⁻) may lead to a
situation in an aqueous electrolyte approximately that of liquid
metal embrittlement. Adsorption models based on this premise have
been proposed for stress corrosion cracking. At any rate, it should
be appreciated that σ and in some cases τ may both be affected by
variations in the charge density in the electrical double layer,
regardless of whether the variations are due, for example, to the
applications of external potential or to the chemisorption of
surface-active species.

    What is lacking in all of the above, of course, is the recog-
nition that in phenomena such as stress corrosion cracking the metal
electrode is likely to be film covered. Indeed, as mentioned earlier,
the alloys which are most resistant to general corrosion, due to the
presence of passive films, are also among the most susceptible to
stress corrosion cracking. In effect, as shown in Figure 9, we are
not dealing with a metal-electrolyte interface, but a metal/film/
electrolyte interface. It is interesting that the mechanical behavior of semiconductors and insulator solids is environment sensitive as well. Such chemomechanical effects are described in detail
elsewhere[48,49].

    In brief, there are two schools of thought regarding the
mechanisms by which environments (particularly adsorbates) influence
the mechanical behavior of such solids. The view developed by the
discover of such effects, Rebinder, and his colleagues[49] is that
adsorption-induced softening and strength reduction occur as a

result of the lowering of the specific free surface surface energy
of the solid, i.e., the work of formation of new surfaces during
deformation and fracture.

Alternatively, the view of Westwood and his co-workers is that
many examples of such phenomena may be better understood from considerations of the influence of adsorption-dependent, surface potential induced, redistributions of the charge carriers in the solid,
on the generation, motion and interaction of near-surface dislocations. Perhaps the most obvious demonstration of the dependence
of the mechanical behavior of nonmetallic solids on surface potential
is the correlation between ζ-potential (a measure of the surface
charge density), near-surface hardness and dislocation mobility
indicated generally in Figure 9. This correlation, which shows that
hardness passes through a maximum when $\zeta=0$, has been observed on
solids such as MgO, Al_2O_3, SiO_2, and others. While an explanation
for the symmetry observed in this correlation is a matter of
discussion[53,54] the point of interest here is the suggestion by
Westwood that chemomechanical effects (i.e., environmentally-induced
changes in the charge distribution) on surface films may, in turn,
affect the near-surface mechanical behavior of the underlying metal
substrate, and, hence, may well play a role in embrittlement phenomena.

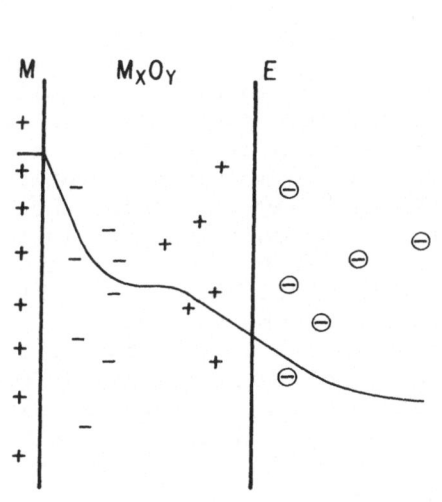

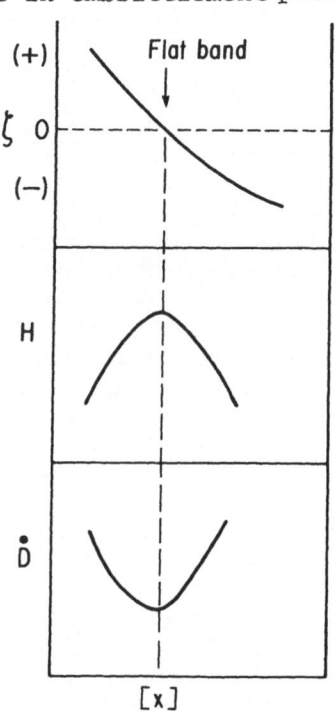

Figure 9. Charge and potential distribution at a metal/oxide/
           electrolyte interface and corresponding mechanical
           behavior effects.

B.   Grain Boundary Phenomena

The importance of impurity segregation at internal interfaces, particularly grain boundaries, has been discussed in terms of phenomenology of temper embrittlement[55], stress corrosion cracking[56], liquid metal embrittlement[6] and hydrogen embrittlement[57]. In each case, intergranular crack morphology is often, if not exclusively, observed. For example, temper embrittlement of steels appears related to the grain boundary accumulation of alloying elements such as Ni and Cr and impurity elements (metalloids) such as P, Sb, Sn, etc.[59-62]. Intergranular embrittlement of W[63], Cu[64], and phosphor bronze[65] have been associated with segregated P, Bi, and P, respectively. In many cases embrittlement has been attributed to a reduction in (grain boundary) surface energy, a consequence of the segregation of solutes to the grain boundary, or to galvanic effects in electrolytes arising out of chemical inhomogeniety between grain boundaries and contiguous grains[56]. Of course, depending upon the nature of the impurity element, increased or decreased reactivity may be expected to occur as a consequence of segregation. In the following, attention will be focussed on the part played by segregated impurities in the intergranular hydrogen embrittlement of polycrystalline metals and alloys. In addition, we will consider the effectiveness of grain boundary and lattice diffusion as well as of dislocation transport of hydrogen in the embrittlement process.

Impurity – Environment Interactions in Electrolytes. Impurities may play an important role in embrittlement induced by cathodically produced hydrogen, particularly if the segregated species happen to by metalloids such as those mentioned above which are known to be effective catalytic poisons for the hydrogen recombination reaction in electrolytes[58]. Perhaps the most effective way to examine the influence expected of metalloid elements on cathodic kinetics is to consider Figure 10 which shows the exchange current density for hydrogen evolution for some of the elements in the Periodic Chart. The exchange density may be considered a measure of the catalytic efficiency of a given element for the hydrogen evolution reaction at the reversible potential. Notice that the exchange current density for the hydrogen reaction is orders of magnitude higher on noble metals (typically ~10^{-3} A/cm^2) or other transition metal surfaces (~ 10^{-6} A/cm^2) than on metalloid surfaces (~ 10^{-13} A/cm^2). The overall hydrogen evolution reaction may be broken down into, for example, a proton reduction step, $H^+ + e = H_{ads}$, and a hydrogen adatom – adatom combination step. Metalloid elements effectively poison the reaction $H_{ads} + H_{ads} = H_2$ thereby increasing the population of adsorbed uncombined hydrogen on the electrode surface and, consequently, the probability that atomic hydrogen will be absorbed by the metal also increases. The influence of soluble metalloid salts in increasing the absorption of hydrogen from electrolytes at cathodic potentials is well documented[67]. Hence, as suggested by Latanision and

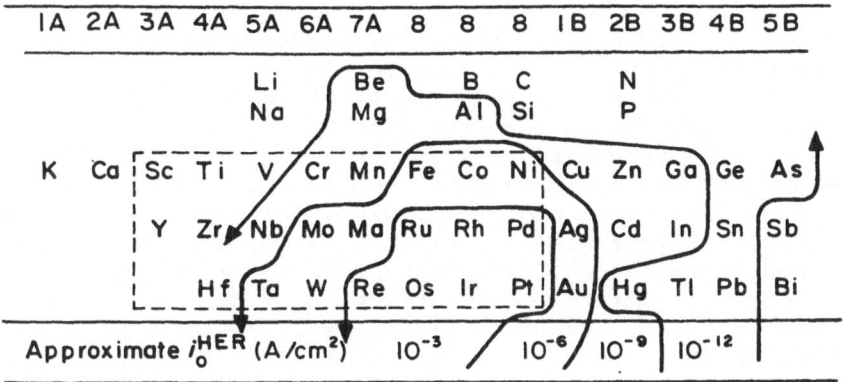

Figure 10. The exchange current density for the hydrogen evolution reaction, i_o^{HER}, for various elements in the Periodic Chart (from West[66]).

Opperhauser[57], one might expect metalloid-segregated grain boundaries to act as preferential sites for the absorption of cathodic hydrogen into polycrystalline metals, Figure 11. In essence, it was argued that the entry of hydrogen into nickel occurred preferentially in the proximity of the grain boundary intersections with the free surface due to the presence therein of segregates which act to poison the combination of hydrogen atoms formed by the discharge of protons from the electrolyte. At locations remote from the grain boundary, protons are reduced forming hydrogen adatoms which in the absence of a poison have a high probability of combining to form molecular hydrogen and are, thus, subsequently evolved. In the vicinity of a grain boundary, in contrast, rather than evolving from the electrode surface in molecular form, uncombined hydrogen adatoms increase in number at the interface and the probability of their absorption into the metal lattice increases. While metalloids may be expected to have the effect of increasing the rate of atomic hydrogen absorption, noble elements such as Ru, Pt, Pd, Rh, etc., would have the opposite influence with regard to cathodic charging. Hence, perferential segregation of noble metals to the grain boundaries of alloys otherwise embrittled intergranularly during cathodic charging may be beneficial. Indeed, the work reported by Pickering and coworkers[68] demonstrates very nicely that Pt introduced into the surface by ion implantation decreases the entry and permeation of electrolytic hydrogen into iron. Somewhat different but related considerations apply if the source of hydrogen is molecular as has been discussed by Berkowitz et al[25]. In this case, noble metals (hydrogen dissociation catalysts) segregated to the grain boundaries would be expected to stimulate hydrogen absorption and subsequent cracking. On the other hand, catalytic poisons should retard cracking as shown in the interesting work of Liu et al[69].

It should be appreciated that the embrittlement of nickel described above[57-58] is not the result of segregation of metalloids
alone as indicated by the fact that identical tensile specimens deformed in the absence of hydrogen were not embrittled.   Indeed embrittlement is associated with the interaction between segregated
species and the surrounding environment.   Such impurity-microchemis-
try interactions may apply to other metals and alloys as well.
Grain boundary segregation of phosphorus, for example, has been observed in thermally treated nickel-base alloys such as Inconel
600[70,71] and Hastelloy C-276[24], both of which are subject to intergranular hydrogen embrittlement.   Likewise, recent extensive work
by Bruemmer, et al[72], has shown a similar fracture mode transition
in hydrogen charged iron as a function of grain boundary sulfur segregation.   In addition, of course, it has been observed that the
tempering temperature range producing shortest life in steels exposed to cathodic hydrogen corresponds also to the typical temper

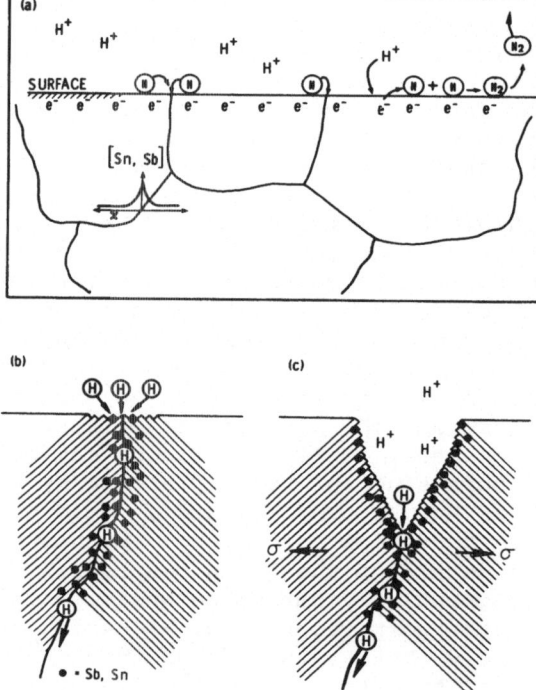

Figure 11.   Schematic indicating the preferential absorption of
             atomic hydrogen along metalloid-segregated grain bounda-
             ries and subsequent intergranular embrittlement of poly-
             crystalline metals and alloys (after Latanision and
             Opperhauser[57]).

embrittlement range[61]. Moreover, it is known, as pointed out earlier, that temper embrittlement is associated with grain boundary accumulation of Sb, P, Sn, etc., all of which are very effective hydrogen recombination poisons. Recognizing that water and water vapor are damaging to high strength steels[28] -- presumably due to the presence of hydrogen produced as the cathodic equivalent of the oxidation or corrosion of iron -- it has been suggested[58] that temper embrittlement may be due not solely to the accumulation of impurities at the grain boundaries but, as in the case of nickel, in part as well to the local rate of absorption of hydrogen by the matrix. It should be mentioned that Yoshino and McMahon[61] attribute the increased hydrogen sensitivity of temper embrittled steels to the combined effects of impurities and hydrogen in reducing the intergranular cohesive strength. The point of the present discussion is to suggest that the accumulated impurities may be responsible as well for the presence of hydrogen in the grain boundaries.

        Given all of the above, some of the elements which may be involved in the intergranular embrittlement of nickel are shown in the sequence in Figure 12. Recent studies (see reference 1 for a review) suggest that yielding begins in the surface grains of polycrystals through the action of dislocation sources near the free surface. The result is that one expects some hydrogen is dragged into the interior along with mobile dislocations which may then interact with grain boundaries, Figures 12(a) and (b). Some hydrogen is likely to enter the solid at other than poisoned grain boundaries, evidence for which is the fact that serrated yielding has been observed in large-grained polycrystals and in similar experiments with monocrystals[73] cathodically charged and deformed simultaneously. The latter suggests that dislocation-solute (hydrogen) interactions occur. There is evidence[74], as well, to show that hydrogen induces softening of nickel under certain circumstances and this should, of course, be considered. Likewise, atomic hydrogen which presumably enters the solid preferentially at grain boundary intersections with the free surface may diffuse via the grain boundaries into the solid. In the later stages of deformation, internal dislocation sources become operaticnal and the incidence of dislocatons interactions with the grain boundaries increases. It is conceivable, for example, that dislocations generated by sources located at grain boundaries may sweep hydrogen into the bulk. The attendant stress and the presence of hydrogen in the vicinity of the grain boundaries may subsequently lead to embrittlement, Figure 12(c), perhaps as a result of the chemisorption-induced reduction in the cohesive strength of atomic bonds at regions of stress concentration.

        While I do not wish to treat the question of hydrogen transport by dislocations as well as by grain boundary or lattice diffusion in detail here (our views in this regard have been presented else-

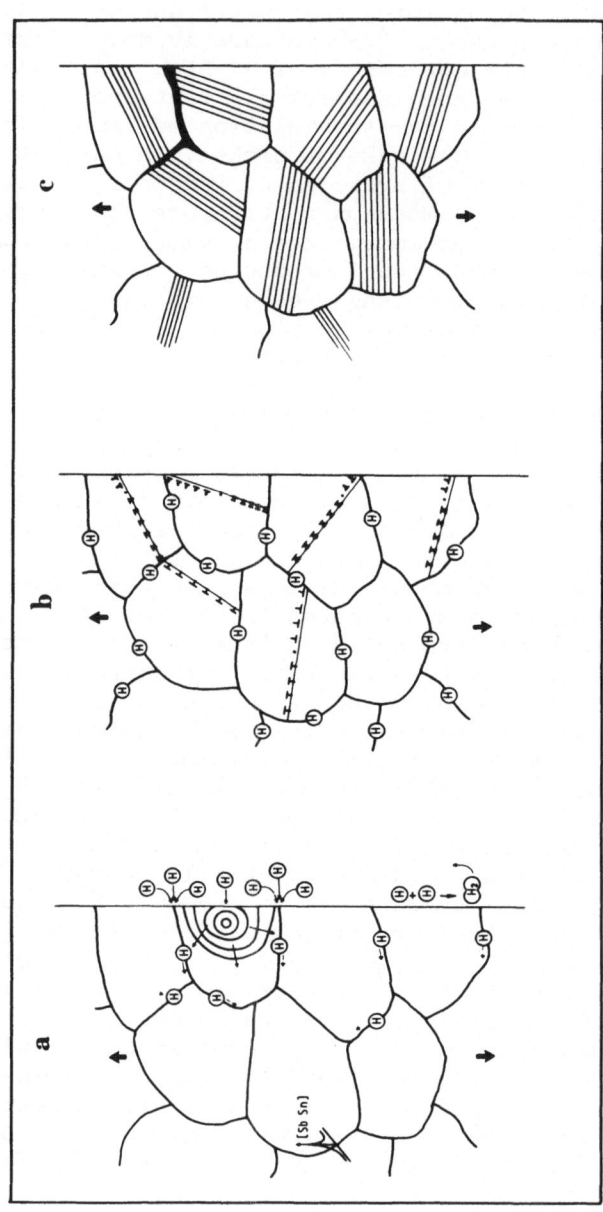

Figure 12.   Schematic showing sequence perhaps involved in intergranular hydrogen embrittlement (after Latanision and Opperhauser[58]).

where recently[58]), I do wish to make two points. Firstly, although
one would like to examine the absorption and transport of electrolytic hydrogen in metalloid-segregated grain boundaries, this is
experimentally difficult to do in situ. On the other hand, we have
recently explored the use of metallic glasses as a structural and
chemical analog of segregated grain boundaries in electrolytic
hydrogen permeation studies[75,76]. Structurally, Ashby et al.[77] have
pointed out that grain boundaries may be described on an atomic
scale as a packing of polyhedra, a model which has been used as well
to effectively characterize the structure of metallic glasses. In
effect, both grain boundaries and metallic glasses may be considered
to possess a certain short range order on an atomic scale, but not
the long range periodicity typical of perfect crystals. Chemically,
the transition metal-metalloid type glasses (typically 80 atomic per
cent transition metals, 20 atomic per cent metalloid compositions)
are of compositions that are good approximations of the chemistry
of solute segregated grain boundaries in polycrystalline metals
and alloys. As mentioned earlier, grain boundary segregation of
phosphorus has been observed in thermally treated nickel-base alloys
such as Inconel 600 and Hastelloy C-276. Hence, Ni-P binary glasses
may be considered to be good structural and chemical analog of
grain boundaries in thermally treated nickel-base alloys. Without
elaborating in detail, using electrolytic hydrogen permeation techniques[78], we find[76] that metalloid-containing glassy alloys do
absorb considerable amounts of hydrogen, Figure 13, presumably
catalytically stimulated as described earlier. On the other hand
in a $Ni_{81}P_{19}$ glassy alloy, which simulates P segregated grain
boundaries in alloys such as Inconel 600 or Hastelloy C-276, the
diffusivity of hydrogen is found to be ~ 10^{-10} cm²/sec, or roughly
of the order of lattice diffusion in nickel. This is significant
since it suggests, given the grain boundary-metallic glass analog,
that fast grain boundary diffusion of hydrogen may be unlikely in
nickel-base alloys. Still, considering that the permeation

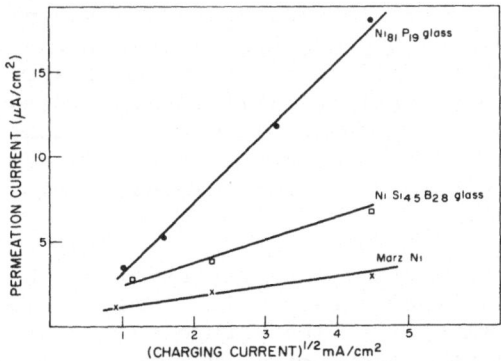

Figure 13.   Steady state permeation flux as a function of charging
             current for various crystalline and glassy nickel-base
             materials (after Latanision, et al.[76]).

flux, if not the diffusivity, of hydrogen in metalloid-containing glasses is higher than in polycrystalline nickel, one may consider that the introduction of substantially more hydrogen occurs along segregated grain boundaries than via the bulk. Work is in progress now using means other than electrochemical to explore the possibility that considerable hydrogen transport may occur along grain boundaries (i.e., a large local flux) but not at significantly greater rates than by lattice diffusion.

The second point I wish to make concerns dislocation transport of hydrogen, which will receive considerably more discussion during the workshop sessions. Observations of dislocation transport have been reported in terms of tritium release measurements performed during deformation and by autoradiographic techniques[79-81] as well as in some recent electrolytic permeation studies by Kurkela and Latanision[82]. The latter work demonstrates that mobile dislocations transport hydrogen at rates much higher than lattice diffusion in polycrystalline nickel. Typical permeation build-up and decay transients under plastic deformation are shown in Figure 14. Hydrogen permeates through the 0.1mm specimen in less than 10 seconds and a steadystate is reached in less than a minute. On switching off the charging current, the anodic current decays to the value observed prior to charging. When deformation is stopped the anodic current decays back to the original background value. Increasing strain rate increases the observed permeation flux. The largest strain rate (4.2×10^{-4}/s) was below the critical strain rate, given by $\sim 10^{-11} p_H$ (p_H = density of hydrogen-carrying dislocations), above which hydrogen transport does not occur. The effective diffusivity of hydrogen in such experiments is found to be linearly dependent upon the strain rate and is of the order of 10^{-5} cm^2/sec in contrast to 10^{-10} cm^2/sec in unstrained nickel at room temperature. From these results, it is evident that in plastically deformed nickel dislocation transport of hydrogen is the predominant mechanism and that the transport rates are several orders of magnitude higher than in unstrained nickel where lattice diffusion is the predominant mechanism. It should be appreciated, however, that dislocation transport or any other rapid transport process does not guarantee in itself that the host material will be embrittled. Such transport mechanisms will only be important in materials that are sensitive to hydrogen. For example, hydrogen is very permeable in palladium, but embrittlement is not[83] anything like as serious as in nickel or some iron-base alloys.

In short, identification and/or control of the partitioning of solutes - alloying elements as well as metalloid impurities - may well provide a clue as to the nature of a particular intergranular embrittlement phenomenon (i.e., stress corrosion cracking vs. hydrogen embrittlement) as well as suggest means of reducing susceptibility by metallurgical treatment. Of course, we need at this stage to know much more about the details of such segregation

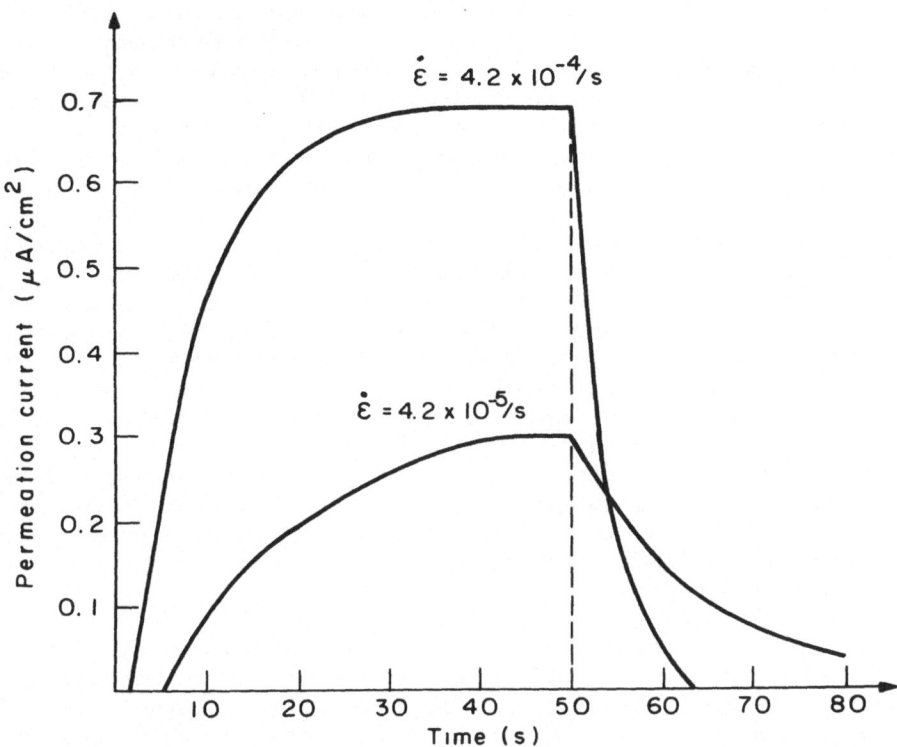

Figure 14. Effect of Strain Rate on the Decay Transient I_c=50mA (after Kurkela and Latanision[82]).

phenomena – how might one treat an alloy to induce selective segregation; what interactions might one expect between segregated solutes; etc.? It is also clear that despite years of activity we still do not understand on an atomic scale how and what embrittling species (hydrogen atoms, chloride ions, liquid metal atoms, etc.) do to induce decohesion at a crack tip. This remains to be answered, I believe, by means other than traditional mechanical testing. Detailed studies of the surface structure, electronic structure, and bonding of adsorbed hydrogen, for example, on strained metal surfaces, are just beginning to proceed but are crucial, we believe, to understanding such phenomena. Here is an area where our coordination chemist and quantum chemist and other friends may begin to help us.

C. Embrittlement Analogs in Chemistry and Physics

I believe it is clear that embrittlement phenomena such as stress corrosion cracking, hydrogen embrittlement, liquid metal embrittlement and temper embrittlement are academically exciting and, even more clearly, technically important. These problems are also, I believe, poorly understood on an atomistic level. In short, we know quite a lot about the phenomenology and control of environmentally-induced embrittlement, but very little about atomistics. In some respects, this should not be surprising. As I have mentioned earlier, these are multidisciplinary problems involving not just materials science and mechanics but chemistry as well. The latter, it seems to me, is the area where we have been least effective in attempting to understand the atomistics of embrittlement, and probably, the area where some remarkably meaningful contributions may be made in the near future, provided that we properly cultivate the interest of this discipline. In the following I wish to present a short description of some aspects of chemistry which may be helpful. Many of these views are based on discussions with my colleagues Traugott Fischer, Keith Johnson, and George Whitesides, who are far more expert in these matters than I, and will fortunately follow me in the program with discussions of far greater depth. I hasten to add that the following is not intended to represent a complete or exhaustive indication of the input which might be forthcoming, but, rather, a starting point.

While it is possible to describe liquid metal embrittlement, hydrogen embrittlement, etc., on the basis of environmentally induced decohesion, the atomic scale interactions between the adsorbate (Hg, H, etc.) and strained metal surfaces which cause decohesion are not well understood. In this context, one interesting electronic model of hydrogen embrittlement was proposed in 1960 by Troiano[32]. In this model, Troiano presumes that hydrogen behaves in an electropositive sense and donates its electron to the unfilled d-bands of the metallic cores as shown schematically in Figure 15. The increase in electron density leads to an increase in the repulsive force between adjacent metal cores or, in other words, a

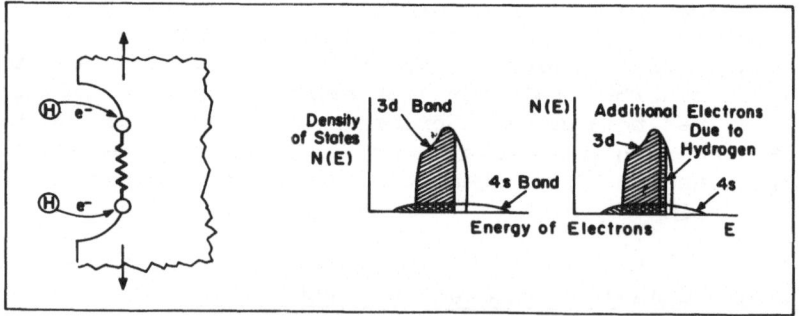

Figure 15.   An electronic model for hydrogen embrittlement of transition metals (after Troiano [32]).

decrease in the cohesive strength of the lattice, i.e., in not unlike
the manner described in Figure 8. Recognizing that the great cohesive
strength of the transition metals may be associated with their band
structures, particularly the filling of d-orbitals, the above model
has a certain appeal.

Of course, there are alternate approaches to considering the
nature and significance of the charge transfer processes which occur
and which may be associated with the atomistics of hydrogen embrittlement as well as other (at least potentially) adsorption-induced
embrittlement phenomena such as liquid metal embrittlement. For
example, if we look into the literature of catalysis we find that
there is considerable interest and understanding of one process
which is of interest to us, namely the dissociative adsorption of
molecular hydrogen. We can examine this problem in elementary
terms with the aid of Figure 16 which shows the energeties
involved when two hydrogen atoms are made to approach one another.
When the spins of the electrons on the two atoms are opposed, attractive or bonding interactions occur between the atoms, with a pronounced minimum at r_o , whereas repulsive or antibonding interactions occur if the hydrogen 1s electrons have parallel spins.
In effect, the two original hydrogen 1s orbitals are transformed
into two new orbitals: one of these is a bonding molecular orbital
and has a lower energy than the parent atomic orbitals while the

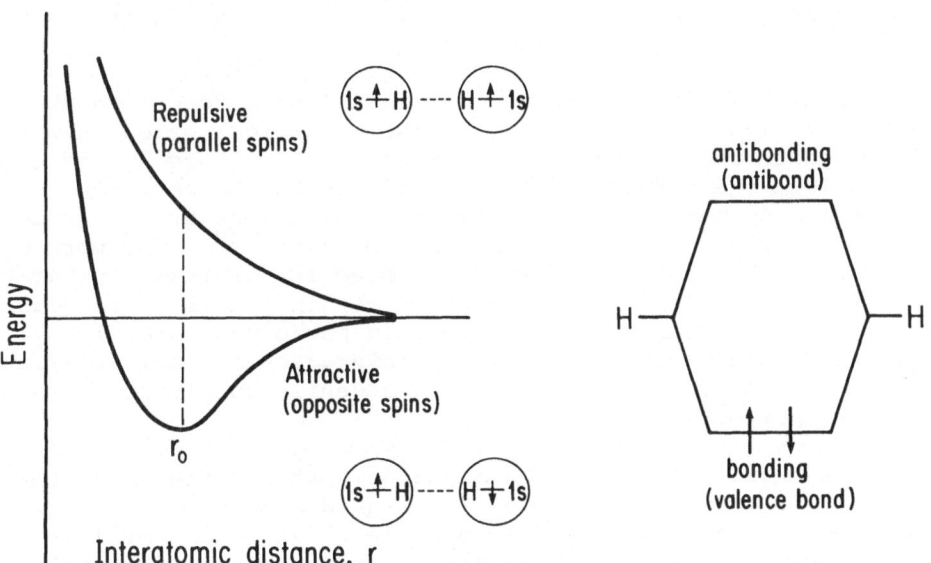

Figure 16.   The hydrogen molecule.

other is an antibonding orbital and has a higher energy than the
original atomic orbitals. A stable hydrogen molecule, at a
separation r_O, would just fill the bonding orbital. Indeed,
on this basis we can understand why a stable He_2 molecule does not
exist. In dissociative chemisorption, it is envisioned that electrons
from the catalyst (for example, Fe, Ni, or other transition metals)
are dumped into the antibonding orbitals of the hydrogen molecule.
In such circumstances, the hydrogen molecule begins to dissociate.
The efficiency of this dissociation depends on the amount of overlap
between the metal orbital which is involved and the antibonding
orbital of the hydrogen molecule. (See, for example, references
84 and 85.)

Of course, in the above description we have concerned ourselves
with the dissociation of the adsorbate, hydrogen. What happens to
the metal atoms in the surface of the transition metal catalyst?
It is, of course, possible to identify bonding and antibonding
orbitals in the density of state diagram for the metals involved,
as described in Figure 17. On this basis it would seem that one
might consider the stability of the metal surface in terms of the
relative location and population of the bonding and antibonding
orbitals in both the metal and adsorbate. Indeed, as pointed out
by Fischer[85-86], dissociative adsorption (catalysis) is the mirror
image of chemisorption-induced embrittlement (and vice versa). On
this basis one may consider that some metals, those which are embrittled by hydrogen, may transfer electrons from metal bonding
orbitals to antibonding orbitals in the hydrogen molecule, while
nonembrittled metals may dissociate hydrogen by transfer of charge
from metal antibonding orbitals to hydrogen antibonding orbitals.
In both cases hydrogen is dissociated.

Certainly, matters are more complicated than described above.
There is, for instance, the issue of the mechanical stresses which
are involved in the embrittlement situation. How does stress (or
strain) affect the relation between orbitals in the metal and
adsorbate. I do not believe that we have an answer at this stage,
but some work along these lines is in progress[87,88]. The work of
Eberhart et al.[88], based on SCF-Xα-SW molecular orbital cluster
calculations, includes the influence of strain and is particularly
exciting. Similar calculations have been performed recently by
Briant and Messmer[89] in assessing the intergranular embrittlement
of nickel by sulfur. Losch[90] has also recently considered a chemical
model of temper embrittlement.

Although calculations which might demonstrate the above embrittlement of metals by hydrogen (or liquid metals, etc.), have not
yet been performed, there is some experience in organometallic
chemistry which by analogy with embrittlement phenomena is quite
intriguing. For example, transition carbonyl molecules are known to

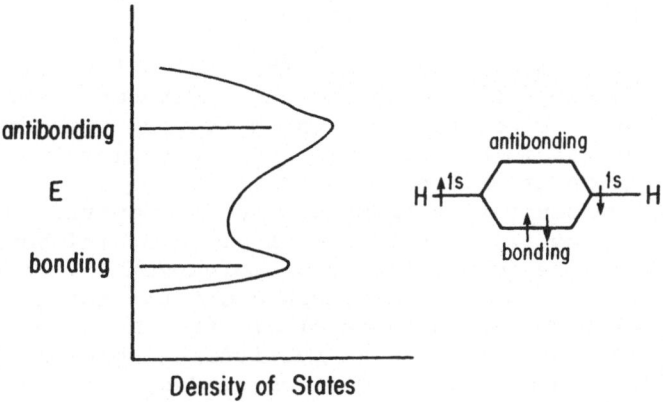

Figure 17.   Adsorption of hydrogen on metal surfaces.

dissociate in the presence of molecular hydrogen; for example, as in the following[91,92]:

$$Co_2(CO)_8 + H_2 = 2HCo(CO)_4$$

In effect the hydrogen molecule splits, but so too does the bond between the cobalt atoms in the carbonyl molecule. While the Co-Co bond in this carbonyl molecule may not be metallic in the usual sense, such reactions are, on an atomic scale, an interesting analog of the hydrogen embrittlement of transition metals.

At any rate, it seems to me that there is great potential for fruitful interactions between those involved in research in embrittlement phenomena and in catalysis.

CONCLUDING REMARKS

In keeping with my initial objectives, I have attempted in
this paper to provide a general overview of some of the phenomenology related to the atomistics of (particularly, environmentally-
induced) fracture and of those areas where interdisciplinary
activity involving materials science, mechanics, physical chemistry,
surface physics and chemistry and atomic modeling might provide
valuable insight. May of the details which I have omitted will,
I suspect, appear in our workshops.

In all of the above phenomena, the action typically begins at
the interface between the solid and the environment which surrounds
it, and it should not be too surprising to find that many of the
current mechanistic models for these phenomena involve film formation, adsorption of critical species from the environment, interfacial solute concentration gradients, etc. However, we know
little about the possible influence of yielding initiation in the
surface layer on the embrittlement of polycrystals. Nor, for
example, do we seem to know very much about the influence of surface
charge on crack propagation in metal electrodes, about the detailed
atomic-order interactions between embrittling species and strained
crystal surfaces, about the atomistics of decohesion or about the
electron distribution at a crack tip. We have made some progress
in recent years in these directions, and the potential for sizeable
advances is enormous, provided, I believe, that the concerted
activity of the disciplines represented here is stimulated.

We have learned quite a lot about the phenomenology of embrittlement, but it is safe to say that in an atomistic sense we
still do not know in detail what hydrogen (but not nitrogen or
oxygen, etc.) does on an atomic scale to induce this degradation.
The same applies to other examples of environmentally-induced
fracture; what is it about the ubiquitous chloride ion that
induces premature catastrophic fracture (stress corrosion cracking)
of ordinarily ductile austenitic stainless steels? Why, moreover,
are chloride ions troublesome but chemically similar iodides
inhibitors for SCC in such stainless steels? In short, despite
all that we may know about the phenomenology of fracture on a
macroscopic scale, we know precious little about the atomistics of
embrittlement phenomena such as those described above.

On the other hand, it is interesting to note that physical
chemists and surface chemists also have interest in the same kind
of interactions that occur on an atomic scale when metals such as
nickel or platinum, for example, are used as catalysts for chemical
reactions. Such metals are very effective catalysts for the dissociation of molecular hydrogen. Indeed, considerable experimental study has been directed toward the surface chemistry of
hydrogen adsorption, etc., on transition metal surfaces. But much

of the same uncertainty in terms of fundamental understanding pervades this area of science. Can we predict which metals will be effective catalysts and for which chemical reaction? It would seem that the development of fundamental understanding of the kind implicit in the above discussion of the catalytic dissociation of hydrogen (embrittlement of the hydrogen molecule!) would impact, as well, understanding of the fundamentals of the embrittlement of the metal surfaces (dissociation of metal atomic bonds). One might wonder in a converse sense if such interdisciplinary inter- action might benefit the catalysis community as well. For example, is it possible that the effectiveness of a catalyst might be sig- nificantly increased if the catalyst were allowed to plastically deform while performing its chemical function?

At any rate, I believe that more direct and frequent commun- ications must be stimulated between the kinds of disciplines described above. Indeed, it seems very clear now that much is known by physical chemists and quantum chemists that bears directly on problems of embrittlement and fracture of interest to materials scientists and those involved with the mechanics of solids. This information is often, however, not transferred. Development of the atomic scale understanding which is sought is not only of academic significance but may well lead to technological advances in chemistry and materials science ranging from catalysts for use in energy conversion and storage, for example, to the improved lifetime and reliability of complex and expensive structures in aggressive environments. In this context, in fact, it seems to me that the technological significance of the interactions of environments such as hydrogen, liquid metals, hot (perhaps molten) salts, etc., with materials will become increasingly more important in the decades to come. There is, for example, the issue of a hydrogen economy -- i.e., the use of hydrogen as an energy medium for this planet.[93] This is not unlikely, particularly if effec- tive (semiconductor) photoelectrodes can be developed to split water, since water and sunlight are both plentiful and, at the moment, free! In addition, of course, questions such as materials for storage and transmission or distribution of hydrogen (perhaps in this case materials that are already in place) will become increasingly important. I suspect that if the previous decades of research and engineering on environmentally-induced embrittle- ment were considered exciting, the next decades will be even more so and the impact of this effort will be of even greater conse- quence.

ACKNOWLEDGMENTS

The paper reflects work performed in the Corrosion Laboratory at MIT on programs sponsored largely by the Office of Naval Research, Contract No. N00014-78-C-0002, NR 036-127 (the work on

hydrogen embrittlement) and by the National Science Foundation,
Grant No. DMR 78-05712 (the work on metallic glasses). I am
grateful for the continued encouragement of Dr. P.A. Clarkin and
Dr. R.J. Reynik. Likewise, I am pleased to acknowledge many exciting
and instructive discussions over a period of several years with
T.E. Fischer, A.R.C. Westwood, R.W. Staehle, K.H. Johnson, N.H.
Macmillan, H.H. Uhlig and my outrageously provoking graduate
students at MIT.

REFERENCES

1. R.M. Latanision and J.T. Fourie, eds., "Surface Effects in
   Crystal Plasticity", Noordhoff, Leyden (1977).
2. J.C. Scully, ed., "Theory of Stress Corrosion Cracking", NATO
   Scientific Affairs Division, Brussels (1971).
3. R.W. Staehle, A.J. Forty and D. Van Rooyen, eds., "Fundamental
   Aspects of Stress Corrosion Cracking", NACE, Houston (1969).
4. A.W. Thompson and I.M. Bernstein, eds., "Hydrogen Effects on
   Behavior of Materials", AIME, New York (1976).
5. A.R.C. Westwood and R.M. Latanision, in "Corrosion by Liquid
   Metals", Plenum Press, New York (1979), p. 405.
6. N.S. Stoloff in "Environment-Sensitive Fracture of Engineering
   Materials", ed. Z.A. Foroulis, AIME, New York (1975), p. 486.
7. A.R.C. Westwood and J.R. Pickens, These Proceedings.
8. A.S. Tetelman and A.J. McEvily, Jr., "Fracture of Structural
   Materials", Chapter 9, Wiley, New York (1967).
9. R.W. Hertzberg, "Deformation and Fracture Mechanics of
   Engineering Materials", Chaper 11, Wiley, New York (1976).
10. R.M. Latanision, O.H. Gastine and C.R. Compeau, in "Environment
    Sensitive Fracture of Engineering Materials", ed., Z.A. Foroulis,
    AIME, New York (1979), p. 48.
11. R.M. Latanision, M. Kurkela andF. Lee, in "Effect of Hydrogen
    on Behavior of Materials", AIME, in press.
12. A.W. Thompson and I.M. Bernstein, in "Advances in Corrosion
    Science and Technology", volume 7, eds., R.W. Staehle and
    M.G. Fontana, Plenum Press, New York (1980).
13. T.P. Hoar and J.G. Hines in "Stress Corrosion Cracking and
    Embrittlement", Wiley, New York (1956), p. 107.
14. F.A. Champion, in "Symposium on Internal Stresses in Metals and
    Alloys", Institute of Metals, London (1948), p. 468.
15. H.L. Logan, J. Res. N. B. S., 48: 99 (1952).
16. A.J. Forty, in "Physical Metallurgy of Stress Corrosion
    Cracking", Interscience, New York (1959), p. 99.
17. N.J. Petch, Phil. Mag., 1: 331 (1956).
18. H.H. Uhlig, "Corrosion and Corrosion Control", 2nd Edition,
    Wiley, New York (1971).
19. T.J. Smith and R.W. Staehle, Corrosion, 23: 117 (1967).
20. R.W. Staehle in Ref. 3.
21. O. Reynolds, Manchester Lit. Phil. Soc., 13: 93 (1874).

22. D.E. Hughes, Scientific American Supplement, 237: July 17, (1880).

23. R.M. Latanision and H. Opperhauser, Jr., Met. Trans., 5: 483 (1974).

24. B.J. Berkowitz and R.D. Kane, Corrosion, 36: 24 (1980).

25. B.J. Berkowitz, J.J. Burton, C.R. Helms and R.S. Polizzotti, Scripta Met., 10: 871 (1976).

26. B.E. Wilde, Corrosion, 27: 326 (1971).

27. M. Pourbaix, These Proceedings.

28. G.G. Hancock and H.H. Johnson, Trans. AIME, 236: 513 (1966).

29. D.O. Hayward and B.M.W. Trapnell, "Chemisorption", 2nd Edition, Butterworths, London (1964).

30. M.R. Louthan and R.P. McNitt, in Ref. 3, p. 496.

31. C.A. Zapffe and C.E. Sims, Trans AIME, 145: 225 (1941).

32. A.R. Troiano, Trans. ASM, 52: 54 (1960).

33. R.A. Oriani, Berichte der Bunsenges fur Phys. Chem., 76: 848 (1972).

34. C.D. Beachem, Met. Trans., 3: 437 (1972).

35. J.A. Clum, Scripta Met., 9: 51 (1975).

36. S.P. Lynch, Metals Forum, 2: 189 (1979).

37. D.G. Westlake, Trans. ASM, 62: 1000 (1969).

38. J.J. Gilman, Phil. Mag., 26: 801 (1972).

39. P. Bastien and P. Azou, Proc. 1st World Metallurgical Congress, ASM, Cleveland (1951), p. 535.

40. J.K. Tien, A.W. Thompson, I.M. Bernstein and R.J. Richards, Met. Trans., 7A: 821 (1976).

41. A.R.C. Westwood, C.M. Preece, and M.H. Kamdar, in "Fracture, An Advanced Treatise", Academic Press, New York (1971), p. 589.

42. A.R.C. Westwood, C.M. Preece, and M.H. Kamdar, Trans. ASM., 60: 763 (1967).

43. A. Kelly, W.R. Tyson, and A.H. Cottrell, Phil. Mag., 15: 567 (1967).

44. S.P. Lynch, in "Proc. Fourth Intl. Conf. on Fracture", Permagon, New York (1977), p. 859.

45. P.J. Birdseye and D.A. Smith, Surface Science, 23: 198 (1970).

46. R.M. Latanision, N.H. Macmillan and R.G. Lye, Corrosion Science, 13: 387 (1973).

47. G. Lippmann, Ann. Chem. Phys., 5: 494 (1875).

48. N.H. Macmillan, in Ref. 1, p. 629.

49. E.D. Shchukin, in Ref. 1, p. 701.

50. A. Pfutzenreuter and G. Mazing, Z. Metallk., 42: 361 (1951).

51. V.I. Likhtmann, L.A. Kochanova, D.T. Leykis and E.D. Shchukin, Elektrokhimiya, 5: 729 (1969).

52. R.M. Latanision, H. Opperhauser, Jr., and A.R.C. Westwood, Scripta Met., 12: 475 (1978).

53. R.M. Latanision, in Ref. 1, p. 3.

54. J.S. Ahearn, J.J. Mills, and A.R.C. Westwood, J. Appl. Phys., 49: 614 (1978).

55. C.J. McMahon, Materials Science and Engineering, 25: 233 (1976).

56.   R. Speiser and J.W. Spretnak, in "Stress Corrosion Cracking
       and Embrittlement", Wiley, N.Y. (1956), p. 52.
57.   R.M. Latanision and H. Opperhauser, Jr., Met Trans., 5: 483
       (1974)
58.   R.M. Latanision, M. Kurkela and F. Lee, in "Effect of
       Hydrogen on Behavior of Materials",in press.
59.   H.L. Marcus and P.W. Palmberg, Trans. AIME, 245: 1664 (1969).
60.   D.F. Stein, A. Joshi and R.P. Laforce, Trans. ASM, 62: 736
       (1969).
61.   K. Yoshino and C.J. McMahon, Jr., Met Trans., 5: 363 (1974).
62.   R.O. Ritchie, Met. Trans., 8A: 1131 (1977).
63.   A. Joshi and D.F. Stein, Met Trans., 1: 2543 (1970)
64.   A. Joshi and D.F. Stein, J. Insti. Metal, 1: 178 (1971)
65.   J.M. Popplewell and J.A. Ford, Met. Trans., 5: 2600 (1974)
66.   J.M. West, "Electrodeposition and Corrosion Processes",
       Van Nostrand-Reinhold, London (1971).
67.   F. Zakroczymski, Z. Szklarska-Smialowska and M. Smialowski,
       Werkstoffe und Korrosion, 27: 625   (1976).
68.   M. Zamanzedeh, A. Allam, H. Pickering, and G.K. Hubler, J.
       Electrochem. Soc., 127:1688 (1980).
69.   H.W. Liu, Y.L. Hu and P.J. Ficalora, Engineering Fracture
       Mechanics, 5: 281 (1973).
70.   N. Pessall, G.P. Airey and B.P. Lingenfelter, Corrosion 35:
       100   (1979).
71.   G.S. Was, Sc. D. Thesis, M.I.T., June 1980.
72.   S.M. Bruemmer, R.H. Jones, M.T. Thomas and D.R. Baer, Scripta
       Met., 14: 137   (1980).
73.   R.M. Latanision and R.M. Staehle, Scripta Met., 2: 667 (1967)
74.   E. Heubaum and H. Birnbaum, ONR Tech. Rept., March 1981.
75.   R.M. Latanision, J.C. Turn and R.C. Compeau, in "Proc. Third
       Int'l Conference on Mechanical Behavior of Metals," vol. 2,
       Pergamon, Toronto (1979), p. 475.
76.   R.M. Latanision, C.R. Compeau and M. Kurkela, in "Proc.
       Alexander R. Troiano Symposium on Hydrogen Embrittlement
       and Stress Corrosion Cracking", in press.
77.   M.F. Ashby, F. Spaepen and S. Williams, Acta Met., 26: 1647
       (1978).
78.   M.A. Devanathan and A. Stachurski, Proc. Roy. Soc., A270:90
       (1962).
79.   M.R. Louthan, G.R. Caskey, J.A. Donovan and D.E. Rawl, Mat.
       Sci. Eng., 10:357   (1972)
80.   J.A. Donovan, Met Trans., 7A:145   (1976).
81.   L.M. Foster, T.H. Jack and W.W. Hill, Met Trans., 1:3117
       (1970).
82.   M. Kurkela and R.M. Latanision, Scripta Met., 13:927   (1979).
83.   J.O'M. Bockris, M.A. Genshaw, and M. Fullenwider, Electrochem.
       Acta, 15: 47   (1970).
84.   J.C. Slater and K.H. Johnson, Physics Today, 34 (Oct. 1974).
85.   T.E. Fischer, J. Vac. Sci. Technol., 11: 252   (1974).

86.   T.E. Fischer, in Ref. 1, p. 127.
87.   M. Ciftan and E. Saibel, Int. J. Engng. Sci., 17: 175 (1979).
88.   M. Eberhart, R. O'Malley, A. Abdelmassih, and E. Johns,
      "A New View of Hydrogen Embrittlement," Project report in
      MIT Course 3.22J/10.613J Solid State Surface Science, May,
      1981.
89.   C.L. Briant and R.P. Messmer, Phil. Mag. B, 42: 569 (1980).
90.   W. Losch, Acta Met., 27: 1885 (1979).
91.   E.D. Muetterties, in "Transition Metal Hydrides", Dekker, N.Y.
      (1971).
92.   R. Ugo, Catal. Rev. - Sci. Eng., 11: 225 (1975).
93.   J. O'M. Bockris, "Energy Options," Halsted, New York (1980).

DISCUSSION

Comment by S.P. Lynch:

Table II in your paper indicates that steels are not embrittled
by mercury. However, observations show that some high-strength steels
are embrittled by mercury and that the susceptibility to embrittlement
decreases with decreasing strength. Other solid-metal/liquid-metal
combinations listed as nonembrittling could also be embrittling in
some circumstances.

Reply:

As I have mentioned in the text of my presentation, the specificity of LME is not as restrictive as once thought. Table II is
intended only as a rough guide to common solid metal-liquid metal
couples. I agree with your comments, of course.

Comment by J. O'M. Bockris:

The atomistics of fracture divides itself sharply into two groups
of phenomena. In the one (e.g., liquid metal embrittlement) water
(or moisture) does not play a part. In the other (e.g., stress
corrosion cracking), it does. There seems little doubt that stress
corrosion cracking (i.e., most corrosion - there does not have to be
an outside applied macro-stress) has the greatest financial implication.

Given this situation, it should be said that there is not much
mystery. Both in a recently published Russian book (Cherapanov) and
in an international meeting at Minnakami in 1980, there can be seen
to be a consensus, thus: corrosion cracks have two motions. The
first and the usual is slow (e.g., 0.1 mm per hour). It arises from
the coupled action of two charge transfer reactions, metal dissolution
and hydrogen evolution. There is sometimes reported to be a second

motion: the crack undergoes a rapid sudden tearing motion of perhaps
a micron.

Thus, in respect to the first type of movement, bond breaking
is secondary, and non-rate determining. In respect to the tearing
movement, bond breaking must occur (as in all cracking) but there
is evidence for some systems that the rate-determining step is the
aggregation of impurities (e.g., H) to the crack surface.

These remarks indicate the areas of knowledge where increased
investigation is needed if one is to get progress. Bond theory and
cohesion calculations (which at first would seem the heart of fracture
theory) may be of less relevance than calculations concerning the
rate-determining step. Quantum mechanics has not developed in
respect to calculations of rates. Here, absolute reaction rate theory
would yield much more than quantum mechanical approaches.

Reply:

I am not certain that I agree with your comments entirely,
particularly regarding the relevance of quantum mechanical calculations of fracture-related processes. Perhaps we could consider
this question again after we've heard from the people performing
such calculations later in the program.

Note Added in Proof:

Subsequent work[1-4] has demonstrated that the data shown in Figure
14 is complicated by what appears to be a thermal or Joule heating
effect originating in the relatively high charging current densities
used in the experiments described. Work is in progress[4] to attempt to
distinguish thermal or Joule effects from dislocation transport in
cases where both are significant transient phenomena.

1.   R. Otsuka and M. Isaji, Scripta Met., 15, 1153 (1981).
2.   M. Kurkela and R.M. Latanision, Scripta Met., 15, 1157 (1981).
3.   R.W. Lin and H.H. Johnson, Scripta Met., 16, (1982) in press.
4.   G.S. Frankel and R.M. Latanision, unpublished research, 1982.

GENERAL OVERVIEW:

ATOMISTICS OF SURFACE REACTIONS

T. E. Fischer

Exxon Research and Engineering Company
P. O. Box 45
Linden, New Jersey

INTRODUCTION

The adsorption of a molecule is the first step in the modification of the mechanical properties of a metal by the environment. Once adsorbed, the molecule can play a number of different roles that are relevant to the phenomena of fracture. It can react with the surface to form a coherent and stable layer that protects the solid from further influence from the environment; it can react with the solid to corrode it. This reaction can be activated by mechanical energy stored in locally enhanced stresses and lead to stress corrosion cracking. Hydrogen can dissociate and penetrate the metal and lead to hydrogen embrittlement. The molecule can displace from the surface, or react with, a potentially harmful adsorbate and stop the environmental degradation of the solid. Thus, the study of the physical and chemical interaction of fluids with solid surfaces is of importance to the understanding of environment effects on the fracture of a material under stress.

There are further reasons for the usefulness of surface chemistry. Fracture is a complex phenomenon that includes the rupture of the chemical bond between neighboring atoms and the creation of a new surface. Progress in the mastery of this complex field will be made if we get a better understanding of the cohesive forces in metals and their modification by the adsorption of various species. Surface science is developing experimental and theoretical techniques that have shed much light on this problem from a very funda-

mental viewpoint. Thus, we believe it is in order, at the beginning of this Institute, to reacquaint ourselves with the methods of
surface science, with the kind of information they provide and the
kind of questions we attempt to answer. In this, we shall emphasize
particularly the progress that took place since the previous meeting
in Hobegeiss, six years ago. Having done that, we shall examine in
some detail a number of surface reactions that take place in environmental effects on fracture, and close this review with some surface science considerations that, we believe, have some relevance to
the fundamental understanding of the phenomena of cohesion and fracture.

PROGRESS IN SURFACE SCIENCE

Great advances have been made in the last twenty years in the
observation and theoretical understanding of properties of solid
surfaces and their interaction with gases, especially at low pressures. The reason is that gases, at least in the pressures below
10^{-4} Torr (10^{-2} Pascal) represent so little matter per unit volume
that the space in front of the solid allows the movement of light,
electrons, ions and molecules without noticeable interference. Such
low pressures obviously represent a limitation in the simulation of
technological situations, but they offer the scientific advantage of
presenting stable surfaces and chemical reactions that are simple
and slow enough to be amenable to experimental observation. Above
all, these low pressures permit to probe the surface by the absorption, scattering or emission of photons, x-rays, electrons and ions.
The depth of the layer from the surface over which a particle aver-
ages its information is equal to the mean free path of that particle before it is absorbed or scattered with random energy loss and
therefore rejected by the measuring instrument. This depth is about
two atomic layers for electrons of 50 eV and ten layers for electrons and ions with energy near 1 keV; it is a fraction of a micro-
meter for light and high energy ions (hundreds of keV). Thus, by
using the right probing particle with appropriate energy, the exper-
imenter can investigate the solid surface to the depth of interest
to him. Electron and ion beams can be directed and focused and can,
therefore, analyze the inhomogeneities of a given material. This
is particularly useful in the study of technological materials; it
can be used as well for the study of transport properties on homogeneous surfaces.

Established surface analysis methods have been described in the
previous conference, in what follows, we present improvements in
these methods and techniques that have been developed or more widely
applied since. In the second half of this paper, several of these
methods will be presented again, together with their results that
are of relevance to hydrogen embrittlement.

Chemical Composition

The chemical composition of the surface can be determined by energy- or wavelength- dispersive analysis of x-rays, Auger spectrometry, x-ray photoemission (also called ESCA) and secondary ion mass spectrometry (SIMS). Each technique has its specific characteristics making it the preferred one for a given situation. X-ray analysis (EDAX) is normally used in the scanning microscope. When the electron beam of the microscope (with 20 to 30 keV energy) enters the sample, one of its effects is to eject electrons from the core levels of atoms. These atoms return to their ground state by ejecting an x-ray of characteristic wavelength. These x-rays are analyzed and reveal the identity and concentration of the elements present. X-ray emission is sufficiently efficient only for elements heavier than neon. (The metallurgically important H, C, N, are not detected). Since x-rays travel through relatively thick solids, the spacial resolution is determined by the volume into which incoming electrons are scattered, a depth of about 1-2μm and a radius of ∿1μm.

In Auger spectrometry,[1] the action of the exciting beam is the same as with EDAX: it ejects core electrons, but the core hole is filled by the Auger process: an electron of higher energy drops into the hole and loses its energy by giving it to another electron that is ejected from the atom. The Auger process is most effective at low atom numbers, all elements except hydrogen and helium can be detected (the sensitivity is lower at the higher element numbers where it is best for EDAX). Auger is also more surface sensitive than EDAX; its depth of measurement is determined by the energy of the emitted electrons, it is about 0.3 nanometer (nm) at 20-80 eV energy and increases to 3 nm for 1 keV electrons. Lateral resolution is about 0.1μm at present state of the art.

Secondary Ion Mass Spectroscopy (SIMS) is extraordinarily sensitive: it can detect concentrations as low as 0.1 parts per million. Since the exciting beam consists of ions, it can be focused and directed to scan the surface of interest. The lateral resolution is about 1μm, the depth resolution is usually considered to be the first monolayer, but recent insight[2,3] into the sputtering process lets us think that it is somewhat deeper, namely about 1.2 nanometer per kilovolt energy of the incoming ions. The disadvantage of the method is that it is not quantitative. The ionization probability of secondary atoms changes, not only from one element to the other but also among the chemical states of one element, and these differences in sensitivity range over several orders of magnitude. SIMS is therefore the method of choice when one looks for trace elements. Birnbaum and coworkers,[4] for instance, have used this method to investigate grain boundary segregation in a nondestructive manner, by analyzing the composition of grainboundaries intersected by the surface of the material. SIMS is sensitive to all elements, including

hydrogen.   X-ray photo- emission spectroscopy (also called ESCA) is
probably the most widely used technique in chemistry.   It consists
of exciting electrons out of core levels by x-ray absorption and
measuring their energy.   The energy of a given peak in the energy
distribution identifies the element and its intensity provides the
concentration of the element; fine structure and small energy shifts
of the lines are used to determine the valence state of the element
under consideration.   Similar chemical information is increasingly
retrieved from the line-shape in Auger spectroscopy.

Further techniques of chemical analysis employ high-energy particles.   Rutherford backscattering[5] is based on the elastic collision of a high-energy nucleus with the atoms of the sample.   A simple elastic collision formula determines the mass of the scattering
atom:

$$E_{scatt} = E_{inc} (M-m)^2/(M+m)^2$$

Where m is the mass of the scattered ion, M the mass of the scattering atom in the material, E_{inc} is the energy of the incoming
ion and E_{scatt} is the energy of the scattered ion.   An example is
shown in Figure (1).   The method is quantitative; furthermore, the
variation of the amount of a given element as function of depth

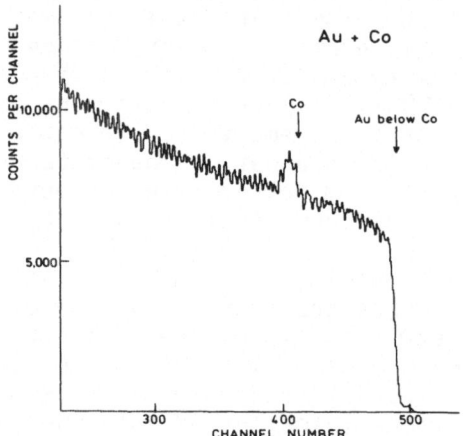

Figure 1.   Rutherford backscattering of 2 MeV $^4He^+$ ions from gold
covered by a thin layer of cobalt.   The ions scattered by cobalt
have less energy because Co is lighter than Au.   (Dearnaley[6]).

under the surface can be obtained by using the well-known energy
loss of the particle per unit distance traveled in the solid.   Certain elements (including hydrogen) can also be detected by nuclear
resonance reactions.[7]   Since the incoming particle must have a
precise energy to react, one can analyze the surface and any chosen

depth (up to 1μm) by giving the incoming nucleus just enough energy
to react with the atom of interest after having lost the amount of
energy corresponding to the chosen length of travel under the surface. Both methods require accelerators; they do not allow scanning
the surface but give quantitative, nondestructive information on the
depth-profile of the composition.

Depth profiling of the composition is done with the other
methods (Auger, XPS, SIMS) by sputter etching of the surface and
continuously analyzing the new exposed surface. Progress has been
made recently in the understanding of the kind of information one
obtains.[2] It has been found that the etching beam rapidly homoge-
nises the concentration in a layer corresponding to its projected
depth of penetration, namely about 1.2 nanometer per kilovolt energy of the incoming ions. Thus, a monolayer of a given element deposited on a solid would appear in the depth profile as 30 to 40% of
a monolayer (because the electrons from the underlying layers give a
reading of the elements belonging to the substrate) followed by an
exponential decay with decay constant equal to the depth of the
mixing layer.

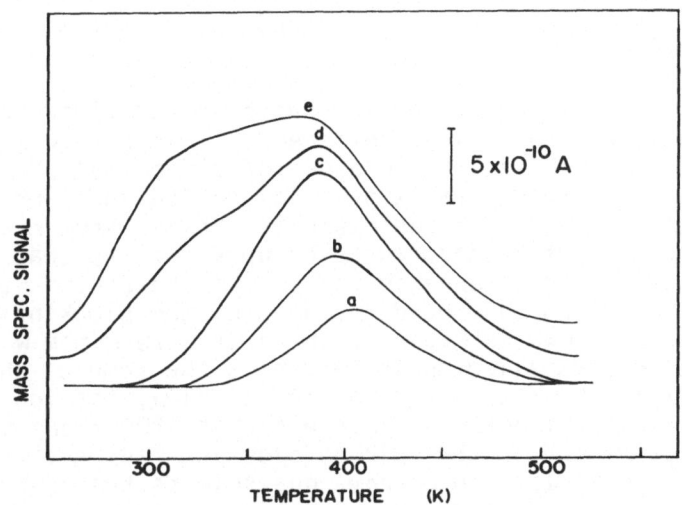

Figure 2. H_2 desorption from Fe (100) - coverage variation at 250K.
Exposures are (a) 1.10^{-6} torr sec. (b) $2.5\ 10^{-6}$, (c) 10^{-5}, (d) 2.5
10^{-5}, (e) 10^{-4} torr sec (Benziger and Madix[37]).

All the techniques can be used to determine the amount of adsorbed molecules in solid-gas interactions. By heating the surface
with a chosen rate (usually linear with time), one can detect the
heat and kinetics of desorption which often allow to deduce what
surface reaction is taking place.[8,9] Desorption can be measured

by the decrease in surface concentration with the methods described
above or by detecting the evolved molecules with a mass spectrometer
(Fig. 2).

Structure

Information on the atomic structure of the surface, on the identity of molecules and the modifications to their electronic structure due to adsorption forces, and finally on the vibration frequencies of surface atoms and adsorbed molecules, can be obtained by a
number of techniques. The quality of the information obtained depends, of course, on the type of sample that are analyzed. A sample
that has been simplified to the maximum extent, such as a pure single crystal with a flat and clean surface allows techniques using
nonfocused excitation beams and measuring the anistropy of the response.

Low Energy Electron Diffraction (LEED) allows a determination
of the surface structure including the position of adsorbed species.
The interpretation of LEED data still requires rather extensive
theoretical treatment because of the occurrence of multiple scattering caused by the high scattering cross section that makes LEED
surface sensitive.

Excellent information on surface and bulk structure can be obtained in single crystals by angle-dependent Rutherford scattering,
in particular by exploiting the channeling effect.[5,7,10] In the
latter, ions whose trajectory is parallel to low-index crystal axes
can travel long distances in the channels between atom rows without
being scattered. Surface structures or bulk defects that place
atoms into these channels enhance the scattering at the given angle.
Surface structures can also be determined by the blocking of back-
scattered ions as shown in Figure 3 where it determines whether the
distance between crystallographic layers is the same at the surface
as in the bulk.[11,12] The surface plane is (110); the ion beams im-
pinges in the [110] direction. Scattering at 120^0 along the [01-1]
direction is prevented in the bulk by blocking. By analyzing only
those ions that have lost no energy, one selects the ions scattered
in the first layer. If there is relaxation (Figure 3a), the blocking angle must be different from its bulk value. Since this angle
is 120^0 as for the bulk (Figure 3b top), one concludes that the interplanar distance at the surface and in the bulk are indeed the same.

Samples that are more complex do not allow such extensive characterization. EXAFS (Extended X-ray Absorption Fine Structure Spectroscopy) is a powerful new technique for structure measurements.[13]
It makes use of the fact that the cross section for the absorption
of x-rays by electron excitation out of core levels is proportional
to the amplitude of the excited electron wave at the site it occu-

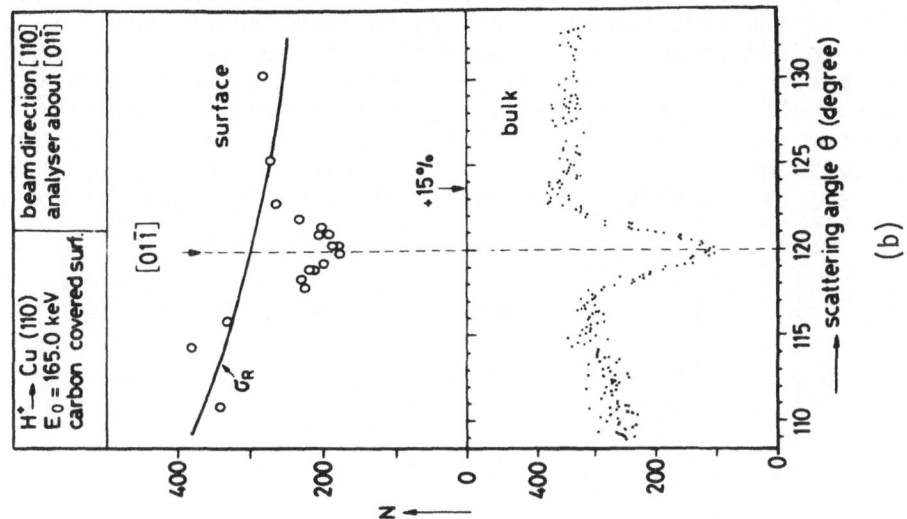

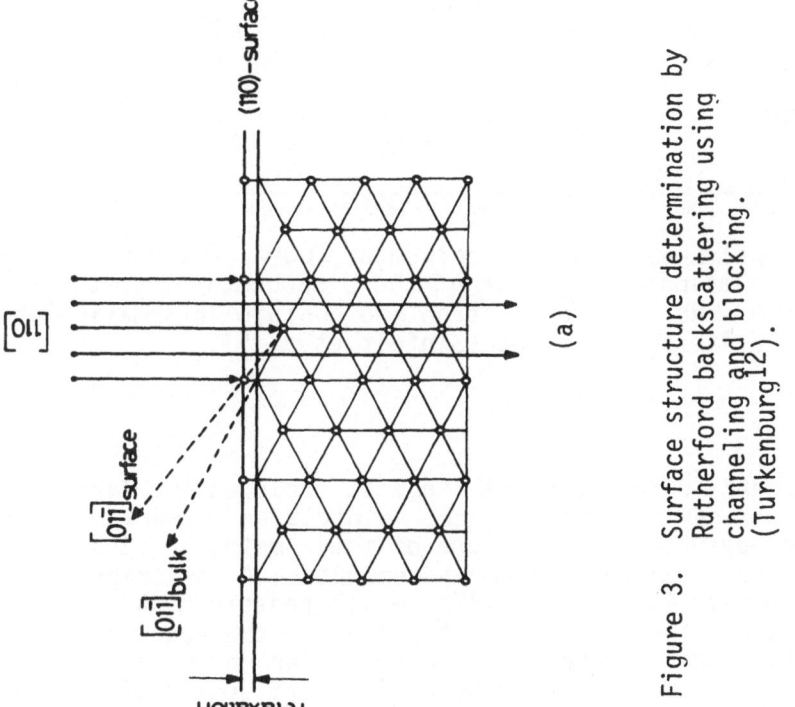

Figure 3.   Surface structure determination by
            Rutherford backscattering using
            channeling and blocking.
            (Turkenburg[12]).

pied before excitation. Since the excited electron propagates out
of its atom of origin, it is reflected by the neighbor atoms. The
reflection of the electron wave sets up an interference pattern.
Because of this diffraction, the amplitude of the excited electron
in the atom core will have maxima and minima as its wavelength and
energy vary. These oscillations are present in the absorption coefficient for x-rays near an absorption edge (Fig. 4). This method
has two advantages. First, it does not require a single crystal, it
works for any configuration. Secondly, it probes the environment of

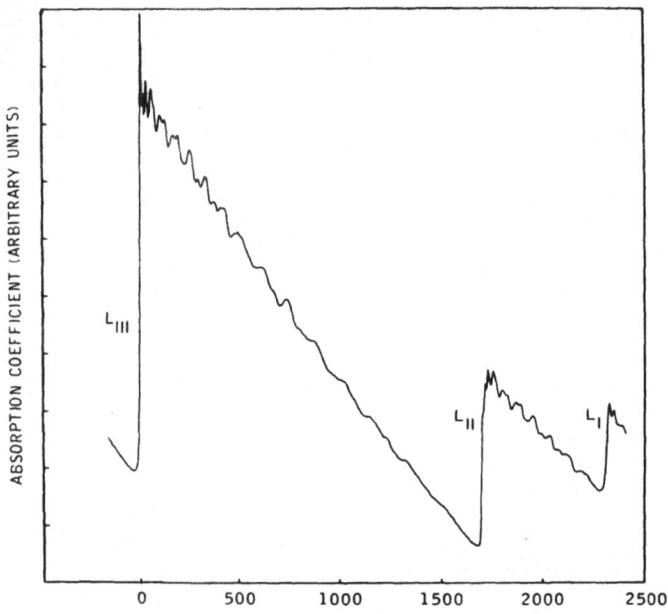

Figure 4. Typical L-absorption spectrum at 100°K of a 2.5μ platinum
foil. The oscillations above each of the L absorption edges are due
to diffraction of the excited electron and contain information about
the structure of its surroundings (Sinfelt et al[14]).

any chosen element in the material since it explores the fine structure at a given x-ray absorption edge. Sinfelt and coworkers,[14] for
instance, have measured the structure of small metallic clusters
dispersed in porous silica or alumina supports. A variation of this
technique is surface EXAFS (or SEXAFS) which rejects bulk events
because it measures the absorption of x-rays by the Auger electrons
the excited atoms emit in their return to the ground state.[15] These
techniques utilize a continuously turnable x-ray source of sufficient intensity. This requires either a specially developed rota-
ting anode or access to a synchrotron.

Spectroscopy

Angle-resolved ultraviolet photoemission (ARUPS) excited by polarized uv light has now been developed to the point that it allows
a rather complete mapping of the band structure of a clean metal,
including its bands of surface states.[16] It also permits the measurement of the electronic structure of adsorbed molecules, including
the identification and characterization of the molecular orbitals
participating in the chemisorption bond and the distortion of the
molecule due to adsorption. We are thus at the point of having a
new discipline at our disposal, namely experimental molecular orbital chemistry. An important recent development[17] is the introduction of low electron energy loss spectroscopy (LEELS) that allows
to measure the vibrational energy of adsorbed molecules with great
sensitivity (0.01 monolayer). This technique gives the surface
scientist the ability to probe the stiffness of chemical bonds and
its variation with adsorption or surface reaction and thereby to
gain information on the identity and properties of the adsorbed
molecules; it is the analog of infrared spectroscopy which has been
so useful to the chemist. Electron loss and optical infrared spectroscopy are now both used for adsorption studies;[18] the electronic
method is more sensitive (0.01 vs .1 monolayer) but the optical
method has much better energy resolution (0.1 meV vs 10 meV). Since
very low energy electrons are used, the resolution of the electronic
method presents problems with ferromagnetic materials.

Materials Modification

Finally, we should consider materials modification techniques,
for instance, ion implantation.[19,20] They are not analytical tools
as the ones we have just described, but these techniques, especially
ion implantation, produce such characteristic and well controllable
changes in composition and structure, and these changes have such
profound effects in the fracture occurring near the surface (such as
wear and fatigue life), that their thoughtful use for the purpose of
better understanding what facilitates or inhibits fracture under
stress might well be fruitful.

ADVANCES IN THE ELECTRONIC THEORY OF SURFACES AND ADSORPTION

Together with advances in experimental methods, we have wit-
nessed significant advances in the theory of surfaces and their interactions with the environment. One of the difficulties in the
theoretical treatment of surfaces lies in the breakdown of symmetry.
The problem of the solid is usually simplified by the periodicity
of the lattice and is most conveniently treated in momentum space;
hence the theory of the band structure of solids. The atom and the
molecule present the converse situation of a very local problem usu-

ally with spherical symmetry, or at least the superposition of atoms with spherical symmetry. When treating adsorption, one is forced to treat together these two problems with different symmetries; this is too complex to be treated exactly. Without attempting a complete overview we distinguish four essential directions from which this problem is approached. All theories are concerned with the properties of the valence electrons and how they determine the total energy, the cohesion of metals, their surface energy and the bonding of adsorbates.

One approach is an extension of solid state physics.[21] It treats the surface region as a two-dimensionally periodic system and adsorption is treated in terms of a full, periodic monolayer of atoms or molecules.[22] Another approach is more akin to chemistry: the system under study is approximated by a small cluster of atoms. This approach allows more flexibility in the surface structure and permits to explore, for instance, the bonding of any atom onto surface sites of different symmetries. In the cluster method, there are two main approaches: the self-consistent field (SCF-Xα) methods[23,24] permit a good approximation to the electron orbitals in a relatively complex system composed of transition metals with a large (≈ 10) number of d-electrons each, but does not yield good approximations for the bond strength; the valence-bond method[25,26] calculates the total energy with more rigor but is not as adept at treating large systems. A third approach[27-29] concentrates on many body aspects of the interaction of the electrons; the positive cores of the atoms are approximated by a constant space charge. This "jellium" model treats work functions[30] and their modification by adsorbates as well as surface energies. The fourth approach is empirical. It does not attempt to calculate the properties of a given system from first principles but endeavors, with a somewhat simplified treatment, to understand trends in the periodic table.[31,32] Such methods treat a number of quantities, such as metallic cohesion,[32] surface energy, work function, chemisorption bonds,[33] etc. Other theories treat the mechanisms that underlie the different measurement techniques we have reviewed.

SURFACE REACTIONS OF METALLURGICAL INTEREST

With the methods we have described, a considerable amount of knowledge on the properties of metallic surfaces and gas-surface interactions has been accumulated. This research is pertinent mostly to catalysis and electronic devices. A fair amount of this work has produced insight that is of relevance to the problems of fracture, the strength of materials and the way it is modified by the environment.

Let us look, for instance, at hydrogen embrittlement. It is well known that many structural materials, steels in particular, lose part of their ductility when exposed to hydrogen gas, that this effect is much stronger in hydrogen sulfide and that it is decreased or disappears when oxygen or SO_2 is mixed with the gas. We also know that hydrogen embrittlement is a room temperature phenomenon, that it disappears at low (T < 200 K) and moderately high (T ~400 K) temperatures.

From the studies of interaction of hydrogen and hydrogen sulfide with transition metals, we have learned the following: the heat of adsorption of hydrogen varies relatively little from one transition

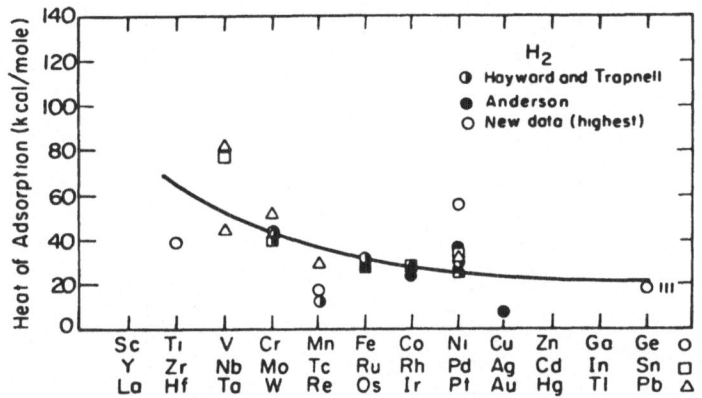

Figure 5. Heat of adsorption of hydrogen on transition metals (Toyoshima and Somorjai[34]).

metal to the other[34] (Fig. 5). It is largest (50 kcal/mole) at the left side of the periodic table (Ti, Zr, Hf) and decreases as the d-band is more populated to reach about 20 kcal for Ni, Pd, Pt. The strength of the chemisorption bond is equal to one-half the dissociation energy (110 kcal/mole) and one-half the adsorption energy (Figure 5) of hydrogen. This amounts to 80 kcal/mole at the left side and 65 kcal/mole at the right side of the transition metals. This variation is too small to explain by itself why certain materials are embrittled by hydrogen and others are not.

The sticking probability of hydrogen on many transition metals is relatively low (S_0 0.2). It appears that the presence of unoccupied d-orbitals just above the Fermi level (E_F) of the metal is essential for the efficient dissociation of molecular hydrogen. This has been shown recently by El-Batanouny and coworkers[35] who deposited thin layers of palladium on niobium (Fig. 6). At very low coverages, Pd forms an overlayer that is commensurate with the Nb substrate; photoemission shows that its d-band is fully occupied and lies at least 1.5 eV below E_F. At higher coverages, Pd forms an incommensurate (111) phase and the Fermi level moves into the d-

band. In this state, the surface adsorbs and dissociates hydrogen
with near unity probability on impact. In the commensurate Pd (110)
state with full d-band, the Pd fails to increase the sticking probability of hydrogen over its value for clean niobium.

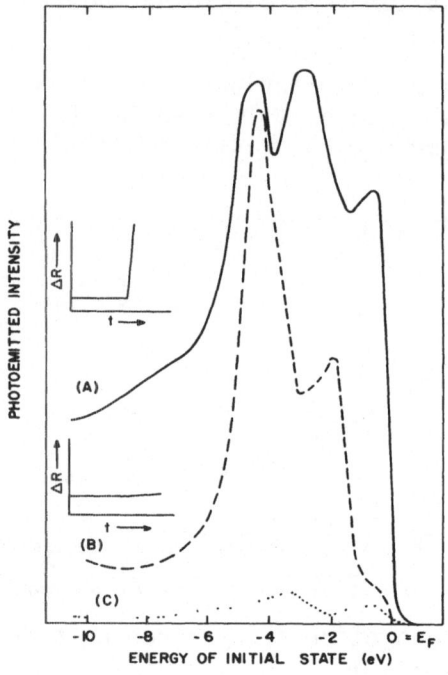

Figure 6. Effect of the electronic structure of palladium on the
kinetics of hydrogen dissociation. Pd was evaporated on a Nb (110)
surface. Hydrogen uptake is measured by the increase in resistivity
of the film. (a) Multilayer Pd film with (111) surface, Fermi level
in the d band, (b) Commensurate Pd (110) film (submonolayer, Fermi
level above d band, (c) Niobium. (El-Batanouny et al[35]).

     The prevention of hydrogen embrittlement when O_2 or SO_2 are mixed
with the hydrogen corresponds with experience in surface science.
To begin with, oxygen is much more strongly bound on transition
metals than hydrogen[34] (Fig. 7) and easily displaces the latter,
often through the synthesis of water. The temperature dependence of
hydrogen embrittlement (Fig. 8) presents an interesting problem.[36]
Surface measurements have shown that hydrogen is desorbed from iron
surfaces at relatively low temperatures;[37,38] (Fig. 2). But this
effect is not a satisfactory explanation for the temperature dependence of embrittlement since the disappearance of hydrogen embrittlement is very abrupt and occurs at temperatures that are too low for
desorption.[36] The disappearance of hydrogen embrittlement below
200 K could be attributed to the decrease of hydrogen diffusion.
Recent observations with photoelectron spectroscopy suggest an al-

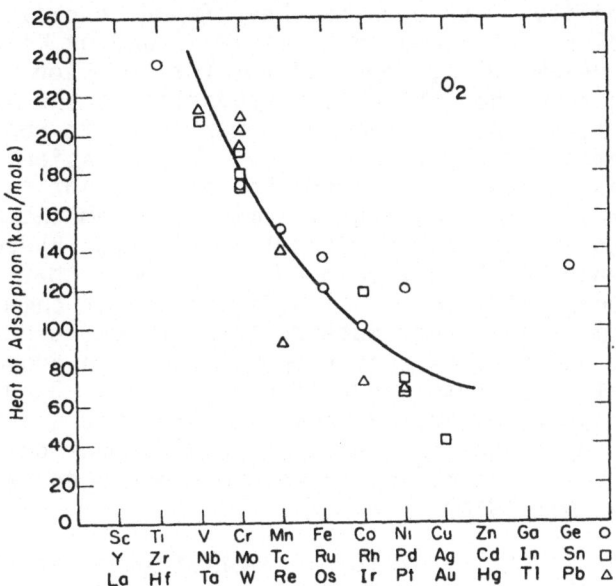

Figure 7. Heat of adsorption of oxygen on transition metals (Toyoshima and Somorjai[34]).

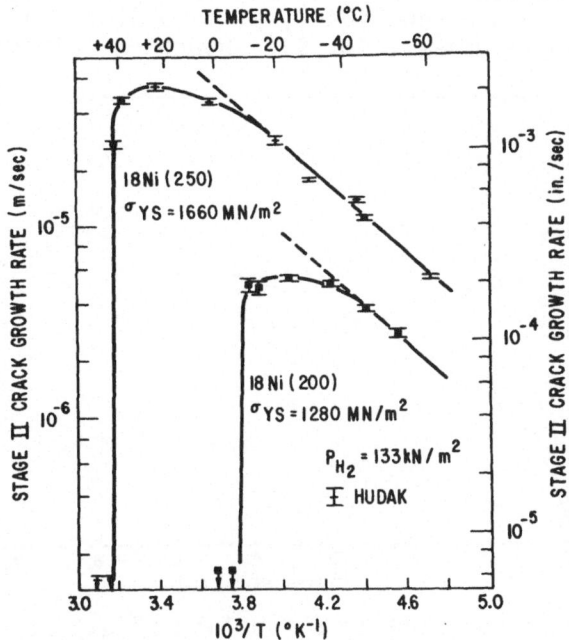

Figure 8. Temperature dependence of hydrogen embrittlement for two high strength steels (Gangloff and Wei[36]).

ternative possibility. Hydrogen has been found to exist in two
electronically different forms on nickel[39-41] and on Ni-Ti alloys[42].
At temperatures below 200 K, one observes the emission of electrons
that are excited from the orbital corresponding to the hydrogen
bound to the surface. As the sample is warmed up to room temperature, the photoemission spectrum changes. The emission from the hydrogen orbital is no longer observed; the only deviation from a
clean surface is a slight decrease in emission from the d-band of
the metal. These changes occur without any loss of hydrogen from
the metallic sample.[42] It has been hypothesized[41] that the low temperature form corresponds to the classic picture of chemisorption:
the hydrogen atom resides just outside the surface, probably on top
of a metal atom, and that at room temperature the hydrogen has moved
into or underneath the first layer of metal atoms. Because of the
similarity of transition temperatures, we raise the question whether
the temperature-dependence of hydrogen embrittlement does not correspond to this surface transition in adsorbed hydrogen rather than to
the onset of bulk diffusion.

The difference in the severity of embrittlement between H_2S and
H_2 is likewise reflected in surface chemistry. In general, a hydrogen molecule impacting on a transition metal has a small probability
of being adsorbed, while H_2S adsorbs or decomposes on transition
metals with unit probability. On iron, for instance, the sticking
probability of hydrogen is 0.16 in the most favorable case; namely,
a clean (110) surface below 140K[38], it is 0.03 on the Fe(100) surface at room temperature[37] (Fig. 9). Hydrogen sulfide interacts

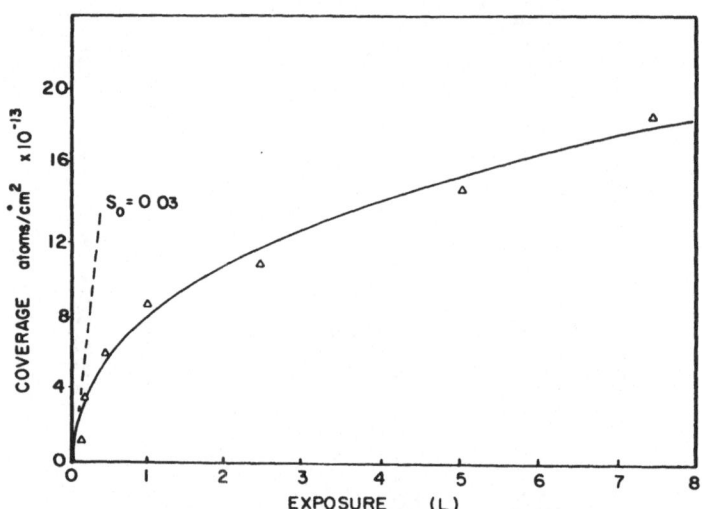

Figure 9. Hydrogen coverage as a function of exposure for H_2 on
Fe (100). The slope at the origin gives a probability $S_0 = 0.03$
for a molecule impinging on a clean surface to be adsorbed.
(Benziger and Madix[37]).

with the iron surface with unity probability on impact[43]. It is chemisorbed at low temperatures (150 K) and dissociates above 200 K to chemisorbed sulfur and hydrogen. The latter is desorbed as H_2. Considering the different reactivities on H_2 and H_2S on clean surfaces, we are lead to hypothesize that the difference between the two gases is even more pronounced with oxidized surfaces. No such studies have been undertaken yet at low gas pressures with the methods described above but the effect was clearly shown by Berkowitz and Horowitz[44] in an electrolytic system.

It has been argued[45] that a poison deposited on a metal surface has opposite effects on hydrogen penetration from the gas phase and from the electrolyte. In the gas phase, sulfur prevents the dissociation of H_2 and H penetration; in the electrolyte, sulfur allows the adsorption of protons but prevents their recombination and evolution as H_2 and thereby increases H penetration into the metal.

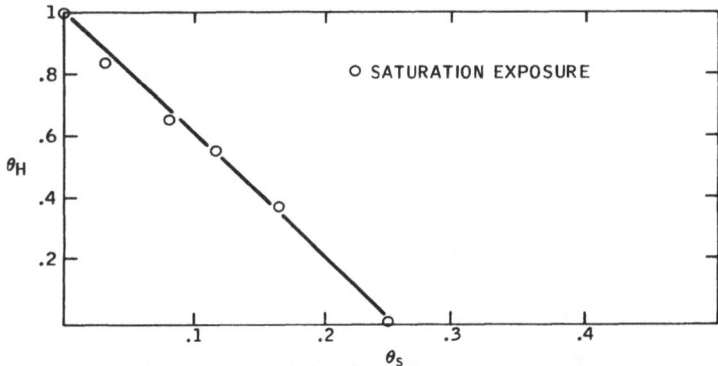

Figure 10. Maximum amount of hydrogen that can be adsorbed on a Ru(001) surface covered with partial monolayers of sulfur. θ_H:ratio of H to Ru atoms, θ_S:ratio of S to Ru atoms on the surface (Schwarz[47]).

It is often argued that H_2S is more effective in embrittling metals than hydrogen because it deposits sulfur onto the surface which poisons the hydrogen recombination reaction. Model experiments[46] of surface reaction poisoning have identified three poisoning mechanisms: (a) the contaminated surface prevents the adsorption of reagents, (b) the reagents adsorb but the weakened interaction with the surface increases the reaction temperature and lowers the desorption temperature so that desorption replaces reaction, (c) the strongly bonded and immobile poison atoms prevent the surface migration that is necessary for reaction. The suppression of H_2 adsorption by a sulfur-saturated iron surface has been reported[37]. The more detailed results available on Ru are instruc-

tive: The total amount of hydrogen that can be adsorbed rapidly decreases with increasing sulfur concentration (Fig. 10); sulfur acts
to block dissociation sites for hydrogen adsorption and recombination sites for hydrogen desorption[47] (Fig. 11), it also decreases
the binding energy of adsorbed hydrogen.   No hydrogen is adsorbed on
Ru when the sulfur concentration is one-fourth that of Ru surface
atoms or more.   H₂S, by contrast, is still adsorbed until the totally inert surface with a S/Ru ratio of one-half is achieved.   Above
a S/Ru ratio of 1/4, the surface sulfur inhibits the recombination

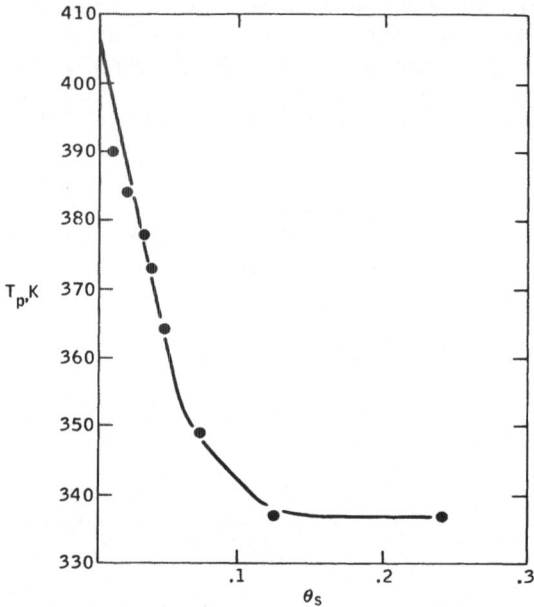

Figure 11.   Temperature of the desorption peak of hydrogen as a
function of the amount of sulfur present on the Ru(001) surface.
The temperature of this peak is proportional to the heat of desorption.   (J. Schwarz[47]).

of the H atoms deposited by H₂S so that higher temperatures are required for the evolution of H₂ and the penetration of H into the
metal is favored[48].   Figure 12 shows the change in work function of
a Ru(001) surface as the amount of sulfur (measured by Auger spectroscopy) on the surface increases.   Above 70% of the maximum sulfur
coverage, one obtains a decrease in work function due to the presence of co-adsorbed hydrogen.   Heating to 500K results in evolution
of H₂ and the increase in work function to the linear plot.   The
question then arises why embrittlement by H₂S should not stop once
the surface is saturated with sulfur.   Compatibility with surface
science experiments suggests that the adsorption responsible for em-

brittlement takes place at the crack tip itself where fresh surface
is exposed by the fracture.

   Much progress has been achieved in the theory of chemisorption,

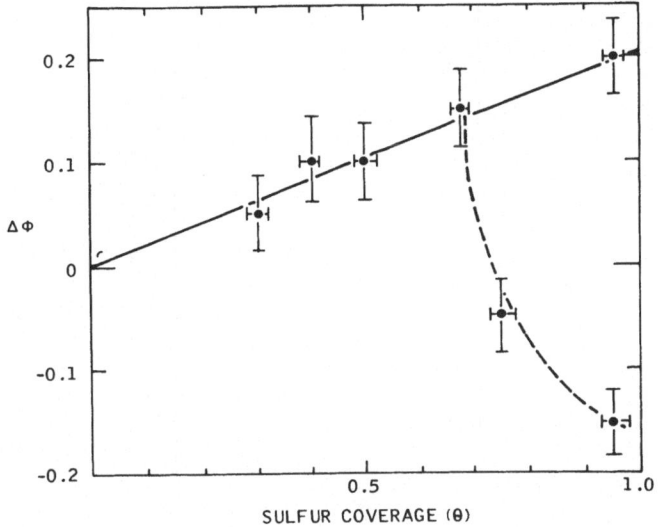

SULFUR COVERAGE (θ)

Figure 12.   Change in work function of the Ru(001) surface as a
function of amount of sulfur deposited by reaction of H_2S.   Dotted
line:values obtained at room temperature.   Full line:values due to
the presence of sulfur only. (Kelemen and Fischer[48]).

but controversy still remains. We note, in particular, the calculations of Briant and Messmer[24] who provided an explanation of temper
embrittlement by showing that a sulfur atom in a grain boundary
forms such strong bonds with its nickel neighbors that the nickel
electrons normally providing cohesion between the two neighboring
grains are entirely diverted towards the sulfur with the consequence of decreased intergrain cohesion. The bonding of hydrogen
to nickel, for instance, is attributed to the s-p electrons of the
metal by some,[26,49] and to the d electrons by others.[22]   Experiments
favor involvement of d electrons[39-41]. We are not aware of any calculation attempting to explore the basic assumption of the Oriani
model of hydrogen embrittlement, namely, that bonding of a transition metal to a hydrogen atom decreases the cohesion between metal
atoms. It seems that such an endeavor is within the capabilities
of present theories. Neither are there any measurements on solid
surfaces that are capable of exploring directly the effect adsorption has on the metallic cohesion in the vicinity of the adsorbate.
Examples abound that show qualitatively the influence of adsorption
on the forces between metal atoms. On the (100) surface of platinum, for instance, we know that the surface is reconstructed: the

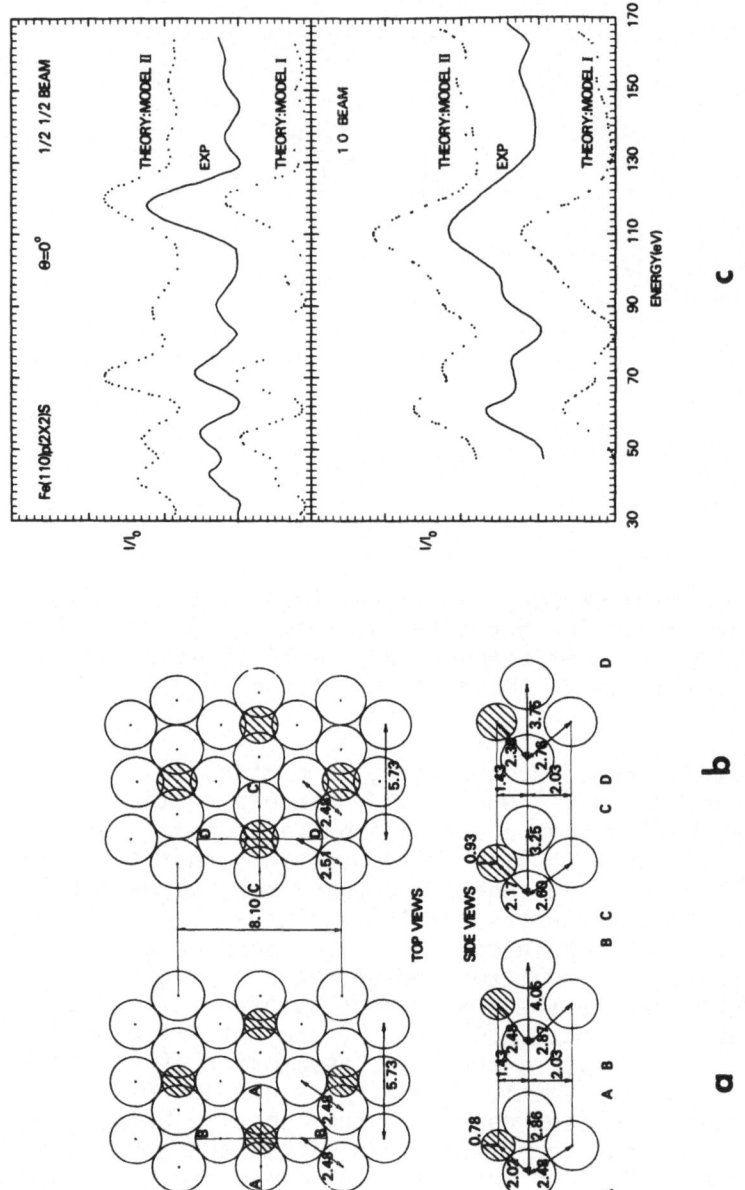

Figure 13. Changes in the positions of iron atoms on an Fe (110) surface induced by the adsorption of sulfur: a) Model of surface structure with unreconstructed substrate, b) model of surface structure with displaced Fe atoms, c) comparison of calculated and measured diffraction beam intensities showing that model b) is correct (after Shih, et. al[52]).

equilibrium structure of the top layer is not that of a parallel layer in the bulk;[51] LEED measurements are compatible with a hexagonal instead of the square arrangement of atoms. The adsorption of most gases pushes the surface platinum atoms back into positions corresponding to the bulk structure. Adsorption of CO and NO and its low temperature reaction to CO_2 and N_2 which both desorb leave us with a surface layer where the Pt atoms occupy the same positions as in the bulk. Another case is the adsorption of sulfur on Fe(100) which causes a displacement of the iron atoms of the topmost layer[52] (Fig. 13).

Surface scientists often compare the adsorption onto metallic surfaces with the corresponding reactions of metallorganic molecules.[53] These molecules do show a direct analog of the Oriani model of hydrogen embrittlement: when dicobalt octocarbonyl

$$
\begin{array}{c}
CO \diagdown \\
CO \diagdown \\
CO - Co - Co \\
CO \diagup \\
\end{array}
\begin{array}{c}
\diagup CO \\
\diagup CO \\
\diagdown CO \\
\diagdown CO \\
\end{array}
$$

is exposed to hydrogen and heated, one observes that hydrogen is adsorbed on the cobalt and dissociated: one H atom is attached to each Co.[54] This bond weakens the intermetallic bond which breaks when the temperature is sufficient. One observes the reaction

$$Co_2(CO)_8 + H_2 \rightleftharpoons 2\ HCo(CO)_4$$

$X\alpha$ calculations by Johnson[55] explain this phenomenon by population of antibonding orbitals due to adsorption of hydrogen.

A theoretical or experimental treatment of the modification of metallic cohesion by chemisorption of various gases could help us understand why hydrogen embrittles certain metals and not others despite the fact that its chemisorption bond varies little from one transition metal to the other and why for instance sulfur which creates a very strong bond with all transition metals[56] and causes temper embrittlement when dissolved in the solid, does not make an embrittling contribution of its own when present as H_2S environment but serves merely to increase the hydrogen activity.

The problem of the fracture of metals and its modification by the environment is not related to metallic cohesion in a trivial manner. To begin with, the total cohesive energy of a metal actually represents the heat of sublimation. The elastic constants and the phonon spectra are reflections of the curvature at the minimum of a plot of the cohesive energy as a function of interatomic distance. The force an isolated bond can withstand before the atoms are separated represents the largest slope of the binding potential.

So the calculation of the bond energy does not, per se, provide the
strength of a bond. Furthermore, a metal is a three dimensional aggregate of atoms which experiences local stress concentrations due
to variations in plastic deformation that are caused by the resistance to dislocation movement offered by the inhomogeneity of the
material. Most fracture processes are much more complex than those
described by the Griffith theory of newly created surface energy
because they are inevitably accompanied by plastic deformation.
Modern attempts to understand fracture on the atomic scale make use
of computer simulation. In these calculations, the cohesion of the
metal is simulated by a two-center potential between any two neighboring atoms. However, most electronic theories teach that such a
potential does not adequately describe cohesion in the metallic
bond. It seems that progress could be achieved in the understanding
of fracture if an approximation to metallic cohesion that is acceptable from the viewpoint of fundamental theory and tractable in the
computer modeling of fracture could be devised.

REFERENCES

1. T. E. Fischer in "Surface Effects in Crystal Plasticity,"
    R. M. Latanision and J. T. Fourie eds. Noordhoff Leyden,
    (1977), p. 127.
2. Z. L. Liau, B. Y. Tsaur and J. W. Mayer, J. Vac. Sci. Technol,
    16:121 (1979).
3. D. M. Heyes, M. Barber and J. H. R. Clarke, Surface Sci.
    105:225 (1981).
4. P. Williams, C. A. Evans Jr., M. L. Grosbeck and H. K.
    Birnbaum, Phys. Stat. Sol. A34:K97 (1976).
5. See for instance: "Methods and Phenomena: Methods of Surface
    Analysis," S. P. Wolsky and A. W. Czanderna eds., McGraw Hill,
    New York (1975).
6. G. Dearnaley, in "Ion Implantation Metallurgy," C. M. Preece
    and J. K. Hirvonen eds., AIME, New York (1980).
7. "Ion Beam Handbook for Material Analysis," J. W. Mayer and
    E. Rimini eds., Academic Press (1977).
8. P. A. Redhead, Vacuum 12:208 (1962).
9. J. L. Falconer and R. J. Madix, J. Cat. 48:262 (1977).
10. "Channeling," D. V. Morgan ed., J. Wiley and Sons, New York,
    (1973).
11. W. C. Turkenburg, W. Soszka, F. W. Saris, H. H. Kersten and
    B. G. Colenbranner, Nucl. Inst. and Meth. 132:587 (1976).
12. W. C. Turkenburg Doctoral Thesis, Univ. Amsterdam (1976).
13. D. E. Sayers, F. W. Lytle and E. A. Stern, Phys. Rev. Lett.
    27:1204 (1971), D. E. Sayers, E. A. Stern and F. W. Lytle,
    Phys. Rev. B35:584 (1975).
14. J. H. Sinfelt, G. H. Via and F. W. Lytle, J. Chem. Phys.
    72:4832 (1980).
15. P. H. Citrin, P. Eisenberger and R. C. Hewitt, Phys. Rev.

Lett. 45: 1948 (1980).

16. "Handbook of X-Ray and Ultraviolet Photoelectron Spectroscopy" D. Briggs ed., Heyden, London (1977).

17. H. Ibach in: Advances in Solid State Physics, Vol. 11 Pergamon (1971). H. Ibach, H. Hopster and B. Sexton, Appl. Surface Sci 1:1 (1977), H. Ibach, J. Vacuum Sci. Technol 15:407 (1978).

18. A. M. Bradshaw and F. M. Hoffmann, Surface Sci 72:513 (1978).

19. "Ion Implantation Metallurgy," C. M. Preece and J. K. Hirvonen eds., The Metall. Soc. of AIME, New York, (1980).

20. "Treatise on Materials Science and Technology, Vol. 18 Ion Implantation" J. K. Hirvonen ed., Academic Press, New York (1980).

21. K. M. Ho, M. L. Cohen and M. Schlüter, Phys. Rev. B15:3999 (1977).

22. S. G. Louie, Phys. Rev. Lett. 42:476 (1979).

23. K. H. Johnson, Int. J. Quantum Chem. 11:39 (1977), also these proceedings.

24. C. L. Briant and R. P. Messmer, Phil. Mag. 342:569 (1980).

25. C. F. Melius, B. D. Olafson and W. A. Goddard III., Chem. Phys. Lett. 28:457 (1974).

26. T. H. Upton and W. A. Goddard, III, Phys. Rev. Lett. 42:472 (1979).

27. W. Kohn and L. J. Sham, Phys. Rev. 140, A 1133 (1965); P. Hohenberg and W. Kohn, Phys. Rev. B136:864 (1964).

28. V. L. Moruzzi, A. R. Williams and J. F. Janak, Phys. Rev. B15:2854 (1977).

29. C. D. Gelatt, E. Ehrenreich and R. E. Watson, Phys. Rev. B15:1613 (1977).

30. N.D. Lang, Solid State Comm. 7, 1047 (1969).

31. A. R. Miedema, Philips Tech. Rev. 36:217 (1976).

32. D. G. Pettifor CALPHAD 4:305 (1977).

33. A. J. Wilson and C. M. Varma, Phys. Rev. B22:3805 (1980).

34. J. Toyoshima and G. A. Somorjai, Catal. Rev. Sci., Eng. 19:105, (1979).

35. M. El-Batanouny, M. Strongin, G. P. Williams and J. Colbert, Phys. Rev. Lett. 46:269 (1981).

36. R. P. Gangloff and R. P. Wei, Metallurgical Trans. 8A:1043 (1977).

37. J. Benziger and R. J. Madix, Surface Sci. 94:119 (1980).

38. F. Boszo, G. Ertl, M. Grunze and M. Weiss, Appl. of Surf. Sci. 1:103 (1977).

39. J. E. Demuth, Surf. Sci. 65:369 (1977).

40. F. J. Himpsel, J. A. Knapp and D. E. Eastman, Phys. Rev. B19:2873 (1979).

41. W. Eberhardt, F. Greuter and E. W. Plummer, Phys. Rev. Lett. 46:1085 (1981).

42. T. E. Fischer, S. R. Kelemen and R. S. Polizzotti, J. Catal. in press.

43. S. R. Kelemen, private communication.

44. B. J. Berkowitz and H. H. Horowitz, J. Electrochem Soc. in
     press.
45. B. J. Berkowitz, J. J. Burton, C. R. Helms and R. S.
     Polizzotti, Scripta Met. 10:871 (1976).
46. T. E. Fischer and S. R. Kelemen, J. Catal. 53:24 (1978).
47. J. A. Schwarz, Surface Sci. 87:525 (1979).
48. S. R. Kelemen and T. E. Fischer, Surface Sci. 87:53 (1979).
49. P. Cremaschi and J. L. Whitten, Phys. Rev. Lett. 46:1242
     (1981).
50. R. A. Oriani, Ber. d. Bunsen, Ges. f. Phys. Chem. 76:301
     (1971), Ann. Rev. Mater. Sci. 8:327 (1978).
51. C. R. Helms, H. P. Bonzel and S. Kelemen, J. Chem. Phys.
     65:1773 (1976).
52. H. D. Shih, F. Jona, D. W. Jepsen and P. M. Marcus, Phys. Rev.
     Lett. 46:731 (1981).
53. R. Ugo, Catal. Rev. Sci. Eng. 11:225 (1975).
54. E. D. Muettertis, Transition Metal Hydrides, (Dekker, New York
     1971).
55. K. H. Johnson, unpublished.
56. J. Oudar, Mater. Sci. and Eng. 42:101 (1980).

DISCUSSION

Comment by J.W. Gadzuk:

As discussed by Dr. Fischer, the past decade has been characterized by the development of many new techniques for probing the
electronic and structural properties of well-defined solid surfaces.
EXAFS shows particular promise in the case of surfaces in which long
range order is not present, possibly even for fracture surfaces.
However, a cautionary warning note must be added. As it stands now,
in order to analyze the absorption fine structure in terms of local
geometry, all the absorbing atoms must be in equivalent geometrical
sites; otherwise the fine structure from different sites will average
out to a smooth curve. This condition may be quite hard to satisfy
for ill-defined fracture surfaces in engineering situations.

Reply:

Dr. Gadzuk is quite right. The restrictions for the applicability and interpretation of EXAFS (and SEXAFS), in fact, are the
same as for classic X-ray diffraction except that the structure
around any chosen element (even if only in 1% concentration) can
be studied by EXAFS and not by classical diffraction. The latter,
for instance, would be unable to study the structure of finely
dispersed metal particles on a porous silica support (which EXAFS
can), or the crystallographic position of sulfur in a fracture
surface (amenable to SEXAFS). As Dr. Gadzuk points out, if these
metal particles or the adsorbate were present in several dissimilar
structures, interpretation would be difficult, or impossible. Nearest
neighbor distance, if everywhere the same, could still be obtained.

Comment by J.O'M. Bockris:

Fischer's list of spectroscopic methods for looking at surfaces is useful but its applicability to the wet interfaces we are looking at is doubtful. One must wash the surface, dry it, reduce to low temperature and go to high vacuum. Much could happen, e.g., to the water in an oxide film on the way. Ellipsometric methods are surely better for the relevant wet interfaces. By means of a combination of ellipsometry and reflectance methods, one could evaluate the absorption coefficient of the surface layer; U.V. and visible ranges would be available in situ. Raman resonance radiation may also prove to be a tool superior to electron spectroscopy for surfaces in solution.

Reply:

This is certainly true. The methods that are described in this paper have been selected because they were developed recently and add their exploratory capacity to our tools. The whole set of tools now available for the investigation of the interface of a solid with a low pressure gas allows to measure almost every quantity of interest and is particularly well suited to interact with the new electronic theories. Thus, we feel that the gaseous case is ready for rapid progress. It would be wrong to imagine that the wet interfaces are the only relevant ones. Interaction of metals with gaseous hydrogen, H_2S or other gases will present an important challenge, for example, to the new energy technologies that must be developed.

Comment by I.M. Bernstein:

Can you comment on progress made in relating the behavior of external crystal surfaces to internal surfaces such as grain boundaries and interfaces. In particular, I wonder how and if we can extrapolate the behavior of hydrogen and sulfur on external, clean, crystallographic surfaces to that of an internal interface. Is, for example, the valence state of sulfur the same in both cases?

Reply:

There are some qualitative analogies. For instance, it is generally accepted that heats of segregation of elements to grain boundaries are a fraction of their value to free surfaces, but the trends and kinetics are the same. I am not aware of an experimental or theoretical determination of the valence state of sulfur at grain boundaries.

Comment by K. Christmann:

In your paper you reported the existence of two electronically different forms of hydrogen adsorbed on nickel, and you referred to a very recent paper from Plummer's group (PRL 46, 1085 (1981)). The authors interpret their UV photoemission spectra (which do not show any alteration as compared to the clean surfaces) as evidence for hydrogen atoms that have moved below the surface and are therefore no longer seen with UPS. In my opinion this interpretation cannot be correct: Our isotherm data for all the systems Ni (111)/H, Pd (111)/H and Pt (111)/H clearly show that under the temperature and pressure conditions used by Plummer et al., H desorption is the dominating process and there would hardly be any hydrogen left in or on the surface.

Reply:

I agree with Dr. Christmann for the case of platinum, at least. We have not repeated the work of Plummer and coworkers[41], but we tend to believe the work of the IBM group[39,40]. Also, our own work[42] shows two electronic forms of hydrogen at low and room temperatures, and we know precisely how much hydrogen is desorbed (about 1/2) and how much remains when the temperature is raised. We are positive that hydrogen is present in the room temperature form on Ni_3Ti. We have no reason to doubt the same to be true for Ni.

Comment by R.A. Oriani:

It is, of course, interesting and scientifically important to investigate the electronic change accompanying the adsorption of hydrogen upon a clean metal surface, the interaction of hydrogen with a metal cluster, or the reaction of hydrogen with a molecule. However, what basis does one have for hoping that such calculations will give one insight into the electronic interactions of hydrogen in the lattice of a metal, or at a grain boundary, leading to decohesion and/or a change in the resistance to dislocation motion? After all, it is the details that one must be interested in, since the very fact that adsorption occurs with a lowering of the surface free energy necessarily means that the average of the force for separation of two portions of a body to create free surfaces is lowered by adsorption. Adsorption of oxygen must also have this effect, yet oxygen does not cause cracking.

Reply:

In order to master the mechanisms by which hydrogen embrittles materials, we must develop further experimental and theoretical methods. One of the possible scenarios starts with surface phenomena, because that can be measured more completely than others, being accessible to probes. The theoretical chemistry of electron orbitals and bonding forces is being developed rapidly but it needs frequent testing by experiment because of the many approximations that must be made. Once these theories will have reached a sufficient level of confidence and predictive power they should be able to treat the local effect of hydrogen in the bulk and at grain boundaries without insurmountable obstacles, provided of course that we have a good idea of the structure and composition of these boundaries. An understanding the differences between the action of oxygen and hydrogen will undoubtedly be a natural result of this work.

Comment by S.P. Lynch:

There does not seem to be general agreement on whether hydrogen embrittlement is associated with adsorbed hydrogen atoms or absorbed hydrogen (solute) atoms (or both). However, the close similarities between HE and LME (where only adsorption occurs) suggest that the effects of adsorption are important.

Reply:

The difficulty of deciding whether hydrogen embrittlement results from the decrease of the cohesion force at the surface or in the bulk results from the fact that a case can be made for either. Surface science studies are relevant for both possibilities because adsorption, dissociation, and penetration of hydrogen and their modification by other gases present (e.g., O_2) are surface effects. The work of Vehoff, Rothe, and Neumann (these proceedings) is certainly compatible with a pure surface effect, but not exclusively.

Comment by J.P. Fidelle:

The abrupt decrease of H.G.E. found in maraging steels as temperature increases is repeatedly quoted, but has not been repeatedly demonstrated by independent experiments since its first publication.

In fact, it is extremely surprising considering the particularly high sensitivity of maraging steels to HE. The possibility of spurious (interfering) effects remains and unless unambiguously confirmed by future experiments, the above result should be considered with care... at least for industrial applications.

Reply (by R.P. Gangloff for T.E. Fischer):

The following facts suggest that the abrupt decrease in the susceptibility of precracked specimens of 18Ni maraging steel to gaseous hydrogen embrittlement is real and reproducible.

1. The phenomenon has been observed for 18Ni (300)[1], 18Ni (250)[2,3] and 18Ni (200)[3] steels by three investigators working independently over a period of almost ten years.

2. The critical temperature for the abrupt transition increased systematically with increasing material strength and with

increasing hydrogen pressure. The latter phenomenon was related to the critical temperature-pressure characteristic of a two dimensional phase transition[3,4].

3. The low to high temperature transition in crack growth rate was correlated [5] with an intergranular to transgranular crack path transition, indicative of a true temperature effect on the hydrogen-steel interaction. (The transgranular mode was chemically assisted and clearly not dimpled rupture.)

4. As stated in reference 3, propagating cracks were arrested by small (<10°C) temperature increases, and stationary cracks were re-initiated through small temperature decreases. This temperature cycling sequence, performed on the same sample, suggests that the transition effect is not due to adsorption of impurities from the environment and in competition with H_2.

5. The experimental apparatus employed to investigate the abrupt transition in maraging steel was used to reproduce the more gradual high temperature decrease in hydrogen susceptibility for 4130 alloy steel as reported by Nelson and Williams.

Many highly hydrogen sensitive steels are not embrittled in H_2/O_2 mixtures. Analogously, the abrupt decrease in sensitivity for maraging steels in pure H_2 is not surprising if one envisions a model based on precluding hydrogen from adsorbing on the steel above a critical temperature. The exact details of such a process remain, however, a mystery[4].

The author agrees that any industrial application of a maraging steel in a hydrogeneous environment should be approached with extreme caution.

References in Reply:

(1) P.S. Pao and R.P. Wei, Scripta Met., Vol. 11, p. 515 (1977).
(2) S.J. Hudak and R.P. Wei, Met. Trans., Vol. 7A, p. 235 (1976).
(3) R.P. Gangloff and R.P. Wei, Met. Trans., Vol. 8A, p. 1043 (1977).
(4) N.H. Chan, K. Klier and R.P. Wei, Scripta Met., Vol. 12, p. 1043 (1978).
(5) R.P. Gangloff and R.P. Wei, ASTM STP 645, p. 87 (1978).

FRACTURE -- APPLYING THE BREAKS

A.R.C. Westwood and J.R. Pickens

Martin Marietta Laboratories
1450 South Rolling Road
Baltimore, MD   21227 U.S.A.

INTRODUCTION

"Quake kills at least 35 in California," reads the headline in
the Baltimore Sun for 10 February 1971. "Grain elevator explodes,
causing 30 million dollars worth of damage" is the news from Corpus
Christi, Texas, for 7 April 1981 (Fig. 1). These examples represent one way in which fracture enters our daily lives. To most
people the word fracture connotes calamities and disasters; the
roof of a public building collapsed by the weight of ice and snow
(Fig. 2); a window smashed by a carelessly kicked ball; a heli-
copter brought down by a fatigued rotor; a finger broken in a
basketball game. Yet a world without fracture would be quite
different from that in which we live. Unbreakable eggshells
would mean not only no omelettes or crepes but also, presumably,
no birds. Unbreakable ice would leave many northern ports inac-
cessible through much of the winter. And, though many of us
would willingly forgo mowing the lawn if grass were uncuttable,
being unable to bite into a crisp apple or a succulent pear would
certainly be a pleasure sorely missed.

In fact, fracture has served man well through the ages. Over
two and a half million years ago, prehistoric man relied on flint
knapping to produce the tools he needed to protect himself, to
obtain food, and to hew wood for fires.[1] Modern man relies on
industrial-scale fracture processes to blast the ores from which
he extracts metals; to crush the rocks he uses in construction; to
drill for oil, water and gas; to grind cement clinker, cocoa and
cosmetics; and to machine materials into useful shapes.

Figure 1.   Damage resulting from grain elevator explosion at Corpus
            Christi, Texas.   UPI.

Figure 2.   Roof of auditorium collapsed by snow load.   UPI.

In this paper, fracture in the home, art, and industry are
briefly surveyed; fracture in blasting, rock drilling, comminution and machining discussed; and some opportunities for research
and application suggested.

FRACTURE...IN EVERYDAY LIFE

Fracture is immensely useful around the home. We rely on
it to crack nuts, grind coffee, crush ice, slice potatoes, chew
cereals and cut bread. Without fracture we would be forced to
subsist largely on water-soluble materials, severely limiting any
tendency towards gourmandise. Indeed, food-"fracturing" teeth
were critical to survival in earlier times, as recounted by the
novelist Michener in "The Covenant."[2]

Simple fracturing tools are commonplace, such as scissors,
razors and shears. In recent years, small motors have been attached to these to speed their action and reduce manual labor. So
nowadays many of us use electric shavers on our chins (or legs),
power mowers to cut the lawn, electric scissors to cut out the
components of a suit or dress, and electric knives to slice the
roast beef or turkey. We may also use a blending machine, whose
whirling blades can reduce an orange or a tomato to pulp in just a
few seconds...an extremely useful tool for the cook whose dinner
menu ranges from vichyssoise to chocolate mousse.

Our pursuit of the life of ease is much dependent on our capability for causing and controlling fracture. Cans of sterilized
food require steel cutting can-openers. Some containers are even
designed with built-in paths of easy fracture...the aluminum pop-
top beverage can. The "safety glass" auto windshield that shat-
ters on impact into small, relatively innocuous pieces instead of
life-threatening pointed slivers, is another example of usefully
controlled fracture.

Consider also the many roles of fracture in medical care, from
the grinding operations used to produce small and readily dissolved
particles of aspirin to the high-speed, water-jet-driven drills of
the dentist. Surgery usually involves cutting with a scalpel,
while brain surgery requires an initial rotary trepanning of the
skull. Physicians are also called upon to repair fractures, and
quite ingenious techniques are being developed for this, including
the use of comminuted and reconstituted bone to remake the ball
joint on the femur.

Sharpening a pencil involves fracture, and so does making a
pencil mark. And when the written words do not convey one's
thoughts as explicitly as desired, so does tearing up the paper.

Grinding lenses so that we can see sufficiently well to be productive is another example of fracture usefully impacting our daily lives. And grinding appropriately scented or colored particles to a sufficient fineness that they can be incorporated in cosmetics represents another socially accepted application, helping both actors and housewives come closer to the images they wish to project.

On a finer scale, the fading of color in fabrics is an example of photo-assisted fracture, in this case of bonds in complex organic molecules. To illustrate, when daylight (or a detergent) causes the Br and CH$_3$ radicals to break off the molecule shown in Fig. 3, the color changes from violet to red.

Earthquakes are at the opposite end of the scale. In California, for instance, quakes of destructive magnitude have occurred about once each year for the past fifty years. Fortunately, they are usually located some distance from a population center, but when they are not, as in San Francisco in 1906 (Fig. 4),[3] or in Avilignia, Italy, in 1980, they can be terrible and devastating.

Because fracture so routinely impacts our lives, its consequences...breaks or cracks...have entered the English language in a variety of usages. Many popular songs recount sagas of broken hearts, and actors going on stage wish each other luck with the phrase, "Break a Leg." There are daybreaks and nervous breakdowns; the crack of the whip, and the wise-crack. Even songwriter Neil Sedaka has commented that "Breaking Up Is Hard To Do." No doubt he was thinking of the crushed stone industry at the time. And, as is well known, the language of international science is "broken" English.

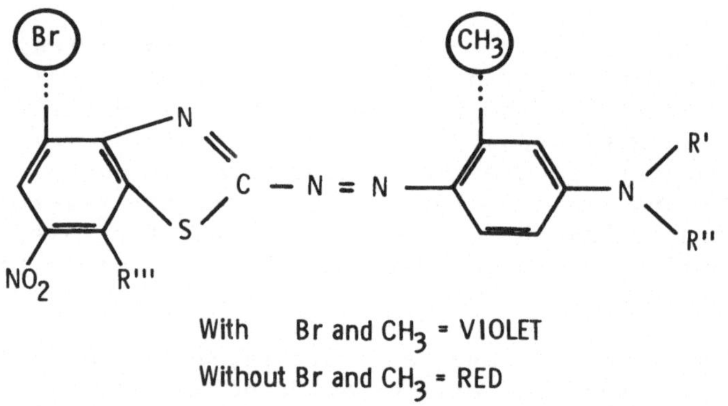

With   Br and CH$_3$ = VIOLET

Without Br and CH$_3$ = RED

Figure 3. Fracturing the Br and CH$_3$ ligands changes the color of this dye molecule from violet to red.

Figure 4.   City Hall, San Francisco, after the 1906 earthquake.[3]

...IN ART

     The most elegant examples of fracture applied to art occur in
sculpting. On the grandiose scale, there is the Great Sphinx of
Egypt, and the heads of four U.S. presidents carved into Mount
Rushmore. Nearer life size, Michaelangelo's masterpiece "David",
the Venus de Milo, and the superb traceries of many European
cathedrals are examples of the classic approach. The sculpting of
large wooden figureheads for ships was popular in the 19th century.
Exquisite smaller pieces are carved in jade, ivory and other more
precious materials.   In fact, almost every conceivable material
has been used for sculpting at some time...from lava to plastics.
In general, however, a fine-grained, homogeneous substance, with
no inherent flaws, is preferred. Thus, for example, marble from
Carrara, and mahogany from Cuba are considered materials of choice.

     Fracture is also used to reveal the intrinsic beauty and op-
tical symmetry of gemstones although, contrary to popular belief,

diamonds are cleaved only as a last resort. More often a diamond
is sawn to shape, and its facets (58 in the standard "brilliant"
cut) are produced by grinding and polishing. Grinding is also
used in the preparation of pigments, paints and inks, and in the
manufacture of cut glass pieces, such as Waterford crystal.

Sometimes, a fracture pattern is itself considered aesthet-
ically appealing, as in the crackle glazes employed in pottery
(Fig. 5). Alternatively, a solid may be fractured along some
predetermined path and then reconstructed to produce a work of
art, as in stained glass windows and mosaics.

...IN INDUSTRY

Fracture is a disaster for a few industries, an instrument
for others, and the basis of several. The catastrophe at the
Flixborough chemical plant, in June 1974, appears to have involved
such fracture-facilitating phenomena as embrittlement of steel by
liquid zinc, and creep cavitation.[5] The Sea Gem oil drilling
rig that collapsed in the North Sea in December 1965 failed because of fractures in the bars joining the legs and platform.[5]
Such large-scale fractures impact the public dramatically via
newspaper headlines. However, tiny fractures also are important,
especially the hosts of small and essentially uncontrolled fractures involved in mining and manufacturing processes. Indeed, the
survival and profitability of a crushed stone company depends
almost entirely on the relative efficiency with which it uses
fracture. It drills, blasts, crushes and comminutes to produce

Figure 5.   Earthenware bowl with rutile and iron glazes entitled
            "Desert Sun."[4]

large rocks for breakwaters, stone for road foundations, aggregates
for concrete, gravel for road surfacing, pulverized limestone for
agriculture and pollution control, and sand for making glass.
Cement companies grind raw materials to a powder before feeding
them to the kiln, and then grind the resulting clinker to generate
portland cement of sufficient fineness to produce adequate strength
within a few hours of being mixed with water. The finer the particles, the more rapidly is strength developed.

    Fracture by comminution is used in the food industry to grind
cereals and spices, the dyestuff industry to produce particles
small enough to dissolve quickly and so provide adequate dyeing
rates, the power-generating industry to pulverize coal, and the
extractive metallurgy industry to separate useful minerals from
gangue.

    Fracturing by means of a rotating or reciprocating bits is
used to drill holes to emplace explosives, or to extract gas and
oil. Jackhammers are used to break up degraded roads, while small
buildings are efficiently fractured by impacting with a heavy
ball swung by a crane.

    By far the largest users of fracture, however, are the manu-
facturing industries, where the cutting, machining, grinding and
polishing of metals, semiconductors, ceramics, and glasses are
involved in producing the components, devices, equipment and uten-
sils of modern man. The cost of such fracturing operations in the
U.S. alone amounts to about 5% of the Gross National Product,
i.e., for 1980, ~ $125 x 10^9.

APPLIED FRACTURE

    Much scientific effort has been profitably expended develop-
ing mechanistic understanding of fracture phenomena to provide
logical bases for selecting the materials and design parameters
to build safe load-bearing structures. Much less attention has
been devoted to clarifying factors which facilitate fracture,
e.g., tool shape, the application of chemically active environments, optimized fragmentation systems, etc., when fracture is the
desired objective. This area of study simply has not attracted
the interest of many of the best scientific minds, in spite of its
technical challenge and economic importance.

    The potential for improving the efficiency of most industrial
fracturing processes is substantial. Comminution, for example, is
extremely inefficient, with only a small percentage of the energy
consumed being expended in the generation of new surface area.
The financial returns from modest improvements in efficiency would
certainly be worthwhile...a few percent increase in the efficiency

of machining operations, utilized worldwide, would save many bil-
lions of dollars each year. Unfortunately, such efforts as have
been made to exploit, for example, environment-sensitive fracture
behavior in practical engineering situations have been both limited
and not too successful. This is largely because the fundamental
mechanisms were incompletely or incorrectly understood. Classic
examples of this are the Rebinder-type effects (adsorption-induced
changes in near-surface mechanical behavior); discovered in 1928,
yet still not widely exploited in industry despite their great potential. Such practical uses as have occurred have been pioneered
almost exclusively by Soviet workers. But, even in the U.S.S.R.,
the level of effort applied has been small in relation to the
conceivable returns.

     Let us now turn our attention to the role of fracture in a
variety of important industrial processes.

... IN BLASTING

     Blasting is an effective means of fracturing the earth's sur-
face so as to remove useful ores and minerals in fragment sizes
appropriate for subsequent economical processing. In the past,
blasting parameters were established largely by trial and error.
Recently, however, the mechanisms of explosively-induced fracture
have begun to be studied (see Fig. 6) with a view to predictably
controlling fragment size and distribution, minimizing ground
vibration levels, and reducing dangerous fly rock.[6-8] Two basic
theories have been developed. The first is that fracture results
from interactions between the rapidly propagating stress waves
produced by the explosion and the structure of the rock being
blasted.[6-8] This "stress-wave" theory, an outgrowth of work by
Sharpe,[9] postulated in its original conception that spallation
generated by reflection of intense stress waves (the compressional
or "P" waves) from a free surface is very important in fragmentation.[10,11] The problem with this theory as initially stated is
that only 3-10% of the energy of the explosive is transmitted in
such stress waves.[12,13] A second theory, developed from the work
of Porter,[14] proposes that the pressure of gases produced by the
explosive charge propagates the radial cracks formed initially by
the shock wave generated on detonation (Fig. 6a).[13] More recently,
Fourney and co-workers[6] have undertaken high-speed photoelastic
studies of explosively stressed polymer samples containing structural discontinuities simulating those found in rock, and identified another mechanism of fracture consistent with the stress
wave theory...a mode II shear failure that occurs when the P wave
interacts with simulated rock joints (Fig. 6b). Now, it is important to know which of these various possibilities, or some
combination of them, is valid for naturally occurring rocks,
because the blasting strategies that would be derived as a consequence are quite different.

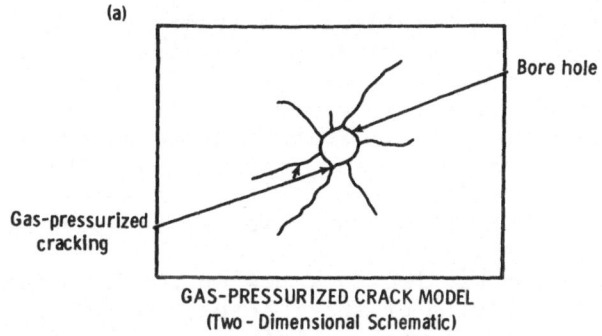

GAS-PRESSURIZED CRACK MODEL
(Two - Dimensional Schematic)

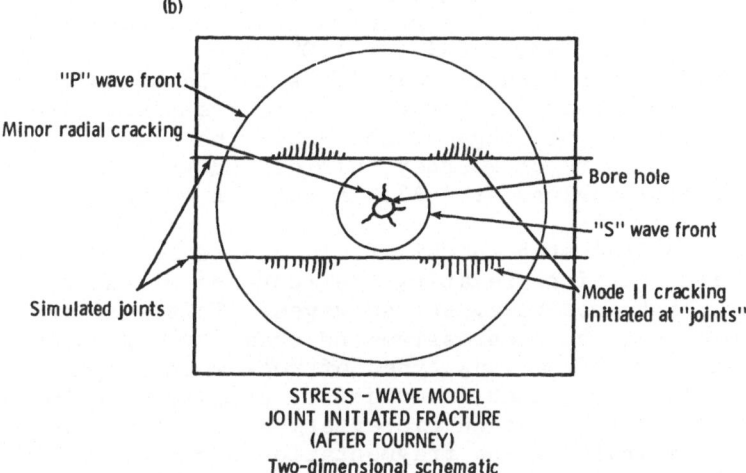

STRESS - WAVE MODEL
JOINT INITIATED FRACTURE
(AFTER FOURNEY)
Two-dimensional schematic

Figure 6.   Possible models of fracture in explosively stressed
            rock.

Accordingly, Winzer et al.[7,8] have begun to conduct in-
strumented blasting experiments in rock for which the size and
distribution of structural discontinuities has been precisely
determined.   The progress of each blast is followed by means of
high-speed cinematography.   The movies reveal surface fractures
opening too soon after the blast for gas-propelled radial cracks
to have arrived at the free surface, no venting of gases through
surface cracks formed shortly after the blast, continued fracture
of fragments after detachment from the main body of the rock, and
many fractures formed in a manner consistent with Fourney's model.

Clearly, the gas-pressure hypothesis is not consistent with
these observations.   On the other hand, it is observed that fracture at the free surface occurs at a time consistent with the
arrival of compressive stress waves.   This is taken as evidence
supporting the stress-wave theory of explosive-induced fracture.

Accordingly, Winzer et al.[8] rationalize their observations as
follows: the interaction of stress waves (compressive, shear and
tensile) with discontinuities in natural rock formations generates
both shear and tensile stresses at such flaws, and these both nucleate and propagate fractures. Fragments are generated by intersecting (coalescing) fractures, and fragments detached from the
rock mass a few hundred microseconds after the initial explosion
continue to fracture because stress waves are trapped within them.

This understanding has led to a new approach to the placement
of charges and to the timing of their detonation, based in part on
the structure of the rock formation being blasted. The improvements in fragmentation efficiency and distribution of the "muck-
pile," and the reduction in fly rock so produced, have been dramatic (cf. Figs. 7a and 7b), leading to 10% reductions in the costs
of production.[15] This result exemplifies the benefits of coupled
basic and applied research in fracture. If transferred to the
crushed stone industry throughout the U.S., the blasting strategies
developed from the work of Fourney, Winzer and their associates
would provide cost savings of > $250 x 10^6 per year.

The next logical step in this field should be the development
of constitutive equations relating the response of the rock mass
to transient, high amplitude, stress waves. This will require data
on the dynamic tensile, compressive and shear strengths, and the
fracture toughness, of various types of rock when responding to
high rates of loading. Such information, coupled with improved
knowledge of the nature and characteristics of those discontinuities that participate in the fragmentation process, should
greatly facilitate the mathematical modelling, and hence the pre-
dictibility, of blasting behavior.

...IN ROTARY ROCK DRILLING

The mechanisms involved in the rotary drilling of hard substances, such as rocks and coal-bearing minerals, have been succinctly summarized by Hustrulid,[16] and this synopsis is based on
his account. When the face of a bit first contacts a rock asperity, some crushing occurs (Fig. 8). Then, as the bit penetrates
further, counter forces...the magnitude of which depend on the
strength and degree of perfection of the rock, and frictional
forces at the bit-rock interface...increase until the cutting edge
can advance no longer. The drill stem continues to rotate, however, and this builds up a torque that eventually is sufficient to
generate stress concentrations at the bit-rock interface that
exceed the local fracture stress, whereupon a chip is generated.
The bit then accelerates forward, and whether it stops or not
before contacting the next asperity depends on such factors as
shaft rotation speed, size of chip produced, and amount of energy
stored in the system.

Figure 7. (a) A badly designed quarry blast, with poor fragmentation and much fly rock; (b) A good shot: no fly rock,
good fragmentation, and an excellent muck-pile. Bench
height about 33 feet. (Courtesy of S.R. Winzer).

If the system is very stiff, drilling tends to be smooth, and
bit life is determined largely by its rate of abrasive wear. If
the system is soft, then penetration may be rough, and bit life is
then influenced also by the fatigue induced by repeated impacting.

The rate at which a rock can be drilled is strongly dependent
on the amount of torque and thrust that can be supplied to the
bit. Generally, the thrust contribution is relatively small,
however, and is often neglected in analytical considerations.

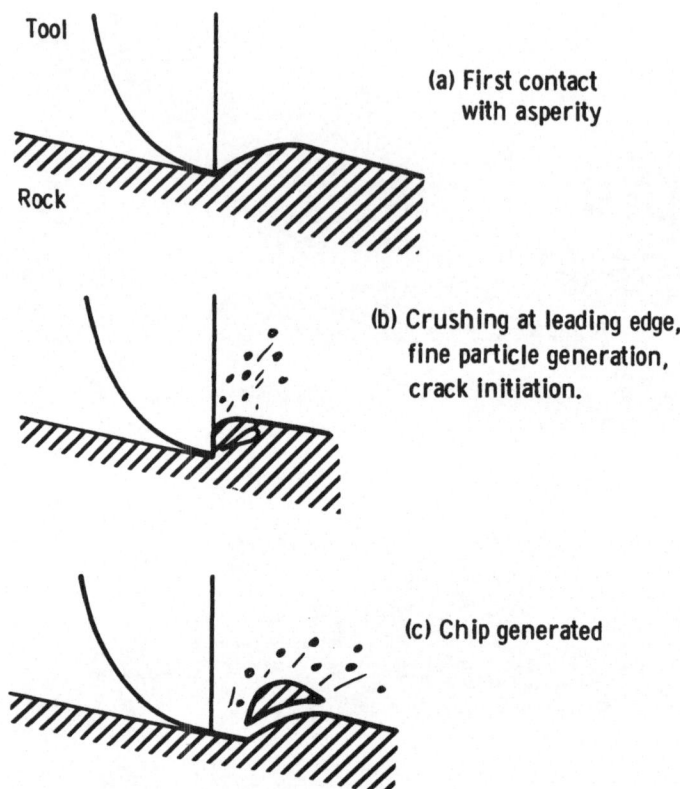

Figure 8. Representation of the probable sequence of events at
the drill tip during cutting.[16]

Drilling efficiency is usually discussed in terms of the energy
required to remove unit volume of material, E_v. For rotary
drilling, the penetration rate, R, is related to E_v by:

$$R \simeq 2\pi n\tau/E_v A$$

where n is bit rotation rate, τ the applied torque, and A the cross
sectional area of the hole. In principle, one tries to maximize
R and total penetration while staying within the torque and rotational capabilities of the drilling machine. In practice, another
critical objective is to minimize bit wear which, as might be imagined, has a substantial effect on penetration rate.

Bailey and Dean[17] have provided the following values of E_v
for removing ice by various means:

Method	E_v (in. lb/in.^{3})
Drill and blast	1740
Auger drill	1450
Manual use of pick axe	23
Melting	42,000

The merit of human intelligence is evident...leading Bailey and
Dean to suggest that "smarter" drilling machines should be developed! In reality, though, energy is a minor expense (< 1%) in deep
drilling, much less important than, say, the cost of downtime
associated with changing the bit.

   Very little is known about the specific mechanisms of bit
wear, and this would be a very fruitful area for research. It is
appreciated, however, that one effective way of reducing the rate
of bit wear is through the use of chemomechanically active fluids.
Indeed, it has been suggested recently[18-20] that this is the predominant mechanism by which Rebinder-type effects influence the
rates of hard rock drilling with diamond bits in field-scale
situations. Specifically, it has been proposed[19,20] that when a
rock is exposed to a fluid that reduces its ζ-potential* to zero,
its hardness is maximized. The coefficient of ploughing friction
between the rock surface and the diamonds in the bit is then minimized. For this reason, the bit runs cooler and the diamonds stay
sharp longer. This enhances both the penetration rate and the
total penetration, and also reduces downtime for bit changing.

   The application of Rebinder - "Chemomechanical" effects to
rock drilling has been nothing if not controversial, with scarcely
two investigators reproducing the same results. The reasons for
this are now becoming clear. In essence they relate to misunder-
standings regarding the mechanisms of the effect, and these have

---

* The concept of the ζ-potential can be briefly explained as
  follows: When a non-metallic solid is immersed in an electrolyte, it acquires a surface potential caused by either 1) absorption of ions from the electrolyte, 2) a change in the distribution of mobile defects in the surface of the solid, or 3) ion
  transfer to or from the electrolyte resulting from the equilibrium established between ions in the solid and those in the electrolyte. A charge-balancing distribution of counter-ions is
  established in the electrolyte that decays exponentially with
  distance from the solid-electrolyte/interface. Relative motion
  between the solid and the electrolyte can shear off the outer
  ions from those at the interface. The potential at the plane
  of shearing is defined as the ζ-potential.

led to inappropriate approaches to testing and application. For
example, testing strategies must recognize that beneficial chemomechanical effects occur primarily as a consequence of reductions
in the rate of bit wear.[18-20] Thus, experiments that involve
comparative measurements of drilling rates over only small fractions of the total bit life will not reliably reveal differences
in the effectiveness of various environments. Not appreciating
this factor has been responsible, we believe, for much of the
present confusion in the literature. Tests must be conducted over
a significant fraction of the life of the bit if they are to be
meaningful. When such is the case, and the experimental data are
plotted in terms of penetration rate vs penetration distance, the
results reveal that appropriately selected surfactant solutions
can increase bit life by factors of three or four.

A second important factor in full-scale drilling is the dynamics of adsorption. For example, it is quite possible to select
a surfactant concentration producing a zero ζ-potential on the
rock, to carry out an appropriate testing strategy for the specific
type of bit, and yet still not observe any significant improvement
in drilling performance. This can come about either because the
surface active molecules did not have time to arrive at the freshly
fractured surfaces before the next cutting edge came along, or because, having arrived, they did not chemisorb quickly enough.

The influence of adsorption dynamics has now been studied[20]
by drilling Westerly granite with bits of three sizes, at different rotational speeds, and under various concentrations of the
surfactant Aerosol C-61. It was found that when bit life parameter
R* (R* = bit life in surfactant solution ÷ bit life in water) was
plotted against the peripheral speed of the bit, a "master" curve
was obtained for each surfactant concentration, Fig. 9.

These data reveal the importance of bit rotation speed in
optimizing the use of chemomechanical effects in diamond-bit
drilling. They also demonstrate that lab scale screening tests,
using small diameter bits, may be used to simulate full-scale
drilling tests in a reasonably reliable fashion. Thus, surfactant
solutions that can economically produce a zero ζ-potential on the
rock to be penetrated, and also meet other criteria concerning
toxicity, corrosivity, and biodegradability, can be rapidly and
inexpensively screened for effectiveness in the field.

Other factors that are now seen to be important in scaling
up laboratory findings include: diamond failure mode, type of bit
(i.e., impregnated versus surface set), redressing effects, bit
load, and particle removal efficiency.[18-20]

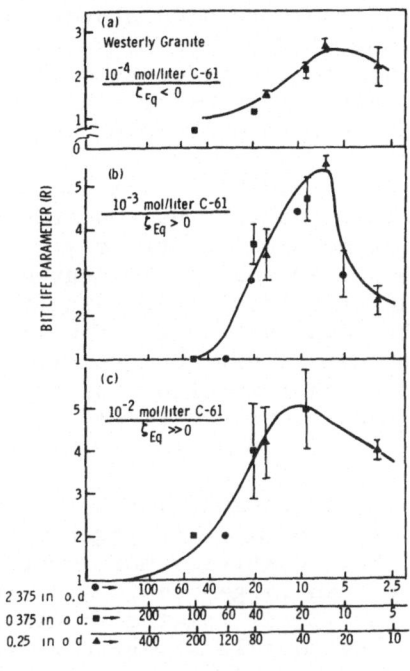

Figure 9.   Variation in bit life parameter, R, for diamond bits
            drilling into Westerly granite induced by three dif-
            ferent concentrations of surfactant Aerosol C-61 as
            a function of bit rotation speed.[20]

It is hoped that this recent, deeper appreciation of the
factors involved in chemomechanically assisted rock drilling will
provide an incentive for workers to pursue both an understanding
of the mechanism(s) by which bit wear is actually reduced by such
active fluids, and the application of these effects in practice.
Economic analyses indicate that the observed 3-4 fold increases
in bit life can reduce the cost of drilling a 3,000-foot hole by
as much as 50%.[21]

...IN COMMINUTION

Schönert[22] has noted that to understand comminution as
involved in industrial crushing and grinding operations it is
necessary to consider several sub-processes. These are: transportation of the particles to the stressing zone, loading the
particles to fracture, removing those particles that are smaller
than the desired size from the grinding zone, and preventing subsequent agglomeration of the fine fragments. Of these, the act of

fracture is the most difficult to understand. Some of the factors involved are: the stress fields caused by the crushing jaws or grinding media, residual elastic stresses due to mechanical or thermal pretreatments, and environmental influences on crack initiation and propagation...as well as on agglomeration.

The specific energy consumed in comminution, E_c, is always substantially greater than the intrinsic surface-free energy, γ, and for metals can be as much as 10^5 times γ. The value of E_c depends on the stressing system and on the degree of size reduction. Regarding the first factor, tensile stressing is more effective if one wishes to break the substance into a few large pieces...e.g., for ease of removal in tunnelling. Compressive stressing, on the other hand, usually produces a multitude of small particles, useful if one is interested in extracting a finely dispersed metal ore from its parent rock.[23]

A number of theories of comminution have been proposed over the years; those by Kick, Bond and Rittinger being best known. All of these indicate that relatively more energy is required to produce small particles than large ores. A recent study by Oka and Majima[24] suggests that E_c is related to particle size d, Young's modulus Y, and the tensile strength of the material σ_T, by

$$E_c = A\sigma_T^2 \, d^{3[1-2/\beta]}/Y$$

where A is a factor dependent on the angle of loading and the value of Poisson's ratio, varying from about 4 to 16, and β is Weibull's coefficient of uniformity. When $\beta \longrightarrow \infty$, the above relationship is equivalent to Kick's theory; when $\beta \simeq 12$ (as it might be for coal or cement), it is equivalent to Bond's theory; and when $\beta \sim 6$ (as it might be for marble or granite), it is equivalent to Rittinger's theory.

That relatively more energy is required to fracture smaller particles can be appreciated by recognizing that they are produced from larger particles, so depleting the supply and reducing the size of intrinsic flaws. Moreover, the capacity for storing the elastic energy needed to propagate a crack is proportional to the volume of the particle.

Schönert's elegant work[22] on single-particle crushing has provided a number of quantitative insights into the fundamentals of comminution. Figure 10, for example, shows how the stress required to compressively fracture glass spheres increases from $\sim$ 40 Kg/mm^2 to 300 Kg/mm^2 as sphere diameter decreases from 100 μm to 2 μm. Figure 11 also demonstrates that higher velocities of impact, and so higher specific energies (proportional

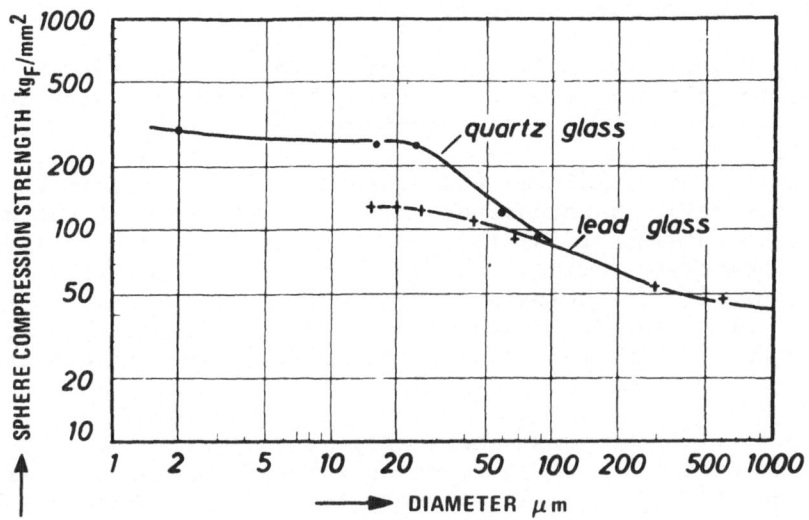

Figure 10. Compressive strength of glass spheres as a function of diameter.[22]

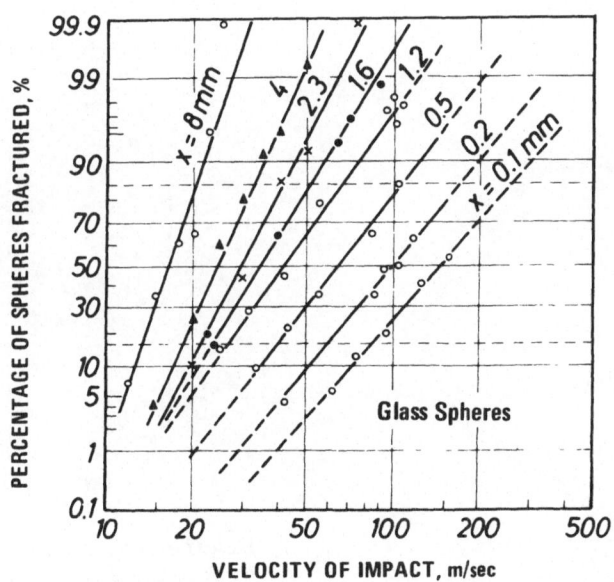

Figure 11. Fracture probability of glass spheres in impact crushing.[22]

to velocity squared), are required to achieve the same probability of breakage with smaller spheres.

At the higher stress levels required to fracture smaller particles, the plastic response of the substance becomes an increasingly important factor in determining its behavior. For example, Fig. 12 shows a 20 μm quartz sphere after loading. This normally brittle material has undergone densification near the points of stressing, a phenomenon that will influence stress distributions on subsequent loading, and also the nature of residual stresses. Sometimes residual stresses result in breakage during unloading, usually producing much coarser fragments, as illustrated in Fig. 13. In general, energy usage in ball mill grinding is about an order of magnitude less efficient than in single particle breakage.[22]

The potential for using chemically active environments to increase the efficiency of comminution is still unclear. There is little question that small concentrations of appropriate surfactants can prevent the re-agglomeration of already fragmented particles, and that this not only improves comminution efficiency,

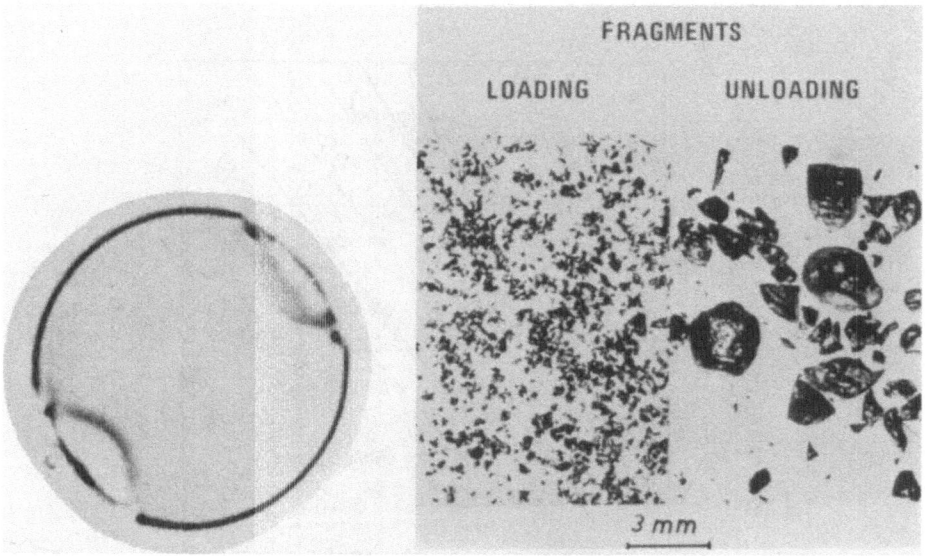

Figure 12. Quartz sphere, 21 μm in diameter, deformed irreversibly in compression.[22]

Figure 13. Fragments produced from 3-mm glass spheres broken either during loading or unloading.[22]

but also the flowability of the powdered product.[25] This is extemely important in practice. The problem of "pack set," which can make it almost impossible to unload a silo or a transportation container containing materials such as ground portland cement, dye particles, or cocoa, can be eliminated by the use of surfactants during comminution. But whether such "grinding aids" actually facilitate the grinding step per se has yet to be unequivocally established. It is known that chemomechanically active fluids can change the extent of plastic deformation and of subsurface cracking for semi-brittle solids like MgO under single impact-loading conditions.[26] But, to date, no correlation between such factors and ball milling efficiency have been clearly demonstrated. On the other hand, under circumstances where agglomeration is not an issue, as in the surface grinding of hard substances such as alumina, it is known that material is removed by plastic deformation and flow-dependent fracture processes,[27] and that material removal rates are maximized by the use of fluids that minimize surface hardness (maximize ζ-potential).[28] The opportunities for useful fundamental work in this area are excellent.

Another aspect of comminution not widely appreciated is that mechanical attrition can markedly influence subsequent chemical behavior, especially reactivity. Thus, fine particles dissolve more quickly (Fig. 14),[29] or sinter more readily...out of proportion to any increase in surface area. It has even been suggested

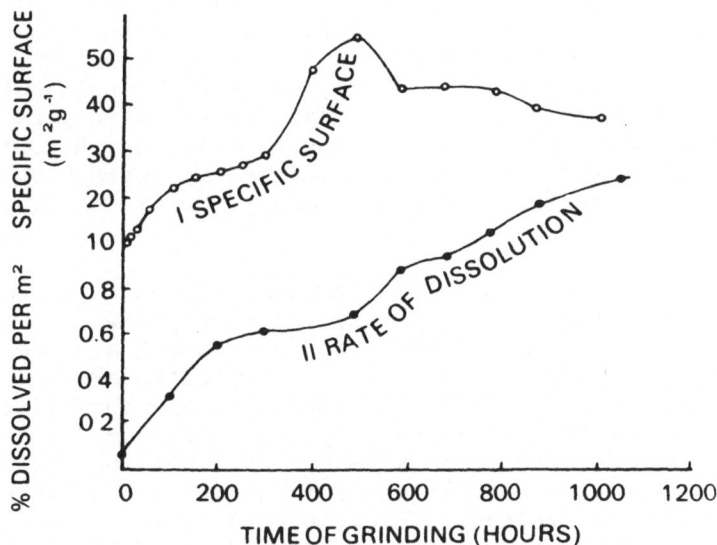

Figure 14. Illustrating marked increase in rate of dissolution of kaolin following grinding, apparently unrelated to increase in specific surface area.[29]

that the "reactive" quartz produced by ball milling induces sil-
icosis more rapidly than normal quartz of the same particle size!
Such "mechano-chemical effects," comprehensively reviewed by Fox[30]
and Lin et al.,[31] can be caused by a variety of phenomena, including stored elastic energy, surface phase changes, oxidation,
enhanced catalytic activity (e.g., of NiO[32]), high densities of
dislocations that (possibly) provide stereo-active sites for
reactions, and exo-electron emission. One oft-observed effect, a
cyclic influence of grinding time on chemical reactivity, may be
related to periodic recovery or recrystallization phenomena.
Such effects are of much importance in the mineral extraction and
processing industries.[33]

One of the more interesting recent developments in metallurgy
has been the use of comminution technology to produce novel powder
metallurgy (P/M) alloys of superior corrosion resistance and
strength. This work originated in that of Irmann[34] in the 1940's,
which led to the "SAP" (sintered aluminum powder) alloys. These
alloys derive their useful properties from a fine dispersion of
interground oxides. The art of producing such materials lies in
developing in the ball mill an appropriate balance between particle
fracture, surface oxidation, and particle rewelding.[35] In the
early 70's, Benjamin and coworkers[36,37] found that by increasing
the energy of interparticle collisions through the use of a mechanical attritor, and by the selection of suitable grinding
atmospheres, alloying additions and surface-active additives (such
as alcohols, formic acid or acetone), he could not only markedly
reduce processing time, but also produce outstanding mechanical
properties. This came about because of reduced dispersoid size
and mean-free path between particles, and improved uniformity of
dispersion. For example, the yield and ultimate strengths of
commercial extrusions of aluminum alloy IN 9051 (which contains ~ 4%
Mg) are typically, respectively, 545 and 586 MPa with ~ 5% elongation.[38] This is more than 200 MPa stronger than Al-4% Mg made by
ingot metallurgy.

The nature of the dispersoid formed as a result of the repeated fracturing, chemical reaction, and re-welding of particles
depends on the composition of both the environment and the matrix material. For aluminum in air, the particles are likely
to be $\gamma-Al_2O_3$;[35] but for IN 9051 in air, they are more likely to
be MgO.[39] In other circumstances, Al_4C_3 is produced.[40,41] The
specific dispersoid formed can markedly influence the grain size
which develops when the powders are subsequently consolidated and
thermomechanically processed. This, in turn, affects the temperature dependence of mechanical properties and resistance to stress
corrosion cracking (SCC). In the latter case, recently developed
P/M aluminum alloys, such as IN 9051, are essentially immune to
SCC at stresses up to the macroscopic yield stress. This may be

either because the grain boundaries are predominantly of low angle,[39] or because they are relatively free from segregated magnesium or impurities.

This field contains many problems worthy of study, the pursuit of which may well lead to the development of many superior, special purpose alloys.

...IN MACHINING

Increases in machining rate can be realized either by improving the efficiency of the cutting tool, or by facilitating the fracture process in the workpiece. In most cases, machining parameters (cutting speed; tool material, rake angle and appropriate geometry for chip breakage, etc.) are selected on the basis of mechanical engineering considerations. Only rarely is the "physical metallurgy" of the workpiece considered. In a few cases, however, modifying the composition and structure of the workpiece material has led to substantial increases in machining productivity, the best known examples of this being the leaded-brasses and the free-machining steels.[42,43] Additives that enhance the machinability of steel include sulphur, tellurium, selenium and lead; apparently, inclusions formed by these elements, e.g., MnS, strongly influence chip formation and breakage.

One explanation for such effects is that MnS inclusions, for example, decohere from the matrix in the shear band formed in the metal ahead of the tool, initiating fractures. Unfortunately, the hydrostatic compressive stresses produced during cutting can cause the MnS-matrix interface to re-weld. Recently, Iwata and Ueda[43] have noted that the morphology of the MnS inclusions affects their decohesion behavior; long ellipsoidal inclusions fracture without forming voids, whereas globular inclusions decohere. Steels containing spherical inclusions machine easier, with better chip behavior. Iwata and Ueda note further that additions of lead not only help lubricate the tool-chip interface, but also reduce re-welding of the voids, providing still better machining behavior.

Machining at very high speeds (> 5,000 sfm) is a technology that could benefit greatly from an improved science base.[44] Apart from the obvious advantages of rapid rates of material removal, certain metals that are quite difficult to machine at conventional rates become much more tractable at higher rates,[45] possibly because of the high local temperatures produced.[46] To clarify some of the factors involved, Komanduri and von Turkovich[46] are investigating the machining of titanium at various rates. They observe that, when machining at moderately high speeds, a concentrated shear band of very large strain is generated periodically. This

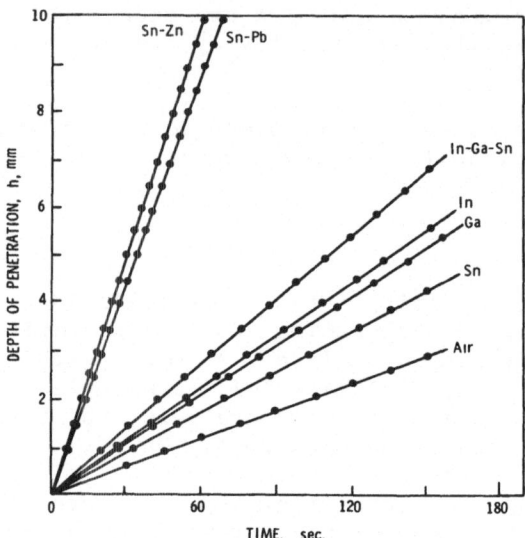

Figure 15.   Penetration of stainless steel as a function of time
             when drilling in various liquid metals.[48]

leads to plastic instability and rapid shear failure. Komanduri[47]
suspects that failure involves adiabatic shear, especially since
titanium is a relatively poor thermal conductor. Much more work
is required in this area.

    Another interesting technique for enhancing machining rates
is to use active liquid metal environments, as demonstrated by
Shchukin and co-workers.[48,49] For example, they have found that
the rate of penetration of a bit into stainless steel can be increased eight fold over that in air by drilling under a liquid
Sn-Zn eutectic at 200°C (Fig. 15). In-Ga-Sn eutectics have also
been used to increase the rates of penetration of bits into aluminum and copper by up to two orders of magnitude. Pertsov[49]
reports that drilling a tempered U8 steel in conventional lubricants is extremely difficult, even using a tungsten carbide drill.
But quite acceptable drilling rates can be produced using a liquid
Sn-Zn eutectic as the working medium.

    Of course, liquid metal-facilitated machining (LMFM) is not
without its difficulties. Care must be exercised to select the
tool and liquid metal such that it is the workpiece that is embrittled and not the tool! The liquid metal must also be carefully removed from the workpiece so that the finished product does
not fail in service because of liquid metal embrittlement. Pertsov[49] claims, however, that with care and sensible practice,
finished products produced by LMFM are as reliable as parts produced more conventionally.

A recent development in this area is the incorporation of tiny particles of active, low-melting-point alloys in the standard lubricating fluids used in metal cutting.[50] At the elevated temperatures in the vicinity of the tool-workpiece interface, these particles melt and facilitate cutting. Apparently, this imaginative, new approach to machining markedly increases both rates of production and tool life, again with no detrimental effects on the subsequent mechanical reliability of the machined parts.

Clearly the opportunities for innovative use of surface active environments in machining...formulated to provide lubrication, cooling, and fracture-facilitating capabilities...are enormous. Interdisciplinary cooperation between metallurgists and organic and surface chemists could be both scientifically stimulating and technically productive.

...IN CONCLUSION

In general, fracture is still perceived negatively, and the objective of most scientific endeavors in this field is fracture prevention. Yet, in reality, fracture is an exceedingly useful phenomenon, serving mankind in a multiplicity of ways. In our view, the positive aspects of fracture deserve much more scientific attention than they have received traditionally. With this in mind, some tasks that might be profitably pursued are:

(i)   conduct theoretical and experimental analyses of the dynamic fracture behavior of rocks, especially under explosive stressing,

(ii)   determine the mechanisms by which surface-active fluids reduce bit wear, and develop reliable procedures to utilize chemomechanical effects in cutting, grinding and drilling operations,

(iii)   establish the mechanisms and relative merits of liquid-metal-facilitated machining; confirm that parts made in this way do not remain embrittled after machining; examine the effectiveness of low-melting point particle-containing lubricants,

(iv)   obtain a better appreciation of the mechanisms and engineering potential of ultra-high speed machining,

(v)   establish whether or not surfactants influence particle fracture (in addition to agglomeration) in ball milling.

These represent just a few of the ways in which we can ensure that our breaks are well and truly applied.

ACKNOWLEDGEMENTS

     It a pleasure to acknowledge particularly useful discussions
with Professor S. Ramalingham and Drs. R. Komanduri, T. McCall,
and S.R. Winzer. The editorial advice and assistance of Mr. R.V.
Bartlett also is much appreciated.

REFERENCES

1.   See Science, 211:806 (1981).
2.   J.A. Michener, "The Covenant," Random House, New York (1980).
3.   From "Earthquake Country," Lane Books, Menlo Park, New York,
       105 (1971).
4.   Pottery by Glen Lukens, illustrated in Creative Horizons,
       28:24 (1968).
5.   V.F. Bignell, Mat'ls. in Eng. Appl., 1:310 (1979).
6.   D.B. Barker, W.L. Fourney, and D.C. Holloway, Proc. 20th U.S.
       Symp. on Rock Mech., 119 (1979).
7.   S.R. Winzer, V.I. Montenyohl, and A.P. Ritter, Mining Congress
       J., 10:46 (1979).
8.   S.R. Winzer and A.P. Ritter, Proc. 21st U.S. Symp. on Rock
       Mech., 362 (1980).
9.   J.A. Sharpe, Geophys., 7:144 (1942).
10. K. Hino, Quarterly Colorado Sch. Mines, 191 (1956).
11. W.F. Duvall and T.C. Atchinson, U.S. Bureau of Mines, Report
       of Invest. 5356 (1957).
12. D.E. Fogelson, W.I. Duvall, and T.C. Atchinson, U.S. Bureau
       of Mines, Report of Invest. 5514 (1959).
13. P.A. Persson, N. Lundborg, and C.H. Johannson, Proc. 2nd
       Cong. of Intl. Soc. for Rock Mech., Belgrade, 1 (1970).
14. D.D. Porter, MS thesis, Colorado Sch. Mines, (1961).
15. S.R. Winzer, D.A. Anderson, and A.P. Ritter, Proc. 22nd
       U.S. Symp. on Rock. Mech., (1981), to be published.
16. W. Hustrulid, Terratek Inc., Report to U.S. Bureau of Mines
       on Contract HO-242-062, July (1974).
17. J.J. Bailey and R.C. Dean, Failure and Breakage of Rocks,
       AIME, New York, 396 (1967).
18. G.A. Cooper and J. Berlie, J. Mater. Sci., 11:1771 (1976).
19. J.J. Mills and A.R.C. Westwood, J. Mater. Sci., 13:2712
       (1978).
20. J.J. Mills and A.R.C. Westwood, J. Mater. Sci., 15:3010
       (1980).
21. J.J. Mills and A.R.C. Westwood, Martin Marietta Laboratories,
       Final Report to N.S.F. on Grant APR-73-07787, February
       (1978).
22. K. Schönert, Preprint Number 71-8-115, AIME Meeting,
       New York, February (1971).
23. C. Fairhurst, Proc. 2nd Cong. of Intl. Soc. for Rock Mech.,
       Belgrade, 413 (1970).
24. Y. Oka and H. Majima, Canad. Metall. Quarterly, 9:429 (1970).

25. F.W. Locher and H.M. v. Seebach, Ind. Eng. Chem. Proc. Prod.
    Dev.), 11:190 (1972).
26. J.S. Ahearn, J.J. Mills, and A.R.C. Westwood, J. Appl. Phys.,
    50:3699 (1979).
27. B.G. Koepke, Science of Ceramic Machining and Surface Finish-
    ing, NBS Spec. Publ. 348, Washington, D.C., 317 (1972).
28. M.V. Swain, R.M. Latanision, and A.R.C. Westwood, J. Am.
    Ceram. Soc., 58:372 (1975).
29. S.J. Gregg, J. Appl. Chem., 4:666 (1954).
30. P.G. Fox, J. Mater. Sci., 10:340 (1975).
31. I.J. Lin, S. Nadiv, and D.J.M. Grodzian, Mineral Sci. & Eng.,
    7:313 (1975).
32. Y. Sadahiro and K. Shimidzu, J. Chem. Soc. Jap., 71:1874
    (1968).
33. V.V. Boldyrev and E.G. Avvakumov, Soviet Chem. Revs., 40:847
    (1971). (Engl. Trans.)
34. R. Irmann, Metallurgia, 125, September (1952).
35. E.A. Bloch, Metallurgical Reviews, 6:22 (1961)
36. J.S. Benjamin, Scientific American, 234:40 (1976).
37. M.J. Bomford and J.S. Benjamin, U.S. Patent 3,816,080, June
    (1974).
38. Data sheets on IN 9051, Novamet, Inc., Dept. N.S., Wyckoff,
    New Jersey.
39. Y.W. Kim, Scripta Met., (1981), to be published.
40. G. Jangg and F. Kutner, PMI 9:24, 1977.
41. J.R. Pickens and R.D. Schelleng, Int. Nickel Co., Report
    on AFML Contract Number F33615-76-C-5227 (1979).
42. M.C. Shaw et al., J. of Eng. for Industry, 83:163, 175, and
    181 (1961).
43. K. Iwata and K. Ueda, Proc. Fourth N.A. Metalworking Res.
    Conf., 326 (1976).
44. Report Number 355, U.S. Nat. Matls. Adv. Board, Washington,
    D.C., November (1979).
45. G. Arndt, Proc. Inst. Mech. Eng., 187:625 (1973).
46. R. Komanduri and B.F. von Turkovich, Wear (1981), in press.
47. R. Komanduri, G.E. Res. Labs., Schenectady, New York, private
    communication (1981).
48. E.D. Shchukin, Z.M. Polukyrova, V.S. Yushchenko, L.S. Bryuk-
    hanova, Dokl. Akad. Nauk., 86:205 (1972).
49. N.V. Pertsov, Surface Effects in Crystal Plasticity, Nordhoff,
    Holland, 863 (1977).
50. I.A. Andreeva, L.S. Bryukhanova, V.M. Miroshnichenko, and E.D.
    Shchukin, Sov. Mat. Sci., 16:250 (1980).

DISCUSSION

Comment by J.O'M. Bockris:

One of the less mentioned influences on the cracking process is the electrical component of the surfaces stress, which tends to reduce the surface tension. Such electrical components arise in all situations except at the zero charge potential. They arise, for example, with the surfaces of semi-conductors and insulators in contact with mositure and, hence, solutions. The average stress can be measured by appropriate integration of the capacity-potential relation.

Reply:

Yes. I think that some studies by Latanision, et al., on the mechanical behavior of zinc monocrystals in 1N Na_2SO_4 as a function of applied potential (Scripta Met., 12, 475 (1978)) illustrate this point.

Comment by N.S. Stoloff:

Embrittlement by hydrogen or by liquid metals readily occurs in single crystals of crystalline metals, as well as in samples of amorphorus Fe-Ni or Fe-base alloys. I question, therefore, the importance of grain boundary fracture in models of environmental cracking. The grain boundaries are sites for impurity segregation, precipitate nucleation and other phenomena which can obscure the basic embrittlement mechanism(s).

Comment by J.O'M. Bockris:

In respect to the wearing effect of rock-drills, and the effect of additives, it appears that electrical effects across the double layer are of vital importance. The zeta potential is relatively small in value (~100 mV) compared with the overall value of the metal-solution p.d. It is the latter which controls the interfacial field strength which interacts with the double-layer at the interface of the rock-solution. The friction depends on the potentials at the interface parabolically (max. at the potential of zero-charge). It seems likely that part of a model for the effect of organic additives would involve the effect they have on this interfacial field.

Reply:

Yes. The zeta potential is simply an experimentally convenient (though sometimes misleading) measure of the magnitude and sign of the true surface potential. A rather more sophisticated approach to the problem of explaining chemisorption-induced changes in the hardness (and, hence, frictional behavior) of non-metallic solids has been conducted recently by Ahearn et al., (see, e.g., J. Appl. Phys. 49, 96 and 614 (1978)), a summary of which is contained in Westwood et al., (Colloids and Surfaces, 2, 1 (1981)).

Comment by P. Haasen:

I am wondering whether the influence of diffusive transport has been taken into account in machining alloys. There is a temperature gradient at the tool/alloy interface as well as a potential gradient which could be manipulated.

Reply:

I am not aware of any work in which the possibility of solute diffusion in the solid has been considered. Presumably, the times available would be very short so that the extent of such diffusion would be quite limited. On the other hand, of course, the rates of arrival of surface active species from the cutting environment to the cutting zone determine whether or not they are able to play any role in influencing cutting efficiency. For any given concentration of surfactant there will be a material removal rate above which the surfactant's usefulness is markedly reduced. Finally, regarding the tool, diffusive transport is quite important; cemented carbide tools are sometimes fabricated with greater carbon contents than "optimum" to compensate for carbon loss during use and consequent loss of toughness.

TUTORIAL LECTURES ON

FRACTURE OF MATERIALS

Session Chairmen: A.S. Argon and P.R. Pratt

THE IDEAL STRENGTH OF SOLIDS

N. H. Macmillan

The Pennsylvania State University

Materials Research Laboratory
University Park, PA 16802, USA

ABSTRACT

There is an upper limit to the mechanical strength of any
material under any particular test conditions. This limit is known
as the ideal strength. The present paper attempts to define the
ideal strength as closely as is presently possible. To this end, a
brief account is first given of some of the highest strengths
recorded experimentally. Then follows a short account of the
published theoretical calculations of the ideal strength. Finally,
the theoretical and experimental values are compared, and it is
shown that in several instances these differ by less than a factor
2x.

INTRODUCTION

This paper seeks to define as closely as is presently possible
what has variously come to be called the ideal, theoretical or
ultimate strength* -- i.e., the maximum strength that can be
obtained from any given solid under any given test conditions. To
this end, short reviews are presented below of (i) some of the very
highest strengths observed experimentally and (ii) the published

---

* The first of these names is to be preferred because the second is
a misnomer for a quantity that appears to have been realized
experimentally and the third leads to ambiguous use of the term
"ultimate tensile strength" in descriptions of the results of
tensile tests.

calculations of the ideal strength. Then, in the final section the
theoretical predictions are compared with the experimental results
where possible, and the question of whether or not the ideal
strength has actually been measured experimentally is considered.

SOME EXPERIMENTAL MEASUREMENTS OF VERY HIGH STRENGTHS

Results

     It is convenient to divide experimental observations of very
high strengths into five categories, as follows: low temperature
bend or tensile tests on very small whisker single crystals (i);
similar tests on larger single crystals of materials such as Ge, Si
and Al_2O_3 (ii), and on rods and fibers of silicate glasses (iii);
tensile tests on C, B, asbestos and other fibers (iv); and indirect
measurements (v). Extensive compilations of such data have been
prepared by Kelly [1] and the present author [2], and many of the
examples cited in Tables 1-5 and discussed below are taken from
these compilations. Note the nomenclature defined in Table 1 and
used throughout this text.

Discussion

     A variety of conclusions can be drawn from the experimental
experience that generated the data listed in Tables 1-5.

     To begin with, it is clear that high strengths are only
obtained from tests in categories (i) - (iii) when the specimens
employed (which, by definition, do not contain grain boundaries)
are both free from gross internal flaws -- i.e., voids, internal
fractures and inclusions -- and have step- and crack-free surfaces
that are at least optically perfect. In the case of whiskers,
which often are prepared from the vapor phase, surfaces of the
requisite quality frequently occur naturally; but in the other two
cases such surfaces must be prepared, either by etching, or by
flame or chemical polishing, or by electro-polishing if the
material is sufficiently conductive.

     A further requirement for the attainment of high strength in
the case of crystalline specimens (categories (i) and (ii)) is that
dislocation glide must be prevented. Some whiskers, and even an
occasional larger crystal [53], satisfy this requirement by virtue
of being dislocation free; and other whiskers contain only a single
axial screw dislocation which, by virtue of its orientation, has no
tendency to glide when the whisker is loaded in uniaxial tension.
But in the case of larger crystalline specimens it is normally only
those composed of materials such as Si, Ge or Al_2O_3 -- which have a

Table 1. Strengths of Whiskers

Material	Maximum Tensile Stress GPa	$\times E^{-1}$ ($\approx$ strain)	Reference	Remarks (stresses in GPa)
NaCl	1.6	0.039	[3]	<100> tension, under water $\tau\{110\}<1\bar{1}0> = 0.8 = 0.044G$
NaCl	1.08	0.026	[4]	<100> tension $\tau\{110\}<1\bar{1}0> = 0.54 = 0.030G$
Mg$_2$SiO$_4$	–	0.12	[5]	Whisker bent in electron microscope between two grids
ZnO	40	0.15	[6]	Cantilever bending. Data reported as published, despite internal inconsistency (40 GPa $\approx$ 0.4E)
AlN	7.6	0.02	[7]	Tension
AlN	7.3	0.022	[8]	Bending. Similar results in tension
Si$_3$N$_4$	13.5	0.036	[9]	---
BeO	19.6	0.057	[10]	---
SiO$_2$ (quartz)	–	0.052	[11]	Tension
Cr$_2$N	–	0.058	[12]	Bending
Fe$_3$C	–	0.047	[12]	Bending
Ge	2.06	0.038	[13]	Torsion. Stress quoted is $\sigma<111>$.
Ge	4.3	0.079	[14]	Bending
SiC	20.3	0.03	[9]	---
Si	6.7	0.04	[9]	---
Si	7.6	0.045	[14]	Bending. Comparable strength obtained from Si-Ge alloy.
Al$_2$O$_3$	15.2	0.03	[11]	Tension
Al$_2$O$_3$	22.3	–	[15]	<11$\bar{2}$0> tension. 10.9 and 15.3 parallel to <0001> and <10$\bar{1}$0>, respectively.
Cr	–	0.038	[12]	Bending
Ag	1.73	0.040	[16]	<100> tension. $\tau\{111\}<1\bar{1}0> = 0.71 = 0.031G$
Cu	2.94	0.015	[16]	<111> tension. $\tau\{111\}<1\bar{1}0> = 0.80 = 0.022G$
Cu	$\approx$ 1.50	0.028	[11]	<100> tension. $\tau\{111\}<1\bar{1}0> = 0.61 = 0.017G$
Cu	1.74	0.030	[17]	<100> tension. $\tau\{111\}<1\bar{1}0> = 0.71 = 0.020G$
Cu	1.59	0.016	[17]	<110> tension $\tau\{111\}<1\bar{1}0> = 0.65 = 0.018G$
Cu	1.71	0.012	[17]	<111> tension. $\tau\{111\}<1\bar{1}0> = 0.46 = 0.013G$
Fe	13.10	0.049	[16]	<111> tension. $\tau\{110\}<111> = 3.58 = 0.060G$
Cd	2.80	0.042	[18]	<11$\bar{2}$0> tension. $\tau\{hkl\}<uvw>$ = 0.033 - 0.069G for various $\{hkl\}<uvw>$. $\sigma<10\bar{1}0> = 0.026E$.
C (graphite)	20.7	0.02	[19,20]	Tension

E, E<uvw> = Young's modulus; G, G{hkl}<uvw> = shear modulus; σ_{max} = ideal uniaxial tensile or compressive strength; τ_{max} = ideal shear strength; $\sigma<uvw>$ = largest tensile or compressive component of the applied stress in any direction <uvw>; $\tau\{hkl\}<uvw>$ = largest shear component of the applied stress on any plane {hkl} in any direction <uvw>.
When σ_{max} or $\sigma<uvw>$ is compared with Young's modulus, or τ_{max} or $\tau\{hkl\}<uvw>$ with the shear modulus, for the same crystallographic orientation and subject to the same constraints, then the symbols E and G are used unqualified. Likewise, when reference is made to either modulus and it is obvious from the context which mode of deformation is under consideration, the same unqualified symbols are employed.
The qualified symbols are used only when it is desired to compare (say) a stress value corresponding to one mode of deformation with an elastic modulus corresponding to another.

Table 2. Strengths of Larger Single Crystals

Material	Maximum Tensile Stress GPa	$xE^{-1}(\approx$ strain)	Reference	Remarks
Ge	3.82	$\approx$ 0.02	[21]	Bending at 78°K. 0.7 x 1.5 mm bar.
Ge	2.94	$\approx$ 0.015	[22]	Bending at room temperature. Etched bar.
Si	4.14	$\approx$ 0.02	[23]	Elongated ring, ~4 cm long with cross section area of ~0.4 cm^2, pulled in tension.
Al$_2$O$_3$	6.85	$\approx$ 0.02	[24-26]	Bending. 1 mm diameter rods, flame polished or etched.

Table 3. Strengths of Glass Rods and Fibers

Material	Maximum Tensile Stress GPa	xE^{-1} (≈ strain)	Reference	Remarks (stresses in GPa)
E-glass	3.59	---	[27]	Strength independent of fiber diameter in range 5-20 μm.
Soda-lime glass	3.38	≈ 0.05	[28]	Tension. Mean value of many tests = 255.
Soda boro-silicate glass	---	0.12	[29]	Bending at 78°K. 125 μm diameter fiber with 50 μm diameter graded-index core and 75 μm thick silicone coating.
SiO$_2$ (silica)	---	0.21	[29]	Bending at 78°K.
SiO$_2$ (silica)	38.42	0.063	[30]	Tension. 3.3 μm diameter fiber. Result inconsistent with other data.
SiO$_2$ (silica)	13.1	≈ 0.18	[31,32]	Bending at 78°K. 0.5-3.0 mm diameter rods.
SiO$_2$ (silica)	13.8	≈ 0.19	[28,32]	Tension at 78°K. 30 μm diameter fibers.
SiO$_2$ (silica)	5.68	≈ 0.08	[33]	Bending. 1 mm diameter, flame-polished rods.

Table 4. Strengths of Other Fibers

Material	Maximum Tensile Stress GPa	xE^{-1} ($\simeq$ strain)	Reference	Remarks
MgO	23.7	$\simeq$ 0.10	[34]	Bending. <100> cleavage filaments.
LiF	---	0.03	[35]	Bending. <100> cleavage filaments.
Asbestos (crocidolite, $Na_2Fe_3Si_8O_{22}(OH)_2$)	6.04	0.03	[36]	Tension.
Asbestos (chrysotile, $Mg_3Si_2O_5(OH)_4$)	5.65	0.024	[36]	Tension.
Asbestos (chrysotile, $Mg_3Si_2O_5(OH)_4$)	4.39	0.028	[37]	Tension.
Al	2.27	$\simeq$ 0.06	[38]	Bending. 0.7 x 13 μm cross-section filament etched from larger crystal.
Fe-C (cast iron)	10.45	---	[39]	Thin wires drawn down by Taylor method [40].
SiC	6.0	$\simeq$ 0.01	[41]	Tension. Reference cites earlier work by same author on other SiC fibers.
C	3.8	$\simeq$ 0.005	[42]	Tension. Also see [1].
B	4.5	$\simeq$ 0.015	[1,43]	Tension. Average strength = 2.3 GPa.
Polydiacetylene	2.0	$\simeq$ 0.04	[44]	Deformation of 0.5 μm thick crystals in electron microscope.
Polyethylene	3.04	0.060	[44]	Tension. Solution spun/drawn filaments.

Table 5. Indirect Measurements of High Strengths

Material	Maximum Tensile Stress GPa	xE^{-1} ($\simeq$ strain)	Reference	Remarks
Au, Al, Cu	---	---	[46,47]	Microindentation. Maximum shear stress $= 0.05G\{111\}<1\bar{1}0>$.
Ge	7-12	0.05-0.08	[48]	Microindentation. Maximum shear stress $= 10-15 = 0.2-0.3G\{111\}<1\bar{1}0>$.
MgO	5-9	0.02-0.035	[48]	Microindentation. Maximum shear stress $= 9-15 = 0.09-0.15G\{110\}<1\bar{1}0>$.
Cu	---	---	[49,50]	Prismatic punching. Maximum shear stress $\simeq 0.039G*$ at 873°K.
W	14	0.034	[51]	Two-layer electropolishing. Maximum shear stress $\simeq 0.044G*$
W	20	0.049	[52]	Two-layer electropolishing. Maximum shear stress $\simeq 0.063G*$
Ni	10	0.05	[52]	Two-layer electropolishing. Maximum shear stress $\simeq 0.063G*$.

* Calculation assumes G to be isotropic.

very high Peierls-Nabarro stress, at least at low homologous
temperatures -- which can meet this last requirement.

Whether either silicate or metallic glasses contain line-
defects in any way analogous to dislocations in crystals, as Gilman
[54-58] and others [59-65] have suggested, remains a matter of
conjecture at the present time. It should be noted, however, that
the suggestion is by no means as illogical as it may at first
sound. The essential property of a dislocation is to correlate in
space and time the unit (atomic-scale) processes by which the atoms
in a solid shear past one another -- i.e., to raise the probability
that, if such a shear event occurs at some place in a solid, the
next such event will occur at an immediately adjacent site rather
than elsewhere in the solid. Thus, it follows that when there is a
high degree of correlation between the unit shear events -- and
there is good evidence that this is the case in metallic glasses,
at least [66] -- the deformation can by definition be described by
a dislocation, even though this dislocation would have rather
different properties from a dislocation in a crystalline material.
For example, the high yield strengths of many metallic glasses [66]
suggests that dislocations in such materials would have higher
Peierls-Nabarro stresses than their counterparts in metal crystals.
Likewise, though care needs to be taken when comparing the maximum
tensile strains reported for whiskers of quartz and forsterite
(Mg_2SiO_4), Table 1, and for silicate glass rods and fibers, Table 3,
because the elastic stiffnesses of the former tend to decrease and
those of the latter to increase [67] at large strains, it is evident
from the numbers given in Table 3 that any dislocations present in
the latter materials also would have to have very high Peierls-
Nabarro stresses.

The third requirement for the attainment of very high strength
is that the specimen must be gripped, mounted in appropriate test
apparatus, and deformed under the chosen test conditions without
introducing any extraneous damage to mar its perfection. It is
unfortunate that these practical problems of handling and testing
are particularly acute for whisker specimens, because most solids
have so far only been prepared with any degree of perfection in
this form.

Because of the various practical difficulties involved in
specimen preparation and testing, the strengths measured in bend
and tensile tests usually show considerable scatter. For example,
only 10% of the NaCl whiskers tested by Gyulai [4] had strengths
greater than 0.2 GPa, whereas the strongest of their number broke
at a stress ~5x greater than this (1.08 GPa). Indeed, of all the
strength values listed in Tables 1-3 -- each of which represents
the highest strength recorded in a series of tests -- only the
value listed for Cd was obtained in a significant fraction of the
total number of tests performed. With this one exception,

therefore, it remains unclear whether the highest strength recorded really is the highest possible, or whether further tests would produce a greater strength.

In the case of Cd, Crump and Mitchell [18] extended thin blade-shaped whisker crystals in $<11\bar{2}0>$ tension, using a pneumatic tensile device fitted in a transmission electron microscope. These crystals had {0001} major surfaces, {10$\bar{1}$0} edge facets, and uniform rectangular cross-sections, and exhibited no evidence of any surface defects of any kind when viewed in either an optical or an electron microscope. The tensile strain produced by a given applied load was determined from the reduction in the spacing of the appropriate spots in the electron diffraction pattern of the strained specimen. Electron microscopic examination following testing showed that only 13 of 72 crystals tested in this way were free of internal growth defects and of damage introduced during mounting in the tensile apparatus. These 13, however, all failed at a tensile strain of 0.042 ± 0.003, significantly larger than the fracture strain exhibited by the other 59. An interesting feature of the experiments was that the crystals did not shatter into many fragments upon fracture, as is usual at very high strengths, but broke into two halves [68]. Thus, it could be determined that half the strong crystals failed by cleavage across a {10$\bar{1}$0} plane, yielding apparently atomically smooth fracture surfaces and leaving no traces of any dislocations in the broken parts. The other strong crystals failed by shear on {10$\bar{1}$1} planes, yielding slightly ragged fracture surfaces and leaving a few dislocations of various Burgers vectors in the fracture zone. The strength values quoted in Table 1 for Cd were calculated from Crump and Mitchell's data by the present author. Both their magnitude and their consistency, and also the extreme localization of the corresponding fracture processes, suggest strongly that these failures took place at, or close to, the ideal strength of Cd.

Table 4 presents "high strength" data obtained from a variety of fibers of unknown or less than optimum perfection, and is included primarily to show that many so-called ultra-high strength materials still have strengths far below the ideal strength. Thus, high strength C and SiC fibers, Table 4, have strengths no more than 30% and 18%, respectively, of those of the corresponding whiskers, Table 1. And even in the most highly aligned polymers, Table 4, C-C chains break at a stress an order of magnitude smaller than the strength of a graphite whisker, Table 1. Similarly, asbestos fibers, Table 4, fail at strains ~0.25-0.5x the breaking strain of quartz and forsterite whiskers, Table 1. Nor is the maximum strength obtained from B fibers remarkable, Table 4, although these are nonetheless useful because of the surprising -- and as yet unexplained -- insensitivity of their strength to handling damage. More interesting in the present context are the considerable strengths realized from an Fe-C wire drawn down and solidified

inside a glass tube [39,40], from an Al filament etched from a
larger crystal [38], and from an MgO cleavage filament [34] --
although the latter result could not be reproduced with LiF [35].
In none of these cases was any attempt made to characterize the
defect content of the specimen, however, so it is again impossible
to be certain that the strengths obtained are those characteristic
of the perfect crystal lattice.

Because of the experimental difficulties associated with tests
in categories (i) - (iv), a few alternative methods of measuring
the ideal strength have been devised. These indirect methods (v)
accept that real solids usually contain defects, and apply high
stresses only in regions small enough to have a high probability of
being defect-free. This greatly reduces specimen preparation and
handling problems, but usually means that the applied stresses are
distributed in a complex and inhomogeneous manner. Hence, it may
be difficult to calculate these stresses accurately.

One such indirect technique is applicable to two-phase
materials -- particularly to metallic alloy systems containing
coherent spherical second-phase particles. In such materials
coherency strains may generate large local shear stresses in the
matrix immediately adjacent to the particle/matrix interface.
Thermal fluctuations will result in spontaneous dislocation
nucleation in this region if (a) these local shear stresses reach
the ideal shear strength and (b) the total energy of the system is
thereby reduced. Alternatively, if aided by deformation or
irradiation, for example, dislocation nucleation may occur even
though the interfacial shear stresses are less than the ideal shear
strength, provided of course that condition (b) above is obeyed
[49]. Hence, electron microscopic investigations of the loss of
coherency of precipitates under different conditions can provide
indirect estimates of the ideal shear strength. In particular, if
elastic isotropy is assumed, studies of the Cu-Co system [50]
suggest that, under the particular loading conditions occurring in
the interface region, the ideal shear strength of Cu is $> 0.039G$ at
873°K, Table 5.

A second indirect method developed by Gane et al. [46-48]
makes use of a very small-scale microhardness indentation technique.
These studies employed very small indenters (field-ion microscope
specimens having radii of curvature < 1000 Å) to indent etched
surfaces of Au, Al, Cu, Ge and MgO at points far from any dislocation etch pits. Both the specimen and the indenter were mounted in
a scanning electron microscope, and a continuously increasing load
was applied to the indenter by an electrical method. The first
damage produced was thus immediately revealed at high magnification,
and the load at which it occurred could be monitored. For each
material, a critical load was found at which sudden yielding
occurred and an appreciable indentation was formed. This observation

was interpreted as meaning that the ideal strength for the particular mode of deformation occurring beneath the indenter had been exceeded. The stresses beneath the indenter at this critical load have been variously estimated from the Hertzian theory of elastic contact [48] and from plasticity theory [46,47] as upwards of 0.05x the appropriate elastic stiffness. Gane [47] comments that similar behavior is shown also by Ag, Ni, Fe and Zn. In the same paper he also reports that comparable stresses are sometimes necessary to blunt similar Au indenters against a hard flat surface, particularly if an appreciable amorphous layer of the polymeric contaminant common to electron microscopes is present to suppress shear stress concentrations at the surface of contact. In similar fashion, Ruoff and Wanagel [69] report that 2 μm radius diamond indenters fail at a contact pressure of 160 GPa.

The remaining data listed in Table 5 were obtained by Smith et al. [51,52] by means of the two-layer electropolishing technique developed originally by Ralph [70] for preparing field-ion microscope specimens. This involves applying a potential between a vertically-mounted rod-shaped specimen and a concentric cylindrical electrode, both of which project both above and below a layer of electrolyte floating on a heavy inert liquid (such as CCl_4), and then polishing that portion of the specimen in the electrolyte down to a narrow neck. Failure occurs when the neck can no longer support the weight of the lower part of the specimen, at a stress that is defined by the ratio of the weight of the lower half of the broken specimen to the area of the fracture cross-section. Typically, the area of this cross-section is small enough that there is a high probability of it being defect-free, which fact can be confirmed when the neck region is examined in a transmission electron microscope to determine its dimensions.

For completeness, it should also be noted that, since a typical field-ion microscope specimen is small enough to have a high probability of being defect-free, it is in principle possible to use such a microscope to measure the ideal strength of any reasonably refractory metal. The procedure is to first image the specimen to determine its radius of curvature and confirm its freedom from defects, and then measure the voltage necessary to cause field evaporation -- i.e., rupture in what is to a first approximation a state of hydrostatic tension. A few such measurements have been made, but they are not reported here because there is still considerable disagreement about how to calculate the stresses produced in the specimen by the applied field [71-74].

It is thus apparent that, despite the difference in atomic density, the interatomic cohesive forces in monocrystalline and glassy materials are of comparable strength, and that this strength is very large. This similarity in strength of dissimilar structures arises because glassy materials retain considerable short range

order -- i.e., they contain continuous three-dimensional networks
of nearest neighbor bonds of similar geometry and comparable
strength to the nearest neighbor bonds of the corresponding crystalline materials. And it is precisely this same feature of glassy
materials that must account for their resistance to the motion of
dislocations, if such defects they are proven to contain.

It is also clear that, in the absence of grain boundaries* and
grosser defects, it is the presence of glissile dislocations and/or
tiny cracks (the latter often initiated at surface inhomogeneities)
that is the main cause of the relative weakness of most specimens
of most materials. Further, because even a single such defect can
drastically reduce the strength of a specimen, and because such
defects occur on a more or less statistical basis, it is also
explained why very high strength, in all materials, is normally
such a structure-sensitive and non-reproducible property, and why
it is most frequently observed in very small specimens.

Such an understanding of mechanical properties suggests that
the strength of any material will increase with increasing structural perfection. It follows that the ideal strength will only be
realized when the greatest possible perfection is attained. There
are, however, certain limits to the degree of structural perfection
attainable in a given specimen.

For crystalline specimens a limit is imposed first by thermodynamic considerations, for even at thermodynamic equilibrium an
unstressed crystalline solid contains such defects as lattice
vibrations, a certain concentration of vacancies, etc. These
defects are necessarily present in any test specimen. In contrast,
defects such as dislocations are not expected to occur under the
same conditions, even after very long times. However, when the
solid is deformed both its thermodynamic potential and the energies
of formation of each kind of defect are changed, and at strains
corresponding to the highest possible strengths these changes could
conceivably be quite large. In the deformed material, therefore,
the equilibrium concentrations of the various species of defects
already present will change, and it is also possible that the
presence of additional species of defects could become thermodynamically favorable. The extent to which changes in defect
content will actually occur in a given test depends on such test
conditions as temperature, loading time and strain rate. These
kinetic factors therefore impose a second limit on the degree of
structural perfection that may be attained in a test specimen.

----

* Note, in passing, that the detrimental effect of grain boundaries
  on strength is much greater in ceramic materials than in metals,
  simply because the non-directional nature of the bonding in the
  latter makes them better able to accommodate lattice mismatch.

It is thus apparent that the ideal strength, like the strength measured in any experiment, is a function not only of the particular material under test, but also of the testing procedure. Hence, the ideal strength is a function of specimen geometry, mode of deformation, temperature, strain rate, type of loading (e.g., dead loading, pressure loading, etc.) imposed by the testing machine, and so on.

In practice, the size and shape, the method of preparation, and the environment of a test specimen determine the area, the crystallographic orientation, and the condition of its surface. Regardless, however, of how perfect this surface may be -- i.e., how nearly it approaches thermodynamic equilibrium -- it remains a region in which the regular atomic array characteristic of the bulk solid is disrupted. In this sense the surface may be regarded as a defect -- albeit one that is necessarily present. This raises the question of whether the surface of a solid is inherently weaker than the interior -- i.e., of whether the ideal strength is ultimately controlled by the surface. The extreme sensitivity of high strength measurements to surface condition suggests that this might be the case.

For amorphous specimens, which by their very nature are never in thermodynamic equilibrium, the details of the preceding arguments must differ. Nevertheless, the same general conclusion can be drawn -- namely, that the requirements of a practical test procedure are likely to influence the attainable degree of specimen perfection, and hence influence the ideal strength.

It may also be relevant to note that, since the attainment of the ideal strength requires the absence of dislocation glide and ductility requires the occurrence of extensive dislocation glide, very strong solids of any kind are necessarily going to be brittle -- and therefore vulnerable to impact loading and surface damage. One spectacular consequence of this absence of ductility is the violent shattering that normally occurs at fracture as very strong specimens try to dissipate the strain energy they contain via the creation of new surface and the acceleration of the fractured parts.

Further, because the rate at which a crystal can deform by dislocation glide depends on the number and the velocities of the mobile dislocations it contains, and because neither of these quantities can increase indefinitely, it can be argued that the very high strengths calculated from some shock loading experiments are measures of the ideal strength -- usually in shear or in uniaxial compression. The essence of this argument is that the crystal will deform homogeneously when it can no longer do so heterogeneously by dislocation glide. There are, however, so many uncertainties about the way in which the stresses and temperatures

generated by shock loading are estimated, and about the way in
which the specimen responds to such loading, that none of the
strength data obtained from such experiments have been included in
this review.

CALCULATIONS OF THE IDEAL STRENGTH

General Considerations

     A test specimen consists of a finite three-dimensional array
of mutually interacting atoms in oscillatory motion about their
equilibrium positions and acted upon by some distribution of
external forces. In principle, the methods of quantum mechanics
provide the means to calculate any particular thermodynamic potential, and hence the ideal strength under any given test conditions,
from the description of this dynamic atomic system. For the system
to be in stable equilibrium the appropriate thermodynamic potential
must be a minimum. Alternatively, if the kinetic energy of the
moving atoms is ignored and the stressed crystal is treated as a
static mechanical system, the conditions for equilibrium and for
stability are that there be no net force on each atom and that the
total system potential energy be a minimum. In each case the
minimum condition is imposed on some potential characteristic of
the whole system, and this potential includes a contribution from
the work done by the external forces in any virtual displacement.
The thermodynamic potential, and hence the ideal strength, is thus
again seen to depend on the test conditions.

     In practice, lack of a sufficiently detailed knowledge of the
interatomic forces and the complexity of the calculations involved
have to date prevented anything other than much simplified calculations of the ideal strength from being carried out. Indeed, it is
only in a few recent molecular dynamic studies that the dynamic
nature of the atomic system has even been taken into account. More
often, the static lattice potential has simply been evaluated from
a knowledge of the equilibrium positions of the atoms in the interior of the solid and then equated to some thermodynamic potential -- though which one it is usually not clear, since the test
conditions have rarely been specified. Thus, neither surface
effects nor the influences of thermal displacements, lattice vacancies, etc. have been taken into account. Consequently, the various
calculations reported below represent best approximations at $0°K$,
when zero point motion is the only internal defect neglected. Nor
can these calculations throw any light on the question of whether
the surface of a solid is weaker than the interior. Indeed, only
in the cases specifically indicated has any attempt even been made
to study the stability of the deformed solid. In all other cases

it is tacitly assumed that the deformation remains stable until the stress component of interest reaches a maximum.

As in the case of the experimental work discussed above, it is convenient to divide the theoretical calculations into several categories as follows: (i) calculations relating the ideal strength to other physical properties; (ii) "bond counting" calculations of the ideal strength; (iii) calculations of the ideal strength from simple central-force interatomic potentials; (iv) calculations of the ideal strength from more sophisticated interatomic potentials; (v) calculations of the ideal strength from higher-order elastic stiffnesses; and (vi) molecular dynamics calculations. The calculations in each of these categories are discussed in turn in the following six sub-sections.

Calculations Relating the Ideal Strength to Other Physical Properties

One way to circumvent both lack of knowledge about the cohesive forces and computational difficulties is to try to relate the ideal strength directly to other macroscopic physical properties.

Polanyi [75] and Orowan [76,77], and subsequently several other authors [1,23,78,79], have adopted this approach in a widely quoted calculation of the ideal uniaxial tensile strength σ_{max}. This calculation assumes that the stress-displacement relationship for any solid pulled in uniaxial tension may be represented by the first half cycle of a sine curve having an initial slope consistent with Young's modulus E. It further assumes that the work done per unit area of atomic plane normal to the tensile axis in raising the stress to the peak value of the sine curve -- i.e., to σ_{max} -- is equal to the surface energy γ. These assumptions lead to

$$\sigma_{max} = \left(\frac{E\gamma}{a_o}\right)^{\frac{1}{2}} , \tag{1}$$

where a_o is the interplanar spacing in the unstressed state of the planes perpendicular to the tensile axis. The σ_{max} values given in Table 6 were obtained from this equation by Kelly [1]. The data used are appropriate to 293°K, except for those in brackets which refer to 0°K [80,81]. For silica glass σ_{max} is obtained from the same equation as 28 GPa, or ~0.4 E [79].

It is clear from Table 6 that the Orowan-Polanyi calculation implies comparatively little dependence of σ_{max} on temperature. Also, although no explicit mention has yet been made of the constraints imposed during the deformation, the calculation implies that σ_{max} will vary with such constraints in the same manner as E.

Table 6. Calculations of the Ideal Uniaxial Tensile Strength σ_{max} by the Orowan–Polanyi Method

Material	Tensile Direction	E GPa	γ mJ m^{-2}	σ_{max} GPa	σ_{max}/E
Ag	<111>	121	1130	24	0.20
Au	<111>	110	1350	27	0.25
Cu	<111>	192	1650	39	0.20
Cu	<100>	67 (75)	1650 (3630)	25 (38.7)	0.37 (0.51)
W	<100>	390 (405)	3000 (6415)	61 (90.8)	0.16 (0.22)
α–Fe	<100>	132 (143)	2000 (4520)	30 (47.9)	0.23 (0.34)
α–Fe	<111>	260	2000	46	0.18
Si	<111>	188	1200	32	0.17
C (diamond)	<111>	1210	5400	205	0.17
SiO$_2$ (silica)	-	73	560	16	0.22
NaCl	<100>	44	115	4.3	0.10
MgO	<100>	245	1200	37	0.15
Al$_2$O$_3$	<0001>	460	1000	46	0.10

Later calculations discussed below, however, show that an increase in E corresponds to an increase in σ_{max} for some solids, but a decrease for others.

The U.S. National Academy of Sciences [79] has generalized the Orowan-Polanyi calculation, assuming σ_{max} to be the maximum value of a more general stress-displacement function containing three adjustable parameters calculated from appropriate surface energy, thermal expansion and elastic stiffness values. Different plausible functions were found to give σ_{max} values varying by about a factor 2x. A similar range of variation was found by Gilman [23] when he used several different functions to relate the stress-displacement curve to one or more of such variables as E, γ, a_o, the coefficient of thermal expansion α_o, the "range" a of the interatomic forces, and the atomic volume v_o. The results reported in Table 7 suggest that, for diamond and NaCl at least, the Orowan-Polanyi approach gives the most conservative estimate of σ_{max}. Other authors [82-84] have obtained estimates of the ideal strength in triaxial tension (i.e., volume expansion) from empirical expressions for the strain energy. For silica glass, for example, Demishev and Bartenev [83,84] estimate the ideal strength in this mode of deformation as 78.3 GPa by this method.

Frenkel [85] first applied the Orowan-Polanyi approach to the calculation of the ideal shear strength τ_{max} of a solid deformed in simple shear. He assumed that for any solid the shear stress τ needed to shear any atomic plane a distance x over its neighbor was given by

$$\tau = K \sin \frac{2\pi x}{b} , \qquad (2)$$

where b is the appropriate repeat distance in the direction of shear. Frenkel also assumed the plane of shear to remain undistorted during the deformation. Then, if K is chosen to be consistent with the shear modulus G, it follows that

$$\tau_{max} = \frac{Gb}{2\pi d} , \qquad (3)$$

where d is the interplanar spacing of the shearing planes. Hence, for an fcc metal in {111}<11$\overline{2}$> shear, $\tau_{max} = G/2\pi\sqrt{2}$.

Equation (2) may also be written as

$$\tau = \frac{\partial U(x)}{\partial x} , \qquad (4)$$

Table 7. Calculations of the Ideal Uniaxial Tensile Strength σ_{max} by the Orowan-Polanyi and Related Methods

Expression for σ_{max}	σ_{max} GPa	
	Diamond	NaCl
$E/10$	120	4.9
$E a/\pi a_o$	140	7.8
$0.52 \, (E\gamma/a_o)^{\frac{1}{2}}$	94	2.7
$3k/8\alpha_o v_o$	270	5.7

k = Boltzmann's constant

where

$$U(x) = \frac{Gb^2}{4\pi^2 d} \left(1 - \cos \frac{2\pi x}{b} \right) . \tag{5}$$

$U(x)$ thus represents the stored elastic strain energy of the bulk crystal per unit area of shear plane. Mackenzie [86] recognized that Equation (5) represented the first two terms of a Fourier series, and suggested that incorporation of higher-order terms in the series would give a better approximation to the stored elastic strain energy. He showed how to deduce the coefficients of the next two terms for the particular cases of an fcc crystal deformed homogeneously in {111}<11$\bar{2}$> shear and {111}<1$\bar{1}$0> shear. This analysis by Mackenzie is the origin of the often quoted result that $\tau_{max} = G/30$ for an fcc metal. If it is assumed that $U(x)$ is determined by first nearest neighbor interactions alone, then Mackenzie's arguments apply also to basal shear of cph crystals of ideal axial ratio $\sqrt{(8/3)}$. Kelly [1] has further extended Mackenzie's ideas to bcc metals (for {110}<1$\bar{1}$1> shear), to cph metals with non-ideal axial ratios, to graphite, and to Al_2O_3. In all cases the estimate of τ_{max} obtained is very sensitive to the coefficients of the additional higher-order terms. Table 8 lists some results obtained by Kelly [1]. Except in the one designated instance these results are computed from room temperature data. Kelly takes $G = 1/4S_{1313}$, where the 81 quantities S_{ijkl} are the conventional "small-strain" second-order elastic compliances or moduli (having dimensions stress^{-1}) [87] and axes 1 and 3 are the shear plane normal and the shear direction, respectively, thereby allowing for any distortion of the shear plane that occurs during the shear process. Note that this is inconsistent with the strict geometric arguments used to obtain the coefficients of the extra terms of $U(x)$ in the first place. Note too that, just as for σ_{max}, this method of calculation suggests τ_{max} does not vary rapidly with temperature.

These simple calculations may be criticized on the grounds that they are all based ultimately on some empirical description of the strain energy function, or of some derivative of this function, rather than on a description of the atomic system. In no case is any model of the interatomic forces taken into account explicitly and, with the exception of the work of Mackenzie [86] and its development by Kelly [1], neither is crystal structure an explicit factor in the calculations. Nevertheless, despite the fact that the results of these calculations can only be regarded as order of magnitude estimates, they are widely quoted. Also, the early work of Polanyi [75] and Frenkel [76] was historically important because it first emphasized the difference between the actual and the ideal strengths of solids. It was the need to reconcile this disparity that lead to the recognition of the role of defects in determining mechanical behavior in most circumstances.

Table 8. Calculations of the Ideal Shear Strength τ_{max} by the Frenkel Method

Material	Shear Plane and Direction	G GPa	τ_{max}/G	τ_{max} GPa
Cu (10°K)	{111} <11$\bar{2}$>	33.2	0.028–0.039	0.93–1.29
Cu	{111} <11$\bar{2}$>	30.8	0.028–0.039	0.86–1.20
Au	{111} <11$\bar{2}$>	19.0	0.028–0.039	0.53–0.74
Ag	{111} <11$\bar{2}$>	19.7	0.028–0.039	0.55–0.77
Al	{111} <11$\bar{2}$>	23.0	0.028–0.039	0.65–0.90
Al	{111} <1$\bar{1}$0>	23.0	0.114	2.62
Fe	{110} <1$\bar{1}$1>	60.0	0.11–0.13	6.6–7.8
W	{110} <1$\bar{1}$1>	150.0	0.11–0.13	16.5–19.5
Al$_2$O$_3$	{0001} <11$\bar{2}$0>	147.0	0.115	16.9
Zn	{0001} <10$\bar{1}$0>	38.0	0.034*	1.3
C (graphite)	{0001} <10$\bar{1}$0>	23.0	0.05*	0.0115

* These values of τ_{max}/G differ from 0.028–0.039 because Zn and graphite have non-ideal axial ratios.

Bond Counting Calculations of the Ideal Strength

Another very simple approach to calculating the ideal strength of a solid is that adopted first by De Boer [88]. This treatment attempts to remove one criticism of the work already mentioned in that it makes use of an explicit, albeit crude, model of the interatomic forces in a given material. As before, however, the results should only be regarded as order of magnitude estimates.

De Boer assumed that the energy of an aliphatic C-C bond was related to the C-C separation by a Morse function* and calculated the force required to break one such bond from the maximum value of the derivative of this function with respect to bond length. The values of the adjustable parameters D, α and R_0 contained in the Morse function he deduced from measurements of the bond length, the bond energy, and an appropriate absorption frequency -- the latter determined spectroscopically. In this way De Boer obtained a value of 5.64×10^{-5} Pa for the breaking strength of one aliphatic C-C bond. Then, to obtain the ideal uniaxial tensile strength σ_{max} of any structure composed of continuous straight C-C chains aligned parallel to the tensile axis, he simply multipled this breaking strength by the number of C-C chains per unit area normal to the tensile axis. This gives σ_{max} = 43 GPa and 38 GPa for phenol formaldehyde and m-cresol formaldehyde, respectively. Tyson [89] similarly obtained σ_{max} = 106 GPa for the <111> direction in diamond. Likewise, Ladik and Náray-Szabó [90] estimated σ_{max} as 24.2 GPa for silica glass by applying a similar analysis to the "pseudo-diatomic" molecule SiO_3-O.

This bond counting approach has also been extended [79] to zig-zag C-C chains, where simultaneous bending and stretching of the bonds occurs. In this case the Morse function model of the stretching of the bonds was combined with an equally empirical model of the bending of the bonds to obtain a function relating bond energy to bond geometry. This function was selected to give a smooth stress-strain curve, and to be consistent with the binding energy and equilibrium stereo-chemistry of the chain, the inversion energy, and the spectroscopically determined force constants for bending and stretching the bonds of the chain. For polymethylene, σ_{max} parallel to the C-C chains was estimated from this model as 30 GPa (at a strain ~0.3).

The same authors [79] also made a few qualitative attempts to extend their argument to more complex structures based on C-C bonds. One example of interest is graphite, where σ_{max} in any direction in the basal plane was estimated to be ~170 GPa. Note that due to its layer structure graphite may also be expected to show high strength in biaxial loading, in contrast to the chain structures.

---

* Defined in the next sub-section.

Tyson [89] employed a similar model to calculate τ_{max} for diamond in {111}<110> simple shear. He considered the change in energy of a C atom during such a shear due both to the bending and to the stretching of the bonds with its nearest seven neighbors. As in his tensile calculations for this material, Tyson assumed the stretching to be described by a Morse function. The bending he described by an empirical function that took into account the fact that the resistance to bending decreased with increasing bond length. He then calculated τ_{max} as the product of the number of C atoms per unit area in the plane of shear and the maximum value of the derivative of the energy per atom with respect to the shear strain, obtaining 91.6 GPa at a shear strain of 0.29.

The bond counting approach has also been used by Perepelkin to calculate σ_{max} (i) parallel to the C-C chains of a wide variety of uniaxially oriented polymers [91], (ii) in the plane of an oriented polyethylene film [92], and (iii) in the basal planes of graphite, hexagonal BN and hexagonal SiC [92], the last four of which structures he treated as consisting of hexagonal arrays of covalent bonds. In each case he modeled the covalent bond by a Morse function fitted to the bond dissociation energy, the equilibrium interatomic spacing, and appropriate spectroscopic data; and he derived σ_{max} either from the force required to break simultaneously all the bonds in unit area of the load-bearing cross-section [92] or from the ratio of the activation energy for bond rupture to a "structure coefficient" [91], which the latter can either be calculated from a model of the chain and its interaction with neighboring chains [93] or deduced from the observed stress-dependence of the time-to-rupture [94]. Some of Perepelkin's results are listed in Table 9. Note the surprising difference in the values for fibers and films of polyethylene.

Because the notion of individual directional bonds only retains even qualitative validity for dominantly covalent solids, this bond counting method of calculating the ideal strength is limited to this class of materials. Even for covalent solids, however, there is considerable uncertainty about the accuracy of the ideal strength values obtained, for none of the functions employed to describe bond bending and stretching has any theoretical justification. Rather, the Morse function and the various angular functions involved were chosen because they seemed physically plausible and were mathematically convenient.

Calculations of the Ideal Strength From Simple Central-Force Interatomic Potentials

Mathematical Formalism. By far the greatest number of ideal strength calculations have employed a radially-symmetric central-force two-body interatomic potential of the Morse, Mie,

Table 9. Bond Counting Calculations of the Ideal Uniaxial Tensile Strength σ_{max}

Material and Form	Tensile Direction	σ_{max} GPa	xE^{-1} ($\simeq$ strain)
Polyethylene (fiber)	Fiber axis	19.5 – 23.0	0.093 – 0.105
Polyvinyl chloride (fiber)	Fiber axis	9.6 – 9.9	0.127 – 0.133
Polycaproamide (fiber)	Fiber axis	18.0 – 19.6	0.11 – 0.126
Polypropylene (fiber)	Fiber axis	9.0 – 9.1	0.086 – 0.108
Polystyrene (fiber)	Fiber axis	5.4	
Polyethylene (film)	In plane of film	3.1 – 3.6	
C (graphite)	In basal plane	12.0 – 13.5	
BN (hexagonal)	In basal plane	11.6 – 13.2	
SiC (hexagonal)	In basal plane	9.5 – 11.6	

Lennard-Jones, Born, or Born-Mayer type, Table 10. In all of these
calculations the kinetic energy of atomic motion has been ignored,
and the potential energy of the crystal evaluated directly from the
model interatomic potential by (static) lattice summation methods.
This potential energy has in each case been assumed to adequately
represent the appropriate thermodynamic potential (e.g., the Helmholtz free energy F in the case of isothermal deformation under
conditions of thermodynamic reversibility). It is therefore convenient to commence this section with a brief outline of the mathematical formalism needed to describe the homogeneous deformation of
a crystal within this approximation. The discussion is further
restricted to crystals in which all atoms are situated at centres
of symmetry, for in this case each sub-lattice deforms similarly
and there is no internal strain in the unit cell [95].

For such a solid, the most general homogeneous deformation
that might be imposed by a hypothetical testing machine can be
written as

$$X_i^{(\alpha)} = x_i^{(\alpha)} + g_{ij} x_j^{(\alpha)} = (\delta_{ij} + g_{ij}) x_j^{(\alpha)}, \tag{12}*$$

where $x_i^{(\alpha)}$ and $X_i^{(\alpha)}$ are the positions of atom α in the initial
(unstressed and unstrained) and current (stressed and strained)
configurations, respectively, δ_{ij} is the Kronecker delta and g_{ij}
is a non-symmetric second-rank tensor. g_{ij} and $(\delta_{ij} + g_{ij})$ are
often called the displacement gradient and the deformation gradient,
respectively. It follows that the separation $R_{\alpha\beta}$ of two atoms
α and β in the strained crystal is given by

$$R_{\alpha\beta} = \left[\left(X_i^{(\alpha)} - X_i^{(\beta)} \right) \left(X_i^{(\alpha)} - X_i^{(\beta)} \right) \right]^{\frac{1}{2}}. \tag{13}*$$

Also, if the crystal contains N formula units $A_m B_n$ . . . per unit
cell and the energy of interaction of atoms α and β at a separation
$R_{\alpha\beta}$ be designated $u_{\alpha\beta}(R_{\alpha\beta})$, Table 10, the potential energy ϕ of the
strained crystal per unit cell is given by

$$\phi = \tfrac{1}{2} \sum_{\alpha} \sum_{\beta} u_{\alpha\beta}(R_{\alpha\beta}). \tag{14}$$

In Equation (14) $\sum_{\alpha}$ runs over the N (m + n + ...) atoms of one unit
cell, and $\sum_{\beta}$ runs over the equilibrium positions of all atoms in a
volume of material surrounding (but excluding) atom α and large
enough to give adequate convergence of ϕ. It is assumed that all
of the atoms involved in both summations are situated far from any

---

* In equations thus marked the Einstein summation convention is
  implied for Roman suffixes.

Table 10. Radially-symmetric central-force two-body potentials

Name	$u_{\alpha\beta}(R_{\alpha\beta})$	Equation Number
Morse	$D\left[e^{-2\alpha(R_{\alpha\beta}-R_o)} - 2e^{-\alpha(R_{\alpha\beta}-R_o)}\right]$	(6)
Generalized Morse	$\dfrac{D}{m-1}\left[e^{-m\alpha(R_{\alpha\beta}-R_o)} - me^{-\alpha(R_{\alpha\beta}-R_o)}\right]$	(7)
Mie	$D\,\dfrac{nm}{n-m}\left|\dfrac{1}{n}\left(\dfrac{R_o}{R_{\alpha\beta}}\right)^n - \dfrac{1}{m}\left(\dfrac{R_o}{R_{\alpha\beta}}\right)^m\right|,\; n > m$	(8)
Lennard-Jones	$A\left(B/R_{\alpha\beta}^{12} - 1/R_{\alpha\beta}^{6}\right)$	(9)
Born	$\pm\, e^2/R_{\alpha\beta} + A/R_{\alpha\beta}^{n}$	(10)
Born-Mayer	$\pm\, e^2/R_{\alpha\beta} - c_{\alpha\beta}/R_{\alpha\beta}^{6} - d_{\alpha\beta}/R_{\alpha\beta}^{8}$ $+ bb_{\alpha}b_{\beta}f_{\alpha\beta}e^{-R_{\alpha\beta}/\rho*}$	(11)

crystal surface, so that they occupy positions characteristic of the perfect bulk crystal.

The homogeneous deformation represented by Equation (12) can be separated by the polar decomposition theorem into a sequence of two such deformations, one of which is a pure strain (i.e., simple extensions or compressions along the three mutually perpendicular directions of the principal axes of the strain), and the other of which is a rigid body rotation [96]. Since ϕ is not changed by a rigid body rotation, it follows that it can also be expressed as some (different) function of the six independent components of the strain, hereinafter designated $\varepsilon_{ij}(= \varepsilon_{ji})$. The preceding description of ϕ in terms of the nine g_{ij} rather than the six independent ε_{ij} is ambiguous, in the sense that many different g_{ij} can give rise to the same ε_{ij} and ϕ, but is nevertheless introduced here because it has been so widely used in the literature. The reasons are several: the g_{ij} are easily visualized and are easily related to the positions of the atoms used in computing lattice sums; and many different ε_{ij} have been used in the literature.

For this latter reason it is useful to continue this discussion by regarding ε_{ij} as some generalized set of six coordinates that provides a measure of the strain relative to the unstressed reference state and introducing the quantity

$$\phi'(\varepsilon_{ij}) = \phi(\varepsilon_{ij})/V_o, \tag{15}$$

where V_o is the volume of one unit cell in the unstressed crystal. A generalized work-conjugate measure of stress t_{ij} and a set of generalized elastic stiffnesses or constants C_{ijkl} can then be defined [98,99] as

$$t_{ij} = \frac{\partial \phi'}{\partial \varepsilon_{ij}} \tag{16}$$

and

$$C_{ijkl} = \frac{\partial t_{ij}}{\partial \varepsilon_{kl}} = \frac{\partial^2 \phi'}{\partial \varepsilon_{ij} \, \partial \varepsilon_{kl}}. \tag{17}$$

It should be noted that each of these quantities is dependent on the definition of ε_{ij}, and that the interpretation of t_{ij} in terms of force and area is in general difficult. It can be shown [2,100], however, that the Cauchy (or true) stress tensor σ_{ij} is related to the g_{ij} by

$$\sigma_{ij} = \frac{1}{V} (\delta_{jk} + g_{jk}) \frac{\partial \phi}{\partial g_{ik}}, \tag{18}*$$

where V is the volume of the unit cell in the stressed crystal; and it also can be shown [99] that the C_{ijkl} become the conventional small-strain second-order elastic stiffnesses in the reference

state when this is unstressed (as in the present case) and the ε_{ij}
are some tensor measure of strain. In the present approximation
for ϕ (or ϕ') there is, of course, no distinction between isothermal
and adiabatic elastic stiffnesses or compliances.

In all of the work reviewed so far in this paper the ideal
strength has been taken as the maximum value of the applied stress
-- i.e., it has been tacitly assumed that the crystals involved
remain stable throughout the deformation, and a maximum stress
criterion for failure has been adopted. However, merely because
the applied stress-strain curve is rising, so that the crystal is
stable against a small increment in the applied stress or strain,
it does not follow that the crystal is also stable against any
arbitrary additional small deformation. Thus, in any ideal strength
calculation, tests should be made of the stability of the deformed
crystal at each increment of stress or strain, and the point of
instability determined. Such checks have been made in a number of
cases (see below), and it is therefore appropriate to conclude this
discussion of mathematical formalism with a description of the
tests of stability that have been employed. For this purpose the
accounts given by Hill [98], Hill and Milstein [99], Milstein [101]
and Parry [102] are followed.

According to the Lagrange-Dirichlet criterion, stability
requires that in any virtual strain $\delta\varepsilon_{ij}$ the increment of crystal
potential energy per unit reference volume $\delta\phi'$ must exceed the work
$\delta\omega$ done by the applied forces -- i.e.,

$$\delta\phi' > \delta\omega \quad \text{for stability.} \tag{19}$$

The condition for equilibrium -- namely, that the potential energy
of the system "crystal plus applied forces" be stationary when all
$\delta\varepsilon_{ij}$ are zero -- demands that $\delta\phi'$ and $\delta\omega$ be equal to first order
when expanded as Taylor series with respect to strain increments
$\delta\varepsilon_{ij}$, and hence application of Inequality (19) requires that both
$\delta\phi'$ and $\delta\omega$ be evaluated to second order. This involves a decision
as to whether to test for the stability of the material under
particular loading conditions or to test for the "intrinsic stability" [98] of the material independent of these conditions. The
former task is much more difficult because in any real mechanical
test the response of the applied loads during any perturbation can
be a function both of the experimental arrangement and of any rigid
body rotation of the specimen that occurs. In the latter case, the
simplifying assumption that the loads are conservative and "follow"
the material during the perturbation is adopted, with the result
that $\delta\omega$ can be expressed as a function of the $\delta\varepsilon_{ij}$ and Inequality
(19) becomes

$$\left(\frac{\partial^2\phi'}{\partial\varepsilon_{ij}\,\partial\varepsilon_{kl}} - \frac{\partial^2\omega}{\partial\varepsilon_{ij}\,\partial\varepsilon_{kl}} \right) \delta\varepsilon_{ij}\,\delta\varepsilon_{kl} > 0 \quad \text{for stability.} \tag{20}*$$

This inequality is, of course, independent of the definition of $\delta\varepsilon_{ij}$.

Prior to the publication of Hill's paper [98], the distinction between intrinsic stability and stability under particular loading conditions was not clearly recognized. Moreover, many writers included only the first order terms in their evaluation of $\delta\omega$, thereby arriving at the criterion

$$\frac{\partial^2\phi'}{\partial\varepsilon_{ij}\,\partial\varepsilon_{kl}}\,\delta\varepsilon_{ij}\,\delta\varepsilon_{kl} > 0 \quad \text{for stability,} \qquad (21)*$$

which is not coordinate invariant. It is therefore useful to follow Hill and Milstein [99] and introduce the concepts of M-strength, S-strength and G-strength, these being the different estimates of the ideal strength obtained from Inequality (21) when the ε_{ij} are the M-, S- and G- variables -- i.e., the unit cell edges and their included angles (a choice which is not tensorially invariant [98]), the elements of the stretch tensor $(\delta_{ij} + g_{ij})$, with $g_{ij} = g_{ji}$, and the squares and scalar products of the edges of the unit cell in the strained crystal (which is equivalent to using the elements of Green's strain tensor $\frac{1}{2}(g_{ij} + g_{ji} + g_{ri}\,g_{rj})*$ [98]), respectively. Note that the preceding complications vanish when the crystal is not stressed because (i) $\delta\omega = 0$ and (ii) C_{ijkl}, Equation (17), becomes the unique tensor of elastic stiffnesses [98], as pointed out above.

It is also important to recognize that ensuring stability against any small homogeneous deformation $\delta\varepsilon_{ij}$ does not, in general, ensure stability against any combination of small atomic displacements that may take place in the loaded crystal, as has been demonstrated in particular cases by Wallace [103]. Thus, what the criteria discussed above do in essence is to test for stability against any lattice vibration having a wavelength large compared to atomic dimensions.

Finally, it should be noted that by diagonalizing the matrix of coefficients of the quadratic form appearing in Inequality (20) or (21), as appropriate, at -- or in practice just beyond -- the point of instability, and then finding the eigenvector(s) corresponding to the negative eigenvalue(s), it is possible to identify the nature of the deformation $\delta\varepsilon_{ij}$ responsible for the loss of stability.

Central-Force Interatomic Potentials. The principal approximation involved in the above analysis is that of identifying ϕ with some particular thermodynamic potential. Some of the consequences of this step have already been outlined above. There is,

---

* The Einstein summation convention is implied in this expression.

however, another aspect to this approximation that must now be
discussed -- namely, how closely a radially-symmetric central-force
two-body interatomic potential can simulate the binding in any real
solid. The several such potentials employed in ideal strength
calculations to date are listed in Table 10 as Equations (6)-(11)
and discussed in turn below.

The Morse potential, Equations (6) and (7), has been used to
calculate the ideal strengths of a variety of metals, even though
it is not designed to simulate the interatomic bonding in any
particular class of solid. Rather, it is used because it is a
simple function that conveniently displays certain of the essential
features of all interatomic cohesive forces. In calculations
involving lattice summations it has the additional advantage of
providing fairly rapid convergence.

The Lennard-Jones potential, Equation (9), is the particular
form of the Mie potential, Equation (8), that provides the best --
albeit still approximate -- description of the cohesive bonding in
a Van der Waals crystal. The $1/R_{\alpha\beta}^{6}$ term correctly represents the
attractive dipole-dipole interaction that accounts for the greatest
part of the binding energy, but the $1/R_{\alpha\beta}^{12}$ repulsive term has no
theoretical justification beyond that it provides the best-fit
inverse power law description of the compressibility of the inert
gases. A and B are adjustable parameters, usually obtained from
the binding energy and lattice parameter. Higher-order (attractive)
multipole interactions, non-central and many-body interactions,
zero-point effects, etc. are all ignored. More detailed accounts
of this model may be found in the reviews by Dobbs and Jones [104],
Pollack [105] and Winterton [106].

The Born-Mayer potential, Equation (11), is an improved version
of the Born model, Equation (10), but is still only a relatively
crude approximation. Both models describe correctly the long-range
electrostatic interactions between the ions that account for by far
the largest part of the binding energy, and the former correctly
accounts for the attractive dipole-dipole and dipole-quadrupole
interactions as well. Experimentally determined values of the
coefficients $c_{\alpha\beta}$ and $d_{\alpha\beta}$ of these latter terms are available for
many ionic crystals [107]. Like the Lennard-Jones model, however,
both the Born and Born-Mayer potentials neglect many-body and non-
central interactions, and employ empirical descriptions of the
short-range repulsive interaction that dominates when the ions are
forced close together. The Born model adopts a simple inverse
power law repulsive interaction, and usually obtains the adjustable
parameters n and A (the latter assumed identical for anion-anion,
anion-cation and cation-cation interactions) from compressibility
data and from the requirement that the binding energy be a minimum
at the observed lattice parameter. The Born-Mayer potential incorporates two improvements, recognizing that the repulsive interaction

differs between pairs of ions of different kinds, and employing an
inverse exponential repulsive interaction which better approximates
to the best quantum mechanical treatment available. For this model
the adjustable parameters incorporated in the repulsive term are
usually obtained from the compressibility, the lattice parameter,
and certain empirical rules about the additivity of ionic radii in
series of alkali halides. A more detailed account is given by Tosi
[107].

Thus, even for ionic and Van der Waals solids, in which the
interatomic forces are best understood and are most nearly two-body
and central-force in nature, the available potential models are
only approximate. In each case the procedure has been to adopt a
reasonable approximation to the analytical form of the major contributions to the interatomic potential, and then to introduce
(say) n adjustable parameters into this approximate function. The
values of these n parameters are then computed from the experimentally measured values of n physical properties, and the resulting
model used to calculate the required $(n + 1)^{th}$ property (i.e., the
ideal strength). In this way it is hoped that the various errors
in the potential model will tend to cancel each other out. Clearly,
the calculation is more likely to yield a reliable result if the
$(n + 1)^{th}$ property is closely related to the other n.

However, easily measurable structure-insensitive physical
properties usually depend on the shape of the bottom of the potential energy well that defines the position of each atom in the
crystal, while the ideal strength tends to depend on the maximum
slope of the side of this well. Thus, the ideal strength is not
closely related to other easily measured properties, and the calculated values of this strength are of uncertain accuracy. Expressed
differently, the radially-symmetric potential models employed are
most reliably applied to structures of high symmetry, and tend to
become increasingly unreliable as the strain increases, symmetry
decreases, and the constituent atoms become polarized. Due to the
relatively much weaker nature of the interatomic forces in a Van der
Waals solid as compared to an ionic solid, it is likely that polarization is rather more serious in the latter case, and that the
calculated ideal strength values accordingly are less reliable.

More extensive discussion of the problems involved in applying
the currently available interatomic potential models to solid state
problems of all kinds are given in the books by Torrens [108] and
Gehlen et al. [109].

Calculations for Van der Waals Solids. The earliest order of
magnitude estimates of the strength of Van der Waals forces were
made by De Boer [88] in the course of his studies, described above,
of m-cresol and phenol formaldehyde polymers. He combined the
dipole-dipole attractive interaction -- i.e., a force of attraction

between atoms varying as the inverse seventh power of their separation -- with the assumption that the repulsive force between atomic planes varies inversely as the tenth power of the interplanar distance, and hence estimated the ideal tensile strength of the Van der Waals forces between the C-C chains of m-cresol and phenol formaldehyde as between 0.06 and 0.4 GPa. De Boer also made a calculation of the contribution of Van der Waals forces to the ideal tensile strength of NaCl (see below).

Subsequently, Mackenzie [86], Tyson [89], Born and Fürth [110], and Macmillan and Kelly [1,2,111,112] have all made estimates of the ideal strength of a monatomic fcc Van der Waals solid. These authors all used the Lennard-Jones 6-12 potential model, Equation (9), and between them have investigated a wide variety of modes of homogeneous deformation. Tyson, like Macmillan and Kelly, used a computer to perform all lattice summations, and the minor discrepancies between the results obtained by these authors appear to be due only to differences in computational procedures. The latter authors typically included several hundred atoms in their summation $\sum_{\beta}$, and obtained a convergence error in calculated stress values of $< 1\%$. The summation procedures used in the earlier work employed various analytical approximations and/or hand calculation methods, and so may be subject to relatively larger convergence errors.

Macmillan and Kelly [1,2,111,112] used values of the adjustable parameters A and B, Equation (9), computed from the equilibrium lattice parameter and the binding energy of unstressed Ar crystals at 0°K [104]. In a series of separate calculations [2] they showed that this model predicts: (i) values of the elastic stiffnesses C_{1111}, C_{1122} and C_{1212} referred to <100> axes of 3.86, 2.21 and 2.21* GPa, respectively, which are within the present experimental uncertainty [113-118]; (ii) {100}, {110} and {111} surface energies of 40, 42 and 38 mJ m^{-2}, respectively, in good agreement with the measured value of 35 mJ m^{-2} for the surface tension of liquid Ar at its triple point [119]; and (iii) that the fcc and cph forms of Ar have binding energies at zero pressure that are virtually identical and much lower than those of putative bcc or sc forms, in accord with the observation that Ar is normally fcc, but can also occur in a cph form that is metastable up to the melting point [120]. These successes of Equation (9) suggest that the Lennard-Jones potential has considerable validity as a model of solid Ar.

Mackenzie [86], Tyson [89] and Born and Fürth [110] all worked in arbitrary units, so only their values of dimensionless geometric parameters or of ideal strengths divided by elastic stiffnesses are directly comparable with the results of Macmillan and Kelly.

---

* For any central-force model of a cubic crystal the Cauchy relations require that $C_{1122} = C_{1212}$.

Table 11 summarizes the most important of the results obtained by these several authors. In this table Ox_3 is chosen as the tensile axis in all cases of uniaxial deformation. Thus, for constrained tension $g_{ij} = g_{33}\delta_{3i}\delta_{3j}$ and for unconstrained tension $\sigma_{ij} = \sigma_{33}\delta_{3i}\delta_{3j}$. When Ox_3 is <100> or <111> the values of g_{11} ($=g_{22}$) corresponding to the unconstrained condition are defined at each value of g_{33} by $\partial\phi/\partial g_{11} = 0$. Unconstrained calculations for less symmetric orientations have not been carried out, because these require minimization of ϕ with respect to more than one parameter to achieve the unconstrained condition. Further details are given elsewhere [2]. Biaxial tension and plane strain is defined by $\sigma_{11} = \sigma_{22} \neq 0$, $g_{33} = 0$, where Ox_i are parallel to <100>. Hence, $g_{11} = g_{22} \neq 0$, $\sigma_{33} \neq 0$, and all off-diagonal g_{ij} and σ_{ij} are zero. Relative to the same Ox_i, triaxial tension is defined by $\sigma_{11} = \sigma_{22} = \sigma_{33} \neq 0$. Hence, $g_{11} = g_{22} = g_{33} \neq 0$, and all off diagonal σ_{ij} and g_{ij} are again zero.

In all of the shear calculations reported in Table 11 the convention was adopted of shearing parallel to Ox_3 on the plane perpendicular to Ox_1 -- i.e., a simple shear g_{31} was imposed. Regardless of the crystallographic orientation of Ox_1 and Ox_3, however, such a shear invariably reduces crystal summetry to such an extent that the unconstrained condition ($\sigma_{13} = \sigma_{31} \neq 0$, all other $\sigma_{ij} = 0$) can never by achieved without minimization of ϕ with respect to several g_{ij} for each value of g_{31}. In all of the calculations listed in Table 11, therefore, the simplification was adopted of constraining the shearing planes to remain undistorted throughout the deformation, and allowing only their separation to vary -- i.e., for each value of g_{31}, ϕ was minimized only with respect to g_{11}. This is the most important relaxation, and has the most marked effect on τ_{max}. For {111}<11$\bar{2}$> shear, for example, $\partial\tau_{max}/\partial\sigma_{11} = -0.172$ [2,80] over a wide range of values of σ_{11}. It is important to recognize, however, that quite large stresses remain unrelaxed in these calculations, and that the effect of these stresses on τ_{max} is unknown.

Throughout Table 11 a maximum applied stress criterion for failure is adopted -- i.e., σ_{max} and τ_{max} are taken as the maximum values of σ_{33} and σ_{31}, respectively, and the ideal strength values quoted for modes of deformation other than uniaxial tension or simple shear are the maximum values of the stress components indicated. In every case the strain values quoted correspond to the appropriate stress maximum.

It is interesting to note that in constrained uniaxial tension σ_{max} and the corresponding value of g_{33} vary comparatively little with orientation. σ_{max} is greatest parallel to <111> and least in a direction <h01> lying in the cube plane about 5° from <110>. Also, the differences between the σ_{max} values for constrained and unconstrained uniaxial tension in particular high symmetry directions

Table 11. The Ideal Strength of the FCC Lennard-Jones Crystal.

Uniaxial Tension	σ_{max}*	g_{33}	$g_{11}=g_{22}$	E*	σ_{max}/E	Reference
<100> constrained	0.320	0.23	0	3.86	0.083	[111]
<110> constrained	0.268	0.20	0	5.24	0.051	[111]
<111> constrained	0.345	0.18	0	5.70	0.061	[111]
<h0l> constrained	0.267	0.20	0	–	–	[111]
<100> unconstrained	0.345	0.26	-0.027	2.24	0.154	[111]
<100> unconstrained	–	$(1+g_{33})/(1+g_{11})$	$\simeq1.25$	–	0.16	[110]
<100> unconstrained	–	0.25	–	–	0.145	[89]
<111> unconstrained	0.352	0.18	-0.006	5.20	0.068	[111]
<111> unconstrained	–	0.18	–	–	0.064	[89]

Simple Shear	τ_{max}*	g_{31}	g_{11}	G*	τ_{max}/G	Reference
{100} <001>	0.243	0.20	0.088	–	–	[111]
{100} <011>	0.196	0.16	0.055	–	–	[111]
{110} <001>	0.425	0.22	0.077	–	–	[111]
{110} <1$\bar{1}$0>	0.261	0.34	0.051	–	–	[111]
{111} <1$\bar{1}$0>	0.137	0.20	0.044	–	–	[111]
{111} <$\bar{1}\bar{1}$2>	0.188	0.20	0.058	–	–	[111]
{111} <11$\bar{2}$>	0.077	0.12	0.015	1.043	0.074	[111]
{111} <11$\bar{2}$>	–	0.13	–	–	0.062	[89]
{111} <11$\bar{2}$>	–	0.13	–	–	0.062	[86]

Biaxial Tension and Plane Strain	$\sigma_{11}=\sigma_{22}=0.256$	$g_{11}=g_{22}=0.13; g_{33}=0$	[111]

Triaxial Tension	$\sigma_{11}=\sigma_{22}=\sigma_{33}=0.254$	$g_{11}=g_{22}=g_{33}=0.09$	[111]

* The values of σ_{max}, τ_{max}; E and G from Reference [111] are in GPa and are appropriate to Ar at 0°K. <h0l> lies in {010} about 5° from <101>.

are small, though the latter are larger in all cases studied. It
therefore seems likely that the ideal unconstrained uniaxial tensile
strength of an fcc Lennard-Jones crystal would not vary by very much
regardless of the orientation of the tensile axis relative to the
symmetry elements of the crystal. The tensile results are also of
interest because they establish that the relaxation of constraint
does not always reduce the ideal strength of a crystal, even though
it necessarily reduces the initial slope of the plot of the σ_{33}
versus g_{33}. Note also that in both the unconstrained calculations
Poisson's ratio was found to decrease steadily with increasing g_{33},
and even to change sign at the very highest stress levels. The
same behavior was also observed in both the Morse potential models
of metals and the Born-Mayer model of NaCl discussed below, so it
would appear that many crystals tend to expand rather than contract
laterally with increasing tensile strain when the applied stress
approaches the ideal strength.

Investigations of the stability of deformed fcc Lennard-Jones
crystals have only been made in the cases of <100> unconstrained
uniaxial tension [110,112], <100> unconstrained uniaxial compression
[110], and <100> biaxial tension and plane strain [112].

In the first case, Born and Fürth [110] determined the
G-strength (which in this case is equal to the M- and S-strengths
[99]) to be ~0.85x the maximum value of the applied stress σ_{33} at
a strain corresponding to $(1+g_{33})/(1+g_{11}) = 1.154$; and Macmillan
and Kelly [112] confirmed that the S-strength was indeed the same.
These latter authors also identified the critical perturbation
responsible for their instability as $\{110\}<1\bar{1}0>$ shear on the $\{110\}$
planes parallel to the tensile axis. On the basis of these stability criteria of failure, therefore, the ideal strength of Ar in
<100> unconstrained uniaxial tension is ~0.29 GPa. The corresponding
strain is $g_{33} = 0.12$, $g_{11} = g_{22} = -0.024$.

In <100> unconstrained uniaxial compression [110], the
G-strength is reached at a strain corresponding to $(1+g_{33})/(1+g_{11})$
$= 0.794$ and is identical to the maximum value of $|\sigma_{33}|$, as deduced
by Hill and Milstein [99]. In this case the G-strength is greater
than the M- and S-strengths, which are equal [99]. It is also
noteworthy that, for this particular crystal and direction of
loading, the absolute value of the G-strength in compression is
only 0.90x the tensile G-strength.

For the case of <100> biaxial tension and plane strain, where
the S-strength is equal to the M-strength and greater than the
G-strength [98], the S-strength is 0.240 GPa, some 6% less than the
maximum value of σ_{33}, and is reached at a strain $g_{11} = g_{22} = 0.09$
[112]. The critical perturbation causing this instability is of a
rather complicated nature, and will not be discussed here.

The calculations in Table 11 for shear modes of deformation reveal the relatively greater anisotropy of τ_{max} as compared to σ_{max} in the case of an fcc Lennard-Jones crystal. τ_{max} is least for $\{111\}<11\bar{2}>$ shear, which mode corresponds to the observed shear due to motion of a partial dislocation in an fcc metal and (probably) also in Ar [121,122]. The values of g_{11} given in Table 11 show also that the increase in interplanar spacing with increasing shear strain g_{31} is least for this orientation of shear. Other calculations [2] reveal further that the unrelaxed stress components are also smaller for this than for the other orientations studied.

To complete this discussion of the ideal strength and stability of Lennard-Jones crystals, it is noted that Thompson [123,124] and Thompson and Shorrock [125,126] have recently begun the task of applying catastrophe theory to the study of the sort of instabilities found by Macmillan and Kelly [112] on the basis of the linear eigenvalue analysis discussed above. Thompson and Shorrock make a variety of simplifying assumptions -- including cutting off the interatomic interaction at distances greater than the nearest neighbor separation -- so their results are not directly comparable with those of either Born and Fürth [110] or Macmillan and Kelly [112]. Nevertheless, it is interesting to note their conclusion -- that the instabilities discovered by these latter authors (which Thompson and Shorrock call points of bifurcation on the primary deformation path) are likely to be non-linearly unstable, and therefore to truly represent the limits of the load-carrying capacity of the crystal in the various modes of deformation involved.

Calculations for Ionic Solids. The first application of the Born potential, Equation (10), to calculation of the ideal strength was made by Zwicky [127], who calculated the ideal unconstrained uniaxial tensile strength of NaCl parallel to <100>. He took n=9 from compressibility data, deduced the constant A from the requirement that the binding energy be a minimum at the observed lattice parameter in the unstressed state, used semi-analytical methods of performing the lattice summations $\overset{\Sigma}{\alpha}$ and $\overset{\Sigma}{\beta}$, found the Poisson contraction by trial and error, and adopted the maximum stress criterion for failure. Tyson [89] subsequently repeated Zwicky's calculation, using a computer to perform all lattice sums and thereby achieving greater accuracy; and De Boer [88] has performed the corresponding calculation in the case of constrained tension. In addition, Bartenev and Koryak-Doronenko [128,129] have used the Born model to calculate the maximum tensile stress for NaCl deformed in <110> and <111> uniaxial tension. They assumed Poisson's ratio to be strain-independent, but to be different in two perpendicular directions. This last assumption is in contradiction to the symmetry requirements in the case of <111> uniaxial tension. Due to their assumptions about Poisson's ratio, Bartenev and Koryak-Doronenko's values of the ideal strength both refer to partly constrained uniaxial tension. All of the above results are

summarized in the upper portion of Table 12. Also included in the
same table is the result of a further calculation by De Boer [88]
of the ideal <100> constrained uniaxial tensile strength of NaCl,
in which he employed a Born potential modified by the addition of a
crude treatment of the Van der Waals contribution to the binding.

The lower half of Table 12 lists the results of two calculations
by Tyson [89] of the ideal {110}<1$\bar{1}$0> shear strength of NaCl, in
both of which the Born potential was employed. The first calculation makes the same assumptions about relaxation and stability as
did the shear calculations for the fcc Lennard-Jones crystal discussed above and listed in Table 11. The second, however, allows
certain distortions of the plane of shear (variation of g_{22} and g_{33})
as well as variation of the interplanar spacing of this plane (i.e.,
variation of g_{11}) as the applied shear strain g_{31} increases.
Comparison of the two results shows that this additional relaxation
results in a reduction ~20% in τ_{max}.

More recently, Tyson [89], Kelly, Tyson and Cottrell [80],
Szomor [130], and Macmillan and Kelly [1,2,111,112] have all employed
the Born-Mayer potential, Equation (11), to extend the calculations
described immediately above. For the adjustable parameters in the
model all of these authors used values appropriate to room temperature.
The ideal strength values they obtained are therefore also appropriate
to this temperature. As has already been pointed out, however, the
potential model represents a best approximation at 0°K. Hence, the
use of room temperature values of the adjustable parameters imparts
a further degree of uncertainty to the ideal strength values
obtained.

Szomor [130] used a computational method similar to that of
Zwicky [127] to investigate the ideal strength of a series of
alkali halides in <100> unconstrained uniaxial tension. Again
following Zwicky, he assumed the deformation to remain stable up to
the maximum applied stress. Szomor's results are summarized in
Table 13, and show that the ideal tensile strength of alkali halides
with the NaCl structure decreases with increasing atomic weight of
the anion, but is relatively insensitive to the atomic weight of
the cation. Also, except for the lithium halides, the ideal strength
of these crystals is a relatively constant fraction (0.064-0.076)
of Young's modulus. For NaCl itself, the ideal <100> unconstrained
uniaxial tensile strength is estimated by Szomor as 2.95 GPa.

Tyson [89], Kelly, Tyson and Cottrell [80], and Macmillan and
Kelly [1,2,111,112] all used values of $c_{\alpha\beta}$ and $d_{\alpha\beta}$ determined by
Mayer [131], and values of $f_{\alpha\beta}$, b_α, b and ρ^* calculated by Huggins
[132,133] (see Equation (11) for definition of these parameters).
Tyson and Kelly, Tyson and Cottrell used the same value of the
electronic charge e as did Huggins, while Macmillan and Kelly
employed a more recent value [134]. All these authors used computer
methods of summation.

Table 12. Born Model Calculations of the Ideal Strength of NaCl.

Uniaxial Tension	σ_{max} GPa	g_{33}	g_{11} and g_{22}	E GPa	σ_{max}/E	Reference
<100> unconstrained	2.00	0.14	−0.023//<100>	−	−	[127]
<100> constrained	2.83	0.17	0	−	−	[88]
<100> constrained	2.62*	−	0	−	−	[88]
<100> unconstrained	2.27	0.14	−	45	0.05	[89]
<110> part constrained	3.86	0.185	{−0.045//<110> −0.030//<100>}	−	−	[128,129]
<111> part constrained	7.15	0.25	{−0.025//<110> −0.031//<112>}	−	−	[128,129]

Simple Shear	τ_{max} GPa	g_{31}	g_{11}	G GPa	τ_{max}/G	Reference
{110}<1$\bar{1}$0>	2.58	0.24	−	20	0.13	[89]
{110}<110>	1.99**	0.20	−	20	0.10	[89]

* This value was obtained from a modified Born model incorporating a crude estimate of the Van der Waals contribution to the binding.

** This calculation allowed distortion in the plane of shear as well as relaxation normal to it.

Table 13.   Born–Mayer Model Calculations of the Ideal Strength of
            Various Alkali Halides in <100> Unconstrained Uniaxial
            Tension.

Material	σ_{max} GPa	E* GPa	σ_{max}/E
LiF	4.73	40.0	0.118
LiCl	2.71	24.2	0.112
LiBr	2.27	19.6	0.116
LiI	1.82	13.0	0.140
NaF	5.00	68.1	0.073
NaCl	2.95	42.1	0.070
NaBr	2.48	34.4	0.072
NaI	1.89	˙26.0	0.073
KF	4.12	56.1	0.073
KCl	2.66	40.6	0.066
KBr	2.28	35.0	0.065
KI	1.81	27.8	0.065
RbF	3.83	50.6	0.076
RbCl	2.56	39.4	0.065
RbBr	2.15	33.1	0.065
RbI	1.76	27.4	0.064
CsF	3.65	50.8	0.072

*E is the reciprocal of the value of S_{3333} calculated from the same
 potential.

Macmillan and Kelly endeavored to assess the accuracy of their version of the Born-Mayer potential by using it to calculate several known physical properties of NaCl. They obtained: (i) values of the elastic constants C_{1111}, C_{1122} and C_{1212} referred to <100> axes of 46.2, 17.1 and 17.1 GPa, respectively, in moderate agreement with the "experimental" values of 48.8±1.1, 12.9±1.0 and 12.5±0.4 GPa obtained by averaging 17 sets of experimental data [135]; (ii) a {100} surface energy of 200 mJ m^{-2}, which is within the present experimental uncertainty [136-142]; and (iii) a binding energy at zero pressure of -7.803 x 10^5 J mol^{-1}, which is within 2% of the experimental value of -7.644 x 10^5 J mol^{-1} [107] and is considerably lower than the binding energy of a putative polymorph having the cesium chloride structure. On the basis of these results Macmillan and Kelly concluded the Born-Mayer potential to be worth using to estimate the ideal strength of NaCl.

The ideal strength estimates obtained by Tyson [89], Kelly, Tyson and Cottrell [80], and Macmillan and Kelly [1,2,111,112] are summarized in Table 14. The notation and conventions employed in this table are entirely analogous to those used in Table 11, but one or two additional results are given. Thus, in the cases of biaxial tension and plane strain and of triaxial tension, both of which are again referred to <100> axes, the elastic stiffnesses computed by Tyson [89] have been listed. Also included in Table 14 is the value of σ_{max} obtained by Macmillan and Kelly [111] for the <110> unconstrained uniaxial extension of NaCl. Note that this latter calculation required the minimization of ϕ with respect to the two independent parameters g_{11} and g_{22} for each value of g_{33} in order to obtain the appropriate Poisson contraction. For all other modes of deformation, the methods of calculation employed for NaCl differed from those used in the case of the fcc Lennard-Jones crystal only in that the summation $\sum_{\alpha}$, which for NaCl runs over one Na$^+$ and one Cl$^-$ ion, could no longer be omitted.

Table 14 reveals that there is an order of magnitude increase in both the elastic constants and the maximum stress values for NaCl as compared to Ar. This reflects the fact that the electrostatic interaction between charged ions is much stronger than the dipole-dipole interaction between neutral atoms. The same table reveals also that for NaCl the ideal strength in both constrained and unconstrained uniaxial tension is least parallel to <100> and greatest parallel to <111>, and that the anisotropy in both modes of tension is far greater than in the case of the fcc Lennard-Jones crystal. In NaCl the high <111> strength results from the difficulty of pulling apart alternate {222} layers of oppositely charged anions and cations, and the low <100> strength is due to the relatively greater mutual cancellation of the attractive and repulsive Coulombic forces across {100} than across any other plane [111]. Another noteworthy feature of the results for NaCl is that in all orientations examined the ideal strength is greater in constrained

Table 14. Born-Mayer Model Calculations of the Ideal Strength of NaCl.

Uniaxial Tension	σ_{max} GPa	g_{33}	g_{11} and g_{22}	E GPa	σ_{max}/E	Reference
<100> constrained	2.82	0.17	0	46.16	0.062	[111]
<100> constrained	3.00	–	0	49.80	0.060	[89]
<100> constrained	5.96	0.34	0	48.73	0.122	[111]
<110> constrained	20.09	1.77	0	49.45	0.405	[111]
<111> constrained	2.40	0.18	–0.031	36.82	0.065	[111]
<100> unconstrained	2.66	0.13	–	40.80	0.067	[89]
<110> unconstrained	4.61	0.29	$\left\{\begin{array}{l}-0.083\,//\langle100\rangle\\ 0.005\,//\langle110\rangle\end{array}\right.$	40.78	0.113	[111]
<111> unconstrained	12.40	0.42	–0.024	42.04	0.295	[111]

Simple Shear	τ_{max} GPa	g_{31}	g_{11}	G GPa	τ_{max}/G	Reference
{100}<001>	4.19	0.30	0.075	–	–	[111]
{100}<011>	3.53	0.30	0.056	–	–	[111]
{100}<011>	3.60	0.30	–	16.4	0.217	[89]
{110}<001>	17.62	0.60	–0.051	–	–	[111]
{110}$\langle\bar{1}10\rangle$	2.67	0.30	0.091	14.54	0.183	[111]
{110}$\langle\bar{1}10\rangle$	2.84	0.27	–	17.2	0.164	[89]
{111}$\langle\bar{1}10\rangle$	3.20	0.44	0.109	–	–	[111]
{111}$\langle11\bar{2}\rangle$	2.29	0.28	0.054	–	–	[111]
{111}$\langle11\bar{2}\rangle$	5.10	0.60	0.021	–	–	[111]

Biaxial Tension and Plane Strain*	$\sigma_{11}=\sigma_{22}=3.50$		$g_{11}=g_{22}=0.16$; $g_{33}=0$			[111]
Biaxial Tension and Plane Strain*	$\sigma_{11}=\sigma_{22}=3.63$	–	$C_{1111}+C_{1122}=65.5$			[89]
Triaxial Tension*	$\sigma_{11}=\sigma_{22}=\sigma_{33}=4.28$		$g_{11}=g_{22}=g_{33}=0.16$			[111]
Triaxial Tension*	$\sigma_{11}=\sigma_{22}=\sigma_{33}=4.34$	–	$(C_{1111}+2C_{1122})/3=80.4$			[89]

* Stresses and elastic stiffnesses are in GPa.

than unconstrained tension, whereas the reverse is the case for Ar.
The features of the Born-Mayer and Lennard-Jones potentials responsible for this reversal are discussed in detail elsewhere [111].
Comparison of Tables 11 and 14 shows that in many modes of deformation the effect of relaxation is greater in the case of NaCl than
in the case of Ar. For {110}<$1\bar{1}0$> shear of NaCl, for example,
which corresponds to the observed glide elements in this material,
$\partial\tau_{max}/\partial\sigma_{11}$ = -0.5 [2,80]. In contrast, when shearing Ar parallel
to its probable glide elements, $\partial\tau_{max}/\partial\sigma_{11}$ = -0.172 [2,80]. The
shear calculations performed for NaCl reveal that τ_{max} has a lower
value for {110}<$1\bar{1}0$> shear than for any other shear that regenerates
the crystal structure, but that the lowest value of all occurs for
{111}<$11\bar{2}$> shear. It is not possible to judge whether the results
of the fully relaxed shear calculations would display a different
anisotropy. In the case of {110}<001> shear the interplanar spacing
of the shearing planes decreases with increasing g_{31} when g_{31} is
small, because each ion initially moves towards one of the opposite
sign in the layer below. And, although the interplanar spacing
does eventually start to increase with increasing g_{31}, σ_{31} reaches
its maximum value before the original spacing is regained.

Macmillan and Kelly [112] also investigated the S-strength of
NaCl for a few modes of deformation, for all of which the M- and
S-strengths are equal to one another and greater than the G-strength
[99]. These studies revealed that in <100> tension the first
instability occurs at an applied stress ~2% less than the maximum
in the constrained case, and only ~0.1% less in the unconstrained
case. The accompanying extension g_{33} was correspondingly reduced
from 0.17 to 0.14 and from 0.18 to 0.17, respectively. And in both
cases the critical perturbation responsible for instability was
found to be continued extension parallel to the applied stress. A
similar analysis of the case of <100> biaxial tension and plane
strain showed that the first instability developed in this case
when the applied stress had risen to ~92% of its peak value. The
corresponding reduction in the accompanying strain g_{11} (=g_{22}) was
from 0.16 to 0.10. It was shown also that the critical perturbation responsible for this latter instability could be represented
by two {110}<$1\bar{1}0$> simple shears whose rotational components cancel.
These shears occurred on the {110} planes parallel to the direction
of zero strain -- i.e., on planes which had no applied shear stress
acting on them. Finally, Macmillan and Kelly made a parallel
examination of the stability of NaCl deformed in {110}<$1\bar{1}0$> shear.
In this case they found the first instability to occur at a value
of σ_{31} only ~0.5% below the maximum, and showed that the accompanying shear strain g_{31} was correspondingly reduced from 0.30 to 0.28.
The critical perturbation responsible for this instability proved
to be complicated, and is discussed elsewhere [2]. It is apparent
from these studies that, as was found for Ar, the S-strength values
for NaCl obtained from Inequality (21) by setting the ε_{ij} equal to

the elements of the stretch tensor are not very different from
those derived from the maximum stress criterion of failure. In
both cases it seems probable that the difference is small compared
with the uncertainty introduced into either calculation by the
design of the interatomic potential employed.

   Calculations for Metals. Milstein et al. [101,143-157],
Thakur and Jenkins [158], and Macmillan and Kelly [159] have all
investigated the application of either the Morse function, Equation (6), or the generalized Morse function, Equation (7), to
calculation of the mechanical behavior of ideal fcc, bcc, sc
(simple cubic) and cph metal crystals. Some of these papers are
concerned primarily with the development of the potential models
themselves [146] and with the questions of whether they can (i)
predict correctly the observed crystal structure and its response
to hydrostatic pressure [101,143,145,149,150,152,153,159] and (ii)
provide a unified understanding of the transformations between the
three cubic structures that can be produced by deforming these
uniaxially [101,154,155,156]. Born et al. [95,110,160-167] some
years ago made (without computers!) what is in several respects a
comparable study of the Mie and Lennard-Jones potentials (Equations
(8) and (9), respectively), and it is instructive to compare the
many similarities between the results reported in the two series of
papers. This review, however, focusses only on one aspect of these
Morse potential studies -- namely, the discussion by Hill and
Milstein [149,150,152,153] of the differences in the stabilities of
the cubic structures predicted by Inequalities (19) and (21) in the
case of constant hydrostatic pressure loading (either positive or
negative) which does not vary during the perturbation $\delta\varepsilon_{ij}$ and
thus allows $\delta\omega$ to be evaluated exactly. As before, Inequality (21)
is applied in the three different forms obtained from the M-, S-
and G-variables.

   Hill and Milstein's results are summarized in Figures 1(a)-(c)
for the fcc, bcc and sc structures, respectively. In each figure,
λ is the (principal) stretch -- i.e., the ratio of the lattice
parameters in the stressed and unstressed states -- and

$$\beta = e^{-\alpha R_o} \tag{22}$$

is a measure of the "effective range" of the potential, since a
larger value of β makes the potential well represented by the Morse
function both narrower and deeper. When the Morse function is
fitted to experimental values of the elastic stiffnesses, binding
energies and lattice parameters of different bcc and fcc metals,
log β is typically found to vary from 3 to 8 [143,168]. Figures
1(a)-(c) show that for each structure the range of stability predicted by each of the four stability criteria is quite different,
although it is in all cases uniquely dependent on the value of β.
This considerable dependence of the range of stability under

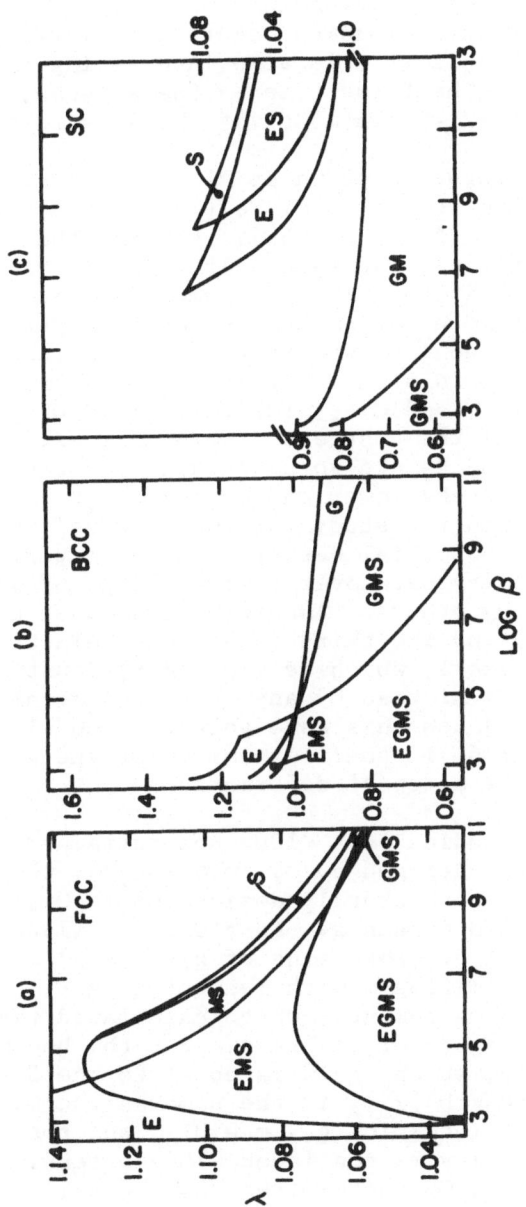

Figure 1. Regions of stability of Morse function models of the fcc (a), bcc (b) and sc (c) structures under constant hydrostatic pressure loading according to (i) Inequality (19) (E) and
(ii) Inequality (21) using G-, M- and S- variables (G, M and S, respectively).

triaxial loading on the choice of stability criterion is in marked
contrast to the behavior of the same crystal models in uniaxial
loading (see below), and also to the behavior of both the fcc
Lennard-Jones crystal and the Born-Mayer model of NaCl in this
latter sort of modes of deformation (see above). This finding is
significant because triaxial loading occurs immediately ahead of
the tip of a crack. Figure 1(c) is also interesting in that it
shows -- apparently for the first time -- the existence of a region
of stability for a central-force model of the sc structure.

In their other papers, Milstein et al. investigated the ideal
strengths of the bcc and fcc forms of Fe [144], of (fcc) Ni [147,
148,151,155,157], and of (fcc) Cu [157]. For Fe, they used the
conventional Morse function, Equation (6), with the values of D, α
and R_O chosen [168] to fit the measured compressibility, binding
energy and lattice parameter of an unstressed Fe crystal; and for
Ni and Cu they used the generalized form of this potential, Equation
(7), varying m from 1.25 to 6 and determining D, α and R_O from the
elastic stiffnesses C_{1111} and C_{1122} (referred to <100> axes) and
the lattice parameter of the unstressed structure extrapolated to
0°K [146]. When D, α and R_O are so constrained, $u_{\alpha\beta}(R_{\alpha\beta})$ is found
to become simultaneously shallower and longer-range in nature as m
increases. Ni was chosen for study for two reasons. First, because
this generalized Morse potential better fits the experimentally
determined compressibility data over a wider range of pressure for
this metal than for many others. And second, because the Cauchy
equality between C_{1122} and the third independent elastic stiffness
(C_{1212} in the present axes), which is implied by Equation (7), is
more nearly satisfied by Ni than by any other fcc metal [146].
Thakur and Jenkins [158] use this same potential model, though only
with m equal to 1.25 or 6, in their extension of the work of
Milstein et al. to other modes of deformation.

For Fe, Milstein studied only <100> unconstrained uniaxial
tension and compression, defined by $\sigma_{ij} = \sigma_{33}\delta_{3i}\delta_{3j}$ when the tensile
(or compressive) axis is arbitrarily designated as Ox_3; and for
each value of the imposed strain g_{33} he found the (isotropic)
Poisson contraction -- i.e., the value of g_{11} (=g_{22}) corresponding
to $\sigma_{11} = \sigma_{22} = 0$ by minimizing ϕ with respect to g_{11}. These calculations were all done on a computer which calculated the lattice
summation $\sum_{\beta}$ to a high degree of accuracy. For the bcc polymorph in
tension, the M-strength, which again is equal to the S-strength and
greater than the G-strength [99], is the same as the maximum value
of σ_{33} (0.17 GPa at an elongation g_{33} = 0.017) and the perturbation
responsible for instability is continued <100> extension. And in
compression, when the M-, G- and S-strengths are all equal [99],
this polymorph exhibited instability with respect to {110}<1$\bar{1}$0>
shear on the {110} planes parallel to the axis of compression at an

M-strength (0.505 GPa) of no more than 5% of the maximum value of $|\sigma_{33}|$. This instability occurred at an axial contraction g_{33} = -0.016. Conversely, the fcc polymorph exhibited an M-strength in compression (3.22 GPa at a contraction g_{33} = -0.074) essentially equal to the maximum value of $|\sigma_{33}|$, but instability in tension with respect to {100}<010> shear on the cube planes parallel to the tensile axis in a direction perpendicular to the same axis at an M-strength of 8.90 GPa, which is only about 0.20x the maximum value of σ_{33} (44.4 GPa). The corresponding values of g_{33} were 0.071 and 0.55, respectively.

The results obtained by Milstein et al. for Cu and Ni in uniaxial tension and for Ni in pure shear (and not simple shear as the title of Reference [148] suggests) are summarized in Tables 15 and 16, respectively. Perhaps the most surprising revelation is that neither the range of stability nor the stress-strain curves obtained vary much with m. For example, as m increases from 1.25 to 6, the G-, M- and S-strengths in <100> unconstrained uniaxial tension decrease only from 17.04 to 15.55 GPa, and the corresponding elongation g_{33} decreases only from 0.108 to 0.103. In each case failure is the result of a {110}<1$\bar{1}$0> shear perturbation on the {110} planes parallel to the tensile axis, and occurs well before the maximum values of σ_{33} (~25 GPa) and g_{33} (~0.26) are reached. Similarly, in <100> unconstrained uniaxial compression, the M- and S-strengths changed only from 7.39 GPa at g_{33} = -0.093 to 7.22 GPa at g_{33} = -0.090 as m increased from 1.25 to 6; and in unconstrained {100}<010> shear (which, when the shear plane normal is designated as Ox_1 and the shear direction as Ox_3, is defined by σ_{13} = σ_{31} $\neq$ 0, all other σ_{ij} = 0) both the M-strength -- which is almost equal to the maximum value of σ_{31} -- and the shear strain at failure are equally insensitive to m. These minor (<10%) changes in the ideal strength and the corresponding strain values when m changes from 1.25 to 6 may be compared with corresponding changes of ~100% and ~250%, respectively, in the maximum value of the force between atoms and in the value of the force between nearest neighbor atoms in the unstressed equilibrium fcc structure [147]. Hence, for the Morse potential model of an fcc metal, the ideal strength is remarkably insensitive to the depth and the range of the inter-atomic potential. This insensitivity presumably results in part from the fact that the adjustable parameters in the potential model were chosen in each case to be consistent with the elastic constants at small strains rather than with any other physical property.

The main thrust of the work of Thakur and Jenkins [158] has been to extend Milstein's studies of Ni to a wide variety of states of biaxial and triaxial loading. Their results are thus difficult to compare with experiment, and for this reason are not reviewed here.

Table 15. Morse Function Calculations of the Ideal Uniaxial Tensile Strengths of Cu and Ni

Metal	m Equation (7)	Mode of Deformation	Maximum $\lvert\sigma_{33}\rvert$ GPa	g_{33}	g_{11} and g_{22}	G-, S- or M- Strength GPa	g_{33}	g_{11} and g_{22}	Reference
Cu	2	<100> constrained tension	32.77	0.450	0	-	-	-	[157]
Cu	2	<100> unconstrained tension*	37.96	0.451	≈-0.04	6.97 (G,M,S)	0.069	≈-0.02	[157]
Cu	2	<100> unconstrained compression	2.51	-0.07	≈0.03	-	-	-	[157]
Ni	2	<100> constrained tension	23.65	0.234	0	-	-	-	[157]
Ni	1.5	<100> unconstrained tension	≈26	≈0.25	-	17.04 (G,M,S)	0.108	≈-0.025	[147,157]
Ni	2	<100> unconstrained tension	25.78	0.256	≈-0.03	16.78 (G,M,S)	0.107	≈-0.025	[147]
Ni	3	<100> unconstrained tension	-	-	-	16.29 (G,M,S)	0.105	-	[147]
Ni	4	<100> unconstrained tension	-	-	-	15.94 (G,M,S)	0.104	-	[147]
Ni	6	<100> unconstrained tension	≈24	≈0.275	-	15.55 (G,M,S)	0.103	-	[147]
Ni	1.5	<100> unconstrained compression**	-	-	-	7.39 (M,S)	-0.093	-	[147]
Ni	2	<100> unconstrained compression	7.39***	-0.088***	≈0.04	7.39*** (M,S)	-0.088***	≈0.04	[147,157]
Ni	3	<100> unconstrained compression	-	-	-	7.33 (M,S)	-0.092	-	[147]
Ni	4	<100> unconstrained compression	-	-	-	7.30 (M,S)	-0.091	-	[147]
Ni	6	<100> unconstrained compression	-	-	-	7.22 (M,S)	-0.090	-	[147]
Ni	2	<110> unconstrained tension	≈9	≈0.10	-	8.78 (M)	0.075	-	[101,151]
Ni	2	<110> unconstrained compression	-	-	-	768 (M)	0.251	-	[151]
Ni	2	<111> unconstrained tension	25.64	0.173	≈-0.01	24-25 (G,M,S)	0.117 (G,M) / 0.135 (S)	-	[101,155]
Ni	2	<111> unconstrained compression	45.87	-0.130	≈0.07	≈45 (G,M,S)	-0.129 (G,M) / -0.125 (S)	-	[155]

*   In <100> unconstrained tension G = M = S [99].

**   In <100> unconstrained compression G > M = S [99].

*** The values quoted are from [157]. [147] gives 7.37 and -0.093, respectively.

Table 16. Morse Function Calculations of the Ideal {100}<010> Shear Strength of Ni

m Equation (7)	Maximum σ_{31}	M-strength	Shear Strain $[(90-\theta)/90] \times 100\%*$
1.25	9.305	≈ 9.3	6.08
2	9.124	≈ 9.1	6.00
6	8.530	≈ 8.5	5.74

* θ = angle in degrees at point of failure between lines that were parallel to the direction of shear (Ox_3) and the shear plane normal (Ox_1) in the unstressed state.

Calculations of the Ideal Strength from More Sophisticated Interatomic Potentials

Attempts to apply more sophisticated potentials to the calculation of the ideal strength appear to have begun with the work of Zhdanov et al. [169-172] in the Soviet Union. Initially, Zhdanov and Konusov [169-171] calculated the G-strength of bcc and putative fcc forms of Li, Na and K in a variety of modes of deformation from a potential of the form

$$\phi = N \left(A \, V^{-k} - V^{-1/3} \sum_\alpha \sum_\beta z^2 e^2 \right) \, , \tag{23}$$

where N and V are the number of atoms in and the volume of the unit cell, respectively, e is the charge on the electron, and the adjustable parameters A, k and z were chosen to best fit the elastic stiffnesses C_{1111}, C_{1122} and C_{1212} (referred to $<100>$ axes) and the latt parameter in the unstressed state. Several of their results are listed in Table 17, with the stresses given as fractions of the measured value of C_{1212} for the bcc polymorph of each metal. Also given in the same table in the same units are the results of a subsequent calculation by Zhdanov and Dobrotvorskii [172] of the ideal strengths of the same metals plus Rb and Cs in triaxial tension. Beyond the information that it is based on the Krasko-Gurskii model pseudopotential [173,174], which is adjusted to fit the lattice parameter of the unstressed crystal and the ionization potential of the free ions, these latter authors provide few details of their calculation, so it is not clear what failure criterion they invoked.

Several features of the results cited in Table 17 are of interest. First, the two approaches are remarkably consistent; and second, neither the strength in triaxial tension nor the strain at failure varies much from one alkali metal to another. Third, it will be noted that in the case of the putative fcc structures the ideal strength is less in pure unconstrained $\{111\}<\overline{11}2>$ shear than in pure unconstrained $\{111\}<11\overline{2}>$ shear -- i.e., there is an "easy" and a "hard" direction of shear in this particular orientation. The same phenomenon can also be observed in Tables 11 and 14 for the Lennard-Jones model of fcc Ar and for the Born-Mayer model of NaCl, respectively. In the case of the alkali metals, however, the easy direction of shear is that for which each atom moves directly towards an atom in the close packed layer immediately below it, and the hard direction corresponds to each atom moving towards a "saddle-point" between two atoms in the same layer, whereas for argon the hard and easy directions are reversed with respect to the atomic motions. In NaCl the situation is more complicated, because the structure consists not of identical close-packed layers of atoms, but of alternating close-packed layers of positive and negative ions. And in this case the easy direction of shear, which

Table 17. Calculations of the Ideal Strength of Metals from the Soviet Literature.

Metal	Structure	Mode of Deformation	Ideal Tensile Strength*	g_{33}	Ideal Shear Strength*	Reference
Li	Bcc	Triaxial tension**	0.115(G)	0.040	—	[171]
Na	Bcc	Triaxial tension**	0.108(G)	0.056	—	[171]
K	Bcc	Triaxial tension**	0.118(G)	0.032	—	[171]
Li	Fcc	Triaxial tension**	0.100(G)	0.032	—	[171]
Na	Fcc	Triaxial tension**	0.094(G)	0.048	—	[171]
K	Fcc	Triaxial tension**	0.102(G)	0.028	—	[171]
Li	Bcc	Pure unconstrained $\{110\}\langle 1\bar{1}0\rangle$ shear***	—	—	0.100(G)	[171]
Na	Bcc	Pure unconstrained $\{110\}\langle 1\bar{1}0\rangle$ shear***	—	—	0.100(G)	[171]
K	Bcc	Pure unconstrained $\{110\}\langle 1\bar{1}0\rangle$ shear***	—	—	0.100(G)	[171]
Li	Fcc	Pure unconstrained $\{111\}\langle 1\bar{1}2\rangle$ shear***	—	—	0.0060	[171]
Na	Fcc	Pure unconstrained $\{111\}\langle 1\bar{1}2\rangle$ shear***	—	—	0.0060	[171]
K	Fcc	Pure unconstrained $\{111\}\langle 1\bar{1}2\rangle$ shear***	—	—	0.0060	[171]
Li	Fcc	Pure unconstrained $\{111\}\langle 1\bar{1}2\rangle$ shear***	—	—	0.0270	[171]
Na	Fcc	Pure unconstrained $\{111\}\langle 1\bar{1}2\rangle$ shear***	—	—	0.0270	[171]
K	Fcc	Pure unconstrained $\{111\}\langle 1\bar{1}2\rangle$ shear***	—	—	0.0270	[171]
Li	Bcc	Triaxial tension	0.136	0.052	—	[172]
Na	Bcc	Triaxial tension	0.162	0.060	—	[172]
K	Bcc	Triaxial tension	0.164	0.056	—	[172]
Rb	Bcc	Triaxial tension	0.160	0.052	—	[172]
Cs	Bcc	Triaxial tension	0.163	0.052	—	[172]

*   Expressed as fractions of the measured value of C_{1212} for the bcc polymorph.

**   Defined as $\sigma_{ij} = \sigma\delta_{ij}$.

***   Defined as $\sigma_{13} = \sigma_{31} \neq 0$, all other $\sigma_{ij} = 0$ when the shears take place on the planes perpendicular to Ox_1 and Ox_3 in the Ox_3 and Ox_1 directions, respectively.

is the same as for the alkali metals, is that for which each ion
moves towards the point midway between two of the opposite sign one
layer below and directly towards one of the same sign two layers
below, while the hard direction is the one for which each ion moves
directly towards one of the opposite sign one layer below and
towards the saddle-point midway between two of the same sign two
layers below [111].

    Basinski, Duesbery and Taylor [175] also have made a series of
calculations of what appears, from the brief account given, to be
the S-strength of Na for a variety of modes of pure shear and
uniaxial deformation. They use a two-part potential [176] consisting of: (i) a central-force ion-ion interaction made up of a
Coulombic term and a decaying oscillatory term of the form (sin
$kR_{\alpha\beta})/kR_{\alpha\beta}$; and (ii) a volume-dependent term that takes account of
the deviations from the Cauchy relations [95]. Derivation of and
justification for this model is given elsewhere [176,177]. For
present purposes it will suffice merely to note a few of its
properties. In particular, note that this potential generates bcc,
fcc and cph polymorphs of Na which all have very similar cohesive
energies at 90°K, and which are all stable at this temperature
against any arbitrary infinitesimal inhomogeneous perturbation in
the unstressed state [177]. This is reasonably consistent with the
observations [178] that Na partially transforms from bcc to cph on
cooling below 36°K, and that below this temperature the two phases
co-exist. Note also that for bcc Na at 90°K the potential predicts
values of the elastic stiffnesses C_{1111}, C_{1122} and C_{1212} referred
to <100> axes of 8.2, 6.9 and 5.8 GPa, respectively [177], in
excellent agreement with the most recently determined experimental
values of 8.62, 7.14 and 5.95 GPa [179].

    In all of the ideal strength calculations reviewed so far, the
crystal has been subjected to successive increments of strain, and
the resultant stress has been calculated after each increment.
Basinski, Duesbery and Taylor proceed rather differently, however,
first calculating the elastic compliances S_{ijkl}, and then using
these to compute the increment of strain resulting from a given
small increment of stress. They then calculate new compliances for
the strained crystal, and iterate the whole procedure until they
reach the point at which instability occurs. The lattice summations
involved in these calculations were all performed on a computer,
and the ion-ion interaction was in all cases cut off at a distance
of three times the lattice parameter in the appropriate phase.
Basinski, Duesbery and Taylor's method has the advantage that fully
unconstrained modes of deformation are achieved directly, without
recourse to the relaxation procedures used by other authors.

    Table 18 presents a summary of Basinski, Duesbery and Taylor's
results for bcc Na. The absolute values of σ_{max}, τ_{max}, E and G
quoted therein have been calculated from the published data assuming

Table 18. Calculations of the Ideal Strength of BCC Na.

Unconstrained Uniaxial Tension and Compression	Ideal Tensile Strength GPa	E GPa	σ_{max}/E	g_{33}	$(1+g_{11})(1+g_{22})$ -1	Remarks*
<100> tension	0.036	2.03	0.018	0.045	-0.042	$\tau\{110\}<1\bar{1}0> = 0.0177$
<110> tension	0.191	5.64	0.034	0.026	-0.016	$\tau\{110\}<1\bar{1}0> = 0.0485$
<111> tension	0.848	13.90	0.061	0.080	-0.031	$\begin{cases}\tau\{110\}<1\bar{1}0> = 0.0485\\ \tau\{110\}<1\bar{1}\bar{1}> = 0.230\end{cases}$
<100> compression	0.102	2.03	0.050	-0.030	0.030	$\tau\{110\}<1\bar{1}0> = 0.0509$
<110> compression	0.073	5.64	0.013	-0.023	0.020	$\tau\{110\}<1\bar{1}0> = 0.0184$
<111> compression	1.502	13.90	0.108	-0.097	0.038	$\tau\{110\}<1\bar{1}0> = 0.0339$

Pure Shear	Ideal Shear Strength GPa	G GPa	τ_{max}/G			
{110}<1̄10>	0.025	0.70	0.036	-	-	$\begin{cases}\tau\{110\}<1\bar{1}\bar{1}> = 0.020\\ G\{110\}<1\bar{1}\bar{1}> = 2.42\end{cases}$

* Stresses in GPa

$G\{110\}<1\bar{1}1> = (C_{1111} - C_{1122} + C_{1212})/3$ and $G\{110\}<1\bar{1}0> = (C_{1111} - C_{1122})/2$, where C_{1111}, C_{1122} and C_{1212} are referred to $<100>$ axes and have values of 8.409, 7.016 and 5.877 GPa, respectively [180]. Note that these values of the elastic stiffnesses differ somewhat from the values given in Reference [177] as appropriate to 90°K. The temperature to which the present calculations apply is not specified, however, so it is not clear whether this difference should be attributed to a change of temperature or to some other cause.

The data in the upper part of Table 18 reveal clearly the great anisotropy of the ideal strength of bcc Na in both uniaxial tension and compression. In both cases the ideal strength σ_{max} is greatest parallel to $<111>$, and in both cases the value in this direction is ~20x the minimum value. However, the weak direction is $<100>$ in tension and $<110>$ in compression. Now, Na shows considerable elastic anisotropy, having both calculated [177,180] and measured [179] values of $2C_{1212}/(C_{1111} - C_{1122}) > 8$. This anisotropy, however, produces only about a 7-fold difference between the minimum ($<100>$) and maximum ($<111>$) values of Young's modulus. Hence, the anisotropy of the ideal strength values shown in Table 18 is not due to elastic anisotropy alone. This is reflected both in the ~3- and ~8-fold variations between the maximum and minimum values of σ_{max}/E in tension and compression, respectively, and also in the different anisotropy in compression of E as compared to σ_{max} or σ_{max}/E.

For bcc Na in pure shear, Basinski, Duesbery and Taylor find τ_{max} to be lowest in the $\{110\}<1\bar{1}0>$ case. Correspondingly, for this orientation of shear the elastic anisotropy results in a particularly low shear modulus. Note that the value of τ_{max} obtained in $\{110\}<1\bar{1}0>$ pure shear is somewhat greater than the resolved shear stress on the most highly stressed $\{110\}<1\bar{1}0>$ glide system at the point of first instability in $<100>$ tension and $<110>$ compression, but is only about half this stress in the cases of $<100>$ compression and $<110>$ tension. Curiously, and for no obvious reason, the influence on this maximum resolved shear stress of a component of stress normal to the plane of shear appears to depend on the ratio of the normal and shear stresses. When these stresses are equal --i.e., the ratio is 1 -- Na behaves like the fcc Lennard-Jones crystal and NaCl, in that a tensile normal stress reduces the maximum resolved shear stress. Conversely, when the ratio is $\sqrt{3}$, it is a compressive normal stress that reduces the maximum resolved shear stress. Basinski, Duesbery and Taylor's investigations of the nature of the particular small (homogeneous) perturbation causing instability in $<100>$ and $<110>$ uniaxial tension and compression show that these two different values of the maximum resolved $<110><1\bar{1}0>$ shear stress correspond to two different modes of failure. Thus, the lower value results when failure is by $\{110\}<1\bar{1}0>$ shear on those $\{110\}$ planes inclined to (and therefore stressed by) the

applied stress, and the higher value occurs when failure is by
{110}<1$\bar{1}$0> shear on the unstressed {110} planes parallel to the
applied stress. Basinski, Duesbery and Taylor do not define the
"complex shear instability" occurring in <111> uniaxial tension and
compression, where the ratio of the stress normal to {110} to the
resolved {110}<1$\bar{1}$0> shear stress is very large (infinite at zero
strain). However, it is interesting to note that in this case both
tensile and compressive normal stresses raise the maximum resolved
{110}<1$\bar{1}$0> shear stress to a value significantly greater than
σ_{max} for {110}<1$\bar{1}$0> pure shear before instability occurs. Clearly,
this aspect of the ideal shear strength of bcc Na merits further
investigation.

Basinski, Duesbery and Taylor also report the results of
several parallel calculations of the ideal strength of putative fcc
and cph forms of Na in various modes of deformation. These are
summarized in Table 19 for the sake of completeness, but will not
be discussed further here except to point out that, in contrast to
the bcc form, fcc Na is stronger in <100> tension than <100>
compression.

Finally, note that Basinski, Duesbery and Taylor did not
calculate any of their stress-strain curves beyond the point of
first instability. It is therefore not possible to determine
whether or not the maximum applied stress criterion of failure
would lead to significantly different values of the ideal strength
in the case of Na.

The most recent attempt to apply more sophisticated potentials
to the calculation of the ideal strength of metals is that of
Esposito et al. [181]. They use a maximum applied stress criterion
of failure to calculate the <100> constrained uniaxial tensile
strength of Cu by five different methods: from an ab initio pair
potential based on electronic structure calculations of the cohesive
energy as a function of volume and the assumption that this energy
can be written as a sum of pairwise interaction energies [182] (i);
via non self-consistent Korringa-Kohn-Rostoker (KKR) band-theoretic
calculations [183] using the muffin-tin approximation (ii); from
self-consistent augmented spherical wave band-theoretic calculations
with (iii) and without (iv) the Ewald correction [184]; and from
the empirical Johnson pair potential [185] (v). All five calculations give surprisingly large values of σ_{max}, Table 20.

Calculations of the Ideal Strength from Higher-Order Elastic Stiffnesses

With the development of precision techniques for measuring
ultrasonic elastic wave velocities in particular directions in highly
stressed crystals, it has recently become possible to determine

Table 19. Calculations of the Ideal Strength of Close-Packed Forms
           of Na.

Structure	Mode of Deformation	τ_{max} or σ_{max}	Remarks
Cph	$\{10\bar{1}0\}<11\bar{2}0>$ pure shear	0.039G	Minimum value of τ_{max}
Cph	$\{0001\}<11\bar{2}0>$ pure shear	0.066G	–
Cph	$<11\bar{2}0>$ unconstrained tension	–	Failure by $\{10\bar{1}0\}<11\bar{2}0>$ shear when $\tau\{10\bar{1}0\}<11\bar{2}0>$ = 0.026G
Fcc	$\{111\}<1\bar{1}0>$ pure shear	0.023±0.005G	–
Fcc	$<100>$ unconstrained tension	–	Failure by shear when $\tau\{111\}<1\bar{1}0>$ = 0.033G
Fcc	$<100>$ unconstrained compression	–	Failure by shear when $\tau\{111\}<1\bar{1}0>$= 0.014G

$G = C_{3131}$, where Ox_1 and Ox_3 are the shear plane normal and the
shear direction, respectively [180].

Table 20. Calculations of the Ideal <100> Constrained Uniaxial Tensile Strength of Cu [181].

Method	σ_{max} GPa	σ_{max}/E^*	g_{33}
Non self-consistent KKR band-theoretic calculations	55	0.43	0.5
Self-consistent augmented spherical wave band-theoretic calculations, with Ewald correction	32	0 25	0.55
Self-consistent augmented spherical wave band-theoretic calculations, without Ewald correction	18	0.14	0.3
From ab initio pair potential	36	0.28	0.5
From Johnson potential	41	0.32	0.3

* E is the measured Young's modulus of polycrystalline Cu.

experimentally the third -- and sometimes the fourth -- order
elastic stiffnesses. As a result, it is now possible, by expanding
the strain energy function as a Taylor series in some suitable
measure of the strain, to obtain an analytical expression for this
function that is correct through the third -- and sometimes through
the fourth -- order terms. And it is then a straightforward matter
to evaluate the first and second derivatives of this expression
with respect to strain at whatever state of strain is desired, and
so calculate the corresponding state of stress and test whether
this is a stable or an unstable state. The technique was pioneered
by Barsch [186] and his students [187,188] in the mid-1960's, but
has only recently attracted wider interest.

Hollinger [188] made a detailed theoretical study of the
procedure that was intended to ascertain how rapidly the calculated
values of different properties converged as higher and higher-order
elastic stiffnesses were added to the Taylor expansion of the
strain energy. For this purpose she studied NaCl and KCl, using
the Born-Mayer potential, Equation (11), with the adjustable parameters fitted to the lattice parameter of the unstrained crystal,
the second-order elastic stiffnesses C_{1111}, C_{1122} and C_{1212} (referred
to <100> axes) and their pressure derivatives, and cutting off
all lattice sums beyond the third nearest neighbors. Hollinger
also assumed dead loading -- i.e., that the first Piola-Kirchoff
stress tensor remained constant during any perturbation of the
stressed crystal -- because $\delta\omega$ can in this case be evaluated
exactly provided it is assumed that the applied loads are conservative and "follow" the perturbation [189]. Her results show that
the strain energy, the Cauchy stress tensor, and the eigenvalues of
the "effective elastic constant matrix" that must be positive-
definite for stability, Inequality (20), all converge more rapidly
with the addition of higher order terms to the Taylor expansion
when this is evaluated in terms of the modified Eulerian strain
tensor ε'_{ij} rather than the more usually adopted Lagrangian
(Green's) strain tensor $\varepsilon_{ij} = \frac{1}{2}(g_{ij} + g_{ji} + g_{ri}g_{rj})^{*}$. Hollinger's
best estimates of the ideal <100> unconstrained uniaxial tensile
strengths of NaCl and KCl are 2.91 and 2.26 GPa, respectively.

Gieske [187] applied the same methods to calculate the
orientation dependence of the ideal unconstrained uniaxial tensile
strength of Al_2O_3 from the experimentally determined second- and
third-order elastic constants of this material. His values of 98,
87 and 87 GPa parallel to <0001>, <11$\overline{2}$0> and <10$\overline{1}$0> may be compared
with the strengths of 10.9, 22.3 and 15.3 GPa obtained by Soltis
[15] from whiskers of the corresponding orientations.

---

* $\varepsilon'_{ij} = \frac{1}{2}(\delta_{ij} - (\delta_{ij} - 2\varepsilon_{ij})^{-1})$ [188], and thus is not easily
  expressed in terms of g_{ij}.

More recently, Ruoff [190] and Nelson and Ruoff [191] have used the same approach to calculate the ideal strengths of Si and Ge in <100> unconstrained uniaxial tension and compression and <111> unconstrained uniaxial compression, using a maximum applied stress criterion of failure. And they have extended their results to diamond by arguing that all three materials should fail in tension and compression, respectively, at the same fraction of their Young's moduli and their shear moduli for simple shear on the {111} plane. These results are summarized in Table 21.

Atomistic Calculations of the Ideal Strength

The application of molecular dynamics techniques to the study of the ideal strength and of the homogeneous nucleation of defects in highly stressed crystals has recently begun in the Soviet Union. Such work as has been done, however, is all of a qualitative nature, so no estimates of the ideal strength have yet emerged. Shchukin and Yushchenko [192] have recently reviewed this activity.

At the same time, what might be called "molecular statics" calculations of the mechanical behavior of amorphous Fe in pure shear [193] and unconstrained uniaxial tension [194] have begun in Japan. The same short-range pairwise Johnson potential [195] was used in both studies -- namely,

$$
\begin{aligned}
u_{\alpha\beta}(R_{\alpha\beta}) &= -2.195976 \ (R_{\alpha\beta}- 3.097910)^3 + 2.704060 \ R_{\alpha\beta} \quad 7.436448 \ \text{eV} \\
&\quad \text{for } R_{\alpha\beta} \leq 2.4 \ \text{Å}, \\
&= -0.639230 \ (R_{\alpha\beta}- 3.115829)^3 + 0.477871 \ R_{\alpha\beta}- 1.581570 \ \text{eV} \\
&\quad \text{for } 2.4 < R_{\alpha\beta} \leq 3.0 \ \text{Å}, \\
&= -1.115035 \ (R_{\alpha\beta}- 3.066403)^3 + 0.466892 \ R_{\alpha\beta}- 1.547967 \ \text{eV} \\
&\quad \text{for } 3.0 < R_{\alpha\beta} \leq 3.44 \ \text{Å}.
\end{aligned}
$$

The cut-off distance of 3.44 Å falls between the second and third nearest neighbor distances for crystalline α-Fe, and the potential is adjusted to reproduce the elastic properties of the same material.

Maeda and Takeuchi [193] considered the pure shear of a two-dimensional array of 254 atoms subject to periodic boundary conditions and strained incrementally. Between each strain increment they relaxed the structure, thereby assuring that the deformation proceeded along the path of minimum energy. This model of amorphous iron had a shear modulus 0.58 ± 0.04x that of the corresponding two-dimensional crystalline array, and yielded in a jerky manner to the accompaniment of substantial yield drops at a shear stress 0.07 to 0.10x its shear modulus (i.e., 0.04 to 0.06x the shear modulus of its crystalline analog). Each yield drop was accompanied by substantial plastic (i.e., irreversible) deformation, which deformation appeared to result from the sequential collapse of holes somewhat smaller than the vacancies in the crystalline

Table 21. Ideal Tensile and Compressive Strengths of Ge, Si and Diamond.

Material	Mode of Deformation	σ_{max} GPa	g_{33}	σ_{max}/E
Ge	<100> unconstrained uniaxial tension	15.2	0.181	0.148
Ge	<100> unconstrained uniaxial compression	11.3	-0.185	0.110
Si	<100> unconstrained uniaxial tension	22.4	0.203	0.172
Si	<100> unconstrained uniaxial compression	16.3	-0.221	0.125
C (diamond)	<100> unconstrained uniaxial tension	168	--	0.16
C (diamond)	<100> unconstrained uniaxial compression	132	--	0.126
Ge	<111> unconstrained uniaxial compression	37.1	-0.26	--
Si	<111> unconstrained uniaxial compression	47.3	0.31	--
C (diamond)	<111> unconstrained uniaxial compression	410	--	--

analog at intervals of 5 to 8 atom spacings along the line of maximum shear. This process resulted in slip nucleation and propagation on a macroscopic scale at constant density, because the collapse of one hole favored the creation of another in the same vicinity.

Yamamoto et al. [194] studied the extension parallel to its axis of a cylinder initially some 30 Å in length and 20 Å in diameter that contained 567 atoms. A total of 103 of the atoms situated at the two ends of the cylinder served as "grips" through which the applied force was transmitted to the remaining 464. An iterative procedure was adopted to deform the crystal, with each iteration consisting of three steps: (i) application of a small increment of tensile strain by shifting the grips parallel to the axis of the cylinder; (ii) proportional displacement of the middle 464 atoms to match in order to improve the convergence; and (iii) minimization of the energy of the array by allowing the atoms of the grips to move in the plane perpendicular to the tensile axis and the remainder to move in all three dimensions. After an initial -- and probably spurious -- decrease in energy for the first 2.6% of elongation, this model produced a steadily rising stress-strain curve until it yielded to the accompaniment of a major yield drop at an elongation of 6.4%. This yield drop was accompanied by a major increase in the average atomic displacement per unit strain increment, which strongly suggests that yield involves some long-range correlated motion of the atoms; and subsequent examination of the atomic displacements at yield confirmed this view. It was also noted that the surface atoms were considerably more mobile than those in the interior. Neither in this case, nor in the work of Maeda and Takeuchi [193], however, did these atomic movements manifest themselves in any significant change in the pair distribution function.

The Temperature Dependence of the Ideal Strength

It has already been pointed out that the potential models used in the calculations described above do not take explicit account of the kinetic energy arising from thermal motion of the atoms when evaluating the total energy of the system "crystal plus applied forces". Strictly, therefore, these models should only be used to calculate the ideal strength at 0°K if the best results are to be obtained. Likewise, for consistency, the adjustable parameters in the potential models should be calculated from experimental data appropriate to 0°K, although data obtained at other temperatures have, in fact, been used in some cases.

It is therefore not surprising to find that very little attention has been given to determining the temperature dependence of the ideal strength. An approximate analytical treatment by Zwicky [127] indicates a reduction in the ideal <100> unconstrained uniaxial tensile strength of NaCl proportional to the root mean square amplitude of (small) thermal vibrations of the atoms. And another indication of the influence of temperature on σ_{max} and τ_{max} may be obtained by inserting into the Orowan–Polanyi and Frenkel equations (Equations (1) and (3), respectively) values of the elastic stiffness, lattice parameter and, in the former case only, surface energy appropriate to different temperatures. Tables 6 and 8 include a very few results obtained in this way by Kelly [1]. These suggest that for Cu, W and α–Fe σ_{max} is reduced by ~30% between 0 and 293°K, while τ_{max} for Cu is reduced by < 10% over the same interval. Due to uncertainty about the surface energy values employed, however, this estimate of the temperature dependence of σ_{max} cannot be regarded as very reliable.

A more important criticism of the calculations outlined above, however, is that they ignore the finite possibility of a dislocation loop or a tiny crack being nucleated at the surface or in the interior of the crystal at any temperature above 0°K as a result of the combined influences of the applied stress and thermal fluctuations. In a highly stressed material such an occurrence would very probably cause fracture. The energetics of the nucleation of a dislocation loop at the surface of a stressed crystal was first considered by Frank [196]. Recently, Kelly [1] has made use of Frank's analysis to determine the temperature dependence of τ_{max}. Kelly concludes that for any material, if only total dislocations were to be nucleated, then the ideal shear strength would show little temperature dependence due to this effect. He also calculated that for Cu, if partial dislocations were to be nucleated, then the ideal shear strength might be halved over the temperature interval 300 to 1173°K. The accuracy of these estimates is also difficult to assess, because parameters such as the stacking fault energy and the elastic stiffnesses, which are treated as constants in Kelly's calculation, may actually vary widely with both temperature and applied stress in some unknown manner.

Accordingly, it has to be concluded that no accurate estimates of the temperature dependence of the ideal strength have yet been made.

COMPARISON OF CACULATED AND EXPERIMENTAL RESULTS

Despite the limitations of the various ideal strength calculations reviewed above and the paucity of consistent experimental measurements of high strengths, it is interesting to make such comparisons of theory and experiment as are possible.

In the case of the fcc Lennard-Jones crystal there is good
agreement between the various published calculations, Table 11, and
such discrepancies as do occur almost certainly result only from
differences in computational procedures. There is also good reason
to believe that the potential employed is more realistic in this
case than in any other. Hence, it is unfortunate that there appears
to be no experimental strength data available for anything but
highly imperfect polycrystalline specimens of purely Van der Waals
bonded materials, and that consequently no comparison between
theory and experiment is possible in this case.

Meaningful comparisons can, however, be made in the case of
NaCl, for which a <100> unconstrained uniaxial tensile strength as
high as 1.6 GPa has been recorded [3], Table 1. In comparison, if
the maximum stress criterion for failure is assumed, the Born model
predicts [89,127] σ_{max} = 2.27 and 2 GPa for this mode of deformation, Table 12, and the various calculations using the Born-Mayer
model [89,111,130], Tables 13 and 14, predict 2.66, 2.40 and 2.95
GPa. Nor is the value of σ_{max} significantly altered if the ideal
strength is instead defined as the stress at which the deformation
first becomes unstable with respect to some infinitesimal homogeneous perturbation [2,112,188]. In this case also, therefore,
the several published calculations are in excellent agreement with
one another. Further, these calculations differ from experiment
only by a factor of 1.5-2x. There is also excellent agreement
between the several calculations of the published ideal <100> constrained uniaxial tensile strength of NaCl, regardless of whether
the Born [88] or the Born-Mayer [89,111] potential is used, Tables
12 and 14. And in this mode of deformation, for the Born-Mayer
model at least, the M- and S-strengths are virtually the same as
the strength derived from the maximum applied stress criterion of
failure [112].

For NaCl extended uniaxially parallel to <110> and, more
particularly, parallel to <111>, the agreement between the Born-
Mayer model calculations of Macmillan and Kelly [2,111], Table 14,
and the Born model calculations of Bartenev and Koryak-Doronenko
[128,129], Table 12, is less good. The discrepancies appear to
arise not only from the differences in potential, but also from the
different constraints imposed and from differences in computational
procedures.

In both their Born-Mayer model calculations for NaCl and their
Lennard-Jones model calculations for Ar, Macmillan and Kelly [111]
included the equilibrium lattice parameter of the unstressed crystal
among the physical properties from which they determined the adjustable parameters in the potential. They then used the resultant
model to calculate not only σ_{max}, but also Young's modulus and the
surface energy, for a variety of orientations. The values of these

latter two quantities, together with the value of the lattice
parameter, may be substituted in the Orowan-Polanyi equation,
Equation (1), to obtain another value of σ_{max}. Comparison of these
two values of σ_{max} affords a test of the validity of the Orowan-
Polanyi equation. Note that the value of this test is independent
of the merits of the Born-Mayer and Lennard-Jones potentials as
descriptions of NaCl and Ar, respectively. Macmillan and Kelly
have performed this test, and find [197] that in every case the
Orowan-Polanyi equation, which is the most conservative of the
equations of its genre [23], Table 7, overestimates σ_{max} by 2-4x as
compared to the value obtained directly (i.e., by using Equation (18)
and adopting the maximum stress criterion for failure). For
example, for the <100> uniaxial extension of NaCl, the Orowan-
Polanyi equation predicts σ_{max} = 5.77 and 5.15 GPa in the constrained and unconstrained cases, respectively [197], whereas the
direct calculation gives corresponding values of 2.82 and 2.40 GPa
[111], Table 14.

   In the case of SiO_2 glass, tensile strengths of 13-14 GPa have
been measured experimentally [28,31,32], Table 3. In comparison,
the Orowan-Polanyi equation leads to estimates [1,79] of σ_{max} of
16 GPa, Table 6, and 28 GPa, while Ladik and Náray-Szabó's bond
counting calculation [90] predicts a value of 24 GPa. In this case
also, therefore, theory and experiment agree to within a factor 2x,
even if no allowance is made for the possibility that the Orowan-
Polanyi equation again leads to an overestimate of σ_{max}.

   The maximum <111> tensile strength obtained by Brenner [16] in
his experiments with α-Fe whiskers was 13.1 GPa at an elongation
close to 0.05, Table 1. These values are almost one third of those
obtained by Kelly [1] from the Orowan-Polanyi equation, Table 6.
This equation also suggests that the ideal uniaxial tensile strength
of α-Fe parallel to <100> should be ~0.65x that parallel to <111>,
and is thus in significant disagreement with Milstein's Morse function calculation [144], which latter estimates σ_{max} as only 0.17 GPa
in this mode of deformation. This latter value is far less than
the component of stress acting normal to {100} immediately prior to
failure in the case of Brenner's strongest whisker, resulting in
the unusual situation of the experimental strength exceeding that
predicted theoretically. This discrepancy can almost certainly be
attributed to the inadequacy of the two-body central-force Morse
potential as a model of the bonding in bcc α-Fe. It is therefore
encouraging to note that the more sophisticated potential employed
by Basinski et al. [175] to describe another bcc metal, Na, predicts a maximum <111> tensile elongation to failure of 0.080,
Table 19, that remedies this situation. It is likewise encouraging
that both the volume-dependent potential, Equation (23), employed
by Zhdanov and Konusov [169-171] and the model pseudopotential
employed by Zhdanov and Dobrotvorskii [172] predict elongations to
failure of 0.03 to 0.06 for all of the (bcc) alkali metals in

triaxial tension, Table 17. Evidently, there is some prospect that improved potential models will lead to a better understanding of the ideal strengths of bcc metals than has so far been gained from either the Orowan-Polanyi approach or the use of the Morse potential.

In the case of fcc metals, the Orowan-Polanyi equation, Table 6, and both the band-theoretic and pair potential calculations of Esposito et al. [181] (which employed a maximum applied stress criterion of failure) give estimates of the ideal uniaxial tensile strength up to an order of magnitude larger than the highest strengths obtained experimentally [11,16,17,38,46,47,50,52], Tables 1, 4 and 5. And Milstein's Morse potential calculations, Table 15, predict an ideal tensile strength for Cu in <100> unconstrained tension that is between 3 and 4x the tensile strength of the strongest <100> Cu whiskers [11,17], Table 1. A comparable discrepancy is also apparent between the results of the same author's Morse potential calculations for Ni, Table 15, and the strength of this metal as measured by the two-layer electropolishing method [52], Table 5. It would therefore seem to be worth investigating whether application of a stability criterion of failure would bring the band-theoretic calculations of Esposito et al. more into line with experiment.

In the case of more-or-less covalent materials with a high Peierls-Nabarro stress, such as Si, Ge and Al_2O_3, which typically exhibit only about a two-fold spread in strength between micron diameter whiskers and millimeter diameter rods, there also remains a substantial gap between the highest experimental values of the tensile strength and the calculated values of the ideal strength. For example, the highest strengths obtained from whiskers and bulk specimens of Si are 7.6 GPa [14], Table 1, and 4.14 GPa [23], Table 2, respectively; yet the Orowan-Polanyi calculation gives the ideal <111> uniaxial tensile strength of Si as 32 GPa [1], Table 6, and the corresponding figure for the <100> direction obtained from the third-order elastic stiffnesses is 22.4 GPa [190,191], Table 21. Similarly, although the tensile strength of bulk Al_2O_3 can be as high as 6.85 GPa [24-26], Table 2, and tensile strengths ranging up to 22.3 GPa have been obtained from whiskers [15], Table 1, Gieske [187] obtained estimates of 98, 87 and 87 GPa for the ideal uniaxial tensile strength parallel to <0001>, <11$\bar{2}$0> and <10$\bar{1}$0>, respectively, from the third-order elastic stiffnesses; and the Orowan-Polanyi equation gives a value of 46 GPa for the ideal <0001> uniaxial tensile strength [1], Table 6.

Comparison of the calculated values of the ideal shear strength presented above with experiment is complicated by the fact that the calculations were performed for pure or simple shear but high strength measurements are almost always made in bending or tension. In the experiments, therefore, there normally are both shear and tensile stresses acting across any plane, and the effect of these

normal stresses on the shear behavior can vary markedly from
material to material (compare, for example, the effects of a tensile
stress perpendicular to observed glide plane on τ_{max} for shear on
this plane in the observed glide direction in the cases of the fcc
Lennard-Jones crystal [2,80], NaCl [2,80] and Na [175]).

Nevertheless, at the stress levels at which failure occurred
in the strongest of Crump and Mitchell's Cd whiskers [18], of
Brenner's Cu and Ag whiskers [11,16] and of Kobayashi and Hiki's Cu
whiskers [17], Table 1, the ratio of the resolved shear stress on
known or likely glide systems to the corresponding shear modulus
was comparable to, and in some cases greater than, the values of
τ_{max}/G calculated by Kelly [1] and Mackenzie [86]. In these cases,
therefore, the gap between theory and experiment is closed to
within the uncertainty of the calculations. The fcc Lennard-Jones
crystal, however, exhibits a somewhat larger value of τ_{max}/G for
$\{111\}<11\bar{2}>$ shear (i.e., for shear on the likely glide plane in the
easy direction) of 0.062-0.074 [86,89,111], Table 11.

For the bcc metals, just as in the tensile calculations, the
simple approach exemplified by the Frenkel calculation, Equation (3),
gives values of τ_{max}/G larger than experiment, as do the calculations of Zdhanov and Konusov [171] with the volume-dependent potential described by Equation (23); but the calculations of Basinski
et al. [175] give values in quite good agreement with experiment.

For the non-metals, the values of τ_{max}/G obtained from any
calculation generally are several times larger than those observed
experimentally, regardless of the orientation of shear considered.

One interesting conclusion may be drawn from this comparison
of theory and experiment. Because experiment, which stresses both
the surface and the interior of the solid, in several instances
more of less agrees with theory, which "stresses" only the interior,
it follows that the surface of a solid cannot be very much weaker
than the interior.

Finally, it should be noted that the calculations of ideal
strengths reviewed here were all made on an assumption of homogeneous deformation. Consequently, the results should strictly
only be applied to problems of inhomogeneous deformation in which
stress and strain vary but little over distances comparable with
the range of the interatomic forces. This is not the case at the
tip of a crack, at the core of a dislocation, or at a particle-
matrix interface, for example, so that due caution must be exercised in applying the results of these calculations to such problems.

ACKNOWLEDGEMENT

This review draws heavily on the author's two previous reviews
of the same subject [198,199].

REFERENCES

1. A. Kelly, Strong Solids, 2nd edn., Oxford University Press, Oxford (1973).
2. N. H. Macmillan, Ph.D. Thesis, Cambridge University, Cambridge (1969).
3. A. F. Ioffe, The Physics of Crystals, McGraw-Hill, New York (1928).
4. Z. Gyulai, Z. Phys. 138:317 (1954).
5. G. W. Sears, J. Chem. Phys. 36:862 (1962).
6. R. B. Sharma, J. Appl. Phys. 41:3371 (1970).
7. F. N. Tavadze, G. G. Surmava, A. A. Nikolaishvili and S. E. Makovets, Sov. Phys. Solid State 15:901 (1973).
8. T. J. Davies and P. E. Evans, Nature 207:254 (1965).
9. C. C. Evans, J. E. Gordon, D. M. Marsh and J. N. Parratt, Report No. 133, Tube Investments Research Lab., Hinxton Hall, Cambridge (1961).
10. E. Ryshkewitz, Sci. and Tech. p. 54, Feb. (1962).
11. S. S. Brenner, in: "Growth and Perfection of Crystals", R. H. Doremus, B. W. Roberts and D. Turnbull, eds., Wiley, New York (1958).
12. W. W. Webb and W. D. Forgeng, Acta Met. 6:462 (1958).
13. V. S. Postnikov, S. A. Ammer and A. I. Drozhzhin, Sov. Phys. Solid State 14:2636 (1973).
14. A. V. Sandulova, P. S. Bogoyavlenskii and M. E. Dronyuk, Sov. Phys. Solid State 5:1883 (1964).
15. P. J. Soltis, Bull. Am. Phys. Soc. 10:163 (1965).
16. S. S. Brenner, J. Appl. Phys. 27:1484 (1956).
17. H. Kobayashi and Y. Hiki, Phys. Rev. B7:594 (1973).
18. J. C. Crump and J. W. Mitchell, J. Appl. Phys. 41:717 (1970).
19. R. Bacon, J. Appl. Phys. 31:283 (1960).
20. O. L. Blakslee, D. G. Proctor, E. J. Seldin, G. B. Spence and T. Weng. J. Appl. Phys. 41:3373 (1970).
21. O. W. Johnson and P. Gibbs, J. Appl. Phys. 34:2852 (1963).
22. T. L. Johnston, R. J. Stokes and C. H. Li, Acta Met. 6:713 (1958).
23. J. J. Gilman, in: Proc. Symposium on the Physics and Chemistry of Ceramics, C. Klingsberg, ed., Gordon and Breach, New York (1963). (Also published in: Mechanical Behavior of Crystalline Solids, U.S. National Bureau of Standards, Gaithersburg, Maryland (1963)).
24. J. G. Morley and B. A. Proctor, Nature 196:1082 (1962).
25. F. P. Mallinder and B. A. Proctor, Phil. Mag. 13:197 (1966).
26. F. P. Mallinder and B. A. Proctor, Proc. Brit. Ceram. Soc. 6:9 (1966).
27. W. F. Thomas, Phys. and Chem. Glasses 1:4 (1960).
28. J. G. Morley, P. Andrews and I. Whitney, Phys. and Chem. Glasses 5:1 (1964).
29. P. W. France, M. J. Paradine, M. H. Reeve and G. R. Newns, J. Mater. Sci. 15:825 (1980).

30. F. O. Anderegg, Ind. and Eng. Chem. 31:290 (1939).

31. W. B. Hillig, J. Appl. Phys. 32:741 (1961).

32. W. B. Hillig, in: C. R. Symposium sur la Résistance Mécanique du Verre et les Moyens de l'Améliorer, Union Scientifique Continentale du Verre, Charleroi, Belgium (1962).

33. B. A. Proctor, I. Whitney and J. W. Johnson, Proc. Roy. Soc. A297:534 (1967).

34. C. O. Hulse, J. Am. Ceram. Soc. 44:572 (1961).

35. T. L. Johnston, Reported in Reference 11.

36. R. Zukowski and R. Gaze, Nature 183:35 (1959).

37. J. Aveston, J. Mater. Sci. 4:625 (1969).

38. N. Gane, Private communication (1969).

39. A. V. Ulitovsky, Pribory Tekh. Eksp. 3:115 (1957).

40. G. F. Taylor, Phys. Rev. 23:655 (1924).

41. S. Yajima, Phil. Trans. Roy. Soc. A294:419 (1980).

42. W. S. Williams, D. A. Steffens and R. Bacon, J. Appl. Phys. 41:4893 (1970).

43. C. P. Talley, J. Appl. Phys. 30:1144 (1959).

44. R. J. Young, Private communication (1979).

45. P. Smith and P. J. Lemstra, J. Mater. Sci. 15:505 (1980).

46. N. Gane and F. P. Bowden, J. Appl. Phys. 39:1432 (1968).

47. N. Gane, Proc. Roy. Soc. A317:367 (1970).

48. N. H. Macmillan and N. Gane, J. Appl. Phys. 41:672 (1970).

49. L. M. Brown, G. R. Woolhouse and U. Valdré, Phil. Mag. 17:781 (1968).

50. L. M. Brown and G. R. Woolhouse, Phil. Mag. 21:329 (1970).

51. D. A. Smith and K. M. Bowkett, in: Proc. 4th European Regional Conf. on Electron Microscopy, Vol. 1, Tipografia Poliglott Vaticana, Rome (1968).

52. G. D. W. Smith, D. A. Smith, G. S. Taylor, M. J. Goringe and K. Easterling, in: High Voltage Electron Microscopy (Proc. 3rd Intl. Conf. on High Voltage Electron Microscopy Oxford (1973)), P. R. Swann, C. J. Humphreys and M. J. Goringe, eds., Academic Press, New York (1974).

53. J. R. Patel and A. R. Chaudhuri, J. Appl. Phys. 34: 2788 (1963)

54. J. J. Gilman, J. Appl. Phys. 44:675 (1973).

55. J. J. Gilman, Physics Today 28:46 (1975).

56. J. J. Gilman, J. Appl. Phys. 46:1625 (1975).

57. J. J. Gilman, in: Physics of Strength and Plasticity, A. S. Argon, ed., The M.I.T. Press, Cambridge, Massachusetts (1969).

58. J. J. Gilman, in: Dislocation Dynamics, A. R. Rosenfeld, G. T. Hahn, A. L. Bement and R. T. Jaffee, eds., McGraw-Hill, New York (1968).

59. W. C. Leavengood and T. S. Vong, J. Appl. Phys. 31:1416 (1960).

60. M. F. Ashby and J. Logan, Scripta Met. 7:513 (1973).

61. N. Rivier, Phil. Mag. A 40:859 (1979).

62. J. P. Hirth, J. Mater. Sci. 12:2540 (1977).

63. J. C. M. Li, in: Metallic Glasses, Am. Soc. Metals, Metals Park Ohio (1978).

64. L. A. Davis, in: Rapidly Quenched Metals, N. J. Grant and B. C. Geissen, eds., The M.I.T. Press, Cambridge, Massachusetts (1976).
65. A. S. Argon, Acta Met. 27:47 (1979).
66. C. A. Pampillo, J. Mater. Sci. 10:1194 (1975).
67. F. P. Mallinder and B. A. Proctor, Phys. and Chem. Glasses 5:91 (1964).
68. J. W. Mitchell, Private communication (1975).
69. A. L. Ruoff and J. Wanagel, Science 198:1037 (1977).
70. B. Ralph, Ph.D. Thesis, Cambridge University, Cambridge (1964).
71. P. J. Smith and D. A. Smith, Phil. Mag. 21:907 (1970).
72. P. J. Birdseye and D. A. Smith, Surf. Sci. 23:198 (1970).
73. H. C. Eaton and R. J. Bayuzick, Surf. Sci. 70:408 (1978).
74. D. K. Bowen and V. Vitek, phys. stat. sol. (b)92:645 (1979).
75. M. Polanyi, Z. Phys. 7:323 (1921).
76. E. Orowan, Z. Krist. A89:327 (1934).
77. E. Orowan, Rep. Prog. Phys. 12:185 (1949).
78. T. L. Johnston, in: Mechanical Behavior of Crystalline Solids, U.S. National Bureau of Standards, Gaithersburg, Maryland (1963).
79. U.S. National Academy of Sciences (Materials Advisory Board), Report MAB-221-M, Washington (1966).
80. A. Kelly, W. R. Tyson and A. H. Cottrell, Phil. Mag. 15:567 (1967).
81. A. Kelly, W. R. Tyson and A. H. Cottrell, Can. J. Phys. 45:883 (1967).
82. F. A. McClintock and W. R. O'Day, Unpublished work (1965).
83. G. K. Demishev and G. M. Bartenev, Sillikattechnik 17:215 (1966).
84. G. K. Demishev and G. M. Bartenev, Sillikattechnik 17:344 (1966).
85. J. Frenkel, Z. Phys. 37:572 (1926).
86. J. K. Mackenzie, Ph.D. Thesis, Bristol University, Bristol (1949).
87. J. F. Nye, Physical Properties of Crystals, Oxford University Press, Oxford (1960).
88. J. H. de Boer, Trans. Faraday Soc. 32:10 (1936).
89. W. R. Tyson, Phil. Mag. 14:925 (1966).
90. J. Ladik and I. Náray-Szabó, Nature 188:226 (1960).
91. K. E. Perepelkin, Sov. Mater. Sci. 6(2):213 (1970).
92. K. E. Perepelkin, Sov. Mater. Sci. 8(2):198 (1972).
93. A. I. Gubanov and A. D. Chevychelov, Sov. Phys. Solid State 5:1898 (1964).
94. A. I. Gubanov and A. D. Chevychelov, Sov. Phys. Solid State 4:681 (1962).
95. M. Born and K. Huang, Dynamical Theory of Crystal Lattices, Oxford University Press, Oxford (1966).
96. A. E. H. Love, A Treatise on the Mathematical Theory of Elasticity, Dover, New York (1944).

97. R. Hill, J. Mech. Phys. Sol. 16:229 (1968).
98. R. Hill, Math. Proc. Camb. Phil. Soc. 77:225 (1975).
99. R. Hill and F. Milstein, Phys. Rev. B15:3087 (1977).
100. R. N. Thurston, in: Physical Acoustics, Vol. 1, Part A, W. P.
       Mason, ed., Academic Press, New York (1964).
101. F. Milstein, J. Mater. Sci. 15:1071 (1980).
102. G. P. Parry, Quart. J. Mech. Appl. Math. 31:1 (1978).
103. D. C. Wallace, Adv. Mater. Res. 3:331 (1968).
104. E. R. Dobbs and G. O. Jones, Rep. Prog. Phys. 20:516 (1967).
105. G. L. Pollack, Rev. Mod. Phys. 36:748 (1964).
106. R. H. S. Winterton, Contemp. Phys. 11:559 (1970).
107. M. P. Tosi, Solid State Physics 16:1 (1964).
108. I. M. Torrens, Interatomic Potentials, Academic Press,
       New York (1972).
109. Interatomic Potentials and Simulation of Lattice Defects,
       P. C. Gehlen, J. R. Beeler, Jr., and R. I. Jaffee, eds.,
       Plenum Press, New York (1972).
110. M. Born and R. Fürth, Math. Proc. Camb. Phil. Soc. 36:454 (1940).
111. N. H. Macmillan and A. Kelly, Proc. Roy. Soc. A330:291 (1972).
112. N. H. Macmillan and A. Kelly, Proc. Roy. Soc. A330:309 (1972).
113. H. R. Moeller and C. F. Squire, Phys. Rev. 151:689 (1966).
114. M. Gsänger, H. Egger and E. Lüscher, Phys. Lett. 27A:695 (1968).
115. H. Egger, M. Gsänger, E. Lüscher and B. Dorner, Phys. Lett.
       28A:433 (1968).
116. D. N. Batchelder, M. F. Collins, B. C. G. Haywood and G. R.
       Sidey, J. Phys. C3:249 (1970).
117. G. J. Keeler and D. N. Batchelder, J. Phys. C3:510 (1970).
118. B. Dorner and H. Egger, phys. stat. sol. (b)43:611 (1971).
119. D. Stansfield, Proc. Phys. Soc. 72:854 (1958).
120. L. Meyer, C. S. Barrett and P. Haasen, J. Chem. Phys. 40:2744
       (1964).
121. R. Bullough, H. R. Glyde and J. A. Venables, Phys. Rev. Lett.
       17:249 (1966).
122. J. A. Venables and D. J. Ball, in: Electron Microscopy 1966,
       Proc. 6th Intl. Congr. Electron Microscopy, R. Uyeda,
       ed., Maruzen, Tokyo (1966).
123. J. M. T. Thompson, Nature 254:392 (1975).
124. J. M. T. Thompson, Ann. New York Acad. Sci. 316:553 (1978).
125. J. M. T. Thompson and P. A. Shorrock, Nature 260:598 (1976).
126. J. M. T. Thompson and P. A. Shorrock, J. Mech. Phys. Sol.
       23:21 (1975).
127. F. Zwicky, Phys. Z. 24:131 (1923).
128. G. M. Bartenev and E. G. Koryak-Doronenko, phys. stat. sol.
       24:443 (1967).
129. G. M. Bartenev and E. G. Koryak-Doronenko, Sov. Phys. J.
       11(3):57 (1968).
130. P. O. Szomor, phys. stat. sol. 28:529 (1968).
131. J. E. Mayer, J. Chem. Phys. 1:270 (1933).
132. M. L. Huggins, J. Chem. Phys. 5:143 (1937).
133. M. L. Huggins, J. Chem. Phys. 15:212 (1947).

134. E. R. Cohen, J. W. R. DuMond, T. W. Layton and J. S. Rollett, Phys. Rev. 27:363 (1955).

135. G. Simmons, J. Grad. Res. Center 34:1 (1965) (Since updated and republished as G. Simmons and H. Wang, Single Crystal Elastic Constants and Calculated Aggregate Properties, The M.I.T. Press, Cambridge, Massachusetts (1971)).

136. S. G. Lipsett, F. M. G. Johnson and O. Maass, J. Am. Chem. Soc. 49:1940 (1927).

137. E. Hutchinson and K. E. Manchester, Rev. Sci. Instr. 26:364 (1955).

138. G. C. Benson, H. P. Schreiber and F. van Zeggeren, Can. J. Chem. 34:1553 (1956).

139. V. D. Kuznetsov and P. Teterin, Surface Energy of Solids, H.M.S.O., London (1957).

140. W. H. Class, Ph.D. Thesis, Columbia University, New York (1964).

141. P. L. Gutshall and G. E. Gross, J. Appl. Phys. 36:2459 (1965).

142. S. M. Wiederhorn, R. L. Moses and B. L. Bean, J. Am. Ceram. Soc. 53:18 (1970).

143. F. Milstein, Phys. Rev. B2:512 (1970).

144. F. Milstein, Phys. Rev. B3:1130 (1971).

145. F. Milstein, phys. stat. sol. 48:681 (1971).

146. F. Milstein, J. Appl. Phys. 44:3825 (1973).

147. F. Milstein, J. Appl. Phys. 44:3833 (1973).

148. K. Huang, F. Milstein and J. A. Baldwin Jr., Phys. Rev. B10:3635 (1974).

149. F. Milstein and R. Hill, J. Mech. Phys. Sol. 25:457 (1977).

150. F. Milstein and R. Hill, J. Mech. Phys. Sol. 26:213 (1978).

151. F. Milstein and K. Huang, Phys. Rev. B18:2529 (1978).

152. F. Milstein and R. Hill, J. Mech. Phys. Sol. 27:255 (1979).

153. F. Milstein and R. Hill, Phys. Rev. Lett. 43:1411 (1979).

154. F. Milstein and K. Huang, Phys. Rev. B19:2030 (1979).

155. F. Milstein and R. Hill and K. Huang, Phys. Rev. B21:4282 (1980).

156. F. Milstein and B. Farber, Phys. Rev. Lett. 44:277 (1980).

157. F. Milstein and B. Farber, Phil. Mag. A42:19 (1980).

158. K. P. Thakur and H. D. B. Jenkins, J. Phys. F10:1925 (1980).

159. N. H. Macmillan and A. Kelly, Mater. Sci. and Eng. 12:79 (1973).

160. M. Born, Math. Proc. Camb. Phil. Soc. 36:160 (1940).

161. R. D. Misra, Math. Proc. Camb. Phil. Soc. 36:173 (1940).

162. M. Born and R. D. Misra, Math. Proc. Camb. Phil. Soc. 36:466 (1940).

163. R. Fürth, Math. Proc. Camb. Phil. Soc. 37:34 (1941).

164. R. Fürth, Math. Proc. Camb. Phil. Soc. 37:177 (1941).

165. S. C. Power, Math. Proc. Camb. Phil. Soc. 38:61 (1942).

166. H. W. Peng and S. C. Power, Math. Proc. Camb. Phil. Soc. 38:67 (1942).

167. M. Born, Math. Proc. Camb. Phil. Soc. 38:82 (1942).

168. L. A. Girifalco and V. G. Weizer, Phys. Rev. 114:687 (1959).

169. V. A. Zdhanov and V. F. Konusov, Sov. Phys. J. 8(4):23 (1965).

170. V. A. Zdhanov and V. F. Konusov, Sov. Phys. J. 9(6):51 (1966).

171. V. A. Zhdanov and V. F. Konusov, <u>Sov. Phys. J.</u> 10(3):64 (1967).
172. V. A. Zhdanov and V. V. Dobrotvorskii, <u>Sov. Phys. Solid State</u> 15:381 (1973).
173. G. L. Krasko and Z. A. Gurskii, <u>JETP Letts.</u> 9:363 (1969).
174. Z. A. Gurskii and G. L. Krasko, <u>Sov. Phys. Doklady</u> 16:298 (1971).
175. Z. S. Basinski, M. S. Duesbery and R. Taylor, <u>in</u>: Proc. 2nd Intl. Conf. on the Strength of Metals and Alloys, Am. Soc. Metals, Cleveland (1971).
176. M. S. Duesbery and R. Taylor, <u>Phys. Lett.</u> 30A:496 (1969).
177. Z. S. Basinski, M. S. Duesbery, A. Pogany, R. Taylor, Y. P. Varshni, <u>Can. J. Phys.</u> 48:1480 (1970).
178. C. S. Barrett, <u>Acta Cryst.</u> 9:671 (1956).
179. R. H. Martinson, <u>Phys. Rev.</u> 178:902 (1969).
180. A. Kelly, Private communication from M. S. Duesbery (1972).
181. E. Esposito, A. E. Carlsson, D. D. Ling, H. Ehrenreich and C. D. Gelatt Jr., <u>Phil. Mag.</u> A41:251 (1980).
182. A. E. Carlsson, C. D. Gelatt Jr., and H. Ehrenreich, <u>Phil. Mag.</u> A41:241 (1980).
183. J. M. Ziman, <u>Solid State Physics</u> 26:1 (1971).
184. A. R. Williams, J. Kübler and C. D. Gelatt, <u>Phys. Rev.</u> B19:6094 (1979).
185. R. A. Johnson and W. D. Wilson, Reference 109, p. 301.
186. G. R. Barsch, <u>J. Metals</u> 17:1028 (1965).
187. J. H. Gieske, Ph.D. Thesis, The Pennsylvania State University (1968).
188. R. C. Hollinger, Ph.D. Thesis, The Pennsylvania State University (1970).
189. B. P. Coleman and W. Noll, <u>Arch. Rat. Mech. Anal.</u> 4:97 (1959).
190. A. L. Ruoff, <u>J. Appl. Phys.</u> 49:197 (1978).
191. D. A. Nelson and A. L. Ruoff, <u>J. Appl. Phys.</u> 50:2763 (1979).
192. E. D. Shchukin and V. S. Yushchenko, <u>J. Mater. Sci.</u> 16:313 (1981).
193. K. Maeda and S. Takeuchi, <u>phys. stat. sol.</u> (a)49:685 (1978).
194. R. Yamamoto, H. Matsuoka and M. Doyama, <u>phys. stat. sol.</u> (a)51:163 (1979).
195. R. A. Johnson, <u>Phys. Rev.</u> A134:1329 (1964).
196. F. C. Frank, <u>in</u>: Symposium on Plastic Deformation of Crystalline Solids, Carnegie Inst. Tech., Pittsburgh (1950).
197. N. H. Macmillan and A. Kelly, <u>Mater. Sci. and Eng.</u> 10:139 (1972).
198. N. H. Macmillan, <u>J. Mater. Sci.</u> 7:239 (1972).
199. N. H. Macmillan, <u>Can. Met. Qty.</u> 13:555 (1974).

DISCUSSION

Comment by R. Bullough:

I am not in the least surprised that the various empirical potentials you cited failed to provide any agreement with mechanical strength, since no strength data was put into them and they are empirical. Some published potentials have, however, been fitted to non-harmonic properties and therefore should have a better hope of achieving a reasonable correlation.

Reply:

I was careful in my talk to point out that the process of adjusting any more-or-less empirical potential to fit harmonic properties and then using it to calculate an anharmonic property such as the ideal strength is fraught with danger because it involves a long extrapolation along the interatomic force-distance curve. The corollary, which I stated explicitly in my paper, though not in my talk, is -- as you correctly point out -- that a potential fitted to anharmonic properties would be more likely to lead to a reliable estimate of the ideal strength. To the best of my knowledge, however, no-one has yet calculated the ideal strength from an empirical potential fitted to anharmonic data, although such potentials have been widely used to study defects (see Refs. 108 and 109 in my paper).

You might also note that I was careful when comparing theory and experiment merely to point out the discrepancy and not to say that the the theory was wrong and the experiment right. The fact is that in no case can we yet be certain that the highest strength recorded experimentally really is the ideal strength, although there is good reason to hope it might be in the cases of Cd and SiO_2; and until we gain such certainty we cannot be sure just how good or bad the calculations published so far really are.

PHYSICS OF FRACTURE

Robb Thomson

Center for Materials Science
National Bureau of Standards
Washington, D.C.   20234

INTRODUCTION

In this paper my task is to set a perspective for the physical problems posed by fracture. Fracture is not a subject often mentioned among solid state physicists, but the field of fracture research is extraordinary in the range of disciplines it spans and draws from, and it will be my purpose to demonstrate that there is an important physical dimension to this subject.

The title of the conference in its focus on atomistics certainly hints at the physical dimension, and it gives us our point of departure. However, before diving directly into a full-scale discussion of discrete lattices, I will first provide in the next section a very succinct background statement of the elastic description of a brittle crack. In the third section, discrete lattice theories are addressed directly, but first in one dimension for the purpose of emphasizing the kinds of phenomena for which an atomistic theory is important. This discussion points to the application of discrete lattice theories to the rates of atomic and chemical processes at the crack tip. In the fourth section we lay out the statistical mechanical framework for the thermal equilibrium of a crack and for thermally activated crack growth both for intrinsic lattices, and including interactions with external chemical environments. Then follows in section five a more detailed presentation and critique of the theoretical techniques for calculating the structure of lattices containing cracks in two and three dimensions. A short statement of the current status of the quantum theory of binding as it relates to defects will be included. Finally, in the sixth section, the question of stability of the crack in the lattice with respect to the emission of dislocations is discussed. A concluding short section summarizes the crucial points where future progress looks most promising.

II.   THE CONTINUUM BACKGROUND.   ANTI-PLANE STRAIN[1]

The simplest crack possesses a planar cleavage surface and a
straight crack line. It is thus analyzable as a defect in two
dimensions. The simplest two-dimensional elastic problems are those
in isotropic anti-plane strain conditions, and the crack in this case
is called a Mode III crack (see Fig. 1b). In anti-plane strain, the
displacement is solely in the x_3 direction. The elastic equilibrium
equations (the field equations) are then simply

$$\frac{\partial^2 u_3}{\partial x_1^2} + \frac{\partial^2 u_3}{\partial x_2^2} = \nabla^2 u_3 = 0 \tag{1}$$

This equation (Laplace's equation) immediately suggests that we
invoke complex functions, and we introduce as the complete solution
of (1), the complex function $\Omega(z)$ from which u_3 is defined as its
imaginary part.

$$u_3(x,y) = \text{Im}(\Omega(z))$$

$$z = x+iy = x_1+ix_2 \tag{2}$$

Hookes law has the form

$$\sigma(z) = \sigma_{32}(x_1,x_2) + i\sigma_{31}(x_1,x_2) = \mu \frac{d\Omega(z)}{dz} \tag{3}$$

from which the strain energy density, W, is

$$W = |\sigma^2| \, 2\mu \tag{4}$$

The simplest basic solutions to the crack problem which describe
the stress field in the immediate (singular) vicinity of the crack are
found by looking for functions $\Omega(z)$ which satisfy the boundary conditions on the cleavage plane, and are simple powers of z, from which
power series solutions can be constructed. On any open surface, the
boundary condition is that no force is transmitted across the surface,
so in general $\sigma_{ij} n_j = 0$ where $\underline{n}$ is the surface normal. In our case
with the cleavage plane on the negative x-axis, Re $\sigma(z) = 0$ for $\theta=\pi$.
The lowest order singularity in stress for this is given by

$$\Omega = \frac{K}{\mu} \sqrt{(2z/\pi)}$$

$$\sigma = \frac{K}{\sqrt{(2\pi z)}} \tag{5}$$

The constant of proportionality, K, in (5) is chosen to have its
traditional form, and K is called the stress intensity factor.

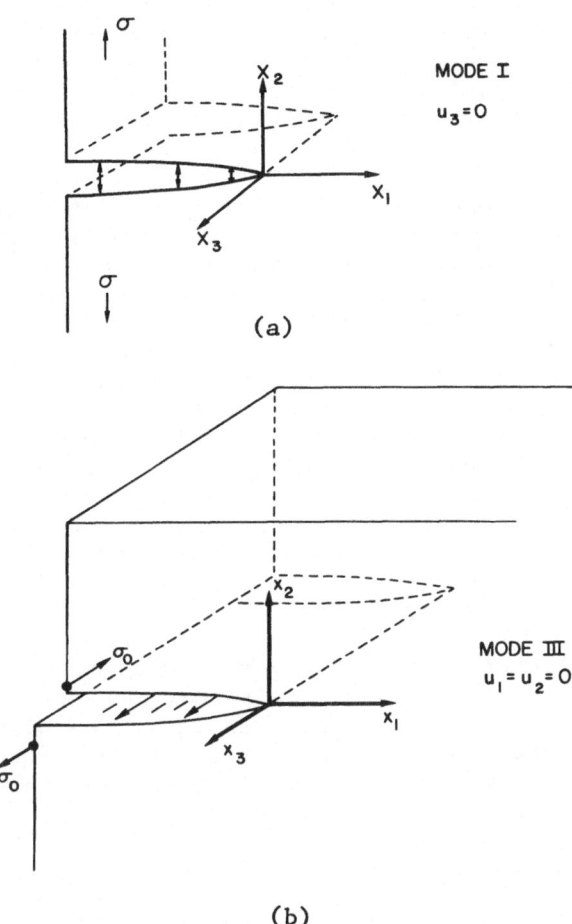

Fig. 1. Two types of simple cracks. (a) in mode I, in response to
tensile stress, a crack opens up on a cleavage plane normal
to the principal stress direction. The cleavage plane is a
mathematical cut in the medium across which no stresses are
transmitted. This type of crack with the cleavage plane displacements normal to the plane is the prototype of a physical
crack; (b) in mode III the stress is a pure shear with the
displacement lying entirely within the cleavage plane in the
direction of the crack line. The crack line is chosen to be
the x_3 axis, and the cleavage plane is the negative $x_1 x_2$ half
plane. Since in a physical crack of this type, atoms facing
one another across the cleavage plane would continue to be
within one another's force fields, such cracks serve principally
as mathematical models.

That (5) is only the first term in a Laurent series expansion is easily seen from the physical fact that a crack will close, and the defect in the solid will disappear completely if the external stress is removed. The crack is indeed the only primitive defect in a lattice which requires an external influence for its existence. Hence, in the simple problem described, the stress at ∞ must approach a finite value, not zero as implied by (5).

We describe the more complete problem by imposing a finite constant stress at infinity, and ask for the modification to the constant stress in the vicinity of the crack caused by the fact that making a cut on the cleavage plane requires that Re $\sigma(z) = o$ on that surface. We also think in terms of a finite crack (Fig. 2). We obtain the result by means of the standard Greens function argument if we add a force (positive on the upper surface and negative on the lower) which exactly cancels the external stress at the point of application on the cleavage surface. By integrating such forces over the whole crack surface, we obtain the desired solution. A requirement of this technique, of course, is knowledge of the appropriate Greens function for a dipole force exerted on the cleavage surface of the crack. The Greens function for the stress is

$$G(z,x') = \frac{1}{i\pi} \sqrt{\frac{(a^2-x'^2)}{(a^2-z^2)}} \frac{1}{z-x'} \tag{6}$$

The term $1/(z-x')$ when integrated over a small half circle on the positive cleavage surface gives a unit force there, and a negative unit force on the negative cleavage surface. The square root functions make $1/\sqrt{z}$ type singularities appear at the crack tip, but also ensure that near z', the crack-like singularities cancel. The total stress is thus

$$\sigma = \sigma_o -\sigma_o \int_{-a}^{a} G(z,x')dx' \tag{7}$$

The general integration of (7) is not possible analytically, but it can be expressed in asymptotic form near one end of the crack. The result can be given the form familiar from (5), $\sigma = K/\sqrt{2\pi z}$ with

$$K = \sigma_o \sqrt{(\pi a)} \tag{8}$$

We will develop the concept of the equilibrium crack by means of a theorem of Eshelby[2] which states that the force on a defect in the body is given by the surface integral

$$f_i = \oint P_{ij} \, dS_i$$

$$P_{ij} = \frac{1}{2} \sigma_{km} u_{k,m} \delta_{ij} -\sigma_{kj} u_{k,i} \tag{9}$$

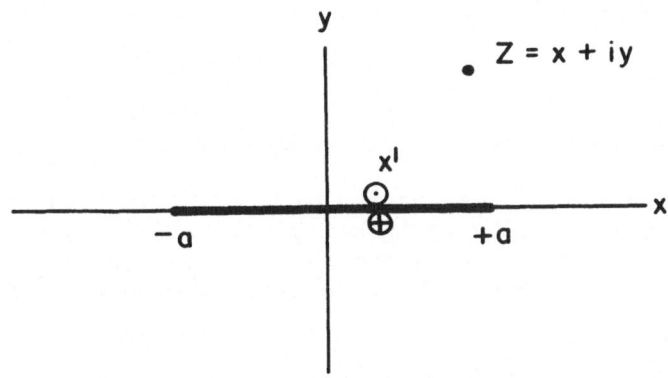

Fig. 2.   A mode III crack of finite length in two dimensions.   The
cleavage plane lies in the xy plane from x=-a to x=+a.
Complex notation is used so that z=x+iy.   Equal and opposite
point forces are exerted on the two faces of the cleavage
plane as shown, whose values on the planes just compensate
for a pre-existing stress in the medium.

P_{ij} is the energy momentum tensor of the elastic body, and the surface
is any surface enclosing the defect or defects.   The defect appears in
this theory as a singularity in the stress field.

In two-dimensional anti-plane strain, this theorem for the force
per unit length of a line defect takes the form

$$\bar{f} = \frac{1}{4\pi i} \oint \sigma^2 \, ds \tag{10}$$

which by the residue theorem becomes

$$\bar{f} = \frac{1}{2\mu} \sum_{s} \mathrm{Res}(\sigma^2) \tag{11}$$

In this form, the force may be exerted on several singularities, but
we shall apply it only to a crack.   The force in the positive x-
direction is the real part of f and the force in the negative y
direction is the imaginary part of f.   For a single crack, (5) gives

$$\bar{f} = K^2/2\mu \tag{12}$$

This force, which is that tending to extend the crack, and caused by
the external weights and springs which exert stress on the body is
balanced by any internal forces exerted by the material.   Since to
extend the crack means that energy must be stored in the cleavage
surfaces, the force (12) will be balanced by the surface tension of

the two cleavage surfaces, which tends to close the crack. Thus in equilibrium,

$$K^2 = 4\mu\gamma \tag{13}$$

where γ is the intrinsic surface tension of the solid.

Finally, the work done on the system as a function of crack length can be obtained by integrating (12) with K given by (8). If we write the energy of the system as the negative of the work done on it, plus the increase of surface energy, we obtain the Griffith result

$$E = -\frac{\sigma_o^2 \pi a^2}{2\mu} + 4\gamma a \tag{14}$$

In writing (14) we remember the system has two crack ends a distance 2a apart, where $\bar{f}$ gives only the force opening one end at a time. Equation (14) includes both the energy stored in the form of elastic strain of the body and the decrease in the potential energy of external weights which arises because the external stress system is a necessary part of the existence of the crack. The external work term is included automatically because it is an explicit part of Eshelby's derivation of the force on the elastic singularity. We note the form of (14) is an inverted parabola, so the equilibrium crack described by (13) is metastable.

III. DISCRETE LATTICES – GRAPHICAL SOLUTION

The most important physical effect of introducing a lattice into the problem of fracture is that the lattice interferes with the free motion of the crack. That is, the lattice introduces energy barriers on top of the smooth energy parabola of Eqn. (14). The easiest way to demonstrate this effect is from a simplified one-dimensional model of a crack. This model is shown in Fig. 3. It is composed of two chains of atoms. The atoms of each chain are tied together with horizontal flexible springs, and the chains are then attached by vertical stretchable springs (we shall term them bonds) for j=n. The first n-1 atom of the chains are unattached, and when external forces are applied at the ends (j=o), the stress is concentrated on the tip atom in the same way as stress is concentrated at the tip of a true crack in a crystal.

The solution of the onedimensional crack model can be obtained in closed form[3], but its character is best seen by considering the compliance of the system to the applied force, F, in a graphical solution. We consider all springs initially to be linear, with displacements proportional to the local force on the springs. Since the entire system is assumed to be linear, the displacement response of the atom at j=o is also linear in the force, F, and we can write

$$u_o = \lambda(n) F \tag{15}$$

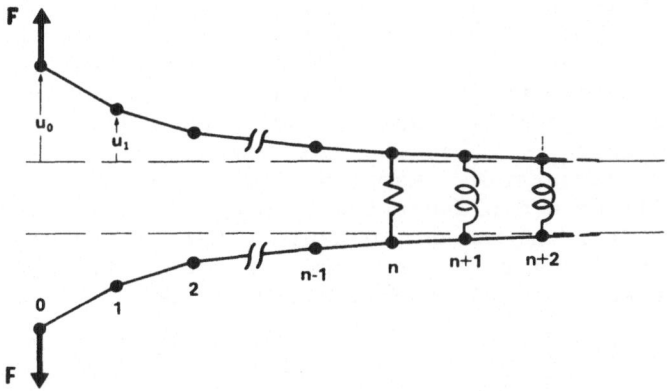

Fig. 3. One-dimensional model of a crack. Two semi-infinite chains composed of atoms tied to one another by bendable springs are attached by stretchable springs beginning at atom n. The first stretchable spring at n is nonlinear. All other stretchable springs $j \geq n+1$ are linear. Equal and opposite forces are exerted at the end to open the chains.

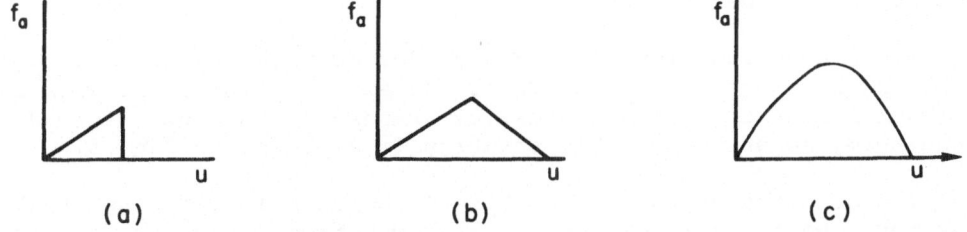

Fig. 4. Examples of nonlinear atomic forces at the crack tip bond. (a) bond snapping force: the nonlinear bond at the crack tip is linear up to some critical value beyond which the force drops discontinuously to zero; (b) bilinear force: the force linearly increases up to a critical value after which it drops linearly to zero; (c) quadratic force: the force is an inverted parabola.

F is the force shown in Fig. 3, and 2 u_o is the total displacement of the two end atoms as shown in Fig. 3. λ is the compliance, and depends upon the "length of the crack", n.

We are, however, most interested in nonlinear problems where at least some atom bonds in the crack tip region have been stretched into their nonlinear regimes. These atoms constitute what is termed the core or cohesive region of the crack. We shall incorporate non-linearity into our simple model by adding an additional arbitrary attractive force pair, f, at the atom pair (j=n) tending to close the crack. Representative examples of this force are shown in Fig. 4. This force will be a nonlinear function of the displacement of the atom pair, but for the moment we neglect this functional relation-ship and think of the force as a specified quantity independent of all u_j. Then we can write the linear relation

$$u_o = \lambda(n) \, F + \eta(n) \, f \tag{16}$$

$\eta(n)$ is a compliance function which relates u_o to f. We can in the same way write for the displacement of the atom pair at n

$$u_n = \lambda'(n) \, F + \eta'(n) \, f$$

or

$$F = \frac{1}{\lambda'} \, u_n - \frac{\eta'}{\lambda'} \, f(u_n) \tag{17}$$

where we have now put in the functional relation between f and u_n. Since $f(u_n)$ is an attractive force, by definition in (16) and (17), f is normally a negative quantity.

The second equation of (17) provides us with a graphical solution of the problem because it says that the external force, F, with the relevant linear compliance terms, is composed of a linear term and bond force term, separately. This solution is depicted in Fig. 5.

The solution shown in Fig. 5 is incomplete in the sense that it does not repeat itself during crack growth from one value of n to the next after the bond at n is broken. To be consistent, after the bond at n is broken, the bond at n+1 becomes stretched into its nonlinear regime, and the process repeats. In Fig. 6, we have sketched the solution for a crack moving continuously from one lattice position to the next. The figure shows that for a given value of F, there are many solutions for various values of n, showing that the crack is stable for a number of lattice positions. This is in direct contrast to the result in the continuum where for a given external stress, there is only one equilibrium position for the crack. Figure 6 is drawn for the functional dependence $F(u_o)$, not $F(u_n)$ as in Fig. 5. We do this for the purposes of the next paragraph. One can go back and forth from $F(u_n)$ to $F(u_o)$ by means of (16) and (17).

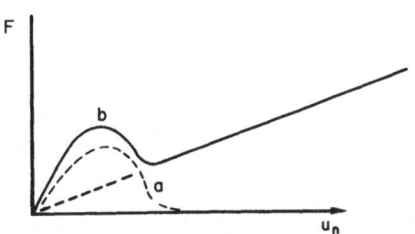

Fig. 5.   Crack solution. According to Eqn. (17), the external force
          is a linear function of displacement plus a linear term in
          the force law function at the tip.

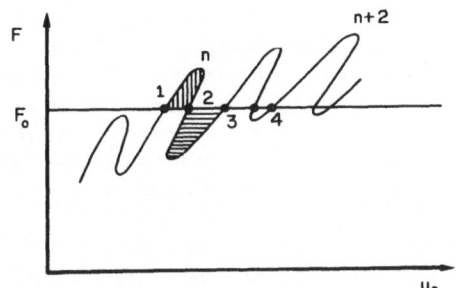

Fig. 6.   Crack solution. When the force is expressed as a function
          of the displacement, u_o, one of the branches of the curve
          shown here results. Other branches for various crack lengths,
          n, are shown superposed self consistently. For a given value
          of F, several solutions are possible corresponding to the
          points 1,2,3,...The forward activation energy for going to
          2 from 1 is shown vertically hatched. Reverse activation
          energy is shown horizontally hatched. Point 2 is a saddle
          point.

Since separate solutions exist for neighboring lattice positions,
these configurations should be separated by an energy barrier. The
energy of the system is obtained by examining the graph of F vs u_o.
This graph is similar to that of F vs u_n, and is what is plotted in
Fig. 6 as noted above. We can compute the energy changes of the
system at the various mechanical equilibrium positions, (1), (2),
(3)...by traversing the system over the path shown from (1) to (2)
to (3), etc. The total energy change is the sum of the work done on
the pair at j=o by the external force over the path (this work is
stored in the springs of the chain), plus the change in potential
energy of the external force machine. These contributions are given
respectively in the equation

$$\Delta E(1,2) = \int_1^2 F \, du_o - F \, \Delta u \tag{18}$$

If we eliminate f from (16) and (17) to find u_o in terms of u_n and substitute for du_o in (18) we find

$$\Delta E = A \int_1^2 f \, du_n + B \int_1^2 u_n \, du_n - f \, \Delta u_o \tag{19}$$

where A and B are expressions containing the compliance constants. We can write this as

$$\Delta E = A \, \Delta \gamma + \frac{B}{2} \Delta(u_n^2) - F \Delta u_o \tag{20}$$

In writing this expression, we note that $\int_1^2 F dF = o$ because F at (1) is equal to F at (2).

First, we note that graphically, (18) is given by the vertically hatched area shown in Fig. 6. The forward activation barrier is given by the area shown from (1) to (2), while the reverse barrier is the area shown from (3) to (2). The configuration at (2) is a saddle point in the configuration space of the system, while (1) and (3) are energy minima. Equation (20) illustrates the decomposition theorem promised above, which follows from the fact that Eqn. (17) is also a simple decomposition into a linear term and a nonlinear bond at the crack tip when it ruptures. Since the second term is a quadratic term in displacement, it is the change in stored energy in the linear spring system when the crack moves forward to the saddle point. The third term is the change in potential energy of the external weights. Thus, the second and third terms are properties of the linear macroscopic lattice, while the first term is a local core energy contribution from the crack tip pair when the bond at the tip ruptures. As we mentioned earlier, if external chemical influences are present, they modify the bond at the crack tip, and the form of the second and third terms is unaffected.

The reader will note in deriving the equations of this section, we have dealt with general properties of a linear spring system which possesses a single nonlinear bond and with an external force applied at one point. The qualitative results are thus very general, and though derived with the one-dimensional crack in mind, they apply equally well to a general crack system in any dimension satisfying our basic assumptions. The values of the compliance constants and the nonlinear bond function will depend, of course, on the particular system, and it is in order to find these values that one must solve the equations for mechanical equilibrium. And, of course, from these compliance constants, one may find particular expressions for the activation energy, size of the trapping regime, etc.

We complete the discussion of the intrinsic crack by noting in Fig. 6 that as the crack extends from n to n+1...the forward activation energy decreases and the reverse activation energy increases. At some value of the crack length, the forward barrier disappears, and the crack expands dynamically. Thus we have shown that the parabolic energy function of the continuum theory must take on the "dinosaur back" features depicted in Fig. 7, where we have plotted the energy of the system in configuration space along a coordinate which threads its way through the energy extrema, representing stable states and unstable saddle point states. This reaction coordinate serves as a measure of the "length of the crack" for comparison with the smooth parabola of the continuum theory as drawn on the same figure. The ends of the lattice trapping zone are shown as the points where the barriers in one direction disappear.

Finally, we turn to a simplified description of how reactions with external chemical species modify these general results[3]. Consider, for example, a diatomic gas in which a single chemical bond exists between the two atoms of the molecule. Let the external molecule be B_2 and let the bonded crystal atoms at the tip be A_2. We assume that the reaction

$$B_2 + A_2 \rightarrow 2AB \tag{21}$$

corresponds to bond rupture and subsequent chemisorption on a one-to-one basis (see Fig. 8). A plot of the energy of the system as the A-A bond comes apart, with the B atoms taking their most favorable configuration is shown in Fig. 9 for a variety of different conceivable cases, including comparison with the pure A-A bond. We recall that the bond energy in our simple model is equal to twice the surface energy per atom pair, and chemisorption on the basis of Fig. 9 thus corresponds to a lowering of the energy to form the surface. On this basis, chemisorption during fracture therefore should enhance the ease of fracture. In other terms, the equilibrium γ value in Eqn. (13) is lowered (see Fig. 10).

Gaseous reactions of the form (21) normally exhibit activation energy barriers, and in the restricted region of a crack tip, we would expect these barriers to be enhanced by what chemists call steric hindrance. The general rule in chemical reactions is that molecules with saturated bonds react with activation energy barriers of the order of one ev. Thus, even though chemisorption reactions assist fracture in the final state, the barriers to the reaction may be significant, and the reaction slow. In the end result, we cannot distinguish contributions made to the barriers from the intrinsic lattice trapping from the contribution of the extrinsic chemistry. The two are clearly intermixed. Graphically, this is shown in Fig. 10a and 10b. A chemical reaction with a barrier will graph as a force which has a <u>negative</u> range corresponding to the negative slope on the far side of the energy hump. This exaggerates the hump in F vs u_n, and extends the lattice trapping regime.

An extreme example of a chemical reaction at the crack tip which is so slow as not to be observable (at the equilibrium crack length

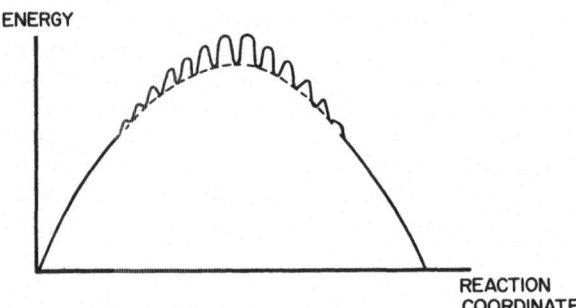

Fig. 7.   Energy of a crack as a function of reaction coordinate or
           length.   The dotted curve is a graph of the continuum energy
           given in Eqn. (14).   The dinosaur-back function results from
           the lattice activation barriers.   The trapping region is
           restricted to a portion of the curve roughly centered about
           the maximum.

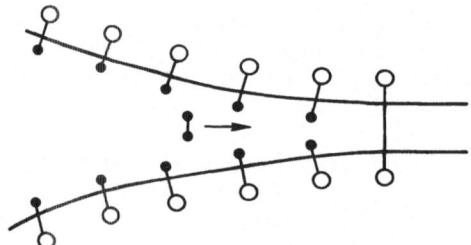

Fig. 8.   Diagram of a chemically assisted fracture.   The external
           molecules chemisorb on the surface and interact with the
           crack tip to decrease the bond strength.

calculated on the basis of the chemisorbed surface energy) is that
of a large molecule interacting with a crack in a lattice of small
lattice spacing.   In this case, the external gaseous species simply
cannot penetrate into the tip region because of its extreme size,
even though it may have a large adsorption energy on a flat surface,
and no enhanced crack growth can occur.   Fracture in this case will
be governed by the intrinsic properties of the lattice.

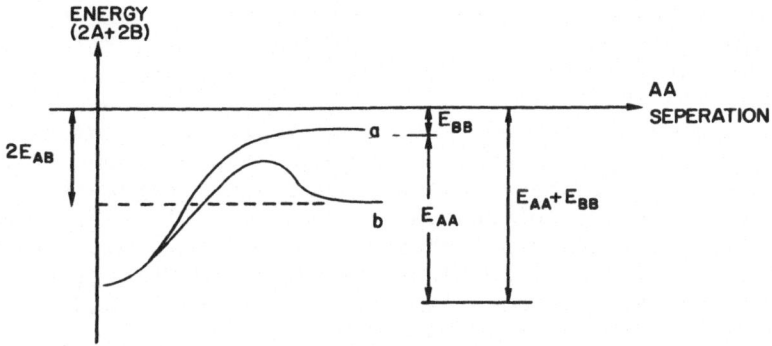

Fig. 9. Morse function diagram for chemically assisted fracture.
The endothermic reaction AA+BB→2AB (path b) requires less
energy than the intrinsic bond rupture AA→2A (path a) in
the fracture process. Hence, fracture is assisted, but a
chemical activation barrier may be present in the reaction
path.

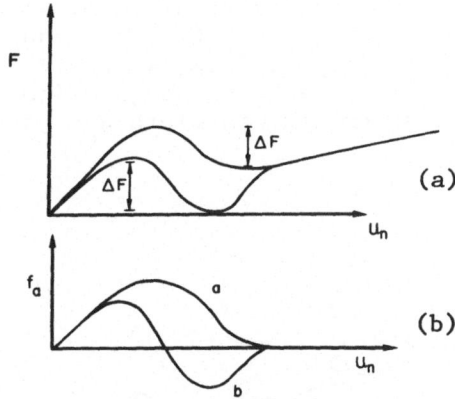

Fig. 10. Crack solution for chemically assisted fracture. The
derivative of the Morse function of Fig. 9 yields an
effective force law for the crack tip bond as shown in
(b). This force law when substituted into Eqn. (17) yields
the crack solution (a). Curves a correspond to the intrinsic
solution of Fig. 5, and curves b show the modification
resulting from chemical reactions.

The example of chemically enhanced fracture we have given was a deliberately simple one, and a variety of more complex phenomena can be imagined. For example, if the bonding state of the chemisorbed species is different in valence from that between the two lattice atoms, then the atom from the external environment does not simply replace the lattice atom in a one-for-one exchange, and a multiple bond rupture in the lattice may be required for each molecular reaction. An even more interesting possibility is for bridging bonds to develop between the atom from the external environment and the opening fracture surface. In that case, the lattice is strengthened by chemical reaction, not weakened. See Fig. 11. The theory thus suggests that the variety of phenomena associated with chemically assisted fracture should be rich, and that specific predictions will depend upon the type of chemical species involved, and the structure and chemistry of the lattice as well.

The experimental situation, briefly described, is that chemically assisted fracture is ubiquitous, but there are no systems understood so well that the preceding ideas have been applied in a quantitative way.

IV. THERMODYNAMICS AND CHEMICAL RATE EFFECTS[4]

A prime feature of the physical picture developed in the previous section is the appearance of activation barriers in the lattice trapping regime, and the idea that external chemical species may also interact with the bond at the crack tip with additional chemically induced barriers. These factors lead immediately to questions of thermal equilibrium of the crack and to questions of the rate with which the crack can overcome the barriers and grow.

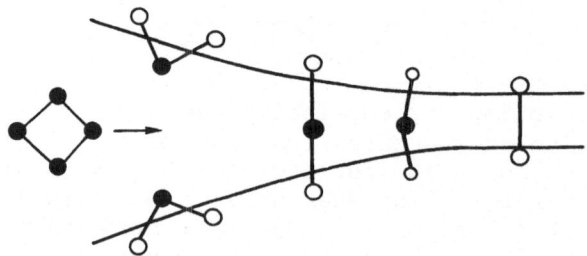

Fig. 11.   Bridging reactions at crack tip. Instead of bonding only to one of the opening surfaces, atoms can attach in a bridging bond across the opening crack, thus resisting fracture.

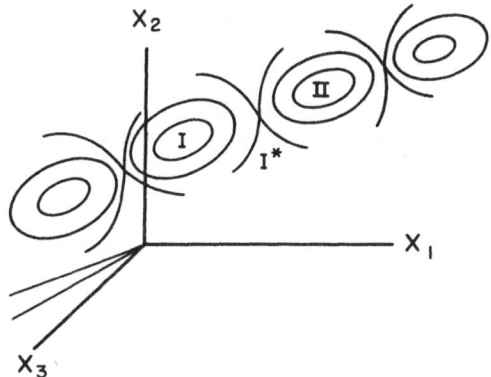

Fig. 12.   Energy function of a crack in configuration space. As the
crack moves from one stable lattice position, I, to another,
II, it must pass through saddle points in configuration
space, I*.

The physical ideas of the previous section imply that the configuration space of the system have the appearance shown in Fig. 12, which
applies either to intrinsic fracture or to that with external environment. When the configuration space of the interacting gas is included,
of course, the minima become level valleys in those directions corresponding to translation of a gaseous molecule in space. However, for
all directions in which the crystal changes configuration, the energy
function has the typical dimpled surface shown.

To calculate the rate of passage in the forward direction from
one minimum position over a saddle to another we can apply classical
reaction rate theory to obtain

$$\nu_f^* = \frac{kT}{h} \frac{Z(I^*)}{Z(I)} \tag{22}$$

where ν_f^* is the forward rate, k is Boltzmann's constant, h is Planck's
constant, T is the temperature, and $Z(I^*)$ and $Z(I)$ are the partition
sums of the system calculated at the saddle point and initial position,
respectively. $Z(I)$ is related to the free energy, A, of the system by
the standard equation

$$A = -kT \ln Z \tag{23}$$

If the system is in equilibrium, then by definition the forward
reactions, I → II, and backward reactions, II → I, are equal, and by
substitution into (22),

$$A_{II} = A_I \tag{24}$$

Thus, A is a minimum relative to its accessible states, a standard
result.

For the system composed of a simple vibrating lattice of N_L
modes without external environment, the partition sums are immediately
calculable, and we obtain

$$\nu^* = \frac{\displaystyle\prod_{i=1}^{N_L} \nu_i}{\displaystyle\prod_{i=2}^{N_L} \nu_i^*} \ e^{-E/kT} \tag{25}$$

where the ν_i are the vibration frequencies of the lattice, and ν_i^* are
the vibration frequencies of the system at the saddle point. At the
saddle point, for the direction of steepest descent, the frequency is
imaginary and only the frequencies in the hyper surface normal to this
direction are counted. This result is the same with similar interpretation to Vineyard's theory of atomic diffusion in lattices. The
energy, E, is precisely the activation energy calculated in the
previous section.

Equilibrium for the crack problem involves an interesting
subtlety. In the usual applications of (25) to the diffusion problem
in solids, equilibrium implies that E(I) = E(II) because the vibration states at I and II are equivalent. However, in the crack problem,
when the crack advances one lattice spacing, two atomic areas of surfac
are added, and the vibration structures are not equal in I and II.
In fact, when the crack advances, it converts one or more bulk vibration modes to surface vibration modes, and these frequencies are not
in general equal to one another. That is, even though in the two
states the total number of modes is the same, their partition between
bulk and surface categories is different. Thus for the crack, equilibrium is not that state where E(I) = E(II), because the entropies of
the states are unequal. By following up this line of reasoning it is
possible to show, however, that in equilibrium

$$\bar{f} = 2\gamma \tag{26}$$

where $\bar{f}$ is the force on the crack in the sense derived in Eqn. (12),
and γ is the thermodynamic surface tension defined in the Gibbs sense.
In (26), γ is not simply half a bond energy, but has its normal
temperature dependent component as well. All this follows from the
fact that strictly E(I) $\neq$ E(II), though in practice the difference
is small when measured in electron volts per atom.

When an external chemical environment is present, a straight-forward application of the same equation for the rate of bond breaking at the tip yields

$$\nu^* = rp \frac{\prod\limits_{i=1}^{N_L} \nu_i}{\prod\limits_{i=2}^{N_L+6} \nu_i^*} \, e^{-E/kT}$$

$$r = \frac{h^5 (1-e^{-h\nu_g/kT})}{64\pi^{7/2} I m_g^{3/2} (kT)^{7/2}\lambda} \tag{27}$$

In deriving (27) the external gas is assumed to satisfy perfect gas laws for a diatomic molecule. p is the pressure of the gas, and r contains the various factors associated with the partition sum of a diatomic gas. ν_g is the vibration frequency of the diatomic molecule, I is its moment of inertia, $2m_g$ is the mass of the molecule, and λ is the nuclear spin factor, which is of order unity. We have assumed in (27) that the adsorbed gas contributes 6 additional modes per molecule to the solid when adsorption occurs, and that one molecule is adsorbed for each bond-breaking event. These modes will be added, of course, to the surface modes. By making various assumptions on the mass of the adsorbing species, etc., numerical estimates can be made from (27).

From its derivation, we see that E in (27) represents the total activation energy of the system, and in general it will not be pos-sible to separate this total into a chemical reaction part and lattice part, because the two are so closely intertwined.

Equilibrium considerations similar to those of a previous paragraph show that at equilibrium

$$\bar{f} = 2\gamma \tag{28}$$

where again γ is the thermodynamic temperature dependent surface tension defined by Gibbs for the surface in equilibrium with the external gas. At equilibrium, again $E(I) \neq E(II)$ except at zero temperature, but we will often assume equality in loose speech because of the slight practical difference.

V. CRACKS IN TWO AND THREE DIMENSIONS

Introduction

With the general discussion of the previous two sections, we are ready to turn to the detailed mathematics of two and three dimen-sions, and the comparison of theory with experiment. Although the

main overall qualitative picture is established by the previous
sections, we are left with such matters as the role of lattice
symmetry, the form of the activation energy as a function of external
force, the effect of the strength and range of the force laws on the
results, the extent to which two-dimensional results (easier to
calculate) can be applied to three dimensions (much harder to calculate), the condition for stability of the crack in the lattice
relative to intrinsic blunting at the crack tip, and detailed
predictions of chemical reaction parameters. Needless to say,
this broad agenda has hardly been touched by the theory, and the
experimental picture is very inadequate.

We begin with a short survey of the theoretical techniques which
have been applied to the fracture problem, together with an associated
bibliography. Atomic force considerations at the crack tip can fairly
say to have begun seriously with work by Barenblatt[5], who investigated
the consequences of postulating a nonlinear force-displacement law for
the atoms at the tip. This was a continuum theory, and he showed that
the shape of the crack near the tip is strongly modified near the tip
by nonlinearity, and in general he was able to deal successfully with
the requirement that no real crack can sustain the singularity at the
tip predicted by the linear theory.

The first discrete studies of cracks were performed in the thesis
of Kaninin[6], followed over a period of years by computer simulations
of iron with his subsequent coworkers, Gehlen and Markworth at Battelle,
using the empirical force law developed for iron by Johnson[7]. The
later work by the Battelle group has been an attempt to simulate iron
and hydrogen in iron. In these calculations an inner core of atoms is
relaxed atom by atom, and then fitted to an infinite continuum.
Sinclair and Lawn, and later Sinclair[8], applied a more sophisticated
version of this technique to simulated Si atoms. The iron calculations
were plagued by a series of problems associated with matching appropriate boundary conditions to the core, but seemed to show that iron
was only marginally stable relative to dislocation production. In
comparison between Si and Fe, the crack in Si was long and narrow
with a small crack opening displacement.

Simulations of a crack in a finite lattice without attempting to
match to an infinite continuum were done earliest by Sanders[9], and by
Weiner and Pear[10], in dynamic studies of a constrained system of atoms
with rigid walls. Ashurst and Hoover[11] presented similar calculations
on a triangular lattice. The most complete calculations have been performed recently by Paskin, Dienes and coworkers[12] with relaxed boundaries
and a 6/12 potential function. In all these calculations, dislocation
formation and crack breakdown was a feature displayed by the crack
under various conditions.

A quite different approach using a lattice statics theory has
been taken by the present author with Hsieh[13] and by Esterling[14] with

a technique complementary to both computer simulation techniques
described above. The lattice statics theory has a partially analytic
form for simple square lattices. It also rigorously addresses the
infinite lattice, but has been developed only for finite length cracks.
Its main application has been to demonstrate lattice trapping and the
analytic form of the activation barriers for various kinds of force
laws and for external chemical reactions.

Lattice Statics

We turn now to a presentation of the mathematics of lattice
theory as applied to cracks, and a comparison of the theoretical
predictions with the sparce experimental findings. The equations
which specify mechanical equilibrium for a general lattice are

$$\sum_{x,j} f_{ij}(\underline{x}-\underline{x}')\, u_j(\underline{x}') = F_i(\underline{x}) + \sum_{\substack{\text{crack} \\ \text{plus} \\ \text{core}}} \Lambda_{ij}(\underline{x},\underline{x}')\, u_j(\underline{x}')$$

$$+ \sum_{\text{core}} g_{ij} \qquad \text{core} \tag{29}$$

or in operator notation

$$Fu = F + \Lambda u + g \tag{30}$$

The f_{ij} is the force on the atom at $\underline{x}$ in the i direction due to a
unit displacement of the atom at $\underline{x}'$ displaced in the j direction.
The matrix operator F depends only on the distance $\underline{x}-\underline{x}'$ in a perfect
lattice. F_i are the external forces applied on the atoms of the
lattice in the i direction. The operator $\Lambda_{ij}(\underline{x},\underline{x}')$ provides the
means by which we describe the defect lattice. When the crack is
formed, we must snip the atom bonds which cross the cleavage plane.
The Λ operator is the annihilation operator for these bonds.
Specifically, $\Lambda_{ij}(\underline{x},\underline{x}')\, u_j(\underline{x}')$ is the bond force which is annihilated between the $\underline{x}$ and $\underline{x}'$ atoms. It is summed only over the atom
pairs which are snipped because of the crack formation and cancels
the appropriate terms in Fu. Its values are the same as the corres-
ponding $f_{ij}(\underline{x}-\underline{x}')$ values. Λ also serves an added function, however.
When the atoms of the crack cohesive zone are stretched into their
nonlinear ranges, we incorporate this fact into the formalism by
again annihilating the linear bond by the Λ matrix, but then we add
the nonlinear function g for that atom. Thus Λ is summed over the
atoms connected by bonds across the cleavage plane, and over the
cohesive core region.

The function, g_{ij} is the force on the ith atom due to its bond
to the jth atom, and depends nonlinearly on the bond distance. It
thus depends upon both atom positions at x_i and x_j for a pair-wise
bond. In the two-dimensional crack application, however, these nonlinear bonds will be considered to act only across the cleavage plane,
which normally exhibits sufficient symmetry to allow g_{ij} to be written
as a function only of u_i, a considerable simplification. For kinks,

and for the dislocation production problems, however, this simplification will not generally apply. The function g is, of course, summed only over the atoms of the core.

The problem is solved by the linearization strategy employed for the one-dimensional problem. We first assume that the g-forces on the atoms are constants independent of any displacement functions. That is, they are treated as given external forces, not functions of atom displacements in order to derive the appropriate compliance functions for the linear portion of the system.

To this end, we define a Green's function conjugate to F by

$$F^{-1} = G \tag{31}$$

so that a formal solution becomes

$$u = G\Lambda u + G(F+g) \tag{32}$$

which we rewrite

$$(1-G\Lambda) \; u = G(F+g) \tag{33}$$

F is an infinite but banded matrix (i.e., it has non-zero elements only in a band adjacent to the diagonal), G is also an infinite matrix. Since Λ is finite, the product, $G\Lambda$, in (33) leads to a decomposition. That is, for the subspace in (33) in which u is a displacement of an atom in the core or on the cleavage surface, the matrix $G\Lambda$ does not couple to terms outside the subspace. That means, if the crack and its core region is finite, we solve the infinite lattice problem by solving a set of linear algebraic equations of finite order. In fact, if the cleavage plane is sufficiently symmetric, the order of the problem can be as low as (n+c), where 2n+1 is the total length of the crack, and c is the number of nonlinear bonds in the core region at each end of the crack. The major problem, of course, in this solution is the determination of the static lattice Green's functions, G, but there is fairly extensive literature for the computation of these quantities[15].

The formal solution is then obtained by writing (33) as

$$u = (1-G\Lambda)^{-1} \; G(F+g) \tag{34}$$

where the quantities $(1-G\Lambda)^{-1} G$ become the linear compliances. At thi stage we substitute the nonlinear functional forms into g and solve the actual nonlinear problem. Since F can be a single force exerted on the center pair of atoms on the cleavage plane in a two-dimensional lattice, and g may only be exerted at the pair of atoms at the tip of

the crack, (34) can be as simple as the expressions we have already
used in (16). Then the full discussion leading up to activation
energy expressions such as (20) still follow. Thus, Eqn. (34) may
be onerous to carry through to an actual numerical solution in any
given case, but its form as a generalization of (16) is clear.

It would be inappropriate for us to enter here into a detailed
discussion of the mathematical anatomy of the general solution we have
outlined above. However, the reader might be interested to see it
applied to a particularly simple case. From that example, the current
status of the method, and its limitations will become apparent. Thus
we consider the two-dimensional square lattice shown in Fig. 13.

In this problem, we restrict displacement to be in the y direction, only, and take forces to be only between nearest neighbors.
Thus

$$F_u = A(u_y(\ell_x+1,\ell_y) - 2u_y(\ell_x,\ell_y) + u_y(\ell_x-1, \ell_y))$$

$$+ B(u_y(\ell_x,\ell_y+1) - 2u_y(\ell_x,\ell_y) + u_y(\ell_x,\ell_y-1)) \tag{35}$$

A and B are the spring constants for shear and longitudinal displacements, respectively.

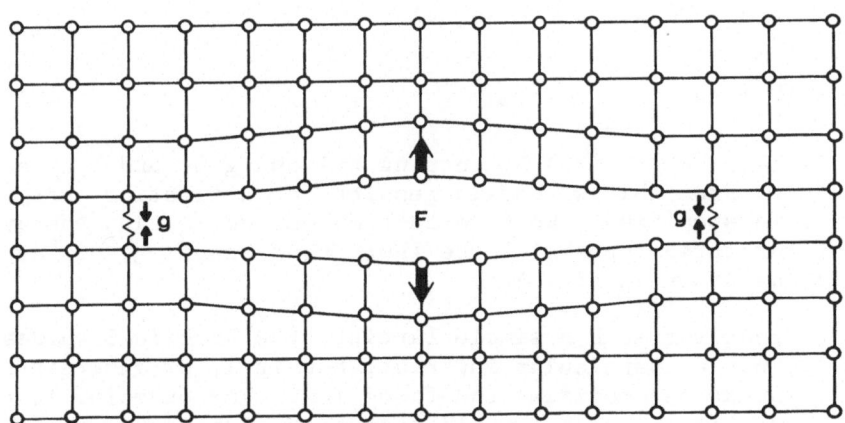

Fig. 13.   Diagram of a crack in a two-dimensional square lattice with
           external forces exerted at the center of the crack. Bonds
           at the two crack tips are nonlinear.

The Green's function solution for this case is most easily given not in terms of the source function for a unit force, but for a dipole created by exerting equal and opposite forces in the y direction at an atom pair facing each other across the cleavage surface. This Green's function is

$$G(\ell-k) = G(m) = \frac{1}{B}\{\delta_{mo} - \frac{C}{\pi} \int_{o}^{\pi} \frac{\cos mq \sin q/2}{[1+C^2 \sin^2 q/2]^{1/2}} dq\}$$

$$C^2 = A/B \tag{36}$$

It has an asymptotic form for large m

$$G(m) = \frac{C}{2\pi m^2 B} \tag{37}$$

Although the asymptotic solution diverges for m=0, it is in error by only about 15% for m=1.

After substitution of G into (33), assuming that F is exerted only at the center pair on the cleavage surface, and that g is exerted only at the two end pairs, we have a set of algebraic equations for the 2n+1 displacements, u_j, j=0, ± 1, ...±n. 2n+1 is the total length of the crack plus core, i.e., 2n+1 includes the two nonlinear bonds at the crack tips. We write the solution in the form

$$u_o = \lambda F + \eta g \tag{38}$$

where λ is obtained in (33) by setting F=1 and g=0, and η by setting F=0 and g=1. From this we obtain results for activation energy, etc., when $g(u_n)$ is specified. We have carried out activation energy calculations for three types of force law: bond snapping, bilinear and quadratic, as shown in Fig. 4.

Of course, for such a simple lattice, the numerical values of the results are not of particular interest, but it is of interest to observe the variation as one modifies the force law. For example, in the bond snapping limit (Fig. 4a) one would expect the lattice trapping effect to be larger than for the softer quadratic force law of Fig. 4c. In Table 1, we show these results where the activation energy is normalized to the bond energy, E_b, and with A/B=1. For comparison we give the range of values calculated by Sinclair[8b] for several force laws for silicon. The activation energies are calculated at the top of the energy function, Fig. 7.

Table 1 – Activation Energy in a Square Two-Dimensional Lattice and
        in Si.   A/B=1.

	Bond Snapping Force	Symmetric Bilinear Force	Quadratic Force	Si[(8b)]
Activation Energy $(\Delta E / E_b)$	0.22	0.047	0.026	.064-.14

In the table, the symmetric bilinear force is one in which the up
and down slopes of the force in Fig. 4b are equal.

The table shows the expected results, that the activation energy
is a strong function of the softness of the force function. Geometric
and symmetry considerations are also expected to play a role, and in
this vein, Esterling[14] and Sinclair[8b] have shown that the motion activation energy for a kink in a crack is much smaller than the energy to
move a two-dimensional crack.

Of equal or perhaps even more interest in these results is the
variation of the activation energy with external stress. We have
noted that a fundamental requirement is that the activation barriers
must disappear at the ends of the lattice trapping regimes. At the
same time, this variation is immediately observable, because it leads
to the stress-crack velocity relation through the rate expressions of
the previous section.

We obtain

$$(\Delta E_+)_{bilinear} = E_o \ (1-F/F_+)^2$$

$$(\Delta E_+)_{quadratic} = E_o \ (1-F/F_+)^{3/2} \tag{39}$$

The parameter F_+ is the value of F where the barriers disappear. We
also find in the case of the bilinear force law that the point where
the forward and backward barriers are equal is given by

$$F_G = \sqrt{(F_+F_-)} \tag{40}$$

F_- is the force at which the reverse barriers disappear. Thus the
Griffith point in this case is the geometric mean, rather than the
arithmetic mean between these two points. From the graphical form
of the solution, Fig. 6, we can see that the quadratic law in (39)
depends basically upon a triangular F/u_o dependence, and the 3/2 law
reflects a rounded F/u_o dependence (the one displayed). A linear

$\Delta E_+(F)$ would follow if the "loading branch" from F to F_+ was parallel to the "unloading branch" from F_+ to the saddle point in Fig. 6, an extreme limiting case. If we think of the variable $(1-F/F_+) = \delta F$ as the decrement force relative to the Griffith value, then the expected variation of ΔE should be near

$$\frac{d\ln(\Delta E_+)}{d\ln \delta F} \overset{\sim}{=} 3/2$$

$$\delta F = 1-F/F_+ \tag{41}$$

for all reasonably continuous force laws.

The functional dependence of ΔE on δF is closely related to the concept of an activation volume in chemical rate and solid state diffusion theory. If an activated process is related to an external pressure or other stress by the law

$$\Delta E = \Delta E_o + \sigma_{ij} E_{ij} \tag{42}$$

then the parameters E_{ij} have the dimension of volume. In our model, the quantity $\partial \Delta E/\partial F$ has the dimension of a <u>distance</u>, and in particular, it is some kind of measure of the displacement induced in the system at the tip by the activation process. Indeed, we find from the various expressions for the compliance that

$$\left(\frac{\partial \Delta E}{\partial F}\right)_{bilinear} = \frac{\alpha}{\sqrt{n}} \; \delta u_n$$

$$(\delta u_n)_{bilinear} = \beta \delta F \tag{43}$$

and

$$\left(\frac{\partial \Delta E}{\partial F}\right)_{quadratic} = \frac{\alpha'}{\sqrt{n}} \; \delta u_n$$

$$\delta u_n = \beta' \; \sqrt{\delta F} \tag{44}$$

where α, α', β, β' are various combinations of the compliance constants, n is the half length of the crack, and the activation displacement δu_n is the change in displacement of the crack tip atoms during the activation process. Because of the factors δF and $\sqrt{\delta F}$, the activation displacements go to zero at F_+ as they should, but again, we note a characteristic difference between the bilinear force laws and the quadratic force laws.

Results from the detailed calculations for the activation displacement, δu_n at the Griffith force are given below. The displacements are surprisingly large fractions of the maximum atomic displacements u_{max}, where u_{max} is defined as the displacement at which the restoring force goes to zero. These values are also in general agreement with the expected values as calculated by the quasichemical method by Wiederhorn et al[18].

Table 2 – Activation Displacements for Various Force Laws

	Bond Snapping	Symmetric Bilinear	Quadratic
$(\dfrac{\delta u_n}{u_{max}})$Griffith	0.4	0.3	0.3

Three-Dimensional Cracks

The behavior of real cracks in solids is yet one step beyond the two-dimensional models delineated to this point, because a crack is actually a three-dimensional object due to the ability of the crack line to assume a curved path even if it remains within its cleavage plane, to say nothing of its freedom to wander from one cleavage plane to another. From the lattice standpoint, a moderate curvature of the crack line is equivalent to a collection of kinks on the otherwise crystallographically oriented crack line. Further, motion of the three-dimensional crack on its cleavage plane is equivalent to nucleation of pairs of double kinks on the line, and the transport of these kinks along the line. Figure 14 is a diagram of atom displacements of atoms at a crack facing each other across the (111) cleavage surface in Si, and showing a single kink in the crack line as calculated by Sinclair[8b].

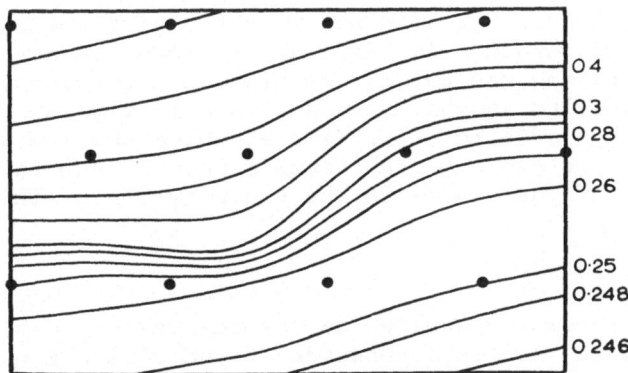

Fig. 14.   Crack kink configuration. Shown is the (111) plane in silicon with atom positions shown as dots. Contour lines are the lines of equal displacement. The kink is highly localized at one site. After Sinclair[8b].

Obviously, theoretical treatment of the kink problem is enormously more difficult than the two-dimensional problem discussed so far. The computer requirement for both direct computer solutions and for the lattice static theory is the limiting factor. Nevertheless, first Esterling[14] using lattice statics, and then Sinclair[8] using the direct approach have addressed single kinks and found that single kink trapping barriers are much smaller than the two-dimensional barriers. This result is the same as the analogous finding that kink barriers for motion on dislocations are much smaller than the two-dimensional Peierls barrier. It points to the proposition that the rate limiting step in slow crack growth is the rate of nucleation of double kinks in the crack line.

Since the double kink is composed of several parallel planes of two-dimensional configurations, the formation energy of the double kink will be of the order of several two-dimensional energy barriers. This result is based on the intuition supplied by the quasi-continuum treatments of the dislocation double kink with the Frenkel-Kontorova model[16]. Unlike the dislocation analogue, however, the stress dependence of the formation energy will be supplied in part by the two-dimensional barrier dependence, which we have already discussed, as well, as partly by the interaction between two close lying kinks. Lin and Hirth[26] have looked at this problem from the point of view of crack dislocations, and made estimates not only of the stress dependence, but the energy of double kinks. Their estimate, in lowest order, is that apart from the stress dependence, the double kink nucleation energy is approximately twice the single kink energy. In the case of Si, their estimate of 1.5 ev for the double kink is reasonable in view of Sinclair's[8b] estimate of .5 to 1.5 ev for twice the double kink energy. We shall thus assume for the total activation energy of a double kink to lowest order

$$\Delta E_+ = 2(\Delta E_+)_{2D} \, w \tag{45}$$

where $(\Delta E_+)_{2D}$ is our 2D result and w is the single kink width. The overall stress dependence is difficult to guess because both the kink-kink interaction and the 2D lattice trapping contribute to it. We are tempted, however, to suggest that a simple linear law might suffice because the range of stress (or K) over which the law will be applied will be small, and experimental results do not seem to be very sensitive to using various powers in the vicinity of 1. Thus, we assume that

$$\Delta E_+ = E_o \, (1-F/F_+) \tag{46}$$

A similar argument applies to the activation displacement. Again, we do not know the stress dependence of the double kink activated state, and shall simply write for the total activation displacement

$$\delta u = 2(\delta w)_{2D} w \tag{47}$$

Thus, on the basis of a very approximate but general argument, we expect that both activation displacement and activation energy should be roughly linear to the kink width, w, for intrinsic crack growth.

The result (47) is important, because of the comparison between the intrinsic crack and the chemically assisted crack. In the latter case, we should expect a class of interactions where both size effects and bonding energetics are favorable in which double kink formation is accomplished by a single bimolecular reaction at the crack tip which does not require cooperative activity in several adjacent atomic planes. Experimentally, this should be associated with first-order chemical kinetics in which the rate is proportional to the first power of the pressure (see Eqn. (27)). Kink formation requiring more than one external molecule would be correlated to a quadratic or higher power pressure dependence.

Such a chemically assisted process should be distinguishable from the intrinsic process outlined above by a smaller activation displacement, because of the smaller number of participating atoms. It may also show a smaller activation energy barrier as well, although the energy barriers probably will be dominated by basic chemical bonding considerations rather than by lattice trapping effects. In any case, the lattice contribution will be low because only one two-dimensional atom plane is involved.

The picture sketched above seems to be consistent with the small amount of experimental data which are available. Figure 15 shows the general experimental findings by Wiederhorn[17] typical of a material (in this case glass) undergoing slow crack growth which is influenced by an external atmosphere (in this case H_2O vapor). The behavior is characterized by three stages: Stage I corresponds to an enhanced velocity proportional to the partial pressure of H_2O; II corresponds to a diffusion limited region where the crack runs ahead of the external atmosphere; and, III corresponds to the intrinsic crack velocity which is independent of external chemistry and is the same as that measured in vacuum. Region III has a much steeper force-velocity relationship than Region I. Table 3 shows the collected results by Wiederhorn et al[18] on various glasses when experimental slow crack velocity measurements are plotted according to the expression

$$v = v_o \exp \{(E+bK)/RT\} \tag{48}$$

K is the stress intensity, R is the gas constant, and v_o, b, and E are determined from the K-v plots.

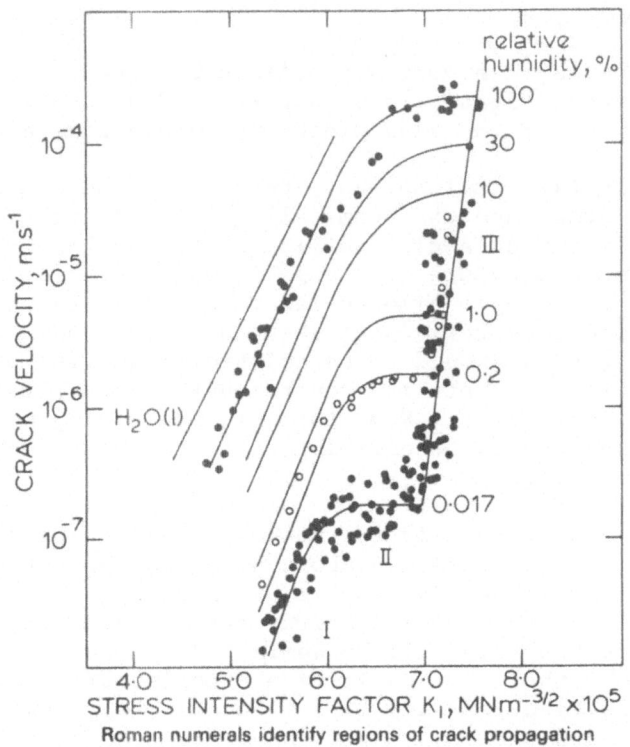

Fig. 15.   Experimentally observed chemically assisted fracture.
Experimental values are those for glass at various pressures
of water vapor.   Three stages of crack growth were observed.
I and II are H_2O sensitive, while III is the intrinsic crack
velocity.   Stage I has the form of Eqn. (48), while II is a
diffusion limited zone where the crack outruns the external
environment.   After Wiederhorn[17].

    The results from the table show that both activation displacement
and activation energy tend to be considerably larger for intrinsic
crack growth in vacuum than in the presence of a corroding atmosphere.

    Wiederhorn et al[18] interpreted the results of Table 3, however,
in different terms.   They noted that electrostrictive effects
associated with charge displacements generated during fracture would
also have the effect of enlarging the cooperative region during
activation.   When chemical species are active in water solution, the
large dielectric constant of water would damp such effects leading to
smaller activation displacements.   Additional experimental investigation will be required to determine if electrostrictive effects are
indeed involved in addition to the kink width effects discussed above.

Table 3. Summary of Stress Corrosion Data in Glass[18]

Environment	Glass	E,kcal mol^{-1} (kJ mol^{-1})	b,m$^{5/2}$mol^{-1}	ln v$_o$
Water	Silica	33.1 (139)	0.216	−1.32
	Aluminosilicate I	29.0 (121)	0.139	5.5
	Aluminosilicate II	30.1 (126)	0.164	7.9
	Borosilicate	30.8 (129)	0.200	3.5
	Lead alkali	25.2 (106)	0.144	6.7
	Soda-lime-silica	26.0	0.110	10.3
Vacuum	61% lead	83.1 (348)	0.51	5.1
	Aluminosilicate III	176 (705)	0.77	14.1
	Borosilicate crown	65.5 (275)	0.26	6.6
	Soda-lime-silica	144 (605)	0.88	−5.7

Force Laws

The lattice statics theories have all been constructed on the basis of very simple ad hoc force laws which are not intended to simulate any particular material. In the computer simulation studies, the Battelle group used the Johnson potential for iron[7], Sinclair[8] used an empirical 3-body force to simulate the covalent bond of silicon, and the 6-12 force law of Paskin et al[12] simulates the rare gas solids. In all these cases, with the possible exception of the rare gas solid, the particular force laws should not be taken too seriously. In particular, the Johnson force was constructed from equilibrium properties of iron, and is not expected to simulate surface effects, of which the crack is a highly distorted example. Sinclair[8b] used a number of modifications for the Si bond force to test the sensitivity of crack parameters to the form of the law, which we believe is in the appropriate spirit for this kind of theory. Silicon is probably an excellent material to model because of the short range of the force, and because silicon is one of the best examples where a covalent bond model might be used successfully. Yet even here, the hybridization which occurs in the surface bonds, and which leads to surface reconstruction, points to complications which could be important in the detailed structure of the highly

distorted core of the crack. Also, it is known that the electronic
structure and formation energy of simple defects in the semiconductors
has been difficult to predict theoretically because of the difficulties
associated with dealing with the structure dependent terms in the
electron binding problem. A crack in a metal is probably the most
difficult defect to simulate because of the very large distortion at
the crack tip and the great difficulty in constructing adequate
effective force laws for such distortions. Indeed, it is probably
wise to consider the crack problem as the most severe test for those
who construct force laws to be used in defect calculations.

        In other examples of strong perturbations in solids, cluster
calculations have been developed with varying success. It is hoped
that one of the results of this conference will be a better perspective
regarding the application of these techniques to the fracture problem.

        In the face of all these fundamental difficulties regarding the
problem of atomic binding in solids, one should ask what, after all,
is the significance of _any_ theory of the structure of cracks built
on such ad hoc force laws? The view of this author is: 1) a truly
fundamental approach to the problem of fracture, for example by the
use of a cluster calculation, is likely to be possible only for a
small number of canonical cases. It will probably be limited to the
treatment of a few atoms at the tip; 2) matching small clusters to
the full lattice will require an understanding of the "lattice problem";
3) over and beyond the fundamental approach, it will undoubtedly remain
necessary to bridge the gap from the fundamental theory of simple
systems to the full gamut of materials and chemical reactions of
interest to fracture problems. For these problems, classes of
"effective force" laws based on rough rules supplied by the fundamental studies would provide the theorist with a very powerful tool
for the study of physical problems in fracture. I believe the
construction of such effective force laws is the most important goal
for fundamental theory; and 4) we should, indeed, beware of extrapolating
fracture theory too far, and keep a close weather eye on experimental
observation.

VI. CRACK STABILITY

        The final major issue remaining in our topic is whether the sharp
cracks we have been describing play any significant role in material
failure. Indeed, many materials undergo ultimate failure not by growth
of a crack, but by purely ductile or shear deformation means. Butter
is an obvious example, but other materials such as steel also can fail
by ductile processes on a microscale. The general role of ductility
in fracture will be dealt with in Knott's lecture at this conference,
but the question for us is whether a sharp crack is in fact stable in
the lattice against shear processes which emit dislocations and blunt
the tip of the crack. Such a dislocation emission may be the first
stage in a fully ductile breakdown of the crack. Stability against

such shear breakdown is, of course, a necessary condition for brittle fracture (see Fig. 16).

The first study of this question was by Kelly, Tyson, and Cottrell[19] who expressed the condition for stability as

$$\left(\frac{\sigma_{shear}}{\sigma_{tensile}}\right)_{max} < \left(\frac{\sigma_{shear}}{\sigma_{tensile}}\right)_{theoretical} \tag{49}$$

The left side relates to the maximum values of shear and tensile stress in the vicinity of the crack, and the right side relates to the maximum sustainable theoretical shear and tensile stresses of the material. The physical argument of Kelly et al was that by definition of a sharp crack, the bond at the tip must be stretched nearly to its maximum force. As this force is transmitted to its neighbors, the tensile stress will be converted at least partially to a high shear stress. Since the theoretical tensile strength is larger than the theoretical shear strength for all materials, there is an excellent chance that a shear stress will be generated at the crack tip of sufficient magnitude as to blunt the tip. Hence, (49) follows. Kelly et al applied it to a number of materials and obtained a general consistency between (49) and the materials known to be ductile or brittle.

Rice and Thomson[20] recast the stability condition in terms of the ability of the material to emit a dislocation. They reasoned that the highly inhomogeneous stress in the vicinity of the crack means that the stability condition must be stated in terms of Fourier components of displacement at the top of the Brillouin zone instead of k=0 as in (49). Such a stability condition is equivalent to production of a dislocation. From this standpoint, these authors showed that a rough criterion for stability, which neglects lattice geometry and three-dimensional effects, is

$$\frac{\mu b}{\gamma} \mathrel{\substack{>\\ \sim}} 10 \tag{50}$$

μ is the shear modulus, b the dislocation burgers vector, and γ the intrinsic surface energy. They concluded that cracks in the face-centered metals and the body-centered metals of the 1st row of the periodic chart should be unstable. Except for iron, brittle cracks in the bcc transition metals, hexagonal metals, ionic crystals, and the covalent diamond lattice insulators were predicted to be stable. Iron was an interesting borderline case which seems to be neither strongly brittle nor strongly ductile.

Both these earlier studies have been based on essentially continuum considerations with cutoffs invoked where necessary. Also, the formation of a dislocation in the vicinity of a crack is a complex three-dimensional problem which involves the nonlinear and atomistic

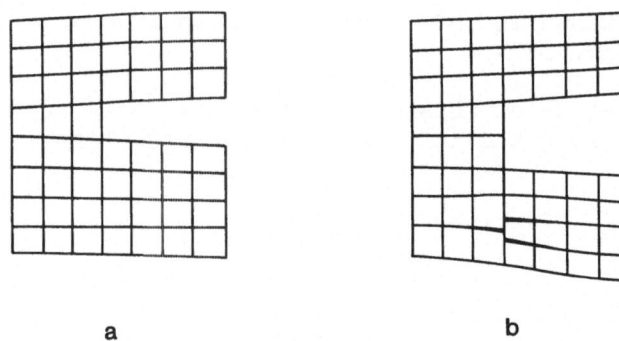

a   b

Fig. 16.   Dislocation emission causing crack blunting.

features of the core regions of both crack and dislocation. For this
reason, both these approaches should be viewed as having only qualitative significance. Nevertheless, criteria like (49) or (50) are
valuable for comparing one type of material with another, and they
both express the underlying physics of the problem, although in
somewhat different ways.

Comparison of (49) and (50) with computer calculations is particularly useful. In spite of the problems of identifying the computer
calculations of the Battelle group with actual iron, it is interesting
to note that those workers have already commented[6c] on the appearance
of a shear-like breakdown in the core of the crack in the two-dimensional
configuration. Tyson[21] has looked at a number of extant computer simulations using criteria (49) and (50) and feels that (49) is a somewhat
more reliable criterion than (50). Finally, Paskin et al[12] in their
calculations for the 6/12 potential have concluded that (50) accurately
predicts the brittle character of cracks in their case, but Dienes will
give a more complete description of these results in his paper.

Howard[22] has made an exceedingly interesting application of (50)
to hydrogen embrittlement. Noting that (50) is a criterion for pure
mode I fracture, he finds that for a given value of γ, there will be
a critical cleavage plane angle above which fracture will be ductile,
and below which it will be brittle. Using hydrogen pressure as a
parameter which controls γ, he then calculates the fraction of facets
of a grain boundary which will be ductile, and that which will be
brittle. From this, in an elegant way, he calculates the toughness
as a function of the H-pressure. In spite of the ingenuity of the
model, before taking it seriously as a hydrogen embrittlement model,
further more direct experimental confirmation is desirable, however,

So far, we have only considered a static crack, and the emission
of only a single dislocation. Jokl, Vitek, and McMahon[23] have suggested
that dislocation emission may be a property of a moving crack, and

indeed, the computer solutions of Paskin[12] et al, and Wiener and
Pear[10] indicate such features. In any case, since the stresses in
the core are nearly sufficient to form dislocations when the stress
is raised sufficiently to make the crack run, possibilities exist
for the core to explore dislocation-like configurations as it moves
from one lattice site to the next. We believe that the two-dimensional
models are not sufficiently powerful to demonstrate these possibilities,
especially when one portion of the crack gets out of phase with another
because of local inhomogeneities. Under these circumstances, local
accelerations and asymmetries will allow a much greater variety of
configurations than the simple two-dimensional models are likely to
suggest without special arrangements.

From these comments, it is clear that the author does not believe
the last word has been said regarding dislocation emission from cracks.
Of course, dislocation emission is only one aspect of the much broader
question of dislocation interactions with cracks in general, but it is
crucial in determining the source of those dislocations.

It would be useful to have direct observational evidence regarding
dislocation emission. Early etch pit experiments of Gilman et al[24]
have already suggested that when cracks are arrested in LiF, dislocations are produced in the absence of other observable sources. More
recently, the experiments of Lawn et al[25] have been the most unambiguous. They have demonstrated that cracks in a variety of brittle
materials (Si, Ge, Al_2O_3, SiC) can be completely sharp and devoid of
dislocation atmospheres, and with no evidence of dislocation emission,
even when the crack is completely static for long periods at low
temperatures. On the other hand, at high temperatures when the
dislocations are mobile, these cracks have dislocations associated
with them. In this author's view, at high T, these cracks can carry
dislocation atmospheres with them in the form of nonblunting dislocations whose slip planes are skew to the crack line (see Fig. 17).
Also, there is some evidence even in Si, which is expected to be among
the most brittle of materials by (50), that dislocations can be emitted
at inhomogeneities, corners, or branching in cracks, in the vein of the
remarks of the last paragraph.

Thus, in the light of very incomplete evidence, we believe that
some materials like the face-centered soft metals which satisfy the
ductility criteria expressed by (49) or (50) will not be able to sustain
a stable sharp crack in the lattice for the physical reasons underlying
(49) and (50). Other materials which satisfy the brittle criteria
appear to be sufficiently close to dislocation emission that inhomogeneities in the crack or in the lattice will permit dislocations to
be formed, perhaps in the blunting configuration, and perhaps in a
nonblunting slip plane. What consequences such emission will have
for the subsequent mobility of the crack are still speculative at this
point.

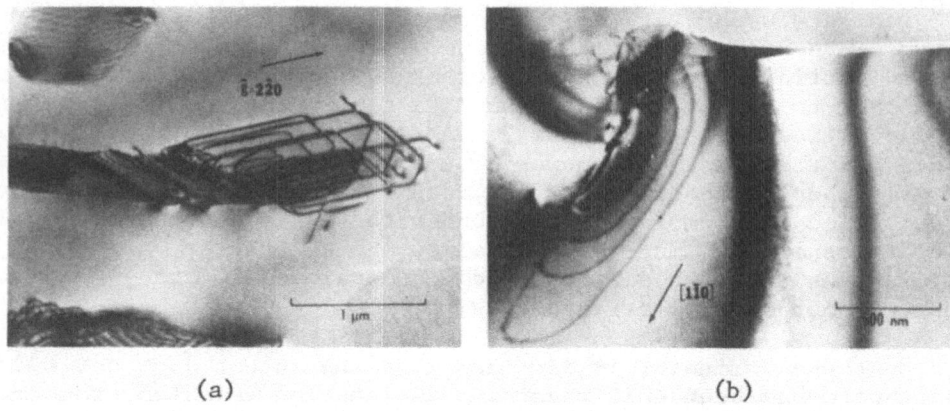

(a)   (b)

Fig. 17.   Dislocation atmospheres associated with cracks in Si.   In
           (a) the dislocations are in slip planes skew to the straight
           crack line, and are in nonblunting configurations.   In (b)
           the crack appears to have a corner, and dislocations appear
           to be emanating or otherwise associated with the crack line
           inhomogeneity.   After Lawn et al[25].

    Even though we have suggested that the problem of dislocation
emission is a three-dimensional phenomenon, perhaps involving crack or
lattice inhomogeneities, we believe that additional two-dimensional
theoretical work is desirable.   A systematic study of how various
classes of force laws, crystal symmetries, and crack geometries affect
the ease of dislocation emission has not been carried out in two
dimensions, and would be valuable for its qualitative insight.
Likewise, the effect of dislocation motion on ease of emission, and
the amount of pinning of a crack caused by an emission event are both
unknown, and susceptible of calculation by current techniques.

VII.   SUMMARY AND PERSPECTIVE

    The overall views of the author on atomic studies of crack
structure can be summarized in the statement that these studies have
their primary applications in understanding chemical effects and the
dislocation interaction and emission aspects of fracture.   In these
terms, the major challenges as seen by the author for the future are:

Crack Instability Associated with Dislocation Emission--This problem
is perhaps the one most nearly approachable at this time, both theoretically and experimentally.   The primary theoretical question to be
explored is how various crystal symmetries and different classes of
force laws affect dislocation emission.   The full three-dimensional
treatment of this problem is not now accessible in these terms, but
much of value could be learned from a comparative study of the effects
of different types of force laws and symmetry in the context of the
two-dimensional models.   Three-dimensional effects can be addressed

in a very qualitative fashion on the continuum level with much help
from direct observation. Our working hypothesis here is that inhomogeneities in either the crack line or the crystal lattice will lead
to the occasional dislocation emission which may accompany a basically
brittle crack, but twodimensional instability will prevent brittle
behavior all together.

Activation Energy and Activation Displacement for Materials of
Differing Classes--An adequate treatment of this problem in our
view would couple the two-dimensional methods with a semi-
continuum model analogous to the Frenkel-Kontorova model for dislocations. In two dimensions, the aim would be to explore the varieties
of behavior and structure which flow from different types and classes
of force laws and crystal symmetry. Lin and Hirth[26] have recently
taken a major step in linking continuum theory to the discrete theories by developing a continuum theory of crack kinks in terms of
continuum dislocations. We view this problem area as having less
significance for the important questions in fracture than the other
two areas, but it could well prove to be an excellent testing ground
for developing mathematical techniques applicable to the more important
crack structure problems.

Classification Schemes for Different Classes of Chemically Assisted
Fracture--This problem represents the major long-range challenge in
the area of physics of fracture, and will require the development of
new theoretical and experimental capabilities. I believe, however,
that we can hope for a level of understanding which correlates chemical
effects in fracture with the more ordinary chemistry of gas phase,
liquid phase or surface reactions, together with sorting out the roles
of bridging reactions at the crack tip site, steric effects in the
restricted geometry afforded by the crack tip, etc. Theory in this
area may become accessible by application of the decomposition theorem
discussed in Section V which may allow small cluster quantum calculations for the molecules at the crack tip to be fitted at the boundary
to the rest of the lattice by means of the lattice compliances. But
it will also be necessary to carry out systematic experimental studies
of chemical effects which give us a catalogue of the chemical reactions
which occur. At the very least, experiments should include crack growth
measurements in controlled atmospheres, such as those by Wei and
colleagues at Lehigh in metals[27], and Wiederhorn and his colleagues
at NBS for glasses[18]. But additional techniques which probe the
phenomena at crack tips directly are badly needed.

To conclude this section and the paper, we return to considering
the theoretical tools described in the earlier sections and attempt
to make a comparative assessment in the light of the challenges laid
out just above.

Of the two broad classes of atomistic theories available (direct
computer simulation, and lattice theory), we have emphasized the latter

in this paper. The lattice theory is a rigorous theory closely akin to lattice dynamical theory, but it incorporates a splitting of the problem into linear and nonlinear parts. The nonlinear part is solved separately as a system of coupled nonlinear equations in which the lattice appears in terms of a set of compliance coefficients. The advantage of the lattice theory is that its solution can be displayed graphically, from which a great deal of physical insight can be abstracted. Also, from it, one can derive the important decomposition theorem, which allows the crack tip processes to be separated from the remainder of the problem, leading to insights of particular power when only one bond at the tip is nonlinear. On the other hand, when applying it to lattices more general than the simple cubic, the method loses some of its analytical simplicity, and the lattice Green's functions must be computed completely by numerical means. A second weakness which probably can be overcome, is that results for a given crack size, 2n+1, require the inversion of a matrix of order n or larger, and asymptotic solutions for very long cracks have not yet been derived. For a complex problem such as a crack stability study, solving the set of coupled nonlinear equations requires first the calculation of the linear compliances for the rest of the lattice. These are calculable and need to be calculated only once for a given lattice and crack problem. Second, the nonlinear equations for the core must be solved, and the difficulty of this stage is probably equivalent to the similar stage in the direct computer simulations. However, a careful assessment of the overall efficiency of carrying out solutions to complex problems by the lattice theory has not been made in comparison to direct computer simulations. Such problems in the lattice theory would be made much easier by an adequate paradigm for handling the asymptotically long crack, however.

The direct computer simulation approach has two subclasses. In the first, a finite core is matched to an infinite medium, and in the second the computer simply deals with as large a block of atoms as possible, hoping for fast convergence to the bulk solution as the block size grows. Because of the later lecture by Dienes at this conference, I will only comment on the first. That technique, called "flex", in the hands of Sinclair et al[28], has become a very powerful technique because of the schemes for setting flexible boundary conditions on the interface between inner core and continuum. Flex has been applied very successfully to the crack and dislocation problems. It can address a limited three-dimensional problem such as a single kink, but not yet the double kink. For simple systems it has been possible to find saddlepoint configurations, but the problem of finding saddlepoints in a complex configuration space has as yet no general solution.

Considerable work will be necessary to push any of these techniques forward toward addressing the major challenges we have listed. We believe the various techniques will complement each other. Lattice theory has been valuable because of the physical intuition it has provided, Flex has proved itself capable of dealing with real lattices and a variety of two- and three-body forces in two and three dimensions, and the computer dynamics techniques of Paskin and Dienes leads very naturally to treatment of dislocation emission and other nonstatic processes.

REFERENCES

1. For general texts on fracture see:
   J.F. Knott, Fundamentals of Fracture Mechanics, Butterworths, 1973.
   B.R. Lawn and T.R. Wilshaw, Fracture of Brittle Solids, Cambridge
      Univ. Press, 1975.
   Two excellent monographs of a more advanced nature are:
   B.A. Bilby and J.D. Eshelby, Fracture, Vol. 1, p99, Ed. by
      H. Liebowitz (1968), Academic Press.
   J.R. Rice, Ibid, Vol. 2, p191.
2. J.D. Eshelby, Solid State Phys., 3, 79; Ed. by F. Seitz and
   D. Turnbull, Academic Press, N.Y., 1956.
3. E.R. Fuller and R. Thomson, Fracture Mechanics of Ceramics, Vol. 4,
   p507, Ed. by R. Bradt et al, Plenum, N.Y., 1978.
4. R. Thomson, J. Matls. Sci., 15, 1027 (1980).
5. G. Barenblatt, Adv. Appl. Mech., 2, 55 (1962).
6. (a) J. Goodier and M. Kanninen, Tech. Rept. 165, Div. of Eng. Mech.,
      Stanford Univ. (1966).
   (b) P. Gehlen and M. Kanninen, Inelastic Behavior of Solids,
      M. Kanninen et al (Eds.), Plenum (1972).
   (c) M. Kanninen and P. Gehlen, Interatomic Potentials and Simula-
      tion of Lattice Defects, Ed. P. Gehlen et al, Plenum (1972).
   (d) M. Kanninen and P. Gehlen, Int. J. Fract. Mech., 7, 471 (1971).
   (e) P. Gehlen, G. Hahn, and M. Kanninen, Scripta Met., 6, 1087 (1972.
   (f) A. Markworth, M. Kanninen, and P. Gehlen, Stress Corrosion
      Cracking and Hydrogen Embrittlement of Iron Base Alloys,
      NACE (1974).
7. R.A. Johnson, Phys. Rev. 145, 423 (1966).
8. (a) J. Sinclair and B. Lawn, Proc. Roy. Soc. A 329, 83 (1972).
   (b) J. Sinclair, Phil. Mag. 31, 647 (1975).
9. W. Sanders. Eng. Fract. Mech., 4, 145 (1972).
10. J. Wiener and M. Pear, J. Appl. Phys., 46, 2398 (1975).
11. W. Ashurst and W. Hoover, Phys. Rev., B, 14, 1465 (1976).
12. A. Paskin, A. Gohar, and G.J. Dienes, Phys. Rev. Let., 44, 940 (1980)
    A. Paskin, D.K. Som, and G.J. Dienes, J. Phys. C., 14, L171 (1981).
13. C. Hsieh and R. Thomson, J. Appl. Phys. 44, 2051 (1973).
14. D. Esterling, J. Appl. Phys., 47, 486 (1976).
15. R. Bullough and J. Hardy, Phil. Mag., 17, 833 (1968).
16. J. Hirth and J. Lothe, Theory of Dislocations, p484, McGraw-Hill,
    N.Y. (1968).
17. S. Wiederhorn, J. Amer. Cer. Soc., 50, 407 (1967).
18. S. Wiederhorn, E. Fuller, and R. Thomson, Metal. Sci., 14, 450 (1980)
19. A. Kelly, W. Tyson, and A. Cottrell, Phil. Mag., 15, 567 (1967).
20. J. Rice and R. Thomson, Phil. Mag., 29, 73 (1974).
21. W. Tyson, Fracture 1977, 2, 159, 1CF4 (1977).

22. I. Howard, Proc. Int. Conf. on Mechanical Properties of Materials
    (ICM3), Cambridge, England, 2, 462 (1979).
23. M. Jokl, V. Vitek, and C. McMahon, to be published.
24. J. Gilman, C. Knudsen, and W. Walsh, J. Appl. Phys., 29, 747 (1958).
25. B. Lawn, B. Hockey, and S. Wiederhorn, J. Matls. Sci., 15, 207 (1980
26. I. Lin and J. Hirth, to be published.
27. R. Wei, K. Klein, G. Simmons, and Y Chou, Symp. on Hydrogen
    Embrittlement and Stress Corrosion Cracking, June 2-3, 1980, Case
    Western Reserve Univ., Proceedings to be published.
28. J. Sinclair, P. Gehlen, V. Hoagland, and J. Hirth, J. Appl. Phys.,
    49, 3890 (1978).

DISCUSSION

Comment by R. Bullough:

    You said you thought the Flex II method and your Augmented
Lattice Green's Function (ALGF) method may be equivalent. Could you
perhaps explain why you hold this view on their equivalence? Flex
II was a linear combination of the elastic solutions in the region
II to represent the field there due to the anharmonic sources in
region I. The ALGF, however, appears to actually represent the
anharmonic sources in the lattice region in quite a different way.

Reply:

    Formally, of course, the ALGF and Flex II are different. The
point of my comment was that both have a similarity in the way they
separate the medium into two parts, one non-linear and one linear.
The question I posed is then which is most convenient or more
efficient, and whether combinations of the two can be taken which
would improve the efficiency of doing complex problems.

Comment by H. Vehoff:

    Is it now possible to simulate the crack tip opening angle of
a stable growing crack, because this value can be used as a measure
for the ratio of sliding to bond breaking?

Reply:

    In a partially ductile crack, the dislocation effects would be
an essential part of the problem. Some attempts along this line have
been made by myself and by Weertman, but the specific deformation
character reported in your own paper would be very interesting to
work out theoretically. The mobility problem in this case also
involves the dislocation rate effects, and has been considered in
recent work by E. Hart.

Comment by A.W. Thompson:

A question and a comment. First, you probably noticed the
collective gasp in the audience when you asserted that fracture
theory is now ahead of the experiments. What metal and what test
conditions would you recommend to test your calculations? The comment
is that, although you may need to neglect plasticity near the crack
in order to do your calculations, in real materials we certainly
cannot neglect events on the scale of the plastic zone (including the
contribution of microstructural elements like inclusions) if we wish
to understand their fracture. So I suspect additional modeling on a
somewhat larger scale will also be needed eventually.

Reply:

My point was that to gain fundamental knowledge about such a
complex subject as fracture requires one to separate the various
aspects of the problem for individual study to the extent possible.
Thus, in my opinion, there is a class of fracture problems, in some
metals in some circumstances where the properties of a brittle crack
are relevant, even though simultaneous dislocation activity is present.
Thus the first task for theory is to work out those simpler components
of the complete system, and in this regard theory is well ahead of
experiment. The question of whether there are any metals in which
the brittle aspects of fracture are paramount (in the way I have
talked about it) is of course an experimental question, but how about
zinc? My final comment is that Hart has found it necessary to include
the intrinsic velocity law of the underlying brittle crack in his
theory of moving quasi-ductile cracks, and in my own work the
equilibrium properties of sharp cracks are crucial to the theory of
dislocation shielded cracks.

Comment by H. Birnbaum:

Your criterion for emission of dislocations from crack tips
predicts that bcc metals such as Nb and Fe are brittle in the sense
that the crack will propagate before dislocations are emitted. Isn't
it troublesome that these metals are ductile down to low temperatures
when they are sufficiently pure? In Nb, at least, the ductility is
observed even in the presence of very sharp notches.

Reply:

Ultimately, the test of any theory is the experimental verifi-
cation. Also, as I stated in my lecture, the theory is still based
on continuum ideas. However, you do not say what the scale of your
observations is, and in a ductile material external sources may be
very close to the crack tip. Also, cracks may carry dislocations
with them, and again the distance of approach may be small. Finally,
it appears that unsteady motion of a crack, and inhomogenities in

the crack or matrix may lead to dislocation formation. In this
latter case, I would expect the dislocation production to be less
than for intrinsic ductility as given by the theoretical criterion.
For all these reasons, the precise conditions under which a crack
may emit dislocations is still very vague.

Comment by A.W. Thompson:

Listening to Howard Birnbaum's question, about the ductility at
very low temperature of pure Nb and Fe despite your prediction of
brittleness, reminds me that the same is true of pure Cd and Ti, the
former having been tested at 1 K. Why do you think that these two
hcp metals, unlike Zn, seem "inherently ductile"? Is it possible that
a brittleness criterion based on or derived from crystal structure
is oversimplified? Could you refine your calculation in foreseeable
ways?

Reply:

As stated in my reply to Birnbaum, one should remember that
"ductility" is composed of two parts so far as fracture is concerned.
The intrinsic emission of a dislocation from the crack tip is all our
calculation addresses. Thus the crack will induce plasticity in its
neighborhood, and drag dislocations with it, and these processes will
take place very close to the crack tip. Further, these dislocations
will blunt the "cleavage plane" behind the crack tip and give it the
appearance, macroscopically, of a ductile crack. However, down underneath all this dislocation activity may be an atomically sharp crack
if that crack is intrinsically brittle in the way I discussed. The
physical description of such a fracture will then be different from
that of a crack growing by true plastic processes as in hole growth.
So the question of whether a crack has a sharp or blunt tip underlying its plastic regime is important both for the theory and as an
experimental challenge. The theory can be refined (see Argon's
comments in the concluding paper), but it will be necessary, probably,
to have better force laws before those refinements will yield a theory
with strong predictive power beyond the current theory. At this
conference, however, we have seen modifications of the theory by both
Haasen and by Sinclair and Finnis which yield very interesting
results.

Comment by J. O'M. Bockris:

Dr. Thomson's calculation depends upon crystal parameters and
thus it seems undesirable to deal with a glass. Wouldn't zinc be a
better system on which to try his approach?

Perhaps I try too much to judge the utility of an approach by
it's relevance to corrosion cracking. Many aspects of the growth
of such cracks are interpreted quasi-quantitatively in terms of

electrochemical kinetics with the rate-controlling step being far
from the crack tip. Does Dr. Thomson think his approach would have
a usefulness for such actual situations comparable with direct
calculation of the crack velocity in terms of these interfacial
electrical processes (e.g., Cherapanov's work)?

Reply:

The atomic calculations I have reported first assumed that the
rate limiting processes are those at the crack tip. There are, of
course, other possibilities such as those mentioned by Bockris, and
even in the gaseous environment, stage II of slow crack growth is
believed to be a velocity regime limted by the transport of the gas
molecules to the crack tip. But in a certain sense, these other
processes are not fracture theory, but something else. The theory
as presented was also limited to the gaseous case in order to focus
simply on atomic processes at the tip. Modifications relating to
the electrochemical case would be interesting to work out.

MECHANICS OF FRACTURE

J.F. Knott

Department of Metallurgy and Materials Science
University of Cambridge

INTRODUCTION

Calculations of the theoretical strengths of crystalline solids are usually based on idealised forms of atomic force-displacement curves, in which the force is defined as the differential with respect to distance of the inter-atomic or inter-ionic energy. The energy curve is similar to that for a diatomic molecule in that it represents the resultant of inter-atomic repulsions and attractions; the nature and strength of the attractive forces depending on the bond type: ionic, covalent, metallic, or Van der Waals. Differences in character between a lattice and a molecule occur at separations of order one lattice spacing, when Friedel oscillations in the screening charge cause the long range component of the interaction potential in the lattice to undergo a damped oscillation about zero. For small displacements, the atomic force/displacement curve is linear, having a slope equivalent to Young's modulus, E. A lattice also has shear stiffness, denoted by the shear modulus μ. Both E and μ are defined macroscopically, usually for randomly-oriented polycrystals which are assumed to be isotropic. In single crystals, both the tensile stiffness and the shear stiffness vary with the orientation of the crystal with respect to the tensile axis and these variations can be substantial: in iron, for example, the minimum value of E is in the [100] direction (132 GPa at room temperature) and the maximum value is in the [111] direction (260 GPa).

In a simple calculation of ideal fracture strength, it is conventional to convert the atomic force/displacement curve to a stress/strain curve, by defining (for a simple cubic lattice) the "atomic area" as a_0^2 and the "original gauge length" as a_0 , where a_0 is the equilibrium lattice spacing. The stress/strain curve is then approximated to half a sine-wave of the form:

$$\sigma = \sigma_{max} \sin (2\pi u/\lambda) \qquad\qquad ... 1)$$

where u is the displacement from the rest position and λ is the wave-length; i.e. when $u = \lambda/4$, $\sigma = \sigma_{max}$. The total area under this curve represents the work done when the atomic plane is fractured and this is often equated to the surface energy, 2γ, of the two fracture surfaces produced. Hence:

$$(\lambda/2\pi)\ \sigma_{max} \left[- \cos(2\pi u/\lambda) \right]_0^{\lambda/2} = 2\gamma \qquad\qquad ... 2)$$

Assuming linear elasticity, for small displacements, we have:

$$\sigma = \sigma_{max}\ (2\pi u/\lambda) = Eu/a_0 \qquad\qquad ... 3)$$

because (u/a_0) is the strain. Substitution for λ from 3) into 2) then gives

$$\sigma_{max} = (E\gamma/a_0)^{\frac{1}{2}} \qquad\qquad ... 4)$$

Typical values for iron subjected to tension in the [100] direction are E = 132 GPa, $\gamma = 2\mathrm{Jm}^{-2}$, $a_0 = 0.25$ nm; hence σ_{max} = 32.5 GPa = 0.25 E. This is very much larger than experimentally-measured values of the cleavage fracture stress of iron (commonly of order 0.5 - 2.0 GPa). More sophisticated forms for the energy curve have been employed, and are treated elsewhere, but equation 4) has been used in discussions of behaviour at crack tips and serves well for illustrative purposes. It is arguable that iron cleaves on {100} planes because E is a minimum in the [100] direction, thus minimising σ_{max}.

An interesting point is that the *shape* of the atomic stress-strain curve is specified by its maximum *amplitude*, σ_{max}, and its initial *slope at low displacements*, E, through equation 3). If σ_{max} is lowered, for example, by the presence of hydrogen, equation 3) indicates that two limiting possibilities may occur. From the equation $(\lambda/2\pi) = a_0 (\sigma_{max}/E)$, we see that at one extreme, E decreases with σ_{max} so as to keep λ (and hence $\lambda/4$, the strain at σ_{max}) constant: at the other, E remains constant, so that λ decreases with σ_{max}. The former case implies a decrease in the tensile stiffness of the bond in the presence of hydrogen, which is not implausible if the general "strength" of 3d electron bonding in iron is affected by the loss or gain of electrons to or from electron levels localised around protons in the lattice. The latter implies a smaller displacement $(\lambda/4)$ to reach σ_{max} (which has itself been lowered) and hence an atomic "embrittlement" of the bond. In both cases, the work of fracture, 2γ, is reduced.

We shall see later that differences between theoretically

calculated values and experimental values of the fracture strengths
of common structural materials are due to the presence in real
materials of micro-cracks or crack-like defects. In some materials,
these can concentrate stress at their tips so as to produce the local
forces for bond separation in a perfectly elastic manner, but, in
others, the material's shear strength is exceeded and local plastic
deformation is produced before there is any tensile fracture of bonds.
Since plastic deformation is so closely associated with crack propagation (and also with crack nucleation) the following section is
devoted to a brief survey of concepts from the theory of plastic
deformation relevant to subsequent discussion.

DISLOCATION THEORY

    As for the ideal tensile fracture strength, calculation of the
ideal shear strength of a crystal (i.e. the shear stress required to
slip one plane of atoms over another) gives values which are significant fractions of the (shear) modulus, typically 0.1μ. Such values
are some two-to-three orders-of-magnitude greater than those observed
experimentally for pure metal crystals. The anomaly is resolved by
invoking the concept of a *dislocation*[1-3], which is a line defect, of
line vector ℓ and Burgers vector $\underline{b}$, defined as the closure failure
when a clockwise circuit of lattice translation vectors, around a
dislocation line is repeated in perfect, undislocated lattice (RH/
FS convention). An *edge* dislocation has $\underline{b} \perp \ell$; a *screw* dislocation
has $\underline{b} \parallel \ell$; any other dislocation is *mixed*, with edge and screw
components. If a dislocation moves in a direction $\underline{m}$, it is possible
to define a plane of movement with normal $\underline{n} = \ell \times \underline{m}$: this plane is
a *slip-plane* if $\underline{n}.\underline{b} = 0$. In a continuum, screw dislocations can slip
in all planes: edges can slip only in the plane containing $\underline{b}$ and ℓ.
The components of $\underline{b}$ along the coordinate axes $X_1X_2X_3, b_1b_2b_3$, are
conventionally written in crystallographic form; $\underline{b} = a_0/2\ [1\bar{1}0]$ has
components $b_1 = a_0/2$ $b_2 = -a_0/2$ $b_3 = 0$. This is the primitive
lattice translation vector for a cubic close-packed (ccp) crystal
and is the Burgers vector for a *perfect* dislocation in that lattice.
The slip-planes in a c.c.p. lattice are of the form {111} . These
are the close-packed, and hence most widely-separated, planes. A
dislocation with $\underline{b} = a_0/2\ [1\bar{1}0]$ can slip in the planes (111) and
$(11\bar{1})$: in cubic crystals, these have normals [111] and $[11\bar{1}]$ and
the cosines of angles $[1\bar{1}0]\char`^[111]$ and $[1\bar{1}0]\char`^[11\bar{1}]$ in the calculation
of $\underline{n}.\underline{b}$ are both equal to zero. In non-cubic crystals, a direction,
such as $\underline{b}$, having indices [uvw] lies in a plane of indices (hkl) if:

$$hu\ +\ kv\ +\ lw\ =\ 0 \qquad\qquad\qquad \dots\ 5)$$

Note that [hkl] is not normal to (hkl) in a non-cubic lattice.

    The stress field around a screw dislocation is of particularly
simple form in cylindrical coordinates r,θ,z, (with $z \parallel \ell$), being,

to first order, a pair of equal shears of magnitude:

$$\sigma_{z\theta} = \sigma_{\theta z} = \mu b / 2\pi r \qquad \qquad \dots 6)$$

where $\sigma_{z\theta}$ and $\sigma_{\theta z}$ are (shear) components of the stress tensor σ_{ij} (the suffices can take values independently of r, θ or z: i indicates the direction of a stress component, j the normal to the face on which it acts: i = j implies tensile stress i $\neq$ j implies shear stress). For an edge dislocation, plane-strain deformation is assumed and the stress-field involves tensions and compressions, in addition to shears. With $D = \mu b / 2\pi (1 - \nu)$ the stresses, in polar coordinates, become:

$$\sigma_{\theta\theta} = \sigma_{rr} = - D\cos\theta / r; \quad \sigma_{r\theta} = \sigma_{\theta r} = D\sin\theta / r \qquad \dots 7)$$

where θ is measured in the anticlockwise sense, starting from the line of the "extra half-plane". The stress-fields are subject to a cut-off at the edge of the dislocation "core", inside which the strains are so high that linear elasticity cannot be applied. A simple estimate of the core radius r_c, is made by setting the strain at the periphery of the core, $b/2\pi r_c$, equal to 0.1, as a reasonable limit to linear behaviour: hence, $r_c = b/0.2\pi \simeq 1.6b$. The elastic energies per unit length of screw and edge dislocations respectively (ignoring core energy) are

$$\varepsilon_{screw} = (\mu b^2 / 4\pi) \ln (r_1 / r_c) ; \qquad \dots 8a)$$

$$\varepsilon_{edge} = (\mu b^2 / 4\pi (1 - \nu)) \ln (r_1 / r_c) \qquad \dots 8b)$$

where r_1 is the outer radius of the "cylinder of influence" and may be of order 1μm in a well-annealed crystal.

In many calculations, it is convenient to use the concept of a "force on a dislocation", $\underline{F}$. This is derived from a work argument. Suppose that a crystal containing a dislocation of line vector $\underline{\ell}$ and Burger's vector $\underline{b}$ is subjected to a general stress field σ_{ij} (with i and j independently taking values of 1, 2 and 3 in right-handed orthogonal Cartesian coordinates $X_1 X_2$ and X_3). The dislocation moves by $\underline{m}$, defining the slip-plane normal $\underline{n} = \underline{\ell} \times \underline{m}$, so that the components of traction on the slip-plane are $\sigma_{ij} n_j$. Writing these as $\underline{\sigma}\underline{n}$, the work done by the applied stress is $(\underline{\sigma}\ \underline{n}).\underline{b}$ which, for $\underline{\sigma}$ symmetrical, can be shown to be equal to $(\underline{\sigma}\ \underline{b}).\underline{n}$, or, since $\underline{n} = \underline{\ell} \times \underline{m}$, to $(\underline{\sigma}\ \underline{b})$. $\underline{\ell} \times \underline{m}$. This is a set of (scalar) triple products, so that $(\underline{\sigma}\ \underline{b})$. $\underline{\ell} \times \underline{m} = (\underline{\sigma}\ \underline{b}) \times \underline{\ell}.\underline{m}$. The work done by the applied stress is then equated to the work done by the "force on the dislocation", $\underline{F}$, as the dislocation moves through $\underline{m}$, i.e. to $\underline{F}.\underline{m}$. Hence, $\underline{F}$ may be written as (the Peach-Köhler relationship):

$$\underline{F} = (\underline{\sigma}\ \underline{b}) \times \underline{\ell} \qquad \qquad \dots 9)$$

The "$x\underline{\ell}$" term should be noted: this emphasises that the force acts at right angles to the dislocation line. A corollary is that $(\underline{\sigma}\ \underline{b})$ is normal to the direction of $\underline{F}$ and this can be achieved if $(\underline{\sigma}\ \underline{b})$ is interpreted as a matrix multiplication. Suppose, for example, that an edge dislocation of line vector $\underline{\ell} = \underline{\ell}_2$ and Burgers vector $\underline{b} = -\ b_1$ (RH/FS convention) is subjected to an applied shear stress σ_{13} (which tends to move the dislocation in the direction X_1). Equation 9) becomes

$$\underline{F} \quad = \quad \begin{vmatrix} 0 & 0 & \sigma_{13} \\ 0 & 0 & 0 \\ \sigma_{31} & 0 & 0 \end{vmatrix} \begin{vmatrix} -b_1 \\ 0 \\ 0 \end{vmatrix} \text{ x } \ell_2 \qquad\qquad \text{... 10)}$$

because the complementary shear σ_{31} must be introduced to maintain equilibrium. Evaluation of the matrix product gives the only non-zero term as $-\sigma_{31}b_1$ (acting along $-X_3$), which, when crossed with ℓ_2 gives $\underline{F}$ as having magnitude $\sigma_{31}b_1 = \sigma_{13}b_1$ and acting in the positive X_1 direction. Other archetypal examples are for a screw dislocation with $\underline{\ell} = \underline{\ell}_2$, $\underline{b} = b_2$ subjected to a shear stress σ_{23}: $\underline{F}$ has magnitude $\sigma_{32}b_2 = \sigma_{23}b_2$ and acts along $-\ X_1$; and for an edge dislocation, with $\underline{\ell} = \underline{\ell}_2$, $\underline{b} = -\ b_1$, subjected to a tensile stress σ_{11}: $\underline{F}$ has magnitude $\sigma_{11}b_1$ and acts along $-X_3$ (a "downward climb" force).

In addition to these definitions in terms of applied stress fields, the concept of "force on a dislocation" may be used to calculate forces between dislocations or between dislocations and free surfaces. If, for example, two parallel screw dislocations of the same sign (relationship of $\underline{b}$ to $\underline{\ell}$) have Burgers vectors b_1 and b_2 and lie at a distance r apart, the mutual repulsion between them is given by:

$$F \quad = \quad \mu b_1 b_2 / 2\pi r \qquad\qquad \text{... 11)}$$

From equation 6) the stress field of dislocation 1 is $\sigma_{z\theta} = \mu b_1 / 2\pi r$ and this produces a force on dislocation 2 of magnitude $\sigma_{\theta z} b_2 = \sigma_{z\theta} b_2$ directed along r. There is an equal and opposite force exerted on dislocation 1 by dislocation 2. The interaction between a screw dislocation and a free surface follows a similar line of argument. Use is made of the concept of an "image" dislocation, to reduce all surface stresses to zero. The attraction between a screw dislocation and the surface is then given by:

$$F \quad = \quad \mu b^2 / 4\pi r \qquad\qquad \text{... 12)}$$

Surface films of modulus higher than that of the matrix produce repulsive forces.

MICROCRACKS AND STRESS DISTRIBUTIONS AROUND BLUNT CRACKS

The main reason why structural materials are so much weaker than the calculation of ideal fracture strength predicts is that they contain distributions of crack-like defects, which serve to produce high stresses locally. The defects can arise in various ways: either naturally, or as part of processing and fabrication procedures. Good examples are to be found in sintered ceramic materials: here, the size of any inherent defect is that of unbonded material, which is directly related to the initial grain size. Metals may be fabricated by powder-processes, but here, the defects are related not so much to the grain size of the metal (assuming atomisation in vacuum or argon) as to the size of any non-metallic inclusions which may come through the final sieving. Such inclusions might limit, for example, the fatigue-life of a superalloy, gas-turbine disc. The fracture of sedimentary rocks may be similar to that of sintered ceramics: certainly, the fractured-surface of an oolitic limestone shows clearly that the fracture proceeds around near-spherical "grains" so that it may have been triggered-off by an inter-granular microcrack. Cast metals may also contain defects, in the form of pores due to gas bubbles, or as inter-dendritic shrinkage cavities. These are even more likely in as-deposited weld metal, because tensile stresses due to shrinkage are set up as the weld solidifies. It is possible that similar effects produce micro-cracking in igneous rocks. Fabrication by welding can also introduce cracks in the heat-affected-zone (HAZ) as a result of hydrogen uptake, or in the parent plate, if the material is subject to "lamellar tearing". Quenching and "stress-relief" heat-treatments in steel products may sometimes produce cracks, particularly if the steel contains high levels of "tramp" impurity elements, such as P, Sn or Sb. Even in fully wrought material, the non-metallic inclusions are likely to act as cracks *ab initio* and brittle particles in the heat-treated microstructure may also be significant, although, in this last case, it is usually necessary to nucleate sharp cracks by dislocation arrays.

The stress-concentrating effect of a blunt crack is given by the expression, due to Inglis[4], for an ellipse of semi-major axis a and semi-minor axis b, oriented with its major axis (along X_2) normal to a uniform tensile stress, σ_{app} (parallel to X_1) in an infinite body. The maximum local tensile stress, σ_{11}, at the ends of the ellipse ($x_2 = \pm a$) is found to be:

$$\sigma_{11 (max)} = \sigma_{app} (1 + 2a/b) \qquad \ldots 13)$$

or, if the radius of the tip is ρ ($\rho = b^2/a$ for an ellipse), $\sigma_{11 (max)}$ is given by:

$$\sigma_{11 (max)} = \sigma_{app} (1 + 2 \{a/\rho\}^{\frac{1}{2}}) \qquad \ldots 14)$$

The stress σ_{11} decreases with distance, r, from the crack tip and varies significantly with angle, θ, around the tip. For rather sharp cracks, $2(a/\rho)^{\frac{1}{2}} \gg 1$, and we may write:

$$\sigma_{11(max)} \simeq 2\sigma_{app}(a/\rho)^{\frac{1}{2}} \qquad \dots 15)$$

It is not uncommon to find expressions in the literature in which the stress-concentrating effect of a crack-like defect has been estimated first by writing $\sigma_{11(max)}$ in equation 15), with ρ equal to a lattice spacing, a_0, and then equating $\sigma_{11(max)}$ to the ideal fracture strength, σ_{max}, as given by equation 4). The result is:

$$\sigma_{app} = (E\gamma/4a)^{\frac{1}{2}} \qquad \dots 16)$$

from which it may be deduced that a microcrack of total length 2a = 0.2μm would be sufficient to reduce σ_{app} to 1 GPa, if the fracture process were perfectly elastic. Although the form of 16) turns out to be generally correct, it has been derived in an improper manner, because 15) assumes linear elastic behaviour, whereas 4) is based on a sinusoidal force-displacement curve. Fortunately, a different approach enables macroscopic fracture stresses to be calculated, without reference to the details of crack-tip separation. This is essentially the thermodynamic approach of Griffith[5], later extended to a variety of quasi-brittle materials and termed fracture mechanics [6,7]. A suitable starting point is to consider the stress distribution around a mathematically sharp crack.

STRESS DISTRIBUTIONS AROUND SHARP CRACKS

From 15) we see that, as $\rho \to o$, so $\sigma_{11(max)}$, the stress *at* the tip of the ellipse, $\to \infty$, so that we are assured that enough stress will be available to break atomic bonds, if such fracture is energetically favourable. To decide this latter point, it is necessary to consider the stress *distribution* ahead of the crack-tip. Measuring r as a distance from the tip and θ as the angle measured in an anticlockwise sense from the X_2 axis, it is found that, *close to the tip*, the stresses may be written as:

$$\sigma_{11} = K(2\pi r)^{-\frac{1}{2}}\cos(\theta/2)\{1+\sin(\theta/2)\sin(3\theta/2)\} \qquad \dots 17a)$$

$$\sigma_{22} = K(2\pi r)^{-\frac{1}{2}}\cos(\theta/2)\{1-\sin(\theta/2)\sin(3\theta/2)\} \qquad \dots 17b)$$

$$\sigma_{12} = \sigma_{21} = K(2\pi r)^{-\frac{1}{2}}\sin(\theta/2)\cos(\theta/2)\cos(3\theta/2) \qquad \dots 17c)$$

$$\sigma_{r\theta} = \sigma_{\theta r} = K(8\pi r)^{-\frac{1}{2}}\sin(\theta)\cos(\theta/2) \qquad \dots 17d)$$

Along the X_2 axis (for $|x_2| > |a|$) and near the tip, σ_{11} varies with distance as $K(2\pi r)^{-\frac{1}{2}}$. The strength of the *singularity*, K, is termed the *stress-intensity factor* and, for a through-thickness crack of length 2a, lying normal to a uniform applied stress, σ_{app}, in an infinite body, it has the value σ_{app} $(\pi a)^{\frac{1}{2}}$. A similar form would hold for an internally-pressurized crack in an infinite body, but the presence of external boundaries (normal to which, stress-components must be zero) alters the expression for K to a general form $K = \sigma_{app}$ f(a/W), where W is the width of the body. Tabulations are available and "Modes" I (tension) II (shear) and III (antiplane strain) are recognised. Over larger distances, the stress σ_{11} is given by:

$$\sigma_{11} = \sigma_{app} (1 - \{a^2/x_2^2\})^{\frac{1}{2}} \qquad \ldots 18)$$

for $x_1 = 0$, $|x_2| > a$, and the displacement within the crack, u_1, is given by:

$$u_{1(x_1 = 0)} = 2(1-\nu^2) (\sigma_{app}/E) (a^2 - x_2^2)^{\frac{1}{2}} \qquad \ldots 19)$$

for $x_1 = 0$, $|x_2| < a$ where ν is Poisson's ratio. This corresponds to plane-strain deformation ($\varepsilon_{33} = 0$). The $(1-\nu^2)$ term is absent in plane stress ($\sigma_{33} = 0$).

Griffith's thermodynamic approach to crack propagation[5] assumed that there would always be sufficient stress available at the crack tip to break atomic bonds and that the occurrence or non-occurrence of catastrophic propagation would then depend on whether the energy of the system as a whole was decreased or increased by propagation. He recognised two energy terms: the surface energy of the two surfaces created by fracture, $S = 2\gamma a$ for a half-crack of unit thickness and length a, and the strain-energy or potential-energy released from the body, U. Using Inglis' results[4], U was calculated by integrating the product of stress x strain for all elements of the infinite body to give (for a half-crack of length a) $U = - \sigma_{app}^2 \pi a^2/2E$. The total energy is then given as a function of crack length by

$$\varepsilon = S + U = -(\sigma_{app}^2 \pi a^2/2E) + 2\gamma a \qquad \ldots 20)$$

This has a maximum when $(d\varepsilon/da) = 0$, i.e. when:

$$\sigma_{app} = (2E\gamma/\pi a)^{\frac{1}{2}} \qquad \ldots 21)$$

The implication is that, for a given crack (half-) length a, there is a critical value of σ_{app}; above which, the energy is lowered by catastrophic propagation: below which, it is lowered by re-healing of the crack faces. The critical value of σ_{app} is the fracture stress σ_F. It is clear that a microcrack of length 2a $\approx$ 0.4μm would be sufficient to lower σ_F to approx. 1 GPa if the fracture process were elastic. Most samples of commercial steels contain brittle carbide particles of thickness greater than 0.4μm, but it is possible to purify iron (e.g. by zone-refining) to levels such that no particles of this size could be contemplated, and yet find that brittle fracture occurs at stresses of order 1 GPa. In zinc, the corresponding microcrack size is several mm, but it is possible to obtain brittle fracture at relatively low stress in zinc crystals which have cross-sections smaller than the critical size. Experiments show that local fractures are then produced by dislocation arrays.

IRWIN'S VIRTUAL-WORK CALCULATION AND FRACTURE MECHANICS

A second method of deriving the energy release is based on the virtual-work theorem and was introduced by Irwin[6]. In the derivation of 21) it was necessary to calculate (dU/da): the change in strain-energy for crack extension da. In fracture mechanics terminology[7], this is defined as the "strain-energy (or potential-energy) release-rate", (per unit thickness) and given the letter G. Mathematically, it is possible to extend a crack notionally from length a to length $(a + \delta a)$ and to calculate the work done by the surface tractions, σ_{11}, over δa, as they reduce to zero by moving through the displacements, u_{11}, corresponding to the extended crack. This is then equal to the change in strain-energy stored in the body. Hence

$$G\delta a = \int_o^{\delta a} \sigma_{11}\, u_{11}\, dr \qquad \qquad \dots 22)$$

From 17a) with $\theta = 0, \sigma_{11} = K(2\pi r)^{-\frac{1}{2}} = \sigma_{app}\, (a/2r)^{\frac{1}{2}}$
From 19), substituting $(a + \delta a)$ for a and $r = (x_2 - a)$,

$$u_{11} = 2\sqrt{2}(1 - \nu^2)\, (\sigma_{app}/E)\, a^{\frac{1}{2}}\, (\delta a - r)^{\frac{1}{2}}, \text{ and}$$

$$G\delta a = 2(1 - \nu^2)(\sigma_{app}^2 a/E) \int_o^{\delta a} (\delta a - r)^{\frac{1}{2}} r^{-\frac{1}{2}}\, dr \qquad \dots 23)$$

By substitution of $r = \delta a \sin^2 \omega$ this can be evaluated as

$$G\delta a = (\sigma_{app}^2 \pi a/E)\, (1 - \nu^2)\, \delta a \qquad \qquad \dots 24)$$

With $K = \sigma_{app}\, (\pi a)^{\frac{1}{2}}$, we obtain

$$G \equiv dU/da = K^2 (1 - \nu^2)/E \qquad \qquad \dots 25)$$

or, in plane stress

$$G \equiv dU/da = K^2/E \qquad \qquad \dots 26)$$

A convenient mode of writing is to use E^1 to stand for E in plane stress and $E/(1 - \nu^2)$ in plane strain, so that 25) and 26) may be combined to give:

$$K^2 \equiv E^1 G \qquad \qquad \dots 27)$$

This is the basic identity that makes fracture mechanics such a powerful tool for the analysis of crack propagation, because it reduces complicated calculations of energy release to much easier calculations of stress intensities, many of which are tabulated[8]. Moreover, the K-values add in a simple algebraic manner, so that the effect of combinations of applied stresses may be deduced in a straightforward manner.

If the energy associated with fracture is composed not only of surface energy, but also of additional terms, associated perhaps with plastic deformation or visco-elastic deformation, it may still be possible to use 27) provided that the additional terms are small and do not seriously affect the elastic virtual-work calculation. Under these conditions, we might find, for example, that fracture is still *characterised* by a critical value of G, G_{crit}, which is larger than 2γ, but, for "quasi-brittle" behaviour, still write

$$K^2 = E^1 G$$

$$K_{crit}^2 = E^1 G_{crit} \qquad \qquad \dots 28)$$

or, for the infinite body, with $K = \sigma_{app}(\pi a)^{\frac{1}{2}}$,

$$\sigma_{app}^2 \pi a = E^1 G$$

$$\sigma_F = (E^1 G_{crit}/\pi a)^{\frac{1}{2}} \qquad \qquad \dots 29)$$

The *interpretation* of G_{crit} resides in the micromechanisms of crack extension: in metals, this involves plastic flow; in polymers, visco-elastic flow and crazing. The *engineering use* of 29) is immediately apparent: an experimental (if uninterpreted) value of *fracture toughness* (G_{crit} or K_{crit}) enables failure stress, σ_F, to be related to defect size, a. In design, σ_{app} is usually determined by considerations such as elastic buckling or plastic collapse and 29) is then employed to set NDT limits on permissible defect size. In defect assessment, the size of crack is known and 29) is used to calculate the fracture stress.

ELASTICITY IN CRYSTALS

If the aim is eventually to focus on crack-tip behaviour at the atomistic level, it is relevant to make the point that equation 27) has been derived in an "engineering" manner, using the averaged, macroscopic value for Young's Modulus, E, corrected for plane strain by the macroscopic value of Poisson's Ratio, ν. In results dealing with single-crystal deformation, the same principle of virtual work holds, but the moduli should strictly be replaced by values of crystal constants for the appropriate orientation of crack[9]. In the most general form of elastic stress/strain relationship, the nine terms of the symmetrical stress tensor, σ_{ij}, are related to the nine terms of the symmetrical strain tensor, ε_{ij}, through the fourth-order tensor of crystal constants, c_{ijkl} (81 terms):

$$\sigma_{ij} = c_{ijkl} \, \varepsilon_{kl} \qquad \dots 30)$$

With $\sigma_{ij} = \sigma_{ji}$ and $\varepsilon_{kl} = \varepsilon_{lk}$, there are 36 independent components to c_{ijkl} and it is conventional to write σ_{ij} and ε_{kl} in matrix form σ_i (i = 1 to 6, $\sigma_{23} = \sigma_{32} = \sigma_4$ etc.) and ε_j (j = 1 to 6). Then the relationships become:

$$\sigma_i = c_{ij} \, \varepsilon_j \qquad \dots 31)$$

Further reduction of terms is possible, if it is recalled that strain-energy, W, is a thermodynamic function of state, its value being dependent only on the instantaneous values of strain and not on the path of integration:

$$dW = \sigma_i d\varepsilon_i = c_{ij} \, \varepsilon_j \, d\varepsilon_i \quad or \quad (\partial W/\partial \varepsilon_i) = c_{ij} \, \varepsilon_j$$

$$(\partial/\partial \varepsilon_j)(\partial W/\partial \varepsilon_i) = c_{ij} \qquad \dots 32)$$

However, since W is simply a function of strain, the order of differentiation is immaterial and the left-hand side of equation 32) is symmetrical with respect to i and j.

Hence $\quad c_{ij} = c_{ji} \qquad \dots 33)$

This reduces the array to 21 independent terms and these must all be evaluated for crystals of the lowest symmetry class (triclinic). Crystal symmetry, however, ensures that certain coefficients become equal to each other: in the tetragonal system for example, the values of tensile stiffness along the a and b axes are the same ($c_{11} = c_{22}$) whilst that along the c axis is different (c_{33}). In the cubic system, only three independent coefficients remain: c_{11}, c_{12} and c_{44}. For an *isotropic* material $c_{11} - c_{12} = 2c_{44}$ and then it is possible to relate the values of the crystal constants to their macroscopic equivalents: Lame's constants, λ and μ:

$$c_{11} = \lambda + 2\mu, \quad c_{12} = \lambda, \quad c_{44} = \mu \qquad \dots 34)$$

where $\lambda = E\nu/(1+\nu)(1-2\nu)$ and μ is the shear modulus $= E/2(1+\nu)$. It is also possible to relate strain to stress *via* the elastic compliances, s_{ij}:

$$\varepsilon_i = s_{ij} \, \sigma_j \qquad \dots 35)$$

and equivalent arguments may be used to reduce these to three independent components for the cubic system. The ratio $2c_{44}/(c_{11}-c_{12})$

is the *elastic anisotropy factor*. It has a value close to unity for
tungsten and 1.2 for aluminium, but differs significantly from unity
for other cubic materials. For iron, values of c_{11} = 237 GPa, c_{12} =
141 GPa, c_{44} = 116 GPa give rise to a factor of 2.4. This is
consistent with the wide variations in Young's modulus with orientation of iron crystals described in the introduction. In polycrystalline materials, these differences are usually averaged-out,
but there are situations in "textured" materials where the individual
grains are all oriented in more-or-less the same crystallographic
orientation. This is done deliberately in silicon-iron, to reduce
eddy-current losses when the material is used as a transformer-core
material, but may arise adventitiously in castings or weld-deposits
because the solidification is such that cells or dendrites grow
preferentially in one orientation, usually <100> in cubic metals.
This solidification texturing is used in single-crystal, gas-turbine
blades to align a [100] direction along the length of the blade, to
minimize the modulus and reduce the stress level for a given thermal
strain. As mentioned in the introduction, the fact that E is a
minimum in the [100] direction may be related to the fact that iron
cleaves on {100} planes. However, tungsten, which is bcc, but isotropic, also cleaves on {100} planes.

BRITTLE AND DUCTILE BEHAVIOUR

     Griffith's equation 21) is strictly a (static) thermodynamic
balance of energy changes with crack length, ignoring entropy changes
or kinetic energy terms, and postulating that all the release of
elastic energy is absorbed in the creation of two surfaces. The
critical crack length is that length above which catastrophic propagation occurs; below which, the crack heals. The separation is
completely elastic, with no (irreversible) plastic flow at the crack
tip. Griffith's initial experiments were made, using glass, but
there is evidence that the results which he obtained were influenced
by stress-corrosion due to water-vapour and that, had these effects
been absent, the value of G_{crit} obtained would have been significantly
greater than 2γ. Marsh[10] envisages the need for some local process
such as "plastic" (or visco-elastic) flow before failure in the crack-
tip region if chemical interactions are absent.

     Better evidence for the simple Griffith energy balance is
perhaps to be found in the "cleavage" of minerals and some metals.
It is claimed that thin mica sheets can be cleaved in a thermodynamically reversible manner in vacuum, if the testing system is
such that the crack growth is stable, e.g. wedge loading. This means
that if the applied load or displacement is reduced, the crack faces
re-bond. It is interesting that the word "CLEAVE" itself has meanings
both of "splitting apart"* and of "binding together"** so that perhaps

*   "split" from O.E. cleofan cf. Ger klieben
**   "bind" from O.E. clifian cf. Ger kleben

we should reserve the term "true cleavage" for this reversible
process. It is experiments such as these that lead to definitions
of what is meant by brittle or ductile (tough) behaviour. Consider
a mode I crack subjected to applied tensile stress and examine the
stresses around its tip (equations 17a)-d)). In particular, there
is a tensile stress, σ_{11} (equation 17a)) which acts directly across
the plane in which the crack lies and a shear stress, $\sigma_{r\theta}$, (equation
17d)) which may be resolved onto an appropriate slip plane. The
general proposal has been made[11], that, if σ_{11} exceeds the tensile
strength of the bond at the crack tip ($\sim$0.2-0.3E) before $\sigma_{r\theta}$ exceeds
the shear strength ($\sim$0.1μ), the material is brittle: if not, it is
ductile, because the crack blunts by spontaneous emission of dislocations before it can propagate in an elastic manner. In a broad
sense, this is an extremely useful definition, because it clearly
separates materials which are "obviously" brittle from those which
are "obviously" ductile. The problems with infinities in the
stresses in equations 17a)-d) were initially overcome by comparing
the *ratio* of $\sigma_{r\theta}$ (resolved onto an appropriate slip-plane) to σ_{11}
(across a cleavage plane) with the *ratio* of (estimated) theoretical
shear strength to theoretical cleavage strength. Later, more
detailed calculations were made for the emission of dislocations
from the crack-tip, through the steep shear-stress gradient, at the
Griffith stress[12]. Allowance was also made for the "image force"
due to the presence of a free surface (equation 12)).

Mica is a layer-silicate structure, possessing strong bonding
within the layers, but weak bonding normal to the layers. If a
tensile stress is applied normal to the layers it is easy to break
the weak inter-layer bonds and very difficult to produce shear at
any angle to this "cleavage plane": the fracture is therefore
almost ideally elastic and, in a suitable loading system, reversible.
Graphite is a simpler structure composed of (hexagonal) layers,
although here the more familiar mode of behaviour is the *sliding* of
the densely-packed layers with respect to each other, in response to
an applied shear parallel to the layers. This enables graphite to
be used as a solid lubricant. Zinc, which is hexagonal, with large
c/a axial ratio (1.86) slips on its (0001) basal plane in response
to shear stress, but cleaves on (0001) in what appears to be an
ideally elastic manner, if a tensile stress is applied normal to
(0001).

A very clear contrast between "obviously" brittle and "obviously"
ductile behaviour is provided by *diamond* and *gold*, both of which have
a cubic F (fcc) lattice. On the one hand, *diamond* has strong,
directed, covalent bonding, so that very high shear stresses are
required to create dislocations at a crack tip and move them away to
blunt the crack, (i.e. it has a high lattice friction stress, or
Peierl's-Nabarro stress). Cleavage cracks in diamond therefore
propagate easily and cleavage is used, prior to final polishing, as

a technique to produce gem-stones. Similar brittle behaviour occurs
in other covalent or ionic solids in which the Peierl's-Nabarro
stress is high. In *gold*, on the other hand, dislocations move
extremely easily and any incipient crack nucleus is blunted
completely. Under external stress systems which do not engender
tensile necking instability, gold can withstand extremely large
plastic strains without failure. In wire-drawing, it can be drawn
into extremely thin wires: 1 gm. of gold has been drawn to a length
of 2 miles (∿3km)[13]. In the production of gold leaf, gold can be
hammered down to thicknesses of order 10nm. The following description has been given of the beating of gold by a hammer[13]: "this
causes gaping cracks upon the edges of the leaves, the sides of which
readily coalesce and unite without leaving any trace of the union
after being beaten upon". Gold's extreme "malleability" therefore
seems to reside partly in its ability to undergo pressure-welding of
cracks, but these cracks form as a result of shear, not cleavage.
Gold may be regarded as the extreme example of the typically ductile
behaviour of all c.c.p. metals (with the possible exception of
iridium) when fractured in air, although {001} cleavage-type fractures
have been observed in aluminium when subjected to liquid-metal
embrittlement (LME) and in a variety of c.c.p. alloys when tested in
fatigue. Definitions of cleavage will be discussed in the following
section.

A final important class of materials is that of b.c.c. metals.
In these, the Peierl's-Nabarro stress increases markedly with decrease
in temperature and, typically, a *transition* is observed, from brittle
behaviour at low temperatures to ductile behaviour at high temperatures. The transition-temperature is a function, not only of the
particular metal tested and the applied strain-rate (which controls
the time available for a thermal fluctuation to overcome energy
barriers to dislocation movement at any given stress level), but also
of the *stress-state*, being higher in stress-states in which the ratio
of tensile-stress to shear-stress is high. This demonstrates the
importance of applied tensile stress and the crack-propagation stage
in the fracture process. Cleavage planes are usually {100} although
these are not easily identified in commercial steel microstructures
and may be altered by, for example, the local elastic strain fields
due to precipitates or other second-phase particles. Details of the
fracture process in steel will be treated in a later section, but it
is of interest first to note that, in calculations of the emission
of crack-tip dislocations, iron must be regarded as very much a
"borderline" material, which the calculations are really too crude
to be able to state categorically whether cracks will blunt or
propagate with sharp tips.

DEFINITIONS OF CLEAVAGE

In the preceding section, the term "cleavage" was used freely
to denote a number of fracture processes, each of which results in
a macroscopically flat surface, usually corresponding to a close-
packed, or low-index, plane of the crystal's lattice. In detail,
this "plane" may exhibit steps (in the limit, of atomic dimensions,
if the crack front intersects a screw dislocation), slip-traces or
striations (in fatigue) or even fine-scale dimples (in LME). A more
precise definition of "cleavage" is difficult, even if we try to
relate this definition to the most common engineering experience of
cleavage, which is found in the fracture of iron and steel at low
temperatures. Two main features would appear to characterise such
cleavage failure:

a)   the *catastrophic* nature of crack propagation, under the
     influence of the applied tensile stress
b)   the *crystallographic* nature of the {100} fracture surfaces.

Contrast these two criteria now with observations on aluminium
alloys, broken in fatigue or LME. These exhibit macroscopic {100}
fracture planes, but covered with fine striations or dimples.
a)   Strictly, *catastrophy* is a matter of degree, because it is
possible, at least conceptually, to devise a testing system,
(perhaps of the wedge-loading type but with a triangular cross-
section, so that the stress-intensity decreases markedly with crack-
extension) in which virtually any "brittle" crack can be made to
extend in a stable manner. It becomes more difficult to control
elastic energy release as the crack-tip displacements associated with
crack advance decrease. What criterion a) is really saying is that,
in the particular system of uniform tension or simple bending, {100}
cleavage cracks in iron require so little crack-tip displacement that
they become unstable, whereas the {100} fatigue crack in an aluminium-
alloy is associated with so much crack-tip plasticity per cycle that
it does not become unstable in response to the applied K_{max} in the
cycle, even though the crack-tip opening may be of a similar nature
to that in iron. It has been suggested[14] that a macroscopic [001]
opening in zone-hardened Al-Cu in fatigue can be achieved by two
simultaneous, equal shears on appropriate {111} planes, i.e.

$$a_0[001] \ = \ a_0/2[011]_{(1\bar{1}1)} \ + \ a_0/2[0\bar{1}1]_{(\bar{1}11)} \qquad \ldots 36)$$

giving a crack front along [110] and a crack propagation direction
along [1$\bar{1}$0]. This is similar in form to the relaxation (or, in
reverse, the nucleation[15]) of a cleavage crack in iron, assuming slip
on {110} planes, and a crack front along [010]:

$$a_0[001] \ = \ a_0/2[111]_{(\bar{1}01)} \ + \ a_0/2[\bar{1}\bar{1}1]_{(101)} \qquad \ldots 37)$$

The crack propagation direction is now [100]. Such cleavage cracks
in iron are, however, extremely sensitive to the value of K_{max} in a
fatigue cycle[16], suggesting a much more inherently unstable mode of
propagation.

b)   It could also be argued that the *crystallography* is only a
matter of degree and that the fact that the Al-4Cu shows striations,
whereas iron does not, is simply due to the fact that the slip-steps
in iron are on such a fine scale that they cannot be resolved. In
plane strain tension, however, such a mode of crack opening by
dislocation emission would be *non-cumulative*[17] and it would be common
to see, in iron, arrested microcracks of less than a grain size in
length. The fact that microcracks always apread completely across
grains, once they have broken-out from carbide nuclei, is good
evidence for an inherently "brittle" mode of crack propagation
through the ferrite grain, i.e. without "significant" crack-tip
dislocation generation and propagation.

As a first approximation to what is meant by "significant"
dislocation movement, a possible limit might be the size of the "end-
region" associated with a crack[18]. This is equivalent to the core-
width of a dislocation and may be estimated by assuming a stress-
strain law, in the region of the crack-tip, equivalent to the
sinusoidal atomic force/displacement curve, rising to σ_{max} at $u = \lambda/4$
(equation 1)). The sinusoidal law could be used directly but it is
easier to replace it by a rectangular curve, of uniform height σ^*
and extent u^*, where u^* is a "cut-off displacement" at which total
fracture is deemed to have occurred ($u^* \sim 0.5-1.0a_0$). The model
then becomes that of Bilby, Cottrell and Swinden (BCS)[19] for a
yielded crack, and, at low stresses, fracture occurs at the critical
displacement u^*, when

$$\sigma_{app} = (E'\sigma^* u^*/\pi a)^{\frac{1}{2}} \qquad\qquad \ldots 38)$$

This is identical to the Griffith equation 21) if the product
$\sigma^* u^*$ (the area under the rectangular curve) is equated to 2γ (the
area under the sinusoidal curve, equation 2)). The length of the
"end-region", R, is then given by:

$$R = (\pi E'/8)(u^*/\sigma^*) \qquad\qquad \ldots 39)$$

For a relatively stiff force law, taking $\sigma^* = 0.2E'$, $u^* = 0.5a_0$
we find that $R \simeq a_0$; for a softer law, say, $\sigma^* = 0.1E'$, $u^* = a_0$,
$R = 5a_0$. Within such "end regions", it is possible that atom movements, equivalent to the generation of crack-tip dislocations, may
occur, but the crack may still propagate in a brittle manner: larger
dislocation movements may give rise to "cleavage-like" separation,
but are slip-separations, non-cumulative and essentially stable in
type. The calculation of R is approximate, but an "end region" of
magnitude one-to-five atomic spacings seems not unreasonable by
comparison with dislocation core widths.

A refinement of the criteria for judging "cleavage", as associated with low temperature fracture in iron or with fracture in covalent or ionic solids, therefore defines:

a)   *catastrophy* as being unstable, plane-strain propagation, controlled by tensile stress or K_{max} in mode I loading, for stress systems in which K increases with crack length.

b)   a *cleavage surface* as being *planar*, apart from intersections of the crack front with *pre-existing* imperfections in the lattice (network dislocations, forest walls, dislocated lath or packet boundaries). No slip traces associated with dislocation emission from the crack tip should be observed, although local movements over distances less than five atomic spacings may have to be conceded, even for brittle crack propagation. These two criteria hold for most "genuine" cleavage fractures, although, for zinc, a) is only partially true, because the fracture is controlled by nucleation[20] and this therefore dependent on the shear stress, not the tensile stress. The observations in aluminium alloys do not then satisfy either criterion and the fracture mode might be referred to as "cleavage-like"; "quasi-cleavage" being used already to describe genuine cleavage fractures in quenched-and-tempered steels, where there are so many interactions of the crack with features of the substructure that the fracture surface is relatively rough.

THE FRACTURE TOUGHNESS OF STEEL

Bounds to the fracture toughness of precracked testpieces of low strength steel, fractured in a cleavage mode, are of order $K_{IC} = 20\text{-}200\text{MPam}^{\frac{1}{2}}$, i.e. from 28). $G_{crit} = 2\text{-}200\text{kJm}^{-2}$, compared with a true surface energy of 2Jm^{-2}. These very large apparent works-of-fracture, which cause steel structures generally to be so tolerant to the presence of defects, are attributed[21] to the need for a large plastic zone to form ahead of the crack tip *before* it is possible to generate local tensile stresses, sufficient to propagate microcrack nuclei, at the positions where microcrack nuclei exist. The incorporation of stress *and* distance is essential, because stress intensity has dimensions of stress (distance)$^{\frac{1}{2}}$. From equation 17a) with $\theta = 0$, $\sigma_{11} = K(2\pi r)^{-\frac{1}{2}}$; the critical value of K is of order ...

$$K \sim \sigma_F \, (2\pi r_{crit})^{\frac{1}{2}} \qquad \qquad \ldots 40)$$

where σ_F is the (statistically averaged) stress required for microcrack propagation and r_{crit} is the (statistically averaged) distance to a microcrack nucleus of size corresponding to the propagation stress, σ_F . For quantitative predictions[22], a full elastic/plastic stress analysis is necessary, but this becomes asymptotic to the elastic solution at distances a few crack-openings away from the crack-tip. The elastic/plastic behaviour may also be treated analytically, assuming power-law hardening[23].

Values of σ_F may be measured most conveniently, using blunt-notched bend specimens[7], for which finite-element elastic/plastic stress analyses are available. It appears that, for many mild steels at moderately low temperatures, the critical stage in the fracture process is the propagation of a microcrack, which has formed in a brittle carbide particle, into the ferrite matrix. Nucleation occurs as the result of high local stresses exerted on the carbide, by dislocation pile-ups or by general fibre-loading of plate-like carbides from the yielded matrix. Nucleation in the ferrite matrix by a pile-up mechanism seems to be ruled out by the fact that fracture is propagation (tensile stress) controlled and dislocation interactions, of the form of equation 37) in reverse, have not been unambiguously observed, although, at very low temperatures, microcracks can be nucleated by twin-twin interactions. Note that equation 37) in reverse does, however, predict that the cleavage plane is (001). Propagation of a microcrack from a grain-boundary carbide of thickness C_O has been modelled[24] as for the mode I propagation of a Griffith crack of length $2a = C_O$, under the action of a simple tensile stress, σ_{app}, but a more detailed treatment[25] includes the further stress-field due to the dislocation pile-up which has nucleated the microcrack. The expression for unstable propagation then becomes:

$$(C_O/d)\sigma_F^2 + \tau_{eff}^2 \{1+(4/\pi)(C_O/d)^{\frac{1}{2}}\tau_i/\tau_{eff}\}^2 > 4E\gamma_p/\pi(1-\nu^2)d \qquad \ldots 41)$$

where d is the grain diameter, τ_i is the lattice "friction-stress", $\tau_{eff} = (\tau_{app} - \tau_i)$ is the "effective" shear stress acting on the pile-up, and γ_p is the work-of-fracture required to propagate the microcrack into the ferrite matrix. For most of the normalised or annealed mild steel microstructures for which σ_F values have been measured, it turns out that the dislocation pile-up term is of relatively little importance, contributing, perhaps, 10-20% to the total stress field available to cause the microcrack to propagate. In a similar fashion, values of σ_F have been obtained for spheroidal carbide microstructures[26]. Here, the system has been modelled as that of Griffith propagation of penny-shaped microcracks, of radius X_O.

Two extremely important findings follow from the experimental values of σ_F. On the one hand, it is possible, particularly for spheroidal particles, to provide a statistical basis[27] for the plane strain fracture toughness, K_{IC}, rather than assign unique values to σ_F and r_{crit} in an expression such as equation 40). For a given microstructure, it is possible, first to determine a frequency distribution of carbide sizes, so that the probability $P(X = X_O)$ is known. This distribution will then be present in any volume element, located at (r,θ) ahead of the precrack. Secondly, the fracture model enables one to determine the fracture stress $\sigma_{F(X_O)}$ required to propagate a microcrack of radius, X_O. At a given temperature (or

matrix yield stress, σ_Y) it is possible to compute, using the
elastic/plastic finite-element stress analysis or its power-law
hardening analytical equivalent, the probability, $P(\sigma_{11(r,\theta)} > \sigma_{F(X_O)})$
that the stress, σ_{11}, in the element (r,θ) is greater than $\sigma_{F(X_O)}$,
for each value of X_O. The probability of failure, $P(F)$, is then
given by the product $P(\sigma_{11(r\theta)} > \sigma_{F(X_O)}) \times P(X = X_O)$ summed for all
elements (r,θ) in the plastic zone. Simple calculations, based
on histograms, give good predictions of K_{IC} in steel and a similar
approach, but using a Weibull distribution of microcracks, seems to
work well in describing the toughness of two-phase ceramics[28].

The second important use of σ_F, and the one pertinent to the
question of whether the microcrack propagates as a sharp crack
through the ferrite matrix or whether it blunts, to form a void, is
that values of γ_p, the work of fracture in the matrix, may be
derived. The following main points can be made:
a)   The calculated *value of* γ_p (for grain-boundary carbides or
spheroidal carbides) centres around $14 Jm^{-2}$, with a possible spread
in the range $11-17 Jm^{-2}$. The *surface energy* of iron γ_{SE} is $2 Jm^{-2}$.
b)   If the critical stage is microcrack-propagation, *controlled by
tensile stress* (which explains the effects of triaxial stress fields
around notches), an examination[25] of the energetics of propagating
cracks initiated by dislocations shows that γ *must increase* as the
crack extends (if it did not, fracture would be nucleation-controlled,
i.e. as soon as a microcrack was nucleated, it would propagate
catastrophically to failure). This implies that γ_p for the *iron*
matrix must be *greater than* γ_c, the surface energy of the carbide.
Relative values of γ_c and γ_{SE} are not known, but it is arguable
that γ_{SE} is less than γ_c. The usual inference is that γ_p *includes*
a *plastic work* term, and this inference is reinforced by the
experimental observation that arrested microcracks in carbides have
clearly blunted by plastic flow in the ferrite.
c)   It is intriguing that σ_F, for slip-initiated cleavage fracture,
is *independent of temperature* and, for the simple Griffith model,
this implies that γ_p is also temperature-independent. From 41) if
τ_{eff} is written as $k_Y^S d^{-\frac{1}{2}}$, where k_Y^S is the slope of the Hall-Petch
plot in shear, (derived later in equations 43) and 44) we have

$$\sigma_F^2 C_O + (k_Y^S)^2 \{1 + (4/\pi) C_O^{\frac{1}{2}} (\tau_i/k_Y^S)^2\} > 4 E \gamma_p/\pi(1-\nu^2) \qquad \ldots 42)$$

For annealed or normalised steel, k_Y^S is independent of temperature
if yielding occurs by slip processes[29]. Although τ_i is quite
strongly dependent on temperature, its effect on the calculation of
γ_p from σ_F is minor, because the dislocation contribution as a whole
is rather small. Consider, for example, as an "average" material,
a steel of grain size $50\mu m$, and carbide size, $C_O = 2.5\mu m$. At
moderately low temperature, its tensile yield stress might be
$600 MPa$ and k_Y^T in tension is $0.7 MPam^{\frac{1}{2}}$. Subtraction of the $k_Y^T d^{-\frac{1}{2}}$
term gives a σ_i value of $\sim 500 MPa$. At this stage, conversion to

shear is unnecessary, because $(\sigma_i/k_y^T) = (\tau_i/k_y^S) = 715$ m$^{-\frac{1}{2}}$. The term $(4/\pi)C_o^{\frac{1}{2}}(\tau_i/k_y^S)$ becomes equal to 1.43, and the squared term in parentheses is then 5.9. The $(k_y^S)^2$ term must be converted to shear, so k_y^T is divided by a suitable Taylor factor, which allows for the variety of slip systems in the b.c.c. lattice. A value of 2.2 has been suggested. With this substitution, the whole dislocation term is equal to 0.6 MPa^{2}m. Corresponding to $\gamma_p = 14$Jm^{-2}, the simple Griffith expression gives the first term $\sigma_F^2 C_o$ equal to 3.92 MPa^{2}m. A decrease of σ_i to 300 MPa reduces the dislocation term to ~ 0.35 MPa^{2}m, so that the percentage change in γ_p for a 40% decrease in τ_i is only of order (0.25/4.32) x 100, i.e. approx. 6%.

The points made in b) and c) above are interesting because b) suggests that the value of γ_p of ~ 14Jm^{-2} is associated with plastic deformation, whilst c) suggests that γ_p is independent of temperature in the range over which slip-initiated cleavage is occurring, even though the yield stress is varying markedly with temperature. The final section treats and attempts to reconcile these apparently contradictory conclusions.

THE "WORK OF FRACTURE" IN IRON

Consider the modes by which plastic relaxation might occur ahead of a crack tip. One mode is the operation of sources, at relatively low stress, at distances remote ($\sim 1\mu$m) from the crack tip: the other is the generation of dislocations from the tip (at a stress of $\sim 0.1\mu$) and their propagation through the steep stress gradient. The former process may depend strongly on temperature and crack-tip velocity (strain-rate): the latter, in its early stages, may be virtually athermal, because thermal fluctuations at low temperatures are not sufficient significantly to affect the local stress required to generate dislocations from the tip. It should be noted that the two modes of relaxation are somewhat analogous to those treated by Cottrell[29] in his analysis of the interpretation of k_y during Lüders band propagation.

Here the slip-band in the yielded grain is modelled as a Mode II shear crack, of length $2a =$ grain diameter, d, but subjected to a shear stress, $\tau_{eff} = (\tau_{app} - \tau_i)$. Then, the (shear) stress at a distance r ahead of the slip-band (shear crack tip) is given by:

$$\tau = (\tau_{app} - \tau_i)(d/4r)^{\frac{1}{2}} \qquad \qquad ... 43)$$

This is equivalent to equations 17a)-d). To spread yield into an unyielded grain, it is assumed that dislocations, located at a (statistically averaged) distance r*, move when the shear stress reaches a critical, (statistically averaged) value $\tau*$. Rearrangement then gives a value for the applied shear stress, $\tau_{app} = \tau_y$, required to propagate the Lüders band from grain to grain:

$$\tau_Y = \tau_i + \tau^* (4r^*)^{\frac{1}{2}} d^{-\frac{1}{2}} \qquad \ldots 44a)$$

$$= \tau_i + k_Y^S d^{-\frac{1}{2}} \qquad \ldots 44b)$$

Multiplication by the Taylor factor of 2.2 then produces the Hall-Petch equation, with $k_Y^T = 2.2k_Y^S$ as used in the previous section. The value of k_Y^S clearly depends on the values assumed by τ^* and r^*. For lightly-pinned dislocations in very low-carbon ferrite quenched from annealing temperatures of order $720^\circ C$ (to retain all carbon in interstitial solid solution), a dislocation density of $10^{12}m^{-2}$ would imply $r^* \sim 1\mu m$ and, for simple bowing of network dislocations, $\tau^* \sim \mu b/\ell$, where $b = 0.25nm$ and $\ell = 1\mu m$, i.e. $\tau^* \sim 0.25 \times 10^{-3}\mu = 20MPa$; $k_Y^S = 20 (4 \times 10^{-6})^{\frac{1}{2}} = 0.04$ MPam$^{\frac{1}{2}}$; $k_Y^T = 0.09$ MPam$^{\frac{1}{2}}$. As the amount of pinning is increased (e.g. by ageing, to allow carbon atoms to diffuse to the dislocations) so k_Y increases, but a limit is reached, at which it is easier to *create dislocations* from the *slip-band (shear crack) tip,* i.e. at the common grain-boundary, than it is to operate heavily pinned, remote sources. Now, we may take $\tau^* \sim 0.1\mu = 8GPa$ and $r^* \sim b = 0.25nm$. Hence $k_Y^S = 8 (10^{-9})^{\frac{1}{2}} = 0.25$ MPam$^{\frac{1}{2}}$, $k_Y^T = 0.55$ MPam$^{\frac{1}{2}}$, in close agreement with the experimental value of 0.7 MPam$^{\frac{1}{2}}$, which, it should be noted, is *independent of temperature* for heavily aged material, provided that yield proceeds by slip, not twinning.

Analogies between this model and the tensile crack are close, particularly if the lack of temperature dependence of k_Y^T is compared with the lack of temperature dependence of γ_p. An extremely useful conclusion is that, taking k_Y^T as its experimental value, 0.7 MPam$^{\frac{1}{2}}$, a *minimum* value can be set on τ^* for the operation of remote sources, since $\tau^* (4r^*)^{\frac{1}{2}}$ for remote sources must be greater than $\tau^* (4r^*)^{\frac{1}{2}}$ for grain-boundary creation, if dislocation creation is the mode of yielding. If $\tau^*(4 \times 10^{-6})^{\frac{1}{2}} > 0.7$, $\tau^* > 350$ MPa. This would apply throughout the whole low-temperature range until twinning became the operative mode of deformation.

If we now consider a microcrack under tension, the shear stress distribution is given by (17d))

$$\sigma_{r\theta} = \sigma_{\theta r} = K(8\pi r)^{-\frac{1}{2}} \sin\theta \cos(\theta/2) \qquad \ldots 17d)$$

The *maximum* shear stress in a continuum is at $70^\circ 32'$ to the line of crack extension, whence, setting $r = 1\mu m$, we derive $K = 6.5 \times 10^{-35} \sigma_{r\theta}$, which, with $\sigma_{r\theta} = \tau^* = 350$ MPa gives $K > 2.3$ MPam$^{\frac{1}{2}}$ to produce any dislocation movement from sources located at $1\mu m$ from the crack tip. If the dislocation has to move on a plane of the $(\bar{1}01)$ type (e.g. equation 37)) to relax a crack on the (001) plane, θ is 45° and $K = 7.7 \times 10^{-3}\sigma_{r\theta}$; i.e. with $\sigma_{r\theta} = \tau^*$, $K > 2.7$ MPam$^{\frac{1}{2}}$.

The actual values of K attained during the microcrack propagation stage may be derived from equation 42), since $K = \sigma_F(C_0/2)^{\frac{1}{2}}$. Ignoring the dislocation term, $K^2 = 2E\gamma_p(1-\nu^2)$ which, with E = 200 GPa and $\gamma_p = 14Jm^{-2}$ gives $K = 2.5$ MPam$^{\frac{1}{2}}$. This value may be reduced slightly, by the order of $(15\%)^{\frac{1}{2}} = 7.5\%$, if the dislocation contribution is taken into account.

The calculations are approximate, but indicate that the balance between operation or non-operation of remote dislocation sources is fine. Mathematically, the calculation may be refined by examining all components of the stress tensor around the microcrack tip (equations 17)) and invoking the Peach-Köhler relationship (equation 9)) to calculate the forces on dislocations of Burgers vector $(a_0/2)<111>$ resolved onto slip-planes of forms $\{1\bar{1}0\}$ or $\{11\bar{2}\}$. For the crack in equation 37) the references axes are simply $X_1 = [100]$, $X_2 = [010]$ and $X_3 = [001]$, which form a right-handed set. Detailed studies[30] of plastic relaxation around cracks produced by hydrogen charging in Fe-3%Si suggest however that the crack is on (001) plane in the $[1\bar{1}0]$ direction, with a crack front along [110], so that $X_1 = [1\bar{1}0]$, $X_2 = [110]$ and $X_3 = [001]$. Different models for the stress distributions predict the movement of $(a_0/2)[\bar{1}1\bar{1}]$ dislocations on $(\bar{1}12)$ slip-planes, $(a_0/2)[\bar{1}11]$ dislocations on $(11\bar{2})$ slip-planes or $(a_0/2)[111]$ dislocations on $(\bar{1}10)$ slip-planes, and these modes of relaxation were observed experimentally, using etch-pits. The general analysis could be re-worked to ensure fully self-consistent solutions, but the values to be substituted for τ^* or r^* are subject to so much error that refinements of this nature are of second-order importance. The conclusion reached in the hydrogen studies, that high resistance to crack propagation is due to easy operation of remote sources is not in dispute. The velocity dependence of dislocations on applied stress is also clearly important[31]. The present discussion is aimed at the possibility that, *when microcracks propagate by catastrophic cleavage,* a small packet of dislocations is continuously generated from the crack tip and travels *en suite* with the crack.

Reverting to the earlier discussion of brittle and ductile behaviour, we recall that the definitions refer to the relative ease of breaking the crack-tip bond in tension or blunting the crack by the generation of dislocations. Calculations, made either in terms of the ratios of tensile strengths and shear strengths[11] or in terms of the generation of dislocations at the Griffith stress[12], indicate that iron is very much a "borderline" material, for which the calculations are not sufficiently precise. Consider now the basic assumption, i.e. that the crack-tip bond "fractures" at the stress, $\sigma_{max} = (E\gamma/a_0)^{\frac{1}{2}}$ equation 4) or, indeed, the postulate that the work-of-fracture is 2γ, which is assumed in integrating the atomic force/ displacement curve and in Griffith's energy balance[5], equation 16). The former assumption is clearly true if the model is simply that of

two atoms, pulled under force-control, as a simple chain. Once the
force-maximum has been exceeded, the system has negative stiffness
and catastrophic separation occurs. The situation is analogous to
catastrophic failure of a tensile specimen at its "U.T.S." in a
force-controlled testing machine. If these same two atoms are now
embedded at a crack tip, however, it is not clear that they will
fly apart when σ_{max} is reached, because they are surrounded by a
"cage" of other atoms, all of positive stiffness, and these atoms
must continue to be displaced, to allow the crack-tip bond to
"separate" i.e. to reach a displacement, perhaps of a_0, rather than
$(a_0/2)$ at σ_{max}. The analogy is now with a tensile specimen deforming
in a stable manner under decreasing load in a displacement-controlled
machine. In terms of energy, the final state of 2γ is achieved, but
through a *mechanism* , of cooperative crack-tip atomic movements,
involving the expenditure of extra work, which provides an *activation
energy barrier* between initial and final states. Movements of this
sort, but accomplished by visco-elastic flow, may be implicit in the
reasoning applied by Marsh[10] to explain values of works of fracture
greater than $2\gamma_{SE}$ for glass.

     In iron, it is tempting[21,32] to try to describe the atom move-
ments as if crack-tip dislocations had been generated and moved
through distances of a few atom spacings, of the order of the size
of the "end-region" around a crack-tip (cf equation 39) et seq.).
If *each* successive severance of the crack-tip bond required such
dislocations to move three spacings on each side of the crack, the
work done would be of order $6\underline{F}.\underline{b}$ where $\underline{F} \sim 0.1\mu b$. This would have to
be expended for each crack advance of magnitude $\underline{b}$, so that the work
per unit area would be of order $0.6\mu b$, i.e. with $\mu = 80$GPa, b =
0.25nm, some $12 Jm^{-2}$. This is quite sufficient to enhance the surface
energy term, 2γ, to the value of $14 Jm^{-2}$ observed in the cleavage
experiments. Again, refinement of the calculation to include effects
of crystallography, such as differences in spacing for crack advance
and dislocation movement, are probably not warranted, although it is
worth noting that if $C_{44}(= 116$ GPa, following equation 35)) is used
instead of μ, the dislocations only have to move 2b rather than 3b.
Such movements could be easily contemplated within the "end-region"
(equation 39) et seq.). The dislocations might be regarded as
"lattice-trapped" during the operation of the mechanism of crack-tip
bond separation, so that they can run back to the surface under the
action of the image-force component[32] (cf equation 12)) when the
applied stresses decrease as the crack-tip moves onwards. This
situation has been recently analysed[32] for a mode-II crack, in
general agreement with the reasoning followed above.

     The interaction of foreign species with this crack-tip separa-
tion process may be mentioned briefly. Two possible limiting effects
were mentioned in the introduction: either a reduction in modulus,
with constant strain at σ_{max}, or an "embrittlement" of the bond, in
which the modulus remains constant, but the strain to σ_{max} decreases.

The second possibility does not seem to be widely favoured in descriptions of embrittlement due to hydrogen or impurity elements such as P, Sn or Sb, although a reduction in the *total* displacement required to sever a crack-tip bond would lead to a marked reduction in γ_p, because much less atomic rearrangement at the crack-tip would be required. On the other hand, if both E and μ were reduced proportionately by the foreign species, it would appear that γ_p would not be reduced so greatly, because the balance between bond-separation and dislocation movement would be the same. An important new approach, however, suggests that tensile modulus and shear modulus may be affected in fundamentally different ways by "embrittling" species. Studies[34] of sulphur segregation to "delta-hedron" interstices in nickel grain-boundaries show that the *bonding is strengthened in the grain-boundary plane,* by the formation of planar hybridised complexes of sulphur 3p orbitals with nickel 3d/4s electrons, whilst *bonding normal to the plane is weakened* by the loss of these 3d/4s electrons from bonding orbitals. In terms of crude analogy, a 3-D silica network has become more "mica-like", and it is then plausible that the balance between tensile strength and shear strength has been so altered that a much more inherently "brittle" material has been produced.

The main attention of this paper has been paid to brittle fracture in iron and steel, as the world's most common and widely-used structural metals and ones in which brittle fracture is a cause of major engineering concern. General techniques of macroscopic fracture mechanics exist which enable fracture control to be exerted in engineering design and these are increasingly being linked with similar techniques which treat steel microstructures as miniature engineering configurations, mainly using continuum mechanics, with some crystallography, applied to defects such as dislocations or microcracks. It is intriguing - literally, "ironic", - that of all materials available for structural use, iron is the one material for which broad generalisations on brittle and ductile behaviour are most inadequate and it is clear that for this material, if for no other, a proper understanding of the way in which it breaks must be treated, not just by the "Mechanics of Fracture" but, in detail, by the "Atomistics of Fracture".

ACKNOWLEDGEMENTS

The author wishes to thank Professor R.W.K. Honeycombe for the provision of research facilities and for his continued and unfailing support for fracture research at Cambridge. Whilst the opinions expressed in this paper are idiosyncratic, they have been strongly influenced by past and present members of my research group, of whom Dr. D.A. Curry, Prof. G.G. Garrett and Prof. R.O. Ritchie warrant specific mention. Discussions on the general topic with Dr. C.Briant, Prof. J.R. Rice and Prof. E. Smith are also gratefully acknowledged.

REFERENCES

1. Hirth J. and Lothe J. "Theory of Dislocations" McGraw-Hill
      (New York) 1968.
2. Nabarro F.R.N. "Theory of Crystal Dislocations" Oxford
      (Clarendon Press) 1967.
3. Cottrell A.H. "Theory of Crystal Dislocations" Blackie and Son
      Ltd. 1964.
4. Inglis N. Trans. Inst. Naval Arch. London 1913, 55, p.219.
5. Griffith A.A. Phil Trans. Roy. Soc. 1920, A221, p.163.
6. Irwin G. Trans. ASME Jnl. Appl. Mech. 1957, 24, p.361.
7. Knott J.F. "Fundamentals of Fracture Mechanics" Butterworths
      (London) revised impression 1979.
8. Rooke D.P. and Cartwright D.J. "Compendium of Stress Intensity
      Factors" HMSO (London) 1976.
9. Nye J.F. "Physical Properties of Crystals" Oxford University
      Press 1972.
10. Marsh D. Proc. Roy. Soc. 1964, A282, p.33 and Tube Investments
      Internal Report 1964 no.161.
11. Kelly A., Tyson W. and Cottrell A.H. Phil. Mag. 1967, 15, p.567.
12. Rice J.R. and Thomson R. Phil. Mag. 1974, 29, p.73.
13. Encyclopaedia Brittanica 1950, 10,   p.480.
14. Garrett G.G. and Knott J.F. Acta Met. 1975 23 p.841.
15. Cottrell A.H. Trans. Amer. Inst. Min. Metall. Petrol. Engrs.
      1958 212 p.192.
16. Ritchie R.O. and Knott J.F. Mat. Sci. and Eng. 1974, 14, p.7.
17. Cottrell A.H. Proc. Roy. Soc. 1963 A276 p.1.
18. Cottrell A.H. Tewksbury Symposium on Fracture, University of
      Melbourne 1963, p.1.
19. Bilby B.A., Cottrell A.H. and Swinden K.H. Proc. Roy. Soc. 1963,
      A272, p.304.
20. Curry D.A., King J.E. and Knott J.F. Met. Sci. 1978, 12, p.247.
21. Knott J.F. Phil. Trans. Roy. Soc. 1981, A299, p.45.
22. Ritchie R.O., Knott J.F. and Rice J.R. Jnl. Mech. Phys. Solids
      1973, 21, p.395.
23. Curry D.A. Nature (London) 1978, 276, p.50.
24. McMahon C.J. Jnr. and Cohen M. Acta Met. 1965, 13, p.591.
25. Smith E. Proc. Conf. on Physical Basis of Yield and Fracture,
      Inst. Phys. and Phys. Soc. Oxford 1966, p.36.
26. Curry D.A. and Knott J.F. Met. Sci. 1978, 12, p.511.
27. Curry D.A. and Knott J.F. Met. Sci. 1979, 13, p.341.
28. Evans A.G., Heuer A.H. and Porter D.L. Proc. 4th Intl. Cong.
      on Fracture, Waterloo, Pergamon 1978, 1, p.529.
29. Cottrell A.H. Symposium on the Relation Between the Structural
      and Mechanical Properties of Metals, Nat. Phys. Lab. HMSO
      (London) 1963, p.456.
30. Tetelman A.S. and Robertson W.D. Acta Met. 1963, 11, p.415.
31. Jokl M.L., Kameda Jun, McMahon C.J. Jnr. and Vitek V. Met. Sci.
      1980 14, p.375.

32.   Knott J.F.   Proc. 4th Intl. Cong. on Fracture, Waterloo,
          Pergamon 1978, 1, p.61.
33.   Smith E.   Matl. Sci. Eng. 1981, 47, 133.
34.   Briant C.L. and Messmer R., Phil Mag B 42 p.569. (1980).

DISCUSSION

Comment by R.H. Jones:

Does the size of the plastic zone at the tip of a crack change
with crack velocity, and does the plastic zone of stable and unstable
cracks differ in size or character?

Reply:

The size of the plastic zone varies with crack velocity, as a
result of the strain-rate dependence of the yield stress. Slip is a
thermally-activated process, and, if more time is available (at low
strain rate), there is greater probability of a dislocation
surmounting an obstacle of given size at a fixed stress level. In
iron, at high strain rates, there is a change in the mechanism of
plastic flow from slip to twinning and it appears that the strain
rate dependence is then lost. Again, in iron, stable crack growth
is associated with large plastic zones (with dislocations generated
from sources relatively distant from the crack tip). Unstable
cleavage nucleated in carbides is, I believe, associated with a few
dislocations emitted from the crack tip. Unstable, fast running
cleavage (moving at say two-thirds the speed of sound) is associated
with a plastic zone produced by mechanical twinning, extending over
a region of the order of one grain diameter.

Comment by R. Bullough:

The question was raised whether twinning associated with dynamic
crack propagation can be observed in single crystals. Such a
correlation between fracture and twinning was observed by Demuyterre
and Greenough by studying the cleavage faces of zinc single crystals
fractured in liquid N_2. The results were analyzed by Bilby and me
and both are published in Phil. Mag. (1954). Mechanical twins were
observed on one cleavage face and kink bands on the other in complete
agreement with the respective initially resolved shear stresses, from
the moving crack tip, for twinning and kinking.

Comment by J.P. Hirth:

With regard to the interesting discussion of γ_p as deduced from
the model of crack initiation from carbides, there is a factor which
should favor flow initiation from the grain or interface boundary
rather than from a source within the grain ahead of the crack. This
is the presence of elastic incompatibility stresses arising from

elastic anisotropy. These stresses decay exponentially with distance
from the boundary and have maximum magnitudes as large as one-half
the applied stress in tension. Thermal activation could still be
accounted for in such a case by adsorption of impurities and pinning
of grain boundary dislocations which could act as the grain boundary
sources.

Reply:

    This is an interesting point which, I hope, can be developed.
It clearly would affect Cottrell's calculation of k_y. In my own
discussion, I have merely used experimental values of k_y to set
an upper limit to the stress required to operate relatively distant
sources, but stresses of the sort described would also alter the
basic equation used to derive the energy balance for a microcrack
emerging from a carbide.

Comment by J.W. Gadzuk:

    Rather surprisingly, in the surface science world from which I
come, the property "surface energy" has played a relatively minor
role. Nonetheless, quite a bit of theoretical activity was generated
in the mid 70's as a result of a controversial theory due to Schmidt
and Lucas for metal surfaces energies which was based upon surface
plasmon effects. A problem arose in resolving this conflict due to
an apparent lack of experimentally determined surface energies for
well-defined solid surfaces. Is this an accurate reflection of the
current state of affairs on data, and if so, how do those people
concerned with crack theories in which surface energies are an
integral input parameter deal with this lack of key input data?

Reply:

    I would be the first to agree that there is a dearth of experi-
mental surface energy values for structural materials of interest.
For iron, the values are essentially obtained using "zero-creep"
experiments on δ-iron (b.c.c.) at high temperatures, with corrections
made for the effect of temperature. For example, Kelly, Tyson and
Cottrell (see reference in my paper) extrapolated to a surface energy
value of $4 Jm^{-2}$ at °K. I believe that some values have been obtained,
using cleavage in the Gilman test, but this begs the question, and it
should be noted that the original analysis of this test was refuted
by Maitland and Chadwick (Phil. Mag., approx. 1969) in their work on
cleavage fracture energies in zinc. My main point, however, was that
one should not necessarily expect a work-of-fracture to be equal to
2γ, except for a fully reversible thermodynamic process with no
activation-energy barrier. I am suggesting that it may be necessary,
in iron, to "shuffle" atoms in the region of the crack tip to provide
a mechanism which allows the two atoms comprising the crack tip bond

to come fully apart. This "shuffle" would entail extra work, which
I have modelled tentatively as the localized movement of crack tip
dislocations over distances of a few atomic spacings.

Comment by J.O'M. Bockris:

Your result (γ_P) is one-half to one order of magnitude away from
the (admittedly doubtful) experimental result. In calculation of a
rate, such a discrepancy might be tolerable. In calculation of an
energy quantity, it suggests that some important feature has been
neglected in the analysis.

Reply:

There is, indeed, a discrepancy of some one-half to one order of
magnitude difference between γ_P (say, 11-17 Jm^{-2}) and γ_S (2 Jm^{-2}).
As I explained, γ_P, must be greater than γ_S for fracture to be
growth-controlled, rather than nucleation-controlled. I have been
suggesting that, for iron, the simple Griffith energy balance may
not be sufficient, because it is necessary to devise a mechanism for
separation of the crack tip bond by a "shuffling" of atoms in the
crack tip region, modeled in my paper by localized crack tip dislocation movement. This provides an "activation energy" barrier to
the fracture process so that the work-of-fracture (γ_P) is greater
than twice the surface energy.

Comment by A.W. Thompson:

In your talk you pointed out that the effective energy term in
fracture, γ_P, can exceed the surface energy, γ_S by factors of perhaps
5 to 1000. This depends, I suppose, on temperature, crack size, etc..
Since one presumes that γ_S is fixed, the deformation contribution
varies from moderate to very large. Could this be connected to
Thomson's and Macmillan's talks (especially the former) by considering
whether they refer to "O.K." circumstances; or for example, Thomson's
categorization of "ductile" and "brittle" materials?

Reply:

The categorization of "ductile" and "brittle" materials is, of
course, a fairly broad-brush treatment, based mainly on behaviour at
room temperature, under static loading (calculations being made for
the "quiescent state", i.e., the precise Griffith energy balance).
All deformation and fracture processes are thermally activated and
behaviour at high temperatures may be drastically different from that
at low temperatures: glass, for example, may be drawn into fine
filaments when heated above its softening temperature. For sub-zero
temperature, however, kT is so small that it is difficult to see how
thermal activation could significantly affect the stress (~0.1μm)

required to emit dislocations from the crack tip. It is on these grounds that I tend to support the crack tip dislocation model in iron, at low temperatures, to explain the constancy of γ_p between say -150°C and -50°C. As the temperature is increased, more dislocations may be emitted or, in my own view, relatively distant sources (at 1μm from the tip) are activated, and the toughness increases dramatically. The behaviour is then, of course, highly rate sensitive. In brief, Thomson's categorization should hold at sub-zero temperatures, but is unlikely to hold at high temperatures.

Comment by D.J. Duquette:

While it is useful to consider atomically sharp carcks and single dislocation sources and/or sinks near cracks, our non-metallurgical colleagues should understand that the true state of stress and/or strain in the vicinity of a crack in a metal is in fact quite complex. For example, direct observations of crack tips in iron by HVEM show dislocation cells, tangles and other complex configurations. From this point of view, our current approaches of either continuum considerations or di-atomic considerations are, at best, attempts at modelling otherwise currently intractable processes.

Reply:

I agree entirely, but would point out that the observations referred to above pertain to stopped cracks. Fast-moving cleavage cracks in iron may have sharp tips, at the most a few atom-spacings wide, with only two dislocations at the tip to provide the mechanism for separation.

Comment by S.P. Lynch:

It should be noted that there are ASTM standards regarding the definition of fracture terms (although they do not seem to be generally adhered to!). ASTM Standards, Part 10, 1977, E616-77 defines crystallographic cleavage as follows:

> crystallographic cleavage - the separation of a crystal along a plane of fixed orientation relative to the three-dimensional crystal structure within which the separation process occurs, with the separation process causing the newly formed surfaces to move away from one another in directions containing major components of motion perpendicular to the fixed plane.

There are a number of notes following this definition including one which essentially states that an atomic mechanism of fracture is not implied. This seems sensible since detailed atomic movements during fracture are not well-established at the present time.

Reply:

I suppose that the first point to make is the rather facile
comment that the presence of definitions in ASTM Standards does not
automatically lead to universal acceptance of these definitions, or,
indeed, imply that these are the best definitions that could be used!
More seriously, in the present case, I have no grumbles at all with
the definition as a definition of <u>crystallographic cleavage</u>, which
is what it defines, although I would point out that it specifies a
plane of separation whilst simultaneously noting (note 4) that "this
form of cleavage is usually...recognizable by a number of fracture
surface features, such as flat facets, steps between parallel facets,
river patterns formed by the joining of steps and tongues and herring-
bone patterns cause by crystallographic twins at the crack". It
should be obvious that such features are clearly incompatible with
the definition of a (i.e., one) plane of separation, but it is
equally clear that those in the field "know what they mean" and no
practical man (in Britain, "the man in the Clapman omnibus") would
disagree with the definition. There is, however, another subset of
fracture workers, concerned more with the testing of materials, who
recognize cleavage (not "crystallographic cleavage") in terms of
mechanical instabilities rather than fractography and may even use
such descriptions as "intergranular cleavage" for the case of smooth
separation of grain boundaries. In such a community, insistence on
the use of the ASTM definition of "cleavage" would be, frankly, silly.
In my paper, I have tried to reconcile the "catastrophic" and
"crystallographic" aspects and I wish that Dr. Lynch had tried to
make constructive criticism of the points that I raised there, because
this is clearly a problem that concerns him greatly. I would have to
return my use of the term "cleavage-like", however, for the fatigue
fracture of Al-Cu alloys, because then the macroscopic [001] opening
was achieved by <u>sliding</u> movements on <u>two</u> (111) and (11$\bar{1}$) planes.

Comment by I.M. Bernstein:

In an apparently illustrative way a number of the speakers,
including yourself, have suggested that hydrogen can affect the
fracture stress of bcc metals by directly modifying the force distance
characteristics, across the cleavage plane, of iron-iron bonds at a
crack tip. It appears to me that such an approach implies that
hydrogen only affects the magnitude of the fracture stress (or perhaps,
cohesive energy) but not the crystallography of the cleavage process.
However, there is now considerable evidence, in both purified iron and
high strength steels, that both the microcracking and the macroscopic
fracture surfaces are the {110} (or possibly the {112} slip planes)
rather than the "expected" {100} cleavage plane. Can you comment on
this apparent anomaly?

Reply:

These comments are of great interest, and contrast strongly with those made by Tetelmann and Robertson (as referenced in my paper), where hydrogen cracking in iron - 3% silicon produced cleavage on the (001) plane, with the crack length along [110] and the crack front along [1$\bar{1}$0]. The "expected" (001) plane is only expected, because it is what is observed for cleavage fracture in the absence of hydrogen. In my introduction, I put forward the very tentative point that Young's modulus is a minimum in the <001> directions (which are normal to {001} planes) and that this might be connected with the choice of {001} cleavage planes. Perhaps the presence of hydrogen in the bcc unit cell strengthens bonding in the <001> direction whilst weakening bonding at right angles (along [110] or [1$\bar{1}$0]in a manner similar to that achieved by sulphur in the grain-boundary deltahedron coordination polyhedra in nickel. On a related point, I have referred to "cleavage-like" facets in some fcc metals in fatigue. Here macroscopic {001} fracture-planes were associated with slip on {111} slip-planes and, in some cases, the facets themselves were completely of the {111} form. This fracture process is, however, strain-controlled, dependent on ΔK and insensitive to K_{max}. From the question, I am not sure whether the {110} hydrogen fractures are strain-controlled, in which case, slip dislocations might transport and localize the hydrogen concentration, or a tensile-stress controlled phenomenon, when a more complicated atomistic mechanism may have to be involved. The length of hydrogen crack "jump" under constant plastic strain but different applied tensile stress (e.g., in the region of triaxial tension below notches of different angles) might be a useful diagnostic technique.

Comment by R.H. Jones:

George Whitesides in the session on surface reactivity and bonding suggested that chemical rate theory be applied to fracture and crack growth rates. For a subcritical crack growing in some corrosive medium can you suggest what material parameters and features might be identified with the activated state and activation energy. Are there similar arguments for the transition from subcritical to critical crack growth?

Reply:

If I may tackle the question in reverse order and restrict my comments to iron and steel, the first point to take is whether the material is above or below its cleavage/fibrous transition temperature. For cleavage fractures, the fracture toughness is a function of yield stress, which is itself a thermally-activated process (see my answer to your other question), but almost all the work input is necessary to set up conditions for the propagation of microcracks. When the microcracks go unstable, I believe that the process is associated

with local, crack tip dislocations, which require such high stresses
for activation (0.1μ) that kT at sub-zero temperatures cannot produce
any significant thermal activation.   The initiation of fibrous
fracture is associated directly with void growth and crack tip
geometry changes induced by slip, generated from sources relatively
distant from the crack tip, so that the rate of coalescence is a
function of the rate dependence of the flow stress.   This is
significant, at room temperature, particularly for lightly-pinned
dislocations.   There is, of course, no guarantee that initiation of
fibrous fracture leads to unstable crack propagation.

In any stress corrosion process, there is a plethora of possible
rate controlling steps, and it is very important to realise that the
rate controlling step in a process may not be directly associated
with the mechanism of crack advance.   It is, for example, quite
possible to contemplate a system which is under cathodic control
(e.g., the rate controlling step is evolution of hydrogen), but in
which the mechanism of advance is anodic dissolution.   For anodic
dissolution, electrons are liberated at the anode, through the
reaction $M \rightarrow M^{R+} + Re$, and if these electrons cannot be consumed at
the cathode the reaction must cease.   Similarly, the reaction will
be stopped if the M^{R+} cations cannot be removed from the environment
of the anode (concentration polarization).   In the model proposed
by Scully, the rate of crack advance is attributed to the balance
between passivation rate and rate of film rupture by thermally
activated flow (somewhat misleadingly termed "creep").   Here is one
example of a relationship between rates-of-reaction in the presence
and absence of an environment, but it relates primarily to near
ambient temperatures, where the mode of fracture in the absence of
environment is fibrous and does not necessarily become unstable unless
the test specimen undergoes plastic collapse.

FRACTOGRAPHY

Regis M. N. Pelloux

Department of Materials Science and Engineering
Massachusetts Institute of Technology
Cambridge, Massachusetts 02139, U.S.A.

INTRODUCTION

The word "fractography" describes the science of studying the characteristic features of fracture surfaces.[1] Visual fractography is an old science which has expanded rapidly in the last fifty years with the use of the optical microscope (1930), the transmission electron microscope (TEM)(1950) and the scanning electron microscope (SEM) (1965). Thus fractography is often referred to as microfractography or electron fractography.

The basic principles to be used in the study and identification of fracture surface features depend upon the user's goals. For instance, the scientist wants to observe the influence of simple, well-defined test conditions and microstructural features on the mechanisms of crack initiation and propagation. The metallurgist would like to be able to predict the fracture toughness benefits to be gained from new heat treatments or new alloys. The consulting engineer wants to determine the causes and conditions of failure from the interpretation of the sequence of the fracture surface markings created by a growing crack during the service life of a component.

Typical fractographic features will range in size from the atomic scale, ($\sim 3\text{Å}$) for instance for an atomic step, to the cm scale for a shear lip or a chevron marking. The challenging problem is to record and analyze all this topographic information and to relate the fracture features to the fracture mechanisms, to the microstructure of the material and to the local stresses and displacements.

The depth of field and the high magnification of the TEM (with replicas) and the SEM make the electron microscope ideally suitable

for fractography. The SEM provides the best quantitative information
by using the techniques of stereography and of diffraction orientation
analysis. Quantitative fractography is fast replacing qualitative
fractography[2]. Point counting and line intercepts can be used to
determine projected and true area fractions of different fracture
features.

Fracture mechanisms

     Figure 1 gives a schematic representation of six (6) basic
fracture mechanisms which are commonly observed in polycrystalline
materials under monotonic tensile stresses.

     Cleavage fracture is defined as a separation that occurs along
well-defined crystallographic planes. It is limited to BCC and HCP
metals. Cleavage cracks originate at locations where crystallographic
slip is inhibited such as at grain or twin boundaries, inclusions,
or carbide films at grain boundaries. Figure 2 gives a good example
of cleavage fracture in low carbon steel. A cleavage facet is never
a single, perfect plane across a crystal or a grain. River markings
represent steps between separate cleavage planes. River steps are
either secondary cleavage planes or tearing (shear) lines. The river
patterns indicate the local directions of crack propagation. The
smallest river step height corresponds theoretically to the Burgers
vector of a screw dislocation intersecting the cleavage plane. Thus
a cleavage plane must contain a large number of river steps due to
its interaction with dislocation networks. However, river step
heights of less than 50Å are not observed because of the lack of
resolution of the SEM. Consequently the local microplasticity associated with the initiation and propagation of a cleavage crack is not
usually resolved as a fractographic feature. It can be demonstrated
that a statistically random distribution of screw dislocations of
density Λ cutting through the cleavage plane will lead to a river

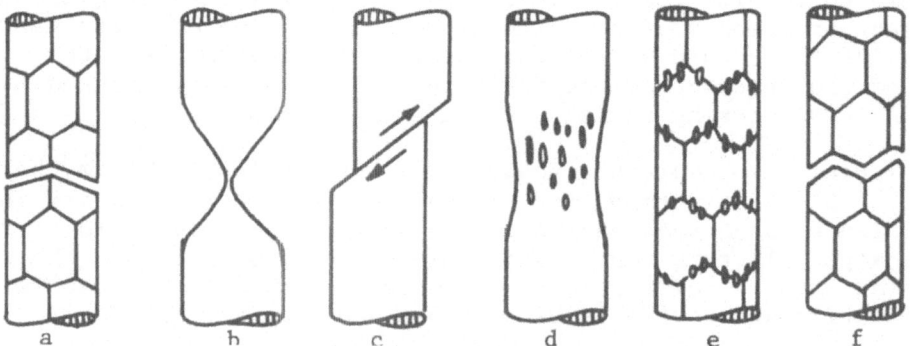

Fig. 1.   Fracture mechanisms. (a) cleavage; (b) rupture by necking;
          (c) rupture by shear; (d) microvoids; (e) intergranular
          microvoids; (f) intergranular cleavage.[7]

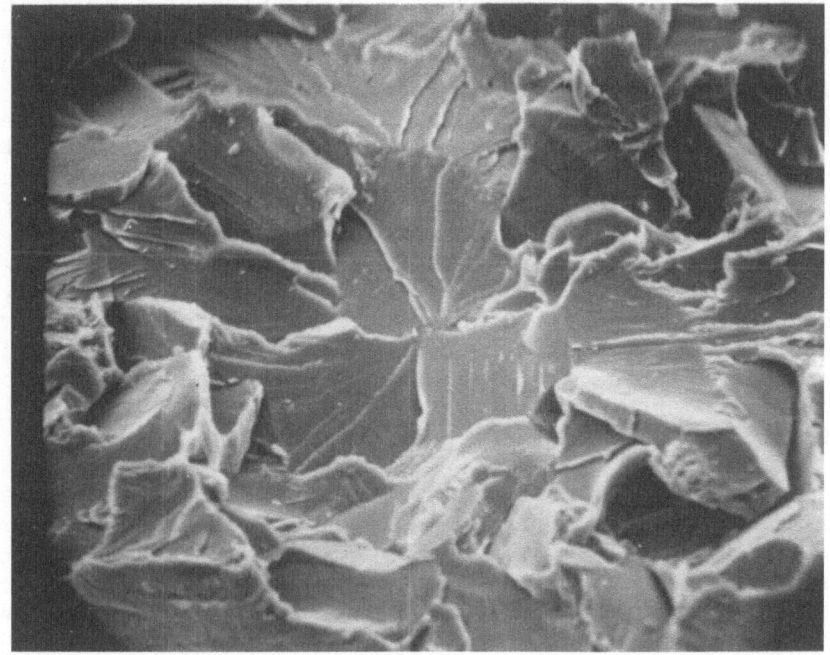

Fig. 2.   Cleavage fracture in low carbon steel (1300 X).

marking step height proportional to $\sqrt{\Lambda}$ and increasing proportionally with the square root of distance[3].

Another feature associated with cleavage is the formation of deformation twins in the small plastic zone ahead of the cleavage crack tip.   Secondary cleavage fracture of the deformation twins leads to cleavage tongues; the cleavage tongues may also be associated with twin-matrix interface brittle decohesion[4].

In the case of complex microstructures such as quenched and tempered high strength steels which contain a high density of small carbides, it is difficult for a crack to follow a well-defined cleavage plane.   The fracture surfaces are made up of small facets and tear ridges associated with the complex crystallographic orientations of the martensite plates and the dense network of carbides.   These complex cleavage fracture features, which may include some microvoid decohesion, are referred to as a quasi-cleavage fracture.

In HCP alloys (Zi and Zr for instance) cleavage planes with complex crystallographic indices (i.e., planes different from the 0002 plane) are observed in the case of slow crack growth by stress corrosion cracking.   This mode of cleavage of brittle decohesion associated with an environmental embrittlement is not understood.

Ductile rupture is defined as the separation of a solid in a
continuous manner by a process of shear or sliding off. In Figure 1,
sketches 1b and 1c illustrate two different mechanisms of rupture by
shear. In Figure 1c shear along a single shear plane leads to complete separation of the tensile bar. Note that the shear plane can
be either a crystallographic plane for a single crystal or a plane
of maximum shear stress for a continuum solid. In general, shear
deformation proceeds along alternating shear planes with either intersecting slip bands or non-intersecting slip bands[5]. Figure 3a shows
the process of crossing shear bands in a plane strain test bar. In
this case the shear strain of the deformed metal is equal to 2.
Figure 3b shows the process of crack extension by shear bands which
do not intersect, with a shear strain equal to 1. The fracture features associated with ductile rupture are slip lines and slip band
traces. These two processes of shear play a very important role in
ductile fracture and in fatigue crack growth.

Ductile fracture by microvoid initiation, growth and coalescence
is a well-known process consisting essentially of debonding of the
matrix-particle interfaces followed by tensile ductile rupture of the
ligaments between the cavities (see sketch 3a). In general the process of tensile ductile rupture involves many slip systems, thus
leading to a fairly isotropic ellipsoid cavity down to a scale of the
order of 500Å. Figure 4 shows a typical example of ductile fracture
by microvoids. When there is a limited number of slip systems available (HCP crystals), a strong crystallographic anisotropy and a slow
strain rate, the voids may take the shape of elongated channels, as
described by the plane strain sketch, Figure 3a. Figure 5 is an

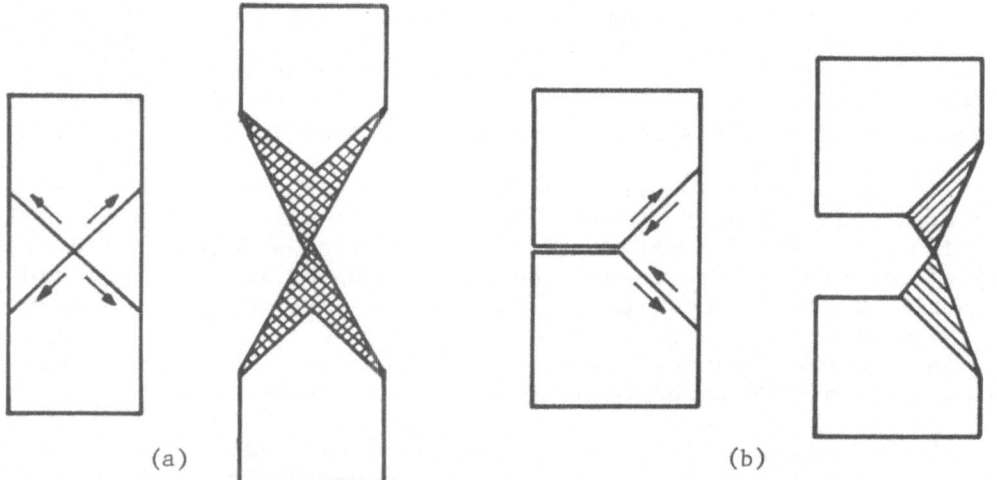

(a)   (b)

Fig. 3.   Ductile rupture by alternating shear. (a) with intersecting
          shear bands; (b) with non-intersecting shear bands.

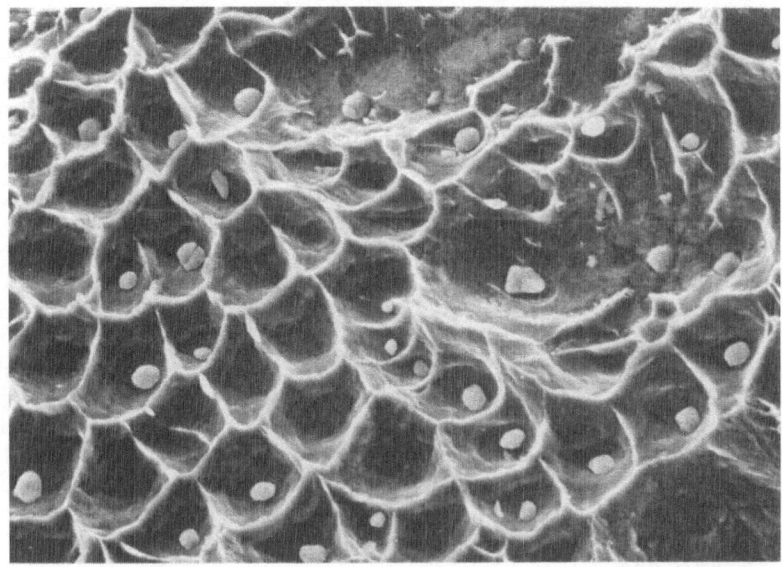

Fig. 4.   Ductile fracture by microvoids (1400 X).

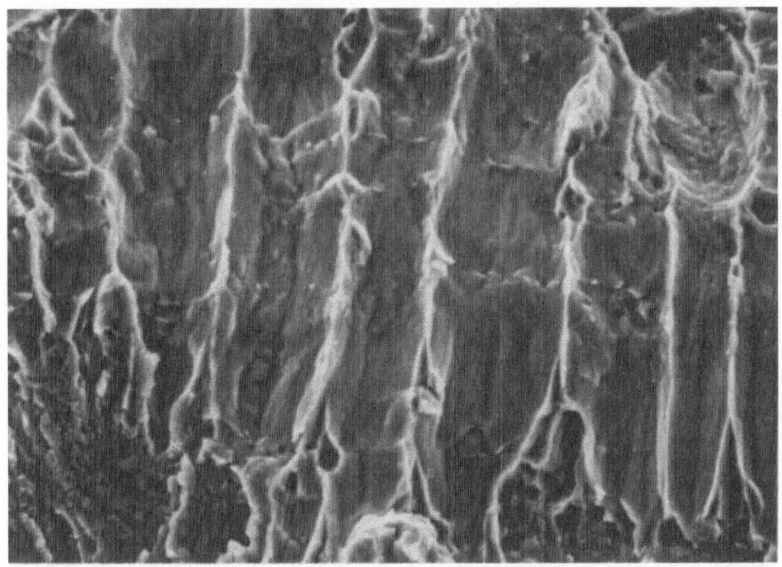

Fig. 5.   Ductile fracture channels in Zircaloy.   The channels which
are called 'flutes' are parallel to [0002].   (1800 X)

example of planar ductile fracture in Zircaloy 2 giving rise to
channels which are referred to as fluting.

Interface decohesion is a common mode of fracture in polycrystalline materials and polyphase structures. Typical examples are
brittle grain boundary decohesion resulting from grain boundary surface embrittlement such as mercury in aluminum or sulfur and phosphorus segregation at grain boundaries in low alloy high strength
steels. Another example of interface decohesion is the debonding of
particle-matrix interfaces which leads to the initiation of microvoids and ductile fracture in multiphase systems. This decohesion
may be due to the coalescence of a few edge dislocations to form a
crack at the interface; thus, it is difficult for particle debonding
to differentiate between a perfect brittle interface debonding and
debonding resulting from localized plasticity. A common mode of
interface decohesion is the intergranular mode of fracture resulting
from ductile voids and/or creep cavitation along grain boundaries.
This process of fracture also involves void growth by ductile rupture
of tensile ligaments between the grain boundary voids along the grain
boundary interfaces.

Fracture in Fatigue

Crack growth in fatigue is especially interesting because of
the repeated resharpening of the crack tip and a general condition

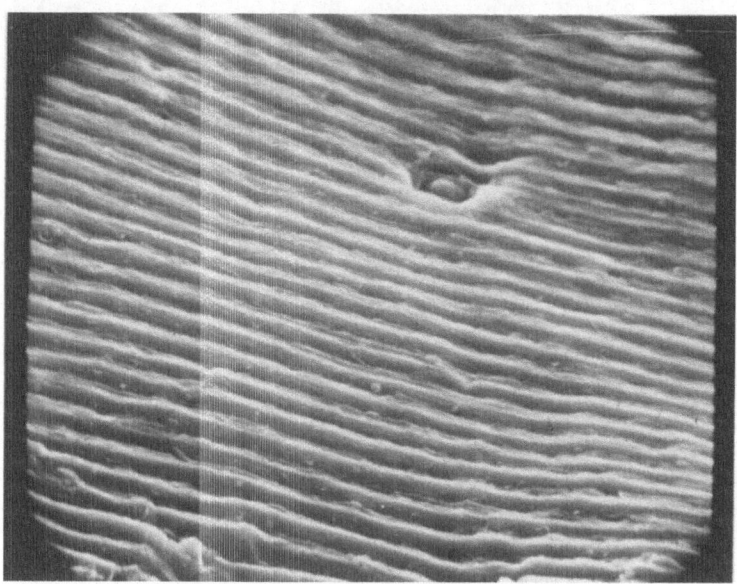

Figure 6.   Typical fatigue striations resulting from alternating
            sliding off at the crack tip.

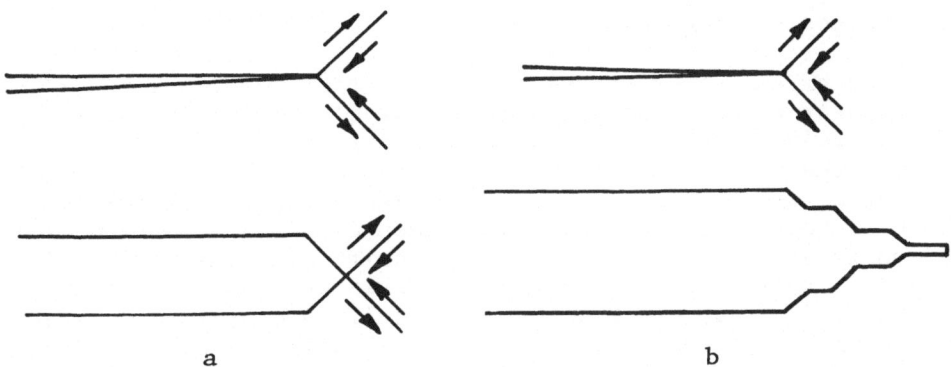

Figure 7.   (a) Crack extension by fatigue by alternating shear;
(b) Crack extension by combined shear and cleavage.

of highly localized shear deformation under cyclic strains.   In Stage
I of fatigue crack growth, short cracks proceed by shear along single
(slip) planes in each grain (Figure 1c).   This Stage I mode of fatigue
cracking is extended to long cracks in alloy systems with shearable
precipitates such as the $\gamma-\gamma'$ nickel base superalloys.   Stage II
cracking corresponds to crack growth in a plane perpendicular to the
direction of maximum stress.   In most alloy systems, Stage II fatigue
cracks grow by repeated alternating shear or sliding off at the crack
tip[6] (see Figure 7a).   The irreversibility of the sliding off process
during closure of the crack is due to oxidation of the crack tip.   In
continuum plasticity the planes of maximum shear stress at the crack
tip make an angle of ±45° with the crack plane.   In single crystals
or in large grains, the shear planes will be crystallographic slip
planes symmetrically oriented with respect to the plane of the crack.
Figure 6 illustrates the fatigue striations which are formed by the
repeated process of sliding off sketched in Figure 7a.

In the case of fatigue crack growth in a corrosive or embrittling
environment, the process of crack tip decohesion can combine alternating shear and cleavage as shown in Figure 7b.   It can be seen that the
ratio of crack advance due to shear versus cleavage is reflected in
the sharpness of the angle of the crack tip.   In general this angle is
difficult if not impossible to determine.   However, measurements of
the crack tip angle as a function of the agressiveness of the environment have been performed recently by Neumann (see these proceedings).

Stress Corrosion Cracking

        Crack extension by stress corrosion cracking occurs by either
intergranular decohesion on transgranular cleavage accompanied by
some microvoid growth and coalescence. The relative amount of
fracture area that fractures by microvoid coalescence is related
to the velocity of the SCC crack. At high crack growth velocity
the SCC component of the fracture surface area may be as low as 10
or 5% while at low crack velocity the entire fracture surface is
due to SCC cracking. The process of stress corrosion cracking is
discontinuous. Each crack arrest is usually accompanied by plastic
deformation due to relaxation at the crack tip by single or alternating shear (see Fig. 7a, b). The complex 3D profiles of SCC
crack fronts make it difficult to assess quantitatively the contribution of crack tip plasticity to SCC cracking. Tear ridges between brittle SCC fracture facets should be clearly differentiated
from crack front slip steps due to crack arrest plasticity. The
tear ridges will run parallel to the local direction of crack
growth. SCC fracture surfaces contain a large number of secondary
cracks and/or crevices which will complicate any detailed fractographic analysis.

CONCLUSIONS

        Fractography is a powerful tool to study the micromechanisms
of fracture. All the fracture features result from one of three
basic modes of separation which are cleavage, rupture by alternating intersecting or single slip and interface decohesion.

        The 5nm resolution limit of the SEM restricts our ability to
differentiate between cleavage and shear separation on an atomistic
scale. However, the clear identification of fractographic features
such as cleavage facets, ductile rupture, microvoid coalescence and
interface decohesion has led to a greater understanding of the correlation between the micromechanisms of fracture, microstructure,
test and environmental variables.

REFERENCES

1.   Fractography and Atlas of Fractographs, in "Metals Handbook",
     V. 9, 8th ed., American Society for Metals, Metals Park, Ohio.
2.   F.N. Rhines, Application of Quantitative Microscopy in Metal
     Science in "Fourth International Congress for Stereology",
     National Bureau of Standards Special Publication 431, Washington,
     D.C. (1976), p. 233.

3. J. Friedel, "Propagation of Cracks and Work Hardening Fracture",
    p. 498, Swampscott Conference, 1959. J. Wiley, New York.
4. C. Henry and J. Plateau, "La microfractographie", IRSID, Saint
    Germain-en-Laye, France.
5. P. Neumann, "Ductile Fracture", Materials Science and Engineering,
    25, 1976, p. 217-223.
6. R. Pelloux, "Mechanisms of Formation of Ductile Fatigue
    Striations", ASM Transactions, 62, (1969), p. 281.
7. M. Ashby, Notes of Lecture, MIT, September 1976.

DISCUSSION

Comment by S.P. Lynch:

I think it is important to note that fractography does not
resolve details on the atomic scale and, hence, it is sometimes
difficult to deduce atomic mechanisms of fracture from fractographic
observations. For example, fracture surfaces which appear to be flat
when examined by SEM may not be atomically flat and fracture may not
be brittle on the atomic scale.

Reply:

The best resolution of the fractographic techniques (SEM or TEM
with replicas) is of the order of 50 to 100 Angstroms. This is one
order of magnitude greater than the "atomistic features". However,
one is not prohibited from intuitive interpolations between 50Å and
3Å to assume the fracture features resulting from micro mechanisms
of fracture.

Comment by A.W. Thompson:

You didn't mention a fracture mode which I think of as both
common and important: quasi-cleavage. Admittedly we don't know as
much as we would like to about the nucleation and propagation of this
kind of fracture, but I believe it is a distinctive mode that should
not be ignored. Would you comment?

Reply:

Quasi-cleavage has been extensively described by Beachem. It
is a mode of fracture where a small cleavage crack grows around a

microvoid. The size of the cleavage crack is limited by the size
of the crystallographic unit (martensite plate). Quasi-cleavage is
often accompanied by small tear ridges which are sites of ductile
tearing.

Comment by H. Vehoff:

Is there any experimental evidence that gaseous hydrogen influences the microvoid nucleation rate in pure metals? A smaller crack
tip opening angle by microvoid coalescence can only be explained if
there is an increase in the nucleation rate. If that is not the
case it must be microcleavage.

Reply:

The work by Vehoff and Neumann appears to be void independent.
The small crack tip angle seems to be due to a continuous combined
mode of sliding off and cleavage.

Comment by H. Birnbaum:

Microvoids are observed on the fracture surfaces of high purity,
single phase metals. Can you comment on the mechanism of microvoid
formation in these cases? Is it necessary to have second phase
particles to nucleate the microvoid?

Reply:

Microvoid initiation requires a local enhancement of the tensile
stress. If a dislocation sub-boundary node will lead to a high
tensile stress it may be the source of a void initiation site. In
general microvoid initiation sites are always linked with the
interfaces of small intermetallic particles.

Comment by R.H. Jones:

There was some discussion about whether the microvoid coalesence
process requires inclusions or non-deformable particles. One way to
evaluate this question is to determine if a plot of microvoid spacing
versus the inverse of the inclusion spacing can be extrapolated to
the origin or if there is a finite void spacing size at infinite
paritcle spacing. Is such data available?

Reply:

This type of work was done by Palmer and Smith (Cambridge
University - Phil. Mag. 1966). There is not a one to one
correlation between width and particles. Voids form only for
$d \geqslant d_{critical\ particle\ size}$. Voids do not form on each particle.

Other authors have performed similar experiments since then.

Comment by R. Bullough:

   I was interested to hear that you think that H_2 will enhance the formation of microvoids by decohering the small particles from the matrix. Similar effects of H_2 are found in M316 stainless steel when subjected to irradiation in the presence of H_2; the presence of H_2 causes many voids to nucleate preferentially on coherent $M_{23}C_6$ precipitate particles. Maybe this preferential void formation occurs because the coherence of the $M_{23}C_6$ precipitate-matrix interface is destroyed by the hydrogen, rather than, as we had thought, that bubble nucleation occurs on the interfaces and the bubbles grow into large voids that eventually engulf the original precipitates.

Reply:

   I agree with the author of the question.

Comment by G. Garrett:

   With reference to the relationship you observe between the "zig-zag" wave length obtained during ductile crack propagation and the square of the fracture toughness, is there any mechanistic basis for this correlation, particularly in view of the fact that the fracture toughness is more conventionally considered to be a parameter describing the initiation of fracture?

Reply:

   The zig-zag waves are observed from the end of the fatigue pre-cracked zone through the whole fast fracture surface. The zig-zags are observed with Charpy bars as well as with slow K_{IC} fracture specimens. The $\lambda - (K_{IC})^2$ correlation appears to be related to a $\lambda - CTOD$ correlation. At this time, the correlation between λ and K_{IC} is purely empirical. There is no mechanistic basis for the correlation.

TUTORIAL LECTURES ON

SURFACE REACTIVITY AND BONDING

Session Chairmen: J. Oudar, M.T. Thomas, and
H.P. Bonzel

MOLECULAR ORBITALS AND THE ATOMISTICS OF FRACTURE

M. E. Eberhart and K. H. Johnson

Department of Materials Science and Engineering
Massachusetts Institute of Technology
Cambridge, MA 02139

R. P. Messmer and C. L. Briant

General Electric Company
Corporate Research and Development
Schenectady, NY 12301

INTRODUCTION

If a metal contains very few impurities and is mechanically
tested in an inert environment, its fracture mode will be either
transgranular cleavage or transgranular ductile microvoid coalescence. If, however, the metal is of commercial purity, the brittle
fracture mode can switch from cleavage to intergranular separation.
This transition is undesirable because intergranular fracture is
generally a very low energy process.

The occurrence of intergranular fracture in a specific situation will depend on a number of metallurgical variables: grain
size, yield strength, test temperature, and notch radius. However,
in all reported cases, one observation is common. Impurities which
have a very low solubility in the bulk have segregated to the grain
boundaries and weakened them. The bulk concentration of these
impurities need only be several hundred ppm for grain boundary
concentrations of 5-10 atomic percent to be observed. Examples of
this impurity-induced weakening of grain boundaries include sulfur
in iron and nickel, phosphorus, tin, and antimony in steel, bismuth
in copper, and oxygen in refractory metals.

If the environment is not inert, intergranular separation may
occur even though the fracture mode in an inert environment would

be transgranular. One much studied example of this is hydrogen
embrittlement. If a sample is tested either in hydrogen gas, or is
cathodically charged with hydrogen and then tested in an inert
environment, one finds that the mechanical properties of the mate-
tial are often degraded and that the fracture mode has become more
intergranular. As with impurity-induced embrittlement, many additional variables can affect the degree to which hydrogen will cause
embrittlement, and one of the worst possible cases is to have both
hydrogen and impurities segregated at the grain boundaries. However,
it is clear that hydrogen must segregate at interfaces and reduce
the cohesive strength across them

    We believe that an important contribution to understanding
some aspects of these complex problems of embrittlement in metals
can be made through theoretical studies of bonding in model systems.
The philosophy is to isolate groups of atoms in clusters which
simulate a possible local environment of interest and to perform
detailed quantum mechanical calculations using molecular orbital
theory to probe the nature of the chemical bonding.

    After discussing molecular orbital theory and the Xα-scattered-
wave technique in the next section, we consider in the following
sections two recent applications of this approach to models of
potential metallurgical interest.

MOLECULAR ORBITAL THEORY AND THE SCF-Xα-SW CLUSTER METHOD

    The most characteristic feature of molecular orbital theory is
that it provides an independent-particle interpretation of the
electronic structure of a molecule or solid. That is, each electron
of the system has associated with it a characteristic spin-orbital
and energy level which are determined by solving Schrödinger's
equation for an electron in the average potential energy of all the
other electrons in the system. The most familiar of such techniques
is the Hartree-Fock self-consistent field method, in which each
molecular orbital is traditionally represented by a linear combination of atomic orbitals (the HF-SCF-LCAO method). For all but the
simplest molecules one must, at least for the present, choose
atomic-orbital basis sets which are fairly far from the correct
solutions of the Hartree-Fock Schrödinger equations. We refer to
this type of approximate solution of the Hartree-Fock equations
using the LCAO representation as the AHF-SCF-LCAO method.

    It is well known that the Hartree-Fock theory provides a poor
description of the electronic structure of a molecule or solid at
large internuclear distances and that it leads to an improper
description of the dissociation of such systems. In order to correct for this behavior, it is necessary for one to work with linear

combinations of Slater determinants of molecular orbitals, the so-called "configuration-interaction" (CI) procedure, which is not only complicated computationally, but also destroys the independent-electron interpretation. A further difficulty is that in performing numerical calculations using the LCAO approximation at either the AHF or CI levels of sophistication, the number of two-electron multicenter integrals which must be calculated even for moderately sized systems becomes enormous. For example, in a recent calculation on a molecule consisting of 26 atoms, all of atomic numbers less than ten, 10^{11} integrals had to be evaluated and the calculations required the equivalent of 192 h on an IBM 360/195 computer, one of the fastest machines available.

A much more practical independent-electron theory, which avoids most of the difficulties associated with the AHF-SCF-LCAO and CI methods, is the recently developed SCF-Xα-SW cluster method. Based on the combined use of Slater's Xα statistical theory of exchange correlation and Johnson's multiple-scattered-wave formalism, this technique allows one to calculate accurate self-consistent spin-orbital energies, wave functions, charge densities, and total energies for complex molecules and clusters, yet requires only relatively small amounts of computer time. Like the AHF-SCF-LCAO approach, the SCF-Xα-SW method is an approximate theory of many-electron systems, but it is not an approximation to Hartree-Fock theory. First of all, in contrast to HF theory, the Xα statistical expression for the total many-electron energy automatically goes to the proper separated-atom limit as the internuclear distances are increased to infinity, i.e., as the system dissociates. Second, it has been shown that the Xα theory satisfies Fermi statistics, thereby ensuring the proper ordering and occupation of electronic energy levels, while the HF theory does not. Third, it has been proven that, under the proper conditions, the Xα statistical total energy rigorously satisfies both the virial and Hellmann-Feynman theorems, which facilitates the calculation of the equilibrium cohesive energies of molecules and solids. Finally, the SCF-Xα-SW method, in conjunction with the "transition-state" concept, leads to an accurate approximate description of the excited electronic states and charge distributions of polyatomic systems, including the effects of orbital relaxation which are usually neglected in applying Koopmans theorem in the AHF-SCF-LCAO approach.

The SCF-Xα-SW method and transition-state procedure have already been applied successfully to a wide range of molecules and clusters in solids. Because the required computational effort does not increase inordinately with the number of electrons per atom or with the number of atoms in the molecule or cluster, the SCF-Xα-SW method is ideally suited for describing the electronic structures of transition-metal complexes and clusters representing the local molecular environment of a grain boundary.

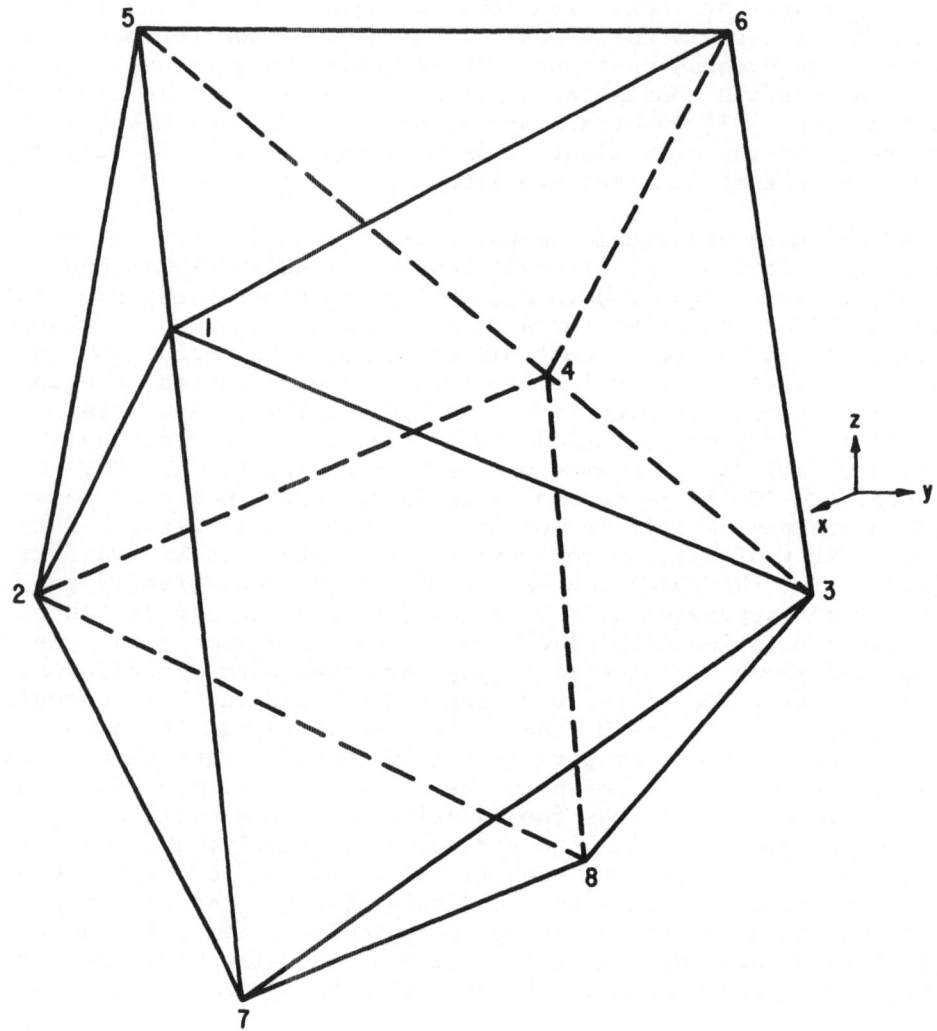

Figure 1

MOLECULAR ORBITAL MODEL FOR GRAIN BOUNDARY EMBRITTLEMENT OF NICKEL
BY SULFUR

As discussed in the last section, the theoretical method we
use to investigate the electronic structure of the local environment at a grain boundary consists of two parts. The first is to
use a cluster of atoms to represent the local environment and the
second is to use molecular orbital theory to solve for the electronic structure of the cluster of atoms.

To apply this technique to grain boundaries one must first
choose a cluster that resembles in some way the local environment
an impurity atom might have were it at the grain boundary. The
choice of such clusters is considerably simplified due to the
recent work of Ashby, Spaepen, and Williams, who have shown that
the structure of grain boundaries in fcc solids can be described in
terms of atoms at the vertices of Bernal deltahedra. The individual
deltahedra are small clusters containing four to thirteen atoms and
thus provide a natural initial model of local atomic structure with
which to investigate the electronic properties by SCF-Xα-SW calculations. In order to investigate the feasibility of applying such
methods to this problem we have chosen to use one particular cluster
in this work.

The particular cluster we chose is the tetragonal dodecahedron
shown in Fig. 1. It contains eight atoms and has D_{2d} point group
symmetry. We do not claim that this structure is necessarily a
prevalent one in grain boundaries. However, it does contain four
atoms (numbered 1, 2, 3, and 4) slightly but symmetrically displaced
from the x-y plane and four other atoms further displaced from this
plane. One can roughly consider that the four atoms nearest the
x-y plane are grain boundary atoms and that the other four are in
the first atomic layers away from the precise boundary.

It is well established experimentally that sulfur embrittles
nickel. In order to investigate the electronic effects of a sulfur
atom at a grain boundary in Ni, we have carried out two calculations
on a model system in which the Ni atoms were placed at each of the
eight sites of the cluster shown in Fig. 1 with an internuclear
separation of 2·49 Å. In the first calculation, we considered the
cluster containing only the eight nickel atoms. For the second
calculation, a sulfur atom was placed at the center of the cluster.
An interstitial site of this type is a possible position which a
sulfur atom would choose at a grain boundary. For these two configurations we have determined the molecular orbitals for the
clusters, their corresponding energy levels, and the valence charge
densities.

The energy levels for each cluster are shown in Fig. 2. We
note that in the Ni_8S cluster several of the energy levels have

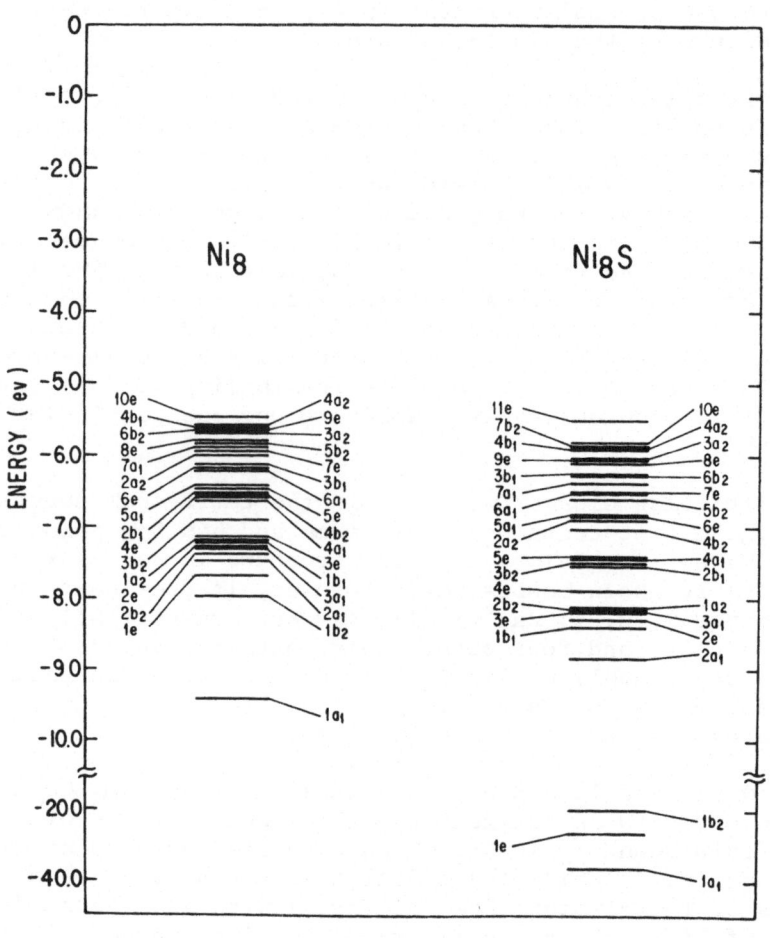

Figure 2

been lowered substantially. Those that have been lowered have
orbital wave functions which are highly localized on the sulfur
atom and its four nearest neighbors, i.e., the four nickel atoms
closest to the x-y plane. Based on this simple analysis one might
assume that some Ni-Ni bonds have been weakened.

The bonding may be better understood if one analyzes the
molecular orbitals of the clusters in more detail; however, in
this particular case a more succinct picture of the bonding in the
cluster can be obtained by summing the orbital charge densities
over all the occupied valence orbitals. One can then plot contours
of the total valence charge density which exists on any plane.
These are shown in Fig. 3 for the x-z plane. In this figure we
have plotted the same set of contours for both clusters. The three
outermost contours (numbered 1, 2, and 3) in both clusters surround
all atoms. However, contour number 4 (which is shown with a
heavier line for clarity) is quite different in the two clusters.
In Ni_8, it shows bonding between the pairs of nickel atoms 1 and 7,
7 and 8, and 4 and 8, where the characteristic "neck" regions
between the pairs of atoms is responsible for bonding. In Ni_8S,
the shape of contour number 4 is quite different. The characteristic neck bonding region between atoms 1 and 7, and 4 and 8 is
gone and consequently so is the concomitant strong bonding between
the two types of nickel atoms. Furthermore, the size of the neck
region between atoms 7 and 8 is reduced from that in Ni_8, implying
a weakening of this metal-metal bond as well. Contour 4 in Ni_8S
is split into two distinct regions. The upper region of contour 4
and an inner contour of the nickel atoms nearest the sulfur atom
enclose both the nickel atoms and the sulfur. This would lead to
very strong bonds between this set of nickel atoms and the sulfur.
However, there is no such bonding interaction between the sulfur
atoms and the other nickel atoms.

From these results a clear picture begins to emerge. Adding
sulfur to the cluster of nickel atoms causes strong NiS bonds to
be formed between the nickel atoms nearest the sulfur. Yet this
weakens bonds which the nickel atoms would ordinarily have with
the other nickel atoms that are not as close to the sulfur atom.
In this cluster the strong Ni-S bonds involve the sulfur atom and
the four nickel atoms nearest the x-y plane. We suggested earlier
that these atoms might be thought of as being in the grain boundary.
However, the bonds between the nickel atoms nearest the x-y plane
and those farther away are weakened. These bonds might represent
those from the boundary out to the first layer of the bulk. Clearly,
if those are weakened, the stress required for fracture would be
less.

Therefore, based on the calculations we can suggest an electronic mechanism by which grain boundary embrittlement could occur.
We stress that these are the results from calculations on only one

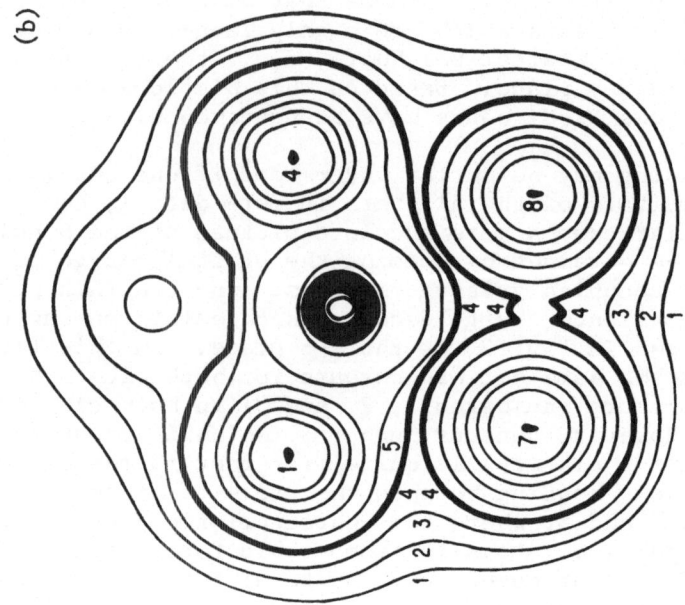

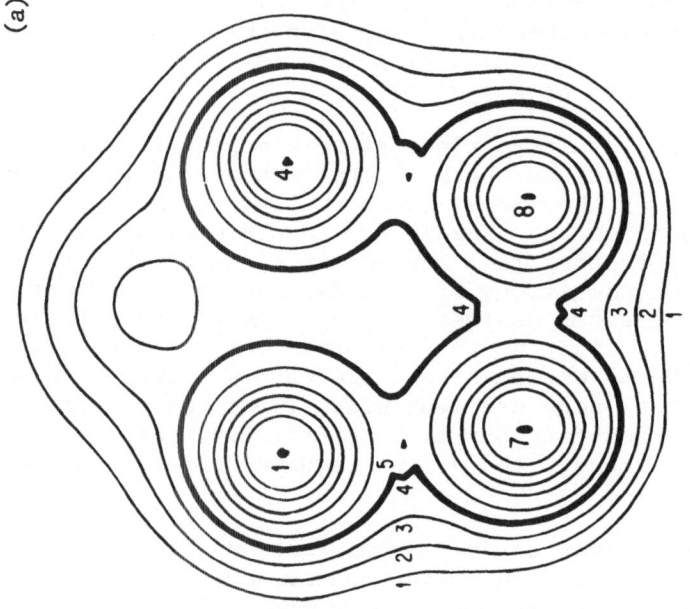

Figure 3

system. Furthermore, we have chosen an embrittling element which
probably occupies an interstitial site at the grain boundary.
Other embrittling elements would clearly be substitutional. Also,
many other types of geometrical configuration are possible at the
boundary. Yet, with these results, we believe that it is not pre-
sumptuous to propose that the above-described mechanism is one
mechanism by which segregated elements embrittle the grain boundary.

MOLECULAR ORBITAL MODEL FOR HYDROGEN EMBRITTLEMENT OF IRON

     The causes and mechanisms of hydrogen embrittlement have been
and will continue to be the subjects of detailed investigation.
As a result, several models have been offered in explanation of the
embrittlement phenomenon. These models, however, cannot account
for embrittlement at the atomic level, as a chemical interaction
between hydrogen and the embrittled metal. It is the purpose of
this paper to present the results of SCF-Xα-SW molecular-orbital
calculations on iron-hydrogen clusters in order to explain hydrogen
embrittlement of ferrous alloys at the most fundamental level.

     Hydrogen atoms are most frequently found in octahedral holes
of transition metals. If molecular hydrogen were adsorbed within
tetrahedral holes then concerted dissociation of molecular hydrogen
would result in placing atomic hydrogen in octahedral holes.
Accordingly, the four-atom iron cluster shown in Fig. 4 will be used
to model the tetrahedral hole and the site for hydrogen adsorption
in bcc iron. (Both the tetrahedral and octahedral holes of bcc
iron are misnamed since they actually present a D_2d and tetragonal
symmetry respectively.) In an earlier paper, the results of SCF-
Xα-SW molecular-orbital calculations of four-, nine-, and fifteen-
atom clusters of iron were presented. It was shown that these
clusters provide a model for the local electronic structure of bcc
iron. A comparison of these results indicates that both the four-
atom and fifteen-atom clusters exhibit the large exchange splitting
and high magnetic moment characteristic of bulk iron. Net spin
electron number per atom for Fe_{15} is 2.7 and for the Fe_4 cluster
2.5 compared with 2.12 for the bulk. The density of states in the
four-atom cluster is sufficiently similar to the density of states
in the 15-atom cluster to justify the use of Fe_4 as a model for a
site of hydrogen adsorption. It is important to note that the
fifteen-atom cluster provides three environments for the atoms of
the cluster. In bulk iron (neglecting surface effects) each iron
atom is in the same environment. It is this fact which is responsible for the variation of properties predicted by the cluster
method with those measured in bulk iron. In a calculation of this
type, however, where an impurity is being modeled, the impurity
breaks the symmetry of the bulk and there are a number of environments around the impurity site. As a result, the cluster method is
ideally suited for the study of dilute impurity states and

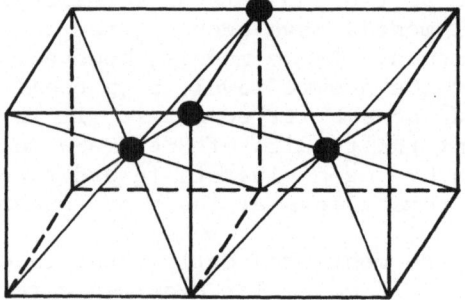

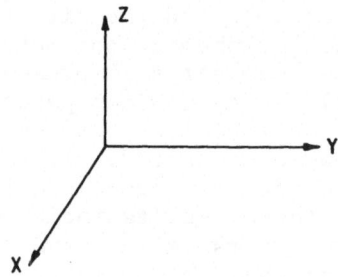

Figure 4

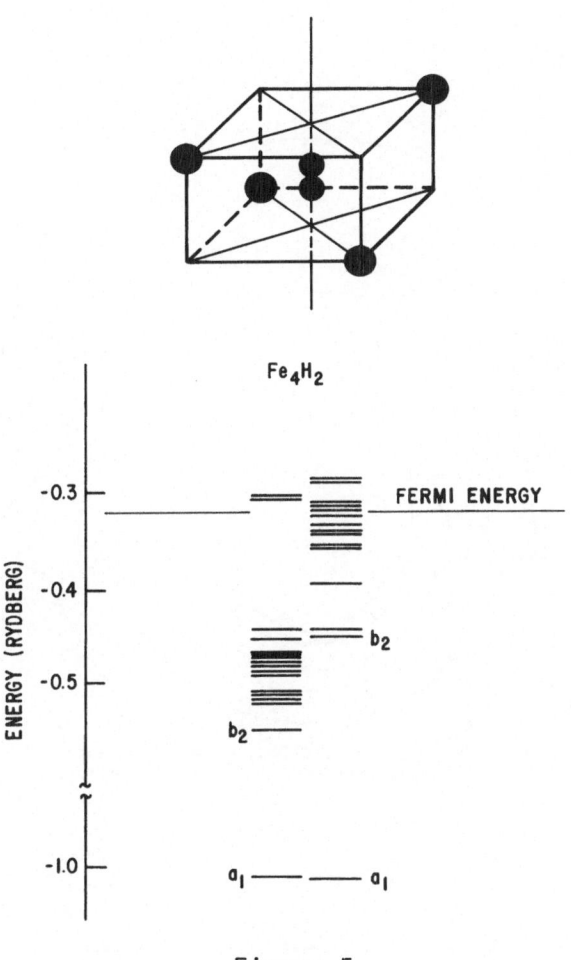

Figure 5

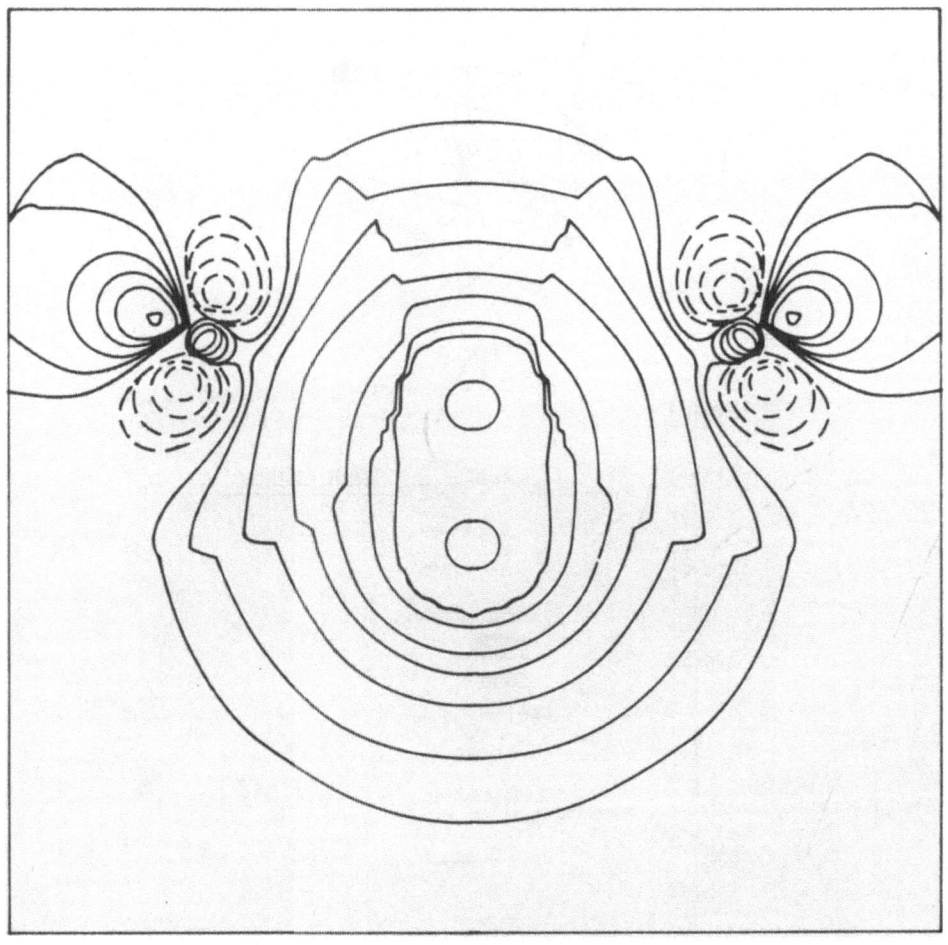

Figure 6

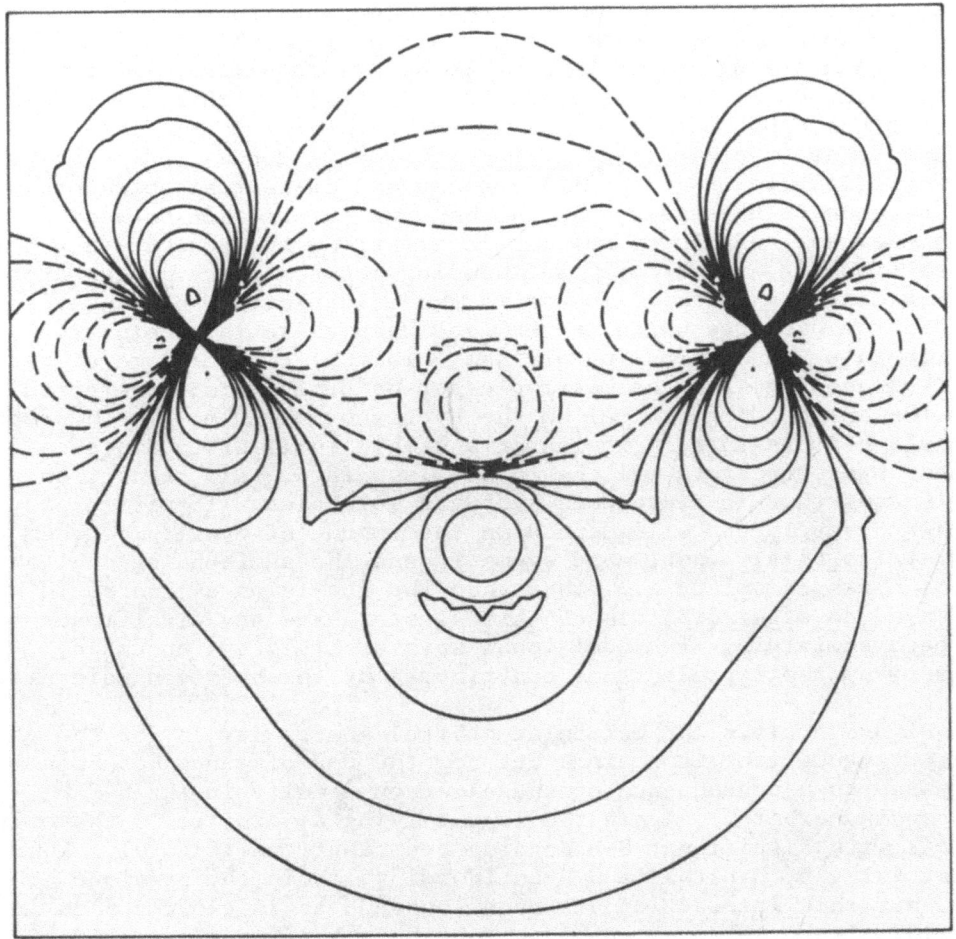

Figure 7

consequently smaller clusters than those used in the study of bulk
properties can be used.

The molecular-orbital energy levels of a hydrogen molecule
inserted into the four-atom cluster is shown in Fig. 5. The a_1
and b_2 orbitals of the four-atom iron cluster have the same symmetry as the hydrogen molecule bonding and antibonding orbitals.
These hydrogen molecular orbitals couple with the lowest lying
a_1 and b_2 orbitals of the Fe_4 cluster to produce the molecular
orbitals shown in Figs. 6 and 7. As a result of this coupling the
lowest lying a_1 and b_2 orbitals of Fe_4H_2 are stabilized relative
to the position of these orbitals in Fe_4. The a_1 orbital is predominantly a H-H bonding orbital the H-Fe interaction is non-
bonding. The b_2 orbital in the Fe_4 cluster is d-d π-bonding between
nearest neighbors and δ-bonding between second-nearest neighbors.
The addition of H_2 strengthens the bonding interaction between
second-nearest neighbors, has little effect on the nearest-neighbor
interaction, and is strongly antibonding between the hydrogen atoms.
The H-H distance in this cluster is the molecular H-H distance.
This is the distance which is realized when there is no electron
density in a H-H antibonding orbital. In this cluster some of the
electron density from the metal cluster b_2 orbital is transferred
into the antibonding orbital of the hydrogen molecule. The hydrogen
molecule H-H distance is no longer a stable equilibrium distance
and the hydrogen atoms will begin to dissociate. All transition
metals are known to dissociate hydrogen molecules. The efficiency
of this dissociation will depend on the amount of overlap between
the metal orbital of suitable symmetry and the antibonding orbital
of the hydrogen molecule. Other than the low-lying a_1 and b_2
orbitals, no other orbitals of this cluster have any significant
hydrogen admixture. Two additional spin orbitals are occupied,
however, by the two electrons contributed by the hydrogen molecule.

Figure 8 gives the molecular orbital energy levels for the
same arrangement of iron atoms but now the H-H distance has been
increased as a consequence of the electron density in the H-H
antibonding orbital. Again the lowest lying a_1 orbital is the only
orbital with significant H-H bonding contributions (Fig. 9). The
lowest lying b_2 orbital is H-H antibonding, as in the previous
case, but this interaction has been lessened as is clearly seen in
Fig. 10. There is a reasonably strong H-Fe bonding interaction
within this b_2 orbital, and this interaction should be maximized
when the hydrogen atom is sitting between the iron second-nearest
neighbors, i.e., when the hydrogen atoms are in the tetragonal
holes of the bcc structure.

While hydrogen adsorption and dissociation may be explained
in a fairly straightforward manner by molecular-orbital theory,
attempts to explain embrittlement at the molecular level must be
assisted by considering embrittlement as a competitive process

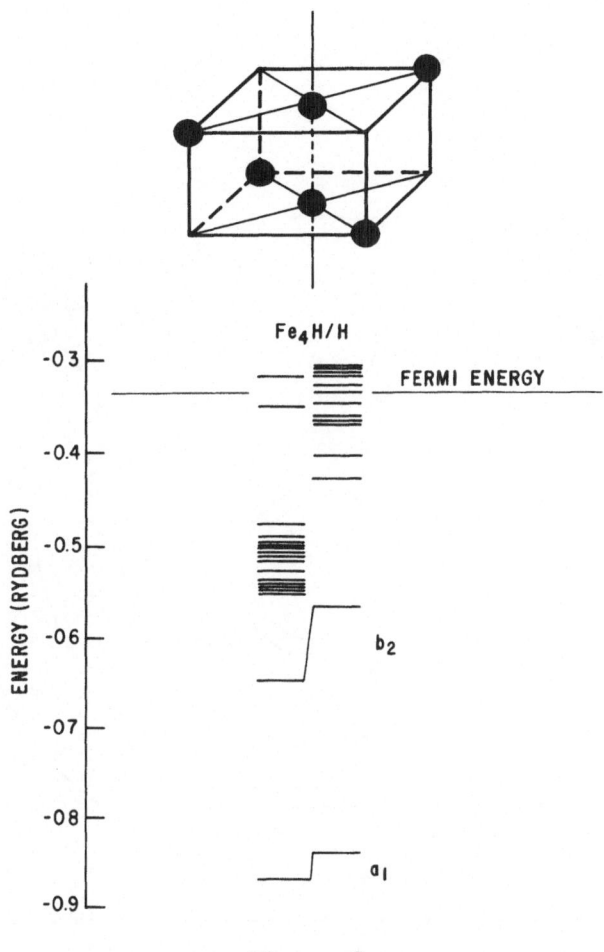

Figure 8

Figure 9

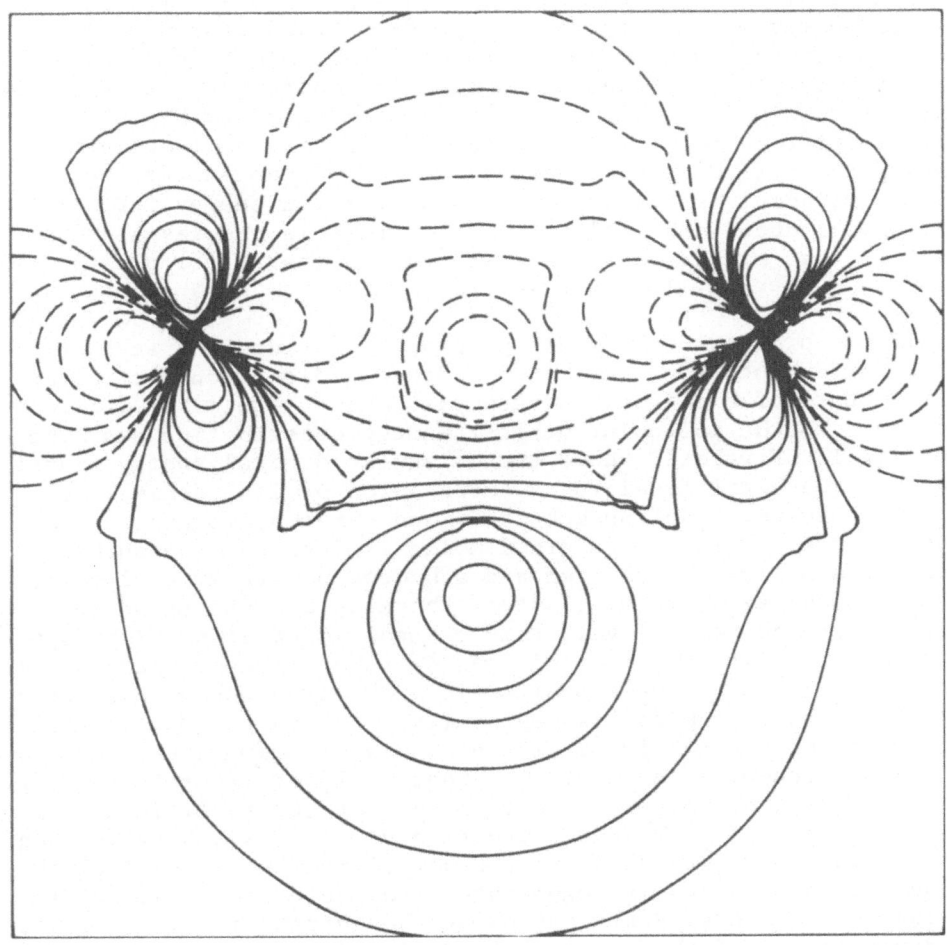

Figure 10

between two mechanisms. The first of these is the process of plastic flow facilitated by dislocation movement. The second is characterized by nuclear motion in a direction which will counter the strain energy of the crystal, a purely elastic response. An energy of activation may be associated with each of these processes. The activation energy for the first process will depend predominately upon the structure of the crystal, the Burgers vector, etc., while the activation energy of the second process is simply the fracture energy of the crystal. A ductile metal responds to excessive strain through the first mechanism, and this may be attributed to the fact that the activation energy for plastic flow is significantly less than the activation energy for brittle fracture. A brittle material will respond to strain by fracturing before plastic flow begins because the activation energy for dislocation movement is higher than the activation energy for brittle fracture. A ductile-brittle transition must be the result of a mechanism that either raises the activation energy for plastic flow or lowers the energy for brittle fracture.

If the process of hydrogen adsorption affects the activation energy for dislocation flow, then hydrogen adsorption must alter the structure of bcc iron. Structural changes are often manifest as changes in the molecular orbital energy distribution near the Fermi level. Figure 11 shows the molecular orbital energy distribution near the Fermi level for a tetrahedron of iron atoms, the four-atom cluster modeling bcc iron, and three-hydrogenated-iron cluster corresponding to adsorbed molecular hydrogen, adsorbed dissociated molecular hydrogen and a hydrogen saturated cluster, Fe_4H_6. In the tetrahedral cluster, there is an unoccupied t_2 orbital near the Fermi level, this orbital is split when the symmetry of the cluster is reduced from tetrahedral to D_2d as in the 4-atom bcc cluster. The e orbital is stabilized and becomes occupied while the b_2 orbital is destabilized. This process is reminiscent of the Jahn-Teller effect of inorganic complexes, where a complex spontaneously distorts in order to lower its total energy. This leads to the conjecture that the bcc structure of iron is a consequence of a Jahn-Teller distortion of a fcc structure. When hydrogen molecules are added to the bcc cluster the e orbital is depopulated and is nearly degenerate with the b_2 orbital, and is similar to the energy distribution of the tetrahedral iron cluster. The dissociation of the hydrogen molecule stabilizes the b_2 orbital but the addition of more hydrogen seems once again to depopulate this orbital. Hydrogen adsorption would appear to promote a close-packed phase transition near the site of adsorption. There is evidence that hydrogen adsorption does promote phase transformations. In HCP iron hydrogen adsorption is known to lower the stacking fault energy indicating that hydrogen is promoting the fcc-hcp phase transition.

This is just the type of effect that would raise the

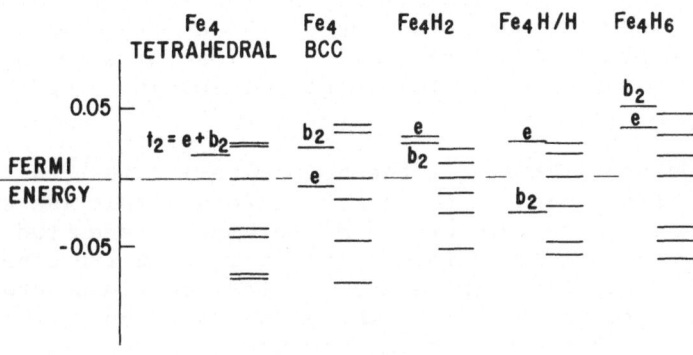

Figure 11

activation energy for dislocation movement around the sites of
hydrogen adsorption. However, no surface reconstruction of iron
due to hydrogen adsorption has ever been noted and fcc metals,
notably nickle, embrittle in a hydrogen environment. While hydrogen
undoubtedly has an effect on the structure of bcc iron, it is unlikely that this alone can account for hydrogen embrittlement.

    This leaves the possibility that hydrogen adsorption lowers
the activation energy of brittle fracture. If this is the case
then hydrogen-adsorbed clusters should have molecular orbitals that
show similarities to metal clusters that have been strained, for
such clusters require less additional energy to fracture, i.e., the
activation energy for brittle fracture has been lowered. Accordingly,
the molecular orbital energy levels for the four-atom bcc cluster
in which the iron atoms have been strained 20% along the twofold
rotation axis was calculated. The relative energy distribution of
the majority spin states for this cluster is shown in Fig. 12,
along with the four-atom model of bcc iron and the dissociated
molecular hydrogen-iron clusters. (The minority spin states for
all clusters show the same ordering of the energy levels as do the
majority spin states. No information is lost by studying only the
majority spin states, while the forest of levels is considerably
thinned.)

    The process of straining the bulk material results in con-
siderable rearrangement of the energy levels within the energy
manifold. A rough measure of the difference between two energy
manifolds is the number of intersections between the connecting
lines of the manifolds. There are ten intersections between the
bulk and the strained material, while there are only two intersections connecting the strained and the hydrogenated-iron energy
manifolds. Saturation of the Fe_4 cluster with hydrogen results in
almost no rearrangement of the energy manifold. What rearrangement
occurs is confined to the lowest five orbitals of the manifold.

    The direction in which the molecular-orbital energy levels move
upon straining can be seen to be the result of removing electron
density from between nearest neighbors, and replacing it in the
region perpendicular to the direction of strain, i.e., between
second-nearest neighbors. The result should be a contraction of
the cluster perpendicular to the direction of strain. This is the
Poisson's effect. Within this analysis, it is clear that hydrogen
atoms in tetragonal holes will mimic the behavior of strain by increasing the electron density between second-nearest neighbors. It is
worth noting that only the two lowest lying orbitals of the hydrogenated-iron cluster have any hydrogen admixture; therefore, the rearrangement of the other orbitals must be the result of a relaxation
effect due to the stabilization of the lowest a_1 and b_2 orbitals.

    It seems reasonable then that if the lowest lying a_1 and b_2

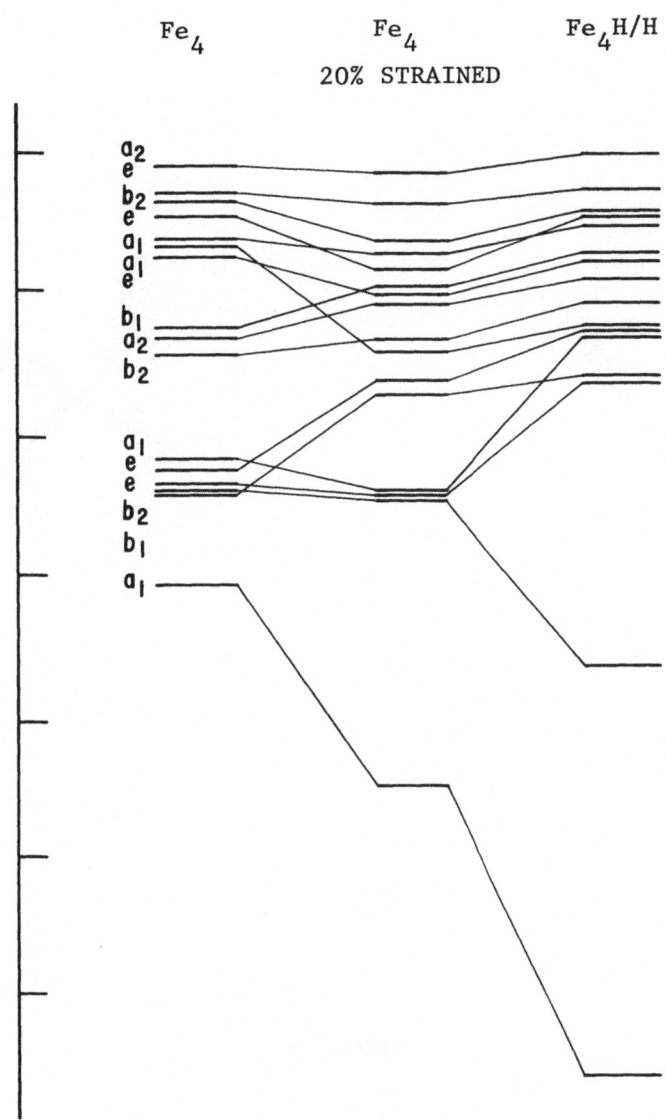

Figure 12

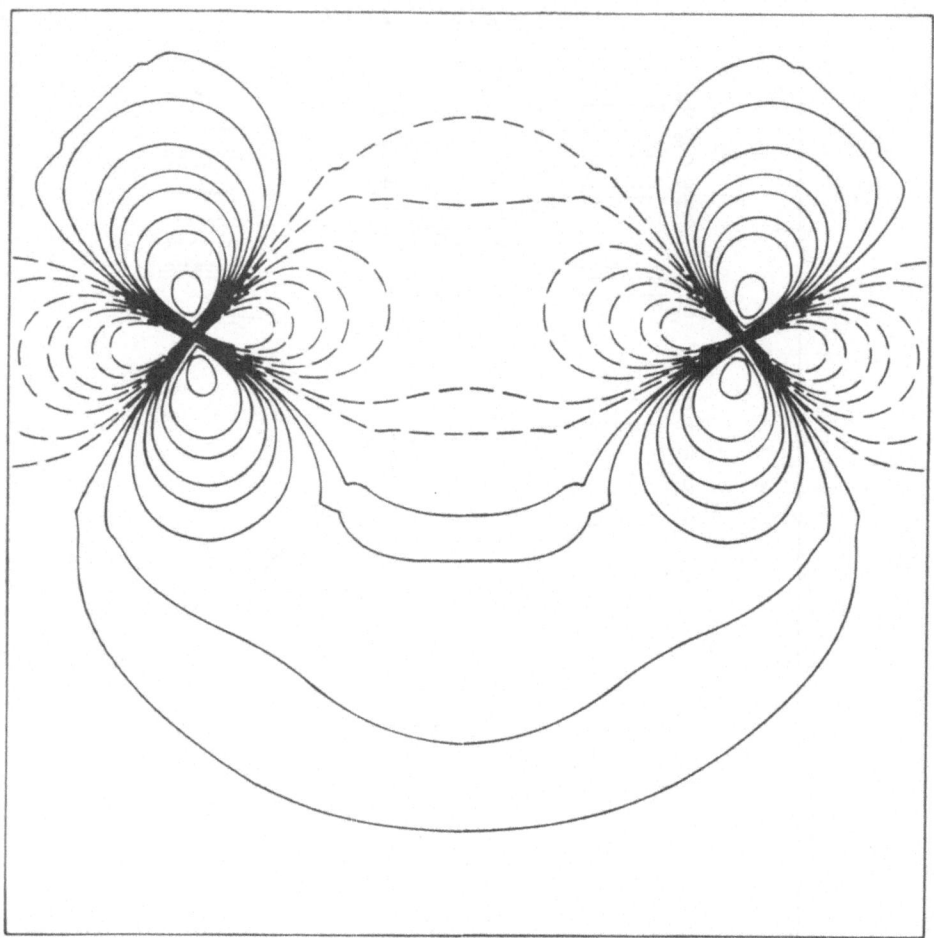

Figure 13

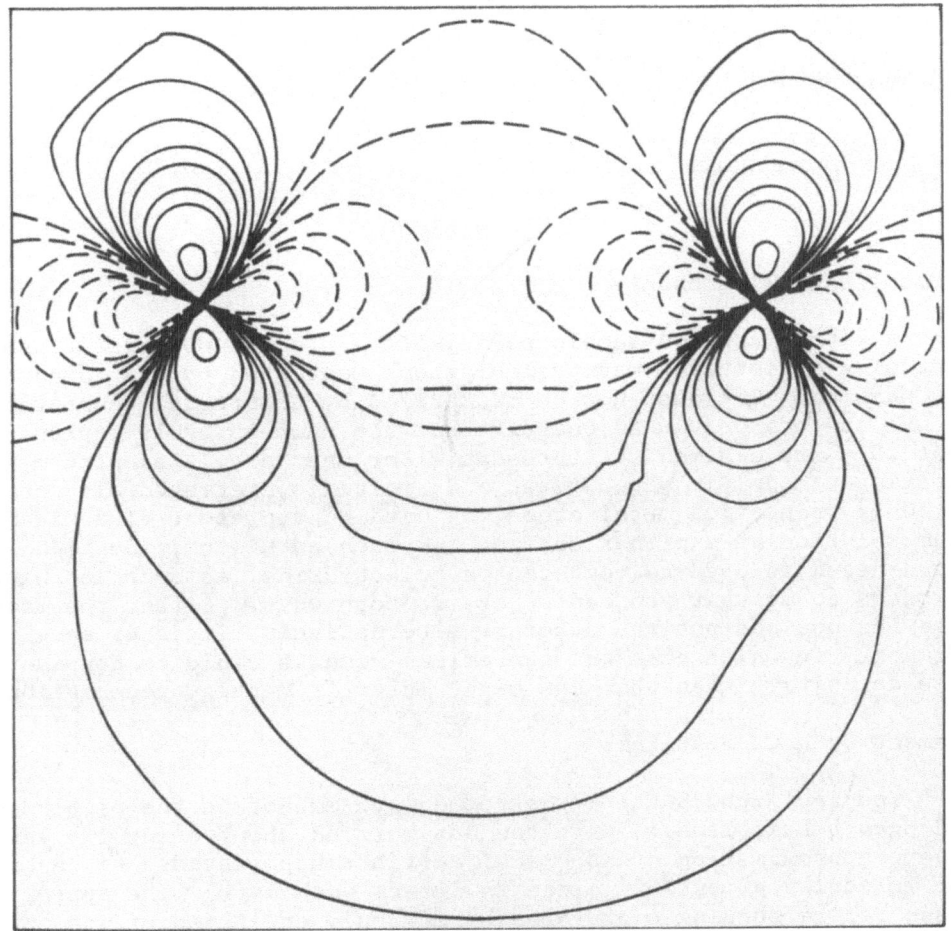

Figure 14

orbitals of the strained and hydrogenated cluster are similar in
composition, the other orbitals of the manifold should also show
similarities. Figures 13 and 14 compare the lowest b_2 orbitals in
an xz plane containing the second neighbors for the unstrained and
strained (20%) Fe_4 clusters, respectively. Again, both the hydrogen-
adsorbed (Fig. 10) and strained cluster (Fig. 14) show greater
electron density between second neighbors and even similar electron
distributions in areas away from the hydrogen atom.

ACKNOWLEDGMENTS

     This work was supported in its entirety at M.I.T. and in part
at G.E. by research grants provided by the Office of Naval Research.

DISCUSSION

Comment by J.W. Gadzuk:

     A bit of caution should be injected into the optimistic assess-
ment of x-α scattered wave calculations suggested by Drs. Johnson
and Messmer. One must not be mesmerized by theoretical numbers
before one has convinced oneself that the numbers truly represent
what they are claimed to represent. For instance, a skeptic might
question a model in which first a H_2 molecule, tetrahedrally coordinated by transition metal atoms, is used to represent dissociative
chemisorption at a planar surface and then an H atom, identically
coordinated is used to represent a bulk hydride, as done in this talk.
It seems to me that you can't have it both ways. Either the model is
good for one and not the other or alternatively, it is of equal merit
for both, in which case the numberical results could be considerably
less definitive than what one might infer from this presentation.

Comment by D.G. Pettifor:

     You apply the SCF x-α scattered wave method to the problem of
hydrogen embrittlement. Are you not worried about using the very
severe approximation of spherical muffin tin potentials to describe
the molecular potential, since in dimers such as H_2^+ the approximation
leads to the bonding eigenvalue having only one-third of its correct
bonding strength? You also mentioned that iron is not superconducting
because its Fermi level falls in the antibonding region. Why then
are Tc and Re, whose Fermi levels lie in this region, such good
superconductors?

Comment by F. Cyrot-Lackmann:

     I am not sure that one can assess as definitive and noncontro-
versial the results of electronic structure of transition metal
clusters as it seems that different theoretical approaches give

different answers concerning either the energy levels distribution or the role of s or d electrons for hydrogen chemisorption (see, for example, Veillard et al. for Cu by ab initio H.F., Baetgold by EH methods, discussion of H/Ni by Goddard, Bagus, and others). Concerning your interpretation of magnetic properties of transition metals through the antibonding character of the state at E_F, how do you differentiate from Goodenough's approach who, many years ago, tried to understand the magnetism from this kind of criteria through a localized scheme. How do you discriminate within this scheme between the first series and the 2nd and 3rd transition metal series whose magnetic properties are completely different. Moreover, the scientific community now, I think, has a whole agreement that magnetism in transition metals is a collective phenomena due to the itinerant character of the valence electrons.

Reply:

1. Hartree-Fock Theory and semiempirical (e.g., tight-binding) theories are in gross error in the treatment of transition-metal systems.

2. Goodenough is not good enough.

3. The Xα method has already been shown (in the literature) to describe magnetic properties accurately in both the limits of transition-metal complexes and extended transition-metal systems.

Comment by H. Birnbaum:

1. In estimating the effects of H on the stability of the b.c.c. Fe should not the energy of nearest <u>and</u> second nearest neighbor bonds be considered since at least second nearest neighbor bonding is necessary to stabilize the b.c.c. structure?

2. Can you estimate the relative energies of the deformed 4 atoms cluster and the deformed cluster with 2H atoms? Does the addition of H stabilize (decrease the energy of) the strained cluster?

3. Since the partial molar volume of solution of H in Fe is large (about 28% of the atomic volume of Fe) should not the effect of H on the cohesive energy of the Fe in the cluster be calculated for a cluster configuration where the Fe-Fe distances are expanded to increase the volume by the volume of solution?

Reply:

Reply:

    1. Yes.

    2. Yes.

    3. Perhaps.

Comment by R. Bullough:

    Have you relaxed the rigid ion cluster to see if the presence of the hydrogen molecule causes the cluster to relax to a morphology at all resembling that present at a crack tip.

Reply:

    Not yet, but we will in future work.

Comment by T.E. Fischer:

    Can your approach already suggest or demonstrate why hydrogen embrittles metals and oxygen does not?

Reply:

    Yes--work in progress.

Comment by R.H. Jones:

    Hydrogen embrittlement of high strength steels has been shown by several authors as requiring only a few atomic parts per million ($C_H \simeq 10^{-6}$). Since this quantity is considerably smaller than your calculations with four iron and 2 hydrogen atoms, what are the possibilities and difficulties of extrapolating your calculations to these hydrogen concentrations?

Reply:

    No special difficulties--work in progress.

Comment by P. Haasen:

    Concerning your cluster calculation of the superconducting transition temperature: What is the equivalent of the electron-phonon interaction of the BCS theory? Can you account for the isotope-effect on Tc?

Reply:

    The dynamic Jahn-Teller effect. Yes.

COHESION AND DECOHESION IN THE METALLIC BOND

D.G. Pettifor

Department of Mathematics
Imperial College
London

INTRODUCTION

It has taken nearly fifty years since the classic work of
Wigner and Seitz[1] on the cohesion of monovalent sodium for their
10% accuracy in the cohesive energy, equilibrium lattice constant
and bulk modulus to be attained amongst the polyvalent metals of
the periodic table. Ironically in the light of their much cited
quotation[2] beginning "If one had a great calculating machine....
it is not clear a great deal would be gained", it has been the
advent of fast computers which has sparked off the rapid progress
made during the past few years. In particular, the computations
have placed the one-electron picture on a sound foundation by
demonstrating the surprising accuracy of the local density functional (LDF) theory of Hohenberg, Kohn, and Sham[3] in evaluating the
energies of atoms, molecules, and solids[4]. This theory has all
the elegant simplicity of the independent particle Hartree approximation, but with exchange and correlation explicitly included
within a local approximation. The success of the LDF scheme is
illustrated in fig.1. by the results of Moruzzi, Williams, and
Janak[5] for the cohesive properties of the elemental metals across
the 3d and 4d transition series. We see that for the non-magnetic
4d series the equilibrium Wigner-Seitz radius (or lattice constant),
cohesive energy, and bulk modulus are given to better than 10%.
The large deviations in lattice constant and bulk modulus observed
amongst the 3d series is due to the presence of magnetism and is
removed by generalizing the LDF theory to include spin polarization[6].
It must be stressed that there are no arbitrary parameters in the
theory, the only input being the nuclear charge and crystal structure.

This success of the LDF theory in describing the bonding
between atoms allows the interpretation of the results within a band
framework, since the motion of a given electron is governed

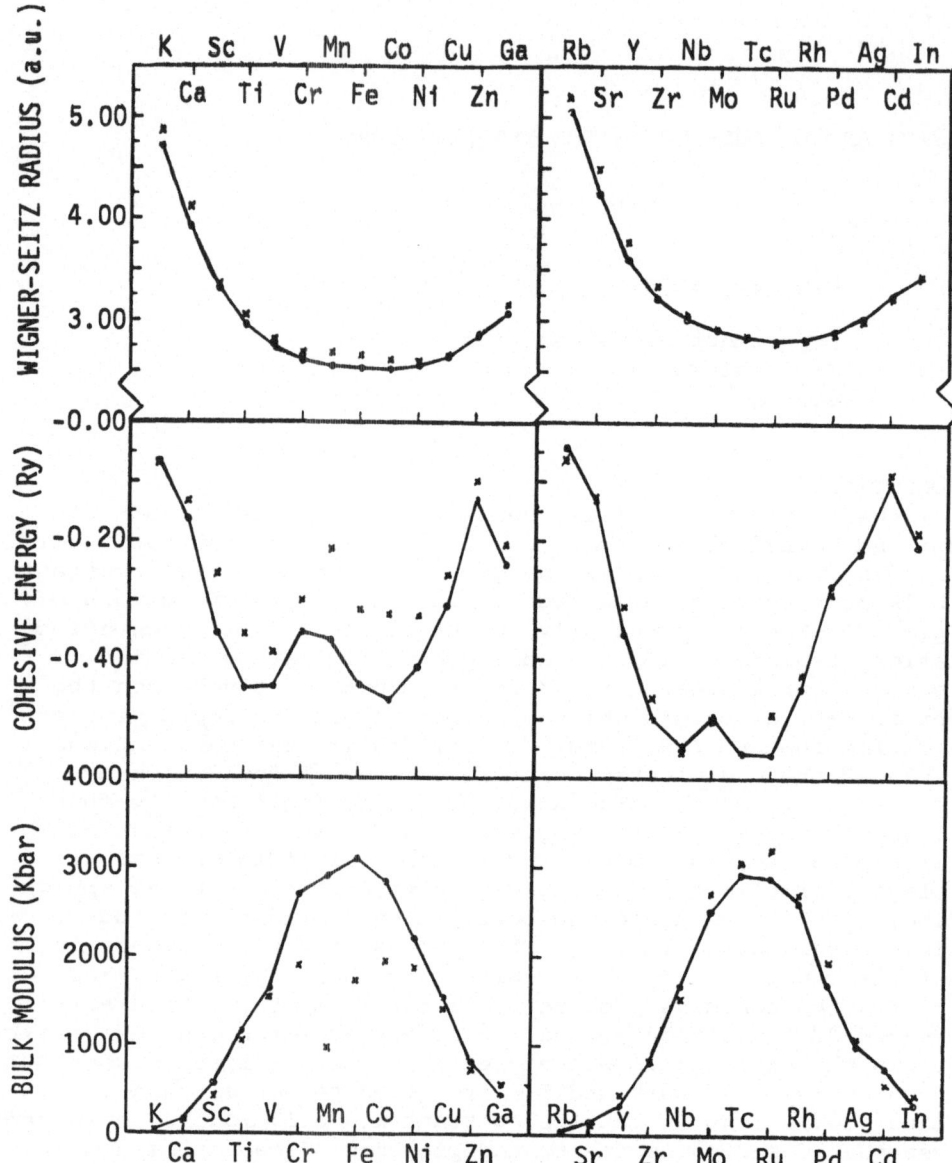

Fig. 1.   The equilibrium Wigner–Seitz radii, cohesive energies,
          and bulk moduli of the 3d and 4d transition series.
          Experimental values are indicated by crosses and the
          computed LDF values by the connected points (after
          Moruzzi, Williams, and Janak[5]).

by the one-electron Schrodinger equation[7]

$$\left[-\nabla^2 + v(\underline{r})\right] \psi(\underline{r}) = E \psi(r),$$ (1)

where the potential $v(\underline{r})$ includes the local exchange and correlation contribution in addition to the usual average coulomb field of all the other electrons (and ions). As is well known the eigenvalues E of the free atom broaden out into bands of states as the atoms are brought together to form the solid. In this paper I will discuss the nature of these energy bands in simple metals, transition metals, and their binary alloys respectively, thereby unraveling the microscopic origin[2] of the attractive and repulsive forces in the metallic bond. Unfortunately, it will be seen that the behaviour of the metallic bond during deformation or fracture is not describable by an interatomic pair potential, so that the problem still remains of finding accurate, yet simple models that avoid the extremely time consuming and expensive solution of the Schrodinger eq. (1) at each step.

SIMPLE METAL BONDING

The sp bonded metals are called simple because their bands are nearly-free-electron-like, as is illustrated schematically in fig. 2(a). Wigner and Seitz[1] showed that the cohesion of the monovalent alkali metals is determined by two competing terms:

(i) the <u>attractive</u> contribution from the formation of the bonding state Γ at the bottom of the band and (ii) the <u>repulsive</u> contribution from the increase in the kinetic energy on filling up the band with electrons to the Fermi level E_F.
The bottom of the conduction band Γ satisfies the bonding boundary condition over the surface of the Wigner-Seitz sphere of radius S, where the electron is assumed to see the potential of the ion,[7] namely

$$v(S) = -2Z/S,$$ (2)

where $Z = 1$ for the monovalent alkali metals. Γ is, therefore, well-described by the Frohlich-Bardeen expression

$$\Gamma = -3Z/S + ZR_c^2/S^3,$$ (3)

where R_c, a constant of intergration, can be identified with the ionic core radius (of an optimized model potential). Wigner and Seitz[1] found an attractive bonding contribution of 3.1eV and a repulsive kinetic energy contribution of 1.9eV, giving a cohesive energy within 10%[8] of the experimental 1.1eV/atom for Na.
Generalizing[8] these Wigner-Seitz ideas to the polyvalent simple metals by including the free electron gas exchange and correlation contributions, we can write the binding energy per atom as

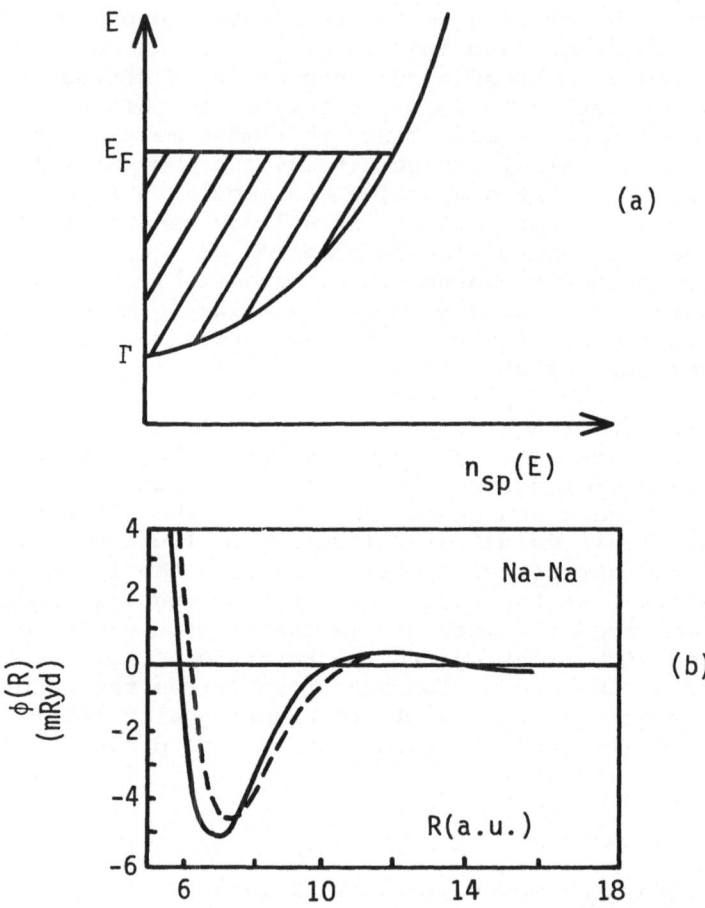

Fig. 2.   (a) The free electron density of states $n_{sp}(E)$.

(b) The Na-Na interatomic pair potential in the two
alloys Na_2K (solid curve) and NaK (dashed curve)
at their equilibrium volumes (after Hafner[12]).

$$U = Z\Gamma_{eff} + U_{ke}, \tag{4}$$

where
$$U_{ke} = 2.21 Z^{5/3} / S^2 \tag{5}$$

Γ_{eff} takes care of the usual[9] double counting contributions that arise in band theory and is given by

$$\Gamma_{eff} = - A/S + Z R_c^2 / S^3, \tag{6}$$

with
$$A = 1.8 Z + 0.916 Z^{1/3} + [0.112 - 0.0335 \ln(S/Z^{1/3})] S.$$

The three terms in A are the Ewald, exchange, and correlation contributions respectively. For Na with $Z = 1$ and $S = 4$ a.u., $A = 2.98$ which compares very well with the value of 3 in eq. (3) arising from Wigner and Seitz's[1] treatment of exchange and correlation in monovalent metals. It is quite clear from the presence of the repulsive core contribution in Γ_{eff} that the bulk properties of the sp metals are related directly to the size of the ionic core. For example, the increase in core radius R_c in going down the alkali metal group is responsible[8,10] for the increase in equilibrium lattice constant and decrease in binding energy observed experimentally.

Although the broad trends in binding energy, equilibrium lattice constant and bulk modulus can be understood[8,10] within this Wigner Seitz picture, structural problems can only be tackled by going beyond the first order expression (4) and perturbing the free electron gas to at least second order in the pseudopotential. The resulting expression for the binding energy per atom may be written in the real space representation[11] as

$$U = U_{eg} - V\kappa_{eg}^{-1} + \frac{1}{2} \phi(\underset{\sim}{R} = 0; \rho) + \frac{1}{2} \sum_{\underset{\sim}{R} \neq 0} \phi(\underset{\sim}{R}; \rho) \tag{7}$$

where U_{eg} is the kinetic and exchange - correlation energy of the free electron gas of density ρ and κ_{eg} is the free electron gas compressibility that enters through the compressibility sum rule on transforming from reciprocal space to real space[11]. $\phi(\underset{\sim}{R} = 0; \rho)$ represents the electrostatic interaction between an ion and its own screening cloud of electrons.[8] $\phi(\underset{\sim}{R}; \rho)$ is the usual central pair potential that arises within second order pseudopotential theory and is illustrated in fig. 2(b) for the case of Na. As has been pointed out by Finnis[11], the dominant contribution to the binding energy, $\frac{1}{2} \phi(\underset{\sim}{R} = 0)$, can be approximated by the very simple expression $-Z^2/R_c$, so that

$$\text{Binding energy / electron} \underset{\sim}{\sim} - Z/R_c . \tag{8}$$

This gives the observed binding energies of the sp bonded elements to within about 20% (see fig. 3) and shows at once why the binding

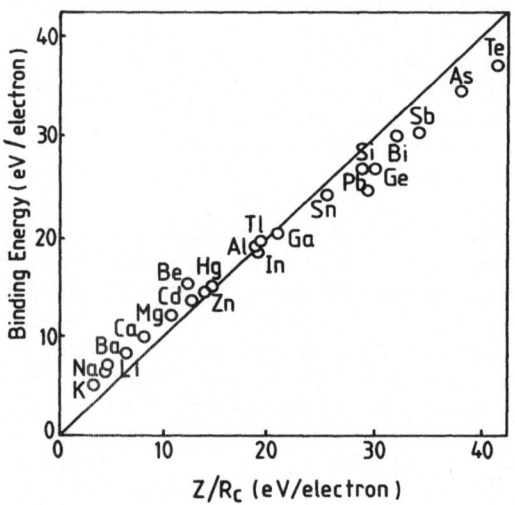

Fig. 3.   The binding energy per electron versus Z/R_C, where R_C is the optimized model potential core radius (after Finnis[11]).

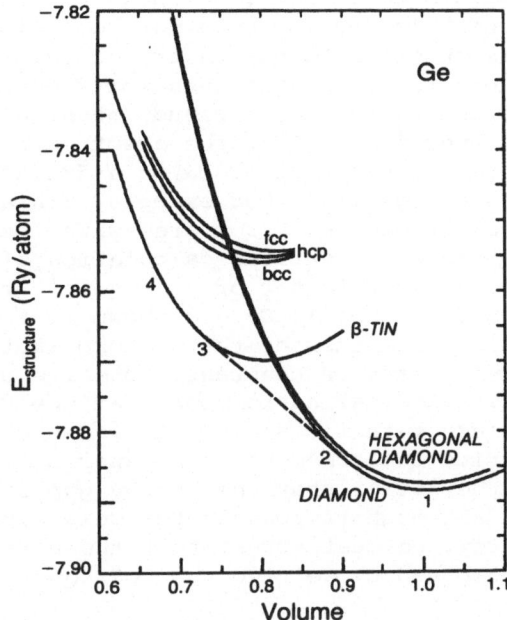

Fig. 4. The energy of the diamond, hexagonal diamond, β-tin,
hcp, bcc and fcc structures as a function of the atomic
volume (normalized to the measured equilibrium volume)
for Ge (after Yin and Cohen[15]).

energies decrease with increasing core size.

Two points must be made concerning the use of interatomic pair potentials for describing the energetics of simple metals. Firstly, it is clear from fig. 2(b) that the pair potential contribution to the binding energy is about 0.4eV/atom which is only about a third of the observed cohesive energy of sodium, namely 1.1eV/atom. Thus, there is no microscopic justification for describing the bonding in simple metals with pair potentials alone. Secondly, it is clear from fig. 2(b) that the pair potential is density dependent, which immediately raises problems [11,13] of the applicability of equilibrium lattice pair potentials to defects, surfaces, fracture, etc.

However, to end this section on a more positive note, the zoo of pseudopotentials on offer to the unwary scientist (usually with ensuing contradictory results) has been whittled down in the last few years by constraining the ionic pseudopotentials to reproduce not only the ionic spectra but also the electronic charge density[14] outside the ion core. The result has been a dramatic increase in the accuracy of the energetics. For example, Yin and Cohen [15] find a cohesive energy of Si of 4.67eV compared with the experimental value of 4.63eV, whereas the earlier calculation of Wendel and Martin[16] was too large by a factor of 2. Their[15] results for the structural energy differences in Ge are shown in fig. 4, where the equilibrium bulk properites of the diamond lattice are within 4% of the experimental lattice constant, cohesive energy, and bulk modulus. It is interesting to note that the β-tin, hcp, bcc, and fcc structures are metallic and lie at least 0.25eV above that of the diamond lattice in agreement with experiment[34]. This is consistent[15] with the notion that the energy gap stabilizes the diamond structure, although it contributes only about 5% to the total cohesive energy. We must stress that these results are parameter free being computed by solving the LDF equation (1) self-consistently.

TRANSISTION METAL BONDING

The transistion metals are characterized by a relatively narrow partially filled d band that overlaps and hybridizes with the broad nearly-free-electron-like sp band, as is illustrated schematically in fig. 5(a). C is the centre of gravity of the d band and W its width. An unambiguous[17] description of the role played by the valence sp and d electrons in transition metal bonding can be obtained[18] by working not with the total LDF energy but with the first order change in energy on changing the volume V of the Wigner Seitz sphere i.e. the pressure. We find[18] that the sp and d partial pressures can be written

$$3P_{sp}V = 3N_{sp}(\Gamma - \varepsilon_{xc}) + 2U_{sp}^{ke} \tag{9}$$

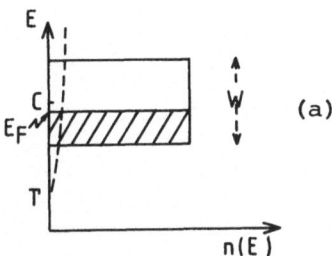

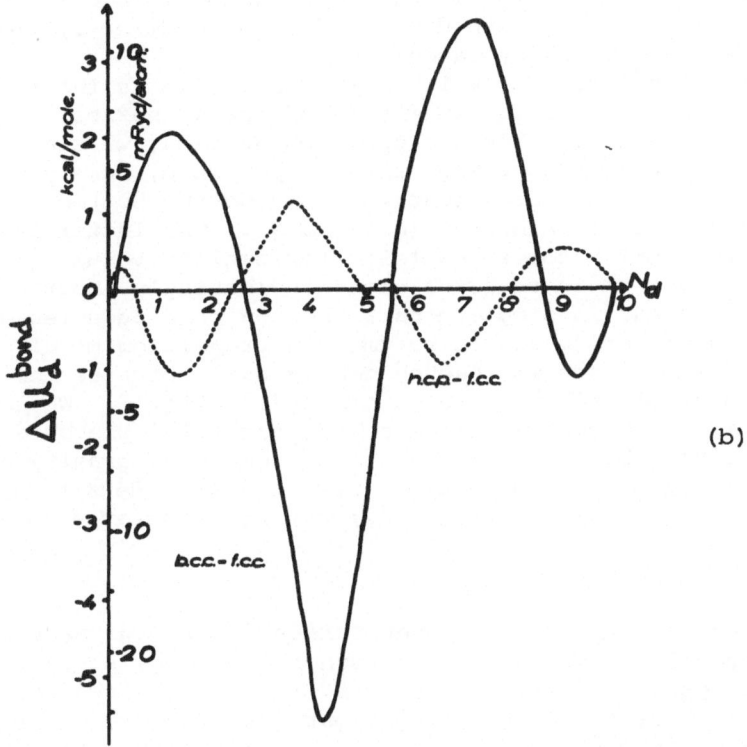

Fig. 5.   (a) Schematic illustration of the transition metal
density of states, neglecting the hybridization between
the d band (full curve) and the sp band (dashed curve).

(b) The difference in d bond energy between the bcc,
hcp, and fcc structure as a function of band filling
(after Pettifor[24]).

$$3 P_d V = 2 N_d (C - \varepsilon_{xc})/m_d + 5 U_d^{bond} \tag{10}$$

where
$$U_{sp}^{ke} = \int^E F(E - \Gamma) \, n_{sp}(E) \, dE \tag{11}$$

and
$$U_d^{bond} = \int^E F(E - C) \, n_d(E) \, dE . \tag{12}$$

N_d and N_{sp} are the number of $\ell = 2$ and $\ell \neq 2$ electrons within the Wigner Seitz spere, where ℓ is the angular momentum quantum number. ε_{xc} is the value of the exchange and correlation energy density at the Wigner Seitz radius S and in the absence of correlation equals $3/4\, v(S)$, where $v(S)$ is the value of the potential at the Wigner Seitz sphere boundary. m_d is the effective mass of the d band which is related [19] to the width through $W = 25/(m_d S^2)$. Eqq. (9) – (12), therefore, allow us to write down the pressure directly from a knowledge about the energy bands.

The results[18] for Tc (which has been chosen as the illustrative example because it lies in the middle of the 4d series) are shown in fig.6. As expected from the previous section, we see that there are two contributions to the partial sp pressure, namely the repulsive sp kinetic energy contribution and the bottom of the conduction band Γ contribution (c.f. eq.9). The latter is attractive for $S > 3$ a.u., but becomes repulsive for smaller volumes as the repulsive core term in eq.(3) becomes increasingly important (see fig. 6a). Whereas in simple metals the ion core occupies only about 10% of the equilibrium atomic volume, in transition metals with nearly half-full bands it reaches nearly 35%[20] because their equilibrium atomic volumes are so small (see fig.1). We see that there are also two contributions to the partial d pressure, namely the attractive d bond and the repulsive centre of gravity contributions respectively. If we assume a rectangular d density of states as drawn in fig. 5(a), then eq. (12) can be integrated to yield

$$U_d^{bond} = - \frac{W}{20} N_d (10 - N_d) . \tag{13}$$

This is just the contribution that Friedel[21] argued was responsible for the parabolic behaviour of the cohesive energy across the non-magnetic transition metal series (c.f. fig. 1). Since the d band width varies[22] as $V^{-5/3}$, the d bond partial pressure becomes increasingly attractive as the metal is compressed. This is countered by a repulsive contribution as the d band centre of gravity C moves up with respect to the potential $v(S)$ at the Wigner Seitz boundary (c.f. eq. 10 and fig. 6a). The net d contribution in Tc remains attractive in the region of the equilibrium atomic volume (see fig. 6b).

Thus, the picture that emerges of bulk stability in transition metals is that

(i)   the atoms are being pulled inwards under the <u>attractive</u>

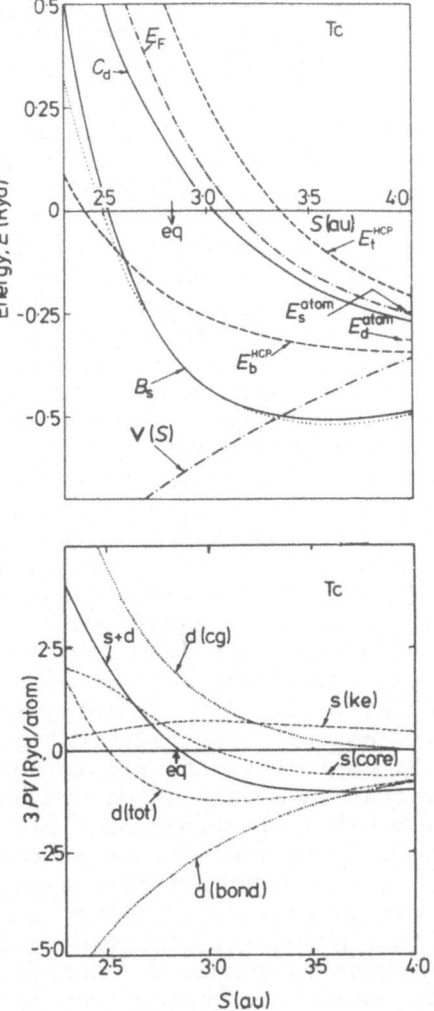

Fig. 6.   (a) The LDF energy bands of Tc as a function of the
Wigner-Seitz radius S; eq marks the equilibrium Wigner-
Seitz radius.   v(S) is the potential at the Wigner-
Seitz radius, which would be zero in the absence of
exchange-correlation.   The dotted curve gives the
Frohlich-Bardeen fit (eq. 3) to the bottom of the
conduction band at r. E_t and E_b mark the top and
bottom of the d band respectively.   sp-d hybridization
has been neglected (after Pettifor[18]).

(b) The sp and d contributions to 3PV as a function of the
Wigner-Seitz radius S for T_c.   s(core) labels the
contribution from the first term in eq.(9). (after
Pettifor[18]).

influence of the bonding d electrons, and (ii) they are
being kept apart by the <u>repulsive</u> pressure arising from the
sp electrons being squeezed unwillingly into the core regions.

Moreover, a careful analysis of the sp and d contributions to the
<u>cohesive energy</u> by Gelatt, Ehrenreich and Watson[23] shows that the
d bond contribution is by far the dominant factor, thereby
supporting Friedel's [21] argument. The renormalization in the
position of the d-band, which is so important in maintaining the
<u>equilibrium volume</u> (see fig. 6) does not give a large contribution
to the cohesive energy as can be shown directly[18] by intergrating
eq.(10). It, therefore, comes as no surprise that ten years ago
the d bond contribution was found[24] to determine which crystal
structure amongst fcc, bcc, or hcp is the most favoured as the
d band is progressively filled with electrons. Fig. 5(b) shows
that the correct trend from hcp → bcc → hcp → fcc was computed,
although predicting incorrectly[25] a return to the bcc structure
near the noble metal end of the series. Moreover, spin polarized
band theory is able to account [26] for the anomalous bcc structure
of iron and predict its transition to hcp under pressure. Finally,
although the equilibrium atomic volume is maintained by a balance
of the repulsive sp and attractive d pressures, the <u>bulk modulus</u>
arises almost entirely [18] from the sp electrons being excluded
from the core region, as can be seen from fig. 6(b).

The metallic bond in transition metals cannot be described by
interatomic pair potentials. The d bond energy is proportional to[27]
the band width W. (c.f. eq.(3)). But within tight binding theory
W is proportional to $z^{\frac{1}{2}}$, where z is the atomic co-ordination number.
This reflects the fact that in metals we do not have saturated bonds
between the neighbours but the electrons must resonate between the
z nearest neighbour atoms. The limitations of describing transistion
metal energetic's with central pair potentials has been illustrated
very recently by the work of Carlsson, Gelatt, and Ehrenreich[28],
who obtained a unique pair potential $\emptyset$(R) from their LDF curve of the
total energy U as a function of volume V. Because these potentials
have been defined to reproduce the correct <u>volume</u> dependence of the
total energy for a given crystal structure, they give good values
of the cohesive energy, equilibrium atomic volume and bulk modulus
(the errors being a measure of the accuracy of the LDF approximation;
c.f. fig. 1). However, if one looks at the dependence of the total
energy on atomic rearrangement or <u>structure</u> keeping the volume fixed,
then this potential is totally unreliable. For example, it predicts
the close packed structures of Mo to be more stable[28] than the
observed bcc phase. This is not surprising, because it is well
known (see fig. 5(b) and fig. 8 of ref. 20) that the symmetry of the
d orbitals can maximize the bonding for a half full band if the local
arrangement of atoms is bcc rather than close packed. The angular
correlation of the neighbours is ignored in the two body potential,
so that it can lead to the wrong crystal structure. A similar
conclusion has recently been reached concerning twin and stacking

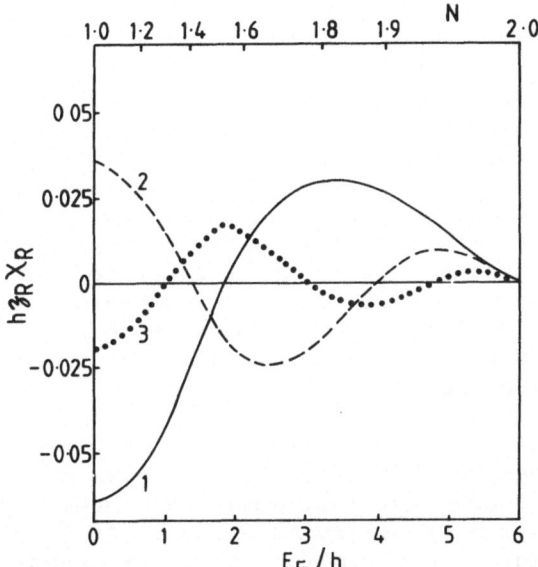

Fig. 7. The non-local susceptibility χ_R as a function of the band
filling for a simple cubic s band. h is the tight binding
hopping integral and $\mathscr{z}_R$ the co-ordination number of the
R^{th} atomic shell (after Pettifor[32]).

fault stabilities[29] in bcc transition metals.

However, if one is interested in perturbations with respect to
a given starting lattice, then second order perturbation theory may
be used which will result in 2 body interactions, as, for example,
occurs in lattice dynamics. Gautier, Ducastelle, and Giner[30] have
defined effective pair interactions for discussing ordering energies
in binary transition metal alloys by taking the disordered state as
the reference and Moriya and Hasegawa[31] have performed similar
calculations for discussing the energetics of local spin fluctuations
about the disordered paramagnetic state, for example. These pair
interactions depend on the non-local susceptibility χ_R which measures
the effect on the atom at the origin due to a local perturbation at
the site R. Formally, χ_R is written

$$\chi_R (E_F) = - \frac{1}{\pi} \mathcal{I}_m \int^{E_F} G_R^2 (E) \, dE \quad , \tag{14}$$

where $G_R (E)$ is the appropriate lattice Green function. Explicit
algebraic expressions for the first three nearest neighbour χ_R
may be derived [32] for the tight binding s band in the simple
cubic lattice (by starting with an analytic expression for the
s band density of states). They are illustrated in fig. 7 and
display the well-known [30,31] oscillatory behaviour with band filling
or distance. Such oscillatory behaviour has been beautifully observed
and documented by Powell, Martel, and Woods[33] amongst the interplanar
force constants of Nb Mo alloys.

HEATS OF FORMATION

Recently Miedema[34] and his colleagues have developed a very
simple and highly successful semi-empirical scheme for calculating
the heats of formation of binary AB alloys. It is a natural extension
of Pauling's[35] original description of electronegativity in which
the heat of formation ΔH was was proposed to be
proportional to the square of the electronegativity difference ΔX.
Miedema[34], in order to overcome the problem that Pauling's scheme
always predicts a negative ΔH, included a repulsive contribution
proportional to the square of the mismatch in the cube root of the
charge density at the Wigner Seitz sphere boundary $\Delta \rho^{1/3}$. Thus,
(in its simplest form)

$$\Delta H_{Miedema} = - P(\Delta \phi^*)^2 + Q(\Delta \rho^{1/3})^2 + R \quad , \tag{15}$$

where the co-ordinate ϕ^* was adjusted to give the correct sign of
the heat of formation of about 500 binary AB alloys. (It assumes a
value close to Pauling's electronegativity X). R is an additional
contribution required for transition metal sp metal alloys in which
p d hybridization effects are important. In this section the microscopic origin of the attractive and repulsive contributions to ΔH
will be explored.

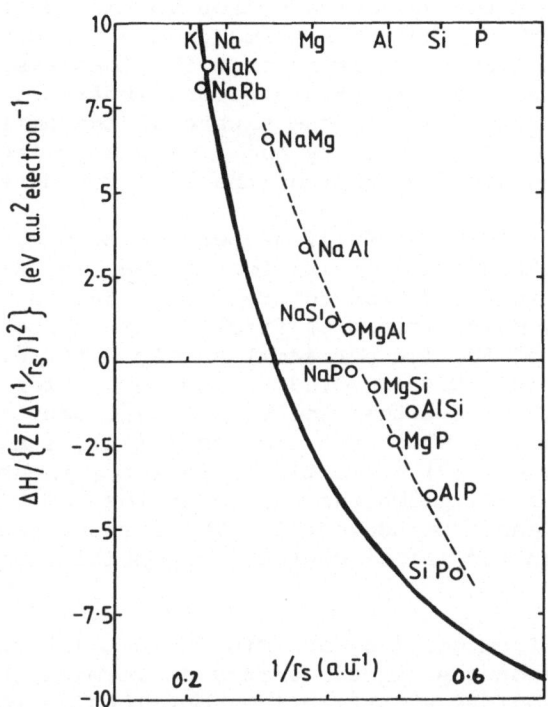

Fig. 8. The heat of formation per electron for the Na series normalized by $[\Delta(1/r_s)]^2$ as a function of $1/r_s$ (after Pettifor and Gelatt[36]).

Simple metal-simple metal

    Consider two pure metals A and B with equilibrium charge
densities ρ_A and ρ_B respectively being brought together to form
the AB alloy with charge density ρ_{AB} and volume $V_{AB}=\frac{1}{2}(V_A+V_B)$. The
second order change in energy per atom resulting from the
two electron gas terms in eq. (7) can be written[36]

$$\Delta H_{eg} = -\frac{5}{9}\bar{Z}(2.210 - 0.641\,\bar{r}_s - 0.023\bar{r}_s^2)[\Delta(1/r_s)]^2. \qquad (16)$$

r_s is the radius of the average electron volume, so that $1/r_s$ is
proportional to $\rho^{1/3}$. The three terms inside the bracket are the
kinetic, exchange, and correlation contributions respectively
(c.f. eq. 6). Thus, the electron gas contribution to the heat of
formation per electron goes as the square of the difference in the
cube root of the electron density $\Delta\rho^{1/3}$ with a prefactor that
depends on the average cube root of the electron density.

    In fig. 8 the LDF values[36] of ΔH per electron normalized by
$[\Delta(1/r_s)]^2$ are plotted against $1/\bar{r}_s$ for the Na, Mg, Al, Si, P
binary AB alloys in the CsCl structure. They fall approximately on
a single curve[37], which is not at first apparent looking at the
numerical values of ΔH, (eV per atom) namely 0.15, 0.35, 0.28, -0.12
for NaMg, NaAl, NaSi, NaP and 0.03, -0.08, -0.46 for MgAl, MgSi,
MgP respectively. We see that the volume-dependent electron gas
term ΔH_{eg} (eq. 16), in which the compressibility sum-rule
contribution from eq. (7) is critical, reflects the broad behaviour
of the LDF results, although the change in the contributions[36] from
$\phi(\underline{R})$ in eq. (7) cannot be neglected. The structure-dependant
potential contribution is, of course, responsible for the ordering
energy.

    It is clear from eq. (16) and Fig. 8 that (i) the relaxation
in the free electron gas kinetic energy on forming the alloy lowers
the energy and provides an attractive contribution to ΔH, whereas
(ii) the loss in exchange and correlation energy on forming the
alloy provides a repulsive contribution, which dominates at low
electron densities.

Transition metal-transition metal

    Consider two transition metals A and B with free atom d levels
C_A and C_B respectively being brought together to form the AB alloy
with volume $V_{AB} = \frac{1}{2}(V_A+V_B)$. By analogy with Friedel's[21] treatment
of pure transition metal cohesion, the AB alloy band may also be
approximated[38] by a rectangular density of states of width W_{AB},
where

$$\frac{1}{12}W_{AB}^2 = \frac{1}{12}W^2 + \frac{1}{4}(\Delta C)^2. \qquad (17)$$

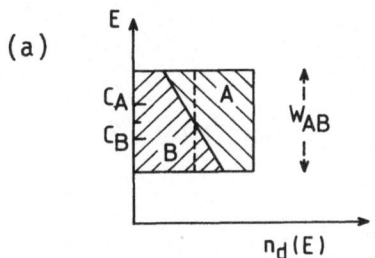

(a)

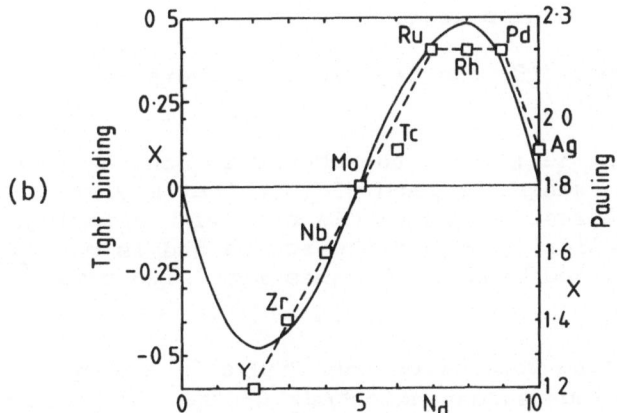

(b)

Fig. 9. (a) The AB alloy density of states $n_d(E)$. The shaded skew densities of states represent the individual A and B site densities (after Pettifor[42]).

(b) The tight binding electronegativity X_d across the 4d series compared with Pauling's values (squares).

The first term is the contribution to the square of the alloy band width that arises from nearest neighbour bonding, whereas the second term reflects the increase[39] in bonding that arises from the chemical differences in the constituent atoms measured by $\Delta C = C_B - C_A$. Filling up this alloy band in fig. 9(a) with the average number of d electrons per atom $\bar{N}_d$ and comparing the resulting band energy with that obtained from the pure metal bands in fig. 5(a), we find[39] for the 4d series with a band width W of 10eV that

$$\Delta H = \Delta H_O + \Delta H_1 \quad , \tag{18}$$

where (in eV/atom)

$$\Delta H_O = \frac{1}{4}[1 - \frac{3}{100} \bar{N}_d (10 - \bar{N}_d)] \, (\Delta N_\alpha)^2 \tag{19}$$

and

$$\Delta H_1 = \frac{5}{6} \alpha \left[\frac{(5 - \bar{N}_d)(N_O - \bar{N}_d)}{1 + \alpha (\bar{N}_d - N_O)^2} \right] (\Delta N_d)^2 \quad , \tag{20}$$

with $\alpha = 0.03o8$ and $N_O = 7.3$ being parameters fitting the observed equilibrium atomic volumes (c.f. fig 1).

We see from eq. (18) and fig. 10 that there are two contributions to ΔH:

(i)   The first term gives the change in bonding energy ignoring any changes in band width W that may arise from the changes in nearest neighbour distance with alloying. It varies parabolically across the series and is most <u>attractive</u> for a half-filled band and becomes repulsive towards the band edges.

(ii)   The second term gives the change in energy that results from the nearest neighbour distance dependence of the band width, assuming[22] $W \propto R^{-5} \propto V^{-5/3}$. It is strongly <u>repulsive</u> near the beginning of the series where there are large volume differences between neighbouring atoms (see fig 1).

The results of the simple band theory agree reasonably well with Miedema's[34] semi-empirical values of ΔH obtained from eq. (15), the main deviation occurring for $N_d = 8.5$ where the experimental value for RhPd[40] favours the more repulsive theoretical value. The more attractive values of ΔH found by Miedema[34] near $\bar{N}_d = 5$ reflects structural bonding effects[41] which stabilize the CsCl (bcc) structure in this region (c.f. fig. 56). The semi-empirical values in fig. 10 also demonstrate that the bonding saturates as ΔN_d increases for a constant average band filling $\bar{N}_d$ (For ex., consider NbTc, ZrRu, YRh corresponding to $\bar{N}_d = 5$). This is consistent[38] with eq. (17) which gives

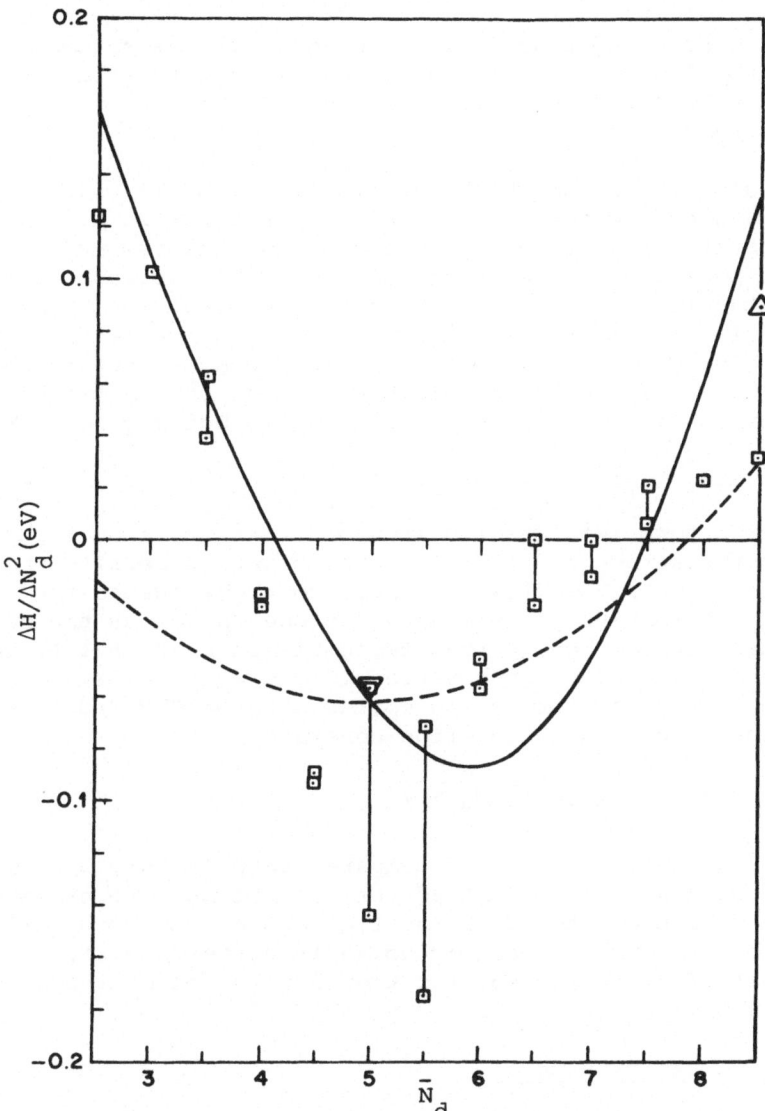

Fig. 10. The heat of formation ΔH for the 4d series divided by the square of the difference in number of valence d electrons $(\Delta N_d)^2$ as a function of the average number of valence d electrons $\bar{N}_d$. The full curve is the theoretical $\Delta H/(\Delta N_d)^2$, whereas the dashed curve is the $\Delta H_O/(\Delta N_d)^2$ contribution. Miedema's values for 4d alloys with $\Delta N \leqslant 4$ are represented by squares with the points for common $\bar{N}_d$ connected by lines (after Pettifor[38]).

$$W_{AB} = W[1+3(\Delta C/W)^2]^{\frac{1}{2}} \tag{21}$$

so that the second order approximation eq. (19) overestimates the bonding as ΔN_d (and hence ΔC) increases. This saturation is illustrated very clearly by fig. 1 of the LDF results of Williams, Gelatt, and Moruzzi[41].

Miedema's picture on which eq. (15) is based is essentially classical[41], since he regards the individual atoms as macroscopic pieces of metal. His attractive contribution was assumed to reflect the lowering in energy brought about by equilibrating the chemical potentials through a flow of electrons at the Fermi level. However, as is illustrated in fig. 9(a), all the d electrons in the occupied band are affected on alloying because they are bound quantum mechanically to their neighbours. The charge transfer Q_d is, therefore, not simply proportional to ΔE_F but may be written[42]

$$Q_d = \frac{1}{2}\Delta N_d + \frac{3}{10}\bar{N}_d(10-\bar{N}_d)(\Delta C/W). \tag{22}$$

The first term reflects the flow of electrons from right to left across the series due to increasing electron density and the second term reflects the flow from left to right due to the increasingly attractive d level as one proceeds across the series. If we assume that the charge transfer Q_d is proportional to the difference in the electronegativities ΔX and write $Q=\Delta X$, then eq. (22) may be integrated for the 4d series with $W=10eV$ and $\Delta C=-\Delta N_d eV$ to yield (apart from an integration constant)

$$X_d = -\frac{1}{2}N_d[1-\frac{1}{50}N_d(15-N_d)] . \tag{23}$$

This is plotted in fig. 9(b) and compares surprisingly well with Pauling's[35] values of X for the 4d series. It must be stressed, however, that although the d charge flow appears to be described accurately by Pauling's electro-negativity difference ΔX, the metallic heat of formation contribution ΔH_0 is not of the simple ionic form proportional to Q_d^2.

Simple metal-transition metal

The LDF results of Gelatt, Williams, and Moruzzi[43] for the heats of formation of 4d transition metals with Li row elements are shown in fig. 11 for the NaCl structure. They have interpreted these curves in terms of the self-consistent energy bands, which are illustrated schematically in fig. 12 for PdLi, PdB, and PdF respectively. Fig. 12(a) shows the bands of the elemental metals, where we see that the sp band of the non-transition metal drops with respect of the Pd d band as one moves across the Li series. Moreover, the width of the sp band reflects the variation in equilibrium atomic volume, being largest for the smallest atom B.

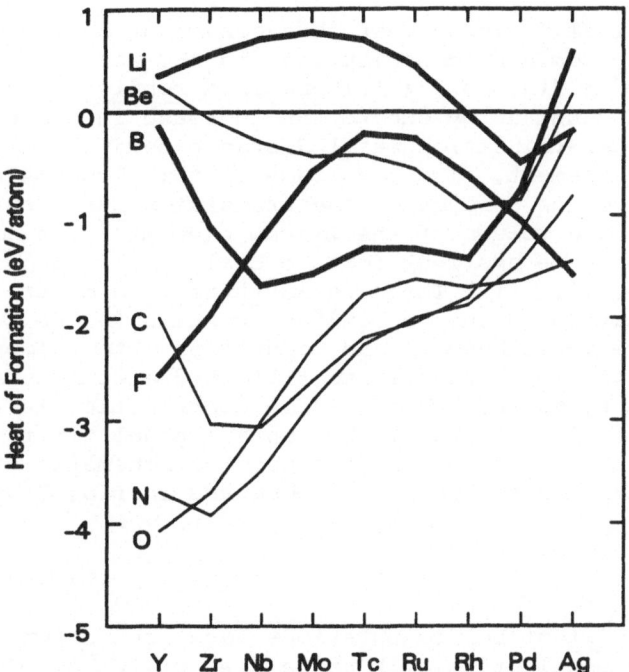

Fig. 11. The heats of formation of 4d transition metals with Li
row elements in the NaCl structure (after Gelatt,
Williams, and Moruzzi[43]).

Fig. 12(b) shows the resulting bands of the Pd alloy. For PdLi
there is little mixing between the Pd d and the Li sp bands, the
resultant alloy bands comprising a narrower d band (because the
presence of the Li atoms increases the Pd-Pd distance in the alloy)
and the usual broad nearly free electron sp band. In the case of
ρdB, the sp band is lower and there is strong pd hybridization, which
is reflected in the band of six pd bonding states separated by a set
of four d non-bonding states from the corresponding six pd antibonding states. In PdF the sp band has dropped below the d band and
there is little hybridization, but a strong ionic bond from the
transfer of one electron from the Pd d band to the F p band.

Gelatt et al.[43] argue that this behaviour of the energy bands
results in two dominant contributions to ΔH, which are illustrated
schematically in fig. 13. (i) There is a _repulsive_ contribution from
the loss of d bond energy due to the increase in nearest neighbour
transition metal-transition metal distance because of the presence
of the Li row element. The magnitude of this loss is roughly
proportional to the product of the transition metal bulk modulus by
the square of the volume of the Li row element. Its contribution
will, therefore, be smallest for the B alloys, as shown in the upper
panels of fig. 13. (ii) There is an _attractive_ contribution resulting
from the formation of the pd bonding orbitals in the transition metal
borides or the ionic bond in the transition metal fluorides. Since
the non-bonding orbitals start to fill at 8 electrons and are full
by 12 electrons, the pd hybridization contribution is constant in
this region as illustrated by the central panel for the borides in
fig. 13. We see that the sum of these two schematic contributions
in fig. 13 reflects the behaviour observed in fig. 11 for the LDF
results.

CONCLUSION

Self-consistent LDF computations have been shown to give accurate
values of the equilibrium bulk properties of the elemental metals and
their alloys. These results have been interpreted in terms of simple
band models, in particular the nearly-free-electron approximation for
the sp bonded metals and the tight binding approximation for d bonded
metals. During the next decade we will see these computations
extended to areas which do not contain the high symmetry of the
problems considered in this paper - namely to defects, surfaces, and
more complicated alloys. At the same time we can expect a development of the models outlined in this paper - in particular, the
development of effective potentials for describing the energetics of
local atomic arrangements that avoids the direct solution of the
Schrodinger equation (1) and which would, therefore, be of direct
applicability to the problems of fracture considered at this
conference.

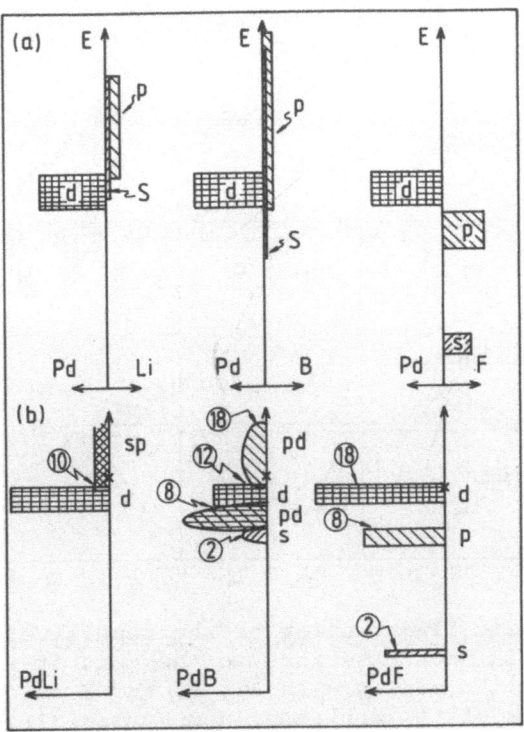

Fig. 12. (a) and (b). The densities of states of the pure A,B constituents and the AB alloy respectively (after Gelatt, Williams, and Moruzzi[43]).

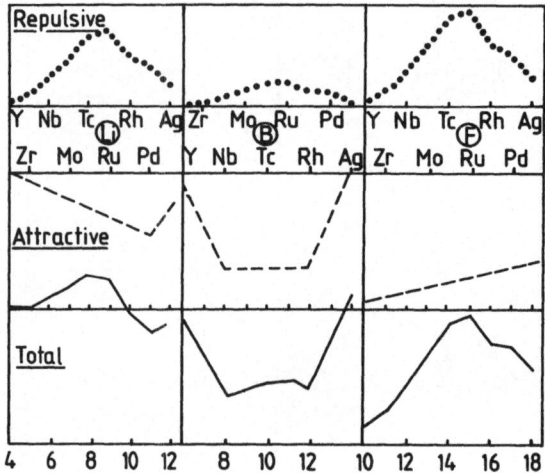

Fig. 13.   Schematic illustration of the repulsive, attractive,
            and total contributions to the heat of formation of 4d
            transition metlas with Li, B, and F respectively
            (after Gelatt, Williams, and Moruzzi[43]).

REFERENCES

1. E. P. Wigner and F. Seitz, Phys. Rev. 43, 804 (1933).
2. E. P. Wigner and F. Seitz, Sol. St. Phys. 1, 97 (1955).
3. P. Hohenberg and W. Kohn, Phys. Rev. 136, B864 (1964);
    W. Kohn and L. J. Sham, Phys. Rev. 140, A1133 (1965).
4. O. Gunnarsson and B. I. Lundguist, Phys. Rev. B13, 4274 (1976).
5. V. L. Moruzzi, A. R. Williams and J. F. Janak, Phys. Rev. B15,
    2854 (1977); V. L. Moruzzi, J. F. Janak, and A. R. Williams,
    Calculated Electronic Properties of Metals (Pergamon, New
    York, 1978).
6. J. F. Janak and A. R. Williams, Phys. Rev. B14, 4199 (1976).
7. We use atomic units throughout i.e. 1 a.u.=0.529 10^{-10}m,
    1 Ryd.=13.6eV. (0.0leV/atom=1.0 kJ/mole=0.23 kcal/mole).
8. V. Heine and D. Weaire, Sol. St. Phys. 24, 250 (1970).
9. c.f. U(Hartree)$\neq\Sigma_i E_i$.
10. N. W. Ashcroft and D. C. Langreth, Phys. Rev. 155, 682 (1967).
11. M. W. Finnis, J. Phys. F4, 1645 (1974).
12. J. Hafner, Phys. Rev. B15, 617 (1977).
13. R. Taylor, Electrons in Disordered Metals and at Metallic
    Surfaces (Ed. by P. Phariseau, B. L. Gyorffy, and
    L. Scheire; Plenum) 473 (1979).
14. see, for example, J. Harris and R. O. Jones, Phys. Rev. Lett.
    41, 191 (1978); see also ref. 13.
15. M. T. Yin and M. L. Cohen, Phys. Rev. Lett. 45, 1004 (1980);
    and to be published.
16. H. Wendel and R. M. Martin, Phys. Rev. Lett. 40, 950 (1978).
17. Provided the Wigner-Seitz or Atomic Sphere are good
    approximations.
18. D. G. Pettifor, Commun. Phys. 1, 141 (1976); J. Phys F7, 613
    (1977); J. Phys. F8, 219 (1978).
19. O. K. Andersen, Phys. Rev. B12, 3060 (1975).
20. D. G. Pettifor, Calphad 1, 305 (1977).
21. J. Friedel, The Physics of Metals (Ed. J. M. Ziman, C.U.P.)
    Ch. 8 (1969).
22. V. Heine, Phys. Rev. 153, 673 (1967).
23. C. D. Gelatt, H. Ehrenreich, and R. E. Watson, Phys. Rev. B15,
    1613 (1977).
24. D. G. Pettifor, Metallurgical Chemistry (Ed. O. Kubaschewski,
    London H.M.S.O.), 191 (1972); rare earth crystal structure
    sequence: with J. C. Duthie, Phys. Rev. Let. 38 564 (1977).
25. The LDF muffin tin KKR calculations yield the same incorrect
    structure for Pd (J. F. Janak, private communication) and
    illustrate the importance of going beyond the approximation
    of spherically averaged charge density, as, for example,
    the pseudopotential calculations of ref. 15 are now able
    to do.
26. J. Madsen, O. K. Andersen, U. K. Poulsen, and O. Jepsen, Proc.
    Conf. on Magnetism and Mag. Materials (New York, A.I.P.)
    327 (1976).

27. F. Ducastelle and F. Cyrot-Lackman, J. Phys. Chem. Solids 31,
     1295 (1970); V. Heine, Sol. St. Phys. 35, 82 (1980).
28. A. E. Carlsson, C. D. Gelatt, and H. Ehrenreich,
     Phil. Mag. 41A, 241 (1980).
29. A. M. Papon, J. P. Simon, P. Guyot, and M. C. Desjonqueres,
     Phil. Mag. B40 159 (1979).
30. F. Gautier, F. Ducastelle, and J. Giner, Phil. Mag. 31, 1373
     (1975); F. Ducastelle and F. Gautier, J. Phys. F6, 2039
     (1976).
31. T. Moriya and H. Hasegawa, J. Phys. Soc. Japan 48, 1490 (1980).
32. D. G. Pettifor (to be published).
33. B. M. Powell, P. Martel, and A. D. B. Woods, Phys. Rev. 171,
     727 (1968).
34. A. R. Miedema, P. F. de Chatel, and F. R. de Boer, Physica
     100B, 1 (1980) and references therein.
35. L. Pauling, The Nature of the Chemical Bond (Cornell Univ.
     Press, N.Y.; 1960).
36. D. G. Pettifor and C. D. Gelatt (to be published).
37. $\Delta H/(\Delta Z)^2$ plotted against $\bar{Z}$ also falls on a single curve showing
     little evidence for the importance of terms beyond second
     order (unlike the transition metal alloys; c.f. refs. 38
     and 41).
38. D. G. Pettifor, Phys. Rev. Lett. 42, 846 (1979); Sol. St.
     Commun. 28, 621 (1978); The Physics of Transition Metals
     (Inst. Phys. Conf. Ser. No.55) 383 (1981).
39. c.f. the parallel with the AB diatomic molecule where the
     bonding-anti-bonding separation h_{AB} is given by
     $$h_{AB}^2 = h^2 + (\Delta C)^2.$$
40. K. M. Myles, Trans. Met. Soc. 242, 1523 (1968).
41. A. R. Williams, C. D. Gelatt, and V. L. Moruzzi, Phys. Rev.
     Lett. 44, 429 (1980).
42. D. G. Pettifor, J. Mag. Mag. Mat. 15-18, 847 (1980).
43. C. D. Gelatt, A. R. Williams, and V. L. Moruzzi, The Physics
     of Transition Metals (Inst. Phys. Conf. Ser. No. 55)
     392 (1981).

DISCUSSION

Comment by T.E. Fischer:

These results are impressive for the perfect crystals. What
is the prognosis that they can be extended to cracks, grain boundaries, dislocations and other defects relevant to fracture?

Reply:

It will be necessary to develop further the simple models outlined for perfect crystals, as it will not be possible to solve
computationally the Schrodinger equation for large collections of
atoms relevant to fracture. I believe, however, that the advances

made during the past ten years will be continued and that by the
next Latanision meeting we will have a more realistic potential for
treating the metallic bond.

Comment by R.A. Oriani:

    In view of the length of time that probably must pass before
theoreticians can provide sufficiently comprehensive and detailed
calculations relevant to the problem of how hydrogen in solid
solution affects the metal-metal bonds in the bulk, it would be
very helpful if theoreticians would consider what experiments should
be done to illuminate that question, and what data should be obtained
that would be useful to the theoreticians in their calculations.

THEORY OF CHEMISORPTION ON TRANSITION METALS

IN RELATION WITH HETEROGENOUS CATALYSIS

F. Cyrot-Lackmann

Groupe des Transition de Phases
C.N.R.S.
B.P. 166
38042 Grenoble Cedex
France

ABSTRACT

The strength of the chemisorptive bond is shown to be closely correlated to the local electronic structure of the metal catalyst depending on roughness and size. First, one emphasis is that surface atomic sites with a low coordination number have striking modifications from the bulk in their electronic properties. Surface relaxation effects are also studied. Surface bond lengths are more shortened for rougher surfaces. Similar conclusions are obtained for transition metal clusters which also show a contraction, larger for the smaller clusters. The influence of spin orbit is also discussed and is shown to be important for the third series. Then trends and order of magnitude of the binding energies of simple gases, such as hydrogen, oxygen and sulphur are calculated and results explain the experimental facts. Finally, dissociative and molecular adsorption of diatomic molecules are discussed. Dissociative adsorption is a thermodynamic phenomena and appears only at finite temperature.

I. INTRODUCTION

While a first-principles theory of heterogeneous catalysis appears to be some years away even for simple reactions, there has been considerable advance within the past few years in the chemisorption theoretical concepts and calculations which will lay the groundwork for such a theory. Most catalysts have a complex structure showing heterogeneities and sites with different neighborhood. A key problem is the identification of active sites for adsorption. One thus has to study chemisorption as a function of the structure of the surface. Progress in the comprehension has been associated with the development of surface science involving experimental studies on well characterized surfaces and the study of microscopic parameters. Interpretation of surface electron structure associated with the adsorption and reaction of molecules on metals is more complicated than atomic adsorption but more inter-esting. It gives indeed information on the fundamental question of the competition between intramolecular and intermolecular forces.

A lot of experimental data exist now on chemisorption of simple gases on transition metals and they have been recently reviewed by many authors. [1,2] Although large scattering exists in the data, some systematic trends have been observed and there exist a great body of analytic effort on the conditions of molecular adsorption. This will help us to construct simple theory leading to an understanding of the observed trends. One can sum up the main trends in the following way.

a. Diatomic gases like CO, H_2, NO, O_2 can chemisorb associa-tively or dissociatively on a number of transition metals. Broden et al.[3] have noted the existence of a borderline in the periodic table as shown in Fig. 1 on the basis of available experimental evidence at room temperature. Molec-ular adsorption can be expected to the right of this line and dissociative adsorption is likely to exist to its left. The approximate borderline moves to the left of the series on going down from 3d to 5d metals and when passing of adsorption of NO to CO. The borderline depends, of course, strongly on the experimental conditions (temperature, pressure of the gases reactants) and increasing temperature favors the dissociation.

b. The heat of adsorption is of the order of the heat of forma-tion of the corresponding compounds. The bond energy is also of the order of the dissociation energy of the corres-ponding molecules. Table 1 gives some numbers for the hydrogen adsorption. This is a very old finding, and Langmuir in 1918 was probably the first one to express explicitly the similarities between adsorption on solids and usual chemical bonds. One can also note that the strength of the bond follows with the adsorbate the order

     N > O > H > CO > NO

c. There is a smooth variation of heats of adsorption with the number of d electrons of the metal: it decreases from the early to late transition metals. Fig. 2 shows, for example, such behavior for oxygen and hydrogen.

Sc	Ti	V	Cr	Mn	Fe D	Co	Ni M	Cu
Y	Zr	Nb	Mo	Tc	Ru M	Rh	Pd M	Ag
La	Hf	Ta	W D,M	Re	Os	Ir M	Pt M	Au

CO

Sc	Ti	V	Cr	Mn	Fe	Co	Ni D+M	Cu
Y	Zr	Nb	Mo	Tc	Ru D	Rh	Pd M	Ag
La	Hf	Ta	W	Re	Os	Ir D+M	Pt	Au

NO

Fig. 1.   Adsorption behavior at room temperature of CO and NO.
M denotes molecular adsorption and D dissociative adsorption.

Table 1.   Hydrogen Adsorption

	Ni-H	Cu-H	Ag-H	Pt-H
$(M-A)_{mol}$	60	66	53	83
$(M-A)_{chim}$ in Kcal/mole	63	56	52	57

d.   The heat of adsorption is not very sensitive to the crystallographic plane. One finds a change of a variation of 1 to 10 kcal/mole between various surface planes compared to 20 to 100 kcal/mole for the heat itself. No systematic trends exist for the more adsorbing plane, as there are a large scattering of the existing data. One can in some cases observe multiple binding states, particularly for stepped surfaces.

e.   Interaction between adsorbates has been observed. It leads to a decrease of the heat of adsorption when the coverage is increased. In some cases it permits the formation of ordered adsorbed phases and one can observe as a function of the temperature and coverage an order-disorder transition with typical phase diagram. The energies of interaction are of the order of 0.5 to 1 kcal/mole.

f.   There is a high surface mobility for small adsorbed species (O, H, N, CO...). Indeed the interaction energy and the variation of heat of adsorption between different atomic sites are of the same order of magnitude as the thermal energy at room temperature (kT $\simeq$ 0.5 kcal). One can thus question the validity of the usual assumptions of localized adsorption and independent adsorbates on which the main approaches of the kinetics of chemisorption are widely based.

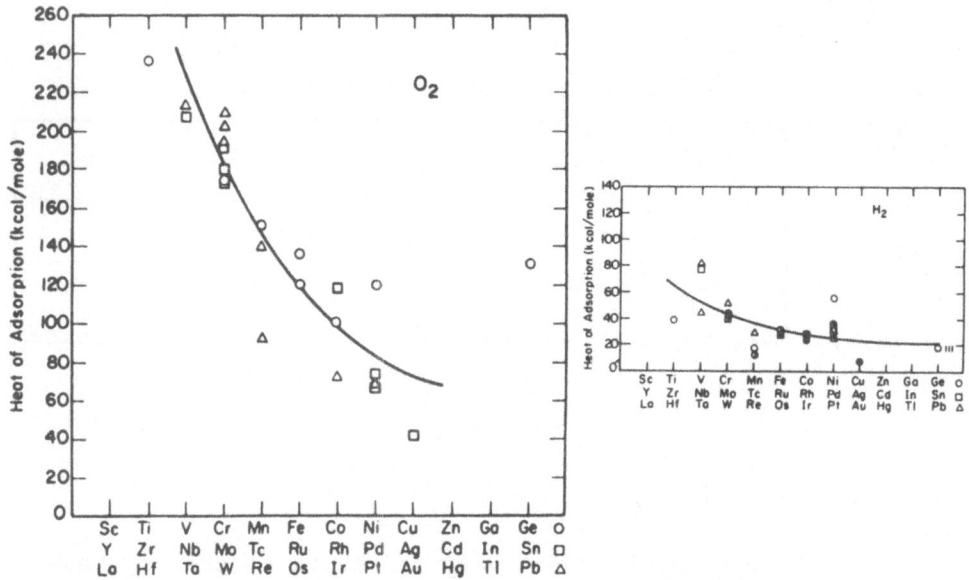

Fig. 2.   Heats of adsorption of oxygen and hydrogen on poly-
          crystalline transition metal surfaces from (2).

g. One generally observes a small change of the work function
    when adsorbing gases, leading to a negligible charge
    transfer.

h. Results of surface crystallography, although tentative
    in some cases due to the limited data base, indicate that
    the adsorption sites usually maximize the number of
    nearest metal neighbors: the adatom chooses the hollow
    site. The adsorption bond lengths fall mostly within the
    range of known bond lengths for the same pair of atoms in
    molecules or bulk solids. Contraction of metal-metal bond
    length usually found for clusters or surfaces are also
    usually canceled by absorption.

As a conclusion, the large heat of adsorption is relatively easy to understand with the concept of the chemical
bond. But the range of energy of the order of thermal energy
(i.e., 0.1 to 1 kcal), which is crucial to understanding the
different behaviors of various cluster metals for chemi-sorption
and catalysis, need a precise microscopic approach on the atomic
and electronic scale.

The qualitative description of chemisorption is well known.
In the process of chemisorption, when an adsorbate
approaches the surface substrate, there is a broadening
of its atomic levels due to the hybridization with the
substrate electronic levels. Schematically, there are two
limiting cases:

    - For a weak adsorbate-metal interaction, the adsorbate
      keeps its atomic character and its atomic level and
      is slightly broadened forming a virtual bound state.

    - For a strong adsorbate-metal interation, the system
      adsorbate plus its neighboring surface atoms of the
      substrate behaves like a surface molecule coupled
      to the substrate. The atomic level of the adsorbate
      will thus form two broadened states, a bonding and
      an antibonding one. One expects that these two
      regimes will give different energetics of chemi-
      sorption, the second case leading to larger binding
      energies. These two regimes are two extreme limits
      and many of the physical interesting cases are
      in between.

For a more quantitative approach, two ways exist in the
literature: one based on a cluster model theory, and one

on the alloy analogy of the adsorption based on concepts
more familiar respectively in quantum chemistry or solid
state physics.

The first approach pictures the metal substrate by a
small cluster of metal atoms which cannot exceed a few
tens of atoms due to numerical difficulties. This
approach emphasizes the role of the surface molecule
picture and treats the electrons in a localized scheme.
One thus immediately obtains a bond energy of the order
of magnitude of the dissociation energy of the corresponding molecule. However, detailed results are not
reliable, as such small clusters cannot simulate real
solid surfaces. For instance, the ionization potential
and the electron affinity of such clusters are different
and are also different from the work function of bulk
solids. Similarly, the band width, the cohesive energy
are different from the bulk or from large clusters. For
example, we calculate that for a cubooctahedron of 55
atoms, the cohesive energy is 80% for a cluster of 147
atoms, i.e., within 0.5 and 1 kcal. [13,14] Difficulties
are also encountered in this model to treat the interaction between adsorbates and all problems related to
coverage.

The second approach describes the metal substrate through
a band structure model theory with itinerant electrons
similar to that used in bulk solids. It uses similar
techniques to those developed for treating impurities
in metals. The two situations are very similar and many
known results of solid state physics can be translated
for chemisorption. The strength of the interaction
between the adsorbate and the substrate will therefore
not only depend on the surface molecule directly formed
when adsorbing but also on the whole band structure of
the metal due to an indirect interaction through the
rest of the substrate. In such an approach the order
of magnitude of the chemisorptive bond is not so
obvious to predict. However we will show in the following that in a tight binding model, we recover the
order of magnitude of the dissociation of molecules.
But the interest of this approach stems from the possibility to handle problems directly related to the
detail atomic description of the solid surface and
which are outside the possibilities of the cluster
model theory such as the size effect; the heterogeneity
of surfaces; possible magnetic effects; the importance
of spin orbit coupling which enables understanding
differences between Ni, Pd, Pt; the effects of coverage.

II. ELECTRONIC STRUCTURE OF TRANSITION METALS: TIGHT BINDING SCHEME

The band electronic structure of a transition metal is formed by a narrow d band with a high density of states overlapping with a broad sp band. Many of the peculiar properties of transition metals have been shown to be related to the d character of their valence states and the sd hybridization can be neglected as a first approximation. The smallness of the overlap between the d atomic orbitals on neighboring atoms justifies the use of the simple LCAO molecular orbital theory (the tight binding scheme analog to the Huckel one) to describe the d band of transition metals.[4]

In the LCAO method, the wave function Ψ of an electron is expressed in terms of linear combinations of atomic orbitals $\phi_\lambda(\underset{\sim}{r}-R_i)$

$$H\ \Psi\ =\ E\ \Psi$$

$$\Psi\ =\ \sum_{i,\lambda}A_{i\lambda}\ \phi_\lambda(\underset{\sim}{r}-\underset{\sim}{R_i})\ =\ \sum_{i,\lambda}\ A_{i\lambda}\ \phi_{\lambda i} \tag{1}$$

where i labels the site and λ the orbital degeneracy

$$H\ =\ T\ +\ \sum_i\ V(\underset{\sim}{r}-\underset{\sim}{R_i})\ =\ T\ +\ \sum_i\ V_i \tag{2}$$

$$\tag{3}$$

$$(T+V_i)\ \phi_{\lambda i}\ =\ E_{0\lambda}\ \phi_{\lambda i}$$

$E_{0\lambda}$ being the atomic orbital level.

Equations (1,2,3) applied to the Schroedinger equation lead to the secular determinant

$$\det\ |\ H_{ij}\ -\ E\ S_{ij}|\ =\ 0 \tag{4}$$

for the energy values E, where S_{ij} and H_{ij} are the overlap and hamiltonian matrices

$$S_{ij}\ =\ <\ \phi_i\ |\ \phi_j\ >$$

$$H_{ij}\ =\ <\ \phi_i\ |\ H\ |\ \phi_j\ >$$

In general orbitals on different sites are not mutually orthogonal, so that $S_{ij} \neq 0$ for $i \neq j$, although we assure the orbitals normalized, i.e., $S_{ii} = 1$. But the overlap is usually small and can be neglected as a first approximation. H_{ij} can be expressed in terms of two center integrals, respectively of the type

$$<\ \phi_i\ |\ V_i\ |\ \phi_j\ >\ =\ t_{ij} \tag{5}$$

called hopping (or resonance) integrals and leading to the width of
the band (W $\sim$ zt, z = number of coordination) and
$< \phi_i \mid V_j \mid \phi_i > = \alpha_i^\gamma$, called crystalline field (or Coulomb)
integrals leading to the shift of the band respective to the atomic
level E_0.

   Several recent works permit putting the tight binding method
onto a proper quantitative basis. These works have been guided by
the success of the semiempirical fitting or interpolation scheme.
If some information is known from experiment or accurate ab initio
computation about the energy levels of a molecule or a crystal,
e.g., energies at points of high symmetry in the Brillouin zone,
then one can fit a few parameters in (4) and hence interpolate the
remainder of the energy bands. This is the way mostly used to get
the tight binding parameters. But one can also compute them
directly from the atomic parameters of the free atoms (potential and
wave functions) obtained through ad initio calculations.[18] This
success has thus raised theoretical questions, and recent works
throw light on the justification of the tight binding scheme by
discussing the effect of the overlap between the atomic orbitals,
the suitability of atomic orbitals as a basis function and the
treatment of electron correlation. This leads to the understanding
of the physical ideas of the tight binding scheme. The H_{ij} are
interatomic quantities but are given by an atomic description in
terms of intraatomic quantities characteristic of the nature of the
free atom. The theory defines thus a standard transition metal
with a canonical shape for band structure which can be used for
scaling between different elements, crystal structures and interatomic spacings.[4,5,6]

   The theory can be extended to alloys and to compounds with s
or p bands of tight binding form. One can for example apply
parameters from the metal-oxygen bond to oxides or to chemisorption
of oxygen on metals. The parameters are thus transferable
quantities which permit transfering information in various situations. One recovers thus the idea of the chemical bond associated
with delocalized (or itinerant) electrons.

   Finally, let us point out that while in a periodic solid the
use of Bloch's theorem allows simplification of Equation (4), for
an arbitrary cluster, the size of the matrices H and S equals the
total number of orbitals. For small clusters (up to some tens of
atoms), one can diagonalize directly, but for larger clusters or
solids with rough surfaces, one cannot calculate the eigenstates.
We will use thus a moments methods associated with a continued
fraction expansion so as to calculate directly the electronic
density of states without having to know the eighenstates.[7]

III.   ELECTRONIC PROPERTIES OF TRANSITION METAL SURVACES AND CLUSTERS

The study of the change of the electronic properties of a transition metal when going from the bulk to semiinfinite crystals and to clusters of various size and geometry is of fundamental importance for the understanding of the mechanism of chemisorption. The following questions are of particular interest:

   - At what size does a cluster behave as a bulk or a semiinfinite surface?

   - What are the essential features of the electronic properties of active sites for adsorption?

   - What are the stable geometry of clusters, and how does this change with adsorption?

A useful tool to study the electronic properties of a peculiar atomic site R_i is the local density of states (L.D.S.) $n_i(E)$ on the site R_i

$$n_i(E) = \sum_{\lambda,n} [A_{i\lambda}(E_n)]^2 \delta(E-E_n)$$

where E_n are the eigenstates of the system and the $A_{i\lambda}$ are defined by Equation (1).

The L.D.S. are sensitive to the localization of the electrons near the site R_i and give information on the spatial distribution of electrons in the various orbitals λ.[4]

III.1 Semiinfinite Transition Metal Surfaces

One can resume the detailed results obtained for various cleavage planes and various crystallographic structures in the following way.[8]

   - For a dense cleavage plane (i.e., (111) for an FCC lattice, (110) for a BCC one), the L.D.S. on the surface behaves like the bulk (Fig. 3a).

   - For a nondense cleavage plane such as the (110) for a FCC lattice or the (110) for a BCC lattice, there exists a virtual bound state peak in the middle of the band. This peak is due to the resonance of the surface atoms ineracting weakly with the bulk ones.[9] There is correlatively a weakening in the shape of the L.D.S. at the top of the band (Fig. 3b).

- When one enters in the crystal, the L.D.S. tend to have
a behavior similar to the bulk one.

These features of the L.D.S. on surfaces exist for all the
transition metals and have been observed by spectroscopy. They
lead to a slight increase in the asphericity of the charge density
os the d electrons when going from a bulk to a surface site giving
indications on the possible adsorption sites.[10]

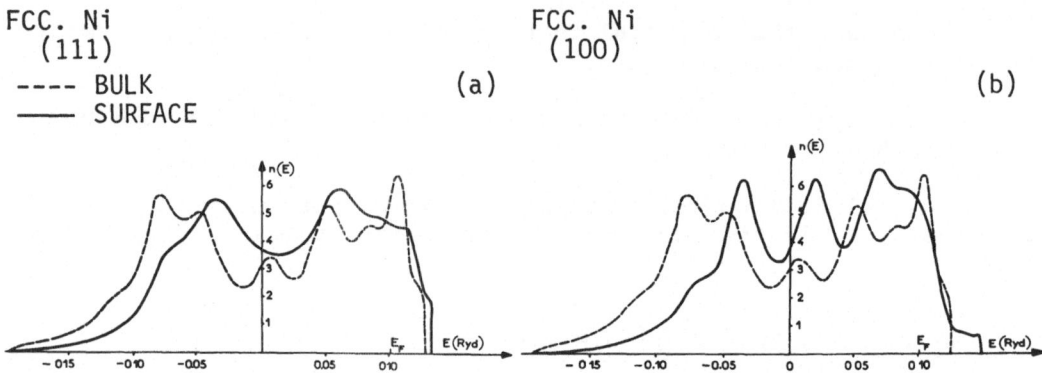

Fig. 3.   Local density of states on the surface of nickel compared
          with the bulk one (broken curve)   a) (111) plane;
          b) (100) plane.

III.2.   Stepped Surfaces

Some dramatic differences of reactivity between flat surfaces
and stepped surfaces (i.e., high index surfaces) were reported,
particularly in the case of Pt.[11] This suggests that electronic
properties may differ markedly in the various surfaces sites of an
heterogeneous surface. This is indeed what we have found in a calculation concerning L.D.S. on various sites of stepped Ni and
Pt.[12,13,14]

A sharp surface virtual bound state peak is found in the L.D.S.
at the protruding edge of the stepped surfaces of Pt and the
symmetry of states near the Fermi level is found to be rather
dependent on the geometry of the surface[12] (Fig. 4).

In the case of Ni. the conclusions are rather different and
stepped surfaces behave like the non dense low index planes.[13]
Spin orbit coupling $\xi.L.S.$ play thus an important role in the case

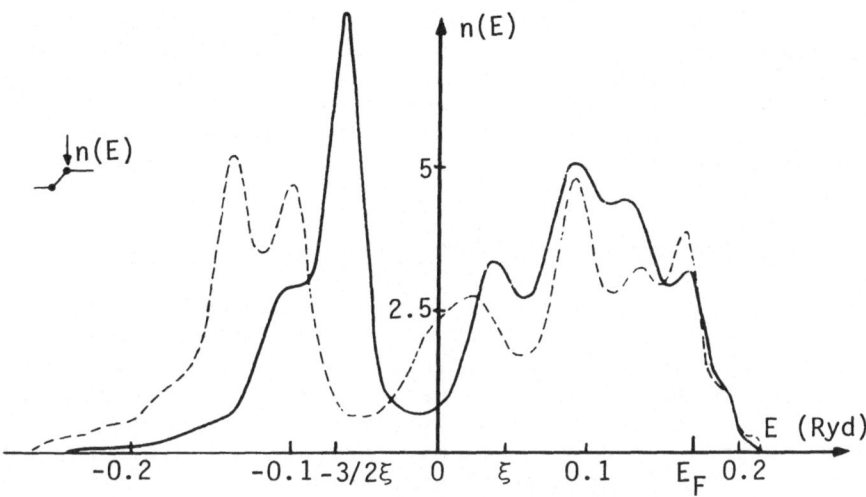

Fig. 4. The L.D.S. on the first atomic ion of stepped Pt|6(111)x (100) | compared to the bulk one (broken curve).

of Pt (where ξ = 0.42 eV) in contrast to the case of Ni (ξ = 0.07 eV) where it is neglectible with respect of the width of the d band.

III.3. Clusters

To ensure a maximum surface area effect, catalysts are clusters usually made of powders supported on an inert holder and a major problem in heterogeneous catalysis is characterization of the active sites. A way to get solution, is to find a correlation between particle size and some catalytic properties, selectivity or specific rates. It is thus of particular interest to study the electronic factors associated with change in the size and geometry of clusters.

We have studied Ni and Pt clusters of size varying between 13 atoms to 2089 atoms (i.e. dispersion rate of 0.92 to 0.30 and a diameter of 5 to 40 Å for Ni) and of different geometries of cubic and five fold symmetry, such as icosahedron and cubooctahedron[13,14,15].

Our results for the L.D.S. described in detail in refs. 13, 14, and 15 show the following trends:

- the band features of clusters approach those of bulk for large size, i.e., clusters having a few hundreds atoms

- it appears extra peaks of surface resonant states in the middle of the band for some atomic surface sites having a low coordination number. These features are similar to those existing in the case of rough surfaces, but enhanced by the

size effect. In particular, spin orbit coupling has a spectacular effect in the case of Pt clusters (Fig. 5). The L.D.S.
on the corner atoms show indeed in that case the opening of a
pseudo-gap in the band, contrary to the case of Ni clusters
where there exists a central peak. This has interesting consequences on the chemisorptive properties of these clusters, and
is currently studied.

III.4: Relaxation and Stability of Surfaces and Clusters

The geometrical location of atoms at a surface is a fundamental
quantity in the description of chemisorption and the geometry is a
clue to the understanding of bonding. Surface crystallography begin
to be widely studied, however, experimental precise results are still
very scarce.[1] For transition metal surfaces, the systematic seems to
show that contraction always occurs, contraction larger for rough
surfaces and which is at most of some percent. For clusters, the
scarce measurements of interatomic distances show also that they are
shorter than in bulk.[17]

We have calculated this surface relaxation effect in our tight
binding model. The energy is made of two terms: the electronic band
energy attractive one E_B and the repulsive energy contribution E_R,

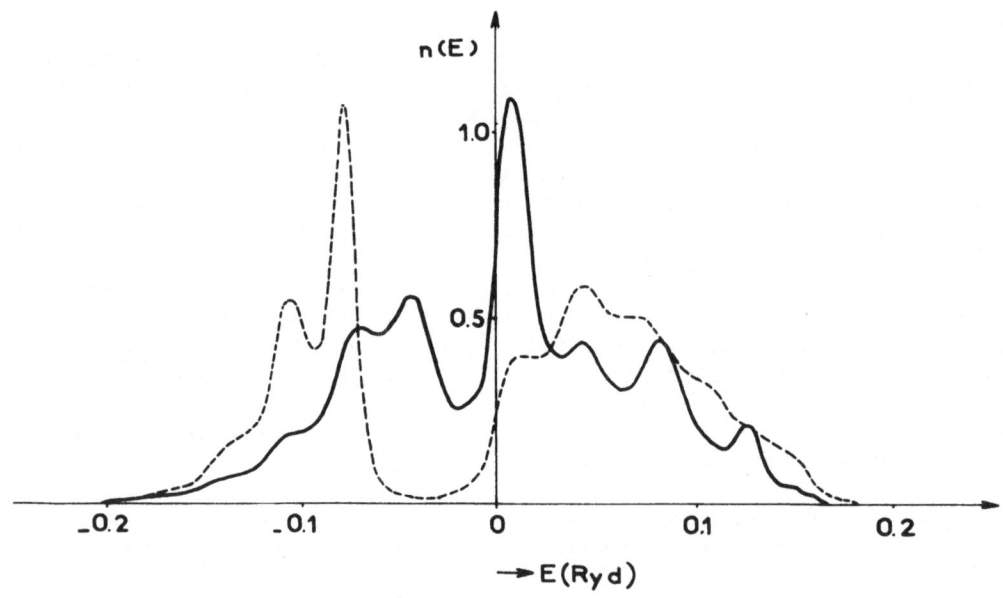

Fig. 5.   Comparison of L.D.S. on the corner atom of the cubooctahedra
          of 309 atoms in the absence (———) and presence (----) of
          spin orbit.

due to the repulsion between core shells and the compression of valence states due to orthogonalisation. This last term E_R can be represented by a two body repulsive interaction between nearest neighbors i, j

$$E_R = \sum_{ij} Ce^{-pR_{ij}}$$

the band contribution E_B writes

$$E_B = \int^{E_F} En(E) \, dE$$

where n(E) is the electronic density of states and E_F the Fermi level. The dependence[18] of the two center integrals as a function of the distance can be taken as $e^{-qR_{ij}}$. The minimization of the total energy as a function of the R_{ij} gives the interatomic distance in the bulk and permits also to calculate the elastic properties.[4,19]

We use the same method at the surface and we assume that only the first surface plane relaxes. The interplane distance becomes $R_0(1 + \delta)$, with R_0 being the bulk interatomic distance. We find that

$$\delta R_0 = -\left(\sqrt{\frac{Z}{Z-Z_s}} -1 \right) \frac{1}{2p-q\left(1+ \sqrt{\frac{Z}{Z-Z_s}}\right)}$$

where Z_s is the number of broken bonds; p, q, are two parameters which can be obtained from the bulk elastic constants, and which have nearly constant values along the series, with $\frac{p}{q} \sim 3$. Thus we always obtain a contraction which is larger the smaller R_0 is, i.e., which increases along the series as R_0 is a decreasing function of the number of d electrons. $|\delta|$ also increases with the number of broken bonds, i.e., with the roughness of the surface plane.

We have also done a similar calculation for clusters[14] and we have studied the relative stability of these clusters by minimizing their total energy as a function of the interatomic distance. Results show that there is a contraction of the bond lengths larger for smaller clusters size and larger for icosahedron than for cubooctahedron. For example, for a 15 Å Ni cluster, the contraction is of the order of 3% for fcc and 5% for icosahedron. These results are in agreement with recent experimental work on Ni and Pt catalysts by EXAFS technique.[17] The isosahedra structure is slightly favoured for small clusters size with respect to fcc, up to Ni clusters of 15 Å but difference of energies between both structures is small, i.e., $\Delta E \sim 500$ cal/mole so that they can both exist (Fig. 6).

I wish to point out that electronic configuration of the atomic surface sites will change from the bulk, and that a full selfconsistent calculation should take into account also this effect. Indeed

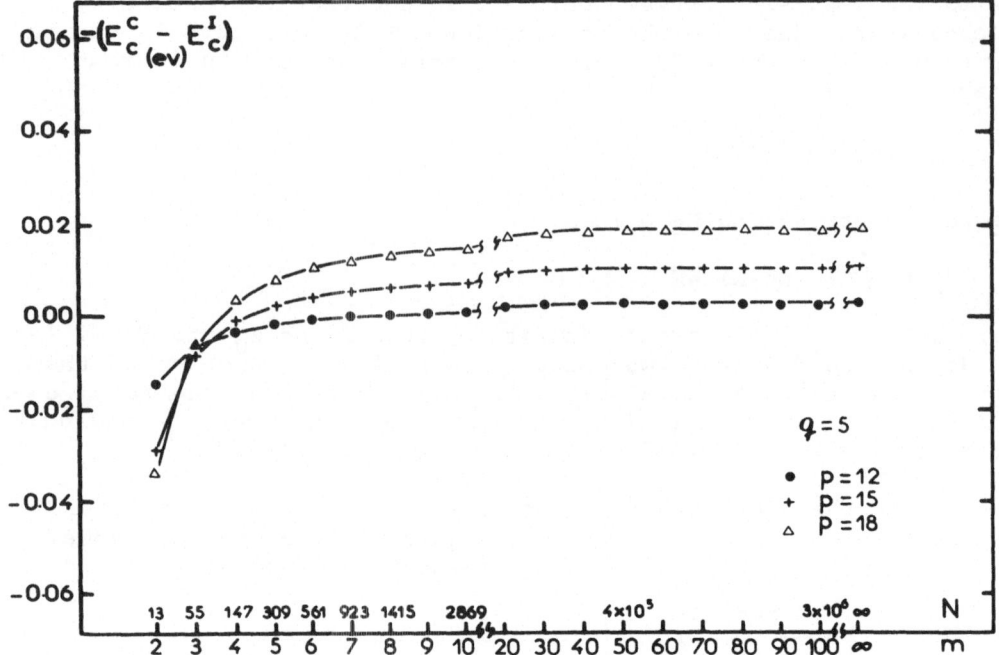

Fig. 6.   Difference of cohesive energies of cubooctahedral and
          icosahedral Ni clusters as a function of size.

one expects for sites having a low coordination number or for small
clusters, a change from the bulk electronic configuration towards the
atomic one, due to a relative displacement of the s and d bands.
Thus, for example, the LDS on corner atoms of small palladium clusters will have a full d band instead of 0.4 holes in the bulk d band.
This explains the results of measurements showing that small Pd
clusters have a diamagnetic surface contribution to the susceptibility.[20]

IV.   CHEMISORPTION

Many attempts have been made to correlate the electronic factors
of transition metals, such as their empty d orbitals, with their
power to chemisorb, including for example the application of Pauling's
valence bond theory or more recent quantum mechanical treatments of
clusters or surfaces. But these approaches have some serious drawbacks, being generally qualitative models representing very roughly
the complexity of the surfaces and of the chemisorptive bond.

However, any attempt at an understanding of adsorption must take
into account the reality of the electronic structure of the surface
of the catalyst. The important changes occurring in the electronic

properties when going from the bulk to surfaces or to clusters, described in the above section will affect the chemisorption process.

IV.1: General Description of the Chemisorptive Bond

When an adsorbate approaches a metal surface, there is an hybridization of its discrete electronic levels with the bond of the metal leading to a broadening in virtual or resonant states. To describe the adsorption of an atom A on a substrate, one can write the total himiltonian as a sum of three terms

$$H = H_a + H_m + H_c$$

where H_a describes the free adsorbate characterized by an atomic level ε_a with z_a electrons, H_m the substrate characterized by its band structure, i.e., the LDS at the surface $n_s(E)$, and H_c the coupling between both. This coupling can be described by the effective overlap B between the adsorbate and the Z_s corresponding atomic sites of the substrate and by the charge transfer occurring between the adsorbate and the substrate leading to a change in the atomic level $\varepsilon_a \rightarrow \varepsilon_a^x$ of the adsorbate.

The usual approaches are based on a one electron shceme, either a tight binding model or a Hartree Fock model hamiltonian of Anderson-Friedel type. These two approaches are very similar and differ only in the way of treating the charge transfer, i.e., of calculating ε_a^x. In the tight binding model that we are using in the following of this paper, the charge z_a brought by the adsorbate is equal to the charge variation induced on the substrate and the adsorbate when coupled. In fact, we are using a separate charge neutrality condition for the adsorbate and the substrate as experimentally the change in work function indicates a neglectible charge transfer $(0.1 \ e^-)^1$, but one could in principle take into account some ionicity of the bond. In the Hartree Fock treatment, the change in the position of the adsorbate level is due to the Coulomb repulsion U between two electrons, i.e., $\varepsilon_a^x = \varepsilon_a + U < n >$ where n is the number of electrons. But the values one can deduce for the charge transfer are quite important $(1-2 \ e^-)$ and this approach seems to suffer from serious and unrealistic drawbacks.

The LDS on the adsorbate is given by an equation of similar type for both approaches

$$n_a(E) = \frac{\Gamma(E)}{[E - \varepsilon_a^x - \Omega(E)]^2 + \Gamma^2(E)} \qquad (6)$$

with $\Omega(E) = \frac{1}{\pi} p \int \frac{\Gamma(E')}{E-E'} dE'$

$\Gamma(E) = \pi Z_s B^2 n_s(E)$ is the chemisorption function, $n_s(E)$ represents the LDS at the surface only for simple adsorbate geometries, otherwise one has to consider the orbital group density of states. One finds easily on this expression for $n_a(E)$ the two limiting cases, virtual bound state regime or surface molecule regime for respectively a weak or a strong coupling, i.e., depending on the strength of $\Gamma(E)$.

The strength of the coupling will depend obviously on the effective overlap between the adsorbate and the corresponding substrate site, on its coordination number and also on its local density of states. This last term will be particularly important when there exist resonant surface states close to the Fermi level.

IV.2: Adsorption and Magnetism

The understanding of surface magnetic properties of transition metals or their change when they participate in an adsorptive process still remains a challenging problem. However, magnetic techniques are often used to characterize the state of dispersion of supported metal catalysts, and to throw some light on the intermediate states produced in the adsorption process.[21,22] Any model description of the adsorption process must then permit to explain these magnetic properties. When adsorbing a gas such as H_2, O_2, CO, on a nickel substrate, there will be a broadening of the adsorbate electronic levels and a displacement of the relative positions of the s and d bands of the nickel. These shifts of the s and d bands will cause a reduction of the local density of states at the Fermi level on the active sites of nickel for adsorption and on its first neighbors.

These conclusions are confirmed for example by a calculation for oxygen overlayers on Ni (100).[23] Experimentally oxygen atoms are found to be in fourfold coordinated sites, forming p(2x2) or c(2x2) overlayer. Their distance to the substrate is rather well determined, and various spectroscopic investigations have also been performed. Our results show very similar trends in the change of the electronic structure of Ni (100) when adsorbing oxygen with various converages (Fig. 7). There is a shift of the d band towards the negative energies and consequently a drastic reduction of the LDS at the Fermi level. A simple reasoning using the Stoner criterion relating the existence of ferromagnetism to a high value of the density of states at the Fermi level will thus lead to a demagnetization of the active sites of nickel. The results for a very low coverage ($\theta \to 0$) indicate thus that one atom of oxygen demagnetizes 4 atoms of nickel. For a p(2x2) oxygen overlayer ($\theta = 1/4$) the surface layer of nickel is likely to become non-magnetic. This is in agreement with experiments which have revealed a decrease of the saturation magnetization of Ni when adsorbing oxygen.[24]

Another case of interest is the correlative change in the magnetic properties of nickel and nickel-copper catalysts when chemisorbing hydrogen, oxygen and in the binding energies which have been shown to depend on the structure of the catalyst.[22,24,25]. This effect

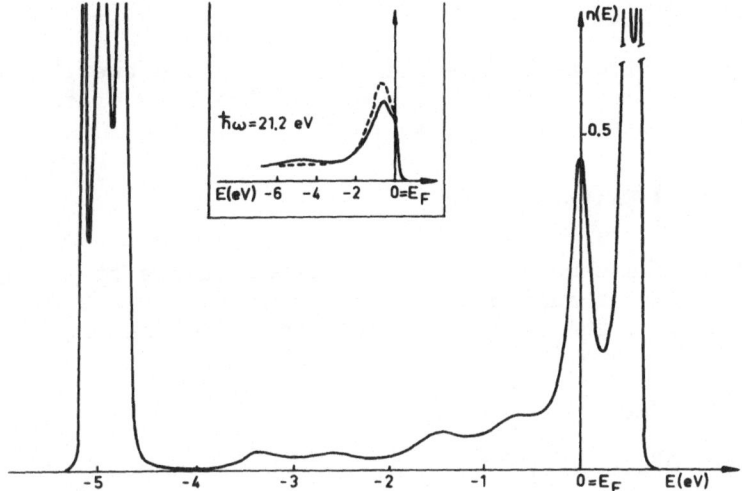

Fig. 7a. LDS on the p(2x2) oxygen overlayer on Ni(100). In the
         insert, the photoemission spectra[16] (hω = 21.2 eV);
         (---) clean surface, (——) surface with 0 overlayer.

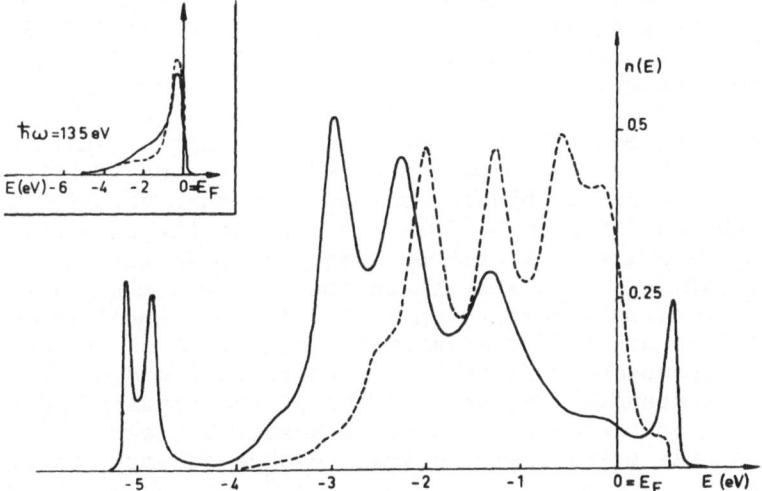

Fig. 7b. LDS(——) on the first Ni(100) plane with an adsorbed p(2x2)
         oxygen overlayer compared with the clean surface (---). In
         the insert, the photoemission spectra[16] (hω = 13.5 eV)m
         (---) clean surface, (——) surface with 0 overlayer.

can be explained as due to the difference in the electronic properties of surface nickel atoms with the roughness of surface or
alloy composition.[6]

IV.3: Binding Energy

In a tight binding scheme the adatom binding energy can be written as a sum of two terms, the attractive band width dependent part and the repulsive one E_R, so that

$$U(M-A) = \int^{E_F} E \delta n(E) dE + E_R \tag{7}$$

where $\delta n(E)$ is the change in the density of states of the system due to the adsorption of adatom A on metal M and E_F is the Fermi level. Let us remind that the binding energy can be deduced from the heat of adsorption ΔH_{ads} through

$$U(M-A) = \Delta H_{ads}$$

or $$U(M-A) = \Delta H_{ads} + \frac{D(A_2)}{2}$$

for the case of atomic and molecular adsorptions respectively, where D is the dissociation energy of the A_2 gas molecule. $\delta n(E)$ will be composed on two parts: the first one is the local density of states (LDS) on the adsorbate, given in a one electron scheme by Eq. (6); the second one is the change of the LDS on the atoms of the metal. We can reasonably assume that this change is non negligible only on the atoms of the substrate directly linked to the adsorbate and on their nearest neighbors.

The binding energy is evaluated through Eq. (7) using the one electron tight binding scheme described in Section IV. In principle, one should minimize the binding energy as a function of distance to calculate the equilibrium bond length between the adsorbate and the metal, following the same method developed in Section III.4 for surface relaxation. But results on the relaxation of clean surfaces[14] or of an adsorbate on a half-filled band[26] show that the relaxation does not affect greatly the magnitude of the energies. We will here consider thus the bond length as fixed as we are interested in understanding the main trends of the binding energy. We assume also that there is no charge transfer between the adatom and the metal. This stems from the small change of work function during adsorption.[1]

Let us focus on the band contribution to the binding energy. We have seen that the strength of the coupling depends on three factors: the adsorption site geometry, the LDS on the substrate adsorption sites and the direct coupling B between the electronic states of the adsorbate and the substrate. This last term writes as a function of the hopping integrals.

$$<\phi_A|V_A|\phi_M> = t_{AM}$$
$$<\phi_M|V_M|\phi_A> = t_{MA} \tag{8}$$

where ϕ_A and ϕ_M are respectively the atomic orbitals on the adsorbate and the metal atoms directly linked, and V_A and V_M the corresponding potentials. Let us remind here that the LDS on the substrate will be given in terms of the t_{MM} hopping integrals between two metal atoms (see section 2 and Eq. 5). These hopping integrals between orbitals on adjacent atoms are usually defined in terms of Slater-Koster formulas[31] writing the s, p, d orbitals as a linear combination of σ, π, δ functions refering to the component of angular momentum around the axis joining the atoms. Thus, for example, for a transition metal, the t_{MM} are defined by three parameters ddσ, ddπ, ddδ. For the coupling between hydrogen and a transition metal, t_{HM} is given by one parameter sdσ and for the coupling between oxygen (or sulfur) and a transition metal, t_{OM} (or t_{SM}) is given by two parameters pdσ and pdπ.

In order to describe the main trends of variation of the binding energy, we will use a simplified description of the LDS through its first few moments and a continued fraction expansion. This has been proved to be rather accurate as energy quantities involve integrals over the density of states and are therefore not sensitive to its fine structure. We introduce the exact second moment in the first step of the continued fraction and sum up then the fraction.[19] The second moment of the LDS on the adsorbate writes as

$$b_a = Z_s (sd\sigma)^2 \text{ for hydrogen}$$

(9)

$$b_a = Z_s \frac{1}{3} [(pd\sigma)^2 + 2(pd\pi)^2] \text{ for oxygen or sulfur}$$

where Z_s is the number of coordination of the adsorption site ($Z_s=1$ for top site; $Z_s=2$ for bridge site; $Z_s=3$ or 4 for centered site). The second moment of the bulk substrate LDS writes as

$$b = 12 \frac{1}{5} [(dd\sigma)^2 + 2(dd\pi)^2 + 2(dd\delta)^2]$$

(10)

The first contribution U_1 to the band part of the binding energy comes from the broadening of the adsorbate levels which is given in this model by:

$$n_a(E) = \frac{b_a}{2\pi} \frac{(4b-E^2)^{1/2}}{b_a^2 + (b-b_a)E^2} \quad \text{if } b_a < 2b$$

If $b_a > 2b$, i.e., if the coupling is strong, we also get two bound states which emerge from the band at energy $\pm b_a/(b_a-b)^{1/2}$ with a weight $(b_a-2b)/2(b_a-b)$. This simple model permits thus to recover splits off states well-known to appear in the strong coupling limits.[6,28] In the usual cases of adsorption, we are generally in the strong coupling limit (see Table II). One gets then, for

TABLE II

	$d(\overset{\circ}{A})$	$b_{a(eV)^2}$	U_{1eV}	U_{2eV}	$U_{tot_{eV}}$	$U_{expl_{eV}}$
Ni–H	1.86	12	3	0.5	3.4	2.8
Ni–O	2.08	4.37	3.6	0.7	4.2	4.3
Ni–S	2.28	4.64	4.5	0.7	5.2	4.5

with b equal to $1.54(eV)^2$ for pure Ni with an interatomic distance of 2–46 Å.

example, in the case of hydrogen adsorption

$$U_1 = -2\left[\frac{b_a(b_a-2b)}{2(b_a-b)^{3/2}} + \frac{b_a}{2\pi(b_a-b)}\left(2b_a^{1/2} - \frac{b_a-2b}{(b_a-b)^{1/2}}\text{arctg }2b_a^{1/2}\frac{(b_a-b)^{1/2}}{b_a-2b}\right)\right]$$

(11)

For oxygen or sulfur adsorption, the formula is more complicated due to the partial filling of the adsorbate p band.[32]

The second contribution U_2 to the band part of the binding energy comes from the adsorption sites of the metal. It can be simply evaluated by the change of the second moment of surface atoms. From the formula relating the change δb of second moment b of the density of states to a change δE_c in the cohesive energy,[7,14] which writes

$$\frac{\delta E_c}{E_c} = \frac{1}{2}\frac{\delta b}{b}$$

We thus get for the change of energy of the adsorption site, for example for hydrogen adsorption

$$U_2 = \frac{1}{2}\frac{3b_a}{20b}E_c'$$

(12)

where E_c' is the cohesive energy of a surface atom.

A calculation of the binding energy needs the knowledge of the second moments b_a and b given by Eqs. (9) and (10); it is thus necessary to evaluate the hopping parameters t_{AM} as a function of the adsorption bond length. Up to now, no such quantities have been yet computed and the data used were obtained by a fit to the corresponding

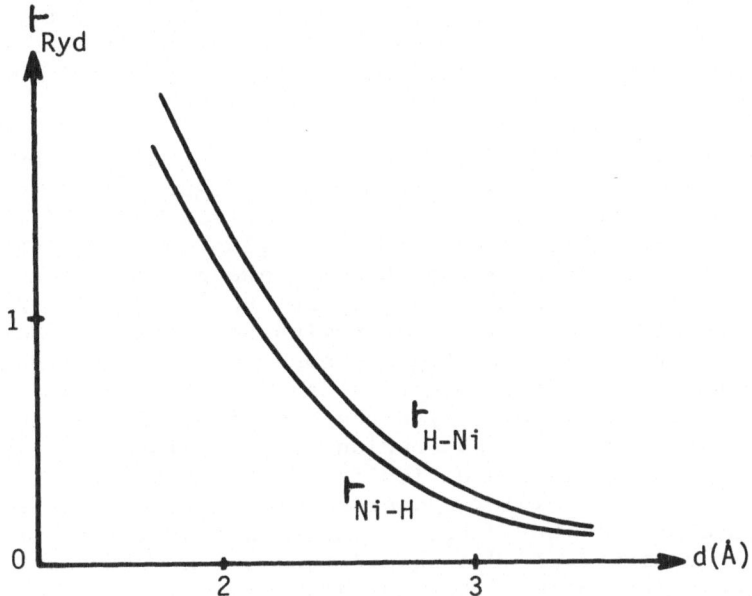

Fig. 8. Hopping integrals for a Ni-H bond as a function of Ni-H
         distance.

bulk compounds band structure calculation, when these exist through
an interpolation scheme. This was the way we proceed for the calculation of the electronic spectra of oxygen adsorbed on Ni (110).[23]
But this is not satisfying especially for energy calculations, as one
has an estimation of these parameters for only a fixed distance A-M
corresponding to the bulk compound. We have thus done a calculation[18]
of these hopping parameters from first principles using the atomic
structure calculations of Herman and Skillman done through a Hartree-
Fock-Slater approach. The detailed results are described in ref. 18.
A typical curve giving the variation of these parameters as a function of the distance between A and M is given in Fig. 8 for Ni-H.
Some results are shown in Table II for adsorption on Ni for a coordination $Z_s = 3$ of adsorption site.

    The variation of the binding energy within the transition metal
series will be given by the band contribution, as the repulsive
contribution is nearly independent of the filling of the d band.
It will only cause a small lowering of the magnitude of the binding
energy. One should, therefore, point out that the large variation
of the hopping parameters t_{AM} with the adsorbate-metal distance[18]
(Fig. 8) and the uncertainty on the bond adsorption length make difficult a precise determination of the magnitude of the binding energy.
As already mentioned above, the binding energy should be minimized as
a function of the bond length to determine the adsorption geometry,

in a way similar to that developed in Section III.4, for clean surfaces. In particular we could expect some relation between the adsorption bond length as a function of the rugosity of the surface
plane, and of the adsorption coordination number.

For metals at the end of the series, such as Ni, Pd, Pt, due
to their small cohesive energy, the order of magnitude of the binding
energy is given by the first contribution U_1 (see Table II). This
term varies slightly when going from the beginning of the series to
the end. But the variation of the binding energy along the series is
dominated by the second term U_2, mainly due to the large variation of
parabolic shape of the cohesive energy with the series.[4] Moreover,
the hopping parameters t_{AM} and t_{MN} decreases with the filling of the
series.[18] Both effects will lead to a decrease of the binding energy
along the series, in agreement with the experimental results particularly for oxygen or sulfur adsorption. The variation of the binding
energy of hydrogen within the series is smoother than for the other
simple gases. Such a behavior cannot be accounted by this simple
model and could be attributed to the importance of correlation effects
in the case of hydrogen.[27] One should then use a perturbation
approach through the ratio of the correlation energy to the band
width starting either from the band limit (small ratio) or from the
atomic limit (large ratio).[32]

The magnitude of the binding energy can be affected by resonant
surface states close to the Fermi level.[6] In that case, one can find
examples where one goes from a weak coupling regime to a strong
coupling regime when adsorbing respectively on a dense plane and a
rough plane of a transition metal of the middle of the series. The
surface resonances are thus quenched by the effect of metallic
adsorbates.[30] This will increase the anisotropy of the binding
energy. This may be an explanation of the large discrepancies one
observes in the experimental results of heats of adsorption of simple
gases on single crystal surfaces of transition metals of the middle
of the series.[2] The spin orbit coupling will also play an important
role as shown in III for metals of the third series such as Pt and
illustrated by the peculiar effects of steps. Experiments show also
that there can be many binding states of an adsorbed small molecule
on a given surface indicating that even a single crystal plane may
be heterogeneous when viewed by the adsorbed species. This effect
is enhanced in the case of adsorption on small aggregates such as
those used in catalysis. In that case, the way of preparation of
the catalyst or the metal particle size induce important changes in
the catalytic reactions which can be related to difference in
electronic properties with the surface structure.[6,24,25,29]

IV.4: Dissociative versus Molecular Adsorption

We will now show how one can understand the dissociation behaviors of the diatomic molecules such as CO, N_2, NO, H_2 on transition

metal surfaces. The adsorption has been studied by various experimental techniques and, for example, dissociative adsorption has been clearly demonstrated on a variety of transition metal surfaces. Correlations have been proposed which divide the transition metals up into groups of those metals which dissociatively adsorb the molecules and those metals which only exhibit molecular adsorption in comparable temperature and pressure conditions. Broden et al.[3] have thus noted some general trends. For a given series, the further to the left the substrate metal lies, the greater is the tendency to dissociate (Fig. 1). Secondly, the borderline moves to the right as the bond strength in the free molecules decreases. One can also remark that molecular adsorption is preferred at low temperature and high pressure.

The naive idea is to compare the energy gained in molecular and dissociative adsorption. However, we first emphasize that adsorption is a thermodynamical process and one has to consider also the entropy change in both cases for finite temperature. We point out that at zero temperature, molecular adsorption or more generally coupled adsorption has always a lower energy. Indeed when introducing any small coupling between the two adsorbate atoms, one always lowers the energy. Dissociation occurs only at finite temperature when the energy lost by the zero coupling between the adsorbed atoms is compensated by the entropy gained as the two atoms move independently. Dissociative adsorption will occur if the two adsorbed species without interaction has its state of energy level just above the ground state. On the contrary, we will speak of molecular adsorption if it is the energy state of the molecule in the gas phase which is just above. Let us illustrate this by a very crude model. A molecule A B with an intramolecular coupling V is adsorbed on two atoms 1 and 2 of the metal substrate. The coupling between the adsorbate and the metal surface atoms is B. We will discuss only the case in the limit of small band width compared to B and V that is mostly the case. We can thus reasonably assume that the two atoms represent the solid as the adsorption is usually strong (see Fig. 7 and Table II). We take one electron on each atom and the same atomic level for the few atoms. The energy of the system is $- 2(V^2+4B^2)^{1/2}$, the energy of the molecule A B in the gas phase being $- 2V$, and the energy of the dissociated adsorbed molecule being $- 4B$. The coupling B always lowers the free enrgy of the molecule A B. Two possibilities occur for the energy level distribution as shown in Fig. 9.

At finite temperature one gains entropy in both cases, and if the entropy gained is of the same order of magnitude, the case (a) will correspond to molecular adsorption and case (b) to dissociative adsorption. The dissociation patterns of diatomic molecules on transition metal surfaces can then be understood within this crude moderl. For a given molecule, we have shown that the binding energy of one of its species is greater at the beginning of the series, and thus the dissociative process occurs more likely at the left of

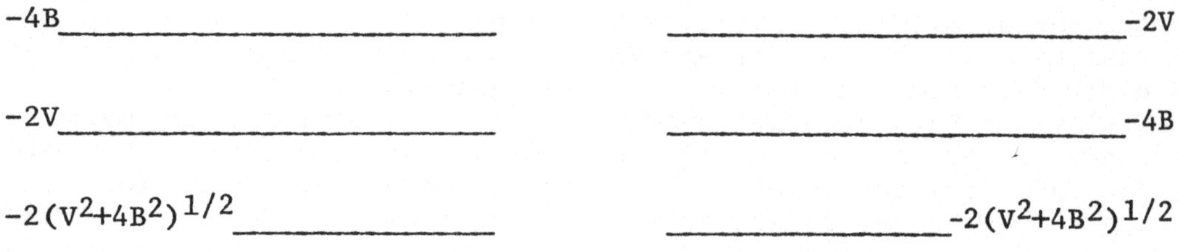

-4B _____   _____ -2V

-2V _____   _____ -4B

$-2(V^2+4B^2)^{1/2}$ _____   _____ $-2(V^2+4B^2)^{1/2}$

                    (a)   (b)

Fig. 9.   Scematic model of energy levels of a molecule A B adsorbed
          on a metal through a bond B.

the series.   Secondly, as the strength in the free molecule increases,
the molecular adsorption is more likely and the borderline moves to
the left.   Let us remind here that the bond molecule strength is
increasing from H_2 to O_2, NO, N_2 and CO (from 4.5 eV to 11 eV).

This simple model cannot help us in the case $V \sim 2B$ and one has
thus to do a detailed calculation of the entropy and compare   the
free energy to get more precise results.   To take into account a
finite coverage, one has also to perform a detailed thermodynamical
analysis.

REFERENCES

1.   Somorjai G.A., Principles of Surface Chemistry, Prentice Hall,
     1972.
     Rhodin T.H. and Ertl G., The Nature of the Surface Chemical
     Bond, North Holland, 1979.
     Smith J.R., Theory of Chemisorption, Springer Verlag, 1980.
     Oudar, J., Catal. Rev. 22, 1980, pp. 171.
2.   Toyoshima I. and Somarjai G.A., Catal. Rev. 19, 1979, pp. 105.
3.   Broden G., Rhodin T.H., Brucker C.F., Benbon R. and Hurych Z.,
     Surf. Sci. 59, 1976, pp. 593.
4.   Friedel J., The Physics of Metals, Ed. Ziman J.M., 1969;
     Heine V., Solid State Physics 35, 1980, pp. 1.
5.   Anderson P.W., Pys. Rev. Lett. 21, 1968, pp. 3;
     Fricker H.S. and Anderson P.W., J. Chem. Phys. 55, 1971, pp. 5028;
     Cyrot-Lackmann F. and Del Re G., to be published.
6.   Cyrot-Lackmann F., Studies in Surf. Sci. Catal. 4, 1980, pp. 241.
7.   Cyrot-Lackmann F., J. de Pysique C1, 1970, pp. 67;
     Gaspard J.P. and Cyrot-Lackmann F., J. Phys. C6, 1973, pp. 3077.
8.   Desjonqueres M.C. and Cyrot-Lackmann F., J. Phys. F5, 1975, pp.
     1968.
     Desjonqueres M.C., and Cyrot-Lackmann F., Solid St. Comm. 20,
     1976, pp. 855.
9.   Friedel J., J. de Physique 37, 1976, pp. 883.

10. Desjonqueres M.C. and Cyrot-Lackmann F., J. Chem. Phys., <u>64</u>, 1976, pp. 3707.
11. Somorjai G.A., Adv. in Catal. <u>36</u>, 1977, pp. 1.
12. Desjonqueres M.C. and Cyrot-Lackmann F., Solid St. Comm. <u>18</u>, 1976, pp. 1127.
13. Gordon M.B., These de 3e Cycle, Grenoble, 1978.
14. Gordon M.B., Cyrot-Lackmann F., Desjonqueres M.C., Surf. Sci. <u>80</u>, 1979, pp. 159; Surf. Sci. <u>68</u>, 1977, pp. 359.
15. Khanna S.N., Cyrot Lackmann F., Boudeville Y., Rousseau J., Surf. Sci., to be published.
16. Ngugen T.T.A., Cintri R.C. and Avignon M., Proc. 3rd Int. Conf. on Solid Surfaces, Vienna, 1977, pp. 493.
17. Renouprez A., Fouilloux P. and Moraweck B., Studies in Surf. Sci. Catal. <u>4</u>, 1980, pp. 421.
    Apai G., Hamilton J.F., Stohr J. and Thompson A., Phys. Rev. Lett. <u>43</u>, 1979, pp. 165.
18. Boudeville Y., Rousseau J., Cyrot-Lackmann F., and Khanna S.N., Solid State Comm. <u>39</u>, 1981, pp. 253 and to be published.
19. Cyrot-Lackmann F., Phys. Rev. B<u>22</u>, 1980, pp. 2744.
20. Ladas S., Dalla Betta R.A. and Boudart M., J. Catal. <u>53</u>, 1978, pp. 356.
21. Selwood P.W., Chemisorption and Magnetization, Academic Press, 1975.
22. Derouane E.G., Simoens A., Colin C., Martin G.A., Dalmon J.A. and Vedrine J.C., J. Catal. <u>52</u>, 1978, pp. 50.
    Martin G.A., Dalmon J.A. and Dutartre R., Studies in Surf. Sci. Catal. <u>4</u>, 1980, pp. 467.
23. Desjonqueres M.C. and Cyrot-Lackmann F., Surf. Sci. 80, 1979, pp. 208.
24. Dalmon J.A., Martin G.A. and Imelik B., "Thermochimie", Ed. CNRS, n° 201, 1971, pp. 593.
25. Prinsloo J.J. and Gravelle P.C., J.C.S. Faraday <u>76</u>, 1980, pp. 2221.
26. Sawyers C.M., Surf. Sci. <u>99</u>, 1980, pp. 471.
27. Friedel J., private communication.
28. Cyrot-Lackmann F., Desjonqueres M.C. and Gaspard J.P., J. Phys. C<u>7</u>, 1974, pp. 925.
29. Gault F., Garin F. and Maire G., Studies Surf. Sci. Catal. <u>4</u>, 1980, pp. 451.
30. Richter L. and Gomer R., Surf. Sci. <u>83</u>, 1979, pp. 93.
31. Slater J.C. and Koster G.F., Phys. Res. <u>94</u>, 1954, pp. 1498.
32. Cyrot-Lackmann F., to be published.

DISCUSSION

Comment by J.O'M. Bockris:

I have one comment and one question.

1. I find Ms. Cyrot-Lackmann's starting point--that adsorption
    of H makes a negligible difference to the work function--
    inadmissable and hence her basic model to be in doubt.
    Mignolet, and earlier Suhrmann and Czesch, found that the
    surface potential due to adsorption of H on metals was of
    the order of 1 volt.   Moreover, it was +eV or -eV depending
    on the metal.   A trivial electrostatic calculation shows
    that, assuming the surfaces concerned to be fully covered,
    $\mu \simeq 1.6$ Debye.   This surely implies a rather considerable
    charge transfer.

2. My question concerns the type of adsorption heat calculation
    which would approach a degree of usefulness for a real
    situation, e.g., that in catalytic hydrogenation.   Depending
    upon the rate-determining step, those hydrogen atoms will
    be relevant which adsorb on the tight-binding sites, or the
    low-bound states.   Hence, an important aspect of useful
    calculations would be that of calculating:

$$\frac{\partial \ \Delta H}{\partial \ \Theta}$$

Do Ms. Cyrot-Lackmann's students attack this point?   Again,
the difference involves about 1eV.

    Thus, these matters cast a 2eV shadow on calculations reported
in this paper.   But heats of adsorption of H on M are 2-3 eV.

Reply:

    I don't agree at all with your comment upon the magnitude of the
change in work function when adsorbing hydrogen, oxygen, or sulfur,
on a transition metal.   Indeed, as shown by, experimental results
(see, for example, the paper by K. Christmann (JCP, 70, 4168 (1979)
and ref. 1 of my paper).   For example, the change of work function
of H on close-packed surfaces is at most on the order of 0.3eV
inducing an experimental dipole moment of about 0.05D, i.e., a charge
transfer of at most 0.1 e$^-$, which thus can be negligible at first.
The hydrogen-metal bond has thus a covalent character.

    Concerning your question about the variation of the heat of
adsorption with the coverage θ, the results mentioned in my paper
have been obtained for a coverage $\rightarrow$ 0. The heat of adsorption will

also depend on the nature of the adsorbed site of the metal and its environment as shown through its coordination number, its bond-length, and its local density of states. We can also calculate the variation of the heat of adsorption with Θ in the way we did for explaining the entropy of mixing of liquid transition metal alloys of the type Fe-C and Fe-Si, and we are currently looking at that point. But as I emphasized in my talk, a precise quantitative calculation would need a complete self-consistent calculation, minimizing the binding energy as a function of bond length, and taking into account the change in electronic configuration on the adsorption sites. This is possible in principle but would need a fairly large amount of computer time.

Comment by R.A. Oriani:

Since you have said that there is very little charge transfer between an adsorbate hydrogen and the substrate, whereas one believes that a dissolved hydrogen contributes its electron to the Fermi distribution, it would seem to me that one would not expect that the manner in which adsorbed hydrogen affects the bonding between substrate atoms is similar to how the dissolved hydrogen affects the bonding between the lattice atoms of the metal. What is your comment on this thought?

Reply:

Very little change transfer does not mean that the electron of hydrogen does not mix with the valence electron states of the metal to form a chemisorptive bond, i.e., modifying its own electron distribution and that of the local electronic structure on the metal adsorption sites and their neighbors. It means only that each atom remains approximatively neutral with a covalent bond between H and metal. Therefore, I do not expect that the bonding of metal atoms will be affected in the same way for dissolved hydrogen in bulk and chemisorbed hydrogen at metal surfaces. It will depend, of course, on the location of H atoms compared to the neighbor of metal atoms. One would think a priori that in surface H can be adsorbed without causing large deformation of the lattice, which reduces the metal-metal bond strength. But to get a precise answer one should do a calculation in the way explained in this paper for the study of stability of various types of clusters, to examine how the strength of the metal-metal bond is affected by hydrogen on surface and dissolved in bulk. This could be done, but to my knowledge has not yet been achieved. It is surely something to do.

RELATIONS BETWEEN FRACTURE

AND COORDINATION CHEMISTRY

George M. Whitesides

Department of Chemistry
Massachusetts Institute of Technology
Cambridge, MA, 02139

INTRODUCTION

This paper summarizes certain mechanistic techniques used in inorganic chemistry to study processes occurring at metal centers, and suggests applications of these techniques to the study of the atomistic mechanism(s) of fracture of metals. Coordination chemistry, organometallic chemistry, and catalysis also offer a number of kinds of structural and mechanistic information pertinent to the study of fracture.

In considering the process of fracture of materials, as in considering other processes, the questions of interest are questions of rates: Under a given set of conditions, how rapidly will a material fracture? Under what conditions will this material fracture rapidly? To begin to answer these questions in atomic or molecular terms, it is necessary to be able to identify the slowest step in the fracture process: that is, the elementary step whose rate determines the overall rate of fracture. This elementary step might be any of a large number of possibilities: breaking individual metal-metal bonds as the fracture advances; rearranging bonds within the bulk metal close to the fracture zone as part of plastic deformation of the metal; breaking metal-oxygen or metal-sulfur bonds in a passivating surface layer; removal of products from the cathodic or anodic regions of a local electrochemical cell; rearrangement of groups of bonds (slipping), either individually or in a concerted manner; formation or migration of surface metal hydrides; many others. When the slow step has been identified, one can profitably consider in detail the interactions of the participating atoms in the step, and perhaps develop rational strategies for slowing (or accelerating) it.

337

In many (perhaps at present most) cases it is difficult or impossible to rigorously identify the slow step in a fracture process. In the absence of firm information, one is restricted to discussion of probable, plausible, or possible cases, and of phenomena which seem generally likely to be important or relevant to fracture a priori. Here we consider one such general topic which is relevant to fracture: viz., the qualitative coordination chemistry of the fracture surface, and methods for exploring this type of coordination chemistry. This topic is a very broad one, and this manuscript concentrates on approaches drawn from organometallic, mechanistic, and catalytic chemistry. Other approaches based on surface physics are not discussed.[1-3]

Fracture of a metal creates new surface. The free-energy change accompanying this process has contributions from many sources, of which several are the structure and composition of the bulk material which is cleaved, the energy of breaking metal-metal (or metal-hydrogen or metal-oxygen) bonds, the energy of forming bonds between the new surface atoms and any adsorbed species, and the energy of reconstructing the surface (that is, the energy of modifying the bonding of the newly-formed surface atoms with their immediate neighbors and of these immediate neighbors with more distant centers. Consideration of factors influencing the rate of formation of new surface is clearly relevant to the fracture process, but formation of new surface is not necessarily the slow step in fracture. Nonetheless, the formation of new surface, and its concurrent or subsequent modification by adsorption of species the environment is undoubtedly an important process in many fractures.

Adsorption on a metal surface is, of course, also critical to other areas of science -- heterogeneous catalysis, corrosion, adhesion, friction, wear -- and one might hope to find useful parallels between these areas. In catalysis, in particular, reactants adsorb on the surface of a metal or metal salt. After adsorption they undergo combinations of migrations, reactions with the surface, and reactions with one another which lead ultimately to products. The products must then desorb to create room on the surface for fresh reactants. Again, the adsorption processes are often not overall rate limiting, but they are important in determining the energetics of the overall transformation of reactants to products.

It is, of course, difficult to answer questions of detailed mechanism concerning any surface reaction, in part because it is so difficult to obtain detailed information concerning the structure of surfaces. Fortunately, many of the same questions concerning adsorption at metal centers, and about the making and breaking of bonds among metals and between metals and adsorbates arises in another, related field: homogeneous catalysis.[4] Here the metallic species which are involved are (at least initially) in structurally well-defined environments (howbeit in environments in which the nearest-neighbor groups are usually organic phosphines, carbon monoxides, and olefins instead of other metal atoms). The structures of these catalytic groups provide models for surface species which are at least stimulating and provocative (and which may even be relevant to fracture), and it is often possible to follow the course of catalytic reactions in very great detail at the molecular level.

RATES OF REACTIONS: TRANSITION STATE THEORY

To provide some background for discussion of these soluble organometallic analogs of surface structures and surface processes, it is useful to mention several details of the formalism which is almost universally used in chemistry in discussing rates. This theory is the so-called "transition state theory". This theory is, of course, well-known in materials science, but the aspects of the theory which are emphasized in materials science and chemistry are surprisingly different. The utility of transition state theory in chemistry is that it limits problems in rates to the consideration of only two structures: that of the starting material(s) in their ground state, and that of the species having the highest free energy between starting material and product. The rate of the process is then given by an equation of the familiar form 1 which relates this rate to the differences in energy, $\Delta G^{\ddagger}$, between these two states (figure 1). The vertical axis in this figure is the Gibbs free

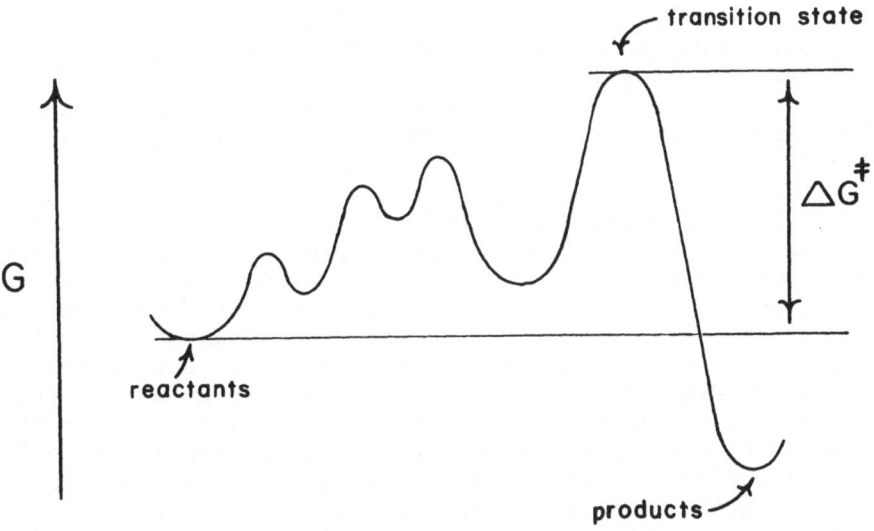

Fig.1. Schematic reaction coordinate (lowest-energy section through a potential surface) connecting reactants and products.

energy (which <u>includes</u> contributions from entropy: equation 1).

$$\Delta G^{\ddagger} = \Delta H^{\ddagger} - T\Delta S^{\ddagger} \tag{1}$$

The horizontal axis is a not-very-well-defined quantity which represents progress along the potential surface connecting reactants and products. The rate of the process being considered is given by the familiar equation 2

$$\text{rate} = \kappa \frac{kT}{h} e^{-\Delta G^{\ddagger}/RT} \tag{2}$$

(In equation 2, the constants k (Boltzman), h (Planck) and R (Gas) have their usual meaning, and $\kappa \simeq 0.5$ is an adjustment factor included for not very satisfying reasons and often ignored.) The general form of this treatment is familiar to almost everyone with technical interests, but two of its implications deserve emphasis in the particular context of fracture.

• <u>Only</u> two structures are of interest in considering the rate of a process: the ground state and the transition state. Other intermediate states which may seem scientifically interesting or which may appear to be high energy are of no importance to calculation of rates, if their energy (G) is not the in fact <u>highest</u> along the reaction coordinate.

The great virtue of transition state theory is just its simplicity: one need only know $\Delta G^{\ddagger}$ to calculate an approximate rate. <u>If</u> one knows the elemental composition and structure of the transition state, one <u>may</u> be able to calculate its energy and estimate the rate of the process. If one does <u>not</u> know what the transition state is, discussion of rates in atomic and molecular detail is meaningless.

• The Gibbs free energy, $\Delta G^{\ddagger} = \Delta H^{\ddagger} - T\Delta S^{\ddagger}$, includes contributions from both enthalpy and entropy. This simple fact is widely ignored by theoreticians interested in estimating rates. Such individuals calculate <u>energy</u> (that is, a quantity which can be roughly equated with enthalpy); they almost never consider entropies. There is, as a result, a critical weakness in their methods and conclusions. Until it can be demonstrated that entropy terms are not important, or until some method is available for calculating them, calculations of $\Delta H^{\ddagger}$ should be considered at best a poor approximation to $\Delta G^{\ddagger}$. This caution holds particularly true for any process involving water or other solvents (for example water solvating a newly created fracture surface), adsorption of components on the new surface, or release of atoms from the surface into solution: in all of these cases entropy terms may be large.[5]

Consider the application of transition state theory to consideration of the rate of propagation of a fracture along a crystal plane,
with the fracture tip in contact with a medium which contains a
species L which adsorbs on the metal surface (Figure 2). To repeat
a point for emphasis: the utility of the transition state theory
is in focusing attention on the highest energy state (i.e. the transition state). What is the elemental composition of this state? What
is the character of the bonds connecting the atoms in the transition
state? What is its energy, relative to reactants? In the case at
hand, one can imagine two limiting cases. In one case, the metal-
metal bonds break completely before the species L adsorbs at the
resulting, newly formed surface metal atoms (Figure 2A); in the
second, L adsorbs as the metal-metal bonds are breaking, and lowers

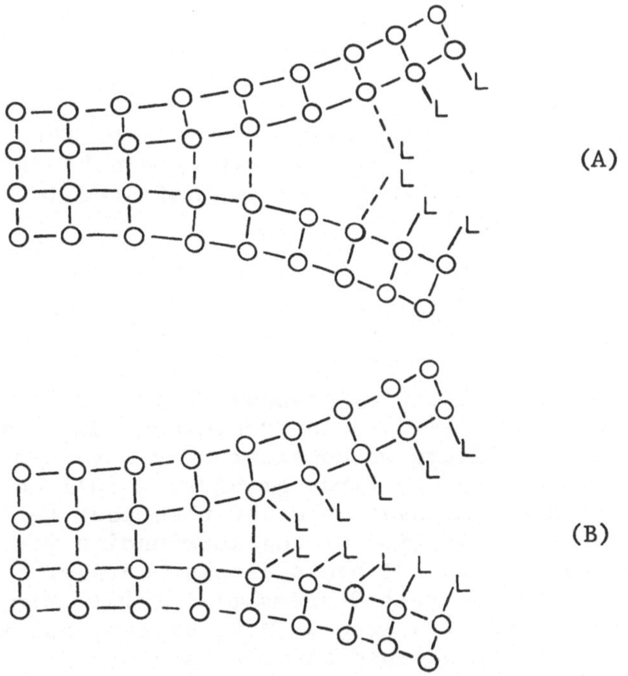

(A)

(B)

Fig.2. Schematic representation of a fracture tip propagating
        by a process in which a component L from the medium
        does not (A) and does (B) participate actively in
        breaking metal-metal bonds.

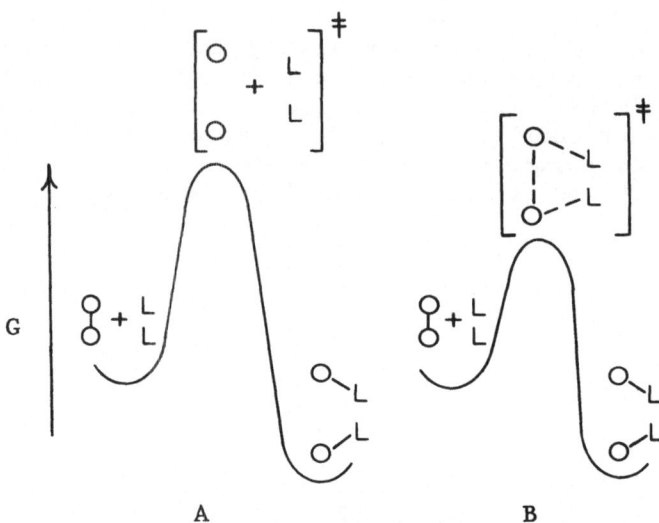

Fig.3.   Simplified reaction coordinates for the
         cleavage of metal-metal single bonds in
         the fracture processes indicated in
         Figure 2.

the energy of this bond breaking process (Figure 2B). These two
alternatives would be represented in free energy diagrams as shown
in Figure 3.   In this figure we consider only one pair of metal
atoms for simplicity.   In the first process, a full metal-metal bond
is broken without any compensation other than that reflecting some
small structural reorganization of the surrounding metal atoms.   In
the second, as the metal-metal bonds breaks, metal-L bonds form.
The sequence of events in path B seems preferable from the vantage
of the energetics of the four-center M_2L_2 system, but is probably
emtropically less favorable than path A, and may also have a
significantly unfavorable energetic contribution due to non-bonding
interactions between the L group.   One does not know at present which
is actually important in a real instance of fracture, but the inference that surface-active species (in this context, hydrogen, liquid
metals, perhaps species such as chloride ion) accelerate some
fracture processes suggests that path B may be important in at
least some instances.

The type of analytical exercise represented by this example is not idle: it indicates the <u>minimum</u> information required to rationalize the rate of a process. One <u>must</u> know the elementary composition of the transition state, and some elements of the structure of this state. Without this information, theoretical calculations of energetics are argueably irrelevant. The burden of taking the first steps in understanding the mechanism of fracture thus presently rest with the experimentalist, since, at present, only experimental work can provide reliable information on composition and structure.

It is easy to emphasize the importance of this type of detailed information in rationalizing rates, but much more difficult to obtain it in practice. In no instance involving fracture is the information available. In a few cases in heterogeneous catalysis it is now being developed, and a significant body of information is available for homogeneous reactions involving metals. In the following sections, I touch on examples of homogeneous and heterogeneous reactions involving metals, and sketch some of the chemical concepts and techniques which have proved useful in studying these reactions. The major focus of this discussion will be that of coordination of ligands to metals. This subject is relevant to the influence of species coordinating to a newly created fracture surface on the energy of the steps creating that surface.

MECHANISMS OF METAL-CATALYZED REACTIONS: THE IMPORTANCE OF VACANT
COORDINATION SITES

Consider, by way of background, two important metal-catalyzed reactions: hydrogenation of ethylene using platinum metal as catalyst (Figure 4), and hydrogenation of ethylene using a soluble rhodium(I) complex as catalyst (Figure 5). Very little is known with great certainty concerning the heterogeneous catalytic reaction. The current interpretation of the available experimental data is that the catalyst adsorbs ethylene strongly at vacant coordination sites (to yield one or several surface complex(es) whose structure(s), especially under the conditions encountered during catalysis, is (are) not well known).[6] Dihydrogen subsequently adsorbs on remaining vacant coordination sites with dissociation into surface metal hydrides. The adsorbed ethylene and this surface hydride react and form surface ethyl groups. These surface ethyls react subsequently with more surface hydride and generate ethane. When the ethane desorbs from the surface, vacant coordination sites are regenerated which reenter the catalytic cycle.

The mechanism for hydrogenation of ethylene by the soluble rhodium complex is conceptually similar.[7] Ethylene coordinates at a vacant coordination site on the metal, dihydrogen adsorbs dissociatively, the coordinated ethylene and a metal hydride react and form an intermediate ethylrhodium intermediate, and ethane is produced by reaction of this species with a second equivalent of metal hydride.

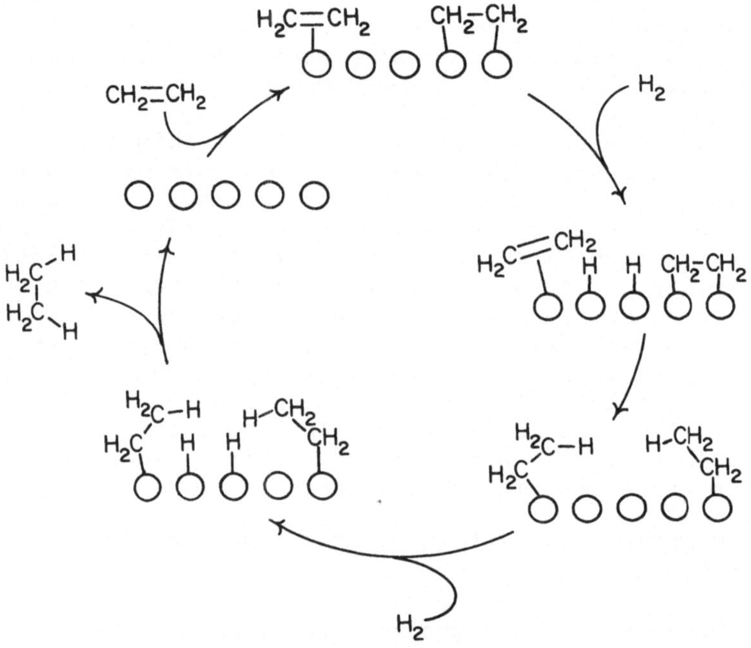

Fig.4. Schematic mechanisms for hydrogenation of ethylene over platinum.

    In both of these mechanisms there remain substantial ambiguities: Does dihydrogen or ethylene adsorb first? Which adsorbs more strongly? What are the detailed structures of the reaction inter- mediates? What is the overall rate-limiting step? Some of these questions can be answered tentatively (for example, formation of ethane by reaction of ethylmetal and metal hydride is probably rate- limiting in both reactions under at least some circumstances); others cannot. Nonetheless, from the vantage of a discussion of fracture, these details are irrelevant. The important aspects of these reactions is their dependence on the availability of a vacant coordination site on the metal: if no vacant coordination site is available, no reaction occurs. Thus, adsorption of the reactants at vacant metal coordination sites is a critical element of coordination catalysts, and an understanding of the factors influencing this

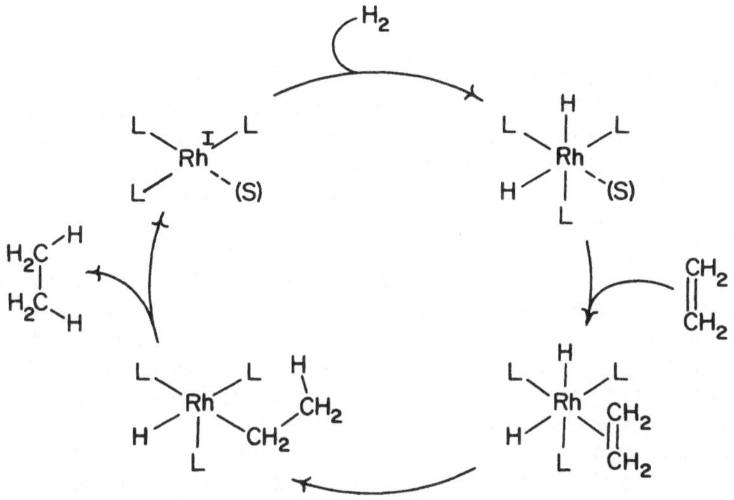

Fig.5. Schematic mechanism for hydrogenation of ethylene by dihydrogen catalyzed by a soluble rhodium(I) complex: L = $(C_6H_5)_3P$, S = solvent typically CH_2Cl_2 or CH_3COCH_3.

adsorption is essential to understanding catalysis. Similar adsorption phenomina are certainly involved in adhesion, lubrication, and corrosion, and in adsorption at freshly created fracture surfaces.

COORDINATION TO METALS: USEFUL QUALITATIVE CONCEPTS

Coordination chemistry is an area of great sophistication, but much of the information in it which is immediately useful to considerations of the (atomistically) less sophisticated area of fracture can be summarized in a limited number of simple empirical models for the interaction of metals and metal ions with ligands, and for the interaction of ligands with one another. I outline two of these here: the theory of "Soft and Hard Acids and Bases", developed by Pearson (and related to an earlier classification by Chatt and Ahrland) and Tolman's classification of the sizes of ligands by "cone angles". In addition, I mention the Hammet equation as an example of a Linear Free Energy Relation (LFER), an intellectual construct which has proved invaluable in mechanistic chemistry. Finally, I touch

briefly on the subject of entropies. Other useful models used to
rationalize the reactivities of metal ions in solution -- especially
classical ligand field theory and its mathematical developments --
are not discussed, since these have not proved particularly useful
in discussing organometallic chemistry or catalysis, and do seem
likely to be useful in fracture at this stage in its development.

Hard and Soft Acids and Bases (HSAB).[8] The most commonly used
model used in classifying the relative strengths of coordination of
metal ions and ligands is based on the idea that there are two major
types of bonding: ionic and covalent. Ionic bonding (for example,
in Na^+F^-) is familiar. Covalent bonding is most commonly discussed
using a model developed by Dewar and Chatt (illustrated for a complex
of silver(II) and ethylene in Figure 6).[9] In this model the covalent
bond between silver and ethylene is considered to be composed of two
components: one component (the σ-bond component) reflecting overlap
between the filled π orbital of ethylene and a vacant s orbital on
silver, and another component (the π-bond component) reflecting
overlap between a filled silver d orbital and the vacant π* orbital
on ethylene. The σ-component is believed to be the more important in
determining the energy of most metal-ligand bonds, and the π-component
serves the primary function of preventing a large separation of
charge by transfer from olefin to metal by allowing back-donation of
charge from metal to olefin.

The SHAB classification is based on the idea that ions which
are small, highly charged, and non-polarizable ("hard" ions) can
interact with one another strongly ionically, but will not interact
strongly with large, polarizable species with low charges ("soft"
ions or molecules). The latter, in turn can interact with one

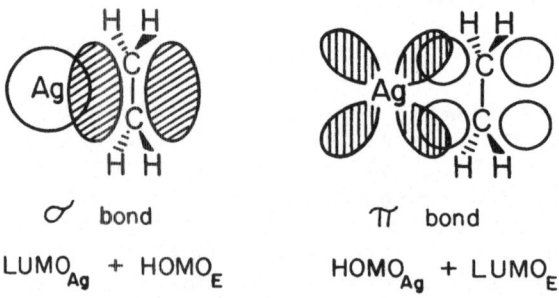

σ bond

$LUMO_{Ag}$ + $HOMO_E$

π bond

$HOMO_{Ag}$ + $LUMO_E$

Fig.6.   Dewar-Chatt model for the bonding of silver
          ion to ethylene. LUMO = lowest unoccupied
          molecular orbital; HOMO = highest occupied
          molecular orbital.

Table I. Examples of Hard and Soft Species

Hard	Soft
H_2O	H_2S, H_2Se
H_3N	H_3P, H_3As
Li^+, Na^+, K^+	M^0, bulk metals
Mg^{+2}, Ca^{+2}	M^0L
Al^{+3}	
Cu^{+2}	Cu^{+1}
	CO, $CH_3N\equiv C$
	$CH_2=CH_2$
	$HC\equiv CH$
F^-, OH^-	Br^-, I^-
$SO_4{}^{2-}$, $ClO_4{}^-$	$CH_3CS_2{}^-$, CN^-
H^+	H^-

another covalently. Table I gives examples of species classified as "hard" and "soft". Empirically, hard ions coordinate one another more strongly than soft ions, and _vice versa_: like prefers like.

The usefulness of this classification for discussions of fracture, freshly created fracture surfaces of metals) are "soft", zero-valent metals (that is, in the context of a discussion of fracture freshly created fracture surfaces of metals) are "soft", and would thus be expected to coordinate especially strongly with soft species (liquid sodium and mercury metals, sulfur- and phosphorous-containing species, hydrogen, olefins, carbon monoxide). After these metal surfaces have reacted oxidatively (with O_2 or H_2O), they are usually covered with "hard" species (high-valent ions, metal oxides and hydroxides), and would be expected to coordinate more strongly with hard ligands.

It is difficult to argue the empirical usefulness of the SHAB classification, and it in no way detracts from its usefulness to point out that this classification may be incorrect in its fundamental assumptions. It is a theory which entirely ignores the interaction of the coordinating species with solvent, and there is a growing suspicion that solvation may play a dominant role in

determining the strengths these bonding interations. In any event,
in certain reactions carried out in parallel in polar solution and
in the vapor phase, suspicious charges in selectivity are observed.
Equation 3 gives an example drawn from organic chemistry;[10]

$$NH_2^- + CH_3O\overset{\overset{\displaystyle O}{\|}}{P}(OCH_3)_2 \quad \begin{cases} \xrightarrow{\text{solution}} CH_3O^- + H_2N\overset{\overset{\displaystyle O}{\|}}{P}(OCH_3)_2 \\ \\ \xrightarrow{\text{vapor}} H_2NCH_3 + {}^-O\overset{\overset{\displaystyle O}{\|}}{P}(OCH_3)_2 \end{cases} \qquad (3)$$

The upper reaction is considered to be typical of a "hard" interaction between the charged nucleophile (NH_2^-) and phosphorous; the
bottom is a "soft" interaction. The fact that the characteristic
reactivity changes with the character of the medium suggests an
important role for the medium in determining the interaction type.

The relevance of concerns about the origin of the SHAB classification to considerations of fracture is twofold: First, since the
process of fracture may not expose the new metal surface to a
polar solvent, classification of ligand-binding affinities taken
from polar solvents should be used with some caution. Second, the
suggestion that a major contribution to these binding affinities
may come from interaction of the component with solvent rather than
with one another provides another general caution concerning the
indiscriminate use of the products of current efforts in theory,
since these are based entirely on calculations which ignore medium
effects

Hammett Equation: Linear Free Energy Relations. The study of
"so-called" linear free energy relations (LFER) has proved one of
the most useful and productive areas of modern mechanistic chemistry.[11]
A LFER postulates a relation between changes in free energies
(either $\Delta G°$ values, for equilibria, or ΔG values, for rates for
two reactions -- one a simpler reaction considered to be well-
understood (typically the ionization of a proton acid) and the
second a more complex reaction of interest. Consider the two
reactions shown in equations 4 and 5. One can easily measure the
influence of substituents X on the ease of ionization of benzoic
acids (that is, on the acidities of these acids). This type of
reaction is as simple as one is likely to find, and study of the
response of the acidity of the acids to the structure of X provides
a way of studying and defining the mechanisms of interaction of these
substituents with the $-CO_2H$ and $-CO_2^-$ groups. The information
obtained from this study can be used to characterize other reactions
(for example, chelation of copper(II) by the substituted

$$X-C_6H_4-\overset{O}{\underset{O-H}{C}} \rightleftharpoons X-C_6H_4-\overset{O}{\underset{O^-}{C}} + H^+ \qquad K_a \qquad (4)$$

$$X-\underset{H-C=O}{\overset{O}{\underset{}{C}}}Cu(II) \rightleftharpoons X-\underset{H-C=O}{\overset{O^-}{}} + Cu(II) \qquad K_I \qquad (5)$$

$$\Delta(\Delta G^\circ)_x^{\text{reaction I}} \; \propto \; \Delta(\Delta G^\circ)_x^{\text{reaction 2}} \qquad (6)$$

$$\log \frac{(K_I)_x}{(K_I)_{x=H}} = \rho \log \frac{(K_a)_x}{(K_a)_{x=H}} \qquad (7)$$

$$= \rho \sigma_x \qquad (8)$$

salisaldehydes; equation 5). One postulates a LFER (equation 6): the <u>change</u> in the free energy of this reaction which accompanies a particular change in substituent in one reaction is assumed to be linearly proportional to that accompanying the same change in a second reaction. Equation 7, which relates the observed equilibrium constants for the two types of reaction to one another, follows from this assumption. In equation 7, or its alternative form equation 8, there are two types of parameters. The so-called σ parameters characterize the substituents X; the ρ parameters are characteristic of the reactions. Table 2 gives a short list of σ constants; much more complete lists can be found elsewhere, together with lists of ρ values.

The usefulness of LFER's is in the study of complex reactions and processes: it provides a way of establishing empirical analogues between reactions. For example, Figure 7 gives a plot of values of log k for the coordination reaction shown in equation 5 versus values of the ionization constants of the corresponding salicy-aldehydes (equation 9)[12]

$$X-\underset{H-C=O}{\overset{OH}{}} \rightleftharpoons X-\underset{H-C=O}{\overset{O^-}{}} + H^+ \qquad (9)$$

Table 2.   Substituent Constants σ.[a]

X	σ	X	σ
CH_3	−0.17	OH	−0.37
CH_2CH_3	−0.15	OCH_3	−0.27
C_6H_5	−0.01	F	0.06
CHO	0.44	Cl	0.23
$CO_2CH_2CH_3$	0.45	Br	0.23
CN	0.66	I	0.28
CF_3	0.54	SH	0.15
NH_2	−0.66	SO_2NH_2	0.57
$N(CH_3)_3^+$	0.82	SO_3^-	0.09
N_2^+	1.91	PO_3H^-	0.26
NO_2	0.78	$Si(CH_3)_3$	−0.07

[a]From ref. 11, p 66.

The fact that this plot is (roughly) a straight line -- that is, that a LFER relates the two reactions -- makes it possible to estimate the details of bonding in the copper chelate with some precision. Although this approach has not been applied to the study of mechanisms in fracture, it holds great promise. For example, if one were to study the rate of fracture of copper metal in contact with an aqueous solution of the substituted salicy-aldehydes used in Figure 7, and were to find that the rate of fracture correlated with association constants with copper(III) (equation 5), one would immediately be in a position to postulate that the salicylaldehyde molecules were involved in the rate-limiting step for fracture. If the ρ values for the fracture and for association with copper(II) were similar, one might also be able to postulate that copper(II) was involved in the fracture process.

In brief, LFER relations have proved to be an invaluable technique in very complex problems in mechanistic chemistry, and should also be applicable to problems in fracture. They provide a highly-developed method for establishing analogies between reactions and/or processes.

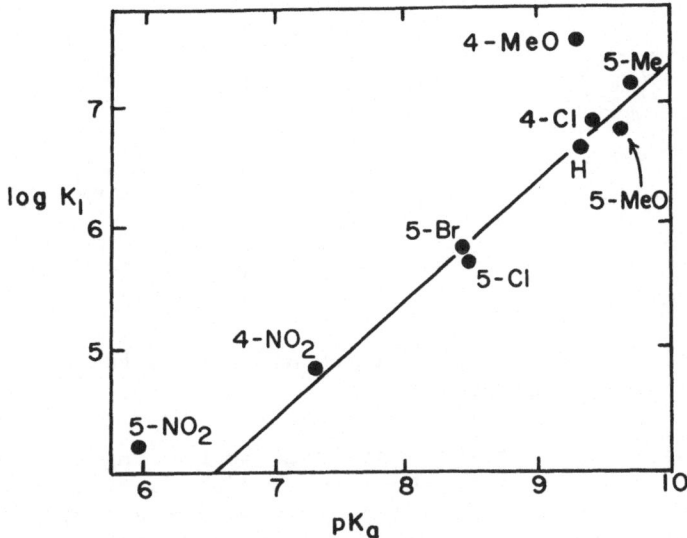

Fig.7. Plot of the logarithms of the equilibrium constants
        for coordination of substituted salicylaldehyde anion
        with copper(II) versus those for ionization of these
        salicylaldehydes.[12]

    Cone Angles. The HSAB classification and LFER are useful
primarily for exploring electronic effects on bonding. For many
reactions of metals with ligands, purely steric effects (that is,
effects due to size and non-bonded interactions) dominate. A
simple approach which has proved very useful is that of Tolman.[13]
In this approach, the "size" of ligands is estimated using
molecular models, by measuring the angle subtended by the groups
attached to the coordinating atom (Figure 8). This figure also
gives data for a typical coordination reaction: competition among
different phosphines and phosphines (L,L') for a coordination site
at nickel(0) (equation 10).

$$L \; + \; L'Ni(O)(CO)_3 \; \rightleftharpoons \; L' \; + \; LNi(O)(CO)_3 \qquad (10)$$

This approach shares with many other techniques in mechanistic
chemistry the fact that it is very empirical. It is, nonetheless,
a useful way of thinking about the coordination of ligands at
metals in solution, and probably also on metal surfaces.

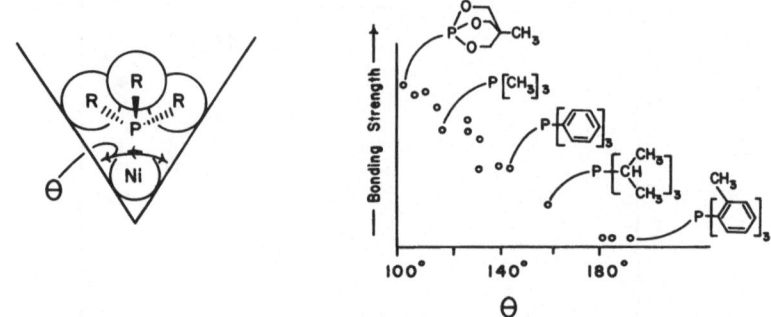

Fig.8.   Cone angles and relative affinities for Ni(CO)$_3$
         for several phosphines.

Entropies.   All of these approaches to bonding concentrate on
energies.   It is essential to keep in mind the very large potential
magnitudes of entropic terms.   For example, for a simple dissociation
of one particle into two, the change in translational entropy can be
large (equation 11).[14]

$$A\text{-}B \longrightarrow A + B \qquad\qquad -T\Delta S \sim 8 - 12 \text{ kcal/mole} \qquad (11)$$

Since this contribution to $\Delta G°$ or $\Delta G^{\ddagger}$ for a reaction may be comparable
to or larger than contributions from enthalpy, it must be estimated
in any serious quantitative consideration of rates or equilibria.

THE COORDINATION OF FRACTURE SURFACES

A number of distinct types of chemistry are relevant to
problems in fracture (Table 3).   The chemistry of fracture and
corrosion is more complex and more difficult to understand than the
types of chemistry which have been studied with greatest success in
mechanistic chemistry.   Chemistry cannot offer easy solutions to the
complex problems of fracture, but it can offer detailed information
on simpler systems which seem certain to be relevant to certain
elementary processes in fracture.   In what follows I simply touch
on several areas which illustrate representative areas in which
inorganic chemistry might be able to contribute to fracture.   Much
of the discussion in this section will be speculation.

Corrosion Inhibition by Adsorption.   A number of types of
superficially different species (carbon monoxide, phosphonium
salts, sulfides, acetylenes, long-chain amines) inhibit corrosion
and influence the rate of crack formation and fracture initiation.
Can one draw inferences about the probable modes of action of these
materials from inorganic chemistry?   We consider these in several
groups.

Table 3. Problems in Fracture, and Related Areas of Inorganic
Chemistry.

Problem	Area of Inorganic Chemistry
<u>Chemistry of Passivating Films</u>: Metal oxides and halides; formation constants and solubilities <u>Corrosion Inhibition</u>: Surface oxide films; chemistry of metal phosphates, borates, and chromium oxides	Coordination Chemistry
<u>Metal Surface Chemistry</u> Corrosion inhibition by adsorption; interaction of metals with R_2S, R_4P^+, CO, $HC{\equiv}CC(CH_3)_2OH$ Surface Electrochemistry: $M + H_2O \rightarrow H_2 + M_nO_m$ $M + Ph_4P^+ \rightarrow M_nO_m + Ph_3P\ Ph\cdot$ $M + RBr \rightarrow M^{n+} + R\cdot\ +\ Br^-$	Organometallic Chemistry Electron Transfer Chemistry
<u>Hydrogen Embrittlement</u> Metal hydride and dihydrogen formation Hydrogen migration (surface, bulk)	Metal Hydride Chemistry

Bulk metals (i.e. freshly-formed metal surfaces) are "soft" using the SHAB classification, and are expected and observed[15] to coordinate CO, phosphines, sulfides, and acetylenes strongly, but not phosphnium salts, sulfonium salts, or amines or ammonium salts. Equations 12-15 show relevant examples of stable complexes formed by interaction of metals or metal complexes with soft ligands.

Soft ligands might thus be expected to form surface organometallic compounds by adsorption. If these compounds are stable, they might provide barrier films, although the chemistry of these barrier films might be much more complex than suggested by the relatively simple structures shown in these equations. In particular, acetylenes may polymerize over clean metal surfaces, and phosphines and (bi)sulfides are reduced to metal phosphines and metal sulfides.

What is the mechanism of action of phosphonium salts and high molecular weight alkyl amines? These materials are not able to coordinate to low-valent metals. There seem several possibilities: Metals readily reduce phosphonium salts to phosphines under some conditions, so phosphonium salts may simply be precursors for phosphines (equations 16). Alternatively, the phosphosium salts,

$$Ni \ + \ CO \ \longrightarrow \ Ni(CO)_4 \qquad\qquad (12)$$

$$Ph_3P \ + \ Fe(CO)_5 \ \longrightarrow \ Ph_3PFe(CO)_4 \qquad (13)$$

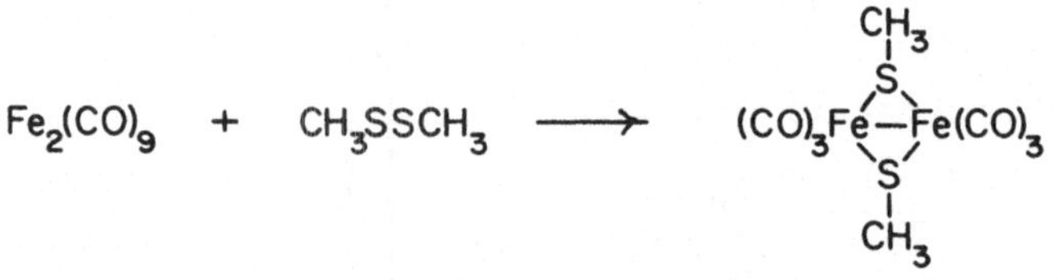

$$Fe_2(CO)_9 \ + \ CH_3SSCH_3 \ \longrightarrow \ (CO)_3Fe\!-\!Fe(CO)_3 \qquad (14)$$

$$\begin{array}{c} Fe_3(CO)_{12} \\ + \\ CH_3C\!\equiv\!CCH_3 \end{array} \ \longrightarrow \qquad\qquad (15)$$

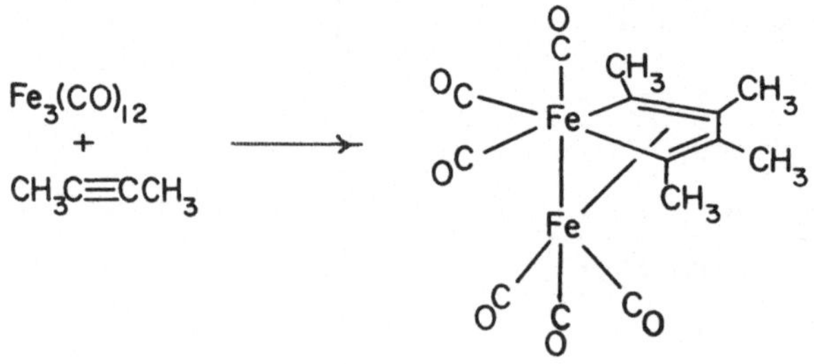

$$Ph_4P^+ + M(0) \longrightarrow Ph_3P + Ph\bullet + M^{n+} \qquad (16)$$

$$Fe(CO)_5 + RNH_2 \longrightarrow \left[Fe(NH_2R)_6\right]^{2+} \left[Fe(CO)_4\right]^{2-} \qquad (17)$$

or transformation products of the ammonium salts (of the type illustrated by equation 17) may adsorb or precipitate on an <u>oxidized</u> metal surface and modify its properties in a way that render it resistant to corrosion.

In any event, the initial steps in reactions of soft ligands with metal surfaces probably involve <u>reduced</u> metals, while those of hard ligands probably involve oxidized metals.

<u>Corrosion By Organic Solvents</u>. A particular but important type of corrosion is that of electropositive metals (magnesium, aluminum) in contact with halogenated organic solvents. The probable mechanism for this type of corrosion is now well established from studies of a closely related system important in organic synthesis: that is the formation of organomagnesium species (Figure 9).[16] The <u>overall</u> rate-limiting step is often breaking an oxide film on the surface. After this film is broken, the rate of reaction is limited by the rate of mass transport of RBr to the magnesium surface. The first step is a single-electron transfer from magnesium metal to alkyl halide. This step generates

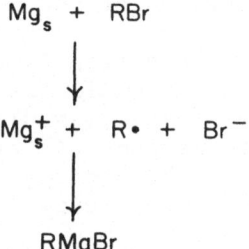

Fig.9.   Mechanism of Reaction of Magnesium with
         an Alkyl Bromide (RBr) in Diethyl Ether
         Solution. Mg_s is a magnesium surface
         Atom.

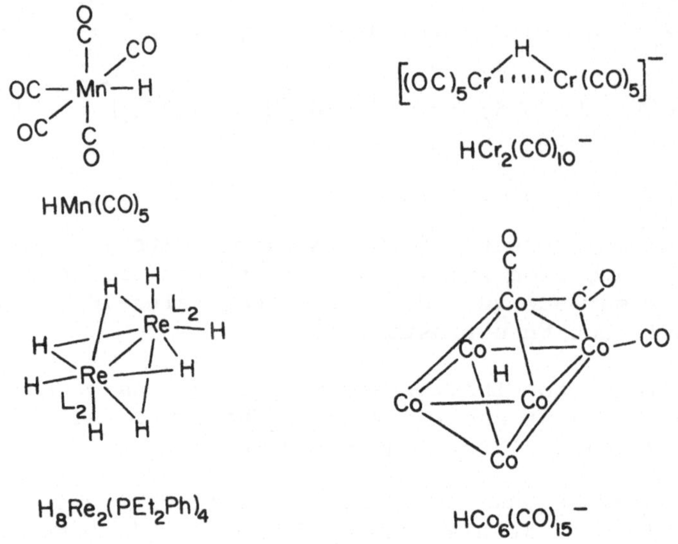

Fig.10.   Examples of organometallic hydrides. The positions
of most of the carbon monoxide ligands of $HCO_6(CO)_{15}^-$
are omitted to show the hydrogen and the directly-
bonded cobalt atoms.

an alkyl radical (R·) as an intermediate. A subsequent step
reduces this species to an alkylmagnesium species, and abstracts
a magnesium atom from the surface of the metal.

Hydrogen Embrittlement. Hydrogen is unique as a ligand in
organometallic chemistry. It is small (it has probably the
smallest steric constraints of any ligand); it is mobile; it forms
stong bonds to metals; it forms many different types of bonds.
Figure 10 illustrates some of the bonding arrangements observed
for hydrogen in organometallic complexes.[17]

One interesting speculation concerning the mechanism of
hydrogen embrittlement of metals is that dihydrogen may cleave
metal-metal bonds during fracture. A schematic mechanism would
be that shown in equation 18. The bond energies necessary to
estimate the thermodynamics of this process are not known, but the
approximate values given in equation 18 suggest that it is not
impossible. Moreover, a number of good organometallic analogs

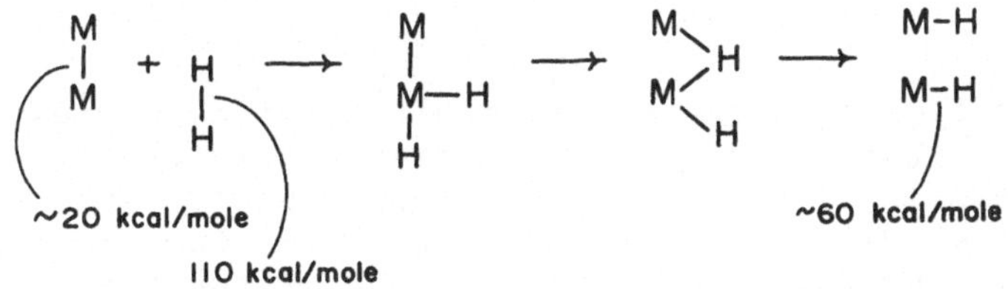

exist for cleavage of metal-metal bonds by dihydrogen (Figure 11).[18]
In no case is the mechanism of one of these reactions known, but a
study of the reverse reaction (formation of an osmium-osmium bond
with expulsion of dihydrogen) suggests that an Os-H-Os group may
be an intermediate.[18]

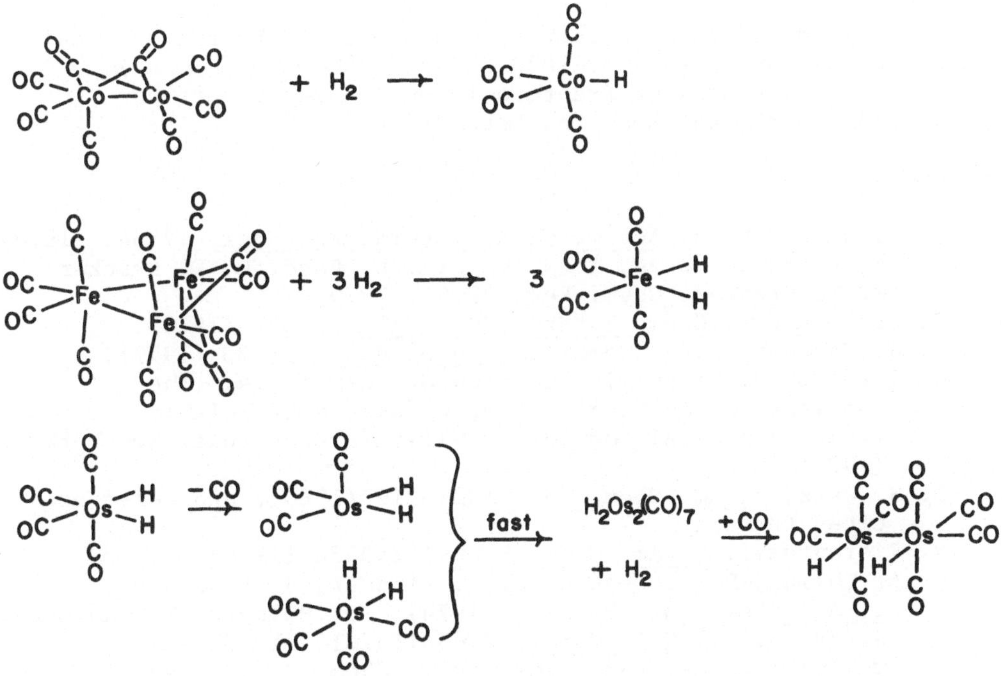

Fig.11.   Cleavage of Metal-Metal Bonds by Dihydrogen.

CONCLUSIONS

Fracture is a complex and difficult-to-study series of processes. Chemistry may have procedures and information to offer which will help in understanding these processes. The techniques developed for studying rates in chemistry are based on transition state theory. The emphasis on the concept of the transition state -- that is, the highest energy species or configuration of atoms along the reaction coordinate -- may have application in analysis of fracture mechanisms. A series of qualitative or semi-qualitative techniques and classifications (SHAB theory, linear free energy relations) have proved useful in chemistry for studying complex processes. In addition, organometallic chemistry and catalysis offer a wealth of structural and mechanistic information concerning bonding of non-metal species to metals. This information should be useful in considering in atomic detail the processes which occur in creation of fresh fracture surfaces.

ACKNOWLEDGEMENT

The writing of this manuscript was supported by the National Science Foundation (Grants 80-12722CHE and DMR 78-24185   ). My interest in the area of fracture is due in major part to conversations with Professor Ron Latanision.

REFERENCES

1.   E. L. Muetterties, *Angew. Chem. Intern. Ed. Engl.* 17:545 (1978); E. L. Muetterties, T. N. Rhodin, E. Band, C. F. Brucker, W. R. Pretzer, *Chem. Rev.* 79:91 (1979).
2.   G. Ertl, *Angew. Chem. Intern. Ed. Engl.* 15:391 (1976).
3.   D. G. Castner, G. A. Somorjai, *Chem. Rev.* 79:233 (1979).
4.   Review: G. W. Parshall, "Homogeneous Catalysis: The Application and Chemistry of Catalysis by Soluble Transition Metal Complexes" Wiley-Interscience, New York (1980).
5.   D. H. Wetz, *J. Am. Chem. Soc.* 102:5316 (1980), and references cited therein.
6.   R. Pierantozzi, *J. Am. Chem. Soc.* 101:5436 (1979).
7.   C. A. Tolman, P. Z. Meakin, D. L. Lindner, J.-P. Jesson, *J. Am. Chem. Soc.* 96:2762 (1974). J. Halpern, T. Okainoto, J. Zakhariev, *Mol. Catal.* 2:65 (1977). Y. Ohtani, M. Fujimoto, A. Yamagishi, *Bull. Chem. Soc. Japan* 51:2562 (1978).
8.   R. G. Pearson, *Survey Prog. Chem.* 5:1 (1969).
9.   M. J. S. Dewar, *Bull. Soc. Chim. Fr.* 18:C71 (1951). J. Chatt, L. A. Duncanson, J. Chem. Soc. 2939 (1953).
10. R. V. Hodges, S. A. Sullivan, J. L. Beauchamp, *J. Am. Chem. Soc.* 102:935 (1980).

11. J. Hine, "Structural Effects on Equilibria in Organic Chemistry"
       Wiley-Interscience, New York (1975).
12. J. G. Jones, J. B. Poole, Tomkinson, R. J. P. Williams,
       J. Chem. Soc. 2001 (1958). K. Clark, R. A. Cowen,
       G. W. Gray, E. H. Osborne, ibid. 245 (1963).
13. C. A. Tolman, Chem. Rev. 77:313 (1977).
14. S. W. Benson, "Thermochemical Kinetics" Wiley-Interscience,
       New York (1968).
15. F. A. Cotton, G. Wilkinson, "Advanced Inorganic Chemistry",
       John Wiley & Sons, New York (1980).
16. L. M. Lawrence, G. M. Whitesides, J. Am. Chem. Soc. 102:2493
       (1980), and references cited therein.
17. R. Ban, R. G. Teller, S. W. Kirtley, T. F. Koetzle, Accounts
       Chem. Res. 12:176 (1979).
18. J. R. Norton, Accounts Chem. Res. 12:139 (1979).

DISCUSSION

Comment by J.O'M. Bockris:

I share Professor Whitesides' opinion that an approach to cracking most likely to yield fruit is in terms of chemical mechanism determination. The key process in this approach is the determination of the rate-determining step. Thus, in general, $A \rightarrow B \rightarrow C \underline{\quad} \rightarrow$ R.D.S. $\underline{\quad} \rightarrow E$ for the process $A \rightarrow E$. If the R.D.S. is known, the rate of $A \rightarrow E$ is immediately calculable by standard formalism.

Many persons unfamiliar with the ubiquitous approach of the theory of absolute reaction rates may think that the identification of the nature of the activated complex may be a difficult matter, particularly for a complex process involving many partial steps, some of which may be parallel and simultaneous. Whilst I do not wish to project the image of a utopian pathway to clarity, my experience of the method (applied to reactions in liquids) suggests that fairly complex processes have sometimes quite simple rate-determining steps. If the situation were that one had to compute the step without the aid of experiment, I think we should be no better off than the physicists. However, a cautious and well thought out campaign of experiments often gives a strong indication of what the r.d.s. may be. (At the worst, it greatly reduces the possibilities.)

It so happens that I can give a strikingly relevant example. My co-worker, P.K. Subramaniam, determined the rate of hydrogen permeation through a series of Fe-Ni alloys. He found a P_H- composition relation in which the values varied by some 5 orders of magnitude. There was much structure (following phase changes in the alloy). By happenstance, we came across the work of K. Nobe, who had determined the corrosion rate under stress of a similar alloy

series. The corrosion rate and permeation (rate) had identical
structure and degree of variation. I can surely conclude that the
r.d.s. in the corrosion and permeation are identical. But now I
face a totally different and much easier problem: it is likely that
the r.d.s. in this case is $H_3O + e_{(m)} \rightarrow MH$, proton discharge from
solution across the double layer to the metal surface.

Two minor points:

1. The distinction between $H_{ads} + H_{ads} \rightarrow H_2$ and $H^+ + H_{ads} + e \rightarrow H_2$
    in removal processes for H on a metal is most important when
    it comes to cracking. Thus, for the first, the coverage is
    low and for the second, high. Other matters being equal, the
    latter is more likely to constitute a danger--promote crack-
    ing--than the former.

2. In respect to the effect of solution-bourne inhibitors and
    promoters of permeation, one mechanism for promotion is
    simply bond weakening by means of lateral electrostatic
    interaction between M-H bonds and the organic adsorbate.

Comment by R. Bullough:

    You began your talk by drawing an analogy between fracture and
polymerization. I presume this analogy only applies to the thermally
activated propagation of an ideally brittle crack in a crystalline
lattice. In any real fracture process there is some plastic deformation with the emission of dislocations or the stimulation of dislocation sources at the crack tip. In the work of fracture this
plastic work usually dominates the energetics of the crack propagation
and defines the rate of crack propagation. Can the chemical rate
theory approach with molecular analogs give any insight into such
dislocation dominated processes?

Reply:

    Yes, in the sense that one can construct an analogy to the
formalism in which only chemically activated reactions are employed
in which activation energies contain a significant contribution from
potential energy terms originating in strain. The essential ideas
remain the same. There are also useful procedures analogous to
transition state theory which are used to analyze photochemical
reactions and others in which the transition state energy is attained
by some process other than thermal fluctuation. In any event, the
value of the transition state theory in fracture is less for its
particular utility in quantitative analysis of rates and more for
its qualitative emphasis on the elemental composition and structure
of the transition state.

Comment by A.R.C. Westwood:

You have noted that certain organic reactions require (or involve) a reorientation of the ligands of one or other of the participating species. Is it likely that an equivalent function is played by lattice strain at the tip of a crack, i.e., is strain-assisted (dependent) chemisorption of an active species at a crack tip a likely situation in embrittlement. As an example, note that in order to constitute a liquid metal embrittlement "couple" the participating metals usually exhibit low mutual solubility, and little tendency to form intermetallic compounds. In other words, in equilibrium conditions, they tend not to associate. Hence the surmise that, perhaps, non-equilibrium configurations may be involved, e.g., a strained crack tip.

Reply:

A very similar idea has been used to rationalize rates observed in enzymatic catalysis. It is certainly possible that strain may assist certain reactions, but no exact model is known.

Comment by D.J. Duquette:

I would like to support the speaker in emphasizing the importance of rate determining steps in the fracture problem. It is a factor often ignored by structural scientists who study beginning and end states and ignore transitional states which may, in fact, control the process of interest. A minor point of contention, however, related to the role of chromate on corrosion resistance. While the model suggested by the speaker, involving transitional states of Cr, may have relevance to Cr containing alloys where anodic dissolution is still not thoroughly understood, passivity of Fe by chromate is generally considered to be a simple process involving a compound with a noble redox potential and low activation energy for reduction which accordingly stimulates anodic dissolution of the metal thus forming a protective film. Identical results can be obtained for iron using molybdates, tungstates, and other heavy metal oxide complexes as well as with nitric acid and, in some cases, simply dissolved oxygen.

Reply:

I stand corrected.

Comment by R.M. Latanision:

Almost without exception, as far as I am aware, catalysts are used in an essentially unstressed (mechanical) condition. On the other hand, if the reactivity of a metal catalyst is related to the presence of exposed metal atoms, I would expect that the action of a catalyst might be changed significantly if it were strained plastically (i.e., dynamically strained rather than prestrained or statically stressed) while performing as a catalyst. Plastic deformation generates surface steps by virtue of the egress of dislocations. Are you aware of any attempt to examine the reactivity of dynamically deforming catalysts?

Reply:

No, I am not aware of any such effort. A practical problem with an attempt to carry out this sort of experiment lies in reporting it: Catalytic rate constants are usually reported as turnover numbers (moles product/mole of catalyst surface atoms/time). It is not clear how one would measure surface areas under dynamic conditions.

HYDROGEN ADSORPTION ON METAL SURFACES

Klaus Christmann

Institut für Physikalische Chemie
der Universität München, Sophienstraße 11
D-8000 München 2, FRG

INTRODUCTION

For a variety of reasons, the interaction of gaseous hydrogen with solid surfaces of metals and semiconductors has found widespread interest in the past, and countless experimental and theoretical investigations have been performed in this field. The term 'interaction' primarily focuses on the adsorption of the hydrogen molecule on the metal surface, but in principle it also includes a possible penetration of the hydrogen into the bulk crystal as well as a reaction with the bulk material to form the respective metal hydride. Both these interaction phenomena play a major role in practical chemistry and physics: The hydrogen adsorption step is considered a central reaction in heterogeneously catalyzed hydrogenation reactions. In forming an adsorptive bond to the metal certain other bonds of an adsorbed molecule may become weakened or even cleaved, and new species are easily formed by the subsequent surface reaction which normally is of a Langmuir-Hinshelwood type, i.e. both partners react from the adsorbed state. The desired (gaseous) products leave the surface by thermal desorption under the reaction conditions and can be separated from the reaction mixture.

Sometimes it occurs that hydrogenation reactions (e.g. the well-known nitrogen hydrogenation to form ammonia) require high temperatures and high pressures of hydrogen. In these cases serious material problems can arise in that the hydrogen is able to permeate the reactor walls or to undergo chemical reactions with the constituents of the vessel material, e.g. carbon. As a result, the mechanical properties of the material get altered and the reactor can no longer withstand the reaction conditions. In general, hydrogen induced embrittlement and fracture phenomena are known to be a se-

riously limiting factor when handling certain materials like steel
or titanium in a hydrogen atmosphere at elevated temperatures and
pressures, but only little is known so far about the details of the
mechanism of the underlying chemical reactions.

　　　The present paper will mainly focus on the hydrogen adsorption
step and is almost exclusively devoted to ultra-high vacuum studies
under well-defined conditions using single crystalline materials.
It will be organized as follows: First, a somewhat general concept
is offered to understand the mechanism of molecular and dissociative
hydrogen adsorption, followed by a major experimental section in
which recent results on the energetics, structure and electronic
properties of hydrogen adsorbed layers will be highlighted including
experimental evidence for H induced surface reconstruction and hydrogen penetration into surface-near regions of the metal crystal lattice.
Finally, we will conclude our results and give some comments on the
importance of a detailed understanding of the metal - hydrogen interaction in view of catalysis, metallurgy, and energy storage problems.

THE MECHANISM OF THE HYDROGEN ADSORPTION

　　　The physical system that we have to concentrate on consists of
a diatomic hydrogen molecule in its ground state infinitely far away
from an unstructured ideal metal surface. The molecule approaches
the surface along the reaction coordinate, z, which is directed normal to the metal surface. According to Lennard-Jones [1], the energetic situation can be described by a simple potential energy diagram which is shown in Fig. 1. The hydrogen molecule at infinity is
given the potential energy zero. As it gets closer to the surface,
weak van-der-Waals attractive interaction forces become perceptible

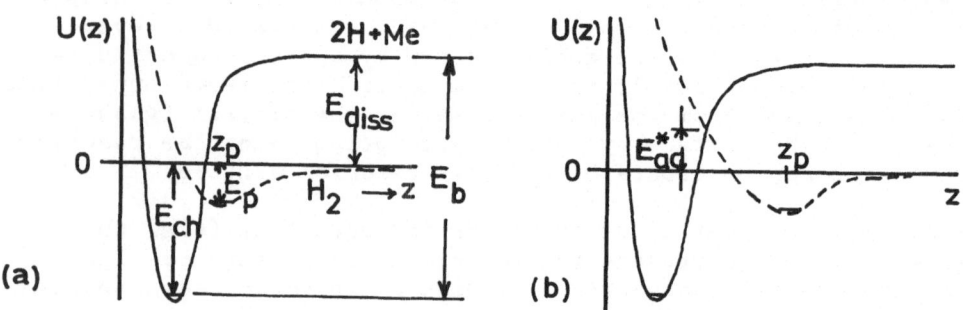

Fig. 1. Potential energy diagram for hydrogen adsorption according
　　　　　to Lennard-Jones [1]. Example (a) refers to non-activated
　　　　　adsorption with the point of intersection of the two poten-
　　　　　tial curves below E = 0, example (b) illustrates the case of
　　　　　activated adsorption with the intersection above E = 0.

first, and the H_2 molecule moves into a shallow potential energy minimum with the depth E_p located at a distance z_p in front of the surface. If it would be forced even closer to the surface, the molecule would experience massive repulsive interaction forces as characterized by large positive potential energy values. The minimum therefore represents the equilibrium position of the molecule which performs vibrations around z_p. The corresponding energy E_p commonly is referred to as physisorption energy and is of the order of a few kJoule/mole. This brings it about that physisorbed hydrogen can only be attained at sufficiently low temperatures, where the thermal energy content of the molecule is quite low. The reason though why hydrogen nevertheless adsorbs quite strongly on many metal surfaces is easily elucidated by Fig. 1 (a): Suppose the case that the H_2 molecule is cleaved into atoms first by supplying it with an external energy of 432 kJoule/mole (which is just the dissociation energy of the hydrogen molecule, E_{diss}), and thereafter the H atoms are brought to the surface separately. It appears that comparatively strong chemical interaction forces can now operate between the hydrogen 1s orbital and the metal electron wavefunctions, and a relatively large energy, namely twice the Ni-H bond energy) is obtained. This reaction is called chemisorption, and the respective energies typically range between 500 and 600 kJoule/mole, depending on the chemical nature of the metal. The position of the chemisorption minimum as the new equilibrium location of the H atom is given by the distance z_{ch} and is much closer to the surface (about 1 - 2 Å) than the physisorption minimum mentioned before. Due to the considerable depth of the chemisorption well, even at room temperature hydrogen atoms will only occasionally be able to escape from this potential.

The real situation now can be considered as a superposition of the two processes: At the cross-over point of the two respective potential energy curves the hydrogen molecule will fall apart into two H atoms spontaneously, and the chemical interaction energy will be released which clearly overcompensates the heat of dissociation. This reaction is known as 'dissociative adsorption', and it can be written as

$$H_{2(g)} + 2\,Me \rightarrow 2\,H_{(ad)} + 2\,Me \rightarrow 2\,Me\text{-}H$$

with the energy balance (E_{ch} = heat of adsorption, viz. the net energy

$$E_{ch} + E_{diss} = 2\,E_{b,\,Me\text{-}H} \tag{1}$$

gain; E_b = binding energy of a single metal - hydrogen bond). Depending on whether the cross-over point of the physisorption and chemisorption potential energy curves will lie above or below the potential energy zero line, a small extra activation energy will be required to enable the hydrogen molecule to overcome the barrier with the height E_{ad}^{*} which situation is illustrated in Fig. 1b, or the dissociation into the atoms will occur spontaneously (shown in Fig.1a).

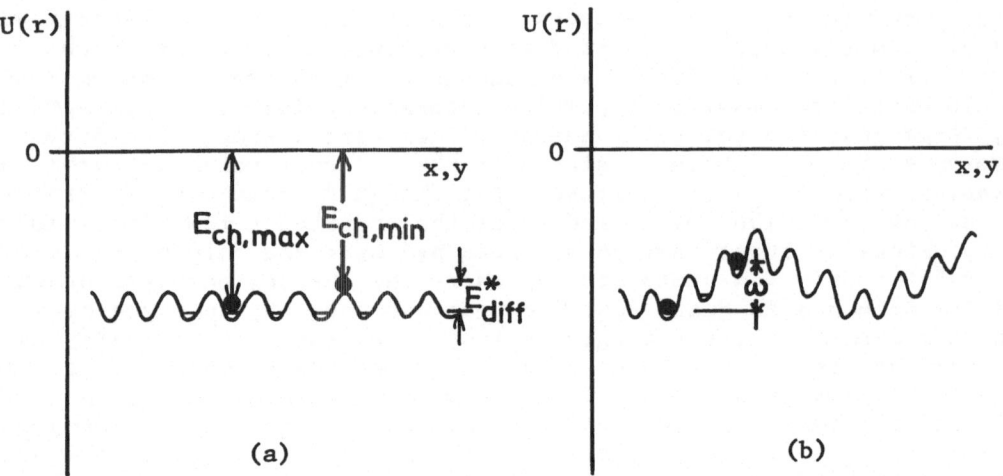

Fig. 2. Lateral variation of the potential energy of a metal surface:
(a) for a single adsorbed particle, (b) for adjacent adsorbed
particles with repulsive mutual interaction.

These two cases is referred to as 'activated' and 'non-activated' adsorption, respectively. With most of the transition metals, the hydrogen adsorption is non-activated as the chemisorption reaction takes
place readily. There are, however, few examples where the existence
of an activation barrier of several kJoule/mole has been reported.

Once a H atom has entered the chemisorption potential well its
position on the surface is rather fixed, i.e. it will not easily re-
evaporate into the gas phase. It just vibrates around its equilibrium
distance z_{ch} and has a residence time τ that is given by the so-
-called   Frenkel equation:

$$\tau = \tau_o \exp\left(+\frac{E_{ch}}{kT}\right) \tag{2}$$

(with the preexponential factor τ_o being of the order of a vibration
period).

In order to really understand the behavior of a chemisorbed hydrogen layer we have to explicitly consider the atomistic structure
of a periodic solid surface, i.e. we can no longer argue with a completely smooth and perfectly even surface but rather have to take into account the lateral variation of the potential energy which evidently depends on the location of the metal surface atoms, i.e. on
the translation vector $r(x,y)$. Consequently, also E_{ch} varies periodically along the surface as is schematically illustrated in Fig. 2a
for a single adsorbed particle. The difference between $E_{ch, max}$ and
$E_{ch, min}$ is equal to the activation energy for surface diffusion, E_{diff}^*, and typically one order of magnitude smaller than
E_{ch} itself.

Within an alternative (but equivalent) description the H adsorbed layer is given sort of a two-dimensional band structure for the motion parallel to the surface, based on the assumption that the H atom (not the electrons) is located in a two-dimensional box the dimensions of which are governed by the lattice constant, i.e. the atom positions of the metal surface. Like in electron band structure bands and band gaps occur with the band width becoming larger as the H atoms gain thermal energy [2].

It has to be pointed out that it is the magnitude of the surface diffusion energy, E_{diff}^*, that determines whether the adsorption is localized or mobile. Accordingly, we can define another residence time, τ', for a H atom on a certain adsorption site of the metal surface:

$$\tau' = \tau'_o \exp \left(+ \frac{E_{diff}^*}{kT} \right) . \qquad (3)$$

For a hydrogen adsorbed layer, E_{diff}^* and therefore also τ' usually is fairly small which accounts for the experimentally observed high mobility of adsorbed hydrogen atoms.

The situation gets rather different if a whole ensemble of adsorbed particles is considered instead of a single atom. It can be rationalized immediately that, no matter what the distance of two adjacent possible adsorption sites is, two H atoms cannot be brought together over any distance. There is a nearest neighbor interaction potential U(r) with a strong repulsive part at small H - H distances r which modulates the periodic potential of the metal surface mentioned before. This is demonstrated by Fig. 2 (b), and we may state that an adsorbed hydrogen atom will "feel" whether or not an adjacent site is occupied. In other words: The hydrogen - metal binding energy E_b (and therefore also the heat of adsorption, E_{ch}) depends on the concentration and configuration of the totally adsorbed particles (E_{ch} normally decreases continuously as the coverage increases). In principle, one has to distinguish two different types of interaction potentials: Direct (i.e. dipole - dipole repulsion or orbital overlap) and indirect (through-bond) interaction forces can occur. For hydrogen adsorption, the by far more important contribution stems from the indirect interactions. As was shown theoretically [3] these forces act over 1 - 3 lattice constants and often have oscillatory character, i.e. they may be repulsive at a nearest neighbor site but attractive at a next-nearest neighbor site. The typical order of magnitude of these indirect interaction forces is about one tenth of the heat of adsorption; for hydrogen values between 2 and 5 kJoule/mole have been derived.

It has to be kept in mind that it is these interaction forces that are responsible for the formation of long-range order within an adsorbed layer. With strong mutual repulsive interactions, for example,

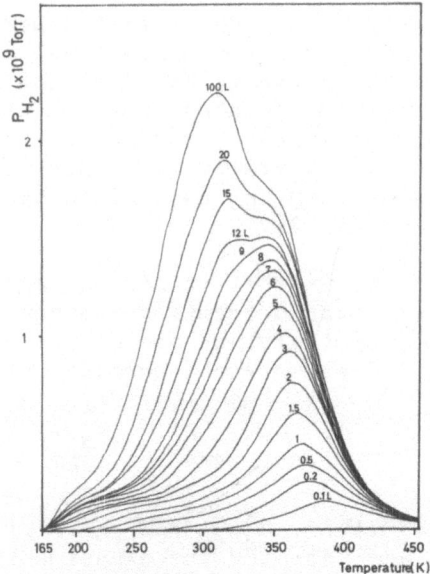

Fig.3.   Series of hydrogen thermal desorption spectra from a Ni(100)
         surface with the H$_2$ exposure as parameter. The adsorption
         temperature was 150 K, the heating rate ß was 10 K/s. After
         ref. [5].

the adsorbed atoms will be forced on certain periodic adsorption
sites even at fairly small overall coverages, whereas noticeable
attractive interactions will favor cluster growth or island formation. It will be shown later how careful experimental investigations
of the temperature and coverage dependence of ordering phenomena
will reveal the nature and magnitude of possible mutual interactions
between the adsorbed hydrogen atoms.

RECENT EXPERIMENTAL RESULTS

a) The energetics of the hydrogen - metal bond

     As was pointed out in the foregoing section, the energy of the
hydrogen - metal bond follows directly from the heat of adsorption
via Equ. (1). Experimentally there are basicly two ways to determine
this quantity.

     1. Thermal desorption experiments, where a gas-covered surface
is heated uniformly and linearly in a pumped vacuum system and the
hydrogen partial pressure is monitored as a function of the sample
temperature, yield the activation energy for desorption, E_{des}^* (i.e.
the energy to remove a hydrogen atom from the bottom of the chemisorption potential well to just outside) as a function of the hydrogen coverage Θ. Actually, E_{des}^* reflects E_{ch}^* only, if no adsorption activation barrier exists, i.e. $E_{ad} = 0$, and the desorption

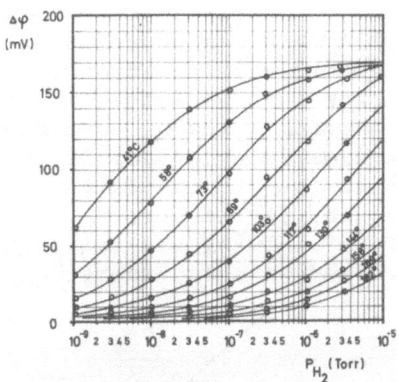

Fig. 4.   Hydrogen adsorption isotherms for a Ni(100) surface, after
ref. [8]. The hydrogen-induced work function change, $\Delta\varphi$,
was used as a coverage monitor.

process can be described by simple rate equations [4]. Then the position of the hydrogen pressure maximum is a fairly reliable measure
for the activation energy of desorption and thus for the heat of adsorption, E_{ch}. Moreover, the area under a thermal desorption curve
is proportional to the respective hydrogen surface coverage prior to
the desorption experiment. An example is given in Fig. 3 which displays a set of H desorption curves for a Ni(100) single crystal surface. The spectra exhibit two features typically observed in hydrogen desorption: The development of two desorption maxima ("β_1 and
β_2 states") with roughly 50 - 100 K temperature separation (the β_1
state usually starts to develop around $\Theta_H = 0.5$), and a pronounced
shift of the high-temperature (β_2) desorption maximum towards lower
temperature values as the hydrogen surface coverage increases. This
is indicative of a second-order rate process as caused by the recombination of two isolated hydrogen atoms.

ii. Isosteric heat (q_{st}) data based on thermodynamic equilibrium
measurements give a still clearer information about the energetics of
the chemisorption process than the thermal desorption data, since the
kinetic parameters do not enter explicitly. q_{st} follows immediately
from adsorption isotherms, i.e. the variation of the coverage Θ_H as
a function of the pressure P at constant temperature T: $\Theta_H = \Theta_H(P)_T$,
by applying the phase equilibrium thermodynamics and solving the
Clapeyron differential equation for q_{st}:

$$\left(\frac{dP}{dT}\right)_{\Theta=const} = \frac{q_{st}}{T\left(V_g - V_{ad}\right)} \tag{4}$$

(with V_g and V_{ad} being the molar volumes of the gas phase and the adsorbed phase, respectively). Adsorption isotherms for the Ni(100)/H system are shown in Fig. 4 and can be quite conveniently measured by using the hydrogen-induced work function change, $\Delta\varphi$, as a hydrogen coverage indicator. For this purpose, of course, one has to determine the relation between $\Delta\varphi$ and Θ_H separately, for instance by a thermal desorption peak area integration. It is evident from Fig. 4 that a given H coverage can be attained with various hydrogen pressure/temperature values, and if the term V_{ad} is neglected in Equ.(4) and V_g replaced by RT/P by applying the ideal gas law, and moreover q_{st} is assumed not to depend on the temperature, Equ. (4) can be given the form:

$$\left(\frac{d\ln P}{d\left(\frac{1}{T}\right)}\right)_{\Theta=const} = -\frac{q_{st}}{R} \tag{5}$$

A plot of ln P versus 1/T then should yield a straight line for a given coverage, the slope of which contains the 'isosteric differential heat of adsorption' q_{st} for that particular coverage Θ_H as indicated by the corresponding work function change $\Delta\varphi$.

In the manner just described isosteric heats have been determined for many hydrogen - transition metal adsorption systems, and the following table (Table I) lists some of these q_{st} values, extrapolated for zero hydrogen coverage, as well as E_b values (as calculated after Equ. (1)) for the adsorptive Me - H bond, along with the respective data for the diatomic metal hydride molecules.

Table I.   Selected experimental data for Me - H adsorptive bond
           strengths and metal hydride molecule bond dissociation
           energies (after ref. [6]). All values in kJoule/mole

investigated system	q_{st}	metal - hydrogen adsorptive bond energy and reference		bond dissociation energy of the diatomic hydride molecule
Fe(110)/H	104	268	[7]	
Ni(111)/H	96	264	[2, 8]	251
Cu(111)/H	36	234	[9]	276
Pd(100)/H	102	267	[10]	
Ag films	8	< 220	[11]	222
Pt(111)/H	46	239	[12]	347

Interestingly, the E_b values for the Me - H adsorptive bond lie within a rather narrow range and show reasonable agreement with the corresponding hydride dissociation energies. Two somewhat more general conclusions may then be allowed at this point: Not only appears the hydrogen - metal adsorptive binding mechanism rather similar irre - spective of the metal substrate, but there is also a fairly localized character of the H - Me bond established in that the overwhelming contribution to the bond energy seems to be provided by the metal atom(s) directly underneath the hydrogen adatom. This does not mean, however, that only one or two metal atoms are responsible for trapping and adsorbing a hydrogen molecule. As will be shown later, there is clear evidence from experiments with bimetallic surfaces [13] that ensembles of at least 4 adjacent metal surface atoms are required to capture and dissociate a hydrogen molecule arriving from the gas phase before the H atoms formed can enter the chemisorption potential well .

This brings us to a point that some emphasis has to be put on: All isosteric heat data and also the E_b data listed in Table I refer to the chemical equilibrium situation - possible kinetic effects are not included and do not show up. In practise, however, the existence of kinetic limitations can slow down a chemisorption process considerably. A good example is the interaction of hydrogen with a Cu(111) surface, where the adsorption is indeed activated by about 12 kJoule/ mole, and only fairly large pressures of hydrogen and appreciable surface temperatures enable a sufficiently large number of H_2 molecules to pass the activation barrier and undergo subsequent dissociation [6]. Once this process has occurred, the Cu-H chemisorption bond is formed readily, and the strength of this bond does not deviate too much from the average of the listed E_b values.

Turning to the coverage (Θ) dependence of q_{st} which is shown for various surfaces in Fig. 5, according to ref. (14), it is

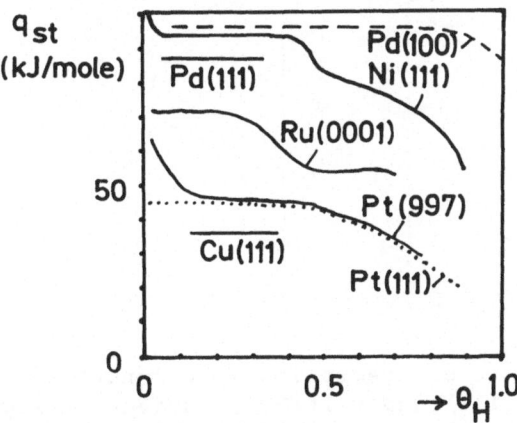

Fig. 5.   Comparison of some $E_{ch}(\Theta) = q_{st}(\Theta)$ relationships for various metal - hydrogen adsorption systems. After ref. [14].

quite obvious that q_{st} falls as the coverage increases thereby pointing to a pronounced energetic heterogeneity of the respective surfaces. From Fig. 5 we can readily delineate between an a priori heterogeneity (caused by inherent crystallographic imperfections) and
an a posteriori heterogeneity (as induced by the presence of the adsorbate). The variation of q_{st} at very small coverages certainly has
to be attributed to preferential adsorption of hydrogen atoms on particular surface defect sites, whereas the decrease of q_{st} beginning
at medium coverages clearly reflects the onset of the operation of
mutual repulsive H - H interactions which come into play as the H
surface concentration increases. Both types of energetic heterogeneity are essential in hydrogen adsorption and deserve to be considered in some more detail.

With regard to the a priori heterogeneity (which not only offers
certain high adsorption energy sites but may also facilitate the dissociation reaction and therefore accelerate the entire adsorption pro-
sess), we have undertaken an attempt to systematicly investigate the
influence of crystallographic surface defects on both the energetics
and kinetics of the hydrogen adsorption on platinum [15]. We prepared
two surfaces with (111) orientation: One surface was cut and polished
so as to give an utmost flat and smooth (111) face with very low defect concentration, the other one was cut from a Pt boule about 6.5
degrees off so as to produce a (997) orientation which exhibits artificial defect sites, i.e. alternating terraces with (111) orientation
nine atoms wide, and monoatomic steps. These surfaces were exposed to
hydrogen under otherwise identical conditions, and the results on both
the energetics (as indicated by the heat of adsorption, q_{st}) and the
velocity of adsorption (as indicated by the coverage dependence of the
hydrogen sticking probability s) are illustrated in the Figs. 6 a and
b.

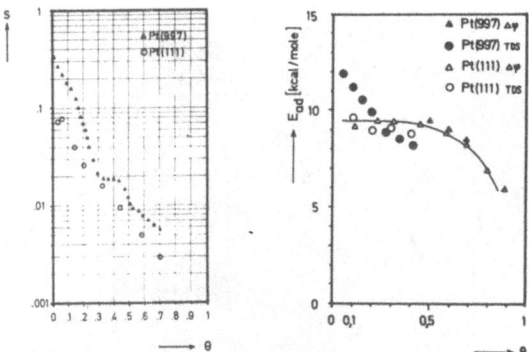

Fig. 6. (a) isosteric heat of adsorption as a function of the H co-
         verage for a Pt(111) and a Pt(997) surface. (b) sticking
         probability s (the ratio between adsorbed and exposed H atoms)
         as a function of the H coverage for a Pt(111) and a Pt(997)
         surface. After ref. [12] and [15].

The effects are quite clear: Not only did we create about 25% of sites with a different coordination on the stepped surface (namely near the top and the bottom of the steps) which apparently provide a 10 - 15 % stronger bond to the hydrogen atom, but also the sticking coefficient initially increased by about a factor of 4 - 5 as compared to the flat surface. Without entering into any further detail we just briefly mention the dramatic effect of this accelerated adsorption on chemical surface reactions, for instance the hydrogen - deuterium isotope exchange, where an increase in the reaction rate on the stepped surface of a factor of 10 was found. This observation is completely in line with findings by Somorjai [16, 17] who reported on molecular beam studies on quite a similar stepped platinum surface and obtained a largely increased exchange reaction rate if the molecular beam was directed perpendicular to the steps, but almost no effect with the beam parallel to the step arrays.

In general, it is believed that quite similar phenomena play an important part also in the course of many heterogeneously catalyzed gas phase hydrogenation reactions in that a high local concentration of reactive species exists at certain defect sites. As soon as those sites become blocked by irreversible adsorption of impurities such as sulfur or carbon, the overall reaction rate will be substantially lowered and the catalyst is regarded as 'poisoned'. From all our previous CO, O_2, N_2 and H_2 adsorption studies it appears that only the hydrogen case (even with well-oriented low-index planes) gives rise to features like the ones described above, namely increased initial binding energy and sticking probability [18]. This makes us believe that the small hydrogen atom is particularly sensitive to crystallographic imperfections whereas the larger oxygen or carbon monoxide molecule apparently is not.

As far as the induced heterogeneity is concerned which always appears at higher coverages in the $q_{st}(\Theta_H)$ relation, cf. Fig. 5, it is quite clear that with the effective diameter of the adsorbed H atom ranging around 1.5 Å [2] and the typical distances of possible H adsorption sites on a metal surface direct repulsive interactions between adjacent adsorbed H atoms ought to be a limiting factor only for H coverages greater than 1, when more H atoms than metal surface atoms are present. Since the H binding energy starts to drop already around $\Theta_H = 0.5$, where only half the possible adsorption sites are occupied, we clearly have to invoke indirect interactions that operate on the nearest and next-nearest neighbor sites. If these interactions were to occur one would expect ordered hydrogen phases within a certain temperature and surface coverage regime, and indeed there are some cases known where low-energy electron diffraction (LEED) investigations have proven the existence of long-range order within the adsorbed hydrogen phase [10, 19, 20]. The considerable mobility of hydrogen at room temperature may obscure the formation of a noticeable order, but cooling to below 200 K often helps to fix the H atoms in periodic adsorption sites by increasing the lifetime τ'

and to produce a LEED 'extra' pattern.

For the sake of simplicity we may confine ourselves to pairwise repulsion forces and assume a regular array of metal surface
atoms so as to provide a sufficient number of equivalent adsorption
sites the adsorbing H atoms can compete for. The problem consists
in determining the pairwise repulsion energy ω and can theoretically treated by methods of statistical mechanics. A related
problem is the two-dimensional Ising model which was originally designed for describing the spin properties of a magnetic alloy [21],
and for which an analytical solution exists [22]. As was e.g. shown
by Hill [23], there is a relation between the critical temperature
T_c of the ordered two-dimensional phase (which would be symmetric
with respect to $\Theta = 0.5$, i.e. for half the adsorption sites filled)
and the pairwise interaction energy ω. For a square lattice the
following expression holds:

$$\omega = 1.76 \, k \, T_c \qquad\qquad (6)$$

Experimentally, phase diagrams can be evaluated by following
the long-range order as a function of the temperature at a fixed
coverage. This can be accomplished by monitoring the intensity of a
LEED beam (which is a good measure for the long-range order [24])
as a function of the temperature so long as no thermal desorption
effects will lead to a loss of adsorbed hydrogen atoms. Quite appropriate in this respect are the Ni(111)/H and the Pd(100)/H adsorption systems where we obtained the phase diagrams reproduced in Fig.
7 a and b.

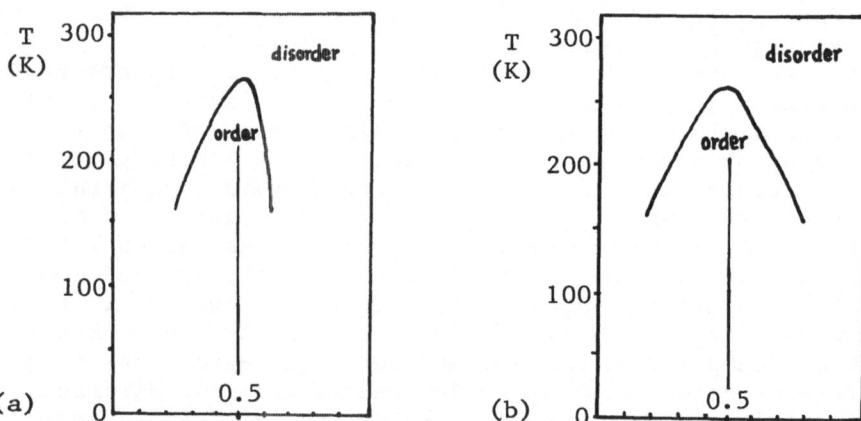

Fig. 7   Experimentally observed phase diagrams for ordered adsorbed
         hydrogen phases. (a) Ni(111)/H (ref.[19]),(b) Pd(100)/H ,
         (ref.[10]).

However, even at a first glance it becomes evident that these diagrams are by no means symmetric with respect to Θ_H = 0.5, opposite
to the predictions of the simple theoretical model. Particularly the
Ni p(2x2)H phase diagram is not, and the reason certainly is that
our assumption of strictly pairwise interactions and equivalent adsorption sites over the entire coverage range is too simple. Instead,
the structure of the hydrogen phase on Ni(111) which will be discussed in some more detail in a moment is such that equivalent adsorption
sites exist only up to Θ_H = 0.5; thereafter equivalent threefold coordinated sites are no longer available. With Pd(100)/H, on the other
hand, we have evidence that besides pairwise forces also three-body
("trio") interactions [25] gain importance as Θ approaches the value
1.0. The extended width of the phase diagrams in both cases as compared to theoretical predictions based on the Ising model can be
qualitatively explained as being due to additional weak attractive
next-nearest neighbor interactions [26].

However even with the admittedly very crude model of pairwise
interactions and equivalent adsorption sites we get at least an idea
of what the magnitude of the repulsive interaction forces is, although the extraction of a more refined information would certainly
require sophisticated statistical calculations. It turns out that ω_{rep}
usually ranges between 2 and 6 kJoule/mole; table II gives a summary
of critical temperatures T_c and pairwise repulsion energies obtained
for the three ordered hydrogen phases on Ni, Pd and Fe (Table II).

To conclude this section on the energetics of H adsorption, we
would like to add that the appearance of a second low-temperature
state in hydrogen thermal desorption experiments mentioned in the
beginning of this section (cf. Fig. 3) nicely parallels our indirect
interaction model in that the difference ΔT between the high and low
temperature desorption state also corresponds to an energy difference
of 2 - 8 kJoule/mole and is therefore another good indicator of the
magnitude of the induced interaction forces.

Table II.   Experimental data on the phase transitions in chemisor-
            bed hydrogen systems, i.e. the critical temperature T_c,
            and the pairwise repulsion interaction energy ω_{rep}
            as determined from the temperature dependence of LEED
            diffraction maxima (corrected for Debye-Waller contribu-
            tions). After ref. [10, 19, 20].

investigated system (observed ordered phase)	critical tempera-ture (K)	pairwise repulsive energy (kJ/mole)
Ni(111) - p(2x2)H	270	4.2
Pd(100) - c(2x2)H	260	2.1
Fe(110) - p(2x1)H	$\sim$ 250	$\sim$ 2 - 3

b) The structure of hydrogen adsorbed layers

Using single crystal samples in vacuo valuable structural information can be obtained by all scattering techniques using e. g.
electrons, ions or neutral atoms. To some extent, also high-reso-
lution electron energy loss methods may contribute to a structural
understanding [27] of adsorbed hydrogen phases. In a typical diffraction experiment the beams scattered off the sample elasticly can be
analyzed with respect to their angular distribution (the 'diffraction pattern') and their intensity - energy variation. Unfortunately,
the diffraction of low-energy electrons (LEED) as one of the most
powerful standard techniques for surface structure determination [28]
is not particularly sensitive with respect to hydrogen: The small
Coulomb field makes the H atom a very weak scatterer for low-energy
electrons, and the major diffraction contribution arises from forward scattering events in which the electrons scattered off a H atom
are directed towards the bulk of the sample to undergo further diffraction on the metal atom electric potentials [29]. Nevertheless
it was possible in some cases to obtain reasonably intense H-induced
LEED 'extra' spots, e. g. with Ni(110) [8, 30], Pd(110) [31], Fe(110)
[20, 32], Ni(111) [19, 33], Pd(100) [10], W(100) [34] and Mo(100) [35].
The observed intensity arises either from direct scattering off the
H adatoms, or is produced by metal substrate atoms that have moved to
new equilibrium positions under the influence of the adsorbed hydrogen, a phenomenon referred to as 'adsorbate-induced surface reconstruction'. Within the framework of plausible structural assumptions
the diffraction intensity can be calculated as a function of the electron energy, and in kind of a trial-and-error procedure compared
with the experimentally obtained intensity - energy curves in order
to determine those geometric parameters that give the best fit between theory and experiment [28, 36]. In principle, this method should
be feasible to solve the surface reconstruction problem, too, however, owing to the large number of surface atoms involved in this
process (large unit cells!) a LEED calculation is quite complex
and not always successful. Whether or not hydrogen induces a surface reconstruction has been a matter of controversy for some time.
Although still no general rule can be offered, there are particular
features that point into the one or the other direction: Very weak
fractional order LEED beams (i. e. 1 - 10 °/o of the intensity of an
integral order beam) observed with close-packed surfaces at low temperatures are believed to reflect the 'true' hydrogen scattering
case, whereas rather high intense beams (of the order of the substrate beam intensity) observed with 'open' surfaces at temperatures around 300 K may point to a hydrogen induced surface reconstruction. Examples of this latter kind are the W and Mo(100) faces and
the Ni and Pd(110) faces, where some arguments in favor of a reconstruction model exist, although a complete LEED analysis is still
lacking. The Ni(111)/H system, on the other hand, to our opinion
represents an example where the observed 'extra' LEED pattern is
due to direct scattering off the adsorbed hydrogen phase [19]. In

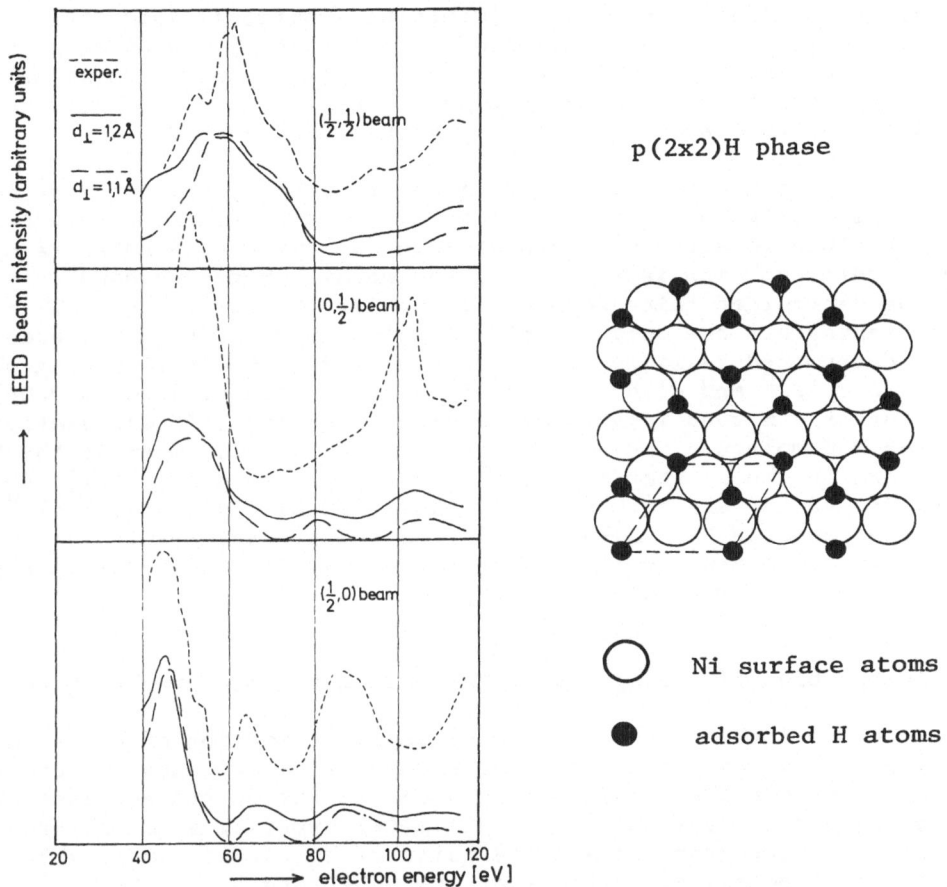

p(2x2)H phase

○ Ni surface atoms

● adsorbed H atoms

Fig. 8. LEED spectra for the p(2x2)H structure on Ni(111). For three beams the experimental (dotted line) and two theoretical curves (full line) are shown. The perpendicular Ni - H layer spacing has been varied from 1.1 Å to 1.2 Å.

Fig. 8 LEED intensity - energy curves ("I(V)-spectra") are reproduced along with the underlying structural model in which the H atoms are located in threefold hollow sites of the Ni(111) surface so as to form a network of hexagonal graphite-like rings [19]. This unique model (with two H atoms in a unit cell) is supported by recent phase diagram calculations [26] or the excitation of phonon losses as observed in ELS experiments by Ibach [37].

There was some progress recently in the determination of the configuration of hydrogen adlayers on Ni(110) by means of atom scattering using thermal molecular beams of helium [38, 39].

Although this method may yield important additional information the uncertainties in the assumption of a realistic atom scattering potential and in the evaluation of the corrugation function as well as the elaborate technique make the He scattering still inferior to the well-established and comparatively simple LEED method.

One of the most interesting quantities that come out a structural analysis is the hydrogen - metal layer distance. If in addition the absolute location of the H atom on the surface is known (a highly coordinated site is the rule), the metal - hydrogen bond length can be determined quite accurately. So far the Ni(111) p(2x2)H structure represents the only example in which the Me-H bond distance has been determined reliably [19]; based on a Ni - H layer spacing of 1.15 ($\pm$ 0.1) $\overset{o}{A}$ and with a threefold site a value of 1.84 ($\pm$ 0.06) $\overset{o}{A}$ for the Ni - H bond length was obtained from the LEED calculation. In the NiH molecule the bond length is 1.47 $\overset{o}{A}$ [40], whereas the Ni-H distance in the cluster compound $Ni_4H_3(C_5H_5)_4$ is 1.69 $\overset{o}{A}$ [41]. (hcp)-bulk Ni-hydride exhibits bond lengths of either 1.62 $\overset{o}{A}$ or 1.87 $\overset{o}{A}$ for the two possible Ni - H bonds [42]. So the present result for the hydrogen chemisorption fits quite reasonably into this relatively wide range of comparable bond lengths.

c) Interaction of hydrogen with the bulk metal and preceding reactions

In the context of a structural characterization of the H adlayer it appears logical to report some experimental evidence for processes which go beyond the simple hydrogen chemisorption and at least involve surface-near regions of the bulk metal crystal, since already these sort of 'precursor' reactions quite often lead to noticeable structural and chemical alterations of the substrate.

Although it cannot be the intention of the present work to give a detailed and complete description of bulk metal - hydrogen interaction phenomena (a vast amount of literature exists in this field [43]), it is deemed useful to make a few remarks on those processes, too, since they are clearly related to their 'precursor' reactions in and near the surface through which the mass transport has to take place. There are basicly three classes of different hydrogen - bulk metal interactions: (a) A simple dissolution of hydrogen gas in the metal crystal, with no stochiometric compound being formed and the H atoms located in interstitial sites (H absorption), (b) a chemical reaction between hydrogen and the metal leading to a stochiometric compound called 'hydride', and (c) a selective chemical reaction between the permeating hydrogen and non-metallic alloy or impurity constituents of the bulk crystal, such as carbon or sulfur, to form gaseous products like methane or hydrogen sulfide which tend to leave the crystal after they have been formed. All three types of reactions are important in the context of hydrogen-induced metal embrittlement and fracture phenomena.

Quite often it is fairly difficult to distinguish the simple hydrogen absorption case from the hydride formation, since only the evaluation of a complete metal - hydrogen phase diagram (which requires careful measurements of the H uptake as a function of temperature and pressure) can reveal the hydride compound existence. H-induced phase transformations or just lattice expansion phenomena as observed by x-ray diffraction alone may not be sufficient to indicate the formation of a stochiometric hydride phase.

Binary hydrides are known to exist for Pd, Zr, Ce, Ta, Th, Ti and V under certain temperature and H_2 pressure conditions. In all of these cases the hydride formation is an exothermic process that should take place readily as long as no kinetic limitations occur. The same reaction, however, is endothermic for Cu and Ni, and for most of the other metals which do not form hydrides but rather dissolve hydrogen gas like Fe, Co, Mo, Ag or Pt. This implies that the uptake of hydrogen should increase as the temperature increases in agreement with experimental observations [44].

If we briefly consider the possible effects of the hydrogen dissolution and hydride formation on the structural and mechanical properties of a metal crystal we can again distinguish basicly three different phenomena. It has to be clearly born in mind that some of these phenomena are also observed in the surface-near regions of the respective metal crystal and are also readily seen with single crystal surfaces under ultra-high vacuum conditions.

(a) H-induced segregation. Hydrogen gas penetrates into the metal lattice and, in getting absorbed, replaces impurity atoms which may either segregate at and near grain boundaries or even move to the metal surface using e. g. screw dislocations as diffusion channels. Reactions of this kind are of some importance in metallurgy as they may create centers that may become sources for fracture processes under mechanical stress conditions.

From a recent study on the hydrogen adsorption on Ni(111) [45] there was strong evidence obtained for a hydrogen-induced segregation of carbon on the surface. The reaction started already around room temperature and with hydrogen pressures in the range of 10^{-5} Pa, and as the temperature of the Ni surface was raised to 300°C, carbon concentrations as high as half a monolayer could be detected by Auger electron spectroscopy. With the iron(110) hydrogen system we have indications that similar carbon segregation phenomena occur under related conditions [20], whereby it is not known whether the H atoms really penetrate into the metal lattice or just use dislocation defects to get inside. As long as there is no suitable means to detect and characterize lattice defects in single crystal surfaces it appears rather hopeless to pursue these interesting questions further.

(b) H-induced lattice expansion. Without changing the overall crystal structure, the uptake of hydrogen can cause considerable (up to 4 °/o) distortions of the crystal lattice with some of the metal atoms being noticeably moved from their former equilibrium positions. It is obvious that a lattice expansion of this kind can give rise to a significant degradation of the mechanical properties of the metal in question as indicated by an alteration of the elastic constants.

There are many interesting examples where quite related phenomena have been observed in hydrogen adsorption studies. One extreme case which was mentioned already is the H-induced surface reconstruction, where metal surface atoms really move under the influence of adsorbed hydrogen. It is not known so far whether part of the hydrogen atoms may even be able to 'dip' into the metal surface, i.e. move beneath the first layer of metal atoms. That these effects occur readily with palladium is not surprising, where even at quite low H pressures (around 10^{-6} Pa) saturation of the adsorbed layer can be attained which represents a reservoir for H atoms to diffuse into the bulk. The structural changes of the surface region (which is exclusively detected by LEED) in the presence of hydrogen can be neatly followed by monitoring LEED spectra as a function of the hydrogen exposure. An example is given in Fig. 9 for a Pd(111) surface interacting with H at room temperature. The pronounced changes in the positions of the intensity maxima as well as in the entire shape of the curves really indicate serious structural alterations of the Pd lattice. Quite related observations have been made with the Pt(111)/H system [12] where not only LEED data but also a significantly reduced preexponential factor in thermal desorption pointed to a considerable hydrogen-induced surface relaxation [47]. Recent ion scattering work revealed a H-induced surface relaxation of Pt(111) by about 1.3 °/o [46], and current experimental LEED work on Fe(110) seems to indicate a surface relaxation due to the presence of hydrogen either [48].

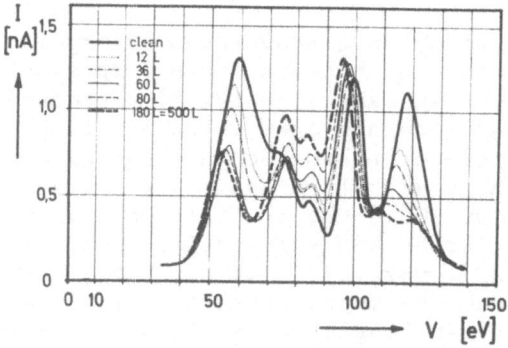

Fig. 9. Variation of the intensity of the specularly reflected (0,0) LEED beam from a Pd(111) surface with the exposure to hydrogen at room temperature. After ref. [49].

(c) H-induced lattice transformation. In this case the entire lattice of the metal under consideration is changed as hydrogen absorption takes place. This process usually requires more vigorous conditions as far as temperature and hydrogen pressure is concerned. Particularly titanium can undergo a H-induced lattice transformation as the titanium hydride is formed. There are many other cases in which alkali or alkaline earth metals react with hydrogen to form the respective hydrides which are no longer of 'metallic' character but are ionic compounds and of course have a lattice structure different from the metal.

d) The electronic interaction between hydrogen and metal surfaces

The presentation of the H - Me potential energy curves in the beginning is nothing but a phenomenological description of the interacting system. It cannot really answer the questions as to why an isolated H atom reacts much stronger with a metal surface than a H_2 molecule, nor why an activated adsorption occurs in the one case and a non-activated adsorption in the other. So far we have always used the term 'chemical' interaction forces, and in order to really elucidate the origin and the nature of these forces one clearly has to treat the hydrogen - metal interaction problem by quantum mechanics, and as a natural consequence also the details of the electronic interaction would be obtained by such a treatment. Within the scope of this paper it is simply impossible to list and explain all the various quantum mechanical theories that have been developed by now in order to describe the hydrogen adsorption problem. It seems reasonable, however, to present just some basic ideas and to refer to a single calculation that has been published recently on the hydrogen interaction with a Ni and a Cu cluster [50].

In Fig. 10 the metal is modelled by a continuum of electronic states that are filled up to the Fermi level. In order to remove an electron from this 'Fermi sea' one has to supply it with the work function Φ which can be assumed to be 5 eV for an average transition

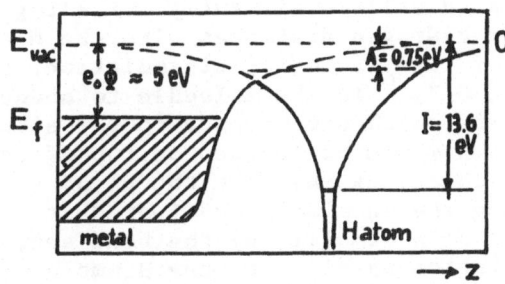

Fig. 10. Schematic illustration of the interaction of a H atom with a transition metal surface

metal like nickel. The H atom is still far away from the surface and
is characterized by a sharp and distinct 1s orbital 13.6 eV below the
vacuum level which acts as an energy reference. Accordingly, the ioni-
sation energy of the H atom is 13.6 eV, however its electron affinity,
i.e. the tendency to fill the 1s orbital with a second electron to
form the H⁻ ion is only 0.75 eV. With respect to the Fermi level of
the metal, the occupied H 1s orbital is about 8 eV lower in energy,
whereas the unoccupied 1s level is approximately located 4 eV above
the Fermi energy. The situation gets altered markedly, if the H atom
now is brought closer to the surface so as to enable its orbitals to
overlap with the sp- or d-wavefunctions of the metal. First of all,
the hydrogen 1s levels get wider and likely change their position on
the energy scale. The coupling between the hydrogenic orbitals and
metallic wavefunctions leads to the formation of two resonance levels,
one with bonding character and located even below the occupied 1s level, the other one anti-bonding and located above the Fermi level.
This quantum mechanical concept was first developed by Newns [51] who
showed that the adsorbing H atom can form sort of a surface molecule
with the metal. The energetic position of the H 1s levels with respect to the metal Fermi energy makes it quite clear that there is no
possibility of a charge exchange neither from the occupied H 1s to
the metal Fermi level nor from the metal Fermi level to the unoccupied H 1s orbital. Therefore the H-Me bond usually is of covalent
character, and experimentally determined dipole moments for adsorbed
H atoms are quite small (about 0.05 D [2]).

Although the relatively simple concept of Newns was able to predict the energetic situation fairly well, it was not able to give a
comprehensive answer to the question as to what the actual role of the
metallic sp- or d- electrons is in forming the bond and how the first
step, namely the dissociation of the H_2 molecule, takes place. In a
recent very promising ab initio calculation by Melius et al. [50]
using a multi-configurational self-consistent field approximation
the H_2 adsorption was modelled by a hydrogen molecule whose molecular
axis was aligned with the axis of a Ni_2 cluster. The Ar-like core of
the Ni atoms was fitted by an effective potential, and the interactions of all the 3d and 4s electrons of the Ni with the H 1s and 2p
orbitals were calculated self-consistently for altogether three different intermolecular hydrogen distances with the Ni-Ni and H-H axis
at a fixed distance of 1.48 Å. During dissociation, the H - H distance has to change from 0.74 Å in the molecule to about 2.49 Å in a
Ni-H - Ni-H arrangement which corresponds to the situation where the
two dissociated H atoms would be located on top of adjacent Ni atoms.
The result of the calculations was that the Ni 3d electrons do not
participate in forming the chemical bond (which is provided mainly by
the 4s electrons) but help stabilizing the bond and, in addition,
provide a lower activation barrier for the H_2 molecule to get into
contact with the 4s electron orbitals by changing their spatial
and spin symmetry. That these results are principally correct is
proven by similar calculations using the Cu_2-H_2 cluster. Whereas a

quite similar binding energy was obtained for the interaction between
the Cu 4s and the H 1s orbital there is an essential difference as
far as the role of the 3d electrons is concerned: With Ni, the spatial
orbital and spin symmetry of the 3d wavefunctions can be changed without a noticeable loss in the total energy of the system and a non-
-activated reaction path for the dissociation of the H_2 molecule can
be offered. With Cu, on the other hand, only a single wavefunction
can be constructed from all the occupied 3d orbitals, and it is not
possible to create an appropriate reaction path for the dissociation.
This is in excellent agreement with the experiment, where just on Cu
surfaces the hydrogen adsorption appears to be activated.

Unfortunately there is evidence that the real situation is more
complicated than the calculation by Melius [50] suggests, at least as
far as the dissociation step is concerned. From recent studies of the
hydrogen adsorption on bimetallic Cu-Ru surfaces in which a Ru(0001)
single crystal surface was partially covered with Cu atoms [52] it
turned out that the uptake of hydrogen of the (normally very active)
Ru surface decreased dramatically as soon as small amounts of Cu
(about 10 $^o/_o$ of a monolayer) were added. Since the Cu was statistically distributed over the Ru surface under the reaction conditions
we had to invoke a pronounced ensemble effect [54] in order to understand the loss in the H uptake. For some reason, the adsorption of
a single hydrogen molecule obviously requires ensembles of about 5-
-10 adjacent Ru atoms, probably with a particular orientation with
respect to each other. It may very well be that only a certain spatial orientation of the impinging hydrogen molecule in conjunction
with a special configuration of the surface atom orbitals allows the
dissociation of the H_2 molecule to occur, and the probability of
having those metal atom configurations in the surface decreases
sharply as the number of blocking Cu atoms increases. It has to be
mentioned here that this is also quite an important issue with respect to heterogeneous catalysis, where it was found [53] that even
small amounts of Cu added to a Ru containing catalyst completely inhibited hydrogenolysis reactions but did almost not influence dehydrogenation reactions. The largely inhibited hydrogen adsorption
offers an elegant way to understand this selectivity effect [52].

CONCLUSIONS

The present survey of the role of hydrogen adsorption in catalysis and metallurgy could only touch some of the problems. It is
nevertheless hoped that some of the ideas presented may contribute
to an additional understanding of the problems that will become even
more challenging in the future. Heterogeneous chemical reactions
that involve a transfer of hydrogen, for example hydrogenation and
dehydrogenation or hydrogenolysis reactions are believed to gain
even more interest in the future as increasingly amounts of fuel

will be produced on the basis of coal (coal liquefication) or natural gas. In order to synthesize light hydrocarbons from fossile fuels
a cracking of complicated organic molecules with high molecular mass
has to be achieved, and in many cases just the hydrogenolysis reaction does the job, particularly in the presence of catalysts. In this
case it would be extremely useful to really know how the catalyst operates, i.e. which species appear in the course of the reaction and
which do not, and which factor is responsible for a certain selectivity that has been found empirically. Although there is still a big
gap between our single crystal work under ultra-high vacuum conditions and the real catalysis as far as pressure, catalyst area and
cleanliness is concerned, there are quite promising attempts to bridge the gap from both sides, and single crystal studies at pressures
in the Pa regime have found a considerable interest in the recent
past [55].

With a given low temperature (T_1) of the condenser, the efficiency of any power plant or reactor can be improved markedly if the
temperature of the high-temperature (T_2) heat reservoir is kept as
high as possible. This holds particularly for the adiabatic chemical reactor where the first reactor stage should operate at high a
temperature as possible in order to get the best conversion for a
given reaction. In principle, most of the processes could be carried
out at high temperatures, if there were no material problems arising
under these conditions. The presence of hydrogen in these cases quite
often is absolutely deleterious to materials like steel or titanium
alloys (which otherwise have excellent properties in that they withstand chemical corrosion even under vigorous conditions). Although
it is felt that our present knowledge on the microscopic processes
that lead to hydrogen-induced embrittlement and fracture is still
poor a cooperation in the field of surface science and metallurgy
with combined surface/bulk studies of hydrogen adsorption, solubility and succeeding reactions will be extremely helpful in getting the
problems solved.

Besides catalysis and metallurgy an increasingly important field
is the use of hydrogen as a working fluid in energy conversion and
storage devices [56]. Special alloys containing e.g. Mg and Ni or La
and Ni which are capable of dissolving large amounts of hydrogen may
be used as absorbers and resorbers in absorption heat pumps with hydrogen acting as a working fluid. They offer particular advantages
at high working temperatures and can thus improve the efficiency of
ordinary steam power plants quite considerably [56]. It has to be re-
membered that also with this kind of application the material problems in handling gaseous hydrogen at elevated temperatures persist.

ACKNOWLEDGEMENTS

It is a pleasure to acknowledge support of this work by the
Deutsche Forschungsgemeinschaft (SFB 128)

REFERENCES

[1]   J.E. Lennard-Jones, Transact. Faraday Soc. 28, 28 (1932)
[2]   K.Christmann, R.J. Behm, G.Ertl, M.A.van Hove and W.H.Wein-
      berg, J. Chem. Phys. 70, 4168 (1979)
[3]   T.B.Grimley and M.Torrini, J.Phys. C 6, 868 (1973);
      T.L. Einstein and J.R.Schrieffer, Phys.Rev. B 7, 3629 (1973)
[4]   P.A.Redhead, Vacuum 12, 203 (1962);
      D.A.King, Surf. Sci. 47, 384 (1975)
[5]   K.Christmann, Z.Naturforsch. 34a, 22 (1979)
[6]   G.A.Goyden, in "Dissociation energies and spectra of diatomic
      molecules", Chapman and Hall, London 1968
[7]   F.Bozso, G.Ertl, M.Grunze and M.Weiß, Appl.Surf.Sci. 1, 103
      (1977)
[8]   K.Christmann, O.Schober, G.Ertl and M.Neumann, J.Chem.Phys.
      60, 4528 (1974)
[9]   C.S.Alexander and J.Pritchard, J.Chem.Soc. Faraday Transact.
      I 68, 202 (1972)
[10] R.J.Behm, K.Christmann and G.Ertl, Surf.Sci. 99, 320 (1980)
[11] B.M.W.Trapnell and D.O.Hayward in "Chemisorption", Butterworth,
      London 1964
[12] K.Christmann, G.Ertl and T.Pignet, Surf. Sci. 54, 365 (1976)
[13] K.Y.Yu, D.T.Ling and W.E.Spicer, J.Catal. 44, 373 (1976)
[14] K.Christmann, Bull.Soc.Chim.Belg. 88, 519 (1979)
[15] K.Christmann and G.Ertl, Surf.Sci. 60, 365 (1976)
[16] R.J.Gale, M.Salmeron and G.A.Somorjai, Phys.Rev.Lett. 38, 1027
      (1977)
[17] S.L.Bernasek and G.A.Somorjai, J.Chem.Phys. 62, 3149 (1975)
[18] K.Christmann, Habilitationsschrift, Universität München 1978
[19] M.A.van Hove, G.Ertl, K.Christmann, R.J.Behm and W.H.Weinberg,
      Solid State Commun. 28, 373 (1978)
[20] R.Imbihl, K.Christmann, R.J.Behm and G.Ertl, to be published
[21] E.Ising, J.Phys. 31, 253 (1925)
[22] L.Onsager, Phys.Rev. 65, 117 (1944)
[23] T.L.Hill, in "Introduction to statistical mechanics", Addison-
      -Wesley, New York 1964
[24] G.Doyen, G.Ertl and M.Plancher, J.Chem.Phys. 62, 2957 (1975)
[25] T.L.Einstein, Surf.Sci. 84, L497 (1979)
[26] E.Domany, M.Schick and J.S.Walker, Solid State Commun. 30,
      331 (1979)
[27] A.M.Baro, H.Ibach and H.D.Bruchmann, Surf.Sci. 88, 384 (1979)
[28] M.A.van Hove, in "The Nature of the Surface Chemical Bond",
      T.N.Rhodin and G.Ertl edts., North Holland Amsterdam 1979
[29] S.C.Ying, J.R.Smith and W.Kohn, Phys.Rev. B 11, 1483 (1975)
[30] J.W.May, Advanc.Catal.Rel.Subj. 21, 219 (1970)
[31] H.Conrad, G.Ertl and E.E.Latta, J.Catal. 35, 363 (1974)
[32] R.Imbihl, Diplomarbeit, Universität München 1980
[33] J.C.Bertolini and G.M.Dalmai-Imelik, Coll.Intern.CNRS, Paris
      1969, pg. 135
[34] P.J.Estrup and J.Anderson, J.Chem.Phys. 45, 2254 (1966)

[35]   M.W.Roberts and C.S.McKee, "Chemistry of the Metal-Gas Inter-
        face", Clarendon Press, Oxford 1978, pg. 396
[36]   J.B.Pendry, "Low energy electron diffraction", Academic Press,
        New York 1974
[37]   H.Ibach and H.D.Bruchmann, Phys.Rev.Lett. 44, 36 (1980)
[38]   K.H.Rieder and T.Engel, Phys.Rev.Lett. 43, 373 (1979)
[39]   T.Engel and K.H.Rieder, Phys.Rev.Lett. 45, 824 (1980)
[40]   A.Heimer, Z.Phys. 105, 56 (1937)
[41]   T.F.Koetzle, R.K.McMullan, R.Bau, D.W.Hart, R.G.Teller, D.L.
        Tipton and R.D.Wilson, Adv.Chem. 167, 61 (1978)
[42]   W.Büssem and F.Gross, Z.Phys. 87, 778 (1934)
[43]   see, for example, "Hydrogen in Metals", edited by I.M.Bern-
        stein and A.W.Thompson, Proceedings of an International Con-
        ference on the Effects of Hydrogen on Material Properties
        and Selection and Structural Design, September 23-27, 1973,
        Champion, Pennsylvania, American Society for Metals (1974)
[44]   F.A.Lewis,"The Palladium Hydrogen System", Academic Press,
        London, New York 1967, pg. 7
[45]   H.Kuipers, Diplomarbeit, Universität München 1977
[46]   J.A. Davies, D.P. Jackson, P.R. Norton, D.E. Posner and
        W.N. Unertl, Solid State Commun. 34, 41 (1980)
[47]   R.J.Madix, G.Ertl and K.Christmann, Chem.Phys.Lett. 62, 38
        (1979)
[48]   R.Imbihl, R.J.Behm, K.Christmann and G.Ertl, in preparation
[49]   K.Christmann, G.Ertl and O.Schober, Surf.Sci. 40, 61 (1973)
[50]   C.F.Melius, J.W.Moskowitz, A.P.Mortola, M.B.Baillie and M.A.
        Ratner, Surf.Sci. 59, 279 (1976)
[51]   D.M.Newns, Phys.Rev. 178, 1123 (1969)
[52]   H.Shimizu, K.Christmann and G.Ertl, J.Catal. 61, 412 (1980)
[53]   J.H.Sinfelt, Y.L.Lam, J.A.Cusamano and E.A.Barnett, J.Catal.
        42, 227 (1976)
[54]   W.M.H.Sachtler, Le Vide 164, 67 (1973)
[55]   H.P.Bonzel, Surf.Sci. 69, 239 (1977)
[56]   G.Alefeld, in "Festkörperprobleme"(Advances in Solid State
        Physics), J.Treusch Edr. Vieweg, Braunschweig 1978, pg.53,
        Vol. XVIII

DISCUSSION

Comment by J.O'M. Bockris:

I think that Dr. Christmann may overstate the case when he says
that the heat of adsorption of H on various metals shows a small
variation. Within the transition metals the variation is some three
times although the points are a bit scattered. But one should add
non-transition metals too. Then, the variation is over around
50 Kcal mole^{-1}.

I suggest that one might be able to see adsorbed H in ellipsometry (Parson and Barret); or in reflectance spectroscopy (Bewick).

Reply:

I did not say that the variation in the heat of adsorption, q_{st}, is small (in fact, it is not as you point out: it ranges from below 30 kJoule/mole for silver up to 150 kJoule/mole for tungsten). What I tried to say is that the variation in the binding energy of a single hydrogen-metal bond is rather small (around 200-250 kJoule/mole), as long as transition metal surfaces are involved. If one wants to include non-transition metals there would be, of course, a much larger variation of q_{st}, in fact on most of these metals hydrogen would not chemisorb at all. The reason for this (and there was too little time to point that out in detail) is believed to be the different electronic structure of these metals. Only transition metals have unfilled d-orbitals which have been shown to be essential in making the hydrogen chemisorb, either by directly participating in the Me-H bond formation, or by just modifying the potential energy surface of the interacting system so as to enable the H_2 molecule to dissociate.

With regard to the optical studies you mention, I am not informed about these activities (there are not too many), but I agree that optical spectroscopies on H adsorbed layers would be an interesting subject to pursue.

Comment by R.A. Oriani:

Has enough work been done to be able to state what are the factors that determine whether or not the adsorption of hydrogen will reconstruct a surface?

Reply:

Theorists try very hard to determine the factors that cause a surface reconstruction with and without hydrogen. The problem, however, is a very complicated one since many atoms are involved. Experimentally, there is some hope that from recent ion backscattering, synchrotron UV photoemission and LEED measurements those factors can be disentangled. Based on LEED studies it is fairly safe to say that appreciable intensities of hydrogen-induced superstructure beams observed around room temperature with crystallographic "open" planes indicate H-induced surface reconstruction. Examples are the Ni (110)/H or W (100)/H systems.

Comment by R.H. Jones:

You showed a difference in H_2 adsorption with changes in surface ledge and kink density. Have similar measurements been made to show the change in H_2 adsorption on crystalline and amorphous material of the same chemical composition?

Reply:

No, I am not aware of such studies. To my knowledge, no metal
can be grown in a really amorphous structure, so one had to try on
glassy materials like certain Pd-Si metallic glasses and on Pd-Si
crystalline alloys of the same surface composition in order to make
the comparison you asked for. Whether those materials, however, will
be homogeneous enough in the surface may be quite doubtful.

Comment by F. Cyrot-Lackmann:

Does your interpretation of thermal desorption spectra take into
account the high surface mobility of hydrogen on the surface. Could
this explain the complicated profiles you find in some cases?

Reply:

Actually, a high surface mobility of a gas in a precursor state
(weakly held molecular state) can give rise to significant effects
on both the desorption peak shape and the desorption rate maximum
(D.A. King, Surf. Sci. 1976). For chemisorbed hydrogen, however,
and in the present situation we have no evidence for any molecular
hydrogen state, the high surface mobility does not affect the interpretation of a TD spectrum, since the hydrogen recombination reaction
is known to be the rate-determining step. The reason for the
occurrence of multiple binding states at higher coverages (β_1 state)
simply is the operation of mutual repulsive interactions between
adjacent adsorbed H atoms which then desorb from lower binding
energy sites.

Comment by H. Vehoff:

For low coverages the work function is proportional to the
coverages. Is there a simple relationship to guess at which
coverages such a proportionality fails?

Reply:

Recent careful work function-coverage studies for various
adsorbates have shown that a strictly linear relationship between
$\Delta\Phi$ and Θ is the exception rather than the rule. Particularly at
medium and higher coverages (and this is known for a long time) the
adsorbed particles interact with each other thereby experiencing not
only a weakening of their bond to the surface, but also dipole-dipole
depolarization forces which lead to an underproportional increase of
$\Delta\Phi$ with respect to Θ. For strongly charged ad-particles there exists
a relationship between $\Delta\Phi$ (or the dipole moment μ which is directly
related to $\Delta\Phi$) and the coverage Θ called the Topping formula which
contains the polarizability α of the adsorbed atom and predicts
a linear relation between

$$\frac{\Theta}{\Delta\Phi} \quad \text{and} \quad \Theta^{2/3}.$$

For hydrogen, however, the charge concentrated on the adatom is so small that no such dipole effects have ever been found. On the other hand, the use of $\Delta\Phi$ as a coverage monitor is allowed in all cases (no matter what the function $\Delta\Phi(\Theta)$ lookes like) where there exists a well-defined functional relationship in which $\Delta\Phi$ depends only on the coverage and not (as may be the case with some ordered adsorbate systems) on the temperature.

ADSORPTION ON METAL SURFACES: SOME KEY ISSUES

J. W. Gadzuk

National Bureau of Standards
Washington, D.C.   20234, U.S.A.

INTRODUCTION

Given that the first experimental observations of the electronic energy levels of an adsorbate-covered metallic surface were published only 15 years ago and that it was not until 1971 that the first photoemission spectrum of an adsorbate was obtained,[1] remarkable progress has occurred over the past decade in understanding, on a microscopic level, the phenomenon of adsorption.[2] The purpose of these notes is not to review the field exhaustively, but instead to provide a perspective on some key issues and their possible resolutions which might be relevant to atomistics-of-fracture-related surface phenomena, in the hope that those readers most interested in fracture will come away with both a feeling for what types of answers the surface-science community can provide and also with a better intuition of the microscopic basis of adsorption.

The phenomenology of static adsorption on metals is illustrated in Fig. 1. Within a one-electron (mean field) picture, the substrate is regarded as an attractive potential well characterised by a broad s-p conduction band of occupied width = E_F the Fermi energy, possibly some narrow, high density of states d-bands centered around ε_d, a workfunction ϕ, and continuum wavefunctions ψ_m constrained to the metal half space if $\varepsilon_m < \phi$. Similarly the adsorbate is characterised by occupied orbitals ψ_a with orbital energy V_i (ε_{ao} with respect to the Fermi level) and unoccupied levels with electron affinity = A. The uncoupled systems are shown in Fig. la. Allowing the adsorbate and substrate to interact results in the formation of total wave-functions ψ_{sys} spread over the complete system. The relative admixture of ψ_m and ψ_a in ψ_{sys} depends upon (or alternatively, is

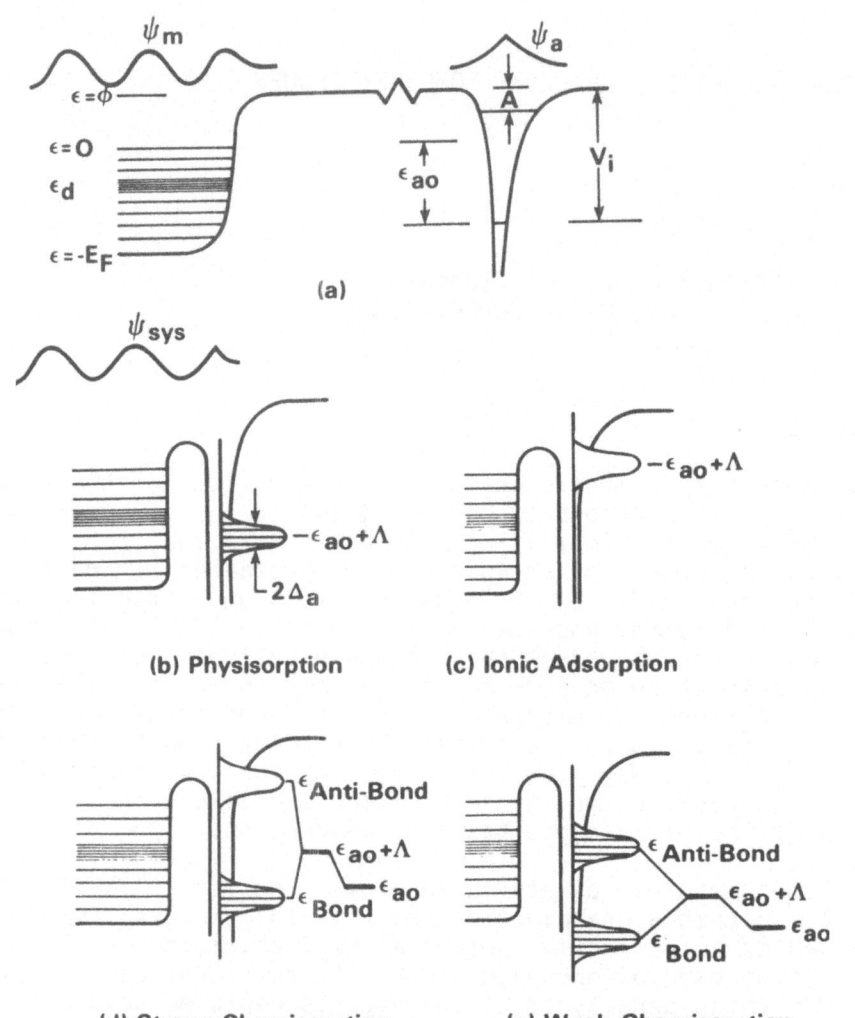

Fig. 1 –   Potentials, energy levels, and wavefunctions for (a)
           non-interacting adparticle and metal, and (b-e)
           various adsorption configurations.

determined by) the nature of the bond formed between the adsorbate and substrate which in turn is intimately tied up with the localised electronic structure in the vicinity of the adparticle. Some possibilities are shown in Figs. 1b-e. The relatively long-range interaction between the substrate s-p electrons and the adsorbate valence electrons effectively shifts the adsorbate level by an amount Λ and spreads the discrete state out into a band of width $2\Delta_a$. The location of the broadened "virtual" level with respect to the Fermi level determines the degree of occupancy which in turn determines the nature of the non-d-electron-involved surface bonding. For instance, as shown in Fig. 1b, the virtual level lies nearly totally below the Fermi level and the adsorbate remains neutral as is characteristic of weakly physisorbed rare gases.[3] In contrast, Fig. 1c shows the case in which the level rises above the Fermi level and electron charge transfer from adsorbate to substrate occurs, as is typical of strongly ionically adsorbed alkali atoms.[4] The intimate relationship between these two regimes has been demonstrated by Flynn and coworkers.[5] If the adsorbate can sit close enough to the surface so that there is significant overlap between the substrate d-electrons and the adsorbate valence levels, bonding and anti-bonding "molecular orbitals" can form. For the strongly interacting case shown in Fig. 1d, only the bonding orbitals are occupied and a strong chemical bond is formed. On the other hand, when both bonding and anti-bonding orbitals are occupied, as shown in Fig. 1e, the resulting bond or chemisorption energy would be small.

One of the goals of adsorption studies is to understand the properties of composite surfaces in terms of the properties of the non-interacting components. From Fig. 1, one would presume that starting input data include substrate band structure, symmetry-classified wavefunctions, crystal structure, and adsorbate electronic structure and wavefunctions. Through judicious interplay between spectroscopic and structural experiment and theory, the electronic and geometrical properties of the coupled system can be determined, bonding situations classified in terms of the scheme outlined in Fig. 1, and from these microscopic properties, macroscopic ones ascertained.

In the remainder of these notes, I will present my view of some key issues and the ways in which the surface science world is dealing with them. After establishing some basic theoretical background, I will concentrate on questions in which experiment and theory are working together. In Section II, the abstract problem of adsorption is elucidated and some comments on single-electron vs. many-body effects are made. Some of the more prevalent "ab initio (ish)" modeling procedures together with computationally simplified approximations to them will be discussed in Section III.

Section IV is devoted to some brief comments on surface-sensitive spectroscopies. The question of non-adiabatic dynamical surface processes (rather than statics) is raised in Section V. Finally the specific case studies of a) alkali adsorption, b) series of transition metals adsorbed on a transition metal, c) O, H, CO adsorbed on a series of transition metals, d) dissociative chemisorption vs. molecular physisorption, and e) relationship between surface electronic properties and hydride kinetics are discussed in Section VI.

II. ADSORPTION IN THE ABSTRACT

Starting with the proposition that "everything" is known about the adsorbate, characterised by a many-body Hamiltonian H_{ads} $(\underset{\sim}{r};\underset{\sim}{R})$ with

$$\underset{\sim}{r} = \underset{i=1,n}{\Sigma} \{\underset{\sim}{r_i}\} \quad \text{and} \quad \underset{\sim}{R} = \underset{j=1,N}{\Sigma} \{\underset{\sim}{R_j}\}$$

the set of coordinates of all n electrons and N nuclei or ion cores of the adsorbate and also about the substrate, similarly characterised by $H_{sub}(\underset{\sim}{r'};\underset{\sim}{R'})$, the eigenstates of the coupled system

$$H = H_{ads}(\underset{\sim}{r};\underset{\sim}{R}) + H_{sub}(\underset{\sim}{r'};\underset{\sim}{R'}) + H_{int}(\underset{\sim}{r},\underset{\sim}{r'};\underset{\sim}{R},\underset{\sim}{R'}) \qquad (1)$$

are desired. In obvious notation, H_{int} denotes the adsorbate-substrate interaction which depends upon the coordinates of both systems. With a Born-Oppenheimer treatment of nuclear coordinates, taken as numerical parameters rather than dynamical variables, the many-body spectral distribution

$$\rho(E;\underset{\sim}{R},\underset{\sim}{R'}) = \frac{1}{\pi} Im< 0|\frac{1}{E - H + i\delta}| 0> \qquad (2)$$

contains all the information we need about the coupled system. Here $|0>$ is the <u>complete</u> electronic ground state for given $\underset{\sim}{R}$, $\underset{\sim}{R'}$. The total energy is expressed as

$$E_{Tot}(\underset{\sim}{R},\underset{\sim}{R'}) = \int E\rho(E;\underset{\sim}{R},\underset{\sim}{R'})dE + E_{elec}(\underset{\sim}{R},\underset{\sim}{R'}) \qquad (3)$$

where E_{elec} represents nuclear electrostatic contributions. The adsorption geometry is given by that set of nuclear coordinates satisfying

$$\delta E_{Tot}(\underset{\sim}{R},\underset{\sim}{R'}) = \frac{\partial E}{\partial \underset{\sim}{R}}Tot \cdot \delta \underset{\sim}{R} + \frac{\partial E}{\partial \underset{\sim}{R'}}Tot \cdot \delta R' = 0 \qquad (4)$$

together with a check that this represents a minimum. The fact that several sets of $\underset{\sim}{R}$, $\underset{\sim}{R'}$ might satisfy Eq. 4 just says that metastable adsorption arrangements are possible. Local minima might occur if the starting geometry of the substrate $R' = \Sigma\{R'_j\}$ is chosen to include surface imperfections, steps, edges, or cracks. Needless to say, carrying out such a program is impossible without the introduction of severe approximations. Chemisorption theory has been concerned with the development of approximations to Eqs. 1-4 which enable feasible but also meaningful calculations which in turn may lead us to a well-grounded intuition of the chemisorption bond and then onto something like Pauling rules for surfaces.

By far the most common simplification is some version of independent-electron, mean-field theory in which each electron is imagined to move in the static, self-consistent field provided by all other electrons plus ion cores frozen into a prescribed geometry; the particular choice is oftentimes prompted by independent surface crystallographic determinations or billiard ball reasoning. Exchange and correlation effects are usually simulated by an effective potential V_{xc} obtained within the local-density approximation, that is $V_{xc}(r)$ is taken to be identical with that of a uniform electron gas of density $n_- = n_-(r)$, the density at point r where V_{xc} is sampled.[6] Within this scheme, the ground-state properties of the coupled system can be obtained from the self-consistently determined charge density which follows from the set of equations

$$[-\frac{\nabla^2}{2} + V_{eff}(r;R,R')]\ \psi_{sys}^i(r;R,R') = \varepsilon_i(R,R')\ \psi_{sys}^i(r;R,R')$$

$$V_{eff}(r;R,R') = V_{elec}\ (r;\ [n_-(r;R,R')];\ R,R') \qquad (5a\text{-}c)$$

$$+\ V_{xc}\ [n_-(r;R,R')],$$

$$n_-(r;R,R') = \sum_{i=i}^{n} |\psi_{sys}^i(r;R,R')|^2$$

where it is to be noted that V_{elec} is the total electrostatic interaction involving electrons, ion cores, and external fields and V_{xc} is assigned a specific functional form[6], usually chosen proportional to $n_-(r)^{1/3}$. Solution of Eqs. 5a-c for the inhomogeneous electron charge density is a standard, albeit difficult and time-consuming, exercise in computational physics. The advantage of focusing on the charge density follows from a theorem due to Hohenberg and Kohn[7] where it is proven that the ground-state wavefunction and thus all ground-state properties are unique functionals of the density $n_-(r;R,R')$. Of all possible self-consistent solutions to Eq. 5, the correct one is obtained by minimizing the energy functional

$$E_{Tot} = E_{Tot}\ [n_-(r;R,R');R,R'] \qquad (6)$$

where E_{Tot} is a universal functional of the electron charge density. Actual execution of this procedure has been well discussed elsewhere.(2b,2e,6,8)

An advantage of one-electron theory is that contact is easily established with intuitive concepts such as orbitals, bands, bonds and their interrelationship.(9) Descriptions of localised adsorption events are often given in terms of various density of states (DOS) such as the local DOS

$$\rho_{LOC}(\underset{\sim}{r},\varepsilon;\underset{\sim}{R},\underset{\sim}{R}') \equiv \Sigma_i |\psi_{sys}^i(\underset{\sim}{r};\underset{\sim}{R},\underset{\sim}{R}')|^2 \; \delta(\varepsilon-\varepsilon_i(\underset{\sim}{R},\underset{\sim}{R}')) \tag{7}$$

where ψ_{sys}^i is the coupled system "wavefunction" (actually a solution to Eq. 5a, which is not exactly the same thing as a one-electron orbital) depending upon the adsorbate-substrate separation and displaying the qualitative features of ψ_{sys} in Fig. 1. Another useful quantity is the projected DOS such as that of the adparticle orbital onto system states:

$$\rho_{ad}(\varepsilon;\underset{\sim}{R},\underset{\sim}{R}') = \sum_i |<\psi_a|\psi_{sys}^i(\underset{\sim}{R},\underset{\sim}{R}')>|^2 \delta(\varepsilon-\varepsilon_i(\underset{\sim}{R},\underset{\sim}{R}')) \; . \tag{8}$$

As an example, the projected adparticle density of states within the virtual resonance limit shown in Fig. 1b can be obtained from Eq. 8 in the form:

$$\rho_{ad}^{vir}(\varepsilon;\underset{\sim}{R},\underset{\sim}{R}') = \frac{1}{\pi} \frac{\Delta_a(\underset{\sim}{R},\underset{\sim}{R}')}{(\varepsilon-\varepsilon_{ao}-\Lambda(\underset{\sim}{R},\underset{\sim}{R}'))^2 + \Delta_a(\underset{\sim}{R},\underset{\sim}{R}')^2}$$

where the level shift and width functions Λ and Δ_a depend upon the ion-core geometry and hence the separation.

It is to be noted that all quantities presented in Eqs. 5-8 depend parametrically on the geometry $\{\underset{\sim}{R},\underset{\sim}{R}'\}$ which we would like to be able to determine from a minimization procedure such as Eq. 4. In practice this has not been possible to date other than in some idealised situations discussed in the next section.

Consideration of many-body effects usually centers on the role of non-local and frequency-dependent electron-electron interactions due to the polarizable nature of the electron gas, plasmon-related phenomena, interpretation of core-level spectroscopies, and non-adiabatic surface processes. Due to space limitations, systematic exploration of their important implications will not be attempted here. When necessary, mention will be made within the text.

III. TRACTABLE ADSORPTION MODELS

Various simplified classes of models have been developed, which enable numerical evaluation of Eqs. 5a-c, under restricted conditions. The simplifications generally include the introduction of some symmetries into the system and oftentimes the approximation of continuous infinite systems by discrete finite ones. The three most prevalent models which permit some sort of ab initio-like calculation, namely clusters, jellium, and films, are shown in Fig. 2.

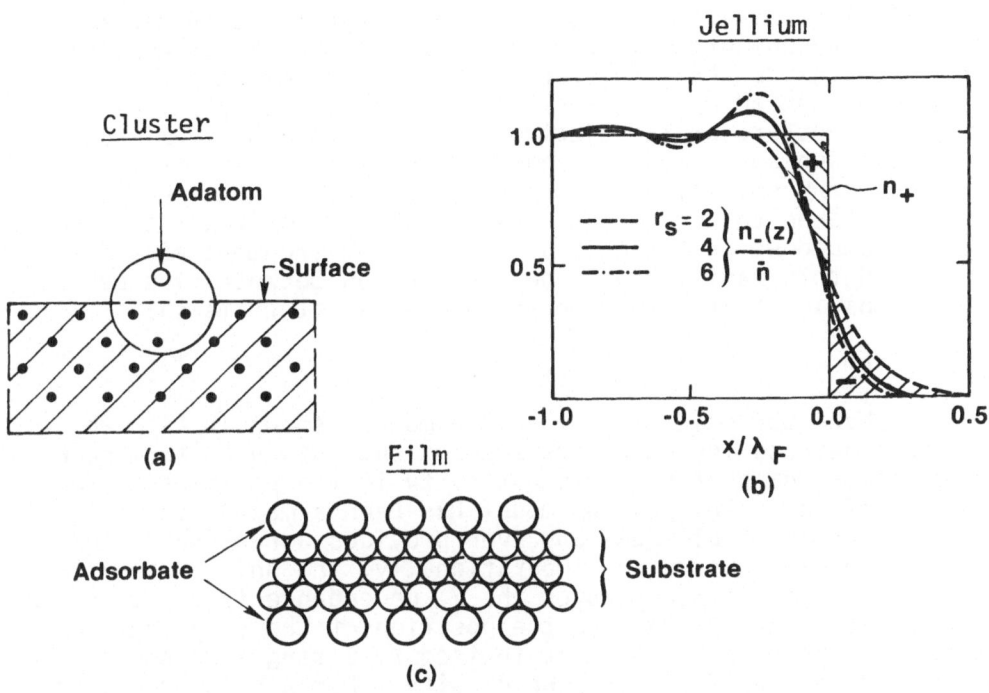

Fig. 2 - Tractable surface models for ab initio calculations.
    a)   Finite cluster embedded in an anisotropic background (From R. O. Jones in Ref. 2f).

    b)   Jellium characterised by smeared out positive background $n_+ = |\bar{n}|$ and self-consistent electron charge density $n_-(z)$. (From O. Gunnarsson in Ref. 2f).

    c)   Finite thickness film with translational invariance in parallel planes.

3.1a Clusters

The adsorption bond is considered to be localised. Thus a single adparticle on a semi-infinite substrate is hoped to have similar properties to that of the same adparticle interacting with a few substrate atoms, for which quantum-chemical calculational procedures of finite systems are possible. In addition to the chapter by Johnson in this volume, other reviews are recommended.[10]

The main advantages of this procedure are that calculations can be done and the results are easily mapped onto intuitive chemical notions. A number of shortcomings exist though. Unless the cluster is taken to be quite large, with several adsorbates within it, the model is at best applicable for isolated single adsorbates and not ordered mono-layers whose properties depend upon adsorbate-adsorbate interactions. Any phenomena (such as inelastic or non-adiabatic processes) which depend upon the continuum structure of the electronic excitation spectrum are poorly represented by cluster models. Most severe is the "outer-sphere-boundary condition" in which the potential outside the cluster is taken to be constant. As shown in Fig. 2a, the outer sphere is embedded in a highly anisotropic environment at a surface and this fact could render comparison between cluster calculations and flat-surface phenomena somewhat whimsical.

3.1b Jellium

The characteristic feature of the model, shown in Fig. 2b, is that the substrate ion cores are smeared out into a uniform background with a sharp step or discontinuity in the positive charge at the "surface". The jellium model has been a major testing ground for theoretical ideas and even provides a reasonable model for simple metal substrates. Operationally, the uniform smearing of the background charge removes the $\underset{\sim}{R}'$ dependence from Eqs. 5 and 6 (other than keeping track of the position of the step edge) and allows the substrate to be characterised by a single parameter, the conduction-band free-electron density. For atomic or diatomic-molecule adsorption in which the diatomic approaches the surface head on, cylindrical symmetry along the surface normal provides essential simplifications in numerical procedures. If the diatomic-bond distance is held fixed (not necessarily justified), then the set of adsorbate $\underset{\sim}{R}$ coordinates reduces to a single parameter z, the adsorbate-jellium step-edge separation which further simplifies calculations. Crystallography is usually dealt with by perturbatively inserting a lattice. Lang and coworkers[11], the Scandanavian group (2b,6,12,13) and Ying(2d) are most closely identified with development of this model.

3.1c Films

The finite-thickness film model, shown in Fig. 2c, has the advantage that much of the machinery developed in band-structure theory can be adapted here. The essence of the model is that periodicity is maintained in planes parallel to the surface so a basis set of two-dimensional Bloch functions with continuous wavevectors $k_{||}$ and band indices can be identified, although these functions do require an additional quantum number to represent the discrete quantization in the z direction. For practical calculations, the adsorbate must be in an ordered layer commensurate with the substrate, thus eliminating the possibilities of describing isolated or low-coverage adsorption events. The film must be sufficiently thick (current calculations have ~13 layers as an upper limit) so that opposing surfaces do not "see" each other and also so that "bulk" results are reasonably simulated within the interior. With these limitations, the ordered-film technique permits the most realistic modeling of surface electronic structure.[14]

3.1d Generalities

A crucial ingredient in calculations of surface electronic structure is self-consistency, which is an attainable end in the three models just presented. There is a simple reason why self-consistency is so much more important at a surface than within the bulk. From the electronic point of view, the electron density can extend out into the vacuum as far as required to form a minimal energy configuration whereas in the bulk, the total charge per atom is constrained to remain within the Wigner-Seitz cell (although energy minimization determines the size and shape of the cell, this is usually regarded as independent input data). The surface dipole potential is a direct manifestation of the electron charge asymmetry. In addition, the ion cores of the surface atoms can rearrange to lower energy configurations, if necessary. Williams has given this reconstruction, a prevalent phenomenon in transition metal surfaces, an intuitively appealing basis.[15] Within a transition metal, attractive bonds are formed amongst the d-electrons on neighboring sites which, on their own would compress the lattice. However, compression of the s-p electrons is repulsive and this resistance balances the d-electron attraction at the equilibrium lattice configuration. At the surface, the s-p electrons can resist the compression by squirting out into the vacuum, thus allowing for tighter d-bonding, as seems to be observed experimentally.

In spite of the achievement of self-consistency, these models do have limitations as already mentioned which restrict applications to highly symmetric situations. Nonetheless, knowing the physical consequences of self-consistency from such ab initio models, approximate models can then be constructed in which those consequences are built in and which allow consideration of more general geometries.

3.2a Effective Medium Theory

The basic idea of this scheme is that a simple relationship should exist between the total energy of a system in which an adsorbate is placed in the exponentially decreasing (inhomogeneous) electron charge-density tail sticking out of a substrate and a system in which the same adsorbate is immersed in a homogeneous jellium characterised by a density equal to that of the local density at the actual adsorption site[16]. Obviously, calculation of the latter quantity is much easier. Thus rather than evaluating Eq. 6 for the inhomogeneous electron gas, one proceeds by first determining $n_-^o(\underset{\sim}{r})$, the free-surface charge density. Next,

$$E_{Tot}^{im} = E_{Tot}^{im}(n_-^o)$$

the total energy for the impurity within the homogeneous electron gas is obtained using standard techniques[16,17] and finally the adsorbate-substrate interaction is taken as

$$E_{Tot}(\underset{\sim}{R}) = E_{Tot}^{im}(n_-^o(\underset{\sim}{r}=\underset{\sim}{R})) + \delta E$$

where the correction δE is hoped to be small. At least for simple adsorbates on jellium, this technique, with $\delta E = 0$, yields bonding energies and equilibrium separations in reasonable agreement with ab initio calculations. The method could be useful for surfaces containing imperfections (cracks and steps) for which first principles adsorption calculations are out of the question but for which unperturbed electron charge densities might be obtained.

3.2.b Parameterised Model Hamiltonians

The conceptual development of adsorption theory has oftentimes been carried out within the framework of a model Hamiltonian which emphasizes the role of a particular selected bonding mechanism. Analytic solutions to quantities such as projected density of states and binding energies are obtained in terms of a few parameters which are in principle calculable, but which can also be deduced from spectroscopic data. These parameters include hopping or resonance

integrals which are a measure of a given electron's desire to move
to a neighboring site, intra-atomic Coulomb integrals, overlap
integrals, and substrate bandwidths. The model-Hamiltonian approach
is really a solid-state equivalent of quantum-chemical schemes such
as CNDO, Huckel, etc. Since the calculated results are so directly
related to the values of the input parameters, this approach is most
meaningful for studying systematic trends rather than absolute numbers,
a task made easy by the analytic structure of the "solutions". The
current status of Model Hamiltonians has been well reviewed by
Einstein, Hertz and Schrieffer in Ref. 2d.

3.2.c Surface-adapted-alloy heat of formation theories

The empirically deduced relation for binary AB alloy heats of
formation, due to Miedema[18] and given by

$$\Delta H(AB) = -P(\Delta\phi_{AB}^{*})^2 + Q(\Delta n_{AB}^{1/3})^2 \qquad (9)$$

has inspired a considerable body of theoretical inquiry[19] in search
of a microscopic justification, as discussed by Pettifor in this
volume. In Eq. 9, $\Delta\phi_{AB}^{*}$ is the difference in workfunctions of the
base materials, presumably directly related to an electronegativity
difference, and thus a measure of charge transfer, $\Delta n_{AB}^{1/3}$ is the
difference in charge-density tails at the respective Wigner-Seitz
cell boundaries (a measure of the intra-cell electronic readjustment
required for boundary matching), and P and Q are universal constants.
To the extent that the physical consequences of metallic bonding
are contained in a simple law such as Eq. 9, it seems reasonable to
believe that the physics uncovered in the microscopic investigations
of its basis[19] would be similar to the physics of a surface
bond formed between A adsorbed on B. This proposed similarity
has already been exploited in an interpretive study of surface
segregation in binary alloys.[20] As will be further detailed
in Section VI.3, Varma and coworkers have constructed a theory
for heats of adsorption on transition metal surfaces based upon
alloy studies.[21] They argue that the general features of the
electronic structure and not the fine details are the significant
factors. In particular, such quantities as low-order moments of local
and projected density of states changes, bandwidths, d-electron
count, and Fermi-level position are most important. Bonding is
taken to occur via d-electron overlap and s-p electrons enter
mainly through screening renormalization of various parameters.
Trends in adsorption data are reasonably reproduced.

Approximate theory of this type, if put on a firm foundation,
should be quite useful in areas such as fracture in which the geometry is so irregular that any promise for ab initio calculations
is hopelessly far into the future.

IV. SURFACE SPECTROSCOPY

Much of our newfound knowledge of surfaces has been a result of the development of surface-sensitive spectroscopic techniques. In general, surface sensitivity is achieved by having either the incident or emitted beam (or both) as a charged particle, usually electrons. Due to the $\sim 10\text{A}$ inelastic mean free path for ~ 5-500 eV electrons, any events which involve non-inelastic electrons necessarily occur within $\lesssim 10\text{A}$ of the surface.[22] The spectroscopic probes are used to study electron valence and core levels [2b,2c,2e,23] and vibrational states[24].

4.1 Valence levels

The relatively weakly bound valence levels are most commonly investigated with ultraviolet and soft x-ray photoemission. Energy-resolved studies yield energy spectra, which when compared with gas-phase spectra, provide detailed information about bonding mechanisms. Angle-resolved spectra in addition yield information on orbital shapes, bonding geometries, and surface band structure (dispersion).

4.2 Core Levels

X-ray photoemission and Auger-electron emission are the most common techniques used for obtaining core-level spectra. Since the reasonably discrete core electron binding energies (~ 100-1000 eV) are, to a good first approximation, a characteristic elemental signature, these techniques are widely used for compositional analysis. In addition, small (~ 1-10eV) shifts in binding energies, when properly interpreted, can yield information on the chemical environment. This interpretation requires the decomposition of an observed shift into so-called initial-state chemical shifts and final-state relaxation shifts.

4.3 Vibrational Spectroscopy

High resolution (~ 1-8 meV) electron energy-loss, infrared absorption, and Raman spectroscopy provide information about the characteristic vibrational modes associated with an adsorbed surface complex. Both the absolute frequencies of the bonding mode as well as surface-induced shifts and widths of perturbed intra-molecular modes are used to ascertain bonding geometries and some electronic properties of the chemical bond.[25]

An interesting proposal by Oelhafen[26] demonstrates in a direct way, possible connections between electron spectroscopies and alloy heats of formation discussed in Section 3.2c, at least for binary alloys with non-overlapping d-bands such as those

between early and late transition metals. Photoelectron spectra
for pure metals A and B and for the AB alloy are schematically
shown in Fig. 3. The point to note is that both individual core-
level shifts and d-band centroid shifts for each constituent can be
identified. If it is assumed that the valence-band spectra is
proportional to the d-band density of states, then it is possible,
within the context of the band theories[19], to obtain the expression

$$\Delta H(AB) = N_d^A(\Delta\epsilon_{core}^A + \Delta E_d^A) + N_d^B(\Delta\epsilon_{core}^B + \Delta E_d^B) \qquad (9)$$

for the heat of formation, where N_d^A and N_d^B are the average number
of d electrons per atom of the constituents and ΔE_d^i is the energy
shift of the d-band centroid shown in Fig. 3. All quantities
appearing in Eq. 9 are direct observables. The values for $\Delta H(AB)$
deduced from Eq. 9 and photoemission spectra for a large number of
alloys are in quite satisfactory agreement with the heats determined
theoretically[18,19] and, when possible, experimentally.

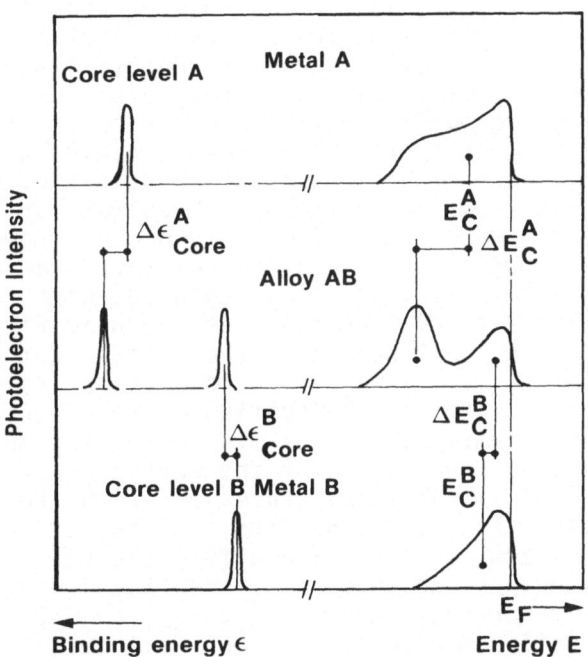

Fig. 3 - Core and valence level photoemission spectrum showing
          the relationship between spectra for pure A and B
          transition metals and for binary AB alloys. (Ref. 26).

V.   FROM STATICS TO DYNAMICS

Actual processes at surfaces occur in real time for which non-adiabatic effects are very important. As discussed in Section 6.5, the kinetics of hydrides is a good example in which non-adiabaticity in surface processes could be an overall rate-limiting factor. Traditionally, substrate phonons have been considered to be the source of any energy dissipation[27], but recently the important role of substrate electron-hole pair excitation has been receiving much attention[27,28]. Due to the energy mismatch between substrate-phonon spectra and energy requirements in surface reactions, the continuum of electronic excitations could be the more significant factor in non-adiabatic processes.

As an illustration consider the situation of an atom or molecule far removed from the metal substrate and which possesses two internal electronic states $|1>$ and $|2>$, each with a "significantly" different charge distribution. For instance, these states could represent A and A^+ (an atom and ion), or H_2 and 2H (a diatomic and dissociated molecule). Allowing the particle to interact with the substrate, subject to the constraint that no substrate-mediated mixing between $|1>$ and $|2>$ occurs, two different particle-surface interaction curves result, as for instance in Fig. 4a. Within the Born-Oppenheimer approximation, these curves, characteristic of the electronic state of the system, serve as potential-energy surfaces over which the nuclei move. As shown, the curves cross at some distance $z=R_c$ from the surface. If now we allow for substrate mixing, this can be represented by adiabatic curves with avoided crossings[29] or by the diabatic curves shown in Fig. 4a in which case a diabatic electronic transition[28c] between internal states $|1>$ and $|2>$, in the region of $z=R_c$, may occur. However, in order that nuclear motion (and thus the reaction) can continue onto curve 2, a vibrational level of 2 must coincide with K_i, the initial nuclear translational energy. As shown in Fig. 4a, this is unlikely and thus the $1 \rightarrow 2$ reaction is blocked. But this picture takes the substrate as a static entity, always in its ground state. If proper account is taken of the continuum of substrate-electronic excitations, then a family of curves over which the nuclei can move is required. Fig. 4b shows a few of the continuum of curves in which the particle is in internal state $|2>$ with the substrate in some excited state. Each curve has its own vibrational spectrum and Franck-Condon factors connecting with incident nuclear states. If pair excitations were not possible, then the $1 \rightarrow 2$ diabatic transition could occur only when $K_i = \varepsilon_{vi}$, ε_{vi+1}... On the other hand, with the continuum of excited electronic curves, there are always vibrational states available, for any energy K_i, and this fact removes one inhibition blocking the diabatic transition. Thus the surface reaction rate certainly depends upon the probability for creating low-energy substrate electron-hole pair excitations, which in turn must be related to the electronic structure at the Fermi level. An intriguing example of this will be presented in the next section.

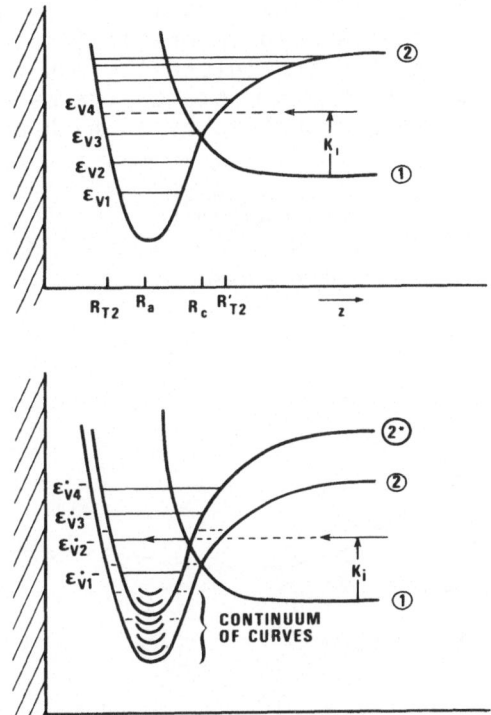

Fig. 4 - Diabatic potential energy curves for a particle initially in some electronic state 1 with kinetic energy K, incident upon a surface where an internal transition to state 2 may occur.

     a)   Two state system with a passive substrate.

     b)   Continuum system when substrate excitation is possible.

VI. CASE STUDIES

Having spent considerable effort in understanding some of the fundamentals of static and dynamic adsorption theory, we are now at a point where it is reasonable to look at some specific applications of these general ideas.

6.1 Alkali Adsorption

Adsorption of alkali films on a variety of substrates has been the focus of both experimental and theoretical inquiry for many decades.[4,30] This longstanding interest stems from the fact that submonolayer alkali adsorption on most substrates can reduce the workfunction of the composite surface by as much as $\sim$ 3 eV, as shown in Fig. 5.[31]

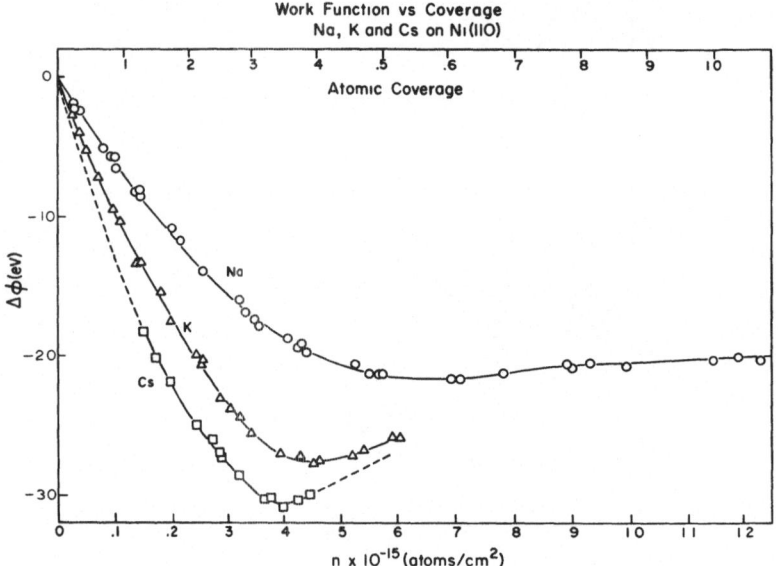

Fig. 5 - Change of surface workfunction vs. coverage for Na, K, and Cs adsorbed on Ni(110) (Ref. 31).

The basis of this behavior is easily understood upon consideration of the ionic adsorption illustration in Fig. 1c. In general the alkali atoms, which to begin with have low ionization potentials, experience a sufficient upward shift in their valence electron level that at the equilibrium adsorption separation, the broadened level lies mainly above the Fermi level. Consequently the valence electron hops into the substrate, leaving a positive ion with a fractional net charge, which in turn induces a screening ("image") charge constrained to remain within the substrate. A net dipole

moment is associated with this charge configuration. A low coverage
array of such dipoles forms an approximately continuous dipole
layer directed opposite that of the pure substrate dipole barrier,
thus initially reducing the workfunction by an amount proportional
to the coverage and to the dipole per alkali, as shown in Fig. 5.
At high coverages, interactions between adsorbates reduce this
effect, leading to a minimum and eventual attainment of a workfunction characteristic of the pure alkali.

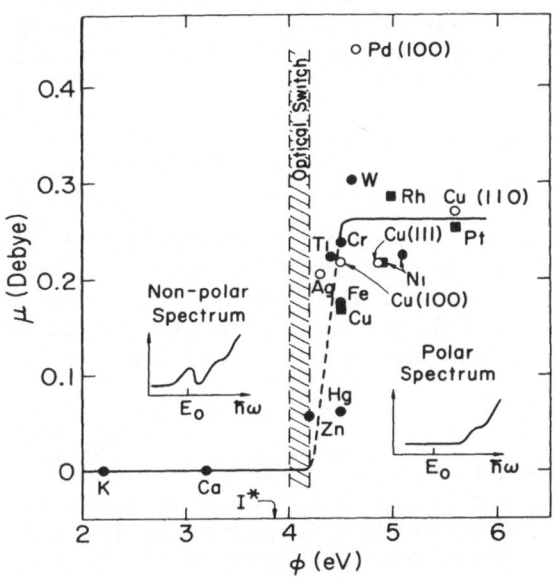

Fig. 6 -   Variation with substrate workfunction of the dipole moment
           per adsorbed xenon atom. The dipole moment emerges at the
           optical charge-transfer instability (hatched band), which
           separates complexes giving polar spectra from those giving
           non-polar spectra. (Ref. 5).

     An intriguing possible demonstration of the interrelationship
between bond polarity, substrate workfunction, and adatom electronic
structure has recently been suggested by Flynn and Chen, within the
context of physisorption (Fig. 1b) of Xe on numerous substrates.
The atomic ground state of Xe is the closed shell $5p^6$ configuration.
Upon adsorption, the substrate perturbation induces a configuration
interaction with excited states, principally the $5p^5 6s^1$ configuration
whose chemical properties are similar to those of alkali atoms.
However, depending upon the substrate workfunction, the Cs-like
component of the adsorbed Xe ground state may "hop" into the substrate, as in ionic adsorption, resulting in a polar "physisorption
bond" for the same reasons as in alkali adsorption. If ϕ is larger
than an effective (substrate independent) "ionization potential" for
the excited $5p^5 6s^1$ configuration, then the bond should display a

noticeable dipole moment. Available Xe dipole moment vs. substrate workfunction data, assembled by Flynn and Chen,[5] is shown in Fig. 6, where the "sharp" transition between non-polar and polar physisorption correlates very well with substrate workfunction, in agreement with the virtual level picture of adsorption.

6.2 Transition Metals on Transition Metals

Transition metal bonding has also received considerable attention due both to the technological importance of transition metals and also to the prototypical role given to the d-electron-determined properties of these metals. The underlying philosophy in the cohesive theories of pure transition metals,[32] alloys,[19] and adsorbed layers[2a,33] is one in which the relevant energy is somehow simply related to the d-orbital occupation number. This is discussed in some detail by Pettifor in this volume.

The essence of the argument is the following. Within a one-electron band picture, the interaction between atoms broadens the discrete d state into a band of width W. The assumed uniform-density-of-states ($\rho(\epsilon)=10/W$) band is filled up to the Fermi energy determined by the number of d-electrons

$$\epsilon_{Fermi} = \frac{W}{2}\left(\frac{N}{5}-1\right)$$

with respect to a band minimum at $-W/2$). Since the bottom (top) 5 states are bonding (anti-bonding), the cohesive energy increases with d-count up to N=5 and then decreases as the anti-bonding states must be filled to accommodate the extra d-electrons. Assuming the cohesive energy to be purely of d-origin, the parabolic form

$$\epsilon_{coh} = \int_{-W/2}^{\epsilon_F} \epsilon\rho(\epsilon)\,d\epsilon = \frac{WN(N-10)}{20} \tag{10}$$

easily follows. Various levels of sophisticated improvement on the quantification of Eq. 10 have appeared, but the basic notion of an approximately symmetric energy expression peaking at N=5 is usually still obtained.

Plummer and Rhodin[34] have measured binding energies of the 5d transition metals adsorbed on various single crystal facets of a W field emission tip and the results are shown in Fig. 7. The general trend of the data is in qualitative accord with an expression of the form of Eq. 10, suggesting that d-orbital mechanisms involved in chemisorption bonds are of the same qualitative origins as in solid bonding. Since no ab initio calculations capable of suggesting d-electron systematics in adsorption have yet been carried out, one is well advised to study the solid heat-of-cohesion theories[19] for additional insights that might pertain to adsorption data such as that shown in Fig. 7.

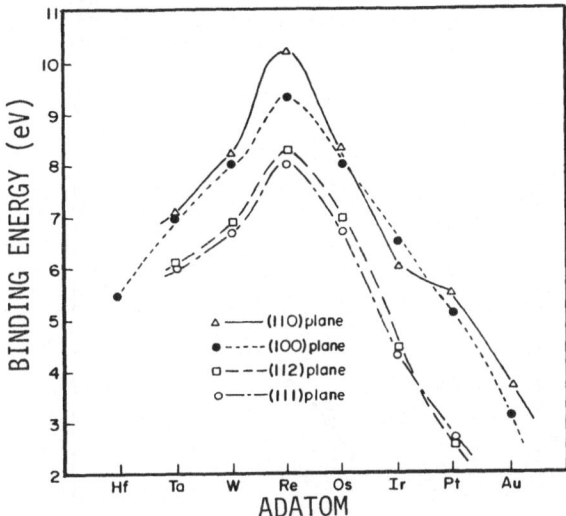

Fig. 7 - Binding energy of the 5d-transition-elements adsorbed
on the four low index planes of tungsten. (Ref. 34).

Note that theoretical work based on model Hamiltonians has been
carried out[33] which does improve upon Eq. 10. Finally, systematic
studies of other properties such as surface energy and activation
barriers for diffusion show trends similar to those of Fig. 7 and
Eq. 10.

6.3 Gases on a transition metal series

In contrast to the N=5 peaking of transition metal binding
energies discussed in 6.2, gases such as O, H, and CO show a
general trend in which the binding energy for all adsorbed species
is maximum at the beginning of the transition-metal series and
decreases as the number of d-electrons increases. The qualitative
aspects of this trend, shown for the case of oxygen in Fig. 8, is
independent of the gas or transition metal series and has stimulated
Varma and coworkers[21] to attempt a semi-empirical modelling in
which the trends are understandable in terms of basic transition-
metal parameters. The hypothesis within the Varma approach include
the following. a) Bonding is principally through the d-orbitals,
somewhat analogous to the picture given in 6.2, with the substrate
s-p electrons mainly screening or renormalising parameters appearing
in a purely substrate d-electron-adatom bond theory; b) Fine
details of the substrate density of states are much less important
than low order moments, namely the mean energy (first moment),
bandwidth (second moment), and the number of d-electrons or equivalently the Fermi energy; c) self-consistency and local charge
neutrality is important; d) the adatom levels, before coupling
with the substrate d electrons, lie near the bottom of the d-band.

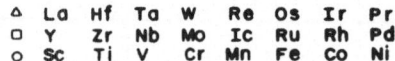

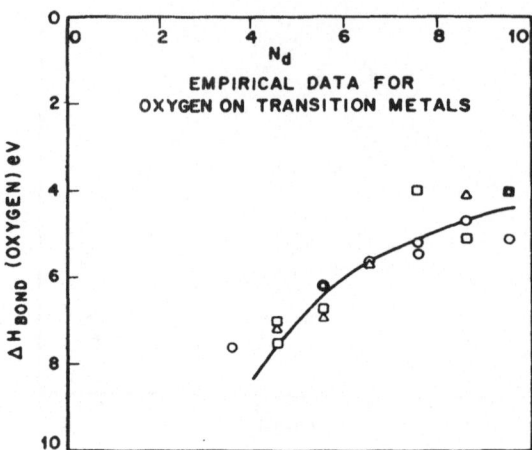

Fig. 8 -   Experimentally deduced binding energy of atomic oxygen
           on polycrystalline transition-metal surfaces. (Ref. 21).

    Some qualitative consequences of the model can be understood
in terms of Figures 1d and e, slightly modified. Imagine that the
d-band of width W straddles the Fermi level with the occupancy
determined in the way discussed in Section 6.$\underline{2}$. The average energy
per d-electron (effective orbital energy) is ε = NW/20 with respect
to a zero at the bottom of the band (for the uniform density of
states model). The s-p renormalised adsorbate orbital energy
$\equiv \varepsilon_g$ (= ε_{ao} + Λ in Fig. 1) is assumed to lie below the band and is
thus a negative number with respect to the same zero. Letting the
effective d-orbital couple to the adatom orbital via some matrix
element V, the perturbation theory energy shift of the adatom
orbital ($\overset{\sim}{\varepsilon}_{bond}$ - ε_{ao} - Λ in Fig. 1d) is

$$\Delta\varepsilon \overset{\sim}{\sim} V^2/(\varepsilon_g - NW/20) \overset{\sim}{\sim} \frac{V^2}{\varepsilon_g}(1 + \frac{NW}{20\varepsilon_g} + \ . \ . \ .)$$

which is most attractive for N=1 (recall, ε_g < 0) and increases
monotonically for N $\leq$ 10. Varma et al. have carried out a much
more detailed and realistic analysis, obtaining results such as
those shown in Fig. 9 for the binding energy of hydrogen adsorbed
on the 4d transition metals. Nonetheless, the principal variation
of ΔH_{bond} with N enters through a change in local density of states

in the vicinity of the adatom and the main mechanism for the N dependence in their model is a more quantitative version of the argument just presented. In effect, as N increases, so does the difference $\varepsilon_g - \bar{\varepsilon}$ appearing as a denominator in the binding energy expression and this change shows up as a decrease in ΔH_{bond} with increasing substrate d-electron count.

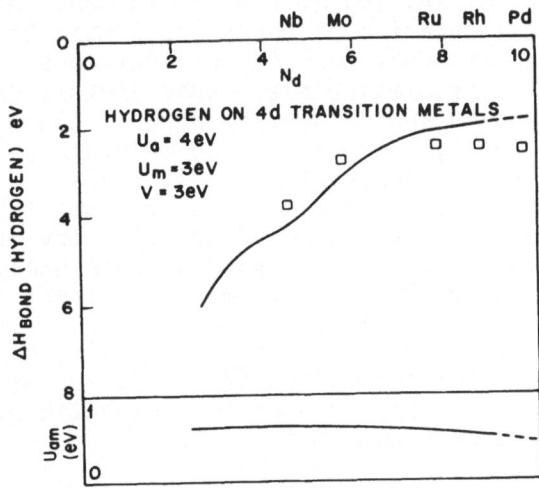

Fig. 9 – Calculated and experimentally obtained binding energy for hydrogen on 4d transition-metal surfaces. Also given is U_{am}, the metal atom-adatom Coulomb parameter, required to obtain local charge neutrality (Ref. 21).

6.4 Molecular dissociation and chemisorption

The study of diatomic molecule adsorption on surfaces has been intense since this is the simplest example illustrating the actual competition going on between various possible alternatives in a surface process in which the bonding within the molecule is pitted against the bonding of the molecule to the surface. The question is, which path is taken (and why) in the reaction

$$nM + AB \begin{cases} pM\text{-}A + (n\text{-}p)M\text{-}B & \text{(dissociative)} \\ nM\text{-}AB & \text{(non-dissociative)} \end{cases}$$

Dissociation can occur if the energy associated with the bonds formed between the metal atoms and the individual components of the diatomic is lower than that of the bond formed between the metal atoms and the intact molecule. Such an idealized one-dimensional situation is shown in Fig. 4, where curve 1 is taken to be

the molecule-metal interaction (a shallow attractive well could
also be allowed for in this curve) and curve 2, the individual
atom-metal interaction (strictly speaking only meaningful as drawn
for homo-nuclear A_2 molecules). Thus the energy cost to dissociate
the molecule (V_2 ($z\to\infty$) - $V_1(z\to\infty)$) is more than made up for by the
gain in forming the atom chemisorption bonds. This would be the
whole story if equilibrium thermodynamics was applicable. For most
experiments, one is not dealing with a gas in equilibrium with a
surface so the actual dynamics of the process must be considered.
In particular, if the thermal beam of incident molecules cannot
undergo an internal dissociative electronic transition in which the
AB bond is broken and AM and BM bonds formed in the vicinity of R_C,
the curve crossing point, then either the molecule is quasielasti-
cally reflected back from the surface or it transfers enough of its
kinetic energy into substrate excitations so that it becomes trapped
in the molecular well. On the other hand, if entry onto the dissociative curve is possible, then sticking is reasonably assured.
The point is that the substrate plays many different roles:

a) Provides a bonding mechanism for deep V_2 wells; b) Provides a
force for dissociation of the molecules; c) Serves as a dissipative
heat bath; d) Provides a continuum of non-adiabatic potential
energy curves for nuclear translations.

We will briefly consider the two most commonly studied diatomics
on surfaces, namely CO and H_2. In the case of CO, the molecular
orbitals include what is called a "lone pair orbital" (5σ derived
mostly from a C p_z atomic orbital) protruding out the oxygen free
end of the carbon which is available for strong bonding with the
metal atoms, without altering much the remaining structure of the
CO. Consequently a strong, non-dissociative bond is formed, as in
Fig. 1d, in which the shifted 5σ orbital (ε_{ap} +Λ) forms a bonding
orbital with the substrate d-electrons and the CO sits upright on
the surface.

H_2 is more interesting in the sense that both types of adsorption are observed and understood on a theoretical basis. For a
complete interpretation, multi-dimensional potential energy surfaces
(as opposed to 1-D curves) such as that one due to Nørskov
et. al(12) for broadside H_2 on Mg(0001) and shown in Fig. 10 are
desirable.

Features of this surface include possibly van der Waals
bonded physisorbed H_2 for $z \gtrsim 7$ a.u., "chemisorbed" H_2 (M) at z~3.5
a.u., an activation barrier for dissociation (D) at $z \sim 3$ a.u., and
chemisorbed H(B) at z $\sim$ 2.2. a.u. Experimentally, if H_2 can be
dissociated, then it will strongly chemisorb. This is a very
important rate limiting factor in hydride kinetics, to be discussed

in the next section. On the other hand, even without dissociation if the heat bath aspects of the substrate are sufficient, H_2 is also known to adsorb weakly as a molecule, sometimes called precursor sticking. Kasemo and Tornqvist[35] have nicely demonstrated the necessity for dissociation in hydrogen uptake by a Ti surface. Pure Ti soaks up H, forming a good hydride. The surface can be passivated by forming an oxide film. However, deposition of Ti atoms on top of the oxide then results in rapid H uptake which suggests that the oxide merely insulates the incident H_2 from the Ti which has the ability to dissociate H_2. The deposited Ti atoms assist in H_2 dissociation, after which the individual H atoms readily diffuse through the oxide.

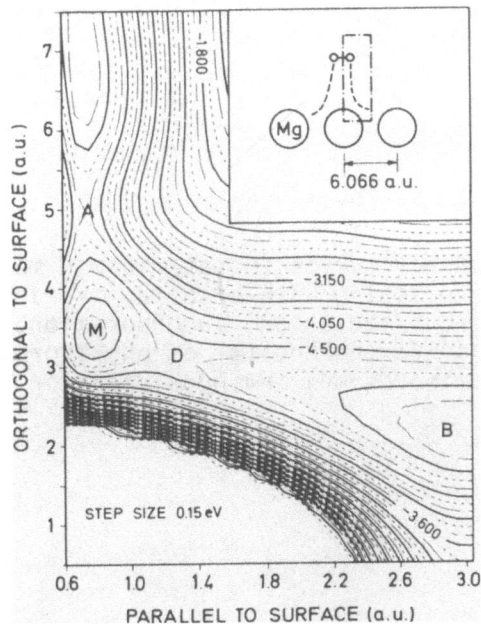

Fig. 10 - Equipotential-energy curves for H_2 dissociating over
         the atop site of a Mg(0001) surface into bridge sites.
         The energies in electron volts are relative to those
         of the free atoms. The molecular adsorption energy
         in M is thus only 0.4 eV relative to a free molecule
         (Ref. 12).

6.5 Hydride Kinetics

The final specific example, namely hydride kinetics, brings together many of the different concepts discussed in this chapter and focuses them onto a problem of intellectual interest, practical significance, and relevance to the hydrogen embrittlement/atomistics of fracture audience. The energetics of hydrogen incorporation is shown in Fig. 11.

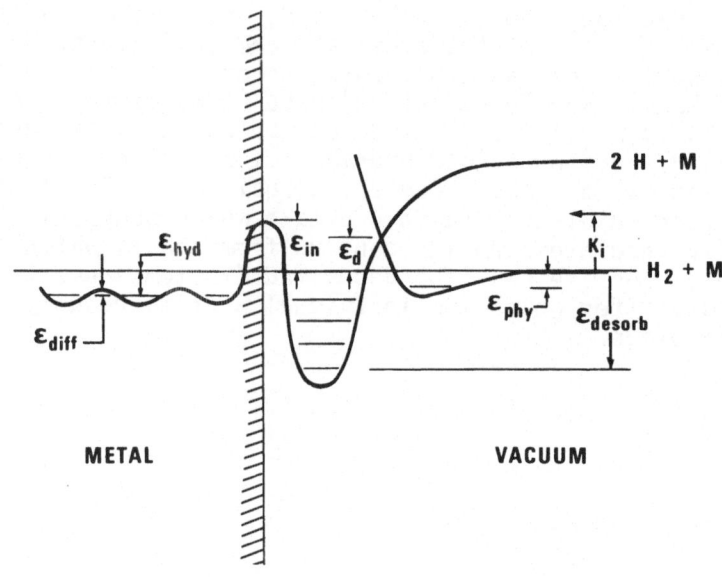

Fig. 11 - Potential energy curves for atomic and molecular
          hydrogen interacting with a solid. Various activation
          barriers for dissociation, incorporation, and diffusion
          are shown as are the "heats" of physisorption, disso-
          ciative chemisorption, and hydridization.

The H_2 incident on the metallic surface can either reflect,
physisorb, or dissociate, in which case subsequent nuclear motion
is on one of the continuum of 2H + M curves which are possible
because of substrate excitations. While in the chemisorption
well, the atoms may either vibrationally relax to the ground state
or move into the metal, where they may or may not diffuse away
from the surface. In considering the rate limiting steps in such
a concerted process, one must know the relative magnitudes and
signs (with respect to a common origin, here taken to be the H_2+M
vacuum zero) of the activation barriers for dissociation (ε_d),
hydrogen incorporation (ε_{in}), and diffusion (ε_{diff}) as well as the
role played by any other non-adiabatic influences. Certainly no
hydride formation is possible though, if the dissociation step
forms a bottleneck.

A beautiful set of experiments by Strongin and coworkers[36]
on hydrogen uptake in niobium illustrates ways in which state-of-
the-art surface science can provide meaningful information on this
problem. The observed facts are the following. The relatively
slow hydrogen uptake for a Nb(110) surface was monitored for
composite surfaces, in which submonolayer films of Pd had been
deposited. From LEED observations it was seen that for up to one
monolayer films, the Pd grows in a commensurate 1 x 1 Pd structure,
labeled Pd*(110) and in this domain, the hydrogen uptake was slow.
For larger Pd coverages, LEED patterns characteristic of the bulk
Pd(111) fcc mesh were observed, consistent with either a structural phase
phase transition within the Pd*(110) or formation of new layers
with the bulk Pd lattice. At the point where the Pd(111) structure
appeared, the hydrogen uptake rate increased dramatically, suggesting
that Pd(111) but not Pd*(110) could dissociate H_2. Photoemission
spectra were then obtained in order to see if there was a significant
change in the electronic structure associated with the structural
change and the results are shown in Fig. 12.

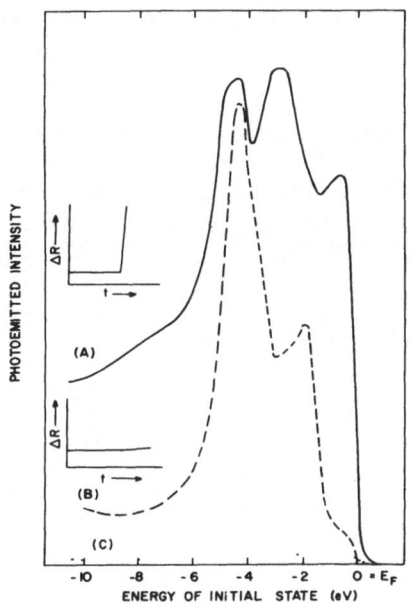

Fig. 12 - Normal-photoemission spectra at hω=90 eV: Curve A,
        a Pd(111) overlayer on Nb(110); curve B, a Pd*(110)
        overlayer on Nb(110); and curve C, Nb(110). The
        insets show schematically the hydrogen-uptake curves
        where the change in resistance (ΔR) of the Nb foil
        is plotted against time. Hydrogen uptake is measured
        by the change in the resistance of the Nb foil with
        hydrogen bulk concentration. (Ref. 36).

The results are quite dramatic. They show a quite low
Fermi-level density of states for the passive Pd*(110) structure,
but a very large one for the active Pd(111) indicating that the
Fermi level d-electron occupation is a very significant factor in
the dissociating powers of a surface. A speculative interpretation
of this observation can be offered, upon reconsidering the curves
shown in Fig. 4. Let Fig. 4a represent the adiabatic situation in
which H_2 is very unlikely to dissociate, as chances of K_i matching
onto a vibrational level of the atom-metal well are small. However,
from Fig. 4b, if the electron-hole pair continuum of the surface
can be excited, then motion can continue on the dissociative curve
and the probability for dissociative chemisorption is greatly
enhanced. From detailed microscopic theory of reactive surface
scattering[28c], it is believed that the probability for such pair
excitation is proportional to the square of the Fermi-level density
of states. Thus, all other things being equal, that substrate
with the largest Fermi level density of states would be expected
to have the greatest propensity for dissociative adsorption. On
the basis of this argument alone and the observed differences in
the photoelectron spectra, the probability for dissociative adsorption of H_2 on Pd(111) should be at least 100 times larger than on
Pd*(110) which is certainly in accord with the observations of
Strongin et. al.[36].

It seems quite likely that studies of this type, bringing
together the tools of a modern surface science lab and intuition
resulting from both static and dynamic model theories of surfaces
could be used to great advantage by researchers involved in other
fields where surface phenomenon play a significant but peripheral
role, such as in the atomistics of fracture. It is with this hope
in mind, that the present overview of some current issues in
adsorption concepts has been offered.

REFERENCES

1. H. D. Hagstrum, Phys. Rev. 150, 495 (1966); E. W. Plummer,
    J. W. Gadzuk and R. D. Young, Solid State Comm. 7, 487 (1969);
    D. E. Eastman and J. K. Cashion, Phys. Rev. Letters 27, 1520
    (1971).
2. a) J. W. Gadzuk in "Surface Physics of Materials", ed. by
    J. M. Blakely (Academic, NY, 1975); b) B. I. Lundqvist,
    H. Hjelmberg, and O. Gunnarsson, in "Photoemission and the
    Electronic Properties of Surfaces," ed. by B. Feuerbacher,
    B. Fitton, and R. F. Willis (Wiley, Chichester, 1978);
    c) "The Nature of the Surface Chemical Bond," ed. by
    T. N. Rhodin and G. Ertl, (North Holland, Amsterdam, 1979);
    d) "Theory of Chemisorption," ed. by J. R. Smith (Springer-
    Verlag, Berlin, 1980); e) M. Scheffler and A. M. Bradshaw in
    "The Chemical Physics of Solid Surfaces and Heterogeneous
    Catalysis, Vol. II Adsorption on Solid Surfaces," ed. by
    D. A. King and D. P. Woodruff (Elsevier, Amsterdam, in press);
    f) "Electrons in Disordered Metals and at Metallic Surfaces,"
    ed. by P. Phariseau, B. L. Györffy, and L. Scheire (Plenum,
    NY, 1979).
3. G. G. Kleiman and U. Landman, Solid State Comm. 18, 819 (1976);
    E. Zaremba and W. Kohn, Phys. Rev. B 15, 1769 (1977); N. D. Lang,
    Phys. Rev. Letters 46, 842 (1981).
4. J. W. Gadzuk, J. K. Hartman, and T. N. Rhodin, Phys. Rev. B4,
    241 (1971); J. P. Muscat and D. M. Newns, Progr. Surf. Sci, 9,
    1 (1978).
5. C. P. Flynn and Y. C. Chen, Phys. Rev. Letters 46 447 (1981).
6. L. Hedin and S. Lundqvist, Solid State Phys. 23, 1 (1969);
    "Theory of the Inhomogeneous Electron Gas," ed. by S. Lundqvist
    and N. H. March (Plenum, in press).
7. P. Hohenberg and W. Kohn, Phys. Rev. 136, B864 (1964).
8. N. D. Lang and A. R. Williams, Phys. Rev. B 18, 616 (1978);
    O. Gunnarsson, in Ref. 2f.
9. J. W. Gadzuk, Surf. Sci. 43, 44 (1974); K. H. Johnson,
    D. D. Vvedensky, and R. P. Messmer, Phys. Rev. B 19, 1519 (1979);
    L. Salem and C. Leforestier, Surf. Sci. 82, 390 (1979).
10. R. O. Jones, in Ref. 2f; A. B. Kunz, in Ref. 2d.
11. N. D. Lang and W. Kohn, Phys. Rev. B 7, 3541, (1973);
    N. D. Lang, Solid State Phys. 28, 225 (1973); A. R. Williams
    and N. D. Lang, Phys. Rev. Letters 40, 954 (1978).
12. J. Nørskov, A. Houmøller, P. K. Johansson, and B. I. Lundqvist,
    Phys. Rev. Letters 46, 257 (1981).
13. O. Gunnarsson in Ref. 2f; B. I. Lundqvist, O. Gunnarsson,
    H. Hjelmberg and J. K. Nørskov, Surf. Sci. 89 196 (1979);
    O. Gunnarsson, H. Hjelmberg, and J. Nørskov, Phys. Scripta
    22, 165 (1980).
14. F. J. Arlinghaus, J. G. Gay, and J. R. Smith, in Ref. 2d.
15. A. R. Williams (personal communications).

16. J. Nørskov and N. D. Lang, Phys. Rev. B 21, 2131 (1980).
17. G. W. Bryant, J. Phys. F 10, 321 (1980), and references therein.
18. A. R. Miedema, Phillips Tech. Rev. 36, 217 (1976).
19. D. G. Pettifor, Phys. Rev. Letters 42, 846 (1979); C. M. Varma, Solid State Comm. 31, 295 (1979); A. R. Williams, C. D. Gelatt, Jr., and V. L. Moruzzi, Phys. Rev. Letters 44, 429 (1980).
20. J. C. Hamilton, Phys. Rev. Letters 42, 989 (1979).
21. C. M. Varma and A. J. Wilson, Phys. Rev. B 22, 3795 (1980); A. J. Wilson and C. M. Varma, Phys. Rev. B 22, 3805 (1980); W. Andreoni and C. M. Varma, Phys. Rev. B 23, 437 (1981).
22. T. N. Rhodin and J. W. Gadzuk in Ref. 2c.
23. "Photoemission in Solids, I and II," ed. by M. Cardona and L. Ley (Springer-Verlag, Berlin, 1978).
24. "Vibrational Spectroscopy of Adsorbates," ed. by R. F. Willis (Springer-Verlag, Berlin, 1980); "Vibrations at Surfaces," ed. by A. A. Lucas (Plenum, NY, 1981).
25. Ref. 2e; P. Hollins and J. Pritchard, Chem. Phys. Letters 75, 378 (1980); J. W. Davenport, ibid. 77, 45 (1981); J. W. Gadzuk, ibid. (in press).
26. P. Oelhafen, J. Phys. F 11, L41 (1981).
27. J. C. Tully, Ann Rev. Phys. Chem. 31, 319 (1980).
28. a) J. K. Nørskov and B. I. Lundqvist, Surf. Sci. 89, 251 (1979); b) K. Schönhammer and O. Gunnarsson, Phys. Rev. B 22, 1629 (1980); c) J. W. Gadzuk and H. Metiu, Phys. Rev. B 22, 2603 (1980) and in press; H. Metiu and J. W. Gadzuk, J. Chem. Phys. 74, 2641 (1981).
29. T. F. O'Malley, Adv. in At. Mol. Phys. 7, 223 (1971).
30. S. A. Lindgren and L. Wallden, Phys. Rev. B 22, 5967 (1980) and references therein.
31. R. L. Gerlach and T. N. Rhodin, Surf. Sci. 19, 403 (1970).
32. J. Friedel, in "The Physics of Metals," ed. by J. M. Ziman (Cambridge, London, 1969), Chap. 8; C. D. Gelatt, Jr., H. Ehrenreich, and R. E. Watson, Phys. Rev. B 15, 1613 (1977).
33. D. M. Newns, Phys. Rev. Letters 25, 1575 (1970); T. B. Grimley and B. J. Thorpe, Phys. Letters 37A, 459 (1971); F. Cyrot-Lackman and F. Ducastelle, Phys. Rev. B 4, 2406 (1971).
34. E. W. Plummer and T. N. Rhodin, J. Chem. Phys. 49, 3479 (1968).
35. B. Kasemo and E. Törnqvist, App. Surf. Sci. 3, 307 (1979).
36. M. Strongin, M. El-Batanouny and M. A. Pick, Phys. Rev. B 22, 3126 (1980); M. El-Batanouny, M. Strongin, G. Williams, and J. Colbert, Phys. Rev. Letters 46, 269 (1981).

DISCUSSION

Comment by R.H. Jones:

Since there is considerable interest in using surface segregation to simulate grain boundary segregation and because many elements which cause intergranular embrittlement may also be involved in surface effects

on embrittlement, it would be informative to compare the band
structures and energies of adsorbates at external surfaces and
internal interfaces. Can you comment on the possibility of making
this comparison?

Reply:

As I mentioned in the discussion of "tractable models", all
current realistic numberical schemes require some sort of symmetry
in order to make calculations feasible. A system with an adsorbate
or impurity at an interfacial boundary, most likely between materials
of different lattice constants or between different crystallographically oriented faces of the same material, does not possess any such
symmetry, as the impurity is embedded within the ill-defined healing
or mismatch region between the well-defined bulk regimes. Consequently, realistic calculations, in my opinion, are not now possible
although I'm sure that Keith Johnson and Dick Messmer would disagree.

Comment by J.F. Knott:

You have described effects of local potential fields in surfaces
on the adsorption process. Would you care to comment on the possible
effects of applied potentials on adsorption of species? I am thinking
particularly of situations as those which occur on field-emission
tips or field-ion tips.

Reply:

This question can be addressed on several levels. From the point
of view of a surface electron spectroscopist, applied fields could
shift electronic energy levels in an observable and informative
manner. This would require the large ($\sim$0.1 - 10 eV/Å) fields
realized in field emission or ion microscopes. Such an effect has
been demonstrated in work at NBS but due to the fact that tunneling
(field emission) spectroscopies only access electronic states within
$\sim$leV of the Fermi level, the usefulness of this type of controlled
experiment has been limited. On a more macroscopic level, field
induced adsorption (due to the free atom polarizability) or
desorption has been much discussed and realized by field ion micro-
scopists. Due to the non-uniform electric fields across a field-
ion tip, whose values and distributions can only be inferred from
experimentally controllable variables, realistic quantification of
such processes is not easy. Consequently, the phenomenon of field
induced adsorption has not provided a great deal of insight or data
relevant to situations outside the field ion world. Finally, for
flat surface - vacuum interfaces, any possible applied potentials
will most likely not be able to generate sufficiently strong fields
to make an observable difference. There is some theoretical work
(T.F. George and coworkers) in which laser-field-induced adsorption/
desorption processes are being considered (beyond the obvious heating

or selective vibrational excitation), but at present time this must
be regarded as very speculative.

Comment by M. Puls:

   We know that in Zr alloys an oxide layer greatly reduces the
H_2 uptake. Is this a general result in other metals and what do you
think is the reason for this?

Reply:

   As mentioned in my written notes (but not in the oral presentation, hence the question) similar effects have been observed by
Bengt Kasemo and Eric Törnqvist for the case of H_2 uptake in pure
and oxide coated Ti. Going one step further, deposition of Ti atoms
on top of the oxide then led to good H_2 uptake suggesting that H_2
interaction with metal atoms is required for dissociation. Once
dissociated, atomic hydrogen readily diffuses through the oxide
layer and the uptake again proceeds readily.

INTERACTIONS BETWEEN ADSORBED SPECIES AND STRAINED CRYSTALS

E. D. Shchukin*

Institute of Physical Chemistry of the Academy of
Sciences and Moscow State University, Moscow, USSR

1. INTRODUCTION

Surface interaction between a solid, particularly strained
or being strained, and a medium which features adsorption can give
rise to failure and rearrangement of interatomic bonds in the solid
(dissolution, adsorption-induced strength reduction, various kinds
of corrosion) or in the adsorbed molecules (heterogeneous catalysis
or certain mechanochemical phenomena). Adsorption interactions may
decrease the potential barrier thus affecting the reaction kinetics,
as is the case with catalysis, or reduce the energy of the final
state and bring about such thermodynamical effects as adsorption-
induced strength reduction and plasticization. The mechanism of
these interactions depends on the chemical nature of the solid and
requires a variety of experimental techniques and theoretical
approaches to be revealed in each specific case.

The failure and rearrangement of bonds involved in the effects
under consideration may often be generally described in terms of a
sum of energy contributions due to (1) mechanical work (of the
applied force or internal stresses), (2) thermal fluctuations and
(3) physico-chemical interactions at the interface. The proportion
of each given factor may be essentially different under different
conditions, and the surface free energy σ ("the deficiency of the
energy of cohesion" owing to the presence of non-compensated bonds)
may be used here as a convenient characteristic parameter. In strain
and fracture of solids the value of σ determines the work to be spent
on the creation of new surface area. Adsorption from the medium is
an important factor here provided its energy contribution is more or
less compatible with σ (though, generally speaking, the bonds with
the adsorbate should not be too strong so that the migration of the

molecules of the medium along the crack walls would not be prevented). This group of effects will be discussed in the two following sections.

2. MOLECULAR DYNAMICS OF ADSORPTION-DEPENDENT FAILURE

This section deals with a theoretical approach. The method of
molecular dynamics consists in observing the "live picture" of
behaviour vs. time of an ensemble including a few tens (hundreds) of
particles (usually atoms or molecules interacting via Lennard-Jones
"6--12" potential) by means of integrating the equations of motion
for all the particles at given initial coordinates and pulses. The
results are not associated with the concrete nature of chemical
bonds and therefore offer certain opportunities for generalization.

Using this method, Yushchenko and the author investigated certain surface phenomena in solids: (a) the durability (time-to-
failure) of an individual interatomic bond under a constant load P
was found to be described by an equation

$$\tau = \tau_0 \exp \left[(U_0 - \gamma_p) / kT \right]$$

which agrees with macroscopic data obtained by Zhurkov et al. It
was shown that the activation energy may decrease substantially
in the presence of foreign (surface active atoms). (b) the effect
of tensile and compressive stress on the diffusion of interstitial
impurities in the crystal and on solubility has been investigated.
(c) The disjoining pressure of adsorption layers on the walls of
the crack (cavity) has been estimated. (d) The stages of lattice
strain, generation and motion of dislocations and other defects,
accumulation of damages, brittle and plastic failure at different
temperatures have been analyzed. These processes appear fundamentally dependent on adsorption of a surface active component which
may cause local dissolution and melting, facilitate crack generation
and propagation, etc. Such investigations unite the macroscopic ideas
on the role played by the decrease in σ with the picture of failure
at the molecular level.

3. MECHANICAL BEHAVIOR OF CATALYSTS

This section deals with typical cases of failure involving
chemisorption and thus relevant to the chemistry of surface interactions. Strong interaction between the surface of a catalyst and
the adsorbed molecules (which arises from the presence of "non-compensated bonds") decreases the potential barriers associated with
the occurrence of one or another intermediate state, thus decreasing
the activation energy and facilitating the entire process. (The
effect may also be due to a process, e.g. plastic deformation, making the surface essentially less equilibrium -- provided the local
distortions are severe enough to ensure energy heterogeneities

compatible with σ).

On the other hand, chemisorption during catalysis may substantially decrease the surface energy of the catalyst and facilitate its failure. This seems to be a phenomenon widespread in practice. Kontorovich, Suzdaltseva, Paransky, the author et al., carried out a systematic investigation of strength and durability of porous diverse materials as affected by adsorption, chemisorption and catalysis. The following problems were studied: (1) Mechanical behavior of MgO granules catalyzing the conversion of isopropyl alcohol over a wide range of conditions of their preparation and thermal treatment, loading and the level of internal stresses. It was found that MgO structures may be strengthened, say, by fine silica additions. (2) The mechanical behavior of cobalt-molybdenum catalyst of incomplete oxidation of propylene. Isotherms of adsorption were obtained for several products involved in the process. Pronounced strength reduction was observed under the effect of chemisorption of acetaldehyde and especially acroleine (probably due to the special role of π-bonds) as well as the decrease in durability. (3) The abrasion resistance of Al--Cr--K catalyst of n-buthane dehydrogenation at different temperatures as depending on the moisture content in the granules, on the reduction of chromium and on the process of catalysis. (4) Several other cases of the effect of adsorption (chemisorption) on the mechanical behavior of catalysts. The strength reduction observed amounts to 10% and even 50%. Accordingly, the decrease in durability (time-to-failure under a constant load) may be equal to several orders of magnitude. These results help understand the premature failure of catalysts and work out the means of strengthening the granules in order to overcome the effect of adsorption-induced strength reduction.

---

* Editor's Note: Professor Shchukin was invited to lecture at this Conference, but was unable to attend. His extended abstract is published here in recognition of his many contributions to environment-sensitive mechanical behavior.

TUTORIAL LECTURES ON

INTERFACES

Session Chairman: M.P. Seah

NONEQUILIBRIUM SURFACE AND INTERFACE THERMODYNAMICS

John W. Cahn*

National Bureau of Standards
Center for Materials Science
Bldg. 223, Rm. A153
Washington, D.C.   20234

Thermodynamics strictly is the science which tells one what is not possible. It is quite powerful for equilibrium for then all changes have become impossible. It is less useful far away from equilibrium. While it still forbids many changes, many remain possible and thermodynamics provides no information about which of those will actually occur. Thermodynamics provides us with inequalities based on the second law which all possible processes must comply with.

In simple homogeneous systems kept isothermally and at a constant pressure, the second law is restated in terms of a Gibbs free energy that can not increase. More complicated systems in which many processes occur or which are inhomogeneous, are difficult for applications of thermodynamics. For example, if a process which by itself would be forbidden occurs in the presence of a highly irreversible process, their combined contributions could satisfy the second law, but unless the processes were somehow coupled only one would occur. Similarly in an inhomogeneous system, a process in one part that by itself would be forbidden could be coupled with another elsewhere. The assumption that the processes are uncoupled or that the different parts of a system are uncoupled is a convenience for it allows us to apply the thermodynamic rules to the individual processes. It is however an assumption whose validity must be demonstrated by theory or experiment.

Fracture is usually a highly irreversible process. Fracture along an internal interface deals with an initially inhomogeneous system. The application of thermodynamics to just the interfacial

parts during fracture is one example where these uncoupling
assumptions can be shown to be invalid. Surfaces and interfaces
can not exist by themselves. They are contacts between two phases.
Many surface thermodynamics textbooks manage to convey the false
impression that equilibrium surfaces are treated in isolation, but
this is decidedly not true. (1) The definition of surface excess
quantities involves the properties of the abutting phases and the
restrictions on the degrees of freedom imposed by the presence of
even a nonabutting phase affects thermodynamic properties of
surfaces. (2) During fracture these phases are nonhydrostati-
cally stressed and usually far from thermodynamic equilibrium.
The elastic energy stored in the stressed phases is often much
larger than the changes in surface and interfacial energy during
fracture. It becomes a question whether equilibrium surface
thermodynamics has a valid application to fracture and whether
the surfaces and interfaces are uncoupled to the extent that
inequalities involving just surface properties can be constructed.

It is appropriate to consider first interface creation as
close to equilibrium as possible. The creation of interfaces in
a system of fluid phases is thermodynamically simpler. The fluids
remain hydrostatically stressed. Unlike the case for solids,
reversibly stretching of a fluid surface is a process identical to
surface creation. As a result the surface tension, which is a
stress tensor isotropic in the surface plane is identical in
magnitude to a well-defined specific excess surface free energy,
which is a scalar. We can create surface by stretching it and
note the work we do against the surface tension. We can let the
surface contract and note the work that can be recovered from the
system. If the system initially is fully equilibrated then the
surface tension of stretching the surface must exceed the surface
tension when this surface is contracting, and the surface tension
of an equilibrated surface must lie in between. This follows
directly from the second law. If the order of surface tension
magnitudes were otherwise, we could invent cycles of expanding
and contracting surfaces in which net work would be done by a
system which initially was fully equilibrated.

As soon as the system initially is no longer in equilibrium
it is capable of doing work. Can this work somehow be coupled
into the surface? It would destroy our ability to use the above
inequalities. Consider a historic example used by Gibbs (3) to
illustrate such a failure. Alcohol is known to lower the surface
tension of water. Place an alcohol solution into a large room
not saturated with alcohol vapor. Create surface by stretching.
Allow the alcohol to evaporate. Allow the surface to contract.
During stretching the surface tension was low, during contraction
it was high. Net work was recovered from the system during this
cycle. This is chemical work that came from the chemical potential differences of alcohol within the system. Here then is a

quite simple example of coupling of a diffusion process and surface creation process which warns one about considering just the
interfacial processes during irreversible surface creation. For
fluids, compositional effects are the major cause of nonequilibrium surface effects. A quickly created fluid surface even in an
equilibrated multicomponent system will not have equilibrium
adsorption and since surface tension can be measured rapidly there
are extensive studies (4) of "dynamic surface tension." In Gibbs'
example. local surface equilibrium might be a good assumption when
the surface is created in a time scale slow compared to diffusional
jump times for molecules but fast compared to the time for diffusional equilibration of the entire system. In that case the elements of the surface sit at definite chemical potentials in a
diffusion profile and the surface properties can be assumed to be
characteristic of those corresponding to the local state variables.

When fluid surfaces are created on time scales short compared
to molecular jump times theory is more difficult. A theoretical
attempt (5) to treat the surface between two phases that are not
in equilibrium with each other, but are assumed to remain completely homogeneous right to the interface during diffusional
transfers is widely used but is of doubtful validity. The assumed
discontinuity in chemical potentials between the phases right at
the interface led to the assumption of a third different chemical
potential for each specie for the interface layer, while the
unrealistic assumption of continued homogeneity of the phase gives
rise to two peculiar quantities for each specie which the authors
call "cross-chemical potentials." With five chemical potentials
the authors claim to have a completely general theory of all
nonequilibrium surfaces, but its validity under any realistic
nonequilibrium conditions is doubtful and in any case requires a
complicated and large data base.

Fracture is usually not done in equilibrated systems. Even
if the system were fully equilibrated before the load was applied,
the application of the load usually destroys conditions for
equilibrium. With careful design experiments may be devised to meet
criteria of equilibrium or of uncoupling of surface processes and
thereby allow rigorous application of thermodynamic principle to
surface creation alone. For solids a nonequilibrium interface
differs from the equilibrated one not just in adsorption, but in
structural ways as well. Consider for example a stacking fault
in an alloy created by the rapid motion of a partial dislocation.
Even if there is no adsorption, nonequilibrium local order has
been created. On free surfaces adsorption is known to lead to
surface reconstruction and this too is not a factor in any published model of nonequilibrium surfaces.

Under nonhydrostatic stresses as encountered in fracture the
chemical potentials of substitutional components of the solid take on

values at surfaces that depend on the normal stress there (6, 7) and thus differ with orientation of the surface, even if the solid is homogeneously stressed.

Several papers dealing with just the thermodynamics of nonequilibrium adsorption during fracture have recently been published (8). Elaborate cycles of fracturing under different chemical paths have led to estimates of nonequilibrium surface properties. The cyles bear some relationship to a topic in fluid surface thermodynamics called disjoining pressure.

The subject of disjoining pressure (9, 10) concerns reversible "fracture" along a fluid interface done by injecting a layer of a different fluid phase along the interface. The pressure of the injected fluid is the disjoining pressure and is a measure of the work of surface creation. A classic example of disjoining pressure is the "fracture" of a "vapor column" containing a soap film. Injecting water into the film converts a single soap film into two liquid-vapor surfaces. Since the energy of a soap film differs from that of two liquid-vapor surfaces, work has to be done to convert the film into two surfaces. The disjoining pressure at any thickness of the film is equal to the derivative with respect to thickness of the surface free energy change.

During slow fracture along a pre-existing interface there is a continuous change in surface energy and adsorption as the interface separates into two surfaces. Two obvious metallurgical examples are analogous to the disjoining process; liquid metal embrittlement and hydrogen embrittlement of grain boundaries. Injecting liquid metal or hydrogen along a grain boundary is the disjoining process. In an ideal case, the phases are not stressed, but gently and reversibly pried apart. The only stressed region is the interface itself. For fluids this process can be made completely reversible and treated thermodynamically including adsorption changes. For fracture, consideration of the disjoining pressure experience may lead to experimental design in which surface thermodynamics can play a rigorous role in promoting understanding.

Thermodynamics of surfaces has played an important role in the development of fracture criteria. It seems clear that many of these classical papers were based on simplifying assumptions whose validity needs to be reexamined. The recent concerns with the subtleties of the effects of adsorption underscore this point (8). The difficulty of finding and proving the validity of simple inequalities for surface creation in nonequilibrium systems and Gibbs' counterexample should serve as warning that these concerns are real and deserve our attention.

REFERENCES

1. R. Defay, I. Prigogine, A. Bellemans and D. H. Everett,
    "Surface Tension and Adsorption," J. Wiley, New York (1966).

2. J. W. Cahn, "Thermodynamics of Solid and Fluid Surfaces," in:
    "Interfacial Segregation," p. 3, ASM (1979).

3. J. W. Gibbs, "The Collected Works," Vol. 1, Yale Univ. Press,
    (1948), footnotes p. 273.

4. A bibliography of methods of measuring dynamic surface
    tension is given in ref. (1), p. 68.

5. Ref. (1), p. 62ff.

6. Ref. (3), p. 196.

7. F. Larché and J. W. Cahn, Acta Met 26:1579 (1978)

8. J. P. Hirth and J. R. Rice, Met. Trans. 11A:1501 (1980).

9. A. D. Read and J. A. Kitchener in "Wetting," p. 300,
    Soc. of Chem. Ind. Monograph #25, London (1967).

10. Extensive literature on disjoining pressure is found the
    many Conference Proceedings edited by B. V. Deryagin,
    "Research in Surface Forces," Transl. by Consultant Bureau,
    New York, (1963) – present.

---

*Editor's Note: Dr. Cahn was unable to attend the conference.
                  His extended abstract is included because of its
                  importance to the general theme of the conference.

ON THE STRUCTURE OF GRAIN BOUNDARIES IN METALS

H. Gleiter

University of Saarbrücken
D-6600 Saarbrücken
West Germany

INTRODUCTION

The purpose of this review is to summarize the present state of understanding of the atomic structure of grain boundaries by highlighting those results that may provide most insight. For the earlier work, the reader is referred to existing review articles and conference reports [1 - 8].

I. EQUILIBRIUM GRAIN BOUNDARY STRUCTURE

1. Dislocation Models

The first dislocation model of a high angle grain boundary was proposed by Read and Shockley [9] for symmetrical high angle tilt boundaries. By analogy to the strain-field energy of a small angle tilt boundary |9| , it was assumed that, if the dislocations are uniformely spaced in the plane of the boundary, a low energy interface results. For all other tilt angles the boundary may be described as a boundary with a uniform dislocation spacing and a superimposed small angle tilt boundary that accounts for the deviation from the tilt angle required for uniformly spaced dislocations. This idea has been used in the subsequent years also in boundary models that are not based on the dislocation concept (cf. section I.2). Two inherent limitations exist in the dislocation model developed by Read and Shockley. First, the singular behaviour of the elastic strain fields near the dislocation centers was removed by an inner "cut-off"

radius. The second deficiency is the linear superposition resulting in a complete neglect of the interactions among the dislocation in the array.

A first attempt to account for such effects was
made by Li |10|. Li proposed a hollow core dislocation
model[+]) of grain boundaries (Fig. 1) and recognized
that as the tilt misorientation increased, with its
attendant decrease in dislocation spacing, the core
shape changed as the traction-free core condition
assumed in Li's model was violated. He, therefore, extended the dislocation model by postulating that shape
and radius of the actual core depended on the boundary
structure. The result was an increase of the core radius
and a change of the core shape with increasing misorientation in the sense indicated in Fig. 1. To
ameliorate some of the inherent deficiencies of the con-
tinuum dislocation array model, Glicksman et al. |12 ,
13| added a phenomenological core-energy term of
thermodynamic origin. This led to the concept of the
equilibrated "heterophase" dislocation model with a
second phase in the core. A finite array of edge dislocations of this type was superimposed to obtain the
elastic energy per dislocation, but, then, included in
a self-consistent manner was a chemical energy term for
a core of circular cross-section. The total energy was
minimized with respect to the core radius.

As was pointed out earlier, the linear superposition techniques used in the above analysis per
definition neglect the elastic interactions occuring

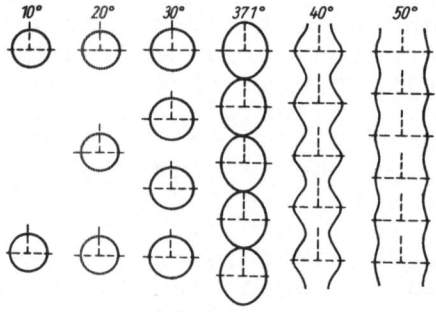

Fig. 1

Calculated shape of the dislocation cores in tilt
boundaries corresponding to different tilt angles.
The tilt angles are indicated by the nos. above the figs.

+) The idea of a hollow core of a dislocation was
   discussed several years before by Frank |11| .

among the grain boundary dislocations. An attempt to include these interactions was made by Masamura and Glicksman |13 , 14| using the heterophase dislocation model and solving the two-dimensional elastostatic boundary value problem for an arbitrary periodic spacing between the dislocations. At temperatures not too far below the equilibrium melting point, liquid-core dislocations should be prevalent in boundaries according to this model. In fact, measurements of the energy of high angle grain boundaries in bismuth at temperatures close to the melting point |12 , 15| seem to support this result (Fig. 2). Some of the inherent difficulties of the method used by Glicksman et al. have been discussed critically by Jones |16|. Recent observations |17| of interphase boundaries and the solid/liquid interphase in carbontetrabromide-hexachloroethane alloys gave little support for any discontinuity of the type shown in Fig. 2. The same conclusion was drawn from measurements of grain boundary energies in copper as a function of temperature |18|. If a phase transition due to the collapse of heterophase dislocations would occur, the boundary energy should become independent of the misorientation angle (Fig. 2) as a slab of liquid is formed in the boundary core. This was not observed.

In addition to the above models based on linear elasticity theory, non-linear dislocation models (Peierls model, Frenkel-Kontorowa model) have been proposed |19 - 21| .The results obtained |21| suggest that the width of the core of a dislocation in a boundary depends on the boundary structure (Fig. 3). In boundaries of low energy the cores of the dislocations are comparable in size to a lattice dislocation whereas with increasing deviation from the low energy structure the core diameter grows and, finally, spreads to

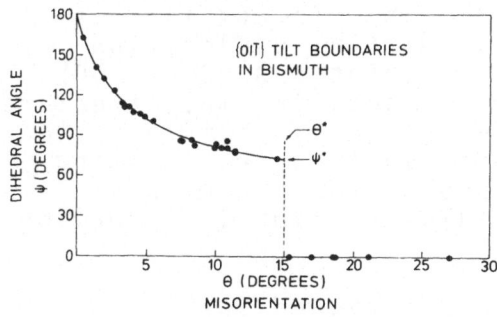

Fig. 2

Dihedral angle (Ψ) at the solid/melt interface at the point where a grain boundary of misorientation θ terminates at that interface. The conspicuous discontinuity occuring at the cordinate (θ+, Ψ+) may be interpreted as a first-order phase transition in the boundary |12|.

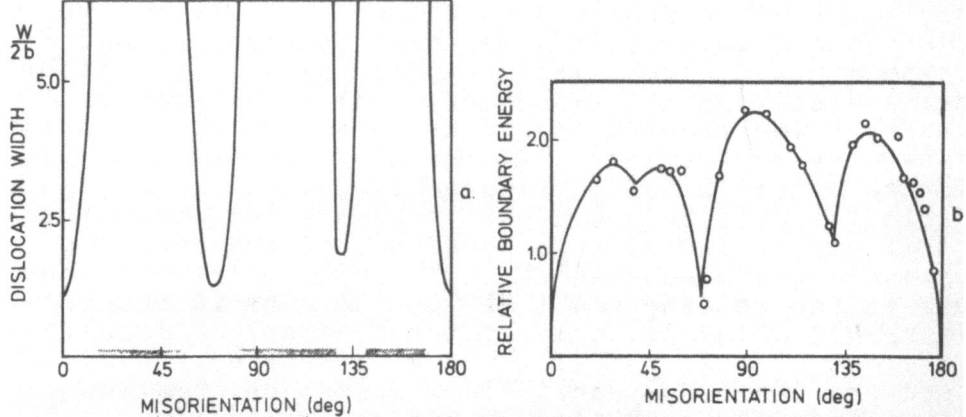

Fig. 3   Calculated widths w (divided by the Burgers-
          vector 2b) of the misfit dislocations in
          symmetric 110 tilt boundaries in aluminium
          (Fig. 3a). The measured energy misorientation
          curve of these boundaries is shown in Fig. 3b.
          The dislocation widths plotted in Fig. 3a are
          calculated for G = 2.5 10^{11} dyn/cm^2,
          2b = 2.86 10^{-8} cm and a unit energy (1.0 in
          Fig. 5b) of 400 erg/cm^2. The energy measure-
          ments are taken from ref. 22.

infinity if the boundary energy is independent of the
misorientation between the two crystals forming the
interface. The physical reason for this effect will be
discussed in section II.2.2. A considerable number of
experimental observations |12 , 23 - 31 , 32 , 33|
support these views. Nevertheless the problem is still
a matter of controverse |34 - 38| . In addition to
non-linear dislocation models, partial dislocation
models |20| and disclination descriptions |39 , 40|
have been considered in order to describe the structure
and energy of grain boundaries (Fig. 4). Similarly to
all dislocation models proposed so far, the existing
disclination models do not account for rigid body
relaxations between the crystals (section I,2) forming
the boundary.

      Recently, this concept was extended further by
using a formalism |47 - 50| that describes the dis-
torsion field, the tensor of Burgers vector density,

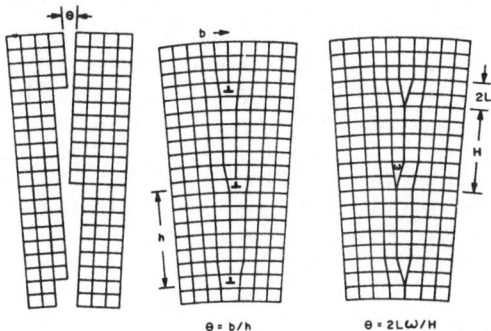

Fig. 4   A simple tilt boundary made of dislocations or
          disclinations. The tilt angle is θ and the
          Burgers vector of the dislocations is b |39| .

the metric strain and the torsion of a grain boundary
by means of differential geometry. In a non-rigid
lattice model of a symmetric tilt boundary of arbitrary
misorientation |51| a grain boundary was portrayed in
terms of an array of voids and/or asymmetric cracks
upon whose surfaces are distributed surface dislocations
|51| . The boundary shown in Fig. 5a contains no elastic
energy and consists of large voids which, in turn,
correspond to a high surface energy. The total amount
of free surface energy can be reduced by a partial
(Fig. 5b) or total (Fig. 5c) coalescence. In the case

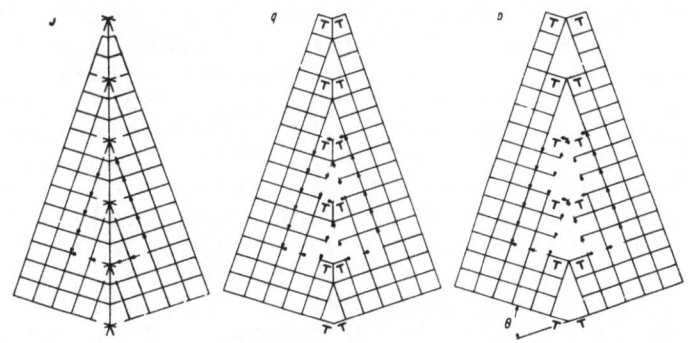

Figs. 5a,b,c   a) Completely relaxed tilt type symmetric
                grain boundary of misorientation angle
                θ = 36.9°; b) partially coalesced
                modification of the grain boundary
                shown in a); c) total coalescence of
                the grain boundary shown in a) |51| .

of total coalescence (Fig. 5c), the elastic distorsion
is a maximum and the surface energy is a minimum. Real
crystals exhibit a behaviour somewhere in between the
two extremes as represented by Fig. 5b. The calculations
indicate that a minimum energy configuration of the
boundary is obtained if the boundary dislocations are
spread out in the plane of the boundary. As was pointed
out above, this delocalization of boundary dislocations
is also suggested by other arguments.

2. Coincidence Models

The pioneering work of Kronberg and Wilson |52|
led to the concept of coincidence sites in grain
boundaries. In the subsequent years this concept was
adopted by several other workers |53 - 56| and
developped in more detail. For example, Fig. 6 shows the
coincidence site lattice formed by two hexagonal arrays
of atoms. Arkharov |57| developped the first model of a
grain boundary in which an attempt was made to interpret
the structure and the properties of grain boundaries on
the basis of the coincidence site concept. He considered
a [100] tilt boundary in a cubic lattice. A more
detailed model was given some years later by Brandon.

Brandon's model |54 , 58| represents an extension
and combination of the coincidence lattice concept |52|
and the dislocation model |9 , 11| (cf. section I,1).
The coincidence theory in the form proposed originally
(Fig. 6) brings out the dependence of the boundary
structure on the orientation relationship between both
crystals but ignores the effect of boundary inclination.
The effect of boundary inclination was incorporated by
suggesting that boundaries constrained to lie at an
angle to the density packed coincidence plane take up
a step structure as indicated in Fig. 7. In the subsequent development of the atmistic models of grain
boundaries, attention was focused essentially on the
following two questions: (i) what characterizes the
structures of boundaries of low energy and (ii) the
development of the general crystallographic theory of
grain boundaries. These two problems led to two separate
lines of development in the understanding of the
structure of grain boundaries. Let us first consider the
ideas put forward to improve our understanding of the
characeristic features of low energy boundary
structures.

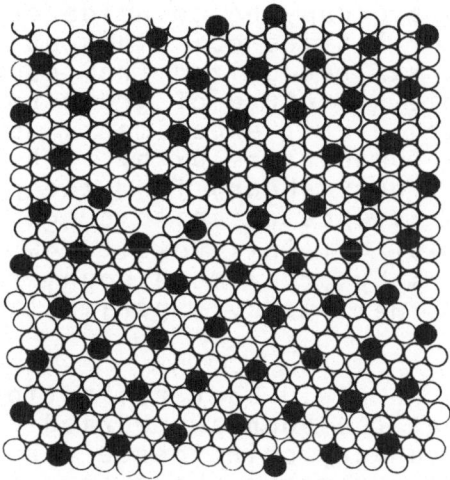

Fig. 6   Coincidence site lattice (black circles) in
         two crystals that are rotated by 38º (22º) about
         an axis normal to the paper |52|.

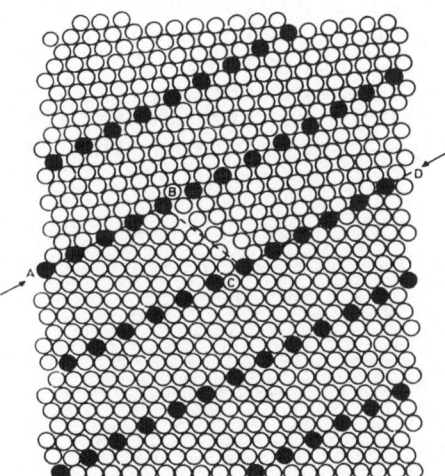

Fig. 7 Structure of a grain boundary according to the
       model of Brandon et al. |54, 58|. The boundary
       shown is a 50.5º [110] tilt boundary between
       two crystals that have b.c.c. structure. The
       sections AB and CD of the boundary lie in a
       plane of high coincidence density. (The atoms
       represented by black circles occupy coincidence
       sites.)

The concept of atoms occupying coincidence sites (in terms of boundary or lattice coincidences) had to be abondoned after it was recognized from computer simulations of the atomic structure of grain boundaries |59 - 61| and from bubble raft experiments |62| that two crystals forming a boundary relax by a shear type displacement (rigid-body shear) from the position required for the existence of coincidence site atoms at the boundary (Fig. 8). This result was confirmed in the subsequent years by other more sophistaceted computer work |63 - 65| as well as by experimental observations |66 - 68|.

The existence of rigid body relaxations led to the conclusion |69| that the boundary periodicity rather than the existence of boundary coincidence per se is the physical meaningful parameter. However, recent measurements |70 , 71| of the orientation relationships corresponding to low energy boundaries in noble metals and noble metal alloys indicate |71 , 72| that the boundary periodicity may not be the only parameter that controls the energy of a grain boundary. The low energy boundaries experimentally observed in the various systems investigated suggested a division of all low energy boundaries in two groups: "electron sensitive" and "electron insensitive" boundaries |70|. The "electron insensitive" boundaries were those, that were observed in all metals and alloys of the same lattice structure irrespective of alloy composition. The "electron

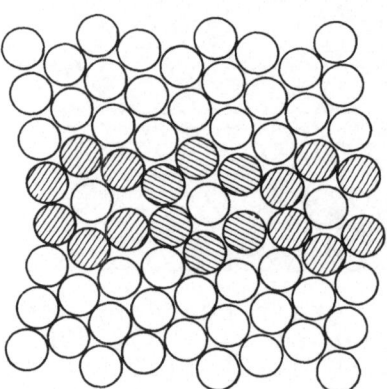

Fig. 8   18° symmetric tilt boundary displaying six-fold corrodinated units. Shaded atoms represent structural units |60|.

sensitive" boundaries were observed only in some pure
metals or in alloys of certain compositions. On the basis
of this result, it was proposed that the energy (and
structure) of the "electron insensitive" boundaries is
controlled primarily by the geometry of the atomic
arrangement in the boundary (atomic packing density
similar to the lattice), whereas the energy of the
second group depends primarily on the electron band
structure of the material. Examples of "electron
insensitive" aries are 70.5° [110] or 50.5° [110]
boundaries. The 59° [110] or the 81° [110] boundaries
belong to the second category. The significance of the
electronic contribution to the energy of grain boundaries
was pointed out by several authors |73 - 74| on the
basis of free electron calculations. In fact, Seeger
and Schottky showed (Fig. 9) that the screening of the

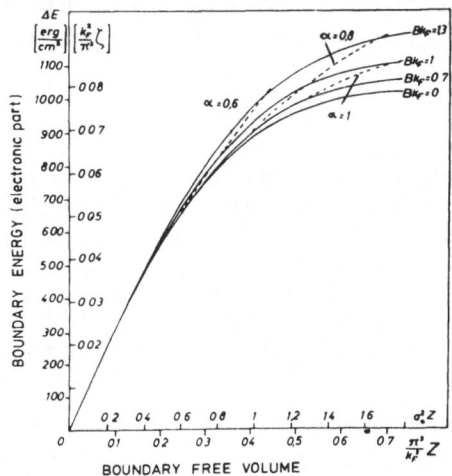

Fig. 9 Electronic contribution (ΔE) to the boundary
        energy due to electron screening effects (of
the positive charge deficit in the boundary) as a
function of the boundary free volume (positive charge
deficit) for a free electron gas model. a is the
lattice constant, Z is the charge per atom, k_F is the
wave vector at the Fermi energy ζ , a^2Z is the positive
charge deficit per unit area of the boundary. The
numbers given on the energy axis are calculated for
silver (a_o = 4.078 10^{-8} cm, k_F = 1.204 10^{-8} cm $^{-1}$,
m^*) = m). The boundary is represented as a square well
potential of width 2B and hight U_o. The broken curves
are computed for various heights of the potential well
(α = U_o/ζ) |73| .

positive charge deficit at a grain boundary by the
conduction electrons may be the dominant part of the
boundary energy.

The considerations so far discussed have neglected
the effect of temperature, pressure etc. on the boundary
structure. There is substantial evidence in the literat-
ure |18 , 75 - 82| suggesting structural phase trans-
formations in grain boundaries similar to the well known
phase transformations of free surfaces |83| . Figs. 10a
and 10b show (hypothetically) how such a transformation
may be visualized. Due to the smaller free volume of the
boundary structure shown in Fig. 10a, the Gibbs free
energy of this structure is lower at high pressures
than the Gibbs free energy of more open structure (Fig.
10b). Therefore, at a certain pressure, the boundary
structure may transform from the more open structure
into the densly packed structure as the pressure is in-
creased |18 , 81| .

The development of the crystallographic theory of
grain boundaries was pioneered by the work of Bollmann
|84 - 86| in form of the O-lattice theory. This
theory is a mathematical method for calculating the
crystallographic structure of interfaces and the permissible Burgers vectors of the perfect interfacial
dislocations between two arbitrary crystals in arbitrary
relative orientation for any position of the boundary
between the two crystals. For the deviation of relevant

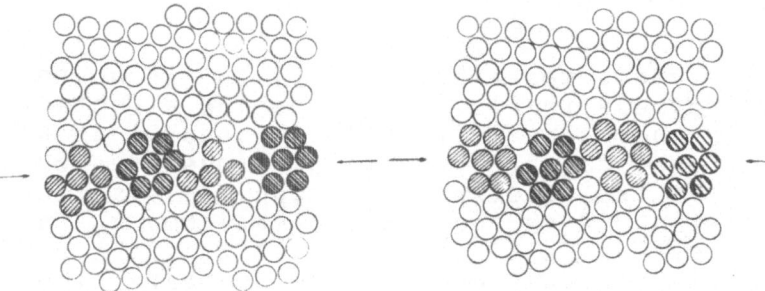

Figs. 10a,b   Schematic model of the atomic arrangement
               in a grain boundary due to a phase trans-
               formation for a boundary between two
               hexagonal arrays of atoms.

equations the reader is referred to the above publications. The question upon the extent to which all grain
boundaries can be represented by low energy structures
and superimposed dislocation networks predicted by the
crystallographic theory has been considered in a number
of experimental |26 , 28 , 32 , 87 , 88 - 93|
and theoretical papers |10 , 12 - 15 , 21 , 94 -
97| . The experimental evidence was recently reviewed
by Goodhew |98| pointing out that still relatively few
examples exist in which the predicted dislocation
structures have been revealed. An example for the
observation of the crystallographically predicted dislocation structures is shown in Fig. 11. Observational
difficulties limit the angular range over which networks
of discrete dislocations can be detected by electron
microscopy. These difficulties may be due to the small
Burgers vectors of grain boundary dislocations or due
to wide cores (and, thus, a weak strain field) as
evidence by the analysis of the contrast width | 24 , 28
- 38| . The importance of the latter effect was
emphasized by observations on misfit dislocations in the
interfaces between ∝/ß brass |26| . The dislocation net-
work in such interfaces became invisible upon small

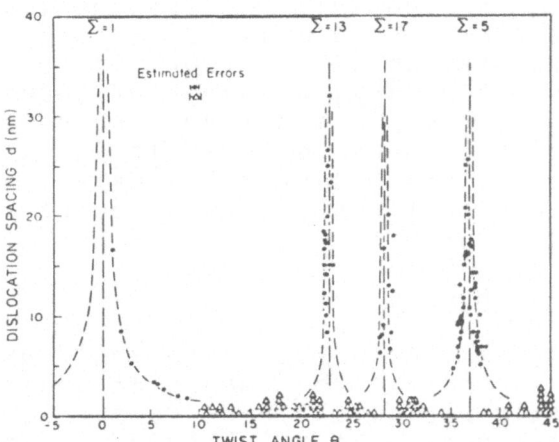

Fig. 11   Spacing d of the secondary grain boundary
          dislocations found in 001 twist boundaries in
          gold. The dashed curves were calculated from
          the crystallographic theory. The filled circles
          represent the measured spacings. The open
          triangles represent cases where dislocation
          networks were not detected |88| .

rotations although the crystallographically predicted
dislocation spacing |170| should have been well above
the resolution of the electron microscope suggesting
that a non-localization of the misfit in the cell walls
may occur under certain conditions. Information about
the structure of boundaries deviating from low energy
orientation relationships may also be derived from
studies of the energy of the boundaries and X-ray
diffraction experiments. Recent measurements on the low
energy misorientation relationships of grain boundaries
in noble metals and noble metal alloys |18 , 70 , 71|
have suggested that the structure of any high energy
boundary is related to (at least one of) the neighbouring
low energy boundaries. This result agrees with a considerable number of other grain boundary energy
measurements and with recent X-ray diffraction experiments
on grain boundaries |98| . The periodicity of [001] twist
boundaries in gold with a reciprocal coincidence site
density of Σ = 13, 85 and 377 showed the periodicity
expected on the basis of the crystallographic theory.

The answer to the question whether or not a given
grain boundary will tend to preserve a structure characteristic of a given coincidence orientation relationship
depends on the energies of alternative possible boundary
structures. Indeed, two examples have been documented
for Σ = 5 and Σ = 9 boundaries, where the manner in
which a coincidence structure is maintained, depended on
the boundary orientation |99 , 100| . In the particular
case of the Σ = 5 boundary, a change of boundary plane
favoured a change from minimum length Burgers vectors
to crystal lattice Burgers vectors. It is, thus, clear
that the simple b^2-rule (Frank-rule) is not generally
applicable for dislocations in grain boundaries as was
done in the literature.

3. Plane Matching Models

The electron microscopy of high angle grain
boundaries has shown that the images of grain boundaries
exhibit frequently periodic linear features, which have
been demonstrated not to be Moiré fringes. These
observations prompted Pumphrey |101| to explain the
periodic line structures on the basis of a plane
matching approach. The situation is shown schematically
in Fig. 12. The overall array of traces shown in Fig. 12
produces a Moiré pattern which possesses parallel bands
of high opacity. These bands are interpreted in terms of
poor atomic fit lying along the dashed lines in Fig. 12.
Between these bands are corresponding bands of high

Fig. 12   Slightly mismatched traces of two sets of
           lattice planes (dark lines) in the boundary.
           The two sets of planes are assumed to impinge
           from the two adjoining crystals |101|.

transparency (associated with good atomic matching).

Pumphrey suggested that atomic relaxation occurs in the
vicinity of these Moiré bands resulting in strain fields
that are imaged in the electron microscope as a single
set of prallel lines which has the same geometry as the
bands. The various defects that may result in grain
boundaries if planar matching is the dominant, have
been worked out by Ralph et al. |102|. Balluffi and
Schober |103| have discussed the relationship between
the plane matching approach and the description of a
grain boundary by means of crystallographic theory
(cf. section I.2). In principle, a plane matching
structure may always be accounted for (at least
geometrically) in terms of the boundary dislocation
network given by the crystallographic theory. Hence,
geometrically the plane matching representation may not
represent an independent model of the boundary structure.
The difference between both approaches lies in the
physical significance which is associated with the
boundary dislocation structure.

     As an example, Figure 13 |104| shows a pattern of
parallel lines in a grain boundary with the misorientation 10.4° [8.5, 9.3, 1.0]. It seems difficult to interpret these lines in terms of grain boundary dislocations.
as the boundary shown deviates more than 10° from the
nearest high density coincidence orientation (Σ = 19).
If the limit Σ = 100 is imposed, the closed CSL is

Σ = 51 resulting in an array of $\frac{a}{102}$ [1,1,10] edge dislocations with a spacing of about one lattice constant
so they are probably physically not significant. It was,
therefore, concluded that boundaries may exist in which
a CSL interface is not maintained but which show dislocation type defects. The observed spacing and orientation of the lines shown in Fig. 13 agrees, however,
with those predicted by the plane matching model from the
mismatch of the (110) planes. Recently, independent
evidence for the possible existence of plane matching
structures in grain boundaries has been reported for
several other systems |105 , 106|.

4. Polyhedral Unit Models

The basic idea of the polyhedral unit models is that
the structure and properties of a grain boundary may be
described in terms of a two-dimensional array of one or
several types of atomic configurations (also termed as
atomic clusters or polyhedral units).

Historically, the idea of describing a boundary by
means of a two-dimensional array of certain atomic configurations apparently has been suggested first on the
basis of observations by field ion microscopy |54|. The
low energy of such a configuration was rationalized in
terms of a two-dimensional array of an atomic configuration containing a coincidence atom plus the surrounding
atoms.

However, with the advent of computer simulations of
the atomic structures of grain boundaries it was recog-
nized |59 - 61| that atoms occupying coincidence sites
may not exist in coincidence grain boundaries (cf. sect-
ion II.2). This result prompted Weins et al. |60| to
suggest that the structure of a grain boundary may be
described by a two-dimensional periodic pattern of
characteristic atomic groups consisting of a central

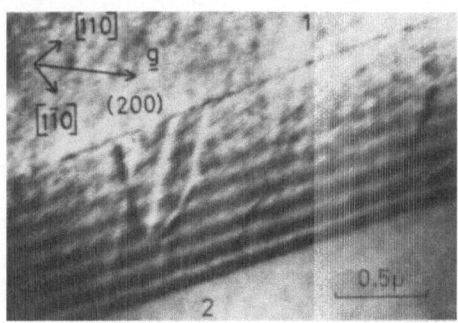

Fig. 13 Grain boundary in
recrystallized Al 0.9^{wt}/o Mg
(cold rolled 92%, annealed
30 min. at 350° C). The
spacing of the periodic
lines is 185 Å |104|.

atom surrounded by five, six or seven atoms (Fig. 14).
In terms of this mode, the structure of a general grain
boundary was interpreted as a combination of several
types of these units. In recent computer calculations
of the structure of [100] symmetrical tilt boundaries
in aluminium, close packed triangles of atoms (which
stack to give trigonal prisms) were found in low energy
relaxed structures |107| . It was, therefore, proposed
that a low energy structure of a boundary may be associated with the existence of planar, close packed groups
of atoms (Fig. 15). There is also based on high resolution microscopy behaviour in a covalently bonded material,
germanium |206| . Tetragonal coordination was found to be
preserved as far as possible across the boundary. The
idea to compare the atomic arrangements in a grain
boundary with the atomic arrangements in amorphous
structures was apparently, first proposed by Potapov
et al. |108| and Aaron and Bolling |109 , 110|. A 38°
[110] boundary was found |207| to consist of periodically
arranged rings formed by five atoms with a central atom
between them (Fig. 16). On the basis of these observations it was concluded that a grain boundary may be represented in terms of atomic configurations existing in
amorphous metals. In the work of Aaron and Bolling |109 ,
 110| the free (excess) volume of a grain boundary was
explicitly calculated for dislocation boundaries and
boundary structures formed by random close packed units.
The idea of describing the structure of a grain boundary
in terms of the packing of certain of the 8 basic
deltahedra that result when equal spheres are packed
to form a shell, such that all spheres touch their
neighbours was developed in detail by Ashby et al.
|112 , 113|. It was found that in all of theses cases

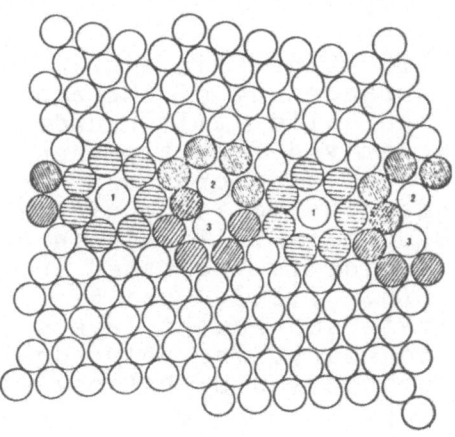

Fig. 14   37.8^o symmetric
tilt boundary displacing
the coordinated (poly-
hedral) units (1 seven-
fold unit, 2 five-fold
unit and 3 five-fold unit).
Shaded atoms represent
the structural units |60| .

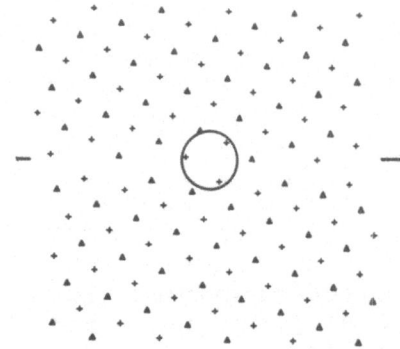

Fig. 15 Triangular group
of atoms (circle) in a (210)
symmetrical tilt boundary
|107| .

the structure may be described completely and uniquely
as nesting stacks of certain of the deltahedra (Fig.
17). As only very special arrangements of such polyhedra can fill space, the requirement of spacefilling
at grain boundaries and the compatibility with the

adjoining grains can in general only be satisfied if
the polyhedra are elastically distorted or if they are
interdispersed with other atomic configurations |114| .
Therefore, the comparison between grain boundary structures and structural elements of amorphous materials is
not without problems as the atoms in a boundary cannot
relax to the same extent as in a glass. This difference
is also born out by positron annihilation experiments
|115| suggesting that the atomic packing in a grain
boundary is more "open" than in glassy structure. This
conclusion is supported by Mössbauer measurements on
grain boundaries in Zn, Fe and Al |116 - 118| .

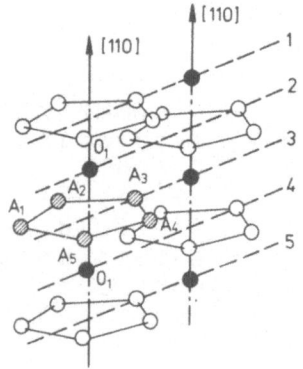

Fig. 16 Schematic digram of
the arrangement of the atoms
in a 40⁰ 110 tilt boundary
in W derived from a sequence
of field ion microscopy
images. The position of the
[1ī0] common tilt axis in the
two grains is indicated in
Fig. 16. 1,2,3,4 and 5
indicate subsequent layers of
the boundary. The polyhedral
rings proposed by Patapov et
al. are labelled by letters |108| .

II. STRUCTURAL (NON-EQUILIBRIUM) DEFECTS IN GRAIN BOUNDARIES

This section is concernd with the atomic structure of point defects and dislocations that are not part of the equilibrium structure of the boundary.

1. Point defects

The attempts that have been made so far to study the atomic structure of vacancies in grain boundaries are based on computer simulation techniques |7 , 119 - 125| and dynamic hard sphere models |7 , 122|. Depending on the atomic structure of the "vacancy-free" boundary, two kinds of vacancy structures were obtained by means of the computation |123 - 125|.

1) Vacancies in short periodic boundaries (Fig. 18) of good fit may be described as an empty site in the boundary (localized vacancy) surrounded by a displacement field that consists of cone-shaped sectors, the tips of which lie approximately at the site of the vacancy (Fig. 18). The relaxation decays in each cone with increasing distance from the vacancy. Due to this decay, the structure of the boundary some lattice constants away from the vacancy, remains essentially unaltered.

2) The generation of a vacancy in such a random boundary induced a displacement field that may be considered as being composed of the following three components:

(a) A displacement field somewhat similar to the cone-shaped sectors observed in short periodic boundaries of good atomic fit (Fig. 18).
(b) A group comprising several boundary atoms in the "vacancy-core" rearranges by about one interatomic spacing, e.g. the group shown in Fig. 19a (marked by heavy arrows) comprises four atoms.

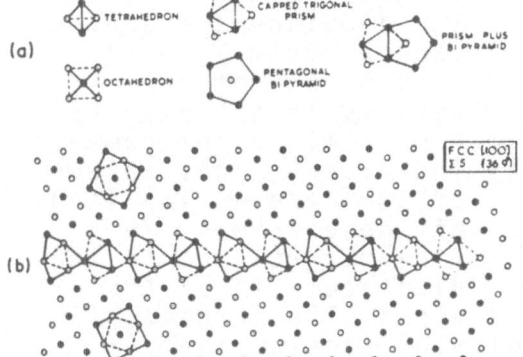

Figs. 17a,b Demonstration of graphically constructed boundaries in terms of polyhedral units |111 , 112|. (a) Undistorted projections of the units appearing in the following diagram. Octahedra and tetrahedra are not shown in subsequent diagram. (b) A Σ = 5 [100] tilt boundary between f.c.c.

(Figure caption cont.)
crystals. It is composed of stacks of capped trigonal
prisms along [100], adjacent stacks staggered by half
the lattice parameter normal to the plane of the drawing.

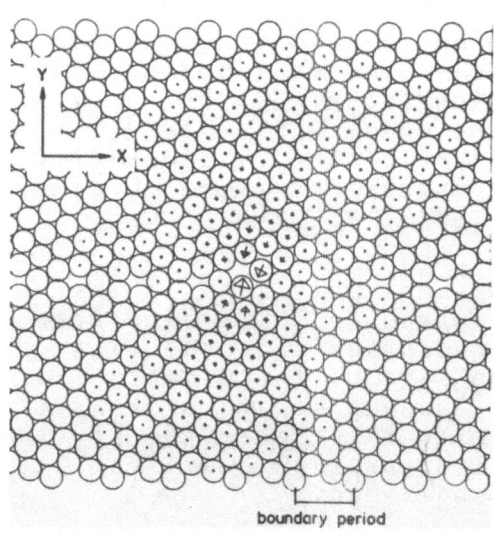

boundary period

Fig. 18 Atomistic
structure of a vacancy in
a symmetrical $\Sigma = 7$ tilt
boundary between two
hexagonal arrays of atoms.
The arrows indicate the
displacement of the atoms
relative to the positions
they occupied before the
vacancy was introduced.
The largest and smallest
arrows correspond to dis-
placements of 25.5% and
0.643% of the interatomic
spacing. The arrows scale
linearly. Third neigh-
bours were included in
the calculation |124|.

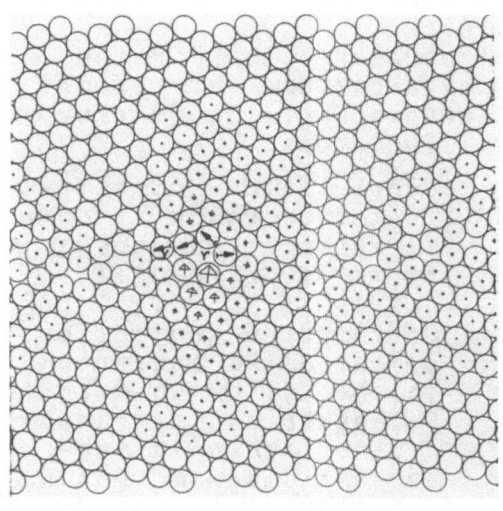

Fig. 19a Vacancy struct-
ure that resulted by
removing an atom at the
site Y from the boundary.
The atomic relaxations
caused by the presence
of the vacancy result in
the displacement of the
vacancy from Y to Y'.
This motion of the
vacancy is achieved by
the atomic displacements
indicated by the arrwos
of the type ➤. The lengths
of these arrows are
linearly related to the
displacements of the
atoms. The largest of
the arrows corresponds to a displacement of 1.04 a;
where a is the interatomic spacing. The long range

(Figure caption cont.)
displacement fields associated with the vacancy at the
site Y' are described by the arrows of the type →.
These arrows indicate the displacements of the atoms
relative to the position they occupied before the
vacancy was introduced. The largest and smallest arrows
correspond to displacements of 0.164a and 0.082a. The
computational conditions are the same as in Figs. 19a
and 19b. The atomic relaxations associated with the
boundary vacancy caused a translational motion of the
two crystals toward one another by 0.028a. The arrows
shown in Fig. 19a do not contain this translation |124| .

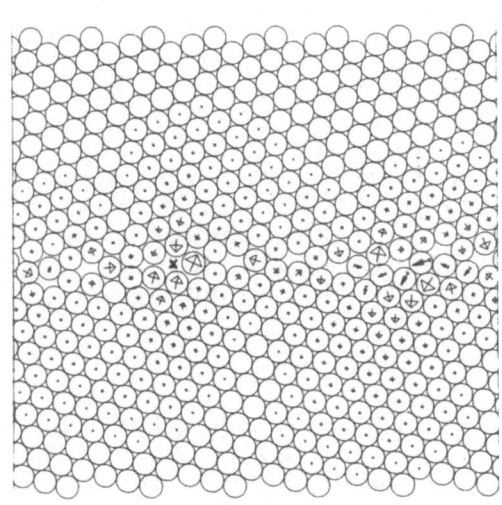

Fig. 19b Computed
boundary structure and
displacement field of
a vacancy generated by
removing the atom X.
The meaning of the
arrows of the types →
and → is the same as in
Fig. 19a. The largest
and smallest arrows of
the type → in the right
part of the figure
indicate atomic dis-
placements of 82.6% and
32.2% interatomic
spacings. The maximum
(minimum) displacements
indicated by the second
type arrows (→)
corresponded to 0.1777a
and 0.082a |124| .

(c) Displacement fields in boundary regions several
    interatomic spacings away from the site at which
    the vacancy was introduced, e.g. the displacement
    fields on the right and left section of Figs. 19a
    and 19b.

Bristow et al. |122| applied a computer simulation
method to study vacancy structures in three-dimensional
crystals. Fig. 20 shows the relaxation around a vacancy
in a twist boundary in tungsten. In this case relatively large displacements occured around the vacancy, and
it became essentially delocalized. A large variety of

other relaxed vacancy structures were found including
cases where the vacancy was "semi-localized" or split
depending upon the boundary structure, the site of the
vacancy in the boundary, and the interatomic potential.
These results agree with observations by Dahl et al.
|121| , Naberezhnykh et al. |126 ,   283| and with the
experimentally observed high efficiency of high energy
grain boundaries as vacancy sources whereas in boundaries of good atomic fit an energy barrier was found to
exist for emission of vacancies |125 ,   127 -   129| .

        In addition to computer simulations, investigations
by bubble raft and hard sphere dynamic model |122| as
well as boundary modelling by colloidal systems |130|
were carried out and seem to support the above results.

        The physical reason for the different types of
vacancy structures observed in boundaries of good and
poor atomic fit may be understood as follows. The free-
volume of a vacancy in a random grain boundary is reduced by a translational motion of the two adjacent
crystals towards the boundary. This motion is made
possible by atomic displacements described in the core
of the vacancy and in boundary areas far away from this
defect. In the case of a short-periodic boundary of
good atomic fit, such displacements would destroy the
good fit (low energy) structure of the boundary.

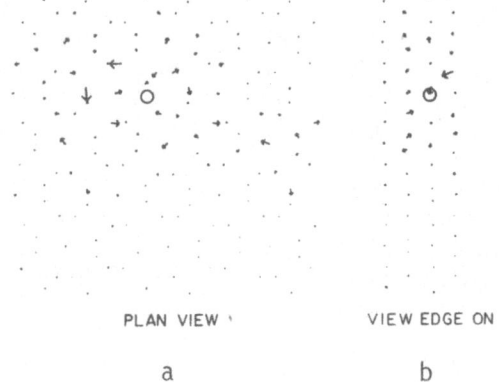

PLAN VIEW   VIEW EDGE ON

a   b

Fig. 20 Displacement
field obtained by com-
puter simulation model
around a vacancy in a
[001] twist boundary
(Σ = 5, θ = 36.87°)
in b.c.c. tungsten.
(a) and (b), plan and
edge-on views of atomic
relaxations around the
vacancy. Each atomic
relaxation represented
by a vector displacement
projected on the plane
of the paper |122| .

2. Dislocations

Essentially, three models for the understanding of the structure of extrinsic dislocation in grain boundaries have been put froward: the "dissociation", the "core spreading", and the "strain sharing" model.

2.1 Dissociation model

This model |84 , 99 , 105 , 131 - 134| describes the extrinsic boundary dislocations in all types of grain boundaries in terms of the coincidence site lattice theory (CSL) and the D.S.C. dislocation model of a boundary |56 , 84|. In the case of exact coincidence boundaries, a lattice dislocation entering a boundary and forming an extrinsic dislocation is envisaged to dissociate into boundary dislocations, the Burgers vectors of which are given by the primitive D.S.C. lattice vectors. The major energy source for driving the dissociation reaction is believed to be the reduction of the elastic energy associated with the strain field of the dislocations.

If a boundary deviates from the exact coincidence lattice orientation, the dissociation model assumes the boundary to contain an (equilibrated) network of (secondary) boundary dislocations. The incorporation of an extrinsic dislocation is assumed to occur again by the dissociation into boundary dislocations, the Burgers vectors of which are given by the D.S.C. lattice. An example is shown in Fig. 21. Despite of the attractive features of the dislocation model, it is not immediately obvious that a dislocation model is the most physically realistic model for all types of boundaries due to the narrow dislocation spacing and the small Burgers vectors.

II.2.2 Core delocalization model

The dislocation core delocalization model proposes the structure of an extrinsic dislocation to depend on the boundary structure. In boundaries with well defined boundary dislocations, the incorporation of an extrinsic dislocation is assumed to occur by a dislocation reaction of the type described in the previous section [II.2.1]. However, in the case of high-energy boundaries the reaction is proposed to occur by the widening (delocalization) of the cores of the dislocations in the plane of the boundary (Fig.22), |21 , 26 , 33 , 135 , 136|.

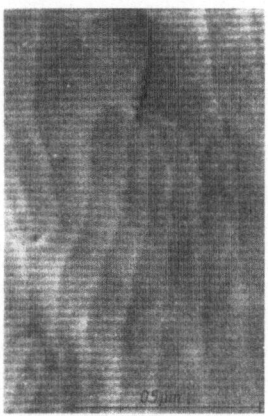

Fig. 21 Splitting of a lattice dislocation into five boundary dislocations in a Σ = 29 boundary in stainless steel according to the equation 1/2 [1,1,0] = 3(1/58) [3,7,0]
+ 2(1/58) [10,4,0].

The physical reason for the spreading of the core of
the extrinsic boundary dislocations seems to be the
reduction of the energy of the extrinsic boundary dislocation. Hence, an extrinsic boundary dislocation in a
highly ordered low energy boundary is expected to have
a core size which is comparable to the core size of a
lattice dislocation as a widely spread core would
destroy the low energy structure of the (highly ordered)
boundary over a large area. However, if an extrinsic
dislocation is introduced in a boundary with little
long range order (high energy boundary), the structural
change associated with the dislocation in general influences the boundary energy insignificantly as the
original boundary structure was already of high energy.[+]

+) The energy density in the core of a high energy
   boundary is in the order of the latent heat of fusion.
   This energy density presents an upper limit, and,
   hence, the introduction of an additional dislocation
   in such a high energy structure can not increase the
   energy significantly further. It has also be argued
   |137 , 138| that the relatively low density 81 ,
   |139| of the material in the boundary should result
   in lower restoring forces there, and, thus, allow
   the long range strain field to reduce its energy by
   widening the boundary dislocation core.

Hence, an increase of the core size of the newly introduced dislocation results in a small or in no increase
of the energy of the boundary. However, the increase of
the dislocation core size reduces the energy stored in
the long range strain field of the dislocation and,
therefore, the total energy of the system is reduced
if the core of the dislocation is widely spread in the
plane of the boundary (Figs. 22a,b).

II.2.3 Strain sharing model

The strain sharing model |87| is based on the
observed visibility of extrinsic dislocations in
symmetrical [110] tilt boundaries in aluminium. The
observed behaviour was interpreted in terms of the
"strain sharing" effect explained in Figs. 23a,b.
An extrinsic dislocation introduced into a grain boundary will repell neighbouring structural dislocations.
If the resulting relaxation or accommodation of the
structural array is extensive, the strain field of the
extrinsic dislocation will be effectively spread out
over a large area and the original extra dislocation
line added will tend to lose its identity. The latter
situation is illustrated in Fig. 23b.

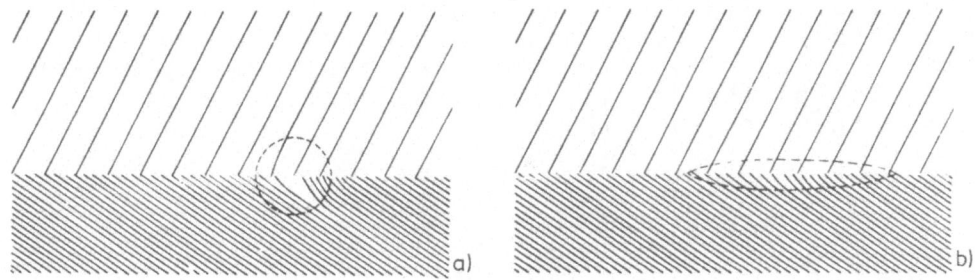

Figs. 22a,b Schematic diagram showing the structure
            of a localized (a) and a delocalized (b)
            boundary dislocation. For simplicity, only
            one set of lattice planes of the two ad-
            joining crystals forming the boundary is
            shown. The approximate size of the disloc-
            ation core is indicated by the broken
            lines |135|.

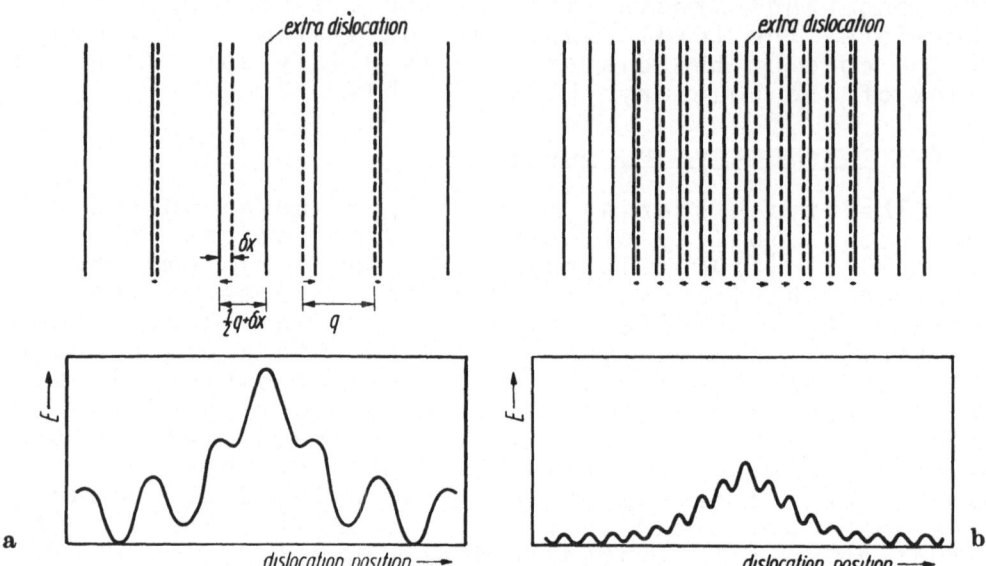

Figs. 23a,b Schematic illustration of the spreading of
the elastic strain field (E) of an extra dislocation
(marked) when added to a network of parallel boundary
dislocations with the same Burgers vector. The added
dislocation/final dislocation positions are shown by
solid vertical lines, the original positions by broken
lines. In (b) the density of structural dislocations is
greater than in (a) |87|.

SUMMARY

The first part of this paper reviews the existing
models (dislocations, coincidence, plane matching, polyhedral units) of grain boundaries. In the second part,
the present understanding of the atomic structure of
defects (vacancies, dislocations) is summarized.

REFERENCES

1. H. Gleiter and B. Chalmers, Progr.Mat.Sci. 16 (1971).
2. H. Gleiter, Phys.Stat.Sol. (b) 45 (1971) 9.
3. H. Hu (ed.), The Nature and Behavior of Grain Boundaries, Plenum Press, New York (1952).
4. P. Chaudhari and J.W. Matthews (eds), Surf. Sci. 31 (1972).
5. G.A. Chadwick and D.A. Smith (eds), Grain Boundary Structure and Properties, Academic Press, New York (1976).
6. W.C. Johnson and J.M. Blakely (eds), Interfacial Segregation, ASM, Metals Park, Ohio (1979).
7. R.W. Balluffi (ed), ASM Materials Science Seminar: Grain Boundary Structure and Kinetics, 15-16 Sept. 1979, Milwaukee, WI, ASM, Metals Park, Ohio (1980).
8. H. Gleiter, Materials Sci. and Eng., in press.
9. W.T. Read and W. Shockley, Phys.Rev. 78 (1950) 275.
10. J.C.M. Li, J.Appl.Phys. 32 (1961) 525.
11. F.C. Frank, Acta Cryst. 4 (1951) 497.
12. M.E. Glicksman and C.L. Vold, Surf.Sci. 31 (1972) 50.
13. R.A. Masamura and M.E. Glicksman, Can. Met.Quart. 13 (1974) 43.
14. R.A. Masamura and M.E. Glicksman, NRL Report 7851, Naval Research Laboratory, Washington, D.C. (1975).
15. C.L. Vold and M.E. Glicksman, in ref. 4 p. 171.
16. D.R.H. Jones, J.Mat.Sci.[4](1974) 1.
17. W.Kaukler, Ph.D.Thesis, University of Toronto (1981).
18. U. Erb and H. Gleiter, Scripta Met. 13 (1979) 61.
19. A. Seeger and R. Hörning, quoted in A. Seeger, Handbuch der Physik, Vol. VII/1, Springer Verlag, Berlin (1955), p. 654.
20. J.H. van der Merwe, Proc.Phys.Soc. A63 (1950) 616.
21. H. Gleiter, Scripta Met. 11 (1977) 305.
22. C.G. Hasson and C. Goux, Scripta Met. 5 (1971) 889.
23. O. Khalfalla and L. Priester, Scripta Met. 14 (1980) 839.
24. T. Mori and K. Tangri, Met.Trans.A10 (1979) 733.
25. R.W. Balluffi and R. Schindler, Proc. V. Polish Conf. on Electron Microscopy, J.A. Kozubowski (ed), Warsaw-Jadwisin (1978) 51.
26. P. Kluge-Weiss and H. Gleiter, Acta Met. 26 (1979) 117.
27. H. Föll and D. Ast, Phil.Mag. 40 (1979) 589.
28. P.H. Pumphrey, H. Gleiter and P. Goodhew, Phil. Mag. 36 (1977) 1099.

29. R.A. Varin, Phys.Stat.Sol. (a) 51 (1979) K 189.
30. R.A. Varin, Phys.Stat.Sol. (a) 52 (1979) 347.
31. R.A. Varin, J. Wyrzykowski, W. Lojkowski and M W.
     Grabski, Phys.Stat.Sol. (a) 45 (1978) 565.
32. H. Kuhn, G. Bäro and H. Gleiter, Acta Met. 27
     (1979) 959.
33. Y. Ishida, T. Hasegawa and F. Nagata, Trans.Jap.
     Inst.Metals 9 (1968) 504.
34. H. Gleiter, Scripta Met., 14 (1980) 569.
35. V. Vitek, A.P. Sutton, D.A. Smith and R.C. Pond,
     Phil.Mag. A39 (1979) 213.
36. A.P. Sutton and V. Vitek, Scripta Met., in press.
37. D.H. Warrington and R.C. Pond, Phil.Mag. 39 (1979)
     821.
38. P.H. Pumphrey and P.J. Goodhew, Phil.Mag. 39 (1979)
     825.
39. J.C.M. Li, Surf.Sci. 31 (1972) 12.
40. K.K. Shih and J.C.M. Li, Surf.Sci. 50 (1975) 109.
41. M. Marcinkowski, K. Sadananda and W.F. Tseng, Phys.
     Stat.Sol. (a) 17 (1973) 423.
42. M. Marcinkowski and K. Sadananda, Phys.Stat.Sol.
     (a) 18 (1973) 361.
43. M. Marcinkowski and E. S. Dwarakadasa, Phys.Stat.
     Sol. (a) 19 (1973) 597.
44. M. Marcinkowski and K. Sadananda, J.Appl.Phys. 45
     (1974) 1521.
45. ibid. 45 (1974) 1533.
46. M.J. Marcinkowski, Archiwum Mechaniki Stosowang
     31 (1979) 763.
47. M.J. Marcinkowski, Unified Theory of Mechanical
     Behavior, J. Wiley and Sons (1979).
48. E. Kröner, Kontinuumstheorie der Versetzungen,
     Springer Verlag (1958).
49. M.J. Marcinkowski, Phys.Stat.Sol. (a) 49 (1978)
     725.
50. M.J. Marcinkowski, J.Mat.Sci. 14 (1979) 205.
51. M.J. Marcinkowski and K. Jagannadham, Phys.Stat.
     Sol. 50 (1978) 601.
52. M.L. Kronberg and F.H. Wilson, Trans AIME 185
     (1949) 501.
53. F.C. Frank, Conf.Plastic Def. of Cryst. Solids,
     Mellon Inst., Pittsburgh (1950) p. 150.
54. D.G. Brandon, B. Ralph, S. Ranganathan and M.S.
     Wald, Acta Met. 12 (1964) 813.
55. S. Ranganathan, Acta Cryst. 21 (1966) 197.
56   W. Bollmann, Phil.Mag. 16 (1967) 363.
57. V.I. Arkharov, Fiz Met i Metall. 12 (1961) 223.
58. D.G. Brandon, Acta Met. 14 (1966) 1479.
59. M. Weins, H. Gleiter and B. Chalmers, Scripta Met.
     4 (1970) 737.

60. M. Weins, B. Chalmers, H. Gleiter and M.F. Ashby, Scripta Met. 3 (1969) 601.
61. M. Weins, H. Gleiter and B. Chalmers, J.Appl.Phys. 42 (1971) 2639.
62. Y. Yamraguchi and V. Vitek, Phil.Mag. 34 (1976) 1.
63. P.D. Bristowe and A.B. Crocker, Acta Met. 25 (1977) 1363.
64. R.C. Pond, D.A. Smith and V. Vitek, Acta Met. 27 (1979) 235.
65. P.D. Bristowe and A.B. Crocker, Phil.Mag. A38 (1978) 487.
66. R.C. Pond and D.A. Smith, Can.Metall. Quart. 13 (1974) 39.
67. P.H. Pumphrey, T.F. Malis and H. Gleiter, Phil.Mag. 34 (1976) 227.
68. K. Marukawa, Phil.Mag. 36 (1977) 1375.
69. B. Chalmers and H. Gleiter, Phil.Mag. 23 (1971) 1541.
70. G. Herrmann, H. Gleiter and G. Bäro, Acta Met. 24 (1976) 353.
71. H. Sautter, H. Gleiter and G. Bäro, Acta Met. 25 (1977) 467.
72. G. Herrmann, H. Sautter, G. Bäro and H. Gleiter, Scripta Met. 9 (1975) 357.
73. A. Seeger and G. Schottky, Acta Met. 7 (1959) 495.
74. M. Kubalski and M.W. Grabski, subm. to Phys.Stat.Sol.
75. K.T. Aust, Can.Met. 9 (1969) 173.
76. C.J. Simpson, K.T. Aust and W.C. Winegard, Met.Trans. 2 (1971) 987.
77. D.W. Demianczuk and K.T. Aust, Acta Met. 23 (1975) 1140.
78. P. Lagarde and M. Biscondi, Can.Met. Quart. 13 (1974) 245.
79. H. Gleiter, Zeitschr.f. Metallk. 61 (1970) 282.
80. E.W. Hart, in ref. [3] p. 155.
81. H. Meisser, H. Gleiter and E. Mirwald, Scripta Met. 14 (1980) 95.
82. H. Gleiter, Radex-Rundschau 1 (1980) 51.
83. J.G. Dash and J. Ruwalds (eds), Phase Transitions in Surface Films, Plenum Press, N.Y. (1980) NATO Advanced Study Institutes Series, Senes B. Physics Vol. 51.
84. W. Bollmann, Crystal Defects and Cryst. Interfaces, Springer Verlag, Berlin (1970).
85. W. Bollmann and A.J. Perry, Phil.Mag. 20 (1969) 33.
86. D.H. Warrington and W. Bollmann, Phil.Mag. 25 (1972) 1195.
87. C.A.P. Horton, J.M. Silcock and G.R. Kegg, Phys.Stat.Sol. (a) 26 (1974) 215.
88. R.W. Balluffi, Y. Komem and T. Schober, Surf.Sci. 31 (1972) 68.
89. P. Goodhew, in ref. [7].
90. C.A.P. Horton and J.M. Silcock, J.Micr. 102 (1974) 334.
91. W.A.T. Clark and D.A. Smith, Phil.Mag. 38 (1978) 367.
92. M. Mori and Y. Ishida, Scripta Met. 12 (1978) 11.
93. R.A. Varin, Phys.Stat.Sol. (a) 51 (1979) K189.
94. W. Lojkowski, H.O.K. Kirchner and M. Grabski, Scripta Met. 11 (1977) 1127.
95. P.H. Pumphrey, Scripta Met. 9 (1975) 151.
96. H. Gleiter, Phil.Mag. 36 (1977) 1109.

97.   W. Bollmann, Acta Cryst. A33 (1977) 730.
98.   W. Gaudig and S.L. Sass, Phil.Mag. A39 (1979) 725.
99.   W. Bollmann, B. Michaut and G. Sanifort, Phys.Stat.Sol. (a)
      13 (1972) 637.
100. D. Vaughan, Phil.Mag. 22 (1970) 1003.
101. P. Pumphrey, Scripta Met. 6 (1972) 107.
102. B. Ralph, P.R. Howell and T.F.Page, Phys.Stat.Sol. (b) 55
      (1977) 641.
103. R.W. Balluffi and T. Schober, Scripta Met. 6 (1972) 697.
104. P. Pumphrey, Scripta Met. 7 (1973) 895.
105. G.R. Kegg, C.A.P. Horton and J. Silcock, Phil.Mag. 27 (1973)
      1041.
106. W. Gronski and G. Thomas, Scripta Met. 11 (1977) 791.
107. D.A. Smith, V. Vitek and R.C. Pond, Acta Met. 25 (1977) 475.
108. L.P. Potapov, B.F. Golovin and P.H. Shiryaev, Fiz Met. Metall.
      32(5) (1971) 227.
109. H.B. Aaron and G.F. Bolling, Surface Sci. 31 (1972) 27.
110. H.B. Aaron and G.F. Bolling, in Grain Boundary Structure and
      Properties, G.A. Chadwick and D.A. Smith (eds), Academic Press,
      New York (1976) p. 107.
111. M.F. Ashby, F. Spaepen and S. Williams, Acta Met. 26 (1978)
      1647.
112. M.F. Ashby and F. Spaepen, Scripta Met. 12 (1978) 113.
113. H.J. Frost, M.F. Ashby and F. Spaepen, Scripta Met. 14
      (1980) 1051.
114. R.C. Pond, D.A. Smith and V. Vitek, Scripta Met. 12 (1978)
      669.
115. H.S. Chen and S.Y. Chang, Phys.Stat.Sol. (a) 25 (1974) 581.
116. T. Ozawa and Y. Ishida, Scripta Met. 11 (1977) 835.
117. Y. Ishida and T. Ozawa, Scripta Met. 9 (1975) 1103.
118. S. Nasu and H. Gleiter, to be published.
119. K.W. Ingle, P.D. Bristowe and A.G. Crocker, Phil.Mag. 33
      (1976) 83.
120. B.A. Faridi and A.G. Crocker, Phil.Mag. A41 (1980) 137.
121. R.E. Dahl, J.R. Beeler, R.D. Bourquin, Interatomic Potential
      and Simulation of Lattice Defects, Plenum Press (1972) p.673.
122. P.D. Bristowe, A. Brokman, F. Spaepen and R.W. Balluffi,
      Scripta Met. 14 (1980) 943.
123. H. Hahn, Diploma Thesis, Univ. of Saarbrücken (1980).
124. H. Hahn and H. Gleiter, Acta Met., 29 (1981) 601.
125. H. Gleiter, Progr.Mat.Sci., in press.
126. V.P. Naberezhnyky, E.P. Feldman and V.M. Iurchenko,
      Metalofizika 2 (1980) 11.
127. R.W. Siegel, S.M. Chang and R.W. Balluffi, Acta Met. 28
      (1980) 249.
128. R.L. Segall, Acta Met. 12 (1964) 117.
129. W. Jaeger and H. Gleiter, Scripta Met. 12 (1978) 675.
130. Y. Ishida, K. Okamato and S. Hachisu, Acta Met. 26 (1978)651.
131. R.C. Pond and D.A. Smith, Phil.Mag. 36 (1977) 353.

132.   D.J. Dingley and R.C. Pond, Acta Met. 27 (1979) 667.
133.   T.P. Darby and R.W. Balluffi, Phil.Mag. A37 (1978) 245.
134.   W.A.T. Clark and D.A. Smith, Journal of Mat.Sci. 14 (1979)776.
135.   P.H. Pumphrey and H. Gleiter, Phil.Mag. 30 (1974) 593.
136.   K. Smidoda and H. Gleiter, Zeitschr.f. Metallk. 89 (1978) 81.
137.   T. Johannesson and A. Tholen, Metal Sci.J. 6 (1972) 189.
138.   M.F. Ashby, Surf.Sci. 31 (1972) 498.
139.   H.B. Aaron and G.F. Bolling, Surf.Sci. 31 (1972) 27.

DISCUSSION

Comment by P.R. Pratt:

Your diagrams showed strong, low energy, grain boundaries of at least two distinct orientations between the stable end orientations after 4 and 40 units of vibration. Did these correspond to twin orientations? If not, to what do you attribute their relative strength?

Reply:

The two strong low energy boundaries were due to twin boundaries. It is well known that, in general, twin boundaries show little embrittlement. I have chosen this particular orientation region because the strong signal of twin boundaries facilitates the correlation between the various diagrams corresponding to different vibrations.

Comment by I.M. Bernstein:

A number of years ago Rath and I studied the orientation dependence of hydrogen-induced grain boundary cracking in purified iron. In support of the results found by Professor Gleiter, we found that boundaries with misorientations ≲15° never cracked independent of the nature of the misorientation (tilt, twist or mixed) or the boundary inclination. However for higher angles, there were broad adjoining misorientation regions of cracking and non-cracking which, except for twin orientations, were not related to regions of geometric coincidence. If this crack susceptibility is related to relative degrees of solute segregation, are there any orientation properties that can be used to predict the degree of segregation?

Reply:

According to our results the general answer would be as follows: a small amount of segregation is to be expected for low energy boundaries. In many cases (but not always) these boundaries correspond to coincidence orientation relationship. In polycrystals the situation is complicated by the structure of the grains. Due to the substructure, the various sections of a grain boundary differ in structure and hence also in the degree of segregation.

Comment by B.J. Berkowitz:

Did you consider that the sintering process may be orientation dependent? For example, it is possible that the neck size and shape will vary with misorientation. Specifically, a smaller neck will not have as high a load bearing capability as a large neck. Also, high curvature of the neck could result in a "notching" effect which will likewise facilitate fracture. If this is so, then the correlation you observe may be controlled by the configuration of the connection between the particle and substrate rather than (or in addition to) effects of orientation in intrinsic bonding strength. Do you have metallographic evidence that show that this is not so?

Reply:

Yes, we have measured the sizes of the necks of spheres that come off at the beginning of the shaking and also of spheres that required a long shaking time to break the boundary. No correlation between the sizes of the necks and the embrittlement was observed. More specifically, the average neck size was independent of the time required to break the boundary by shaking. The "notching" effect - if it were more important - would depend on the relative neck size in comparison to the diameter of the spheres. As no such effect was found metallographically, we were led to conclude that "notching" or other geometrical factors are not the factors controlling the breaking of the boundaries between the spheres and the plate.

Comment by M. Puls:

You suggest that the grain boundaries are embrittled as a result of the segregation of impurities to the dislocation "cores" which are assumed to represent the boundary structure. In the bulk, generally, impurity segregation to dislocations causes strengthening. Why would the opposite occur in the grain boundary case?

Reply:

The suggestion deduced from the experimental observations is that impurities segregate at or near the defects (structural units, poly-hedral units or dislocations) in boundaries. This segregation may well pin these defects just like in the lattice. In the lattice, the impurity interaction with dislocations leads to the strengthening of the material. In the grain boundary, the same may be true. But in addition to the defect pinning, the presence of impurities at the boundary may also weaken the atomic bonding as was suggested recently on the basis of quantum mechanical calculations by Briant and Messmer (Phil. Mag.). Concerning these calculations it may be noted that they were carried out for a low energy boundary structure. The experiments reported here suggest, however, that it is the high energy boundaries which embrittle preferentially whereas the low energy boundaries do not tend to show this effect. So, it may be useful to apply these calculations to defects in low energy boundaries or to high energy boundaries. On the basis of what is known so far, it is to be expected that a similar weakening of the atomic bonds due to the presence of impurity atoms occurs at or in the vicinity of boundary defects or in high energy boundaries.

Comment by H. Vehoff:

If I understood, you have the following sequence: a) a low energy boundary, b) dislocations (no embrittlement) and then segregation (embrittlement). If you have such a sequence it must be possible to calculate what the impurity really does? Are there some ideas? Do you get a little surface layer between the grain boundary and the bulk at an angle of 0.1 degrees?

Reply:

Observations by Auger spectroscopy suggest the existence of a monolayer of impurity atoms at grain boundaries under certain conditions. Whether or not these impurity atoms form a brittle "surface layer" is still an open question and cannot be answered on the basis of our results.

Comment by P. Haasen:

You stated that the diffraction lines from spheres which withstood shaking were sharper than 0.1°. This corresponds to a grain boundary dislocation spacing $\geq$ 500 lattice spacings. To embrittle slightly less perfect boundaries by overlapping Bi-segregation zones as you propose would need then rather big "particles". Has one seen these by TEM or other techniques? Does the segregation kinetics correspond to such big zones?

Reply:

Investigations by TEM on the core width of grain boundary dislocations have led to the result that the cores of these dislocations may be wider than 100b. Such large core widths would explain why relatively large grain boundary dislocation spacings are sufficient to result in embrittled boundaries. For experimental reasons the segregation treatment was carried out at temperatures close to the melting point. Due to the high diffusivity at these temperatures the segregation kinetics observed are consistent with the formation of large zones.

Comment by K. Christmann:

     You showed convincingly that only the presence of impurities
and defects in the grain boundary zone at the same time gives rise
to embrittlement. Is it not possible to think of an impurity which
operates the opposite way; namely, sort of glues the two crystal
grains together?

Reply:

     Yes, this type of process has been discussed several times in
the literature. The major difficulty is to get experimental evidence
because the measurement of the strength of a boundary is a difficult
problem. The only exception is boundary embrittlement.

Comment by D.J. Duquette:

     In view of your very interesting experiments separating plastic
deformation and impurity segregation it is tempting to assume that
phenomena which may deliver species (such as hydrogen) to grain
boundaries by either enhanced diffusion or dislocation transport,
would be very sensitive to grain misorientation. Accordingly the
distribution of impurity rather than the concentration appears to be
the more important variable. Thus, slip planarity, slip plane spacing
and local dislocation density should determine the relative sensitivity
of a given grain boundary to embrittlement. Could you please comment?

Reply:

     It may well be that the distribution of impurities is an important
quantity for the embrittlement. However, the observations reported
suggest a large difference in the impurity concentration between
special and random boundaries. This observation would be difficult
to understand if the impurity distribution would be the only important
parameter. In the experiments described, the mechanism delivering
the impurities to the boundaries was lattice diffusion which is
insensitive to the misorientation in cubic systems.

INTERFACIAL SEGREGATION IN MULTICOMPONENT SYSTEMS

Michel Guttmann

Centre des Matériaux
Ecole Nationale Supérieure des Mines de Paris
B.P. 87, 91003 Evry, France

INTRODUCTION

It is now recognised that many properties of metallic materials are critically affected when segregation occurs to their interfaces. The equilibrium segregation of impurities deteriorates the resistance of the grain boundaries of many structural alloys to various types of intergranular failures, e.g. low-temperature fracture, creep rupture, stress corrosion cracking, fatigue cracks propagation[1,2]. However in these basically multicomponent systems, the alloying elements can also segregate, and the interactions between the various components can alter their segregation pattern.

The first part of this paper is devoted to simple models which allow interactive cosegregation to be rationalised. A second section presents experimental examples of this phenomenon which can be quantified by these models, and the last section shows that fracture properties are controlled by the intergranular coverages of both types of elements.

THE THERMODYNAMICS OF CO-SEGREGATION

Predictions of the Gibbs formalism

In principle the Gibbs adsorption isotherm can be used to demonstrate the influence of interactions between solute atoms on their segregation behaviour in multicomponent systems[3]. The coefficients of interaction $\varepsilon_{ij}^{\alpha}$ between solutes i and j in the bulk solution α are defined as the first order coefficients of the Taylor series expansion of the activity coefficients γ_i^{α} :

$$\varepsilon_{ij}^{\alpha} = \varepsilon_{ji}^{\alpha} = \partial\ln\gamma_i^{\alpha}/\partial y_j^{\alpha} = \partial\ln\gamma_j^{\alpha}/\partial y_i^{\alpha} \tag{1}$$

where the y_i^{α} are the ratios of the bulk concentrations of each solute i to that of the solvent. The adsorptions Γ_I and Γ_M of two solutes in small bulk concentrations are related to the interfacial tension σ according to the equations :

$$\Gamma_I \simeq -\frac{\partial\sigma/\partial\ln y_I^{\alpha}}{RT(1+y_I\varepsilon_{II}^{\alpha})} - y_I^{\alpha}\frac{\varepsilon_{MI}^{\alpha}}{(1+y_I^{\alpha}\varepsilon_{II}^{\alpha})}\Gamma_M = \Gamma_I^{o} - y_I^{\alpha}\frac{\varepsilon_{MI}^{\alpha}}{(1+y_I^{\alpha}\varepsilon_{II}^{\alpha})}\Gamma_M \tag{2}$$

$$\Gamma_M \simeq -\frac{\partial\sigma/\partial\ln y_M^{\alpha}}{RT(1+y_M^{\alpha}\varepsilon_{MM}^{\alpha})} - y_M^{\alpha}\frac{\varepsilon_{MI}^{\alpha}}{(1+y_M^{\alpha}\varepsilon_{MM}^{\alpha})}\Gamma_I = \Gamma_M^{o} - y_M^{\alpha}\frac{\varepsilon_{MI}^{\alpha}}{(1+y_M^{\alpha}\varepsilon_{MM}^{\alpha})}\Gamma_I \tag{3}$$

If the Γ_i^{o}, which are approximately equal to the adsorptions in the respective binary A+i alloys are known, the effect of the MI chemical interaction on the surface behaviour of the ternary system can be anticipated. Since an attractive MI interaction implies a negative $\varepsilon_{MI}^{\alpha}$ coefficient (eqn. 10), it gives rise to mutual enhancement of the adsorptions of I and M. Inversely, a repulsive interaction ($\varepsilon_{MI}^{\alpha}>0$) reduces the adsorptions of both elements if both are strongly surface active (i.e. if $\Gamma_i^o>0$), or reduces that of the weaker segregant while enhancing that of the stronger one if their segregation tendencies are widely different (i.e. $\Gamma_M^o<<\Gamma_I^o$).

Unfortunately the $\varepsilon_{MI}^{\alpha}$ have been measured mainly in liquids (e.g. ref. 4 for the Fe-base system), and their absolute values in the solid state must be appreciably different. In addition, the adsorptions can be calculated using these equations only if the variations of σ are measured experimentally. An alternative consists in resorting to models of the interface, which introduce supplementary equations at the price of narrowing the scope provided by the Gibbs formalism.

The layer model

Simple "layer" models of the interface have been discussed in detail elsewhere[5],[6]. The present paper will rather emphasize the practical character of the zero-order quasi-chemical approximation for interpreting recent segregation data.

The basic hypothesis is that the interface ϕ is a two-dimensional (2-D) "phase" where all the thermodynamic quantities are defined in the same way as in three-dimensional (3-D) ones, but for the introduction of the specific 2-D intensive variable σ (surface tension) and associated extensive variable Ω (area), the product $\sigma\Omega$ being the 2-D homologue of $-pV$ in the expression of the Gibbs free energy, G^{ϕ}. Thus the formal definition of the interfacial chemical potentials, μ_i^{ϕ}, functions of the activities a_i^{ϕ} and interfacial concentrations X_i^{ϕ} is

$$\mu_i^\phi = \mu_i^{o\phi}(p,T,\sigma) + RT\ln a_i^\phi = \left[\zeta_i^{o\phi}(p,T) - \sigma\omega_i\right] + RT\ln\gamma_i^\phi X_i^\phi$$

where $\zeta_i^{o\phi}$, γ_i^ϕ and ω_i are respectively the standard values of the partial Helmoltz free energy F^ϕ, the activity coefficient and the partial area of component i in the interface.

The interfacial concentrations at equilibrium are readily obtained from the equilibrium condition : $\mu_i^\phi = \mu_i^\alpha$ for all i, which yields the general segregation equations :

$$\frac{X_i^\phi}{(X_A^\phi)^{\omega_i/\omega_A}} = \frac{X_i^\alpha}{(X_A^\alpha)^{\omega_i/\omega_A}} \; \exp(\frac{\Delta G_i}{RT}) \tag{4}$$

The segregation energies ΔG_i are functions of the $\zeta_i^{o\phi}$, γ_i^ϕ and ω_i.

The zero-order quasi-chemical (or regular) approximation. Analytical expressions of these parameters can be constructed for various types of atomic arrangement and interaction in the interface using 3-D analogues. The zero-order quasi-chemical approximation is the easiest one to handle analytically in order to emphasize the effects of interactions between segregating atoms of different species. It is a generalisation of Fowler's binary model[7].

In spite of its imperfections, the regular approximation is a powerful tool for the analysis of the thermochemical properties of 3-D alloys, as illustrated by the systematic work of Hillert and coworkers[8,9]. Some of the analytical expressions of the thermochemical functions developed by these authors have been formally transposed to the case of interfaces of different structures[6].

The analogue of the 3-D substitutional solution is an interface where all atomic sites can be occupied by all atomic species (competitive segregation). In a ternary A-M-I system, eqn. 4 reads :

$$\frac{X_i^\phi}{1-X_I^\phi-X_M^\phi} = \frac{X_i^\alpha}{1-X_I^\alpha-X_M^\alpha} \; \exp(\frac{\Delta G_i}{RT}) \qquad i = I, M \tag{5}$$

$$\left.\begin{aligned} \Delta G_I &= \Delta G_I^o - 2(\alpha_{AI}^\phi X_I^\phi - \alpha_{AI}^\alpha X_I^\alpha) + (\alpha_{MI}'^\phi X_M^\phi - \alpha_{MI}'^\alpha X_M^\alpha)\\[4pt] \Delta G_M &= \Delta G_M^o - 2(\alpha_{AM}^\phi X_M^\phi - \alpha_{AM}^\alpha X_M^\alpha) + (\alpha_{MI}'^\phi X_I^\phi - \alpha_{MI}'^\alpha X_I^\alpha) \end{aligned}\right\} \tag{6}$$

The ΔG_i^o are the intrinsic segregation free energies of i and the α_{ij}^τ are the i-j interaction coefficients in each phase τ :

$$\alpha_{ij}^\tau = -Z^\tau N\left[e_{ij}^\tau - (e_{ii}^\tau + e_{jj}^\tau)/2\right] \tag{7}$$

where N is Avogadro's number, Z^τ is the coordination number and the e_{kl} are the pair-wise energies between k and l atoms in τ. $\alpha_{MI}^{'\tau}$ is the preferential ternary interaction coefficient between M and I relative to the binary A-I and A-M interactions :

$$\alpha_{MI}^{'\tau} = \alpha_{MI}^{\tau} - \alpha_{AI}^{\tau} - \alpha_{AM}^{\tau} \qquad (8)$$

The case where M and I do not compete for sites in the interface is the formal analogue of 3-D interstitial-substitutional solutions or non-stoechiometric $(Fe,M)_a(I,Vacancy)_c$ compounds as treated by Hillert and Staffansson[9]. Calling Y^ϕ the concentrations in the respective substitutional (Fe,M) and interstitial (I) "sublattices", the segregation equations read :

$$\frac{Y_i^\phi}{1-Y_i^\phi} = \frac{X_i^\alpha}{1-X_i^\alpha} \exp(\frac{\Delta G_i}{RT}) \qquad i = I,M \qquad (9)$$

$$\Delta G_I = \Delta G_I^O - 2(\frac{\beta_I^\phi}{c^\phi} Y_I^\phi - \frac{\beta_I^\alpha}{c^\alpha} Y_I^\alpha) + (\frac{\beta_{MI}^\phi}{c^\phi} Y_M^\phi - \frac{\beta_{MI}^\alpha}{c^\alpha} Y_M^\alpha) \qquad (10)$$

and the symmetrical equation for ΔG_M is obtained by permuting the M and I indexes. The preferential ternary coefficients β_{MI}^τ in this formalism are equal to the negative standard excess free energy of formation of the saturated $M_a I_c$ compound over that of the $A_a I_c$ one :

$$\beta_{MI}^\tau = - \Delta G_{MI}^{'OT} = -(\Delta G_{M_a I_c}^{OT} - \Delta G_{A_a I_c}^{OT}) \qquad (11)$$

In the interface a^ϕ and c^ϕ are the fractions of sites available for A+M and I atoms in the substitutional and interstitial "sublattices" respectively, $a^\phi + c^\phi = 1$. β_{MI}^τ can be expressed in terms of pair-wise interaction coefficients α_{MI}^τ. The exact relationship[10] reduces to[11] :

$$\beta_{MI}^\tau \simeq \alpha_{MI}^{'\tau} a^\tau c^\tau \qquad (12)$$

when it is assumed that the M-A interaction is negligible.

The effect of excess interfacial interactions will be more easily discussed if it is assumed for simplicity that the interfacial and bulk interactions are comparable, which will turn out to be approximately true for grain boundaries. Then eqns. 6 and 10 become eqns. 13 and 14 respectively :

$$\Delta G_I = \Delta G_I^o - 2\alpha_{AI}(X_I^\phi - X_I^\alpha) + \alpha_{MI}^{'}(X_M^\phi - X_M^\alpha) \qquad (13\text{-}a)$$

$$\Delta G_M = \Delta G_M^o - 2\alpha_{AM}(X_M^\phi - X_M^\alpha) + \alpha_{MI}^{'}(X_I^\phi - X_I^\alpha) \qquad (13\text{-}b)$$

$$\Delta G_I = \Delta G_I^o - 2\frac{\beta_I}{c^\phi}(Y_I^\phi - Y_I^\alpha) + \frac{\beta_{MI}}{c^\phi}(Y_M^\phi - Y_M^\alpha) \qquad (14\text{-}a)$$

$$\Delta G_M = \Delta G_M^o - 2 \frac{\beta_M}{a^\phi} (Y_M^\phi - Y_M^\alpha) + \frac{\beta_{MI}}{a^\phi} (Y_I^\phi - Y_I^\alpha) \qquad (14\text{-}b)$$

The molecular (or "associated solution") approximation. Other models of interactive behaviours can be imagined[6]. For instance, by analogy with 3-D fluids[12], I and M atoms can form quasi-molecules which are in equilibrium with them according to a mass action law. This model is not relevant to the type of bonding existing in bulk metallic alloys and in their internal interfaces, but it could be better adapted to their outer surfaces where segregated or chemisorbed atoms can be involved in localised molecular-like types of bonding[13].

Effects of the interactions between atoms in the interface on their equilibrium segregation. The simplified expressions of ΔG_I and ΔG_M, eqns. 13 or 14, clearly reveal the roles of the chemical interactions between segregated atoms.

The first term ΔG_i^o is the segregation free energy of element i at infinite dilution of all elements in the grain boundaries or in the absence of any interaction (ideal or Henryan system). This behaviour is that described by the formally identical models of McLean and Langmuir for grain boundary segregation and adsorption on free surfaces, respectively.

The second term in these equation shows a dependence of the segregation free energy of each element upon its own interfacial excess. The effect of this well known Fowler's term[7] is to increase segregation, compared to the case of the ideal solution, when the i-i interactions are preferentially attractive in the A-i binary (α_{Ai} or $\beta_i < 0$), and to decrease it in the reverse case.

The third term in eqns. 13 and 14 relates the segregation energy of each solute to the intergranular excess of the other segregant via the relative M-I interaction. When the latter is preferentially attractive, $\alpha_{MI}^{'\phi}$ and β_{MI}^ϕ are positive and the segregation of each element enhances that of the other. If one element, say M, is weakly or not surface active in the A-M binary, ($\Delta G_M^o \simeq 0$), its segregation can still be driven by that of the more surface active element I via the purely chemical term in ΔG_M, and it will in turn amplify the segregation of I via the homologous factor in ΔG_I. This is illustrated in fig. 1-a for the simple case of non-competitive segregation. This figure shows that the co-segregation of both elements and its temperature dependence increase with the interaction, and it also illustrates the effect of increasing the bulk concentration of M.

The same pattern of behaviour is predicted in the case of a competitive segregation mechanism but for the complications arising from the competition which tends to reduce the interfacial coverage of both elements, fig. 1-b. However this effect is appreciable only at large coverages of either solute. In practice, it is often difficult to

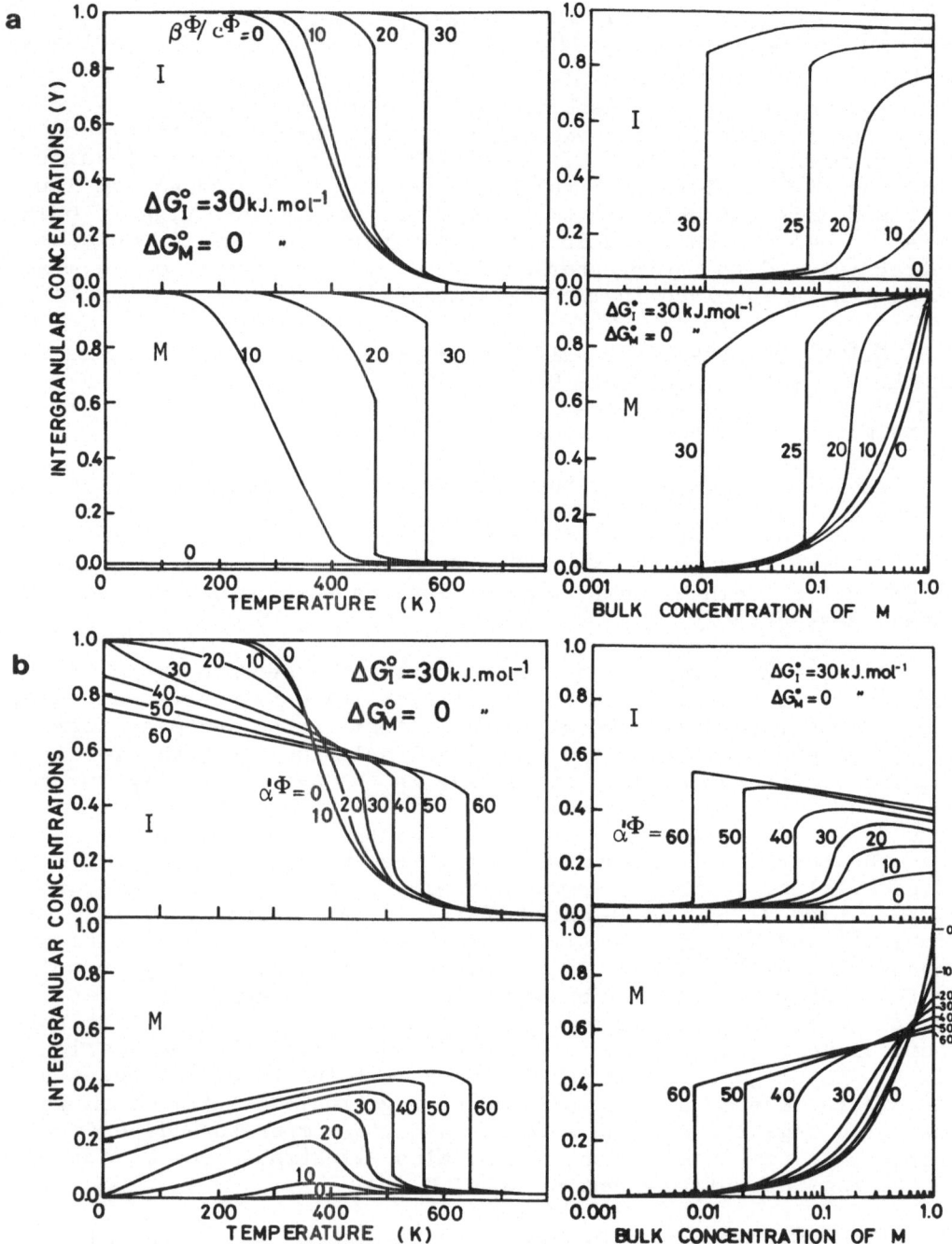

Fig. 1   Influence of attractive M–I interaction on equilibrium
          segregation ; (a) Without site competition; (b) With site
          competition.[6]

identify a competitive segregation with certainty and to differentiate it from a repulsive segregation.

When the relative M-I interaction is repulsive (β_{MI}^{ϕ} and $\alpha_{MI}'^{\phi} < 0$), the segregations of both elements are reduced with respect to those in the respective binaries, if both elements are rather strong segregants. However, a more complex pattern arises when one species is much less surface active than the other, say $\Delta G_M^o << \Delta G_I^o$, fig. 2. This can be seen easily by neglecting the Fowler's term in eqns. 13 or 14. The weaker segregant is eventually expelled by the stronger one, whose segregation is enhanced. The predictions of this model are thus consistent with those of Gibbs formalism, eqns. 2 and 3.

Phase transitions in segregated three-component layers. The problem of 2-D phase transitions on binary free surfaces has been reviewed by Blakely and coworkers[14],[15], and discussed in connection with ternary systems[6] and grain boundaries[16]. It was suggested[16] in particular that the properties of the 2-D compounds in saturated grain boundaries should in some way be related to those of the 3-D compounds existing in each alloy system, the former serving as a precursor state to the nucleation of the latter. This was illustrated by the case of the Fe-Te and Fe-Se binaries[17] and should also be true in Ni-S[18] and Cu-Bi[19]. The appearance of faceting in the grain boundaries, which occurs at saturation coverage in these systems, indicates a drastic decrease in interfacial tension which must arise from the formation of an interfacial 2-D compound[16] as it is the case when the (110) surface of Au becomes saturated with S[20].

It will be sufficient to discuss two simple cases where a preferential M-I interaction can favor or induce 2-D condensation.

Fowler's analysis shows that when the I-I interaction is attractive in a binary A-I interface, i.e. when $\alpha_{AI} < 0$, *clustering* of I atoms occurs below a critical temperature $T_c = -\alpha_{AI}/2R$. Similarly in a ternary system, a preferential M-I attraction produces a 2-D *miscibility gap* between the solvent atoms A and clusters of M+I atoms below a critical temperature $T_c \simeq (\alpha_{MI}' - \alpha_{AI} - \alpha_{AM})/8R$. In the segregation curves, figs. 1, 2, this gives rise to steps which widen as the interaction or the concentration of one component is increased.

The other possible event is the *"precipitation"* of a 2-D "compound" ϕ_p having a negative free energy of formation with respect to the more dilute 2-D "solution". The most simple case is that where the former is a stoichiometric $(A,M)_{a"}I_{c"}$ compound in which the M-I interaction is more attractive than the A-I one, the preferential interaction energy $\Delta G'^{o\phi_p}_{MI} = -\beta^{\phi_p}$ being negative. The relationship between the 2-D "solubilities" (which are the I and M coverages at which 2-D precipitation occurs) and temperature T$\ast$ is formally equivalent to that derived for 3-D precipitation with similar hypotheses, for instance eqn. 16(below) for regular types of interactions in the "compound". Assuming for

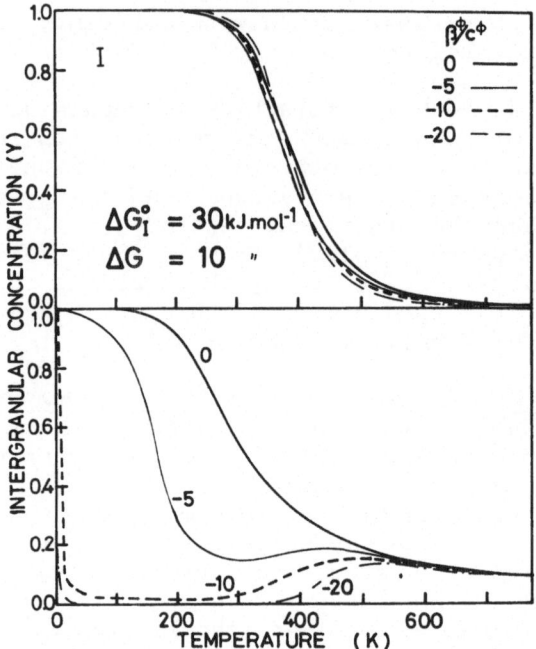

Fig. 2   Influence of repulsive M-I interaction on non-competitive
equilibrium segregation.[6]

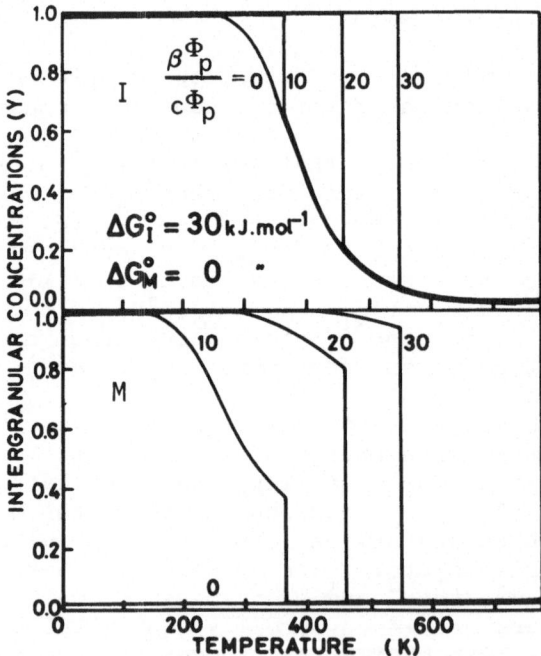

Fig. 3   Equilibrium segregation in the case of the formation of
a 2-D compound with an attractive M-I interaction.

simplicity that segregation above the 2-D "solvus" temperature T^* is of the ideal (McLean) type, it will be described by eqns. 5 or 9 with $\Delta G_I = \Delta G_I^O$, $\Delta G_M = \Delta G_M^O$. Below T^*, the interface is saturated with I, and the segregation free energy of M is increased to :

$$\Delta G_M = \Delta G_M^O + (\beta_{MI}^{\phi} P/a")$$

Fig. 3 shows that this type of interaction also causes the segregations of both elements to enhance each other below the critical temperature at which the step occurs.

Discussion. As pointed out by Jäger[21], the zero-order approximation does not predict the progressive departure from random atomic distribution, which normally arises from the interactions as the coverages increase. In complete analogy with Fowler's binary model, it is merely a rule of thumb to anticipate the occurrence of a 2-D phase transition when the interaction is large enough. More realistic models of segregation have to take into account the environment of the interfacial atoms, their respective sizes, and the interactions between them and those situated in adjacent layers. Sophisticated statistical mechanics calculations based on the "central atom"[22-a] or "surrounded atom"[22-b] models are in remarkable agreement with experimental data, but their complexity is beyond the scope of the present review.

Also recent theoretical work by Szklarz and Wayman[23] has shown that the segregation of a non-magnetic atom in the grain boundaries of a magnetic matrix introduces a large non-linear contribution to the segregation free energy, fig. 4. The influence of a given solute x can be described with the help of a magnetic parameter ΔT_x defined as the effectiveness of x in shifting the Curie temperature T_C^{α} of the bulk solid solution with respect to the pure solvent, say Fe, T_C^{Fe} : $T_C^{\alpha} = T_C^{Fe} + X_x^{\alpha}\Delta T_x$. Fig. 4 shows that a negative ΔT_x leads to a considerable positive contribution to the segregation free energy of x and to its interfacial coverage, especially at low temperatures. It also gives rise to a paramagnetic – ferromagnetic 2-D phase transition in the interface, fig. 4, the Curie temperature of the boundary $T_C^{\phi} = T_C^{Fe} + X_x^{\phi}\Delta T_x$ being lowered by segregation when $\Delta T_x < 0$. In the case of P in αFe, $\Delta T_P = -1200$ K.[23]

The regular solution model, originally developed for dilute 3-D solutions, is perhaps not a quite accurate approximation for interfaces with large solute coverages, but it appears equivalent to a first order Taylor series expansion of the segregation energies :

$$\Delta G_i \simeq \Delta G_i(T=0, \; X_j^{\phi}=X_j^{\alpha}) + \left(\frac{\partial \Delta G_i}{\partial T}\right)_o T + \sum_{j \neq A} \left(\frac{\partial \Delta G_i}{\partial X_j^{\phi}}\right)_o (X_j^{\phi} - X_j^{\alpha}) \qquad (15)$$

where : $\Delta G_i(T=0, \; X_j^{\phi}=X_j^{\alpha}) = \Delta H_i^O \qquad (\partial \Delta G_i/\partial T)_o = -\Delta S_i^O$

Identifying eqn. 15 with e.g. eqn. 13 yields :

$$(\partial \Delta G_i / \partial X_i^\phi)_o = - 2\alpha_{Ai}^\phi \qquad (\partial \Delta G_i / \partial X_j^\phi)_o = \alpha_{ij}^{!\phi} \qquad (j \neq i, A)$$

As will be shown in the next chapter, the linear approximation is sufficient to account for the presently available segregation data. The simplified physical meaning thus conferred to the coefficients of eqn. 15 will allow an easy empirical comparison with bulk thermo-chemical properties. In this procedure, the magnetic contribution will in fact be included in the "chemical" contribution to both the interfacial and bulk quantities.

The effect of bulk interactions on equilibrium segregation

When the M-I interaction is attractive, precipitation of MI compounds may occur in the matrix as the interaction energy or the concentration of either element is increased. For example, a strongly reactive alloying element M may scavenge an impurity I. This reduces the amount of I dissolved in the matrix X_I^α thus hindering its segre-gation to interfaces as is obvious from eqns. 4, 5 or 9, whereas the interaction between interfacial atoms increases their segregation free energies, as discussed in the previous section. In other words, a strong MI affinity considerably reduces the absolute I coverage even though it increases the interfacial enrichment ratio $r_I^\phi = X_I^\phi / X_I^\alpha$. This is the ternary analogue of the inverse relationship between solubil-ities and enrichment ratios in binary systems[24].

A symmetrical situation is that where the scavenging element, having a stronger affinity for another species I' than for I, precip-itates into M-I' compounds, thus *releasing* I into the solid solution and causing its segregation to *increase*.

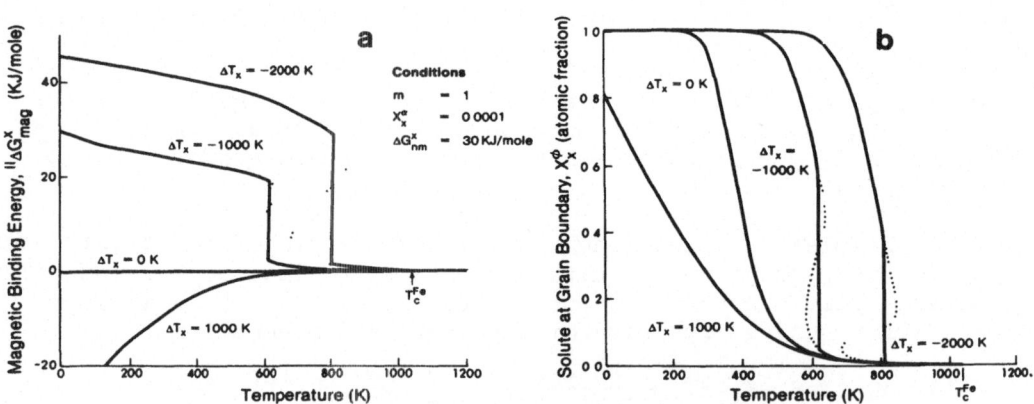

Fig. 4.   Influence of magnetism on equilibrium segregation.[23]
(a) Magnetic contribution to the segregation free energy;
(b) Temperature dependence of segregation (ΔG_{nm}^X is the non-magnetic contribution to ΔG_x).

In many cases, precipitation can be slow compared to interfacial segregation, even though the same bulk diffusion coefficients are involved in both processes. This occurs, for instance, when the nucleation of the 3-D precipitates is slow due to a high activation barrier, or when particle growth is controlled by a slow interface mechanism. In such a situation, the interfacial coverages can be in equilibrium with the instantaneous concentrations of the elements dissolved in the matrix, even though the latter are not the equilibrium values. As bulk precipitation proceeds towards its completion, i.e. towards the state of the $\alpha(3\text{-}D) \rightleftarrows B(3\text{-}D)$ equilibrium, the $\alpha(3\text{-}D) \rightleftarrows \phi(2\text{-}D)$ segregation equilibrium is slowly *displaced*, and the interfacial coverages X_I^ϕ decrease. Segregation therefore reaches a maximum as time elapses. The occurrence of this maximum does not mean that the segregation is of the "non-equilibrium" type, which refers to a *mechanism* of a basically different nature, but that the *state* of equilibrium of the *matrix* is not established. Transient segregations of this type occur in binary A-I alloys where the solubility of I is low. For instance the segregation of Te in Cu exhibits a maximum which has been interpreted in these terms.[25] The same is true of the Ni-S system,[26] where, in addition, the presence of minute amounts of Al or Ca drastically reduces S solubility, but where the kinetics of sulfide precipitation are very slow. The faceting of S-saturated boundaries disappears with prolonged aging i.e. with the progressive depletion of the solid solution in S, because it is reversibly associated with a transient segregation of the equilibrium type, rather than with a non-equilibrium type of segregation as might be inferred from the ambiguous formulation of Loier et al.[18] The same behaviour is expected in interactive multicomponent systems where the precipitation of M-I compounds is slow.

The equilibrium solubility of I, $X_I^{\alpha *}$, in a ternary α matrix can be calculated using the simplified form of Hillert's[8] regular solution model for mutually miscible stoichiometric $(A,M,\ldots)_{a'} I_{c'}$ compounds :

$$X_I^{\alpha *} = \exp(\Delta G_{A_{a'} I_{c'}}^{oB}/a'RT)\{1 + \sum_M X_M^\alpha [\exp(\frac{\beta_{MI}^B}{a'RT}) - 1]\}^{-a'/c'} \qquad (16)$$

The $\Delta G'^{oB} = -\beta_{MI}^B$ are the preferential M-I interaction energies in the complex compound $B = (A,M,\ldots)_{a'} I_{c'}$ as defined by eqns. 11 and 12 with $\tau = B$. The numerator of eqn. 16 is the solubility of I in the binary A-I solution.

This model has been applied with success to the precipitation equilibrium of alloyed carbides in γ-iron,[8] where the Fe-C and M-C interactions are very strong, and in the above simplified form (eqn. 16) to that of alloyed phosphides in α-iron[11] where the Fe-P and M-P interactions are considerably weaker. It should therefore be applicable to any comparable system in which the M- and A- based compounds of I are isomorphic and mutually miscible. If the solubility data in the binary A-I and ternary A-M-I alloys are known, this simple model provides a practical means of evaluating the preferential

M-I interaction in a concentrated phase at equilibrium with the dilute ternary solution. The grain boundary "phase" also has these character-istics, and it will be seen below that the knowledge of the interac-tions in the bulk compound is a powerful tool for assessing those in grain boundaries.

The kinetics of segregation in multicomponent systems

The rate of interfacial build-up is classically analysed in terms of the law derived by McLean for binary alloys[27] under the approximate form valid for the initial linear part of the segregation-$(time)^{1/2}$ plot :

$$\frac{Y_i^{\phi}(t) - Y_i^{\phi}(0)}{Y_i^{\phi} - Y_i^{\phi}(0)} \simeq 2(X_i^{\alpha}/Y_i^{\phi}d)(D_i t/\pi)^{1/2} \tag{17}$$

where $Y_i^{\phi}(t)$ is the interfacial build up of i at time t, $Y_i^{\phi}(0)$ and Y_i^{ϕ} its initial and equilibrium values, respectively. Although Tyson[28] has criticised the validity of McLean's equation and demonstrated that not only the equilibrium but also the kinetics of segregation were very sensitive to M-I interaction effects, eqn. 17 is generally used as a practical approximation. This is justified by the fact, illustrated in the present article, that the dominating effect of a strong M-I interaction on the segregation kinetics of an impurity I occurs via the scavenging of I by M in the matrix. If complex M-I and/ or M-I' precipitation occurs at a rate slower than that of I segrega-tion, the kinetics of this segregation primarily reflect those of the precipitation processes which control the amount of M and I in the solid solution, as discussed in the previous section.

EXPERIMENTAL OBSERVATIONS, EVALUATION OF THE MODELS

Co-segregation to free surfaces

Segregation to free surfaces is studied not only because of its effects on the intrinsic properties of surfaces such as heterogeneous catalysis[13] but also because it provides useful clues for understand-ing intergranular segregation. It must be kept in mind however that the solute-interface attraction is always larger at free surfaces than at grain boundaries. Therefore it can only be said that an ele-ment which does not segregate to the former will certainly not seg-regate to the latter.

Ternary iron-base alloys. Special attention has been devoted to iron-base systems because all the types of segregation-induced inter-granular failures listed in the introduction have been reported to occur in steels. In particuler it is now admitted that reversible temper embrittlement is controlled by the interactions between non-metallic impurities and transition metal alloying elements. Co-

segregation to the surface of high purity ternary Fe-M-I alloys with
I = P, Sb and M = Ni, Cr, Mo, V, Ti was therefore systematically
compared to that in the corresponding binary Fe-I and Fe-M alloys by
means of Auger electron spectroscopy.[29]

Antimony and phosphorus strongly segregate to the surface of
pure iron : when their bulk concentration is of the order of a few
hundred atomic p.p.m., their surface coverages reach saturation
values at low temperatures. Evaporation rather than desegregation
is responsible for the decrease observed at higher temperatures,[30]
fig. 5-a. Therefore only an underestimate of their segregation free
energy can be obtained : $\Delta G^O_{Sb} > 105$ kJ.mol^{-1} and $\Delta G^O_P > 80$ kJ.mol^{-1}.

The metallic elements are much less surface-active in Fe.
Titanium, the comparatively strongest segregant, fig. 6, has a ΔG^O_{Ti}
of 21 kJ.mol^{-1}, followed by nickel with ΔG^O_{Ni} = 15 kJ.mol^{-1}.
Molybdenum, chromium and vanadium could not be made to segregate at
temperatures as low as 500°C for bulk contents of 2 at.%, thus
$\Delta G^O_{Cr} \simeq \Delta G^O_{Mo} \simeq \Delta G^O_V \simeq 0$.

The effect of a moderately attractive Ni-Sb interaction is revealed by the increase of the Ni segregation and of the widening of
the temperature range in which it occurs, which are observed when Sb
is also segregated, fig. 5-b. However the segregation of Sb is not
altered by that of Ni, fig. 5-a, since the surface of the corresponding binary Fe-0.06 at.% Sb alloy is already saturated with Sb below
700°C and since the evaporation rate is apparently not modified above
this temperature.

The Ti-Sb interaction is much stronger and manifests itself by
two effects, fig. 6. The surface coverage of Ti is considerably larger
in the presence of segregated Sb than in the binary Fe-2 at.% Ti, and
remains at a constant value of 0.50 which is apparently indicative of
a saturation coverage. Reciprocally, the segregation of Ti extends
towards much higher temperatures the range in which the evaporation
of Sb is prevented. The Sb coverage thus remains at its saturation
value up to 900°C. This indicates that the Ti-Sb bond is sufficiently
strong to increase the sublimation energy of Sb at the Ti-enriched
surface compared to the unalloyed surface of Fe.

The Ti-P surface interaction is stronger than the Ti-Sb one. The
minute amount of residual P present in the high purity Fe-2 at.% Ti
alloy segregates only after the Ti coverages has reached its equilibrium value (18 at.%), fig. 7. This small P build-up then causes the
Ti surface concentration to rise again with time.

Molybdenum also interacts with phosphorus at the surface since it
segregates in a Fe-1.2at.%Mo-0.073at.%P whereas it does not so in the P-
free alloy, fig. 8-a. However this interaction is much weaker than the
Ti-P one since even a saturated monolayer of P is not able to drive
more than 3at.%Mo to the surface. As for the effect of Mo on P segre-

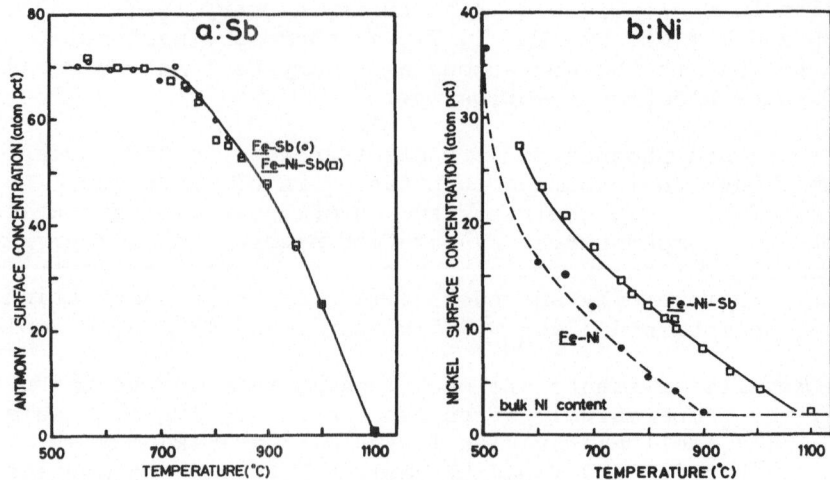

Fig. 5   Steady state segregation of Sb and Ni at the free surface of
Fe-0.06at.%Sb, Fe-2at.%Ni, and Fe-0.06at.%Sb – 2at.%Ni.[29]

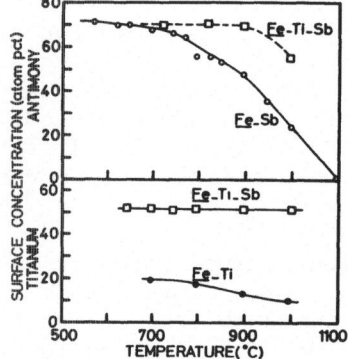

Fig. 6   Steady state segregation of Sb and Ti at the free surface of
Fe-0.06at.%Sb, Fe-2at.%Ti, and Fe-0.06at.%Sb – 2at.%Ti.[29]

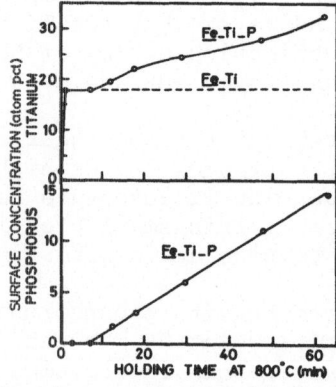

Fig. 7   Segregation kinetics of P and Ti in Fe-2at.%Ti containing
very small amounts of residual P.[29]

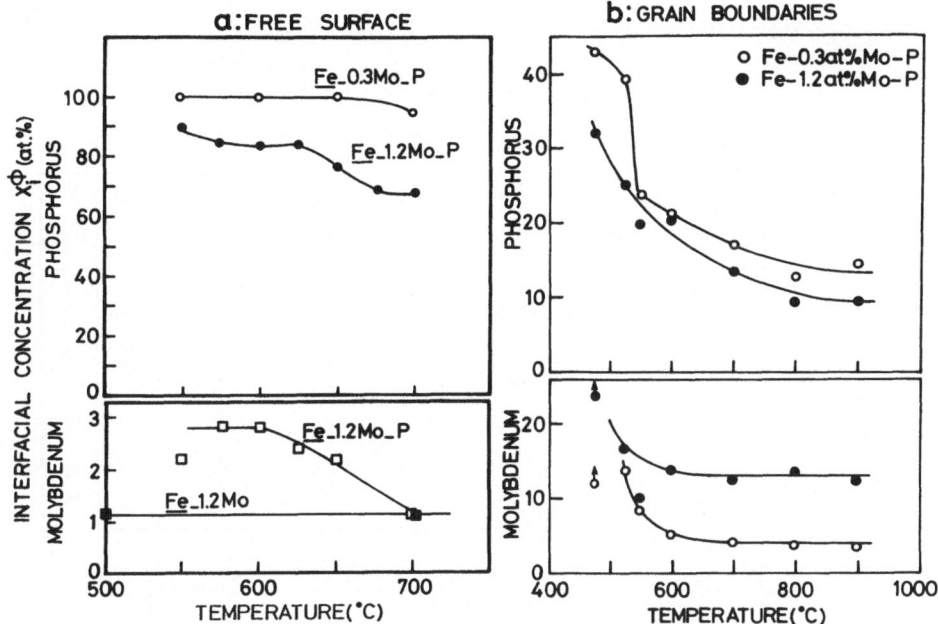

Fig. 8 Steady state segregation of Mo and P in Fe-1.2at.%Mo,
Fe-1.2at.%Mo - 0.073at.%P and Fe-0.3at.%Mo - 0.073at.%P :
(a) Free surface,[29] ; (b) Grain boundaries.[37]

gation, fig. 8-a shows that it is in opposition to the behaviour observed in the previous Fe-M-I systems. The decrease in P coverage with
increasing Mo addition results from the scavenging of P by Mo in the
matrix due to the bulk Mo-P interaction. It will be discussed below.

As for chromium and vanadium, their segregation is totally unaffected by that of Sb which indicates that the Cr-Sb and V-Sb interactions at the surface are virtually zero. Conversely, when the other
residuals such as C,N,S, present in minute amounts in the binary
Fe-M alloys, begin to segregate, the surface concentrations of Mo,
V,Cr, and also of Ti increase concomitantly, in agreement with eqns.
2-3 and the well known M-I affinity for these couples of elements.
Also the saturation coverage of Ti in the presence of segregated C
is larger than in that of Sb, indicating the existence of different
2-D saturated structures depending on the interaction.

In all the above examples where the segregation of the non-
metallic element I has not reached a state of equilibrium because it
is evaporating or because its bulk content is too small, the coverage
of the transition metal M has the time to equilibrate with the
amount of impurity present at the surface. Therefore the equilibrium
equations (eqn. 9) can be applied to M segregation. This yields the
ΔG_M vs. Y_I^S plots shown in fig. 9 and the values of the surface interaction coefficients β_{MI}^S listed in table 1.

Table 1. Comparison between preferential interaction energies β_{MI}^{Φ}
(kJ.mol^{-1}) in bulk compounds (B), in grain boundaries (gb)
and on free surfaces (S) of Fe-base alloys.

M / I		Mn	Cr	Ni	V	Ti	Mo	P	C	Ref.
Bulk Compounds β_{MI}^{B}	Sb	12.5	~0	18.0	2.9	26.3				10
	P	12.5	19-24X_{Cr}^{α}	5.6	27.0	41.4	33.0±1.5			6
Grain Boundaries β_{MI}^{gb}	Sb	>0	0	~15	>>0					6
	P	10.5±3.0	17.1±2.4	<10±2			37.5±6.5	2.2	-9.0	35
	C							-9.0	+8.4	35
Free Surfaces β_{MI}^{S}	Sb		0	<4	0	12				29
	P					42	4			29
	C				44	102	30			29
	S		0			10				29
	N		24		26		11			29

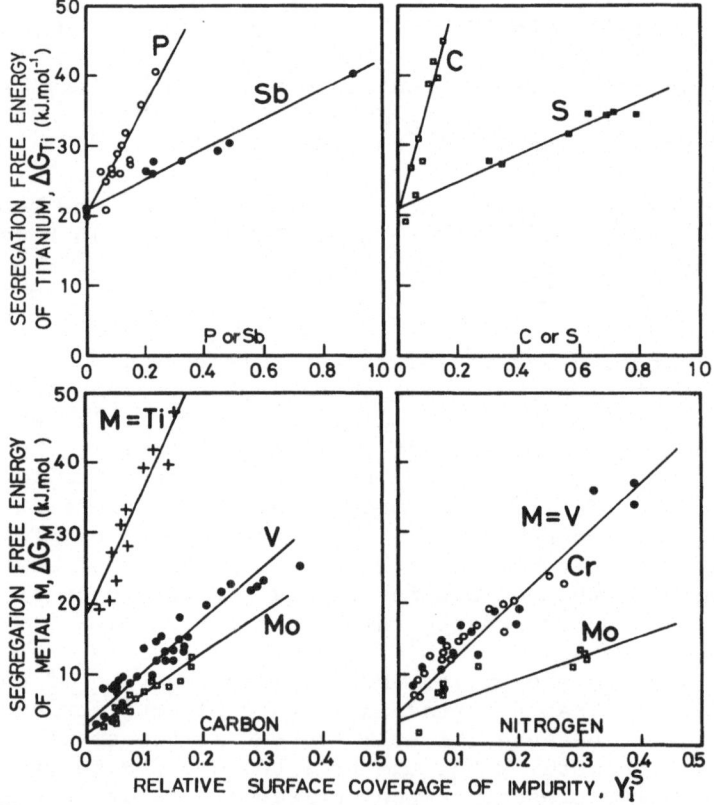

Fig. 9   Influence of the surface coverage of non-metallic elements
on the segregation free energies of Ti,V,Cr,Mo in Fe.[29]

The reactive metals Mo and Ti can also tie up P and Sb atoms
in the matrix. We have seen in fig. 8 that this decreases the *equilibrium* segregation of P in Fe-Mo-P, but it also causes the *rate* of
P and Sb build up to be slower in Fe-Ti-P,[31] Fe-Mo-P and Fe-Ti-Sb.[29]
The concentration of P or Sb in solid solution can be calculated by
fitting eqn. 9 to the equilibrium data and eqn. 17 to the kinetic
data. Then eqn. 16 yields the apparent values of the bulk M-I interaction coefficients β_{MI}^{B} listed in Table 1. They can be compared with
those derived[11,6] from equilibrium solubility data in the Fe-M-P and
Fe-M-Sb systems with the help of eqn. 16. The agreement is good for
Ti-P and Ti-Sb, which suggests that the approximate models used are
realistic. For Mo-P the value deduced from segregation kinetics is
too small, indicating that the Mo-P precipitation equilibrium is not
reached at the temperatures considered. In such a case eqn. 16 is
not valid.

The M-Sb surface interaction coefficients are much smaller than
the corresponding bulk ones, Table 1, but exhibit the same ranking
with respect to the transition metals. This behaviour is qualitatively
consistent with the smaller number of bonds at the surface than in
bulk phases. As for interactions involving P atoms, even though the
Ti-P affinity appears consistently stronger than the Mo-P one both
at the surface and in the matrix, there is a considerable discrepancy between surface and bulk values for Mo-P whereas they are equal
for Ti-P. This observation has not been explained yet. It might be
due to differences in the atomic structures of the respective surfaces which have not been taken into account in the expressions of
the segregation free energies.

Other examples of synergistic co-segregation. Joshi et al.[32] have
reported that Zr an O cosegregate at 1800°C in a Nb-1% Zr alloy,
with a simultaneous maximum between 1300°C and 1400°C. Evaporation
takes place at higher temperatures. Conversely, the segregation of O
in a binary Nb-O alloy ends at 1300°C, oxygen going back into solution. This behaviour is consistent with the stronger affinity of O
for Zr than for Nb, the enthalpy of formation of the Zr oxide, 261
kJ.mol^{-1}, being larger than that of the Nb-one, 184 kJ.mol^{-1}.

The segregations of S and Ca at the surface of commercial nickel
exhibit parallel time depences at all temperatures,[26] fig. 10. At
625°C, they both increase linearly with $\sqrt{t}$, in agreement with eqn. 17,
and do not reach saturation in 5h. At 700°C, S coverage reaches its
saturation value in 80 min., then precipitation of calcium sulfide
at the surface causes the Ca coverage to vanish and that of S to
decrease. However there remains enough S in solution to saturate the
surface again. At 800°C, co-segregation of S and Ca is so rapid that
S build up exceeds saturation for a few minutes. This suggests that
Ca raises the saturation coverage of S compared to the binary Ni-S
surface, as in the case of Ti+C on Fe. These transient enrichments
can occur as long as the equilibrium 3-D compounds are not precipitated

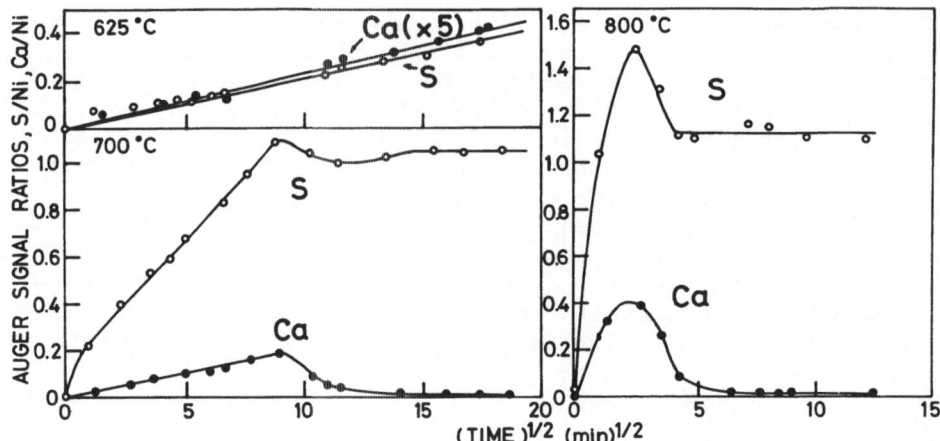

Fig. 10. Kinetics of the co-segregation of S and Ca to the free surface of commercial purity Ni at 625°C, 700°C, and 800°C.[26]

and the matrix remains supersaturated in S and Ca.

Surface segregation in the presence of an adsorbed species

The above theoretical considerations also apply to this case.
The segregation of one component can increase or accelerate gas
adsorption if the interaction between them is preferentially attractive with respect to that between the gas and the other component.
This phenomenon could be responsible for a number of catalytic
effects. It is the other consequence of preferential attractions,
i.e. the enhancement of segregation by gas adsorption, which is the
more frequently observed[13]. This is the case of S in Pd, Pt-Cu and
Mo under an H_2 atmosphere. In binary systems it even frequently
occurs that the component which normally segregates at the clean surface is replaced by the other one when the gas is adsorbed[13] : this
is the case in Au-Pt, Cu-Ni and Pt-Sn, where the adsorption of CO
causes Pt, Ni and Pt to segregate in the place of Au, Cu and Sn
respectively. Similarly, Pd segregates in the place of Au when O_2 is
adsorbed on Pd-Au[13] and the adsorption of S enhances the segregation
of Cu in Cu_3Au.[33]

A quantitative treatment using eqn. 9 has been applied to Sn
segregation in Pb in the presence of adsorbed O_2.[34] The coverage and
the segregation energy of Sn as deduced from Auger data increase
linearly with the imposed O_2 coverage, fig. 11. The small value of
ΔG_{Sn}^0 is consistent with the fact that the segregation of Sn was un-
detectable in vaccuo. Conversely, the adsorption of O_2 lowers the
surface concentration of Bi below bulk concentration in a Pb-Bi alloy,
fig. 11. These results are consistent with the above model in view
of the relative affinities of the various oxides, since the enthalpy

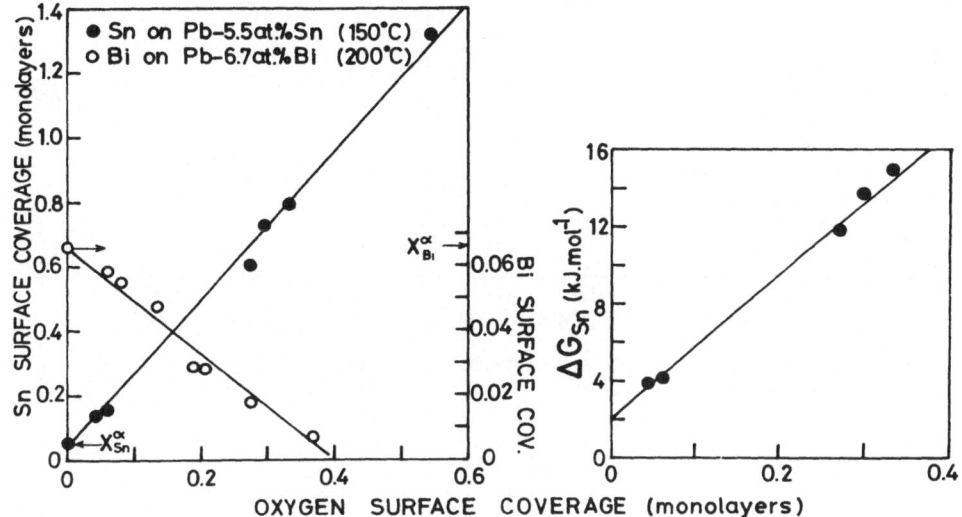

Fig. 11. Influence of adsorbed oxygen on the segregation free
energy of Sn and on the surface depletion of Bi in Pb-Sn
and Pb-Bi alloys.[34]

of formation of SnO_2 (-580 kJ/mole O_2) is larger, and that of Bi_2O_3
(-385 kJ/mole O_2) is smaller, that that of PbO (-435 kJ/mole O_2).[3]

Co-segregation in grain boundaries

Segregation of P in iron and steels. More Auger data being
available for iron-base alloys than for any other, quantitative
analysis of these data has recently allowed the effects of intergranular and bulk interactions to be systematically assessed and
rationalised.[35]

The moderate Ni-P and Mn-P interactions are able to promote co-
segregation but are too weak to induce any observable scavenging in
the matrix. Applying eqn. 9 to the data of Mulford et al.[36-a] about
3.5% Ni-1,7% Cr steels shows, fig. 12, that the segregation free
energies of P and Ni increase linearly with the intergranular concentration of each other with a comparable slope. Least square correlations yield : $\Delta G^o_{Ni} = 3\pm3$ kJ.mol^{-1}, (which indicates that Ni is a very
weak segregant in grain boundaries), $\Delta G^o_P \simeq 46.1$ kJ.mol^{-1}, and $\beta^{gb}_{NiP} =$
(10.5±2.0) kJ.mol^{-1}. However the Ni-P interaction is certainly
overestimated due to the presence of other elements such as Mn and
Cr in the steel studied.

The Mn-P co-segregation results obtained with four low-alloy
steels[35] reveal a comparable value of the Mn-P interaction,

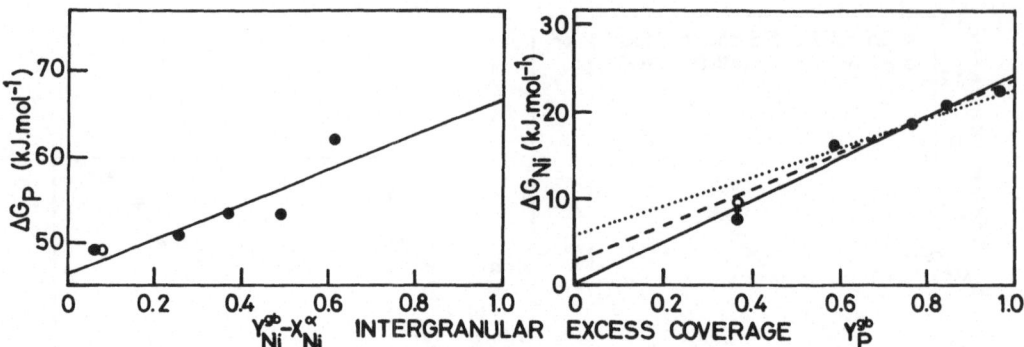

Fig. 12   Variations of the intergranular segregation free energies
          of P and Ni with each other's interfacial coverage in Ni-
          Cr steels.[35]

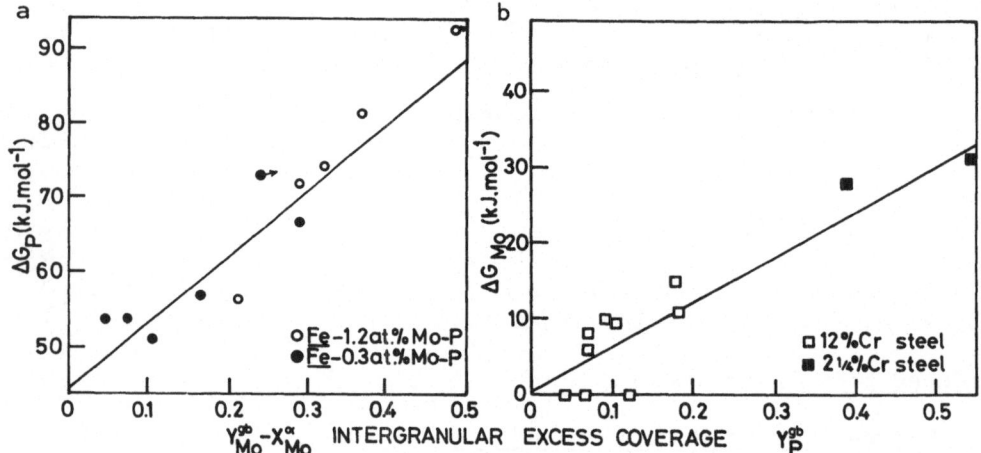

Fig. 13   Variations of the intergranular segregation free energies
          of P and Mo with each other's interfacial coverage :
          (a) ΔG_P in ternary Fe-Mo-P alloys; (b) ΔG_{Mo} in low-C steels.[35]

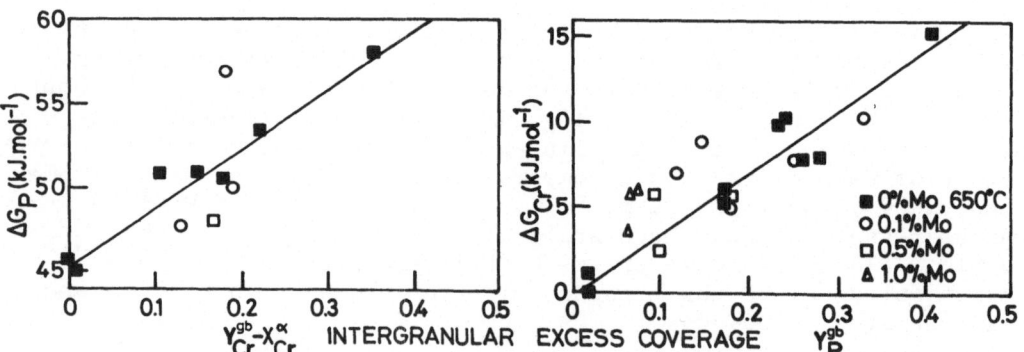

Fig. 14   Variations of the intergranular segregation free energies
          of P and Cr with each other's interfacial coverage in
          12%Cr steels.[38]

β_{MnP}^{gb} = (10.5 ± 3.0)kJ.mol^{-1}. Due to the low concentrations and segregations of other metallic elements in these steels, this value must
be closer to the real one than that found for Ni-P. The intrinsic
segregation free energy of Mn, although small, is definitely positi-
ve, ΔG_{Mn}^{O}= 8 ± 4 kJ.mol^{-1} which shows that Mn is able to segregate alone
in Fe. This is important in view of the suspected embrittling capaci-
ty of this element (see below).

The strong Mo-P interaction affects co-segregation via both its
intergranular (amplifying) and bulk (hindering) effects.

In ternary Fe-Mo-P alloys,[37] the increase in Mo content reduces
the equilibrium intergranular P coverage in spite of the concomitant
increase in Mo segregation, fig. 8-b. This behaviour, which is qualitatively analogous to that observed at the free surface of the same
alloy described in the previous section, shows that the effect of
scavenging P atoms by Mo ones in the bulk overwhelms that of the
intergranular Mo-P interaction. The P segregation kinetics were in
some cases consistent with the theoretical lowering of P solubility
calculated using the equilibrium value of β_{MoP}^{B} (Table 1). In other
cases, the rate of P build up was still considerably slower, but this
was due to the large S segregation and the strong repulsive P-S
interaction.[37] The intergranular interaction could be obtained by
plotting ΔG_{P}^{O} vs. Mo coverage, fig. 13-a, which yielded β_{MoP}^{gb} ≃
44 kJ.mol^{-1}. The symmetrical plot of ΔG_{Mo}^{O} vs. P coverage would have
been biassed by the presence of appreciable amounts of N and S in
the boundaries and the strong affinity of Mo for these residuals.

The segregation free energy of Mo had therefore to be evaluated[35]
in alloys where P was the only non-metallic element present in the
grain boundaries. This was the case in 12%Cr-Mo steels where virtually
all the carbon is precipitated into Cr-carbides and in very low-C
$2^{1}/4$Cr-1Mo steels. The plot in fig. 13-b shows that the intergranular
Mo-P interaction, β_{MoP}^{gb} ≃ 31 kJ.mol^{-1} is comparable to the value deduced from the measurement of ΔG_{P} in fig. 13-a, and that Mo is not
able to segregate alone in Fe, i.e. ΔG_{Mo}^{O} ≃ 0. However if ΔG_{Mo} is calculated from segregation data in a variety of low-alloy steels, much
larger values are found at small P coverages, which apparently implies
a largely positive ΔG_{Mo}^{O} and a much smaller β_{MoP}^{gb}. This behaviour is
certainly due to the segregation of C and to the strong Mo-C affinity.
It may be a controlling factor of grain boundary cohesion in Mo-
bearing steels (see below).

Chromium interacts much less strongly than Mo with P, but it is
able to give rise to comparable effects at much higher concentrations.
[38] Its segregation is quite small in low-alloy steels but becomes
measurable in 12%Cr steels where it increases with P coverage. Fig.14
confirms that the Cr-P co-segregation is synergistic, and the correlations of the data obtained for a single temperature (filled symbols)
yield :

$$\Delta G_P(kJ.mol^{-1}) = 45.7 + 31.8 \ (Y_{Cr}^{gb} - X_{Cr}^{\alpha})$$

$$\Delta G_{Cr}(kJ.mol^{-1}) = -0.4 + 36.5 \ Y_P^{gb} \simeq 36.5 \ Y_P^{gb}$$

The interaction coefficient* $\beta_{CrP}^{gb} = (17.1\pm2.4)$ kJ.mol^{-1} is intermediate between the Mn-P and Mo-P one. These results also demonstrate that Cr alone would not segregate in binary Fe-Cr since $\Delta G_{Cr}^o = 0$.

The segregation kinetics of P, when analysed with the help of eqn. 17, indicate that P was appreciably scavenged by Cr in these steels and that the precipitation equilibrium of Cr and P was nearly established during the segregation process.[38]

These calculations were extended to quaternary co-segregation and precipitation in 12%Cr-Mo-P steels[38] and the observed behaviour was satisfactorily accounted for by the same equations using the same values of the bulk and intergranular interaction coefficients, Table 1.

The comparison of bulk and intergranular interaction coefficients between phosphorus and transition metals, Table 1, shows that they are strikingly similar for each M-P couple, except Ni-P for which the grain boundary value is overestimated. Therefore the relative M-I interactions in bulk compounds in equilibrium with the same solid solution as that in which grain boundary segregation is considered appear as reliable guides to assess intergranular segregation behaviour.

Phosphorus also strongly interacts with non-metallic elements. Its segregation varies inversely with those of S in Fe-Mo-P alloys[37] and of C in Fe-P-C alloys.[39] These observations are better interpreted in terms of repulsive P-S and P-C interactions than of site competition in view of the large positive values of the ε_{PC} and ε_{PS} coefficients (eqn. 1) in liquid Fe at 1500°C.[4] The plot of ΔG_P vs. Y_C^{gb} yields a negative $\beta_{PC}^{gb} = -9.0$ kJ.mol^{-1}, but that of ΔG_C vs. Y_P^{gb} is of opposite sign which is in contradiction with the above model. It could be shown[35] that this apparent discrepancy was due to the existence of a strong repulsive C-C interaction $\beta_{CC}^{gb} = 8.4$ kJ.mol^{-1} in agreement with the large positive ε_{CC} at 1500°C,[4] and ΔG_C reads :

$$\Delta G_C(kJ.mol^{-1}) = 92.4 - 33.4 \ Y_C^{gb} - 18.0 \ Y_P^{gb}$$

Similarly, the existence of a positive ε_{PP} at 1500°C[4] indicates that the P-P interaction is repulsive. A value of $\beta_{PP}^{gb} = 2.2$ kJ.mol^{-1} accounts for the apparent increase of ΔG_P with temperature observed by Erhardt ang Grabke[39] more satisfactorily than the entropy term determined by these authors. The actual segregation entropy must be quite small, and the segregation free energy of P reads :

---

* remembering that in eqns. 14, $a^{gb} = c^{gb} = 1/2$.

$$\Delta G_P(\text{kJ.mol}^{-1}) = 55.5 - 18.0 \ Y_C^{gb} - 8.9 \ Y_P^{gb} + \sum_M 2\beta_{MP}^{gb}(Y_M^{gb} - X_M^{\alpha})$$

taking for β_{MP}^{gb} the β_{MP}^{gb} or rather the equilibrium β_{MP}^{B} values listed in Table 1.[23] The magnetic contribution to ΔG_P, which has been ignored in this treatment must be considered as included in the P-P and P-M "chemical" interaction terms.

The C-P repulsion has an important implication with regard to the role of carbide forming metals : when the addition of Cr for instance suppresses the segregation of C by depressing its solubility, the P coverage can rise to its maximum value.[39] This might be a major mechanism of the deleterious effect of Cr on temper brittleness in low-alloy steels[35] since segregated carbon does not only oppose P segregation but also intrinsically improves boundary cohesion.

Other systems. The co-segregation pattern of Sb, Sn and alloying elements in steels should be qualitatively similar to that described for P. Table 1 shows that the Sb-Ni bulk affinity is stronger than the Sb-Cr one, which explains the larger segregation of Sb in the presence of Ni than of Cr[36-b]. The strong Sn-C repulsion in liquid Fe[4] suggests that the effect of C on Sn and Sb segregations must be even stronger than on that of P. This is certainly responsible for the much more drastic enhancement of Sb than of P segregation by Cr additions to 3.5% Ni steels or by decarburisation of 3.5% Ni-1.7%Cr steels.[36]

In a Ni-base alloy similar to Monel 400, the segregation of P is accompanied by Cu depletion of the grain boundaries.[40] This observation indicates that the relative Cu-P interaction is repulsive in Ni, which is consistent with the formation enthalpy of Cu_3P (—31.6 kJ. mol-1) being less negative than that of Ni_3P (—48.4 kJ.mol^{-1}).

The segregation measurements of S in the grain boundaries of Ni_3Al and $Ni_3(Al,Ti)$[41] suggest that the apparent segregation energy of S increases with temperature. The model proposed by the authors to explain this behaviour, which assumes that the boundary contains a spectrum of sites with a distribution of binding energies, is physically sound but the variation observed might rather arise from the strong affinity of S for Al. Aluminium sulfides are so much more stable than nickel sulfides that the solubility of S must be much smaller in the compound than in pure Ni. In the absence of solubility data in Ni_3Al, only a wild guess can be made. Considering the compound as a virtual Ni alloy containing 23 at.%Al, and taking $\beta_{AlS}^{B} = -[\Delta H_{1/2(AlS)}^{o} - \Delta H_{1/2(NiS)}^{o}] = 31.5 - 11.5 = 20$ kJ.mol^{-1}, eqn. 16 yields a solubility of S of 24 at.ppm. at 1000°C which is well below the nominal contents of the alloys studied. The increase in dissolved S content with increasing temperature is certainly responsible for the apparent increase in its segregation energy as was the case for P when it was scavenged by Mo or Cr additions in steels.[38] The results[41] also suggest that Ti co-segregates with S in $Ni_3(Al,Ti)$ and amplifies its segregation.

THE INFLUENCE OF CO-SEGREGATION ON INTERGRANULAR BRITTLENESS

The M-I interactions obviously affect brittleness by altering
the segregation pattern of the so-called embrittling impurities, but
the direct effect of segregated alloying elements on boundary cohesion
might well be even larger.

In low-alloy steels, the embrittlement associated with a given
intergranular P concentration decreases as the Mo build-up increases
in the grain boundary.[37] Similarly the segregation of Ti reduces the
embrittlement due to Sb.[42] More generally it appears that the embrittling potency of phosphorus EP_P, defined as the increase in impact
transition temperature TT per at.% P segregated in the boundary,
$EP_P = \Delta TT/\Delta X_P^{gb}$, is strongly dependent on the intergranular content of
the transition metal additions. By studying steels where essentially
one metallic element segregates at a time, it has been demonstrated
that segregated Mo depresses EP_P, whereas Mn and especially Ni segregations raise it.[43] Correlating the data obtained on a variety of
steels and heat treatments, fig. 15, it was thus possible to express
EP_P as a linear function of the intergranular M coverages, whose
coefficients indicate the intensity of the beneficial (negative) or
detrimental (positive) effect of M on boundary cohesion[43] :

$$EP_P(°C/at.\%) = 5.43 + 0.11\ X_{Mn}^{gb} + 0.23\ X_{Ni}^{gb} - 0.31\ X_{Mo}^{gb}$$

The intrinsic embrittling potency of P, 5.43°C/at.%P, is divided by
3 when 12 at.% Mo are segregated. Saturation coverages of Mn or Ni
($X_M^{gb} = 0.5$) multiply it by 2 and 3 respectively. These figures emphasize the extraordinarily beneficial effect of Mo. However in view
of the weak segregation potencies ΔG_M^o of these elements, these effects
arise only if there is already an embrittling impurity segregated in
the boundary to drive them there, and it has not been possible to
determine whether these elements intrinsically improved boundary cohesion or merely reduced P-induced embrittlement.

Conversely, carbon is a powerful segregant which intrinsically
improves boundary cohesion. Besides their intrinsically embrittling
effect, the residuals P, Sn, Sb further deteriorate interfacial cohesion by repelling C atoms from the boundary. Similarly, carbide-
forming metals such as Cr are detrimental because they trap C in the
matrix. Another possible type of effect is examplified by the strong
co-segregation of Mo and C : it could be wondered whether the remedial
effect of Mo and Ti should not be at least partly attributed to the
carbon they drive to the boundary.

CONCLUDING REMARKS

In view of the number of parameters which control segregation
and embrittlement in complex systems, it is often difficult to decide

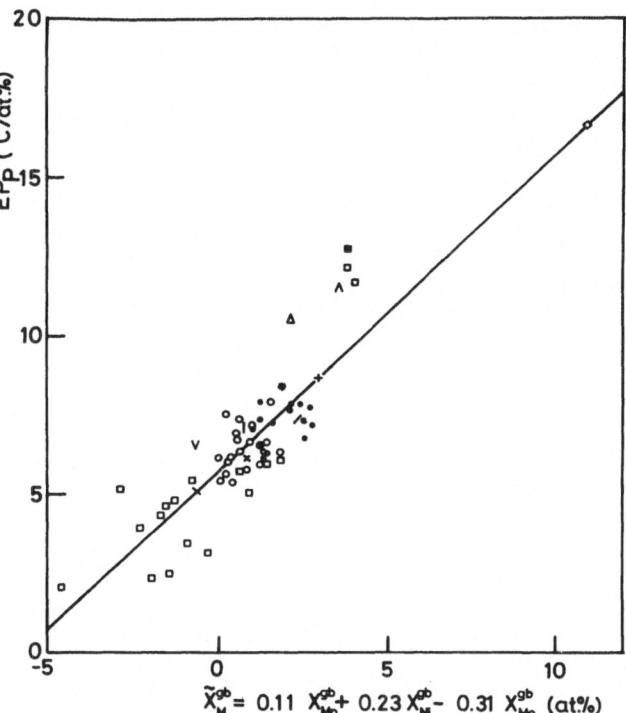

Fig. 15. Influence of intergranular coverages of Ni, Mn and Mo on the embrittling potency of phosphorus EP_P.[43]

which is the one actually responsible for a given effect. Although physically sound, a segregation model may not apply to the particular observation for which it was developed, because a hidden variable has been ignored. For example, an apparent variation of the segregation energy with temperature can be either due to a segregation entropy, or to chemical interactions between segregated atoms of identical or different species, or to magnetic effects, or to interactions between bulk atoms which give rise to variations in their concentration in the solution. These issues can be solved by performing systematic experiments on homologous series of alloys in each particular system, provided that their basic thermodynamic bulk properties are always kept in mind. Also, a given element generally affects segregation-dependent intergranular failure by different mechanisms, directly by segregating at the boundaries and indirectly by modifying the segregation of other detrimental or beneficial elements.

REFERENCES

1. C.L. Briant and S.K. Banerji, <u>Int. Met. Rev.</u>, 23:164 (1978)
2. M. Guttmann, The influence of interfacial segregation in

embrittlement phenomena, in:"Advances in the Mechanics and Physics of Surfaces", R.M. Latanision and R.J. Courtel, eds. Pergamon Press, New York (1981).

3. L.S. Darken and G. Simkovich, Scr. Metall., 13:431 (1979).
4. G.K. Sigworth and J.F. Elliott, Metal Sci., 8:298 (1974).
5. M. Guttmann, Surface Sci., 53:213 (1975).
6. M. Guttmann and D. McLean, Grain boundary segregation in multicomponent systems, in:"Interfacial Segregation", W.C. Johnson and J.M. Blakely eds., ASM, Metals Park (1979).
7. R.H. Fowler and E.A. Guggenheim, Statistical Thermodynamics, University Press, Cambridge (1960).
8. M. Hillert, Calculation of phase equilibria, in:"Phase Transformations",ASM, Metals Park (1970).
9. M. Hillert and L.I. Staffansson, Acta Chem. Scand., 24:3618 (1970).
10. M.C. Cadeville, J.M. Friedt and M. Maurer, Mater. Sci. Eng. in the press (1981).
11. M. Guttmann, Metal Sci., 10:337(1976).
12. I. Prigogine and R. Defay, Thermodynamique Chimique, Desoer, Liège (1951).
13. E.E. Latta and H.P. Bonzel, Surface segregation and gas adsorption, in:"Interfacial Segregation", ASM, Metals Park (1979).
14. J.M. Blakely and J.C. Shelton, Equilibrium adsorption and segregation, in:"Surface Physics of Materials", vol.1, J.M. Blakely ed., Academic Press, New-York (1975).
15. J.M. Blakely and H.V. Thapliyal, Structure and phase transitions of segregated surface layers, in:"Interfacial Segregation", ASM, Metals Park (1979).
16. M. Guttmann, Metall. Trans. A, 8A:1383 (1977).
17. C. Pichard, J. Rieu and C. Goux, Mém. Sci. Rev. Mét., 70:13 (1973).
18. C. Loier and J.Y. Boos, Metall. Trans. A, 12A:129 (1981).
19. A.M. Donald and L.M. Brown, Acta Metall., 27:59 (1979).
20. M. Kostelitz, J.L. Domange and J. Oudar, Surface Sci., 34:431 (1973).
21. I. Jäger, Mater. Sci. Eng., 42:245 (1980).
22. a- C.H.P. Lupis and J.F. Elliott, Acta Metall., 15:265 (1967).
    b- J.C. Joud, J.C. Mathieu, P. Desré and E. Bonnier, J. Chim. Phys., 68:489 (1971); 68:493 (1971); 69:131 (1972). L. Goumiri, Doctorat d'Etat Thesis, Grenoble (1980).
23. K.E. Szklarz and M.L. Wayman, Acta Metall., 29:341 (1981).
24. E.D. Hondros and M.P. Seah, Scr. Metall., 6:1007 (1972).
25. H.L. Marcus and N.E. Paton, Metall. Trans., 5:2135 (1974).
26. A. Larère, C. Roques-Carmes, Ph. Dumoulin and M. Guttmann, submitted to Acta Metall.
27. D. McLean, Grain boundaries in metals, Clarendon Press, Oxford (1957).
28. W.R. Tyson, Acta Metall., 26:1471 (1978).
29. Ph. Dumoulin and M. Guttmann, Mater. Sci. Eng., 42:249 (1980).

30. C. Lea and M.P. Seah, Phil. Mag., 35:213 (1977).
31. W.R. Graham and A.C. Yen, Metall. Trans. A, 9A:1461 (1978).
32. A. Joshi, M.N. Varma and M. Strongin, Metall. Trans.,
    5:861 (1974).
33. R.A. Van Santen, L.H. Toneman and R. Bouwman, Surface Sci.,
    47,n°1:64 (1975).
34. Ph. Dumoulin, J.L. Caillerie and M. Guttmann, Surface Sci.,
    in press (1981).
35. Ph. Dumoulin, M. Guttmann and M.L. Wayman, to be published.
36. R.A. Mulford, C.J. McMahon, D.P. Pope and H.C. Feng
    (a) Metall. Trans. A, A7:1183 (1976); (b) ibidem,
    A7:1269 (1976).
37. Ph. Dumoulin, M. Guttmann, M. Foucault, M. Palmier,
    M. Wayman and M. Biscondi, Metal Sci., 14:1 (1980).
38. R. Guillou, M. Guttmann and Ph. Dumoulin, Metal Sci.,
    15:63 (1981).
39. H. Erhardt and H.J. Grabke, Proc. Int. Conf. on "Residuals
    in Iron and Steels", Ljubljana (1980).
40. L.A. Heldt, A.W. Funkenbusch, D.F. Stein and T.R. Pinchback,
    Corrosion 79, International Corrosion Forum, paper 156,
    Atlanta (1979).
41. C.L. White and D.F. Stein, Metall. Trans. A, 9A:13 (1978).
42. H. Ohtani, H.C. Feng and C.J. McMahon, Metall. Trans. A.,
    7A:1123 (1976).
43. Ph. Dumoulin, M. Guttmann, Ph. Maynier and P. Chevalier,
    Journées d'Automne de la Société Française de Métallurgie,
    Paris,(Oct. 1981). To be published.

DISCUSSION

Comment by A.W. Thompson:

The grain boundaries you examine are very special ones, i.e.,
the ones which break. Do you think the same behavior occurs at all
boundaries, but merely to a slightly different degree, or could there
be qualitative differences which cause the "breakable" boundaries to
be unique? These qualitative differences might include microstructural ones. One might also inquire, as McMahon has done, how segregation
varies even among the boundaries which do break.

Reply:

In principle, the same behavior is expected for segregation at
all boundaries, in the sense that chemical interaction must affect
co-segregation whenever co-segregation occurs. However, quantitative
differences between actual coverages must arise from qualitative
differences in atomic structures among the different types of grain
boundaries, as discussed by Professor Gleiter in this Conference,
and observed by Ogura and McMahon, in the case of temper-brittle
steels. The difference can be thought of, in thermodynamic terms,

as arising mainly from differences in intrinsic segregation free
energies ΔG_x° (gbt), (where x referes to the segregating species and
gbt to the grain boundary type), and saturation coverages, a^ϕ(gbt),
c^ϕ (gbt). Even if the interaction coefficients α_{ij}^ϕ or β_{ij}^ϕ are
assumed independent of gbt, equations 13 and 14 indicate that the
chemical interaction - induced segregation will be greatly gbt-
dependent. In the special (e.g., coherent twin) boundaries for which
all the ΔG_x° are ≈ 0, very little segregation occurs whatever the
chemical interaction is, and these boundaries are not embrittled.

Comment by H. Birnbaum:

    Segregation studies at grain boundaries require that boundaries
be fractured brittlely. Have you examined the composition of both
halves of the specimen? Does the fracture proceed along the boundary
or in the material adjacent to the boundary?

Reply:

    We did not study this pattern experimentally, but we always used
the maximum value of the segregation found among a number of intergranular facets of a given sample. This procedure gave the best
correlation of all properties and the smaller scatter. Grant Rowe
and Clyde Briant at G.E. have addressed that problem and if my
memory is correct they found both types of behavior depending on the
system: either the segregant was evenly distributed between the
two halves of the intergranular fracture surface, or it was randomly
distributed, the sum of the coverages on the corresponding halves
being roughly constant amont the grain boundaries analyzed. This
means that the grain boundary type-dependence (see Thomson's question)
of segregation was smaller (in that case) than the scatter induced
by the uneven distribution of segregated atoms between the two
halves of the fracture face, but it showed that the crack propagated
essentially within a few (2) atomic diameters from the grain boundary
plane.

Comment by J.F. Knott:

    I would like to ensure that we make a clear distinction between
SEGREGATION, as influenced by thermodynamics and kinetics, and
EMBRITTLEMENT, which is a phenomenon observed in mechanical testing.
Segregation is only one aspect of embrittlement, which is influenced
also by the type of test (in particular, the stress-state), by the
material's yield strength and by its microstructure (grain size,
carbide distribution, packet size, etc.). In commercial steels,
variations in the amount of an element such as molybdenum may have
effects on yield strength and microstructure quite separate from
those on segregation. Although I appreciate that these effects have
been assessed and allowed for in some of the results which you

report, the general point must be made that shifts in impact transition temperature alone cannot determine whether a given variable
is affecting segregation or the fracture micromechanisms.

Reply:

   I agree entirely with the comment. Your last sentence, however,
sounds somewhat odd, since I sincerely thought that my paper was
essentially devoted to experimental examples where the effect of a
given variable on segregation had naively been determined by...
measuring the segregation. As to the effect of major additions
(e.g., Mo) on fracture properties, the specific contribution of
their segregation to embrittlement was considered only in cases
where their bulk (i.e., structural) effects had been shown to be
negligible, or even to be opposite to the former (e.g., enhancement
of grain boundary fragility together with softening of the matrix,
etc.).

Comment by P. Haasen:

   What is the width, d, of the segregation zone at the surface or
grain boundary? (You treated this zone as three dimensional in your
thermodynamic model.) Can one estimate d from the $(Dt)^{1/2}$ plots of
segregation kinetics you showed?

Reply:

   In opposition to what you said, the thermodynamic model presented
here basically considers the interface--whatever it is-- as two-
dimensional. Physically, this means 1 to 2 atomic diameters.
Basically, the kinetic model of McLean (eqn. 17) does not allow d
to be deduced from segregation measurements because the segregation
data injected into its equation are already put in terms of interfacial concentrations by assuming a given interface thickness, in
general one monolayer. In other words the quantities actually
measured are $Y_i^\phi d$, i.e., the excess adsorption Γ_i. Under the simplifying assumption that $Y_i^\phi(0)=0$, eqn. 17 becomes:

i.e.
$$Y_i^\phi(t) \cdot d \simeq 2X_i^\alpha (D_i t/\Pi)^{1/2}$$

$$\frac{\Gamma_i(t)}{\Gamma_i^\circ} \simeq 2X_i^\alpha (D_i t/\Pi)$$

where Γ_i° is the absolute saturation adsorption at the interface. The
equation which has been obtained also by an independent derivation
shows that the result is indepenedent of d.

Comment by R.H. Jones:

There is considerable interest in using surface segregation to evaluate grain boundary segregation, because of the ability to do real time experiments; however, while there can be qualitative similarities between them, there are many differences. Examples of the differences include the observation of site competition between Sn and S in grain boundaries of iron but not at the free surface. In view of these differences would you comment on the limitations on the use of surface segregation to simulate grain boundary segregation in multi-component systems? In particular, would you comment on the difference in your phosphorus segregation at the free surface and the grain boundary in the Fe - 0.3 Mo alloy.

Reply:

After trying to use surface segregation experiments as a guide for grain boundary behavior, I think now that the quantitative, and some qualitative, differences between the two types of interfaces are really too large to allow more than qualitative indications as to, for example, the mere existence of interactions. The written paper reports for instance the considerable discrepancy between the surface and intergranular Mo-P interactions. As to the P segregation difference in the 0.3% Mo alloy, they simply reflect the fact that the intrinsic segregation free energy ΔG_P° is, in general, twice as large at free surfaces as at grain boundaries, therefore, saturation coverage of the former is reached at temperatures and bulk contents for which the latter are far from saturated and they are still varying with these two parameters.

The only area of quantitative agreement we found between the two types of experiments was the effect of bulk interaction - i.e., solubility effects - on the kinetics of "normal" interfacial build-up. Beside that, the transient effects observed at the free surface of the Ni-S-Ca system (Fig. 10) were not observed in the grain boundaries. Actually, the segregation of Ca was not even detected in the latter! (See Ref. 26).

TUTORIAL LECTURES ON

SOLUTION CHEMISTRY

Session Chairman: P. Neumann

THE SOLID/ELECTROLYTE INTERFACE

B.E. Conway

Chemistry Department
University of Ottawa
Ottawa, Canada

ABSTRACT

The solid/electrolyte interface is considered at four levels: the interface in the absence of solvent or solution; the interface in relation to adsorbed, oriented solvent dipoles when the solid is in contact with a solution; the interface when adsorbed ions are present, giving rise to an ionic double-layer, and the consequences of interaction of these ions with the oriented solvent co-plane; and the formation of monolayer and submonolayer quantities of oxide species at solid metal/aq. electrolyte interfaces where ion adsorption effects play a major role in the stages of oxide monolayer formation, e.g. at Pt, Ir, Au.

The behavior of solid oxide and semiconductor/solution interfaces is briefly considered.

In the spirit of this Workshop, the presentation will follow the format of a pedagogical review.

INTRODUCTION

The solid/electrolyte interface provides the location where processes occur which affect the electrochemical stability of most materials in contact with electrolytic solutions, or with molten or solid electrolytes, e.g. in corrosion. This interface is also the location of impressed electrode processes which arise in electrocatalysis and in related electrolytic processes, e.g. evolution of Cl_2, O_2 or H_2. Recent interest in solid/electrolyte interfaces is centered on photo-enhancement of electrolytic

processes in aqueous and non-aqueous media in relation to the
semi-conducting properties of the solid and its band structure.

The title of this presentation covers a broad area encompas-
sing many constituent fields. It will therefore be necessary in
this paper to restrict the subject matter to several topics which
are of current and established interest in electrochemistry of
interfaces and materials science, and to which the present author
has made some contributions. Within these confines, the following
matters will be treated:

(a)   The state of metal and oxide interfaces at the solid/
        vacuum interface;
(b)   modifications of the state of solid/vacuum interfaces
        when solvent molecules and electrolyte ions are present;
(c)   the adsorption and orientation of solvent dipoles at
        metal and oxide/electrolyte solution interfaces;
(d)   the specific adsorption of ions at charged interfaces
        and the nature of double-layers at solid/electrolyte
        interfaces;
(e)   ion-solvation effects in ion adsorption and lateral
        interaction in double-layers at solid/electrolyte
        interfaces;
(f)   modification of metal interfaces with solutions upon
        oxide film formation;
and (e)   some aspects of double-layer properties of oxide-
        solution interfaces.

A schematic diagram of the situation at a solid/electrolyte
                       interface is shown on the diagram following.
INTERFACES OF SOLIDS

1.   The Interface with Vacuum or Gases, at Solids

A few preliminary remarks must be made about interfaces of
solids with vacuum or gases before interfaces of solids with a
solvent or with an electrolyte solution are considered. Emphasis
will be given to metals.

The discontinuity of structure of a solid phase at its
interface leaves unused orbitals in various orientations depending
on the index of the crystal face exposed. These orbitals are
responsible for the chemisorptive affinity of most metal surfaces
for various adsorbates and also for the specific surface free
energy of the interface insofar as the unsatisfied surface
orbitals tend to have some mutual interaction. In certain cases,
this can lead to reconstruction of the surface layer of a solid,
e.g. Pt,[1] into a 2-d overlay lattice having a different structure
from that of a parallel section of the underlying bulk material
and consequently different geometrical coordination.

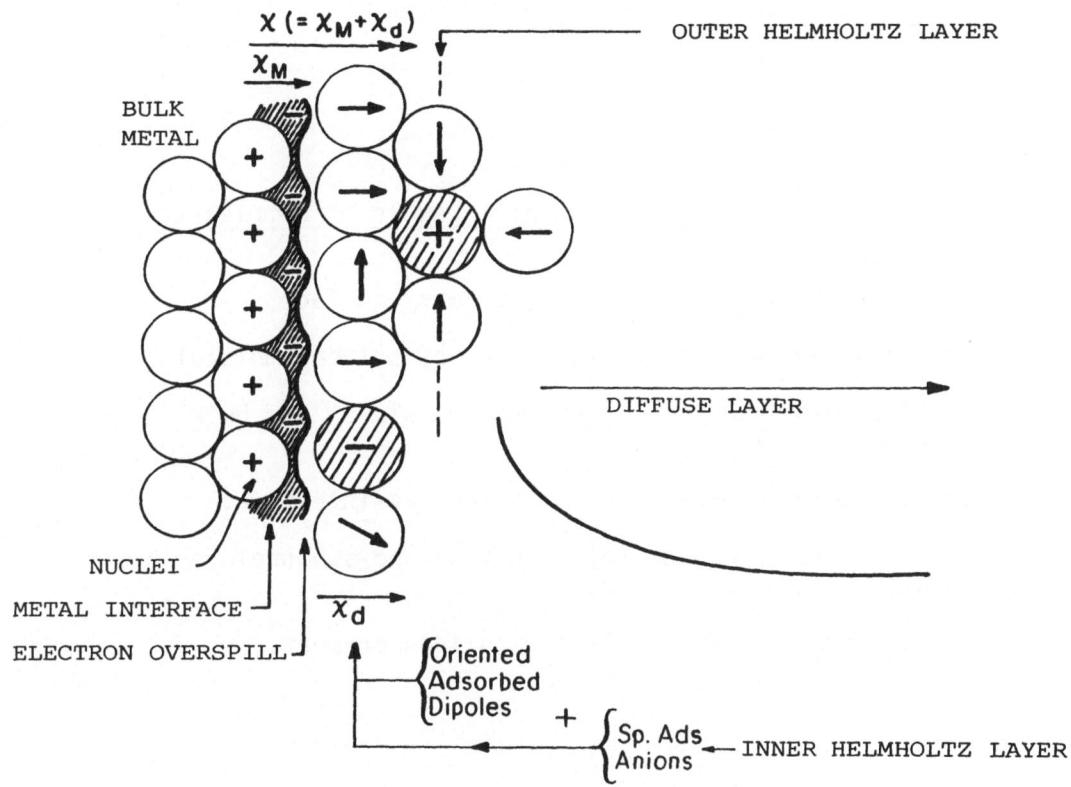

Schematic diagram of the complex interphase at a
solid / electrolyte boundary with
solvent molecules present

The Solid/Electrolyte Interface: Levels of Complexity

(a)   Solid/gas (vacuum) interface - primary surface.

(b)   Solid/vacuum interface + monolayer of solvent molecules.

(c)   Solid/vacuum interface + oriented monolayer and bulk
      solvent molecules.

(d)   Solid/liquid interface (c) with ions at interface.

(e)   Solid/liquid interface (d) with ions and submonolayer or
      monolayer oxide films present.

(f)   Solid oxide/liquid interface with ions present.

At semiconductors, the discontinuity in coordination and bonding at interfaces can lead to "surface states" which are associated with band bending as the surface is approached from within the bulk.

With regard to electron distribution, the discontinuity at a metal interface is associated with some electron "overspill" which, with the geometrical arrangement of the nuclei of surface atoms, gives a contribution to the (unmeasurable) intrinsic surface potential, χ_M, of the metal which is contained in the measurable real potential of electrons, the work function ϕ_M.

That the intrinsic surface potential at the vacuum/metal interface is substantial is indicated by the measured surface potential *changes* that are associated with physical adsorption of noble gases[2] which, by themselves, could contribute no molecular surface dipole moment. The changes $\Delta\chi$ of χ, due to noble gas adsorption, are on the order of $0.5 \sim 0.8$ V and thus correspond to the polarizing influence of a substantial surface field. Taking the polarizability of say Ar as $1.63 \times 10^{-24} cm^3$, the surface field is seen to be <u>ca.</u> $4 \times 10^7 V\ cm^{-1}$. Alternatively, the screeing effect of the noble gas electron shells can be regarded as eliminating effectively the intrinsic χ_M so that $\Delta\chi \simeq -\chi_M$. This surface field will tend to exert a considerable polarizing influence on solvent molecules when a metal is placed in a solution in the zero charge condition (see section (2)).

The density of free electrons near a surface of a metal or semiconductor determines the distance within the solid phase, from its interface, to which an external field vector can penetrate and modulate the electron distribution. This distance, called the Thomas-Fermi screening length, is only of the order of 0.1 nm at metals and has a significance analogous to the Debye ionic atmosphere screening distance, $1/\kappa$, in electrolyte solutions or ionic double-layers where ion charge rather than electron charge distribution is involved. More closely analogous effects to those in ionic solutions arise near semiconductor interfaces and the screening length is then normally much larger than at a metal, depending on the electron or hole density. The Thomas-Fermi screening distance is important in electric modulated reflectivity studies at metal electrode interfaces[3].

The geometrical arrangement of atoms in the surface layer of most crystalline materials is of great importance in determining chemisorptive, catalytic and electrocatalytic properties of the interface, as well as the electron work function of that surface since the χ_M contribution in ϕ_M is substantially influenced by the surface geometry and coordination of the interfacial atomic layer. It is also important in the near-surface mechanical properties of metallic materials.

The geometrical or structural arrangements of surface atoms
on the principal index planes [(111), (110) and (100)] of f.c.c.
crystal lattices are well known[4] and differ substantially. They
are of great importance in electrochemical surface science and
catalysis. At some principal index planes, reconstruction may be
thermodynamically preferred and lead to a more complex overlay
lattice, as mentioned above for the case of Pt[1]; this then determines the surface properties of the interface.

On high index planes, generated by cutting single crystals
just off a crystal axis or just-off parallel to a principal index
plane, a multi-stepped surface can be generated as in the elegant
work of Somorjai[5]. Then specific properties of step-planes and
step edges can be distinguished, e.g. in adsorptive and catalytic
behavior. Similar specificities may be expected for such surfaces
with regard to solvent and anion adsorption at metal / electrolytic-
solution interfaces (see below), but few investigations have yet
been made on this topic except some initial experiments on H
chemisorption by Ross[6] at Pt electrodes.

2.   The Interface of Solids with a Solvent

The interface of a solid with a solvent must next be considered in some detail as it determines in important ways the
behavior of solid/electrolyte interfaces when the electrolyte
is an ionic solution.

(a)   Macroscopic behavior

In a macroscopic thermodynamic sense, the solid/solvent
interface is characterized by the interfacial tension, γ, and
the contact angle. At a liquid metal interface, the partial
derivatives of γ w.r.t. electrode potential, E, and chemical
potential, μ_i, give the surface charge per cm^2, q_M, on the metal
surface and the surface excess of a species i at various potentials
according to the principles of the Gibbs adsorption equation
applied to electrocapillary phenomena.

At Hg electrodes, this approach and the complementary one
of direct double-layer capacitance measurements[7], have given a
very detailed picture of the behavior of the double-layer at Hg
in contact with a variety of aqueous and non-aqueous electrolyte
solutions.

The situation with solid electrodes is unfortunately much
less developed than at Hg owing to the problem of obtaining solid/
solution interfacial tension data. However, some progress has
been made in this direction by means of measurements of interfacial

friction[8], changes of strain at thin plates polarized on one side
and by sensitive extensometric measurements on very thin wires
suspended in a solution and subject to controlled-potential
polarization[9]. Interpretation of results in terms of stress and
surface energy in relation to q_M is not simple.
     A different approach has been made by Morcos[10] who has
obtained electrocapillary information on solid/solution interfaces
(Si, graphite and metals) by following meniscus rise as a function
of potential. Plots similar to electrocapillary curves are
obtained.

     The method is related to studies of the effect of potential
on the wetting of surfaces, previously investigated by Frumkin
and co-workers[11] of his school who measured the contact angle of
a gas bubble on an immersed electrode surface, following an
earlier approach of Möller. This approach is related to Young's
relation for a 3-phase interfacial contact:

$$\cos \theta = (\gamma_{s/g} - \gamma_{s/\ell})/\gamma_{\ell/g} \tag{1}$$

where θ is the contact angle and γ's are the interfacial tensions
at the indicated solid/gas, solid/liquid and liquid/gas interfaces.
This method does not give unambiguous results[11] and most of the
difficulties are avoided if the meniscus height, h, measured[10]
at a vertical plate electrode rather than θ. h and θ are related
by

$$\sin \theta = 1 - dgh^2/2\gamma_{\ell/g} \tag{2}$$

where g is the gravitational constant. The equation relating
the meniscus rise h and the interfacial tension, $\gamma_{s/\ell}$, of the
interface of interest is then

$$\gamma_{s/\ell} = \gamma_{s/g} - \gamma_{\ell/g} (2Kh^2 - K^2h^4)^{\frac{1}{2}} \tag{3}$$

where K is a constant for each liquid equal to $dg/2\gamma_{\ell/g}$.

     Information on electrical properties of the interface is
obtained in the usual way from a Lippmann equation for the charge
density

$$q = -(\partial\gamma_{s/\ell}/\partial E)_\mu \tag{4}$$

or directly from the meniscus rise as f(E)

$$q = \gamma_{\ell/g}(2K)^{\frac{1}{2}} \frac{1 - Kh^2}{(1 - Kh^2/2)^{\frac{1}{2}}} \left(\frac{\partial h}{\partial E}\right) \tag{5}$$

Applications have been made to a number of interfaces of
metals, oxides and semiconductors with electrolytic solutions[10].

Although the meniscus rise method has given some useful
information on solid/electrolyte interfacial behavior, this
approach does not, however, allow of detailed evaluation of the
thermodynamic properties of the double-layer to the extent
afforded by the electrocapillary or capacitance methods at Hg
or Ga.

At polarizable, non-corroding metals, such as Au, the
capacitance method has afforded, in recent years, e.g. at the
C.N.R.S. labs at Bellevue, some of the most detailed information
on the solid metal/solution interphase at solid metals, especially single-crystal surfaces of Au or Ag[12,13]. Problems arise
experimentally in the substantially greater frequency dependence
of capacitance than at Hg and in knowing the potential of zero
charge (pze), e.g. at Au, which depends (owing to anion adsorption) on the solution in which measurements are made and is, in
any case, not at all so reliably determinable (cf. the extensometric results[9]) as at Hg. Pzc values for solids, it is to be
noted, can be derived by the meniscus rise method of Morcos,
described above.

The significance of "electrocapillary" effects at solid metals
in terms of q_M has been examined by Latanision[79].

(b) Microscopic behavior: solvent adsorption and orientation

Thermodynamic studies, especially at metal electrode interfaces (principally Hg and Ga), have given, over the part 40 years,
a detailed picture of ion distribution and adsorption, and the
structure of the double-layer at charged interfaces. The principal concepts and results that have come out of this work since
the time of Grahame's review[7] (1947) are the following:

(a) a distinction between inner and outer Helmholtz planes
     in the double-layer as loci for adsorption of speci-
     fically adsorbed ions (usually anions) and electro-
     statically bound ions (usually cations), respectively;
(b) methods for separating the diffuse (Gouy) layer capaci-
     tance and thence evaluating the inner layer capaci-
     tance;
(c) the evaluation of cation and anion "components of
     charge" in the two Helmholtz planes;
(d) the recognition of the importance of imaging and
     discreteness-of-charge effects;
and, of special importance in this section,
(e) the role of solvent dielectric and dipole orientation
     behavior in the inner layer region of the charged
     solid/solution interphase.

The nature of solvent polarization in the inner layer adjacent to the surface that can bear a net charge density is of fundamental importance for the properties of solid/electrolyte interfaces and is complementary to the specific adsorption behavior of ions in this region (see following main section). Surprizingly little attention was paid to the ubiquitous presence of solvent in the double-layer at charged interfaces in the early work (ca. 1945) but the special role of an adsorbed solvent (H_2O) layer was implicit in Eyring, Glasstone and Laidler's treatment of H_2 over-voltage in 1939 (later proved by Frumkin to be incorrect in some aspects) and was also recognized in Bockris and Potter's work on H_2 overvoltage on Ni in 1952.

The quantitative treatment of the role of solvent in the behavior of double layers was made first from two rather different directions: by MacDonald[14] in 1954 in terms of solvent dielectric polarization and by Watts-Tobin[15] in 1961 in terms of a two-state solvent dipole orientation model. Earlier, Conway, Bockris and Ammar[16], and also Grahame[17], had treated the problem of solvent dielectric saturation (for H_2O) in the double-layer. Together, these treatments led to an improved understanding of the role of solvent polarization in the capacitative properties of charged interfaces (for reviews on this topic, see refs. [18] and [19]). The question of solvent orientation and its role in the capacitative behavior of metal electrode interfaces (especially Hg[7] and Ga[21]) tended to dominate the literature on double-layers and organic molecule adsorption, e.g. the "BDM" theory[20] and treatments of Damaskin[21], during the past 20 years. Especially the significance of the hump on the capacitance vs potential curves for Hg, and its dependence on anions of the electrolyte and properties of the solvent, has attracted much attention. Less work is available on solid metals on this question.

MacDonald[14] gave calculations on the double-layer capacity at the Hg electrode in terms of (a) dielectric constant in the "charge-free" layer of water at the interface and (b) the electro-striction in this layer associated with the tension $\varepsilon E^2/8\pi$ due to the field E. By means of three or four adjustable constants, a good fit to the experimental differential capacitance-potential relation for 0.01 N aq. NaF at 298 K was obtained. The main parameters which determine the fit of the theoretical treatment to the experimental data (of Grahame) are the constants α and β which are defined by

$$\alpha = (z_o^0/z_o - 1)/8\pi p = \beta/8\pi \tag{6}$$

where p is the electrostrictive pressure $\varepsilon E^2/8\pi$ and z_o^0/z_o is a reduced relative thickness of the double layer; b is the cofficient of field squared in the Booth-Grahame[17] relation

Table 1. Models for Inner-Layer Solvent Behavior at Charged
          Interfaces.

Dielectric Polarization Models	Dipole Orientation Models	Cluster/Dipole Models
Continuum dielectric theory with normal dielectric constant	Free dipole orientation	Free dipole + dimer orientation
	Electrostatically interacting dipole orientation in 2 or 3 states	Free dipole + cluster orientation
Continuum dielectric theory with dielectric saturation (a) Debye-Langevin model (b) Kirkwood-Onsager model (Booth-Grahame)	H-bonded dipole orientation* Partial orientation in two-dimensional H-bonded network with varying degrees of H-bond bending	

*In some ways, the case of electrostatically interacting dipoles
is similar to that for H-bonded dipoles. However, directional
factors and some electronic delocalization become important in
the H-bonded model, e.g., for H_2O.

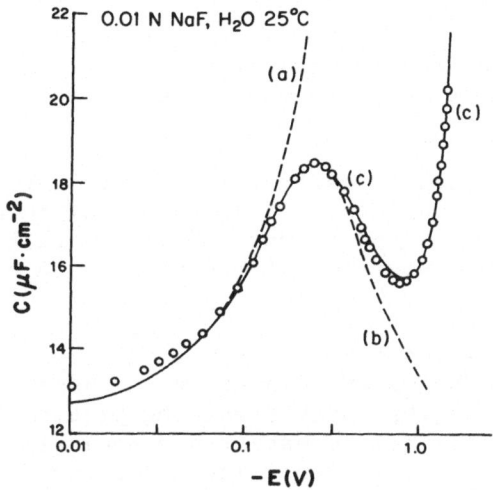

Fig.1
Comparison of calcul-
ated and expt.capacit-
ance for 0.01M NaF at
Hg at 298K according
to the dielectric
polarization model of
McDonald[14].

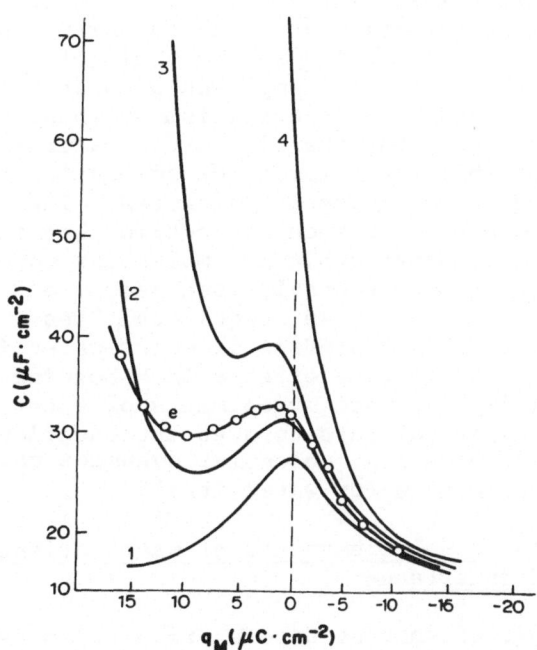

Fig.1a Capacitance curves for the inner layer as a function of
surface charge according to Damaskin and Frumkin for various
values of their characteristic capacitance, K_2.

$$\kappa = c + a(1 + 2bE^2)^{-\frac{1}{2}} \qquad\qquad (7)$$

giving the differential dielectric constant κ as a function of field E. The constant α determines the compression in the double-layer and b the dielectric saturation[22].

A comparison of the calculated results with experiment for $\alpha = 0$, $b = 0$; $\alpha = 0$, $b =$ finite and for finite values of both α and b is shown in Fig. 1. Both α and b must have appropriate finite values in order for the experimental behavior to be simulated adequately.

In relation to other treatments of the double-layer capacity at Hg based on properties of the water solvent in the inner layer, it is to be noted that the theories which treat the problem in terms of a field-dependent dielectric constant are equivalent in some ways to the molecular model theories which allow for field-dependent orientation of H_2O dipoles or clusters of dipoles (see below). In the field-dependent dielectric theories of Conway et al[16], McDonald[14] and Grahame[17], the orientation aspect of water molecule polarization in the double-layer is implicit in the variation of ε with E while interaction effects are implicit in the correlation factor g for orientation in the Kirkwood-Onsager theory of dielectric polarization which is the basis of the Booth treatment[22] of dielectric saturation employed by Grahame[17] and by MacDonald[14]. A frequently made criticism of these dielectric theories is, of course, that the dielectric saturation function can hardly be appliable for a lamina of solvent in the inner layer only one or two water molecules in thickness. The question then resolves itself into whether such dielectric theories are any less realistic than the somewhat arbitrary molecular model treatments which have been given more recently (see below) and will now be discussed. The answer is not an unequivocal "yes"; for example, the angular dependence of H-bonding in water-water interactions is not considered and cooperative effects in H-bonding have received no attention in molecular model treatments of water in the double-layer, except in empirical ways through introduction of cooperative orientation in clusters. Molecular dynamics calculations are required as have been made for water itself.

(c)   Molecular model treatments of water orientation at
      charged interfaces

The treatments of solvent dipole orientation as a function of electrode surface charge q_M seek to evaluate a reciprocal capacitance contribution arising from the dependence of the surface dipole layer potential difference χ_d on q_M. At electrode interfaces, $\Delta\chi_d$ is a function of potential and hence q_M; then the dependence of $\Delta\chi_d$ on q_M can be written as a differential coeffi-

cient $d(\Delta\chi_d)/dq_M$ which is to identified as a reciprocal differential capacitance, $1/C_d$, due to dipole orientation. It combines
in a _series_ relationship with other capacitance contributions of
the overall double-layer due to free charge accumulation. Therefore, only if C_d is sufficiently small will its effect on the
overall measured capacitance be important. Hence its correct
evaluation is a critical matter in modern theories of the double-
layer which properly take into account the solvent layer. For
example, the two-state Watts-Tobin model gives rise to anomalous
behavior with respect to this capacitance (see below).

One of the main questions is whether the capacitance contribution due to the dependence of orientation on q_M can account for
the capacitance hump usually observed[7,14] near the potential of
zero charge (at Hg).

(i) _Two-state dipole orientation treatments_. A useful
molecular model treatment of solvent orientation polarization at
an electrode interface was given by Watts-Tobin and Mott[15]. Two
orientation states of the solvent dipoles, up ↑ and down ↓,
aligned with the electrode field E arising from net surface charge
density q_M were envisaged. The polarization in the interphase at
the electrode surface was calculated in terms of the relative
population N↑/(N↑ + N↓) and N↓/(N↑ + N↓) of the two states of
orientation. N↑ and N↓ are determined (a) by the field; (b) by
temperature and (c) by any lateral interaction forces[20,23] between
the oriented and unoriented dipoles. Interaction effects were
not, however, taken into account in the original treatment.

The Boltzmann distribution function is employed to calculate
the relative populations N↑ and N↓ and hence the local polarization per cm^2 may be calculated. Thus

$$\frac{N\uparrow}{N\uparrow + N\downarrow} = e^{U/kt} ; \frac{N\downarrow}{N\uparrow + N\downarrow} = e^{-U/kt} \tag{8}$$

where U is the net energy of the oriented dipole in the field.
When lateral interactions are significant[20], U has the form

$$U = \mu E + Wn \ N\downarrow/N - Wn \ N\uparrow/N \tag{9}$$

where $N = N\uparrow + N\downarrow$, W is the pairwise lateral interaction energy
between oriented dipoles and 2n the coordination number in the
interphase. An orientation distribution function R can be
defined as (cf. ref. 20)

$$R = \frac{N\uparrow - N\downarrow}{N\uparrow + N\downarrow} = \tanh[U/kT] \tag{10}$$

i.e.

$$R = \tanh \left[\frac{\mu E}{kT} - RWn/kT \right] \tag{11}$$

R measures of the orientation polarization per unit area in the
interphase at the charged surface since $N\uparrow$ + $N\downarrow$ equal to the
total number of orientable dipoles per cm^2, assuming electrostri-
ction in the double-layer[14] does not change this number. The
variation of R towards a saturation orientation value of ± 1 i.e.,
when $N\uparrow \equiv N\downarrow + N\uparrow$ or $N\downarrow \equiv N\uparrow + N\downarrow$ occurs over range of E
dependent on the magnitude of nW. $N\uparrow + N\downarrow$, i.e., $\theta\uparrow$ or $\theta\downarrow \rightarrow 1$,
corresponds to dielectric saturation in the double-layer. E can
be related to the surface charge q_m by $E = 4\pi q_M/\varepsilon$ so that R as
$f(q_M)$ can be obtained. A more developed treatment by Levine,
Bell and Smith[24] took into account (a) the induced dipole moment
μ_i, contribution due to neighboring dipoles of permanent moment
μ_p; (b) the effect of neighboring dipoles on the field at a given
site, viz.,

$$E = 4\pi q_M + (n_e/d^3) \ (\mu_p + \mu_i) \tag{12}$$

where n_e is an effective coordination number for dipoles distant
d between each other; n_e is taken as 11 according to a relation
due to Topping[25]. And (c) the possibility that one orientation
is intrinsically preferred to another due to chemisorption[26]
or image forces[20] (this is found experimentally, e.g. from the
charge at which maximum adsorption of a symmetrical organic
adsorbate of zero net dipole moment arises).

The main weakness of the treatments of Watts-Tobin and Mott,
and Levine et al. is the neglect of H-bond structural effects
in the interphase, i.e., correlated orientation effects. Such
effects are not really taken account of by the lateral interaction
term WnR, since this has no general angular dependence on orientation, only extreme orientations being considered[15,20]. The
limiting 2-state situation is undoubtedly a serious oversimplif-
ication since molecules of the water dielectric in the interphase
will be oriented at various angles and some average polarization
orientation angle θ will be a function of field and Wn_e, and will
also be determined by the angular dependence of H-bond energy
between neighboring molecules[27,28].

(ii) Three-state orientation model

It should be mentioned that a 3-state model of
solvent dipole orientation at electrodes, with configurations
$\uparrow \rightarrow$ or $\leftarrow \downarrow$, was considered in detail by Fawcett[23] and gives
a better account of inner layer capacitance behavior and organic
molecule adsorption than the Watts-Tobin 2-state model.

The 3-state model gives a satisfactory account of the capacitance behavior in ethylene carbonate and methanol and is based on a 2-dimensional layer of interacting but unassociated dipoles. These cases exhibit capacity curves with a simple maximum or minimum.

The inner layer capacitance C_i is treated in terms of a geometrical contribution $1/4\pi d$ and a dipole orientation polarization term, viz.

$$1/C_i = 4\pi d + \frac{4\pi N_T d^3}{C_e} \left(\frac{\partial X}{\partial q}\right) \tag{13}$$

where C_e is the effective coordination number taking account of image effects, q is the surface charge density, and X is the mean field at a dipole site due to surrounding dipoles. $\partial X/\partial q$ is obtained from the charge dependence of fractions of dipoles oriented $\uparrow, \downarrow$ or $\rightarrow$.

(iii) Cluster models for water adsorption and orientation

Experimental studies indicate that some specific direction of orientation of water dipoles tends to occur at Hg[26;29-32], Ga[33] and other metals[34,35] in the absence of an opposing coulombic field due to excess change. Hence it is necessary to take into account both orientation effects due to chemisorption and those due to changing electrode charge. This was, in fact, recognized by Levine, Bell and Smith[24] while Bockris, Devanathan and Müller[20] noted an equivalent preferential orientation effect which they considered would arise because of interactions of the non-central dipole in water with its image charges in the metal (Hg) surface. This is, of course, a different effect from that associated with specific interaction of the lone-pair orbitals on O in H_2O with the atoms in the Hg surface but both are connected with the electron charge distribution in the water molecule and its donor/acceptor properties. At molecules containing an S atom, e.g. thiourea, the specific affinity of the S-end of the molecular dipole for Hg is very much larger.

Frumkin and Damaskin[21] assumed that the dipole potential contribution to the total surface potential could be written $\Delta\chi_d = \Delta\chi_1 + \Delta\chi_2$, where the components 1 and 2 arise from freely oriented dipoles in the electric field due to excess charge and from chemisorbed H_2O dipoles, respectively. The non-chemisorbed dipoles were regarded as being H-bonded in groups ("clusters") which themselves only weakly interact with one another. This conclusion about association was deduced[36] from the shape of the adsorption isotherm for various organic substances[37,38]. However, it must be stated that the shape of the electrostatic isotherm in terms of the dependence of the free energy of adsorption for a

given coverage (isosteric condition) on electrode surface charge
(e.g. the width of the bell-shaped curve at half its height[29])
can be reproduced by a lateral repulsive interaction effect term
of various forms, so it seems unlikely that the shape of the
isotherm can be uniquely indicative of role of the clusters.

Following Damaskin and Frumkin's paper[21], Parsons[39] presented a more detailed treatment of the state of a solvent at an
electrode interface in terms of a "primitive" four-state model.
A similar multi-state treatment was subsequently given by Bockris
and Habib[40].

Parsons[39] took account more quantitatively of the number of
water molecules per cm^2 in the double-layer and avoided some of
the arbitrary assignments of parameters, discussed above, that
were involved in the Damaskin-Frumkin theory. Solvent molecules
were assumed to exist in two states of aggregation at the surface
of the electrode: free molecules and small clusters. The free
molecules could adopt either of two extreme "up" or "down" orientations, as in the Watts-Tobin model, with a dipole moment μ.
The clusters could adopt also one of two extreme "orientations"
depending on the orientation of their resultant perpendicular
dipole moment μ_c w.r.t. the metal surface. μ_c was expressed per
molecule of the cluster and a parameter ρ was defined as the
ratio μ_c/μ to characterize the state of polarization of the cluster.

The potential drop across the inner Helmholtz plane was then
calculated in terms of the various dipole orientation components
derived from the orientation distribution function. Following
usual procedure, the contribution of the solvent dipole capacitance to the inner layer capacitance, was obtained by differentiating the dipole p.d. w.r.t. q_M. A series of "reduced" parameters
were defined to simplify the representation of various terms:
a reduced charge $s = \mu q_M/\varepsilon kT$ and a reduced dipole capacity
$K = \varepsilon^2 kT/\mu N_T K_0$ (where K_0 is the capacity of the inner-layer at a
fixed orientation of solvent dipoles).

Inner-layer capacities were calculated as a function of the
reduced charge s and are shown in Figs. 2a and 2b for two values
of the reduced dipole capacity K with A_- taken as 10^{-8}. A
series of curves is given for various A_+ values varying from 10^{-2}
to 10^{-6} or 10^{-7}. With A_+ taken as 10^{-2} and A_- as 10^{-4} the general
features of the experimental capacitance curve for aq. NaF at $0°C$
[123] are quite closely reproduced. The calculated capacities C
which are plotted in Figs. 2a and 2b are expressed in terms of
the ratio $C = C_i/K_0$ where C_i is the inner-layer capacitance
derivable from analysis of experimental results.

The physical significance of A_+ must be mentioned as it

determines the general shape of the capacitance $\underline{vs}$ s curves (Figs. 2a and 2b), e.g. in changing from a form similar to that for Hg ($A_+ = 10^{-4} - 10^{-5}$ in Fig. 2a, b) to that for Ga($A_+ = 10^{-2}$). A_+ is analogous to Damaskin and Frumkin's parameter[21] K_2; here it measures the energy associated (amongst other things) with binding of the dipole to the metal. A large value of $\mu_{b,+}$ (O-end of H_2O dipole adsorbed at the metal) displaces the sharply rising part of the capacity curve to more positive s values where the field-effect orientation becomes predominant.

In this treatment, the inner-layer capacity behavior is quite well accounted for by H_2O dipole and cluster orientation.

With solvents other than water, different types of inner-layer capacitance behavior can be distinguished, as shown schematically in Fig. 2c. It is possible that type (ii) would manifest behavior like that of water if a wider range of positive q_M could be explored. Type (iii) is, however, different and corresponds to weak association and binding to the surface. The transition from type (i) to type (iii) is qualitatively reproduced by the theoretical calculations as larger values of log A_+ in Figs. 2a,b are taken. In certain solvents, two humps appear, on either side of the p.z.c.

In such cases, it seems that ion interaction in the double-layer as proposed by Bockris, Devanathan and Müller[20] must be invoked to explain at least one of the humps. Also, in water, the development of the hump depends very much on the nature of the electrolyte anion, so the capacitance behavior cannot be treated only in terms of water dipole or cluster orientation without some proper allowance for specific solvent orientation effects caused by the adsorbed anions in addition to that caused by excess surface charge, q_M. Indeed, from the known specific influence of ions on the local state of bulk water, it would be very surprising if such effects were not of major importance in the solvent layer in the interphase[48,49].

In considering the experimental situation, Bockris and Habib[40] concluded that the capacitance hump which occurs in aqueous solutions on the positive side of the p.z.c. cannot be due to a significant dipole layer capacitance $dq_M/d\Delta\chi$. In the treatments of Damaskin and Frumkin[21] and of Parsons[39], on the other hand, it has been seen that such a capacitance contribution is the main basis of the hump and anion adsorption effects are regarded[39] as secondary, although the latter effects still require more consideration. It seems (see below) that the origin of the hump cannot be attributed uniquely to one or other of these effects.

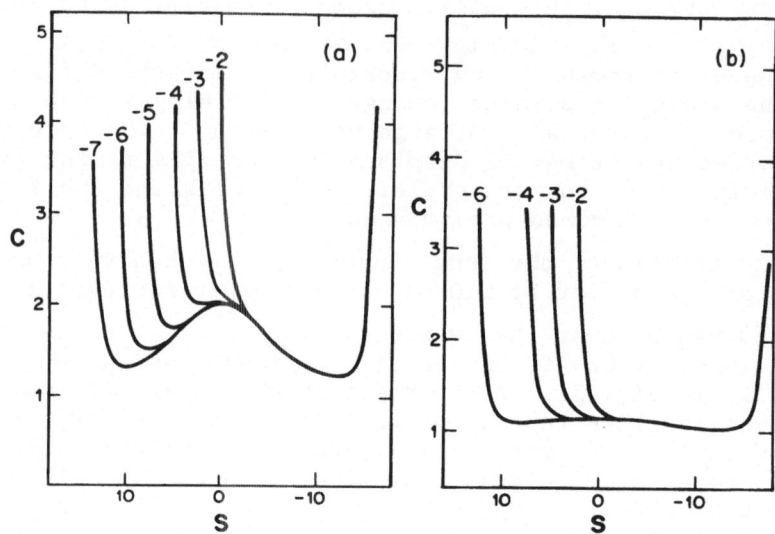

Fig.2a and b. Calculated inner-layer capacity as a function of
reduced charge parameter, s, in cluster model of
Parsons (ref.39). See text for discussion of parameters.

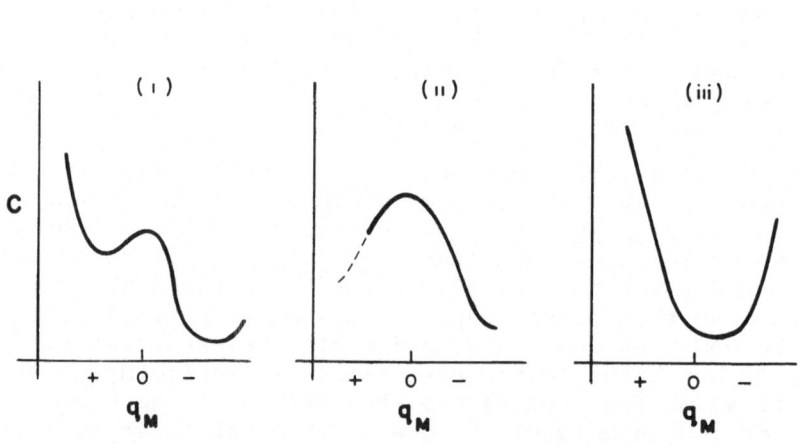

Fig.2c. Limiting types of variation of inner-layer capacitance
with charge (Parsons, ref.39), as observed in various
solvents.

(iv)   "Two-dimensional" H-bonded lattice model

Some of the artificial aspects of two-state orientation models, monomer-dimer treatments and cluster models would be eliminated if the orientation polarization in a two-dimensional interface of water could be treated. In such a model, a lattice of H-bonded water molecules would be considered and their orientation in the double-layer field would be calculated in terms of bending and breaking of the H-bonds between H_2O molecules laterally in the interphase, as well as those connecting such water molecules to the bulk. The potential energy functions for bending of H-bonds between H_2O molecules have been previously evaluated[28];also the intermolecular stretching behavior can be calculated and has been characterized spectroscopically in bulk liquid water. The H_2O dipole orientation in a field due to excess charge at a metal boundary interface would be deduced, in such a treatment, in terms of the bending and breaking of H-bonds in a quasi-2-dimensional lattice rather than in terms of artificial limiting ↑↑ , ↓↓ or ↑↓ orientations and dimer interactions. Such a treatment would be analogous to MacDonald's[14].

In this model, the presence of "monomer" dipoles would arise as a consequence of extremes of H-bond bending in the water lattice; bending of an H-bond beyond ca. 35° corresponds[40] to "dissociation" of the bond, while larger angular excursions correspond to repulsion forces. Molecules with intermolecular bending excursions less than this value are part of the interphasial lattice, the analogue of clusters in the other treatments referred to earlier.

Some indication that this approach may give a more realistic treatment of solvent orientation in the inner-layer at charged metal interfaces is afforded by the fact that Pople's analogous calculations[41] for bulk water gave a good account of the bulk dielectric properties of water, including the temperature coefficient of ε .

(v)   Spontaneous orientation of water due to chemisorption

Electrocapillary curves at Ga in aq. solution show much more asymmetry than at Hg. This is, however, not due to stronger anion adsorption. Anions, such as I⁻, are in fact less strongly adsorbed[21,33]. Thus, the extent of increase of capacitance on Ga with potential changing to more positive values cannot be accounted for quantitatively by increasing adsorption of anions. Stronger specific adsorption and orientation of water dipoles than at Hg is the explanation of the different properties of the double-layer at Ga in comparison with Hg. It is clear that if the specific adsorption of water at metals depends on the chemical identity of the metal, then image interactions[20] alone are not the only factor in preferential orientation of water at Hg and other metals.

The general difference of adsorbability and specific orientation of H_2O dipoles in aq. solution at various metals was noted
by Trasatti [30,31,32,42] in relation to the dependence of the potentials of zero charge of the metals on their electronic work
functions. His conclusions give further evidence of a different
kind that water adsorption at metals is specific for the metal or
families of metals. Trasatti showed that the relation between
electronic work function, Φ, and potential of zero charge $\phi q_M = 0$
for a variety of metals could be best represented by a series of
several, almost parallel lines, rather than a single line (Fig. 3).
The differences are due to the different χ-dipole components of
the surface potential in the equation for ϕ when $q_M = 0$, arising
from different extents of spontaneous orientation of H_2O dipoles
at the metal surface. Two main groups of metals are distinguished
in these plots: one of d-metals (transition series) where H_2O
dipoles are strongly oriented at $q_M=0$ and another of sp metals
at which the binding of water is less strong and the orientation
only approaches that at d-metals when q_M is large and positive.

Some indication of the relative value of $\Delta\chi$ for H_2O
can be derived from charge-potential plots [43] derived by integration of capacitance-potential data. Fig. 5 shows the general
form of such relations. At strong negative charges, the q_M-V
plots should become parallel due to similar orientation of water
dipoles in the coulombic field provided by excess negative
charge. As the condition $q_M \doteq 0$ is approached, the $\Delta\chi$ at the
weaker adsorbing metal drops to zero or a small value while at
the d-metal $\Delta\chi$ remains appreciable. Hence the difference of potential of the two metals at $q_M=0$ (the difference of p.z.c's) becomes larger than that at $-q_M \gg 0$ and includes the $\Delta\chi$ value for
the water-adsorbing metal as well as the work function difference
$\Delta\Phi$.

Trasatti also examined the correlation between the $\Delta\chi$ values
for water at various electrode metals M and the heat of formation
of the oxide MO from M and O_2. While a linear relation seems to
arise, it is doubtful if the forces causing specific adsorption
and orientation of H_2O dipoles at metals can be regarded as similar to those forming oxides "MO" which are usually ionic and in the
form of a three-dimensional lattice. Indeed Trasatti himself has
recently rejected [44,45] such an idea [46] with regard to correlation
of anion specific adsorption with the energy of formation of the
metal-anion salt.   Probably it is more appropriate to plot $\Delta\chi$
against the difference of electronegativity of the metals and the
effective electronegativity of bound O in water. This will give
some measure of the tendency for chemisorption of H_2O as a charge-
transfer complex at the metal surface.

From the above, it is seen that there is definite evidence
for some specific chemical affinity between water dipoles and

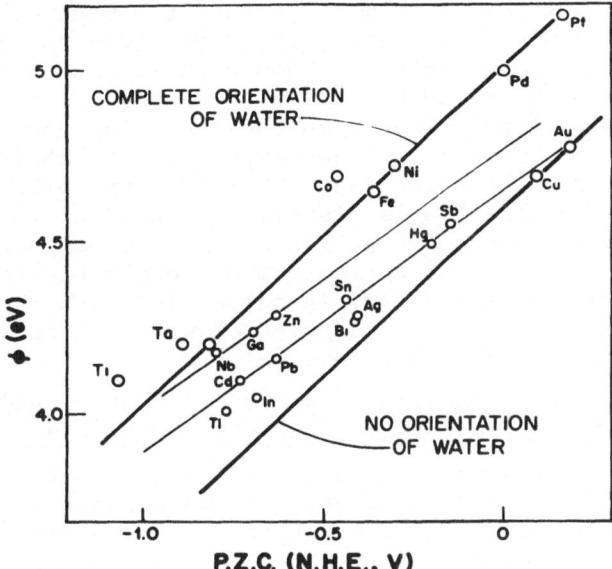

Fig.4. Trasatti's plot for work function vs. pzc for various
metals,showing the effect of the oriented,chemisorbed
layer of water dipoles.

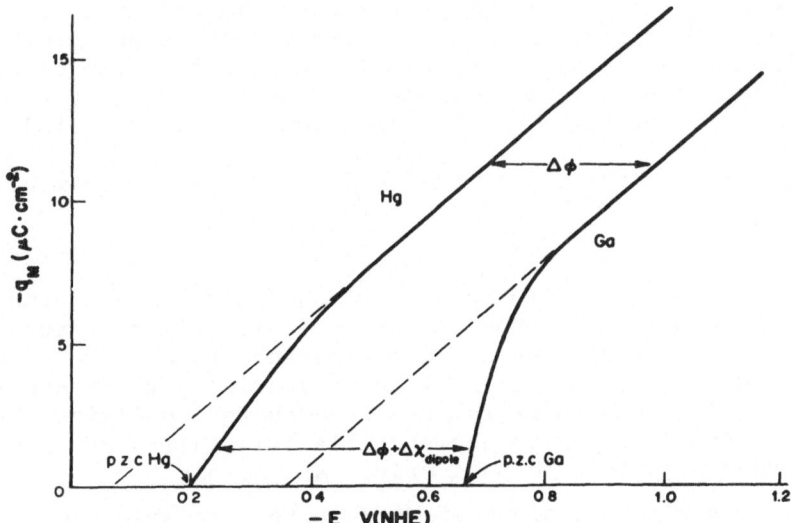

Fig.5. Trasatti's charge vs. potential plot forindicating
role of adsorbed,oriented water layer,depending on
the metal.

metal surfaces. The extent of metal-specific H_2O dipole orientation which this affinity can bring about will depend on the
competing orientation tendency of H-bonds to surrounding water
molecules in the bulk and to specifically adsorbed ions in the
inner layer. It will thus depend on the H-bonding energy which
water molecules at the metal surface experience in their interactions with other water molecules. This in turn, will depend
on temperature. The problem is closely analogous to that of
defining solvation numbers for an ion and the net orientation
of solvent molecules next to an ion. These quantities depend on
the ion-dipole interaction energy in comparison with the dipole-
dipole (H-bonding) interaction energies of water molecules in the
bulk, and, of course, on temperature.

3.   Experimental Determination of Water Orientation and Surface-
     Potential at Charged Interfaces by Dipole Displacement

     When oriented dipoles of water, e.g. at an electrode, can
be displaced by competitive adsorption of an organic substance
which itself generates no surface potential, the observed shift
of potential at a given charge will be equal to the previously
existing (dipole) surface potential. Many organic substances have
themselves a dipole moment, e.g. ethylene glycol or 2-butyne-1-4-
diol which complicates (cf. ref. 47) interpretation of water
displacement experiments. Also, they may exist in field-dependent configurations or conformations, as with ethylene glycol or
dioxane. By using the rigid, flat and overall non-polar molecule
pyrazine N$\ll$_____$\gg$N in water, and naphthalene in CH_3OH solutions
at Hg, Conway and Dhar[29] were able to obtain surface potential
data for these solvents at the Hg electrode as a function of charge q_M free from the above objections. They also calculated the
orientation distribution $(N\uparrow - N\downarrow)/N_T$ from the experimental $\Delta\chi$
changes.

     The surface dipole potential $\Delta\chi_2$ is given by the Helmholtz
relation

$$\Delta\chi_d = 4\pi N\bar{\mu}/\varepsilon_s \qquad (14)$$

where N is the number of dipoles per cm^2, $\bar{\mu}$ their moment normal
to the interface and ε_s is the effective dielectric constant
of the layer. Addition of an organic substance to a Hg/solution
interface causes an observable shift of potential, ΔV, at constant charge, including a shift for the zero charge condition (Esin
and Markov - "EM"- effect) if any specific orientation of solvent
or of the added organic adsorbate exists at the p.z.c.

     Let the experimental slope of an EM effect relation for an
added organic adsorbate of zero dipole moment such as pyrazine
be $m = d(\Delta V)/d\Gamma = \delta(\Delta V)/\delta\Gamma$, where δ represents a finite change
of ΔV or Γ with increasing coverage by the adsorbate, itself remaining unoriented. The number of oriented solvent dipoles can be

represented as $(N\uparrow - N\downarrow)_{q_M}$ cm^{-2} at a given q_M and will be $f(q_M)$. At a given q_M, each pyrazine molecule replaces x adsorbed water molecules (x = 3). Hence, for Γ mole cm^{-2} of adsorbed pyrazine, x mole cm^{-2} water oriented in the ratio $N\uparrow/N_T$ and $N\downarrow/N_T$ will be replaced. Hence a degree of net solvent orientation in the double-layer will have been eliminated by displacement equivalent to $x\Gamma(N\uparrow - N\downarrow)/N_T$ where $N_T = N\uparrow + N\downarrow$. This change of extent of solvent dipole orientation per cm^{-2} gives rise to

$$\Delta V_{q_M} = \frac{x\Gamma(N\uparrow - N\downarrow)}{N_T} 4\pi\mu/\varepsilon_s \qquad (15)$$

The experimental slope of m of the integral $\Delta V - \Gamma$ relation is hence

$$m = d\Delta V_{q_M}/d\Gamma \doteq \delta\Delta V_{q_M}/\delta\Gamma = \frac{x(N\uparrow - N\downarrow)}{N_T} 4\pi\mu/\varepsilon_s \qquad (16)$$

In terms of the BDM/Watts-Tobin treatment for adsorbed solvent dipole orientation

$$\frac{N\uparrow - N\downarrow}{N_T} = \tanh\left[\frac{\mu E}{kT} - \frac{\bar{W n}}{kT}\frac{(N\uparrow - N\downarrow)}{N_T}\right] \qquad (17)$$

which enables $(N\uparrow - N\downarrow)N_T$ to be estimated for various values of $\bar{W n}/kT$ where W is the solvent dipole-dipole interaction energy in a lattice of mean coordination number $2\bar{n}(=6)$ and E is the outer field given by $-4\pi q_M/\varepsilon_s$. W can be obtained from the width of the bell-shaped curve relating energy of adsorption to charge, center- ed about the charge for zero dipole orientation (-2 to -4 μC. cm^{-2} at Hg). Use of eqns. (15) and (16) requires that the ΔE_{q_M} caused by adsorption of an organic does not include a change of potential due to modified anion adsorption. Thus, the necessary experiments must be carried out in a dilute solution of a non-specifically adsorbed supporting electrolyte.

The experimental results[29] for pyrazine at various q_M are shown in Fig. 6. ΔV_{q_M} increases with q_M as $\pm q_M > ca. -2\mu$C. cm^{-2} while an appreciable slope m is observed at $q_M = 0$ indicating, as discussed earlier, significant spontaneous H_2O dipole orient- ation due to chemisorption and image interaction with the surface at $q_M = 0$.

With pyridine, a dipolar adsorbate, the orientation of the adsorbate itself can be distinguished from the displacement of oriented H_2O dipoles in the measurement of ΔV_{q_M}. Then ΔV_{q_M} is made up of two terms:

$$\Delta V_{q_M} = \Delta\chi_{H_2O} + \Delta\chi_{P_y} = x\Gamma_{P_y}\frac{(N\uparrow - N\downarrow)}{N_T} 4\pi\bar{\mu}_{H_2O} + 4\pi\Gamma\bar{\mu}_{P_y} \qquad (18)$$

where $\bar{\mu}_{P_y}$ is the normal component of moment of pyridine when the latter molecules become oriented and ε_s, following Levine et al.,[24] is not included.

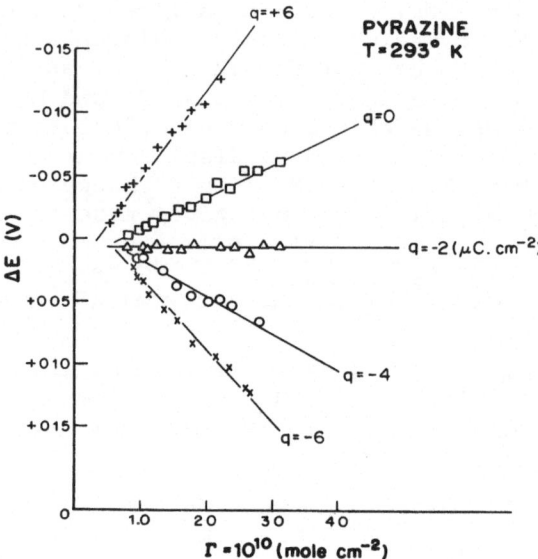

Fig. 6.   Shifts of inner layer potential by displacement of oriented
          water film due to pyrazine adsorption at Hg.

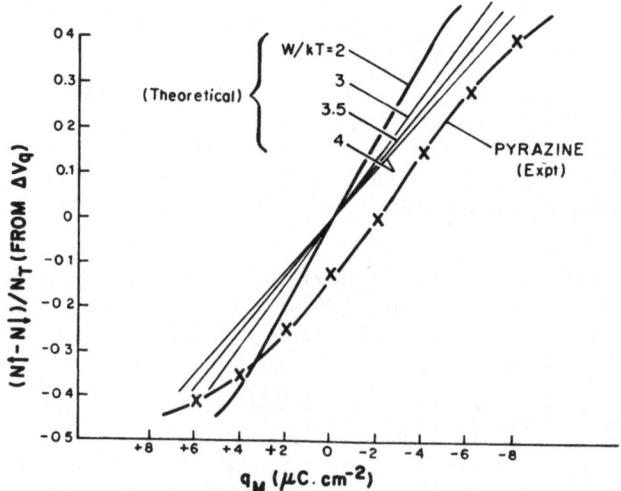

Fig.7.   Comparison of experimental orientation distribution
         function for water dipoles in the inner layer with
         results calculated from BDM theory with various W's.

4.   The Entropy of the Water Layer

The behavior of oriented solvent in the inner layer will
be analogous to that in the solvation spheres of ions[50] except
that the orienting field will be variable in magnitude and sign,
dependent on q_M, the surface charge density on the interface.
The orientation will also depend on the population of specifically
adsorbed ions in or near the inner layer[49].

Derivatives, with respect to temperature, of the surface
free energy give $- \Gamma_s$ the surface excess entropy as treated by
Hills et al.[51,52]. This is a complex quantity[51-54] which includes
the entropy of adsorbed oriented solvent molecules and of water
associated with ions adsorbed at the interface.

Harrison, Randles and Schiffrin[53] made a thermodynamic
analysis of the entropy of formation of a charged interphase,
ΔS_{M-soln}, expressed by

$$\Delta S_{M-soln} = S^\sigma - m_{Hg}^\sigma \, S_{Hg}^\sigma - m_{H_2O}^\sigma \, \bar{S}_{H_2O} - m_+^\sigma \bar{S}_+ - m_-^\sigma \bar{S}_- \qquad (19)$$

where m terms are the numbers of moles of the indicated component
in the interphase, S terms are the entropies and bars indicate
partial molar quantities; S^σ is the total entropy per unit area
of the interphase. ΔS_{M-soln} is the difference of entropy of the
components when in the interphase and when in solution; it is
hence the entropy of formation of the interphase. ΔS_{M-soln} is
related to the surface excess entropy Γ_s and the surface excess
of components by the relation

$$\Delta S_{M-soln} = \Gamma_s - \Gamma_- \bar{S}_- - \Gamma_+ \bar{S}_+ = \Gamma_s - \Gamma_+ \bar{S}_{salt} \pm q_M \bar{S}_\pm F \qquad (20)$$

where q_M is the metal surface charge, $-(\partial \gamma / \partial E)_{T,P,n_1,n_2 \ldots}$ and

$$\Gamma_s = S^\sigma - m_{Hg}^\sigma \, S_{Hg}^\sigma - m_{H_2O}^\sigma (\bar{S}_{H_2O} + \bar{S}_{salt} \, n_{salt}/n_{H_2O}) \qquad (21)$$

The n quantities are the numbers of moles of the indicated
components in the bulk solution. Also,

$$\Gamma_\pm = m_\pm^\sigma - m_{H_2O}^\sigma \, n_\pm/n_{H_2O} \qquad (22)$$

Γ_s and $\Gamma_\pm$ are obtained from the derivatives $-(\partial \gamma / \partial T)_{P,n_1,n_2 \ldots}$
and $-(\partial \gamma / \partial \mu_\pm)_{T,P,}$ etc., respectively.

Experimentally, the entropy of formation of the interphase is obtained at a given surface charge by integration of the temperature coefficient of reciprocal capacity and of the potential of zero charge, with respect to q_M. The method of performing the required calculations for various reference electrodes was considered[51] and the evaluation of inner-and diffuse-layer contributions to ΔS_{M-soln} described[53].

Values of the entropy of formation of the interphase for various solutions as a function of q_M were obtained in relation to the overall temperature coefficient of surface tension of Hg in aq. NaF solution.

An important result is that the entropy of formation of the interphase, and of the inner-layer component, shows a maximum around $q_M = -2$ to $-4\,\mu C.\,cm^{-2}$. This is also the charge where maximum adsorption of non-polar organic substances arises at Hg in a non-specifically adsorbed electrolyte[29]. At this charge, the water dipoles have no net orientation; correspondingly, at the potential of zero charge, the water dipoles must be oriented with O atoms pointing <u>towards</u> the Hg surface[26,29].

The fact that the Hg/water interface exhibits an entropy maximum rather than a minimum at $q_M = 0$, indicates that the surface does not behave simply like an hydrophobic structure-promoting wall, as is the case with the "surface" of large R_4N^+ ions. Presumably, dipole imaging, which occurs "through" the interface, allows the H-bonded water structure to behave, from an electrostatic point of view, as though there was no interface present. At a non-metallic surface of low dielectric constant, the situation would be quite different since the dipole-image interactions are then repulsive. Unfortunately, the Hg/water interface can never by studied with proper electrical control in the absence of ions, so it is not possible to say to what extent the observed entropy maximum at $q_M \lesssim 0$ might be modified if the ionic concentration in the double-layer could be made zero.

A first theory of the entropy of oriented solvent molecules at electrodes was given by Conway and Gordon[54] who took into account the configurational entropy of ↑ and ↓ states in the 2-state model and also the related librational entropy of solvent dipoles in the inner-layer field (cf. ionic solvation entropies[50]), dependent on q_M.

(α) <u>Configurational effects</u>

Solvent dipole orientation can be considered in terms of the model of Watts-Tobin and Mott in which dipoles are regarded as being in an up (↑) and down (↓) directions with respect to the surface but with lateral interactions allowed for[20]. A treatment of the resulting configurational entropy of mixing S_c, gives very simply,

$$S_c = -R[\theta\downarrow \ln \theta\downarrow + \theta\uparrow\ln \theta\uparrow] \tag{23}$$

where the θ terms are the surface fractions (and hence relative coverages) of dipoles oriented in $\uparrow$ or $\downarrow$ directions. In eqn. (23) interaction effects (see below) have not been considered (cf. the Bragg-Williams approximation). The orientation of dipoles will be determined by their interaction with the electrode field E and by their mutual interactions[20]. The resulting distribution function for interacting dipoles is given by

$$\frac{N\uparrow - N\downarrow}{N_T} = \tanh\left[-\frac{\overline{Wn}}{kT}\left(\frac{N\uparrow - N\downarrow}{N_T}\right) + \frac{\mu E}{kT}\right] \tag{24}$$

where W is, as previously, the interaction energy per pair of dipoles in a configuration with mean coordination number $2\overline{n}$ and N, N are the numbers of dipoles in $\uparrow$ and $\downarrow$ orientations, respectively, and $N_T = N\uparrow + N\downarrow$. In terms of the relative coverages ,

$$(2\theta\downarrow) - 1 = \tanh\left[-\frac{Wn}{kT}(2\theta\uparrow - 1) + \frac{\mu E}{kT}\right] \tag{25}$$

which is evaluated below for various values of the interaction factor $f_i = \overline{Wn}/kT$ and the electrostatic field-dipole interaction term $\mu E/kT$. The configurational entropy of the layer of molecules with mixed orientations then follows from eqn. (23). $\theta\uparrow$ and $\theta\downarrow$ may be related to q_M through E given by $E = -4\pi q_M/\varepsilon_s$ where ε_s is the non-orientation dielectric constant component for the surface layer, and may be taken as ca. 6. Results are shown in Fig. 7.

The problem considered here is closely related to that involved in the calculation of magnetization in a lattice of magnetic dipoles. In fact, Cooper and Harrison[55] considered the case of an antiferroelectric arrangement below a critical temperature but concluded it would not arise near the p.z.c. at ordinary temperatures.

(β) Librational entropy

In an electric field, a dipole tends to become oriented but in reality this orientation is rarely complete and the mean degree of orientation at a given T is determined by the field-dipole interaction energy factor $\mu E/kT$, as at ions. Librational entropy is difficult to consider separately from the 2-state configurational orientation entropy which is in some sense a special case here. Evaluation of the librational entropy, which corresponds to a distribution of all possible orientations, appears to be the most realistic way of dealing with the orientation entropy.

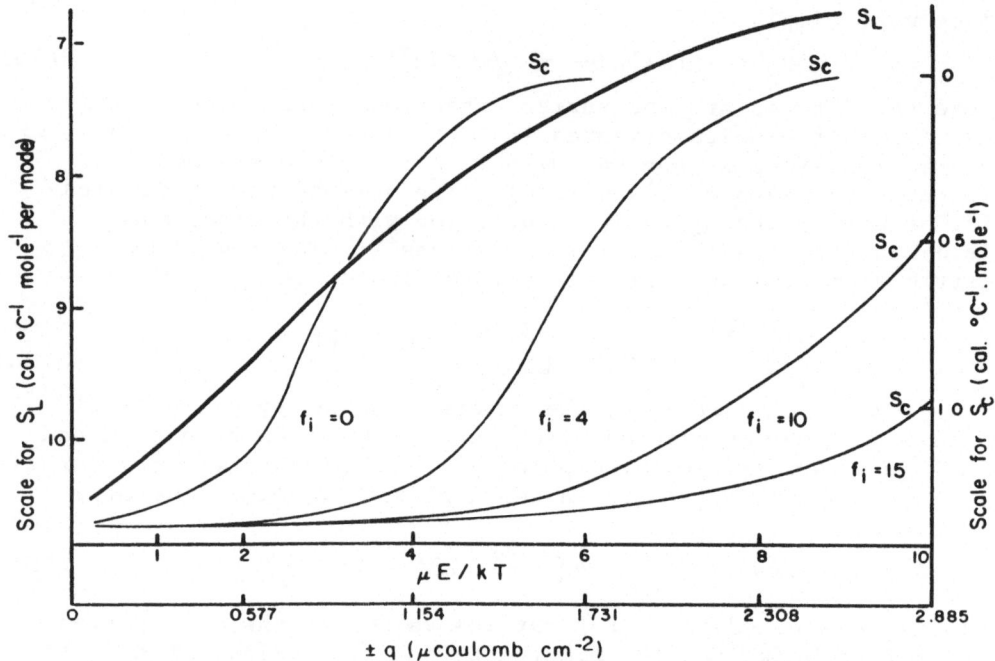

Fig.7 Configurational and librational entropy of the inner
water layer as a function of surface charge,q,according[54]
to the interphasial entropy theory of Conway and Gordon[54].

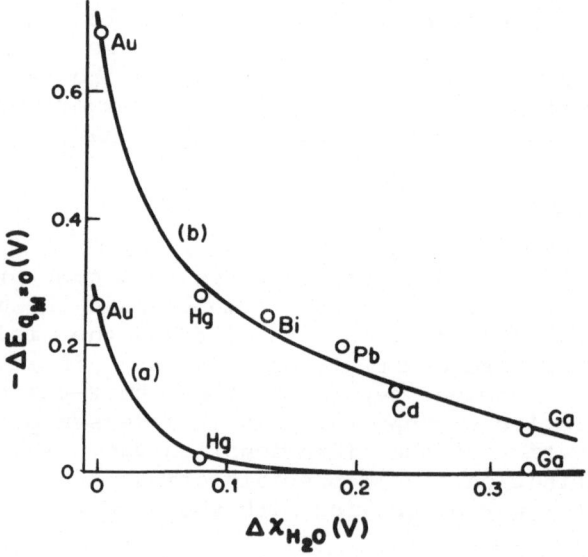

Fig.8 Shift of potential due to I^-adsorption at various metals
in relation to the H_2O surface dipole potential(ref.44).

The dipole executes librative motion in the field and the partition function for the libration mode is

$$f_L = \frac{8\pi^2 (8\pi^3 I_1 I_2 I_3 k^3 T^3)^{\frac{1}{2}}}{\sigma h^3} \cdot \frac{\sinh [U_e/kT]}{U_e/kT} \qquad (26)$$

where U_e is the electrostatic field-dipole interaction energy and I's are the principal moments of inertia of the water dipole. The entropy S_L associated with the librational energy states in the field is then

$$\frac{S_L}{R} = \ln \frac{8\pi^2 (8\pi^3 I_1 I_2 I_3 k^3 T^3)^{\frac{1}{2}}}{\sigma h^3} - \ln U_e/kT +$$

$$\ln \sinh \; U_e/kT - \frac{U_e}{kT} \coth [\frac{U_e}{kT}] + 5/2 \qquad (27)$$

S_L can be evaluated for various values of the inner Helmholtz field $E = -4\pi q_M/\varepsilon$ and is shown in Fig. 7 as $f(q_M)$ for one librational mode. The variation of entropy with q_M is appreciable and larger than that of S_c with q_M. It is also approximately linear with U_e and thus provides a basis for the observed[54] compensation between energy and entropy terms.

(γ) Vibrations

It can be assumed that the internal vibrations of H_2O molecules are little affected by the field at the interface except indirectly by hydrogen bond bending or breaking effects. In any case, the internal vibrational entropy is small at room temperatures since the bend and stretch vibrational quanta are between 8 and 16 kT. However, near the electrode surface, water dipoles will also tend to execute vibrations normal to the surface, as at ions, and the mode is equivalent to restricted translation. This component is not a major one in the field-dependent part of S for the solvent at the interface.

The above treatment, based on the paper of Conway and Gordon[54] for interpretation of entropies of substitutional adsorption at the Hg/water interface, provides a reasonable quantitative explanation for the observed entropies of adsorption of pyridine and pyrazine at Hg. In particular, it gives a basis for the increasing positive entropy of adsorption as surface charge, $\pm q_M$, is increased due to release of low entropy water, previously electrostricted in the double-layer, on account of displacement by the organic adsorbate . The entropy of water itself is found to be a maximum at or near the p.z.c.

Bockris and Habib[56] applied their H_2O monomer/dimer model[40a] to a calculation of the entropy of water at the Hg electrode, using the entropy terms treated by Conway and Gordon[54]. They obtained good agreement with experiment for Γ_s as a $f(q_M)$.

The two-state orientation model[15][55] with interactions[20] was also examined by Cooper and Harrison[55] who showed that it was unable to describe simultaneously both the inner-layer entropy of water and the differential capacity as a $f(q_M)$. They considered the conditions for dipole orientation in relation to those treated in theories of ferro- and antiferro-magnetism for interacting magnetic dipoles. Under certain conditions, the 2-state model leads to the expectation of a negative capacitance ! The reasons for this catastrophe were treated by Parsons[57] and originate from artificialities of the Watts-Tobin model, in particular that only 2 limiting orientational states, ↑ and ↓ are considered.

SPECIFIC ADSORPTION OF IONS

1. Phenomenology of Specific Adsorption

The solid/electrolyte boundary is always associated with a "double"-layer comprising inner and outer Helmholtz layer regions[7], corresponding to distances of closest approach of specifically adsorbed anions and hydrated cations, respectively, and a diffuse-layer ionic atmosphere extending in the direction normal to the interface.

From the point of view of stability of metal surfaces, both mechanically[58] and with respect to Faradaic corrosion reactions, the "specific" or chemisorption of ions, usually anions, is of the greatest importance. Much attention has therefore been devoted in the electrochemical literature to this question. At Hg, a very complete evaluation of the adsorption behavior of chemisorbed ions, e.g. halides, has been made (e.g. ref. 48a) and their effects on kinetics of many electrode processes are well understood, but perhaps least so in corrosion reactions.

Adsorption of ions at electrode surfaces originates on account of (a) the coulombic forces associated with any net surface charge q_M which is held on the interface and (b) any tendency for the ion to be chemisorbed at the electrode surface. The latter effect has been termed specific adsorption and originated in Stern's treatment of the double-layer in 1924. More recently, it has been termed "super-equivalent" adsorption[49], since the charge associated with the presence of the chemisorbed ions often exceeds that necessary simply to balance the excess charge q_M held in the electrode surface - indeed this is a criterion of specific adsorption.

Various forms of adsorption isotherms for ion adsorption have been proposed[48a,60] and tested[48a,60] in applications to experimental results.

The point of significance, to be treated in this section, is the origin of the chemisorptive forces that give rise to specific adsorption of ions. It will be seen that this matter must also involve some discussion of the role of adsorbed solvent molecules (see previous sections) since ions, in becoming specifically adsorbed, must usually displace, or change the interaction of, solvent molecules previously bound to the surface.

2. Chemical Factors in Specific Adsorption of Ions

(i) Replacement of adsorbed solvent molecules

Specific adsorption of ions may be treated as a substitutional adsorption process[49,59] like that for organic substances at electrodes[20,29]. Thus, in the simplest analysis, an anion A^- may become adsorbed at a metal surface* with displacement of n previously adsorbed water or solvent (S) molecules:

$$A^-_b + nS(M) \rightleftharpoons A^-(M) + nS_b \tag{28}$$

where subscript b represents the bulk solution. The overall free energy change of such a substitutional adsorption is obviously made up of terms for the solvation energy of A^- in bulk, the binding energy of S on M and the adsorption energy of A^- on M. Thus, the specific adsorbability of anions can depend not only on the properties of the ion itself, but also in an important way on the chemisorption of water in the inner layer.

In a more detailed consideration of process (28), the change of solvation of A^- upon adsorption, which would determine in part the adsorption energy of A^-, would be considered. Process (28) would then more accurately be represented, for H_2O, by

$$A^-(M \, H_2O)_b + n \, H_2O(M) \rightleftharpoons A^-(m-x)H_2O(M) + (n+x)H_2O_b \tag{29}$$

e.g., for a substitutional adsorption of hydrated ions where a change of hydration number from m to m-x occurs upon adsorption. For most specific adsorptions, some degree of direct contact of the ion with the metal must be envisaged.

Although displacement of adsorbed water is included in the above reaction schemes as an important element of the overall chemisorption process, it appears to have been overlooked in the papers of Barclay[61] and of Vijh[46], although it had already been treated in the relevant paper of Anderson and Bockris[49] .

---

* In some treatments[60], the anion is shown separated from the metal surface by 1 layer of oriented H_2O dipoles.

Experimentally, it seems that chemisorption of water with associated preferential orientation of the O atom of the H_2O molecule towards the metal (i.e., also as a "ligand" to the metal surface) is a major factor in leading to the observed differences of strength of chemisorption of anions at various metals, e.g., Hg, Ga, Au and Pt.

The relation between anion adsorption and displacement of adsorbed water is supported by the good correlation which arises if the shift of p.z.c., at a series of metals due to I^- adsorption at 10^{-1} or 10^{-2} mol 1^{-1} is plotted (Trasatti, ref. 44) against the surface potential of water at these metals. The surface potential of water measures, of course, the tendency for chemisorptive orientation of H_2O molecules at the metal. The shift of p.z.c. decreases from Au through Hg to Ga in the series shown in Fig. 8, as the surface potential ΔX_{H_2O} increases. A relation having a similar trend is obtained if the shift of p.z.c. for I^- adsorption is plotted against the inner layer capacity C_i. This result is rationalized again if it is assumed (but cf. refs.[20,40]) that C is directly governed by the state of water chemisorbed at the metal interface[21,62]. However, other complicating factors evidently enter into this relation as $\Delta V_{q_M} = 0$ changes appreciably from Au to Hg, Bi and Pb without much corresponding change in C_i. This must reflect the different affinities of I^- for the metal surfaces in relation to the affinity of water. It is clear that the affinity of water for metal surfaces is not the only factor determining the relative affinity of chemisorbing ions for various metal surfaces. There is obviously a competitive effort depending on both factors, rather analogous to competitive hydration effects in ion-selective membranes.

(ii) Ion hydration, electrostatic interactions and surface compound formation

In general, the larger the anion the more weakly is it hydrated, so that large anions are more easily desolvated than small ones and can hence approach surfaces more closely. Their lone-pairs are then more accessible for donor-acceptor interactions with the metal.

Increasing anion radius is also correlated with two other properties important for adsorbability: (a) polarizability and (b) electron donicity, the latter being determined by the electronegativity of A in its anionic form. Two other factors that can determine specific adsorbability of anions have been considered in the literature; they are (a) the strength of the image interaction between A^- and the metal surface, which will depend on the effective radius of A^- and (b) the formation of surface compounds[45,46]. The latter possibility is unlikely[49], since, at

Hg, the order of increasing specific adsorption of monatomic halide anions is the opposite of the direction of the strength of Hg-halogen bonds.[49]

The surface compound concept[46] is applicable more to a process such as

$$A_b^- + nS(M) \rightarrow AM + nS_b + e(M) \tag{30}$$

This is the type of process which arises in underpotential deposition and is well known in the case of surface oxide formation at noble metals ($\theta_{A(M)} > 1$) and in Cl$^-$ deposition at Hg[63,64] prior to Hg$_2$Cl$_2$ formation. While such a process has some similarities to anion specific adsorption, the degree of charge transfer[59,64] is more complete and for such a reaction demetallization[46] of the surface is an important result of the reaction. This is not the case in most examples of specific adsorption[49]. The image interaction effect also gives an opposite trend to that observed - the image interaction would be greater for F$^-$ than for I$^-$. However, the image effect must be estimated with due regard to the hydration radius of the adsorbed ion and not its crystallographic radius. Hydration effects can invert the expected direction of image interaction and give a larger image interaction for I$^-$ than for F$^-$.

In practice, it is notoriously difficult to apply an image interaction law quantitatively for distances of approach comparable with atomic dimensions where electron charge density distributions are determined as much by the local atoms of the metal and the solvent as by the presence of a neighboring ion in the inner layer. The limitations of the image law for interaction at short distances (1-3 Å) have been critically discussed by various authors[65,66]. At such distances, the image interaction term must be replaced by one taking into account polarizability in the ionic field and the local electron donor/acceptor properties of the anion/metal system. The latter type of effect has been the subject of a number of recent papers (e.g., 59, 64) concerned with the extent of partial charge transfer in anion chemisorption (electrosorption valency effect) at electrodes and in Faradaic atom deposition where the electrosorption valence factor has been evaluated.

(iii) Anion chemisorption as a "ligand" - metal interaction

A useful departure from the idea that strength of specific adsorption parallels the energy of "surface compound" formation[7,46] was given in recent papers by Barclay[61] and by Conway, Angerstein-Kozlowska and Sharp[64] where the role of the specifically adsorbed ion as a "ligand" to the surface is stressed.

In these treatments, the surface bond is regarded as a coordinate bond as in Werner complexes, rather than as a full covalent or ionic bond in bulk compounds. In Barclay's treatment, the chemical "hardness" or "softness"* which reflect the polarizability and the Lewis electron donor/acceptor properties of the ion/metal system are stressed. In the case of specifically adsorbed anions at the p.z.c. and at positive q_M, the interaction is that of a Lewis-base type ligand with the metal acting as the conjugate Lewis acid. The importance of such interactions over competing effects of hydration of anions was stressed by Barclay in the case of S^{2-} which is strongly hydrated, yet is one of the most strongly chemisorbed ions, e.g., at Hg. Similar considerations apply to SO_4^{2-} which is strongly adsorbed at Pt and Au, so that both d and sp electronic effects are involved. The A^--M bond is regarded by Barclay as not determined by the bond strength in a bulk AM compound (where a three-dimensional lattice is involved and different kinds of bonding may arise) but rather, is governed by the difference in electronegativity of the metal surface (this will in turn depend on q_M since the metal – electrode surface presents an interface of variable Lewis acid/base character determined by surface charge density) and the anion adsorbed.

The correlation shown below, between Pearson's classification for anions and the potential of zero charge at Hg (determined by anion adsorption), was given by Barclay[61]. It is seen that the softest anions give the greatest negative shift of p.z.c., corresponding to strongest adsorption. The relations are, however, only qualitative and are related to polarizability and densities.

The order of strengths of specific adsorption of anions at electrodes (Hg, Pt, Au) is approximately the same as that of their ligand strengths for coordination to common transition metal ions.

TABLE 2:   Relationship Between E_{pzc} and Softness of Anions
(From Barclay, ref. 61)

Anion	$-E_{pzc}$ (mV vs SCE)	Pearson's Classification of anion	Anion	$-E_{pzc}$ (mV vs. SCE)	Pearson's classification of anion
S^{2-}	880	Soft	Cl^-	461	Hard
I^-	693	Soft	OAc^-	456	Hard
CN^-	645	Soft	NO_2^-	450	Intermediate
CNS^-	589	Soft	HCO_3^-	440	Hard
Br^-	535	Inter.	CO_3^{2-}	440	Hard
N_3^-	509	Inter.	SO_4^{2-}	438	Hard
NO_3^-	478	Hard	F^-	437	Hard
ClO_4^-	470	Hard			

* These are the terms used by Basolo and Pearson[67] in their classification of Lewis acid-base interactions.

Klopman[68] has suggested that a scale of ionic softness in solution can be based on the sum of its desolvation energy and valence shell orbital energy. Softness in a base is determined by low desolvation and high orbital energy and in an acid by low desolvation and low orbital energy. Soft-soft interactions involve covalent type bonding between empty and filled orbitals of comparable energies (cf. donor-acceptor behavior). Generally, soft-soft and hard-hard interactions give the best complexation energy.

The most detailed a priori quantitative calculations on ion specific adsorption were given by Anderson and Bockris[49], in terms of ion hydration, dispersion forces repulsions and image effects (cf. Wroblowa and Müller[60]).

While the direct metal/ion interaction energy varies but little (-49.1 kcal mol^{-1} for Na^+ at Hg and -49.2 kcal mol^{-1} for I^-) due to compensating differences between the energy terms, the change of energy in the overall chemisorption process from solution is appreciably ion dependent, as tabulated below:

TABLE 3: $\Delta H°$, $\Delta S°$ and $\Delta G°$ quantities calculated for specific adsorption of ions at Hg

Ion:	Na^+	K^+	Cs^+	F^-	Cl^-	Br^-	I^-
$\Delta H°$ (kcal mol^{-1})	18.5	6.8	-6.8	64.1	-9.6	-12.9	-16.4 (expt. -11.5)
$\Delta S°$ (e.u. mol^{-1})	16.3	13.9	1.6	17.5	-2.1	-5.1	-11.0
$\Delta G°$ (kcal mol^{-1})	13.7	2.7	-7.3	8.7	-9.0	-11.4	-13.1

From Anderson and Bockris, ref. 49

The superequivalent adsorption, e.g. of I^-, increases with positive q_M (coulombic and Lewis acid/base effect) and also with ionic radius. The latter effect is due both to decreasing energy of solvation and increasing polarizability and donicity with increasing radii (of anions). With organic cations such as R_4N^+, hydrophobic adsorption effects[50] are dominant.

With halide ions, contrary to some suppositions, the trend of strengths of specific adsorption is opposite to that of bond strengths in Hg-halide formation (cf. refs. 45,46, 49).

At solid metals, some high degrees of specificity and strength of binding are found. Cl^- and Br^- adsorption at Pt is

detectable at 10^{-7} M levels while SO_4^{2-} is detectably adsorbed
at Au at 10^{-8} M [69]. These effects are important for the stability and surface oxidation behavior of solid metals. Russian workers, e.g. Kazarinov and Balashova, have provided much of the
information on adsorption of anions on solid electrodes through
their method employing radio-labelled ions.

(iv) "Charge-Transfer" effects and "electrosorption valency" in ion adsorption

The state of binding of specifically adsorbed ions
at metals has been referred to earlier in connection with the
role of image forces and coordinative bonding. An important
indication of the role of charge-transfer complexation in
specific adsorption of anions (and also in underpotential deposition of metal atoms) is the so-called "electrosorption valency",
a term employed by Vetter and Schultze[70] but earlier considered
by Lorenz and by Conway and Bockris in metal deposition. This
term is related to the value of the coefficient λ in an electrochemical adsorption equation of the form[59]

$$M H_2O + A_{aq}^z \quad MA_{ads}^{z+} + \lambda e + \nu H_2O \tag{31}$$

where z is a negative integer for anions. λ is a quantity such
that $0 < \lambda < -z$, i.e., $= z_{ads} -z$ where z_{ads} is the charge number
for the ion in the adsorbed state. In terms of experimentally
determinable quantities, Lorenz[71] interpreted the derivative
$(\partial q_M / \partial \Gamma)_E$ as a partial charge transfer, λe, where Γ is the surface
excess of the adsorbed ion.

The formal definition of the electrosorption valency γ is
made through the thermodynamic relation

$$\gamma = -\frac{1}{F} (\frac{\partial q_M}{\partial \Gamma})_E = \frac{1}{F} (\frac{\partial \mu}{\partial E})_\Gamma \tag{32}$$

where μ is the chemical potential of the adsorbing ions in the
bulk. To a good approximation λ is related to γ by the relation

$$\gamma \doteq gz -\lambda(1-g) \tag{33}$$

where g is a parameter measuring the penetration of the adsorbed
ion into the electric field of the double-layer and is to be
identified with the potential $\emptyset_{ads}$ at the adsorption site
relative to the metal/solution p.d.:

$$g = (\emptyset_{ads} - \emptyset_s)/(\emptyset_M - \emptyset_s) \tag{34}$$

where $\emptyset_s$ is the inner potential of the bulk electrolyte. g

is a geometric factor characterizing the location of adsorbed
ions relative to the Helmholtz layer . A value of = 0 represents
purely electrostatic adsorption of ions as was assumed by a number
of authors prior to the consideration of adsorption with partial
charge transfer.

Fig. 9 shows some γ/z values for various electrosorption
systems in water as a function of electronegativity difference
$\chi_M - \chi_A$. A similar relation applies to methanol as solvent.

Evaluations of the electrosorption valency demonstrate the
important role of electronic effects in specific adsorption and
Faradaic atom electrodeposition,and indicate that electrostatic
and hydration effects cannot alone account for specific adsorption
of ions at metal interfaces. In fact, the extent of charge
transfer between I^- and Pt is so complete that iodide is adsorbed
at Pt in the state of an almost neutral I atom[72].

The question of partial charge transfer in specific
adsorption raises several basic problems concerning the nature
of specific adsorption and the interaction with the metal: (a)
what is the role of classical image force interactions[66] at short
distances ? (b) what is the role of displacement of hydration water around the ion ? (c) to what extent is the interaction due
to polarization forces at the interface, or is the apparent relation to ionic polarizability only a secondary one due to other
similarly related properties, e.g., the energy of hydration of
the ion as a function of its radius ? (d) to what extent are
the electron exchange interactions important, especially with the
"softer" anions ?

Recently, Smickler has considered the latter situation and
has made MO calculations for ion adsorption at electrodes,
including the case of alkali metal ions.

Conway, Angerstein-Kozlowska and Sharp suggested[64] that a
useful basis for considering the relationship between adsorption
arising in a Faradaic charge-transfer step at a surface producing
a chemisorbed neutral species and "specific adsorption" of anions
is offered by comparing the situation at a metal surface with that
in ligand complexation, of acceptor metal ions, usually of the
transition metals.

(v) <u>Relation between anion specific adsorption and
     inner layer solvent molecule orientation</u>

In most work on models of the oriented solvent layer
at electrodes, little or nothing is said about the effects of the
presence of specifically adsorbed ions (especially at positive q_M
where anions are involved) on the solvent orientation, though this

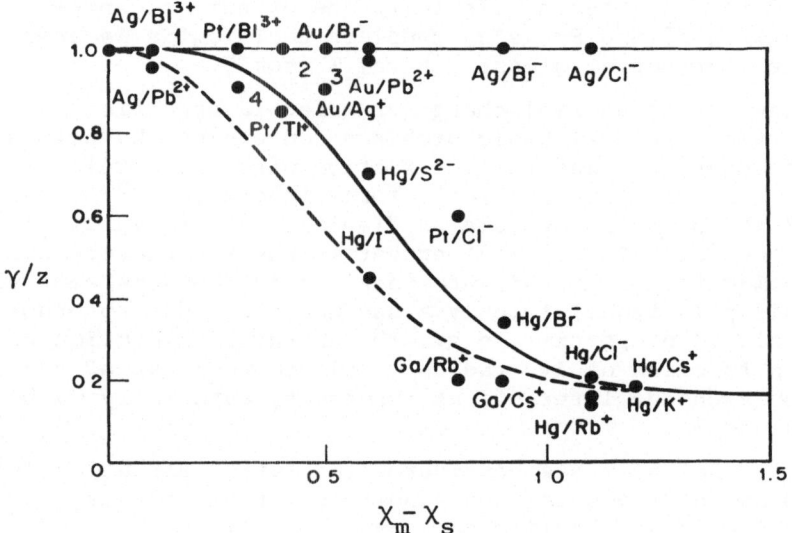

Fig.9   The electrosorption valence factor γ/z plotted against the electronegativity difference $\chi_m-\chi_s$ for various electrosorbed species (from ref.59)

must determine the capacitance behavior and probably accounts for
the anion specificity of the capacity hump in various solutions.

The present author calculated[48] the volume fraction of
solvent molecules in the inner layer which are also involved in
hydration interactions with specifically adsorbed anions at
various surface excesses corresponding to experimentally measured
ionic components of charge, e.g. for Cl^- adsorption. Some
figures for these fractions are given below. Also, the 3-dimens-
ional concentrations of anions, equivalent* to the interionic
distances in ad-ion monolayers at various coverages, were evaluated and show that, in the specifically adsorbed anion layer,
hydration-shell co-sphere interactions must play a substantial
role in determining the energy in the interphasial monolayer ,
as they do 3-dimensionally in electrolyte solutions at
concentrations $> 1 \sim 2$ molar.

TABLE 4: Relation between some Γ_i, q_i and equivalent*
3-dimensional c_i values

Γ_i (mol cm^{-2}) x 10^{11}	5.19	10.37	15.5	20.74	25.9	31.1
$\pm\, q_i$(C m^{-2})	0.05	0.1	0.15	0.2	0.25	0.3*
c_i(mol dm^{-3})	0.50	0.78	1.43	2.20	3.09	4.25
Average interionic distance (nm)	1.79	1.27	1.03	0.89	0.80	0.73

* The maximum $\pm\, q_i$ taken here is 0.3 C m^{-2} which corresponds
to strong specific adsorption. Higher values are rarely encountered except for incipient halide phase formation at Hg or Ag.
Appreciable values of q_i in the range $0.15 \sim 0.25$ Cm^{-2} are
encountered in halide-ion specific adsorption at Hg.

TABLE 5: Volume fraction (%) of ion-influenced co-sphere
water in the inner layer at various extents of ion
adsorption, q_i

$\pm\, q_i$ (C m^{-2})	0.05	0.10	0.15	0.20	0.25	0.30
Γ_i (molecules cm^{-2} x 10^{-13})	3.13	6.26	9.39	12.52	15.65	18.78
Volume fraction (%)	14.8	29.7	44.6	59.5	74.3	(89)

* The relation between 2-d and 3-d concentrations is
$c_i \equiv \Gamma_i^{3/2} N_A^2 \, 100 \text{ mol dm}^{-3} \equiv (q_{i/F})^{3/2} N_A^2 \, 1000 \text{ mol dm}^{-3}$.

It is seen that, even at moderate q_i ($\pm$ 0.15 C m^{-2}) [7],
the volume fraction of ion-influenced water is quite large so
that the inner-layer water dipole orientation can hardly be considered independent of the ions that are also accumulated, when
specifically adsorbed, in the double-layer in response to
electrode surface charge density. The orientation distribution
functions in 2 and 3-state model systems for water in the inner
layer will have to be modified to take these ion effects into
account.

RELATION TO THE MECHANICAL PROPERTIES OF SOLID MATERIALS

Pertinent to the foregoing is the question to what extent
are bulk-type properties such as plasticity, fracture, etc., of
metals dependent on the surface state of the solid, including
effects associated with water adsorption and orientation, and
anion specific adsorption. While there are no direct experimental
indications that the latter factors are primary determinants
of solid state mechanical properties (through surface layer
effects), Latanision[58] has given important critical reviews in
which evidence is examined that surface properties, including the
surface charge density[74] q_M and potential of zero charge conditions, can have significant, if not sometimes substantial, effects
on solid-state mechanical properties. Such properties are those
that involve changes of surface state or creation of new surfaces
such as slip planes[75-79] or deformations introduced by hardness
tests. In particular, the state of metal surfaces may effect
the creation[79] and/or movement of dislocations which are
involved in the mechanical behavior of crystalline solids.

We shall not reiterate here the evidence for surface effects
on apparent bulk properties, nor the basis of controversies which
remain since the material recently published by Latanision[58,73-79]
has covered these topics very thoroughly.

However, the main view appears[75,76] to be that changes in
mechanical properties induced by adsorption at the interfaces of
the material are due to the influence of adsorbed species on the
mobility of dislocations near the surface. Shchukin et al[25] suggest that dislocation mobility (hardness) becomes sensitive to
adsorption at the surface of the material only when (a) surface
steps are formed during dislocation motion and (b) the surface
environment causes appreciable reductions in the energy of
surface slip steps formed during deformation e.g., in hardness
testing or cold working.

Investigations[74] of surface charge density effects on cleavage fracture behavior of Zn single crystals show that the fracture is, in fact, surface charge (potential) dependent. It appears
that the mobility of otherwise crack-blunting dislocations may be
the factor most affected by changes of surface charge density.

There are clear relations between fracture stress and strain, and potential, and time-to-failure as a function of potential, the minima or maximum being at ca. - 1.2 V vs s.c.e. [74].

THE SOLID OXIDE/ELECTROLYTE INTERFACE

1. Monolayer Surface Oxidation of Metals

An important aspect of the properties of metal/electrolyte interfaces and their stability is the tendency for surface oxidation to occur either on account of corrosion or because of an impressed anodic current. Of particular interest is the development of monolayer and sub-monolayer oxide films on the noble, and initially at the almost noble (Cu, Ag), metals prior to thick-film oxide formation.

At the submonolayer level, the specific adsorption of anions plays a major role in competitively determining the onset of surface oxidation process, e.g. at Pt, Ir, Ru, Au, and their continuing development to higher coverages. The surface oxidation chemically satisfies the free orbitals at metal surfaces and causes major modification to the stability of the metal surface and to its catalytic and adsorptive properties[85,86], e.g. for H_2 oxidation, O_2 reduction and Cl_2 evolution.

At the noble metals, the general surface oxidation behavior shows a number of common features[69]. An initial, almost reversible region of submonolayer oxide film formation is first observed, e.g. at Pt, Ir, Ru, but the extent of this region is less at metals (e.g. Pt, Au) for which the onset of surface oxidation occurs at higher potentials, > 0.8 V. It is also less when specific adsorption of anions interferes competitively with the initial stages (submonolayer) of the oxidation.

Further oxidation (increasing coverage towards a monolayer) occurs at Pt and Au in several easily distinguishable states below the monolayer limit which are believed to be associated with development of successive overlay lattices of OH or O species on (or reconstructed with) the underlying metal substrate lattice plane.

When irreversible formation of surface oxide occurs, the reduction of the monolayer or submonolayer film always takes place,of course, at less positive potentials than those required for the formation process but, more interesting, with a much narrower peak (e.g. at Pt, Rh, Au, Pd) in a cyclic-voltammetry experiment than that observed in the oxide formation direction. This difference is observed no matter how slow are the formation and reduction processes conducted. In fact, faster sweeps reveal more reversible behavior, so that the irreversibility is associated with a slow reconstruction (place exchange) of the lattice of deposited oxide species.

The place-exchange process is facilitated by adsorption of
anions on the electrode.- see Fig.12 below.

Fig.10 shows an interesting aspect of these phenomena: at
low temperatures and sufficiently high sweep-rates, the reversible
and irreversible process of Pt (and Au) can be well resolved[80].
The irreversible process apparently becomes predominant only at
the more positive potentials, > ca. 1.0 V E_H. Co-adsorbed anions,
e.g. Cl^-, from the electrolyte modify the extents of reversibly
and irreversibly reduced oxide species below the monolayer
limit.

At Au, the reduction profile is more easily resolved into
component processes. Initially (Fig.11) a reversible O_{A1}/ O_{C1}
stage is observed at low coverage ($\theta_O \simeq 0.1$), dependent
on the anion and pH of the solution. On holding the potential
constant, or on going to somewhat higher potentials (> 1.42 V),
a second reduction peak O_{C2} (Fig.11) is observed which is
associated with diminution of O_{C1}. On longer times of oxidation
(still below monolayer coverage) a third region develops, O_{C3},
which is associated with reduction of the main growing film
to $\theta_O \rightarrow 1$. This region eventually corresponds to the main
reduction peak on a cathodic sweep.

The O_{C2} region that develops at low coverages, saturates
at ca. $\theta_O \cong 0.15$, at a level depending on the anions that can be
co-adsorbed from the electrolyte.

The initially observed peak is associated with reversibly
deposited O species and the O_{C2} peak with anion-stabilized
rearranged, OAu species on the electrode. The main (O_{C3}) peak
is for reduction of rearranged OAu species in the monolayer
at coverages too high to allow significant co-adsorption of
anions. The effects are "general" being seen similarly at Pt or
Au, and are therefore not artefacts associated with oxidation of
difference exposed crystal planes.

2.   The Solid/Electrolyte Interface at Thick Oxide Films

Thick oxide films are more often encountered at solid metal
materials than monolayer films and arise from aerial (thermal) or
aqueous corrosion. Unlike very thin films. they are often
insulators or poor semiconductors. While less is known quantitatively about oxide/electrolyte solution interfaces than about
metal/solution interfaces, oxide/solution interfaces are of major
importance in the stability of materials to corrosion and in photo-
electrochemical effects arising in photo-assisted electrolysis
at semiconducting oxides, e.g., TiO_2, TiO_2-RuO_2.

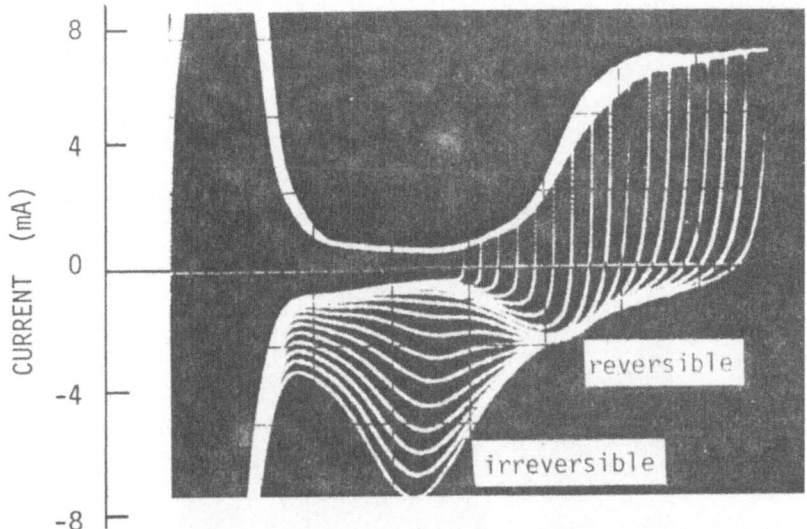

Fig. 10. Resolution of reversible and irreversible components of Pt surface oxidation at low temperature (213K) in 6M H_2SO_4.

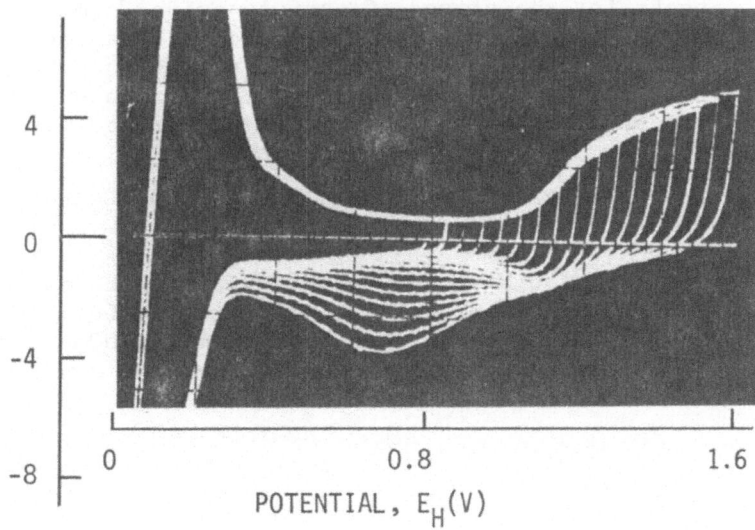

Fig. 10a. As in Fig. 10, but in the presence of 10^{-4}M Cl$^-$ ion. (From Conway and Mozota, in course of publication, Faraday Trans., I).

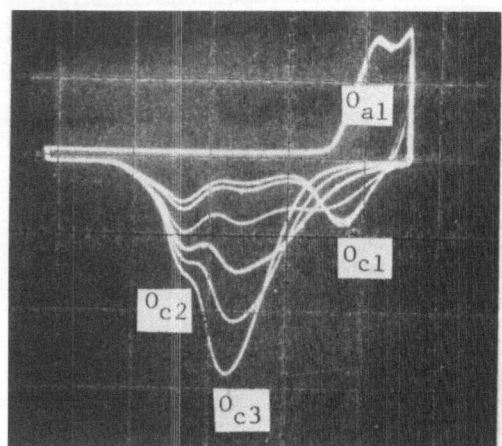

Anodic

currents

Cathodic

Fig.11 Progression of irreversibility in component reduction
       peaks for the stages of surface oxide formation at Au
       at sub-monolayer coverages.Effects are anion and temp-
       erature dependent.Peaks in the reduction profile are
       numbered O_{c1}, O_{c2} and O_{c3} in the order in which they
       appear with increasing degree of sub-monolayer oxid-
       ation.O_{c1}, O_{a1} are almost reversible.

(From Angerstein-Kozlowska,Barnett and Conway,in
course of publication).

DETAILS: 1.0 M aq. $HClO_4$,s=50 V s^{-1};oxide extension
         in sub-monolayer growth at 1.468 V E_H; $_h$ =
         10^{-4} to 10 s at 298 K.Scales:200mV division^{-1}
         on X-axis,current 1mA division^{-1}on Y-axis.

MONOLAYER O FORMATION PLACE-EXCHANGE and ANION ADSORPTION

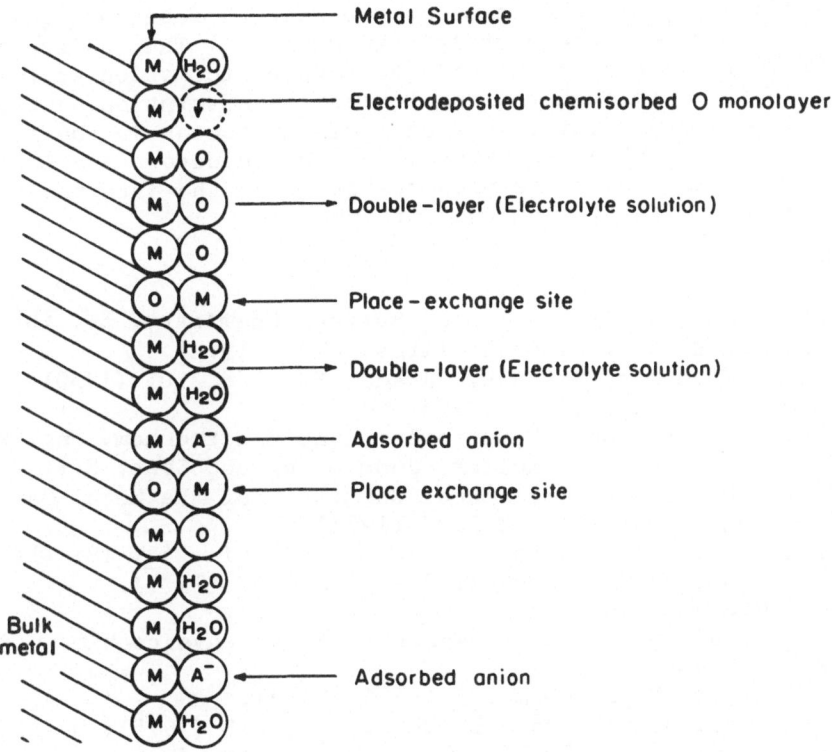

Fig.12 Schematic diagram of course of sub-monolayer oxidation
       of a metal showing co-adsorption of anions and presence
       of adsorbed water molecules at the interface.

The interface of thick insulating oxide films is associated with a double-layer like that at metals except that the charge density arises on account of ionization of OH or O groups,or from ion adsorption. Points of zero charge (isoelectric points) can be evaluated by means of titration experiments coupled with zeta-potential measurements. At oxide interfaces, even simple monovalent cations, as well as anions, can be specifically adsorbed and influence the interface in major ways.

Space prevents a full discussion of the properties of thick-film oxide interfaces here. Further reading references are 81-84, e.g. by Ahmed,who has made some of the most thorough and quantitative studies of oxide/electrolyte interfaces. The ionization behavior is treated in terms of aquo-complex formation at the oxide interface, through various ionization/adsorption equilibria. H-bonding to solvent water is a determining factor in the surface properties[87,88]. When the oxide is a semiconductor, the ionization phenomena influence the band-structure in the interphasial region on the oxide side of the surface.

REFERENCES

1. G.A. Somorjai, Principles of Surface Chemistry, pp. 40-42, Prentice Hall, Englewood Cliffs, N.J. (1972).
2. J.C.P. Mignolet, Discuss. Faraday Soc., $\underline{8}$, 105 (1950); J. Chem. Phys., $\underline{21}$, 1298, (1953).
3. J.D. McIntyre, Adv. Electrochem. and Electrochem. Eng. $\underline{9}$, 62 (1973), Ed. R.H. Muller, John Wiley and Sons, N.Y.
4. M.W. Roberts and C.S. McKee, Chemistry of the Metal Gas Interface, Chapter 3, Oxford (1978).
5. D. Blakely and G.A. Somorjai, J. Catalysis, $\underline{42}$, 181 (1976).
6. P.N. Ross, J. Electrochem. Soc., $\underline{126}$, 67 (1979).
7. D.C. Grahame, Chem. Rev., $\underline{41}$, 441 (1947).
8. P.A. Rehbinder and E.K. Wenström, Acta. Physicochim. U.R.S.S. $\underline{19}$, 36 (1944).
9. T. Beck, J. Phys. Chem., $\underline{73}$, 466 (1969).
10. I. Morcos, J. Electroanal. Chem., $\underline{62}$, 313 (1975); Coll. Czeck. Chem. Comm., $\underline{36}$, 689 (1971); J. Coll. Interface Sci., $\underline{37}$, 410 (1971).
11. A.N. Frumkin, A. Gorodetskaja, B. Kabanov and N. Nekrasov, Phys. Z. Sowjetunion, 1, 225 (1932); $\underline{5}$, 418 (1934).
12. M. Sotto, J. Electroanal. Chem., $\underline{69}$, 229 (1970); $\underline{70}$, 291 (1976);
13. R. Parsons, J. Electroanal. Chem., $\underline{118}$, 3 (1981).
14. J.R. MacDonald, J. Chem. Phys., $\underline{22}$, 1867 (1954).
15. R.J. Watts-Tobin. Phil Mag., $\underline{6}$, 133 (1961).
16. B.E. Conway, J.O'M. Bockris and I.A. Ammar, Trans. Faraday Soc., $\underline{47}$, 256 (1951); B.E. Conway, Ph.D. thesis, London, 1949.
17. D.C. Grahame, J. Chem., Phys., $\underline{18}$, 903 (1950).
18. R.M. Reeves, Chapter 4 in Modern Aspects of Electrochemistry

Vol. 9, Plenum Press, N.Y. (1974).

19. B.E. Conway, Adv. Coll. Interface Sci., 8 91 (1977).

20. J.O'M. Bockris, M.A.V. Devanathan and K. Müller, Proc. Roy. Soc., London, A274, 55 (1963).

21. B.B. Damaskin and A.N. Frumkin, Electrochim. Acta, 19, 173 (1974).

22. F. Booth, J. Chem. Phys., 19, 391; 1327 (1951).

23. R. Fawcett, J. Phys. Chem., 82, 1385 (1978).

24. S. Levine, G.M. Bell and A.L. Smith, J. Phys. Chem., 73, 3534, (1969).

25. J. Topping, Proc. Roy. Soc. London, A114, 67 (1927).

26. S. Trasatti, J. Chim. Phys., 72, 561 (1975).

27. J. Lennard-Jones and J.A. Pople, Proc. Roy. Soc. London, A205, 155; 163 (1951).

28. B.E. Conway, Can. J. Chem. 37, 613 (1959).

29. B.E. Conway and H.P. Dhar, Croat. Chem. Acta, 45, 173 (1973).

30. S. Trasatti, J. Chem. Soc. Faraday Trans., I, 68, 229 (1974).

31. S. Trasatti, J. Electroanal. Chem., 39, 163 (1972); 44, 362 (1973).

32. S. Trasatti, Surface Sci., 32, 735 (1972).

33. A.N. Frumkin, N. Polianovskaya, N. Grigoryev and I. Bagotzkaya, Electrochim. Acta, 10, 793 (1965).

34. S. Trasatti, J. Electroanal. Chem., 33, 351 (1971).

35. J.W. Schültze, Ber. Bunsenges phys. Chem., 73, 483 (1969).

36. B. B. Damaskin, Elektrokhimiya, 1, 63 (1965).

37. R. Parsons, J. Electroanal. Chem., 8, 93 (1964).

38. B.E. Conway, H.P. Dhar and J.M. Joshi, Electrochim. Acta. 18, 789 (1973).

39. R. Parsons, J. Electroanal. Chem., 59, 229 (1975).

40. B.E. Conway, Can. J. Chem. 37, 613 (1959).

40a. J.O'M. Bockris and M. Habib, Electrochim. Acta, 22, 41 (1977).

41. J.A. Pople, Proc. Roy. Soc., London, A202, 323 (1950).

42. S. Trasatti, J. Electroanal. Chem., 54, 437 (1974).

43. S. Trasatti, J. Electroanal. Chem., 64, 128 (1975).

44. S. Trasatti, J. Electroanal. Chem., 65, 815 (1975).

45. S. Trasatti, J. Phys. Chem., 79, 2452 (1975).

46. A. K. Vijh, J. Phys. Chem., 78, 2240 (1974).

47. E. Dutkiewicz, J.D. Garnish and R. Parsons, J. Electroanal. Chem., 16, 505 (1968).

48. B.E. Conway, J. Electroanal. Chem., in press (1981).

48a. R. Parsons, Trans. Faraday Soc., 51, 1518 (1955).

49. T.N. Anderson and J.O'M. Bockris, Electrochim. Acta, 9, 347 (1964).

50. B.E. Conway, Ion Hydration in Chemistry and Biophysics, Elsevier Publ. Co., Amsterdam, (1980).

51. G.J. Hills and R. Payne, Trans. Faraday Soc., 61, 316; 326 (1965).

52. G.J. Hills and S. Hsieh, J. Electroanal. Chem., 58, 289 (1975).

53. J.A. Harrison, J.E. B. Randles and K.J. Schiffrin, J. Electr-
    roanal. Chem., 25, 197 (1970); 48, 359 (1973).
54. B.E. Conway and L.G.M. Gordon, J. Phys. Chem., 73, 3609 (1969).
55. I. L. Cooper and J.A. Harrison, J. Electroanal. Chem., 66,
    85 (1975).
56. J.O.'M. Bockris and M. Habib, J. Electroanal. Chem., 65,
    473 (1975); J. Electrochem. Soc., 24, 123 (1976).
57. R. Parsons, J. Electroanal. Chem., 109, 369 (1980).
58. R.M. Latanision, Surface Effects in Crystal Plasticity: A
    General Overview, in "Surface Effects in Crystal Plasticity",
    Eds. R.M. Latanision and J.F. Fourie, NATO Adv. Study
    Institute, Series E.
59. J.W. Schultze, Electrochim. Acta, 21, 327 (1976).
60. H. Wroblowa and K. Müller, J. Phys. Chem., 73, 3528 (1969).
61. D.J. Barclay, J. Electroanal. Chem., 19, 318 (1968); 28
    443 (1970).
62. R. Parsons, J.E. Electroanal. Chem., 8, 93 (1964).
63. S. Gottesfeld and B.E. Conway, J. Chem. Soc., Faraday Trans.
    I, 70, 1973 (1974).
64. B.E. Conway, H. Angerstein-Kozlowska and W.B.A. Sharp, Z.
    phys. Chem., N.F., 98, 61 (1975).
65. R.H. Sachs and D.L. Dexter, J. Appl. Phys., 21, 1304 (1950).
66. P. H. Cutler and J.C. Davis, Surface Sci., 1, 194 (1964).
67. R.G. Pearson, J. Amer. Chem. Soc., 85, 3533, (1963);
    J. Chem. Educ., 45, 501; 643 (1968).
68. G. Klopman, J. Amer. Chem. Soc., 90, 223 (1968).
69. B.E. Conway, H. Angerstein-Kozlowska, B. Barnett and
    J. Mozota, J. Electroanal. Chem., 100, 417 (1980).
70. K. J. Vetter and J.W. Schultze, J. Electroanal. Chem., 34,
    131; 141 (1972); 53, 67 (1974); Ber. Bunsenges. phys. Chem.,
    76, 920-927 (1972).
71. W. Lorenz, Z.phys. Chem., N.F., 232, 176 (1966); 244, 65 (1970).
72. R.F. Lane and A.T. Hubbard, J. PHys. Chem., 79, 808 (1975).
73. R.M. Latanision and R.W. Staehle, Acta Met., 17, 307 (1969).
74. R.M. Latanision, H. Opperhauser and A.R.C. Westwood, Scripta
    Metallurgica, 12, 475 (1978).
75. F.D. Shchukin, V.I. Savenko, L.A. Kuchanova and P.A. Rehbinder
    Dokl., Akad. Nauk., S.S.S.R. 200 (2), 400 (1971).
76. A.R.C. Westwood, D.L. Goldheim and R.G. Lye, Phil. Mag., 16,
    505, (1967); 17, 951 (1968).
77. A. R.C. Westwood, Molecular Processes on Solid Surfaces,
    McGraw-Hill, N.Y. p. 591 (1969).
78. M. H. MacMillan, R.D. Huntington and A.R.C. Westwood, Phil . .
    Mag., 28, 293, (1973).
79. R.M. Latanision, N.H. MacMillan and R.G. Lye, Corrosion Sci.,
    13, 387 (1973).
80. H. Angerstein-Kozlowska, B.E. Conway and W.B.A. Sharp,
    J. Electroanal. Chem., 43, 9 (1973).
81. S.M. Ahmed, J. Phys. Chem., 73, 3544 (1969).

82. S.M. Ahmed and D. Maksimov, Can. J. Chem., <u>46</u>, 3841 (1968).
83. S.M. Ahmed and D. Maksimov, J. Coll. Interface. Sci., <u>29</u>,
     <u>97</u>, (1969).
84. S.M. Ahmed, in "Oxides and Oxide-Films" (Ed. J.W. Diggle)
     Vol. 1, Marcel Dekker Inc., N.Y. p.319-517.
85. J.W. Schultze and M.A. Habib, J. Applied Electrochem., <u>9</u>,
     255 (1979).
86. B.E. Conway and D.M. Novak, J. Electroanal. Chem., <u>99</u>,
     133, (1979).
87. E. McCafferty and A.C. Zettlemoyer, Disc. Faraday Soc.,
     <u>52</u> , 239 (1971); J. Coll. Interface. Sci., <u>34</u>, 452 (1970);
     Croat. Chem. Acta, <u>45</u>, 173 (1973).
88. D.R. Bassett, E.A. Boucher and A.C. Zettlemoyer, J. Coll.
     Interface Sci., <u>34</u>, 436 (1970).

DISCUSSION

Comment by R.M. Latanision:

I think it may be useful to point out, given the spirit of this
meeting, that the same mathematical formalism which is used to
describe potential and charge distribution at the solid-electrolyte
interface (namely, the Poisson-Boltzmann distribution law) was
developed in the late 1800's and describes, as well, the potential
and charge distribution at a semiconductor surface. The charge
carriers are different (electrons and holes in the case of semiconductors and anions and cations in the electrolyte) but the
formalism is the same. Here, is a good example of one discipline
(solid state physicists) learning or benefiting from the experience
of another (electrochemists).

Reply:

Yes, this is an interesting historical point. The Poisson-
Boltzmann equation was also used three-dimensionally by Debye and
Hückel (1923) in their theory of ion distribution and interactions
in electrolyte solutions. The procedure provides a convenient
averaging method for evaluating the effects and properties of a
distribution of charges such as nowadays can be evaluated by
advanced statistical-mechanical approaches including numerical
methods such as the Monte-Carlo technique.

Comment by D.J. Duquette:

Most PZC and adsorption measurements have been made in renewable
Hg surfaces. Adsorption effects on solid metals and alloys may be
very important to embrittlement processes, however. Do you see any
experimental or theoretical advances which will allow an accurate
model of adsorption on solid metallic surfaces?

Reply:

Such studies are much needed. Some progress is being made in
this area at relatively "polarizable" metal interfaces such as those
of Au and Ag in work at the CNRS Institute at Bellevue, France (R.
Parsons, et al.). A fair degree of surface specific behavior is
found on the (111), (110), and (100) surfaces of these metals for
ion adsorption as deduced by double layer capacitance measurements.
Similarly, work of the Russian School (Kazarimov, Balashova, et
al.) using radio-labelled ions, e.g., SO_4^{2-}, Na^+, have given useful
information on the specificity and potential dependence of adsorption
of these ions at Pt and Au.

Quantitive thermodynamic understanding of the behavior of the
interfaces of solid metals with aqueous solutions, camparable with
that obtainable with renewable Hg interfaces is, however, lacking
at the moment. This is important area requiring continuing work.

Comment by J.O'M. Bockris:

In Professor Conway's model of the interface, he suggests that
water is attached to anions and contributes to double layer properties. However, water which has lost its own translatory motion does
not exist much for anions bigger than F^-, so that the consistency
of the model with data on solvation numbers seems missing. Coordinated water is, of course, present but apparently is not sufficiently
attracted to the ions to orient in the short time in which it remains
in nearest neighbor position with the ion.

Reply:

The purpose of the calculations mentioned (J. Electroanal.
Chem. 1981, in press) is to establish two things: a) the extent
to which ion co-sphere overlap interaction energies influence the
adsorption isotherm for specifically adsorbed anions at Hg; and b)
whether the orientation of solvent dipoles at an electrode interface
can be considered determined by the surface charge density q_M only
or whether the q_M-dependent surface excess of adsorbed anions,
attracted into the inner layer, will itself influence the
solvent dipole orientation.

The first question (a) involves the co-sphere energy U_{co} which
can be calculated for various ions and is determined by the coordination number (4~6 for Cl^-, Br^-) and the time-average water dipole
interactions at the ions. U_{co} is a substantial fraction of the total
solvation energy of the ion and the overlap contribution at q_i~20-
$30\mu C$ cm^{-2} is appreciable, >kT, and ion specific.

The second question can be answered by considering how the

dielectric displacement εE caused by the ion in the inner region of the double layer, $(ze/(r_i+r_{H_2O})^2$, and by the surface charge q_M per cm^2 on the metal compare at a distance say of 1 H_2O radius from the ion and from the electrode surface. For Cl^- ion in the double layer, εE caused by the anion is equal to that caused by q_M when q_M is ca.15.1μC cm^{-2}, while for Br^- ion equal q_M would be ca.12.3μC cm^{-2}.

Whatever may be the phenonmenological situation regarding solvation numbers of anions determined by various methods, it is evident that the presence of anions in the inner layer region of the double layer can cause a dielectric polarization influence on the interphasial water comparable with that of the surface charge. q_M on the metal for quite normal, experimentally accessible surface charges is in the range 10~25μC cm^{-2}. Hence, the orienting influence of such ions must be considered in a complementary way to that of the charged surface itself.

The calculations show that a satisfactory model of the double layer must take into account both the q_M-dependent surface concentration of ions and their influence on solvent orientation in a coupled way to that determined also by q_M.

CONCERNING ADSORBED AND ABSORBED HYDROGEN ON AND IN FERROUS METALS

H. J. Flitt and J. O'M. Bockris

School of physical Sciences, Flinders University of South
Australia, Bedford Park, SA, Australia and Department of
Chemistry, Texas A&M University, College Station, Texas

ABSTRACT

Fe specimens containing 0.2% carbon were chosen to minimize D_H
and maximize c_H. 0.4 M H_2SO_4 solutions were used. Additives were
napthanitrile, napthalene, benzonitrile, H_2S, $Na_2H_2PO_4$.

Measurements: (1) Steady state hydrogen evolution as a function
of potential in the absence and presence of additives; (2) Constant
current transients with the initial potential in the hydrogen
evolution region and a region positive to the corrosion potential;
(3) Laser desorption of hydrogen-a Neodynium-Yag laser was used to
introduce micro-cavities into the metal after its exposure to hydrogen
from solution at various fugacities. The H was measured to hydrogen
from solution at various fugacities. The H was measured in a
quadrupole mass spectrometer, and analyses of the pH/time transient
yields the H concentration (c_H) in the region of metal struck by the
laser.

The results obtained are the Tafel parameters: the fraction (Θ_η)
of the surface occupied by H at various η's (±50%), and the
concentration of absorbed hydrogen (±10%). Tafel parameters in the
addative show b > 2RT/F, (i_o) additive < (i_o) pure. d In Θ/d_η
4RT/F. The c_H is linear with η in the non-additive cases. Extrapolation to $\eta = 0$ gives $c_H \simeq$ that determined by other approaches.
The equilibrium constant, c_H/Θ, depends parabolically upon potential;
inflection occurs at 400 mV NHE. Phosphate and benzonitrile reduced
Θ most. $\Theta \simeq$ 1-10% referred to the whole surface. All additives
increase c_H and lower Θ_η in some potential regions.

The HER mechanism is slow discharge followed by coupled atomic combination. The rate determing step takes place on the regions between carbide and ferritic phases. Results are consistent with Tempkin kinetics. The Tafel slope in the presence of additives is a result of this, together with the effect of the variation of Θ_{org} with potential. Θ/η relations are interpreted on the basis of the hypothesis that $G_{\Theta=\Theta}^{ads} = G_{\Theta=\Theta}^{ads} + r \ln \Theta$. The fugacity/$\eta$ is consistent with Temkin kinetics in the pure case; complexities intervene, in that with additives, the relation of c_H to Θ as a function of η shows fair numerical agreement with experiment, discrepancies of which are due to trapping. What determines a substance to be a promoter or inhibitor of H in the metal is analyzed.

INTRODUCTION

Theoretical relations exist between the surface concentration of hydrogen on a metal in solution, the hydrogen overpotential maintained there, and the tendency of the metal to embrittle.[1] Experiments which show that changes in the interior structure of metals occur with increase of the overpotential in the negative direction have been made.[2]

The measurement of the surface concentration of hydrogen adsorbed on iron is difficult, because the methods used to do this induce a competing dissolution of the substrate, which prevents the anodic current during a sweep being identified with that for faradoic dissolution of the adsorbed hydrogen. Correspondingly, absorbed hydrogen in a metal is usually obtained by vaccum extraction, and this technique is cumbersome. These difficulties of measurement have hitherto prevented the experimental investigation of the relation of adsorbed to absorbed hydrogen as a function of electrode potential and various solution consitiuents. In the present paper, a novel approach has been used to obtain the absorbed hydrogen at various potentials of ingress,[3] and the constant current double pulse method has been used to obtain values, albeit involving several crude approximations uncertainly estimated, of the surface concentration of hydrogen at potentials corresponding to those of the absorption measurements.

EXPERIMENTAL

Systems Studied

The choice of ferrous metal for the study was made, because it was wished to have a high hydrogen solubility, but a low diffusion coefficient to diminish the hydrogen excaping from the specimen. The optimization of such tendencies is obtained by dissolving carbon at a low concentration in iron. No other elements were added to reduce complication in interpretation. The carbon content chosen was 0.2%. The carbon decreases the diffusion coefficient of hydrogen five times compared with that in pure iron.[4]

The electrolyte was 0.05 M sulfuric acid, made up with doubly distilled water.

Measurements were made in the presence of certain additives-- napthanitrile, benzonitrile, napthaline, hydrogen sulfide and sodium dihydrogen phosphate.

The nitriles are known to be corrosion inhibitors.[†] Napthaline has been classed as a promoter.[5] Hydrogen sulfide was chosen because of its well-known effects in promoting corrosion, and sodium phosphate because of the promotion of hydrogen embrittlement in phosphating.[5]

Program of Experiments

The steel, of composition stated above, was made up by the BHP Company. The original specimen was cut into numerous sections, 1 cm^2 each. The specimens were spot welded to an iron wire and sealed to a glass rod. The electrochemical cell is shown in Figure 1. The general approach was to polarize the electrode to a given current density, then apply, at each current density, the constant current double pulse, using the basic anodic current density of 50 mA/cm^2 with a series of cathodic current densities. After these measurements were made, a steady state Tafel line was measured.

The electrodes were polarized for several hours at the same series of constant cathodic current as those used for the measurement of adsorbed hydrogen.

The Tafel lines were measured, using 15 current densities, over the range 1.0-10 mA/cm^{-2}. The anodic transients were carried out for six cathodic current densities covering this range. Laser desorption measurements were made after exposure of specimens at the same current densities for measurements of the internal concentration.

After a specimen had been subject to long-term cathodic polar- ization, it was removed, washed in doubly distilled water and dried, before immersing in liquid nitrogen, where the specimen was maintained until it was transferred to the vacuum system in which the internal hydrogen was to be determined.

---

†The words, inhibitor and promoter, usually refer to the functions of these materials in corrosion. Their relation to the increase or decrease of hydrogen in the metal is discussed below.

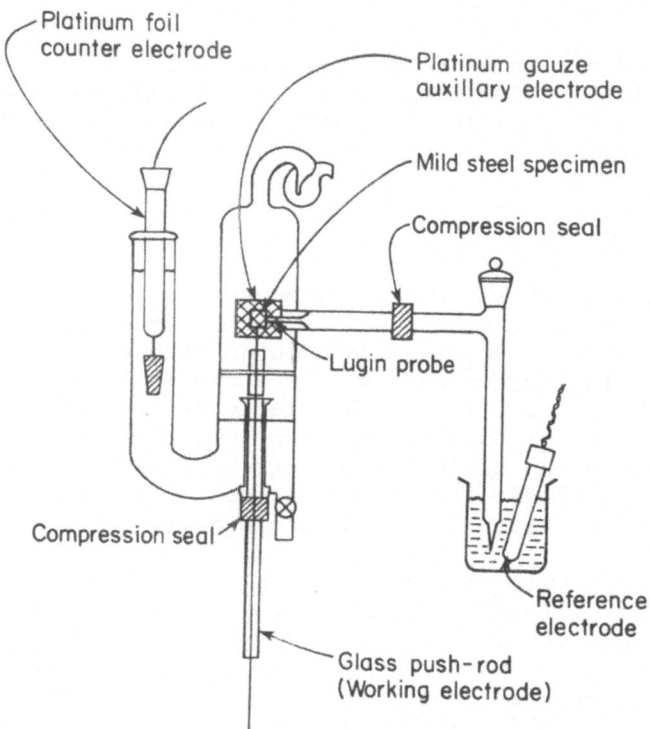

Figure 1   Conventional electrochemical cell.

The specimen was screwed to a push rod, which had been immersed in liquid nitrogen before being attached to the push rod. The cold rod was then transferred to the vacuum system (Fig. 2).

The vacuum system contained a quadrupole mass spectrometer and and ion gage. The principle was to vaporize a section of the metal in the specimen by making spherical cavities of radius around 20 microns in diameter. The hydrogen emitting from each of such cavities was around 10^{-12} moles.

Experiments in the vacuum system were carried out at a base pressure of some 8.10^{-8} Torr. An ion pump was maintained working during the experiments, because degassing from the rest of the system was of comparable rate with that from the vaporization of hydrogen in the metal. The transient concentration of hydrogen was measured by means of the quadrupole mass spectrometer and ion gage. The total emitted gas was calculated from the pressure transients (see below).

After the specimens were inserted into the vacuum system, they were subjected to radiation from the defocussed low power laser beam. According to the technique applied by Levine, Ready and Bernal,[6] the defocussed beam releases only surface hydrogen, without making a cavity which releases the bulk hydrogen. This low powered beam was, then used to clean the surface by the use of a successive series of pulses, until the pressure peaks had become constant and small (red-adsorption from the degassing of the apparatus). The laser was now brought into focus, so that the power level was in a range sufficient to make craters. This procedure was repeated on several spots on the same specimen, and an average for a given specimen and current density at which it had been charged obtained.

We give detail of the two techniques involved below, including a typical table of direct laboratory notes.

The Constant Current Double Pluse Method Applied to Iron

The constant current double pulse method has been described in the literature.[7] It has been used by others, where it was applied to determine the hydrogen concentration on pure Fe.[8]

The method consists of taking two constant current/time transients. In the one, the starting potential is the anodic transient from the cathodic overpotential corresponding to the constant current density concerned. Thus, during this anodic transient, the anodic current is used to dissolve the hydrogen from the surface together with whatever iron dissolves in the relevant potential range.

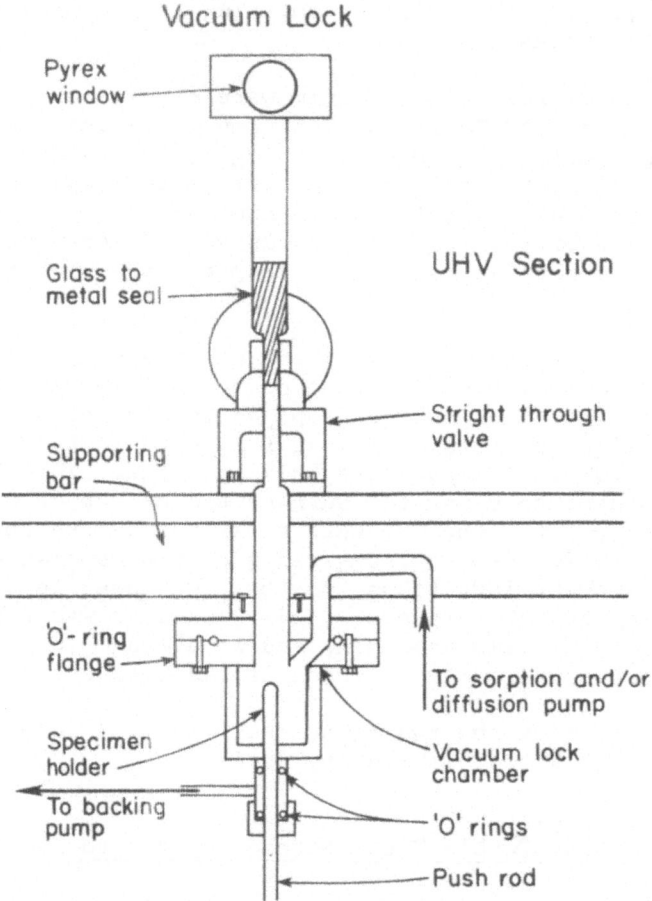

Figure 2   Entry of specimen of ion into vacuum chamber of
quadrupole mass spectrometer.

In the second transient, what is supposed to be done is that the
starting potential is that at which the hydrogen surface concentration
is lowered to zero. During this transient, only metal dissolution
should occur. By the use of a series of simple electrode kinetic
equations[7] involving iteration, it is possible to obtain a value for
surface coverage, Θ, of the hydrogen for any potential. This method
works well where $i_{o,H}/i_{o,Fe}$ is relatively large.[7] It is, however,
associated with a number of approximations. In the following, these
are analyzed to obtain the net uncertainty to which a measurement of
Θ_H on Fe is subject.

Iterative Procedure, $I_{an}\Theta_V$

In the constant current double pulsed method, one obtains a
function $(i_H - I_{an}\Theta_V)$, where i_H is the hydrogen dissolution current
density at a given V, and i_{an} is the corresponding current density
for iron dissolution at the V concerned.

The $(i_H - i_{an}\Theta_V)$ relation is found as a function of potential
from the two different types of constant current transients and
then replotted as a function of time with respect to the hydrogen
desorption transient. Iterations are made. In the first, $i_{an}\Theta$ is
taken as zero and zeroth approximation for the value of Θ_H,
corresponding to a given η, obtained. This Θ is then used to
replot the $(i_H - i_{an}\Theta)$ relation. In this second plot, values of i_H
are calculated by adding to the value of $i_H - i_{an}\Theta$, the zeroeth
approximate values of $i_{an}\Theta$ for various values of the overpotential
corresponding to the potential of measurement. This gives a second
plot of $i_H(\Theta) - t$. A second value of Θ is now obtained, and the
process is continued until (after some six iterations) Θ becomes
constant.

In Figure 3, the iterated value of Θ is shown to converge for
a certain value of V. The asymptotic value is that of Θ for the
overpotential from which the initial transient from the cathodic
region is made.

Error Due to Change of the Double Layer Capacitance with Potential

In the computations reported, the double layer capacitance of
the iron/solution interface has been assumed constant, both with
potential and with the presence of additives. In fact, the shape
of the C_{dl}/potential plot for iron in the solution used was found
to be <u>qualitatively</u> similar to that for mercury. However, the Tafel
lines cover the region of the potential of zero charge, for iron, so
that there is likely to be some variation of double layer capacitance
over the potential range of the measurement.

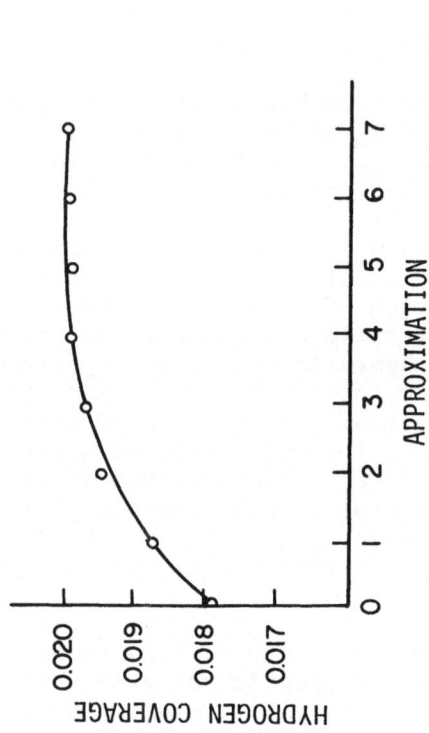

Figure 3   Example of iteration in obtaining hydrogen coverage
from measurements made at 0.5 mA/cm^{-2}.

Let it be assumed that the double layer capacitance doubles over the value measured from the initial slope of a transient ($25.6 \ \mu F/cm^{-2}$). Recalculating the results of the system without additives using the increased capacitance, one finds that the coverage predicted increases from 34.10^{-7} coulombs/cm^{-2} to 40.10^{-7}, i.e., by about 20%. Organic materials on the surface would be expected to lower the double layer capacity, and this might reach an extreme of 50%. With this assumption, a typical computation of Θ_H shows it is diminished by some 40%. Hence, +20% to -40% is the range of error which must be allowed for, on this ground, in respect to Θ_H.

Error Due to Residual Hydrogen on the Surface

The second anodic pulse, that from the less cathodic potential, is supposed to occur when no hydrogen is left on the surface. However, in the present method, the anodic pulse was made for potentials about 0.01 to 0.02 V positive to the corrosion potential. The residual hydrogen on iron at the corrosion potential has been estimated[9] to be 1%.

At the lower limit of the measurements, this quantity is the same order as that of hydrogen measured for less cathodic conditions. At the highest hydrogen coverage estimated by the present procedure (10%), the error would be 10%.

Diffusion of Hydrogen from the Bulk of the Metal

Diffusion of hydrogen from the interior would be given an overestimate of coverage. The equation used for calculation of the rate of hydrogen diffusion from the bulk is:[10]

$$i_p = c_H \left(\frac{D}{\pi t}\right)^{\frac{1}{2}} F \tag{1}$$

where i_p is the permeation current density. Trapped hydrogen can be considered immobilized during the time of the anodic pulse. Only lattice hydrogen is mobile enough to diffuse to the surface during the pulsing. The diffusion coefficient for lattice hydrogen is higher than the apparent value because of the effect of the traps. According to results of Newman and Shreir,[11] there is a factor of about 10 in the diffusion coefficient between the two types of hydrogen. Thus, the value for the lattice hydrogen was taken as 10 times greater than that used with the steels of the type we are dealing with. According to the data of Marandet, the value for 0.2% carbon steel is 1.10^{-7} cm^{-2} sec^{-1}. A value of 1.10^{-6} cm^{-2} sec^{-1} was therefore conservatively taken for calculating diffusion to the surface.

Figure 4 shows a plot of current density corresponding to hydrogen diffusing out of the metal onto the surface. About 9×10^{-8} coulombs per cm^{-2} occurs in the typical time of a pulse. The error due to this cause would correspond to an overestimate of some 10%.

Error Due to Absorbed Organics

Thompson and Hackerman[12] found no electrochemical reactions observable during their use of benzonitriles as inhibitors of the corrosion of steel. Similar arguments have been presumed to apply to napthonitrile.

According to Bockris, Green and Swinkels,[13] napthaline undergoes no electrochemical reaction until a range positive to the hydrogen reversible potential. In the present work, the potentials are never positive to this potential.

In respect to the presence of phosphate on the surface of the electrode for the solution concentrations used, phosphating (with monolayer formation) needs concentrations of several moles in solution. Hence, concentration of phosphate significant in respect to the whole surface is not likely to be adsorbed on the surface of the iron in the present work. In confirmation of this, the Tafel slope did not change in the presence of phosphate compared with that observed in its absence.

In respect to H_2S as a promoter, according to Kawashima, et al.,[14] the reaction of dissolved H_2S with iron does not involve a net electrochemical reaction, i.e., it is:

$$Fe + H_2S \rightarrow Fe^{++} + S^{=} + H_2 \tag{2}$$

Analogously, Kohkhar, et al.,[15] found no anodic oxidation of H_2S on iron.

Conclusion Concerning Errors in the Double Pulse Method

The largest error is that due to the residual hydrogen on the surface. Some of the errors involve both under and some overestimates. The net result is that the surface, Θ, is uncertain to about ±50% over most potential ranges, but up to an overestimate of + 100% at low cathodic overpotentials.

LASER DESORPTION FROM THE SPECIMEN
Cleaning of the Surface

After the specimen has been inserted into the mass spectrometer,

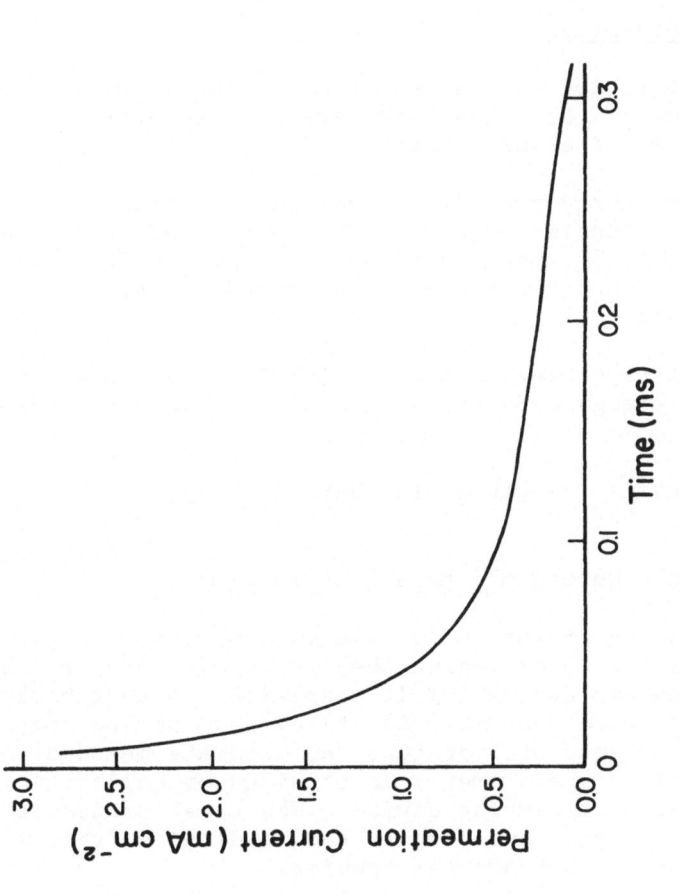

Figure 4   Plot of current density corresponding to hydrogen diffusing out of the metal on to the surface as a function of time.

it has to be cleaned (see above). The defocussed laser beam was
allowed to strike the same area in the specimen six times, the
resulting desorbed gases being monitored by the quadrupole mass
spectrometer and recorded on the storage oscilloscope. When the
evolved gas is constant, the surface is ready for use as a specimen
for the recording of the exiting hydrogen concentration from the
cavity (formed by the focussed laser beam).

Metal/Laser Vaporization

The model used for the calculations of the metal vaporization
was that of Ready.[16] His equations were used to calculate the flux
needed for the vaporization of iron.

The minimum energy needed to cause vaporization in iron is 150
joules per square centimeter for a 300 microsecond pulse. In the
experiments, values of energy used varied from 100-500 joules per
square centimeter, and the crater thus formed varied from 20-100
microns in radius.

The time of vaporization varies with flux, as shown in Figure 5.
The laser pulse energy density as a function of depth is shown in
Figure 6.

Further details are given elsewhere.

Calibration of the Quadrupole Mass Spectrometer

The calibration of the quadrupole mass spectrometer to measure
pressure was made by conditioning the system with hydrogen by several
times introducing and evacuating it therewith, so that background
gettering effects were reduced. All filaments had low emission
currents to the pumping of hydrogen. An accurate measurement of
pressure with a mass spectrometer is attained by calibration against
an absolute pressure-measuring device. The usual procedure is to use
a precalibrated ion gage and compare the readings of this with the
mass spectrometer for a known gas species.

In the present measurements, an ion gage was used for the
hydrogen measurements, applying the manufacturer's factor for the
conversion of his readings, based upon nitrogen (instead of hydrogen).
Because of this, to get hydrogen readings, the ion gage measurements
were multiplied by a conversion factor of 0.4. To compare the mass
spectrometer pressure readings with the ion gage, the dynamic response
principle is used. The hydrogen is introduced gradually at a rate

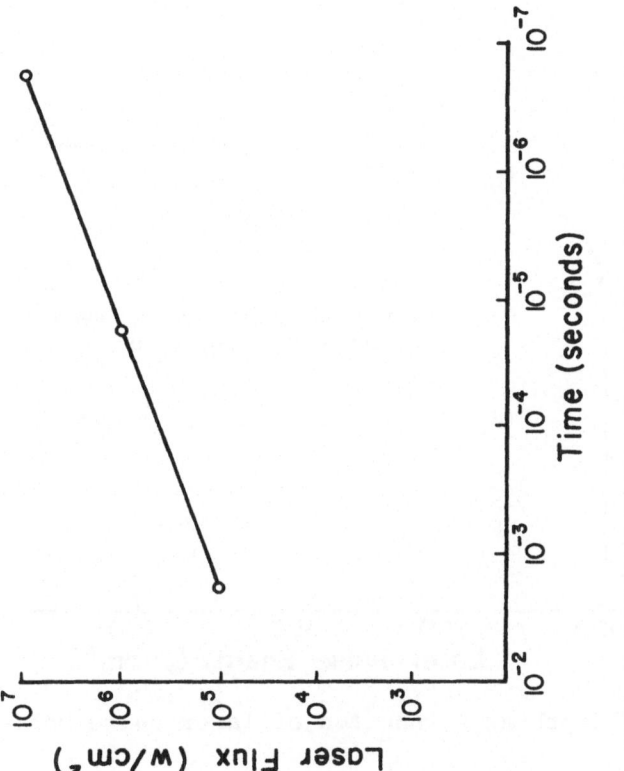

Figure 5 Laser flux and the time of vaporization.

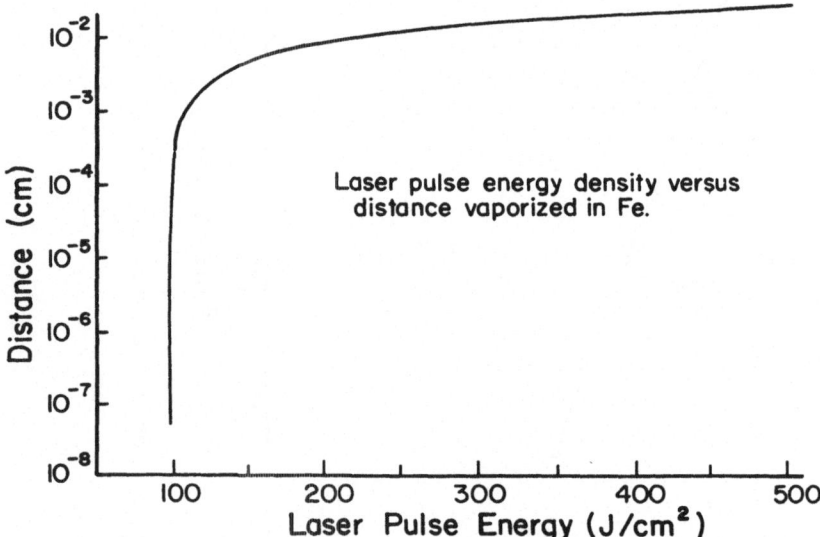

Figure 6   Depth as a function of laser pulse energy.

sufficient to keep the pressure between 10^{-7} and 10^{-5} Torr. The peak height in the mass spectrograph was measured, and three entirely independent series of measurements are shown together in the long line in Figure 7.

These findings show variations from linearity observed at the top and bottom of these lines, but they can be rationalized. Thus, at the upper end of the line, electronic saturation of the electrometer occurs, due to high output. At the lower part, the background pressure becomes significant in comparison with hydrogen pressure. CO and N_2 were the main other gases present.

The justification of these calibration experiments is the work of Winters, Denison and Bills.[17] Further, however, McNobbs[18] was able to get a calibration with reactive gaseous oxygen in the pressure range 10^{-4} to 10^{-6} Torr and obtain a linear relation, which was also justified theoretically (Fig. 7).

Analysis of the Laser-Induced Pressure Transients

The data obtained in the method herein described are in the form of pressure/time transients. However, the system is being pumped out all the time, so that no simple relation exists between the pressure observed at a given moment and the amount of hydrogen released by the vaporization experiment. An analysis must be made of the dynamics of the vacuum system.

A model has been derived which can predict the corrections in the data to compensate for the system pumping, its outgassing and the pressure differences between the region near the specimen and the region near the ion gage. The block diagram of the model included in Figure 9 refers to P_1 as pressure at a time, T, after a vaporization in the quadrupole chamber of volume, V_1. P_2 is the pressure in the sample chamber at a given time. At this time, T_1, in the sample chamber, S, is the conductance (flow rate) of the connecting tube between the quadrupole and the sample chamber, and is in litres sec^{-1}; S_p and Q_2 are, respectively, the gas flow rate between the sample chamber and the ion pump in litres sec^{-1} and litres atmospheres sec^{-1}, respectively (Fig. 8).

Such a system can be represented by three boundary equations:

$$Q_1 = (P_2 - P_1)S = V_1 \frac{dp}{dt} \qquad \qquad \qquad \qquad \qquad (3)$$

$$Q_2 + Q_1 = P_2 S_p + (P_2 - P_1)S = -V_2 \frac{dP_2}{dt} \qquad (4)$$

$$P_{11}V_1 + P_{21} V_2 = P_2 V_2 + \int_o^t P_2 S_p \, dt \qquad \qquad (5)$$

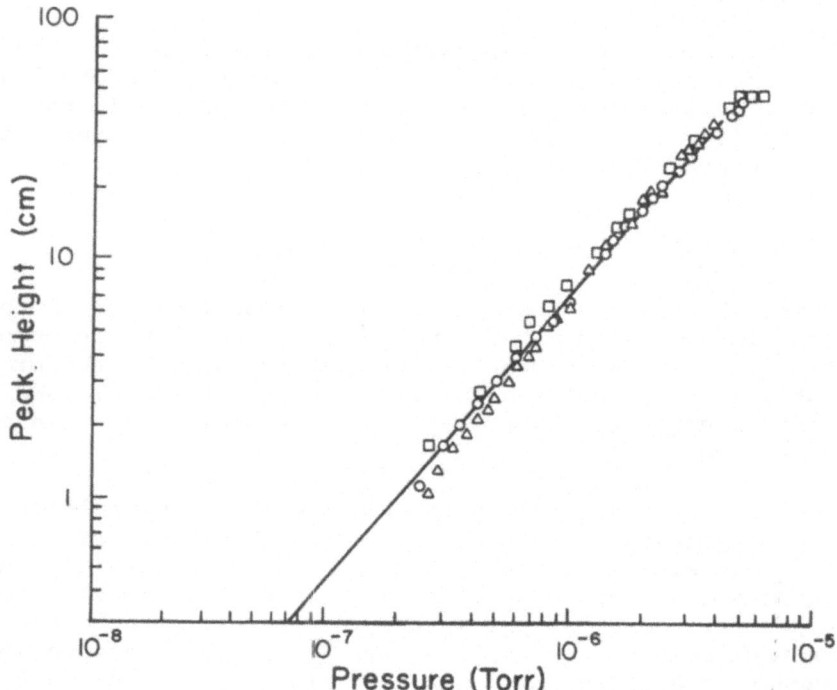

Figure 7   Peak height as a function of pressure in the quadrupole
           mass spectrometer.

The solution to these equations is given elsewhere.[19] The solution for the hydrogen pressure in the ion gage compartment comes out to be:

$$P_2 = C_1 \exp \lambda_1 t_2 + C_2 \exp \lambda_2 t_2 \tag{6}$$

C_1, C_2, λ_1 and λ_2 are constants of the apparatus. The solution of these equations can be found in the case of zero pumping speed. It is necessary to know this, because the solution represents the amount released in unit time into the system by laser vaporization.

We have ignored the effect of outgassing. A correction can be made for this by solving the equation:

$$Q = Q_{outgassing} - \frac{dP}{dt} V \tag{7}$$

Where Q is the quantity of the gas removed from the system in litre atmospheres sec^{-1}.

The application of this to the system involves the block diagram in Figure 8, and a typical transient response is given in Figure 9. The application of the model to a real system involves the block diagram (Fig. 10), and a typical transient response is given in Figure 10.

The voltage traced from the oscilloscope is converted to pressure by means of the calibration curve (see above).

Using the decay of the transient, the pump speed is obtained by differentiating Equation (6) and the maximum on the transient of P against t, it is possible to calculate P_2(initial). After obtaining this parameter, a graph of pressure versus time (Fig. 11) can be re-constructed to give the gas released into the system independent of pumping (cf., the line for S_p = 0 in Fig. 13).

The lower curve in Figure 11 represents the theoretically constructed transient of the oscilloscope trace, and the crosses on it represent the experimental pressures during a decay transient from Figure 10. Agreement is excellent and confirms the analysis and calibration curves.

The pressure obtained from the middle straight line in Figure 10 (that which has compensated for pumping and outgassing) can be con-verted to the amount of gas released from the crater.

By dividing the volume of the crater (dimensions obtained by optical microscopy) and the data from the amount of gas released into the vacuum system, the solubility of the gas in that part of the metal subject to vaporization (and previously subject to evolution

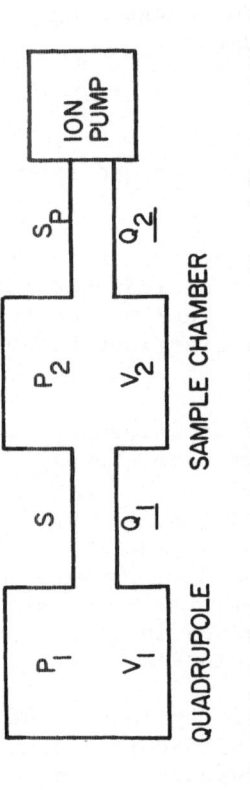

Figure 8   Model for pressure transients: P and V are pressure and
volume in the indicated compartments; S and Q are gas
flow rate and ion pump rate in litres per second and
litres atmosphere per second, respectively.

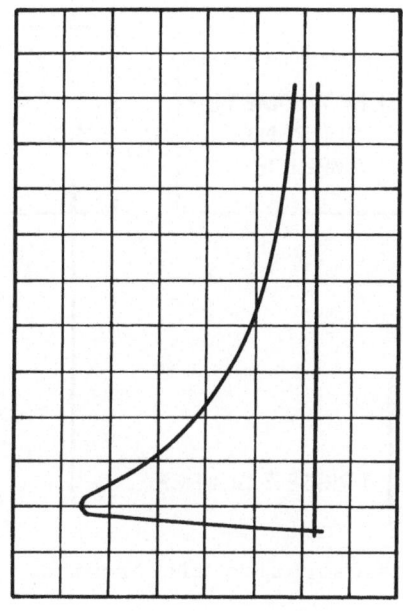

**Transient response of pressure versus time CRO
trace of photomultiplier output**

Figure 9   Typical transient response of pressure vs. time on the
CRO trace of a photo multiplier unit.   X = 0.2 sec/div;
Y = 0.5 V/div.

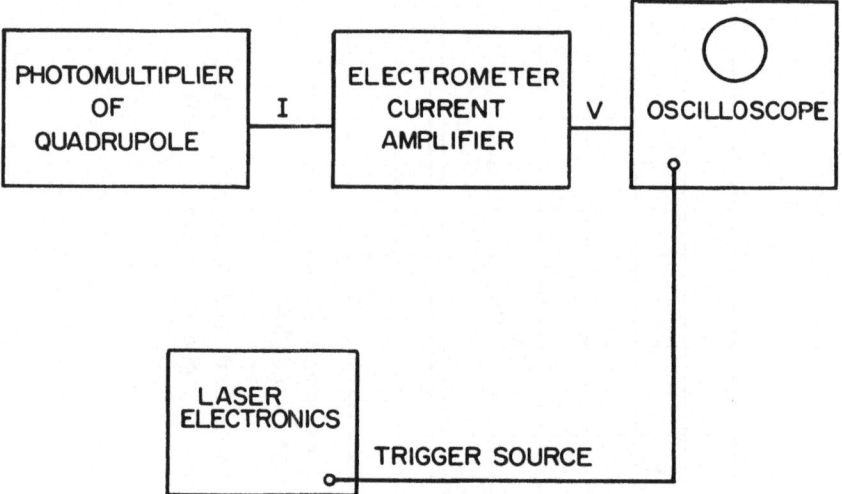

Figure 10   The desorption electronics.

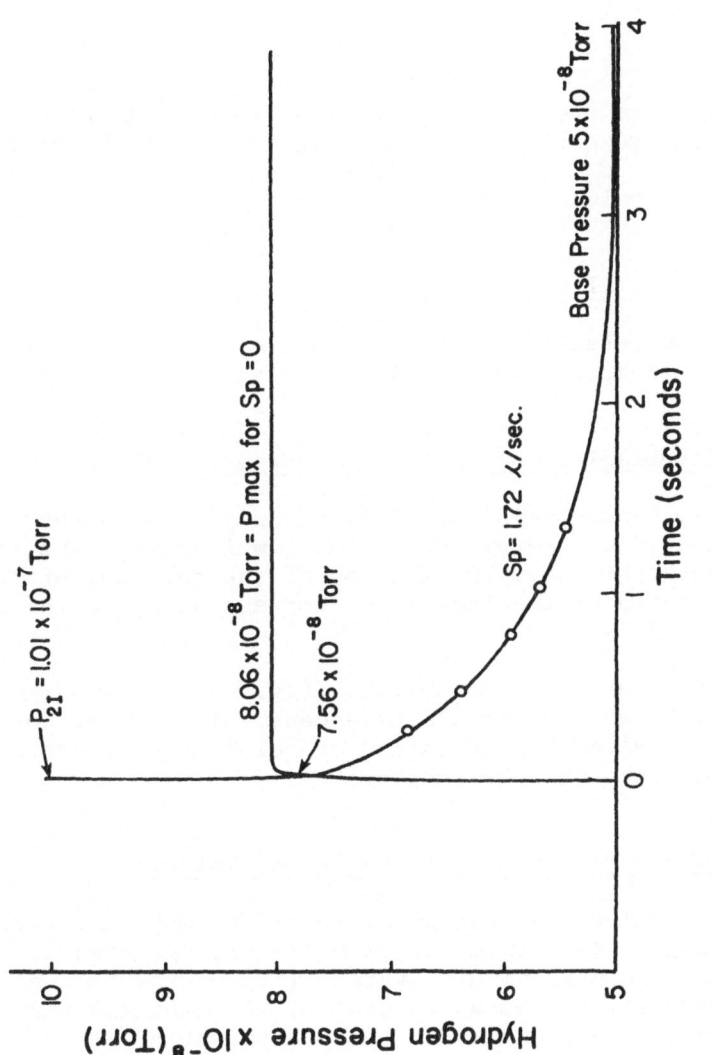

Figure 11 Result of transient analysis (circles are bonds from Figure 9).

of hydrogen at a certain overpotential or hydrogen fugacity) is
obtained. Details of typical calculations are given in Tables 2
and 3.

EXPERIMENTAL RESULTS

Tafel Lines

       Tafel lines are shown in Figure 12. The non-additive case has
the highest exchange current density. The values of b are equal
to or higher than 120 mV, except for the non-additive case near the
p.z.c., where the values were 0.08.

Coverage/Overpotential Relationship

       Plots of the relation of log Θ to η are shown in Figure 13.
There are some linear portions having gradients of > 4RT/F.

Relationship Interfacial Solubility to Overpotential

       The phrase, "interfacial solubility", may be used for the
solubility measured by the present method, which gives the concentration of hydrogen within some 100 microns of the surface, after its
exposure to hydrogen at the fugacities corresponding to the overpotentials applied.

       The log of interfacial solubility (Fig. 14) is linear with
the overpotential for the non-additive case. In systems containing
nitriles and napthaline, the relation of solubility to overpotential
becomes more complex (cf., Discussion).

Hydrogen Solubility Measurements: Degree of Validity

       The interfacial concentration measurements may be tested. If
the values of log C_Θ as a function of η are extrapolated to zero
η, they reprsent a condition of reversibility corresponding to
measurements made for the metal in contact with the gas phase by
non-electrochemical methods, where the overpotential is zero and
the hydrogen gas pressure is 1 atmosphere. Hence, a test of the
results is to find whether the values extrapolated to zero η fit
hydrogen concentration measurements at 1 atmosphere in other systems
at 25°C. To obtain such data, it is important to compare with
corresponding carbon concentrations, and in Figure 16, a plot is
given for five steels. Four points come from the work of others
(recalculated to 25°C). The fifth point, extrapolated from the data

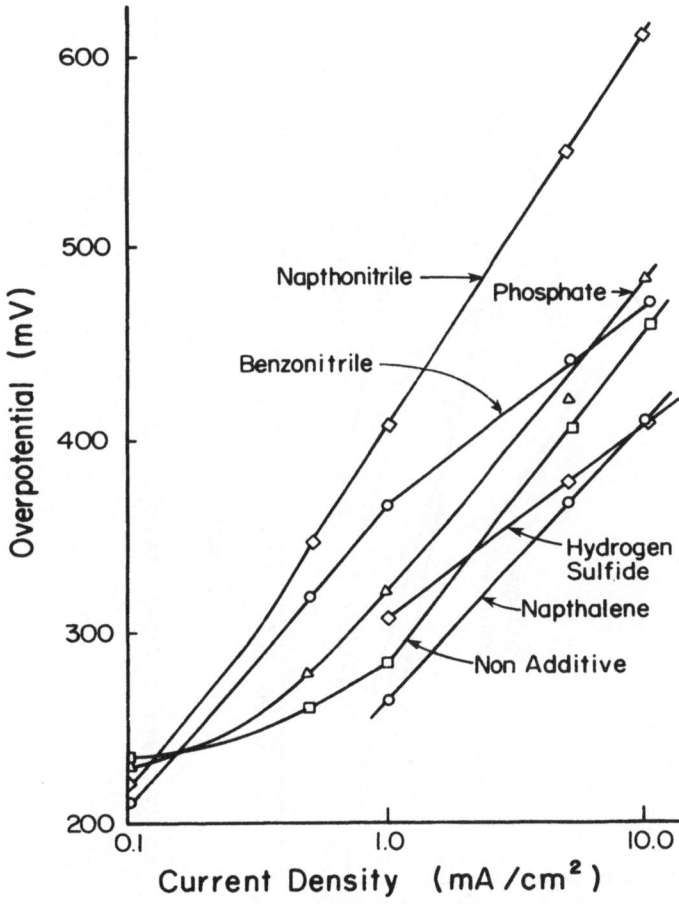

Figure 12   Tafel lines for steady state hydrogen evolution as a
function of current density.

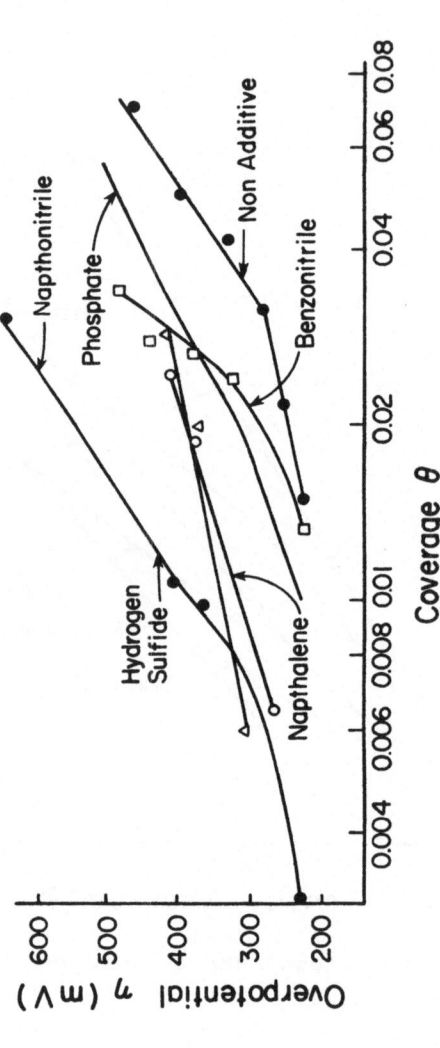

Figure 13   The hydrogen coverage/overpotential relation in various systems.

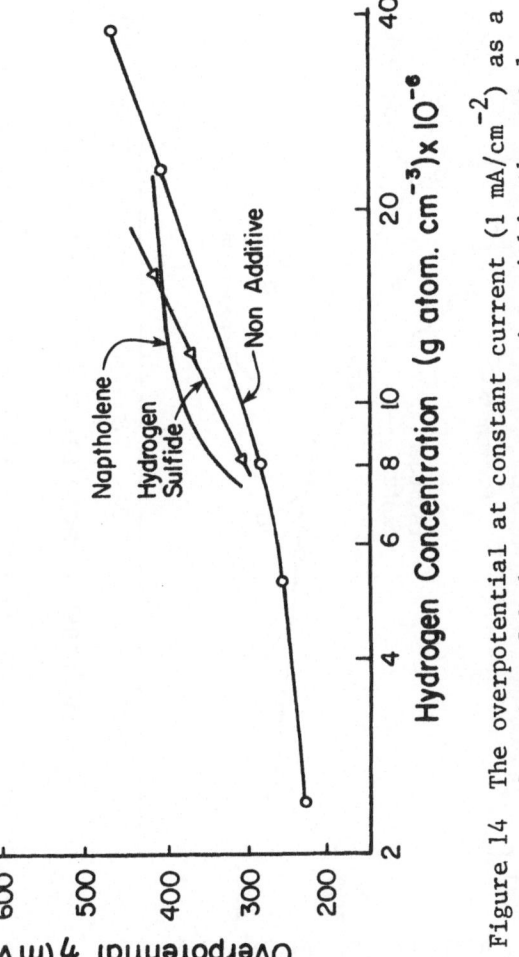

Figure 14 The overpotential at constant current (1 mA/cm^{-2}) as a function of hydrogen concentration within the metal.

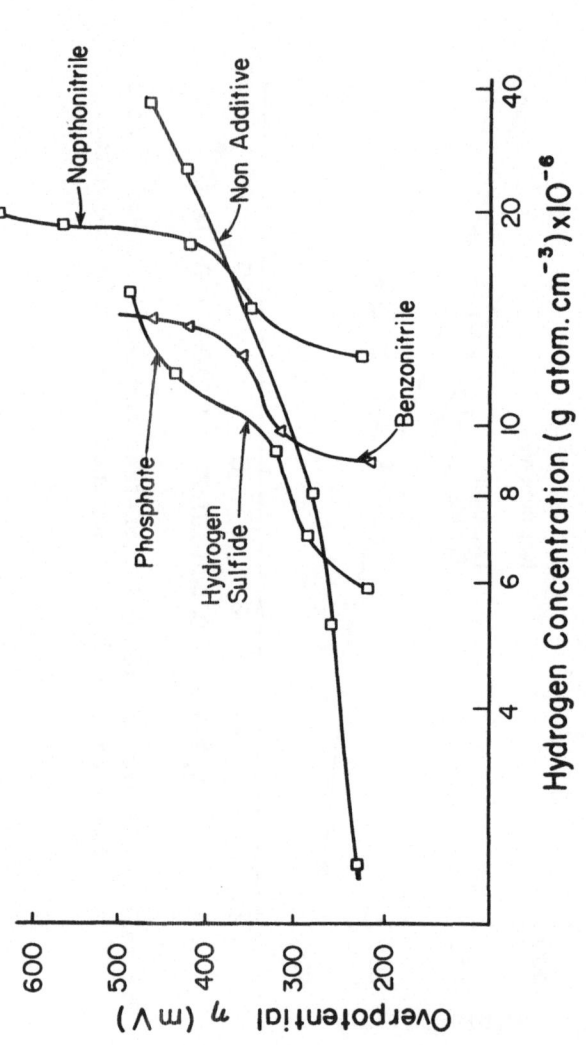

Figure 15   The hydrogen interfacial solubility as a function of overpotential.

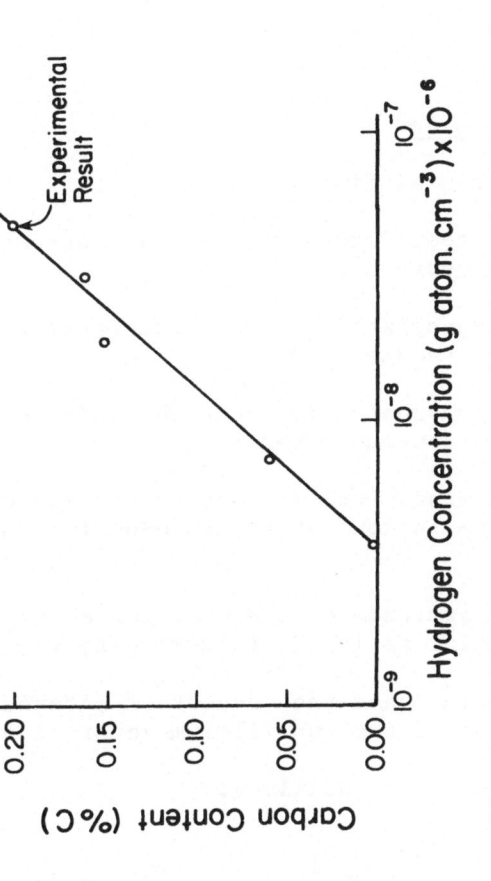

Figure 16 The hydrogen concentration (solubility) at 1 atmosphere and room temperature as a function of carbon content.

of Figure 14 to zero overpotential, falls upon the lines. Considering
that both an extrapolation to zero overpotential was made and an
extrapolation to 25°C, the degree of agreement is very good.

Equilibrium Constant for Adsorbed and Absorbed Hydrogen

Inflections occur in all systems in the region of 400 mV NHE
(Fig. 17).

DISCUSSION

Summary of Results Obtained

Among the experimental findings are:

1) All the additives, independed of their effect upon corrosion,
decrease the exchange current density.

2) Nitriles and Phosphates increase the Tafel slopes, while
sulfide and napthaline reduce them.

3) Nitriles and phosphates increase the $d\eta/d \ln \Theta$ slopes, but
sulfide and napthaline decrease them.

4) Phosphate and benzonitrile reduce the value of Θ to the
greatest extent. There is no consistent reduction due to napthaline
and hydrogen sulfide.

5) The order of magnitude of the hydrogen surface coverage during
its change with potential is 1-10%, referring to the whole surface.

6) All additives increase the content of internal hydrogen. There
is a parabolic relation of the equilibrium constant with overpotential.

7) Θ_η is lowered by all additives.

Mechanism of Hydrogen Evolution in Acid Solution on Steels

In an analysis of the mechanism of the hydrogen evolution
reaction, Conway and Bockris[20] pointed out that the situation in
respect to mechanism with iron is intermediate between the two basic
mechanisms of proton discharge onto empty sites, and proton discharge
onto adsorbed hydrogen. Bockris and Koch[21] measured the ratio of
hydrogen evolution in pure D_2O and pure aqueous solutions and found
evidence of favor of a coupled-discharge reaction.

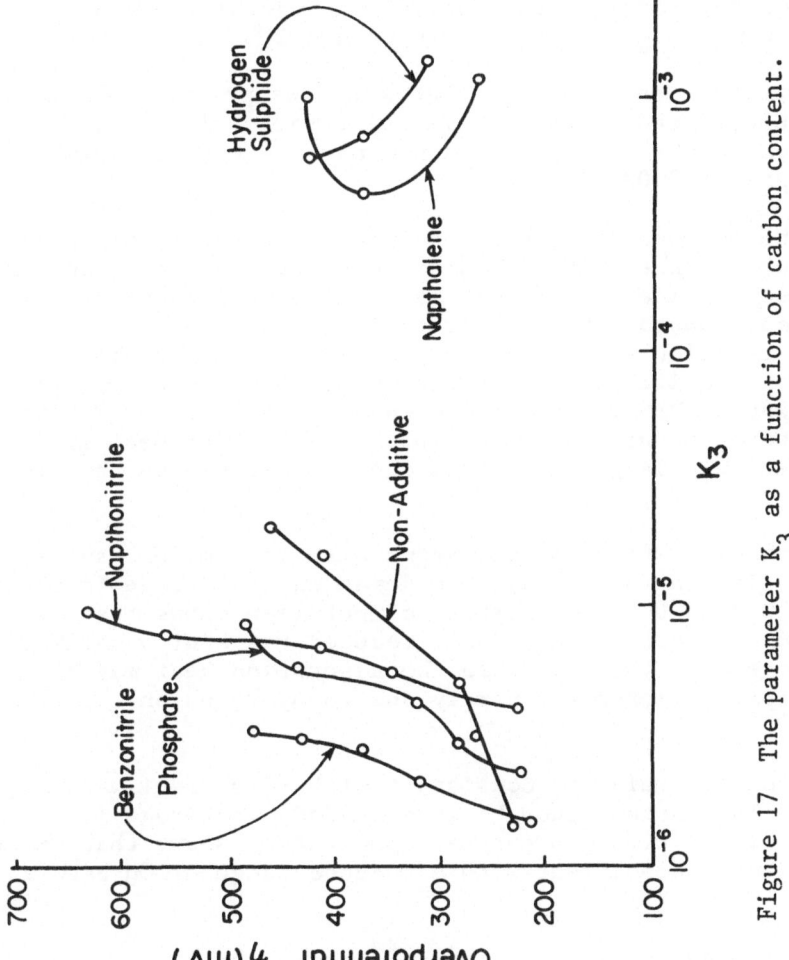

Figure 17 The parameter K_3 as a function of carbon content.

Applying these thoughts to the present mechanism, there is at first a contradiction, for they lead to a Tafel slope of 2RT/F, whereas the measured one is c. 3RT/F.

The previous statements were made for pure iron, and we are here considering a surface which is a ferrite having carbide inclusions. Such a significance of the values of Θ, which have been measured to be in the range of 1-10%. This applies to the overall coverage. However, on the parts of the surface at which the hydrogen evolution reaction occurs, Θ_H may be greater than 0.1.

An attitude analogous to that which we present here was earlier taken by Cleary and Greene.[22] In their view, the hydrogen evolution reaction on steels took place on the carbide phase. They reported Tafel slopes of 0.02 to 0.03.

In the phenomenology presented here, higher slopes have to be interpreted. Hence, the mechanism suggested by Cleary and Greene[22] may be less applicable. Stress is laid upon the heterogeneity of the surface. Suppose that the rate determining step of the hydrogen evolution reaction takes place on the interfacial regions between the carbide and rerritic phases in the model shown in Figure 18. The Θ measured would then represent a composite, due to the Θ which would be on the heterogeneous phases, where the adsorption heat may be assumed to be relatively large and partly due to hydrogen on the ferrite phases.

A mechanism which is consistent with this model was suggested by Thomas.[23] This author thought it demanded a coverage in the active region of 0.2 to 0.8, but further computation shows that the mechanism is only exactly consistent with a reduced range of Θ which would be on the heterogeneous phases, where the adsorption heat may be assumed to be relatively large and partly due to hydrogen on the ferrite phases.

A mechanism which is consistent with this model was suggested by Thomas.[23] This author thought it demanded a coverage in the active region of 0.2 to 0.8, but further computation shows that the mechanism is only exactly consistent with a reduced range of Θ_H between 0.4 and 0.7.

Under these conditions, Thomas[23] shows that:

$$\frac{\partial \eta}{\partial \log i} = 0.180 \tag{8}$$

This Tafel slope fits the results on our surfaces in the absence of additives. The mechanism would be a couple-discharge mechanism occuring principally at interfacial regions on the heterogeneous surface.

Mechanism of Hydrogen Evolution in the Presence of Additives

The Tafel slope in the presence of napthanitrile was 0.21 (Fig. 12). A simple isotherm for the reversible absorption of an organic molecule on the electrode surface is:

$$\frac{\Theta_{org}}{1 - \Theta_{org}} = \frac{C_{org}}{C_w} \exp \frac{-(\Delta G^\circ_{org} - \Delta G^\circ_w)}{RT} \tag{9}$$

where Θ_{org} is the coverage with the organic, and C_{org} and C_w are the solution concentrations of the organic and the water. ΔG°_{org} and ΔG_w are the standard free energies of adsorption for the organic and water, respectively.

Trivial manipulation shows that, with $K_{org} C_{org} \gg K_w C_w$:

$$1 - \Theta_{org} = \frac{C_w}{C_{org}} \exp \frac{-(G^\circ_w - G^\circ_{org})}{RT} \tag{10}$$

Except near the p.z.c., this shows a potential dependence of:

$$1 - \Theta_{org} = \frac{C_w}{C_{org}} \exp \left(\frac{\Delta G^\circ_{org}}{RT}\right) \exp \left(\frac{-\Delta G^c_w}{RT} + \frac{\mu_w \chi}{KT}\right) \tag{11}$$

This expression can be used in the coupled-discharge mechanism, which can be expressed as:

$$Fk_1 c_{H}+ (1 - \Theta_{org} - \Theta_H) \exp \left(\frac{-\beta VF}{RT}\right) \tag{12}$$

where Θ_H is the hydrogen.

We may absorb Θ_H in the constant, because, on the basis of the Tempkin conditions to be assumed below, the variation of Θ_H with potential has a negligible effect on the i, compared with the exponential effect of the potential terms.

$$i_1 = Fk_1 c_H + \frac{C_w}{C_{org}} \exp \left(\frac{\Delta G^\circ_{org}}{RT}\right) \exp \left(\frac{-\Delta G^c_w}{RT} + \frac{\mu \chi}{kT}\right) \exp \left(\frac{-\beta VF}{FT}\right) \tag{13}$$

A similar dependence on potential exists for ΔG°_{org} as for ΔG°_w. Thus, with $\Delta \mu = \mu_{org} - \mu_w$ and $\chi = V_1/\delta$, where V_1 is the electrode potential on the rational scale:

$$i_1 = Fk_1 c_H + \frac{C_w}{C_{org}} \exp\left(\frac{\Delta G°_{org}}{RT}\right) \exp\left(\frac{-\Delta G°_w}{RT}\right) \exp\left(\frac{-\beta VF}{RT} + \frac{\Delta \mu V}{RT\delta}\right) \exp$$

$$\left(\frac{-V_{pzc}}{\delta RT} \Delta \mu N\right) \tag{14}$$

The Thomas analysis for a coupled-discharge mechanism results in an introduction of 2/3 β in place of β, and Δμ/δF also becomes multiplied by 2.3. Then:

$$\frac{\partial \eta}{\partial \log i} = 2.303 \ RT/F \ \left(2/3 \ \beta + \frac{2\Delta \mu N}{3\delta F}\right)^{-1} \tag{15}$$

Utilizing $\delta = 5.10^{-10}$ m, the Tafel slope of 210 mV corresponds to $\Delta \mu = -6.3 \ 10{-30}$ mC. But $\mu_{org} = -6.3 \ 10^{-3}$ mC $+ \mu_w$. As $\mu_w = 6.2 \ 10^{-30}$, $\mu_{org} \approx 0$. The benzonitrile is, therefore, adsorbed in a flat position.

Similar models can be presented for the other additives examined, and they give reasonable accounts of values of the Tafel slopes in their linear portions.

The Coverage/Overpotential Relations

In the coupled-discharge reaction for hydrogen under Tempkin conditions, the expression for the discharge and combination rates gives (from Ref. 23):

$$\frac{\Theta^2}{1-\Theta} = \frac{k_1}{k_2} c_H + \exp\left(\frac{-\beta VF}{RT}\right) \exp\left[\frac{-34f(\Theta)}{2RT}\right] \tag{16}$$

where $f(\Theta)$ is the function which relates the heat of adsorption of increasing Θ to that at zero Θ.

To account for the observed value (Fig. 13) of $\partial \eta/\partial \log \Theta = 0.518$ (cf., the o.240 expected on a Langmuir treatment), it is assumed (cf., Adamson[24]) that:

$$\Delta G_\Theta^{ads} = \Delta G_{\Theta=\Theta}^{ads} + r \ln \Theta \tag{17}$$

Thence, with $(\Theta^2/1 - \Theta \simeq \Theta/1 - \Theta)$:

$$\frac{\partial V}{\partial \log \Theta} = 2.303 \ \left(3/2 \ r/\beta B + \frac{+2RT}{\beta F}\right) \tag{18}$$

This equation is consistent with results of Figure 17, if r = 3.8 kJ mole^{-1}. Gileadi and Conway[25] found 2.5 kJ mole^{-1} for this kind of coefficient.

The approximation, ln $\Theta^2/1 - \Theta = 0$, is consistent with a Tempkin model. At $\Theta \simeq 0.7$ in the local reactive sites, the approximation is exact. Thus, $\partial\eta/\partial$ ln Θ has moved from the Langmuir value of 240 mV to one of 500 to 600 mV because of heterogeneity effects.

Utilizing a treatment similar to that described above, one obtains a series of r values for Θ/η relations with the other additives, which are correspondingly consistent with the range given by Gileadi and Conway.[25]

Overpotential/Fugacity Relationships

The relationship between the fugacity of hydrogen and the corresponding overpotential depends upon the mechanism of the reaction. For a coupled-discharge/chemical combination, the relationship under Langmuir conditions is:

$$F_{H_2} = 10^{1.5} \exp \frac{(-\eta F)}{2RT} \tag{19}$$

When the adsorption energy changes in a Tempkin manner, one obtains the relation:

$$F(\Theta) = RT/r \ln (Af_{H_2}) \tag{20}$$

where $f(\Theta)$ is the functional dependence of the adsorption energy upon coverage.

A corresponding equation to this (Gileadi and Conway[25]) is:

$$C = \exp \left(\frac{-\beta\eta F}{RT}\right) \exp \left[\frac{-3\gamma r f(\Theta)}{RT}\right] \tag{21}$$

where C is a constant. Using this equation for $f(\Theta)$, one obtains, the r = β = $\frac{1}{2}$, from Equation (20) and (21):

$$f_{H_2} = C/A^{-2/3} \exp \left(\frac{-\eta F}{2RT}\right) \tag{22}$$

A test of this equation can be made by relating the fugacity to the internal concentration of hydrogen, determined in the present experiments. Utilizing Sievert's Law applied to the Langmuir case ($f_{H_2} = 10^{1.5} \exp -\eta F/2RT$) and the Tempkin case ($f_{H_2} = C/A^{-2/3} \exp$

$-\eta F/3RT$), one obtains:

$$\frac{\partial \eta}{\partial \lg c_H} = 4 \cdot 2.303 \ RT/F \quad \text{(Langmuir)} \tag{23}$$

$$\frac{\partial \eta}{\partial \lg c_H} = 6 \cdot 2.303 \ RT/F \quad \text{(Tempkin)} \tag{24}$$

Figure 14 shows the dependence of the internal hydrogen concentration on overpotential. The slope is 0.3, in the higher overpotential regions, which agrees with the Tempkin case.

Hill and Johnson[26] preferred the law:

$$f_{H_2} = k \ c_H^n \tag{25}$$

for steel. We have taken $n = 1/2$ and evaluated the comparison between theory and experiment. If n is more than 1/2, the $\partial \eta / \partial \lg c_H$ slope would decrease, in agreement with our observations of 300 V per decade, compared with the theoretical 254. This implies that there is more hydrogen than that which is interstitial, as expected in the presence of traps.

Adsorbed and Absorbed Hydrogen

Hence, from Equation (21) and Sievert's Law:

$$c_H = B^{1/2} \exp \left(\frac{-\eta F}{4RT}\right) \ e^{-3/4 \ r/RT} \tag{26}$$

one may plot the parameter:

$$K_2 = \frac{c_H}{\Theta^{-3/4 \ r/RT}} \tag{27}$$

against η.

If K_2 is divided by Θ, one obtains:

$$K_3 = \frac{c_H}{\Theta^{-3/4 \ r/RT}} \tag{28}$$

When K_3 is plotted as a function of η, one obtains Figure 17. In this plot, promoters (hydrogen ingress sense) are divided from inhibitors.

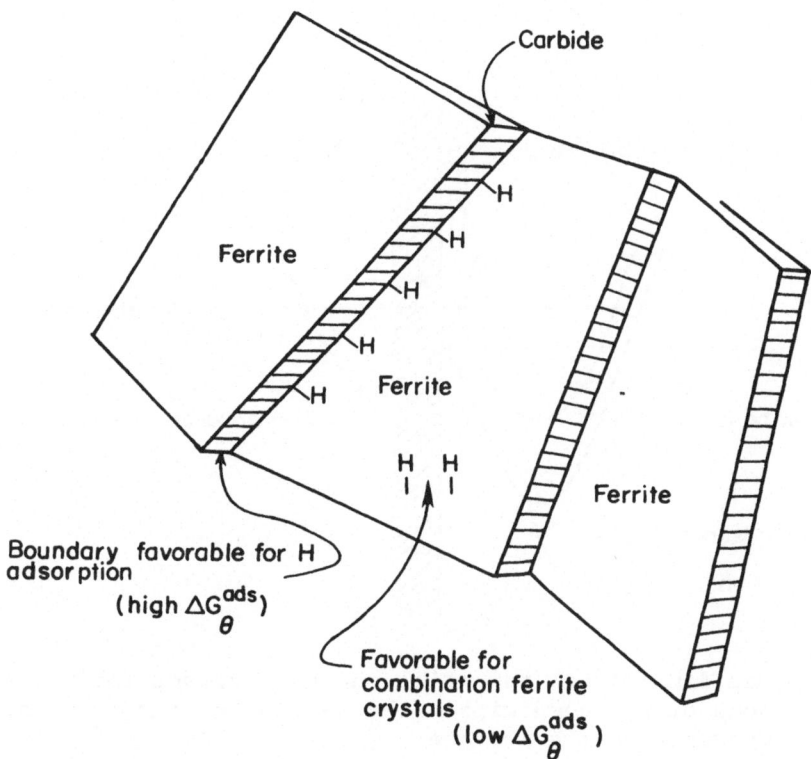

Figure 18   The rate determining step assumed for hydrogen evolution.
It takes place on the interfacial regions between carbide
and ferritic phases.

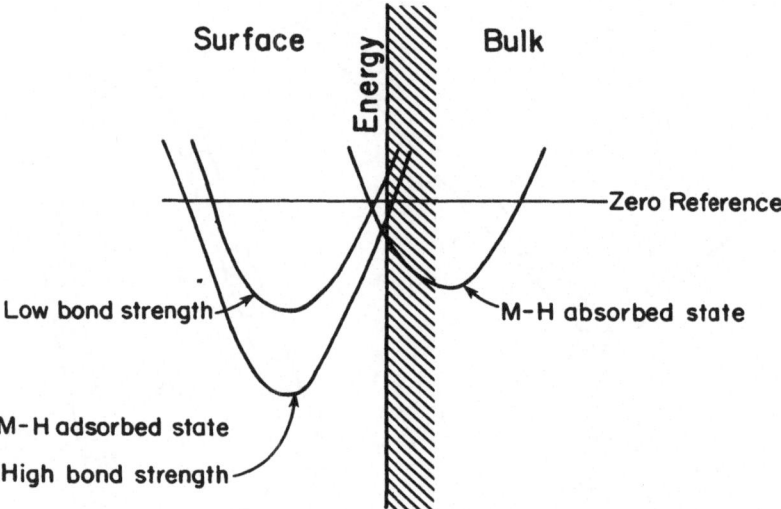

Figure 19   One–dimensional potential energy diagrams for the adsorbed
            bond of hydrogen within the metals and the adsorbed
            hydrogen on the surface.

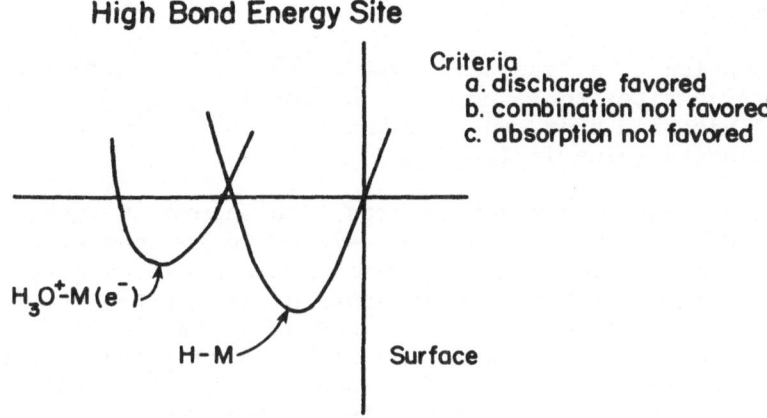

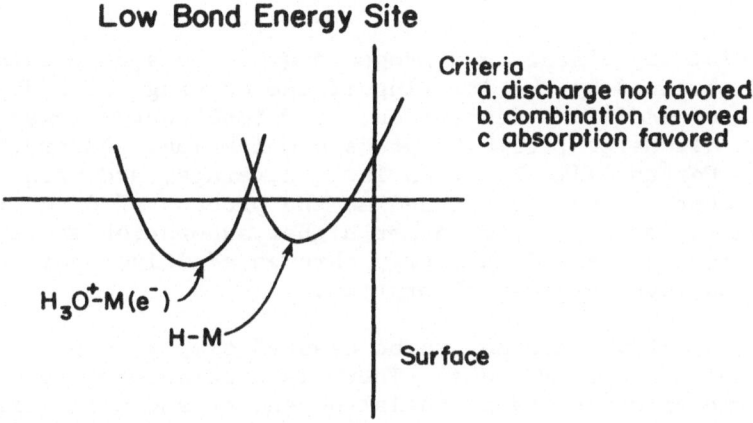

Figure 20   The relative positions of the adsorbed hydrogen bond
(one-dimensional potential energy) brought into
proton/oxygen bond H_3O^+: high bond energy site and
low bond energy site.

Promotion and Inhibition of Hydrogen Ingress into Steel

The paramter which most importantly affects hydrogen coverage on steel is the bond strength of the adsorbed atoms. A low bond energy favors adsorption (Fig. 19). Thus a good promoter of hydrogen ingress is a substance which lowers the bond energy of hydrogen to the electrode surface. On this basis, hydrogen sulfide and napthalene gave promotion.

Adsorption in a "specific" way on the steel surface by additives causes active sites to be covered, giving a lowering of the bond energy of adsorbed hydrogen. Langmuir behavior returns. The exchange current density value is lowered.

However, when the discharge reaction occurs at high energy sites, and the hydrogen coverage does not extend to low energy sites, adsorption of hydrogen is not favored; heterogeneity gives an inhibitory effect on hydrogen absorption. A spill-over of hydrogen to low energy sites would give an increased combination rate of hydrogen atoms. Adsorption on the high energy sites by the additives of napthaline and hydrogen sulfide has two effects: the adsorption bond energy of steel for hydrogen is reduced (exchange current density lowered) and hydrogen coverage is reduced. Figure 20 illustrates this.

An inhibitory effect on hydrogen ingress on a heterogeneous surface is obtained by the lowering of the coverage of hydrogen as a result of additives adsorption. H-M bond energy lowering reduces the discharge rate but speeds combination. Adsorption decreases. Correspondingly, with lower coverage, hydrogen on low adsorption energy sites is reduced as the process of discharge and combination may occur at the former higher adsorption energy sites which have been decreased in energy through additives adsorption and are thus less favorable for absorption.

Hence, additives are not to be classed simply as promoters and inhibitors of hydrogen ingress. There is a balance of several effects, some of which are potential-dependent and will compromise together to give a definite promotion or inhibition.

These effects are tabulated in Table 1.

ACKNOWLEDGEMENTS

The authors wish to acknowledge support from the Flinders University of South Australia and from the Robert A. Welch Foundation at Texas A&M University.

Thanks are due to Lynn McCartney Murphy of Texas A&M University

TABLE 1.   ADSORPTION SITES AND THE MECHANISM OF PROMOTION
AND INHIBITION OF HYDROGEN INTO IRON.

SYSTEM	HIGH ENERGY SITES	LOW ENERGY SITES
Non-additive	are present; medium coverage; favor discharge; do not favor absorption.	are present; low coverage; favor combination; favor absorption.
Promoter	are not present, being covered by additive.	are present; low coverage; favor discharge and combination; favor absorption.

Inhibitor	are present with reduced
	  energy;
	reduced coverage as compared
	  to non-additive;
	favor dischage and combina-
	  tion
	absorption not readily
	  favored.
		are present;
		reduced coverage as
		  compared to non-
		  additive;
		favor combination.

TABLE 2. SURFACE TRANSIENT DATA (0.1 N H_2SO_4).

CATHODIC CURRENT DENSITY	0.1 mA	0.5 mA	1 mA	5 mA	10 mA
Electrometer Voltage for Transient Maximum	0.8 V	1.0 V	1.6 V	2.1 V	0.8 V
Electrometer Voltage for the Base Pressure	0.6 V	0.6 V	0.6 V	0.6 V	0.6 V
Comparison Curve (Voltage to Pressure)					
Electrometer Setting			← 10^{-7} A →		
Pumping Speed of Ion Pump (Sp)	2.3 $1s^{-1}$	0.87 $1s^{-1}$	0.94 $1s^{-1}$	0.60 $1s^{-1}$	0.6 $1s^{-1}$
Surface Transient Pressure Maxima (P_{max})	2.9 x 10^{-7} Torr	3.55 x 10^{-7} Torr	5.08 x 10^{-7} Torr	6.26 x 10^{-7} Torr	2.9 x 10^{-7} Torr
Base Pressure (P_{1I})	2.4 x 10^{-7} Torr	2.4 x 10^{-7} Torr	2.4 x 10^{-7} Torr	2.4 x 10^{-7} Torr	2,4 x 10^{-7} Torr
Surface Transient Peak Time for Maximum (t_{max})	3.2 x 10^{-2} sec	5.14 x 10^{-2} sec	5.67 x 10^{-2} sec	7.16 x 10^{-2} sec	5.3 x 10^{-2} sec
Surface Pressure Pump Speed (zero)	3.09 x 10^{-7} Torr	3.69 x 10^{-7} Torr	5.32 x 10^{-7} Torr	6.81 x 10^{-7} Torr	3.02 x 10^{-7} Torr
Pressure initially in Specimen Chamber (P_{2I})	3.55 x 10^{-7} Torr	4.55 x 10^{-7} Torr	7.27 x 10^{-7} Torr	9.75 x 10^{-7} Torr	3.43 x 10^{-7} Torr
Net Surface Pressure Change (ΔP)	0.69 x 10^{-7} Torr	1.29 x 10^{-7} Torr	2.92 x 10^{-7} Torr	4.4 x 10^{-7} Torr	0.62 x 10^{-7} Torr

TABLE 3. BULK AND SURFACE TRANSIENT DATA (0.1 N H_2SO_4).

CATHODIC CURRENT DENSITY	0.1 mA.cm^{-2}	0.5 mA.cm^{-2}	1.0 mA.cm^{-2}	5.0 mA.cm^{-2}	10 mA.cm^{-2}
Electrometer Voltage for Surface Plus Bulk Transient Maximum	1.0 V	1.6 V	2.6 V	4.6 V	4.1 V
Electrometer Voltage for the Base Pressure	0.6 V	0.6 V	0.6 V	0.6 V	0.6 V
Comparison Curve (Voltage to Pressure)					
Electrometer Setting		←——————— 10^{-7} ———————→			
Pumping Speed of Ion Pump (Sp)	2.3 1s^{-1}	0.87 1s^{-1}	0.94 1s^{-1}	0.60 1s^{-1}	0.60 1s^{-1}
Bulk Plus Surface Transient Height Pressure	3.55 x 10^{-7} Torr	5.08 x 10^{-7} Torr	7.4 x 10^{-7} Torr	1.14 x 10^{-6} Torr	1.04 x 10^{-6} Torr
Base Pressure (P_{1I})	2.4 x 10^{-7} Torr	2.4 x 10^{-7} Torr	2.4 x 10^{-7} Torr	2.4 x 10^{-7} Torr	2.4 x 10^{-7} Torr
Bulk Plus Surface Transient Peak Time for Maxima (T_{max})	2.9 x 10^{-2} sec	5.77 x 10^{-2} sec	6.0 x 10^{-2} sec	7.5 x 10^{-2} sec	7.49 x 10^{-2} sec
Bulk and Surface Pressure for Pump Speed (zero)	3.83 x 10^{-7} Torr	5.3 x 10^{-7} Torr	7.8 x 10^{-7} Torr	1.26 x 10^{-6} Torr	1.15 x 10^{-6} Torr
Pressure nitially in Specimen Chamber after Pulse (P_{2I})	2.79 x 10^{-7} Torr	7.24 x 10^{-7} Torr	1.14 x 10^{-7} Torr	1.94 x 10^{-6} Torr	1.75 x 10^{-6} Torr

(continued)

TABLE 3.   (Continued)

CATHODIC CURRENT DENSITY	0.1 mA.cm^{-2}	0.5 mA.cm^{-2}	1.0 mA.cm^{-2}	5.0 mA.cm^{-2}	10 mA.cm^{-2}
Net Pressure Change	1.43 x 10^{-7} Torr	2.9 x 10^{-7} Torr	5.4 x 10^{-7} Torr	1.14 x 10^{-7} Torr	9.1 x 10^{-7} Torr
Bulk Hydrogen Pressure Change	0.74 x 10^{-7} g.atom	1.61 x 10^{-7} g.atom	2.48 x 10^{-7} g.atom	7.0 x 10^{-7} g.atom	8.48 x 10^{-7} g.atom
Hydrogen Concentration	2.4 x 10^{-6} g.atom.cm^{-3}	5.2 x 10^{-6} g.atom.cm^{-3}	8.0 x 10^{-6} g.atom.cm^{-3}	2.27 x 10^{-5} g.atom.cm^{-3}	2.84 x 10^{-5} g.atom.cm^{-3}

for her assistance in the preparation of this manuscript.

REFERENCES

1. J. O'M. Bockris and P. K. Subramanyan, Electrochim. Acta,
    16:2169 (1971).
2. P. Bastien and P. Azou, Proc. First World Metallurgical Cong.,
    ASM, Cleveland, Ohio (1951).
3. H. J. Flitt and J. O'M. Bockris, in course of publication, cf.,
    Flitt thesis, Flinders University of South Australia (1980).
4. W. Beck, J. O'M. Bockris, J. McBreen and L. Nanis, Proc. Roy.
    Soc., 290:191 (1966).
5. H. H. Uhlig, "Corrosion" (3rd edition), Wiley and Sons, New York
    (1972).
6. L. P. Levine, J. P. Ready and E. G. Bernal, J. Appl. Phys.,
    38:331 (1967).
7. W. Mehl, M. A. V. Devanathan and J. O'M. Bockris, Rev. Sci.
    Inst., 29:180 (1958).
8. C. D. Kim and B. E. Wilde, J. Electrochem. Soc., 118:202 (1971).
9. J. O'M. Bockris and H. Kita, J. Electrochem. Soc., 108:676 (1961).
10. J. O'M. Bockris, M. A. Genshaw and M. Fullenwider, Electrochim.
    Acta., 15:47 (1970).
11. J. F. Newman and L. L. Shreir, Corr. Sci., 11:25 (1971).
12. C. D. Thompson and N. Hackerman, Corr. Sci., 13:317 (1973).
13. J. O'M. Bockris, M. Green and D. A. J. Swinkels, J. Electroanal.
    Chem., 111:743 (1964).
14. A. Kawashima, K. Hashimoto and S. Shimodaira, Corrosion, 32:321
    (1977).
15. M. I. Khokhar, P. Fugassi and E. G. Haney, Corrosion, 27:91 (1971).
16. J. F. Ready, J. Appl. Phys., 36:462 (1965).
17. H. F. Winters, D. R. Denison and D. G. Bills, Rev. Sci. Inst.,
    33:520 (1962).
18. J. McNobbs, Vacuum, 23:391 (1973).
19. H. J. Flitt and J. O'M. Bockris, in course of publication, cf.,
    Flitt thesis, Flinders University of South Australia (1980).
20. B. E. Conway and J. O'M. Bockris, J. Chem. Phys., 26:352 (1957).
21. J. O'M. Bockris and D. F. A. Koch, J. Chem. Phys., 65:1941 (1961).
22. H. J. Cleary and N. D. Greene, Corr. Sci., 7:821 (1969).
23. J. G. N. Thomas, Trans. Faraday Soc., 57:1603 (1961).
24. A. W. Adamson, "Physical Chemistry of Surfaces," (3rd Edition)
    John Wiley and Sons, New York (1976).
25. E. Gileadi and B. E. Conway, "Modern Aspects of Electrochemistry,"
    Vol. 3, Ch. 5, eds. J. O'M. Bockris and B. E. Conway, Butter-
    worths, London (1964).
26. M. L. Hill and E. W. Johnson, Trans. Met. Soc., A.I.M.E., 215:717
    (1959).

DISCUSSION

Comment by A.W. Thompson:

I would just like to point out that although the correlation you
showed between hydrogen permeation and corrosion rate, as functions
of Ni-Fe composition, may be consistent with hydrogen control of
stress corrosion cracking, it certainly cannot prove such control,
and indeed could be interpreted in other ways equally well. I would
also hope it is obvious that the connection between corrosion rate
and stress corrosion cracking (e.g., crack velocity) is at best
tenuous, and in general has no necessary form whatever, so that its
use as a correlation is unconvincing and certainly an unfortunate
precedent.

Reply:

Phenomenon A, with a known rate-determining step x, behaves,
when presented with a certain stimulus, according to the phenomenology P Q R. Another phenomenon, B, with an unknown rate-determining
step, behaves, when presented with the same stimulus, with exactly
a P Q R reaction. Further, of various rate determining steps which
B may have, x is a recognized possibility. Given such circumstances,
one has an indication - a rather strong one - that the r.d.s. of B
is indeed x. Clearly, it is a suggestional situation. Models are
never completely proven. To consider an alternate possibility, a
critic clearly has to suggest one consistent with the facts. Thereafter, diagnostic experiments can be devised in the usual way.

ELECTROCHEMICALLY OBTAINED INFORMATION CONCERNING THE ATOMISTICS OF FRACTURE IN THE PRESENCE OF MOISTURE

J. O'M. Bockris

Department of Chemistry, Texas A&M University

College Station, Texas 77843

INTRODUCTION

There are three types of cracking: (1) pure cracking, e.g., glass; (2) metal-metal embrittlement, e.g., sodium-steel; (3) stress corrosion cracking, e.g., most cracks.

The complexity of fracture is at a minimum in case 1 and a maximum in case 3. Economically, the greatest significance is in the third type.[1,2]

THE DEVANATHAN-STACHURSKI-BOCKRIS (DSB) APPARATUS

The examination of the interaction of hydrogen with metals under corroding conditions was difficult because of estimation of hydrogen within the metal. These difficulties were diminished with the introduction of the DSB Apparatus.[3,4]

The apparatus contains a membrane which acts as a bi-electrode. It separates two solutions. The solution on the left involves the bi-electrode as a cathode, and contains in it a platinum anode. On the cathodic side of the membrane, most of the hydrogen is evolved and escapes to the atmosphere. A small amount of atomic hydrogen on the cathodic side escapes combination and enters the metal.

On the right hand side, the electrode functions as an anode in respect to a platinum cathode which is in the compartment on the right. It may be necessary[5] to add a thin layer of palladium to the surface of the electrode on the right hand side, because under the anodic condistions necessary for the dissolution of hydrogen, there may be an undesirable dissolution or passivation of the iron. It is essential that hydrogen dissolution be the only electrode reaction occurring from the right hand side of the membrane.

A further condition is that, in steady state, the dissolution conditions for the protons, their concentration at the right hand surface of the metal membrane is virtually zero.

When a cathodic current is made to pass across the membrane in contact with the solution, the reaction comes to steady state in time less than 0.01 seconds and hydrogen enters the membrane. The passage through the membrane is under diffusion control and a well-known mathematical analysis allows the profile of the hydrogen[6,7] to be timed to be obtained. Knowledge of the diffusion coefficient (obtainable from the switch-on transient) and the limiting current allows one to obtain the H concentration just inside the metal, c_o. This should not be confused with the solubility of hydrogen in a normal sense. The latter corresponds to the equilibrium solubility of hydrogen throughout a metal corresponding to the hydrogen pressure in the surrounding atmosphere, at a given temperature. The concentration c_o corresponds to the concentration of hydrogen at a certain fugacity corresponding to the overpotential, η, placed upon the electrode by an outside electronic circuit. This η is a bias potential and has a negative value, with respect to the reversible potential. It represents the difference between the potential of the electrode when hydrogen is being evolved from the left-hand circuit at a given rate diminished by the potential which the electrode has at equilibrium.

The basic experiments[8] which can be done with this apparatus involve build up and decay trainsients and an examination of the steady state diffusion conditions.

For a situation corresponding to low overpotential, permeation of hydrogen will be symmetrical in respect to time during the switch-on and switch-off conditions. It should be possible to repeat the transients any number of times.

There are other tests of the apparatus. The permeation must depend upon the thickness of the membrane. Tests for pinholds are essential. The diffusion coefficient must be identical with that obtained by careful gas-metal-contact work.

THE INTERPRETATION OF THE PERMEATION-TIME TRANSIENTS

During switch-on and switch-off conditions in the DSB apparatus, the transients are firstly observed to be symmetrical. This identity continues to be observed, however, only so when the

fugacity of the hydrogen on the surface of the electrode-- and
correspondingly, the overpotential--is low. When it exceeds a
critical value (more negative than -0.3 V for pure iron[4,9]) the
shape of the permeation-time transient differs radically from that
observed at less negative overpotentials. Thus, when one has
exceeded some critical overpotentials, the permeation-time relation
at values of overpotential more -ve than the critical one, no longer
rise and attains a steady state. It rises and falls precipituously.
The decay transient becomes radically dissymetric from the switch-
on transient. Attempts to resuscitate the switch-on permeation-
time transient, result in a different permeation-time transient. The
decay transient from this high fugacity condition[10] contains
"bumps."[11]

     There is one way, however, whereby the earlier permeation time
relation can be reattained. This is to wait for many (e.g., 10)
hours. The permeation eventually rises once more to values expected
by extrapolation from the earlier transients, those obtained before
the critical overpotential was reached.

     The interpretation of these happenings might be made on the
basis of imagined happenings outside the iron itself. During the
cathodic treatment of the left hand side, no films may exist on the
surface of the electrode, hydrogen must be evolving upon an iron
or oxide-free surface.

     Under these conditions, the interpretation of the shape of the
permeation-time line is qualitatively unambiguous. When the overpotential has exceeded a certain (-ve) value, the hydrogen fugacity
on the surface is greater than a critical value, that corresponding
to a molecular hydrogen pressure, which evidently opens traps. The
fall of the permeation means that hydrogen entering the metal fills
traps[12,13] instead of passing through to the right hand side. When
the traps corresponding to a certain overpotential are eventually
filled-- after several hours--hydrogen passes through the metal once
more and so the expected permeation is reattained. The bumps upon
the decay transient are due to hydrogen coming out of various
kinds of traps within the metal.

     The concepts give a reasonable numerical consistency between
the total hydrogen which comes out of the iron and that which would
be expected to be trapped.

FUGACITY AND OVERPOTENTIAL

     There have been several references in the above sections to
fugacity, but also to overpotential, as though they were exchange-
able terms. They are. However, the relations between each

depends upon the mechanism for the hydrogen evolution reaction.
There are many mechanisms.[14] The relation between fugacity and over-
potential is shown in Table I for a case where the hydrogen is
Tempkin adsorbed.

Overpotential on iron is -0.1 to -0.5 V. The fugacity corresponding to these overpotentials would typically be in the range 10-
10^6 atmospheres. The mechanism-dependent relation between fugacity
and pressure must be taken into account in these considerations.
The metal in the voids inside the metal is subject to the pressure
corresponding to the overpotential concerned.

SOLUBILITY AND STRESS: THE SIGNIFICANCE OF THE LARGE $\bar{V}_H$

The realtion between solubility and stress was not developed
until 1965.[4] A simple relationship is:

$$c_\sigma = c_0 e^{\bar{V}_H \sigma / RT} \tag{1}$$

From this relation, an exponential rise of solubility with
stress will begin at a stress of $RT/\bar{V}_H$ so that the higher the value,
$\bar{V}_H$, the lower the stress at which the solubility will increase to a
very large values. The above equation cannot hold over long times
when σ exceeds a value for plastic flow. However, it may be that
for a short time in a metal until the stress is relieved.

Until 1965 it was not known that the partial molar volume for
hydrogen in iron was greater than 2 ccs per mole, it was thought to
be around 0.1 ccs/mole. The actual experimental value first
determined by Beck, Bockris, Nanis and McBreen, changed the situation
thoughts concerning the effect of stress upon hydrogen solubility
and the damage which this could do. The above relation implies that,
at crack tips, and other heterogeneities in the metal, the solubility
may be 10^5-10^6 larger than the saturation solubility of iron at 1
atmosphere. Under these conditions, H atoms will be present in the
same order of magnitude in respect to number per cc as Fe atoms.

SURFACE HETEROGENEITY AND HYDROGEN SOLUBILITY

Surfaces contain heterogeneities. At all of these, there will
be unusually large degrees of stress. Knowing this local stress,
one can find the local solubility of hydrogen. Subramanian[15] has
shown that the increase amounts to several orders of magnitude and
such values make the likelihood that hydrogen enters the metal at
inclusions, and at other heterogeneities present upon the surface
of pure iron a considerable one.

TABLE I. RELATION OF OVERPOTENTIAL AND FUGACITY (TEMPKIN CONDITIONS)

	$\dfrac{F}{RT}\dfrac{d\eta}{d \ln f}$	Fugacity (f_{H_2}) (in atm.) In terms of		Remarks
		η	θ	
Fast discharge–slow recombination	$1/2(1-\theta)$	$\exp\left(\dfrac{+2\eta F}{RT}\right)$	$\left(\dfrac{\theta}{1-\theta} \cdot \dfrac{1-\theta_R}{\theta_R}\right)^2$	Nerst equation valid. Could be embrittling.
Fast discharge–slow electrochemical	$1/(1.5-\theta)$	$\exp\left(\dfrac{+2\eta F}{RT}\right)$	$\left(\dfrac{\theta}{1-\theta} \cdot \dfrac{1-\theta_R}{\theta_R}\right)^2$	Nerst equation valid. Could be embrittling.
Slow discharge–fast recombination	$1/(0.5+\ )$	1	$---$	Non-embrittling
Slow discharge–electrochemical desorption	2	$\exp\left(\dfrac{-2\eta F}{RT}\right)$	$\left(\dfrac{\theta}{1-\theta} \cdot \dfrac{1-\theta_R}{\theta_R}\right)^2$	Non-embrittling
Coupled discharge electrochemical desorption	2	$\exp\left(\dfrac{+2\eta^* F}{RT}\right)$	$\left(\dfrac{\theta}{1-\theta} \cdot \dfrac{1-\theta_R}{\theta_R}\right)^2$	Only predictable if η^* is known experimentally
Coupled discharge–recombination	$\left(\dfrac{2-\theta}{1-\theta}\right)$	$10^{1.5}\exp\left(\dfrac{+\eta F}{2RT}\right)$	$\left(\dfrac{\theta}{1-\theta} \cdot \dfrac{1-\theta_R}{\theta_R}\right)^2$	$\theta \ll 1$. Could be embrittling.

$k_b = 0$, $g = 0$, $\beta = 1/2$, η^* = potential at which fast discharge-slow electrochemical desorption mechansim changes to coupled discharge-electrochemical desorption mechanism.

θ = coverage at the reversible potential; η = positive for cathodic processes.

INITIATION OF A CRACK

The mechanism of the initiation of cracking is uncertain. It may be that the initiation is related to the beginning of pitting. When the surface is in a corroding state, it contains hydrogen inside the surface. However, this hydrogen will be concentrated at points of high stress.

When the hydrogen enters the metal it will reach voids in the interior and therein form hydrogen gas which will be at the fugacity corresponding to the hydrogen overpotential. These voids will expand in regard to meet the surface and eventually form a continuous line on the surface of the metal.

HYDROGEN CONCENTRATION AT THE CRACK TIP

The stress referred to in the above equation refers to the local situation. This is related to the microscopic stress in the whole specimen by the equation:

$$\sigma_{tip} = \sigma_0 \, (2 \, 1/r)^{\frac{1}{2}}$$

Here 1 is the depth of the crack and r is the radius of curvature at the bottom of the crack. If we take ℓ as 1 millimeter and r as 10^{-6} cm, then the value of the term in parenthesis is $10^{2.5}$. Using (2) in (1) makes a very considerable difference to the concentration of hydrogen at the tip bottom. With stresses of the order of 10kg mm^{-2}, the value of σ_{tip}/σ_0 may be as much as 10^5. Taking into account the increase of hydrogen solubility at reasonable over-potentials and the concentration inside the metal given by (1), one obtains values which show that the hydrogen is present at the tip crack to an extent of around 1 atom of H per atom of Fe.

This gives a helpful picture of the crack tip. The hydrogen bonds will be weakened. For this reason, the exchange current density of the iron dissolution reaction at crack tips will be greater than that in its absence. Correspondingly, the crack tip will move faster than had been anticipated.[16,17]

THE ELECTROCHEMISTRY OF THE CRACK TIP

The tip is anodic. Electrons from the solution of it go to positions at the side of the crack which enable protons to be discharged there. Some of the electrons go also to the surface of the metal and there they discharge reduced oxygen or hydrogen according to the pH. Under these conditions, the crack advances slowly, at the order of 1 millimeter per hour for iron.

THE TEARING MECHANISM

From time to time the slow movement of cracks through the metal
is joined by a tearing mechanism when the crack moves rapidly for
about 0.01 millimeters. This movement depends on the fact that a
highly concentrated atmosphere of H extends around the crack bottom,
and when the material has been weakened, the stress near the tip
crack bottom gives a simple non-electrochemcal tearing.

THE RATE DETERMINING STEP IN STRESS CORROSION CRACKING

Once the tip has begun to move, there are four rate determining
steps which may be considered.

1. Breaking open of the passive layer at the botton of the tip.
2. Dissolution of the anodic solution of metal from the crack bottom.
3. Evolution of hydrogen on the sides of the crack tip. 4. The
permeation of hydrogen through the metal to the bottom of the crack.

Sufficient experiments have not been done to make clear which of
these steps is r.d.s. However, there is one case where the evidence
suggests that the fourth case is r.d.s. This is the work by Nobe[18]
who found that the rate of corrosion (probably stress corrosion
cracking) of nickel-iron alloys followed the same course with
composition as the hydrogen permeation. For this system, therefore,
in the conditions used by Nobe, hydrogen transport to the cracked
tips is probably rate determining.

RELEVANCE OF MATERIAL STATED HERE TO THE WORK OF FLITT AND BOCKRIS

The work of Flitt and Bockris (This Symposium) concerns the
determination of hydrogen on and in the metal. The main point of
it is to show how the hydrogen inside the metal varies with the
hydrogen on the surface, the latter being controlled by the fugacity
of the surface hydrogen. This, in turn, depends on the overpotential,
which is related to the rate determining step and overall path for
hydrogen evolution.

THE NEEDED RESEARCH

Among the researches needed to understand cracking under
practical conditions are the following:

(1) The H_2 evolution mechanism. This is a well ploughed field
for Hg, Pt and a few other metals. It is unknown for steels and for
the relevant pH regions. The concentrations of H on and in the metal
are very necessary pieces of information.

(2) Investigation of H near the crack-tip. This must clearly be an extremely important--and difficult--investigation.

(3) Double layer structure at Fe and steel at appropriate solutions. Knowledge on anion adsorption is particularly important.

(4) Partial molar volume of H in steels: Even 15 years after the first determination, few values exist. Were they known, solubility at a given stress would be obtained.

(5) Local stress determination. Such determinations are made by X rays. However, they cannot at present be made sensitive to sufficiently small areas.

(6) Rate determining step evaluation: Knowledge of this is the fulcrum to any practical measures for the control of cracking.

REFERENCES

1. C. F. Barth, A. R. Troiano, Corrosion, 28, 259 (1972).
2. F. G. Hanna, A. R. Troiano, E. A. Steigerwald, Trans. Am. Soc. Metals, 57, 658 (1964).
3. M. Devanathan and Z. Stachurski, Proc. of Roy. Soc., A270, 90 (1962).
4. W. Beck, J. O'M. Bockris, J. McBreen, L. Nanis, Proc. Roy. Soc., A290, 220 (1966).
5. R. Heindersbach, J. Jones, Report for U.S. Army Research Office, University of Rhode Island (1976).
6. J. McBreen, Ph.D. Dissertation University of Pennsylvania (1965).
7. R. M. Barber, "Diffusion in and through Solids," MacMillan Co., New York (1941).
8. J. O'M. Bockris, J. McBreen, L. Nanis, J. Electrochem. Soc., 112, 1025 (1965).
9. J. Chene, J. Galland, P. Azou, 4A4 Proc. Conf. 2nd Int. Congress Hydrogen in metals, Paris, France (1977).
10. J. O'M. Bockris and P. K. Subramanyan, Electrochim. Acta, 16, 2169 (1971).
11. J. O'M. Bockris, P. K. Subramanyan, J. Electrochem. Soc., 118, 114 (1971).
12. B. Marandet (1977), Proc. Conf. Stress Corrosion Cracking and Hydrogen Embrittlement of Iron Base Alloys. NACE-5, 774 (Unieux-Firminy, France (1973)).
13. R. D. McCright (1977) ibid. NACE-5, 306.
14. B. V. Tilak, C. G. Rader, B. E. Conway, Electrochim. Acta, 22, 1167 (1977).
15. P. K. Subramanyan, Medical Technological Publication, Ine. Rev. Sci. Phys. Chem. Ser., One (1973) 6 181-238, ed., J. O'M. Bockris, Butterworth.

16. H. J. Flitt, W. R. Revie, J. O'M. Bockris, Corrosion Australase,
     $\underline{1}$, 3 (1976).
17. H. J. Flitt, Ph.D. Dissertation, Flinders University of South
     Australia.
18. R. R. Sayano and K. Nobe, Corrosion, $\underline{25}$, 260 (1969).

OCCLUDED CORROSION CELLS AND CRACK TIP CHEMISTRY

Marcel Pourbaix

Centre Belge d'Etude de la Corrosion CEBELCOR
Brussels, Belgium

The present lecture will be a follow-up to a lecture entitled "Electrochemical Aspects of Stress Corrosion Cracking"[1], given in 1971 at the Ericeira NATO Science Committee Evaluation Conference on "The Theory of Stress Corrosion Cracking in Alloys", and to a lecture entitled "Electrochemical Corrosion"[2], given in 1975 at the Lyngby NATO Advanced Study Institute on "Stress Corrosion Cracking".

1. THE OCCLUDED CORROSION CELLS O.C.C.

It is well known that most phenomena of localized corrosion in the presence of aqueous solutions take place under conditions where the access of the electrolyte is restricted, due to a particular geometry of the corroding material (structures with interstices such as riveted plates, threaded joints, flange joints, gaskets, ...), to the existence of solid deposits (corrosion products, scales, barnacles, ...) or to other causes (f.i. the narrowness of intergranular or transgranular cracks). In each of these cases, where corrosion occurs under conditions of restricted diffusion, and which B.F. Brown has suggested to group under the wording "Occluded Corrosion Cells O.C.C.", the chemical composition of the corroding solutions inside the more or less occluded cavities may be very different from the composition of the bulk of the solution. And the importance of this composition difference is so great that it <u>must</u> be taken into account for every thorough analysis of the corrosion process. This is valid in any case of localized corrosion: pitting corrosion, crevice corrosion, corrosion under deposits, intergranular corrosion and stress corrosion cracking.

Localized corrosion implies the existence of relatively small local "anodes" where an oxidation (corrosion) is proceeding, and of relatively large "cathodes" where a reduction takes place. The oxidant which is reduced on these cathodes is most of the time dissolved oxygen, coming from the air, so that localized corrosion generally results of a "differential aeration" between large aerated surfaces (where the metal is made more or less passive by a more or less protective oxide film) and small oxygen-free cavities (where the metal is active, free from protective oxide). All phenomena of localized corrosion (including stress-corrosion-cracking) are thus, as said by Montuelle, "sicknesses of the passive state". Localized corrosion does not occur on active, non passive, metals and alloys, which may either remain immune (uncorroded), or suffer general corrosion (without occluded cavities).

A simple method to promote and study in the laboratory localized corrosion of a given metal and alloy in a given solution may thus often consist in the use of a two-compartment device similar to the one shown in Figure 1[3], where a specimen of the considered metal or alloy is immersed in the considered solution, where oxygen (or air) is bubbled in the large compartment and oxygen-free nitrogen (or hydrogen) is bubbled in the small one. By recording, as a function of time, the electrode potential and the pH of the metal in both compartments, one may observe, in the oxygen-free solution, its evolution from the passive to the active state. And, at the end of the experiment, when electrode potentials and pH's have become stabilized, the solution inside the oxygen-free compartment is exactly the same as the solution which exists inside the oxygen-free cavities which really occur in practice.

2.   LOCALIZED CORROSION OF IRON AND CARBON-STEELS IN THE PRESENCE OF AQUEOUS SOLUTIONS CONTAINING CHLORIDE, AT ABOUT 25°C

2.1. Occluded Corrosion Cells

Figure 2 shows notably, as a result of experiments performed in 1969 with such a device by J. Van Muylder et al.[3], the influence of a slow external cathodic polarization on the electrode potential – and pH – conditions inside an iron crevice, in the case of a bulk solution of pH 12.0 containing 0.001 mole NaCl per liter. The two points marked zero correspond to no external polarization; one sees that, inside the oxygen-free crevice, the pH of the solution is then 3.8 and the electrode potential of the iron is about $- 0.33$ volt$_{she}$; under these circumstances, the electrode potential of the iron in the bulk of the solution (pH 10.0) is about $- 0.10$ volt$_{she}$. The potential difference between two identical reference electrodes dipped into the two solutions, which is equal to the electrode potential difference of the iron speciments in the two solutions (about $- 0.10 + 0.30 = 0.20$ volt), is the diffusion potential (plus an

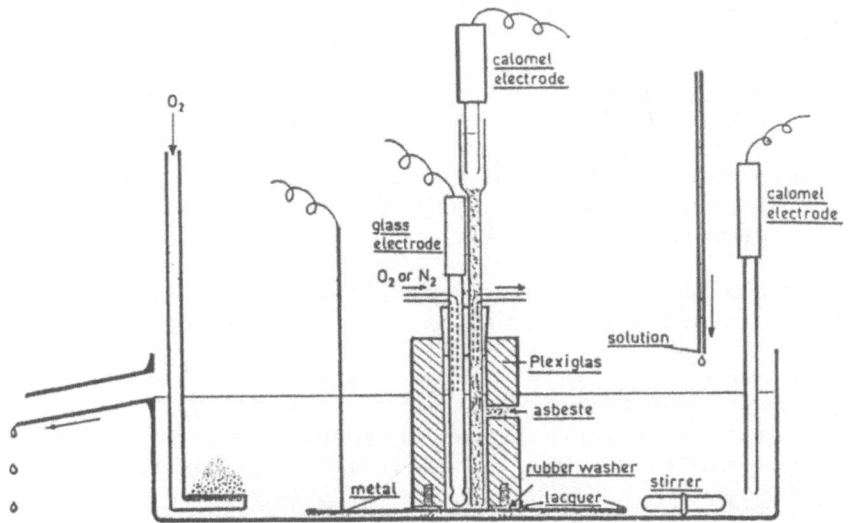

Fig. 1. Device for studying crevice corrosion.

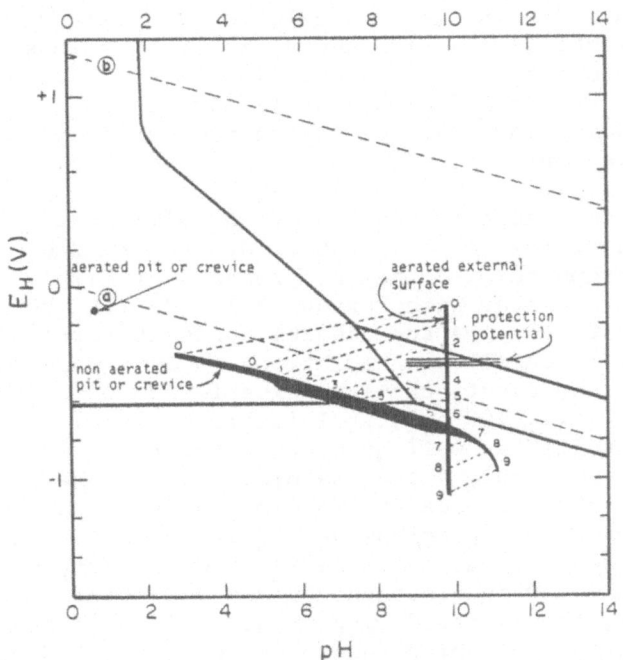

Fig. 2. Influence of a slow cathodic polarization on the potential and pH-conditions inside a crevice (mild steel with an aerated solution of 0.001 M NaOH and 0.001 M NaCl). (After J. Van Muylder, L. Sathler et al.[31]

eventual "ohmic drop") between these two solutions, and is also the "driving force" of the differential aeration process.

Figure 3 shows results obtained by B.F. Brown[4,5] in his famous work on the cathodic polarization of steel cracks. One sees that Brown's curve on cracks is very similar to Van Muylder's curve on crevices; and this confirms the assumption that, for a given metal in a given bulk solution, all forms of O.C.C. have the same chemistry and the same electrochemistry.

In the case of iron in contact with about neutral water, corrosion occurs primarily according to the oxidation

$$Fe \longrightarrow Fe^{++}_{aq} + 2e^-$$

with formation of ferrous ions which hydrolyze into more stable compounds, notably $FeOH^+_{aq}$ according to the reaction

$$Fe^{++}_{aq} + H_2O \longrightarrow FeOH^+_{aq} + H^+_{aq}$$

This reaction promotes an acidification of the solution inside an eventual cavity, and the amount of dissolved iron will stabilize when the saturation in the less soluble iron salt will be reached. In the presence of chloride and at 25°C, this less soluble salt is $FeCl_2.4H_2O$; the saturated solution is 4.6 molar in $FeCl_2$ (i.e. 4.6 molar in Fe and 9.2 molar in Cl) and its pH is 3.8. This is just the pH which was observed in the pioneer's experiments of B.F. Brown and J. Van Muylder.

It thus appears that solutions similar to those existing inside occluded corrosion cells O.C.C. of iron in the presence of water containing some chloride may be prepared by simply saturating pure oxygen-free water with ferrous chloride (in the presence of iron powder for reducing the possibly present ferric ions). One may so obtain very easily such solutions in any quantity suitable for all chemical and electrochemical experiments of scientific or technical interest (f.i. 100 cm^3, or 1 liter, or more). This has been done by C.T. Fujii[6] as well as by L. Sathler in his Brussels thesis[7,8]. As shown by Figure 4 from Sathler's work, the pH of oxygen-free water in the presence of iron, which is about 9.5 in the absence of $FeCl_2$, decreases progressively when $FeCl_2$ is added, and stabilizes at 3.8 when saturation in $FeCl_2.4H_2O$ is reached (log (Fe) = 0.66, i.e. 4.6 molar in dissolved Fe and 9.2 molar in Cl). As shown by Figure 5, the electrode potential of iron in these $FeCl_2$ solutions moves, with increasing concentration, from − 0.59 to − 0.33 volt$_{she}$.

L. Sathler has used so obtained solutions for studying the kinetics of the behavior of iron in oxygen-free solutions of ferrous chloride. Anodic potentiokinetic experiments performed at a scanning rate of 50 mV.mn^{-1} (Figure 6) have shown that, up to + 0.8 volt$_{she}$,

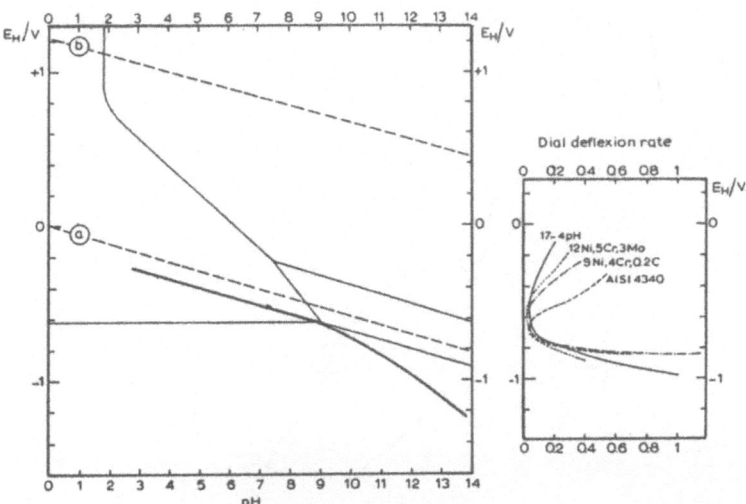

Fig. 3. Influence of a cathodic polarization on the electrochemcial
characteristics and propagation rate of a stress-corrosion
crack (chromium steels in contact with NaCl 3.5% solution).
(After B.F. Brown.[4,5])

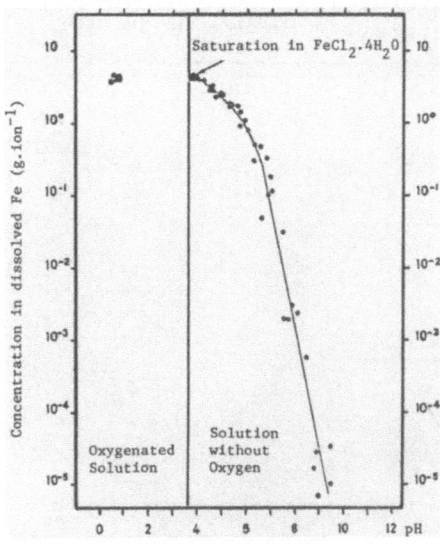

Fig. 4. Relationship between pH and the iron content of solutions of
ferrous chloride in the presence of iron powder. Influence
of oxygen on the characteristics of saturation in ferrous
chloride (25°C). (After L. Sathler.[7, p.98, 8, p.709])

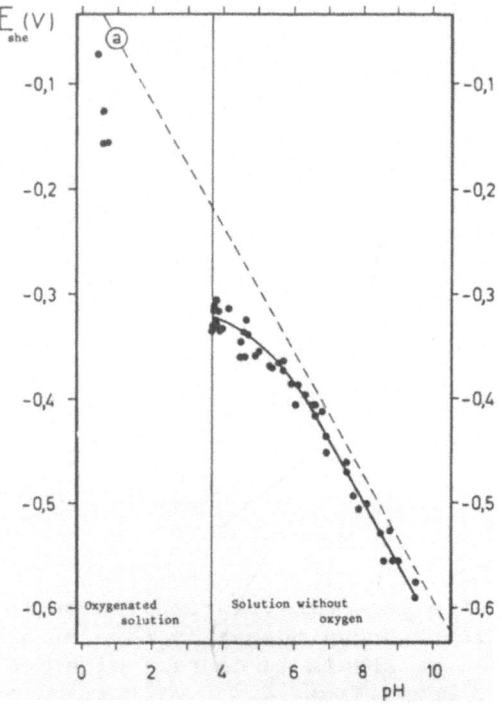

Fig. 5. Relationship between pH and the electrode potential of iron
         in solutions of ferrous chloride. Influence of oxygen on
         these characteristics when saturated in FeCl$_2$.4 H$_2$O.
         (After L. Sathler.[7], p.105, 8, p.710)

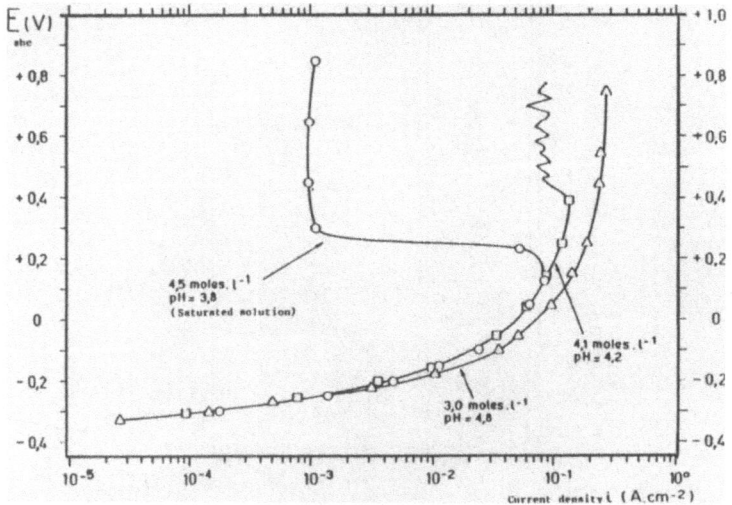

Fig. 6. Influence of the concentration of oxygen-free solutions of
         ferrous chloride on the anodic behavior of iron. Potentio-
         kinetic polarization curves (50 mV.mn^{-1}) (25°C).
         (After L. Sathler.[7], p.113, 8, p.711)

no passivation occurs in non saturated solutions (3.0 moles.l^{-1}, pH 4.8, and 4.1 moles.l^{-1}, pH 4.2), with some pitting in the second of these solutions. But, in the saturated solution (4.6 moles.l^{-1}, pH 3.8) passivation occurs when the electrode potential reaches about $- 0.20$ volt$_{she}$. The existence of this passivation has been confirmed by <u>potentiostatic</u> experiments conducted during about 13 hours at different electrode potentials. Figure 7 shows anodic and cathodic polarization curves obtained by L. Sathler in saturated $FeCl_2$ solutions, as well as an analysis of these curves by superimposition of elementary curves relating to reactions $Fe \rightleftharpoons Fe^{++}_{aq} + 2e^-$ and $2H^+_{aq} + 2e^- \rightleftharpoons H_2$.

This Figure 7 shows notably:

1 - the Tafel line <u>a</u> for the reduction $2H^+_{aq} + 2e^- \longrightarrow H_2$ (equilibrium electrode potential $E_{oa} = - 230$ mV$_{she}$, and exchange current log $i_{o.a} = - 6.82$ (A.cm^{-2}) (*); slope $- 135$ mV per decade). The equation of this Tafel line is $E = - 1146 - 135$ log i (mV).

2 - the Tafel lines <u>c</u> for the oxidation $Fe \longrightarrow Fe^{++}_{aq} + 2e^-$ and for the reduction $Fe^{++} + 2e^- \longrightarrow Fe$ (equilibrium electrode potential $E_{o.c} = - 360$ mV$_{she}$ and exchange current log $i_{o.c} = - 7.10$ (A.cm^{-2}). The anodic line is straight until log i = - 5 (slope + 23 mV per decade), becomes then incurved and shows passivity at - 185 mV$_{she}$ (log i = - 2.28 (A.cm^{-2}), or i = 5.2 mA.cm^{-2}, i.e. 60 mm Fe.year^{-1}). The cathodic line is fully straight in the considered current range (slope - 35 mV per decade). The equations of the two straight lines are $E_{ox} = - 197 + 23$ log i and $E_{red} = - 608 - 35$ log i.

3 - the zero current curve relating to the overall reaction $Fe + 2 H^+_{aq} \longrightarrow Fe^{++}_{aq} + H_2$. At zero current, the electrode potential is - 335 mV$_{she}$ and the corrosion velocity (often called "corrosion current") is log i = - 6.05 (A.cm^{-2}), or i = 0.89 μA. cm^{-2}, i.e. 10 μm Fe.year^{-1} and 290 mg H_2.cm^{-2}.year^{-1}.

On the basis of the presently available informations, we have drawn at Figure 8 an experimental potential-pH diagram intended to present approximately a general picture of the electrochemical behavior of iron in solutions obtained by dissolving increasing amounts

---

(*) We remind that a current density of 1 μ A.cm^{-2} corresponds to:
- 0.0373 μg equivalent per cm2 and per hour, i.e.:
  - 0.0373 μg H_2. cm^{-2}.hour^{-1} = 0.326 mg H_2.cm^{-2}.year^{-1}
  - 1.041 μg Fe. cm^{-2}.hour^{-1} = 9.12 mg Fe.cm^{-2}.year^{-1}
  - 1.32.10^{-3} μm Fe.hour^{-1} = 11.6 μm Fe.year^{-1}

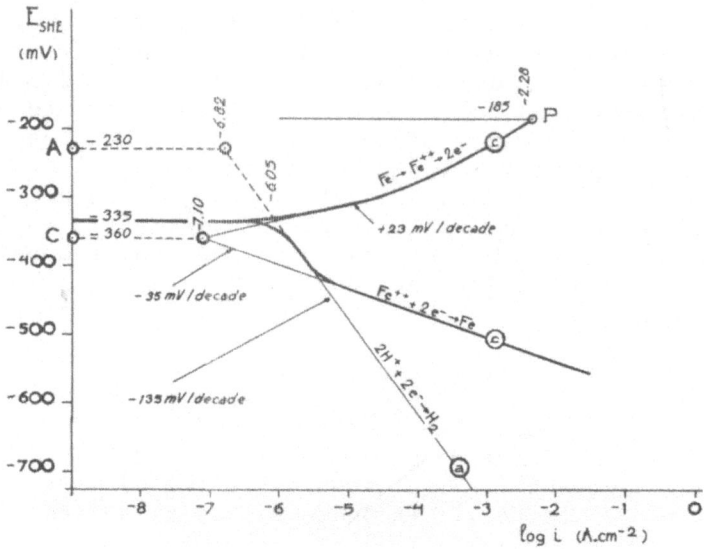

Fig. 7.   Polarization curves on iron in the presence of a saturated
          and stirred solution of ferrous chloride.   (pH 3.8 - 25°C).
          (After L. Sathler.7, p.113, 8, p.711)

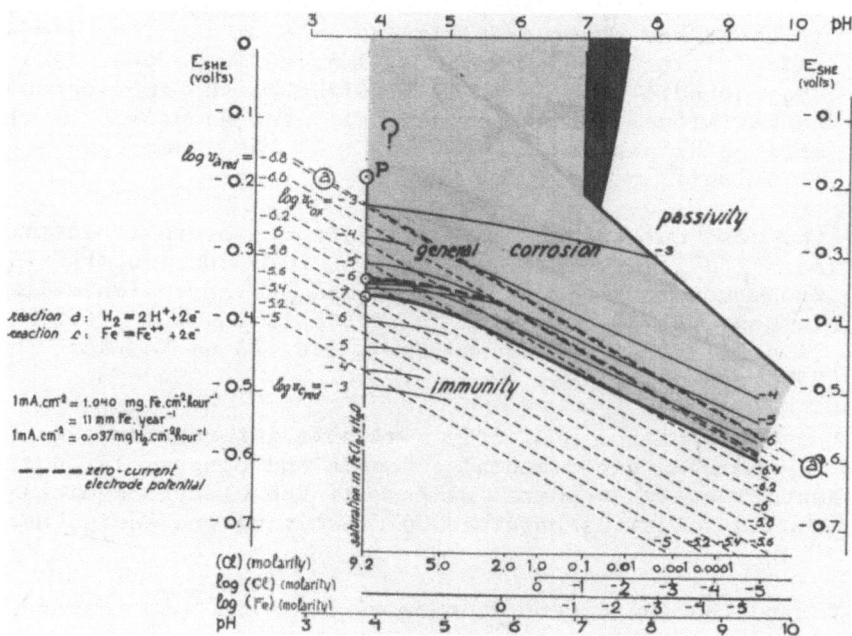

Fig. 8.   Experimental potential-pH diagram for the electrochemical
          behavior of iron in solutions of ferrous chloride (25°C).
          (Schema).

of ferrous chloride in oxygen-free pure water at 25°C, up to saturation. This Figure 8 has been drawn as follows:

1 – line a represents the equilibrium conditions of reaction $2H^+_{aq} + 2e^- = H_2$ for $pH_2 = 1$ atm (equation $E = o - 59.1$ pH (mV_{she}))

2 – below this line a, and parallel to it, is a family of ten lines which relate to given velocities of the hydrogen evolution on iron (log $v_{ared} = -6.8$ to -5.0 (A.cm^{-2})), while assuming that the data of L. Sathler given in Figure 7 for the saturated solutions are valid in the whole considered pH range (3.8 to 9.5). (Equation $E = -921 - 59.1$ pH $- 135$ log i (mV_{she})).

3 – Figures 4 and 5, which reproduce Figures 24 and 26 of L. Sathler's thesis ((7), pp. 98 and 105, (8), pp. 709 and 710), give the following values for the relationship between the pH of oxygen-free solutions of ferrous chloride and the concentration in dissolved Fe (and Cl), and the zero-current electrode potential of iron.

Table I: Relationship between the pH of oxygen-free solutions of ferrous chloride and the concentrations of dissolved Fe and Cl, and the zero-current electrode potential of iron, at 25°C.

pH	$-$ Log (Fe) (mole.1^{-1})	(Fe) (molarity)	(Cl) (molarity)	E_{Fe} mV$_{she}$	
3.8	+ 0.67	4.6	9.2	$-$ 335	(saturation)
4.9	+ 0.40	2.5	5.0	$-$ 345	
5.8	0.00	1.0	2.0	$-$ 375	
6.3	$-$ 0.3	0.5	1.0	$-$ 395	
6.9	$-$ 1.0	0.1	0.2	$-$ 434	
7.0	$-$ 1.3	0.05	0.1	$-$ 439	
7.5	$-$ 2.0	0.01	0.02	$-$ 467	
7.6	$-$ 2.3	0.005	0.01	$-$ 473	
8.1	$-$ 3.0	0.001	0.002	$-$ 502	
8.3	$-$ 3.3	0.0005	0.001	$-$ 520	
8.7	$-$ 4.0	0.0001	0.0002	$-$ 537	
8.9	$-$ 4.3	0.00005	0.0001	$-$ 541	
9.0	$-$ 4.4	0.00003	0.0006	$-$ 554	
9.3	$-$ 5.0	0.00007	0.00002	$-$ 577	
9.5	$-$ 5.3	0.00002	0.00001	$-$ 591	

The data given in Table I have allowed us to indicate the concentrations in iron and in chloride on the pH scale in abscissae of Figure 8, and to draw in this Figure (in a thick dotted line) the zero-current potentials of iron as a function of pH.

4 - On page 116 of his thesis,[7] L. Sathler gives the following data for the equilibrium conditions of reaction $Fe = Fe^{++}_{aq}$ + $2e^-$ (c) in the saturated solution (pH 3.8):

- equilibrium electrode
  potential $E_O = -360$ mV_{she}

- exchange current $i_O = 8.0 \times 10^{-8}$ $A.cm^{-2}$

                    i.e., $\log i_O = -7.10$ $(A.cm^{-2})$.

L. Sathler admits -441 mV_{she} for the standard equilibrium potential of reaction $Fe = Fe^{++}_{aq} + 2e^-$ (c). This leads to the values given in Table II for the influence of the iron concentration (and pH) on the equilibrium potential of iron in solutions of ferrous chloride in oxygen-free pure water, at 25°C. These equilibrium potentials are in fact the "immunity potentials" below which iron is immune (thermo-dynamically stable); these data thus allow to draw in Figure 8 the frontier between the "area of immunity" and the "area of general corrosion" of iron in the considered solution.

5 - According to previous work,[9,10] - Fig. 71 the influence of pH on the passivation potential of iron in stirred chloride-free solutions is given, in the pH range from 1 to 12, by a

Table II: Relationship between the pH of oxygen-free solutions of ferrous chloride and the concentration of dissolved iron, and the equilibrium electrode potential of iron, at 25°C.

(Fe)	log (Fe) (molarity)	pH	E_O Fe (mV_{she})	
4.6	+ 0.67	3.8	− 360	(saturation)
1.0	0	5.8	− 441 ?	
0.1	− 1	6.9	− 470	
0.01	− 2	7.5	− 499	
0.001	− 3	8.1	− 529	
0.0001	− 4	8.7	− 558	
0.00001	− 5	9.3	− 588	

Table III: Relationship between the pH, the passivation current density and the passivation potential of iron, in chloride-free solutions.

pH	$\log i_p$ (A. cm^{-2})	E_p (volt$_{she}$)
5.2	$- 2$	$- 0.056$
8.0	$- 3$	$- 0.296$
11.0	$- 4$	$- 0.553$

straight line which has the following equation:

$$E_p = + 0.389 - 0.0856 \text{ pH (volts}_{she})$$

and the values of the passivation current are given by the relation:

$$\log i_p = - 0.10 - 0.362 \text{ pH (A.cm}^{-2}).$$

This allows to draw in Figure 8 a frontier between the area of general corrosion and an area of passivity (valid for chloride-free solutions) and to mark on this line, as shown in Table III, the points corresponding to several values of the passivation current i_p.

6 - Figure 9 is an ancient experimental potential-pH diagram[11] which represents notably the influence of the chloride concentration (from 10^{-3} to 10^{0} molar) on the passivation - and rupture - potentials of iron in stirred aqueous solutions, with indication of the corresponding areas of pitting. One sees that the presence of chloride does not affect the frontier between the areas of general corrosion and pitting, but may, at low pH's and at high chloride concentrations, damage the protective quality of the passivating film, leading then to pitting or to general corrosion. By plotting on Figure 9 the values of pH and chloride concentrations relating to solutions of ferrous chloride (given in Table I), one obtains the values given in Table IV for the rupture - (or pitting-) potentials E_R of iron in the presence of these solutions. These values have been used for drawing in Figure 8 three lines which separate an area of pitting from the areas of passivity and of general corrosion.

7 - The four lines marked "$\log v_{c \ ox}$" at Figure 8 indicate approximately, in the area of general corrosion, the circumstances where the corrosion reaction $Fe \rightarrow Fe_{aq}^{++} + 2 \ e^{-}$ (c)

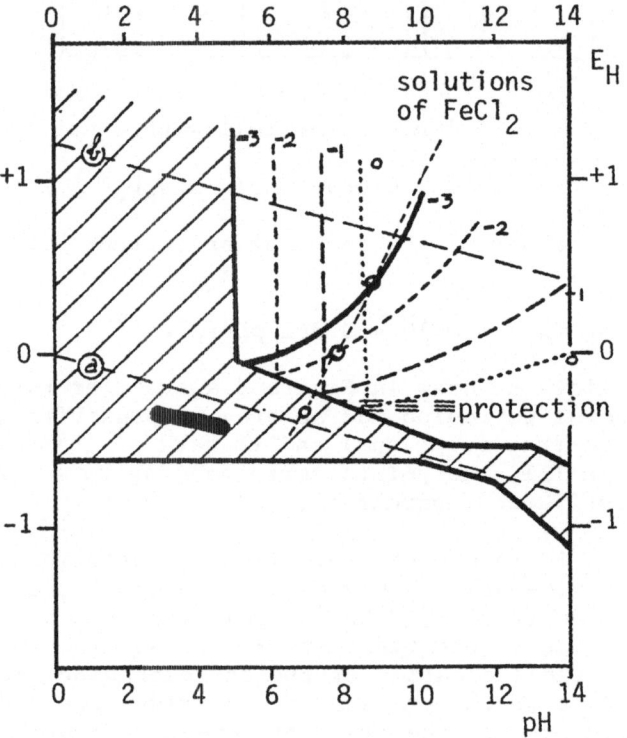

Fig. 9.   Influence of the chloride content on the rupture- and pitting
          potentials of iron. ((Cl⁻) 10^{-3} to $10°$ M).   (Schema).

Table IV: Relationship between the pH, the chloride concentration of solutions of ferrous chloride, and the pitting potential of iron in these solutions.

pH	log Cl ($mole.1^{-1}$)	E_R (volts ($_{she}$))
8.8	- 3	+ 0.4
7.6	- 2	0.0

occurs at a given velocity (from 10^{-6} to 10^{-3} A. cm^2, i.e., about 0.001 to 10 mg Fe. $cm^{-2}.hour^{-1}$, or 0.011 to 110 mm. $year^{-1}$). The lines relating to 10^{-6} and 10^{-5} A. cm^{-2} (index - 6 and - 5) have been drawn under the assumption that the overpotentials of this reaction are then, whatever the concentration of the solution, the same as the overpotentials related to the saturated solution, i.e., that these "isocurrent lines" are parallel to the "immunity line". For the two higher corrosion velocities (10^{-4} and 10^{-3} A. cm^{-2}), we have simply joined with a somewhat straight line the points relating to the considered velocity for the concentrated solutions (9.2 to 5 molar) and for the passivity line. One sees that both methods give the same result for the 10^{-4} A.cm^{-2} line.

8 – In the <u>immunity area</u>, the four lines marked "log v_{cred}" indicate circumstances where the iron electrodeposition reaction $Fe_{aq}^{++} + 2e^- \rightarrow Fe$ occurs at a given velocity (from 10^{-6} to 10^{-3} A.cm^{-2}, i.e., from 0.0104 to 1.04 mg Fe.cm^{-2}. $hour^{-1}$). For drawing these lines, we have assumed that the overpotentials resulting from Sathler's thesis in the saturated solution remain the same until the iron concentration has decreased from 9.2 to 5.0 molar; these portions of isocurrent lines are thus, here again, parallel to the immunity line.

Coming back to Figure 6, we remind that Sathler has observed that:

- in the presence of a 3.0 molar solution of $FeCl_2$ (pH = 4.8), anodic polarization leads to general corrosion without any passivation (in conformity with Figure 8).

- in the presence of a 4.1 molar solution of $FeCl_2$ (pH = 4.2), some pitting appears at an electrode potential of about + 0.4 $volt_{she}$ (where, according to the

approximate schema of Figure 8, general corrosion should
take place).

– in the presence of a 4.6 molar saturated solution
  (pH = 3.8), passivation occurs at – 0.185 volt$_{she}$. As
  stated by Sathler, this passivation potential is very
  close to the metastable equilibrium potential between
  metallic iron Fe and the ferric oxides γFe_2O_3 (maghemite)
  and $\gamma FeOOH$ (lepidocrocite). Sathler has observed that,
  in fact, this passivation is due to the formation,
  below a deposit of crystals of $FeCl_2.4H_2O$, of a duplex
  layer of ferric oxides in the γ form, which consists
  of γFe_2O_3 at its interface with the metal and of γ
  FeOOH at its interface with the saturated solution of
  ferrous chloride. Considering that, in Figure 8, the
  passivation point P is situated largely inside the area
  of general corrosion, it is likely that this passivation
  is only effective when crystals of $FeCl_2.4H_2O$ are present, and ceases if these crystals disappear.

Some of the important results of Sathler's work are summarized in Figure 10. This Figure and the hereabove statment
show notably that:

– when iron corrodes in pure water under oxygen-free
  conditions, the pH stabilizes at about 9.5, the amount
  in dissolved iron is about $10^{-5.3}$ mole.l^{-1} (about 0.3
  p.p.m.), and the zero-current electrode potential of
  iron is about – 0,59 volt$_{she}$, i.e., about 35 mV below
  the "a line" relating to the hydrogen equilibrium electrode potential under 1 atm. hydrogen. The affinity of
  the hydrogen evolution reaction is thus then about 35
  mV per faraday.

– increasing additions of ferrous chloride lead to a decrease of the pH, which stabilizes at 3.8 when saturation in $FeCl_2.4H_2O$ is reached. This highly concentrate
  solution which is, as said by B.F. Brown, like a "syrup",
  is about 4.6 molar in dissolved iron and about 9.2 molar
  in chloride. The zero-current electrode potential of
  iron in this solution is – 0.335 volt$_{she}$, and its corrosion-current is 0.89 μ A. cm^{-2}; iron then corrodes
  with hydrogen evolution with a velocity of 10 μm.year^{-1}.

– if the electrode potential of iron in this saturated
  solution is increased (by anodic polarization), the
  corrosion-current first increases up to a maximum of
  5.2 mA.cm^{-2} (corresponding to a corrosion velocity of
  60000 μm.year^{-1}) when the electrode-potential reaches
  –0.185 volt$_{she}$. Iron then becomes passive due to the

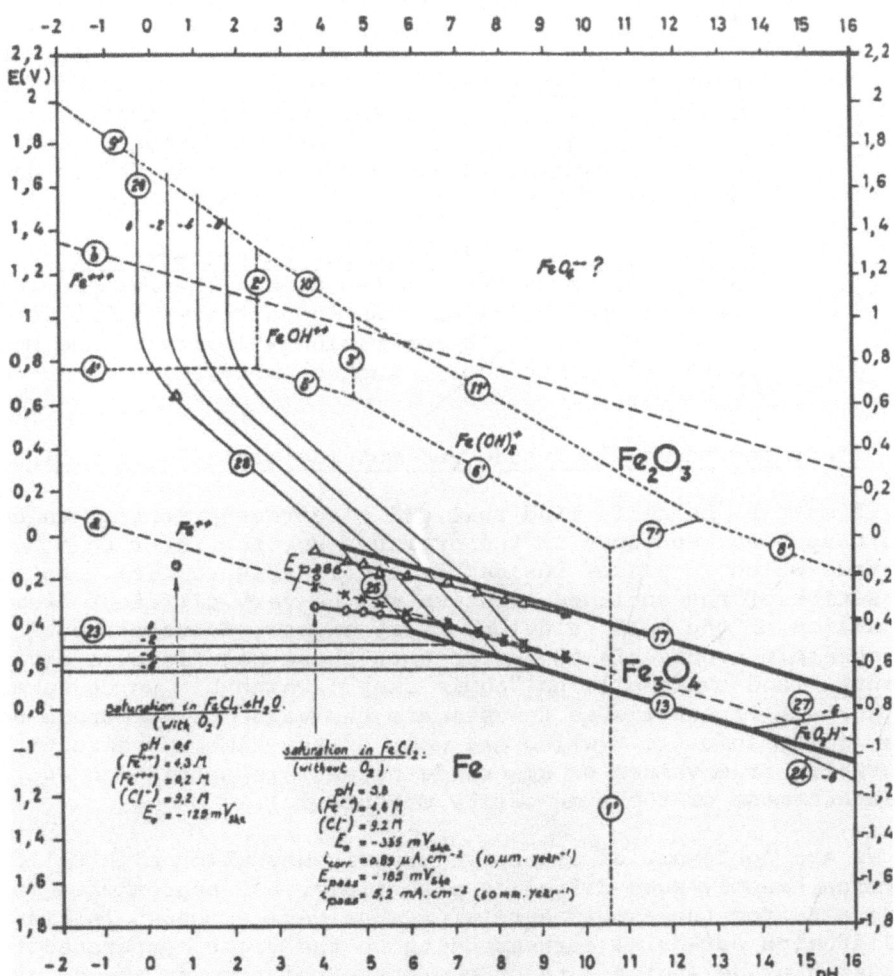

Fig. 10.   Influence of increasing concentrations of oxygen-free solutions of ferrous chloride in the evolution of the pH, and on
the electrode-potentials of iron[o], of bright platinum[Δ] and
of platinized platinum[x].   Effect of the oxygen on the electrode potential of iron and on the pH of the saturated
solution.   (After L. Sathler and J. Van Muylder.[31])

formation, between the metal and a deposit of $FeCl_2.4H_2O$, of a layer of ferric oxides in the γ form (maghemite γFe_2O_3 at its interface with the metal, and lepidocrocite γ FeOOH at its interface with the solution).

- if the electrode potential of iron in this saturated solution is <u>decreased</u> (by cathodic polarization) below -0.360 volt$_{she}$ (which is the immunity potential of iron in this solution), the metal becomes immune. Corrosion then stops and, on the contrary, dissolved iron deposits on the metal.

- if some oxygen is added to the saturated solution, the pH drops from 3.8 to about 0.6, and the zero-current electrode potential of iron jumps from -0.325 to about -0.125 volt$_{she}$. The corrosion velocity of the metal increases tremendously, and no passivation may occur.

2.2. Electrode-Potentials Inside and Outside Occluded Corrosion Cells

It must be borne in mind that all electrode-potential values which have been mentioned in the previous section refer to a reference electrode which is placed <u>inside</u> the o.c.c. (Figure 11a). As the composition of the occluded solution may be very different from the composition of the bulk solution, there exists, between identical reference electrodes placed inside both these solutions, a <u>diffusion potential</u> (and eventually an "ohmic drop") which has to be taken into account if one wishes to estimate the value of electrode potentials inside an o.c.c. (which are most of the time not directly measurable) from values of electrode-potentials outside an o.c.c. (which are most of the time easily measurable).

We are not aware of extensive experiments which might allow to determine exactly such diffusion potentials, but approximate values may result from the experiments of J. Van Muylder shown in Figure 2: the diffusion potential corresponding to the o.c.c. saturated in ferrous chloride (point zero of the curve relating to the non aerated pit of crevice) is then, as there is practically no ohmic drop between the internal parts of the iron speciments, equal to the difference between the electrode-potentials of the metal in the two solutions, i.e., about $-0.12 + 0.33 = +0.21$ volt. Incidentally, assuming that, at 25°C, the diffusion potential between two chloride solutions at different concentrations increases of 59 mV every time that the chloride content increases by a factor of 10, these diffusion potentials versus a bulk solution 0.001 molar in Cl (35 ppm) would be 59 mV if the solution in the o.c.c. is 0.01 molar in Cl, 118 mV if 0.1 molar, 177 mV if 1 molar, and about 227 mV if 9.2 molar. This last value is very close to the here abovementioned value 0.21 volt which has been observed experimentally.

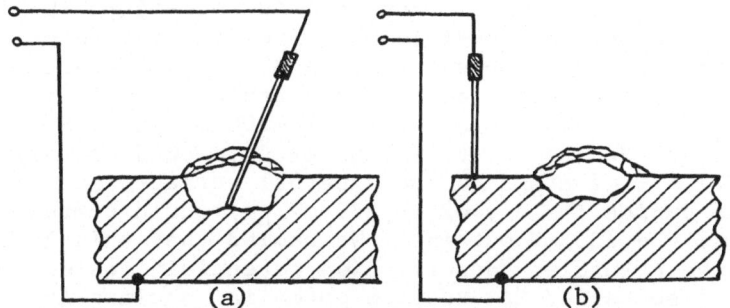

Fig. 11.   The electrode potential of a pitted metal:
           a) as measured within the pit;
           b) as measured outside the pit.

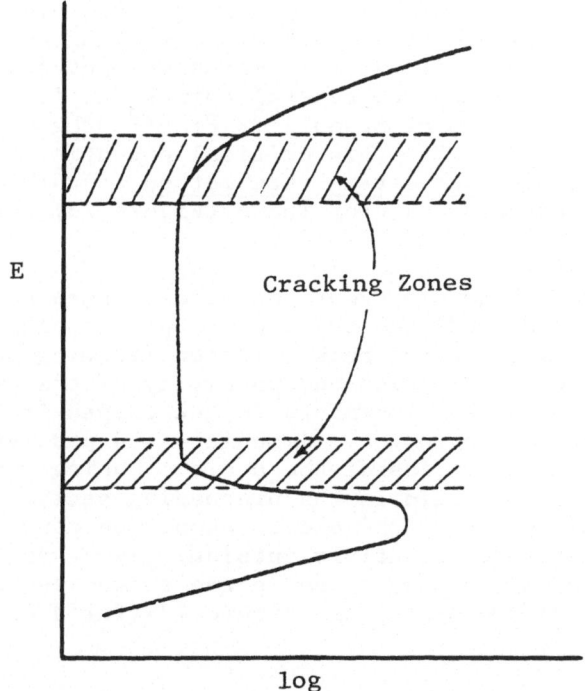

Fig. 12.   Potentiokinetic polarization curve and electrode potential
           areas where stress corrosion cracking occurs.   (Schema).
           (After R.W. Staehle.[22])

This means notably that, in case of a bulk solution 0.001 molar in Cl and of an iron o.c.c. saturated in $FeCl_2.4H_2O$ (9.2 molar in Cl^-), the zero-current conditions of iron inside the o.c.c. where the _internal_ electrode potential is - 335 mV_{she}) would be achieved when the electrode potential of the _external_ surface in the bulk solution would be about - 335 + 227 = -108 mV_{she}. The passivation conditions inside the o.c.c. (- 185 mV_{she}) would correspond to about - 185 + 227 = + 42 mV_{she} on the external surface, and the immunity conditions inside the o.c.c. (- 360 mV_{she}) would correspond to about - 360 + 227 = - 133 mV outside. We remind that this last potential, which is the so-called "deactivation potential" inside the o.c.c., is also the so-called "protection potential against localized corrosion" as measured in the bulk solution.

3. TENTATIVE APPLICATION TO THE STRESS CORROSION CRACKING OF STEELS, AT 25°C*

3.1. General

Thanks to the work performed notably by H.H. Uhlig et al.,[13-21] R.W. Staehle,[22] R.N. Parkins,[23] P.E. Morris[24] and G.J. Theus,[25] it is well known that, when stress corrosion cracking occurs, this happens in a given range of external electrode-potentials, near the frontier between passivity and general corrosion, i.e., in a region where, according to words of Staehle[22], p. 279 "the passivating film shows kinetic instability". This critical potential range may be determined by the constant strain rate method of Parkins (Figure 14) for ordinary temperatures, and of Theus (Figure 13) for high temperatures.

It seems that these dangerous potential ranges may be quantitatively explained and predicted by simply superimposing to polarization curves determined in the bulk solution (showing notably the conditions of general corrosion and passivity of the _external_ surface) polarization curves determined in the oxygen-free occluded solution (showing the corrosion-, immunity- and passivity-conditions of the _internal_ surfaces, inside the o.c.c.), being understood that, as a result of what was said in the preceeding section 2.2., the electrode potentials inside the o.c.c. should be converted into the corresponding electrode potentials outside. Two examples of this method of studying s.c.c. are given in the tentative schemas of Figure 15 (for a carbon steel) and Figure 17 (for a chromium steel).

---

* See[12] - Fig. 8, pp 609, 610

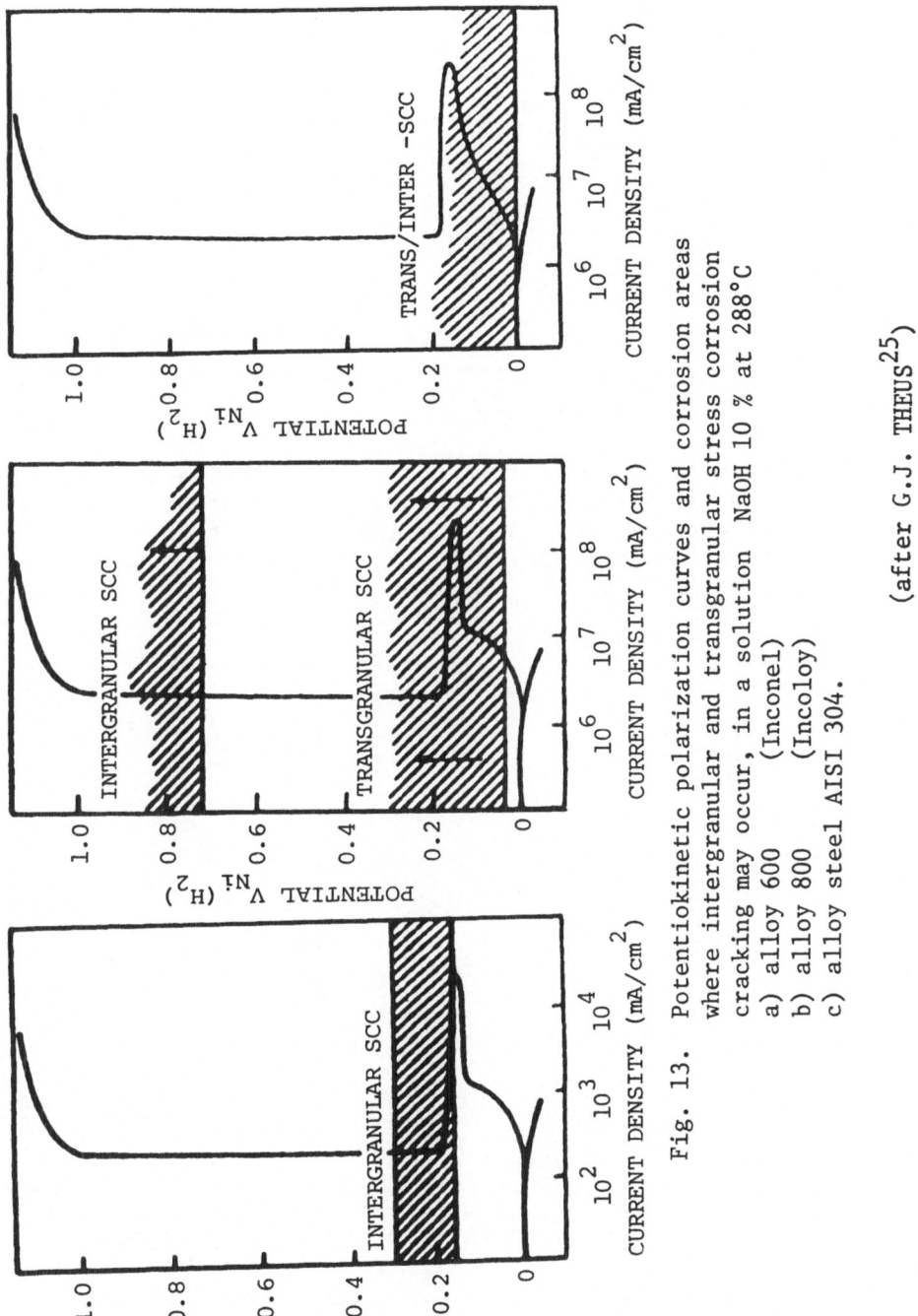

Fig. 13. Potentiokinetic polarization curves and corrosion areas
where intergranular and transgranular stress corrosion
cracking may occur, in a solution NaOH 10 % at 288°C
a) alloy 600 (Inconel)
b) alloy 800 (Incoloy)
c) alloy steel AISI 304.

(after G.J. THEUS[25])

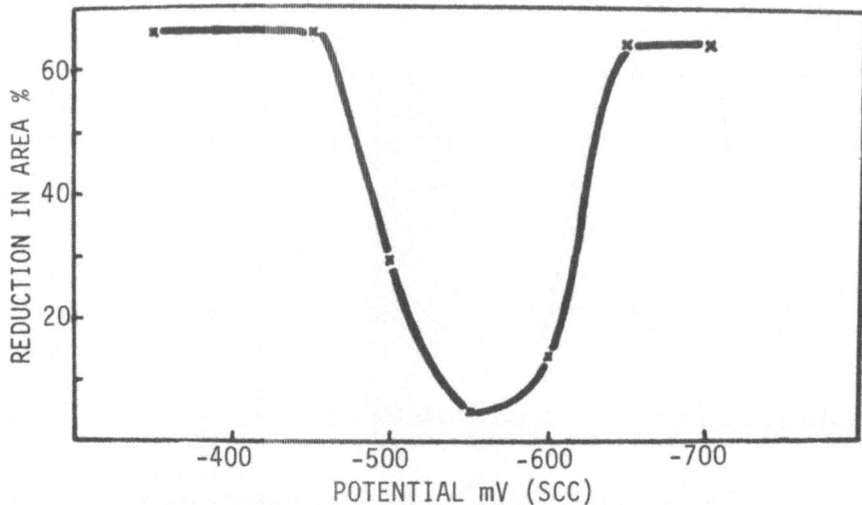

Fig. 14.   Determination of stress corrosion cracking electrode poten-
tials by potentiostatic tests with constant strain rate.
Mild steel in solution 2 N $(NH_4)_2CO_3$, at 75°C.
(After R.N. Parkins.[23])

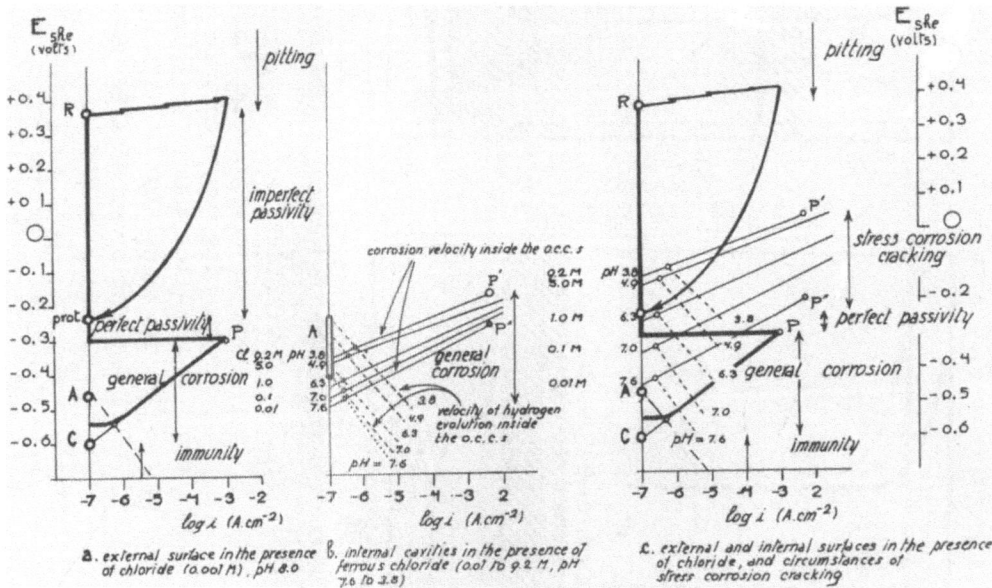

Fig. 15.   Electrochemical behavior of a carbon steel in a solution
10^{-3} molar in Cl polarization curves in oxygen-free solutions.   (Schema.)

3.2. Stress Corrosion Cracking of Carbon Steel

The three diagrams of Figure 15 show schematically three series of polarization curves of a carbon steel in oxygen-free solutions.

Figure 15a relates to the <u>external</u> surface of the steel in a solution of pH 8 containing some chloride (0.001 M in Cl). The plain line is an anodic polarization curve related to the oxidation of the steel. The dotted line relates to the evolution of hydrogen on the steel. This Figure 15a shows notably an "immunity potential" C (about -0.60 volt$_{she}$ in an iron-free solution), a "passivation potential" C (about -0.30 volt$_{she}$), a "rupture potential" (or "pitting potential") R ($+0.37$ volt$_{she}$), and a "protection potential against the propagation of localized corrosion" prot. (about -0.23 volt$_{she}$). These four electrode potentials divide the potential scale in an area of <u>immunity</u> (below C), an area of <u>general corrosion</u> (between C and P), an area of <u>perfect passivity</u> (between P and prot.), an area of <u>imperfect passivity</u> (between prot and R), and an area of <u>pitting</u> (above R).

Figure 15b, which has been drawn according to the data given in Figures 7 and 8, relates to the <u>internal</u> surface of o.c.c.'s, which we shall assume to contain different amounts of dissolved $FeCl_2$, up to saturation (9.2 molar in $FeCl_2$, and pH 3.8). One sees that passivation occurs in the 0.01 molar chloride solution (pH 7.6, point P'), but disappears when the chloride content increases (0.1 molar and pH 7.0). Some passivation exists again in the saturated solution (9.2 molar and pH 3.8), at point P', when the electrode potential reaches -0.185 volt$_{she}$, at an anodic current $10^{-2.28}$ A.cm^{-2}, i.e. 5.2 mA.cm^{-2}, corresponding to a corrosion velocity of 5.2 mg.Fe.cm^{-2}.hour^{-1}, or 1.2 μm.hour^{-1}. The dotted lines show, for the five considered compositions of the solutions inside the o.c.c.'s (pH 7.6 to 3.8) the rate of hydrogen evolution. For instance, in the saturated solution, the rate of hydrogen evolution at the zero-current potential (-0.335 volt$_{she}$) is $10^{-6.05}$ A.cm^{-2}, i.e. 0.89 μA.cm^{-2} = 0.033μg H_2.cm^{-2} hour^{-1}; in this saturated solution, the rate of hydrogen evolution increases to 10^{-5} A.cm^{-2} (0.37 μg H_2.cm^{-2}.hour^{-1}) when the electrode potential drops at -0.46 volt$_{she}$.

At Figure 15c, we have reproduced figure 15a (which relates to the external surface in contact with the bulk solution) and the two series of five lines drawn at Figure 15b (for the corrosion- and hydrogen evolution-reactions inside the o.c.c.'s), each of these lines being moved upwards as said in section 2.2 for taking into account the effect of the diffusion potential between these two solutions.

Table V shows the values admitted in Figures 15b and 15c for the immunity- and passivation-potentials, and for the related current-densities.

Table V. Values of Current Densities and of Electrode Potentials Used for Drawing Figures 15b and 15c. Electrochemical Behavior of Carbon Steel in a Solution 10^{-3} Molar in Cl (pH 8).

pH	Cl (molarity)	Current densities log i ($A \cdot cm^{-2}$)		Figure 15b — Electrode potentials versus a reference electrode placed inside the o.c.c. (mV_{she})		Figure 15c — Electrode potentials versus a reference electrode placed outside the o.c.c. (mV_{she})	
		immunity	passivation	immunity	passivation	immunity	passivation
3.8	9.2	− 7.1	− 2.3	− 360	− 185	−360 + 233 = −127	−185 + 233 = + 48
4.9	5.0	− 7.1	−	− 380	−	−380 + 218 = −162	−
6.3	1.0	− 7.1	−	− 445	−	−445 + 177 = −268	−
7.0	0.1	− 7.1	−	− 475	−	−475 + 118 = −357	−
7.6	0.01	− 7.1	− 1.5 ?	− 505	− 270	−505 + 59 = −446	−270 + 59 = − 211

One sees at Figure 15c that the so-called "protection potential against the propagation of localized corrosion" (about - 0.23 volt$_{she}$) is very close to the external electrode potential where the solutions inside the o.c.c.'s have become about neutral (pH = 6.3) and where the iron has become immune in these solutions. The metal has thus become "immune" or "deactivated". When the external electrode potential of the metal is raised above this protection value, the corrosion rates in the o.c.c.'s increase dramatically and the solutions in the o.c.c.'s become more and more acid, until the external electrode potential has reached + 0.05 volt$_{she}$ and the solutions inside the o.c.c.'s have become saturated in ferrous chloride. If no oxygen is present in the o.c.c.'s, the metal there becomes passive and localized corrosion ceases. Such an absence of oxygen is most of the time not achieved in pits or crevices, but exists automatically in most cases where corrosion occurs within narrow interstices, such as stress-corrosion cracks and intergranular decohesions.

Note that, if o.c.c.'s saturated in ferrous chloride (pH 3.8) are existing, their zero-current potential (measured on the external surface) is - 0.10 volt$_{she}$, and the common value of the corrosion current and of the hydrogen evolution current is log i = -6.05 (A.cm^{-2}), i.e. 0.89 µA.cm^{-2}. This means, inside the o.c.c., a corrosion rate of 0.89 µg Fe cm^{-2}.hour^{-1} and a hydrogen evolution of 0.033 µg H$_2$.cm^{-2}.hour^{-1}. If the external electrode potential is dropped to - 0.25 volt$_{she}$, there occurs in the saturated o.c.c.'s an electrodeposition of iron and an evolution of hydrogen (log i = - 4.8 (A.cm^{-2}), i.e. 16 µA.cm^{-2} or 0.59 µg H$_2$.cm^{-2}.hour^{-1}). The iron - and chloride - concentrations inside the o.c.c. decrease and the pH increases, and all this stabilizes at pH 6.3 (1.0 molar Cl and 0.05 molar Fe).

As a result of this, the six following electrode-potential areas are apparent on Figure 15c (for a carbon steel, in a solution of pH 8.0 0.001 molar in Cl):

- below - 0.60 volt$_{she}$ (potential C), there is complete immunity of the steel, which is however the seat of hydrogen evolution, and may thus sometimes suffer embrittlement.

- between - 0.60 and - 0.30 volt$_{she}$ (potential P), there is general corrosion of the external surface, with some hydrogen evolution on it below - 0.47 volt$_{she}$.

- between - 0.30 and about - 0.23 volt$_{she}$ (protection potential), there may be some "perfect passivation"; i.e. no corrosion occurs, even inside the eventually existing cavities.

- between about - 0.23 and + 0.05 volt$_{she}$ (passivation potential of the o.c.c.'s in the presence of oxygen-free saturated ferrous chloride solution), corrosion increases dramatically inside the eventually existing cavities, where hydrogen is evolved. This is the area of possible stress corrosion cracking, if the metal is susceptible to this form of attack.

- between − 0.25 and 0.37 volt$_{she}$, corrosion increases inside the existing cavities which are not oxygen-free.

- above + 0.37 volt$_{she}$, new pits appear on the external surface.

3.3. Stress Corrosion Cracking of a Chromium Steel

Some work is presently in progress for applying to alloy-steels the method of approach which has been developed for carbon steel in section 3.2 of the present report. Although reliable data are not yet available on this work, we shall present here a tentative analysis of the electrochemical behavior of a chromium steel, based on work performed in the early 70's under sponsorship of the Center for Applied Thermodynamics and Corrosion, in collaboration with CEBELCOR and the University of Florida (E.D. Verink ,[26] R.L. Cusumano,[27] K.K. Starr.[28])(*)

Figure 16 gives, for the pH range from zero to 11, a provisional schema of an experimental potential-pH diagram for the electro-chemical behavior of a 12% Cr steel. The presentation of this figure is similar to the presentation given at Figure 8 relating to iron. However, due to lack of data concerning the composition of the solutions inside the o.c.c.'s of this Fe-Cr alloy, we have provisionally admitted that these solutions are about the same as those existing for iron, i.e. solutions up to 9.2 molar in chloride, where they become saturated in $FeCl_2.4H_2O$. Besides this, due to lack of data concerning the behavior of this chromium steel in such concentrated chloride solutions (of iron- and chromium chlorides), we have used here data available from work performed in dilute solutions. Thus the curvatures observed at Figure 8 for solutions more concentrated than 1 molar in chloride do not appear at Figure 16.

This Figure 16 has been drawn as follows:

1. The family of ten isocurrent lines parallel to the "hydrogen line" a relates to given velocities of the hydrogen evolution on the considered steel ($\log v_{a_{red}}$ = − 5.8 to − 4.0 A.cm^{-2}),

---

(*) The Center for Applied Thermodynamics and Corrosion was created in 1971 for promoting Belgian-American collaboration in these fields. CEBELCOR has brought to this Center its expertise in thermodynamics and electrochemistry, and the University of Florida did the same in physical metallurgy. Details about this Center, which is open to other organizations, have been given elsewhere[29] and are available on request.

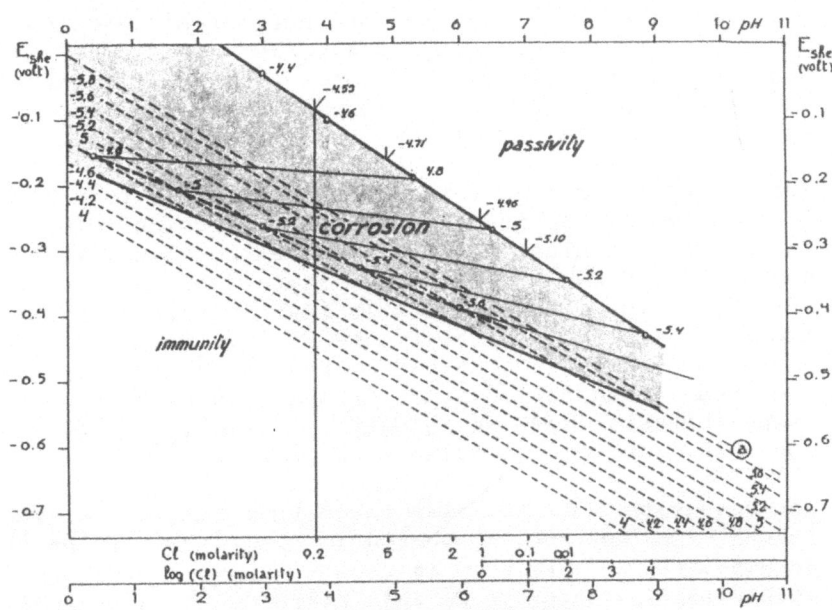

Fig. 16.   Experimental potential-pH diagram for the electrochemical
behavior of a 12 % chromium steel (25°C).   (Schema.)

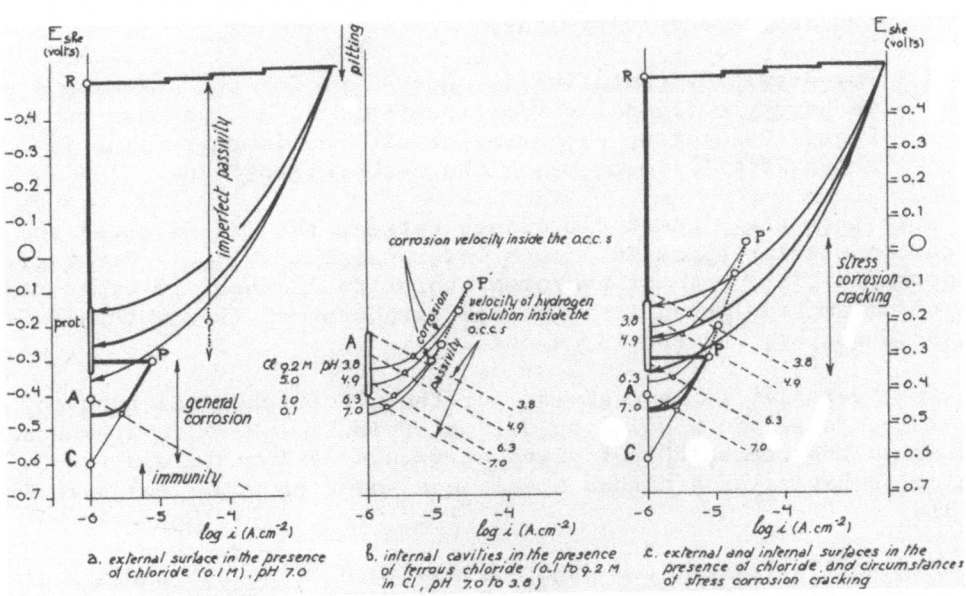

Fig. 17.   Electrocnemical behavior at a 12 % Cr steel in a solution
10^{-1} molar in Cl polarization curves in oxygen-free solutions.   (Schema.)

while assuming that the exchange current of reaction
$2H^+ + 2e = H_2$ (a) on this steel is $10^{-5.85}$ A.cm^{-2}, as already
done in a previous work.[26], Fig. 7   The equation of this
family of lines is E = $-$ 0.720 $-$ 0.059 pH $-$ 0.123 log i
(where E is given in volts$_{she}$ and i in A.cm^{-2}).

2.   The zero-current line for the chromium steel (thick dotted
line, jointing the points where the corrosion veloctiy and
the hydrogen evolution velocity have the same values) is the
one already drawn in this same Figure 7 of reference [26] (for
pH 3.0 to 6.6). Note that this line is about the same as
the one given for iron at Figure 8. According to presently
unpublished results obtained by Yang Wu [30], the zero-current
electrode potential of a 17% Cr steel in a solution saturated in FeCl$_2$.4H$_2$O, at 25°C, is $-$ 250 mV$_{she}$, which is 75 mV
above the zero-current line of Figure 16.

3.   The passivity line has been taken from Figure 8 of reference[26] and has been graduated in logarithmic values of the
passivation current (log i$_p$ from $-$ 5.6 to $-$ 4.4, A.cm^{-2}),
according to Figure 10 of this reference[26]. The equation
of the passivity line is E$_p$ = $-$ 0.210 $-$ 0.030 pH (volt$_{she}$).

4.   Iso-current lines for the corrosion rates of the chromium
steel have been obtained by joining with straight lines
the points of same corrosion rate on the zero-current line
and on the passivity-line.

5.   For drawing a tentative immunity-line for the chromium steel,
we have provisionally admitted that, as is the case of
Figure 8 relating to carbon steel, the immunity-line is
about 25 millivolts below the zero-current line.

Note that the biggest difference between the 12% Cr steel and
the carbon steel resides in a much lower passivity line. For a given
overpotential, the rate of hydrogen evolution is about 10 times less
on a chromium steel than it is on the carbon steel (however this is
perhaps doubtful, and should be checked).

In Figure 17, we have given, for the electrochemical behavior
of a 12% Cr steel in a solution 0.1 molar in Cl (pH 7), a presentation
similar to the one which was given at Figure 15 for the electrochemical behavior of a carbon steel in a solution 0.001 molar in Cl
(pH 8).

Figure 17a relates (as Figure 15a) to the external surface of
the steel. Figure 17b relates to the internal surface of an o.c.c.
supposed to be the same as the one occuring in an o.c.c. of carbon
steel (note that the presence of chromium may perhaps lead to a decrease of the pH below 3.8, for the saturated solution). Figure 17c

Table VI. Values of Current Densities and of Electrode Potentials Used for Drawing Figures 17b and 17c. Electrochemical Behavior of a 12% Cr Steel in a Solution 10^{-1} Molar in Cl (pH 7).

pH	Cl (molarity)	Current densities log i (A.cm^{-2})		Figure 17b — Electrode potentials versus a reference electrode placed inside the o.c.c. (mV$_{she}$)		Figure 17c — Electrode potentials versus a reference electrode placed outside the o.c.c. (mV$_{she}$)	
		immunity	passivation	immunity	passivation	immunity	passivation
3.8	9.2	- 6.0 ?	- 4.53	- 325	- 85	-325 + 115 = -210	- 85 + 115 = + 30
4.9	5.0	- 6.0 ?	- 4.71	- 370	- 160	-370 + 110 = -270	-160 + 100 = - 60
7.0	1.0	- 6.0 ?	- 4.96	- 430	- 255	-430 + 59 = -377	-255 + 59 = -205
7.0	0.1	- 6.0 ?	- 5.10	- 460	- 305	-460 + 0 = -460	-305 + 0 = -305

is a superimposition of the two Figures 17a and 17b, each of the lines of Figure 17b being moved upwards as shown in Table VI.

One sees again at Figure 17c that the so-called "protection potential against the propagation of localized corrosion" of the considered 12% chromium steel (about -0.15 to -0.35 volt$_{she}$, which is not very reproducible[26], Fig. 13) is very close to the external electrode potential where the solutions inside the o.c.c.'s have become neutral, and where the steel has become immune in these solutions. Two big differences with the carbon steel are that the 12% steel becomes much more easily passive in these solutions, even when they are not saturated in any salt, and that the so-formed protective film (which contains ferric and chromic oxides) is of much better quality than the film formed on iron or on carbon steel (which contains only ferric oxides). The propagation of localized corrosion may thus result, according to the circumstances, of a _deactivation_ or of a _repassivation_ of the steel inside the o.c.c.'s.

- between about -0.35 and $+0.03$ volt$_{she}$, there may be inside the o.c.c.'s locally corrosion and locally passivation, with complete and perfect passivation above $+0.03$ volt$_{she}$. Hydrogen evolution may occur inside the o.c.c.'s if the external electrode potential is lower than -0.1 volt$_{she}$. This is the area of possible stress corrosion, if the steel is susceptible to this form of attack.

- between about -0.35 to -0.15 volt$_{she}$ and $+0.51$ volt$_{she}$, corrosion increases inside the existing cavities which are not oxygen-free.

- above $+0.51$ volt$_{she}$, new pits appear on the external surface.

3.4. Tentative Schemas for Stress Corrosion Cracks in the Presence of Aqueous Solutions Containing Chloride

Figure 18 shows, on the basis of Figures 8 and 15, a tentative schema for an _iron crack_ formed in the presence of an aqueous solution containing chloride.

According to this schema, the crack-tip, which has an electrode-potential of about -0.33 volt$_{she}$, acts mainly as an "anode" where ion corrodes with formation of Fe^{++}, $FeOH^{+}$, ($FeCl^{+}$) and H^{+} ions, and of electrons e^{-} which move mostly to the external more or less passive aerated "cathode" (electrode-potential about -0.1 to 0.4 volt$_{she}$) where they reduce oxygen dissolved in the bulk water. Part of the electrons e^{-} remain at the crack-tip where they reduce H^{+} ions to gazeous H_2.

The corrosion velocity of the oxygen-free crack tip reaches a

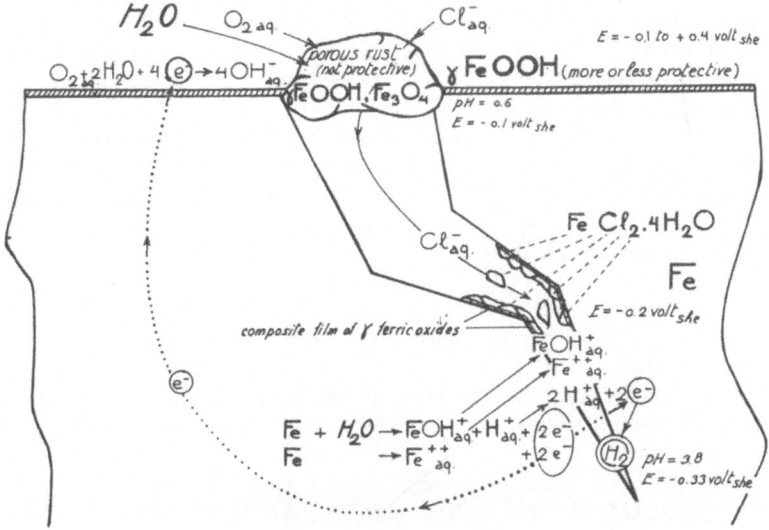

Fig. 18.   Iron crack in the presence of chloride.

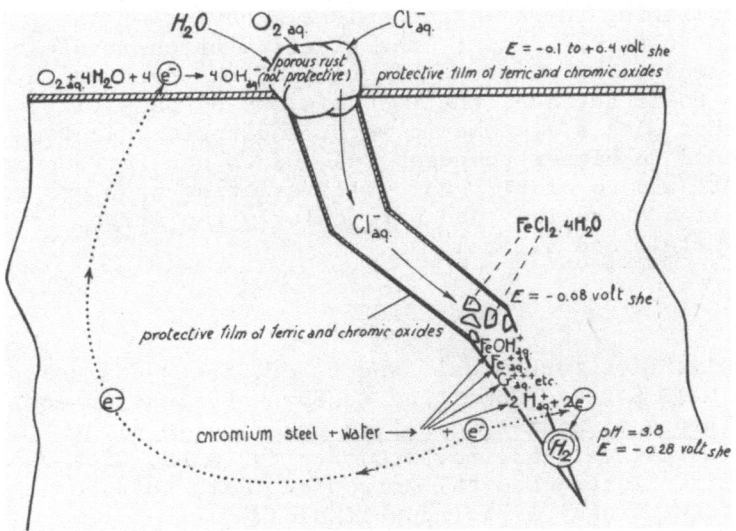

Fig. 19.   Chromium steel crack in the presence of chloride.

maximum of about 60 mm per year (or 5.7 µm per hour) when the solution becomes saturated in ferrous chloride (pH 3.8, 9.2 molar in Cl,
electrode-potential about - 0.2 volt$_{she}$). The metal then becomes
more or less passive due to the formation, under crystals of
$FeCl_2.4 H_2O$, of a composite film of maghemite γFe_2O_3 and of lepidocrocite $\gamma FeOOH$. At the external part of the crack, where some oxygen
may diffuse from the bulk water through a porous layer of rust,
together with chloride ions and water itself, the pH may drop to
about 0.6 (electrode potential about - 0.1 volt$_{she}$) and the corrosion
rate may very much increase.

Figure 19 shows, on the basis of Figures 16 and 17, a similar
tentative schema for a <u>chromium steel crack</u> formed in the presence
of an aqueous solution containing chloride. It is likely that, as
in the case of an iron crack, the corrosion velocity decreases where
the solution becomes saturated in ferrous chloride, but, contrarily
to what happens for iron (or carbon steel), the corrosion stops
completely when the electrode-potential reaches about - 0.08 volt$_{she}$,
due to the formation of a highly protective composite film of ferric
and chromic oxides on the metal.

We wish to point out that Figures 8 and 16 and thus the related
Figures 15, 17, 18 and 19, are based on several assumptions and
should thus be considered as provisional schemas susceptible of
important corrections. The quantitative results of these figures
should therefore be considered with suspicion.

Much more research is needed for clearing up the chemical composition of the solutions inside the o.c.c.'s of metal and alloys in
the presence of different bulk solutions (containing chloride or not),
and for establishing the electrochemical behavior of these metals
and alloys in these solutions: the kinetics of their dissolution-
and redeposition-reactions, the kinetics of the eventual hydrogen
evolution on their surface, their conditions of passivity, etc. ...
We believe that such a systematic work, and better diagrams, which
may be extended to higher temperatures and to non-ferrous metals and
alloys, might help to clarify the still existing problems related
to the localized corrosion, and particularly the stress corrosion
cracking of metals and alloys.

REFERENCES

1. M. Pourbaix, Electrochemical Aspects of Stress Corrosion Cracking,
     Proc. NATO Science Committee Research Evaluation Conference
     "The Theory of Stress Corrosion Cracking in Alloys",
     Ericeira 1971, 17-63, ed. J.C. Scully, NATO, Crussels (1971).
2. M. Pourbaix, Electrochemical Corrosion, NATO Advanced Study
     Institute, Lingby 1975 (unpublished).
3. M. Pourbaix, Fundamentals of Cathodic Protection, and Applications
     (in French), Rapports Techniques CEBELCOR 111, RT. 166 (1969).

4. B.F. Brown, C.T. Fujii, E.Ph. Dahlberg, Methods for Studying the Solution Chemistry within Stress Corrosion Cracks, J. Electrochem. Soc. 116, 218-219 (1969).

5. B.F. Brown, The Role of the Occluded Corrosion Cell in Stress Corrosion Cracking of High Strength Steels, Rapports Techniques CEBELCOR 112, RT. 170 (1970).

6. C.T. Fujii, Electrochemical and Chemical Aspects of Localized Corrosion, Rapports Techniques CEBELCOR 123, RT. 213 (1974).

7. L. Sathler, Contribution to the Study of the Electrochemical Behavior of Iron in the Presence of Solutions of Ferrous Chloride, with Reference to Localized Corrosion, (in French), Thesis Brussels (1978).

8. L. Sathler, J. Van Muylder, R. Winand and M. Pourbaix, Electrochemical Behavior of Iron in Localized Corrosion Cells in the Presence of Chloride, Proc. 7th International Congress on Metallic Corrosion, Rio de Janeiro, 1978, Vol. 2, 705-717, Publ. Abraco (1979).

9. M. Pourbaix, Corrosion, Passivity and Passivation of Iron. The Role of pH and Potential, (in French), Thesis Brussels (part), Bull. Soc. Roy. Belge Ingenieurs Industriels (March 1951), Publ. CEBELCOR F. 21 (1951).

10. M. Pourbaix, "Lectures on Electrochemical Corrosion", Publ. Plenum Press and CEBELCOR (1973).

11. M. Pourbaix, Significance of Protection Potential in Pitting and Intergranular Corrosion, Corrosion 26, 431-438 (1970), Rapports Techniques CEBELCOR 114, RT. 179 (1970).

12. M. Pourbaix, Electrochemistry and Corrosion, (in French, with extended abstract in English), Rapports Techniques CEBELCOR 134, RT. 244 and RT. 245 (1978). See also Proc. 7th International Congress on Metallic Corrosion, Rio de Janeiro, 1978, Vol. 1, 34 - 64, Publ. Abraco (1979).

13. H. Lee, H.H. Uhlig, Effect of Nickel in Cr-Ni Stainless Steels on the Critical Potential for Stress Corrosion Cracking, J. Electrochem. Soc. 117, 18 (1970).

14. R. Newberg, H.H. Uhlig, Stress Corrosion Cracking of 18% Cr Ferritic Stainless Steel, J. Electrochem. Soc. 119, 981 (1972).

15. H. Mazille, H.H. Uhlig, Effect of Temperature and Some Inhibitors on Stress Corrosion Cracking of Carbon Steels in Nitrate and Alkaline Solutions, Corrosion 28, 427 (1972).

16. R. Newberg, H.H. Uhlig, Stress Corrosion Cracking Behavior of Pre-Cracked Stainless Steel, J. Electrochem. Soc. 120, 1629 (1973).

17. A. Asphahani, H.H. Uhlig, Stress Corrosion Cracking of 4140 High Strength Steel in Aqueous Solutions, J. Electrochem. Soc. 122, 174 (1975).

18. H.H. Uhlig, K. Gupta, W. Liang, Critical Potentials for Stress Corrosion Cracking of 63-37 Brass in Ammoniacal and Tartrate Solutions, J. Electrochem. Soc. 122, 343 (1975).

19. D. Hixson, H.H. Uhlig, Stress Corrosion Cracking of Mild Steel in Ammonium Carbonate Solution, Corrosion 32, 56 (1976).

20. A. Asphahani, H.H. Uhlig, Stress Corrosion Cracking of High and Low Strength Carbon Steels in Nitrate Solutions, Corrosion 32, 117 (1976).
21. H.H. Uhlig, Applying Critical Potential Data to Avoid Stress Corrosion Cracking of Metals (to be published in J. Applied Electrochem.)
22. R.W. Staehle, Stress Corrosion Cracking of the Fe-Cr-Ni Alloy System, Pro. NATO Science Committee Research Evaluation Conference "The Theory of Stress Corrosion Cracking in Alloys", Ericeira, 29 March – 2 April, 1971, pp. 223-288, Edited by J.C. Scully, Publ. NATO, Brussels (1971).
23. R.N. Parkins, Stress Corrosion Cracking of Low-Strength Ferritic Steels, Proc. NATO Science Committee Research Evaluation Conference "The Theory of Stress Corrosion Cracking in Alloys", Ericeira, 29 March – 2 April 1971, pp. 449-468, Edited by J.C. Scully, Publ. NATO, Brussels (1971).
24. P.E. Morris, New Electrochemical Techniques for High Temperature Aqueous Environments, Proc. Corrosion 76 "Electrochemical Techniques for Corrosion", Houston, Texas, March 22-26, 1976, pp. 66-72, Publ. NACE, Houston (1977).
25. G.J. Theus, J.R. Cels, Slow Strain Rate Technique: Application to Caustic SCC Studies, Proc. ASTM Symposium on SCC "The Constant Strain Rate Technique", Toronto, May 2-5, 1977, pp. 81-96, Publ. RDTPA (Babcock and Wilcox).
26. E.D. Verink, M. Pourbaix, Use of Potentiokinetic Methods at Successively Increasing and Decreasing Electrode Potentials in Developing Alloys for Saline Exposures, Rapports Techniques CEBELCOR 117, RT. 191 (1971).
27. R.L. Cusumano, Establishment of a k hree-Dimensional Potential-pH-Composition Diagram for Binary Fe-Cr Alloys in 0.1 Molar Chloride Solutions, M.Sc. thesis Gainesville (1971).
28. K.K. Starr, The Significance of the Protection Potential for the Fe-Cr Alloys at Room Temperature, Ph.D. thesis Gainesville (1973).
29. F.N. Rhines, M. Pourbaix, Creation of a Center for Applied Thermodynamics and Corrosion, Rapports Techniques CEBELCOR 117, RT. 192 (1971).
30. Yang Wu, Personal Communication (8 May 1981).
31. L. Sathler, J. Van Muylder, On the Chemical and Electrochemical Nature of the Solutions Inside Occluded Corrosion Cells (in French) – Rapports Techniques CEBELCOR 126, RT. 224 (1975).

DISCUSSION

Comment by R.M. Latanision:

I just want to add a comment to Professor Pourbaix's excellent lecture. Namely, there is a certain symmetry involved in the influence of localized geometries on the potential and pH distribution. In short, there have been observations of hydrogen evolution from crack tips in anodically polarized specimens and, conversely, dissolution on the surface of cracks formed on specimens polarized cathodically. Indeed, depending on the potential and pH in a crack tip relative to the potential and pH at the electrode surface, both reactions may even occur simultaneously in a crack tip.

Reply:

I thank Professor Latanision for his interesting comment.

Comment by D.J. Duquette:

Your results underline one of the basic problems in determining the "mechanism(s)" of stress corrosion cracking or hydrogen embrittlement. That is, you have shown that cracks and pits often contain both hydrogen and anodic dissolution products. Accordingly, the rate determining step for cracking may be controlled by the kinetics of the rate determining step in possibly anodic dissolution, hydrogen production or even diffusion independent of the atomic mechanism of cracking. Would you agree that it may be more productive to study the electrochemical processes involved in cracking until more is known about the physics and chemistry of the solid crack tip?

Reply:

I thank you for your kind comments. I certainly agree that systematic studies on the electrochemical processes involved in cracking may quickly lead to results of great scientific and technical importance, notably by helping to set-up better alloys and convenient protection methods. Such results may be obtained even in the absence of further progress about the physics and the chemistry of the crack tip.

However, in view of the best possible solution of the problem, the best would be that such rapid electrochemical studies be conducted parallel with slower metallurgical ones, by teams having expertise in both fields. This is what we had in mind when creating in 1970, together with F.N. Rhines (University of Florida), the "Center for Applied Thermodynamics and Corrosion" (see CEBELCOR's Rapports Techniques RT 132 (1970), to promote Belgium-American collaboration in these fields.

THE GROWTH OF SURFACE FILMS IN ELECTROLYTES

J.C. Scully

Department of Metallurgy

Leeds University, Leeds LS2 9JT.

INTRODUCTION

The growth of surface films in electrolytes is a subject that covers a wide range of phenomena concerned with many different metals and alloys. At one end of the range are thin passive films (<5nm), at the other end, thicker metal salt films. In discussing the subject in relation to a conference on fracture this paper has had two objectives. The first has been to describe the general phenomenon of film growth by considering briefly the thermodynamic factors, after which the principal theories of passivity are described. Initiation and growth are discussed together with structural and chemical factors. The second objective has been to illustrate the relevance of film formation to environmental aspects of fracture. Inevitably, in covering such a wide subject area only a relatively simple presentation is possible in the space available. It is intended that the cited references will suffice to allow a deeper study of any particular subject area to be made if this is required.

THE THERMODYNAMICS OF PASSIVITY

When a metal reacts with an aqueous electrolyte a number of reactions may occur which can be divided into two basic types depending upon whether the product is soluble or insoluble. Both involve an oxidation process. Thus oxidation of a metal can lead either to active dissolution

$$M \rightleftarrows M^{z+} \text{ (aq.)} + ze \tag{1}$$

for which the equilibrium potential V is

637

$$V = V_O + \frac{RT}{zF} \ln a_M z^+ \text{ (aq.)} \tag{2}$$

where V_O is the standard equilibrium potential, $a_M z^+$ (aq.) is the activity of the solvated cation, and R, T, z and F have the usual meanings, or to oxide formation

$$2M + zH_2O \rightleftarrows + 2zH^+ + 2ze \tag{3}$$

for which V is given by

$$V = V_O - \frac{RT}{F} pH \tag{4}$$

The equilibria between metal and solution can be represented in a diagramatic form, showing the regions of stability of the different phases. Potential - pH diagrams, usually known as Pourbaix diagrams[1], have been constructed for most metals in pure water at 25°C and at atmospheric pressure.

Potential - pH diagrams merely show regions of thermodynamic stability of a compound (oxide or hydroxide). The protective properties of such compounds are ultimately dependent on factors like porosity, crystal structure and growth kinetics. Hence they can be used only as a first approximation in predicting whether passivation is likely to take place.

Although in neutral and alkaline solutions the onset of passivity is in reasonable agreement with the thermodynamically calculated values, in acid solutions passivity can be established at much lower potentials than those predicted by Pourbaix diagrams. These are thermodynamically metastable oxides or hydroxides which dissolve very slowly in solution. Another departure from equilibrium conditions is illustrated by the fact that higher than "normal" oxidation states are often reported in the literature, e.g. Ni_3O_4, Ni_2O_3, NiO_2[2] and Co_3O_4[3]. Also, non-stoichiometric compositions are more the rule than the exception in passive films. Examples are $NiO_{1.5-1.7}$ on nickel[4,5] and the films on stainless steels[6-8]. These and other non-equilibrium states are not surprising considering that very large driving forces are normally used in anodic polarization[9]. From a thermodynamic point of view the potential at which a thin film starts to be stable may be lower than the equilibrium potential for the bulk phase, as shown by Vermilyea[10]. Also, when the composition of the film is not uniform through its thickness, as is the case for stainless steels[6-8], no single equilibrium potential can be expected.

It must also be mentioned that equilibrium diagrams for the most relevant systems in practice, namely alloys in aqueous solutions with ionic additions, are more difficult to construct.

THE MECHANISM OF PASSIVITY

The nature of the surface film that renders a metal surface passive has not been easy to determine. Even the definition of passivity has created problems since, although all metals develop oxides and hydroxides at certain values of electrolyte pH and electrode potential[1], this does not necessarily make them passive. In this section an attempt is made to discuss the nature of the passive film from the viewpoint of why it causes a metal surface to become relatively inert.

The passive state is usually characterized by a low corrosion rate and by a shift in potential into the noble region. These two characteristics have led to two different definitions of passivity. According to the definition based on the corrosion rates[11] "a metal or alloy is passive if it substantially resists corrosion in an environment where thermodynamically there is a large free-energy associated with its passage from the metallic state to appropriate corrosion products". The other definition states that a metal, usually considered to be active according to its position in the e.m.f. series, is passive if it behaves electrochemically like a noble metal.

A more convenient definition of passivity was given by Wagner[12]. The definition, which distinguishes between passivity and inhibition, is as follows: "a metal or alloy is called passive when the amount of at least one of the metallic components con-sumed by a chemical or electrochemical reaction in a given time is significantly lower at a higher affinity than at a lower affinity". This is equivalent to saying that a metal shows passive behaviour when its anodic polarization curve is of the type shown in Fig. 1.

The cause for the loss in reactivity at potentials above the primary passive potential V_{pp} is the presence of a film, the passive film, on the surface of the metal. Although its existence is no longer disputed its nature has not been definitely established yet. Some authors maintain that the film consists of an adsorbed monolayer, while others consider it to be a three-dimensional oxide layer. However, as will be shown below, rather than being irreconcilable, both ideas are complementary in the sense that, in most cases, particularly for the transition metals, adsorption precedes oxide film formation.

The Adsorption Theory

The adsorption theory postulates that the film consists of adsorbed oxygen whose role is to decrease the exchange current density, i_o, or to increase the overpotential, η, for anodic dissolution. According to the oxide-film theory the passive layer consists of a three-dimensional oxide which isolates the metal from the solution; the corrosion rate of the metal is then dependent upon the properties of the passivating oxide.

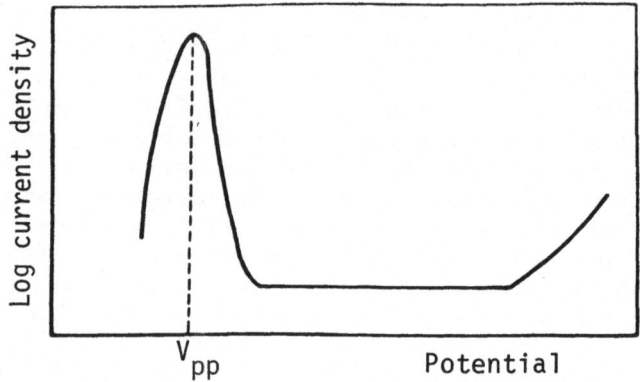

Fig. 1.   Schematic potentiostatic anodic polarization curve of a
           metal exhibiting active-passive behaviour. V_{pp} – primary
           passive potential.

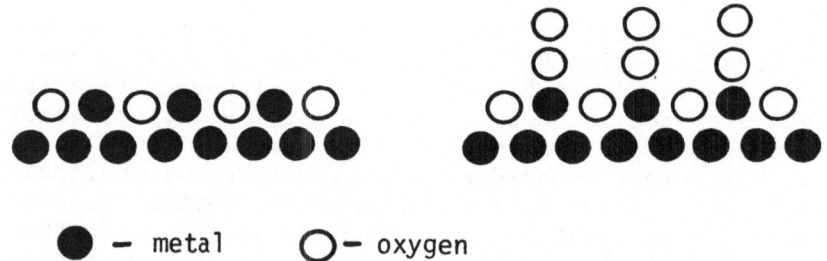

Fig. 2.   Schematic illustration of chemisorbed atomic and
           molecular oxygen making up the passive film.

As far as thick films are concerned there cannot be much argument. Visible films of lead sulphate on lead immersed in H_2SO_4 and
of iron fluoride on steel immersed in aqueous HF are examples of
protective films that successfully isolate the metal from its
environment. It is in the region of thin film thicknesses (<5nm)
that most argument arises. The reason for this is not so much related with the experimental evidence, which gives support to both
theories, but mainly with differences in terminology. This is
particularly relevant when the passive film corresponds to less
than monolayer coverage. Uhlig[13], for instance, has proposed for
the passive film the structure represented in Fig. 2. This is
based on low-energy electron diffraction data obtained by
Mac Rae[14,15] indicating that, for nickel single crystals, the first
adsorbed layer consists of a regular array of oxygen and nickel
ions located in the same approximate plane of the surface.   In
earlier works Uhlig[16,17] had proposed that the passive film on
transition metals consisted of an adsorbed structure, where the
metal ions kept their original positions in the metal lattice, in
contrast with what is suggested in Fig. 2. This was explained on
the basis of the high heats of sublimation of transition metals
compared to those of non-transition metals. When the heat of sublimation is high the metal ions tend to remain in their lattice
positions, resulting in the formation of adsorbed structures.
When the heat of sublimation is low the formation of oxides, which
requires the atoms to leave the lattice, is considered to be the
preferential reaction. A similar distinction between adsorbed and
oxide structures on Pt has been proposed by Damjanovic[18]: "...an
adsorbed oxygen layer is distinguished from oxides in that, in the
latter, platinum atoms have left their regular positions in the
lattice of platinum to enter, together with oxygen atoms, into a
new arrangement of atoms similar to the arrangement in the formation of the nuclei of the respective oxides", whereas in the former
the platinum atoms maintain their positions in the lattice.
According to this definition the structure represented in Fig. 2
would be that of an oxide rather than of an adsorbed film, because
some metal atoms have already abandoned the metal lattice. This
particular example and the multiplicity of terms that can be found
in the literature to characterize very thin passive films, e.g.
monolayer (or less) of oxide, two-dimensional film, adsorbed oxygen
and oxygen-containing species, show that semantic discrepancies can
be a major source of disagreement, perhaps as important as results
obtained under different experimental conditions.

Uhlig is usually associated with the adsorption school but
from as early as 1948[19] he made a distinction between the transition metals, for which the adsorption model would apply, and the
non-transition metals, which would be rendered passive by a
barrier-type film.

The main characteristic of the model proposed in early

works[16,17] is the existence of a chemisorbed monolayer of oxygen
atoms through gaps of which a layer of oxygen molecules is chemisorbed.   In 1967 an important modification was introduced in this
model and the structure shown in Fig. 2 which, as seen before, can
be classified as an oxide, was proposed[13].   Uhlig analysed from a
thermodynamical point of view the transition from an adsorbed
structure to an oxide film.   After the chemisorption of the two
oxygen layers more O_2 molecules are adsorbed, the strength of the
bond corresponding to physical adsorption rather than to chemisorption.   The free-energy for the transformation

$$(nO_2).O.M \text{ (ads.)} + 2nM \rightarrow (2n + 1) \text{ MO} \tag{5}$$

is considered to be largely positive if only the chemisorbed layers
(n=1) are taken into account.   As more O_2 molecules are physically
adsorbed, the formation free-energy of the adsorbed structure becomes less and less negative and the value of ΔG decreases until,
eventually, it becomes negative.   The oxide becomes then the
thermodynamically more stable form of the passive film.

More recently, the presence of metal ions, as well as of
protons, in a "sea of adsorbed negatively charged oxygen ions" has
been proposed by Uhlig[20], according to whom this structure can be
considered as a non-stoichiometric oxide.   The question of stoichiometry is not a major source of disagreement between the two
theories of passivity.   For instance, Bockris and colleagues[4],
who support the oxide-film theory, consider that the essential step
in the passivation of nickel in acid solutions is the formation of
a well conducting, non-stoichiometric NiO oxide.   The main contradiction between both theories is related to the role of the passive
layer in hindering the transfer of metal cations into the solution,
as is now explained.

For Uhlig[13] the adsorbed oxygen layer partially or wholly
replaces the adsorbed (coulombic) water layer, thus increasing the
activation energy for dissolution.   As a result, the exchange
current density i_o decreases and the overvoltage η $\left[\eta = \beta \ \log\left(\dfrac{i}{i_o}\right)\right]$
increases.   The situation is considered as one of
increasing anodic polarization caused by successive additions of
oxygen layers (decreasing i_o).   As will be discussed later, the
oxide theory proposes a mechanism of passivity based on the film
acting as a diffusion barrier.

The adsorption model of Uhlig was developed for transition
metals only and has not been extended to non-transition metals.
The distinction made between these two groups of metals is related
to differences in their respective electronic properties.

Initially, the prediction of the passivating structure was
based on the ratio of the work function ϕ to the heat of sublimation ΔH_s [19].   When the ratio $\phi/\Delta H_s$ is less than 1 (transition

metals) the electrons tend to be removed more easily than the metal atoms; this is considered to favour the adsorption of oxygen. When the ratio is greater than 1 the above tendency is reversed, the oxide becoming the most likely form of the passive film.

In 1967 the problem was treated in a more accurate way[13]. The free-energy change ΔG for the adsorbed-layer to oxide-layer transformation, Eq. (5), was considered. The values of ΔG were calculated for Ni, Zn and Fe. In the case of Ni the adsorbed structure was found to be more stable than the oxide ($\Delta G > 0$), whereas in the case of Zn the calculations predict that the oxide ZnO is more stable than $Zn.O.O_2$($\Delta G < 0$). Fe is in a borderline situation ($\Delta G \approx 0$). The determining factor in the calculations is the heat of adsorption of oxygen, which depends on the electronic configuration. Unfilled d orbitals (transition metals) are expected to lead to strong bonds between oxygen and metal (high heats of adsorption), whereas filled d orbitals (non-transition metals) are likely to lead to weak bonds (low heats of adsorption).

Uhlig and co-workers have extended the electron configuration theory to Cu-Ni[2] and Cr-Ni-Fe[22], alloys. In Cu-Ni alloys passivity is observed only when the nickel content is greater than 35%. This is attributed to the d bonds becoming unfilled at this concentration. The addition of electron donors (Zn, Al, Ga, Ge) increases the critical nickel concentration, whereas the addition of transition metals (Fe, Co) decreases it[21,22]. The authors have interpreted these results as a confirmation of the theory.

The lack of correspondence between the Flade potential and the calculated equilibrium potential for reactions involving any of the known iron oxides is a matter that has attracted much interest over the years. Uhlig has proposed[17] that the passive film forms according to the reaction:

$$3H_2O + Fe \text{ surface} \rightarrow O.O_2(\text{ads. on Fe}) + 6H^+ + 6e \quad (6)$$

The standard equilibrium potential, calculated on the basis of data obtained for chemisorption of gaseous oxygen, is 0.56V(SHE); the experimental values range between 0.58 and 0.68V(SHE). As the adsorbed structure is not thermodynamically more stable than the oxide, it is not very clear why the theoretical value calculated by Uhlig is in reasonable agreement with the experimental data. In the calculations Uhlig always considers the formation of stoichiometric oxides, the thermodynamic properties selected being those of the bulk. The films observed in practice are not necessarily thermodynamically stable and/or stoichiometric. On the other hand, during the nucleation of an anodic film, the surface free-energies of the electrode-solution, electrode-film and film-solution interfaces have to be considered, in addition to the energy change per unit volume accompanying the formation of the film. It has been shown[10] that when the surface free-energies are taken into account a monolayer of film may become more stable than the bulk phase, making possible its formation at potentials negative to the equilibrium potential. As a result, the free-energy change for the reaction[5] would decrease when the surface free-energies are taken into account and the transformation of an adsorbed film into an oxide becomes easier than that which is predicted by Uhlig's calculations.

Passivity is considered by Kolotyrkin[23] as a special case of change in anodic kinetics. The electrochemical reaction on a passive surface consists, as in the active state, of a transfer of cations into the solution. The action of adsorbed oxygen is assumed to be one of increasing the overpotential in the double layer, which results in the hindrance of the dissolution process.

The Oxide-Film Theory

According to this theory the phenomenon of passivity results from the formation of a protective film consisting of the products of the reaction of the metal with the electrolyte. The film is thought to constitute a barrier to the migration of metal ions into the solution under the action of an electric field. The oxide-film theory applies not only to oxides, both stoichiometric and non-stoichiometric, but also to other types of barrier films, namely oxyhydroxides and salt-like films, to which passivity has sometimes been ascribed. Hence, a more appropriate designation for the theory would be "barrier-film" theory. The presence of a film on metal surfaces before the onset of passivity is also observed in some cases, as is indicated below. The theory derives considerable support from electron diffraction, ESCA and Auger analysis, as well as from ellipsometric measurements. For instance, it has been well established that the main constituent of the passive film on iron is $\gamma-Fe_2O_3$ [24-26] and that its thickness ranges from ca. 1 to ca. 5nm, depending on solution composition, applied potential and time[27-31]. The characteristics of this oxide, namely composition, structure and electronic properties, are still a source of disagreement, even among the supporters of the oxide-film theory. However, the most controversial and fundamental question that has to be answered is in what way this oxide can be responsible for the onset of passivity?

Salt films may also occur before or during passive film formation. For example, Mueller[32] observed very thick layers of metallic sulphates precipitated under stagnant conditions. Usually, the films responsible for passivity are much thinner and invisible.

The mechanisms of formation of a salt-layer and of an oxide film, respectively the dissolution-precipitation and the solid-state mechanism, are still the two generally accepted at the present time for the formation of a protective layer. However, it can be difficult to distinguish between both, especially when metal dissolution occurs in parallel with the solid-state reaction. According to Armstrong[33] the most useful experiment to distinguish

between the two mechanisms is the rotation speed dependence of
the steady state current-voltage curve. When the dissolution-
precipitation mechanism is operative the metal should dissolve until
a critical concentration, given by the solubility product of the
film, is reached. The current at which precipitation starts increases with the square-root of the angular velocity[33], the corresponding passivation potential, V_{pp}, being given by the Tafel
equation. It follows from this that V_{pp} should increase as the
angular velocity becomes faster. On the other hand, no dependence
of the active-passive transition is anticipated for a solid-state
mechanism.

The effect of anion adsorption and thin film formation on the
rate of metal dissolution have been investigated by Armstrong and
Thirsk[34]. They found that specific anion adsorption is not a
major cause of passivity and that, in general, a three-dimensional
film is required to prevent metal dissolution. In addition, they
found that two-dimensional films may also have an inhibitory effect,
but in a number of cases they do not[34].

For Vetter[35] the formation of a three-dimensional film is not
essential and he suggests that one oxide monolayer, and sometimes
even less, is sufficient to totally inhibit the anodic dissolution
of noble metals, nickel and chromium. Sato and Okamoto[36]
suggested that passivity was caused by a dissolution-precipitation
mechanism represented by the following reactions:

$$Ni + OH^- \rightarrow NiOH^+ + 2e \tag{7}$$

$$3NiOH^+ + OH^- \rightarrow Ni_3O_4 + 4H^+ + 2e \tag{8}$$

The Ni_3O_4 layer was regarded as having self-repairing properties.
In the event of pores being present in the film, the concentration
of $NiOH^+$ within them would rise rapidly (reaction (7)). This would
lead to an increase in the rate of reaction (8), and consequently
to the rapid blocking of the pores by Ni_3O_4. A similar pore-
plugging phenomenon was invoked by Pavlov and co-workers[37] to
explain the formation of protective layers on the $Pb/PbSO_4$ electrode.

The identification of oxides and the measurement of film
thicknesses corresponding to three-dimensional layers are good
evidence for the oxide-film theory. A third and strong argument
in favour of this theory is given by Flade potential data. For a
considerable number of metals[38] there is a close correspondence
between the Flade potential, which approximates to the passivation
potential, and the equilibrium potential of reactions of the type:

$$M + nH_2O \rightleftarrows MO_n + 2nH^+ + 2ne \tag{9}$$

for which

$$V_F = V_F^o - 0.058pH \text{ (V)} \tag{10}$$

According to the oxide-film theory, the direct hydration of metal ions occurring in the active state ends as soon as the oxide film becomes interposed between the metal and the solution. The large change in the dissolution kinetics observed above the passivation potential must be related to the properties of the film and, in particular, to its chemical resistance and ionic conductivity. In those cases in which the surface of the metal is bare in the active region the beneficial role of a barrier film, formed at potentials greater than V_{pp}, can be easily understood. But when the active metal is covered by some sort of solid product, the onset of passivity has to be explained on the basis of some fundamental change in the properties of this product.

Ellipsometric studies by Bockris et al[4] have shown that a pre-passive film, probably $Ni(OH)_2$, exists on nickel immersed in a $0.01N\ H_2SO_4 + 0.5M\ K_2SO_4$, pH $\stackrel{\sim}{=}$ 3.15 solution. At the passivation potential the hydroxide undergoes a non-stoichiometric change to $Ni\ O_{1.5-1.7}$, accompanied by a sharp increase in the absorption coefficient. This increase was taken as an indication of a change from ionic to electronic conduction. The drop in the corrosion current observed above V_{pp} would be, according to the authors, a mere reflection of the decrease in the rate of migration of the metal ions through the film. The restricted application of the mechanism is perhaps more clearly shown by the fact that Al_2O_3 and Ta_2O_5, good passivating films, are very poor electron conductors[38].

In recent years not only the classical defenders of the adsorption theory, namely Kolotyrkin and Uhlig, have compromised with the oxide-film school, but also the reverse has been taking place. As seen above, Vetter proposed an adsorption step followed by a place-exchange step for the mechanism of film formation[35]. The same sequence, which is represented in Fig. 3, has also been favoured by Bockris et al[29].

Mac Dougall and Cohen[39] have recently shown that for less than monolayer coverage of nickel electrodes in acid sulphate solutions the film is reduced with a current efficiency of 100%. This film is thought to be NiO_{ads}, which subsequently undergoes a transformation into nickel oxide NiO. The latter was found to be much more difficult to reduce than the adsorbed film[39]. The oxide on a Fe-24%Cr alloy has also been reported to have similar characteristics[40]. This oxide was found to exist only at potentials positive to the activation potential, V_z, which was defined as the minimum potential at which complete passivity was observed. In this region an adsorbed film is thought to provide most of the protection. For example in a potentiostatic transient the current is reduced to 1% of its initial value with a surface coverage of only $\theta=0.01$ ($\theta=1$ for complete coverage). In order to explain this high degree of passivation with only a small coverage, the preferential-dissolution/preferential-adsorption mechanism was put forward[41,42]. It assumes that the reactivity of a metal

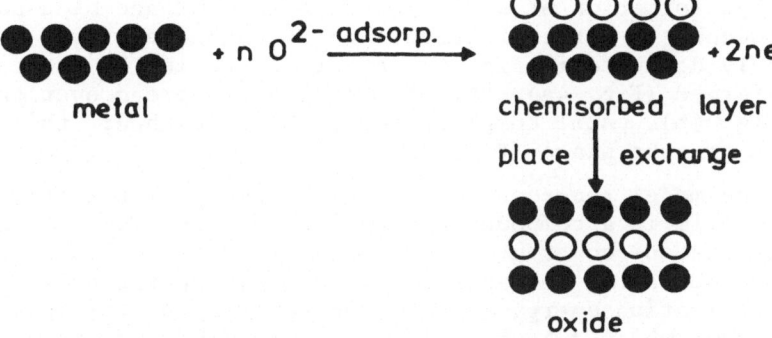

Fig. 3. Formation of the first oxide monolayer by chemisorption followed by place-exchange of the M-O pairs.

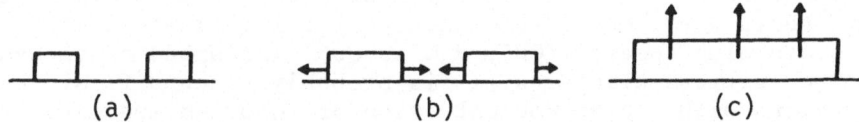

Fig. 4. Schematic sequence of events leading to the formation of a three-dimensional film. After being nucleated (a) the film spreads laterally (b) until monolayer coverage; finally, thickening (c) takes place.

depends on its microscopic surface topography. The order of reactivity of the surface sites, both for dissolution and for adsorption is: kink > ledge > terrace.   The theory conforms fairly well
with experiment in the potential region V_a < V < $(V_a+200$ mV) but
is not valid for higher potentials;   in this case a three-
dimensional oxide layer is required to initiate and maintain passivity[42]. This theory has also been applied to iron in borate
buffer solutions at potentials V_{pp} < V < $(V_{pp}$ + 200 mV).

THE NUCLEATION OF PASSIVE FILMS

        The formation of a new phase on the surface of a bare
electrode occurs in three different and consecutive stages,
schematically depicted in Fig. 4. After the nuclei of the new
phase are formed (Fig. 4a), the film starts to spread over the
surface (Fig. 4b), until complete coverage is reached; thickening
starts then to take place (Fig. 4c).

        The model represented in Fig. 4 assumes implicitly that
the passive film is a compound and not an adsorbed layer. Any
increase in surface coverage by adsorbed particles can only be
attained by increasing the applied potential. On the other hand,
a certain activation energy is required to form the critical nucleus
of any compound which subsequently grows, at constant potential,
until complete surface coverage is reached.

        If a combined adsorption-oxide-film mechanism is adopted
one question concerning the initial adsorbed film may arise. How
does this film influence the nucleation and lateral spreading of
the secondary oxide film? On those sites having a large heat of
adsorption, e.g. kinks[41], an adsorbed structure should form more
or less instantaneously. At this moment metal atoms would begin
to leave their positions in the lattice, forming clusters with the
oxygen containing species (OH^-, H_2O or SO_4^{2-}) adsorbed on the surface. The critical cluster size is probably reached first in
those regions with a high concentration of adsorbed species, namely
kinks. Only after these processes have taken place can the schematic sequence represented in Fig. 4 start.

        Apart from being the precursors of the oxide nuclei, the
adsorbed species are likely to influence the development of the
oxide film in another way. If during the spreading of the film
over the metal surface they are not incorporated into the oxide
lattice, they may hinder the step motion, having an effect similar
to that of impurities[43]. The effect of adsorbed extraneous species
on the growth of oxide films has received very little attention in
the literature.

FILM NUCLEATION

        In the absence of particles that can act as preferential
sites for deposition of the new phase - heterogeneous nucleation -
the atoms or molecules have to rearrange themselves in clusters of

a minimum size before growth can take place. These critical clusters, or nuclei, have equal probability to grow and to dissolve.
Larger clusters will tend to grow whereas smaller clusters have a
greater tendency to dissolve. The formation of a nucleus is similar to a polymerization reaction[44], as shown in Fig. 5. No interaction of subcritical clusters is considered because their concentration is usually very small and, therefore, the probability of
two of them colliding is negligible.

For a circular nucleus the radius r* is given by[45]:

$$r^* = \frac{\lambda s}{zF|\eta|} \qquad (11)$$

where λ is the boundary free-energy per unit length, s is the area
covered by one mole in a mono-atomic layer and z, F and η have the
usual meaning.

In what seems a more correct treatment, Vermilyea[10] has
considered the surface free-energies of the electrode-solution,
σ_1, electrode-film, σ_2, and film-solution, σ_3, interfaces. For a
film patch of height $\bar{a}$ the expression for the critical nucleus
size is:

$$r^* = \frac{Ma\sigma_3}{4.2 \times 10^7 \; azF\rho\eta - M(\sigma_2+\sigma_3-\sigma_1)} \qquad (12)$$

where ρ and M are, respectively, the density and the molecular
weight of the film material.

After the nuclei have formed, either by homogeneous or
heterogeneous nucleation, their growth may occur in a number of
different ways. When metal ions cannot enter the solution, growth
can proceed at the edges or at the top of the nuclei. However,
because of ion migration through the film, growth at the top should
be a slow process. As a result, the centres are likely to acquire
a platelet shape. When metal ions can enter the solution the
nuclei also have a tendency to spread as platelets, although they
may initially grow perpendicularly to the surface, especially on
dirty electrodes[46]. If deposition occurs at a uniform rate, the
nuclei tend to acquire a platelet shape because the increase in
diameter is twice as fast as the increase in height. Experimental
evidence compiled by Vermilyea[46] indicates that while films formed
directly on the metal surface have a tendency to become flat,
those resulting from the transformation of another film grow in
three dimensions.

Although a large number of factors influence the kinetics
of electrodeposition, three processes can be considered as rate
determining[47]: (a) metal dissolution, (b) resistance in the
electrolyte or in the external circuit, and (c) film formation.

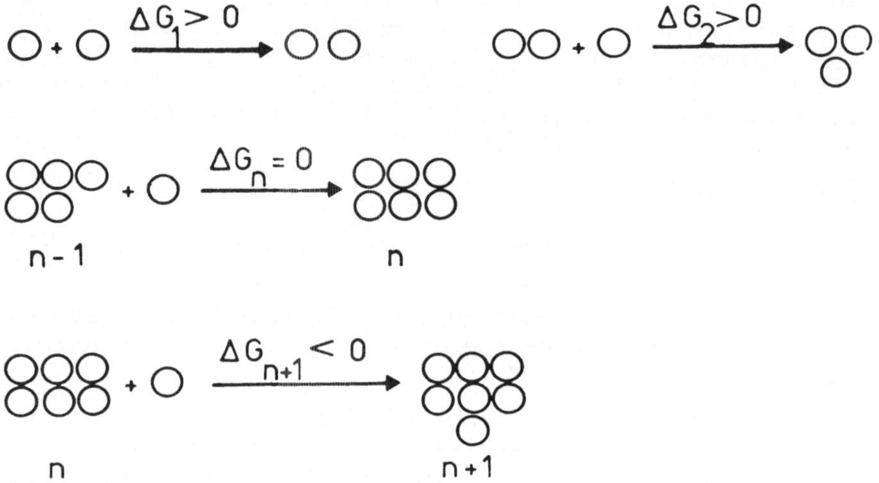

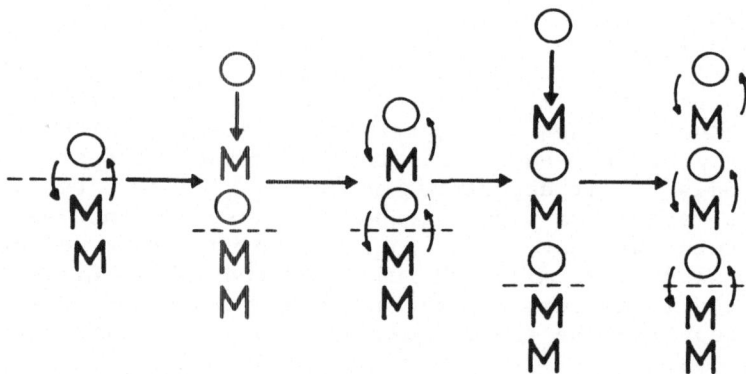

Fig. 5.   Chain of polymerization reactions involved in the formation of the critical cluster (nucleus); large clusters
grow spontaneously ($\Delta G_{n+1} < 0$).

Fig. 6.   Schematic representation of the place-exchange process;
M—metal ion, O—oxygen ion.

When metal dissolution is the slowest process the current density, under potentiostatic conditions, decays according to the following law[47]:

$$i = i_1 \exp \left(- \frac{t}{\tau}\right) \tag{13}$$

where $\tau = zF\rho\delta/Mi_1$, δ being the film thickness and i_1 the current density on the bare surface.

Although a dissolution-precipitation mechanism may apply to the passivation of some systems, a solid-state transformation is the most widely invoked. The theory of electrodeposition of films resulting from a solid-state reaction has been extensively treated by Thirsk and co-workers[46-50]. They have considered several situations: (a) instantaneous and progressive nucleation, (b) one-dimensional, two-dimensional and three-dimensional growth, (c) effect of mass transport through the film patches, (d) interaction of patches only with themselves and interaction with the boundaries of the electrode, (e) effect of surface and bulk diffusion, and (f) galvanostatic and potentiostatic conditions. For more detailed information the reviews published by Thirsk et al[46,47,50] should be consulted.

FILM GROWTH

Film thickening is usually regarded as occurring by one of two mechanisms, viz. ion-migration and place-exchange. A three-dimensional film may also be obtained by the successive deposition of several monolayers, each developing by a combination of nucleation and electrodeposition processes similar to those described in the previous section. It would seem that this type of three-dimensional growth may be applicable to films formed by a dissolution-precipitation mechanism but not to those produced by a solid-state reaction.

Ion Migration

Transport in anodic films has been extensively studied both from a theoretical and an experimental point of view.

During oxide growth metal cations leave the lattice at the metal-oxide interface and migrate towards the solution, whereas anions (oxygen-containing and others) will tend to move in the opposite direction. The combination of cations and anions, to form the film, may occur in principle anywhere between the metal-oxide and the oxide-solution interfaces. Depending on which is the slowest step, the rate of film thickening may be controlled by the movement of ions (a) across the metal-film interface, (b) through the film, and (c) across the film-solution interface.

The first two cases, as well as situations of mixed control, are those usually considered in the literature.

The fundamental equation in electrolytic transport is[51]:

$$j_i(x) = -D_i \left(\frac{dn_i}{dx} + \frac{z_i e n_i}{kT} \frac{d\psi}{dx} \right) \tag{14}$$

where $j_i(x)$ is the flux of species i in the direction of the x-axis, D_i the diffusivity of species i, n the number of particles of i per unit volume and ψ the electrical potential. This equation indicates that the total flux is the sum of two contributions: one due to a concentration gradient and the other associated with the electric field. The solution of this equation may be complicated by the existence of space charges in the film or changes in composition and/or structure during growth.

When the film does not change composition ($dn_i/dx = 0$) and structure during growth and when only one species is mobile, solution of equation (14) is relatively simple. Two limiting cases can be considered: thick films ($\delta > 10^3$nm) and thin films ($\delta < 10$nm) corresponding respectively to low ($E < 10^4$ V/cm) and high ($E > 10^6$V/cm) electric field strengths. The former grow according to a parabolic law

$$\delta^2 = A t \tag{15}$$

whereas the latter follow the inverse logarithmic law

$$\frac{1}{\delta} = C - D \ln t \tag{16}$$

Equation (15) can easily be obtained integrating equation (14), taking into account that the potential gradient, $d\psi/dx$, is approximately constant, because the film is considerably thick compared to the thickness of space charge regions at the film interface.

The inverse logarithmic law of film growth results from the theory developed by Mott and Cabrera for ionic transport[52].

Place-exchange

This mechanism has been proposed by Sato and Cohen[53] to account for results obtained with iron in boric acid-borate solutions. The sequence of events leading to the formation of the passive layer is schematically shown in Fig. 6. Each M-O pair rotates 180° and, while the O atom gets underneath the first metal atomic layer, the M atom reacts with another oxygen ion. This process repeats itself and by successive rotations the M atoms move away from the metal-oxide interface. The theory is then developed on the assumption that all M-O pairs in a row rotate simultaneously. The current density associated with film growth, which is proportional to $d\delta/dt$ (Faraday's law), is expressed as:

$$i_f = K \exp \left(\beta V - \frac{\delta}{B} \right) \tag{17}$$

where K, β and B are constants. Integration of this equation gives the direct logarithmic law of film growth

$$\delta = k_0 \ln \left(\frac{t}{\tau} + 1 \right) \tag{18}$$

sometimes written as

$$\delta = k_1 + k_0 \ln (t + \tau) \tag{19}$$

k_0, k_1 and τ being constants.

Film growth kinetics are usually studied by either coulometric or ellipsometric techniques. A few examples can be cited. Moshtev[54] carried out coulometric experiments with iron in boric acid-borate solutions, in the same conditions of Sato and Cohen's work and found that the results were consistent with the high-field migration mechanism rather than with the place-exchange mechanism. No reasonable explanation has been given yet for this discrepancy.

Results reported by Ord et al[55-57] on Fe, Ta and Pt suggest that a high-field assisted ion migration mechanism is operative for the three metals. The same conclusion was drawn by Goswami and Staehle[30] for Fe, Fe-Ni, Fe-Cr and Fe-Cr-Ni alloys in neutral solutions. Their results, however, also fit the direct logarithmic law. Bulman and Tseung have used coulometry[58] and ellipsometry[59] to study the film growth kinetics on Type AISI 316 stainless steel in 5N H_2SO_4. With both techniques the film has been found to grow according to the direct logarithmic law. Frankenthal[60] has cast some doubts on these results because Bulman and Tseung have assumed that all current goes into film formation and none into dissolution and also because a film-free surface was probably not obtained with their pre-anodic treatment (the open-circuit potential after cathodic reduction was well within the passive region of Fe-24%Cr in 1N H_2SO_4[40] and in 2N H_2SO_4[41]). Smialowska and Mrowczynski[61] have found ellipsometrically that film growth on iron immersed in sodium sulphate solutions follows the direct logarithmic law for potentials greater than 500 mV (SHE), when the pH varies between 6 and 10, and in the whole range of anodic potentials for pH = 12. In the range 6-10 the films formed below 500 mV grow linearly and are not protective. Data obtained by Kruger and Calvert[27] on iron in slightly alkaline solutions fitted either a logarithmic or an inverse logarithmic rate law.

The Time Dependence of i

The expression for the potentiostatic current transient yielded by the direct logarithmic law can be easily obtained by differentiating Eq. (19) and taking into account that

$$\frac{d\delta}{dt} \propto i_f \quad \text{(Faraday's law)}.$$

For values of t considerably larger than τ the current becomes

$$i_f = a \, t^{-1} \tag{20}$$

The same relationship is obtained for the inverse logarithmic law
(Eq.(16)). On the other hand, the parabolic law (Eq. (15)) gives

$$i_f = a \, t^{-0.5} \tag{21}$$

As pointed out by Wagner[51] a continuous transition from the
limiting case of thin films to the limiting case of very thick
films – with transport obeying the classical laws of diffusion –
should be expected. Accordingly, the general expression for the
current transients, contemplating intermediate situations, is

$$i_f = a \, t^{-b} \tag{22}$$

with $0.5 \overline{<} b \overline{<} 1.0$. A log i_f – log t plot of this equation is a
straight line with a slope $-b$. This type of current decay has been
observed for many systems, e.g. Ti6A14V in 1.0N NaX (X=F, Cl, Br,
I)[62] Type AISI 316 stainless steel in 5N H_2SO_4[58], iron in neutral
borate-boric acid solutions[53], cobalt in borate[63] and bismuth in
benzoate, borate, Na_2HPO_4 and NaH_2PO_4[64].

The value of b cannot be used to distinguish between the ion-
migration and the place-exchange mechanism but it may give valuable
information about the protective properties of the film. As b decreases, i.e. as the situation of parabolic growth is approached,
the charge consumed in anodic oxidation increases and the film becomes less protective.

CORROSION IN THE PASSIVE STATE

The formation of a passive layer is accompanied by a large decrease in the rate of oxidation of the metal and, under steady-
state conditions, current densities well below 1 $\mu A.cm^{-2}$ may be
obtained. Usually the establishment of a dynamic equilibrium
corresponding to a true steady-state is a very slow process, last-
ing in some cases days or even weeks. In the absence of film
dissolution, film thickening would proceed indefinitely and no
stationary state would be reached. Since most films dissolve in
acid solutions, a limiting film thickness is often encountered in
practice. Before a constant thickness is reached the events
involved in the passivation of a metal surface are, in the
most general case,

I   -   film thickening

$$M + H_2O \rightarrow MO + 2H^+ + 2e \tag{23}$$

II   -   chemical dissolution of the film

$$MO + 2H^+ \rightarrow M^{2+} \text{ (aq.)} + H_2O \tag{24}$$

III -   electrochemical dissolution of the film

$$MO + 2H^+ + e \rightarrow M^+ \text{ (aq.)} + H_2O \tag{25}$$

IV   -   electrochemical dissolution of the metal by transport of metal ions through the film

$$M \rightarrow M^{2+} \text{ (aq.)} + 2e \tag{26}$$

As chemical dissolution does not involve a change in the oxidation state the total current density is

$$i = i_I - i_{III} + i_{IV} \tag{27}$$

where the subscripts refer to the corresponding reactions.

Although the amount of metal oxidized, and present either in the film or in the solution, depends on the rate of all the four processes, only film thickening is relatively well understood. Very little effort has been devoted to measuring the amount of metal that has gone into solution.

Although a well passivated metal must be covered by a film with a slow rate of chemical dissolution, this condition, although necessary, may be insufficient to ensure a low corrosion rate. Processes III and IV are also likely to play a significant role. Provided the volume of solution is large enough to prevent any changes in concentration, chemical dissolution should be time independent. It should also be potential independent. The available evidence appears to indicate that the passive dissolution rate is neither time nor potential independent.

Whether all the current passing during film formation is consumed in that process is important since this will provide data upon the growth mechanism. There does not appear to be one general mode of behaviour even for one metal. An equivalence between film thickness and measured charge has been found[27,53] but in other cases[65,66,67] dissolution accounts for a major part of the film. The proportion depends upon the environment, nitrite, for example, allowing less current for dissolution than nitrate for mild steel[68].

The dissolution of oxides has been studied by Valverde[69,70].
Dissolution in many cases is predominantly electrochemical in
nature.

Film Structure

There is a general realisation that film structure is important in determining both breakdown e.g. pitting, and ductility[71].
Amorphous glassy oxides appear to be the most resistant to breakdown as illustrated by the behaviour of amorphous alloys when compared to the same alloys in the annealed state and by the improved
resistance of iron alloys as the chromium content is increased[72],
an effect accompanied by a gradual decrease in film crystallinity.
Film ductility is also affected by crystallinity and this feature
is important in relation to stress corrosion and corrosion fatigue.
Models based upon slip step emergence, for example, required the
film to break and thereby reveal underlying metal. The greater the
ductility of the metal the less likely are such events to occur.

Films on Ta have been shown to exhibit deformations as great
as 50% before fracture[73]. Films on Type 304 steel were more
ductile than those on iron[73]. Both types of ductile films were
non-crystalline. It is difficult to draw conclusions since systematic work has not been done. Ductility in films is affected by
crystallinity but other factors play a role also e.g. hydrogen
and water content, film growth rate, etc. In relation to fracture
it is rather surprising that this subject has not been more fully
explored.

FILM FORMATION AND STRESS CORROSION CRACKING

The importance of film formation in the propagation process
has long been recognized. It has been argued[74] that the crucial
event in the cycle of events producing an increment of crack growth
was a critical delay in the repassivation time. If this time was
too short, insufficient corrosion would occur to continue crack
growth whereas, if it was too long, too much corrosion would occur,
resulting, morphologically, in an elongated fissure or pit rather
than a narrow crack. Such ideas had come directly from experiments
at various constant crosshead speeds on a Ti-5Al-2.5 Sn alloy in an
aqueous NaCl solution. These had shown clearly that repassivation
controlled the crack growth process. At this stage[74] the essential
dynamic interaction of quite separate processes was emphasized: a
creep process producing fresh metal area, and a repassivation
process, together allowing crack propagation to occur only if repassivation was 'inadequate'. The propagation process was considered to occur as the result of a specific relationship between the
two processes.

Since that time the subject has been advanced by attempting to
analyze the propagation process as arising from the charge flowing
between slip step emergence events and repassivation events[75,76].

The Constant Charge Criterion[75] has been described in detail.

The term 'repassivation' has been used to describe forming
processes occurring at crack tips. The original paper, for
example, implied that cracking stopped when repassivation was complete, as described above, when the crack tip strain-rate in a constant load test had fallen to a very low value. The lower the repassivation rate the lower would be the strain-rate at which crack
arrest occurred. In corrosive environments instead of crack arrest,
cracking would give way to whatever form of corrosion occurred in
the absence of stress, as is discussed later. Where crack arrest
did occur the value of the crack tip strain-rate was designated[75]
$\dot{\varepsilon}_r$. For any specimen it was proposed that any threshold stress
for stress corrosion cracking, or K_{Iscc} value, corresponded to the
value of the crack tip strain-rate being $\dot{\varepsilon}_r$. From such considerations, film repair could prevent crack growth on a creeping crack
tip surface.

The physical significance of events giving rise to $\dot{\varepsilon}_r$ was not
considered[75]. Whether, for example, the film was ductile and
extended and perhaps thickened so that it never fractured to such
a depth as to reveal bare metal at such low strain-rates was not
discussed. Instead, abundant data were provided that such a concept was realistic whatever might be the precise physical reality
of such an occurrence. In a Ti-O alloy, for example, in 3% aqueous
NaCl solution, cracking occurred at a crosshead speed of 8.3 μm/s
but not at 3.3 μm/s in SEN specimens. If, after crack initiation
at the higher crosshead speed, the crosshead speed was reduced to
the lower value, crack arrest occurred. Continued straining at
the lower crosshead speed resulted in eventual ductile failure.
Stress corrosion cracking was never re-initiated. Similar events
of cracks arresting in specimens at low crack tip strain-rates
have subsequently been demonstrated in many other systems of stress
corrosion cracking.

Over the last 10-15 years, various workers have demonstrated
that the pH of the solutions at the tips of propagating cracks in
Al and Ti alloys and in high strength steels is low and, in each
case, it is in the same range of pH values as that found inside
pits in the same materials. Cracking commonly occurs just below
the pitting potential, as indicated in Fig. 7, at least for passi-
vatable alloys susceptible to stress corrosion cracking in halide
solutions, e.g. austenitic stainless steels. If a similar explanation for the occurrence of low pH values of solutions at the
tips of cracks is to be made as is made for the same occurrence in
pitting, then for hydrolysis to occur at a sufficiently high rate
requires that the solution be provided with metal ions by the
action of the crack tip strain-rate in breaking or disturbing the
growing film and providing a flux of metal ions, since in the absence of the strain-rate the acidic solution will not be produced
nor maintained below the pitting potential. Unless the creep

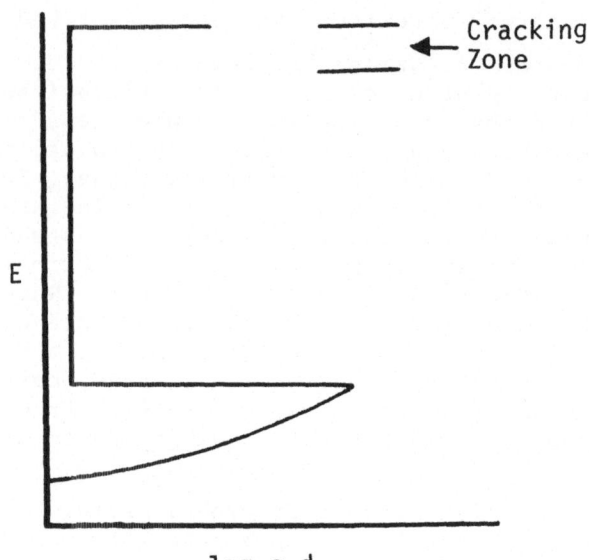

log c.d.

Fig. 7.   The potential zone in which stress corrosion is observed
            just below the pitting potential.

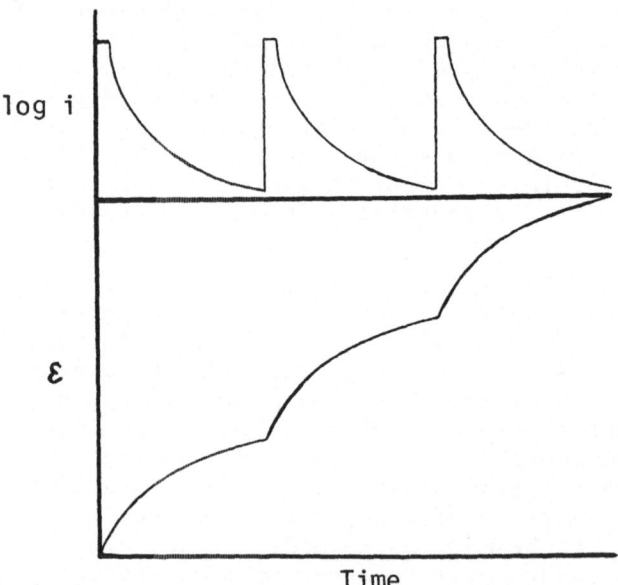

Time

Fig. 8.   Schematic drawing of the effect of repeated identical
            slip transients occurring at the tip of a propagating
            crack on the current/time relationship under a condition
            of constant extension rate. The same charge flows in
            each increment which nucleate each other successively.

process occurs at or above a certain minimum rate ($\geq \dot{\varepsilon}_r$) the surface will repassivate completely and simultaneously the pH of the solution will start to rise towards that of the bulk solution since the hydrolysis reaction rate cannot be maintained. For the same reason the halide ion concentration at the crack tip will start to fall towards that of the bulk solution.

The relatively simple concept being put forward is that cracking is similar to pitting, with the necessary condition that in the cracking process the creep strain-rate is necessary in order to provide dissolving metal ions from newly formed surfaces and thereby maintain the crack tip acidity which otherwise either would not occur or would not be maintained. Since crack velocities can change over several orders of magnitude in a given system, the influence of strain-rate in maintaining crack propagation is best explained by envisaging the crack tip metal surface as being covered by a film of varying composition and, perhaps, morphology. In the extreme case the surface is bare, corresponding to the maximum velocity often observed as a plateau velocity, itself dependent upon pH and solution composition and reaching a maximum at the pitting potential. The idea that the film on the crack tip surface may exhibit a range of properties which determine the crack velocity is not new. The chemical and physical state of this layer remains to be examined. It may, for example, be a mixed layer of oxide and salt layer, with the latter component being the less protective. A direct comparison can be made with the range of current densities observed on growing pit surfaces. While much emphasis has been placed on the maximum values attained in various alloy/electrolyte combinations, a study of pit propagation rates shows that pits can grow at significantly lower rates at potentials close to the protection potential. This reduction is caused by the presence of a film on the pit surface, evidence for which can be deduced from the high values of potential/log i changes that have been recorded for pits[76].

Repassivation experiments with scraping or straining electrodes have been employed in relation to stress corrosion studies. Usually repassivation is rapid, obeying a logarithmic decay behaviour of the form log i α -log t. Such results may have little or no application, however, to crack tip repassivation. At the crack tip the fall in pH can be expected to lower the repassivation rate, as will the increasing Cl$^-$ ion concentration[78]. As already discussed, the type of film forming may be different. Film growth kinetics under such conditions have not been widely investigated although a slowing down as the potential approaches the pitting potential has been observed in aluminium[77] and in austenitic stainless steels[78,79]. A computer simulation of a repassivating scratch on a stainless steel surface[79], however, over a wide range of potentials up to the pitting potential, has shown that near to the pitting potential the log i/log t decay is not a more accurate description of the decay process than an exponential decay (log i α -t) during the early

stages (100 ms) of the film repair process, even with no allowance
being made for the significant amount of hydrolysis occurring in
that region of potential. Practical results[80] for Fe in chloride/
molybdate solutions in the pitting region of potential have indicated a similar conclusion: during the early stages of repassivation
(250 ms) the results fit two exponential decay lines of slightly
different slope and log i/log t decay describes the process only
after the early stage has passed. It is the early stage of repassivation that is of importance since cracking can be viewed as a
series of current transients recurring at short time intervals.
The repassivation rate between such transients is what needs to be
known accurately.

The type of decay law operating during repassivation has a
direct relevance to any electrochemical analysis of crack propagation,
as has been described previously[75]. If an exponential decay is
operative then the current, i, flowing from a surface deforming
at a constant strain-rate, $\dot{\varepsilon}$, is related: log i α $\dot{\varepsilon}$, under constant
potential conditions. If a different decay behaviour is operative
then clearly a different dependency between strain-rate and current
will be observed. This is a very important point since by the
association of the creep rate with the repassivation process what
is being emphasized is that the relationship between i and t during
an increment of crack growth is both mechanically and electrochemically dependent.

The sequence of current transients emanating from a crack tip
and observable hypothetically under ideal experimental conditions
is depicted in Fig. 8 where the conditions are intended to be a
constant extension rate in a specimen under potentiostatic control.
The creep-rate is considered to be constant, giving rise to a
series of identical creep transients. The minimum charge, Q_{min},
flows before a further increment is generated and complete
film repair does not therefore occur before the next increment is
initiated, i.e. $\dot{\varepsilon}$ never falls to $\dot{\varepsilon}_r$. A general formula can be
advanced:

$$v = \text{constant} \ \frac{Q_{min}}{t} \tag{28}$$

where v is the crack velocity and t is the time interval between
successive slip events which are initiated by the propagation of
the crack. If v = 1 mm/h and the slip line separation is 0.1 μm
then 3 transients will occur approximately every second. The
repassivation kinetics over the first 330 ms will determine the
charge flowing during that time.

The value of $\dot{\varepsilon}_r$ determins K_{Iscc} and any other mechanical
threshold between failure and non-failure. It is environment-
sensitive since the competition between metal surface production
rate, hydrolysis and neutralization kinetics will depend, inter
alia, on the nature of the bulk solution. For the same reason it

will be potential-dependent, as is K_{Iscc}, since as the potential
is lowered to more active values the repassivation rate will increase. The establishment and maintenance of the acidic solution
at the crack tip will be more difficult, i.e. it will require a
higher rate of metal surface production. This general effect is
drawn in Fig. 9 which shows the boundaries of cracking in relation
to potential and strain-rate and which also shows how these boundaries are moved by alterations to the solution. The boundaries
indicate the joint role of strain-rate, $\dot{\varepsilon}_r$, and repassivation rate,
$\dot{r}$, dependent in some sense upon the value of the potential, as
already shown for aluminium[77] and austenitic stainless steels,[77,79]
in determining whether cracking occurs in an environment in which
repassivation is possible and in one in which it is not possible.
The lower boundary value of $\dot{\varepsilon}_r$ is considered first.

The dependence of $\dot{\varepsilon}_r$ upon pH can be contrasted with the general independence of pitting potential upon pH. In the former case
the occurrence of acidity is entirely dependent upon the creep rate
generating fresh metal area or disturbing the surface film, while
in the latter case the breakdown of the film and development of
acidity is entirely electrochemical. As the pH is lowered the
boundary falls to lower values of $\dot{\varepsilon}_r$, until when passivity is no
longer possible, it is vertical, corresponding to where the crack
velocity is similar in rate to the corrosion rate. Below that
value of $\dot{\varepsilon}_r$ the cracking will give way to whatever form of corrosion occurs in unstressed specimens. This is frequently intergranular corrosion. It can be expected that the boundary will be
affected by passivating and film-forming inhibitors and, possibly,
by adsorption-type organic inhibitors too, although the effect of
the latter may not be continuous in situations where the crack tip
moves more rapidly than the rate of diffusion of the organic molecules.

In Fig. 9 the upper boundary of $\dot{\varepsilon}_r$ above which cracking does
not occur is drawn. This is sometimes referred to as depassivation-
the implication being that much or even all of the metal surface
undergoes corrosion under these conditions and crack initiation
cannot occur. Ductile failure takes place in a relatively short
time. It is conceivable that cracking might be sustained to a
higher value of $\dot{\varepsilon}$ as $\dot{r}$ increases (with increasingly active potential) in which case the line would not be vertical. The addition
of inhibitors under controlled potential conditions will lower the
corrosion rate and move the boundary to lower values of $\dot{\varepsilon}_r$. Increasing the solution viscosity will have the same effect. Increasing the halide ion will increase the maximum value of $\dot{\varepsilon}_r$.
The upper boundary of $\dot{\varepsilon}_r$ corresponds to the conditions for maximum
crack velocity, observed as a constant velocity in some systems.
It is lowered by lowering the potential over a range of pH values
above a minimum value below which crack velocity is potential-
independent in some systems, e.g. Al and Ti alloys. This is also
drawn on the diagram.

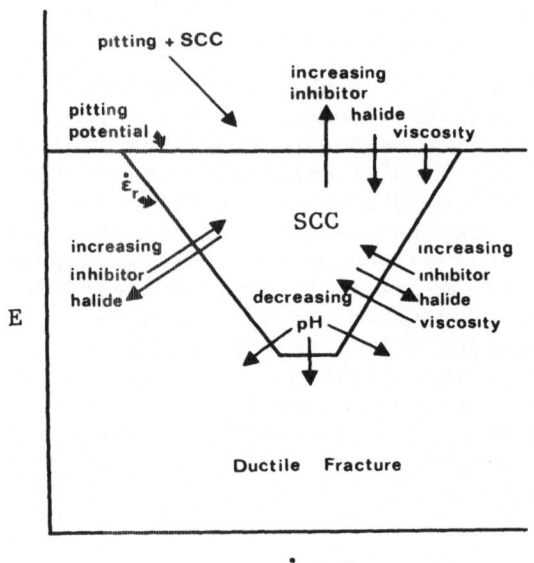

Fig. 9.   Schematic diagram of the relationship between electrode
           potential and crack tip strain-rate, ˙. Below the
           pitting potential the boundaries are determined by solu-
           tion composition, particularly inhibitors, halide
           ion concentration, pH and solution viscosity.

The upper potential boundary in Fig. 9 corresponds to the pitting potential, or possibly the protection potential, if that is different, since below that value pits do not grow. The potential at which pits nucleate and grow in straining aluminium specimens has, however, been designated the pitting potential. At and above the pitting potential there is usually a region of pitting and cracking which gives way to general disintegration as the potential is raised. This may not be the situation for those systems exhibiting a pitting inhibition potential where, perhaps, a region of non-cracking may occur, giving ductile failure. For α-brass in neutral ammoniacal solutions, raising the potential to the value at which the tarnish film dissolves lowers the crack velocity and eventually leads to a region of non-susceptibility. Anodic polarization of mild steels also leads to eventual non-susceptibility; the pitting region is just below the cracking region of potential.

The lower potential boundary in Fig. 9 may be narrow or broad, depending upon the nature of the solution, in particular the ratio of activating to passivating species at the crack tip. As the solution becomes more aggressive the boundary widens until, when passivation is no longer possible, it extends across the full range of $\dot{\varepsilon}$ values and downwards to that value of potential where cathodic protection becomes possible.

Figure 9 illustrates many of the principal features of stress corrosion cracking systems: the cracking range of $\dot{\varepsilon}$ is both environment- and potential-dependent; the highest value of $\dot{\varepsilon}$ depends upon the maximum crack velocity of any system; the cracking range depends upon potential, strain-rate and solution composition.

The overall importance of the repassivation rate in this analysis is clear. A further distinction can be made with respect to cracking morphology. A recent analysis[76] has shown that the repassivation rate determines the fracture path in austenitic stainless steels, mild steel and α-brasses. When the repassivation rate is low cracking is transgranular and when high intergranular, where the difference in rates may be quite small with a narrow range that causes cracking. The conditions at the tip of a propagating stress corrosion crack are finely balanced. Too rapid repassivation gives crack arrest. Too slow repassivation gives blunting and pitting. Somewhere in between the repassivation process covers practically all the crack tip surface, thereby keeping the crack sharp. Within that range differences in the repassivation rates of grain boundaries and slip steps may give rise to different crack morphologies.

REFERENCES

1. M. Pourbaix, "Atlas of Electrochemical Equilibria in Aqueous Solutions", p.312. NACE-CEBELCOR, Houston (1974).
2. T.S. Gromoboy and L.L. Shreir, Electrochim. Acta, 11:895 (1966).
3. N. Sato, "Passivity and its Breakdown on Iron and Iron Base Alloys" eds. Staehle, R.W. and Okada, H. p.1 NACE, Houston (1976).
4. J.O'M Bockris, A.K.N. Reddy and B. Rao, J. Electrochem. Soc., 113:1133 (1966).
5. G.W.D. Briggs, G.W. Stott and W.F.K. Wynne-Jones, Electrochim. Acta 7:249 (1962).
6. J.E. Castle and C.R. Clayton, Corros. Sci., 17:7 (1977).
7. A.E. Yaniv, J.B. Lumsden, and R.W. Stahle, J. Electrochem. Soc., 124:490 (1977).
8. M. Cunha Belo, B. Rondot, F. Pons, J. Le Héricy and J.P. Langeron, J. Electrochem. Soc., 124:1317 (1977).
9. D.A. Vermilyea, "Advances in Electrochemistry and Electrochemical Engineering", Vol. 3, ed. P. Delahay, p.233 Interscience N.Y. (1963).
10. D.A. Vermilyea, Ref. 9, p.223.
11. H.H. Uhlig, "Corrosion and Corrosion Control" (2nd edition), p.61, Wiley (1971).
12. C. Wagner, Corros. Sci., 5:751 (1965).
13. H.H. Uhlig, Corros. Sci., 7:325 (1967).
14. A.U. Mac Rae, Science, 139:379 (1963).
15. A.U. Mac Rae, Surf. Sci., 1:319 (1964).
16. H.H. Uhlig, Z. Elektrochem., 62:626 (1958).
17. H.H. Uhlig and P.F. King, J. Electrochem. Soc. 106:1 (1959).
18. A. Damjanovic, "Modern Aspects of Electrochemistry", eds. J.O'M Bockris and B.E. Conway) Vol. 5, p.369 Plenum Press, N.Y. (1969).
19. H.H. Uhlig, "Corrosion Handbook", p.32, Wiley, N.Y. (1948).
20. H.H. Uhlig, "Passivity and its Breakdown on Iron and Iron Base Alloys" eds. R.W. Stahle and H. Okada, p.19, NACE, Houston (1976).
21. F. Mansfeld and H.H. Uhlig, J. Electrochem. Soc., 117:427, (1970).
22. H.H. Uhlig, Electrochim. Acta, 16:1939 (1971).
23. Ya.M. Kolotyrkin, Z. Elektrochem. 62:664 (1958).
24. G.W. Mellors, M. Cohen and A.F. Beck, J. Electrochem. Soc., 105:332 (1958).
25. C.L. Foley, J. Kruger and C.J. Bechtoldt, J. Electrochem. Soc. 114:994 (1967).
26. N. Sato, K. Kudo and R. Nishimura, J. Electrochem. Soc. 123:1419 (1976).
27. J. Kruger and J.P. Calvert, J. Electrochem. Soc. 114:43 (1967).
28. N. Sato and K. Kudo, Electrochim. Acta. 16:447 (1971).
29. J.O'M Bockris, M.A. Genshaw, V. Brusic and H. Wroblowa, Electrochim. Acta 16:1859 (1971).

30. K.N. Goswami and R.W. Staehle, Electrochim. Acta 16:1895 (1971).
31. C. Lukac, J.B. Lumsden, S. Smialowska and R.W. Staehle, J.
     Electrochem. Soc., 122:1571 (1975).
32. W.J. Mueller, Trans. Faraday Soc. 27:737 (1931).
33. R.D. Armstrong, Corros. Sci., 11:693 (1971).
34. R.D. Armstrong and H.R. Thirsk, Electrochim. Acta 17:171 (1972).
35. K.J. Vetter, Electrochim. Acta, 16:1923 (1971).
36. N. Sato and G. Okamoto, J. Electrochem. Soc. 110:605 (1963).
37. D. Pavlov, C.N. Poulieff, E. Klaja and N. Iordanov, J. Electro-
     chem. Soc. 116:316 (1969).
38. K.J. Vetter, "Electrochemical Kinetics-Theoretical and Experi-
     mental Aspects", p.752. Academic Press, N.Y. (1967).
39. B. Mac Dougall and M. Cohen, J. Electrochem. Soc., 123:1783
     (1976).
40. R.P. Frankenthal, J. Electrochem. Soc., 114:542 (1967).
41. R.P. Frankenthal, J. Electrochem. Soc., 116:580 (1969).
42. R.P. Frankenthal, J. Electrochem. Soc., 116:1646 (1969).
43. D.A. Vermilyea, Ref. 9, p.237.
44. Z. Farkas, Phys. Chem., A125, 236 (1927).
45. T. Erdey-Grúz and M. Volmer, Z. Phys. Chem., 157:165 (1931),
     quoted in Ref. 38, p.316.
46. R.D. Armstrong, J.A. Harrison and H.R. Thirsk, Corros. Sci.,
     10:679 (1970).
47. M. Fleischmann and H.R. Thirsk, "Advances in Electrochemistry
     and Electrochemical Engineering", Vol. 3 ed. P. Delahay,
     p.123 Interscience, N.Y. (1963).
48. R.D. Armstrong, M. Fleischmann and H.R. Thirsk, J. Electroan.
     Chem., 11:208 (1966).
49. R.D. Armstrong and J.A. Harrison, J. Electrochem. Soc.,
     116:328 (1969).
50. H.R. Thirsk and J.A. Harrison, "A Guide to the Study of Elec-
     trode Kinetics", Academic Press, London (1972).
51. C. Wagner, Corros. Sci., 13:23 (1973).
52. N. Cabrera and N.F. Mott, Rep. Prog. Phys., 12:163 (1949).
53. N. Sato and M. Cohen, J. Electrochem. Soc., 111:512 (1964).
54. R.V. Moshtev, Ber. Bunsengesel. Phys. Chem., 71:1079 (1967).
55. J.L. Ord, J. Electrochem. Soc., 113:213 (1966).
56. J.L. Ord and D.J. Desmet, J. Electrochem. Soc. 116:762 (1969).
57. J.L. Ord and F.C. Ho, J. Electrochem. Soc., 118:46 (1971).
58. G.M. Bulman and A.C.C. Tseung, Corros. Sci., 12:415 (1972).
59. G.M. Bulman and A.C.C. Tseung, Corros. Sci., 13:531 (1973).
60. R.P. Frankenthal, "Passivity and its Breakdown on Iron and
     Iron Base Alloys", eds. R.W. Staehle and H. Okada, p.10
     NACE, Houston (1976).
61. Z. Szklarska-Smialowska and G. Mrowczynski, Br. Corros. J.,
     10:187 (1975).
62. H. Buhl, Corros. Sci., 13:639 (1973).
63. G.W. Simmons, E. Kellerman and H. Leidheiser, Jr., "Passivity
     and its Breakdown on Iron and Iron Base Alloys", eds. R.W.
     Staehle and H. Okada, p.65, NACE, Houston (1976).

64. I.A. Ammar, S. Darwish and M.W. Khalil, Corrosion, 32:173
    (1976).
65. K.J. Vetter and F. Gorn, Electrochim. Acta, 18:321 (1973).
66. T.R. Beck, Electrochim. Acta, 18:807 (1973).
67. V. Brusić, Ph.D Thesis, University of Pennsylvania, 1971
    quoted in Ref. 29.
68. J.R. Ambrose and J. Kruger, Corrosion, 28:30 (1972).
69. N. Valverde, Ber. Bunsengesel. Phys. Chem., 80:333 (1976).
70. N. Valverde, Ber. Bunsengesel. Phys. Chem., 81:380 (1977).
71. A.G. Revesz and J. Kruger, "Passivity of Metals" eds. R.P.
    Frankenthal and J. Kruger) p.137, The Electrochemical
    Society, Princeton (1978).
72. C.L. McBee and J. Kruger, Electrochim. Acta, 17:1337 (1972).
73. S.F. Bubar and D.A. Vermilyea, J. Electrochem. Soc. 113:892
    (1966).
74. J.C. Scully, Corros. Sci., 7:197 (1967).
75. J.C. Scully, Corros. Sci., 15:207 (1975).
76. J.C. Scully, Corros. Sci., 20:997 (1980).
77. W.J. Rudd and J.C. Scully, Corros. Sci., 20:611 (1980).
78. P. Engseth and J.C. Scully, Corros. Sci., 15:505 (1975).
79. M. Barbosa and J.C. Scully, "Environment-Sensitive Fracture
    of Engineering Materials" ed. Z.A. Foroulis, p.91, AIME
    Warrendale (1979).
80. T. Kodama and J.R. Ambrose, Corrosion, 33:155 (1979).

DISCUSSION

Comment by D.J. Duquette:

In response to the issue of "creep" at the crack tip, results
in our laboratory have shown that, under conditions of cyclic
deformation, anodic dissolution of Cu and of Fe results in significant amounts of enhanced plastic flow at solid-electrolyte interfaces. Uhlig has also shown that room temperature transient
creep rates in Cu are enhanced by anodic dissolution. I would
prefer, however, that the phenomenon be termed "environment-
enhanced local plasticity" rather than "creep". I would also
like to emphasize that plasticity under anodic dissolution is
in fact enhanced and that a simple model of comparing nominal
"creep" rates with repassivation kinetics is probably not a
realistic model for stress corrosion cracking.

Reply:

I would not disagree with any of these comments. There is
abundant work indicating the influence of dissolution on local
plasticity, in addition to the examples that you cite. The last
point is well worth making. Comparing nominal creep rates with
repassivation kinetics does leave a bit to be desired since, I
have long believed, these two reactions are coupled, as I have

indicated, albeit briefly, in references 75 and 76 and previously.

Comment by J.O'M. Bockris:

There seems even now some doubt about the mechanism of
passivity. Thus, there is much evidence that a monolayer of O^{--}
causes the reduction in dissolution called passivity. However, if
one does Mössbauer, or Auger, on such films there is an indication
that the films themselves (which certainly form after the first
passivating layer) have a special character which is that they are
amorphous and contain H_2O. When Cl^- depassivates iron passive
layers, the films stay but the water goes. Does this mean that the
first O^{--} layer and a seal on top of it are necessary components of
passivation? The answer is relevant to our field. If stress corrosion cracks spread by breaking protecting passive layers at the
bottom of cracks, then that entity which is protecting (passive)
must have a substantial character: one cannot imagine a monolayer
breaking, as a result of the application of stress.

Reply:

It is difficult to carry this discussion further. Little or
nothing is known about the type of film that forms at the tips of
cracks. I have tended to describe it as a mixed oxide/halide entity
(reference 75 and 76), with an oxide-rich type providing more protection than a halide-rich type, the type being determined by the
crack tip solution composition.

I am not certain that I understand the last remark. There is
plenty of evidence that the film is broken periodically at the
crack tip and I cannot see why a monolayer should be difficult to
break. I have discussed the point that breaking thick films may
reveal little metal if the slip step height is the same size or
smaller than the film thickness (reference 76).

NEW CONCEPTS IN
ATOMISTICS OF FRACTURE

Session Chairman: R. Bullough

COMPUTER MODELING OF CRACKS*

G. J. Dienes

Brookhaven National Lab

Upton, NY 11973 USA

Arthur Paskin

Queens College of CUNY

Flushing, NY 11367 USA

ABSTRACT

The theoretical techniques used in modeling cracks in crystalline lattices are reviewed. It is shown that there is generally a trade-off between sample size and realistic interatomic
potentials. Infinite discrete one and two-dimensional lattices
can be handled by the methods of lattice statics, but only with
simple unrealistic potentials. In the hybrid lattice statics
models a very small crystalline region, where the calculations are
done with realistic potentials, is imbedded in an infinite elastic
continuum. In this approach the boundary matching between the two
regions is the difficulty. At the other end of the scale, molecular dynamic techniques can be used on an unconstrained system of
a "large" number of atoms interacting with a reasonably realistic
interatomic potential (this is the only way dynamic simulations
have been done so far). Here, of course, the question is how
"large" is large enough to simulate the behavior of the corresponding infinite system.

The results of a recent molecular dynamic study of a two-
dimensional triangular Lennard-Jones system of about 10,000 atoms
are described. Static calculations on this system show very
little lattice trapping in contrast to the often quoted results
of lattice statics models. Using molecular dynamics, the properties of the system, with and without a crack, are understood
sufficiently quantitatively to allow extrapolation to infinite

---

*This work was supported in part by the U.S. Army Research Office,
and in part by the U.S. Department of Energy under Contract
No. DE-AC02-76CH00016.

size. At 10,000 atoms size effects are small enough to render the
simulation highly reliable. It is also shown that when lattice
statics and hybrid lattice statics calculations are performed using
realistic, long-range potentials, lattice trapping is small.

Dynamic simulations at constant applied stress show an intri-
cate interplay between brittle crack propagation and the tendency
to form dislocations. At low stresses the behavior is brittle
while dislocation generation and crack blunting is observed at
elevated stresses. The velocity of crack propagation has also
been studied using a constant applied strain. A constant velocity
is attained relatively soon after the crack has begun to propagate
at both constant applied stress or strain. At constant strain,
the changes in elastic energy and in the modulus during propagation
are in good agreement with a simplified elastic theory for a nonlinear solid. The change in kinetic energy during propagation has
the functional form suggested by Mott. The credibility of the
various models is discussed.

I. INTRODUCTION

Modeling involves two basic steps: 1) the construction of a
model incorporating the key physical features of the system, and
2) the calculation, analytically or by computer techniques, of the
physical characteristics of interest. If step 2 is carried out to
high accuracy, which can now be done with the availability of
modern computers, then the limitations of the results are those of
the model itself. Modeling can lead to very significant insights,
but it is essential to try to determine which parts of the results
are (or may be) artifacts of the model and which parts are general
and relevant to the real physical system being modeled. There are
always simplifying assumptions and constraints in a model which
need careful scrutiny.

The description of a crack in a solid with an applied external load or strain is rather simple to envision. One has a semi-
infinite periodic three-dimensional array of atoms interacting
with one another via an interatomic potential. A crack is
inserted in the solid and a load is applied externally to appropriate surfaces. If the load is sufficiently large, the crack
grows and the solid ultimately fractures. Why is this phenomenon
so much more difficult to treat than other problems encountered in
solid state physics? In most fundamental phenomena in solids, the
periodicity of the solid is used to reduce the problem to an
examination of a simple representative cell or region. Hence, the
infinite number of atoms is not a serious problem. In the case of
a crack, with its finite extension and the resulting long-range
strain field, such reductions are not possible. Hence, one must
resort to simplified models to try and gain insight into the role
of cracks in brittle fracture.

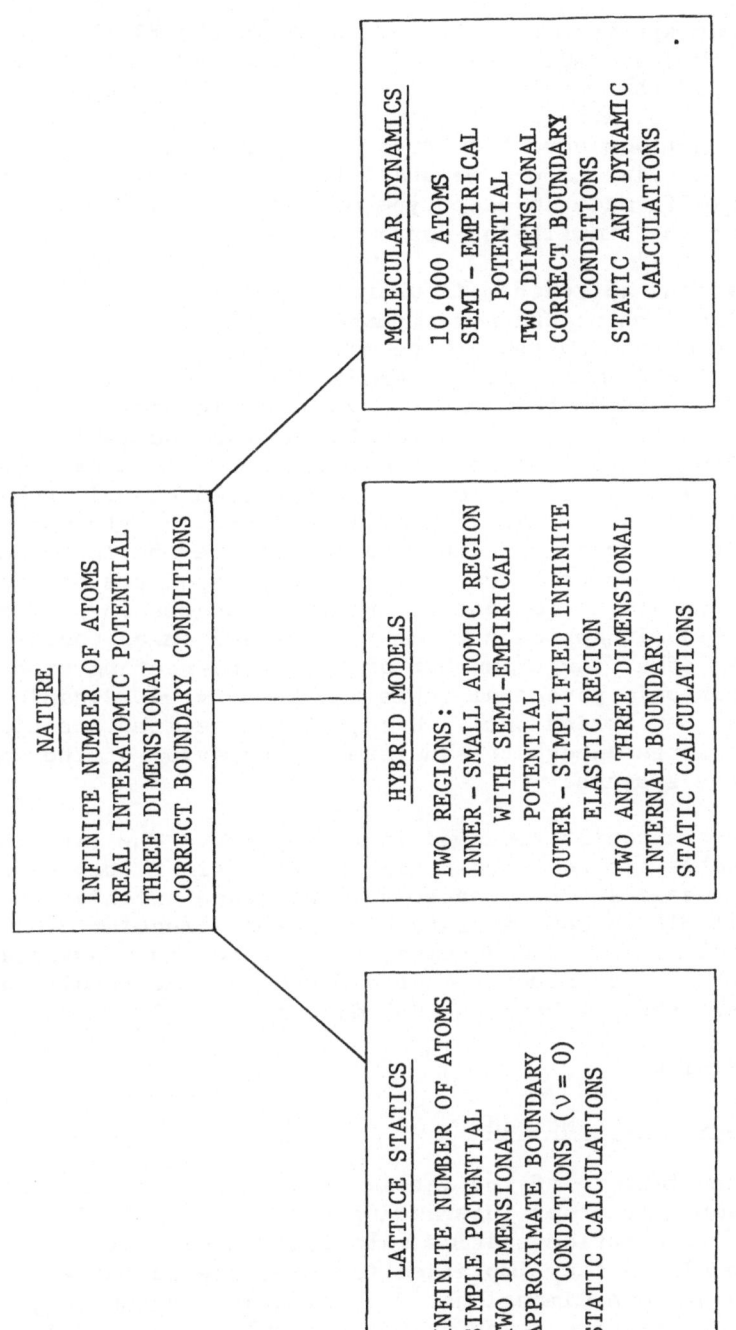

Fig. 1. Illustration of the various current approaches to the modeling of cracks in crystalline solids.

The various approaches are illustrated in the block diagram of Fig. 1. In each case, some trade off has been made between giving up some aspect of the phenomenon in nature for some simplification that enables one to carry out a quantitative study. As the lattice static methods have been treated in some detail by Thomson[1] in an earlier paper, we shall here try to emphasize the kinds of trade-offs made in each type of model and the physical significance of the simplifying approximations.

In the various approaches illustrated in Fig. 1 different facets of the real phenomena are emphasized. If the semi-infinite periodic nature of the system is the major concern then one is led to the lattice static methods.[1-4] Cracks in such lattices can be handled by these methods but only for very simple unrealistic interatomic potentials. In the so-called hybrid models[5-7] a very small crystalline region, where the calculations are done with realistic potentials, is imbedded in a simplified elastic region. The boundary matching, the above simplification, and the appropriateness of the size of the imbedded region are the important approximations in this technique. If the nature of the atomic forces and of the atomic motions is the main concern one is led to the molecular dynamic approach which can be used on an unconstrained system of a "large" number of atoms interacting with a realistic interatomic potential (this is the only way dynamic simulations can be carried out). Here, of course, the question is how "large" is large enough to simulate the behavior of the corresponding infinite systems.

These are the topics reviewed in this paper. The lattice statics approaches are reviewed rather briefly since they have already been treated at this Conference and since excellent reviews are available in the literature.[2-7] The molecular dynamic simulations, both static and dynamic, are treated in considerable detail since these are rather recent and some of the results discussed here have not yet been published.[8-12]

II. LATTICE STATICS

1. Outline of the Method

The key feature of the lattice statics is that it treats an infinite number of atoms interacting via some simplified potential.[1-4] Two lattice statics calculations will be outlined here with emphasis on the simplified forms of the potential. The first is the quasi one-dimensional[13] model which is commonly used to illustrate the lattice trapping which results from lattice statics calculations. As shown in Fig. 2, two semi-infinite chains of atoms are bonded together horizontally by bendable bonds and vertically by stretchable bonds. The zeroth atoms are stretched apart vertically by the forces σ and the first n

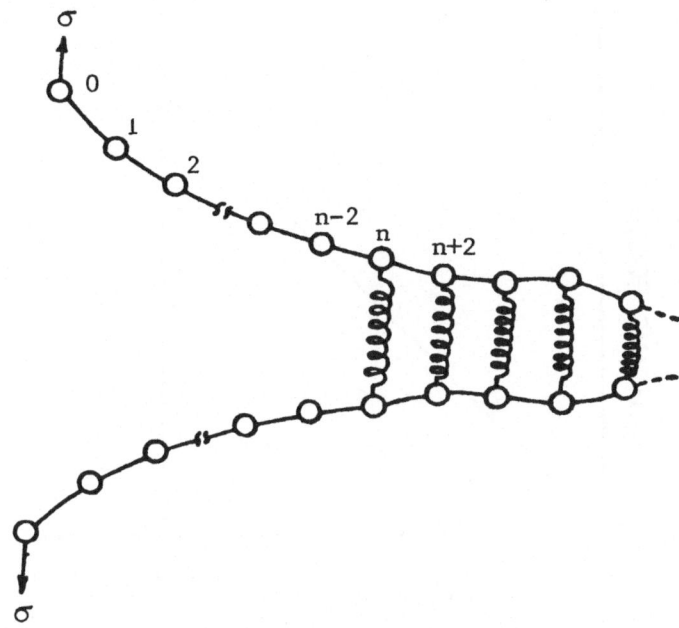

Fig. 2.   One-dimensional model of a crack. Two semi-infinite
          chains of atoms are bonded together horizontally by bend-
          able bonds and vertically by stretchable bonds. The
          zeroth atoms are stretched apart vertically by the forces
          σ and the first n vertical bonds are broken. (After
          ref. 13)

vertical bonds are broken (the transverse springs stretched beyond
their limit) forming a crack of finite length.   In Fig. 3, the
atomic force law is illustrated as a function of the displacement
y.[14] The key feature of the force law is the bond snapping property, i.e. the force abruptly goes to zero at some displacement
y_0. It will be shown that bond snapping considerably enhances
lattice trapping and indeed may be the largest contribution to
lattice trapping. The mathematical details of the lattice statics
calculation will be here outlined.[13-14] The energy u_j in the
lateral bond connecting atoms j with j–1 is

$$u_j = \tfrac{1}{2}\beta(\theta_j - \theta_{j-1})^2; \qquad\qquad j = 1, 2, \ldots \qquad\qquad (\text{II-1})$$

where β is the bendable spring constant. The energy in a stretchable spring is

$$u_j = \tfrac{1}{2}\gamma(2y_j) \qquad\qquad\qquad j = n, n+1. \qquad\qquad (\text{II-2})$$

Making the small angle approximation in (II-1), the total energy
may not be written

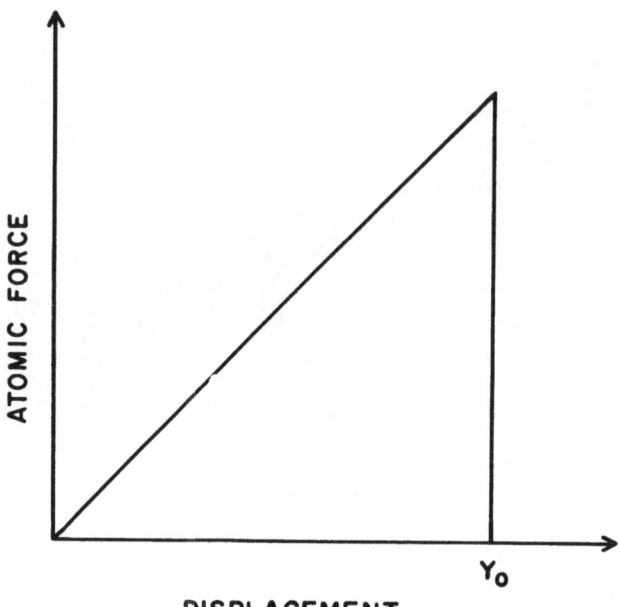

ATOMIC FORCE

Y_0

DISPLACEMENT

Fig. 3.   The assumed force law between the atoms is linear to a
         displacement of y_0 at which point the bond snaps dis-
         continuously and the force falls to zero.   (After ref. 14)

$$u = -2\sigma y_0 + \beta \sum_{j=1}^{\infty} (y_{j+1} - 2y_j + y_{j-1})^2 + 2\gamma \sum_{j=n}^{\infty} y_j^2. \qquad \text{(II-3)}$$

Inserting the equilibrium conditions $(\partial u/\partial y_j) = 0$, conditions are
obtained on a complex vector $\underline{k}$

$$\cos 2\lambda - 2\tanh^2\lambda = 1 + \tfrac{1}{2}k \qquad\qquad\qquad \text{(II-4)}$$

$$2\tan^2\eta + \cos 2\eta = 1 + \tfrac{1}{2}k \qquad\qquad\qquad \text{(II-5)}$$

where $k = \lambda + i\eta$.   The above solution can be used to find the
force σ_+ for which the bond snaps.   The crack is stable for forces
less than σ_+ down to the point where the displacement of atom n-1
is equal to the critical value y_0.   At that point, σ_-, the previously broken bond snaps back together and the crack heals.   The
ratio $R = \sigma_+/\sigma_-$ is the measure of lattice trapping.   As $R \to 1$ there
is no lattice trapping.   At $R \gg 1$, lattice trapping becomes quite
important.   Thomson et al.[13] give three approximate expressions
for R

$$R \to 1+2\lambda \qquad k^{1/4} \ll 1 \qquad\qquad\qquad \text{(II-6)}$$

$$R = 4.3 \qquad k = 1 \qquad\qquad\qquad\qquad \text{(II-7)}$$

$$R = k+9 \qquad k \gg 1 \qquad\qquad\qquad\qquad (\text{II-8})$$

2. Major Results

The expressions (II-6 to II-8) predict the conditions for anything from negligible to large lattice trapping depending on the features of the potential. It is unfortunate that in the earliest lattice statics calculation, a numerical error was made in an independent model calculation, which led Thomson et al.[13] to assume that large lattice trapping was in accord with realistic potentials. To this end, another lattice statics calculation is illustrated, with a different potential used as the interatomic force as shown in Fig. 4. In this calculation the surface energy of the atoms on the crack plane was calculated. The energy was taken to be of the simple form

$$E(n) = \tfrac{1}{2}E_0\left[1 - \tfrac{2}{\pi}\tan^{-1}\tfrac{n}{\alpha}\right]. \qquad\qquad (\text{II-9})$$

The adjustable parameter α is adjusted to fit the results to the actual shape of the crack and to the force law. α can be thought of as the width of a "core region" of a crack. The key feature of the potential relative to the previous model is that there is no bond snapping in this force law. The parameter α contains both the shape of the crack and how rapidly the energy changes from atom to atom in the vicinity of the crack tip. A simple closed

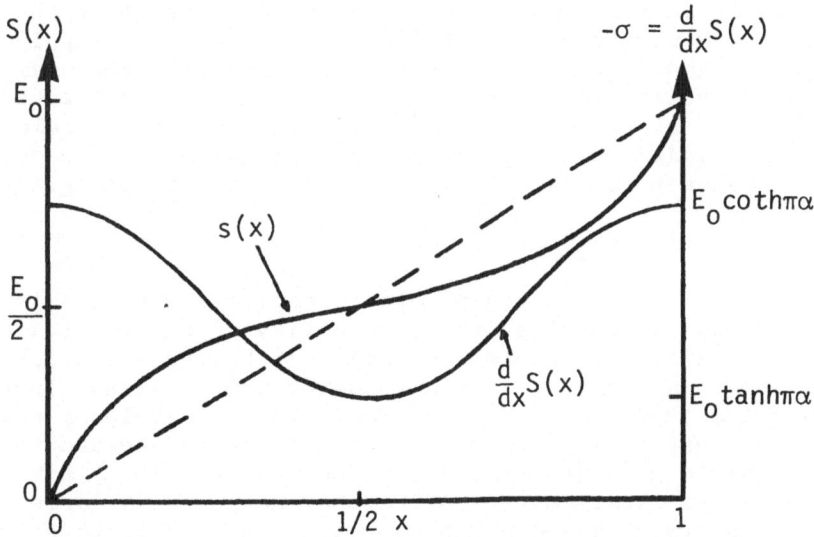

Fig. 4. Functional form of the surface energy for moving a crack $S(x)$ (plotted on the left axis), and the corresponding force $\sigma = -dS/dx$ (plotted on the right). The dashed line represents the average surface energy used in the continuum approximation. (After ref. 13)

form was obtained for R,

$$R = \frac{1+2e^{-2\pi\alpha}+e^{-4\pi\alpha}}{1-2e^{-2\pi\alpha}+e^{-4\pi\alpha}}$$ (II-10)

Thomson et al.[13] performed one of the most important steps in establishing the reliability of any quantitative, unexpected result for a model, since they did the calculation for a very different model in order to see how the results compare. They determined α from the computer calculations of Gehlen and Kanninen[10] using an empirical force law for iron and the computer positions for the atoms near the tip. Unfortunately they made a numerical error in this comparison. Using the data of Gehlen and Kanninen, $\alpha = 0.524$ and $R = 1.15$ indicating small lattice trapping. Thomson et al.[13] quote a value of $R = 2.14$ which led them to conclude that lattice trapping is large. Subsequent lattice static calculations, mainly incorporating bond snapping features, yielded large lattice trapping and have led those working in fracture to conclude that large lattice trapping is a consequence of the discrete nature of the lattice. Clearly, the non-bond snapping lattice static calculation, matching realistic potentials, yields negligible lattice trapping and thus lattice trapping is not a consequence of the discrete nature of the lattice but rather a consequence of the displacements of the atoms near a crack top and the way the potential varies in this regime. Later Esterling[15] used similar lattice static techniques and a variety of force laws and observed that crack healing is sensitive to the long-range portion of the cohesive force. Nevertheless Esterling accepts lattice trapping as non-negligible and goes on to treat some of the thermodynamic consequences. The lattice statics practitioners seem not to take seriously some of the concurrent hybrid lattice static calculations of Sinclair[7] which, prior to Esterlings work, gave evidence that long-range potentials yield little lattice trapping.

3. Credibility of the Results

Thus arises the key problem associated with models and their quantitative results--the credibility of the results. This is the most difficult problem in modeling; how to establish the credibility of a model, and of any unexpected results. There are two general methods to establish credibility: develop a physical picture that explains the results or do other independent model calculations to get confirmation of the results. Both have been tried in lattice statics to confirm the unexpectedly large lattice trapping results obtained. Thomson et al.[3,13-14] tried both but there seem to be serious objections to both the physical description and to the different models.

For the physical explanation of lattice trapping, Thomson et al. draw an analogy to the Peierls stress for dislocation.

This does not seem to be justified for continuous non-truncated interatomic potentials. If the stress that is applied is less than critical, indeed there will be a stress barrier against the crack propagating. However, there will be no stress opposing the crack healing. Any infinitesimal motion to close the crack will result in a lowering of the free energy; hence there will be no Peierls stress to overcome healing. If the stress is slowly increased and the crack does not propagate it should heal. This should be the case until the maximum or critical stress is applied; at this point a slight increase in stress should result in crack propagation. This is essentially the Griffith[16] picture of fracture. Of course, if a material is not entirely brittle and some deformation can take place, then it is possible that the energy is approximately the same for a range of stresses leading to metastability, i.e. a range of stresses at which a crack neither propagates nor heals. This, however, does not seem to be the origin of the lattice trapping calculated by Thomson et al., particularly since even in their two-dimensional calculation,[14] all motion is essentially transverse to the crack surface (i.e. Poisson's ratio is kept at zero). Hence, the Peierls-like barrier does not seem to be a simple consequence of the lattice periodicity independent of the potential and is not a universal phenomenon which yields lattice trapping as a fundamental property of a periodic array of atoms.

The second way to confirm credibility is by using an entirely different model. This was done by Thomson et al. using a surface-energy summation technique, and if it weren't for the numerical mistake made in evaluating R they would have indeed found little lattice trapping in a system that is meant to model iron. This same procedure was used to find R for the two-dimensional array with a Lennard-Jones potential by Paskin et al.[12] A value of $R = 1.02$ was obtained in good agreement with the molecular dynamic results. This suggests that the lattice statics with a realistic value of α, obtained by matching empirical potentials, can give useful insight into the critical properties of fracture. The result also demonstrates the importance of the realistic interatomic potential in any model calculation.

III. HYBRID LATTICE STATICS

1. Outline of Method

The hybrid approach involves separating the infinite solid into two regions. In region I, near the crack tip, both the atomic structure and the atomic interaction are treated explicitly. In region II, the outside region, the force law and also the atomic structure are simplified. In an early Sinclair and Lawn[5] calculation, region II is treated by the continuum approximation and the boundary between regions I and II is fixed. In a later calculation

by Sinclair,[7] the I-II boundary is flexible and higher order terms
are included in the elastic field equations. In the Sinclair and
Lawn hybrid calculations, a suitable semiempirical potential for
the diamond-structure crystals was constructed. The atoms were
then relaxed within region I in such a way that the total potential
energy tended to a minimum. This was systematically done on a computer with the boundary between I and II kept rigid. The boundary
was obtained by solutions of the continuum field equations. While
the results seem to be reasonable, it is difficult to determine
their reliability.

Sinclair[7] thought that the rigid boundary introduced some
error in the hybrid calculations that made them not suitable for a
study of the critical conditions for crack extension or closure.
The reason is that, if the solutions of the inner region are not
compatible or consistent with the solutions of the outer region in
the rigid boundary method, no adjustment can be made for this mismatch in forces. The rigid boundary forces a state of stability
in the system that may be inconsistent with the local stress fields.
If the crack in region I would like to expand (or contract) the
rigid boundary prevents this relaxation. The flexible boundary
model allows some relaxation of the boundary so as to prevent
forces building up at the boundary. Sinclair does this by including
higher order terms in the elastic crack-field equations for region
II; by suitably adjusting the coefficients of these terms a configuration free of residual forces may be obtained for a given
applied stress. The approximation in this procedure is that the
forces are calculated correctly in region I but approximately in
region II.

2. Major Results

a. Rigid Boundary. The results for the rigid boundary
model were mostly qualitative and reasonable. Sinclair and Lawn
found that except for very near the crack tip the continuum and
atomic relaxation calculations showed close agreement, despite the
fact that the elastic nonlinearity and the atomicity render continuum theory strictly invalid throughout the core region. It is
not clear how much of this result depends on the use of rigid
boundaries matched to continuum solutions. In Fig. 5 the bonds in
the vicinity of the crack tip are shown. As illustrated in Fig. 6
the relaxation varies with the size of the atomic region. It is
apparent that the continuum solutions are forcing a larger crack
to exist than the atomic solutions yield. Thus, in this case the
"true" crack is smaller than the "continuum" crack. Unfortunately,
it is difficult to estimate the quantitative errors introduced by
this procedure.

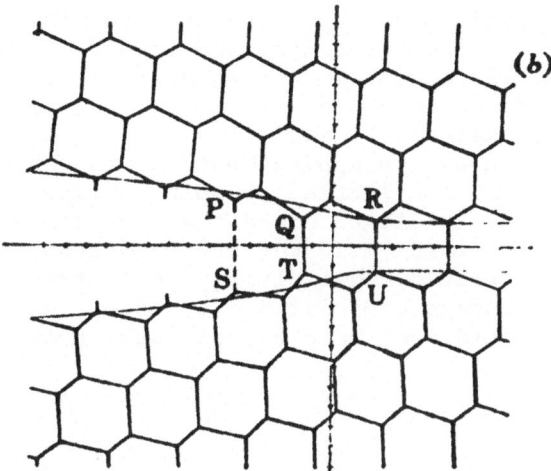

Fig. 5. Computer-relaxed atomic configuration near equilibrium (111) crack tip in silicon. (After ref. 5).

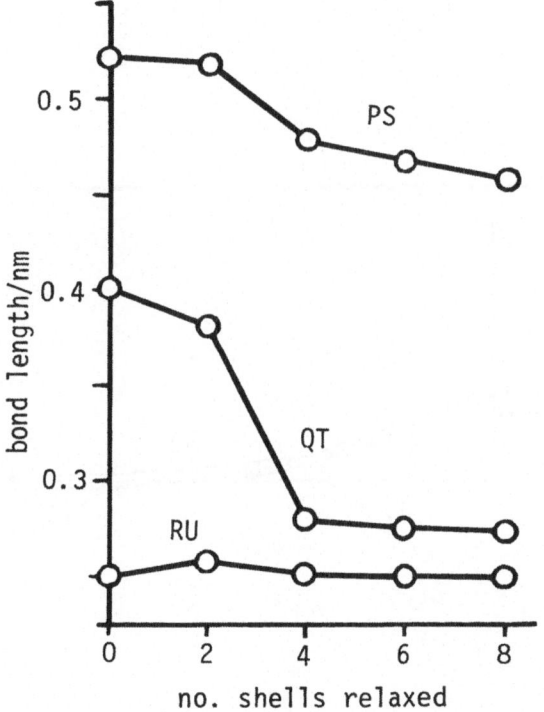

Fig. 6. Dependence of Si-Si bond-lengths PS, QT, RU on size of relaxed region. As region I is increased from 2 to 4 rectangular shells, the bond QT contracts from the "ruptured" to the "stretched" state, thus closing the crack by one atomic spacing. (After ref. 5)

b. <u>Flexible Boundary</u>. The main results of the flexible boundary hybrid treatment were directed towards examining the effect of various potentials on lattice trapping.[7] In Fig. 7, the radical component of the force laws used are shown. The important point to note is that potentials I, II and III are quite long range compared to IV, which has the bond snapping character usually used in lattice statics. In table 1, the lattice trapping ratios for these potentials are given. It should be noted that the long-range potentials yield R ranging from 1.13 - 1.45 but the bond snapping force gives R = 7.3. Thus, the hybrid model clearly demonstrates that bond snapping severely exaggerates lattice trapping, while the long-range potentials yield much more modest lattice trapping.

Table 1. Lattice-trapping ratios for straight [0$\bar{1}$1] edge cracks.[7]

	Pot I	Pot II	Pot III	Pot IV
R	1.13	1.45	1.41	7.3

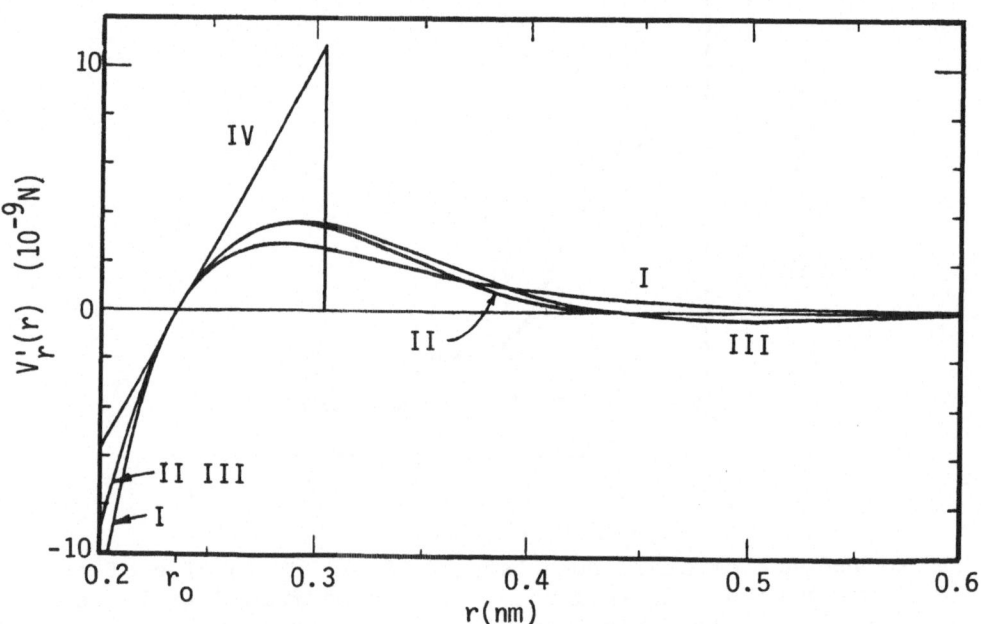

Fig. 7. The radial component of the force laws derived from potentials I to IV. (After ref. 7)

3. Credibility of the Hybrid Model

The fact that the hybrid lattice statics are in relatively good qualitative agreement with molecular dynamic calculations (as discussed in section IV) shows that both models are reasonable. What is more difficult to assess is the accuracy of the hybrid model using flexible boundary conditions. It is not easy to see how one can get a quantitative measure of the accuracy of the model. There must be some artificial mismatch at the boundary as the force laws in regions I and II are not the same. A systematic variation of region I and the resulting variation of the thermodynamic properties to enable an extrapolation to the infinite atomic solid, would lend more credibility to the quantitative results. It is difficult to compare the lattice trapping results for the long-range potentials in the hybrid model with those of the molecular dynamic calculations. The potentials used by Sinclair are non-central whereas the molecular dynamic potentials are central. The non-central nature of the forces may allow nonbrittle distortions not contained in the central force treatment. In any event, hybrid lattice static calculations using semi-empirical potentials predict negligible to little lattice trapping, as opposed to the bond snapping force models which yield large (R > 2) lattice trapping.

IV. MOLECULAR DYNAMICS – STATIC CALCULATIONS

1. Outline of the Method

The computer simulation of atomic motion in solids by the technique of molecular dynamics has been treated extensively in the literature.[17] Basically Newton's differential equations of motion are solved numerically for a large number of atoms interacting with a given potential. The simulations to be described[12] were carried out on a two-dimensional triangular lattice illustrated in Fig. 8 (half sample). The largest sample consisted of 10704 atoms arranged in 79 rows. Mirror symmetry about the central vertical axis was used to reduce the computation time.

For the atomic interaction the Lennard-Jones (6-12) (L-J) potential was used in the form

$$\phi_{ij}(r_{ij}) = \varepsilon_0[(d/r_{ij})^{12} - 2(d/r_{ij})^6] \tag{IV-1}$$

with all energies measured in units of ε_0, and distances r_{ij} in units of d, the equilibrium spacing. The forces were calculated between atoms which were initially nearest neighbors. After the introduction of the crack, forces were calculated between all atoms within 1.6d, well beyond the distance 1.11d where the interatomic force is maximum and slightly less than the second-neighbor distance. The calculations were performed using the Verlet[18] method for solving Newton's equation of motion. The atomic positions,

Fig. 8.   The triangular lattice of 79 rows with a crack of 38
          broken bonds.   The half sample is shown.   The lines
          represent the bonds between atoms.   (After ref. 12).

velocities, forces, total energy, kinetic energy, potential energy,
and modulus of elasticity were monitored throughout. For dynamic
properties the time step in the calculation gave one part in 10^6
accuracy in the energy. For equilibrium (static) calculations a
time step eight times as long was employed to obtain equilibrium
by a less costly procedure.

     In a static simulation the system was brought to equilibrium
under a constant external strain ε before inserting the crack. The
constant external strain was applied by specifying and maintaining
the vertical displacement of the individual atoms in the uppermost
and lowermost rows with all other atomic motions free to adjust.
Thus, there were no additional constraints on the system. The
calculations have also been performed at constant applied stress.
This case will be discussed later. Equilibrium was attained (at
zero temperature) by inserting critical damping into the equations
of motion. A crack was created by cutting some of the bonds across
the central rows of the sample (Fig. 8).

2. Major Results

a. Mechanical Properties of the Triangular 2-D Lennard-Jones Solid: The Perfect Solid. The elastic constants of the 2-D triangular L-J solid are well known, i.e. the response to an infinitesimal stress (σ) (or strain, ε).[9,19] In particular, Young's modulus, M, defined as tensile stress over tensile strain, is 83.14 ε_0/d^2, at an infinitesimal deformation.

As the strain is increased the system shows a highly nonlinear behavior, i.e. M is a function of ε, arising from the highly nonlinear forces of the L-J interaction. For a uniform tensile strain one can calculate M directly[12] from a balance of forces on one deformed triangular cell as shown by the sketch of Fig. 9. With $r_1^0 = r_2^0 = 1$ the pertinent equations are

$$\varepsilon = \Delta/(\sqrt{3/2}) \tag{IV-2}$$

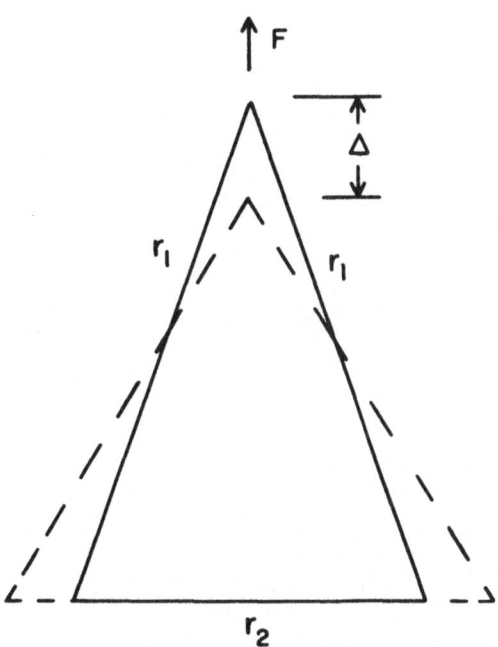

- - - UNDEFORMED TRIANGLE

——— DEFORMED TRIANGLE

Fig. 9. Tensile deformation of the elementary triangle.

$$r_1^2 = \tfrac{3}{4}(1+\varepsilon)^2 + \tfrac{1}{4}(1-\nu\varepsilon)^2$$

$$\tag{IV-3}$$

$$r_2 = 1-\nu\varepsilon$$

where ν = Poisson's ratio. The tensile force, F, is given by

$$F = \frac{2\sqrt{3}}{2}\frac{(1+\varepsilon)}{r_1}\frac{d\phi}{dr} \tag{IV-4}$$

where $\phi(r)$ is the interatomic potential. ν is determined by minimizing the energy with respect to ν, namely

$$2\frac{d\phi}{dr_1}\frac{dr_1}{d\nu} + \frac{d\phi}{dr_2}\frac{dr_2}{d\nu} = 0 \tag{IV-5}$$

with

$$\frac{dr_2}{d\nu} = -\varepsilon$$

$$\tag{IV-6}$$

$$\frac{dr_1}{d\nu} = -\frac{1}{4}\frac{(1-\nu\varepsilon)\varepsilon}{r_1}\ .$$

The modulus, $M(\varepsilon)$ is given by

$$M = \frac{F}{\varepsilon(1-\nu\varepsilon)}\ . \tag{IV-7}$$

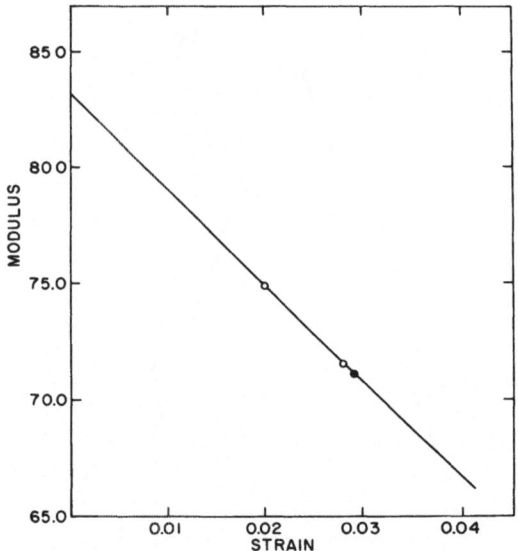

Fig. 10. Young's modulus M against applied strain. The full
circle represents three sample sizes spanning a factor
of 4 in area; the moduli were identical within 0.1%
independent of size. The straight line is the theoretical
behavior of the modulus against strain for a perfect
sample. (After ref. 12)

A series of computer calculations was performed over a range
of applied strains and sample sizes. The data for M are shown in
Fig. 10 together with the theoretical modulus (solid line) calculated by the above equations. The excellent agreement of the theory
and the computer calculations, and the observation of no size dependence in the modulus over a factor of four in the sample area,
show that the strain dependence of the modulus is understood in a
fundamental way for this nonlinear elastic solid. Computer experiments at various applied stresses were also in full agreement with
Fig. 10.

In the region of interest in subsequent calculations the
modulus-strain dependence is accurately given by the linear relation of Fig. 10, namely

$$M = M_0 - \alpha\varepsilon \qquad\qquad\qquad\qquad (IV-8)$$

with $M_0 = 83.14$ and $\alpha = 414$.

b. <u>The Two-dimensional Solid Containing a Crack</u>. When a crack
is introduced into the sample, with a fixed external strain, the
modulus is reduced. This reduction in modulus occurs because the
introduction of a crack, at fixed strain, reduces the stored energy.
As the effective modulus is related to the stored strain energy, a
reduction in modulus results. Again a series of computer calculations of M was carried out at various sample sizes for 38 broken
bonds.[12] The data are shown in the bottom half of Fig. 11.

The size dependence of M for a cracked sample may be analyzed
by a procedure analogous to that given by Berry[20] for a linear
elastic continuum. ΔE, the difference in the elastic stored energy
of a sample with and without a crack, may be written as

$$\Delta E = (A\varepsilon^2/2)\,[(M_0'-M_0) - \tfrac{2}{3}(\alpha'-\alpha)\varepsilon] \qquad\qquad (IV-9)$$

where A is the area and the primes refer to the quantities in the
presence of a crack of half-length ℓ. Making the simplifying
approximation that the nonlinearity of the modulus is the same with
and without a crack, $\alpha = \alpha'$, equation (IV-9) takes on the same form
as the linear case. The computer calculations indicate that this
is a very good approximation. To evaluate M', however, a theoretical estimate of ΔE is needed. As the exact form of the strain
energy of a crack for a non-linear modulus, for small cracks and
large strains, is not known, it was assumed that it is of the usual
form with the linear modulus replaced by M', the effective non-
linear modulus: namely $\Delta E = -\pi\ell^2\sigma^2/M'$ where σ is the applied stress.
By converting from applied stress to applied strain, equation
(IV-9) may be written

$$-\pi\ell^2\varepsilon^2 M' = (A\varepsilon^2/2)(M_0'-M_0) = (A\varepsilon^2/2)(M'-M) \qquad (IV-10)$$

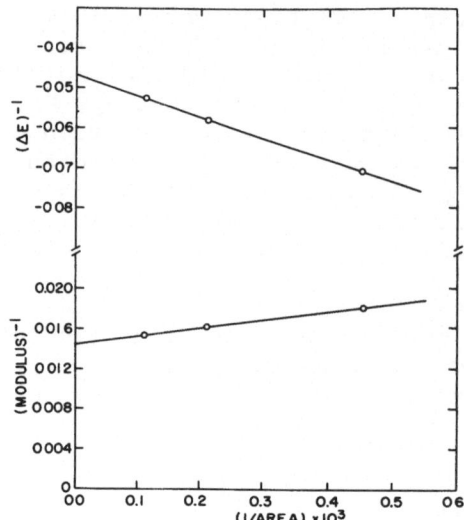

Fig. 11.   The functional dependence of the modulus and the strain
            energy difference ΔE on sample size. Data plotted
            against 1/area as indicated by equations (IV–11) and
            (IV–12) (all values for full sample). The straight
            line is the empirical fit to the data. All samples
            have a crack of 38 broken bonds. (After ref. 12).

and solving for the effective modulus it is found that

$$M' = M/(1 + 2\pi \ell^2/A) . \qquad (IV-11)$$

This is similar to the form obtained by Berry[20] with the linear
modulus M_0 replaced by M. According to these results $(M')^{-1}$ should
be a linear function of 1/A, which is indeed the behavior of the
computer calculations shown in Fig. 11. The behavior of the modulus as a function of sample size is in excellent agreement with
eq. (IV–11).

     The size dependence of the strain energy difference is also
obtained within the same approximation as

$$\Delta E = -\pi \ell^2 \epsilon^2 M'$$

and substitution for M' from eq. (IV–11) gives

$$(\Delta E)^{-1} = -(1 + 2\pi \ell^2/A)/\pi \ell^2 \epsilon^2 M. \qquad (IV-12)$$

Thus, $(\Delta E)^{-1}$ is also expected to be linear in 1/A. As indicated
in the upper half of Fig. 11, the strain energy response also shows
the correct functional dependence, but the slope is larger than
calculated from eq. (IV–12). On the basis of this functional
dependence the extrapolation to infinite size (1/A → 0) can be

arried out. Using this procedure, an infinite sample strain
energy difference of 21.3 ε_0 is obtained, which is about 12% larger
than that for the largest sample (19 ε_0) in the computer simulations.

Thus, the mechanical properties of the triangular 2-D L-J solid
are well understood, both in the presence and the absence of a
crack. It is to be noted for the subsequent discussion that sample
size effects are small for the largest sample used in the computer
simulations.

   c. Lattice Trapping

      Detailed studies were made with a crack consisting of 38
broken bonds (full sample). A series of strains was applied and the
simulation continued until either the system came to equilibrium or
the crack propagated by further bond breakage, or the crack healed
upon restoration of the cut bonds. The results[12] are shown in
table 2, where the applied strain, ε, the length of the first
unbroken bond d_c and the length of the first cut bond d_{c+1} are
listed.

Table 2. Data of lattice trapping study.

Applied Strain ε	d_c*	d_{c+1}[†]	Comments
0.0285		<1.60	crack heals
0.02875	1.510	1.614	stable crack
0.0290	1.549	1.646	stable crack
0.0295	>1.60		crack propagates

*Length of first unbroken bond in units of d.
†Length of first cut bond in units of d.

As shown in the table, the crack was found to be stable at two
strains 0.9% apart. At these equilibrium positions, the interatomic distances of the atoms across the crack surface exceeded
1.6d, the effective range of the potential, while the unbroken bond
was stretched to less than 1.6d. Restoring the interaction, therefore, did not affect the equilibrium. Upon lowering the strain by
0.9%, d_{c+1} became less than 1.6d and the system healed (with and
without damping) upon restoration of the interactions across the
crack. Conversely, upon increasing the strain by 1.7% the unbroken
bond was stretched beyond 1.6d and the crack propagated by bond
breakage (with and without damping). No attempt was made to obtain
closer limits on the strain range of the stable region. It is concluded from these data that the range of lattice trapping is larger
than 1% and less than 3%.

As already pointed out, the strain energy difference in the largest sample differs from that in the infinite sample by only about 10%. As the critical strain is approximately proportional to the square root of the energy, the estimate of any critical strain ratio is believed to be correct to better than 10%. Thus, the size effect would shift the critical strains (an increase with decreasing sample size for a given critical energy) but leave the range in critical strains, and hence lattice trapping, essentially unchanged. A study of lattice trapping at half the size confirmed this analysis. Thus, size effects in this study are quite small and it is concluded that lattice trapping is less than 3% for potentials of the form used.

3. Credibility of the Results

There are major advantages to the molecular dynamic simulation of the equilibrium properties of a crack. A realistic semi-empirical interatomic potential, in this case the Lennard-Jones, can be used. The model system can be completely unconstrained except for the correct boundary conditions. With appropriate attention to the required accuracy[21] in the simulation, the results are exact within the framework of the model.

The major disadvantage of the molecular dynamic technique is the restriction of the system to a finite number of atoms. The basic question is, therefore, how large is large enough to simulate in a relevant way the properties of the corresponding infinite system. In the present case it was shown that the properties of the finite system can be described and understood quantitatively on the basis of simple balance of forces and energies. Thus, extrapolation to infinite size became feasible and the error in using about 10,000 atoms could be estimated quantitatively. As already indicated, the error turned out to be small and did not affect in any significant way the negligible lattice trapping determined for this model.

It should be pointed out that the finite search range of 1.6d used in practice renders the L-J potential a snapping potential but with a small snapping range. The importance of the "cut off" may be illustrated in a simple way. In the sketch of Fig. 12 the extensions of the unbroken bond, d_c, and the first broken bond, d_{c+1}, are shown as a function of the applied strain, ε. Let r_{c1} and r_{c2} be two cut-off bond distances, i.e. the force goes to zero at $r > r_c$. ε_p and ε_h are the applied strains for propagation and healing, respectively. At $\varepsilon > \varepsilon_p$ the bond d_c becomes "broken," while at $\varepsilon < \varepsilon_h$ the bond d_{c+1} "heals" by coming within the range of the potential. In the simple linear sketch of Fig. 12, $\varepsilon_{p1} - \varepsilon_{h1} = \varepsilon_{p2} - \varepsilon_{h2}$. However, the ratio $\varepsilon_p/\varepsilon_h$ clearly increases as r_c is decreased, and vice versa. Thus, a continuous potential, for which $r_c \to \infty$, would be expected to show no lattice trapping.

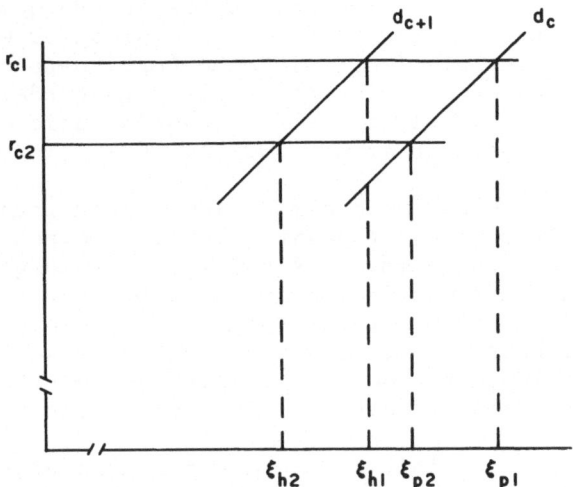

Fig. 12. Sketch of the influence of the "cut-off" in the potential
          on the lattice trapping ratio.

Finally one should mention that all molecular dynamic simulations so far have been restricted to two dimensions as even modern
computers cannot handle a three-dimensional atomic lattice of any
sensible size.

V. MOLECULAR DYNAMICS - DYNAMIC CALCULATIONS

1. Outline of the Method

Weiner and Pear[8] and Ashurst and Hoover[9] made the earliest
calculations of the dynamic properties of fracture. Weiner and
Pear used a quasi one-dimensional geometry and a piecewise linear
tensile force law for atoms in the same column. Ashurst and
Hoover used a two-dimensional triangular lattice constrained so
that Poisson's ratio is zero. They used, for the most part, bond
snapping force laws similar to the Thomson et al.[13] and Esterling[15]
lattice static calculations. It is difficult to assess the quantitative aspects of their results both because of the simplifications in the interatomic potential and the constraints against
motion transverse to the applied stresses. For example, Weiner
and Pear observed supersonic crack velocities at high stress
values, whereas the recent molecular dynamic results yield much
lower velocities. Ashurst and Hoover obtained energies of crack
formation in reasonable agreement with the present molecular dynamic results but also found a lattice trapping ratio of R = 3.7.
Thus, there is no question that molecular dynamic calculations do
yield interesting results but always the problem to be resolved is
how much of the results arise from simplification in the inter-

atomic potential as well as simplifications in the geometry. The
present calculations[11-12] may be thought of as an extension of the
Ashurst and Hoover calculations, with the bond snapping forces replaced by longer-range forces arising from a Lennard-Jones interatomic potential, no artificial zero Poisson's ratio constraint and
the use of larger arrays of atoms (5000 - 10000).

In the present molecular dynamic simulations three methods
were used to initiate crack propagation: (a) A crack was introduced into a prestressed sample. A sufficiently large crack would
propagate. (b) A crack was introduced in a prestressed sample, and
allowed to come to equilibrium and then the applied stress was
raised to a greater than critical value in one large increment.
(c) A crack was allowed to come to equilibrium and the stress (or
strain) increased gradually until crack propagation was obtained.

The method of crack initiation requires some elaboration. A
crack was created by cutting the bonds between a given number of
atoms between the two rows of atoms at the center of the sample.
The interatomic potential between these atoms was set to zero.
These conditions correspond to the insertion of a very thin knife
to create the crack. There was no damping used in the dynamic simulations subsequent to the creation of a crack. When the boundary
condition was that of constant stress a crack would only come to
equilibrium as long as the knife condition was retained. When the
boundary condition was that of constant strain, a crack would come
to equilibrium without the knife condition being retained. Thus,
the different boundary conditions in these dynamic simulations produced quite different results.

In the constant applied stress (c-stress) situation, a constant
external load was applied to the individual atoms of the uppermost
and lowermost rows. All external boundaries were free to move while
the load was kept constant. In the constant applied strain
(c-strain) case, the strain was applied by specifying and maintaining the vertical displacements of the individual atoms in the
uppermost and lowermost rows. All other atomic motions were unconstrained. For the finite samples used, about 5000 atoms, not only
do the surfaces look different (see Figs. 8 and 13a) but the shapes
of the similar size cracks near critical conditions differ. The
c-stress cracks are wider than the c-strain cracks. This shows
that the boundary conditions for finite samples can play an important role, and must be understood in some detail if the results are
to be credible.

Having established the boundary conditions and the method of
introducing the crack and critical stress or strain, the subsequent molecular dynamic calculations are identical to those used
to obtain the static results. The atomic positions, velocities,
and forces as well as the total energy, work, potential energy,

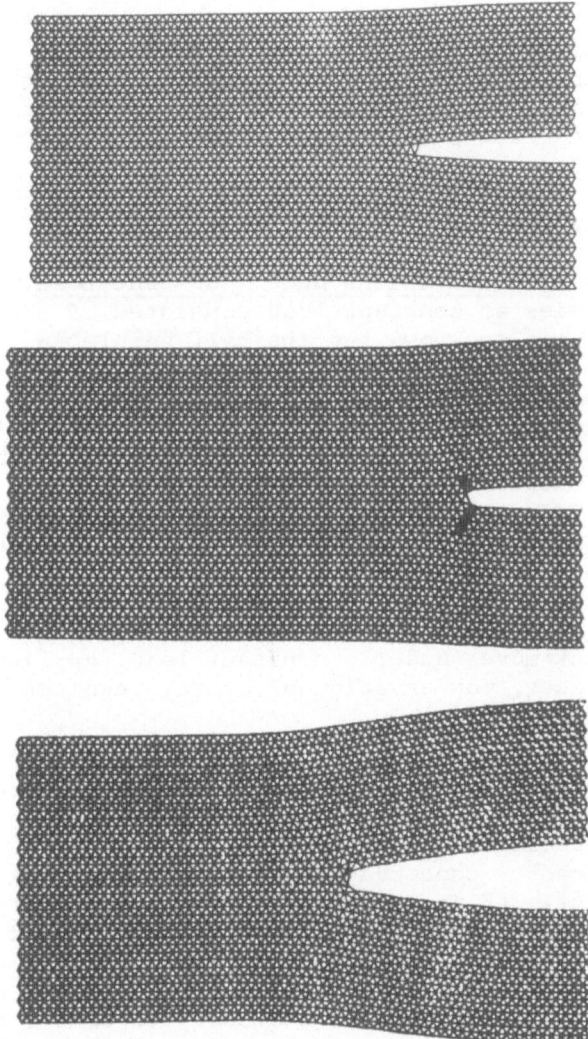

Fig. 13. A two-dimensional triangular lattice with a crack. Because of mirror symmetry the entire sample can be obtained
by reflection about the right-hand side of this figure.
Forces are applied to atoms in the 1st and 39th rows
(applied σ). Lines are drawn between atoms that are less
than 1.6d apart. (a) Clean brittle fracture at $\sigma = 0.5$ and
$L_0 = 39$ (L_0 is the number of initially broken bonds for the
half sample). (b) Incipient dislocation formation at $\sigma =$
1.3 and $L_0 = 19$. The extra lines designate the atoms that
have moved into the range of interaction. (c) Dislocation
formed at the lower end of the crack tip at $\sigma = 1.3$ and
$L_0 = 39$. This dislocation later on propagated to the
surface. (After ref. 11)

kinetic energy, and modulus of elasticity were monitored throughout the calculations. The time step in the calculation was chosen to give one part in 10^6 accuracy in the energy. The results were sometimes tested by increasing the accuracy from 2 to 4 times, to confirm that the results were independent of the accuracy of the numerical procedure.

2. Major Results

a. Dynamic Simulations under Constant Load.[11] The system for dynamic studies at constant load consisted of 5284 atoms arranged in 39 rows as shown for the half sample in Fig. 13a. The mirror symmetry about the central vertical axis was used to reduce the computation. In these calculations the system was brought to equilibrium under a given constant external force, σ (in units of ε/d), before the crack was inserted. A crack was created by cutting the bonds between a given number of atoms in the 19th and 20th rows, i.e. the interatomic potential for these atoms was set at zero. There was no damping subsequent to the creation of the crack. These conditions correspond to the insertion of a very thin knife to create the crack.

As the crack moves under a constant load, applied to both horizontal surfaces, conservation of energy requires that

$$\Delta W = \Delta E_p + \Delta E_k = \Delta E + \Delta E_s + \Delta E_k, \tag{V-1}$$

where ΔW is the work done on the sample, ΔE_p is the change in total potential energy, ΔE_k is the change in total kinetic energy, ΔE is the change in strain energy, and ΔE_s is the change in surface energy of the crack. Conservation of energy was always obeyed in these simulations. ΔE is evaluated as $\Delta E_p - \Delta E_s$, where ΔE_s is taken as two bond energies per pair of interacting atoms, i.e. the depth of the L-J potential, ε_0, per atom. In the Griffith energy treatment ΔE_k is zero. In order to obtain a quantitative criterion for the stress using eq. (V-1) it is necessary to have a relation between ΔW and ΔE. If one assumes macroscopic linear elasticity, $\Delta E = \frac{1}{2} \Delta W$, and therefore, $\Delta E = \Delta E_s$ which is the basic Griffith energy criterion.

In Fig. 14, ΔW, ΔE_s, ΔE, and ΔE_k are plotted versus the relative increase in crack length, $(L-L_0)/L_0$, for two computer simulations at different loads and initial crack lengths. These energy changes were evaluated in the region of early crack motion where ΔE_k is approximately zero and, thus, a direct comparison can be made with the Griffith energy assumption. In these computer simulations $\Delta E/\Delta W$ varies all the way from zero ($\sigma = 0.5$, large crack) to 0.75 ($\sigma = 1.3$, medium crack). While one might say that the Griffith energy criterion ($\Delta E/\Delta W = 0.5$) is approximately true for the latter, clearly the case of zero initial change in strain

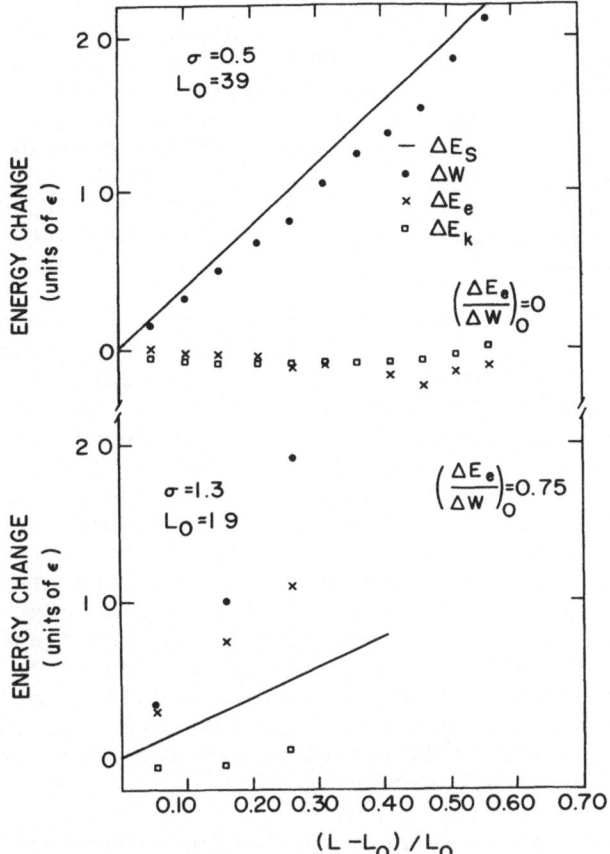

Fig. 14. Changes in energy components as a function of fractional increase in crack length during crack propagation. σ is the applied force (0.5 in the upper figure, 1.3 in the lower) and L_0 is the number of initially broken bonds for the half sample (39 in the upper figure, 19 in the lower). The energy components are ΔW, circles; ΔE_e, crosses; ΔE_k, squares; and ΔE_s as defined in eq. (V-1), solid curves. (After ref. 11)

energy would indicate failure of the elastic mode of brittle fracture.

As a matter of fact, the negligible change in strain energy suggests that a rigid tearing mode is a better description of such a fracture. This fracture (σ = 0.5) exhibited no dislocation formation or bifurcation and was the cleanest example of brittle fracture, with rapid approach to a constant velocity, in these

simulations. One can estimate the critical stress, σ_c, for such a
tearing mode of crack propagation by assuming rigid propagation of
the changing crack shape to the boundary in calculating ΔW. If all
of this work, rather than one half, goes into surface energy, σ_c
turns out to be of the same functional form as the Griffith σ_c but
is lower by a factor of $\sqrt{2}$. This simple estimate yields, using an
effective modulus of $3.0\varepsilon/d^2$, a $\sigma_c = 0.7\varepsilon/d^2$ as compared to the simulation value of 0.5. Thus, even though the energy criterion is
violated for this fracture mode, the critical stress estimate is
approximately valid.

At higher stresses a more complex behavior is observed as shown
by the simulations in Fig. 13a to 13c. Figure 13a represents the
same simulation whose energetics are shown in Fig. 14 (top), i.e.
a large initial crack with an applied σ of 0.5. This crack propagated in a clean brittle manner with very little distortion at the
crack tip. Figure 13b, corresponding to Fig. 14 (bottom), shows a
crack with large directional distortions near the tip indicated by
additional atom-to-atom lines connecting atoms which were originally
not in range but have now come into range. Such lines of distortion,
which may be considered incipient dislocations, are generated and
absorbed as the crack propagates. In Fig. 13c, a large initial
crack with $\sigma = 1.3$, the crack propagated to the configuration shown.
A dislocation can be seen at the lower end of the crack tip. This
dislocation then propagated to the surface and bifurcation occurred.
Raising the stress of the crack of Fig. 13b to $\sigma = 3.0$ resulted in
the immediate formation of dislocations and the subsequent blunting
of the tip, i.e. ductile behavior.

These results can be related to the Rice-Thomson[22] treatment
of crack-dislocation interaction. They consider a material to be
brittle if a dislocation generated in the neighborhood of the crack
tip cannot escape from the tip region. The critical distance, r_c,
at which a straight dislocation is in unstable equilibrium is obtained from considering the following forces: (1) the repulsive
force due to the crack stress field, (2) the attractive image
force of the dislocation in the free surface of the crack, and
(3) another attractive force arising from the additional surface
created by the blunted crack. If r_c is larger than the dislocation
core cutoff, r_0, then the material is brittle. For the present two-
dimensional system application of the Rice-Thomson equations yields
$r_c = 1.8d$. From Esbjorn and Jensen[19] r_0 for this system is 1.7d.
Consequently, this two-dimensional Lennard-Jones system is barely
brittle according to the Rice-Thomson criterion. As indicated
above, crack propagation in this system varies from brittle to
ductile depending on the applied stress and the detailed dynamics.
This suggests that the Rice-Thomson formulation has considerable
quantitative validity.

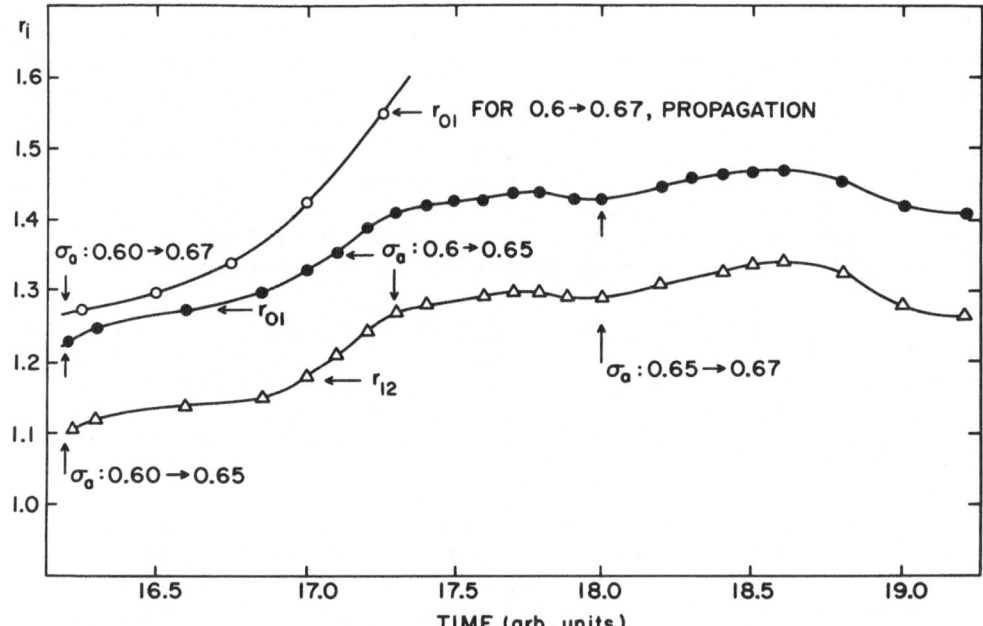

Fig. 15. Healing and propagation of the crack upon various changes in the applied stress.

A more detailed examination of the Rice-Thomson treatment indicates that, as the applied load is increased above critical, the stress at a distance from the crack tip increases. Including this stress dependence in the Rice-Thomson formulation, it is found that r_c is approximately proportional to $(\sigma_c/\sigma)^2$ and thus high applied loads, σ, should reduce r_c and produce a tendency towards dislocation emission. Since the model system is barely brittle at low stresses and low crack velocities, it is not surprising to find dislocation formation and ductility at high applied loads and large crack velocities.

Further information was obtained about the tendency for local deformation by attempting to reach critical stress conditions by very small changes in the applied load for a crack in equilibrium with the bonds cut. By critical damping it was possible to bring the system of Fig. 13a into equilibrium at $\sigma = 0.60$ with the bonds cut (knife present). When the bond interactions across the crack were restored (knife pulled out) the crack healed. When the applied load was raised in small increments the first (r_{01}) and second (r_{12}) unbroken bonds increased slowly as shown in Fig. 15

for the sequence σ_a: $0.60 \to 0.65 \to 0.67$ with large local deformation
but no propagation and with healing upon restoration of the interactions. The two points worth noting in the constant stress static
study are: a stable crack was not possible (with the knife removed
the crack always healed) and over a limited range, where crack propagation was expected from linear extrapolation considerations, the
crack did not propagate. With a more sudden change of σ: $0.60 \to 0.67$
propagation occurred with r_{01} rapidly increasing past 1.6 (the
search range) (Fig. 15). Thus, at constant load, this system is
very sensitive to the rate of change of the applied load.

b. <u>Dynamic Simulations under Constant Strain.</u>[23] As already
pointed out, the boundary condition of constant strain is theoretically advantageous because no external work is done leading to a
simpler energy analysis. Further, crack propagation can be
initiated by a very small change in the applied strain, i.e. from
essentially an equilibrium configuration.

In the dynamic simulations, detailed studies were made on a
sample 59 rows by 95 columns with initially 38 bonds broken. This
system was in equilibrium at an applied strain $\varepsilon = 0.0315$. Upon
increasing ε to 0.0318 the crack propagated and the propagation was
followed until the number of broken bonds increased by a factor of
3 (further propagation would get too near the edge of the sample).
The crack tip velocity, modulus of elasticity and all energies were
monitored during propagation.

The velocity data are shown in Fig. 16b as bonds broken per
time step vs. the step number (i.e. time), and crack tip velocity
vs. ℓ, Fig. 16a. The most striking feature of the data is the
rapid transition to a limiting constant velocity, which turns out
to be 11% of the longitudinal sound velocity. The process starts
with zero kinetic energy which then increases in a very regular
manner. The kinetic energy is discussed later.

The elastic response of the material during propagation is
illustrated in Fig. 17 where the changes in the effective nonlinear
elastic modulus, M', and the elastic energy, ΔE, are shown. Both
of these quantities behave as expected on the basis of the energy
analysis given in section IV for the static simulations. According
to equation IV-11, at a constant ε, $1/M'$ is linear in ℓ^2 with the
slope given by $2\pi/MA$. The area A of this sample is 4772 and M at
$\varepsilon = 0.0318$ is 70 giving a theoretical slope of 1.88×10^{-5}. The
observed slope during dynamical simulation is 1.96×10^{-5} in excellent agreement with the approximate theory for a finite nonlinear
elastic material.

According to equation IV-10 ΔE is linear in $\ell^2 M'$ at a constant
ε with the slope given by $\pi\varepsilon^2$. The data of Fig. 17b, where ΔE is
the half sample, follow this functional form with a slope of

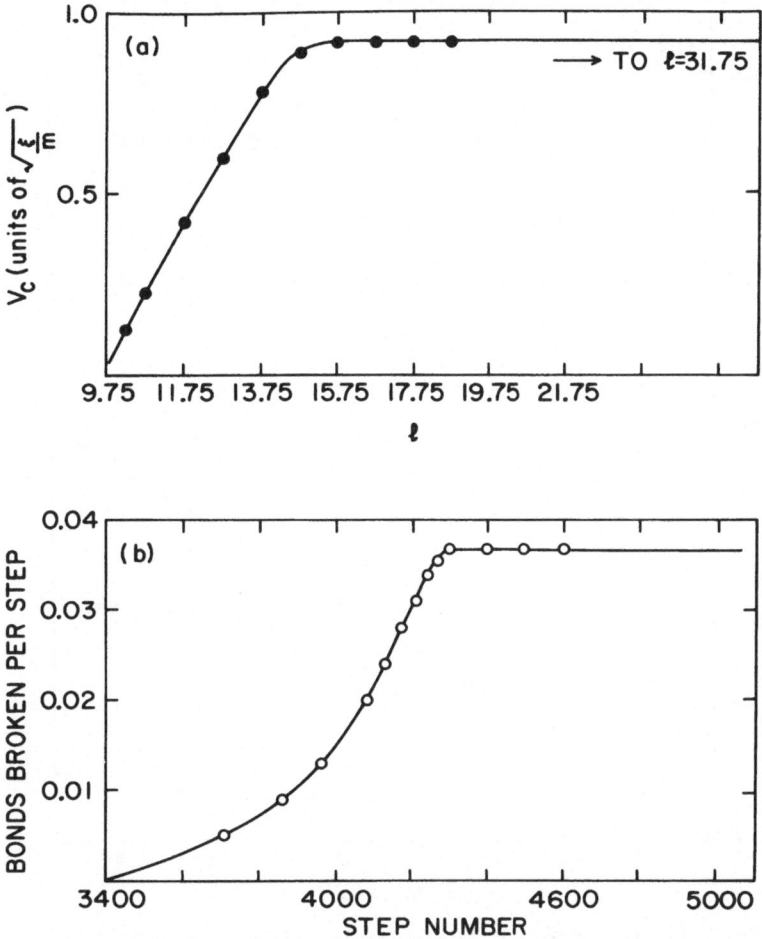

Fig. 16. (a) Crack tip velocity, v_c, vs. crack length, ℓ, during crack propagation at constant strain. (b) Crack tip velocity (number of broken bands per unit time) vs. time during crack propagation at constant strain.

1.8×10^{-3}. The theoretical slope is 1.59×10^{-3} giving a discrepancy of about 12%. It should be noted that there was a similar discrepancy, 16%, in the static ΔE vs. theoretical ΔE in the static simulation. Overall the agreement with the approximate nonlinear elastic treatment is satisfactory. The important observation to be made from the agreement of equilibrium static formulations with the dynamic results, is that the data demonstrates that the dynamic configurations resemble the equilibrium static configurations. This is the basic assumption made in calculations of dynamic crack propagation and it is reassuring to find that it is valid in the

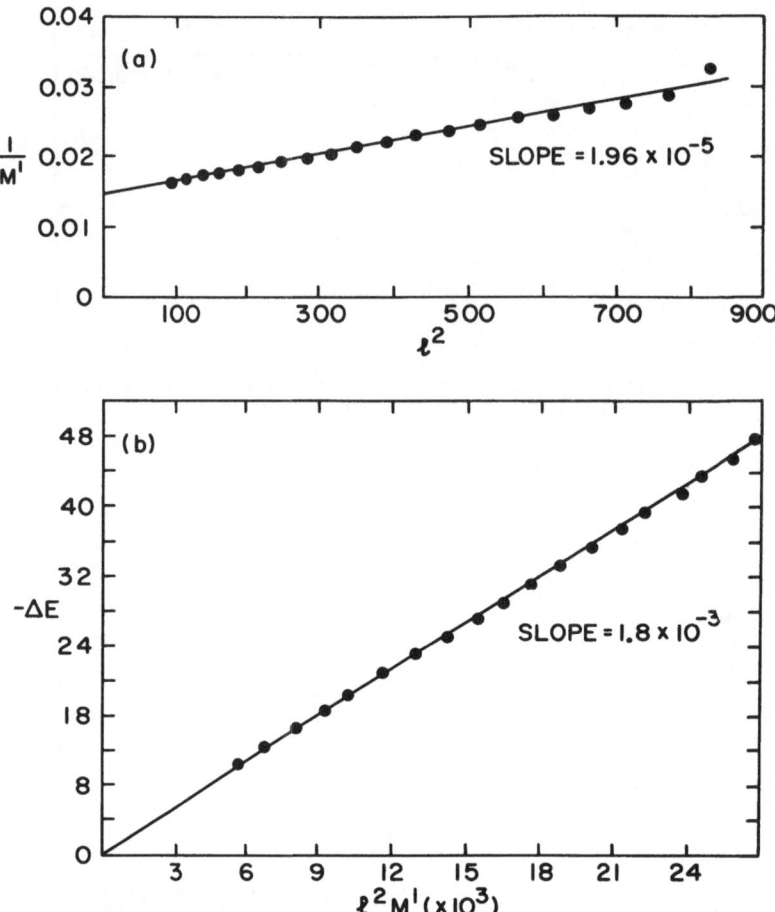

Fig. 17.   (a) The functional dependence of the modulus on crack
            length during crack propagation at constant strain.
            (b) Change in elastic energy vs. $\ell^2 M'$ during crack pro-
            pagation at constant strain.

constant applied strain simulations.

A theoretical approach to dynamic fracture was made by Mott[24]
in an extension of the Griffith thermodynamic approach. Mott added
a kinetic energy term to the static energy and found the configuration which maintains the total energy content as constant. While
this is straightforward, the relationship between the kinetic energy
and the crack velocity is based on a number of simplifying assumptions and dimensional analysis arguments. Quantitatively, Mott
obtains the relationship for ΔE_k, the kinetic energy, as

$$\Delta E_k = \tfrac{1}{2} k \rho v_c^2 \ell^2 \sigma_L^2 / E^2 = \tfrac{1}{2} k \rho v_c^2 \ell^2 \epsilon^2 \tag{V-2}$$

where ρ is the density, v_c the crack tip velocity, σ_L the external stress, E the modulus and k a constant. For constant applied strain, $\epsilon = \sigma_L / E$ and the second form for the kinetic energy is applicable to the present simulations. This is the key Mott result, the rest of the details follow from conservation of energy and must be correct regardless of the form for the kinetic energy. In Fig. 18, the kinetic energy term for the constant external strain case is compared with $v_c^2 \ell^2$; the agreement is striking. It should be noted that initially v_c varies approximately linearly with ℓ and, subsequently, becomes a constant as shown in Fig. 16; nevertheless $v_c^2 \ell^2$ describes the behavior of the kinetic energy throughout.

While the initial behavior of the kinetic energy is somewhat simple to explain, the ℓ^2 dependence, when v_c is a constant, is not so easy to understand nor is the fact that the asymptotic value of v_c is much less than the Mott estimate of ~ 0.4 times the longitudinal sound velocity $(8.49\sqrt{\epsilon/m})$. In the region where v_c varies, $(\ell-\ell_0)/\ell_0 < 1$ and a Taylor expansion can be made of the elastic energy for crack formation as a function of $\ell-\ell_0$. Inserting the conditions that the kinetic energy is zero at $\ell=\ell_0$ and

$$\left. \frac{\partial(\Delta E + \Delta E_s)}{\partial \ell} \right|_{\ell=\ell_0} = 0,$$

one obtains the following expression for the kinetic energy

$$\Delta E_k = -(\Delta E + \Delta E_s) = \frac{(\ell-\ell_0)^2}{2} \, \partial^2 \Delta E / \partial \ell^2 \tag{V-3}$$

With $\ell_0 = 9.75d$, from $\ell=\ell_0$ to 15.75, $\Delta E_k/(\ell-\ell_0)^2$ only varies by about 15% from the average value, while ΔE_k changes by three orders of magnitude. Thus, the initial behavior of v_c can be well described in terms of the initial constraints on the behavior of $(\Delta E + \Delta E_s)$. The behavior where v_c = constant is not so easy to describe. Lawn and Wilshaw[2] use a size dependent relationship for ΔE which would predict that v_c would vary rapidly and vanish much before the crack length had doubled. This does not happen in this simulation. The key difference between the Lawn and Wilshaw analysis and the present one is that they assume (in our notation)

$$\Delta E = -\pi \ell^2 \epsilon^2 (M')^2 / M_0 \tag{V-4}$$

whereas it is found in the simulation that

$$\Delta E = -\pi \ell^2 \epsilon^2 (M'). \tag{V-5}$$

As $M' \propto 1/(1+\alpha\ell^2)$, the Lawn and Wilshaw crack formation energy decreases with increasing ℓ much more rapidly and this makes the kinetic energy vanish at a small ℓ. If the kinetic energy term is calculated using (V-5) it is found to vanish at $\ell \sim 50d$. While this is much larger than the Lawn and Wilshaw estimate, it is still

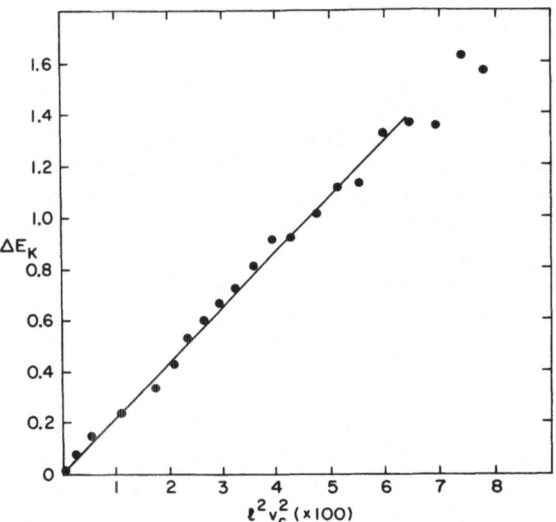

Fig. 18. The functional dependence of the kinetic energy on $\ell^2 v_c^2$ during crack propagation at constant strain.

not in accord with the present results. At first this is quite surprising in that the expression for ΔE is in accord with experiment
and thus the calculation for ΔE_k would also be expected to be valid.
The reason for the discrepancy is that the agreement of ΔE with
experiment is to within several percent, but the ΔE_k is about as
large in the region of interest as the error in ΔE. Hence, the
calculation of ΔE_k has an error of about 100% and thus, the expression for ΔE is just not described accurately enough to calculate ΔE_k
directly from eq. (V-1) with explicit forms for ΔE and ΔE_s.

The constancy of v_c may be approached somewhat differently in
terms of the size effect. The main idea of how the size effect
limits the velocity is that, because of simple size restraints, the
crack height can only grow to some maximum value, δy, limited by
the sample size in the y direction. The v_c may be estimated by
assuming an average acceleration near the crack tip of $F_{max}/2m$, and
thus

$$v_c \sim 2\overline{\left[\frac{F_{max}}{2m}\right]\delta y} \ . \tag{V-6}$$

For the Lennard-Jones potential, $F_{max} = 4.66 \ \varepsilon/d$, for the largest
sample $\delta y \sim 0.5d$, and $v_c \sim 1.5\sqrt{\varepsilon/m}$ compared to the observed value of
$0.9\sqrt{\varepsilon/m}$. This is in the right direction to explain both why v_c
becomes constant at a relatively small fraction of the sound velocity and also why it is a constant whereas a quantitative examina-
tion of $(\Delta E + \Delta E_s)$ suggests a varying value for v_c [i.e. $(\Delta E + \Delta E_s)$ is
not simply proportional to ℓ^2 over any wide range of ℓ]. However,
if v_c is limited by sample size constraints so that it cannot attain

the limiting value of an infinite sample, simply because the available distance for the crack surface to expand is limited, the observed results seem reasonable. This analysis suggests a sample
size effect: the smaller the sample, the smaller the limiting
velocity. This prediction is under investigation.

3. Credibility of the Results

The major argument favoring the credibility of the present
molecular dynamic results is the close agreement of the size dependence of the modulus (in the static calculations) with theory and
the correct functional dependence of the modulus and crack energy of
formation on the sample size and on the crack length during propagation. This suggests that the results for crack propagation are
quantitatively quite good at the early stages of propagation. Of
course, as the crack grows, the size effect becomes more important.
This may introduce some systematic size dependence in the crack
velocity results but such systematics are outside the scope of the
present analysis. The fact that the crack velocities rapidly attain
a constant value less than the sound velocity, in accord with experiment, is another argument for the credibility of the dynamic results.
As the c-strain results produced a number of results in agreement
with our physical understanding of crack propagation and no surprising result inconsistent with this understanding, the c-strain
results are regarded as highly credible.

While similar arguments about size effects can be made about
the c-stress results and their credibility, the one surprising new
result is the crack that propagated with all the applied work
(energy) going to create crack surface energy. Can this result be
true in nature, or is it some artifact of the simulation? As two
out of the three different crack-stress configurations in the
c-stress case propagated in the expected manner, this result is
certainly not a consequence of some grossly spurious constraint in
the simulation. The only feature in the c-stress situation which
differs from the usual mode of initiating fracture is the abrupt
increase in stress. This rapid stress increase may allow the system
to attain a configuration not accessible through the usual equilibrium configurations. Thus, while the unexpected energy sharing in
one of the c-stress cases, which is a violation of the assumption of
the Griffith model, is not believed to be a size effect it may well
be the result of the (artificial) rapid stress increases made in the
calculation. Unfortunately, as crack propagation could not be obtained by a slow increase of the stress, a systematic study of the
origin of this new effect was not possible and its credibility cannot be established. Otherwise, the results are considered quantitatively reasonably (80 - 90%) correct.

The observation of dislocation formation during crack propagation is quite credible for several reasons. From the static mole-

cular dynamic calculations at c-strain and c-stress it is apparent
that the two-dimensional Lennard-Jones solid must be just in the
vicinity of brittle-ductile behavior since small variations in boundary conditions and crack shape can drive the behavior either way.
Further, the Rice and Thomson model for dislocation formation
stresses the importance of the region near the crack tip in determining the brittle-ductile behavior. As the region is quantitatively
well described in the molecular dynamic calculations, the formation
of dislocations should be realistic and the quantitative results
reasonably accurate.

REFERENCES

1.   R. M. Thomson - this conference.
2.   B. R. Lawn and T. R. Wilshaw, "Fracture of Brittle Solids"
      Cambridge University Press, Cambridge (1975).
3.   E. R. Fuller, Jr. and R. M. Thomson in "Fracture Mechanics of
      Ceramics" R. C. Broadt, D. P. H. Hasselman and F. F. Lange,
      ed., Vol. 4, Plenum, New York (1978) pp. 507-548.
4.   D. M. Esterling, Comments Solid State Phys. 9:105 (1979).
5.   J. E. Sinclair and B. R. Lawn, Proc. Roy. Soc. Lond. A329:83
      (1972).
6.   P. C. Gehlen, Scripta Met. 7:1115 (1973).
7.   J. E. Sinclair, Phil. Mag. 31:647 (1975).
8.   J. H. Weiner and W. Pear, J. Appl. Phys. 46:2398 (1975).
9.   W. T. Ashurst and W. G. Hoover, Phys. Rev. B14:1465 (1976).
10.   M. F. Kannimen and P. C. Gehlen in "Interatomic Potentials and
      Simulation of Lattice Defects" P. C. Gehlen et al., ed., Plenum,
      New York (1972) pp. 713-722.
11.   A. Paskin, A. Gohar and G. J. Dienes, Phys. Rev. Lett. 44:940
      (1980).
12.   A. Paskin, D. K. Som and G. J. Dienes, J. Phys. C 14:L171 (1981).
13.   R. M. Thomson, C. Hsieh and V. Rana, J. Appl. Phys. 42:3154
      (1971).
14.   C. Hsieh and R. M. Thomson, J. Appl. Phys. 44:2051 (1973).
15.   D. M. Esterling, J. Appl. Phys. 47:486 (1976).
16.   A. A. Griffith, Phil. Trans. R. Soc. A221-163 (1920).
17.   See, for example, "Computer Simulation for Materials Applica-
      tions" R. J. Arsenault, J. R. Beeler, Jr. and J. A. Simmons,
      eds., Nuclear Metallurgy, Vol. 20 (1976).
18.   L. Verlet, Phys. Rev. 159-98 (1967).
19.   Per-Ole Esbjorn and E. J. Jensen, J. Phys. Chem. Solids 37:1081
      (1976).
20.   J. P. Berry, J. Mech. Phys. Solids 9:105 (1960).
21.   D. O. Welch, G. J. Dienes and A. Paskin, J. Phys. Chem. Solids
      39:589 (1978).
22.   J. R. Rice and Robb Thomson, Philos. Mag. 29:73 (1974).
23.   D. K. Som, A. Paskin and G. J. Dienes, to be published. This
      section is part of a thesis by Dilip K. Som.
24.   N. F. Mott, Engineering 165:16 (1948).

Note Added in Proof:

An interesting paper by E.D. Shchukin and V.S. Yushchenko was published [Journal of Materials Science 16, 313 (1981)] after this review was prepared. It deals with molecular dynamics simulation of mechanical behavior including sensitivity to the environment.

DISCUSSION

Comment by R.A. Oriani:

To simulate the action of an embrittling agent, why not system-atically reduce the strength of the bonds at the tip of a crack at every successive position of the crack tip?

Reply:

A very good suggestion which we already started discussing. We are thinking of a variety of simulation experiments for different locations and time sequences of bond weakening, or perhaps strengthening. In addition to the energetics and the kinetics, we can also obtain the detailed local stress distribution.

ELECTRONIC PROCESSES AT DISLOCATION CORES AND CRACK TIPS

Peter Haasen

Institut für Metallphysik and SFB 126
Universität Göttingen, FRG

INTRODUCTION

It does not seem to be possible yet to present an electronic
theory of dislocation cores and crack tips in bcc transition metals
which are in the center of the fracture problem. Instead the
electronic aspect has become more transparent in semiconductors with
diamond cubic structure which undergo a macroscopic brittle to
ductile transition at about 60% of their absolute melting temperatures[1]. Interestingly enough such crystals like Si and Ge deform
more easily when they are n-doped (say with 10^{19} cm^{-3} As) than in
the intrinsic state, i.e. in a range of substitutional solute concentration which would not affect the brittle ductile transition
of metal single crystals. This is clearly an electronic effect, not
one of the size misfit of the solute, as is shown by the non-
effectiveness of tetravalent solute [2]. The important parameter is
the position of the Fermi level as determined by temperature and
doping relative to the electronic levels (or better: one-dimensional

Dedicated to Prof.Dr.phil., Dr. mont. E.h., Dr.rer.nat.hc. E.Schmid
on the occasion of his 85th birthday.

bands) of the dislocations [3]. This parameter determines the line
charge of the dislocations which in turn has a large and positive)
influence on dislocation mobility. One might say that the line charge
destabilizes a straight dislocation which is its minimum energy
configuration in the Peierls potential. To move it has to create
(and move) kinks which is aided by its charge. The charge can be
changed in a controlled manner by illumination and this leads to a
photomechanical effect [4]. In a region of thickness λ_D (Debye
length) near to a surface covered by an electrolyte the energy
bands may be bent and their electronic occupation changed, leading
to a chemomechanical effect [5,6].

All this depends on where, in the forbidden band of the semiconductor, are the energy levels of the various types of dislocations present, and this in turn is governed by their core structures.
Recently a large theoretical and experimental activity in these
fields is apparent [7-12] taking into account the dissociation of
dislocations into partials. The electronically active site at the
core is the so-called dangling bond, (db) a half-filled orbit of the
sp^3-hybrid at an atom in the diamond structure which has only three
nearest neighbors instead of four. The number of db's in the dislocation core may be changed by core reconstruction. Similar considerations apply to large angle grain boundaries [13] and most
probably also to cracks in the diamond cubic structure. Additional
defect states are normally found due to defect strain, i.e. the
deformation potential, and these are split off the valence and
conduction bands [14]. We will briefly describe dislocation core
structures and energy bands in the next section together with results of powerful electrical (and optical) methods to locate these
defect states and measure their occupation. This will be followed
by a discussion of dislocation mobility vs. temperature, stress
and charging. The mobility of dislocations in turn determines the
extent of the plastic zone which may form at an opening crack tip.
This will be modelled in the last section of the lecture.

CORE STRUCTURES AND ELECTRONIC STATES

It is now well recognized that dislocations in Si and Ge are dissociated into partials which bound a stacking fault of about 40-60 Å width. A screw dislocation splits into two 30° partials, a 60° dislocation into a 30° and a 90° partial, and a 90° dislocation into two 60° partials. These dissociations occur most likely in the so-called glide set of $\{111\}$ planes of the diamond cubic lattice, see the proceedings of a recent conference [15]. An important parameter is the temperature at which the dislocations are introduced into the crystal: At relatively low deformation temperatures $T < 0.6\ T_s \approx 750^\circ C$ for Si, dislocations lie in Peierls valleys, i.e. have 60° or 0° character while after compression at $900^\circ C$ 90° dislocations dominate [1]. Correspondingly 30° and 90° partials are responsible for the changes in electrical properties of the material after low T deformation, 60° partials in the second case. A further difference between crystals deformed above and below a temperature $0.6\ T_s$ is the presence of point defect aggregates in the latter [19]. These mask to a certain extent the properties of the dislocations one wants to observe. Figs 1 and 2 show possible structures of partial dislocations in a form derived from elasticity theory and in a reconstructed form. The reconstructions were performed by Marklund [7] using valence force potentials and computer simulation. Whether reconstruction is energetically favorable or not depends actually on the valence of the dangling bond formation energy E_{db} which is not well known. With a likely estimate $0.25\ eV < E_{db} < 0.85\ eV$ the 30° partial will reconstruct (fig. 2b) while the 90° partial will not (fig. 1a). For the 60° partial the situation is even less clear and there are various plausible reconstructed cores which are energetically degenerate. The electronic states associated with partials were calculated for Si by Marklund applying the tight binding method to periodic arrays of large unit cells containing two dislocations of opposite sign. As a consequence of the translational symmetry in dislocation direction one-dimensional bands occur a) due to broken bonds, b) due to dilatational strain.

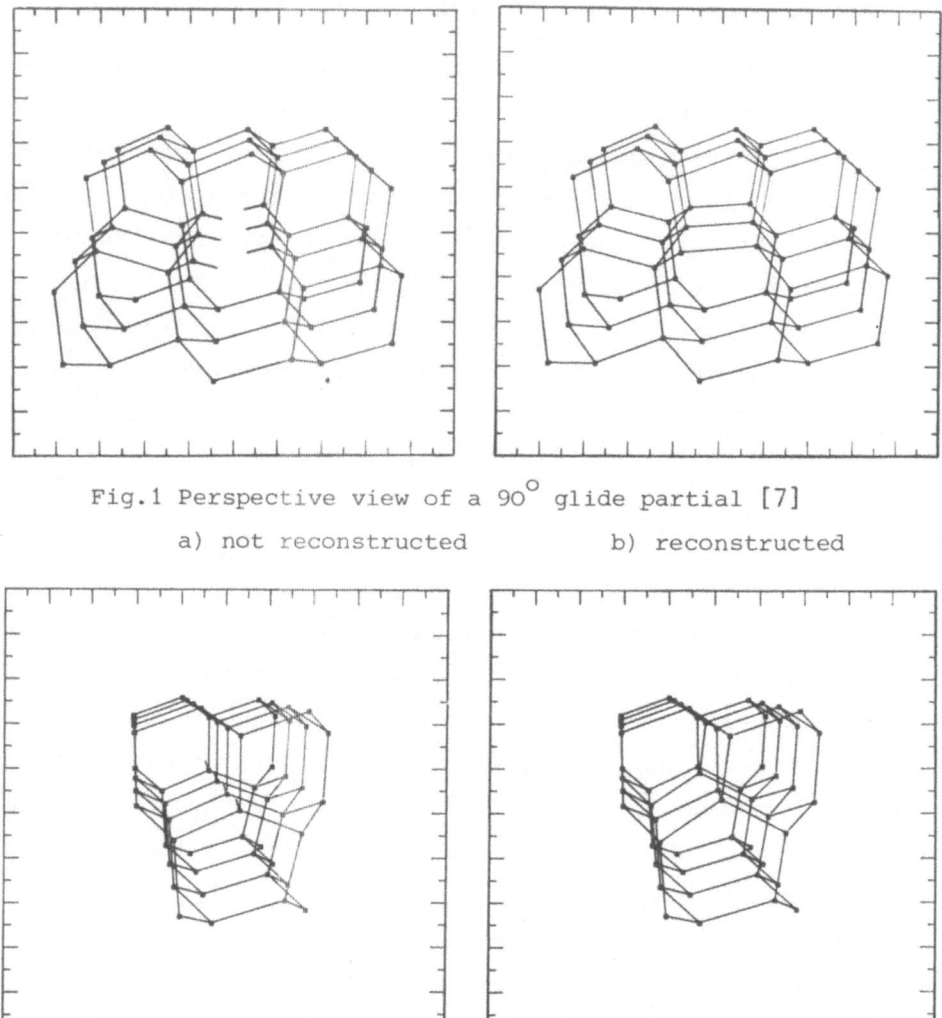

Fig.1 Perspective view of a 90° glide partial [7]

a) not reconstructed   b) reconstructed

Fig. 2 As fig 1, but for 30° glide partial [7]

The former in the core of an non-reconstructed 90° partial
(fig. 1a) are the cause of two overlapping dislocation bands in
the lower half of the gap, a filled and an empty one. The same might
be true for reconstructed 60° partials or perhaps a narrow half-
filled band in the lower half of the gap for another mode of reconstruction [7]. The reconstructed 30° partial should not contain any
dangling bonds. The elastic strain in its core causes however dislocation bands to be split off both band edges (distortion states)

as is found for other types of dislocations [14]. Dislocation states
in Ge are less strongly localized than the corresponding states
in Ge [7]. Fig. 3a shows some of the one-electron spectra calculated
by Marklund for Si, assuming a certain bond-stretching parameter
(ß = 0.4, see [7]).

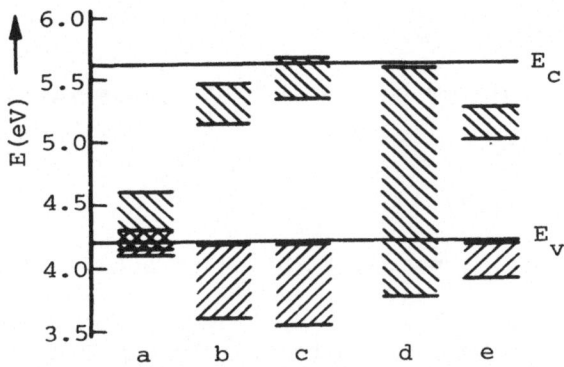

Fig. 3a Energy bands calculated for 90° (a,b,c) and 30° (d,e)
        partials in Si without (a,d) and with reconstruction [7].

CORE CHARGES AND EXPERIMENTAL DETERMINATIONS

       For a given set of one electron levels at the dislocation the
energy of any further electron or hole that is added to the core
depends strongly on the number of charges already present because
of the electrostatic interaction. As the interaction potential varies
only slowly within the half width r_o of the eigenfunctions of the
core states the simple approximation can be made that the whole
spectrum of core states is rigidly shifted by the electrostatic
potential. Then the free energy per site of the dislocation can be
written [16]

$$F = (E_o - E_F) \cdot f + E_{el}(f) - TS(f) \tag{1}$$

where f is the occupation ratio of the dislocation states (f = 0 for
the neutral, f > 0 for the negatively, f < 0 for the positively
charged dislocation). Q = -ef/b is the line charge of the dislocation.

E_O is the lower edge of the empty dislocation band, the upper of the full or the occupation limit in the neutral state of a half-filled dislocation band. E_{el} is the electrostatic interaction energy of all the charges on or around dislocations. S is the entropy of all electrons in the core states. In thermal equilibrium $\frac{\partial F}{\partial f} = 0$ and the Fermi energy

$$E_F = E_O + \frac{\partial E_{el}}{\partial f} - T \frac{\partial S}{\partial f} \tag{2}$$

Read [17] has calculated the configurational entropy for an empty band

$$\frac{\partial S}{\partial f} = k \ln ((1-f)/f) \tag{3}$$

For a full band (-f) replaces f while for a half-filled band $\partial S/\partial f \approx 0$.

The long-range electrostatic potential in the limit eV << kT is

$$V(r) = \frac{2ef}{\varepsilon b} K_O \left(\frac{r}{\lambda}\right) \tag{4}$$

with $\lambda^2 = \dfrac{\varepsilon kT}{e^2 (n_o + p_o) 4\pi}$

(ε = dielectric constant; n_o, p_o = initial concentrations of electrons, holes; K_O = Hankel function $\sim -\ln r/\lambda$ for $r < \lambda$).

Then

$$\frac{\partial E_{el}}{\partial f} = \frac{2e^2 f}{\varepsilon b} \left\{ \ln \frac{\lambda}{r_o} + \frac{1}{4} \left(\frac{N_{cd} - N_{ca}}{<n> - <p>} + 1 \right) \right\} \tag{5}$$

The second term on the right accounts for overlapping potentials and screens of different dislocations (N_{cd}, N_{ca} are the doping concentrations of donors and acceptors).

Fig. 3b shows the theoretical $E_F(f)$ according to eqs. (2,3,5) (without overlap).

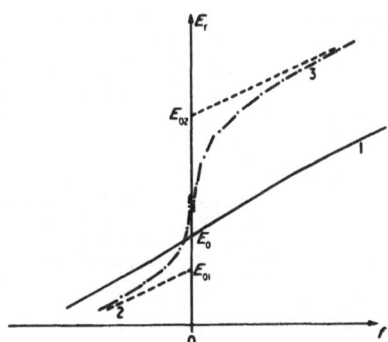

Fig 3b Calculated position of the Fermi level for various occupa-
      tion ratios for (1) a half filled band, (2) a full, (3) an
      empty dislocation band [16].

Experimentally the Hall effect allows to measure average electron
and hole concentrations <n>,<p> which determine the positions of the
Fermi level in the gap (E_g)

in p-material $E_F = kT \ln (C_V T^{3/2} / <p>)$ (6)
in n-material $E_F = E_g - kT \ln (C_c T^{3/2} / <n>)$

where $C_V T^{3/2}$, $C_c T^{3/2}$ are the effective densities of state in
the valence, conduction band. Furthermore f follows from conservation of charges

$$f \cdot \frac{N}{b} = <p> - p_0 \text{ or}$$ (7)
$$= n_0 - <n>$$

with N = dislocation density. Figs 4 and 5a show a comparison of
experimental Hall measurements with the theory, fig. 3b, for high
temperature-deformed p-Ge and p-Si, containing mostly complete 90°
(fig. 4) or screw dislocations (fig. 5a). The dashed "tails" of the
curves in fig. 4 correspond to overlaps of the screens of neighboring dislocations, i.e. in-complete screening, see eq. (5), last
term on right.

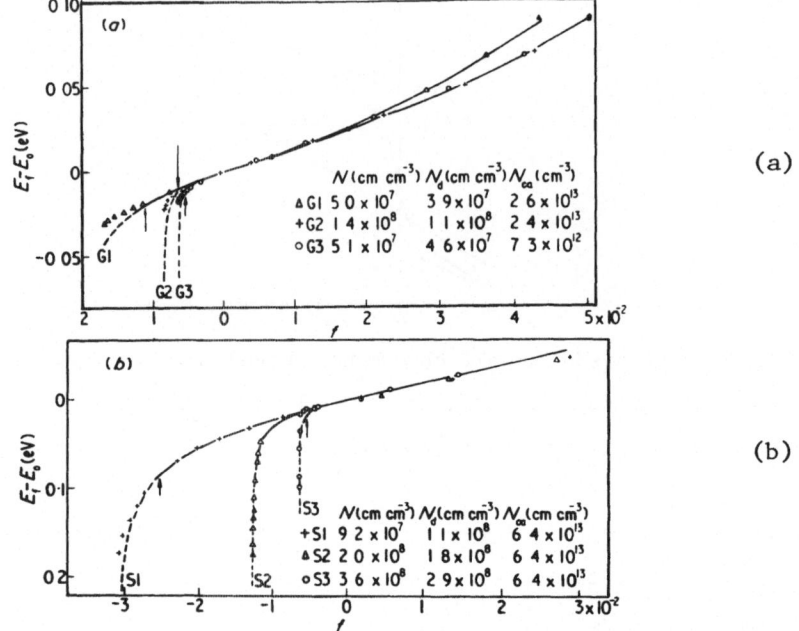

(a)

(b)

Fig. 4 Theoretical (full and dashed lines) and experimental $E_F(f)$
for p-Ge (a), p-Si (b) [16]. N_d is the electrically active
dislocation density.

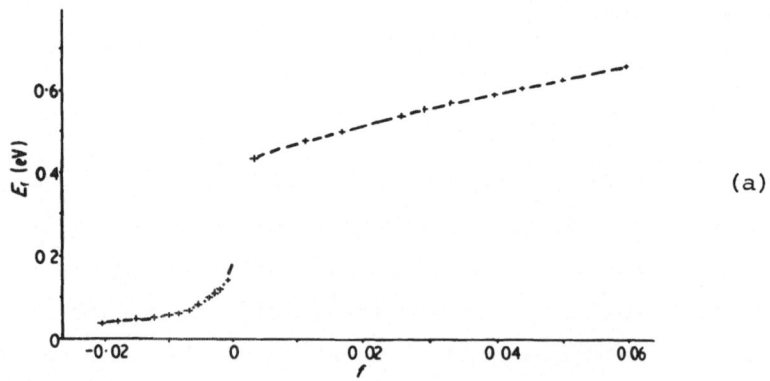

(a)

Fig .5 (a) as fig. 4 for screws in p-Ge [18]; (b) Energy positions
of full screw band, E_{o1}, empty screw band, E_{o2}, and half-
filled edge band, E_o [16], all at 300 K.

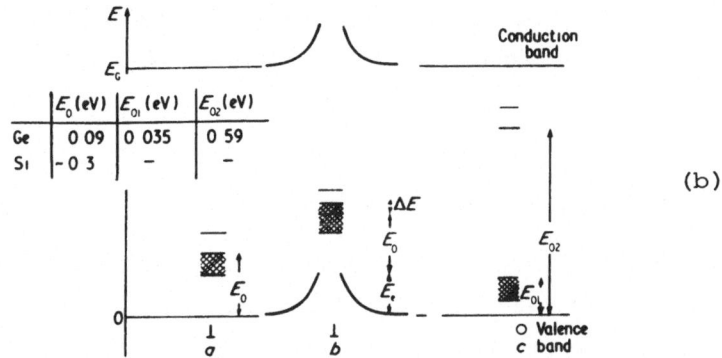

(b)

Fig. 5b indicates the band bending at a negative edge dislocation
while the screw is neutral. For a positive charge the bands
would bend downwards.

Typical occupation ratios are several percent in p-material, up to
30% in n-material (where the linearization used above breaks down).
For grain boundaries (and probably crack surfaces) f-values are
much smaller due to the two-dimensional electrostatic interaction.

The question arises now to which of the partial dislocation
bands the E_{oi} should be attached? In the case of the complete screw
the empty and full bands E_{o2}, E_{o1} must belong to the 30° partial.
The high temperature compressed Si and Ge contain mostly complete
edges; therefore the measured E_o probably belongs to the 60°-partial.
For Ge the best description is here by a half-filled band while for
Si, deformed at $782^\circ C$, Schröter et al [19] can fit their $E_F(f)$ just
as well by a combination of a full and an empty band separated by a
gap of 0.27 eV centered at E_o = 0.31 eV (at 300 K).

Further powerful electrical and optical methods have been introduced in recent years in the study of the electronic energy levels
of dislocations, some of which not only measure the highest occupied
state as does the Hall effect, but also the width of the dislocation
band and full and empty states of the dislocation elsewhere in the
gap. We mention here carrier mobility [3], optical absorption [20],
luminescence [21], photoconductivity [22], DLTS [23], deep level

transient spectroscopy, capacity of a deformed diode [24], EPR
[25,31], microwave conductivity [26]. The results are not always compatible and easy to interpret, especially if specimens deformed at
different temperatures are compared.

DISLOCATION VELOCITY, THERMAL ACTIVATION

The covalent bonding in the diamond structure leads to narrower
dislocation cores than in metals and thus to a high Peierls potential.
Dislocations overcome this by thermally activated formation and
migration of kinks. A dislocation velocity $v \approx b\sqrt{v_K J}$ results if J
double kinks are created per cm dislocation and sec, the kinks
migrate with the velocity v_K and then mutually annihilate.

As usual

$$v_K = v_{KO} \exp (-E_{KW}/kT)$$
$$J = J_O(\tau) \exp (-E_{DF}(\tau)/kT)$$

(8)

Here J_O is a complicated function of the stress [27] because the
dislocations are found to be constricted every $1 \approx 0.5 \ \mu m$ [28] and
a (stress-dependent) neighborhood of these constrictions (= jogs)
is excluded from kink formation. 1 seems to be influenced by thermally activated formation and recovery processes [27,28] according
to

$$1 = 1_O \exp (-\varepsilon/kT)$$

(9)

These constrictions are considered to be obstacles for kink migration, and E_{KW} might be determined by overcoming these. Internal
friction results [29] seem to indicate that the secondary Peierls
potential for kink migration itself is rather small. This is not
clear yet [8]. A further important question is whether kinks are
formed in both partials synchroneously or independently [30]. This
is determined by the stress level τ relative to a critical stress
$\tau_c \approx 3\gamma/4d_O$ (γ = stacking fault energy, d_O = width of dissociation
into partials). For the purpose of the present paper $\tau > \tau_c \approx 10MPa$
and double kinks form independently in both partials. E_{DF} is then

the double kink formation energy on a single partial and depends on stress as the 2 kinks of a pair attract each other

$$E_{DF} = E_{Do} - a\sqrt{\tau - \tau}_c$$

Putting all these factors together the dislocation velocity is [30]

$$
\begin{aligned}
v &= B_o f(\tau) \exp (-(E_{Do} + E_{KW} + \epsilon)/2kT) \\
&= B_o f(\tau) \exp (-U/kT) = Bf(\tau)
\end{aligned}
\tag{10}
$$

where $f(\tau)$ can be approximated in the high stress regime by τ^m, m = 1.35...1.65 [30]. The experimental activation energy is about 1.5 eV in Ge, 2.2 eV for Si ($\tau > \tau_c$). Möller [30] analyses also the case $\tau < \tau_c$ and the formation of single kinks at a surface to separate the various terms in eq (10). A detailed theory has been worked out [1] to calculate the yield and initial creep behaviour of crystals with diamond cubic structure from dislocation dynamics, multiplication and interaction.

CHARGE ENHANCED DISLOCATION MOBILITY

It is now clear that even at the relatively high temperature of deformation dislocations will be charged due to any deviation of theundisturbed Fermi level from the neutral occupation limit of the dislocation states. In the intrinsic range of doping E_F is determined by T alone, in the extrinsic range ($N_c \geq 10^{18} cm^{-3}$ for Ge, 500°C) also by the electrically active solute. Fig 6 shows the relative change in the velocity of 60° dislocations in Si at 600°C with doping, expressed in terms of the position of E_F in the gap [19].

The change is largely due to a change in the activation energy of v: from 2.16 eV for intrinsic Si to 1.5 eV for $1.4 \cdot 10^{19}$ n/cm³ added. But the prefactor B_o in eq (10) then also changes from 10^7 to 10^4 cm/s - not, however, the stress exponent m. For screw dislocations in Si the changes of v with doping appear to be similar to those for 60° dislocations. In Ge however n doping increases

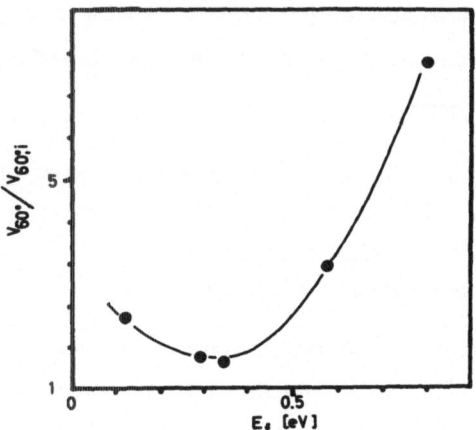

Fig. 6 Velocity of 60° dislocations in doped Si relative to that in
intrinsic Si as a function of the position of the Fermi
level at 600°C [19].

V_{600} while p doping decreases it at 500°C [2]. This difference between Si and Ge is to be expected from the different positions of the
dislocation level $E_O^{60^\circ}$ for the two materials: That in Ge lies so low
in the gap that p doping will discharge it relative to the charged
intrinsic state at 500°C. In Si with $E_O^{600} = 0.34$ eV (300K) charging
in the opposite sense occurs with p doping relative to n in a rather
symmetric way with respect to E_O, see fig. 6. The position of the
screw dislocation bands in Si is not yet known.

A quantitative treatment of the doping effect on the activation
energy for double kink formation has started from the destabilizing
effect of the line charge on the self energy of the dislocation [5].
Following the comparison of Friedel [32] between the elastic self
energies of a wavy and a straight dislocation one finds the line
energy per cm for a charged edge dislocation (shear modulus μ)
using eq (5)

$$E_L = E_L(f=0)\ (1 - \frac{e^2/\varepsilon b}{\mu b^3/8}\ f^2) \tag{11}$$

For Si $e^2/\varepsilon b = 0.32$ eV, $\mu b^3/8 = 1$ eV and $f \approx 0.6$ with

N_{cd} = 1.4·10^{19} cm^{-3} at 700°C [15], so E_L is lowered by 12%. The double kink formation energy is $E_{Do} = \frac{8b}{\pi} \sqrt{E_{PN} \cdot E_L}$ [32] where E_{PN} is the height of the Peierls barrier per unit length. This will be lowered by 6% in our example, in general in proportion to f^2, i.e. according to eq (5) roughly in proportion to the square of the change in E_F. The latter is borne out by the experiment, fig 6, while the absolute magnitude of the change is too small by a factor six. Perhaps E_{PN} also changes with doping, at least in proportional to the change in shear modulus. The latter by itself does not suffice either [1]. A further gain in energy on double kink formation might occur by the occupation of electron states at the kinks themselves [9]. Charged kinks are an independent species in the statistical thermodynamics and their configurational permutation with neutral kinks increases the entropy of the system, leading to a higher equilibrium kink concentration at T > O. Also the kink mobility might be changed by its charge [33] so if E_{KW} is a major contributor to the activation energy of v in eq (10) - which is uncertain at the moment - then ΔE_{KW} will have to be added to the change in E_{Do}. At present there are just too many unknowns to explain quantitatively the dependence of v on doping. The same applies to changes in f produced by illumination (photomechanical effect) or by band bending near a charged surface (chemomechanical effect), see [15].

CRACK TIP AND DISLOCATION EMISSION

After the discussion of the preceeding sections one knows to a certain extent the dislocation dynamics in semiconductors as a function of temperature, stress and position of the Fermi level. The latter in relation to the electron levels at partial dislocations determines the dislocation charge. We have seen that levels and charge depend sensitively on the structure of the dislocation core, whether it has dangling bonds or is reconstructed. Similar arguments apply to surfaces or internal surfaces like grain boundaries and cracks. Model calculations like Marklund's for dislocation pairs [7] are more difficult for cracks because of their long range stresses. There is only one calculation for a crack in the diamond structure by Sinclair [34]. He assumes harmonic bond stretching and bond bending forces of the Morse type while Marklund found it necessary to use the anharmonic Keating potential considering the large core strains [7]. Convergence of the calculations could only be obtained for values of the stress intensity factor K centred around the Griffith value of K_{co} = 0.41 MNm$^{-3/2}$ (concerning the definitions of K and K_{co} see other articles in this volume). There were stable cracks with [011] edge for 5 to 20% larger and smaller K indicating an energy barrier of the Peierls type for crack extension and shrink-age of about 0.1 - 0.3 eV per atomic layer ("lattice trapping of the crack") depending on the choice of potential. Cracks with other edge orientations contain kinks which move when a kink potential of about 1/10 of the above value is overcome. Fig. 7 shows a typical calculated crack tip structure at K = 0.425 MN/m$^{-3/2}$. Qualitatively the structure agrees with that obtained from continuum theory.

The question arises whether the crack is predicted to blunt under stress, i.e. whether dislocations will be emitted from its tip. Experimentally the answer is clearly positive for $T>T_c$ as will be discussed below for initially dislocation-free silicon. A continuum

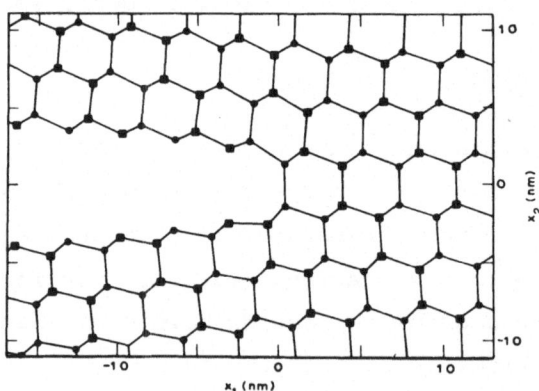

Fig. 7 Calculated crack tip structure for stress intensity
K = 0.425 MN/m$^{3/2}$ for two parallel {011} planes (■,●) [34]

calculation of Rice and Thomson [35] yields for the energy to create
a dislocation half loop within the stress field of a crack tip in
Si 111 eV at a critical loop diameter of 40 b. This would clearly
exclude crack blunting contrary to experiment. Three atomistic
calculations performed so far are reviewed by Tyson [36]. They
assume a simple cubic or square lattice; some include effects of
thermal vibrations. The nature of ductile failure at T = 0 is shown
in Fig. 8.

The typical barrier for the formation of a dislocation at 2.7b
from the crack tip at T = 0 is 0.14 μb^2 high, at 800 K 0.03 μb^2
according to the atomistic simulations. For a critical loop of 40b
depth the activation energies are 6.3 and 1.3 μb^3 or 50 and 11 eV,
respectively, for Si. This is still too high for thermal activation.
If a single partial dislocation with $b_p = \frac{a}{6}$ [112] is nucleated at
the crack, μb^3 is reduced by, a factor $3^{3/2} \approx 5.2$. But the stacking
fault trailing behind the partial must then be added to the energy
balance. This amounts to 0.017 eV/b_p^2 = 0.5 eV for the critical
partial loop of 23 b depth. The activation energy for the formation
of a faulted loop at T = 0 becomes 6.5 eV, at the higher T 1.8 eV.
The latter value put into the usual Arrhenius formula leads to
reasonable nucleation rates at 1100 K where crack tips in Si are in

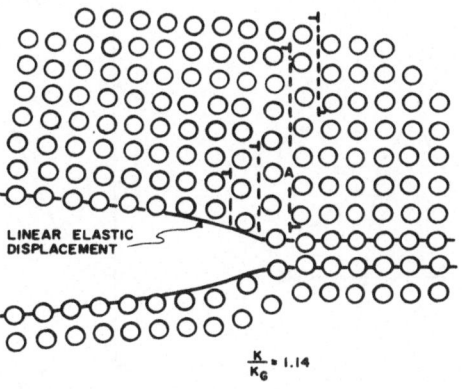

LINEAR ELASTIC
DISPLACEMENT

$\frac{K}{K_G}$ = 1.14

Fig. 8 Atomic positions at a crack tip loaded to 14% above
       Griffith's stress intensity [36]

are in fact observed to emit dislocations. The question now is
whether these move away fast enough to blunt a crack which is
pulled open, i.e. to cause a brittle to ductile transition.

THE BRITTLE TO DUCTILE TRANSITION IN Si

St. John [37] has carried out a series of fracture experiments
on pre-cracked dislocation-free silicon single crystals between
-196° and 1000°C at various strain rates. The trapezoid shape of the
thin Si plate provides on crackline-loading in mode 1 a high K at low
nominal stress and also only a weak dependence of K on crack length.
Fig. 9 shows the crystallography with respect to the crack. (The
crack edge will be kinked, see however [38]).
Fig. 10 shows load deflection curves at three temperatures bracket-
ing a brittle to ductile transition. At the lower temperatures the
specimen fractures at a constant stress intensity $K_{Co} \approx \sqrt{2E\gamma}$ where
E is Young's modulus and $\gamma = 2700$ ergs/cm² is the fracture surface
energy. Crack arrest occurs at $0.9\ K_{Co}$. At higher temperatures K_c
increases sharply while x-ray topography shows the formation of a
limited plastic zone before the specimen fractures. At still a few
degrees higher general plastic yield is observed for K_c values which

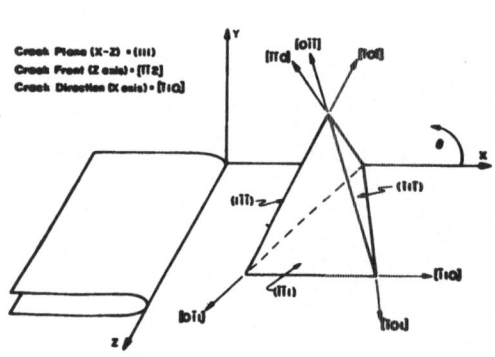

Fig. 9 Slip planes and crack
[37]

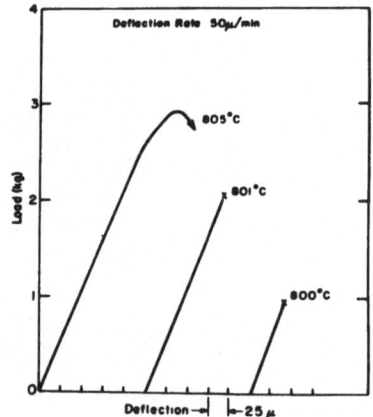

Fig. 10 Load deflection curves
for 3 temperatures [37]

now decrease with increasing T, see fig. 11, as is typical for the yield stress of silicon.

The transition is extremely sharp and depends strongly on the cross head speed $\dot{\delta} = C \cdot \dot{K}$, where $C = 4 \cdot 10^{-6}$ cm$^{5/2} \cdot$ N^{-1} is the machine compliance. The x-ray topographs by St. John [37] and more recently by George et al [38] give the shape of the plastic zone in the region of the original crack tip, the slip systems involved and the dislocation densities estimated. No dislocations were ever observed at the static crack tips in the low temperature plateau of K_c nor in connection with moving crack-tips. In the plastic zone George et al count 10^6 dislocations per cm^2 within a radius of 100 µm and 10^4cm^{-2} somewhat beyond, altogether about several 100 dislocations.

We will now try to give a consistent theoretical description of the dynamics of formation of the plastic zone based in part on ideas of St. John [37] but avoiding some conceptual difficulties (like the "yield stress" of a dislocation-free material) and complications in order to come to a quantitative comparison. The maximum crack tip stress acting perpendicular to the fracture plane depends on the crack tip radius ρ as

$$\sigma_{max} = \frac{2K}{\sqrt{\pi\rho}} \qquad (12)$$

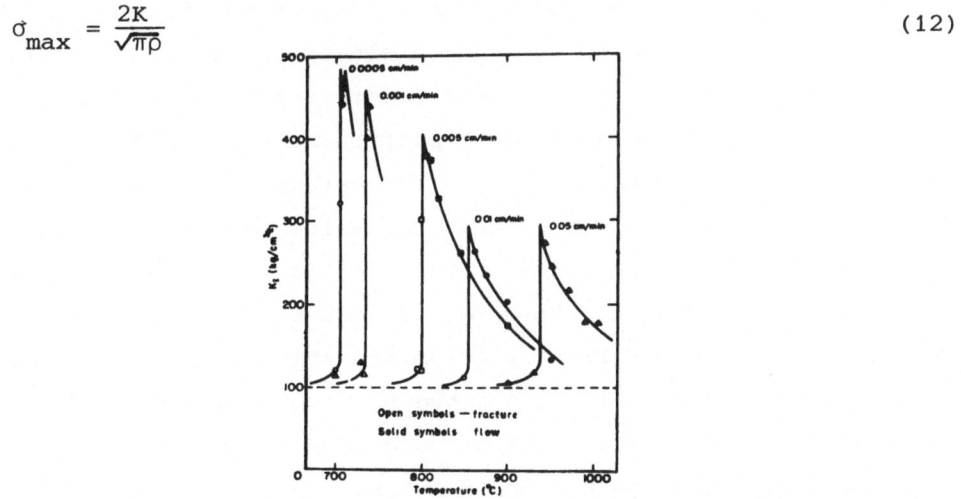

Fig. 11 Critical stress intensity factors for fracture or flow for various cross head speeds [37]

If σ_{max} reaches the cleavage stress E/10 than fracture will occur.
ρ is the sum of an elastic and a plastic component, the latter produced by the crack surface ledges of n dislocation loops emitted
from the tip. The former, ρ_{el}, must be several Burgers vectors large
($\tilde{n}_o b$) judging from the low temperature K_{CO}, so

$$\rho = b(\tilde{n}_o + n(t)) \tag{13}$$

The production rate of new dislocations is assumed to be governed
by the rate at which they move away from the tip

$$\dot{n} = v(\tau^*)/\Delta \tag{14}$$

where Δ is the spacing in the dislocation train and τ^* the stress
that determines their velocity v. τ^* derives from the following
argument: The crack tip stress tensor σ_{ij} in polar coordinates r,Θ
around the tip has the component τ in the slip system considered

$$\sigma_{ij} = \frac{K}{\sqrt{2\pi r}} \psi(\Theta) = \frac{\tau}{\mu} \tag{15}$$

(ψ= trigonometric function, μ = Schmid factor)

Differentiated with respect to time (for constant τ, ψ)

$$\dot{r}\Big|_{\tau,\psi} = 2 \frac{K\dot{K}}{\tau^2} \frac{\mu^2\psi^2}{2\pi} = \frac{K\dot{\delta}}{\tau^2} \frac{\mu^2\psi^2}{\pi C} \tag{16}$$

A constant τ contour moves outward from the tip the faster, the
smaller is τ. On the other hand the dislocation driven by τ moves
more slowly the smaller is τ. These counteracting relations assure
that the dislocation will follow a particular contour τ^* for which
$v(\tau^*) = \dot{r}(\tau^*)$, eqs (10) (16). With the parameters known from refs
[1] [37] one estimates $\tau^* \gtrsim$ 10 MPa. The value appears reasonable
considering the observed width of the plastic zone (500μm) formed
in 120 sec in comparison with measured values for $v(\tau)$ [30]. There
does not seem to be a dynamic pile-up situation for the dislocation
train [39], thus the distance Δ between successive dislocations
should be constant. One can get a rough idea about Δ from the spacing of Frank-Read loops $\Delta = Gb/\tau^* \approx$ 1μm.

Now all the ingredients for a comparison with experiment are
put together to determine the brittle-ductile-transition temperature

T_C where cleavage still occurs in the presence of blunting:

$$\frac{\dot\delta}{C} = \dot K = \frac{E}{40} \cdot \sqrt{\frac{\pi}{\rho}} \quad \dot\rho = \frac{E}{40} \cdot \sqrt{\frac{\pi}{\rho}} \, b\dot n = \frac{K^2_{Co}}{K_C} \frac{B\tau*^m}{2\Delta\tilde n_o}$$

$$= \frac{K^2_{Co}}{K_C} \frac{B}{2\Delta\tilde n_o} (\frac{K_C\dot\delta}{B} \frac{\mu^2\psi^2}{\pi C})^{\frac{m}{m+2}} \tag{17}$$

therefore

$$\frac{\dot\delta \, K_C}{C \, \exp(-\frac{U}{kT_C})} = (\frac{E}{20} \cdot \sqrt{\frac{\pi b}{2\Delta}})^{m+2} \, (\frac{\mu\psi}{\sqrt\pi})^m \cdot B_o \tag{18}$$

Fig 12 shows the proportionality between the strain rate and the dislocation mobility at T_C from the experiments of St. John [37]. For intrinsic Si U = 1.9 eV in fair agreement with the value 2.2 eV determined from direct velocity measurements in the high stress regime [30]. Numerical agreement is found in eq (18) for m = 1,

$\mu = f = \frac{1}{3}$, K_C (1000 K) = 4500 N cm$^{-3/2}$, i.e. n = 250, which are very reasonable values. The only parameter of the theory that has not been directly observed so far is the dislocation spacing Δ. This could be measured in a HVEM as Ohr et al [40] have done it in molybdenum and stainless steel. Perhaps Δ determines the extent of the dislocation-free zone these authors observe near the crack tip. It would also be interesting to repeat St. John's experiments on

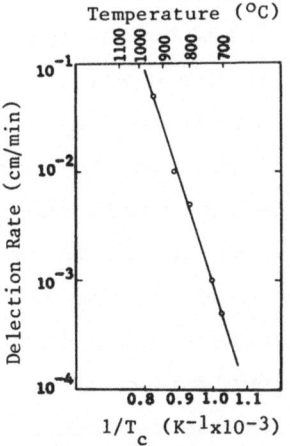

Fig. 12 Cross-head speed vs. inverse transition temperature
according to eq (18) [37]

strongly n-doped silicon where U and T_c should be 30% lower at the same strain rate. It is interesting to note that our plastic zone size of 0.5 mm at a crack length of 10 mm fits the prediciton of the (static) BCS theory [41] assuming n = 250 dislocations and τ^* as the friction stress. In a recent paper [42] Jokl, Vitek and McMahon describe dislocation emission at a crack tip where fracture occurs by bond stretching. They point out that the rupture of the bonds at the very crack tip is only a necessary, not a sufficient condition for fracture. Thermodynamically more strain energy must be released than plastic work is done in crack blunting for fracture to occur. In our model the two criteria lead more or less to the same expression for δ in eq (18) (except for some geometrical factors). Thus this is a necessary and sufficient condition for cleavage at a blunting crack tip.

CONCLUSIONS

The paper should have shown that semiconductors like Si are very intersting materials for the study of the atomistics of fracture. They are available without dislocations but these can be introduced in a reproducible manner. Their mobility is a strong function of temperature and doping, which are well-controlled parameters, but only a weak one of stress which is sometimes difficult to control. Dislocations are slow so the material becomes brittle at normal temperatures where it has a well defined cleavage plane. Electronic processes are well understood in semiconductors, and from this knowledge a clear picture of the charge of a lattice defect can be derived. The problem is the complicated atomic structure of defects, e.g. dislocations, which is not fully understood despite of the relatively simple type of binding. The electronic structure, states and charges of crack surfaces need further atomistic treatment. So does the problem of nucleation of a (partial) dislocation loop at a crack tip in an otherwise perfect lattice at an elevated temperature. The theory of fracture at a blunting crack tip formulated for silicon can in principle be useful also with bcc metals

and ionic crystals of a low density of (mobile) dislocations. However, these normally exhibit a higher stress exponent m of the dislocation velocity (and a lower activation energy) in the range of
their brittle-ductile-transitions which makes basic experiments as
in Si very difficult.

Thanks are due to W. Schröter for a critical reading of the manuscript.

REFERENCES

1. H. Alexander and P. Haasen, Dislocations and Plastic Flow in the
   Diamond Structure, Sol. St. Physics 22:28 (1968)

2. J.R. Patel and A.P. Chaudhuri, Charged impurity Effects on the
   Deformation of dislocation-free Ge, Phys. Rev. 143:601 (1966)

3. R. Labusch and W. Schröter, Electrical properties of dislocations in Semiconductors, in "Dislocation in Solids" Vol. 5
   F.R.N. Nabarro, ed. North-Holland, Amsterdam (1980)

4. E.Y. Gutmanas and P. Haasen, Photoplastic Effect in CdTe
   Jl. de Physique 40:C6-169 (1979)

5. N.H. Macmillan, R.D. Huntington and A.R.C. Westwood, Relationship between ζ Potential and dislocation mobility, Phil. Mag.
   28:923 (1973)

6. P. Haasen, Kinkenbildung in geladenen Versetzungen, phys. stat.
   sol. (a) 28:145 (1975)
   W. Schröter and P. Haasen, The Chemomechanical Effect in Semiconductors in "Surface Effects in Crystal Plasticity", R.M. Latanision et al ed, Nato Adv. Study Inst. Ser E17, p. 681 (1977)

7. S. Marklund, On the core structure of Glide Set 90° and 30°
   Partial Dislocations in Si, phys. stat. sol. (b) 100:77 (1980)
   Electron States associated with Partial dislocations in Si
   phys. stat. sol. (b) 92:83 (1979)

8. R. Jones, Theoretical Calculations of Electron States associated
   with Dislocations, Jl. de Physique 40:C6-33 (1979)

9. P.B. Hirsch, Mechanism for the effect of doping on Dislocation
   Mobility, Jl. de Physique 40:C6-117 (1979)

10. I.L.F. Ray and D.J.H. Cockayne, The dissociation of dislocations
    in Si, Proc. Roy. Soc. A325:543 (1971)

11. H. Alexander, Structure and Mobility of Dislocations in Semi-conductors, in "Dislocations in Solids" F.R.N. Nabarro ed. North-Holland, Amsterdam, in press

12. A. Bourret and C.D. Anterroches, High resolution Studies of Dislocations and Grain Boundaries, 37th Ann. Proc. Electr. Microsc. Soc. Amer., p. 388 (1979)

13. H.J. Möller, Structure and Energy of High Angle Grain Boundaries in Si, Phil. Mag. (1981) in press

14. S. Winter, Bound Electron States close to the Conduction Band in Ge due to 60° dislocations, phys stat sol. (b) $\underline{79}$:637 (1977)

15. Internat. Symposium on Dislocations in Semiconductors, Hünfeld, Colloque 6, Jl. de Phys. $\underline{4o}$:C6 (1979)

16. R. Labusch and W. Schröter, Electrical and Optical properties of dislocations in semiconductors, Inst. Phys. Conf. Ser. $\underline{23}$:56 (1975)

17. W.T. Read, jr., Theory of Dislocations in Ge, Phil. Mag. $\underline{45}$:775 (1954)

18. R. Wagner and P. Haasen, Electronic States at Screw Dislocations in Ge, Inst. Phys. Conf. Ser. $\underline{23}$:387 (1975)

19. W. Schröter, E. Scheibe and H. Schön, Energy Spectra in Si and Ge, Jl. of Microsc. $\underline{118}$:23 (1980); also W. Schröter, to be published

20. H. Schaumburg and F. Willmann, Optical Absorption of Plastically Deformed Ge, phys. stat. sol. (a) $\underline{34}$:K 173 (1976)

21. D. Gwinner and R. Labusch, Luminescence of Screw dislocations in Si, to be published

22. D. Mergel and R. Labusch, Optical Excitation of Dislocation States in Ge, phys. stat. sol. (a) $\underline{41}$:431, $\underline{42}$:165 (1977)

23. J.R. Patel and L.C. Kimerling, Dislocation Defect States in Si, Jl. de Physique $\underline{40}$:C6-67 (1979)

24. S. Mantovani and E. Mazzega, Dislocation Electronic states from Schottky diodes in Si, Jl. de Phys. $\underline{4o}$:C6-63 (1979)

25. V.A. Grazhulis, Application of EPR and electrical measurements to study dislocation energy spectrum in Si, Jl. de Phys. $\underline{4o}$:C6-69 (1979)

26. V.A. Grazhulis, V.V. Kveder and V. Yu. Mukhina, Investigation of the Energy Spectrum in dislocated Si, phys. stat. sol. (a) $\underline{44}$:107 (1977)

27. V.V. Rybin and A.N. Orlov, Theory of dislocation mobility in the low velocity range, Sov. Phys. Sol. St. $\underline{11}$:2635 (1970)

28. G. Packeiser, On the Correlation of Constrictions with jogs in dissociated dislocations in Ge, Phil. Mag. 41:459 (1980)

29. H. Ohori and K. Sumino, Internal Friction in deformed Ge crystals at low temperatures, phys. stat. sol. (a) 14:489 (1972)

30. H.J. Möller, Statistics of kink formation on dissociated dislocations; Jl. de Phys. 40:C6-123 (1979)

31. E. Weber and H. Alexander, EPR of dislocations in Si, Jl. de Physique 40:C6-101

32. J. Friedel, Dislocations, Pergamon Press, Oxford (1964)

33. R. Jones, Structure of kinks on 90° partials in Si and a strained bond model for dislocation motion, Phil. Mag. 42:213 (1980)

34. J.E. Sinclair, Influence of interatomic force law and kinks on the propagation of brittle cracks, Phil. Mag. 31:647 (1975)

35. J. Rice and R. Thomson, Ductile vs. brittle behaviour of crystals, Phil. Mag. 29:73 (1974)

36. W.R. Tyson, Atomistic Simulation of the Ductile-Brittle Transition, in "Fracture 1977", 2a:159, ed. Taplin Pergamon, Oxford (1977)

37. C. St. John, The brittle-ductile transition in pre-cleaved Si single crystals, Phil. Mag. 32:1193 (1975)

38. G. Michot, K. Badawi, A.R. ABD el Halim and A. George, Observation par Topographie aux Rayons X des Configurations de Dislocations developées à l'extremite d'une fissure dans le Si, Phil. Mag. to be published

39. A. R. Rosenfield and G.T. Hahn, Linear arrays of Moving dislocations emitted by a Source, in "Dislocation Dynamics", A.R. Rosenfield et al ed. McGraw Hill, New York (1968) p.255

40. S. Kobayashi and S.M. Ohr, In situ fracture experiments in bcc metals, Phil. Mag. 42:763 (1980)

41. B.A. Bilby, A.H. Cottrell and K.H. Swinden, The Spread of Plastic Yield from a notch, Proc. Roy. Soc. 272A:304 (1963)

42. M.L. Jokl, V. Vitek and C.J. McMahon, A microscopic theory of brittle fracture in deformable solids, Acta Met. 28: 1479 (1980)

The permission to reproduce fig. 8 is acknowledged to Pergamon Press, figs. 1 to 3 to Dr. S. Marklund, figs. 7,9 to 12 to Phil. Mag/Taylor and Francis.

DISCUSSION

Comment by M.P. Puls:

You show large differences in band structure depending on whether your dislocation is reconstructed or not. How sensitive are these bands to the positions of the other atoms in the core as obtained by the pair potential simulation.

Reply:

I must point out that the calculations of Marklund I have referred to are not self-consistent in the sense that the dislocation core structure is obtained from a model using atomic intraction potentials like Keating's, while the electron states are calculated by using the Slater-Koster scheme. Variations in the potential produce quantitative but not qualitative changes in the positions and widths of the dislocation bands.

WORKSHOP SESSION 1

HYDROGEN EMBRITTLEMENT

Session Chairmen: I.M. Bernstein and J.P. Fidelle

HYDROGEN RELATED FRACTURE OF METALS

H. K. Birnbaum

University of Illinois at Urbana-Champaign
Department of Metallurgy and Mining Engineering
Urbana, Il. 61801

INTRODUCTION

The deleterious effects of hydrogen on the mechanical properties of metals are well known and have been extensively studied [1]. Despite the effort which has been placed on understanding these problems our mechanistic understanding is rudimentary at best. Part of the difficulty stems from the fact that the behavior of metals with hydrogen as a solute or in a gaseous hydrogen environment cannot be accounted for on the basis of a single fracture mechanism. A second major difficulty is that the effect of hydrogen on the basic parameters of fracture, the strength of the atomic bonding and the properties of dislocations is poorly understood. Despite these serious handicaps, we have made significant progress in recent years in identifying the important physical parameters of the problem, in adequately establishing the effects of hydrogen on some of these parameters and in beginning to develop a mechanistic understanding of fracture in a number of the metal-hydrogen systems. In this paper we will review some of the recent progress and attempt to high-light some of the outstanding problems which remain to be solved.

The behavior of the metal-hydrogen systems can be classified in various ways; one of the most useful being based on whether the system forms second phases such as hydrides under stress (or as a result of deformation). In the class of hydride forming systems [2,3] we will include those which have thermodynamically stable hydrides as well as systems which have hydride phases which can be stabilized by applied stresses [4,6]. Examples of hydride forming systems include the group IVb and Vb metals and their alloys (Ti,

733

Zr, V, Nb, Ta) as well as a number of other metals such as Mg and Al. All of the hydride formers are severely embrittled by hydrogen when it is present as a solute element and by hydrogen gas when transfer of H between the gas and solid phases can take place. The second category of systems is that of non-hydride formers such as Fe, Mo, W and Cr and their alloys. Some of these systems, such as those based on Fe, are severely embrittled while others such as Mo [7] and W [8] have reduced strain to failure in the presence of a high H fugacity but since they are already relatively brittle materials the effects of H are not as dramatic. Other non-hydride forming systems, such as those based on the noble metals, are not embrittled by hydrogen at the strength levels and hydrogen fugacities to which they have been tested.

The hydrogen related failure mechanisms in these two classes of metals are clearly distinguishable. In hydride forming systems there is clear and increasing evidence [1, 9-14] that the fracture is due to the stress induced formation of a brittle hydride phase the fracture of which then leads to failure. While many questions remain concerning the details of this mechanism, it can no longer be doubted that hydrogen embrittlement results from the stress induced hydrides. Direct evidence for embrittlement by stress induced hydrides has been obtained in the Nb, V, Ti, Zr and Mg based systems. This rather strong statement does not preclude the additional possibility of a second fracture mechanism occurring in the absence of hydride formation. In systems such as the Al-Mg-Zn alloys hydrogen embrittlement has been observed under high H fugacity conditions (stress corrosion cracking) [15,16] but the presence and role of the hydride AlH_3 remains a matter of dispute. In the Ni-H system the hydride can form at high hydrogen fugacity, but under conditions where embrittlement is usually observed the hydride is not stable and fracture occurs by a different mechanism as shall be discussed shortly.

Metals which do not form thermodynamically stable hydrides in the absence of stress may still fail by the stress induced hydride mechanism. The hydride free energy relative to the solid solution can be significantly reduced by a suitable applied stress field and the hydrides may form in the stress field of the crack tip [4-6]. As will be discussed, this stress stabilization of the hydride can cause embrittlement in the Group Vb metals at temperatures where hydrides are not stable in the absence of stress.

The systems based on Fe, Mo, or W do not however fall into this latter category as the hydride free energies are too high to allow stress induced hydride formation. In these non-hydride forming systems the fracture mechanism is poorly understood despite the large amount of work which has been undertaken [1].

In part this stems from the fact that the significant parameters of this problem are only now being clearly established.

HYDRIDE FORMING SYSTEMS

Hydrogen embrittlement of those systems which form hydrides has been extensively studied since the early 1960's [1]. The general behavior of the alloys with H as solute is shown in Fig. 1 in which the fracture ductility is plotted as a function of temperature [10,17,18]. The ductility minimum is characteristic of all of these systems and indeed is observed to characterize the behavior of non-hydride forming systems as well. Another characteristic behavior is the inverse strain rate dependence ie, the increase of the ductility as the strain rate is increased. Both the return of the ductility at low temperatures and the inverse strain rate dependence have been shown to reflect the embrittlement kinetics which requires a flux of hydrogen to the crack tip [10,11]. Hydrogen in random solid solution does not cause embrittlement in these alloys. The ductility of the solid solution is terminated by brittle fracture only under conditions where H can diffuse to stress concentrations to achieve a critical concentration prior to ductile failure. Thus at any strain rate the hydrogen diffusivity determines the temperature below which ductile behavior is observed and at any temperature increasing the strain rate results in a cleavage to ductile failure transition at the strain rate for which the hydrogen diffusivity cannot achieve this critical concentration. The temperature below which the ductility decrease occurs has been shown [10,17-19] to correlate with, but to be higher than, the observed hydride solvus temperatures. Severe embrittlement occurs at temperatures above those at which hydrides form on cooling in the absence of stress, but the high temperature ductile-brittle transition temperatures vary with hydrogen concentration in the same manner as the hydride solvus temperatures. In the temperature range near the high temperature ductile-brittle transition, plastic deformation and ductile failure are competetive with cleavage fracture caused by hydrogen embrittlement. This is meant in the sense that the solid solution plasticity is terminated by ductile fracture (high temperatures) or cleavage fracture (low temperatures). The solid solutions are not inherently embrittled by the presence of hydrogen. Indeed, while the solid solutions may be quite ductile and the strain to failure may be very large above the solvus temperature, the actual fracture process may be a "brittle" process such as cleavage as shown in Fig. 2 [10,11].

The mechanism for fracture in these systems which seems to be able to account for these diverse observations is the stress induced hydride fracture mechanism first proposed by Westlake [9] and shown schematically in Fig. 3. As a result of the distortion field around the H interstitial, ϵ_{ij} its chemical potential in a

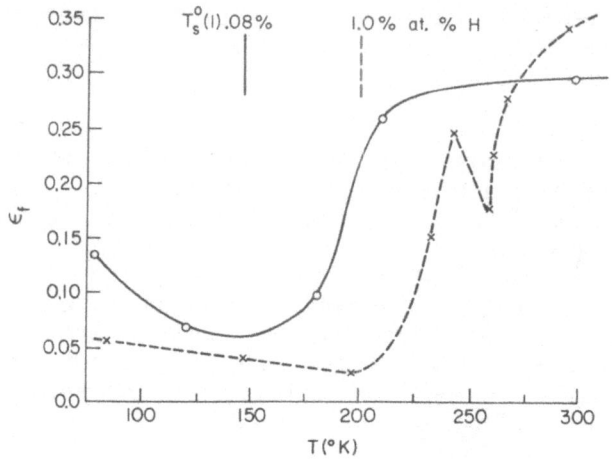

Fig. 1a

Strain to failure
vs temperature for
Nb-H alloys. The
stress free solvus
temperatures are
indicated by the
vertical lines.
Strain rate was
10^{-4} sec^{-1}.

----   1.0 at % H
____   0.08 at % H

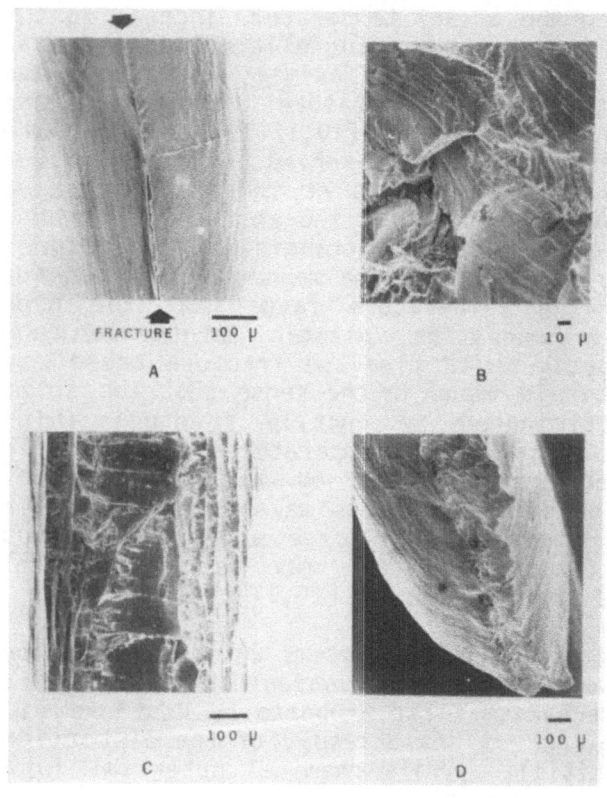

Fig. 1b

Fracture surfaces
of Nb-H alloys
containing 0.08 at
% H.

(A)   210 K
(B)   180 K
(C)   120 K
(D)   77 K

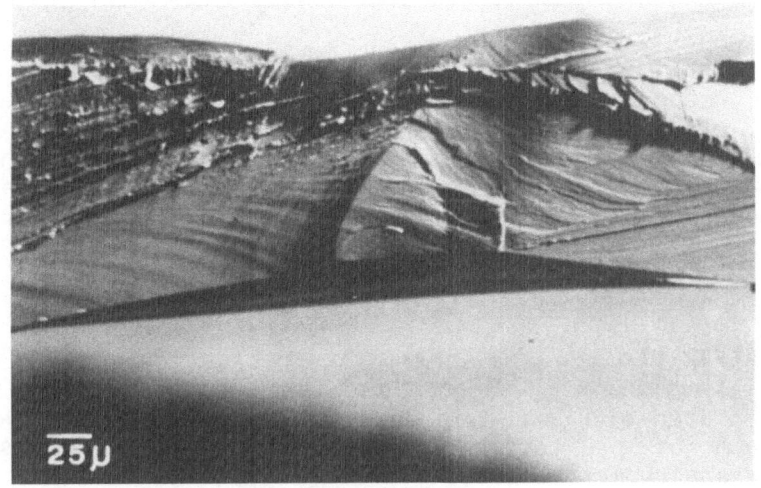

Fig. 2   Fracture surface of niobium specimen with H/Nb=0.01
stressed at 230 K. The initial width of the specimen
was about 500μ. The fracture surface was {110} cleavage
despite the large reduction of area prior to fracture.

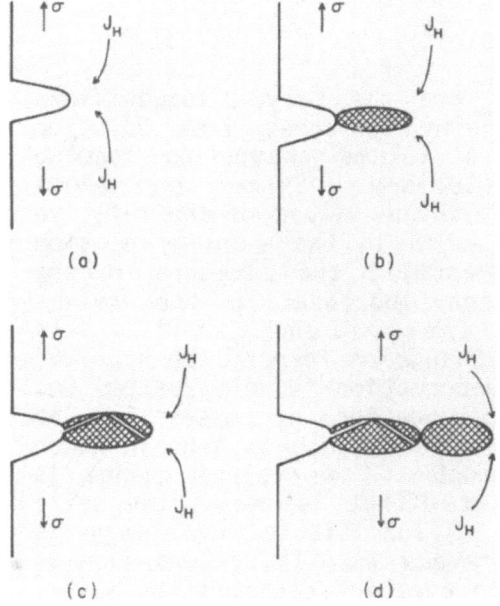

Fig. 3

Schematic representation
of the stress induced
hydride fracture mechanism.
(a) Stress induced flux of
hydrogen to the crack tip.
(b) Stress induced form-
ation of hydrides at the
crack tip. (c) Cleavage of
the hydride. (d) Repeat of
(a) and (b).

stress field, μ_σ can be written [20] as

$$\mu_\sigma - \mu_0 = \int \sigma_{ij} \, \varepsilon_{ij} dV \tag{1}$$

where the reference state chemical potential μ_0 is taken as hydrogen in solid solution at zero stress. The stress gradient which results from the presence of an elastic inhomogeneity such as a crack in a stressed solid, results in an increase in the hydrogen concentration of the crack tip $(C_\sigma - C_0)$ to a value

$$C_\sigma = C_0 \exp \frac{\sigma_s \bar{V}_h}{RT} \tag{2}$$

where $\sigma_s = 1/3\sigma_{ij} \, \delta_{ij}$ is the spherical stress component, $\bar{V}_h$ is the partial molal volume of hydrogen in solution, and only the volume change associated with the H in solution is considered. (Despite its occupancy of the tetrahedral interstitial site, H has been shown to have a cubic distortion field [21]). The increase in the local H concentration occurs at a rate which reflects the diffusion of H in the stress gradient at the crack tip and serves to eliminate the stress induced chemical potential gradient. In addition to this effect, the inhomogenous stress field at the crack tip affects the free energy between the solid solution and the hydride [4-6]; an effect which occurs to a much smaller extent in a uniform stress field [6]. This effect has been treated in various ways but can be viewed as a decrease in the chemical potential of the hydride relative to the solid solution in the stress field of the crack tip resulting in an increase in the solvus temperature to the value given by

$$C_e^\sigma = C_e^0 \exp - (\sigma_s \Delta V_{\alpha\beta}/RT_e^\sigma) . \tag{3}$$

In Eqn. 3 C_e^σ and T_e^σ, and C_e^0 and T_e^0 are the solvus compositions and temperatures under stress and in a stress free solution respectively, and $\Delta V_{\alpha\beta}$ is the molal volume change on forming the β hydride from the α solid solution. Since the values of $\Delta V_{\alpha\beta}$ are in range of 0.1 to 0.25 and the values of the relative partial molal enthalpy of solution of H in the α solid solution relative to the β hydride, $\Delta \bar{H}_H^{\alpha\beta}$ are small, the stresses at the crack tip can cause a significant increase in the solvus temperature (in the case where $(\Delta V_{\alpha\beta} > 0$ and $\sigma_s > 0)$. In addition to this effect, the volume change on forming the hydrides is accommodated in part by plastic deformation in many systems eg. Nb-H [4,22]. These plastic accommodation processes can be assisted by the local stresses. In systems such as V-H, in which the hydride forms as plates and most of the volume change is accommodated elastically, this latter effect is small, the shift of the solvus is given by Eqn. 3, and little hysterisis is observed between precipitation and reversion. In systems such as Nb, Ta and Ti the effect of stress on plastic accommodation must be considered. In addition to causing hydride precipitation above the stress free solvus temperature, the crack tip stress field affects the relative free energies of the various hydride variants and tends to bias the system towards precipitation of those variants which lie in a favorable orientation for crack propagation [13,23].

In hydride forming systems the solid solution remains ductile even at temperatures below the solvus temperature as indicated by direct observation [10-11] and by the fact that the strain to failure remains relatively high even at temperatures within the ductility minimum. Application of a stress at temperatures below the solvus results in fracture of the precipitated hydrides and the crack propagates to the interface with the solid solution. The stress concentration then causes flux of hydrogen to the crack tip and further precipitation of hydride. Cracking of the hydride occurs and the process of stress induced hydride formation and cracking is repeated leading to brittle failure. In those systems (Nb-H, Ta-H) where $\Delta V_{a\beta}$ is accommodated by plastic deformation, a large hysterisis between the hydride formation and resolution temperatures is observed [4-20], as expected from the accommodation free energy terms and the cracked hydride remains along the crack surfaces [10,11,24]. In contrast to this, systems (V-H) in which the accommodation is elastic have a small hysterisis and the hydride will go back into solution once the stress field is removed as the crack propagates [12,13].

This mechanism has been supported by direct observations in the Nb-H [10,11] and V-H [12,13] systems and by appreciable indirect evidence in other systems. In the Nb-H system the fracture surface is covered by hydrides as indicated by electron diffraction and SIMS ion probe studies (Fig. 4) [11,24].

At temperatures somewhat above the solvus this same mechanism will apply [10]. In this case plastic deformation and work hardening occur prior to the initiation of fracture. As the flow stress level increases, the stress induced increase of the hydride solvus temperature also increases as indicated by Eqn. 3. In a sense the plastic deformation of the solid solution is competitive with the onset of stress induced hydrides. If the hydride solvus is shifted to the test temperature by the applied stress, hydride will form at stress concentrations such as dislocation pileups and the plasticity will be terminated by the formation and cleavage of the hydrides. If this does not occur plasticity will lead to the usual ductile fracture. The solid solution is ductile and remains so until the onset of stress induced hydrides.

In the Nb-H system hydrogen embrittlement has been observed [25] at temperatures up to 700 K at compositions greater than about H/Nb = 0.3 (Fig. 5). Since the solid solution extends to

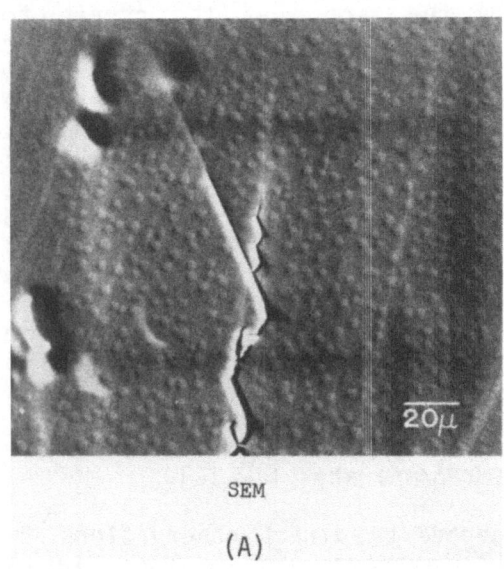

SEM

(A)

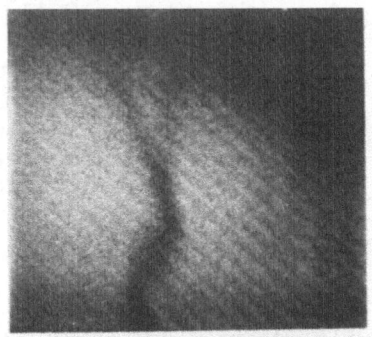

MASS 93 (Nb$^+$)

(B)

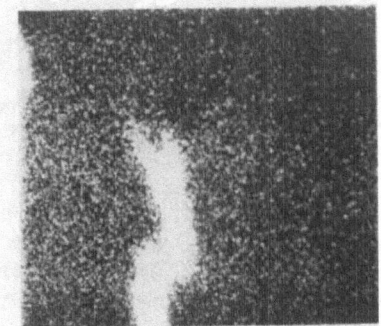

MASS 94 (NbH$^+$)

(C)

Fig. 4   Fracture surface in Nb-H
         alloy showing the formation
         of the NbH hydride along the
         fracture surface.
         (A) SEM image of the fracture
         (B) SIMS image of the
         fracture using mass 93 Nb ions
         (C) SIMS image of the
         fracture using mass 94 NbH
         ions.

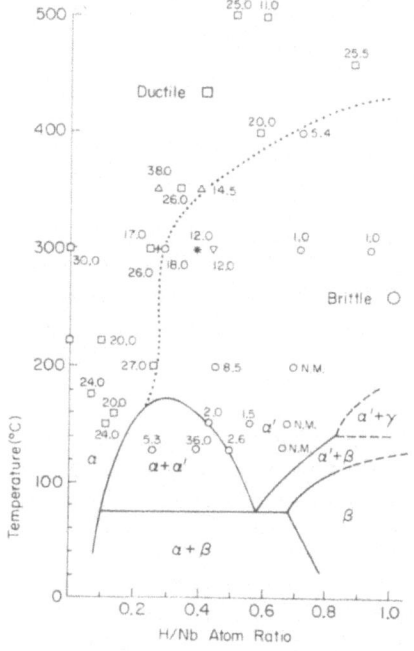

Fig. 5a

High temperature embrittlement of Nb-H alloys. The
phase diagram of the Nb-H
system is shown with the
high temperature ductile
to brittle transition shown
in the solid solution phase
field. The data points
indicate the compositions
and temperatures tested and
the numbers next to each
point indicate the strain
to failure.

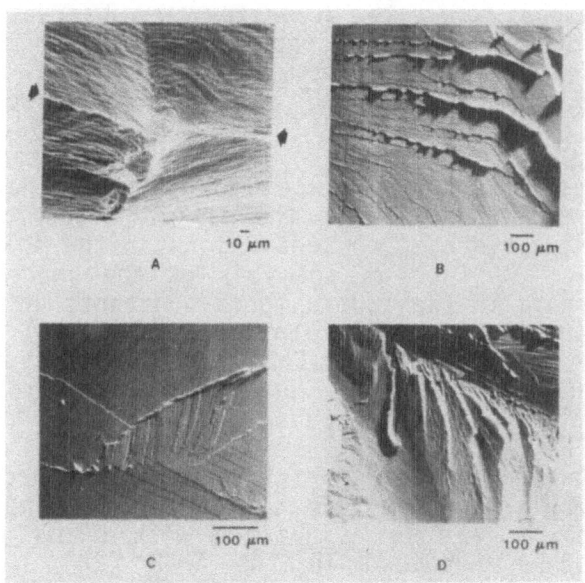

Fig. 5b   Fracture surfaces for specimens tested at 575 K with the
           compositions (a) $NbH_{0.26}$ (b) $NbH_{0.3}$ (c) $NbH_{0.71}$
           (d) $NbH_{0.9}$

above H/Nb = 1.0 at temperatures above about 450 K, it would be expected that the alloys would be completely ductile in the vicinity of H/Nb = 0.3. The ductile-brittle transition occurs on decreasing the temperature or on increasing the H/Nb and is quite sharply defined. This behavior has been shown to be consistent with the stress induced hydride fracture mechanism except that in the high temperature region the induced hydride is the δ , NbH$_2$ dihydride for which the molal volume increase on forming from the α' solid solution is about 25%. The boundary between the ductile solid solutions (H/Nb $\sim$ 0.3 and T $\sim$ 700 K) and the brittle alloys can be shown to correspond to the stress shifted phase boundary between the α' solid solution and the two phase (α' + δ) boundary. The presence of the δ dihydride at the crack tip has not been directly demonstrated.

In this fracture mechanism the stress induced hydride is expected to be a relatively brittle phase as the fracture surfaces are generally cleavage in nature. The fracture surface in the Nb-H system was shown [11] to be the {110}. Single crystals of the β hydride were shown to be extremely brittle, failing by cleavage with {110} and {100} fracture surfaces showing the same features as the low concentration Nb-H alloys [26]. Identical behavior was observed in the α' solid solution. While the V-H alloys have been shown to cleave through stress induced hydrides, single crystals of the β V$_2$H have been reported to be ductile [27]. The reasons for this diversity in observations is not known. While most hydrides are reputed to be extremely brittle, and support for this comes from the difficulty in preparing hydrides in any form other than powder, the reasons for this brittleness are not understood. As will be discussed in a subsequent section, studies of elastic constants, phonon dispersion curves, fracture surface energy, and other properties do not in general indicate that the inherent atomic bonding is reduced by the high hydrogen concentrations in the solid solutions or the hydrides. In fact an analysis of the phonon properties suggests an increase in the atomic force constants in the Group Vb hydrides relative to the solid solutions. TEM [22] and fracture [25] studies of NbH$_x$ have indicated that dislocation motion is very limited in these hydrides and that this, rather than a decrease in atomic bonding, may cause their brittle nature.

The kinetics of hydrogen embrittlement in the hydride forming systems have not been extensively studied. Some indication that the fracture growth rate is associated with H diffusion to the crack tip has been obtained in the Zr-H [28] and Ti-H [29] systems. A recent study [30] of crack propagation kinetics in the Nb-H system has shown definitively that this is the case where H is present as a solute. Several different types of kinetics can be obtained depending on the temperature range relative to the solvus temperature and depending on the type of loading employed (increasing K, decreasing K or constant K type of test where K is the crack tip stress intensity). In decreasing K tests well below the solvus, the crack growth rate is expected to be determined by the flux of H to the stress induced hydride which forms at the crack tip and to be characterized by a linear dependence of $\ln(v)$ vs K behavior as is observed (Fig. 6). The strong isotope dependence in the H diffusivity [31] is reflected in the fracture rates. The ratio of the velocities for Nb-H and Nb-D alloys is observed to be 15 at 77 K in excellent agreement with the expected ratio from the measured H and D diffusivities. This good agreement with the kinetics calculated on the basis of the fracture model is observed over the entire temperature range.

At temperatures closer to the solvus the kinetics of fracture are controlled by both the stress induced formation of hydrides at the crack tip and diffusion of hydrogen to these hydrides from the surrounding solid solutions. This results in a three stage $\ln(v)$ vs K_I curve as shown in Fig. 7. The high stress intensity Stage III corresponds to ductile overload failure which occurs when the fracture is driven at a rate which exceeds the ability of the hydrogen flux to form hydrides at the crack tip. The low K_I Stage I regime corresponds to the threshold stress intensity below which the hydrides are not nucleated and failure does not occur. Stage II behavior is steady state formation and fracture of crack tip hydrides at rates which reflect the flux of hydrogen to the crack tip. The kinetics are described by a modified theory based on Simpson and Puls [32];

$$V = \eta D_H C^0_H \exp \frac{W^{inc}}{RT} \exp \frac{W^a}{RT} \tag{4}$$

where D_H is the hydrogen diffusivity, C^0_H is the concentration of hydrogen in solid solution, W^{inc} is the elastic strain energy of the hydride and matrix due to $\Delta V_{\alpha\beta}$ and W^a is the interaction energy of the hydride with the crack tip stress field. The temperature dependence of Stage II velocities is given by an Arrhenius relation with the activation enthalpy

$$\Delta H = \Delta H_H + W^{inc} + W^a \tag{5}$$

where ΔH_H is the diffusion enthalpy for hydrogen. We can estimate the activation enthalpy to be 6-7 KJ/mole. As seen in Fig. 8 the data fit an Arrhenius relation very well and the measured activation enthalpy is 6.6 ± 1.5 KJ/mole in good agreement with the expected value. A more extensive treatment of the kinetics of crack propagation in the Nb-H system has shown that the stress induced hydride fracture mechanism discussed above is in good quantitative agreement with the measured parameters.

Based on the proposed mechanism several effects of alloying

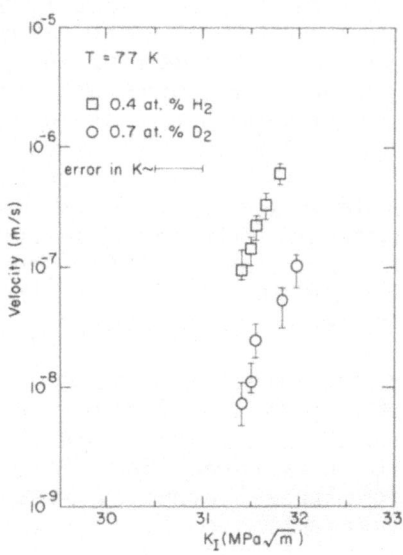

Fig. 6

Crack velocity (v) vs stress
intensity (K_I) for Nb-H and Nb-D
specimens tested at 77 K. The
ratio of the crack velocity at
constant K_I reflects the anom-
olous isotope effect for
diffusion of H to the crack tip.
The actual concentration of H or
D in solid solution is determined
by the solvus.

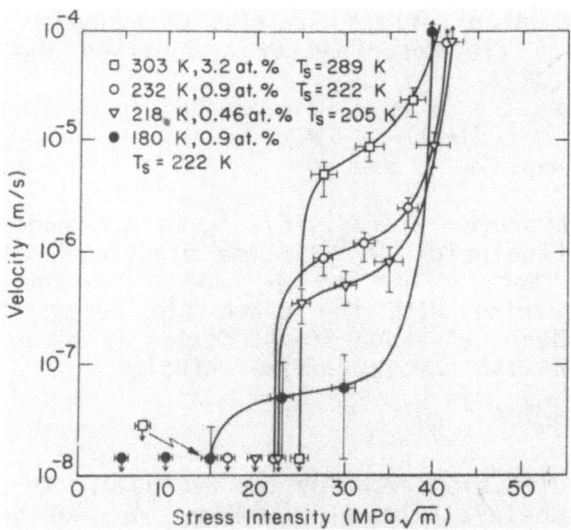

Fig. 7   Crack velocity (v) vs stress intensity (K_I) for Nb
containing the indicated H concentrations and tested at
the indicated temperatures. The tests were carried out
at constant K_I and exhibit three stage behavior. These
tests were carried out near the stress free solvus
temperatures, T_s which are indicated.

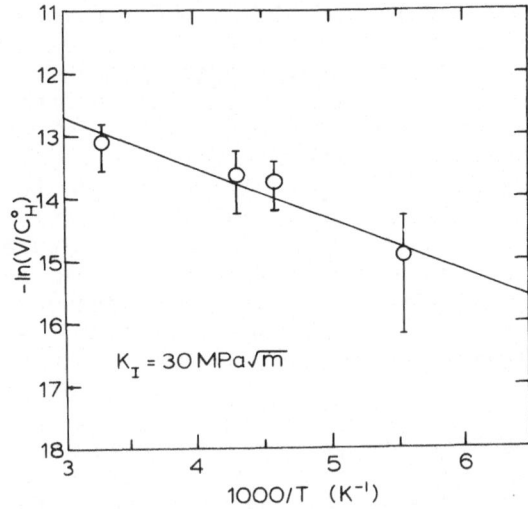

Fig. 8

Arrhenius plot of the temperature dependence of the stage II velocity for the data shown in Fig. 7. This form is in agreement with the theoretical expression, Eqn. 4.

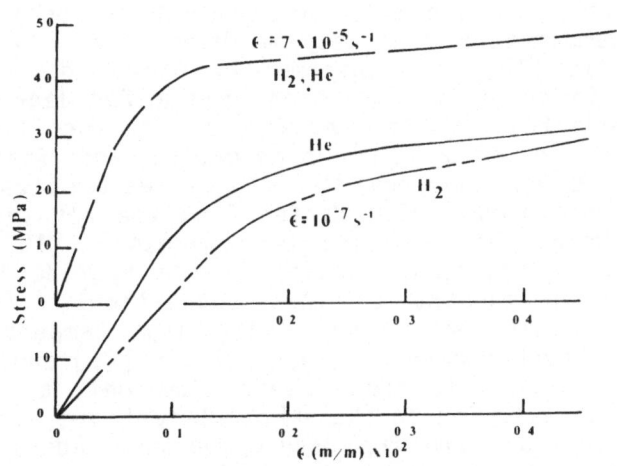

Fig. 9   Initial portions of the stress strain curves for Ni specimens tested in He or H_2 gas environments at "high" and "low" strain rates.

with interstitial and substitutional solutes may be expected. Any solute which causes significant strengthening of the solid solution, such as C, N or O will result in higher stresses at elastic inhomogeneities and hence in larger shifts in the local solvus temperatures. Thus strong alloys should exhibit increases of the temperatures below which the brittle fracture mode is observed as has been noted in Nb-O-N-H alloys [10]. On the other hand, it has recently been shown that substitutional alloys of Nb with Mo, Ti and V have greatly increased solubility for H before the onset of hydride precipitation [33]. In these alloys the increased thermodynamic stability of the solid solution relative to the hydride should result in a decrease in the temperature below which brittle fracture is observed. Trapping of H by solutes may also result in the same effect as this would decrease the rate at which the hydride could form at the crack tip. An inhibition of embrittlement resulting from alloying Nb has been observed [34].

Despite the general understanding of hydrogen embrittlement in the hydride forming systems there are a number of major outstanding problems. One of these is understanding the deformation and fracture of the hydrides themselves. Except for the suggestion [26] that the β NbH is brittle due to the difficulty of dislocation motion and the observation of ductility in V_2H [27], there is little information about the hydride properties. Another observation which is poorly understood is the "high temperature ductility minimum" first reported by McIntyre and Hardie [15] and confirmed by Gahr and Birnbaum [10]. This minimum in ductility, which is only a few degrees in width occurs just above the high temperature loss in ductility in Nb-H and V-H alloys. In situ crack propagation and fractographic studies [10,11] do not show any distinctive features associated with this ductility minimum which differ from the behavior above and below the minimum. While it has been suggested [10] that this minimum reflects an ordering reaction in the hydride there is no strong evidence in support of this suggestion. Another area which deserves additional study is the high temperature embrittlement which has been observed in Nb-H alloys [25] and which may occur in other systems. At the present time only a few studies of the effects of alloying on the embrittlement are available. This area is an important one for developing our understanding as well as possibly developing methods of alleviating the practical problems of using materials in aggressive environments and should be further explored.

NON-HYDRIDE FORMING SYSTEMS

Our mechanistic understanding of hydrogen embrittlement in systems which do not form hydrides (even in the presence of a stress bias at the crack tip) is in a much more rudimentary state than for the hydride forming systems. The behavior of these materials has been extensively characterized but often by investigations which have not been sufficiently well controlled to allow the development of mechanistic understanding. A number of proposed mechanisms have been put forth and as these have recently been reviewed [1] only a few possibilities will be discussed here.

One rather remarkable observation is that the phenomenology of the embrittlement due to solute hydrogen or gaseous H is very similar for both the hydride and non-hydride forming systems. While in the past this has led to some confusion, it is clear that this similarity results from the kinetics of embrittlement rather than the mechanism of fracture. Thus the observation of a ductility minimum vs temperature, of an inverse strain rate effect on the ductility, of three stage ln(v) vs K dependence and many other observations reflect the kinetics of transport of hydrogen to the crack tip and not the mechanism of fracture. One example will be used to illustrate this point. As previously discussed the ductility minimum in the hydride forming systems results from the fact that above a certain temperature the stress is not sufficient to form a hydride and ductile failure occurs. Below a certain temperature the mobility of the hydrogen is limited and within the time scale of the test a sufficiently high concentration does not occur at the crack tip to form the hydride and the ductile failure again occurs. The same general explanation accounts for the behavior in the non-hydride forming systems. A critical concentration of hydrogen, the value of which depends on the local stress, is required for brittle fracture [35]. At low temperatures the hydrogen mobility is limited and within the time of the experiment the critical concentration is not achieved, resulting in ductile failure. Above a certain temperature the stress level at stress singularities is not sufficiently high to increase the local hydrogen concentration to this critical value and again ductile failure occurs. Thus although the detailed phenomena differ in the hydride forming and non-hydride forming cases, the low temperature limit of the ductility minimum is set by the kinetics of hydrogen diffusion to the crack tip and the high temperature limit is set by the thermodynamics at the crack tip. Both types of systems require a critical hydrogen concentration for fracture but for completely different reasons. While understanding the kinetics of fracture is important, the kinetics do not determine the fracture mechanism.

In several important respects the phenomonology differs in the two classes of material. One example is the complexity of morphological effects caused by hydrogen in the fracture of various steels. While the fracture is always transgranular in the hydride forming systems it occurs in both transgranular and intergranular modes in the non-hydride formers. Hydrogen embrittlement of high strength steel generally occurs along prior austenitic boundaries or along martensite lathe interfaces. In many cases the fracture surface is associated with segregated elements such as S, P, As, Sb and others although the relation of this segregation to the process of hydrogen embrittlement is not established [36]. Transgranular cracking of steels, Fe-Si alloys and iron can be obtained as a result of cathodic charging [37,38]. In the case of single phase iron the fracture plane corresponds to the {110} slip plane [38]. Single phase ferrous alloys generally do not have sufficiently high strength to exhibit hydrogen embrittlement under the low fugacity conditions which lead to embrittlement of high strength steels. In an attempt to study hydrogen embrittlement of high strength single crystals of Fe, whiskers were tensile tested in H_2 and H_2S atmospheres which would lead to severe embrittlement of equivalent strength steels [39]. In the case of the whiskers completely ductile fracture was obtained.

The effects of hydrogen on the fracture behavior of steels having a variety of microstructures has been extensively studied by Bernstein and Thompson [40,41]. They have pointed out that the effects of hydrogen can be to alter the mode of fracture from a ductile microvoid to a brittle intergranular mode (high strength steel having a tempered martensite structure), can be to increase the extent of the low ductility cleavage mode in a mixed mode (cleavage and microvoid) type of fracture (lamellar pearlite structure), or can be to have no significant change on the fracture mode (microvoid coalescence) except to affect the microvoid nucleation and growth processes at the later stage of fracture (spheroidal steel). In all of these structures the effect of hydrogen is to markedly decrease the strain to failure and the energy for fracture. This variety of effects of hydrogen on the fracture morphology is much greater than is observed in the hydride forming systems. It would be well to keep this complexity in mind while searching for a mechanistic understanding of the effects of hydrogen on the fracture of metals.

Other non-hydride forming systems have a somewhat less complex dependence of the fracture mode on environment. While the Ni-H system forms a hydride at hydrogen fugacities in excess of about 1.6 GPa there is little possibility of this hydride phase being stable at the low gaseous H fugacity conditions at which embrittlement is observed, even in the stress field at the crack tip. It is possible that such a hydride could be formed during stress corrosion cracking or cathodic charging experiments. Since the hydride is unstable once the H fugacity is decreased it requires some care to detect its presence during these experiments. In straining electrode type of experiments Latanision and coworkers [42,43] have shown that the fracture mode is intergranular and that the segregation of Sb, Sn, and P at grain boundaries acts to enhance the rate of entry of H at these points. While intergranular fracture is the prodominant mode in this type of experiment we have also observed transgranular crack propagation when testing single or bicrystals during cathodic charging [44].

Internally charged Ni generally fractures in an intergranular mode [45,46] whether tested in gaseous H_2 or in an inert atmosphere, while single crystals of nickel which have been internally charged have been reported [47] to fail by a completely ductile mode. Sharply notched single crystals exhibit brittle failure modes when tested in H_2 gas atmospheres with the fracture plane being variously reported as the {113} [48] and the {100} [49]. We have determined the fracture plane to be the {111} by a two surface crystallographic analysis [46,50]. In large grained polycrystalline specimens tested in H_2 gaseous atmospheres the fracture mode was transgranular in specimens heat treated or alloyed to prevent S segregation at grain boundaries and intergranular when such S segregation was allowed to occur [51]. Thus even in nickel, with an inherently simple microstructure, the fracture mode depends on the detailed distribution of the solutes and on the source of the hydrogen (and probably on the hydrogen fugacity all well although this variable has not been systematically studied). The various fracture modes are summarized in Table I.

Table I. FRACTURE MODES IN NICKEL

Source of H	Grain Boundary Segregation	Test Atmosphere	Principal Fracture Mode
Solute H	None	Vacuum or H_2 gas	Intergranular
Solute H	S	Vacuum or H_2 gas	Intergranular
Cathodic Polarization	Sb, Sn	Aqueous	Intergranular
Gaseous H_2	None	H_2 gas	Transgranular {111}
Gaseous H_2	S	H_2 gas	Intergranular
Cathodic Polarization (Formation of hydride)		Aqueous	Intergranular and transgranular {100}

Hydrogen effects on the plastic deformation are as varied as the effects on the fracture paths. In the Group Vb metals, Nb, V, Ta, solute H acts as a dislocation pinning point which retains a high mobility in the lattice. A consequence of this behavior is solid solution strengthening on adding H with the increase in flow stress being larger as the temperature is decreased [52]. This relatively simple behavior is somewhat complicated by the precipitation of the hydride and an interaction with the dislocation strain field leading to a maximum increased flow stress at a temperature near the solvus temperature. In contrast to this behavior, high purity iron exhibits a marked decrease in its flow stress when cathodically charged with H in the temperature range 170 - 300 K and an increase at higher and lower temperatures [53,54]. Low purity iron and steels exhibit solid solution strengthening due to H at all temperatures. While the mechanism for solution softening is not established, it has been suggested [53] that it results from a decrease in the energy to form double kinks on the screw dislocations or from a decrease of the Peierls stress due to presence of H in the dislocation core. These are intrinsic solution softening mechanisms and apply in the cases of strong dislocation - lattice interactions. They are in contrast to the type of mechanisms which are based on the decrease in the effectiveness of solute pinning points as a result of forming small complexes with the H solutes, ie extrinsic solution softening mechanisms [55,56].

Extensive studies of the effects of solute H on the plastic deformation of Ni have shown solution strengthening and serrated yielding near 200 K [57]. The effect of H is however more complex than simple solution strengthening as we have shown in a recent series of experiments [58]. In relatively high purity specimens introduction of solute hydrogen by quenching from an elevated temperature or by testing in H_2 gas atmospheres leads to a small solution strengthening effect at low strains in a low strain rate tensile test . If however, the nickel is solution strengthened by the introduction of C in solid solution, significant solution softening (Fig. 9) occurs during the early stages of slow strain rate testing in H_2 gas or by the introduction of H solutes by quenching. This effect has also been seen directly in the HVEM environmental cell tensile tests (Fig. 10) when a specimen strained in vacuum and held at a constant stress is exposed to H_2 gas atmospheres [58]. At pressures of the order of 10^4 Pa the dislocations exhibit motion and multiplication leading to intense slip and the tangle shown in Fig. 10. This behavior is not characteristic of deformation in vacuum nor does it result from the introduction of other gases. Our results indicate that solution softening of an extrinsic type occurs from the introduction of H into Ni.

It has not been possible to develop the diverse experimental

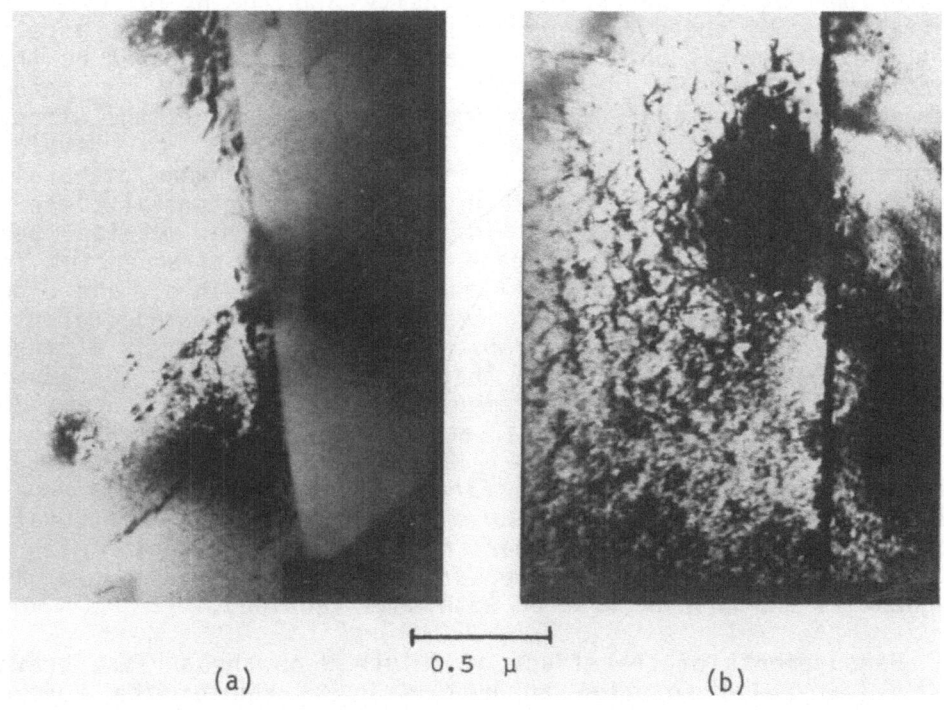

(a)   0.5 μ   (b)

Fig. 10   HVEM micrographs of dislocation structures in Ni formed
during in situ deformation. (a) Specimen stressed in
vacuum. (b) Same as area (a) after the introduction of
H_2 gas while the specimen is held under stress at
constant displacement.

observations made on the non-hydride forming systems into a comprehensive understanding of the fracture mechanism(s). Perhaps the most commonly accepted mechanism of embrittlement in the non-hydride forming metals is "lattice decohesion" which was first suggested by Troiano [59] and developed by Oriani [60] and others. The essence of this mechanism is that solute H decreases the strength of the atomic bonds at the crack tip and this leads to fracture when the H concentration exceeds a critical value. Oriani has shown that this approach leads to a relatively direct connection between the stress intensity and the H_2 fugacity (or equivalently to the hydrogen concentration). The kinetic aspects of the experimental observations can also be accommodated by this approach. Evidence in support of this mechanism is rather indirect as theoretical or experimental treatments of the effects of H on atomic bonding are very difficult to develop. Measurements such as elastic constants and phonon dispersion curves examine the effects of H on the lattice potential close to the equilibrium position. In the group Vb metals these measurements indicate an increase in the phonon frequencies for all of the phonon branches [61] and an increase in c_{44} and B and decrease in c' [62] as the H concentration is increased (recently to concentrations as high as H/Nb = 0.78 [63]). Analysis of these results leads to the conclusion that hydrogen increases the atomic force constants. These measurements do not directly address the effect of H on lattice potential at the large displacements which are of interest in fracture. Measurements of this type are difficult to carry out in the ferrous systems due to the low H solubility. In the only measurement available it was reported that the polycrystalline shear modulus, G decreases with H additions to Fe [64]. (These measurements were carried out at low frequencies and must be treated with some caution).

Measurements of the effect of solute H on the surface energy of Nb (α' solid solution at H/Nb = 0.82) and of the surface energy of the β NbH hydride have been carried out using fracture mechanics techniques and have shown that the solid solution and hydride have about the same surface energy as pure Nb [26]. A similar result was obtained [65] for solid solutions of H in Ni at elevated temperatures where the equilibrium surface energy was determined.

One other class of measurement which may add to our understanding is photoemission studies of adsorbed H on surfaces. On Ni (111) surfaces they indicate bonding of the H to the surface via sp bands [66]. The changes in the surface electronic structure do not clearly establish the effect of H on the Ni bonding and additional work of this sort should be carried out.

The conclusion which must be drawn from the available

experimental evidence, imperfect as it is, is that no evidence exists to support the <u>assumption</u> that H in solid solution decreases the strength of the atomic bonding. While some of the evidence suggests that a strengthening of the bond may occur, this conclusion must be considered extremely tentative. Theoretical treatments are equally inconclusive. Until greater understanding of these problems is in hand it would be best to consider the decohesion type of theory as one of a number of viable suggestions to be tested against experiments.

The above discussion is directed towards the effect of H in solid solution on the energy to create a surface and on atomic bonding. It does not preclude the possibility that the surface energy of these metals is reduced by the segregation of H to the surface or interface. Indeed, it follows from the positive heat of adsorption of H to the surface that there will be an excess concentration at the surface and a decrease in the surface energy. Direct evidence for H segregation at interfaces is relatively meagre due to the difficulty of detecting H on a microanalytic scale. Recent experiments [24,67] using the SIMS ion probe method has provided direct evidence of H (D) segregation at grain boundaries in Nb and in Ni (Fig. 11). As discussed by Petch [68] and more recently by Hirth [69] this surface segregation decreases the fracture energy and as a consequence the stress (or stress intensity) for fracture. Measured values of these parameters lead to fracture energies which are however much greater than the true surface energy of the metals as a result of the plastic deformation which accompanies the propagation of a crack. This problem has been recently discussed by Jokl et al [70] who suggest that the reduction of the surface energy by surface segregation is accompanied by a reduction of the plastic work terms and that this scaling process allows the surface segregation processes to control both energy terms.

The uniqueness of H among the strong surface adsorbers in causing embrittlement remains a difficulty for this point of view. Many segregated species lead to low energy intergranular fracture [71]. It is however, only fugacities of H in the gas phase (H_2 , H_2S, etc) which lead to fracture while other species, (O_2 , H_2O , SO_2 etc.) which have even larger enthalpies of adsorption on clean surfaces do not lead to failure and when present have the effect of stopping crack propogation due to the presence of H_2.

Fracture mechanisms based on the effect of segregated solutes, such as H on the surface energy and on the fracture energy of interfaces are consistent with the bulk of observations in the non-hydride forming systems which indicate that the fracture path is predominantly intergranular. The effect of surface segregation on transgranular fracture is also to reduce

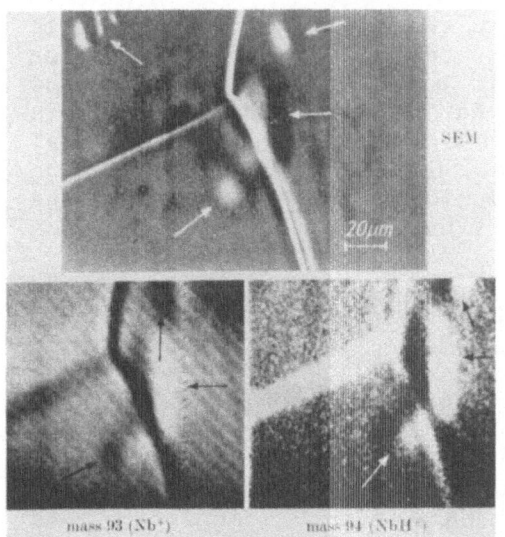

Fig. 11a

SIMS analysis of Nb-H alloys showing H segregation at grain boundaries. The SEM image shows a grain boundary triple point while the SIMS image with Nb^+ ions shows the geometrical contrast at the grain boundaries. The SIMS image with mass 94 shows a high concentration of H at the grain boundaries. Quantitative measurements show that the H concentration does not correspond to the hydride but rather to H in solid solution.

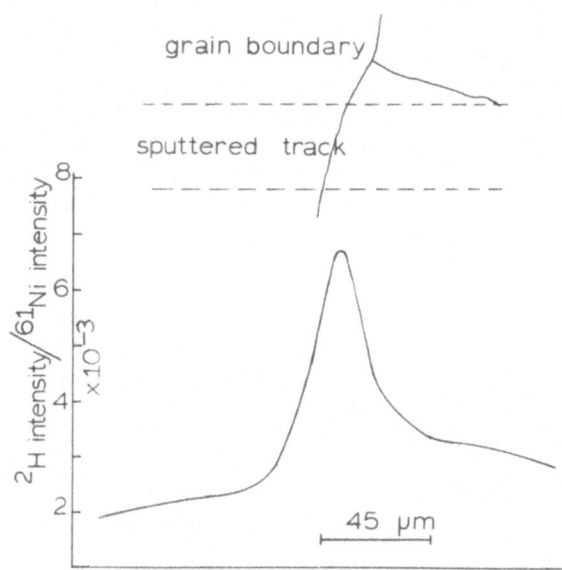

Fig. 11b   SIMS data in the Ni-H system showing the segregation of D at grain boundaries. The grain boundary configuration relative to the linear scan of the ion beam is shown at the top of the figure. The ratio of the $^2H/^{61}Ni$ intensity show a maximum at the grain boundary indicating segregation of D at the grain boundary.

the surface energy if the segregation occurs during the fracture process as may occur with a mobile solute such as H [69]. In the presence of other strongly segregating species, the addition of H apparently leads to a strong predisposition towards intergranular fracture. Segregation of species such as S may weaken the bonding across the grain boundary by forming directional metal - S bonds using electrons which in the absence of the segregant would take part in the metal - metal bonds. There is some recent experimental [72] and theoretical evidence [73] on Ni surfaces that this may indeed occur. Hydrogen, when present as a segregated species at the grain boundaries may act in the same way as these other species to decrease the boundary energy and perhaps the atomic bonding across the grain boundary. It differs from these other segregants in that it retains its high mobility in the lattice at the temperature of the mechanical test and can therefore accummulate at the grain boundary during the stressing in response to stress concentrations resulting from plastic deformation which precedes fracture. Recent SIMS studies of ^{2}H in nickel have shown segregation at grain boundaries both in the presence of segregated S and in its absence [67] (Fig. 11).

The fact that segregation of H at grain boundaries occurs and the expectation that this has a concomitant decrease in the boundary energy does not establish this as the mechanism of embrittlement. Since the fracture energy is dominated by the plastic work terms clearly the coupling between the presence of H at the fracture front and the plastic response of the material must play a dominant role. A coupling between the surface energy and the plastic zone size has been suggested as resulting from the reduction in the local stresses and hence the plastic zone size when the energy to rupture bonds is decreased [70]. Other significant effects may result from the previously discussed effects of H on the behavior of dislocations. In any case the hydrogen related fracture is clearly not a "brittle" process when viewed from the vantage point of the crack tip. These plastic processes have been studied recently in the Ni- H system using in situ HVEM and SEM fracture studies [50] and some examples of crack tips in the Ni-H system are shown in Fig. 12. It is seen that both for intergranular and transgranular fracture there is a very large plastic zone at the crack tip with a large COD or shear displacement along the grain boundaries. The fracture surfaces shown in Fig. 13 exhibit evidence of this local plasticity and are typical of fracture surfaces for which the plasticity is severely confined to regions near the fracture surface despite the fact that they formed in a highly plastic region. The transgranular fracture plane is the {111} slip plane and shows very flat fracture with slip traces intersecting the fracture plane and shallow pyramidal cones and depressions the faces of which consist of the other {111} slip planes. Similarly the intergranular fracture surfaces show slip lines and other evidence of the local

Fig. 12a

100 μ

Fig. 12b

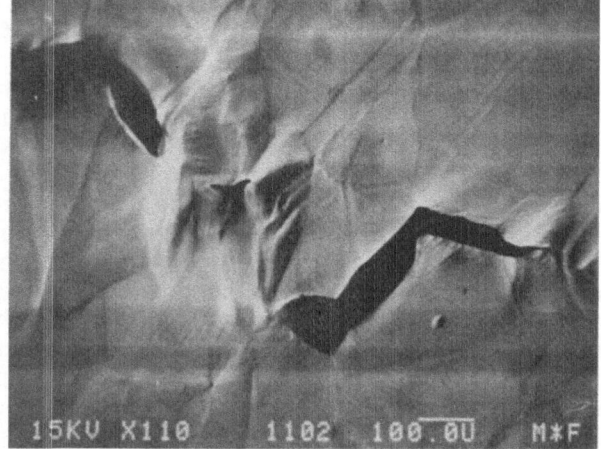

Fig. 12c

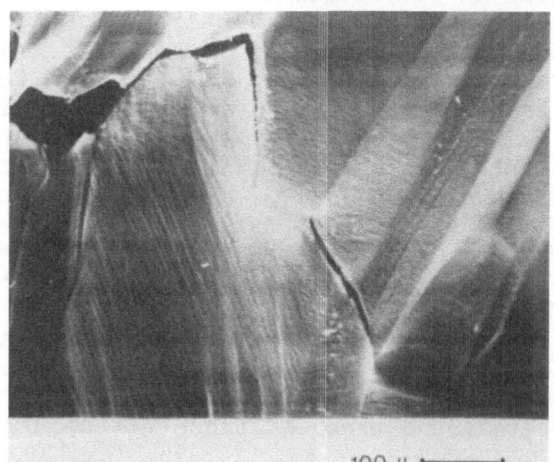

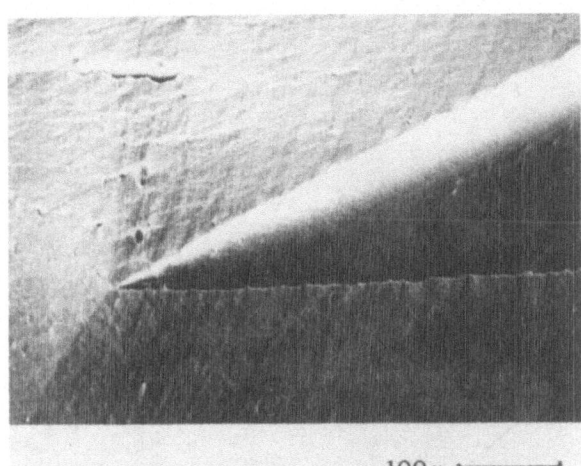

Fig. 12d

Fig. 12 In situ SEM micrographs of propogating cracks in Ni-H alloys. Figures 12a-12c show intergranular fracture in Ni specimens containing solute H. Large crack opening displacements, shear along grain boundaries and general plasticity precedes fracture. Figure 12d shows a transgranular fracture in H_2 gas along the {111} with a large amount of preceding plastic deformation.

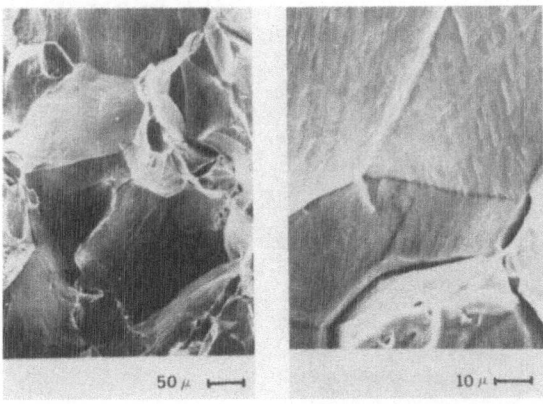

Fig. 13a

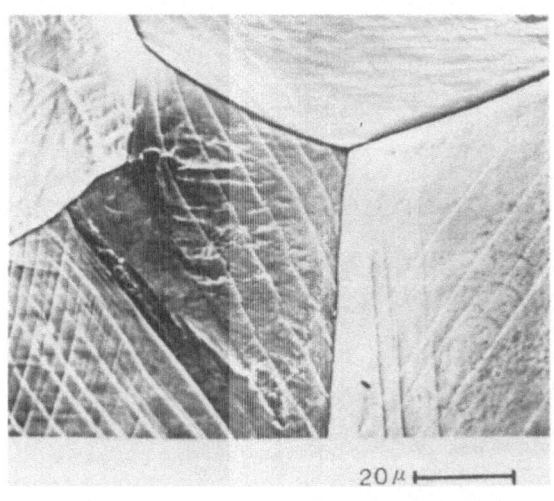

Fig. 13c

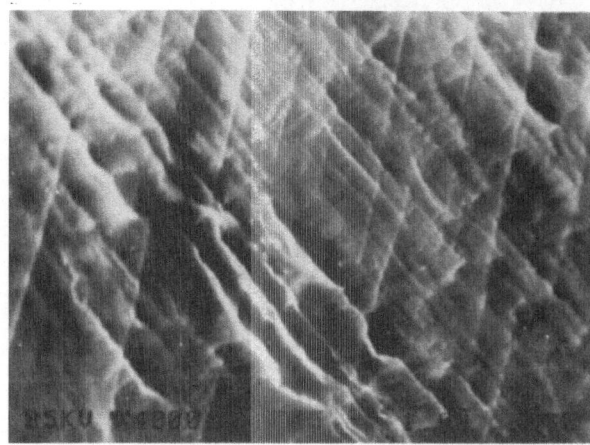

Fig. 13d

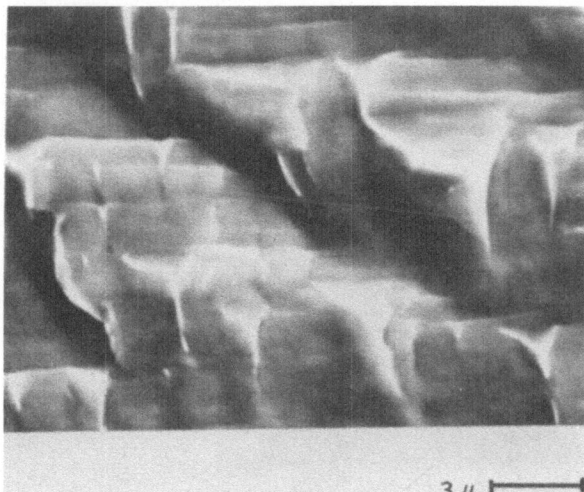

Fig. 13e

3 μ

Fig. 13   Typical fractographs of hydrogen embrittled Ni. Figures
13a-13b show intergranular fracture of Ni containing
H in solid solution. Considerable evidence for plastic
deformation is evident on the fracture surfaces.
Figures 13c-13e show transgranular fracture of Ni
stressed in H_2 gas. The fracture plane is the {111} and
shows evidence for large amounts of plastic deformation
prior to fracture. Figure 13d shows intersecting slip
lines. Figure 13e shows "hillocks" formed by fracture
on intersecting {111} planes.

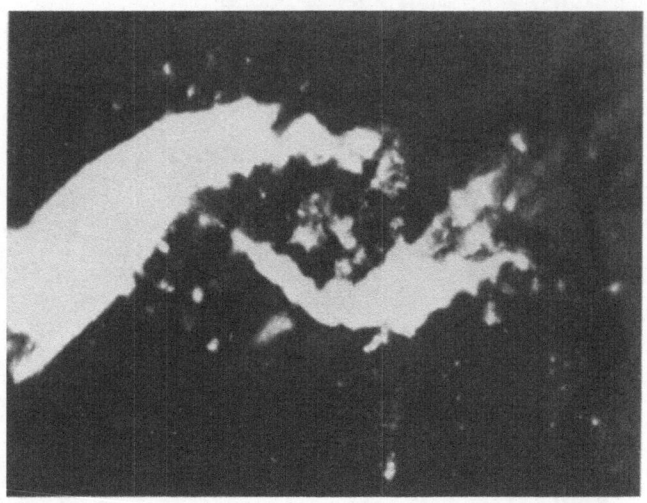

10 μ

Fig. 14a

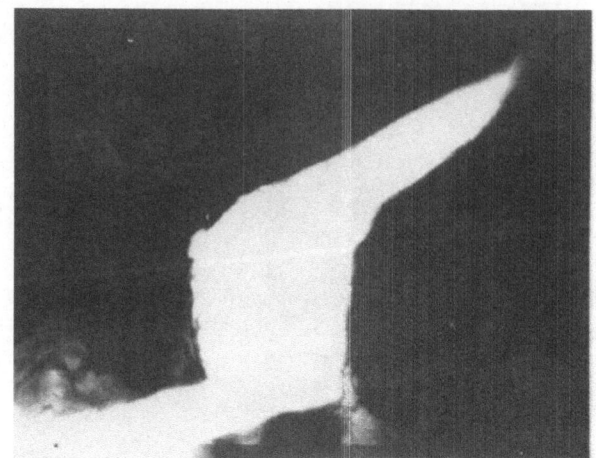

Fig. 14b

10 µ

Fig. 14c

2 µ

Fig. 14d

0.5 µ

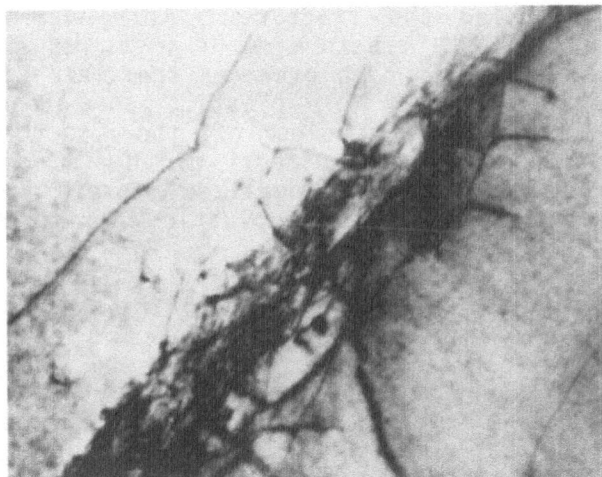

0.5 µ

Fig. 14e

Fig. 14   In situ HVEM micrographs of cracks propogating in Ni.

Fig. 14a   Non-crystallographic fracture of Ni stressed
in vacuum. The cracks propogated transgranularly with
a large plastic zone and nucleation of "holes" in front
of the main crack tip.

Fig. 14b   Crystallographic transgranular fracture of
Ni specimen stressed in H_2 gas. The fracture surfaces
are {111}.

Fig. 14c   High magnification of crack surface showing
the formation of thin sections of the crystal formed
by shear along the {111}. Fracture was in H_2 gas.

Fig. 14d   Crack tip formed in Ni stressed in H_2 gas
and showing the dislocation structure at the crack tip.
The plastic zone was confined to the {111} planes
directly in front of the crack.

Fig. 14e   A higher magnification micrograph of the
region in front of the crack tip shown in Fig. 14d.
The intense plastic zone in front of the crack is shown.

deformation which accompanied the fracture. Corresponding to these features, the in situ HVEM fracture on Ni in H_2 gas shows a very intense slip band (Fig. 14) which precedes the crack tip and a high degree of localization of the plastic zone along the fracture plane. Dynamic HVEM studies of the fracture in the H_2 gas reveal that the major effect of H is on the distribution of the slip into a more localized plastic zone in contrast to the more diffuse and broader plastic zone formed in vacuum.

A number of attempts have been made to model the processes at the crack tip and the effects of H on these processes [74]. In general these have been rather simplistic in nature and while they are interesting exercises they require rather gross assumptions to make them tractable. Modeling of the plastic zone in front of the crack [75] has indicated that a localization of the slip into an intense slip plane in front of the crack can lead to a decrease in the total plastic work and to an increase in the magnitude of the normal stresses. What is required is a theory of the connection between the effects of H on the motion of dislocations and the localization of slip in front of the propagating crack and how such plastic processes lead to "embrittlement". These connections are not well understood. It is worth recalling that while we know a good deal more about the constituent processes, we are not much further along in understanding the hydrogen enhanced plasticity caused embrittlement than when it was first suggested by Beachem [76] (and also worth recalling that his suggestion met with very little acceptance at the time).

In summary, examination of the available experimental and theoretical information does not lead to any definitive conclusion about the mechanism of hydrogen embrittlement of the non-hydride forming metals. The information required for such mechanistic understanding (the effects of H on the atomic bonding, the nature of the interactions of H and other segregants with boundaries, the nature of the interactions between H and these other segregants, the effect of H on the plastic deformation processes) is first being developed and considerable additional work is required. If however, we measure progress by the extent of our recognition of the problems, then appreciable progress has been made.

ACKNOWLEDGEMENT

This work was supported by the Office of Naval Research and the National Science Foundation.

References

1. H. K. Birnbaum, Environment Sensitive Fracture of Engineering Materials (AIME, Warrendale, PA, 1978) ed. by Z. A. Foroulis,

p. 326.

2. G. G. Libowitz, Binary Metal Hydrides (W. A. Benjamin Inc., New York, 1965).

3. Metal Hydrides, (Academic Press, New York, 1968) ed. by W. M. Mueller, J. P. Blackledge and G. C. Libowitz.

4. H. K. Birnbaum, M. L. Grossbeck and M. Amano, J. Less Comm. Met. 49 (1976) 357.

5. R. N. Stevens, R. Dutton and M. P. Puls, Acta. Met. 22 (1974) 629.

6. T. B. Flanagan, N. B. Mason and H. K. Birnbaum, Scripta Met. (In Press).

7. H. K. Birnbaum and H. Wadley, Scripta Met. 9 (1975) 1113.

8. T. M. Devine, Scripta Met. 10 (1976) 447.

9. D. G. Westlake, Trans. A.S.M. 62 (1969) 1000.

10. S. Gahr, M. L. Grossbeck and H. K. Birnbaum, Acta. Met. 25 (1977) 125.

11. M. L. Grossbeck and H. K. Birnbaum, Acta. Met. 25 (1977) 135.

12. S. Takano and T. Suzuki, Acta. Met. 22 (1974) 265.

13. S. Koike and T. Suzuki, Acta. Met. (In Press).

14. D. Cann and E. E. Sexton, Acta. Met. 28 (1980) 1215.

15. S. W. Ciraldi, J. L. Nelson, E. N. Pugh and R. A. Yeske, Third Int. Conf. on Effect of Hydrogen on Behavior of Materials (AIME, New York) (1981) ed. by A. W. Thompson and I. M. Bernstein.

16. See also papers by C.D.S. Tuck, by R. J. Jacko and D. J. Duquette, by G. M. Scamans, and by L. Christodoulou and H. M. Flower, Ibid.

17. D. Hardie and P. McIntyre, Met. Trans. 4 (1973) 1247.

18. C. V. Owen and T. E. Scott, Met. Trans. 3 (1972) 1715.

19. D. G. Westlake, Trans. AIME, 245 (1969) 1969.

20. J.C.M. Li, R. Q. Oriani and L. S. Darken, Z. Physik Chem. Neue Folge., 49 (1966) 271.

21. H. Pfeiffer and H. Peisl, Phys. Lett. 60A (1977) 363.

22. B. J. Makenas and H. K. Birnbaum, Acta. Met. 28 (1980) 979.

23. D. H. Sherman, C. V. Owen and T. E. Scott, Trans. AIME 242 (1968) 1775.

24. P. Williams, C. A. Evans, Jr., M. L. Grossbeck and H. K. Birnbaum, Anal. Chem. 48 (1976) 964; see also Phys. Stat. Sol. (a) 34 (1976) k97.

25. S. Gahr and H. K. Birnbaum, Acta. Met. 26 (1978) 1781.

26. S. Gahr and H. K. Birnbaum, Acta. Met. 28 (1980) 1207.

27. T. Suzuki, Trans. of Japan Inst. of Met. 21 (1980) 17.

28. R. Dutton, K. Nuttall, M. P. Puls and L. A. Simpson, Met. Trans. 8A (1977) 1553.

29. H. G. Nelson, Met. Trans. 7A (1976) 621.

30. B. Hindin and H. K. Birnbaum, (To be published).

31. H. Wipf and G. Alefeld, Phys. Stat. Sol. (a) 23 (1974) 175.

32. L. A. Simpson and M. P. Puls, Met. Trans. 10A (1979) 1093.

33. J. F. Miller and D. G. Westlake, Trans. Jap. Inst. Met. 21 (1980) 153.

34. Y. Sasaki and M. Amano, Proc. of 2nd Int. Conf on Hyd. in Metals, Paris Vol. 1 (Pergamon Press, New York, 1978).

35. R. A. Oriani and P. H. Joesphic, Acta. Met. 22 (1974) 1065.

36. S. K. Banerji, C. J. McMahon and H. C. Feng, Met. Trans. 9A (1978) 237.

37. A. S. Tetelman and W. D. Robertson, Trans. AIME 224 (1962) 775.

38. T. Takeyama and H. Takahasi, U.S. - Japan Seminar on the Characteristics of Deformation and Fracture of B.C.C. Metals (AIME, to be published) ed. by M. Meshii, March 1981.

39. J. J. Au and H. K. Birnbaum, Scripta Met. 12 (1978) 457.

40. I. M. Bernstein and A. W. Thompson, Int. Met. Rev. 21 (1976) 269.

41. R. Garber, I. M. Bernstein and A. W. Thompson, Met. Trans. A 12A (1981) 225.

42. R. M. Latanision and H. Opperhauser, Jr., Met. Trans. 5 (1974) 483.

43. R. M. Latanision, M. Kurkela and F. Lee, Third Int. Conf. on the Effect of Hydrogen on the Behavior of Materials, (AIME, New York) (1981) ed. by A. W. Thompson and I. M. Bernstein.

44. B. Wei and H. K. Birnbaum, Unpublished work.

45. G. C. Smith Hydrogen in Metals (ASM, Ohio) ed. by I. M. Bernstein and A. W. Thompson, (1973) p. 485.

46. F. Heubaum and H. K. Birnbaum, (To be published).

47. A. H. Windle and G. C. Smith, J. Met. Sci. 2 (1968) 187.

48. M. H. Kamdar, Paper 3D10 Proc. of the Sec. Int. Cong. on Hyd. in Met., (Pergamon, Oxford) (1977).

49. S. P. Lynch, Scripta Met. 13 (1979) 1051.

50. J. Eastman, T. Matsumoto, N. Narita, F. Heubaum and H. K. Birnbaum, Third Int. Conf. on the Effect of Hyd. on the Behavior of Materials (AIME, New York) (1981) ed. by A. W. Thompson and I. M. Bernstein; see also R. E. Stoltz and A. J. West, Ibid.

51. T. Matsumoto and H. K. Birnbaum, Trans. Japan Inst. of Met., Suppl. 21 (1980) 493; see also R. H. Stulen, Ibid. p. 501.

52. C. C. Chen and R. J. Arsenault, Acta. Met. 23 (1975) 255.

53. H. Matsui, H. Kimura and S. Moriya, Mat. Sci. and Eng. 40 (1979) 207, 217, 227.

54. C. G. Park, K. S.Shin, J. Nagakawa and M. Meshii, Scripta Met. 14 (1980) 279.

55. R. Gibala and T. E. Mitchell, Scripta Met. 7 (1973) 1143 2.

56. E. Pink and R. J. Arsenault, Prog. in Met. Sci. 24 (1979) 1.

57. A. H. Windle and G. C. Smith, J. Met. Sci. 4 (1970) 136.

58. J. Eastman, F. Heubaum, T. Matsumoto and H. K. Birnbaum, (To be published).

59. A. R. Troiano, BISRA, The Iron and Steel Inst., Harrogate Conf. (1961) p. 1.

60. R. A. Oriani and P. H. Josephic, Acta. Met. 25 (1977) 979.

61. T. Springer, Hydrogen in Metals I (Springer-Verlag, Berlin, 1978) ed. by G. Alefeld and J. Volkl, p. 75.

62. A. Magerl, B. Berre and G. Alefeld, Phys. Stat. Sol. (a) 36 (1976) 161.
63. F. Mazzolai and H. K. Birnbaum, (To be published).
64. E. Lunarska, Z. Zielinski and M. Smialowski, Acta. Met. 25 (1977) 305.
65. E. A. Clark, R. Yeske and H. K. Birnbaum, Met. Trans. 11A (1980) 1903.
66. F. J. Himpsel, J. A. Knapp and D. E. Eastman, Phys. Rev. B 19 (1979) 2872.
67. H. Fukushima, J. Baker, P. Williams and H. K. Birnbaum, (To be published).
68. N. J. Petch and P. Stables Nature 169 (1952) 842; see also N. J. Petch, Phil. Mag. (Eighth Series) 1 (1956) 331.
69. J. P. Hirth, Phil. Trans. Roy. Soc. London A295 (1980) 139.
70. M. L. Jokl, V. Vitek and C. J. McMahon, Jr., Acta. Met. 28 (1980) 1479.
71. M. P. Seah, Acta. Met. 28 (1980) 955.
72. T. N. Rhodin and T. W. Capehart, Surface Science (Netherlands) 89 (1979) 337.
73. C. L. Briant and R. P. Messmer, Phil. Mag. B 42 (1980) 569.
74. E. R. Fuller, Jr., B. R. Lawn and R. M. Thompson, Acta. Met. 28 (1980) 1407.
75. J. R. Wilcox and D. A. Koss, Third Int. Conf. on the Effect of Hydrogen on the Behavior of Materials, (AIME, New York) (1981) ed. by A. W. Thompson and I. M. Bernstein.
76. C. D. Beachem, Met. Trans. 3 (1972) 437.

DISCUSSION

Comment by J.F. Knott:

     You have emphasized the anisotropic nature of β niobium hydride
and its crystallographic orientation with respect to an {001} habit
plane in niobium. The thermodynamic treatment and its prediction
of the effect of stress on the α/(α+β) solvus, however, assumes
isotropic elasticity, with presumably an isotropic dilatation to
describe the formation of β in an α-matrix. By analogy with orientation relationships of G.P. zones in the Al-Cu system, I wonder
whether the calculation might be improved (and, possibly simplified,
if strains in the habit-plane are small) by taking account of the
anisotropic elastic constants or even the crystallographic matching
of product and matrix phase. For GP1 or θ'' in Al-Cu, for example,
the predominant effect would presumably be that of tension normal
to the habit-plane. If this sort of behavior held in the present
case, it might be that the rate of formation of hydride around a
sharp notch could be expressed as a function of slot/crystal
orientation in terms of, for example, the $\sigma_{\theta\theta}$ component of the
crack tip stress field (even if derived assuming an isotropic matrix)
resolved across the appropriate {001} habit plane.

Reply:

     What you suggest can be done and would, of course, improve the
quality of the calculation of the elastic energy of accommodation.
All of the elastic constants of the solid solution and of the hydride
(in Nb-H system) have been measured, so anisotropic elasticity can
be applied. We have not done so because the experimental information
does not yet seem to warrant this theoretical detail. In Nb the
hydride generally forms as a relatively equiaxed precipitate rather
than a plate while in V the hydride forms as a plate. In V the
dominant stress appears to be the stress component normal to the
habit plane. This also seems to be the case in Zr.

Comment by J.P. Fidelle:

     Your experiments on H-charged Nb indicate some ductility
recovery at the lower temperatures. Did you look for the influence
of $\dot{\varepsilon}$?

     Indeed, if you go to very low $\dot{\varepsilon}$ for many metal systems
including steels, the recovery from HE is not as important as
commonly mentioned. According to your kinetic description of stress
induced H diffusion and hydride generation, the same trend should
also apply to Va metals.

Reply:

That is indeed the case. The recovery of the ductility at low temperatures results from the decreased H diffusivity and hence the occurrence of ductile fracture prior to the formation of a hydride. At low strain rates the recovery of ductility occurs at a lower temperature.

Comment by J.R. Pickens:

Do you know of any system in which a stress induced, brittle hydride forms, and that this hydride is only stable under high stress (such as those at the crack tip), yet does not decompose for a time on the order of hours after the stress relaxation caused by fracture?

Reply:

No. Vanadium has a hydride which is established by the crack tip stresses and then reverts to the solid solution on removal of the stress. The reversion is rapid however. Ti-H alloys may do the same.

Comment by M.P. Puls:

Would you care to speculate on why hydrogen in solution in hydride forming metals does not seem to cause embrittlement?

Reply:

Hydrogen may cause embrittlement in a variety of ways and generally the fracture path followed will be the one which is kinetically favorable. We cannot absolutely rule out the possibility that H in solid solution will cause embrittlement. However, our observations and others strongly suggest that solid solutions of the hydride forming metals with H are not brittle even at very high hydrogen concentrations. Hydrogen does not seem to weaken the atomic bonding in these metals as judged from the dependence of the acoustic phonon frequencies, atomic force constants, elastic constants and surface energies on hydrogen concentration. This is discussed in the written text. The hydrides seem to be brittle, not because of reduced atomic bonding, but because of limited dislocation mobility caused by the hydrogen ordering and the fact that dislocation motion disorders the lattice.

Comment by R.M. Latanision:

It seems to me that there are at least two possible explanations of the substantial plasticity observed on intergranular fracture surfaces in your experiments on hydrogenated nickel. One is the notion of environmentally induced injection of dislocations as you

have mentioned. The other, it seems to me, is that dislocation
transport of hydrogen to the grain boundaries would serve the same
purpose. The latter would deliver the embrittling species (hydrogen)
to the grain boundaries and, of course, give rise to the observed
plasticity. The latter was, in fact, originally suggested by Smith
and his coworkers at Cambridge to account for the intergranular
cracking of hydrogenated nickel. Would it not be important to
distinguish between these possibilities?

Reply:

     Yes, it is important to consider dislocation transport as well
as hydrogen enhanced dislocation motion. Both of these processes
probably occur and may, in fact, complement each other. The extent
and form of the plastic zone may be strongly influenced by H transport since the hydrogen carried by the dislocations may also make
it easier for the dislocation to move through the lattice. Hydrogen
transport to grain boundaries via dislocations is a distinct
possibility but does not seem to be necessary for intergranular
fracture. We have observed intergranular fracture in Ni-H alloys
at about 100K with the same features as at the higher temperatures.
At this low temperature H is essentially immobile even with dislocation transport. We believe that the intergranular fracture at
these low temperatures results from H already segregated at the
grain boundaries.

Comment by A.R.C. Westwood:

     With regard to possible influences of hydrogen on dislocation
mobility in Ni, Fe, etc., what can be, or has been, learned from
internal friction studies to clarify the mechanisms by which such
effects could arise?

Reply:

     There has been a large amount of work on the bcc metals using
internal friction by people such as Igata and Mazzolai, and Kronmuller
(with magnetic disaccommodation). Considerably less work has been
done on Ni. I consider the results difficult to interpret in detail
but, in general, they seem to support the concept of H enhanced
dislocation mobility.

Comment by F.R.N. Nabarro:

     How important is the decrease in the line tension of the dis-
location which is produced by the binding of hydrogen? This will
assist the nucleation of dislocations.

Reply:

The changes in line tension due to H at the dislocation may be important in determining the dislocation mobility and arrangements in the plastic zone but at present we do not understand enough of the behavior to be specific. We have not seen evidence of large changes in the line tension in the electron microscope.

Comment by H. Vehoff:

Is there some theoretical or experimental evidence that you can separate a slip plane in hydrogen by tension near a crack tip?

Reply:

We have observed transgranular fracture along the {111} planes (which are the slip planes) in Ni. In addition, we observe secondary cracking along {111} slip bands which intersect the primary fracture plane. In the Fe-H system a number of observations of fracture (cleavage?) along the {110} slip planes have been reported. The theoretical situation is rather more complicated. One possibility is that H alters the plastic zone in front of the crack so as to increase the triaxial stress state and cause fracture in front of the crack (by a bond rupture process perhaps aided by a weakening of the atomic bonds by the presence of H). Sinclair has shown that such an enhancement of the crack tip stress field may result from the self-stress fields of the dislocations in the plastic zones. There are other dislocation models which can be developed to lead to separation along the slip planes but it is too early to be definitive.

ON THE TRANSPORT OF HYDROGEN BY DISLOCATIONS

J.P. Hirth and H.H. Johnson

Ohio State University
Columbus, Ohio 43210

Cornell University
Ithaca, New York 14853

INTRODUCTION

The interaction between dissolved hydrogen and dislocations, especially moving ones, is an intriguing and key metallurgical problem. A resolution would provide a substantial clarification of some of the unresolved mysteries of hydrogen embrittlement as well as a significant contribution to the understanding of defects in crystals.

Bastien and Azou[1] apparently first suggested that moving dislocations might carry hydrogen, and that when hydrogen-laden dislocations annihilated or intersected a void, significant local concentrations might develop. Frank[2] later observed the deformation enhanced hydrogen outgassing of mild steel. More recent experiments which suggest hydrogen transport by dislocations, especially measurements using the isotope tritum, have been discussed in recent reviews.[3-5]

Tien et al.[3] introduced a model which assumed that hydrogen carried by moving dislocations would be stripped from the dislocations and deposited in other traps such as second-phase particles and grain boundaries, and there contribute to decohesion and embrittlement. They specifically suggested that for steel large supersaturations could develop, such that substantial pressures could arise in voids, promoting failure by a pressure mechanism. In contrast, Johnson and Hirth[6] analyzed the supersaturation produced by dislocation annihilation, using a model which produced an upper bound for the supersaturation, and concluded that supersaturation and pressurization were negligible. More recently, Tien et al.[7] reanalyzed the

steel case and affirmed their previous estimate[3] of large supersaturations.

In the present paper, we first present a general physical
discussion of the processes expected to operate when hydrogen is
transported by dislocations. We then consider aspects of earlier
work, showing that several simplifying assumptions in the model of
Tien et al.[3,7] lead to overestimates of supersaturation, and verify
our earlier analysis[6] which predicts negligible supersaturation for
steel at room temperature. In the context of that model, large
supersaturations are possible for steel at low temperatures or at
room temperature for other metals, such as nickel.

Subtleties in the determination of chemical potentials near
internal traps are then considered, along with the appropriate
boundary conditions for diffusional loss from trap sites. Void
growth is treated as an extension of our earlier analysis.[6]
Finally, transport relations are presented for the stepwise motion
of hydrogen by successive dislocation transport and annihilation
events.

OVERVIEW OF HYDROGEN TRANSPORT BY DISLOCATIONS

It is useful to discuss in general two types of behavior in
experiments in which dislocation transport of hydrogen might be
realized. In the first the hydrogen is postulated to be introduced into the material and equilibrated with all traps or defects,
well before plastic deformation is begun. This injection can be
achieved by thermal or thermal/pressure charging or by electrolytic
charging followed by a suitable aging treatment. The surface is
next supposed to be closed to hydrogen egress, so that the system
is closed, consisting of a uniform (on a scale large compared with
the trap spacing) distribution of lattice hydrogen equilibrated
with hydrogen in traps. The total hydrogen concentration and
distribution is governed by the charging conditions and the
trapping parameters.

Experimentally, one may easily close the system after charging.
For some materials, such as niobium, this occurs naturally at
temperatures near 300K by the formation of a thin oxide film
impervious to hydrogen. The niobium-hydrogen system is effectively
a closed system below 250 to 300°C. For other materials, such as
iron and steel, which often leak hydrogen at 300K, a hydrogen-
impervious film can be applied after charging. A thin plate of
copper is quite effective, and there is little doubt that experiments of the type suggested here could be accomplished with iron or
steel hydrogen charged and then plated with copper.

When the closed system is mechanically loaded so that dislocation motion commences, the first occurrence is that some of the

hydrogen trapped at dislocations is returned to the lattice, thus increasing the lattice hydrogen. This happens because the distribution between trap and lattice is altered when the trap is in motion to an extent that is governed by the trap (dislocation) velocity.[8] If the velocity is sufficiently high, all of the hydrogen will be left behind to disperse through the lattice and perhaps be partially absorbed by static traps. For steels, recent experiments suggest that this dumping process may double the lattice hydrogen concentration if all dislocations are moving substantially in excess of the critical breakaway velocity.

For velocities near or less than the critical value, some of the hydrogen will remain with the moving dislocations. Its disposition will depend upon the interactions, including annihilation, between the moving dislocations and on whether they intersect a discontinuity such as a void, inclusion, or decohered interface. As the dislocation slows upon approaching one of these intersections, however, its hydrogen chemical potential, because of the critical dumping process, will be less than that in the surrounding material. Thus, there will be a driving force for hydrogen diffusion back to the dislocation prior to annihilation or intersection. Hence, previously moving dislocations which are stopped are <u>sinks</u> for hydrogen.

The intersection of a hydrogen bearing dislocation with a void is of particular interest. As the dislocation is in essence a moving pattern of atom displacements, one cannot think of the dislocation as directly dumping all of its hydrogen into the void upon intersection: it may remain in the bulk near the interface or at the interface. If the surface topographically resembles the surface where hydrogen enters the system the entry process and time constants would be reciprocal to the dumping process by the principle of microscopic reversibility and entry kinetics could be used to estimate dumping kinetics.

After dislocation intersection with a void, the hydrogen in the remaining dislocation core need not necessarily drain into a preexisting void. Rather, because hydrogen is equilibrated with the void prior to deformation, and because of the emission process described previously, the chemical potential of hydrogen in the void will exceed that in the dislocation core. In this case the net flux would be away from, rather than into, the void.

The second case to be discussed is one where the hydrogen is presented to the surface at the same time that deformation begins. The system is then open, with hydrogen entry accompanying deformation. Dislocations moving in from the specimen surface can be assumed to carry hydrogen at a chemical potential equal to that of the charging medium. There seems little loss in generality in assuming local equilibrium at the charging medium—specimen

interface, consistent with upper bound estimates. Under these
conditions, the dislocation will emit hydrogen for three physical
reasons as it moves into the material; (i) Since the charging
medium chemical potential is greater than that within the material,
hydrogen will tend to diffuse into the material from the moving
dislocation; (ii) the effect of dislocation velocity on trapping
capacity as described earlier will also be present and will lead
to hydrogen emission from the dislocation as it accelerates; and
(iii) dislocation annihilation and intersection can cause its
hydrogen to be dumped. Thus, in contrast to the preequilibrated
case, the dislocations, whether moving or stopped, act as hydrogen
sources.

Evidently, the dislocation transport process contains important aspects which have not been incorporated completely in the
previous models.

PREVIOUS WORK AND MODIFICATIONS

Johnson and Hirth[6] considered the problem of repeated dislocation annihilation at a site, creating a local supersaturation of
H which could be dissipated by diffusion to a free surface. The
upper bound for I_A, the amount of H deposited per unit time in this
model, is

$$I_A = \eta \lambda \dot{\epsilon}/ab \tag{1}$$

in atoms per m per s, where η is the number of H atoms per atom
site and a is the atom spacing along a dislocation, b is the magnitude of the Burgers vector, λ is the mean-free path of a dislocation before it is annihilated and $\dot{\epsilon}$ is the strain rate. The only
condition on Eq. (1) is that the dislocation velocity be less than
the critical velocity for breakaway of the dislocation from its
atmosphere.[8] On the basis of the core binding energy of H to a
non-screw dislocation in iron of $E_B = 58.6$ kJ/mol[9] and the maximum
elastic H-dislocation binding energy of 20 kJ/mol, Fermi-Dirac
calculations[10] indicate that the pertinent atmosphere at room
temperature is a saturated core atmosphere. The value of η appears
to be about 2 on the basis of indirect experimental evidence, in
agreement with the calculations showing core saturation and very
little H in the Cottrell atmosphere.[4,6]

A lower bound on the departure rate is the steady-state diffusion solution, with the assumption of a quasi-equilibrium boundary
condition, which we discuss later. The departure rate of H from a
cylindrical source of radius r_0, which could be a site of dislocation annihilation or a trap, is[6]

$$I_J = \frac{2\pi D(c_1-c_o)}{\ln(2L/r_o)} \tag{2}$$

Here D is the diffusivity of $\underline{H}$, c_1 the matrix concentration in atoms/m^3 at the trap, c_o the equilibrium concentration in the matrix at the free surface, and L the source-surface spacing. Equating (1) and (2) for iron, we found a maximum supersaturation $S = (c_1-c_o)/c_o \cong 10^{-3}$.

Tien et al.[3] considered the arrival and departure of $\underline{H}$ from spherical inclusions or voids. Including an additional quantitative microscopy factor d_1/ℓ, where d_1 is the particle diameter and ℓ the particle, spacing in a square lattice representation of the inclusions, the arrival rate at the particle or void in atoms/s is[4,7]

$$I_a = \dot{\epsilon}d_1^2/b^2 \tag{3}$$

Analogous to the line-source, dislocation-annihilation case, the lower bound to the departure rate is the steady state diffusion flux with quasi-equilibrium boundary conditions in the matrix. With a spherical sink of concentration c_o at diameter d_2, concentric with d_1, the departure flux is[4]

$$I_H = \frac{2\pi d_1^2 D(c_1-c_o)}{(d_1^{-1}-d_2^{-1})d_1^2} \cong 2\pi d_1 D(c_1-c_o) \tag{4}$$

Solving for (c_1-c_o) from Eqs. (3) and (4), we find that it is a fraction $d_1/2\lambda\ln(2L/r_o)$ of the value from Eqs. (1) and (2), thus giving an even smaller predicted supersaturation. This result is typical for a comparison of two and three dimensional diffusion solutions. In both of the above analyses, the supersaturation S or the value (c_1-c_o) is calculated by equating the arrival and departure fluxes, a point of importance for later discussion since it is in contrast to the procedure of Tien et al.[3,7]

Instead of the above estimate of a maximum possible supersaturation by upper and lower bound estimates, Tien et al.[3] approximated the departure rate as a __transient__ one-dimensional flux from the cross-sectional area $\pi d_1^2/4$ of a particle with the concentration gradient approximated by $dc/dx \sim c_1/2\sqrt{2} \ \bar{X}$ where $\bar{X} = \sqrt{2Dt}$ is the r.m.s. diffusion distance in one direction in time t. Moreover, they approximated c_1 by

$$c_1 = c_L \exp(-E_B/RT) \tag{5}$$

where $c_L = a^{-3}$ is the concentration of lattice sites, R is the gas constant and T is absolute temperature. Their result for the departure rate is, then,

$$I_T = \frac{d_1^2 \, c_L}{4} \left(\frac{D}{t}\right)^{1/2} \exp\left(-\frac{E_B}{RT}\right) \tag{6}$$

Hirth[4] considered the ratio of I_H to I_T, even without the approximation of Eq. (5), and showed that for $t = 10^3$ to 10^4s, $D = 10^{-10}$ m²/s (for iron[11]) and $d_1 = 10^{-2}$ to 10^2μm, typical of the range considered by Tien et al.,[3] $I_H/I_T = 10^6$ to 10. Since I_H already predicts a very low value of S, I_T would give an even smaller value. Tien et al.[5] performed the same calculation and found a maximum value of I_H/I_T of 10^2. This difference arises because they considered $t = 10^2$ s and $d_1 \geq 10$μm rather than the broader range of parameters noted above. For large inclusions and low D values $I_H/I_T \sim 1$ and the earlier estimate[3] would apply. For even lower values of D, the ratio I_H/I_T becomes less than unity.[7] These lower values of D would not be characteristic of iron unless other traps were present, and these in turn would provide alternate sinks and lower S even further so they are not relevant for iron or steel. For other metals they would be, but a ratio less than 1 means that transient behavior would be dominant instead of steady state behavior and, as discussed later, this would also lower S provided the equilibrium boundary condition is maintained at r_0.

Since, in their most recent work, Tien et al.[7] consider a modified diffusion calculation for the one-dimensional case, we modify our previous results[6] for this case, which would correspond to a large planar inclusion parallel to a free surface. The arrival flux per unit area is $1/\lambda$ times the result of Eq. (1) or

$$I_{AA} = \eta \dot{\varepsilon}/ab \tag{7}$$

The departure flux at steady state and with quasi-equilibrium boundary conditions is given by Fick's law as

$$I_{HJ} = D(c_1 - c_0)/L \tag{8}$$

The value of S calculated from Eqs. (7) and (8) is a fraction $2L/\lambda \ln(2L/r_0)$ of that given from Eqs. (1) and (2). Since $L \sim \lambda$, we again predict a very small supersaturation. The relaxation time to establish steady-state is $\tau = L^2/\sqrt{2} \, D$ or, with $L = 10^{-7}$m, $\tau \sim 10^{-8}$s. The largest plausible value of L would be 10^{-3} m and even this would give only $\tau \sim 1$ s. Also, in the transient region for any of the above cases the departure flux would be larger than the lower-bound, steady-state estimates so values of S would be even smaller.

Tien et al.[7] cite work to be published in which they find relaxation times for achievement of steady state for the case of one-dimensional diffusion in iron, with $D = 10^{-10}$ m²/s as above, of 79 years when the binding energy is 48 kJ/mol, with an even larger

time for E_B = 58.6 kJ/mol. Moreover, they note that the departure
flux can be less than the steady state value as long as a trap is
incompletely filled. The large relaxation times appear to be
connected with the approximation of Eq. (5) which we consider next.

CONCENTRATIONS AND EQUILIBRIUM BOUNDARY CONDITIONS AT TRAPS

In a Fermi-Dirac treatment of interstitial equilibria, the
standard state can be taken as interstitials in a defect-free
crystal equilibrated with some vapor potential at a free surface.
The standard state chemical potential is then

$$\mu_o = RT \ln [c_o/(c_L-c_o)]$$

$$\cong RT \ln (c_o/c_L) \tag{9}$$

The local value in some supersaturated state is

$$\mu - \mu_o = RT \ln [c_1/(c_L-c_1)]$$

$$\cong RT \ln (c_1/c_L) \tag{10}$$

The local value within a trap site is

$$\mu - \mu_o = -E_B + RT \ln[X/(1-X)] \tag{11}$$

where X is the fractional occupation of trap sites. If quasi-
equilibrium obtains in the matrix adjacent to a trap, as assumed
in our supersaturation estimates, μ must be the same in matrix and
trap, so Eqs. (10) and (11) give

$$\frac{c_1}{c_L} = \frac{X}{1-X} \exp (\frac{-E_B}{RT}) \tag{12}$$

Comparing this to Eq. (5), we see that Eq. (5) when used for the
saturated trap case X → 1 leads to an enormous underestimate of c_1
by a factor $X/(1-X)$ which → ∞ as X → 1. This is the basic reason
why our previous estimates[4,6] of S and $c_1 - c_o$ and those in the
present paper give very small values of S ($\sim 10^{-3}$ or smaller) while
Tien et al.[3,7] find large values of S. The use of Eq. (5) leads to
a great underestimate of S or c_1, hence to a low departure rate of
H from a trap, and hence, for example, to pressurization. This
same approximation would lead to the huge values of relaxation time
found by Tien et al.[7] and discussed in the preceding section. Of
course when X << 1 for a trap, the Boltzmann approximation (5) of
Eq. (12) is accurate, and the assumptions of Tien et al.[7] would be
appropriate.

There remains the issue of whether the equilibrium boundary

condition can be attained. If the sink for dislocation-transported
atomic hydrogen is a void, then recombination to the molecular
state would be expected upon discharge. Hydrogen emission and
diffusion away from the void must then be preceded by molecular
dissociation. It is assumed here that both the recombination and
dissociation rates are sufficiently high that local equilibrium is
established, and it is believed that this is the most probable
situation.

There are, nevertheless, other possibilities. If the H_2 dissociation reaction were selectively impeded, then partial interface
control of the departure process could obtain, and the pressure in
the void would be higher than estimated here. Alternatively, the
H–H recombination process might be poisoned, the entry rate lowered,
and the pressure in a void would be lower than estimated here.
Group V and VI elements are known[12] to be surface active at grain
boundaries and internal interfaces. They slow the recombination
process, enhance dissociation at low surface concentrations, and
slow dissociation at higher, near-monolayer, surface concentrations.
Thus, they could act to either enhance or detract from pressurization.

Also, motion from the void surface or other trap into the
matrix may be a slow process and lead to interface control. The
adsorbed free surface state has a trapping energy relative to the
above standard state of 71 kJ/mol and carbide interfaces give
E_B = 96 kJ/mol typical of large binding energy, or so-called irreversible, traps[4] so that consideration of these cases suffices for
all.

Quasi-equilibrium requires that mass action be satisfied. If
the entire binding energy acts over one atomic spacing, the situation would be that of curve A in Fig. 1. The relaxation time for
equilibration between the trap, site 1, and the matrix, site 2, for
a dilute trap concentration is

$$\tau = g\, \tau_o \, \exp\, [(E_B + Q)/RT] \tag{13}$$

where Q is the activation energy for diffusion, 6.9 kJ/mol,[11] g is
an entropic factor which should be of the order of 0.1 to 1 and
τ_o is the atomic vibrational period $\sim 10^{-13}$s. With E_B = 71 kJ/mol
and Q as above, and g = 1, Eq. (13) gives, τ = 4.5 s. With a value
of E_B of 95 kJ/mol, typical of a carbide interface in iron,
τ = 7 x 10^4s. There are other possibilities, however. Local
relaxations could lead to a value of Q for the first jump which is
larger than that for lattice diffusion, an effect which would
increase τ according to Eq. (13). Alternatively the binding
energy E_B could be relaxed over n atomic distances as in curve B.
This would tend to reduce Q, either the lattice diffusion or
enhanced value, and decrease τ. More importantly, if the trap is

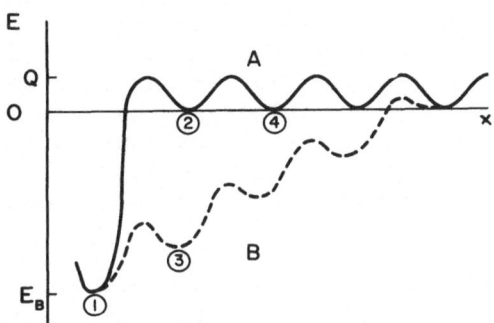

Fig. 1. Possible energy levels at atomic sites 2, 3, 4, as a
         function of distance X from a trap site 1.

saturated, then absolute rate theory would describe equilibrium
between an activated state and a saturated ground state. This
would bring a factor X/(1-X) into Eq. (13) and would greatly
decrease τ. The effect is one such that the saturated trap concentration cannot exceed X = 1. One can think of it as a large driving
force for departure from the trap arising from the entropy of mixing
of the trap vacant site so created. In the absence of specific data
for the local energetics, we proceed on the basis of the above
numerical estimates for τ.

Thus, for a void we conclude that local equilibrium is rapidly
achieved. For a stronger trap the relaxation time to achieve steady
state can be long or short depending on the atomic-scale nature of
the specific problem. For the strongest traps for the present problem, we cannot envision X exceeding 1 without void formation,
though, in which case equilibration would extend to adjacent elastic
binding sites and local-equilibrium should obtain. Hence, the
assumption of local-equilibrium seems justified. Indeed, the
analysis of permeation of iron by $\underline{H}$ from a free surface would imply
that τ did not greatly exceed the maximum value τ ∿ 5s calculated
above for a free void surface.

VOID FORMATION AND GROWTH

The small achievable supersaturations calculated in the preceding analysis predict that internal pressurization of voids in iron cannot occur when the external environment which fixes c_o is at a hydrogen fugacity of 10^5 Pa (1 atm) or less. Thus the work of Hancock and Johnson[13] showing cracking of iron under dry hydrogen pressures of less than 10^5 Pa would conclusively exclude a contribution from internal pressurization. At large fugacities, pressurization can occur, as in blistering,[14] but only to a pressure about equal to the imposed fugacity. The alternatives are that hydrogen is indeed stripped into interface traps and there contributes to void nucleation by lowering the surface energy of the crack nucleus or the bond-breaking cohesive energy in the passage of a nucleus from critical to supercritical size.

The previous results show that for fugacities corresponding to hydrogen pressures less than roughly the yield strength, $\lesssim$ 500 MPa, internal pressurization should not appreciably enhance void growth in iron or steel at room temperature. The supersaturations are all inversely proportional to D so as T decreases S will increase as $S(T)/S(298 \text{ K}) = D(298K)/D(T)$. Using the result of Quick and Johnson[10] for D and our previous result of $S(298) \sim 10^{-3}$, we find for iron

T,K	298	250	200	100
S	10^{-3}	1.7×10^{-3}	3.9×10^{-3}	0.24

Thus, supersaturation remains small at low temperatures for iron as well, provided local-equilibrium obtains at the interface. Whether local-equilibrium holds at low temperatures is questionable, however, so the low temperature steady-state predictions are not as conclusive as those at room temperature. Indeed, Eq. (13) predicts, for the void case of E_B = 71 kJ/mol., that $\tau = 10^{27}$ s. In this case, reentry into the lattice would be impossible. Only if reentry is catalyzed would the steady-state model apply. However, the issue of whether hydrogen could be injected into a void would remain, so the question of pressurization would be an open one.

For other metals, supersaturation and pressurization are possible. As an example, we consider nickel. At room temperature nickel has a diffusivity of 5×10^{-14} m^2/s. Thus, with the other parameters as for iron, we would predict a supersaturation S = 5 at room temperature and larger values at lower temperature.

TRANSPORT KINETICS

Dislocation Velocity

The kinetic rate of hydrogen transport by dislocations

evidently exceeds that by normal diffusion. However, the formal
problem of the dislocation transport is a formidable one, involving
a moving, leaking source with each successive dislocation encount-
ering a less depleted matrix. Only portions of the problem have
been treated in various approximations, but these suffice to give
an estimate of penetration rates.

Tien et al.[3] considered the velocity of a hydrogen interstitial
in the Cottrell atmosphere of a dislocation, simply specified that
the velocity v of the dislocation must be less than the critical
velocity v_c for breakaway and supposed that if $v < v_c$, the dislocation would carry its Cottrell atmosphere until it was stripped.
The maximum penetration rate would be given by the Einstein relation as

$$v_c = \frac{D}{RT} F_c = \frac{D}{RT} \frac{E_B}{30b} \tag{14}$$

where F_c is the critical interaction force of H and a dislocation,
reasonably approximated as $E_B/30$ (a better value for iron would be
$E_B/75$[16]).

There are two limitations in the use of Eq. (14), which has
been used to test experiments on nickel.[17] First it is applicable
only at the critical velocity, which corresponds to the presence of
yield point phenomena, serrated yielding and the like. Second, it
applies for the Cottrell atmosphere and not for the Snoek or core
atmospheres. Thus, while it would be applicable when the Cottrell
atmosphere is dominant, it does not apply for iron at room temperature where the appropriate atmosphere is the core atmosphere and
where $v < v_c$ at ordinary strain rates. Indeed, at ordinary strain
rates, serrated yielding and the cold-work internal friction peak
for iron occur at $\sim$ 200 K so only with the reduced mobility at this
temperature would $v \sim v_c$.

The velocity for drag of a core atmosphere is given in ref.
16 as

$$v = \frac{\sigma b \lambda D_{ok}}{RT} \exp\left(-\frac{Q_k}{RT}\right) \exp\left(-\frac{2F_k^*}{RT}\right) \tag{15}$$

where σ is the resolved shear stress, λ is the dislocation segment
length, $D_{ok} \sim b^2 \nu$ is the preexponential for kink diffusion, ν is
the Debye frequency, Q_k is the activation energy for kink diffusion
and $2F_k^*$ is the free energy of formation of a double-kink on a dislocation. From the analysis[4] of the cold-work internal friction
peak, all of these parameters are known for iron if one takes Q_k as
the activation energy for core diffusion $Q/2$: $\lambda = 1\mu m$, $b = 0.248$ nm,
$\nu = 10^{13}$ s^{-1}, $G = 86$ GPa[8] is the shear modulus, $Q_k = 3.4$ kJ/mol,

and $2F_k^* = 28.4$ kJ/mol.   Thus

$$v = 8.33 \times 10^3 (\sigma/G) \tag{16}$$

in m/s. The greatest uncertainty is in the factor D_{ok}, which might be an order of magnitude smaller if it compares with D_o for lattice diffusion and v would be correspondingly reduced. The critical stress is given by Eq. (18–127) in ref. 8 as

$$\frac{\sigma_c}{G} = \frac{2(E_B - E_s)}{(65RT - Q_k)} \left(\frac{E_B}{Gb^3}\right)^{1/2} \frac{1}{\ln(1/c_o^o)} \tag{17}$$

We have actually added the term $\ln(1/c_o^o)$ since the preexponential term in the solubility, $c_o^o = 0.00185$, is known for $\underline{H}$ in iron.[9,11] Again, all data are known for iron: the energy of solution is $E_s = 28.6$ kJ/mol.[9,11] Thus, $(\sigma_c/G) = 0.016$ and $v_c = 137$ m/s. With the uncertainty in D_{ok} and possible variations in λ, a range for v_c would be $14 \lesssim v_c < 137$ m/s. For typical room temperature tests of iron and mild steel $\sigma < \sigma_c$ and $v < v_c$; only for high strength steels does $\sigma > \sigma_c$.

For nickel, we have $\lambda = 1\mu m$, $Q_k = Q/2 = 17.9$ kJ/mol,[15] $E_s = 42.5$ kJ/mol,[18] $E_B = 60.4$ kJ/mol,[18] $c_o^o = 0.21$,[18] $b = 0.249$ nm, $\nu = 10^{13}$ s^{-1}, and $G = 94.7$ GPa.[8] The data from Stafford and McLellan,[18] giving E_B, give a concentration of traps sites which is too large for dislocations. However, this concentration has a very large scatter and if one takes the lower limit it would be consistent with dislocation core traps in a cold worked metal. Thus, we calculate from Eq. (17) that $\sigma_c = 0.04$ G. The quantity $2F_k^*$, which is needed to estimate v_c is unavailable. If we guess that it has the same value as in iron, we compute $v_c = 1.09$ m/s. In any case, for typical strain rates $\sigma < \sigma_c$ so $v < v_c$. This is consistent with mechanical tests at room temperature,[17,19] which give no indication of breakaway phenomena but do exhibit hardening by $\underline{H}$.

Hydrogen Penetration

The moving, leaking dislocation core problem for a single dislocation has been solved.[20] The results are in the form of infinite sums of sets of Mathieu and Bessel functions, the details of which are beyond the scope of the present work. The results are graphically displayed in reduced coordinates in Fig. 2 for the fractional quantity q of hydrogen remaining in the core relative to its equilibrium amount ($\sim$ zero) with the uncharged bulk. The form is nearly a step function so one can consider the first dislocation of a train to dump its hydrogen into the matrix over a distance $(vt)_1$, increasing the average matrix concentration. The next dislocation would lose less hydrogen over the distance $(vt)_1$ and would

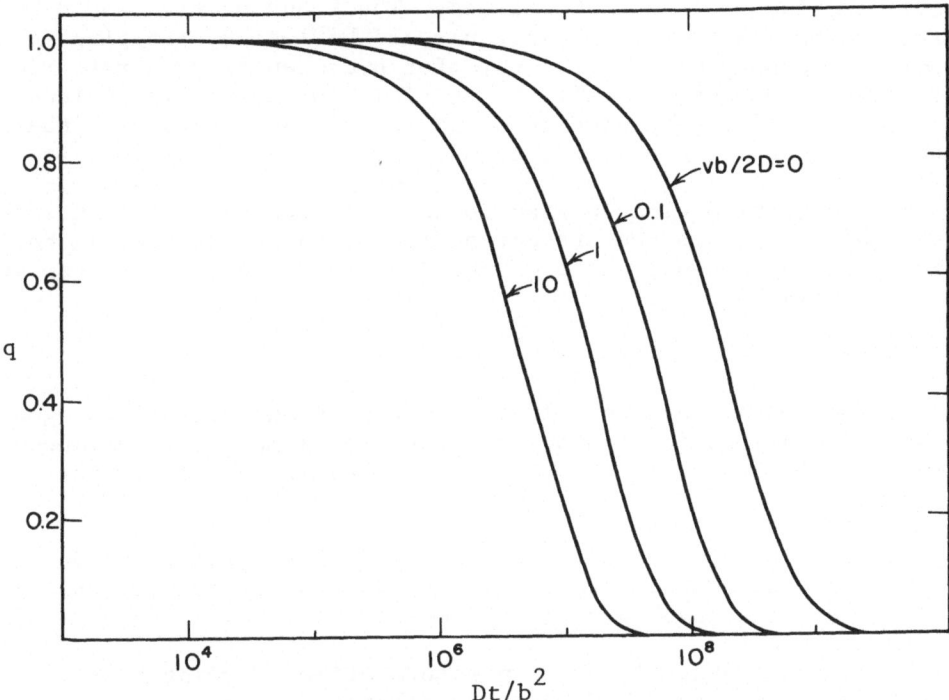

Fig. 2. Fractional draining of a moving dislocation as a function of a reduced time.

lose the remainder over the next incremental distance $(vt)_2$. Thus, the problem can be attacked in a stepwise manner, in general.

For the comparison with experiments on nickel,[17] we adopt the above idea with the incremental distance equal to the cell spacing λ. The mean stress in the experiments was 7.9×10^{-4} G so $\sigma/\sigma_c = v/v_c = 2 \times 10^{-2}$. In the reduced coordinates of Fig. 2 $v_c b/2D = 2700$ so $vb/2D = 54$. The reduced time for which $q = 0.5$ is then $Dt/b^2 = 2 \times 10^6$ giving $t = 2.5s$ to drain the dislocation to $q = 0.5$. This corresponds to a travel distance $vt = 54$ mm at the velocity 21.8 mm/s. However, the specimen thickness was only 0.43 mm. Thus, a single dislocation moving a distance of 0.608 mm on a 45° inclined plane of maximum shear stress would reach the surface in a time of 0.03s. In the actual case, dislocations would annihilate or enter cell walls in passing through the crystal a number of times equal roughly to the distance 0.608 mm divided by the cell spacing of 1 μm or 608 times. The transfer time would be negligible, but part of the hydrogen would be lost by trapping·in the cell wall, so the effective time for hydrogen to penetrate the specimen would be very roughly 608 multiples of the period calculated above or 17s. The prediction is in fair agreement with the observed penetration times of 3 to 8s.[17]

The penetration times appear to decrease somewhat with increasing strain rate.[17] In the present analysis, this effect would be an indirect one, a consequence of the increased mean flow stress at the larger strain rate.

CONCLUSIONS

1.   The theoretical analysis indicates that transport of hydrogen by dislocations in iron or steel at room temperature produces only small supersaturations ∿ 0.001 at voids or other internal traps.

2.   This result would have the consequence that pressurization of voids by dislocation transport of hydrogen is possible only at large external hydrogen fugacities.

3.   Larger supersaturations and pressurization of voids by dislocation transport of hydrogen are predicted to be possible in metals such as nickel which have lower diffusivities for hydrogen.

4.   Kinetic analysis indicates that hydrogen transport occurs in iron and nickel in the form of a core atmosphere at a velocity considerably smaller than the critical velocity for breakaway.

5.   Predictions of penetration times for hydrogen in plastically deforming nickel roughly agree with experiment.

ACKNOWLEDGMENTS

This research was supported by the National Science Foundation in part under Grant DMR 7815735 (JPH) and in part by the Department of Energy under Grant DE-AS02-76-ER-03166 (HHJ).

REFERENCES

1.   P. Bastien and P. Azou, C. R. Acad. Sci. Paris 232: 1845 (1951).
2.   R. C. Frank, in: Internal Stresses and Fatigue in Metals, G. M. Rassweiler and W. L. Grube, eds., Elsevier, New York (1959), p. 411.
3.   J. K. Tien, A. W. Thompson, I. M. Bernstein and R. J. Richards, Met. Trans. A 7:821 (1976).

4. J. P. Hirth, Met. Trans. A 11:861 (1980).
5. A. J. West and M. R. Louthan, Jr., Met. Trans. A 10:1675 (1979).
6. H. H. Johnson and J. P. Hirth, Met. Trans. A 7:1543 (1976).
7. J. K. Tien, S. V. Nair and R. R. Jensen, in: Effect of Hydrogen on Behavior of Materials, I. M. Bernstein and A. W. Thompson, eds., TMS-AIME, Warrendale, PA (1981).
8. J. P. Hirth and J. Lothe, Theory of Dislocations, McGraw-Hill, New York (1968), Chap. 18.
9. A. J. Kumnick and H. H. Johnson, Acta Met. 28:33 (1980).
10. J. P. Hirth and B. Carnahan, Acta Met. 26:1795 (1978).
11. N. R. Quick and H. H. Johnson, Acta Met. 26:903 (1978).
12. R. D. McCright, in: Stress Corrosion Cracking and Hydrogen Embrittlement of Iron-Base Alloys, R. W. Staehle et al., eds., NACE, Houston, TX (1977), p. 306.
13. G. G. Hancock and H. H. Johnson, Trans. TMS-AIME, 236:513 (1965).
14. M. Smialowski, Hydrogen in Steel, Pergamon, Oxford (1962).
15. J. Völkl and G. Alefeld, Topics Appl. Phys. 28:321 (1978).
16. reference 8, p. 606.
17. M. Kurkela and R. M. Latanision, Scripta Met. 13:927 (1979).
18. S. W. Stafford and R. B. McLellan, Acta Met. 21:181 (1973).
19. R. M. Latanision and R. W. Staehle, Acta Met. 17:307 (1969).
20. R. Fuentes-Samaniego and J. P. Hirth, submitted to Phys. Stat. Solidi (a).

DISCUSSION

Comment by P. Neumann:

I did not see a gradient of the fugacity of hydrogen enter your equations. How can you then apply your results to experimental situations (straining of a permeation membrane) in which the same amount of dislocations (of opposite Burgers Vector) travel in either direction, making the net hydrogen flux zero, if fugacity gradients are neglected.

Reply:

In the category of precharged specimens with equilibriated dislocations, the dislocation motion would create a line source of hydrogen producing increased hydrogen fugacity and stopped dislocations would be sinks. Thus, chemical potential (fugacity) gradients would appear, although these are only discussed qualitatively. In the surface entry case, dislocations are annihilated internally by dislocations moving toward the surface creating an internal near surface source with a chemical potential gradient both back toward the surface and toward the interior. With, in the actual case, a variety of sources near the surface, the net effect, except in the near surface region is to create a gradient into the interior. Thus,

as you properly note, conditions exist where with equal fluxes of
dislocations moving in both directions, the net effect would be to
enhance hydrogen transport into the interior. Dislocation micro-
mechanics effects are possible for which the dislocation fluxes in
the two directions would differ, but this would be a second order
effect."

Comment by R.H. Jones:

    Can you imagine a condition where the hydrogen permeation rate
in iron and nickel are equal for equal temperatures, and strain rates
and input fugacities?

Reply:

    Yes, the mean free path for transport in iron would be quite
small and the lattice diffusivity is large, so the enhancement of
the hydrogen flux by dislocations is relatively small, as indicated
by the work reported by Berkowitz. Thus, by adjusting the strain
rate, temperature, fugacity parameters for nickel, one could, in
principle, adjust its permeation rate to one prescribed for iron.

Comment by D.M. Esterling:

    Is there evidence that hydrogen interacts with the core of an
edge dislocation in a different manner than with the core of a screw
dislocation? Are there implications for the mobility of each type
of dislocation?

Reply:

    Yes, there is a strong body of evidence from work on trapping,
internal friction and tensile tests which give a coherent view of
interactions [see J.P. Hirth, Met. Trans. A., 11A (1980) p. 861].
Below about 300°K motion of screw dislocations becomes slow, and
below about 25°K the motion of edge dislocations becomes slow in
pure iron, both results being associated with double kink nucleation.
Hydrogen makes double kink nucleation more difficult for edge dislocations, raising the critical temperature to about 200°K, but easier
for screw dislocations, dropping the critical temperature to the same
value. Thus, hydrogen causes softening from room temperature to
~200°K but hardening below that temperature. Serrated flow, and
associated mechanical behavior, is observed at the critical temperature.

Comment by R.M. Latanision:

    Are the observations of hydrogen-induced dislocation injection
(a la Beachem) and serrated yielding (hydrogen-dislocation interactions leading to hardening) compatible with your modelling?

Reply:

Yes, both are, but neither is necessary. As long as dislocations entering at the surface carry hydrogen in with them, transport occurs whether the injection rate is enhanced or not. There are plausible models for surface sources where, say, hydrogen adsorption at ledges would enhance dislocation injection carrying hydrogen. The Latanision model of subsurface sources would enhance dislocation egress, on the other other hand, and would not tend to enhance entry. With regard to serrated yielding, the phenomenon is associated, with core hydrogen atmospheres as discussed in my response to Esterling.

Comment by J.P. Fidelle:

As a concluding remark, I wish to point out a consequence of the H-dislocations interactions, which is also very important for industrial applied research: straining tests to failure, conducted in the proper strain rate range are faster and always at least as sensitive as delayed failure tests.

Moreover, existing models of H-dislocations interactions have to be enlarged to take into account the recent experimental evidence that embrittlement by dissolved H can take place over more than a 1200°C wide temperature range, e.g.: hydrogen gas embrittlement of TZM Mo alloy, Haynes 188 at +1000°C, HGE of martensitic steels down to at least -160°C, internal HE of stable austenites at 4°K. Thus, the span of the "H dragging by dislocations" mechanism should be approximately indicated in the (likely) case where it would no longer be important at the highest and lowest temperatures.

Reply:

Without a priori atomic calculations, one cannot predict the lowest temperature possible for transport, since it depends on the double-kink formation energy of a dislocation containing a core atmosphere of hydrogen. For the specific case of hydrogen in pure iron, internal friction tests, as manifested by modulus relaxations, show transport by edge dislocations at 4°K, so very low temperature transport is possible. The higher temperature limit can be determined explicitly from the transport theory presented in my lecture. A temperature is reached, depending on the binding energy to the dislocation and the lattice diffusivity, where the hydrogen atmosphere desorbs from the dislocation in such a short time that the effects of dislocation transport are negligible.

HYDRIDE FORMATION AND REDISTRIBUTION IN

Zr-2.5wt% Nb STRESSED IN TORSION

M.P. Puls and A.J. Rogowski

Materials Science Branch
Whiteshell Nuclear Research Establishment
Atomic Energy of Canada - Research Company
Pinawa, Manitoba ROE 1L0 Canada

INTRODUCTION

The mechanism of hydride-induced cracking in zirconium alloys has been investigated extensively in the last few years (see, for example, the reviews by Perryman[1] and Dutton[2]). The mechanism involves the repeated stress-induced redistribution of hydrides to a crack tip and the subsequent fracture of the embrittled crack tip zone. The model for the diffusive redistribution requires expressions for the solubility of hydrides as affected by misfit and external stresses[3,4]. Regarding the latter, earlier models[3,4] have considered only the effect of normal stresses. The present paper deals with the extension of the model to include interaction terms involving applied shear stresses. Experimental results are reported which demonstrate the effect in a Zr-2.5wt% Nb alloy. A more detailed account of the following is given by Puls and Rogowski[5].

MECHANISM OF HYDRIDE PRECIPITATION

Zirconium hydrides precipitate as either the equilibrium fcc-phase ($ZrH_{1.66}$) or as the metastable fct-phase (ZrH). The habit plane is close to the (0001) plane of the zirconium matrix. The hydrides are less dense than the matrix and generate an anisotropic volume misfit as given by Carpenter[6]. Formation of the hydrides requires transformation of the α-Zr hcp lattice to an fcc or fct structure. At the low temperatures of interest, lattice shears appear to be necessary to achieve this phase change. Bradbrook et al.[7] provided experimental evidence suggesting that hydrides may have formed by means of a martensitic reaction. Carpenter[8,9] on the other hand, proposed that fairly simple shears may be responsible for the transformation. Based on Carpenter's analysis, the shear

plane for the transformation is the (0001) plane and the shear
directions are along <10$\bar{1}$0>. The calculated magnitude of the transformation shear S = 0.36. This is an order of magnitude larger than
the dilatational strains. It is likely that the net amount of this
transformation strain transmitted to the surface of the specimen is
considerably reduced, for instance, by twinning of the hydride or
the generation of an incoherent boundary between matrix and precipitate.

EFFECT OF STRESS ON TERMINAL SOLID SOLUBILITY

Detailed derivations of the effect of stress on terminal solid
solubility (TSS) have been given by Puls[10],[11]. The shift from the
stress-free TSS, C_H^S, to the internally constrained, externally
stressed TSS, $C_H^{con}(\sigma^a)$, is given by

$$C_H^{con}(\sigma^a) = C_H^S \exp\left\{ (W_H - \bar{w}_H)/RT \right\} \exp\left\{ (\bar{w}_t^{inc} + \bar{w}_t^a)/xRT \right\} \quad [1]$$

In the first exponential, W_H and $\bar{w}_H$ are, respectively, the work done
and the strain energy per mole addition of hydrogen into solution.
The dominant term usually is W_H giving a positive contribution in
the presence of tensile hydrostatic stresses. The major contribution to the shift in TSS comes from the molal strain energies $\bar{w}_t^{inc}$
and $\bar{w}_t^a$, which are, respectively, the hydride–matrix misfit energy
(always positive, irrespective of the sign of the misfit) and the
misfit-applied stress interaction energy (negative when the stresses
and the misfits are dilatational). The strain energies can be calculated using the methods of Eshelby[12] and are summarized for
hydride plates and needles by Simpson and Puls[4]. We focus our
attention on the interaction term, $\bar{w}_t^a$. For misfitting ellipsoidal
particles having elastic moduli equal to those of the matrix, $\bar{w}_t^a$
has the simple form

$$\bar{w}_t^a = -\bar{V}_{hydride} \sum_{i,j} \sigma_{ij}^a e_{ij}^T \quad [2]$$

where the σ_{ij}^a are the applied stresses (positive when tensile),
e_{ij}^T are the unconstrained misfits and $\bar{V}_{hydride}$ is the molal volume
of hydride assuming a composition per mole of $ZrH_{1.66}$ or ZrH; x in
equation [1] is, then, 1.66 or 1.0, respectively. Note that $\bar{w}_t^a$
is larger than and opposite in sign to W_H. A positive $\sigma_{ij}^a e_{ij}^T$
product thus decreases $C_H^{con}(\sigma^a)$ or, conversely, favours the
formation of hydrides above the stress-free solvus. In a tensile
stress gradient such as near a crack, hydrides are less soluble
at the crack tip than further away. This forms the basis for a
steady-state model for hydride accumulation at the tip of a crack.
In this model, the hydride growth rate is directly proportional
to the product of the amount of hydrogen in solution, the diffusion
coefficient and the difference in solubility, due to the applied

stresses, between the crack tip and bulk hydrides. Although the
effect of hydrostatic stress on the diffusion of hydrogen between
bulk and crack tip hydrides has a minor influence on the kinetics,
it is not the driving force for diffusion. This is determined by
the difference in the boundary solubilities between the crack tip
and the bulk hydrides. Previous efforts have concentrated on the
effect of normal stress gradients on the hydride accumulation rate.
However, the previous section and equations [1] and [2] suggest that
shear stress gradients may also operate in this way. This is demonstrated experimentally in the next section for round notched bar
samples stressed in torsion (anti-plane shear or Mode III). Detailed
expressions for the hydride growth rate at the tip of a Mode III
crack are given by Puls and Rogowski[5].

EXPERIMENTAL RESULTS

    We have taken round notched bar samples of hydrided (50 - 100
μg/g) Zr-2 5 wt% Nb and statically loaded them in torsion (K_{III}
= 15 MPa$\sqrt{m}$), holding them at 150°C - 200°C anywhere from 4 - 1500
hours. Different cooling rates, after hydriding, were used to vary
the hydride sizes. After testing, the samples were sectioned and
examined metallographically. The most common observation,
shown in Fig. 1a, was of a small fan of reoriented hydrides emanating
from the notch tip and roughly following the slip-line pattern
expected in the plastic zone under this type of loading. In a few
cases, the reoriented hydrides had become larger and more distinct,
as shown in Fig. 1b, and some crack growth had occurred.

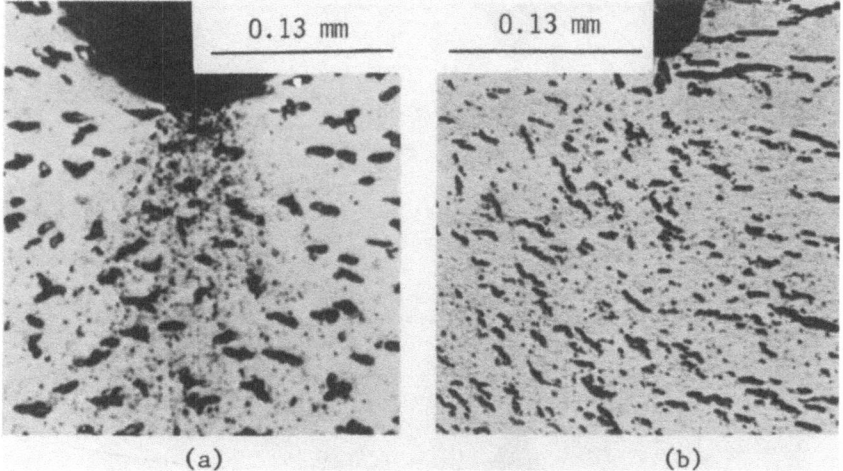

(a)   (b)

Figure 1.   Hydride distribution near the notches of samples stressed
             in torsion: (a) small hydrides formed along slip-lines;
             (b) more extensive reorientation. In both pictures, the
             original hydrides are oriented horizontally.

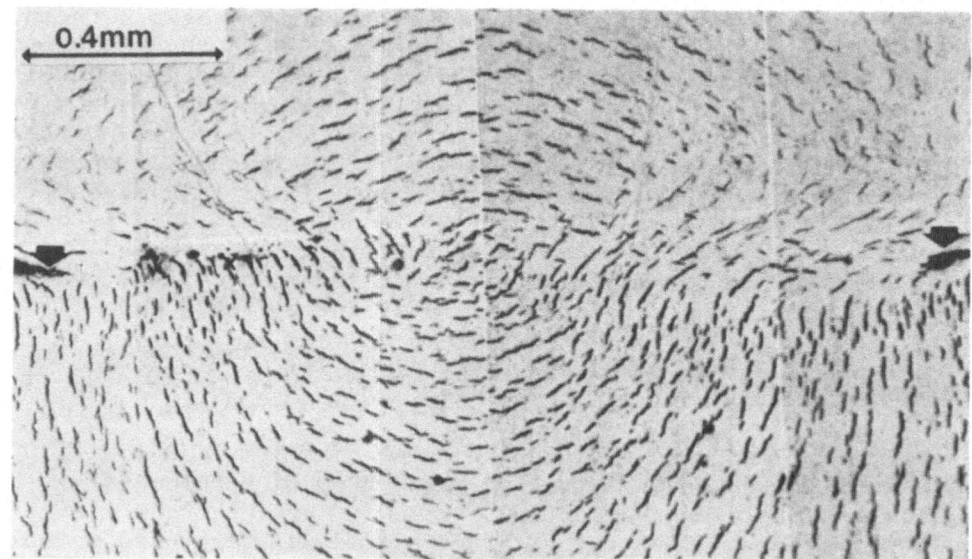

Figure 2. Hydride distribution in a sample stressed in torsion in
which considerable crack growth and reorientation had
occurred. The unreoriented hydrides are vertical. The
cracks (indicated by arrows) have grown in some
distance from the notch.

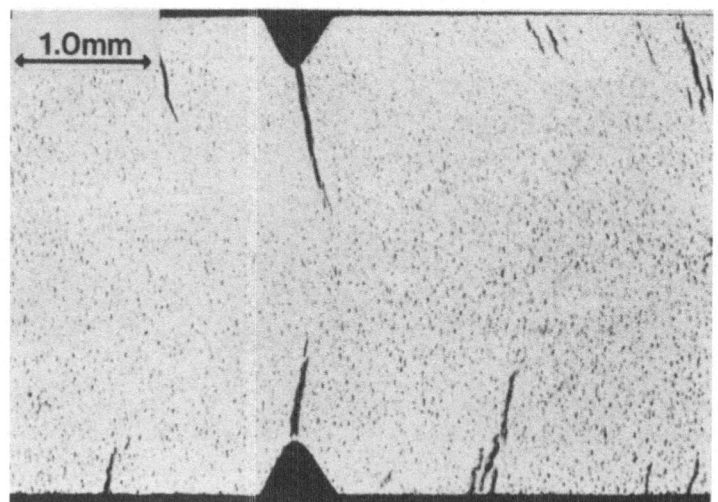

Figure 3. Large reoriented hydrides in a sample stressed in
torsion in which no crack growth had occurred.

In one spectacular case, only 3.5 h under load at 200°C, complete reorientation of hydrides in the region between the notches had occurred (Fig. 2). This reorientation was accompanied by considerable crack growth. In another sample held at 150°C for 1500 h, massive reoriented hydrides were formed (Fig. 3), but no crack growth had occurred. In addition, large reoriented hydrides had also grown in from the outer surface of the specimen away from the notch. The large variability of the results obtained so far may be attributed to variations in the texture and microstructure of the sample and the morphology of the original hydrides. However, an additional problem in these preliminary tests was that, because of friction in the grips, the sample may not always have been free to move in the axial direction. Torsional deformation[13] plus reorientation of the hydrides can result in axial elongation of the sample. This elongation can, if constrained and if large enough, generate inhibiting axial compressive stresses.

Indirect confirming evidence that shear deformations promote hydride precipitation and growth was found by Cann et al.[14] in specimens fractured in tension showing hydride accumulation similar to Fig. 1 in the shear zones near the crack. Another observation (Fig. 4) is the appearance of very narrow long reoriented hydrides near the notch of samples stressed in tension. These hydrides have reoriented to form a pattern which is reminiscent of the slip-line field found near notches under Mode I loading.

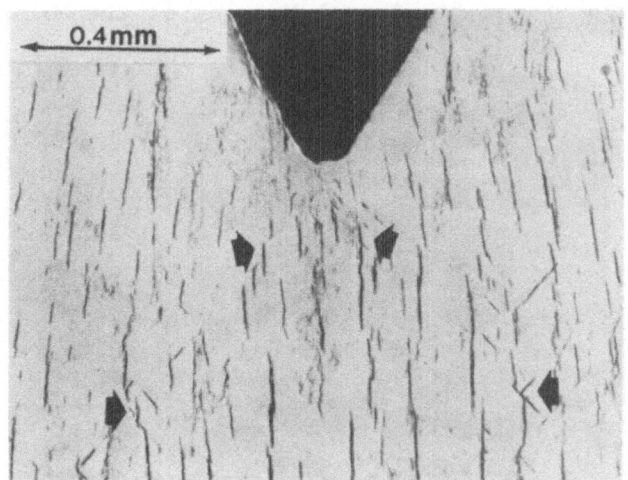

Figure 4. Hydride distribution in a sample stressed in tension showing hydrides (arrowed) which appear to have reoriented along slip-lines. The tensile axis is horizontal.

ACKNOWLEDGEMENT

The authors would like to thank F. Havelock for assistance
with the metallography.

REFERENCES

1. E.C.W. Perryman, Nucl. Energy 17, 95 (1978).
2. R. Dutton, Met. Soc. CIM Annual Volume Featuring: "Hydrogen
    in Metals and Titanium", p. 16, 1978.
3. R. Dutton and M.P. Puls in "Effect of Hydrogen on Behaviour
    of Materials", A.W. Thompson and I,M, Bernstein, eds., Metals
    Society, AIME, New York, N.Y., p. 387, 1976.
4. L.A. Simpson and M.P. Puls, Met. Trans. 10A, 1093 (1979).
5. M.P. Puls and A.J. Rogowski, Atomic Energy of Canada Ltd.
    Report No. AECL-6799 (1980).
6. GJC. Carpenter, J. Nucl. Mat. 48, 264 (1973).
7. J.S. Bradbrook, G.W. Lorimer and N. Ridley, J. Nucl. Mat. 42,
    142 (1972).
8. G.J.C. Carpenter, Acta Met. 26, 1225 (1978).
9. G.J.C. Carpenter, Met. Soc. CIM Annual Volume Featuring:
    "Hydrogen in Metals and Titanium", p. 75 (1978).
10. M.P. Puls, Atomic Energy of Canada Ltd. Report No. AECL-6302
    (1978).
11. M.P. Puls, Acta Met., in press.
12. J.D. Eshelby, Proc. Roy. Soc. A241, 376 (1956).
13. A.M. Freudenthal and M. Ronay, Proc. Roy. Soc. A292, 14 (1966).
14. C.D. Cann, N.S. McIntyre, K. Nuttall and V.R. Deline,
    Proceedings of the 8th Annual Meeting of the Microscopical
    Society of Canada, Montreal, June 1981.

THE HARDENING AND SOFTENING INDUCED BY

HYDROGEN IN CARBON STEELS

R. A. Oriani

Department of Chemical Engineering & Materials
 Science, University of Minnesota
Minneapolis, MN. 55455

This note briefly describes the results of work carried out on
some medium-carbon steels to investigate the plastic response to
the presentation of hydrogen at controlled and measured input
fugacity. The motivation for this work is the realization that
whereas plastic strain can obviously relax stress in a material,
it is also a means of generating stress if the strain is inhomo-
geneously distributed. Therefore, not only is this topic of
interest in its own right but also it is a path to understanding
how the decohesion model, which was developed for high-strength
steels,[1] may be relevant to lower-strength steels.

THE STRESS-STRAIN CURVE OF 1045 STEEL

Tension specimens of AISI-1045 steel, either pearlitic or in
the spheroidized condition, of 0.25 mm thickness, were mounted in
a testing machine after being marked with a photoengraved grid.
Hydrogen was charged cathodically[2] into both sides of the specimen. To investigate the effect of hydrogen on the initiation of
plastic deformation,[3] hydrogen was charged into only one half of
the gage length for two hours prior to straining as well as during
deformation. Straining the specimen produced two segments of
Luders deformation separated from each other by a stress increment.
Measurements of the grid and visual observations showed that for
this steel hydrogen at low fugacity raises the stress both to
generate and to propagate Luders bands, and that the necessary
stress increases with increasing hydrogen fugacity.

To investigate[3] the effect of hydrogen at larger accumulations
of dislocations, tension specimens were deformed in the absence of
hydrogen and then unloaded, after which hydrogen was charged for

two hours prior to reloading and also during additional straining
into only one half of the gage length. Measurements of the photoengraved grids showed that the hydrogen-charged section of the
specimens deforms much less than the hydrogen-free section for
virtually the same stress-time history. Another experiment demonstrated that strain aging does not modify the hardening action of
low-fugacity hydrogen. In addition, in other experiments[3] the
input fugacity of hydrogen presented to the entire gage length of
the specimen was suddenly increased or decreased during straining
at a constant rate of deformation. The effects of the change of
input fugacity were small, and in some instances imperceptible.
When changes were observed it was invariably found that an increase
of hydrogen fugacity produced an increase of flow stress, and that
a decrease produced a lowering of the flow stress. These results
on 1045-steel are the opposite of what Kimura et al.[4] found was
the case for iron of the highest purity.

In another experiment,[3] one half of a 1045 tensile specimen
was hydrogen charged prior to and after deformation to a total of
about 6% plastic strain. The specimen was unloaded and mobile
hydrogen was removed from the charged half. The specimen was
then strained an additional 7% without further hydrogen charging.
The result was that the section which had been initially deformed
with hydrogen and then discharged, subsequently deformed much more
than did the half gage length which had never been exposed to
hydrogen. Thus, plastic deformation with hydrogen at moderate
fugacity leaves a marked softening influence.

RELAXATION EXPERIMENTS

Thin tension specimens of pearlitic 1045 steel were strained[5]
without hydrogen to a desired value of plastic strain, at which
time the crosshead of the testing machine was stopped and the
relaxation of the applied load was followed at room temperature.
At the desired time, cathodic charging at known input fugacities
was begun and maintained for some minutes. If no change was
observed in the character of the load time curve, the input fugacity was increased and held constant for another period of time.
In this way it was found that the logarithm of the load-relaxation
rate is linear in the logarithm of time during the initial relaxation without hydrogen and that the same line is followed for input
fugacities lower than some critical value. However, at and above
that value, a rapid increase in the relaxation rate occurs. The
threshold value of input fugacity is an increasing function of
increasing plastic strain.

The room-temperature creep of spheroidized 1040 steel has also
been measured.[6] The logarithm of the creep rate $\dot{\varepsilon}$ was found to
be linear in the creep strain $\varepsilon - \varepsilon_0$ at constant stress. As in the
stress-relaxation experiments, a value of hydrogen input fugacity

larger than some critical value is required to change the dynamics
of the creep process. There is a threshold fugacity for effecting
a very sudden increase in the creep rate, analogous to the threshold effect found in the stress relaxation experiments. But in
addition, there is also a threshold fugacity for producing an
immediate increase in the absolute value of the slope of $\ln \dot{\epsilon}$ vs.
$(\epsilon-\epsilon_0)$. This means that above a certain critical value of fugacity,
hydrogen causes a greater rate of strain-hardening. Another significant finding is that decreasing the input fugacity into a specimen creeping under a fugacity higher than the afore-mentioned
threshold value causes a rise in the rate of creep. This does not
happen if the initial fugacity had been below the threshold value.
The maximum $\dot{\epsilon}$ attained after decreasing the input fugacity
increases roughly linearly with increasing value of fugacity that
had been applied during the prior creep episode.

DISCUSSION

     The various experimental results on 1045-steel exhibit an
internal consistency, and qualitative understanding in terms of
decohesion theory is possible.[7] Although in very pure iron,
hydrogen facilitates the motion of dislocations, probably by lowering the Peierls barrier,[4] in impure iron and steels the observed
hardening effect of hydrogen has been attributed[6] to the
increased difficulty of cross slip of screw dislocations owing to
the widening of their cores. This is a result of the weakening by
hydrogen of the attractive interactions between iron atoms as
postulated by the decohesion model. The hydrogen-loaded screw
dislocations therefore produce a greater entanglement, and hence a
greater internal stress, σ_i, than when they move without hydrogen.
This is the explanation both of the increase in flow stress produced by hydrogen during dynamic straining, and of the increase of
strain-hardening coefficient during creep.

     The sudden increases in creep rate and in stress-relaxation
rate caused by hydrogen have been interpreted as[6] due to hydrogen-
enhanced nucleation and decohesive growth of microvoids. Such
processes relax the stresses that produced the microvoids in the
first place, so that σ_i is suddenly decreased, leading to a burst
of dislocation activity. The existence of a threshold hydrogen
fugacity is consistent with the basic tenet of the decohesion
model,[1] that bond breaking occurs only above a critical combination of local hydrogen concentration and local stress. The microvoids generated with or without hydrogen, but more easily with
hydrogen, are themselves a softening mechanism which opposes the
intrinsic hardening action of the hydrogen in steels. Therefore,
which effect predominates when a steel is deformed during hydrogen
charging depends on the details of the experiment. The hardening
effect of dissolved hydrogen may generally be overcome by using
very high-fugacity hydrogen such that the corresponding[8]

mechanical pressure of molecular hydrogen gas is larger than the
flow stress of the steel. In these circumstances hydrogen within
microvoids exerts sufficient pressure to expand the microcavities
by causing plastic deformation. Such an induced plastic deformation is artifactual from the point of view of understanding the
mechanisms by which hydrogen affects the plastic properties of
steels.

REFERENCES

1. R. A. Oriani, Berichte Bunsen-Gessellschaft physik. Chem. 76,
    848 (1972).
2. R. A. Oriani and P. H. Josephic, Proc. Symp. Environment-
    Sensitive Fracture of Engineering Materials, AIME, Chicago
    1977, pp. 440-50. Z. A. Foroulis, ed., The Metallurgical
    Society of the AIME, 1979.
3. R. A. Oriani and P. H. Josephic, Met. Trans. 11A, 1809 (1980).
4. H. Kimura, H. Matsui and S. Moriya, Scripta Met. 11, 437 (1977);
    H. Matsui, S. Moriya and H. Kimura, Proc. 4th Int. Conf.
    Strength of Metals and Alloys, Nancy 1976, pp. 291-5.
5. R. A. Oriani and P. H. Josephic, Acta Met. 27, 997 (1979).
6. R. A. Oriani and P. H. Josephic, Acta Met. 29, 669 (1981).
7. R. A. Oriani, Jackson Lake Conference 1980, Proceedings to be
    published.
8. B. Baranowski, Berichte Bunsen-Gessellschaft physik. Chem. 76,
    714 (1972).

DISCUSSION

Comment by S. Lynch:

    There is some evidence that adsorption of hydrogen at surfaces
(internal or external) can produce injection of dislocations. Could
not such an effect contribute to the observed increases in creep
rate due to hydrogen?

Reply:

    It is true that it is possible to hypothesize that the observed
rapid increase of creep rate upon increase of hydrogen fugacity is
due to hydrogen-enhanced injection of dislocations. However, some
special assumptions would have to be made to explain the observed
threshold fugacity, below which creep rate is not increased upon
presentation of hydrogen. Moreover, hydrogen-enhanced dislocation
activity is entirely opposite to what is needed to explain the
observed hydrogen-caused impediment to the formation of Luders bands
in the same steel.

DIFFUSION OF HYDROGEN NEAR AN ELASTO-PLASTICALLY

DEFORMED CRACK TIP

Hideo Kitagawa and Yukio Kojima*

Professor and Post-Doctoral Research Fellow*
Institute of Industrial Science,
University of Tokyo, 22-1, Roppongi, 7-chome
Minato-ku, Tokyo 106 JAPAN

INTRODUCTION (PROBLEMS IN HYDROGEN EMBRITTLEMENTS AND
SELECTION OF APPROACH)

The mechanism of hydrogen embrittlemnt(HE) of high strength
steels at room temperature has been discussed, with no unified
consensus on understanding of its phenomenon itself nor the method
of its quantitative evaluation.

If the phenomenon "embrittlement" can be taken as "low fracture
toughness", the methods of the quantitative evaluation has almost
been established in the field of fracture mechanics(FM). A general
method to measure the occurence or the intensity(or grade) of the
"embrittlement" of materials has not yet been given. "Hydrogen
embrittlement" will be discussed in this study on the basis of FM.
Starting from empirical facts, several problems will be raised, from
which the requirements for the approaches to HE will be introduced.

Problem ① (See Table 1) In FM there have been two opposite
opinions on the HE above. One opinion(A) is that the fracture
toughness (K_{IC}) of the material (steel) itself is decreased by
hydrogen. Another opinion(B) is that the K_{IC} is not decreased, that
is, the embrittlement does not occur due to hydrogen. Stable (subcritical) crack growth affected by hydrogen, however, can occur
under the statical sustained load at the K (stress intensity factor)

---

* Presently, Y. Kojima is a staff member at Nagoya Institute of
  Technology, Department of Production Mechanical Engineering,
  Gokisho-cho, Showa-ku, Nagoya.466 Japan

level below K_{IC} (and above K_{ISCC}). Recent systematic experiments[1] support the opinion (B). Accordingly, it seems most important for practical studies of the HE to clarify the behavior of hydrogen near the tip of an existing crack in a stressed cracked body.

Problem ② What can describe the behavior of hydrogen above? For steels, whose σ_{ys} values are sufficiently .igh, when the average concentration of hydrogen in a cracked steel is higher, lower is its K_{ISCC} and higher its crack growth rate da/dt[2]. The concentration changes time to time and place to place in the cracked body in a particular manner. Variation (with time) of distribution of hydrogen concentration(H.C.) will be taken as the main object to be investigated.

Problem ③ It is believed that any non-destructive experimental determination of the variation(with time)of hydrogen concentration (H.C.) at every sites in a cracked body has not yet succeeded. At present we can only do a study based mainly on some theoretical analysis of a model with given boundary conditions. For the experimental examination of the analysis by using crack growth data, the analysis has to be done for any given boundary conditions(shape of specimens, initial H.C. distribution on surfaces, etc.).

Problem ④ In some reliable HESCC experiments[3] initiation or early growth of cracks were remarkable somewhere along the border of plastic zone at the tip of a large existing crack(i.e., boundary between elastic and plastic zones). This suggests the necessity of an elasto-plastic analysis for the analysis of H.E..

Problem ⑤ There exists a region of K, where H.E. crack growth rate da/dt saturates temporarily and is kept constant, i.e., the region II in K-da/dt relations[4]. If one wants to explain this phenomenon by using H.C., stress or strain, it seems natural to consider the yield strength σ_{ys} as a saturating level of strength or a material non-linearily parameter in the analitical study, where H.C. is taken as a monotonic function of stress or strain.

Problem ⑥ The most important property in H.E. of steels at R.T. is that, when the static strength of unnotched components (e.g.,σ_{ys})is higher, a crack can grow even at lower load(or K)levels, i.e., K_{ISCC} is lower[5]. The above property is mostly applied to the practice for the protection of H.E.. As the analyses which can not explain this property are practically useless, the results of the analysis on H. C., stress or strain are required to include σ_{ys} as far as possible, which can be attained, e.g., by the performance of elasto-plastic analyses.

Problem ⑥' For realizing the H.E. cracking as a time-depending one-side(monotonic) process, introduction of some effects like trapping of hydrogen will be useful, in addition to usual hydrogen diffusion. The σ_{ys}-depending property discussed in Problem 6 can also be analized as an effect of σ_{ys} upon the trapping effect.

Problem ⑦ Other than the problem 6', from the experimental results that the hydrogen diffusion coefficient D below 200°C is much lower than the extrapolation from the D values at higher temperatures[6]

trapping effects by the defects in materials and also their formulation have been proposed.[7,8] In the previous analyses of diffusion
of hydrogen, however, the trapping has scarcely been taken into
consideration.

Problem ⑧   In H.E. cracking, intergranular fracture surfaces are
eminent very often , though it does not seem to be a necessary
condition.[9] This suggests that microscopic local conditions possibly
have important effects upon the fracture process, which requires a
good accuracy of the analysis of H.C., stress and strain even at
the points very close to the crack tip.

     All of the descriptions above are summarized in Table 1. As
clear from the Table 1., an important and available method of study
at present is the follows: that is, "to develop a method to analize
by a calculation the time-dependent variation of the distribution
of hydrogen concentration around an elasto-plastically deformed
crack tip, taking trapping into consideration, accurately even at
the points close to the crack tip." Also it is desirable for the
analysis to get the solution including the effects of various
boundary conditions, for checking the analysis with experiments.
     Because an analysis which satisfies all the requirements above
has not been found previously, an analysis will be developed and
some of the results will be shown.

STRESS AND STRAIN AROUND A NON-LINEARLY DEFORMED CRACK TIP

     Stress and strain in an elasto-plastic state have been ana-
lyzed by FEM. The non-linear equations have been solved by the
Newton-Raphson method[10], using an ealsto-plastic matrix[11].

The models of a cracked body analized are center cracked tension
speci-mens, 50 mm in length, 50 and 100 mm in width and 25 mm in
crack length (=2a). Uniformly distribued stress σ is applied to the
upper and lower ends of the cracked body vertically to the crack
line. The material is assumed to be perfect elasto-plastic (the
work hardening coefficient H'=0). Young's modulus E, Poisson's
ratio ν and the yield strength σ_{ys} are 1.96×10^5 MN/m^2 (20000 kg/mm^2),
0.3 and 1.27×10^3 MN/m^2 (130kg/mm^2) respectively. The diffusion of
hydrogen is controlled by hydrostatic stress $\sigma_h (=(\sigma_x + \sigma_y + \sigma_z)/3)$ as
discussed later. Figure 1 shows the results of the analysis on the
distribution of σ_h in the cracked body along the x-axis ($\theta = 0$), for
two stress states, plane stress and plane strain. In the case of
plane strain, σ_h increases as x approaches 0. In the vicinity of
crack tip, both the absolute value and the gradient of σ_h in plane
strain are extremely large compared with those in plane stress.

     In the case of plane stress as shown in Fig. 1(b), σ_h has a
maximum somewhat distant from the crack tip. In the following analysis, hydrogen assisted cracking will be investigated using the

Table 1.   Selection of the approach to hydrogen embrittlement
of high strength steels at room temperature (from empirical
facts to developed analysis).

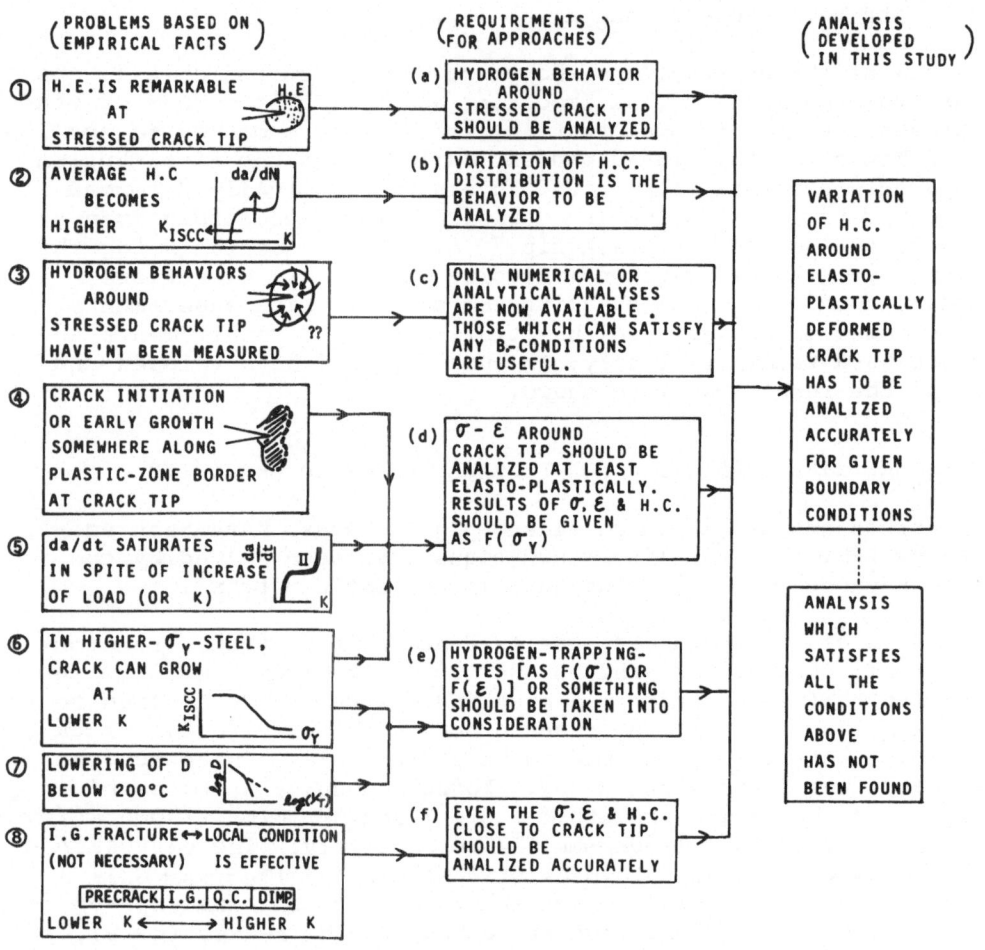

NOTE: H.E.;HYDROGEN EMBRITTLEMENT,H.C.;HYDROGEN CONCENTRATION
       σ ;STRESS, ε ;STRAIN,
       D;DIFFUSION COEFFICIENT OF HYDROGEN

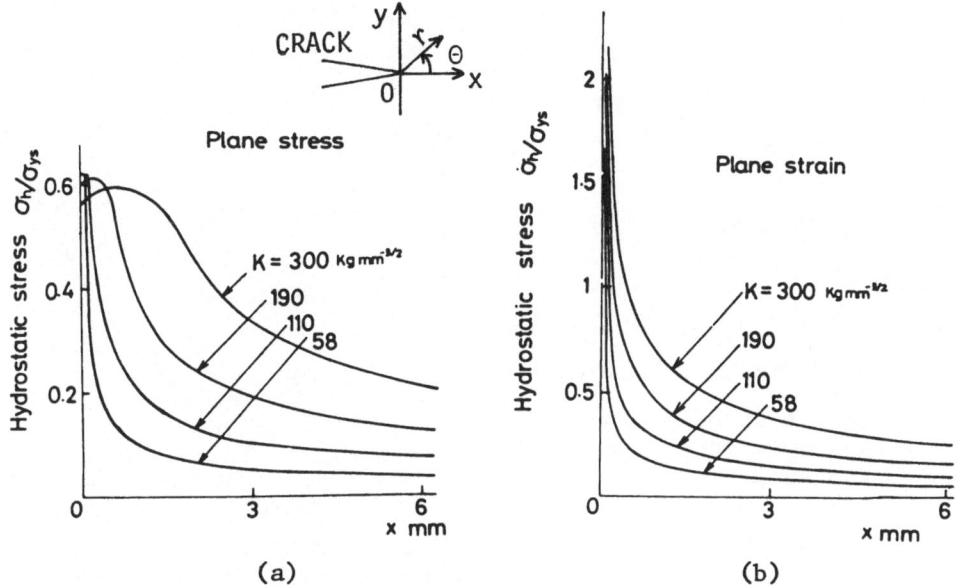

Fig. 1   Distribution of hydrostatic stress σ_h along the x-axis

stress distribution in plane strain, for the reason that these cases are practically important.

The hydrostatic stress σ_h has a maximum $\tilde{\sigma}_h$ at $\theta = 45°$ along the smallest radius for calculation around the crack tip ($r = 0.0176$mm), and the plastic strain ε_p has a maximum $\tilde{\varepsilon}_p$ at $60°$.

Variations of the $\tilde{\sigma}_h$, $\tilde{\sigma}_h/\sigma_{ys}$ and the $\tilde{\varepsilon}$ with respect to K are shown in Fig. 2,3 and 4. The $\tilde{\varepsilon}_p$ continues to increase as K increases, while σ_h or σ_h/σ_{ys} tends to approach each upper- limiting values as K increases. The hydrostatic stress σ_h at $\theta = 0$ has the tendency similar to $\tilde{\sigma}$ at $\theta = 45°$ as shown in Fig. 2. The limiting values of $\tilde{\sigma}_h/\sigma_{ys}$ are the same for different σ_{ys} levels and have been obtained as 2.16 in the present calculation. The $\tilde{\sigma}_h$ can reach higher than $26,000$ kg/cm^2 for a steel with σ_{ys} of 130 kg/mm^2. As the yield strength σ_{ys} increases, σ_h becomes greater, while $\tilde{\varepsilon}_p$ becomes smaller. Considering the fact that the decrease of K_{ISCC} becomes more remarkable as the yield strength increases[5] and the results of the present analysis, it is deduced that stress, rather than strain, controls hydrogen assissted crack growth.

The plastic strain ε_p here is an equivalent plastic strain $\sqrt{2/3} \; \varepsilon_{pij} \; \varepsilon_{pij}$[12].

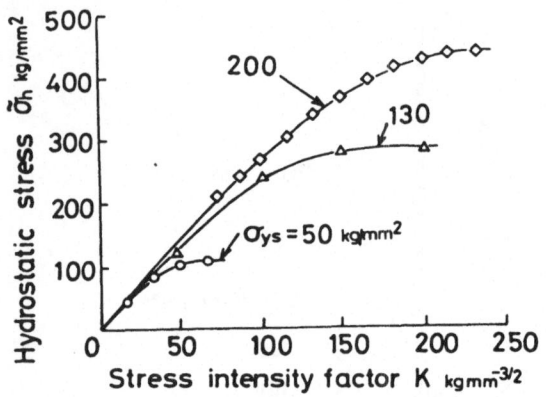

Fig. 2   Variation of hydro-
static stress $\tilde{\sigma}_h$ with stress
intensity factor K.

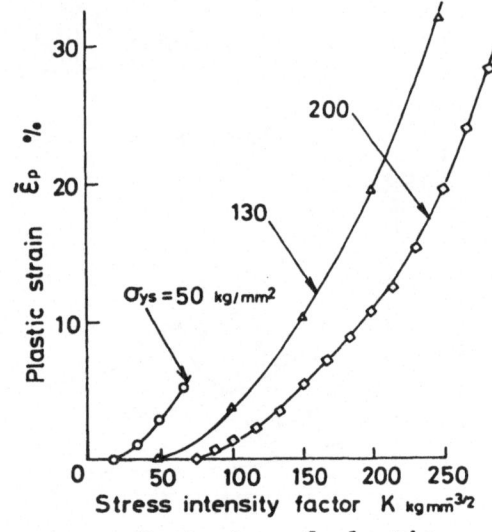

Fig. 4   Variation of plastic
strain $\tilde{\varepsilon}_p$ with stress intensity
factor K.

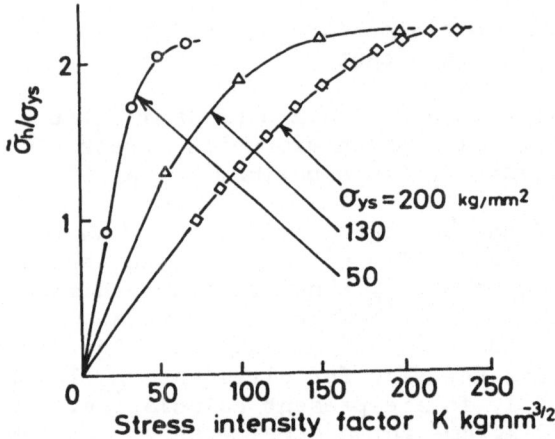

Fig. 3   Variation of nondimensional
hydrostatic stress $\tilde{\sigma}_h/\sigma_{ys}$ with stress
intensity factor K.

DIFFUSION EQUATION OF HYDROGEN COMBINED WITH TRAPPING EFFECT

Diffusion of hydrogen including trapping effect is assumed to
follow the combination of Eqs. 1 and 2.

$$\frac{\partial c}{\partial t} + N \frac{\partial n}{\partial t} = D \, \nabla^2 C - M \nabla C \nabla \sigma_h \qquad (1)$$

$$\frac{\partial n}{\partial t} = kc(1-n) - pn \tag{2}$$

The right side of the Eq. 1 was constructed by referring Liu's and Van Leeuwen's papers[13][14] C and D are the concentration and the diffusion constant of diffusing hydrogen respectively. t is the time. The second term of the right side of Eq. 1 implies the effect of hydrogen diffusion due to the gradient of σ_h and the gradient of C. M is expressed as $M = DV_H/RT$ where V_H is the partial molar volume of hydrogen.[13][14] The second term of the left side of Eq. 1, connected with Eq. 2, which was proposed by McNabb and Foster,[8] represents the effect of hydrogen captured by traps such as defects and voids in the material. N is the density of traps in a unit volume, and n is the fraction of the traps which are occupied by trapped hydrogen. In Eq. 2, p and k are the parameters which control trapping rate. Combined Eqs. 1 and 2 are combined non-linear differential equations, so that they also were solved numerically by use of FEM combined with the weighted residual process.[7]

BEHAVIOR OF HYDROGEN NEAR THE CRACK TIP

In Fig. 5 are shown the variations with time of diffusing hydrogen concentration $\tilde{C}$ and trapped hydrogen concentration $\tilde{n}\tilde{N}$ at the points where the maximum hydrostatic stress occurs. The $\tilde{n}_\infty\tilde{N}$ represents the $\tilde{n}\tilde{N}$ in a steady state after a lapse of sufficiently long time.

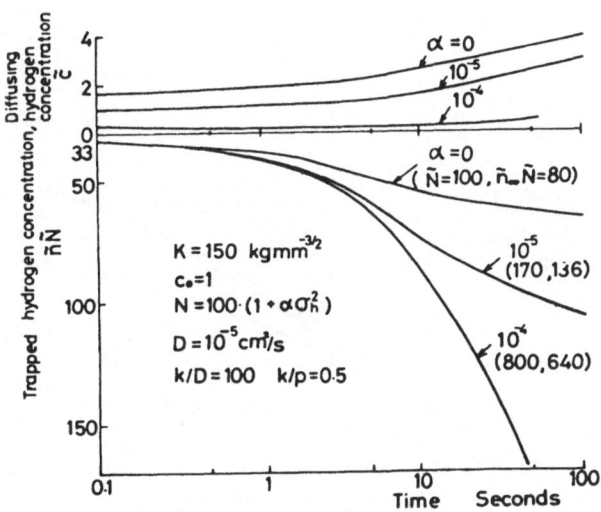

Fig. 5   Variation of diffusing and trapped hydrogen concentration with time.(Center cracked specimen II , Plane strain, $N = 100 \times (1 + \alpha\sigma_h^2)$, $D = 10^{-5}$ cm^2/s, $M/D = 0.00784$ mm^2/kg, $k/D = 100$, $k/p = 0.5$, 50 mm x50 mm specimen)

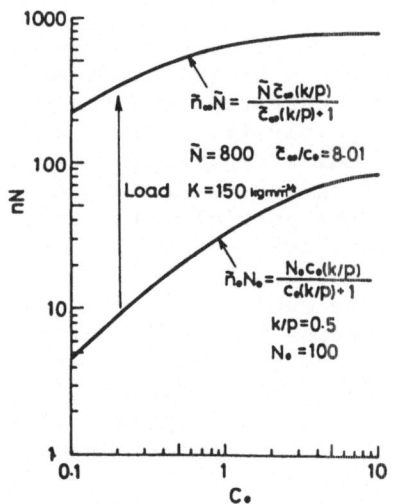

Fig.6 Variation of initial
and maximum trapped hydrogen
concentration $\tilde{n}_o \tilde{N}_o$, $\tilde{n}_\infty \tilde{N}$ with
initial hydrogen concentration
c_o ·(Center cracked specimen II,
Plane strain, $N = 100 \times (1 + \alpha \sigma_h^2)$,
$\alpha = 10^{-4}$, $D = 10^{-5}$ cm²/s,
M/D = 0.00784 mm²/kg, k/D=100,
k/p = 0.5)

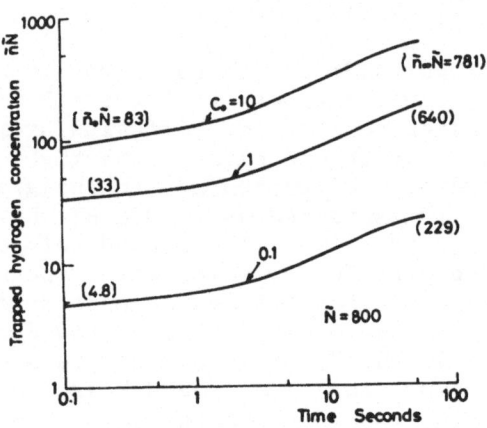

Fig.7 Variation of trapped
hydrogen concentration with
time when initial concentration
c_o varies. (Center cracked
specimen II , Plane strain,
$N = 100 \times (1 + \alpha \sigma_h^2)$, $\alpha = 10^{-4}$,
$D = 10^{-5}$ cm²/s, M/D = 0.00784mm²/kg,
k/D = 100, k/p = 0.5)

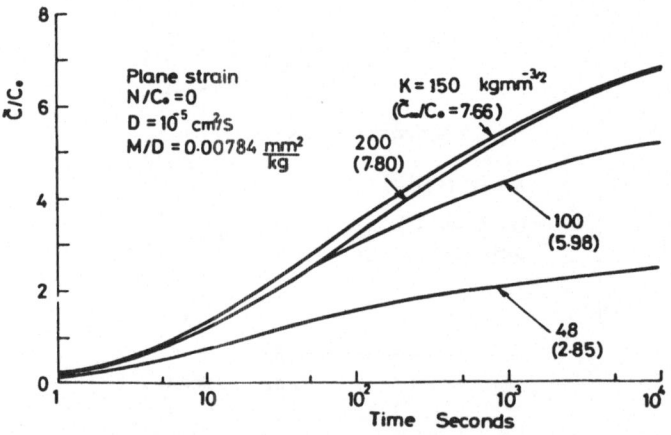

Fig. 8 Variation of diffusing hydrogen concentration $\tilde{C}$
when hydrogen enters from inside of the crack.
(Center cracked specimen II , Plane strain,
  $N = 0$, $D = 2 \times 10^{-5}$ cm²/s, M/D = 0.00784 mm²/kg)

The specimen is assumed to have been charged with uniform concentration of hydrogen C_o before the load is applied.

In the present FEM analysis, the hydrostatic stress σ_h at $r = 0.0176$ mm is assumed to represent σ_h at the crack tip. Since V_H is approximately 2 cc[15], M/D in Eq. 1 is taken as 8×10^{-4} m^2/MN. The diffusion coefficient D in Eq. 1 is taken as 10^{-9} m^2/S[16]. As shown in Appendix the parameters for trapping effect k/p and N are selected to be 0.5 and 100 respectively.

Figure 5 is the results of analyses for the case that before loading there exist initial uniformly distributed traps (N = 100) and initial trapped hydrogen concentration (nN = 33.3) in the material. After loading the number of traps N is assumed to be influenced by stress or strain. It is not certain which the N depends on, stress or strain. Assumption that the N depends on the σ_h is more reasonable for explanation of the higher HE-sensitivity of the steel with higher σ_{ys}. (cf. Figs. 2 and 4) Thus, in the present analysis, it is assumed that N = 100 x $(1+\alpha\sigma_h^2)$. α is a constant. After loading, near the crack tip, the N increases up to a value corresponding to σ_h, as shown in Fig. 5 and diffusive hydrogen is captured by the traps as time goes on. Evidently from Fig. 6, concentration of trapped hydrogen $\tilde{n}\tilde{N}$ increases rapidly with time, when the N strongly depends on σ_h. The concentration of diffusing hydrogen $\tilde{C}$ is low in comparison with trapped hydrogen $\tilde{n}\tilde{N}$, so that total hydrogen near the crack tip becomes nearly equal to $\tilde{n}\tilde{N}$.

Figure 7 shows the variation of $\tilde{n}\tilde{N}$ with time for various values of C_o. As shown in Fig. 7, $\tilde{n}\tilde{N}$ increases with time in proportion to initial concentration $\tilde{n}_o\tilde{N}$ in the early stage of the process.

Since N is assumed to depend on σ_h, the relation between the stress intensity K and trapped hydrogen concentration is suggested in Fig. 2.

In the case of no trapping sites in the material (N = 0 in Eq. 1), hydrogen diffuses only due to the gradient of σ_h produced by the load applied to the crack body. As shown in Fig. 8, the increasing rate of $\tilde{C}/C_o$ becomes higher as $\tilde{\sigma}$ increases with increase of K. For the K level higher than a value 150 kg·mm$^{-3/2}$ (47 MNm$^{-3/2}$) for σ_{ys} =130 kg/mm^2, however, σ_h becomes constant and does not increase (seen in Fig. 2) and $\tilde{C}/C_o$ cannot increase any more shown in Fig. 8.

ADDITIONAL DISCUSSION ON HYDROGEN ASSISTED CRACKING (HAC)

If it can be assumed that hydrogen assisted crack growth occurs when the total hydrogen concentration $\tilde{C} + \tilde{n}\tilde{N}$ or the trapped hydrogen concentration $\tilde{n}\tilde{N}$ reaches a critical value (Ccr), the following discussion can be made on hydrogen assisted cracking $\tilde{C}$, $\tilde{n}\tilde{N}$ or $\tilde{C}+\tilde{n}\tilde{N}$ is a function of K and time, which can be easily understood from the

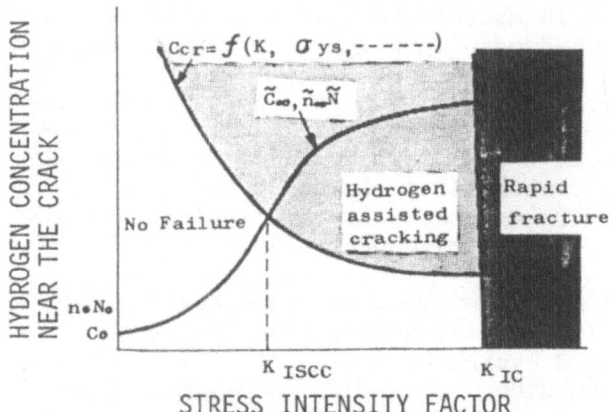

Fig. 9   Schematic representation of relation
between stress intensity factor K and hydrogen
concentration near the crack tip.

results of the preceding analysis. Ccr is a function of K, σ_{ys},etc..
K_{ISCC} can be determined as an intersecting point of two curves,
$\tilde{C}_\infty + \tilde{n}_\infty \tilde{N}$ (or $\tilde{n}_\infty\tilde{N}$) and Ccr, as schematically shown in Fig. 9, where
$\tilde{C}_\infty$ (or $\tilde{n}_\infty\tilde{N}$) is the maximum of hydrogen concentration $\tilde{C}$ (or $\tilde{n}\tilde{N}$)
after the long lapse of time.

At the K level above K_{ISCC} when $\tilde{C} + \tilde{n}\tilde{N}$ (or $\tilde{n}\tilde{N}$) reaches Ccr, the
hydrogen assisted cracking starts to propagate and stops propagating
at a point where $\tilde{C} + \tilde{n}\tilde{N}$ (or $\tilde{n}\tilde{N}$) becomes lower than Ccr and it waits
the next chance of surpassing Ccr by the increasing of hydrogen
concentration. In such a case, the crack will propagate inter-
mittently, as the experimental results so far obtained[17] suggest.
Consider the case where a crack (of length a) extends to a length
(a+ Δa) after incubation time Δt in a discontinuous manner. In the
present analysis, if an increase in crack length Δa is assumed to be
almost constant, crack growth rate da/dt depends mainly on the incubation time Δt, because there is an experimental result[18] that,
in the tests at various temperatures, Δa is almost constant and the
crack propagation rate da/dt ($\fallingdotseq\Delta$a/Δt) is controlled only by variation
of Δt.

At lower stress intensity region, increasing rate of $\tilde{C}$ + $\tilde{n}\tilde{N}$ (or
$\tilde{n}\tilde{N}$) becomes higher as K increases, and accordingly t becomes shorter,
so that da/dt ($\fallingdotseq\Delta$a/Δt) may be dependent on K (Region I behavior). At
higher stress intensity region however, increasing rate of $\tilde{C}$ (or $\tilde{n}\tilde{N}$)
has a tendency of saturation independent of increase of K, and da/dt
becomes almost constant (Region II behavior).

The discussion described above can be applied to any cases
where $\tilde{C}$, $\tilde{n}\tilde{N}$ or $\tilde{C}$ + $\tilde{n}\tilde{N}$ is regarded as hydrogen concentration near the

crack tip. At present, it can not yet be determined which concentration C, nN or $C + nN$, controls the hydrogen assisted crack growth.
As σ_{ys} and/or C_o increase, the C_∞ curve in Fig. 9 shifts upward and
consequently K_{ISCC} decreases. This tendency agrees with experimental
results[5,2] so far obtained.

CONCLUSIONS

By the overview of the states of arts and the problems in the
study of hydrogen embrittlement cracking of high strength steels at
room temperature, a required and available approach has been
selected and composed, following the procedure summarized in Table 1.

Hydrogen behavior and stress-strain distribution near an elasto-
plastically deformed crack tip have been analized by solving the nonlinear hydrogen diffusion equations including trapping effects
combined with an accurately elasto-plastic analysis of stress and
strain near the crack tip.

The computer program composed in this study is applicable to
the specimens of various shapes and materials, and other various
boundary conditions for hydrogen and load. Some of the important
results are the following.

(1) The hydrostatic stress σ_h at the crack tip becomes higher,
as the yield strength σ_{ys} becomes higher, while plastic strain ε_p
becomes lower in contrast to σ_h. From the above results of analysis and the previous experimental results that K_{ISCC} decreases as
the σ_{ys} increases, it is supposed that mainly the stress(rather than
the strain) near the crack tip controls the stress corrosion cracking
of high steels due to hydrogen at room temperature.

(2) The stress near the crack tip (and the corresponding hydrogen concentration) tend to approach each saturated values as K
increases. These results are supposed to be one of necessary
conditions for the occurence of region II behavior in hydrogen
cracking when one wishes to explain the appearance of region II from
the viewpoints of hydrogen behaviors.

(3) By the present analysis, the above stated saturated (maximum)value of the hydrostatic (tensile)stress σ_h near the crack tip
can attain as high as more than twice of the σ_{ys} value(expressed
by the unit of "pressure", atm or Pa). For example, as the σ_{ys}
value of high strength steels can be taken as 100 to $200 \, kg/mm^2$
(about 1,000 to $2,000 \, MN/m^2$),then the $\tilde{\sigma}_h$ value can be at least
20,000 to 40,000 atm (about 2,000 to 4,000 MPa).

(4) After the load is applied to the specimen charged to uniform hydrogen concentration, both the diffusive hydrogen concen-
tration near the crack tip $\tilde{C}$ and trapped hydrogen concentration $\tilde{n}\tilde{N}$
increase with lapse of time, and they approach their saturated
values respectively. However, both the saturated value and the
increasing rate of trapped hydrogen concentration $\tilde{n}\tilde{N}$ are much higher

in comparison with those of diffusive hydrogen concentration $\tilde{C}$, so that the total hydrogen concentration becomes nearly equal to $\tilde{n}\tilde{N}$.

(5) If critical hydrogen concentration Ccr as a requirement for the crack extension can be assumed in addition to the existence of the saturated value of hydrogen $\tilde{C}_{\infty} + \tilde{n}_{\infty}\tilde{N}$ as the function of stress intensity K (assured from the present study), the dependence of K_{ISCC} both on the yield strength of materials and initial concentration of hydrogen and also the existence of region II on the K versus da/dt curve can be qualitatively explained.

REFERENCES

1. Kobayashi, H., Hirano, K., Kayakabe, H. and Nakazawa, H., Proc. 2nd JIM Intern. Symp. Hydrogen in Metals, Minakami, Japan (1980) Trans. Japan Inst. Metals, 21:477 (1979)
2. Yoshizawa, S. and Yamakawa, K., Proc. 7th Intern. Cong. Matallic Corrosion, Rio de Janeiro, 2: 863 (1978)
3. S.Yonezawa, K.Yamakawa and S.Yoshizawa, to be published in BOSHOKU-GIJUTSU(J.Japan Soc. Corrosion Engng.)30:-(1981)
4. Carter, C.S., Corrosion, 27:471 (1971)
5. Sandoz,G., Metall. Trans., 3: 1169 (1972)
6. Johnson, E.W. and Hill, M.L., Trans. AIME, 218:1104 (1960)
7. Darkin, L.S. and Smith, R.P., Corrosion, 5:1 (1949)
8. McNabb, A. and Foster, P.K., Trans. AIME, 227:618 (1963)
9. Beachem, C.D., Metall. Trans., 3:437 (1972)
10. Zienkiewicz, O.C.,"The Finite Element Method in Engineering Science" 374, McGraw-Hill (1971)
11. Yamada,Y. and Yoshimura, N., Intern. J.Mech. Sci.,10:343(1968)
12. Hill, R.,"The Mathematical Theory of Plasticity", The Clarendon Press, Oxford (1950)
13. Liu, H.W.,Trans. ASME, J. Basic Engng., 92:633 (1970)
14. Van Leeuwen, H.P.,Engng. Frac. Mech. 6:141 (1974)
15. Bockris, J. O'M., Beck, W. and Genshaw, M.A., Subramanyan, P.K. and Williams, F.S., Acta Metall., 19:1209 (1971)
16. Ono, K. and Rosales, L.A., Trans. AIME, 242:244 (1968)
17. Steigerwald, E.A., Schaller, F.W. and Troeano, A.R., Trans. AIME, 215:1048 (1959)
18. Aoki,T., et al., Tetsu-to-Hagane (J. Iron and Steel Inst. Japan) 58:S488(1972)
19. Oriani,R.A., Acta Metall., 18:147 (1970)

APPENDIX

Selection of Parameters of Controlling Trapping Included in Diffusion Equations

Let D be a hydrogen diffusion coefficient at room temperature obtained by extrapolating those at high temperatures and Dapp be an observed diffusion coefficient. Since $Dapp \simeq D/[1 + N(k/p)]$[19],D/Dapp at room temperature is taken as about 50^{16}, N and k/p found in Eqs. 1 and 2 as 100 and 0.5 respectively.

A numerical analysis in which hydrogen permeates through membrane of the steel, was carried out to estimate the value of parameter k in Eq. 2. From the above results, k/D was assumed to be a high value of 100.

THE IMPORTANCE OF TRANSIENT EFFECTS RESULTING FROM DISLOCATION

TRANSPORT OF HYDROGEN

I. M. Bernstein and A. W. Thompson

Dept. of Metallurgical Engineering and Materials Science
Carnegie-Mellon University
Pittsburgh, PA 15213

A number of authors have recently considered whether or not dissolved hydrogen, when associated with mobile dislocations, either as a Cottrell atmosphere or as a dislocation core population, can be transported to local imhomogeneities (denoted as traps) in the lattice or grain boundary.[1-3] These considerations have focused on the ability of these moving hydrogen sources[4] to provide a mechanism for either or both local enrichment, or pressurization resulting from molecular hydrogen formation,[5-9] in excess of the amount predicted from thermodynamic equilibrium of the internal hydrogen with the external fugacity. Because of the potential importance of such processes to hydrogen embrittlement, and because much of the ensuing debate has become enmeshed in justifying the validity of the details of a particular theoretical approach, we are concerned that the broader, more important issues have been neglected. Namely, are there materials, or variable ranges in a given material, where dissolved hydrogen can interact with mobile dislocations, can be then transported to local heterogeneities, and be deposited to enrich the local non-equilibrium hydrogen concentration, and possibly recombine to molecular hydrogen, with an associated large volume change and pressurization? Further, can such processes provide an accelerating step for the occurrence or severity of hydrogen embrittlement? We believe that the answer to both queries is yes, but only for very specific conditions, a point emphasized previously.[1,3,8]

Our approach in this paper is to describe three behavior regions; one in which long term supersaturation effects do not appear possible; one in which such effects are both likely and predictable; and the critical third region where the occurrence of stable non-equilibrium processes may or may not occur, depending on material

properties and test conditions. It is this mid-range where sharp
differences of opinion exist, and where a preoccupation with valuing
and justifying one theoretical approach or one set of parametric
terms over another has clouded attempts to establish the usefulness
and generality of the phenomena under consideration.

What are the conditions for which concentration or pressure
levels will not exist in excess of the values dictated by thermodynamic stability? The most obvious case is that of a single phase
material in which there are no defects or heterogeneities capable
of strongly interacting with hydrogen so that local equilibrium conditions cannot be achieved in a realistic time period. Under these
conditions, even if hydrogen can be transported in large quantities
by mobile dislocations, subsequent annihilation with other dislocations or at traps will not lead to enrichment, because there will
be no kinetic barrier to attaining local equilibrium. The only
concentration fluctuations will be those dictated by the local
chemical potential at defects with either a positive or negative
binding energy for hydrogen. If molecular recombination does occur,
say at strain-induced defects, the resultant pressure must be dictated solely by the external fugacity and the size and surface tension of any gas filled voids. If we now extend this first region by
introducing a spectrum of traps of differing strengths and sizes,
such as a population of voids, dispersoids, or precipitates, again
there may be no local enrichment or pressurization resulting from hydrogen transport. So long as there is not a kinetic barrier to the
redistribution of any non-equilibrium concentration or pressure fluctuations due to dumping of hydrogen from their mobile sources, then
again thermodynamic equilibrium should be achieved, usually in a
time scale too short to exacerbate the severity of any hydrogen-
related degradation in properties.

There is the other extremal possibility, namely that kinetics
now completely control the ability of dissolved, trapped, or combined
hydrogen to locally redistribute in response to a driving force
associated with the system's departure from equilibrium. Such a
kinetically dominated situation should exist if traps are very strong,
say with binding energies in excess of 100 kJ/mol. Under these conditions these traps are effectively irreversible, in the sense that
for any realistic time of test the defects will not release their
hydrogen, and such hydrogen is not in local equilibrium with its surroundings.[10] These conditions are reasonably easy to achieve
practically; for example, TiC particles in iron have trap energies
near 100kJ/mol.[10,11] Such enrichment, however, does not have to
lead to pressurization, but in and of itself could promote premature
embrittlement.[1,4,8] However, pressurization is quite possible if
dissociation of the molecular gas and its subsequent leakage away
from the void is again kinetically controlled. For example, this
could be the case if the void traps formed by plastic defor-

mation, or directly by pressurization, have their boundaries poisoned by elements like P and S which can inhibit dissociation,[12] or if
oxygen is adsorbed on the void surfaces to provide a kinetic barrier.[13] Thus in this region, if hydrogen can be transported by
mobile dislocations, enrichment and pressurization at local traps
is not only possible but likely.

The intermediate region between these two extremes is the
one that is potentially the most interesting and the least clear.
It is the case where the strength of the traps are more modest, and
of varying size, and where there are potential variations in temperatures and strain rate, which can profoundly affect the effective
hydrogen diffusivity and the carrying capacity of mobile dislocations
for hydrogen. Under such conditions there is no definitive way to
predict whether local long-term deviations from equilibrium are
possible without being able to fully characterize all the afore-
mentioned parameters, and without having a very detailed knowledge
of the localized and often non-equilibrium reactions between hydrogen
and associated dislocations and traps. Tien et al.[1] have emphasized,
using their pioneering but somewhat simplified model of this behavior,
that very large changes can occur with fairly modest variations in
controlling and critical variables. For example, Figure 1 shows

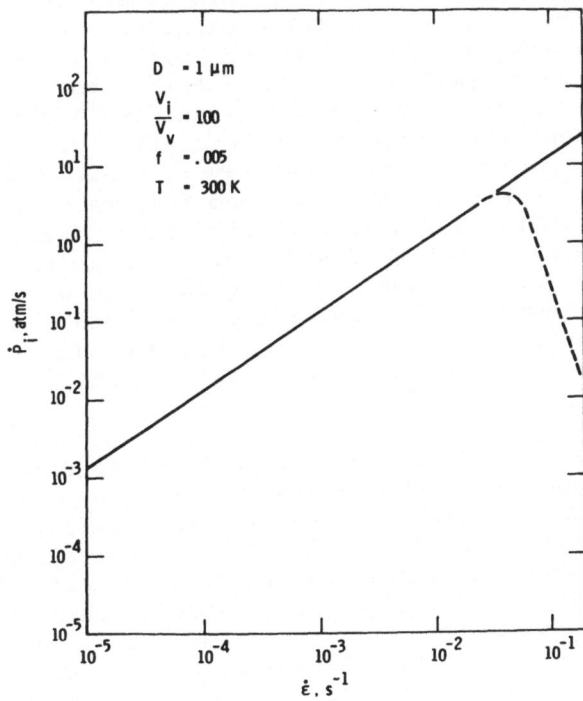

Fig. 1 – Dependence
of pressurization
rate $\dot{P}_i$ on strain
rate, for the values
shown of other para-
meters; dashed line
schematically indi-
cates the onset of
breakaway of hydro-
gen from dislocations;
$\dfrac{V_i}{V_v}$ is the ratio of
inclusion to void
volume; f is the
volume fraction, and
D the diameter of the
inclusions.

that for the set of parameters shown, while local enrichment is
likely, pressurization is only significant at high local strain rates
and for testing times in excess of 10^3s, a value commonly achieved
in tensile tests. The obvious conclusion is that one can always
devise a set of material and testing parameters to exploit the importance of such non-equilibrium processes in hydrogen embrittlement phenomena. Whether these conditions are met in practice depends on the alloy and its total environment, and this can only be
established by combining modelling with carefully designed experiments. In partial support of this, Hirth has considered the possibility of the occurrence of pressurization in a number of alloys of
different crystal structures and temperatures.[7] Using a model requiring an exact knowledge of the statistical nature of the
interaction between hydrogen,and dislocations and traps, and the
exact manner by which hydrogen leakage from traps occurs, he has
concluded that trap enrichment and pressurization are quite possible for nickel at room temperature and for iron at low temperature,
but not for iron at room temperature. While these predictions may
indeed be correct, we submit that the uncertainties inherent in the
model and in the choice of parametric values are as yet too great
for such predictions to be accepted in the absence of supporting
experimental data.

     To illustrate the importance of experimental data in serving
both to direct and test theory, we will cite three different sets
of results on different materials, which, to varying degrees of certainty, support specific hydrogen-dislocation-trap interactions.
The first of these are the results of Kurkela and Latanision[14] who
studied the effect of concurrent straining on the permeation of
hydrogen through membranes of nickel. They conclusively showed that
dislocations can transport hydrogen in nickel at room temperature,
and that the effective diffusivity increased linearly with increasing
strain rate, reaching values of as much as five orders of magnitude
greater than that found in unstrained nickel. The occurrence of a
critical strain rate for breakaway of the dislocation from its
associated hydrogen shows that great care must be taken when de-
vising experiments to search for the occurrence of this effect in
other alloy systems.

     Further support for dislocation transport in another fcc alloy,
and for associated hydrogen enrichment can be found in the results of Albrecht et al.[15] They studied the tensile response of a
high purity equi-axed grained 7075 aluminum alloy, which was tested
to failure either in the hydrogen uncharged condition, after cathodically charging with hydrogen, or after concurrent cathodic charging and
strain. This latter test, the straining electrode test (SET), consisted of charging while the specimen was being strained at a very
slow strain rate to a strain much smaller than the total strain to
failure. Subsequent testing was done in the absence of the hydrogen

charging solution, For these three cases, uncharged, statically
charged and SET, the associated reduction in areas at fracture were
43%, 26% and 3% respectively, vividly demonstrating how the SET condition can lead to severe hydrogen embrittlement of an aluminum
alloy. Associated with this embrittlement was a dramatic change in
fracture mode, as illustrated in Figure 2. Without hydrogen the
failure mode is the expected microvoid coalescence. The introduction
of hydrogen without concurrent strain results in some embrittlement,
and a narrow intergranular zone a few grains deep, a depth doubtless
controlled by the very slow diffusion rate of hydrogen in the aluminum lattice. The introduction of strain during hydrogen entry is
equivalent to the generation of mobile dislocations at the specimen
surface which in turn promotes hydrogen transport, significant embrittlement, and a much deeper intergranular zone. In fact, by carefully adjusting the strain rate and strain during SET, almost the
entire cross-section could be induced to fail intergranularly, with
an associated reduction of area of less than 1%. We can think of
no other self-consistent explanation for these results other than
hydrogen transport by dislocations leading to significant non-
equilibrium hydrogen enrichment at grain boundaries, and to both
an enhanced embrittlement and a change in fracture mode. Clearly
then, hydrogen transport and enrichment can occur in fcc metals and
alloys. Claims that it is not important in austenitic stainless
steels[16] must be viewed with some skepticism since the parameters
examined in that study were narrow in scope and the critical range
for the occurrence of such effects may not have been tested.

   What of bcc metals? There are some suggestive results for the
importance of hydrogen transport and its subsequent role in embrittlement, ranging from autoradiography studies of hydrogen's association
with dislocation,[17] to enhanced embrittlement of steel discs tested
in high pressure hydrogen.[18] While the critical experiments have yet
to be done, there is another set of results which are strongly suggestive of an important role of non-equilibrium pressurization. This
is the room temperature hydrogen embrittlement of spheroidized 1018
and 1080 steel, containing dissolved internal hydrogen. Such systems
can display large changes in reduction in area due to hydrogen, while
the fracture mode of microvoid coalescence is unchanged.[19] Using
data obtained from detailed quantitative metallography, Garber et al.[19]
concluded that hydrogen does not significantly affect the case of void
nucleation or void growth in its early stages. Hydrogen's effect is
only pronounced late in the strain history of the alloy, as shown in
Figure 4.[19]

   In an attempt to rationalize how hydrogen could so profoundly
affect the last stages of void growth, it was concluded that hydrogen
diffusion through the lattice into the microvoids could not adequately
explain the data. Transport calculations did show, however, that sufficient hydrogen could be available in microvoids to exert the required

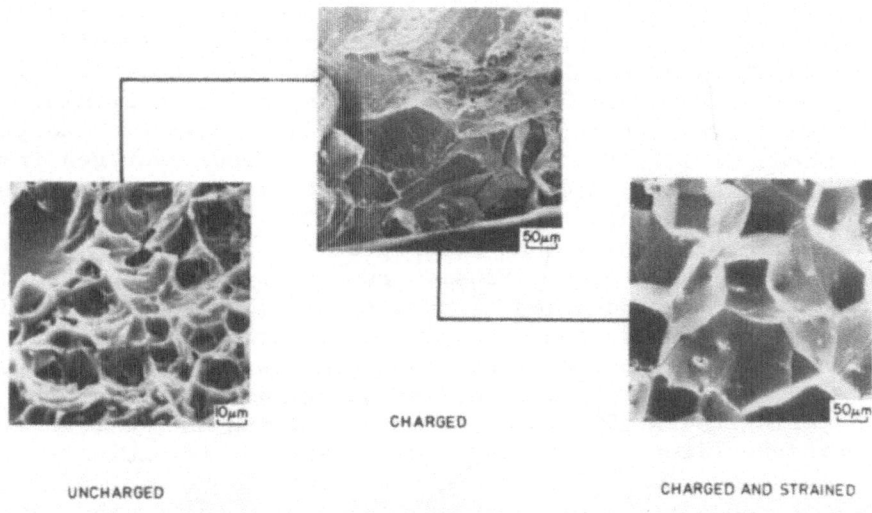

CHARGED

UNCHARGED   CHARGED AND STRAINED

HP-7075;UT

Fig. 2 - Fractographic differences between uncharged, statically charged, and SET in a high purity 7075 aluminum alloy.

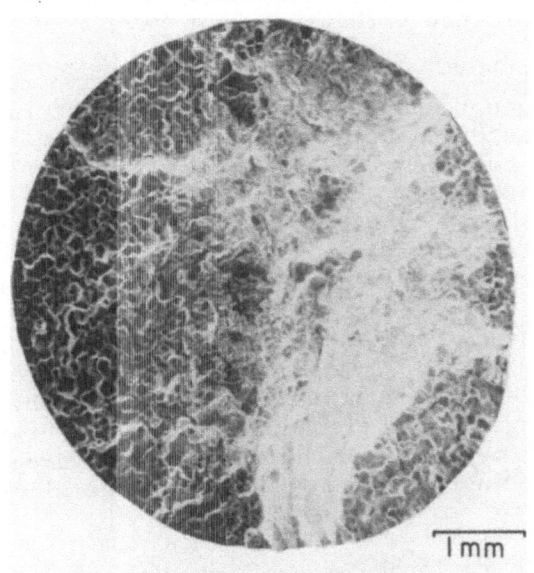

HP 7075;UT; STRAINING EL.

Fig. 3 - The extent of intergranular embrittlement possible using the SET.

pressure to assist growth during link-up, provided that it was transported by mobile dislocations.

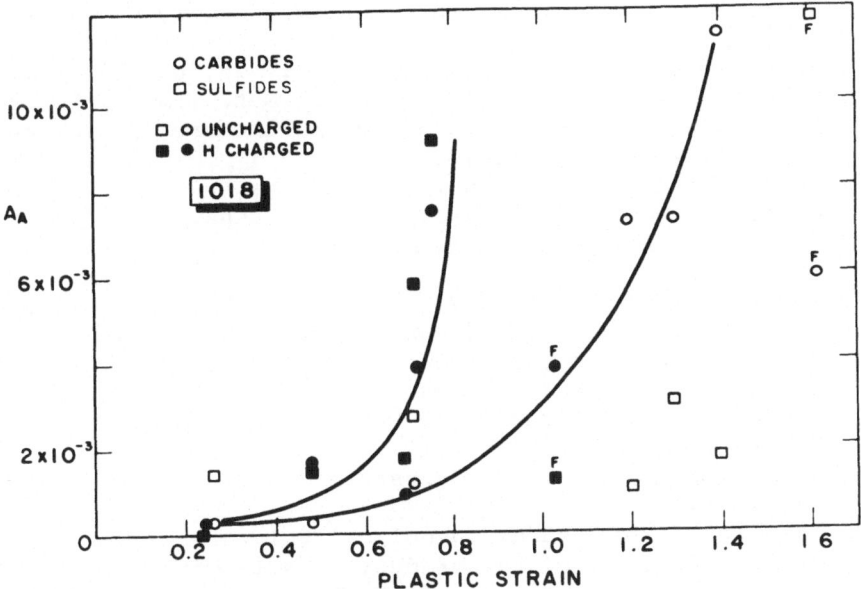

Figure 4 - Area fraction of microvoids A_A as a function of strain for charged and uncharged 1018 steel. Carbide initiated and sulfide initiated voids were separately measured; lines shown are for carbide data only.

In summary, we hope that we have clarified the situation on the role and importance of hydrogen transport, enrichment, and pressurization in hydrogen embrittlement phenomena, particularly under transient conditions. These effects are all possible, but do not necessarily occur in all systems for the reasons discussed. Theoretical attempts to discredit the possible participation of any or all of these processes in specific systems are as yet unconvincing, and do not reflect the spirit and specifics of a number of treatments of this problem.

ACKNOWLEDGMENTS

We wish to acknowledge the many important discussions we have had with Dr. John Tien, who has provided much of the insight on the importance of kinetics on subsequent hydrogen effects. Portions of this work have been supported by the Office of Naval Research, the Air Force Office of Scientific Research and the National Science Foundation.

REFERENCES

1. J. K. Tien, A. W. Thompson, I. M. Bernstein and R. J. Richards: Met. Trans., 1976, vol. 7A, pp 821-29.
2. H. H. Johnson and J. P. Hirth: Met. Trans., 1976, vol. 7A, pp 1543-48.
3. G. M. Pressouyre: Acta Met., 1980, vol. 28, pp 895-911.
4. G. M. Pressouyre and I. M. Bernstein: Met. Trans., in press.
5. J. P. Hirth: 1980 Institute of Metals Lecture, Met. Trans., 1980, vol. 11A, pp 861-90.
6. J. P. Hirth: this Conference Proceedings.
7. J. K. Tien: "Effect of Hydrogen on the Behavior of Materials," eds. A. W. Thompson and I. M. Bernstein, TMS-AIME, NY, 1976, pp 309-26.
8. J. K. Tien, S. V. Nair and R. R. Jensen: "Role of Hydrogen in the Behavior of Metals," eds. I. M. Bernstein and A. W. Thompson, TMS-AIME, Pittsburgh, PA, 1981, in press.
9. J. K. Tien, N. F. Panayotou and R. J. Richards: Proceedings of the Second International Congress on Hydrogen in Metals, Paris, France, 1977, pp 2B11.
10. G. M. Pressouyre and I. M. Bernstein: Met. Trans., 1978, vol 9A, pp 1571-80.
11. T. Asaoka: Thése Docteur-Ingenieur, Univ. Paris-Sud., 1976.
12. H. W. Pickering and M. ZamanZadeh: "Role of Hydrogen in the Behavior of Metals," eds. I. M. Bernstein and A. W. Thompson, TMS-AIME, Pittsburgh, PA, 1981, in press.
13. G. G. Hancock and H. H. Johnson: Trans. AIME, 1966, vol. 236, pp 513-19.
14. M. Kurkela and R.M. Latanision: Scripta. Met. 1979, vol. 13 pp 927-32.
15. J. Albrecht, I. M. Bernstein and A. W. Thompson: Met. Trans., in press.
16. A. J. West and M. R. Louthan, Jr.: Met. Trans., 1979, vol 10A, pp 1675-81.
17. J. Chene and M. Aucouturier: "Role of Hydrogen in the Behavior of Metals," eds. I. M. Bernstein and A. W. Thompson, TMS-AIME, Pittsburgh, PA, 1981, in press.
18. G. M. Pressouyre, J. P. Fidelle and R. A. Laurent: ibid.
19. R. Garber, I. M. Bernstein and A. W. Thompson: Met. Trans., 1981, vol. 12A, pp 225-34.

DISCUSSION

Comment by J.F. Knott:

I would like to raise two points on the 7075 experiments: (a) How can one clearly deduce that the effect of plastic strain is transport of hydrogen by dislocations? It might, for example, be postulated that, for the case of cathodic charging in the absence of strain, the surface was covered by a film which permitted only slow ingress of hydrogen, whereas this film was continuously ruptured

during plastic straining, allowing general access of hydrogen.   (b)
On a related point, what was the heat-treatment of the alloy, and how
might its precipitate structure affect dislocation structure and
hydrogen ingress/transport?

Reply:

With regard to your first point, it seems unlikely that surface
films were important since we know that lattice diffusivity is extremely slow and therefore hydrogen cannot penetrate the Al deeply
even if the surface film is absent. To further test this, we did
a series of tests where we varied the strain and strain rate during
cathodic polarization and found that the depth of intergranular
fracture from the surface scaled with these parameters, consistent
with dislocation transport. These alloys were heat treated to an
underaged temper (24 hours at 100°C). This produced a microstructure
of predominately 20Å coherent precipitates, most likely G.P. zones.
These were uniformly distributed throughout the matrix with no
tendency for preferential grain boundary precipitate free zones.

DISLOCATION TRANSPORT OF HYDROGEN IN STEEL

B. J. Berkowitz and F. H. Heubaum

Exxon Research and Engineering Company
Linden
New Jersey 07036

INTRODUCTION

It has been proposed that the transport of hydrogen by dislocations is an important factor in the embrittlement of alloys. For example, it has been shown that plastic deformation increases the absorption of hydrogen from the environment due to the motion of dislocations during plastic deformation.[1] In addition, there are experiments which show increased solubility and decreased diffusivity of hydrogen for samples that have been cold worked prior to permeation experiments.[2,3] These results indicate that there is an interaction between hydrogen and dislocations and therefore should be considered when evaluating the mechanism of hydrogen embrittlement. Considerable controversy has arisen with respect to the exact role which dislocation transport plays in the embrittlement process. Some workers have suggested that the dislocations transport hydrogen to fracture initiation sites such as inclusions, microvoids or grain boundaries where eventually sufficient pressures develop to promote embrittlement.[4] Others have proposed that transport is a necessary but not sufficient condition for embrittlement in which kinetic factors limit the hydrogen supersaturation and pressures at discontinuities.[5]

The key to understanding hydrogen embrittlement lies in the ability to monitor dislocation/hydrogen interaction during the embrittlement process itself rather than after cold work. This has been successfully done in the case of nickel[6] and a nickel-chromium alloy.[7] Both results show extraordinarily large increases in effective diffusivity as a result of plastic deformation during hydrogen charging. In fact, the diffusivity appears to increase by approximately 5 orders of magnitude from $\sim 10^{-10}$ cm^2/sec to $\sim 10^{-5}$

cm^2/sec. The diffusivity in iron and steel is considerably higher than in nickel base alloys (typically in the range of 10^{-5} to 10^{-6} cm^2/sec). It is interesting to note the similarities in embrittlement behavior of iron and nickel base alloys in spite of the differences in their diffusivity. Therefore, it is our purpose to measure the effect of dynamic plasticity on the hydrogen transport in iron base alloys to compare that with nickel.

EXPERIMENTAL PROCEDURE

Hydrogen permeation measurements were carried out in a modified version of the electrochemical cell originally developed by Devanathan and Starchurski.[8] The cell was mounted between the crosshead of an Instron tensile testing machine allowing determination of hydrogen permeation while simultaneously straining the metal membrane. The strain rate chosen for this preliminary series of experiments was 1.67×10^{-5} sec^{-1}.

Specimens of 4130 steel were machined from cold rolled 0.2 mm thick strips (1.54 cm x 15.24 cm). Gauge length and width were 2.54 cm and 1.90 cm respectively. These specimens were austenitized in vacuum ($\sim 10^{-6}$ torr) at 875°C for 2 hours and water quenched. After mechanical polishing they were tempered at 650°C for one hour and nitrogen cooled. The samples were repolished using alumina powders to remove any temper formed oxide films which have been shown to artificially enhance hydrogen permeation.[9] One side of each sample (anodic side) was electroplated with a thin film of palladium. After insertion into the electrochemical cell, the output side containing 0.1N NaOH was potentiostated anodically at −300 mv (SCE) while the input (cathodic) side, containing 0.5% acetic acid + 5% NaCl solution was galvanostatically controlled during hydrogen charging. Hydrogen permeation currents and load vs time curves were simultaneously recorded during straining. Tests were conducted at room temperature.

RESULTS

Figure 1 shows the effect of elastic and plastic strain on hydrogen transport in 4130 steel. Prior to any deformation, a steady state permeation current was achieved through the membrane. As the stress is first applied to the sample, one notes that the permeation current increases with elastic strain which is consistent with theory.[10]

At the yield point of this steel, in contrast to nickel, the hydrogen permeation decreases. In fact, throughout the plastic region of the stress strain curve, there appears to be an inverse relationship between the amount of hydrogen that permeates and the flow stress. A number of stress strain curves and corresponding

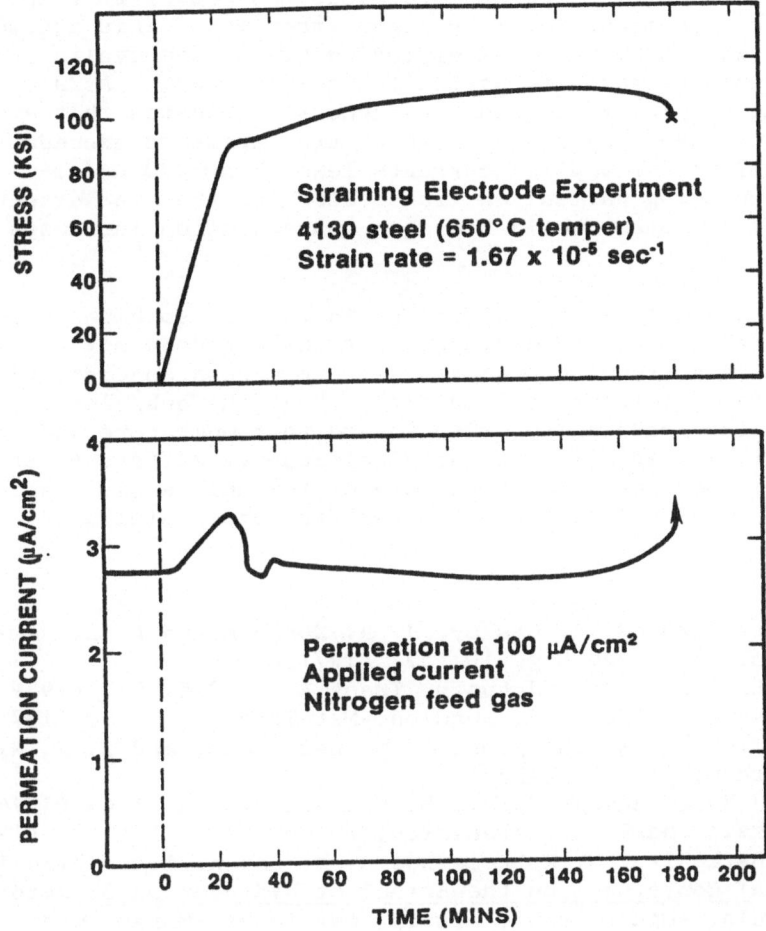

Figure 1.   Hydrogen Permeation During Simultaneous
            Mechanical Deformation

permeation transients for various input currents were recorded and
the general behavior was similar in all cases.

DISCUSSION

    The substantial drop in hydrogen permeation accompanying yield-
ing suggests that the increase in dislocation density (and possible

microcracks at matrix-carbide interfaces) is acting to trap hydrogen and therefore decrease total hydrogen transport. This argument is consistent with previous work employing static and cyclic stresses.[11] The continuous decrease in permeation during plastic straining (until necking or plastic instability occurs) indicates that even at this low crosshead speed, the rate of trap creation exceeds the rate of trap filling for applied currents less than $\sim$200 $\mu A/cm^2$. Simple calculations using permeation transients show that the effective diffusivity decreases while hydrogen concentration increases during straining.

Since the diffusion coefficient in iron is so high to begin with compared to nickel, dislocation transport does not significantly change the rate at which hydrogen gets to stress concentration and other points of fracture nucleation. Thus, the behavior of iron base and nickel base alloys with regard to temperature and strain rate are similar in spite of the significantly different lattice transport properties. It is because of the unique dislocation transports in nickel that the two tend to behave similarly.

REFERENCES

1. M. R. Louthan, G. R. Caskey, J. A. Donovan and D. E. Rawl, Materials Sci. and Eng. 10, 357 (1972).
2. M. L. Hill and E. W. Johnson, Trans AIME, 215, 717 (1959).
3. A. J. Kumnick and H. H. Johnson, Met Trans, 5, 1199 (1974).
4. J. K. Tien, A. W. Thompson, I. M. Bernstein, and R. J. Richards, Met. Trans, 7A, 821 (1976).
5. H. H. Johnson and J. P. Hirth, Met. Trans. 7A, 1543 (1976).
6. M. Kurkela and R. M. Latanision, Scripta Met, 13, 927 (1979).
7. B. J. Berkowitz, M. Kurkela and R. M. Latanision, Third International Conference on the Effect of Hydrogen on Behavior of Materials, August 1980, Jackson Lake Lodge, Moran, Wyoming, A. W. Thompson and I. M. Berstein, eds., New York, Metallurgical Society of AIME (c. 1981).
8. M. A. V. Devanathan and Z. Stachurski, Proc. Roy. Soc., A270, 90 (1962).
9. B. J. Berkowitz and H. H. Horowitz, submitted for publication JECS (c. 1981).
10. W. Beck, J. O'M. Bockris, J. McBreen, and L. Nanis, Proc. Roy. Soc., A290, 220 (1966).
11. R. F. Blundy, R. Royce, P, Poole, and L. L. Shrier, International Corrosion Conference, Stress Corrosion Cracking and Hydrogen Embrittlement of Iron Alloys, R. W. Staehle, J. Hochmann, R. D. McEnright and J. E. Slater, eds, Houston, Texas NACE-5, 636 (1977).

DISCUSSION

Comment by J.O'M. Bockris:

I am concerned with the reliability of permeation results obtained through membranes when the input side was undefined in respect to oxide film. These may be significant time effects on the permeation. What is known about the oxide on the input side? What about the output side? Was a Pd membrane used?

Reply:

It is certainly true that the presence of an oxide on the input side of the membrane could have a significant effect on the permeation results. An extensive investigation of the effect of oxides was carried out and will be reported shortly in the Journal of the Electrochemical Society. Briefly, it was found that the oxide acts as a catalytic poison for the hydrogen recombination reaction and therefore facilitates entry of hydrogen atoms into the steel. With the oxide present, as much as 70% of the hydrogen arriving at the surface is permeated. (Compared to 0.1% to 1% on clean iron surfaces.)

Care was therefore taken during these straining electrode experiments to ensure that the oxide was not playing a role. Prior to charging, the samples were polished to remove the gross oxides formed during tempering. Since oxides are not thermodynamically stable in the electrolyte used, the very thin (50-100 Å) air formed Fe_2O_3 is quickly reduced at cathodic potentials. Auger spectroscopy was used to confirm these results.

With regard to the output side, a Pd plating was applied to reduce Fe dissolution, to keep the background current low. Tests run with only dry N_2 on the input side produced no detectable current at the output side indicating that straining did not enhance dissolution on the output side.

Comment by A.W. Thompson:

I think we need an experiment on pure iron rather than on 4130 steel to answer whether dislocation transport is important in iron. One would also get decreased permeation in a steel like 4130 by cold-work followed by static permeation, so it is tempting to conclude that changes in trap density with strain are dominating the permeation response.

Reply:

Yes. This was the point stated in the presentation and will be covered more fully in the manuscript.

HYDROGEN ASSISTED CRACK GROWTH IN HIGH PURITY LOW PRESSURE HYDROGEN GAS

Horst Vehoff, Wolfgang Rothe, and P. Neumann

Max-Planck-Institut für Eisenforschung GmbH

D - 4000 Düsseldorf, Western Germany

INTRODUCTION

At low partial pressures of hydrogen (less than 0.1 MPa), hydrogen interacts with the metal and lowers the cohesive strength of the lattice (Oriani and Josephic, 1977). The decohesion theory postulates that regions exist near the crack tip with hydrogen concentrations, which are several orders of magnitude larger than the lattice concentration of hydrogen. The microstructure strongly influences the embrittlement process and the crack growth kinetics. (Kumnick and Johnson, 1980). Thus it can be expected, that the experimental study of the mechanism of gaseous hydrogen embrittlement is facilitated by studying the process at the crack tip in its simplest form under the most simple boundary conditions. For these reasons FeSi-single crystals are used, since the micro-processes for ductile and brittle fracture are well known in this alloy (Vehoff and Neumann, 1979,1980). Tests are carried out at different temperatures, hydrogen pressures and plastic crack tip opening rates in order to investigate the influence of the hydrogen concentration at the crack tip on the brittleness of the fracture process (Wei and co-workers, 1980).

EXPERIMENTAL PROCEDURE

Tensile tests with pre-cracked and properly oriented Fe-2.6%-Si single crystals are carried out in high purity low pressure hydrogen gas under plastic strain control. The ratio

$$a_n = 2\Delta a/\Delta\delta = ctg(\alpha/2)$$

(Δa is the crack growth increment, $\Delta \delta$ is the plastic crack tip opening increment, α is the crack tip angle) is a measure of the brittleness of the fracture process. $a_n = \sqrt{2}$ characterizes perfectly ductile rupture by alternating slip for the orientation used. $a_n \to \infty (\alpha \to 0)$ characterizes perfectly brittle fracture. All tests in the hydrogen environment were carried out at small enough crack tip opening rates, $\dot{\delta}$, which would yield perfectly ductile fracture in vacuum (Vehoff and Neumann, 1980). Therefore all embrittling effects characterized by $a_n > a_n^{duct} = \sqrt{2}$ are due to the hydrogen environment. For more details see Vehoff and co-workers (1981).

RESULTS AND DISCUSSION

For pre-cracked single crystals a high concentration of hydrogen can only be expected at the crack tip surface or nearby dislocation cores (Hirth and Carnahan, 1978). If the brittleness of fracture is controlled by the fractional surface coverage, θ, ($0 \leqq \theta \leqq 1$) of hydrogen at the crack tip, the normalized crack growth rate a_n, as a local measure of embrittlement, must be directly correlated with θ, $a_n = f(\theta)$. The simplest possible relationship concerning $f(\theta)$ is a linear relationship.

$$a_n - a_n^{duct} = \beta(\theta - \theta_{thr}) \tag{1}$$

where θ_{thr} is the critical surface coverage for which the fracture mode changes from ductile to brittle. For $a_n \gg a_n^{duct}$ and neglecting θ_{thr} equation (1) reduces to

$$a_n \approx \beta \theta \tag{2}$$

This implies that for crack growth rates for which the hydrogen gas is in local equilibrium with θ and for which the stresses are high enough to initiate brittle fracture a_n must be independent of $\dot{\delta}$ and a function of T and P_{H_2} only. Therefore the $\dot{\delta}$-dependence of a_n is examined for low growth rates and different hydrogen pressures. Fig. 1 shows the results. For the rates used a_n is found to be independent of $\dot{\delta}$ but strongly dependent on P_{H_2}. Therefore it is assumed that for these rates θ is in local equilibrium with the hydrogen gas at the crack tip.

Hydrogen chemisorbs and desorps dissociatively on (100)-Fe surfaces (Boszo and co-workers). For thermal equilibrium θ can be described by a Langmuir isotherm,

$$P_{H_2} = f(T) \cdot \theta^2/(1-\theta)^2 \tag{3}$$

with $f(T) = k_d/\sigma \cdot (2\pi m k T)^{1/2} \cdot \exp(U/kT)$,

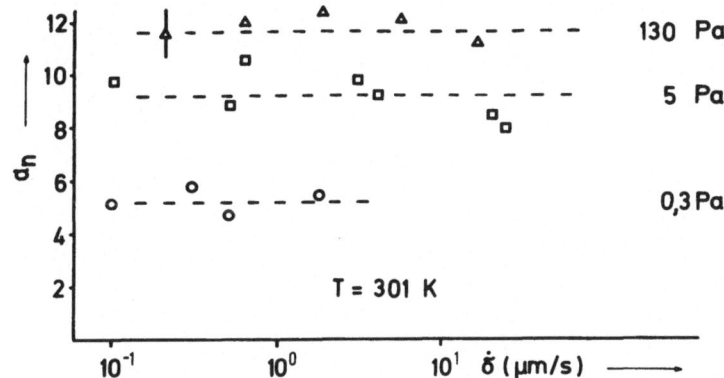

Fig. 1. The $\dot{\delta}$ dependence of a_n for three different
hydrogen pressures.

where k_d is the evaporation coefficient, σ the condensation coefficient and $U = E_a - E_d$ the heat of desorption (Clark, 1970). For
local equilibrium, combining equation (2) and (3) yields

$$1/\sqrt{P_{H_2}} = (1/\sqrt{f(T)}) \cdot (\beta/a_n - 1) \tag{4}$$

Plotting $1/\sqrt{P_{H_2}}$ versus $1/a_n$ for constant temperatures must yield
straight lines. Fig. 2 shows the results. The postulated relationship is obeyed quite well. Hence the pressure and temperature
dependence of a_n is of the same functional form as the pressure
and temperature dependence of the hydrogen surface coverage, θ, on
(100)Fe-surfaces. The 6 straight lines intersect the abscissa at
approximately the same value. This proves that a_n is proportional
to θ with the same proportional factor β at all temperatures examined. (Eq. (4)).

    If the isosteric heat of desorption

$$U = (\partial \ln P_{H_2} / \partial(1/kT))_{\theta = \text{const}} \tag{5}$$

of a (100)FeSi-surface is constant and the proportionality between
θ and a_n (equation (2)) is valid, a plot of $\log(P_{H_2})$ against $1/T$
for constant a_n must give a straight line. The slope of this line
must agree with the heat of desorption from a FeSi-(100) surface.

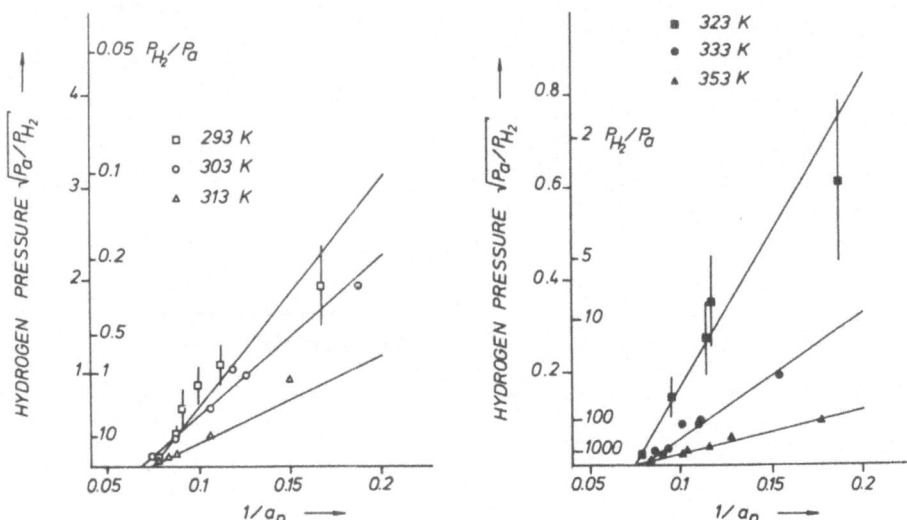

Fig. 2. $1/\sqrt{P_{H_2}}$ versus $1/a_n$ for different temperature.
The straight lines are fitted from equation (6).

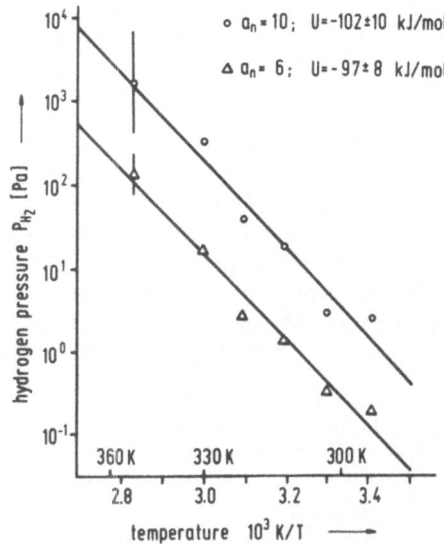

Fig. 3. The logarithms of the hydrogen pressures $\log(P_{H_2})$ for
which a constant a_n is obtained as a function of the
invers temperature $1/T$.

Fig. 3 shows experimental results for two different a_n values. The data are obtained from the fitted lines in Fig. 2. The slopes of the lines in Fig. 3 give an activation energy U = -102±10 kJ/mol, which agrees with the desorption energy of hydrogen from Fe(100)-surfaces $E_d(\beta_2)$ = 100 kJ/mol, measured by Boszos and co-workers (1978) from thermal desorption spectra. These results favour the view that the embrittlement of FeSi characterized quantitatively by an increasing a_n is controlled by the surfaces coverage of hydrogen at the crack tip.

SUMMARY

1. Hydrogen assisted crack growth is measured in single crystals down to hydrogen pressures of 10^{-2} Pa.

2. From the pressure and temperature dependence of a_n it is concludet that the embrittlement is controlled by the surface coverage of hydrogen at the crack tip.

ACKNOWLEDGEMENTS

The authors are grateful for financial support from the Deutsche Forschungsgemeinschaft (Ne 193/7).

REFERENCES

Boszo, F., G. Ertl, M. Grunze, and M. Weiss (1977). Appl.Surf.Sci., 1:103.
Clark, A. (1970). in: The Theory of Adsorption and Catalysis, Academic Press, New York, Chap. 9:209.
Hirth, J.P., and B. Carnahan (1978). Acta Metall., 26:1795.
Kumnick, A.J., and H.H. Johnson (1980). Acta Metall.,28:33.
Oriani, R.., and P.H. Josephic (1977). Acta Metall.,25:979.
Vehoff, H., and P. Neumann (1979). Acta Metall., 27:915.
Vehoff, H., and P. Neumann (1980). Acta Metall., 28:265.
Vehoff, H., W. Rothe, and P. Neumann (1981). Fracture Research, Vol. 1, ICF5, 265.
Wei, R.P., P.S. Pao, R.G. Hart, T.W. Weir and G.W. Simmons (1980). Met.Trans.A., Vol. 11A:151.

DISCUSSION

Comment by R.H. Jones:

What is the thickness of your samples? Are your samples loaded in a plane stress or plane strain mode? Is there any evidence for dislocations in the direction perpendicular to the applied load and in the direction of the fracture as Birnbaum showed in nickel samples fractured in the electron microscope?

Reply:

The samples have a 6x6 mm cross-section and are diagonally
notched. They are loaded in tension. The shape of the specimen
together with the orientation of the crystal assure a straight crack
front from surface to surface. Grown in dislocations are, of course,
everywhere, but there is no evidence of slip lines in the SEM directly
ahead of the crack tip. Averaging over distances of about 0.1μm
(resolution of SEM) the plane of fracture is the bisecting plane of
the two active slip planes, thus being ~45° away from either slip
plane. The difference in comparison with Birnbaum's results are
most likely due to the differences expected to exist in bulk
specimens as compared to foils.

EFFECTS OF CRACK FLANK OXIDE DEBRIS AND FRACTURE SURFACE ROUGHNESS ON NEAR-THRESHOLD CORROSION FATIGUE

R. O. Ritchie* and S. Suresh*

Department of Mechanical Engineering
Massachusetts Institute of Technology
Cambridge, MA   02139, U.S.A.

INTRODUCTION

Over the years, many models have been proposed for the environmental contribution to cracking associated with sub-critical flaw growth by corrosion fatigue and stress corrosion. In general, these models are based on mechanisms involving hydrogen embrittlement and active path corrosion (i.e. metal dissolution) processes.[1] At ultralow growth rates in fatigue (<10^{-6} mm/cycle), however, approaching the cyclic stress intensity threshold ΔK_O for no detectable growth, recent studies in steels have revealed behavior totally inconsistent with such mechanisms.[2,3] For example, near-threshold growth rates in ultrahigh strength steels are found to be *decelerated* in hydrogen gas compared to air,[4] whereas in lower strength steels growth rates are *accelerated* in helium gas and *decelerated* in water, again compared to air.[5] To provide alternative descriptions of environmentally-affected fatigue crack growth at such near-threshold levels, new models have recently been proposed based on the role of oxide deposits, formed within the crack, in promoting crack closure and crack tip blunting in moist, as opposed to dry, environments.[2,3,5] Where thicknesses of the crack flank corrosion deposits are comparable with crack tip opening displacements, e.g. at near-threshold levels, such deposits have been shown to markedly affect crack growth behavior, consistent with observed effects of load ratio, strength level, and environment.[5]

*Currently with the Department of Materials Science and Mineral Engineering, and Lawrence Berkeley Laboratory, University of California, Berkeley, CA   94720, U.S.A.

The object of the present note is to briefly review the concept of oxide-induced crack closure, and to examine the implications of this and similar models to environmentally-affected near-
threshold fatigue behavior in lower strength steels (yield
strengths below 800 MPa).

OXIDE-INDUCED CRACK CLOSURE

The occurrence of enlarged and visible oxide deposits on
near-threshold fatigue surfaces (Fig. 1) has been known for some
time, although its significance was not totally appreciated.[6,7]
Measurements of the surface layers using ESCA, Auger and SIMS
analyses   in $2\frac{1}{4}$Cr-1Mo and 316 stainless steels, however, revealed maximum oxide thicknesses of between three to twenty times
larger than background levels from oxide formed naturally on fresh-
ly-bared samples exposed to the same environment for the same
length of time.[5,8] In moist air environments, the excess oxide
thickness was found to be inversely related to the crack growth
rate, to reach a maximum close to ΔK_o , and to be independent of
the time from crack initiation to final failure (Fig. 2). Deposits
were much smaller in dry environments, such as hydrogen and helium
gases, and were severely reduced to thicknesses comparable with
naturally-formed oxide at high load ratios (R = $K_{min}/K_{max} \gtrsim 0.5$).

The origin of these enlarged oxide deposits at low stress
intensities arises principally from fracture surface abrasion. At
low load ratios as a result of residual plastic deformation left
in the wake of a growing fatigue crack, some closure of the crack
surfaces can occur at positive loads.[9] In oxidizing environments,
where crack tip displacements are small, such "plasticity-induced"
closure can enhance oxide formation within the crack by repeated
breaking and compacting of the oxide. This "fretting-oxidation"[8]
is further aided by Mode II crack tip displacements characteristic
of near-threshold growth.[10] At high load ratios (R $\gtrsim$ 0.5) where
plasticity-induced closure, Mode II displacements and hence fretting oxidation are significantly reduced, no such enlarged oxide
layers are generated. Similarly, oxide deposits are also reduced
in dry atmospheres such as gaseous hydrogen (Fig. 2).

The significance of this oxide debris is that where its size
is comparable with cyclic crack tip displacements, its presence
within the crack will lead to earlier contact between crack surfaces, thereby effectively raising K_{min} and thus reducing the
effective ΔK at the crack tip (Fig. 3). Where excess oxide debris is limited, e.g. in dry environments and at high load ratios,
such oxide-induced crack closure will be much less significant.

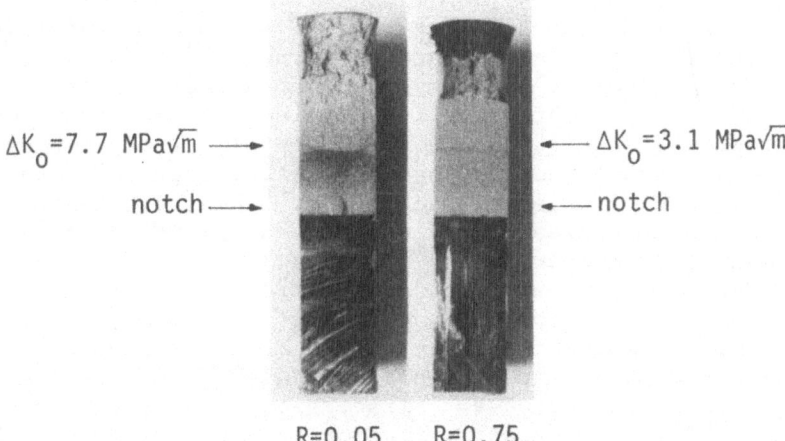

Fig. 1. Bands of corrosion deposits on near-threshold fatigue
fracture surfaces tested in moist air at R = 0.05 and 0.75.

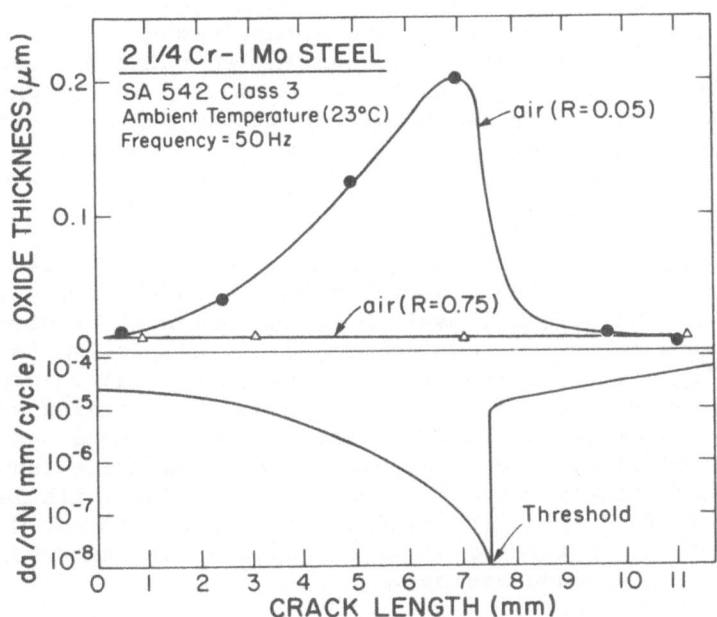

Fig. 2. Variation of crack flank oxide thickness with crack length
and crack growth rate for 2¼Cr-1Mo steel in air.

FATIGUE CRACK CLOSURE AND ENVIRONMENTAL EFFECTS

In many steels, two distinct regions of environmentally-influenced growth rate behavior are often seen,[11] as shown in Fig. 4 for the effect of gaseous hydrogen on 2¼Cr-1Mo.steel. Above typically 10^{-5} mm/cycle at lower frequencies (<5 Hz), hydrogen-assisted growth, generally with a predominately intergranular fracture mode, is observed, and has been attributed to"conventional" hydrogen embrittlement mechanisms.[11] However, below 10^{-6} mm/cycle as the threshold is approached, growth rates are two orders of magnitude higher in dry hydrogen (at 50 Hz) compared to moist air without any apparent change in fracture mechanism. Furthermore, *this acceleration is seen only at low load ratios;* no effect of hydrogen is seen at R = 0.75. Mechanisms of such hydrogen-assisted growth at near-threshold levels can be traced to oxide-induced closure concepts. First, growth rates are expected to be reduced in the moist air, since oxide deposits are thicker (see Fig. 2), thereby raising closure loads and reducing the effective alternating stress intensity at the crack tip. Second, no effect is expected at R = 0.75, since closure effects (and enhanced oxide formation) are largely absent at such high load ratios. Third, since growth rates are "accelerated" in hydrogen because of limited oxide formation, any dry environment should behave similarly. In Fig. 5, near-threshold growth rates at R = 0.05 (50 Hz) in 2¼Cr-1Mo steel are shown to be *accelerated in dry gaseous helium* compared to moist air (no effect was seen at R = 0.75). Moreover, tests in wet hydrogen and distilled water show behavior similar to moist air. Thus, as traces of oxygen in hydrogen gas can reduce susceptibility to hydrogen-assisted cracking arising from hydrogen embrittlement, traces of moisture can reduce susceptibility where the effect arises from oxide-induced crack closure.

Estimates of the extent of closure, derived from ultrasonic measurements, confirm that closure loads are indeed lower in dry environments. Furthermore, maximum excess oxide thicknesses are comparable with pulsating (cyclic) crack tip displacements at the threshold (Fig. 6). Despite experimental difficulties in precise measurements, this result is clearly physically appealing since it infers that the crack cannot propagate at the threshold as it is "wedged-closed" with oxide. Further, since plasticity-induced closure and fretting/abrasion effects are minimized with increasing strength level (and load ratio), the model is consistent with observations[7] of lower threshold values measured in higher strength steels (and at high load ratios).

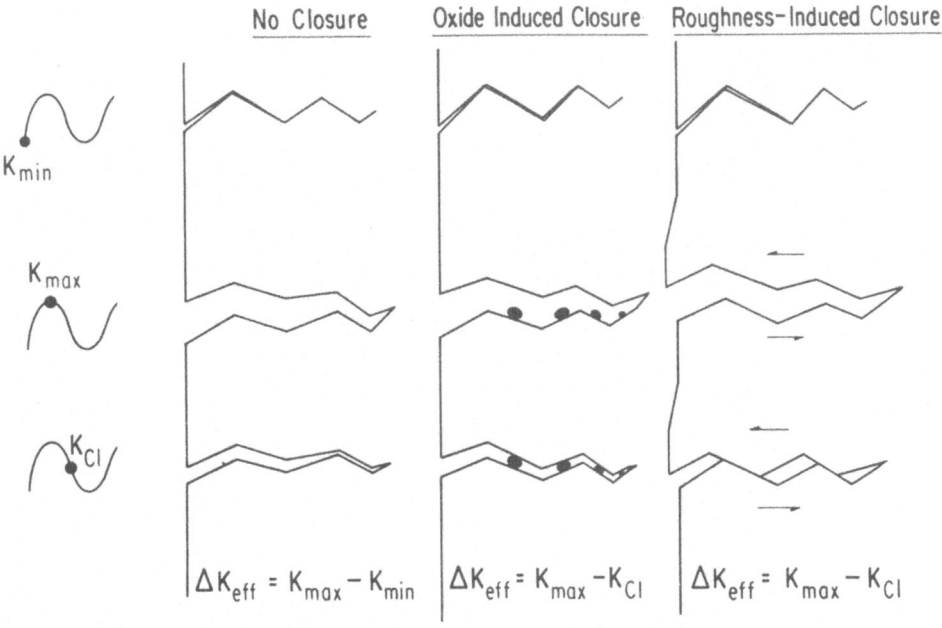

Fig. 3.   Possible mechanisms of near-threshold crack closure.

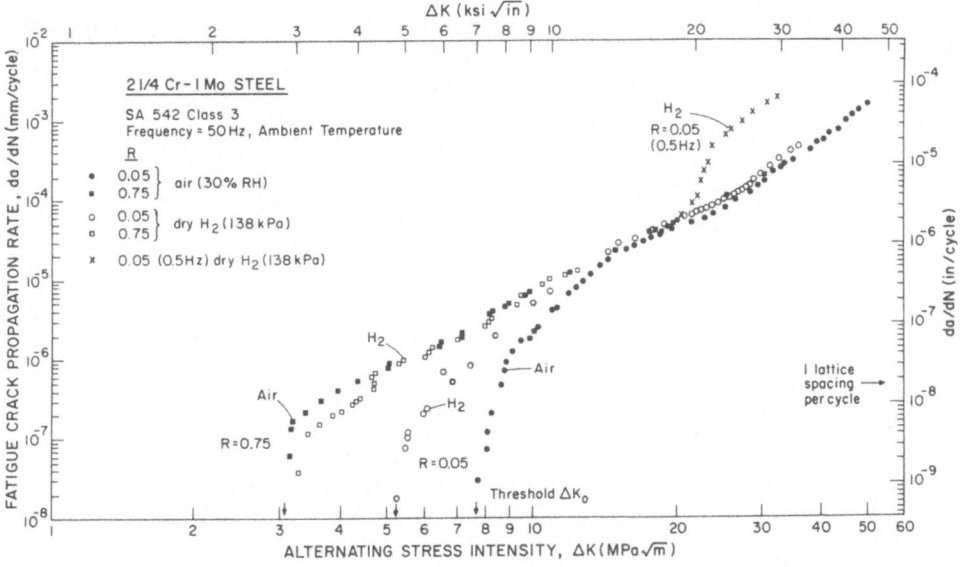

Fig. 4.   Fatigue crack propagation in bainitic 2¼Cr-1Mo steel
          tested in moist air and dry hydrogen gas, showing two re-
          gimes of hydrogen-assisted growth.

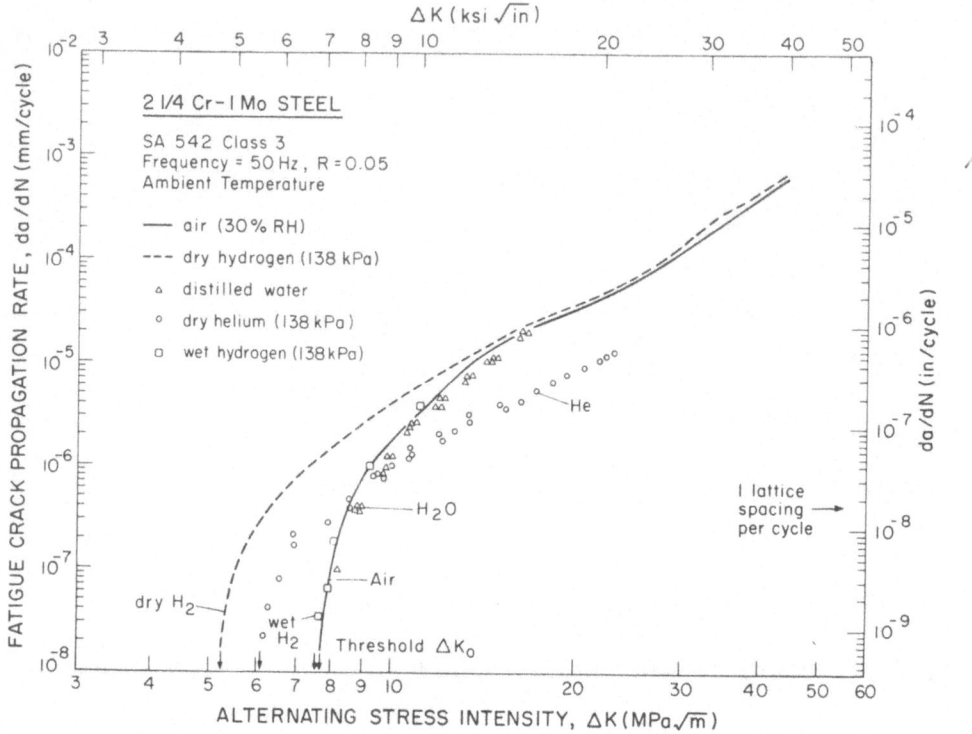

Fig. 5.   Fatigue crack propagation in bainitic 2¼Cr-1Mo steel
           tested at R = 0.05 (50 Hz) showing effect of environment.

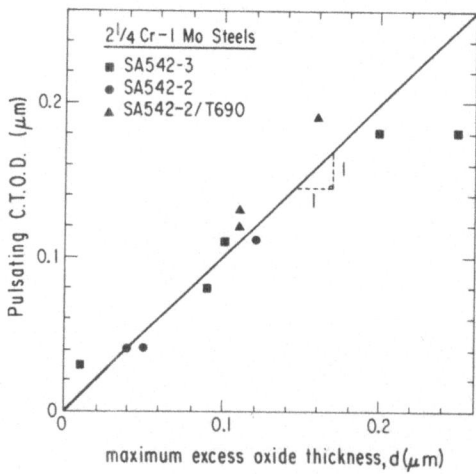

Fig. 6.   Showing correspondence of maximum exess oxide thickness
          with pulsating (cyclic) crack tip opening displacement
          at the threshold.

FATIGUE CRACK CLOSURE AND MICROSTRUCTURAL EFFECTS

Apart from the role of foreign debris inside the crack, additional sources of closure can result where the size-scale of the fracture surface roughness or morphology is of the order of cyclic crack tip displacements.[12] Given the small crack tip displacements and the occurrence of significant Mode II displacements at near-threshold levels, a more irregular fracture surface morphology implies enhanced closure loads by wedging open the crack at discrete contact points along crack faces (Fig. 3). This mechanism of closure is useful in rationalizing certain microstructural effects which are specific to the near-threshold regime.[5,13,14]

Coarsening grain size or decreasing cyclic strength, for example, are known to markedly improve near-threshold crack growth resistance.[7] Similar to the influence of dry environment, such beneficial effects are only seen at low load ratios, again suggesting a dominant role of closure. It seems very feasible to suggest that coarse microstructures may simply result in rougher fracture surfaces, thus promoting "roughness-induced" closure at low load ratios. Such notions are consistent with the metallographic studies of Walker and Beevers[13], where near-threshold cracks growing in coarse α-titanium microstructures displayed a large "out of registry" between mating fracture surfaces.

SUMMARY AND CONCLUSIONS

It is apparent from this work that the phenomenon of fatigue crack closure is particularly significant in steels at near-threshold growth rates. Mechanisms of this closure, however, appear somewhat different to the plasticity-induced closure first envisioned by Elber.[9] At low stress intensities, sources of closure originate from the fact that cyclic crack tip displacements approach the size-scale of the fracture surface morphology and crack flank corrosion debris. It thus follows that factors which tend to increase these size-scales, such as rougher fracture surfaces from coarse microstructures or thicker oxide layers from moist environments, will tend to decrease growth rates and in general raise the near-threshold crack growth resistance. Unfortunately, such beneficial effects of microstructure and environment cannot be maintained for applications involving high mean stresses, since at such high load ratios, the effect of any mechanism of crack closure will be minimal.

ACKNOWLEDGEMENT

The work was supported by the Department of Energy, under a contract with the Office of Basic Energy Sciences.

REFERENCES

1. A. W. Thompson and I. M. Bernstein, The Role of Metallurgical Variables in Hydrogen-Assisted Environmental Fracture, in "Advances in Corrosion Science and Technology," M. G. Fontana and R. W. Staehle, ed., Plenum Press, New York, 7:53 (1979).

2. R. O. Ritchie, S. Suresh, and C. M. Moss, Near-Threshold Fatigue Crack Growth in 2¼Cr-1Mo Pressure Vessel Steel in Air and Hydrogen, J. Eng. Mat. Tech., Trans. ASME Ser. H, 102:293 (1980).

3. A. T. Stewart, The Influence of Environment and Stress Ratio on Fatigue Crack Growth at Near Threshold Stress Intensities in Low-Alloy Steels, Eng. Fract. Mech., 13:463 (1980).

4. J. Toplosky and R. O. Ritchie, On the Influence of Gaseous Hydrogen in Decelerating Fatigue Crack Growth Rates in Ultrahigh Strength Steels, Scripta Met., 15 (1981).

5. S. Suresh, G. F. Zamiski, and R. O. Ritchie, Oxide-Induced Crack Closure: An Explanation for Near-Threshold Corrosion Fatigue Crack Growth Behavior, Met. Trans. A, 12A (1981).

6. R. J. Cooke and C. J. Beevers, Slow Fatigue Crack Propagation in Pearlitic Steels, Mater. Sci. Eng., 13:201 (1974).

7. R. O. Ritchie, Near-Threshold Fatigue Crack Propagation in Steels, Intl. Metals Reviews, 20:205 (1979).

8. D. Benoit, R. Namdar-Tixier, and R. Tixier, Oxidation of Fatigue Fracture Surfaces at Low Crack Growth Rates, Mater. Sci. Eng., 45:1 (1980).

9. W. Elber, The Significance of Fatigue Crack Closure, ASTM STP 486, 230 (1971).

10. D. L. Davidson, Incorporating Threshold and Environmental Effects into the Damage Accumulation Model for Fatigue Crack Growth, Fat. Eng. Mtls. Struct., 3:229 (1981).

11. S. Suresh, C. M. Moss, and R. O. Ritchie, Hydrogen-Assisted Fatigue Crack Growth in 2¼Cr-1Mo Low Strength Steels, Trans. Japan Inst. Metals, 21:481 (1980).

12. S. Purushothaman and J. K. Tien, A Model for Fatigue Crack Closure in Ductile Materials, in "Strength of Metals and Alloys", ICSMA5 Conference Proceedings, P. Haasen, V. Gerold, and G. Kostorz, eds., Pergamon Press, New York, 2:1267 (1979).

13. I. C. Mayes and T. J. Baker, Observations of the Transference of Load across Crack Faces during Fatigue Crack Growth at Positive Values of R in the Low Growth Regime, in "Micromechanisms of Crack Extension," The Metals Society, London (1981).

14. N. Walker and C.J. Beevers, A Fatigue Crack Closure Mechanism in Titanium, Fat. Eng. Matls. Struct., 1:135 (1979).

DISCUSSION

Comment by J.P. Fidelle:

You have beautifully shown that, at odds with static fatigue
(delayed failure tests), H does not decrease the fatigue limit (en-
durance) during dynamic fatigue tests. This is consistent with more
limited pressurization-depressurization tests I have conducted with
He and H_2 gas on embedded disks.

As fatigue is a cumulative process of plastic damage at a microscale, your results confirm the need of a minimum of a kind of plastic
deformation to bring hydrogen embrittlement to evidence.

Reply:

As this is a comment, no reply is necessary.

Comment by R.H. Jones:

There are many practical circumstances where there are reversible
stresses, i.e., steam turbine blades and rotors. What might be the
effect of crack flank oxide debris for the case where the R ratio is
negative?

Reply:

The question of negative load ratios on near-threshold fatigue
crack propagation is certainly both interesting and practically
relevant as Dr. Jones suggests. We, in fact, have not examined this
situation as yet, but we expect that fretting-oxidation effects would
promote enhanced oxide growth within the crack and would, if anything,
be more severe, provided the debris is not extruded out from the sides
of the crack. Accordingly, one might expect that oxide-induced
closure would be promoted at negative load ratios, leading to reduced
ΔK_{eff} values and lower growth rates, compared to say R = 0. We will
certainly attempt such measurements in the near future.

Comment by G.G. Garrett:

How does your interpretation of your results on low strength
steels, leading to your model for environmental influences at fatigue
crack growth thresholds, extend to other alloy systems, for example,
high strength steels or aluminium alloys. In the latter case, in
particular, fatigue crack propagation experiments very often produce
oxide debris in the early stages of crack growth from a machined notch.

Reply:

The majority of our studies have been performed in lower strength
steels of yield strength below 800 MPa where for ferritic-bainitic,
bainitic and martensitic 2 1/4 Cr - 1 Mo steels, the mechanism of
oxide-induced crack closure provides a realistic model for environmentally-influenced near-threshold fatigue behavior. In high and
ultrahigh strength steels, however, the enhancement of oxide formation
within near-threshold cracks is considerably smaller. This is to be
expected since as yield strength is raised, the extent of plasticity-
induced closure is lowered, the abrasive affect of oxide debris on
the harder steel matrix to promote more fretting-oxidation is decreased, and further the maximum thickness of oxide layers is limited
by the smaller crack tip opening displacements. We feel that this
effect of decreased oxide-induced closure with increasing strength
level may in part be the source of the very marked effect of strength
level on the threshold ΔK_o, since when strength is raised threshold
values are invariably lowered.

In aluminum alloys, as you correctly state, enhanced oxide growth
is still evident. Our studies and similiar work at Alcoa, however,
have suggested that the picture may be somewhat more complex than in
steels since it appears that certain microstructures (independent of
strength) are more prone to enhanced oxide formation than others.
Since corrosion deposits are still of a size comparable with near-
threshold crack tip opening displacements, the phenomenon of oxide-
induced crack closure still occurs but its precise significance to
the role of environment in corrosion fatigue for these alloys is as
yet unclear. We are currently examining this effect for aluminum
alloys in our laboratory.

Comment by G.G. Garrett:

At threshold levels in fatigue, you clearly show the influence
of oxide films in slowing down fatigue crack growth. Indeed, this
effect appears to be the dominant one. How then do you interpret
the further acceleration, i.e., lower ΔK_o in vacuum compared with dry
hydrogen or helium, where presumably the scope for "oxide" formation,
or indeed any other "mechanical" surface film is severely limited?

Reply:

The data for dry inert helium do show distinct thresholds, as
shown in the paper, for cyclic crack tip displacements approximately
equal to the maximum oxide thicknes. It appears that even with
rigorous purification and dehumidification of either hydrogen or
helium, there are still sufficient traces of moisture to allow
enhanced oxide formation <u>at low load ratios where accelerated oxidation occurs via the fretting mechanisms</u>. In vacuum superior dehumidification appears to restrict the oxide films to even smaller

sizes, which suggest that no well-defined thresholds should be observed until lower growth rates, as shown by the data. We would predict that the degree of moisture content should control the value of
ΔK_o at low laod ratios in a given microstructure simply from the
model of a threshold ΔK_o being consistent with the crack being wedged-
closed with oxide films (i.e., when oxide thickness $\sim \Delta$CTOD). However,
it is conceivable that in the total absence of such oxide films, a
threshold would still exist, say, at a lower crack opening displacement comparable in size with the fracture surface roughness (roughness-
induced closure) or perhaps at the onset of dislocation generation
ahead of the crack tip. We would conclude than that the difference
between dry helium gas and vacuum in our experiments is a function
of purity (moisture content) in limiting fretting oxide deposits.
However, there is always an added complication in high vacuum environments with the possibility of the occurrence of fracture surface re-
welding. Unfortunately, the mechanics and mechanisms of this particular phenomenum have never been well-defined, and so it is not
possible at this stage to assess the significance of this effect.

WORKSHOP SUMMARY: HYDROGEN EMBRITTLEMENT

Prepared by: I.M. Bernstein and J.P. Fidelle

Discussion
     Leaders: I.M. Bernstein, R.M.N. Pelloux, T.E. Fischer,
               P. Azou, J.P. Fidelle, G.M. Whitesides,
               A.W. Thompson, R.O. Ritchie

     Recorders: H. Vehoff, R.H. Jones, F. Cyrot-Lackmann,
                 M.W. Finnis, S. Lynch, K. Christmann,
                 I.C. Howard, J.R. Pickens

There were four main themes that developed in the eight workshops devoted to this subject, the first two of which were in fact
quite general to many of the aspects covered in the Institute:
1. The applicability of using fundamental chemical and physi-
    cal-based calculations which usually examine the specific
    interactions between a small number of atoms, to study the
    more macroscopic, often phenomenological processes of
    brittle and ductile fracture. And as a corollary;
2. An elucidation of the kinds of critical experiments needed
    to develop a mechanistic interpretation of the large number
    of phenomena that characterize the behavior of hydrogen in
    materials.
         The next two themes are of a different scale since
    they concern hydrogen's interaction with microstructural
    units of a size of the order of 100 times larger than that
    treated in atomistic calculations;
3. The nature of hydrogen transport, in particular, its de-
    pendence on electrochemistry, stress, temperature and of
    most interest in this institute, the possible role of
    dislocations as mobile carriers of hydrogen;
4. The interaction of hydrogen with internal heterogeneities,
    denoted as traps, and the consequences of this for hydrogen-
    assisted cracking and embrittlement.

The recommendations for each of these will be summarized, followed by some general conclusions.

Themes 1 and 2: The solution of real problems in fracture in
atomistic detail will require an information base concerning elementary processes. Specifically, any useful theory of the effect of

847

hydrogen on bonding, will require a description of the most relevant
model clusters, and quantitative calculations of orbitals energies,
the result of which can only, at the present, provide qualitative
information to the metallurgist. Solid state theorists suggest the
following broad systematic influence of bonding on ductility: σ bonds
(covalent or ionic) are associated with brittleness; the more delocalized π and δ are associated with ductility. But this can only be a
general indication, as the local environment of atoms has a major
influence on various elastic properties.

While no kinetic information should be expected from quantum
physical calculation in the near future, there should be value in
studying a somewhat larger scale problem, namely the dislocation core.
It has been suggested that H (or any solute) might favor kink nucleation but inhibit kink migration, and perhaps "soften" screw dislocations and "harden" edge dislocations with respect to glide. To
understand such things would require a theoretical understanding of
the geometry and charge state of H in the core, which cannot be
deduced from any simple theoretical model of the bulk, but which
requires a study of the highly strained and anisotropic situation
of the core itself. If progress can be made in such areas, the
studies could lead to a predictive capability of the role of hydrogen
in modifying the Peierls stress as well as the line energy of a
dislocation.

Perhaps a somewhat more promising approach at this stage will
be the development of a cohesion force field that is reasonably
realistic and usable in computer simulations of fracture and the
modification of this force field by hydrogen. Such computer modeling
of atomic behavior at crack tips (slip vs. decohesion) could likely
be improved by matching the local σ-ε constitutive equations to the
far field loads and displacement. If these approaches are successful,
potential calculations should be done for grain and interphase
boundaries with and without impurities, specifically hydrogen.

In order for the theorists to assess the reasonableness of the
kind of modeling described, they will require a list of metals that
embrittle and of those that do not, as well as a determination of
the most probably mechanisms of embrittlement (decohesion vs. shear)
and its possible variation with temperature, hydrogen pressure and
other relevant variables. There was general agreement that these
requirements have yet to be fulfilled.

A large number of possibly critical experiments and approaches
were suggested to provide a data base to support theoretical calculations as well as to test such mechanistic alternatives as the role
of hydrogen on reducing cohesive energy or the critical stress for
localized shear. The following constitute the consensus on the
most important of these types of experiments:

a.   There was general interest in determining the effect of H
      on bulk properties of metals (e.g., on Young's and higher-
      order elastic moduli and on phonon spectra), since there are
      already correlations in such areas between experimental and
      theoretical work. For example, photoemission experiments
      and band-structure calculations have already provided infor-
      mation on the state of H in bulk metals.

b.   The experimental approaches of surface physicists and
      chemists should be exploited. In particular, the observed
      H-induced reconstructions of metal surfaces certainly
      suggests that H will drastically alter the metallic bonding
      within a dislocation core, but does not suggest how. More
      work is clearly indicated.

c.   Experiments are needed on systems which are metallurgically
      fairly simple, and are themselves well characterized. These
      might include whiskers, amorphous metals, and high purity
      single crystals. An important, but much more complicating
      extension, will be to study the role of hydrogen in poly-
      crystalline materials; in iron with one additional soluble
      component (Fe-P, Fe-S, Fe-Si); and in iron containing a
      second phase. This last system is obviously important in
      studying the influence of solid inclusions in embrittlement,
      with the prototype being iron/cementite (Fe_3C).

Turning to theme (3) there appear to be a number of critical
electrochemical questions that remain to be answered. For example,
there is a need for information and further understanding of the
following subjects:

a.   The nature of electrochemical and chemical deposition of
      hydrogen on surfaces (especially surfaces of base metals)

b.   The critical steps in dihydrogen (H_2) desorption, adsorption,
      and absorption by metals and metal oxides

c.   Rate controlling steps for hydrogen diffusion through
      materials and along surfaces

d.   Factors influencing relative rates of desorption (as H_2)
      and diffusion into bulk metal of surface-adsorbed hydrogen

3.   Experimental methods for determing H concentration in the
      bulk, at interfaces and at the tip of the crack (adsorbed
      and dissolved) should be developed

The question of hydrogen kinetics was extensively considered,
and was dominated by the possibility and implications of hydrogen

transport by dislocations. While it was accepted that this was a
particularly significant phenomenon, it was clear that the nature
of this interaction was not at all understood. It was agreed that
a major difficulty is that of characterization, since the largest
effects in dislocation transport occur essentially as transients.
Since the same is true of the imposed plasticity the complexity of
the interaction is considerable. Dislocation transport thus
increases the difficulty of interpreting kinetic embrittlement data,
since it potentially widens the range of possible transport rates
for hydrogen. It was generally agreed that while dislocation
transport in many systems clearly occurs experimentally, the present
theoretical approaches are less satisfactory, to date relying on
sample calculations in the absence of well-established values of
parameters in the theory. The remaining issue seems to be one of
whether supersaturations occur in particular cases, not the existence
of transport itself. Not only is it not known in which systems
transport occurs most readily but what effect it has on subsequent
plasticity and/or embrittlement has rarely been established.
Suggested experiments included:

a.   Dynamic permeation studies of pure metals and alloys;

b.   Separation of the contribution, if any, of edge and screw
      components to hydrogen transport;

c.   Transmission electron microscopy studies of how hydrogen
      carried by dislocations interacts with particles;

d.   A determination whether and under what circumstances
      hydrogen diffusion along grain boundaries is more rapid
      than in the bulk;

e.   Continued studies of how hydrogen affects plastic behavior;
      these should ideally be carried out on oriented single
      crystals.

The final theme, and the most important practically, was how
hydrogen modifies or accelerates the fracture process. Such
accelerations can occur both in systems which form stable hydrides
and those which do not.

Considering the former case first, the role of hydride stability
appears quite important. In addition, although all hydride-forming
metals can be embrittled by a hydride mechanism (usually cracking
of the hydrides) this is not necessary for all embrittlement cases.
Future studies should include:

a.   Establishing which hydrides are intrinsically brittle;

    b.   Stability of hydrides which, while not expected thermo-
dyamically, form in the high stress region near the crack
tip;

    c.   Effect of strain on hydride formation and stability.

In the non-hydride systems the most interest was how hydrogen
interacted with the crack tip and the microstructural path through
which the crack is running.

A critical issue discussed was whether the accumulation of
hydrogen at points of triaxial stress is either necessary or
sufficient for hydrogen embrittlement. Evidence is apparently quite
diverse, but it does appear that cases exist in which such accumulation is not necessary and in fact need not be sufficient. This
issue, and the complex interaction of hydrogen with temper embrittlement and "brittle" types of microstructure, are problems yet to be
resolved. Some suggested experiments include the following, many
of which are already ongoing:

    a.   The relative rates of hydrogen-assisted cracking in steels
generated from gaseous hydrogen as opposed to moist
environments;

    b.   Continuation of model experiments on specimens with a
rounded notch instead of a sharp crack in order to widen
the plastic zone and thus reduce the demands on spatial
resolution;

    c.   Further studies on hydrogen accumulation at the crack tip
as well as hydrogen effects on crack tip opening angles;

    d.   Experiments to clarify the role of a reduced flow stress
due to hydrogen on enhanced embrittlement.

In summary, while it is not yet appropriate to be optimistic
about the progress made in understanding and predicting hydrogen
effects in metals, from either a theoretical or experimental point
of view, some advances have been made, perhaps most importantly in
establishing a potentially fruitful dialogue between theoreticians
and experimentalists. It was generally agreed that the most critical
issue is how hydrogen embrittles metals and that the most likely
route to arrive at a predictive capability is to first determine if
a single broad-ranging mechanism is responsible. The current
position is that no single mechanism can yet be preferred, and that
moreover each of the five or so existant ones can be demonstrated
experimentally by appropriate choice of material and test conditions.
This state of events probably reflects the complexity of the
phenomenon and our inability to understand all it implications.
For example, evidence for hydrogen-weakened Fe-Fe bonds might not

be decisive unless it were known whether flow stresses were also reduced, and by how much, relative to tensile bond strengths. The ability to perform calculations which can differentiate between these possibilities is hampered by the current inability of theorists to transfer information about interatomic forces from highly symmetrical cases (e.g., the perfect lattice) to highly non-symmetrical ones like opening cracks. Equally, the metallurgist is not yet able to devise the proper kind of experiments to elucidate such critical questions as: the nature of the H-impurity and H-interface interactions, and how such trapping affects subsequent embrittlement; the interaction of hydrogen with plasticity, both localized and general; the rate-determining kinetic steps in crack initiation and propagation; and the whole nature of why and how some alloys are highly susceptible while some are immune to hydrogen-assisted failure.

The problems remaining while daunting are not insurmountable; while progress has been made, much remains to be done.

WORKSHOP SESSION 2

INTERGRANULAR EMBRITTLEMENT

Session Chairman: D.J. Duquette

ATOMISTIC MECHANISMS OF INTERGRANULAR EMBRITTLEMENT

M. P. Seah and E. D. Hondros

Division of Materials Applications
National Physical Laboratory
Teddington, Middlesex, UK

ABSTRACT

Since the composition at grain boundaries generally differs
considerably from that in the grain interior, all considerations
of intergranular fracture must take this into account. In this
survey we consider the historical development of the theories of
low temperature intergranular fracture with particular reference
to temper brittleness. The theory is followed through Rice and
Thompson's analysis of the energy balance between ductility and
brittle fracture with Rice's analysis of the effects of segregants
on the ideal work of fracture term. As developed by Mason, Hirth
and Rice, and Seah the atomistic term involved in this process is
the bond energy of the individual atoms across the fracture plane.
These terms can be calculated simply, to a first approximation,
from tabulated thermodynamic data.

Other important forms of low temperature intergranular
brittleness, hydrogen embrittlement, liquid metal embrittlement
and intergranular stress corrosion cracking are also affected by
grain boundary segregation. In hydrogen embrittlement and liquid
metal embrittlement, the segregant-environment atom bonding is
important whereas in intergranular stress corrosion cracking the
emphasis is placed on the electrochemical potential of the
segregant atoms.

The high temperature intergranular failures in stress relief
annealing and creep embrittlement are similarly affected by
segregants but here the terms involve the effects on surface
energy and grain boundary diffusivity. Each intergranular
fracture phenomenon involves a different atomic parameter and

hence a different hierarchy of segregants is observed in each case.
Recognition of this is important and should be built into future
specification codes.

INTRODUCTION

Material failures that occur along any interface or simple
path without considerable deformation and yielding may be termed
brittle failures. The ability to deform and take up energy
during fracture is an essential property of metals-it is this
quality of toughness which surely led man to forsake stone tools
in favour of metallic ones. Ever since the early metal cultures,
the problem of intergranular embrittlement has existed, cropping
up here and there in various forms: indeed, recorded problems go
back 4000 years[1], although the detailed study of intergranular
embrittlement has, of course, a much briefer history. The first
detailed atomistic study of embrittlement is probably that of
Roberts-Austen in 1888[2], in his work on the effects of impurities
on the fracture properties of gold. Some impurities completely
removed the ductility whilst others had a lesser effect. Of the
impurities which promoted embrittlement, we note his remarkable
discovery of a correlation between the embrittlement and the
atomic volume of the impurity species.

This problem, of why certain impurities promote intergranular
fracture considered at the atomistic level, is the central theme
of the present survey lecture. Part of Roberts-Austen's view of
the problem was solved when phase diagram data became available,
but it was still found that impurities, well within the solid
solubility range, could cause embrittlement and that the fractures
were clearly intergranular. In iron and low alloy steels,
impurities such as P, Sn, Sb and As caused embrittlement, especially
after certain heat-treatments. The elements Bi, Cu, Ge, S, Se, Si
and Te have since been added to the list of embrittlers in iron
and steel[3] although in some cases, for instance S, the effect is
normally nullified by the addition of manganese. The next major
step forward in understanding the micromechanism of impurity
induced intergranular fracture was taken by McLean in 1957[4].

McLean crystallised many of the observations of the time into
a theoretical relation for the equilibrium grain boundary segregation
of solute species, McLean's relation is of the same form as the Langmuir adsorption equation for free surfaces. The study of segregation
has progressed very fruitfully and, as is discussed in Guttmann's
lecture, it is at a stage today where multicomponent effects are
well understood and the segregation levels of important impurities
can be predicted as a function of alloy composition and heat treatment. Thus, the segregation level of different species can now be
directly related to the material and its thermomechanical history.
The problem remains to relate the fracture properties to the

chemical state of the grain boundaries. Modern electron microscopy and surface analytical techniques demonstrate unequivocally that this fracture occurs precisely along the grain boundary plane and across the bonds formed between the segregated impurity and matrix atoms. Thus, it is these precise bonds that must attract our attention.

The above notions apply in particular to temper brittleness[5,6,7] and low temperature intergranular fracture[3]. These fractures occur by impact or tensile forces at temperatures below 200 °C and in non-aggressive or inert environments. The self-same impurity segregation at grain boundaries is now known to exacerbate failure in fatigue[8], hydrogen embrittlement[9], stress corrosion cracking[10,11] and liquid metal embrittlement[12]. These fractures are also associated with the precise atomic path of the grain boundary, but they occur according to different atomic micromechanisms and so in each case, a different atomic property of the segregant is involved and a different effective segregant hierarchy is found. In many of these cases, treatments that reduce the segregation, such as altering the material's heat treatment schedule or purity, lead directly to improved ductility.

Intergranular failures also occur at high temperatures in such phenomena as overheating and hot shortness[13] and also in the well-known hot ductility trough problem[14]. Some failures at high temperature are associated with the grain boundary but not with de-bonding along the precise atomic plane of segregation. In this category, we find the stress relief cracking[15] and creep embrittlement[16] of low alloy steels. These failures occur by the formation of cavities along the grain boundaries at high temperature. The cavities are thought to be nucleated on grain boundary precipitates, such as MnS, and grow by vacancy supply from the grain boundary to reduce the overall strain energy. Ultimate failure occurs when the cavity size and density is such that the net section cannot support the applied load. The time to failure depends on the cavity nucleation and growth rates which are both affected by the grain boundary composition[17]. Thus, a further range of atomic micromechanisms must be invoked although further experimental evidence is required before each mechanism's regime can be properly established.

The common theme of most of the above fractures is the chemical state of the grain boundary and hence the material's purity. It is true that the fracture only occurs as a result of a shift in the balance between the relative properties of the grain and the grain boundary and that modification of the grain microstructure can remove embrittlement, however, this is usually at the expense of the material's specified properties and is not generally practicable. Therefore, much of the work in this survey concerns

changes in grain boundary properties with the grain interior
properties kept constant.

    In the following sections we shall outline, briefly, the
experimental evidence available in each of a number of intergranular fracture phenomena and present some of the ideas that
have been proposed to describe the atomistics of the embrittlement.
To an extent, all of these ideas are relevant in one way or another
to the problem but many do not lead to quantifiable predictions and
so are difficult to test. Of course, only those predictions that
are at least semiquantitative can be ultimately implemented as
procedures to improve the performance of materials. We consider
this aspect of predictability as the important criterion in
assigning value to any theory. Many of the routes to improvements
are evident from the start. For instance, in those problems
involving grain boundary purity the impurities may be removed from
the bulk directly or, where this is uneconomical, they may be
fixed by precipitation reactions with additives such as the rare
earth metals[10,18,19]. However, there are alternatives which
include modifications in the heat treatment schedule, microstructure
and alloying additions.

    A full understanding of the atomistics of the fracture
processes and of the grain boundary composition will ensure that
the most economic route to avoid embrittlement can be maintained
and that, as new materials are developed to meet the increased
demands for efficiency in modern plant, the problems can be dealt
with in the material's design stage and not, as appears to be the
case in many of today's turbine power plants[19], as an on-going
maintenance problem throuthout the life of the plant.

LOW TEMPERATURE BRITTLENESS AND TEMPER BRITTLENESS

    Although temper brittleness is a specific term relating to
alloy steels, we shall consider here that the main differences
from the low temperature brittleness of unalloyed iron and
other metals lies in the segregation terms, discussed in Guttmann's
lecture, and not in the atomistics of the fracture process.
Guttmann deals with the kinetics of the segregation and also the
inter-element couplings which give rise to the observed segregation
in alloys. This aspect is only important here in so far as the
inter-element couplings result in more than one segregant being
present instead of the one appropriate to the unalloyed material.

    Over the years a number of theories have been developed to
explain why the segregants reduce cohesion. In 1957 McLean[4]
developed his theory of segregation and supposed that the segregant
atoms are disposed exactly along the grain boundary plane. McLean
argued that the segregants affect the ideal work of fracture γ,
and that this causes the loss of grain boundary cohesion.

Following Griffith's[20] and Polanyi's[21] theories for pure solids he writes, for rapid fracture:

$$\gamma = 2\gamma'_s - \gamma_b \qquad\qquad (1)$$

where γ_b is the grain boundary energy per unit area immediately before fracture and γ'_s is the surface energy for the surface composition immediately after fracture, emphasising that this is not the equilibrium value, γ_s, appropriate to that composition and temperature. McLean argued further that even if γ were small compared with the plastic deformation work during brittle fracture, γ_p, then γ would still be the determining factor. At a crack tip, if the energy of the surfaces along which fracture is to spread is small, the tensile stress concentration required at the tip of the crack to start or continue fracture is also small. The associated shear stress concentration involved in the total work of fracture is therefore also small and γ_p is small. Hence γ_p is determined by γ.

McLean therefore established a programme of work to measure the experimental dependence of γ_s and γ_b, at equilibrium, on temperature and bulk solute content. The results of this work underpinned the early development of many of the concepts of equilibrium grain boundary segregation. Following Guggenheim[22], McLean[4] derived a number of relationships for segregantion, one of which is the Gibbs adsorption formula for grain boundary segregation from dilute solids with a solute content, X_c:

$$\frac{d\gamma_b}{dX_c} \equiv - \Gamma_b \frac{RT}{X_c} \qquad\qquad (2)$$

where Γ_b is the grain boundary segregation level in moles m^{-2} and RT has the usual meaning. Using a statistical mechanics approach McLean also derived the adsorption relation:

$$\frac{\Gamma_b}{\Gamma_b^0 - \Gamma_b} = \frac{X_c}{1 - X_c} \exp(- \Delta G_b/RT) \qquad\qquad (3)$$

where ΔG_b is the free energy of segregation to grain boundaries. From equations (2) and (3) the reduction in γ_b for a given bulk solute content is defined once Γ_b^0 and ΔG_b are known. Γ_b^0 is generally taken as $(Na^2)^{-1}$ where N is Avogadro's number and a^3 is the atomic volume[23]. Typical results which show how the grain boundary energy of iron alters with phosphorus segregation are given in Fig 1, together with the predictions using equations (1) and (2) and a free energy of segregation of -95 kJ mol^{-1}. Compilation of many values of $d\gamma_b/dX_c$ for different systems are given by

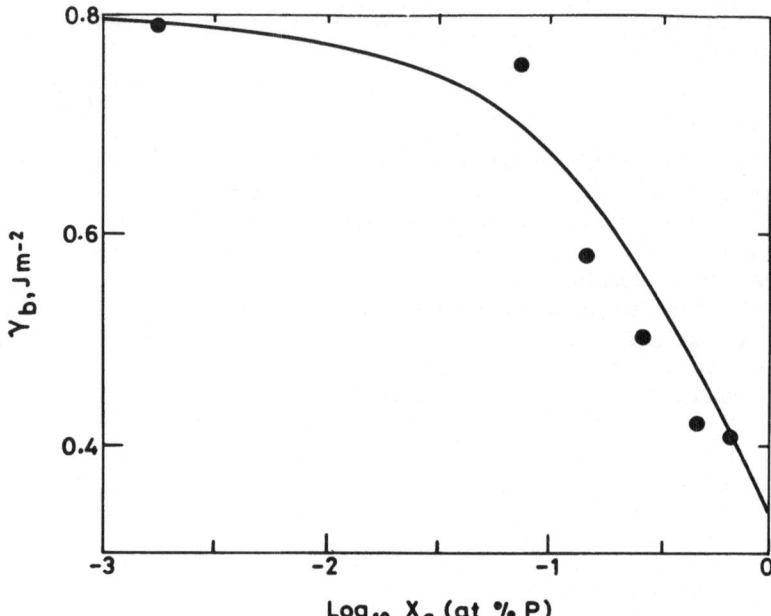

Fig 1   The isothermal variation in grain boundary energy in
         iron at 1450 °C, as a function of phosphorus content,
         after Hondros[24], together with the theoretical
         prediction[23].

Hondros[25], and the counterpart for free surfaces, $d\gamma_s/dX_c$, by
Hondros and McLean[26]. Similar analyses, using equations (1) and
(2), have been shown to be valid for tin segregation to both free
surfaces and grain boundaries in iron[27]. These results and a
general compilation show that $d\gamma_s/dX_c$ is typically 2.5 times[27]
as large as $d\gamma_b/dX_c$ so that the surface energy falls much more
rapidly than the grain boundary energy as the solute bulk content
rises. It should be noted that, in the dilute case, the interfacial activities, $(-d\gamma_b/dX_c)$, $(-d\gamma_s/dX_c)$ are directly related
to the free energies of segregation through:

$$- \frac{d\gamma_b}{dX_c} = \Gamma_b^o \ RT \ \exp(- \Delta G_b/RT) \tag{4}$$

An analysis of many systems by Hondros[25] showed that many of the
systems with a high interfacial activity display intergranular
embrittlement. Thus, Hondros[25] correlated the change in grain
boundary cohesion with ΔG_b.

At this point we summarise the main experimental results with
which the various theories must agree. The degree of embrittlement,

as measured by ductile to brittle transition temperatures, is
linearly proportional to the degree of segregation of each species,
at least over a limited range. The kinetics of the embrittlement
follows the kinetics of segregation[28]. The fracture path follows
the precise grain boundary and the atom bonds ruptured are those
between the segregant atom and the matrix. This point has been
amply verified by surface analysis and field ion microscopy[29].
Because of problems in quantifying the analysis methods exactly[29]
the precise relative embrittling effects of different segregants
are not known accurately, however, it is generally accepted that
for a given level of segregation, Bi, S, Sb, Se, Sn, and Te are
all potent grain boundary weakeners in iron whereas As, Cu, Ge and
Si are moderate and P is somewhere intermediate. Less work has
been done for other matrices but S and Bi are known to embrittle
copper, P and K to embrittle tungsten and S to embrittle nickel[30].
All observations are compatible with a general view that ductility
gives way to brittleness as the test temperature falls and
dislocation movement is impeded. Larger grain sizes and increased
hardness exacerbate the effect and raise the ductile-brittle
transition temperature.

The next development was by Hondros and McLean[31] in 1974. Here
measurements were made of the effect of the grain boundary segregation of Bi in copper on the stress/strain behaviour in uniaxial
tension. They showed that for alloys equilibrated at 550 °C, both
γ_b and the fracture stress rapidly reduced as the Bi bulk and
grain boundary contents rose, as shown in Fig 2. In relation to
equation (1) they note that both γ_s and γ_b are rarely reduced
more than 50% by segregation from the values, γ_s^o and γ_b^o, for
the clean metal. With $\gamma_s^o = 1.64$ Jm^{-2} and $\gamma_b^o = 0.885$ Jm^{-2}, they
find $\gamma_s = 0.885$ Jm^{-2} and $\gamma_b = 0.218$ Jm^{-2} at saturation segregation.
For the most brittle alloy, 2-3 monolayers of bismuth grain
boundary segregation were thought not unreasonable so that the
fracture surfaces, containing 1-1$\frac{1}{2}$ monolayers, would be near
saturation and γ_s' should not be very different from γ_s. For
this alloy, therefore, the ideal work of fracture falls from
2.82 Jm^{-2} to 1.552 Jm^{-2}, a very large reduction. An outline of
their micromechanistic model of the fracture conditions is given
below.

For brittle fracture Hondros and McLean[31] point out that two
conditions must be fulfilled. For condition A, the local tensile
stress normal to the grain boundary at the crack tip, σ_y, must
be greater than the grain boundary cohesive stress, σ_m. Hence
condition A is $\sigma_y > \sigma_m$. The other condition concerns the
balance between brittle fracture and ductility, the latter assumed
to arise through slip on a plane intersecting the crack front. In
this case, the ratio of tensile component to shear component along
this line must be large enough for cracking to occur rather than

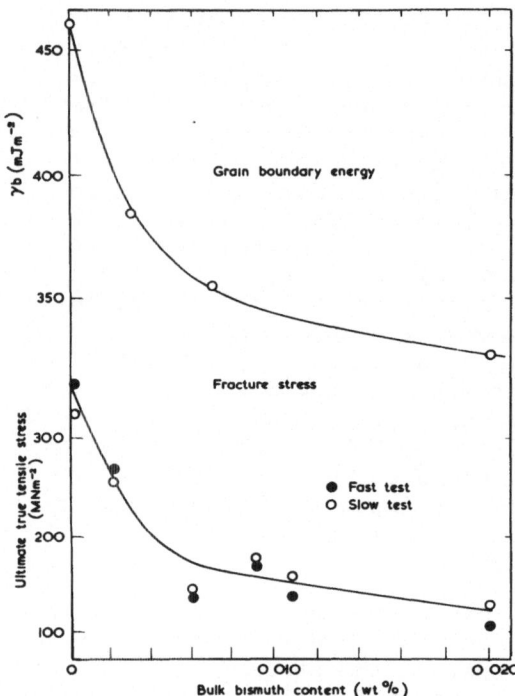

Fig 2   The dependence of γ_b and ultimate tensile strength on
         bismuth content in copper, after Hondros and McLean.[31]

slip[32]. If the relevant measure of the shear component is τ_T and
the local shear resistance is τ_m, the second condition B becomes
$\sigma_y/\tau_T > \sigma_m/\tau_m$ for brittle fracture to occur.

    Hondros and McLean evaluate condition A using Smith and
Barnby's[33] analysis of crack nucleation due to a pile-up of edge
dislocations. The applied tensile stress, σ_a, for brittle
fracture is thus:

$$\sigma_a > \alpha((\gamma_s - \gamma_b/2)\eta/L)^{\frac{1}{2}} \qquad (5)$$

where $\alpha = 3.8$, η is the shear modulus and L the grain size.
Using values for pure copper ($\eta = 60000$ MNm^{-2}, $L = 0.02$ mm), σ_a
is greater than 247 MNm^{-2}, whereas for the maximum segregation
σ_a is greater than 183 MNm^{-2}. Since the experimental results were
one half this value, Hondros and McLean invoked a slightly more
complex situation involving the coincidence of two slip bands
impinging on the crack front. In this case α and σ_a above are
reduced by a factor of 8 and well encompass the experimental
results, ie the two slip band configuration satisfies condition A
by a factor around 8 whereas the single slip band only just
satisfies it.

Condition B was not so easily fulfilled. Estimates of the ratio σ_m/τ_m in the copper matrix give a value of 30. At the grain boundary, assuming that the cohesive stress is proportional to the work of fracture, this is reduced by $(2\gamma_s - \gamma_b)/2\gamma_s$ to give 26 for clean boundaries and 14 for the fully embrittled case. Hondros and McLean estimate σ_y/τ_T to be 9.5 in the two slip plane model and, of course, less in the single slip plane case. Thus, condition A is easily satisfied, condition B is sufficiently fulfilled that a full calculation may show it to be satisfied and, because it is the marginal condition, will usually control the ductility. Hondros and Seah in their review[30] note that the Kelly, Tyson and Cottrell criterion[32] for condition B, has been improved by Rice and Thompson[34] who treat the problem of shear in terms of dislocation production and annihilation at the crack tip and so avoid the problem that the shear stress would have to be uniform across the entire slip plane to allow the atoms to move uniformly past each other. We shall return to the incorporation of Rice and Thompson's approach into the above shortly when we deal with Mason's[35] analysis below.

In 1975, Seah[36,37], considering the changes in the important term above, the work of fracture γ , emphasised that the appropriate value of γ_s , after fracturing a grain boundary with Γ_b moles of solute per unit area, was not the equilibrium value but that appropriate to a surface segregation of only $\frac{1}{2}\Gamma_b$ per unit area. Using thermodynamic arguments he showed that this greatly reduced the effect of segregation on the work of fracture. In 1976 Rice[38] took the thermodynamic analysis further and considered the effect of segregation on γ in two limiting cases; (a) fast fracture in which the segregant is immobile and which corresponds to most room temperature intergranular fracture, and (b) slow fracture in which the segregant can diffuse sufficiently to cover all interfaces and surfaces at equilibrium throughout the fracture process. In the same paper Rice gives a detailed analysis of the ductile/brittle balance approached above by Hondros and McLean[28].

Rice[38] bases his analysis on the earlier paper of Rice and Thompson[34] but reminds us, however, that the critical sizes and distances involved are comparable with lattice dimensions and the problem will only be fully answered by lattice theory. Rice's theory considers the possibility of blunting an atomically sharp crack by nucleating a dislocation at its tip. If the activation energy for this, U_{act} , is negative there is spontaneous crack blunting, and hence ductility, whereas brittle fracture involves a positive value of U_{act}. In turn, U_{act} is the sum of three terms, (a) the dislocation loop self energy, here assumed to be for a semicircular dislocation line of radius rb (b = Burgers vector) expanding on a slip plane at an angle ϕ to the crack plane, (b) the energy of the ledge created by this loop, and

(c) the energy released by creating this dislocation in the stress field of the crack tip. Thus, Rice[38] writes:

$$U_{act} = \frac{2-\nu}{8(1-\nu)} \eta b^3 r \ln \frac{8r}{e^2 \xi_o} + \frac{2}{\beta'} \gamma_{step} b^2 (r - \xi_o)$$

$$- \frac{1.395}{\beta} \frac{b^2}{\sqrt{1-\nu}} \sqrt{\eta b \gamma_{int}} (r^{3/2} - \xi_o^{3/2}) \qquad (6)$$

where η is the shear modulus, ν = Poissons ratio, $\xi_o b$ is the dislocation core cut-off radius, β' = cosec ϕ sec ψ where ψ is the angle between the Burger's vector and the resolved direction of the crack propagation on the slip plane, $\beta = \beta'$ sec$(\phi/2)$, $2\gamma_{int} = \gamma$ the work of reversible separation of the grain boundary and γ_{step} is the energy associated with the step that forms at the crack tip when the dislocation is nucleated, per unit area. The general scheme of this model is shown in Fig 3. Rice rearranges the terms in dimensionless parameters:

$$u_{act} = \left\{ \frac{8(1-\nu)\beta'}{(2-\nu)^2 \beta^2} \right\} \frac{U_{act}}{\eta b^3} \frac{\gamma_{int}}{\gamma_{step}} \qquad (7)$$

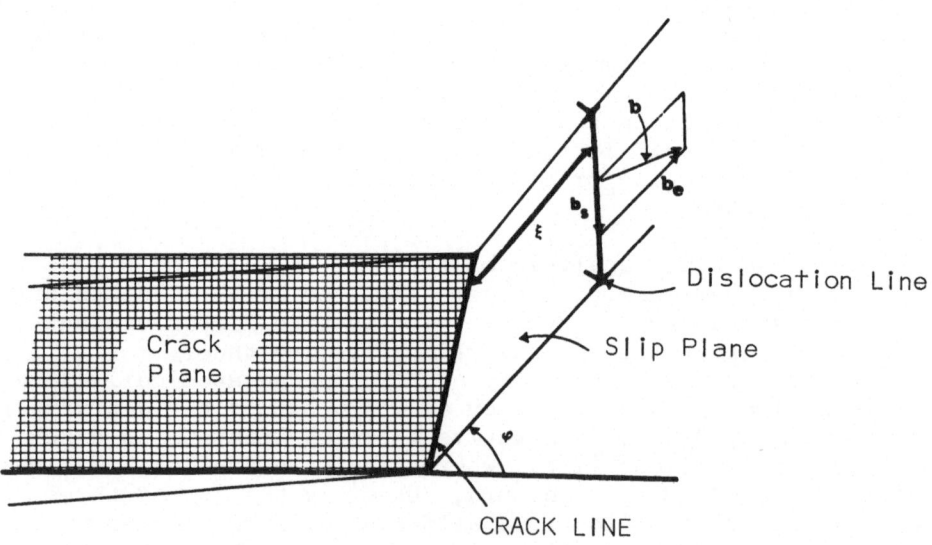

Fig 3   Geometry of the nucleation of a dislocation at the crack tip, after Rice and Thompson[34].

$$S = \left\{ \frac{16(1-\nu)}{5\beta'(2-\nu)} \right\} \frac{\gamma_{step}}{\eta b}, \qquad R_o = \left\{ \frac{16\beta'}{5(2-\nu)\beta^2} \right\} \xi_o \frac{\gamma_{int}}{\gamma_{step}} \qquad (8)$$

and differentiates u_{act} with respect to r to find the maximum value of u_{act}, u_{act}^m. This represents the potential barrier to dislocation blunting of the crack. Rice then plots u_{act}^m as a function of S for various values of R_o, as shown in Fig 4 and, as suggested in the labelling, the bracketed terms are generally close to unity. If u_{act}^m is negative the dislocations nucleate spontaneously and the crack will blunt whereas, if positive, the crack will continue to propagate. These regimes are clear in Fig 4.

Following suggestions by Rice, Mason[35] extended these ideas a little further. Mason basically uses equation (6) and considers, as the crack is loaded and the stress intensity factor is increased, which condition is first reached, a K corresponding to the Griffiths load (brittle fracture) or a K corresponding to a load at which dislocations may be nucleated (crack blunting). Equation (6) may now be written with $2\gamma_{int}$ replaced by $G \equiv (1-\nu)K^2/2\eta$. To calculate the value of K sufficient to nucleate a metastable dislocation loop, the force du/dr acting on the loop is zero and

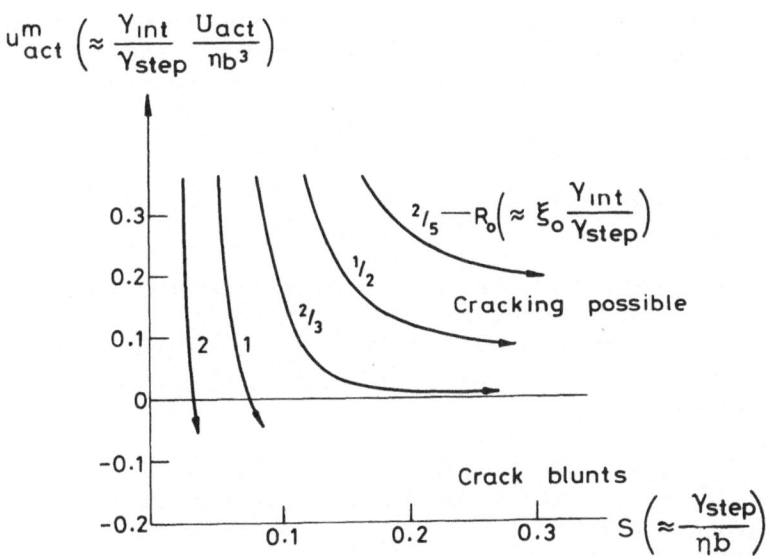

Fig 4   Activation energy for dislocation nucleation, u_{act}^m, as a function of the reduced variables S and R_o, after Rice[38].

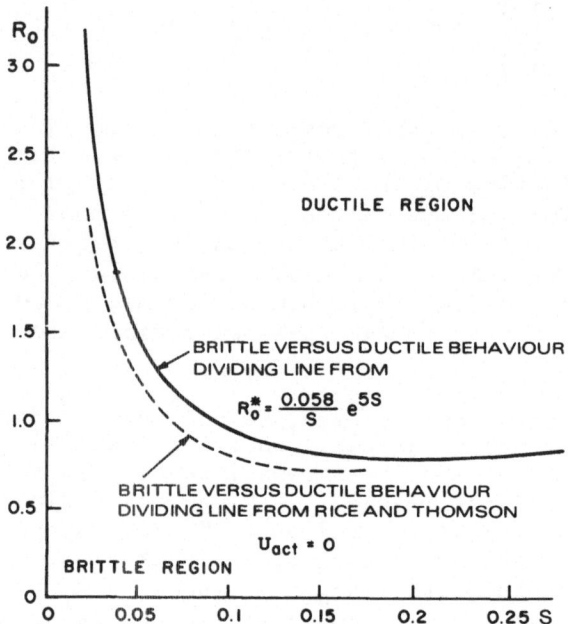

Fig 5   Regions of brittle and ductile fracture on the R_oS plane, after Mason[35].

the critical radius of this loop is defined where the differential of this force is zero. At this critical radius, the ratio of the energy release rate to the energy for brittle crack growth is given by:

$$\frac{G_{int}}{2\gamma_{int}} = \frac{0.058}{R_oS} \exp{(5S)} \qquad (9)$$

If G_{int} is greater than $2\gamma_{int}$, brittle fracture occurs before dislocation nucleation and vice versa. The brittle and ductile regimes may therefore be mapped in R_oS space as shown in Fig 5. Also shown in Fig 5 is the similar plot for $u_{act}^m = 0$ from Rice's[38] earlier analysis.

        To see how the above theory may be realised in practice we shall follow Mason's analysis of Hondros and McLean's results for bismuth segregation in copper[28]. γ_{step} is interpreted as γ_s and $2\gamma_{int}$ as $2\gamma_s - \gamma_b$, the numerical values used are $\eta = 40500$ MNm^{-2}, $b = 0.255$ nm, $\nu = 0.324$ and $\xi_o = 1$. The results, plotted in R_oS space, are shown in Fig 6. The dotted zones show

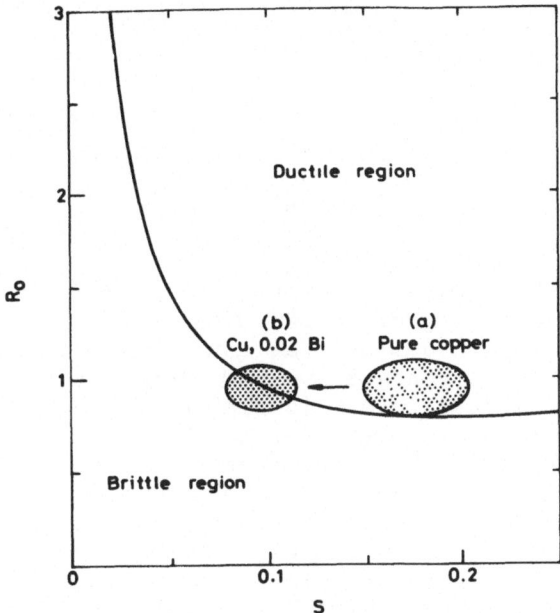

Fig 6 R_O and S values for prospective crack orientations in
         (a) pure copper and (b) copper, 0.02 wt% Bi.

the distribution of R_OS co-ordinates for a range of crack-plane
orientations in the copper lattice. This geometry defines the β
and β' terms which in turn define R_O and S via equation (8).
As may be seen in Fig 6 the distribution shifts from the ductile
region for pure copper, (a), to being marginally brittle for the
most brittle copper-bismuth alloy, (b). These shifts have been
calculated on the basis that γ_{int} and γ_{step} are both reduced
to 54% as a result of the bismuth segregation.

     The above change in the work of fracture, γ_{int}, has also been
the subject of a detailed analysis by Rice[38], as discussed above,
using a thermodynamic approach. Rice considers the interfacial
stress and work of fracture dependence on the interface separation
either as a function of chemical potential, μ, for a constant
total segregation, Γ_b , throughout, or as a function of
segregation, Γ , for a constant chemical potential. By elegant
thermodynamical analysis, he shows that the work of fracture at
constant segregation, γ_Γ , and at constant chemical potential of
the segregant, γ_μ , are given by:

$$\frac{d\gamma_\Gamma}{d\Gamma} = \mu_s(\tfrac{1}{2}\Gamma_b) - \mu_b(\Gamma_b) \quad \text{(constant } \Gamma\text{)} \tag{10}$$

and

$$-\frac{d\gamma_\mu}{d\mu} = 2\Gamma_s(\mu) - \Gamma_b(\mu) \quad \text{(constant } \mu\text{)} \tag{11}$$

The work of fracture at constant μ is appropriate to slow fractures in which the segregant can diffuse sufficiently fast to cover all interfaces and surfaces at equilibrium throughout the fracture process. This analysis was developed to analyse hydrogen embrittlement but is equally applicable to wedge-type slow inter-granular cracking at high temperatures. We shall return to this later but, of more immediate importance is the work of fracture for a constant segregation level. This latter is the condition appropriate to the fast intergranular failure discussed above in which the segregant atoms are immobile during fracture. The problem has been considered by Asaro[39] who extends Rice's work to interphase interfaces and by Hirth[40] who discusses a number of irreversible phenomena which can cause deviations in work from that predicted for the limiting case. More recently Hirth and Rice[41] have presented the thermodynamic analyses of many of the possible work cycles in a clear and stimulating manner.

Hirth and Rice[41] show that in any general separation path which may be involved in a work cycle described in $\mu\Gamma$ space, as shown in Fig 7, the work done in any closed cycle is simply $\oint \Gamma d\mu$ or $-\oint \mu d\Gamma$. Thus, if in Fig 7 curve (a) represents the Γ, μ relation for the grain boundary and curve (b) that for the free surface, we see that Rice's[38] slow fracture, constant μ, represents a trajectory from K to L and his fast fracture, constant Γ, represents a trajectory from K to M. Now, as $\mu \to -\infty$, $\Gamma_s \to 0$ and $\Gamma_b \to 0$, ie there is no segregation without a solute being present. In this limit

$$\gamma_\mu = \gamma_\Gamma = 2\gamma_s^o - \gamma_b^o = \gamma_o \tag{12}$$

where γ_s^o and γ_b^o are the energies for the pure metal. γ_μ and γ_Γ may be simply evaluated using work cycles in Fig 7 as follows. For γ_μ use slow fracture from K to L, with work γ_μ, desorb the free surfaces along the isotherm to $\mu = -\infty$, rejoin the surfaces to form the grain boundary at $\mu = -\infty$ with work γ_o and adsorb along the grain boundary isotherm to μ_K. Thus:

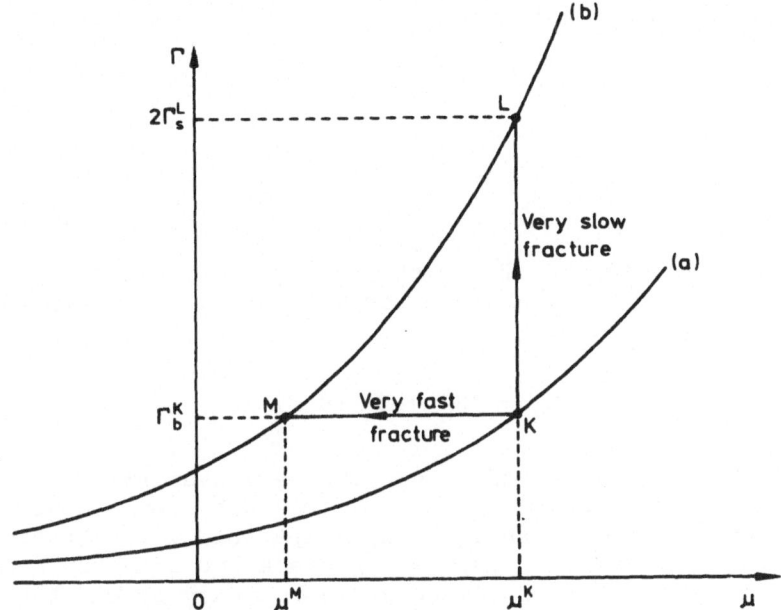

Fig 7   The μ dependence of Γ for solute atoms on (a) the grain
boundaries and (b) the fracture surfaces, after Seah[3].

$$\gamma_\mu = \gamma_0 - \int_{-\infty}^{\mu^K} [2\Gamma_s(\mu) - \Gamma_b(\mu)]d\mu \qquad \text{(constant } \mu) \qquad (13)$$

and in a similar manner

$$\gamma_\Gamma = \gamma_0 - \int_0^{\Gamma_b^K} [(\mu_b(\Gamma) - \mu_s(\Gamma/2)]d\Gamma \qquad \textbf{(constant } \Gamma) \qquad (14)$$

which are the integrals of equations (11) and (10) respectively.

We shall now concentrate on γ_Γ , the work of fracture for
low temperature brittleness and which is the parameter central to
the evaluation of any of the above approaches. In the final
analysis, whether it be Hondros and McLean's theory or that of
Rice or Mason, the actual fracture appears to be governed by the
divergence of γ_Γ from γ_0 . Both Mason[42] and Seah[3] have
evaluated equation (14) in the dilute approximation to obtain a
better feel for the effect. In this approximation, in the usual
way:

$$\mu_c^{K,M} = \mu_o + RT \ln X_c^{K,M} \tag{15}$$

$$\Gamma_b = \Gamma_o X_c \exp(-\Delta G_b/RT) \tag{16}$$

$$\Gamma_s = \Gamma_o X_c \exp(-\Delta G_s/RT) \tag{17}$$

where Γ_o is the value of $\Gamma_{b,s}$ at one monolayer coverage and the 's' suffixes are the surface equivalents of terms otherwise with a 'b' suffix. Thus

$$\gamma_\Gamma = 2\gamma_s^o - \gamma_b^o + \Gamma_b(\Delta G_s - \Delta G_b - RT \ln 2) \tag{18}$$

Only a few values for ΔG_s and ΔG_b are available[30], hence Seah[3] approached the ideal work of fast fracture from a quasi-chemical viewpoint, but which is fully compatible with the above approach. Seah considers a propagating crack and asks how changes in the chemical constitution of the grain boundary affect the ideal work of fracture for an incremental advance of the crack front. Using pair bonding theory for a binary system of solute B in solvent A and nearest neighbour bond energies (negative values), ε_{AA}, ε_{AB} and ε_{BB} assigned to AA, AB and BB neighbours respectively, the energy of the bonds broken in the fracture process may be summed. A schematic view of fracture before and after is shown in Fig 8. The fracture energy per unit area of the clean grain boundary, FE(0), is simply:

$$FE(0) = - \left(\frac{\dfrac{Z_g}{2}}{a_A^2} \right) \varepsilon_{AA} \tag{19}$$

where a_A^2 is the area of an A atom in the surface, Z_g is the co-ordination of atoms in the layer on one side of the boundary to those in the adjacent layer on the other side. On average $Z_g = (1 - \gamma_b^o/2\gamma_s^o) Z_v$ where Z_v is the value in a perfect lattice and γ_b^o/γ_s^o is typically $1/3$[27]. In a similar way, for a segregated boundary

$$FE(\Gamma_b) - FE(0) = \frac{Z_g}{Z} \frac{\Gamma_b}{\Gamma_o} \left[H_B^{sub*} - H_A^{sub*} - \frac{Z\omega}{a_B^2} \right] \tag{20}$$

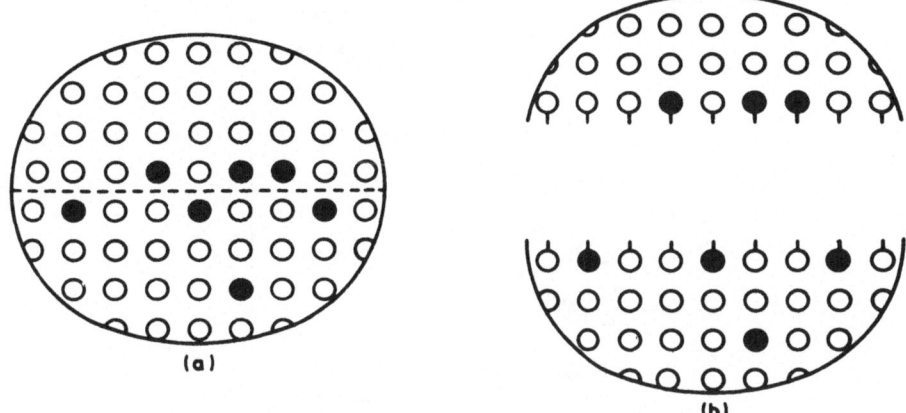

Fig 8   Schematic representation of the atoms (a) either side of
         the unbroken grain boundary and (b) after fracture with
         the ruptured bonds left dangling. The matrix A atoms are
         denoted by open circles and the solute B atoms by filled
         circles, after Seah[3].

where

$$H_i^{sub*} = - \frac{Z\epsilon_{ii}}{a_i^2} \qquad i = A,B \tag{21}$$

is the sublimation enthalpy of pure A or B per unit area, Z is the
total co-ordination of each atom and ω is the regular solution
parameter defined by:

$$\omega = \epsilon_{AB} - \tfrac{1}{2}(\epsilon_{AA} + \epsilon_{BB}) \tag{22}$$

In the above, $FE(\Gamma_b)$ and $FE(0)$ may be identified with
γ_Γ and γ_0. Using the above regular solution approximation,
values of ΔG_s and ΔG_b in equation (18) may be written as a
summation of pair bond and strain release terms[3,43]. Assuming
that the strain contributions are largely satisfied in both forms
of segregation, equations (18) and (20) may be shown to give the
same basic description, ie there is no intrinsic difference between
the thermodynamic and atomic bond approaches. The advantage of
equation (20) is that the values of H_A^{sub*}, B and ω may be derived
from standard texts of thermodynamic data[44,45]. Using this
approach, Seah calculated the term in brackets in equation (20)
for a range of segregants in iron, as shown in Fig 9. In general,
the regular solution term in equation (20) contributes only around

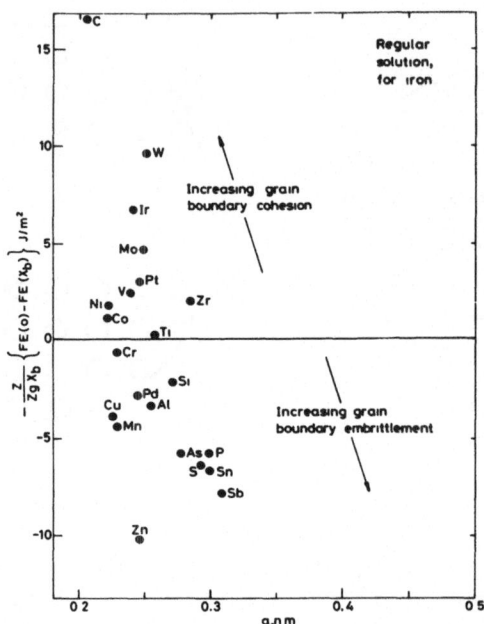

Fig 9   Remedial/embrittlement plot for segregants in iron in the
          regular solution approximation, after Seah[3].

1 Jm^{-2} so that, to a first approximation, it may be ignored. In
this case the ideal solution approximation may be used and from
equation (20), with the ω term removed, the reduction in ideal
fracture energy per unit $Z_g\Gamma_b/Z\Gamma_o$ is just the difference in H^{sub*}
(the bond energy density) between the matrix and segregant species.
On this basis a general remedial/embrittlement plot for all
species may be drawn, as shown in Fig. 10, in which the relative
positions of the species on the H^{sub*} ordinate axis denotes the
degree of embrittlement for different segregants in a given matrix.
For instance, if iron is taken as matrix, all points above Fe will
improve the grain boundary cohesion whilst those below cause
embrittlement. As noted earlier in both Figs 9 and 10, Bi, P, S,
Sb, Se, Sn and Te are predicted to be very embrittling whereas
As, Cu, Ge, Si are all less embrittling and C, B and N are
remedial. Again P and K will embrittle tungsten, S and Bi will
embrittle copper and S embrittle nickel. Thus equation (20)
provides a reasonably accurate first order calculation for
$\gamma_\Gamma - \gamma_o$.

     A recent assessment of measurements of the combined effects
of molybdenum and phosphorus segregation in a low alloy steel by

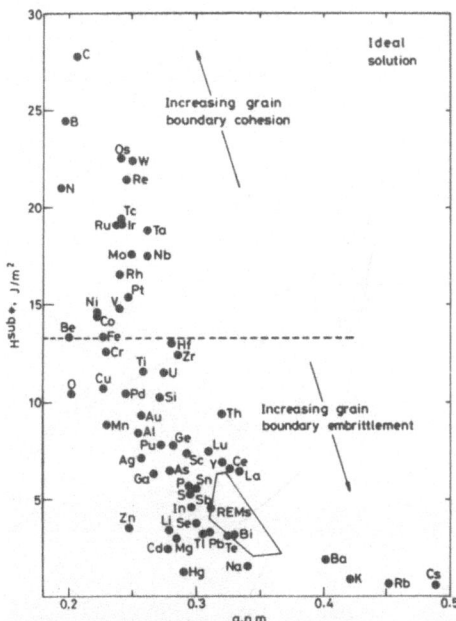

Fig 10   Ideal solution general remedial/embrittlement plot for
          matrix and segregant elements based on a compilation of
          H^{sub*} values as a function of atom size, a, after Seah[3].

Dumoulin, Guttmann, Foucoult, Palmier, Wayman and Biscondi[46], is
given by Seah[47]. The embrittlement at a fixed level of segregation
of phosphorus is seen to reduce linearly as molybdenum segregates,
as shown in Fig 11. This result is already predicted in Fig 10
where it is seen that the Mo atoms are about 0.7 times as remedial
as the P atoms are detrimental, per unit area of grain boundary.
This prediction is shown by the line in Fig 11. The closeness of
the prediction and results is a good illustration of the quantitative predictive power of this approach.

At this stage it is worth looking back to the Rice[38] and Mason[35]
analysis of embrittlement in copper due to bismuth segregation,
presented in Fig 6. The question is how R_O and S from equation (8)
are affected by segregation. From the above calculations, for the
most brittle alloy with 2 monolayers of bismuth segregated, in the
ideal solution approximation

$$\frac{\gamma_{int}(Cu/Bi)}{\gamma_{int}(Cu)} = \frac{H^{sub*}_{Bi}}{H^{sub*}_{Cu}} = 0.3 \tag{23}$$

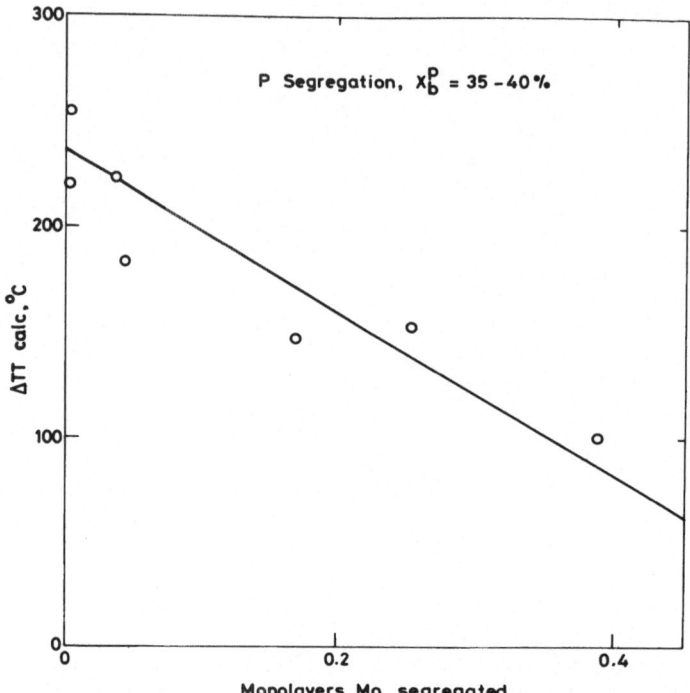

Fig 11   The remedial effect of Mo atoms segregated to a grain
         boundary in steel with 30-40% of a monolayer of P,
         deduced from the data of Dumoulin et al[46], after Seah[47].
         The line is the theoretical prediction.

A simple broken bond model for a step formed by slip at 45°,
using a cubic lattice, as exampled in Fig 8, shows that

$$\frac{\gamma_{step} \ (Cu/Bi)}{\gamma_{step} \ (Cu)} = \frac{H_{Bi}^{sub*} + H_{Cu}^{sub*}}{2H_{Cu}^{sub*}} = 0.65 \qquad (24)$$

This result changes the predictions of Mason[35] shown in Fig 6 to
that shown in Fig 12. Clearly, now, Rice[38] and Mason's[35] theory
agrees very closely with the experimental measurements. Careful
quantification of the Auger electron spectroscopic data from
these alloys[48] shows that, even for the dilute 0.009% bismuth
alloy, the grain boundary content is about 70% of a monolayer
(here we take into account the relevant relative sensitivity
factors, electron inelastic mean free paths, analyser geometry
and assumed equal partitioning of the segregant on fracture) and
this is sufficient to shift the predictions into the brittle

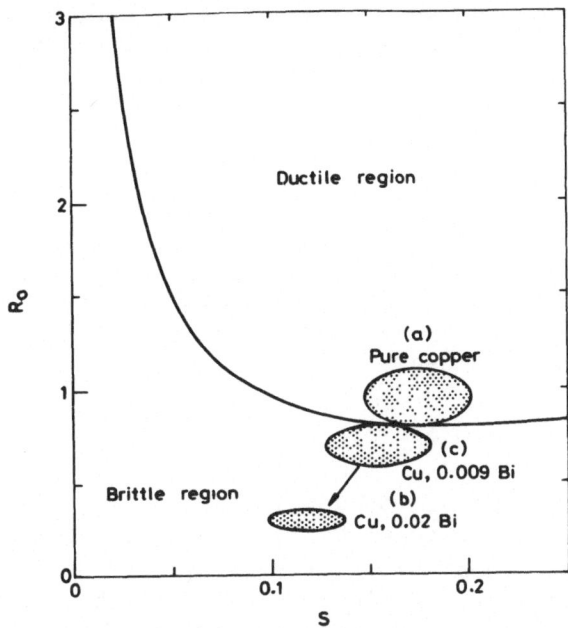

Fig 12 R_0 and S values for prospective crack orientations in
(a) pure copper (b) copper, 0.02 wt% Bi and (c) copper,
0.009 wt% Bi, using equations (23) and (24) to compare
with Fig 6.

region.

The above prediction[38] that the segregation alters the
dislocation nucleation and propagation term is rarely appreciated.
Supportive experimental evidence that this is a real and important
effect is seen in work originating in the very different field of
rock drilling. Here, adsorption from specific liquid environments
has been shown to alter the behaviour of dislocations nucleated at
the surface, dramatically, in freshly cleaved MgO single crystals[49].

The above analyses give a fairly clear atomistic view of the
central phenomena that cause only certain segregants to promote
intergranular fracture and also to understand the relative embrittling
effects of different segregants. More recently McMahon and his
co-workers have tackled the important question of how do these
effects relate to the macroscopic observations of behaviour, ie how
precisely is γ_p related to γ since it is, after all, γ_p that
dominates the macroscopic behaviour. McLean[4] had noted that the
critical stress for failure at the crack tip, σ_c , was

proportional to γ and at the same time was the source of generation of γ_p. No explicit relationship was available until the analysis of McMahon and Vitek[50]. They argue that γ_p may be written:

$$\gamma_p = \sigma_{eff} \, V_p \dot{\varepsilon}_p / \nu_c \qquad\qquad (25)$$

where σ_{eff} is the effective stress at the crack tip, V_p is the volume of the plastic region associated with the advancing crack per unit length of crack, ε_p is the plastic strain in that region and ν_c is the crack front velocity. Using the data of Tomalin and McMahon[51] then

$$\dot{\varepsilon}_p \, \alpha \, \sigma^m, \qquad m = 6.5 \pm 2.2 \qquad\qquad (26)$$

Assuming that V_p and ν_c do not depend on σ_{eff}, then $\gamma_p \, \alpha \, \sigma_{eff}^{7.5}$. Since σ_{eff} is proportional to γ the final result is that $\gamma_p \, \alpha \, \gamma^{7.5}$, ie the overall work of fracture falls rapidly as γ is reduced, leading to a rapid loss in ductility. More recent work[52,53] based on a time-dependent Dugdale-Bilby-Cottrell-Swinden model, gives somewhat weaker effects with the power dependence of γ_p on γ reduced to around 1.5.

Before leaving the topic of low temperature intergranular fracture, some additional concepts should be mentioned as these, to a large extent, represent alternative views of evaluating some of the aspects described above. Jokl et al[52] consider that in transition metals much of the cohesion derives from the d-band electrons and that elements which segregate and dilute the d-band concentration would lead to reduction in the cohesion. They also suggest that segregants which promote bonding in the grain boundary plane rather than normal to it would also reduce cohesion. Jokl et al also note that the relative embrittling effects of different segregants increases as $-\Delta G_b$ increases, as first suggested by Hondros[25]. Since ΔG_s and ΔG_b are related in their chemical bonding and strain release terms[54] this type of effect is generally expected from equation (18).

Two dimensional grain boundary phases have also been proposed[55,56] as well as reconstruction[57], to cause intergranular weakness. Transmission electron microscopy clearly shows that reconstruction of the grain boundary can occur as a result of bismuth segregation in copper where facetting has been observed with typical facet sizes being in the range 50-500 nm[58]. Dingley and Biggin[59] and Pond et al[60] both point out that dislocations moving from one grain to the next require the grain boundary to absorb dislocations in order to satisfy the conservation of

Burger's vectors.   These grain boundary dislocations pile up in
the boundary as yielding occurs. The way this occurs must be
affected by the composition of the boundary but as yet predictions
are difficult. However, the accumulation of dislocations in the
grain boundary will inevitably lead to its weakening.

The above theories and descriptions all apply to low temperature
brittle fractures which include phenomena such as temper brittleness and 350 °C embrittlement. Intergranular fractures occur by
quite separate mechanisms in other failure categories and hence
other segregated impurities become more important than those
predicted in the above work. The phenomena to be considered, in
outline, concern intergranular failures in hydrogen embrittlement,
stress corrosion cracking, liquid metal embrittlement, stress
relief cracking and creep embrittlement. All but the last two
items are also the subjects of other survey lectures.

HYDROGEN EMBRITTLEMENT

Hydrogen embrittlement is a well established field but the
intergranular mode of failure has been considered only in recent
years in relation to the prior grain boundary composition. In
1974 Latanision and Oppenhauser[61] proposed that hydrogen entry into
the grain boundary region, which exacerbates grain boundary embrittlement, should be promoted by hydrogen recombination poisons. These
poisons are well known in their action at the free surface of
metals by adsorption from electrolytes[62] and an equivalent effect
was attributed to the same elements when at grain boundaries,
acting as the crack front proceeds down the boundary. Typical
elements thought of as recombination poisons are, in descending
order of effect[62] in groups V and VI, P, S, As, Sb, Se, Te and Bi.
These are all elements which, in very simple terms, have a stronger
bonding with hydrogen than does iron, so that monatomic hydrogen,
from dissociative adsorption of H_2 or from solution, resides at
these sites much longer than at the normal iron sites and so has a
greater chance of being taken into the lattice. The increased
take-up of hydrogen here is simply proportional to the ratio of the
hydrogen atom residence times at sites involving "poison" atoms to
those at a normal sites together with the appropriate Boltzmann
factors. This approximation is only valid for relatively weak bonds
between the poison atom and hydrogen. For strongly bonding species,
such as nitrogen and oxygen, the residence times become so long
that the hydrogen atom prefers to stay there rather than enter the
iron lattice. Perhaps the term recombination poison is incorrect
in that in this field the poison atom is really an adsorption
promoter. Not only does the promoter increase the rate of hydrogen
take-up but it also increases the total amount that can be taken
up. The above order of elements appears to be the same for both of
these processes[62]. Thus general predictions may be made that these
grain boundary segregants will promote hydrogen embrittlement in

the above order if the fracture process concerns the hydrogen concentration in the metal just ahead of the crack tip.

It is not the purpose here to review the mechanisms of hydrogen embrittlement but merely to indicate how grain boundary segregation can affect the predictions of the main mechanistic models. A model, originally proposed by Troiano[63] envisages that the hydrogen embrittlement arises through the hydrogen that has entered the metal collecting in the lattice at points of maximum triaxial stress near the crack tip. Additionally, enhancement of hydrogen at the grain boundary over that in the bulk will occur as a result of attraction to the poison atoms on the unfractured boundary and this adsorption lowers the work of fracture, γ further, as described earlier. Since the hydrogen can diffuse into the metal ahead of the crack tip it may be considered sufficiently mobile to obey the constant μ work of fracture, γ_μ, of equation (13) which has been shown to be even more affected by adsorption than γ_Γ[38,41]. In this way the hydrogen can reduce the work of fracture considerably. A second model of hydrogen embrittlement[64] considers adsorption of hydrogen directly from the gaseous environment and thus lowering of the work of fracture directly through the γ_μ process. Since the recombination poisons cited above all have stronger bonds with hydrogen than does iron, segregants of this type will not compete with hydrogen for the adsorption sites but will add sites of a higher free energy of adsorption. Thus, in this case as well as that above, a co-operative effect of segregation and hydrogen embrittlement is expected that is greater than a simple addition of the two separate embrittlements.

A number of publications have appeared linking temper brittleness with hydrogen embrittlement but only one[52] provides detailed experimental evidence of hydrogen embrittlement as a function of the segregation of different impurity species. Jokl et al[52] measure both the threshold stress intensity for cracking in 1.7 bar of H_2 at 23 °C, K_{th}, and the K_{1c} in Ni-Cr steel with a range of segregation levels of each of Sb, Sn and P separately. If the hydrogen embrittlement were merely added to the temper brittleness and the segregants played no hydrogen enhancing role, K_{th} and K_{1c} would have a one-to-one correspondence, independent of segregant. Fig 13 shows Jokl et al's[52] results replotted to relate K_{th} and K_{1c}. It is clear that K_{th} does not depend uniquely on K_{1c}, as it would if hydrogen embrittlement were added simply to temper embrittlement, but on the segregant type as well. The effect of Sb is greater in temper brittleness than in hydrogen embrittlement compared with the action of P. This is in accordance with the predictions for hydrogen recombination poisons as described by Smialowski[62]. Typical error bars are shown on the curves and the results look convincing enough at this stage.

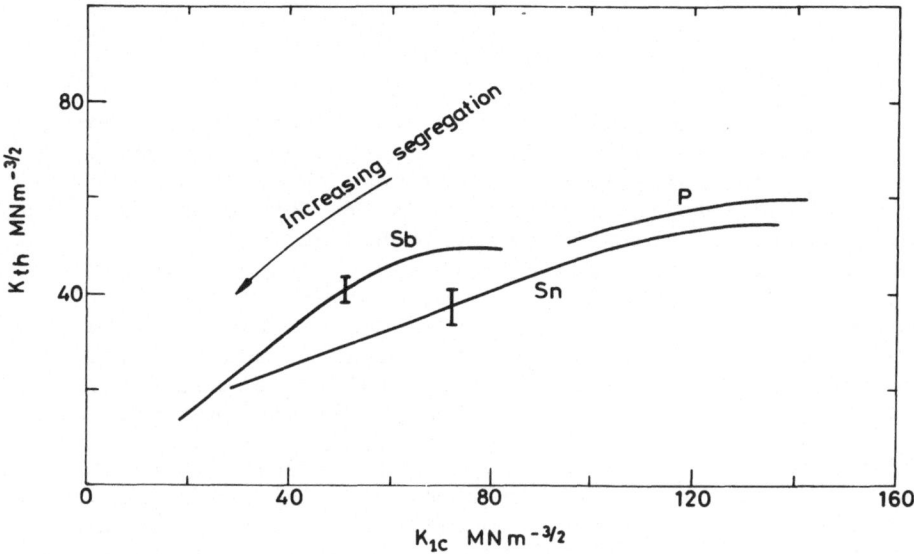

Fig 13   Correlation of K_{th} from fracture data of NiCr steel in
          hydrogen and K_{1c} for various segregation levels of one of
          P, Sb and Sn, after Jokl et al[52].

LIQUID METAL EMBRITTLEMENT

        Liquid metal embrittlement (LME) often occurs by an intergranular
fracture path. LME itself is covered in another survey lecture,
but for the intergranular mode, the prior grain boundary composition
will be important and we should include some general remarks here.
Embrittlement couples are often characterised by low mutual
solubility and a lack of intermetallic formation. These rules are
not always adhered to but suggest that the liquid metal by mildly
wetting and have, in regular solution terms a small positive or
negative value for ω in equation (22).

        It should be noted that many of the criteria for fracture
discussed in section 2 are exactly the same in LME. Thus Kamdar[66]
considers the effects of the liquid metal on the ideal work of
fracture term and also the ratio of σ/τ. We have also seen above
how powerful the quasi-chemical bond term theory can be, and
Kelley and Stoloff[67] show its predictive power for embrittling
couples in practice. In very simple terms, if we consider the LME
of a matrix A with segregant B in a liquid metal C, embrittlement
of the pure metal goes by replacing AA and CC bonds by AC bonds.
Using the pair-bond theory described earlier, the ideal work of
fracture with no segregation is given by:

$$FE(0) = \left(\frac{Zg}{a_A^2}\right) 2\omega_{AC} \tag{27}$$

in contrast to equation (19) for fracture without an environment. If ω is strongly negative, $FE(0)$ is negative, the crack spontaneously propagates in all directions, the matrix bonds are all ruptured and intermetallic compounds are formed on the surface. This interlayer then controls further behaviour. If ω is zero, dissolution easily occurs and rapidly increases the freezing temperature. Embrittlement will be most severe if the liquid environment is retained, and the crack does not blunt. Also $FE(0)$ must not be too high. Thus ω_{AC} will be small and positive. Embrittlement of the segregated metal involves replacement of ω_{AC} in equation (27) by $\{(1-\frac{1}{2}X_b)\omega_{AC} + \frac{1}{2}X_b\omega_{BC}\}$ if the segregant atoms remain on the metal surface. Thus if $\omega_{BC}<\omega_{AC}$ greater embrittlement ensues and if $\omega_{BC}>\omega_{AB}$ the segregant is remedial to embrittlement. This simple argument explains why segregated P is remedial to both the mercury embrittlement of copper[68] and the lead embrittlement of steel[69] whereas Sb and Sn segregation are detrimental to the lead embrittlement of steel[69].

INTERGRANULAR STRESS CORROSION CRACKING

Again, in stress corrosion cracking (SCC) intergranular fractures often occur and the prior grain boundary composition must be taken into account. Two extreme conditions of ISCC may be identified, hydrogen embrittlement cracking, where the stress is very important, and intergranular corrosion, where the action of the stress is primarily in opening the crack to allow ingress of the active liquid environment. A number of articles has been published showing that the grain boundary segregation, as monitored by temper brittleness, has a detrimental effect on SCC[70,71], but it is only recently that the relative effects of different segregants have been measured[72,73], permitting the genuine test of a micro-mechanistic model. The measurements for a range of segregants in mild steel, in both agressive and neutral electrolytes during tensile and bend tests, allow the precise effect of the segregant/agressive electrolyte interaction to be identified. The results for the tensile tests in ammonium nitrate solution and an inert environment (paraffin) are shown in Fig 14 for the pure mild steel base and for the same material with each of ten possible elements segregated to the grain boundary. Factors of three reduction in life are observed. The results are interpreted in terms of a dissolution model in which the divergence of the segregants' electrochemical potential from that of iron, ΔE^o, drives the dissolution. If the segregant has a more positive potential, the segregant atoms themselves are

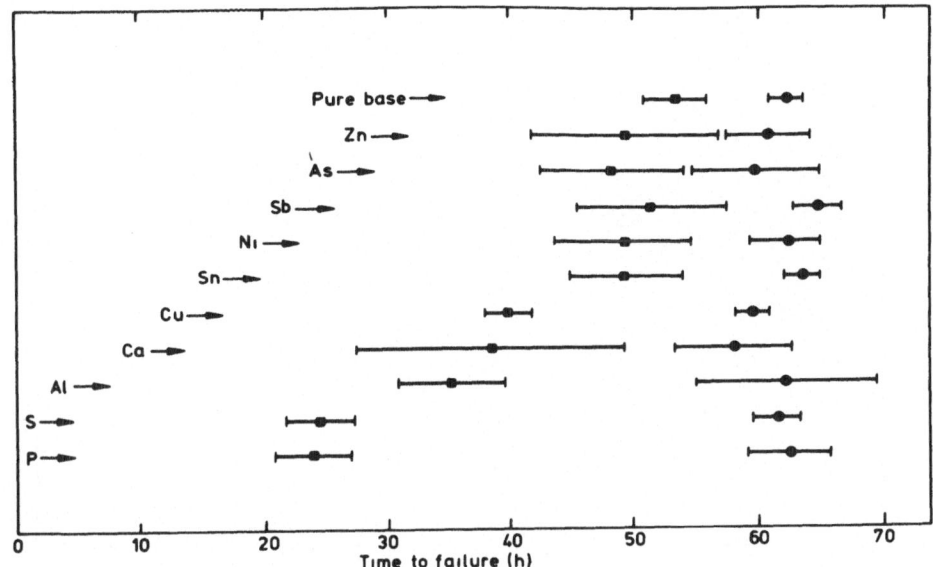

Fig 14   The fracture times of 11 mild steel alloys corroded at a
constant strain rate, ● in paraffin, ■ in 5 M ammonium
nitrate, after Hondros and Lea[11]. The grain boundary
segregants are shown to the left.

dissolved more rapidly and a very sharp grain boundary crack
develops. On the other hand, a more negative potential promotes
dissolution of the iron atoms adjacent to the boundary, enhancing
the intergranular corrosion, but with a somewhat broader crack
than in the former case. Thus, the enhanced SCC is proportional to
$|\Delta E^O|$ but the relative proportionality for the positive and
negative values is undefined. The experimental results shown in
Fig 14 may be replotted to show the relative effect on life for
each segregant, normalised to a grain boundary content of one
monolayer. This is shown in Fig 15 plotted as a function of ΔE^O
It is clear that a general proportionality to ΔE^O exists but
that the effects of positive and negative values of ΔE^O are of
the same order.

STRESS RELIEF CRACKING AND CREEP EMBRITTLEMENT

The above fracture phenomena all have a large degree of
commonality in that the grain boundary separates progressively as
a crack progresses under an applied stress. The chemical
conditions of each phenomenon are quite different but a similar
type of analysis can be applied. This is not the case for the

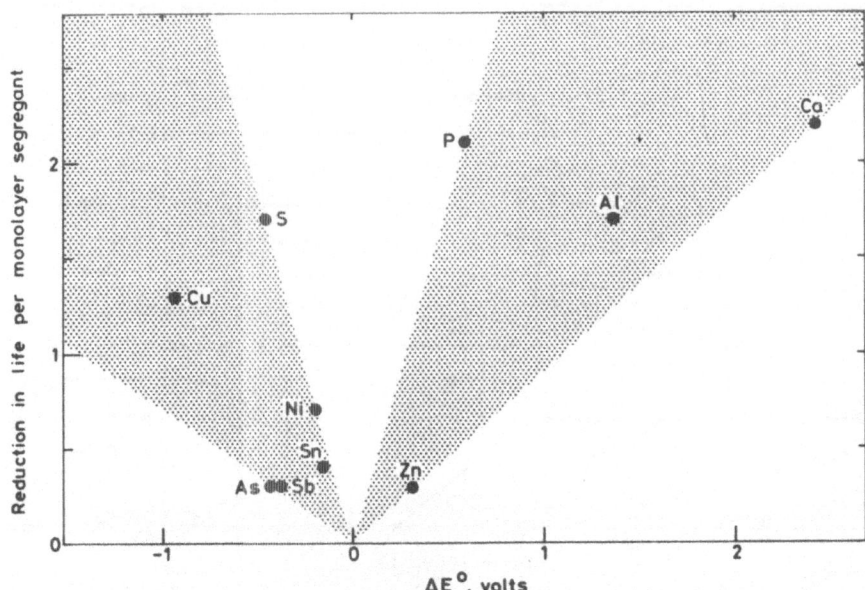

Fig 15   Correlation of the detrimental effects of segregants on
           the ISCC of mild steel in 5M ammonium nitrate solution
           with the difference in electro-chemical potential of the
           segregant and iron matrix atoms in this electrolyte, after
           Lea and Hondros[73].

stress relief cracking and creep embrittlement of creep resistant
low alloy steels. Stress relief cracking occurs in the heat
affected zones of weldments during the stress relief annealing
cycle, typically 1 hour at 700 °C. Creep embrittlement occurs in
the body of the material during service and may only be evident
after some years. Both of these fracture processes occur when the
steels are stressed at high temperatures. Under the applied
stress cavities are nucleated on transverse grain boundaries.
These cavities grow by the accumulation of vacancies until their
density on the grain boundary is sufficiently high that the
remaining net section is not sufficient to withstand the applied
load. Fracture then occurs and the fracture surfaces are seen to
be intergranular but covered with a high population of cavities
with sizes of the order of 1 μm. The effect of impurities such as
P, Sn and Sb in these steels is to lower the rupture ductility and
produce fractures which are characterised by a higher density of
smaller cavities than is the case for purer materials[15,16,73].

     The situation of cavity growth on grain boundaries, eventually
leading to intergranular failures, has been analysed in different

approximations many times[74,75,76]. For the purpose of this discussion it is enough to note that these analyses are generalised in the equation for rupture strain, ε_r , given by Skelton[77] for a model involving a continuous creation of cavities at a rate N per unit area of grain boundary and growing by accumulation of vacancies arriving by grain boundary self diffusion D_{gb} :

$$\varepsilon_r \alpha \ (D_{gb}/N)^{1/5} \tag{28}$$

Seah[17] proposed that impurities could affect the fracture properties through their effects on D_{gb} or N or both together. If it is envisaged that embryo cavities are formed, for instance by slip blocked at a grain boundary precipitate, a range of embryo sizes will occur. Those with radii less than $2\gamma_s/\sigma$, where σ is the local tensile stress, will sinter and disappear, whereas those of larger size become part of the process eventually leading to intergranular failure. Thus reduction of γ_s leads to an increase in the cavity nucleation rate, N, and a reduction of the rupture ductility. That segregants reduce γ_s dramatically is well established[26] and indeed much of the work on surface energies was completed with the creep problem in mind[78]. The extent of the reduction may be predicted from the surface analogues to equations (2) and (3) in terms of the free energy for surface segregation, ΔG_s. In this way Seah[17] showed that the predicted relative effects for P, Sn, Sb, As and Cu in a low alloy steel agreed fairly well with the experimental data for rupture ductilities in stress relief cracking[15,79].

In the lower stress, lower temperature and longer time conditions of long-term creep, a different result was obtained[16] and this was rationalised in terms of the effect of grain boundary segregants on D_{gb}. The much higher value of D_{gb} compared with the lattice diffusion, D_v , arises through the disorder at the boundary. As segregation occurs, the degree of order in the boundary improves, its free energy is reduced and also the grain boundary diffusivity is reduced. This effect was first rationalised by Borisov et al[80] who related D_{gb} and γ_b by a theoretical model:

$$\gamma_b = \left(\frac{kT}{\psi a^2}\right) m \left[\ln \frac{D_{gb}\delta}{aD_v} - \ln m\right] \tag{29}$$

where the symbols k and T have their usual meaning, $\psi = 1$ for interstitial and 2 for vacancy diffusion, a is the matrix lattice spacing and δ is the grain boundary width equal to ma. This relation was derived to relate γ_b and D_{gb} for single component systems but the theory has been shown[8] to be equally valid where γ_b is controlled by segregation in a binary system[81]. Bernardini et al rearrange the terms in equation (29) to show that the grain boundary self diffusivity with segregation, D_{gb}, is related to the value in the pure material, D_{gb}^o, by:

$$D_{gb} = D_{gb}^o \{1 - (2\alpha - b_v)X_c\} \tag{30}$$

where α is the ratio of the segregant atom density at the grain boundary to that in the bulk and b_v is the fractional change in the bulk diffusivity, D_v, per unit change in solute molar content. The α values are related directly to the grain boundary enrichment ratios tabulated elsewhere[30] and, for impurity segregation in low alloy steels, generally dominates the b_v term[82]. In this case, for a given X_c, the stronger the grain boundary segregation the greater the reduction in D_{gb} and, from equation (28), the greater the reduction in rupture ductility. In this way Seah[17] showed that the creep ductility measurements at 550 °C, compiled by the British Steelmakers Creep Committee[83], exhibited an impurity dependence and that this dependence was as expected for a model involving the grain boundary self diffusivity.

CONCLUSIONS

     In the above analysis it is shown that many types of inter-granular fracture can occur and that all are modified by the grain boundary segregation of adventitious impurities. Through an understanding of the atomic micromechanisms it is shown that different parameters of the segregants promote each of the fracture phenomena and that therefore a different hierarchy of problem elements exists in each case. In low temperature fracture the embrittling sensitivity of each element concerns its direct bonding to the matrix lattice, in hydrogen embrittlement it concerns the metastable bond with hydrogen, in liquid metal embrittlement it seems as though the heat of mixing of the segregant and liquid metal is important, in stress corrosion cracking the divergence from the lattice electrochemical potential, in stress relief cracking it is the effect on surface energy and in creep embrittle-ment, the effect on the grain boundary self diffusivity.

     Remedial procedures for intergranular embrittlement have so far concentrated on reducing the level of segregation. Thus, in temper brittleness[84], stress corrosion cracking[10] and stress relief cracking[19], rare earth metal additions have been used with full effect. Realistic proposals for remedial procedures through an understanding of the atomistics have yet to be made.

ACKNOWLEDGEMENTS

The authors would like to thank Drs C Lea and M T Anthony for stimulating discussions.

REFERENCES

1. J H Westbrook, Phil Trans R Soc Lond A 295: 25 (1980).
2. W C Roberts-Austen, Phil Trans R Soc Lond A 179: 339 (1888).
3. M P Seah, Acta Met 28: 955 (1980).
4. D McLean, Grain Boundaries in Metals, Oxford University Press, (1957).
5. C J McMahon, Temper Embrittlement in Steel ASTM STP 407: 127 (1968).
6. M P Seah, Surface Sci 53: 168 (1975).
7. C L Briant and S K Banerji, Int Met Revs 23: 164 (1978).
8. P E Irving, M P Seah and A Kurzfeld, Proc 2nd Int Conf on Mech Behav of Materials, Boston, ASM p563 (1976).
9. S K Banerji, C J McMahon and H C Feng, Met Trans 9A: 237 (1978).
10. C Lea, Met Sci 14: 107 (1980).
11. E D Hondros and C Lea, Nature 289: 663 (1981).
12. M G Nicholas and C F Old, J Mat Sci 14: 1 (1979).
13. D A Melford, Phil Trans R Soc Lond A 295: 89 (1980).
14. C Lea, R Sawle and C M Sellars, Phil Trans R Soc Lond A 295: 121 (1980).
15. B L King, Phil Trans R Soc Lond A 295: 235 (1980).
16. H R Tipler, Phil Trans R Soc Lond A 295: 213 (1980).
17. M P Seah, Phil Trans R Soc Lond A 295: 265 (1980).
18. M P Seah, P J Spencer and E D Hondros, Met Sci 13: 307 (1979).
19. C J Middleton, Phil Trans R Soc Lond A 295: 305 (1980).
20. A A Griffith, Phil Trans R Soc Lond A 221: 163 (1920).
21. M Polanyi, Z Phys 7: 323 (1921).
22. E A Guggenheim, Modern Thermodynamics, London p25 (1933).
23. E D Hondros and M P Seah, Met Trans 8A: 1363 (1977).
24. E D Hondros, Proc Roy Soc A 286: 479 (1965).
25. E D Hondros, Proc Melbourne Conf on Interfaces, Ed R C Gifkins, Butterworths p77 (1969).
26. E D Hondros and D McLean, Monograph No 28, Society of Chemical Industry: London p39 (1968).
27. M P Seah and C Lea, Phil Mag 31: 627 (1975).
28. M P Seah, Acta Met 25: 345 (1977).
29. M P Seah, J Vac Sci Technol 17: 16 (1980).
30. E D Hondros and M P Seah, Int Met Revs 22: 262 (1977).
31. E D Hondros and D McLean, Phil Mag 29: 771 (1974).
32. A Kelly, W R Tyson and A H Cottrell, Phil Mag 15: 567 (1967).
33. E Smith and J T Barnby, Met Sci J 1: 56 (1967).
34. J R Rice and R Thompson, Phil Mag 29: 73 (1974).
35. D D Mason, Phil Mag A 39: 455 (1979).
36. M P Seah, Proc R Soc Lond A 349: 535 (1976).
37. M P Seah, Surface Sci 53: 168 (1975).

38.   J R Rice, Effect of Hydrogen on Behaviour of Materials,
      Ed A W Thompson and I M Bernstein, Met Soc AIME p455 (1976).
39.   R J Asaro, Phil Trans R Soc Lond A 295: 150 (1980).
40.   J P Hirth, Phil Trans R Soc Lond A 295: 139 (1980).
41.   J P Hirth and J R Rice, Met Trans 11A: 1501 (1980).
42.   D D Mason, Segregation Induced Embrittlement of Grain
      Interfaces, ScM Thesis, Brown University, May (1977).
43.   M P Seah, J Catal 57: 450 (1979).
44.   R Hultgren, P A Desai, D T Hawkins, M Gleiser and K K Kelly,
      selected Values of the Thermodynamic Properties of Elements,
      and Selected Values of the Thermodynamic Properties of
      Binary Alloys, ASM, Metals Park (1973).
45.   O Kubaschewski and C B Alcock, Metallurgical Thermochemistry,
      5th Edn Pergamon, Oxford (1979).
46.   Ph Dumoulin, M Guttmann, M Foucoult, M Palmier, M Wayman and
      M Biscondi, Met Sci 14: 1 (1980).
47.   M P Seah, Scripta Met, 15: to be published.
48.   B D Powell and H Mykura, Acta Met 21: 1151 (1973).
49.   A R C Westwood and R M Latanision, NBS Special Pub 348: 141
      (1972).
50.   C J McMahon and V Vitek, Acta Met 27: 507 (1979).
51.   D S Tomalin and C J McMahon, Acta Met 21: 1189 (1973).
52.   M L Jokl, J Kameda, C J McMahon and V Vitek, Met Sci 14: 375
      (1980).
53.   M L Jokl, V Vitek and C J McMahon, Acta Met 28: 1479 (1980).
54.   M P Seah, J Phys F 10: 1043 (1980).
55.   M Guttmann, Phil Trans R Soc Lond A 295: 169 (1980).
56.   E S Machlin, Scrip Met 12: 111 (1978).
57.   W G Hartweck, Scrip Met, 15: to be published.
58.   A M Donald and L M Brown, Acta Met 27: 59 (1979).
59.   D J Dingley and S Biggin, Phil Trans R Soc Lond A 295: 165
      (1980).
60.   R C Pond, D A Smith and R H Wagoner, In Fracture 1977, Vol 2,
      p155, ICF4, Waterloo, Canada (1977).
61.   R M Latanision and H Opperhauser, Met Trans 5: 483 (1974).
62.   M Smialowski, Hydrogen in Steel, Pergamon, London (1962).
63.   A R Troiano, Trans Am Soc Metals 52: 54 (1959).
64.   N J Petch and P Stables, Nature 169: 842 (1952).
65.   M G Nicholas and C F Old, J Mat Sci 14: 1 (1979).
66.   M H Kamdar, In Fracture 1977, Vol 1, p387 ICF4, Waterloo,
      Canada (1977).
67.   M J Kelley and N S Stoloff, Met Trans 6A: 159 (1975).
68.   L P Costas, Corrosion, 31: 91 (1975).
69.   S Dinda and W R Warke, Mat Sci Eng 24: 199 (1976).
70.   U Q Cabral, A Mache and A Constant, C R Acad Sci 260: 6887
      (1965).
71.   R P Harrison, D de G Jones and J F Newman, Proc Int Conf on
      Stress Corrosion Cracking and Hydrogen Embrittlement of
      Iron-Based Alloys, Firminy, France, June 1973, NACE, Houston
      (1977).

72. C Lea and E D Hondros, Proc Roy Soc Lond A, to be published.
73. R Bruscato, Weld J Res Suppl 49: 148S (1970).
74. D Hull and D E Rimmer, Phil Mag 4: 673 (1959).
75. R Raj and M J Ashby, Acta Met 23: 653 (1975).
76. M V Speight and W Beere, Met Sci 9: 190 (1975).
77. R P Skelton, Met Sci 9: 192 (1975).
78. H R Tipler and D McLean, Met Sci J 4: 103 (1970).
79. R C Miller and A D Batte, Met Constr 7: 550 (1975).
80. V T Borisov, V M Golikov and G V Scherbedinskiy, Phys Met
     Metallog 17: 80 (1964).
81. J Bernardini, P Gas, E D Hondros and M P Seah, Proc Roy Soc
     Lond A: to be published.
82. A D LeClaire, Thin Solid Films 25: 1 (1975).
83. British Steelmakers Creep Committee High Temperature Data,
     Iron Steel Inst Spec Pub 156: 257 (1973).
84. M P Seah, P J Spencer and E D Hondros, Met Sci 13: 307 (1979).

DISCUSSION

Comment by R.A. Oriani:

What is the present experimental situation on the effect of specific grain boundary structure and orientation upon the embrittlement by some one solute?

Reply:

The present position is poor. Professor Gleiter has shown the first efficient study of this sort in his presentation at this meeting. It seems highly likely that significant measurements in this area will be obtained by his approach using the microspheres coupled with the new generation of Scanning Auger Microprobes.

Comment by F. Mazza:

The detrimental effect of elements, such as As and P, on the susceptibility of alloys to intergranular attack or to transgranular stress corrosion cracking may assume great practical significance expecially for those alloys in which addition of these elements is usually made in order to increase the resistance of the alloys to dealloying processes in chemical environments.

Experiments carried on in our laboratory gave results in good accord with those now presented by Dr. Seah.

Reply:

These results are very interesting and increase the weight of evidence being accumulated by recent research that impurity segregation can be very important in intergranular stress corrosion crackin

ABOUT INTERGRANULAR FRACTURE OF HYDROGENATED PURE METALS

S. Talbot-Besnard

CNRS,
Ecole Centrale 92290 Chatenay-Malabry, France

There are numerous examples of hydrogenated materials which
fail intergranularly when they are pure and only in this case. In
some cases, carbon, nitrogen or phosphorus were added in solid
solution and it resulted in transgranular cracks and failure.
Grain boundaries were strengthened.

These phenomenons were observed in pure iron (1) in Fe 0,15%
Ti (2) in Fe Cr (3) in austenitic stain less steels (4) in nickel
base alloys (5).

For example, in pure iron (1) the detail of the aspect of
failure is a function of the quantity of hydrogen which was
introduced in the metal and of strain rate (Fig 1-2-) . Craters
are formed and have a very high propagation rate. If the grain
size of the same material decreases, the aspect of the failure
changes : It can be mixt (ductile, inter and transgranular) or
wholly ductile in function of the quantity of hydrogen and strain
rate. Then, it seems that there is a geometrical factor, big
crystals making an easier intergranular failure. But when nitrogen
is added in solid solution transgranular cracks and failure occur.

In the same way, C and P additions decrease the rate of
embrittlement of the pure inconel type alloy (Fig. 3). In hydrogenated high purity 76% Ni - 16% Cr - 8% Fe alloy, the failure is
always intergranular. The addition of metallic elements such as
Sb or Sn has relatively a little effect on intergranular embrittlement. But carbon or phosphorus additions have an influence
which depends on the thermal treatment applied to the alloy.
With 300 ppm C in solid solution, the brittle fracture is always

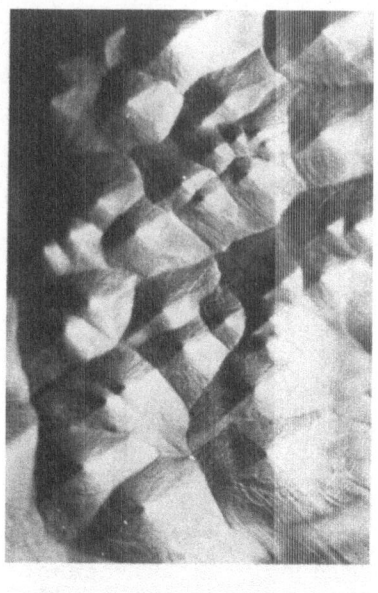

x180

Fig. 1. Intergranular crater
$\varepsilon = 3 \cdot 10^{-4} \text{s}^{-1}$ $\sqrt{I} = 5 \text{mA}^{1/2} \text{cm}^{-1}$.

x180

Fig. 2. Intergranular crater
$\varepsilon = 3 \cdot 10^{-4} \text{s}^{-1}$ $\sqrt{I} = 1 \text{mA}^{1/2} \text{cm}^{-1}$.

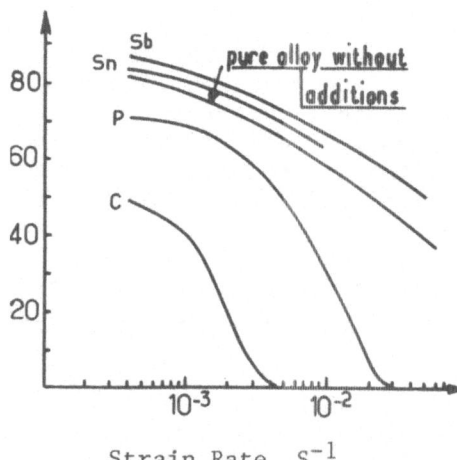
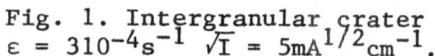

Strain Rate, s^{-1}

Fig. 3. Hydrogen embrittlement of
alloys that have undergone tensile
strain with simultaneous cathodic
hydrogen charging (A%) and (A%)$_H$
are the strain to fracture without
and with hydrogen charging.

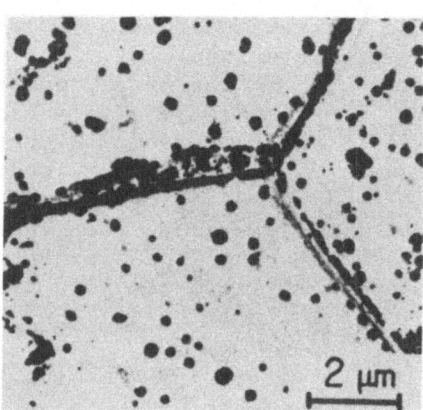

Fig. 4. Tritium trapped on a
grain on a grain boundary of
Fe Ti 0,15%. (2)

intergranular. But a grain boundary segregation of carbon (without a formation of carbides) considerably increases their cohesion and the rupture surface is both inter and transcrystalline.

To understand the action of addition elements it should be necessary to know the reason of the boundary weakness of pure materials.

Several explanations were proposed.

1) It could come from a preferential segregation of hydrogen in grain boundaries. Such a segregation has been shown after a cathodic charging with tritium followed by an observation by microautoradiography on Fe 0,15% Ti (Fig. 4). The study was done after different thermal treatments. The biggest segregations of tritium on grain boundaries were obtained after recrystallisation and quench. In this case the study of the localisation of 14 carbon shows that carbon is in small precipitates or in solid solution in the grain boundaries. The study of cracking of this alloy shows that in this case grains boundaries are strengthened. Then embrittlement of grain boundaries can't be attributed to a segregation of hydrogen. Different results of embrittlement of grain boundaries by carbon can be found (6). They can be explained by the fact that these studies were made on commercial metals and the behavior of other elements as sulfur was not controlled and considered.

2) It could come from a segregation of ultimate impurities in grain boundaries different from carbon and nitrogen. This hypothesis was checked in zone melted iron. A notched sample was broken in an Auger spectrometer by a knock under a vacuum of 10^{-6} torr. The sample was brought at low temperature in liquid nitrogen. The failure was intergranular. Auger analysis is reported fig.5 . Q_i is calculated from h_i corrected height of derivated Auger peaks.

$$Q_i = \frac{h_i x 100}{\Sigma_i \, h_i}$$

i beeing the considered element of impurity.

Embrittling impurities as P, As,Te, Sn, Sb, Se, N, Ni, Cr, are not seen. The peak of S is low. Cl also is present in very low quantity.

However, oxygen peak is high. In fig 6 we have put the peak height versus time after failure. We see that oxygen comes from residual atmosphere. For $1 <t< 5$ mn the peak height varies like $\sqrt{t}$ with a correlation coefficient r = 0,99993. Extrapolation to t = 0 gives a peak height that equals zero and we can conclude that oxygen is absent at the moment of the failure. Carbon is inequaly present in various samples.

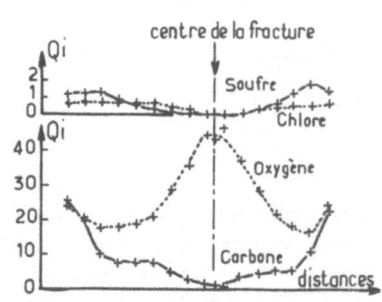

Fig.5. Auger analysis. Q_i is measured along a diameter of intergranular surface of failure.

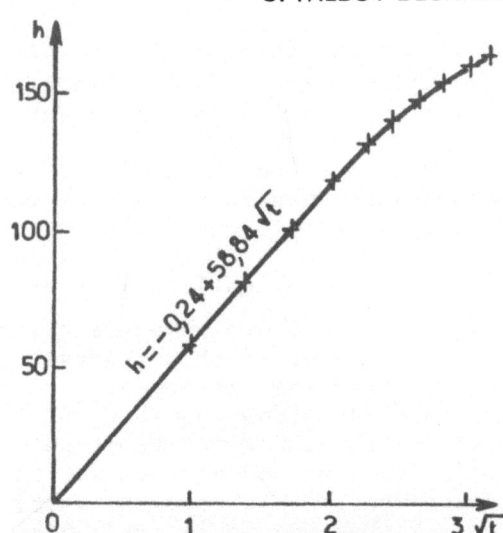

Fig.6. The height oxygen peak in the center of failure versus time after failure.

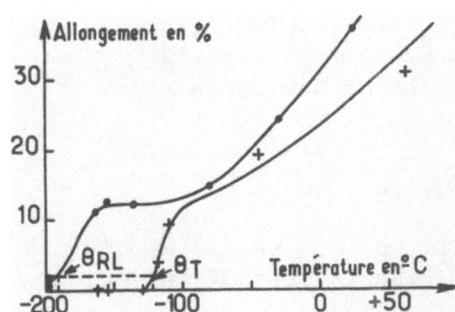

Fig.7. Strain versus tensile test temperature for high purity iron
. Slowly cooled
+ Quenched.

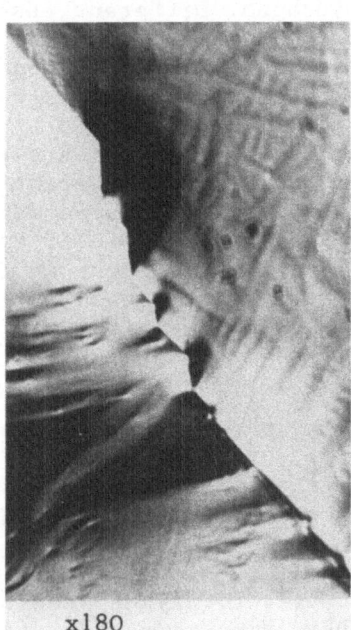

x180

Fig.8. Steps on a grain boundary of high purity iron observed on the intergranular failure.
$\varepsilon = 3 10^{-4} s^{-1}$, $\sqrt{I} = 0,7 mA^{1/2} cm^{-1}$

It is in a bigger quantity near the edge of the failure. This element comes from a massic diffusion after mechanical treatments of wire-drawing and lathe work in spite of pickling before annealing. The same result was obtained in Mo which fails intergranularly when it is pure without H_2 (8).

3) It could come from an intrinsec weakness of the clean grain boundary.

This hypothesis was retained by four authors (1-2-4-5). In the case of Fe-Cr 7-9,4% alloys it was said (3) that failure in liquid nitrogen beeing transgranular it could not be an intrinsec weakness, I do not think so. In high purity iron, the failure in liquid nitrogen is intergranular only when iron is quenched as we see on Fig. 7. If the metal is slowly cooled, ultimate impurities migrate towards boundaries and strenghthen them. The transition temperatures of iron after both thermal treatments are on each side of liquid nitrogen temperature (Fig. 7). Perhaps, the alloy Fe Cr, previously considered, has the same behavior.

It seems the weakness of a clean boundary may be a chemical and geometrical effect. A pure metal has big crystals. It results few orientations for boundary planes. However accomodation is necessary as we see Fig. 8 (1). Cracking increases with grain size as it was seen in iron and iron titanium alloys (2). Accomodation is still reduced by hydrogen that decreases the number of active slip planes. We see that on the aspect of the craters of intergranular failure when a small quantity of hydrogen is introduced slips occur (Fig. 2), but when there is much hydrogen, slips do not occur any more (Fig. 1).

It should be interesting to work on bicrystals for a better understanding of this phenomenon.

References.

1 A. Gourmelon, Thèse Docteur Ingénieur Paris VI 1974 AO 10737.

2 T. Asaoka, Thèse Docteur Ingénieur Paris Sud 1976.

3 C. Paes de Oliveira, M. Aucouturier and P. Lacombe, Corrosion 1980, 36 (2) p. 53-59.

4 D. Nejem, Thèse Docteur 3ème cycle Paris VI 1980.

5 M. Cornet, C. Bertrand, Da Cunha Belo, Hydrogen embrittlement of ultra pure alloys of the inconel 600 type : influence of the additions of elements (C, P, Sn, Sb). A paraître Met. Trans. A 1981.

6 B.B. Rath et I.M. Bernstein, Met. Trans. 1971, 2 2845.

7 M. Cornet, S. Talbot-Besnard, Symposium Environment sensitive fracture- Chicago Oct. 1977 Publ. AIME 1979 p 411-430.

8 G. Lorang, Thèse Docteur d'Etat 12 décembre 1980.

9 F. Faudot, J. Bigot et S. Talbot-Besnard C.R. 1971 273 C p. 1318-1321

RELATIVE EFFECTS OF S, Sb AND P ON THE INTERGRANULAR

FRACTURE OF IRON AND NICKEL TESTED AT CATHODIC POTENTIALS

R. H. Jones, S. M. Bruemmer,
M. T. Thomas and D. R. Baer

Pacific Northwest Laboratory
Richland, Washington 99352

INTRODUCTION

It is generally believed that grain boundary impurities contribute to intergranular fracture in the presence of hydrogen by
either; 1) decreasing the intergranular cohesion; 2) enhancing
hydrogen absorption along grain boundaries; or 3) a combination of
both processes. A comparison of the relative effects of S, Sb and
P on the intergranular fracture of iron and nickel has been completed[1,2] and will be summarized in this paper. With sulfur as a
bench-mark, the effect of S, Sb and P on the intergranular fracture of iron and nickel at cathodic potentials can be expressed by
the following equivalent grain boundary sulfur concentrations,
C_s :

$$C_s^{eq} \text{ (Fe)} = 5 \text{ Sb} + S + \gamma_p P \quad \gamma_p > 0 \qquad (1)$$

$$C_s^{eq} \text{ (Ni)} = \text{Sb} + S + 0.07 P \qquad (2)$$

It can be seen from equation 1 that antimony was five times
more effective when present in the grain boundaries of iron than
was sulfur while phosphorus had a negligible effect. In nickel,
equation 2, antimony and sulfur had similar effects while phosphorus was 1/15 as effective as sulfur. Comparisons can be made
between iron and nickel since it was observed that C_s^{eq} (Fe) =
C_s^{eq}(Ni). Therefore, antimony was five times more effective in iron
than nickel while phosphorus had a negligible effect in both iron
and nickel.

RELATIVE EFFECTS OF SULFUR, ANTIMONY AND PHOSPHORUS IN IRON

The relative effects of S, Sb and P on the intergranular fracture of iron tested at cathodic potentials is shown in Fig. 1a where
it can be seen that a grain boundary P concentration of 0.2 monolayers was insufficient to cause intergranular fracture while 0.2
monolayers of S or 0.08 monolayers of S with 0.03 monolayers of Sb
were sufficient to cause 100% intergranular fracture at slightly
cathodic potentials. Comparison of the results at cathodic potentials with those for embrittlement and hydrogen permeation are made
in Table I. This comparison shows that the effect of S, Sb and P on
intergranular fracture at cathodic potentials and at low temperatures
is very consistent while their effect on hydrogen permeation is only
similar to the fracture results at small (10^{-5} moles/ℓ) promoter
concentrations. Therefore, it is very likely that the major role of
Sb and perhaps S in iron tested at cathodic potentials is to reduce
the critical stress for intergranular fracture.

The small contribution of P to the fracture of iron at cathodic
potentials is qualitatively consistent with its effect on fracture
in the absence of hydrogen. Differences in the magnitude of the P
contribution arise from the contribution of other grain boundary
segregants such as C, N or Ni. It has been proposed that the presence of C or N at the grain boundaries of iron can counteract the
embrittling effect of P. Phosphorus was observed by Smialowski et
al.[5] to enhance hydrogen uptake in iron wires less than S but
greater than Sb and therefore it is concluded that P has a negligible effect on both the critical stress for intergranular fracture
and hydrogen permeation when present in the grain boundaries of
iron.

COMPARISON OF SULFUR IN IRON AND NICKEL

Sulfur segregated to the grain boundaries of iron and nickel
has been observed to have a very similar effect when they are
tested at cathodic potentials. This similarity is shown in Fig. 1b
where the percent intergranular fracture is plotted relative to the
grain boundary sulfur concentration in iron and nickel tested at
equal cathodic current densities. At 50% intergranular fracture,
the grain boundary sulfur concentration is 0.12 and 0.09 monolayers
in iron and nickel, respectively. This similarity suggests that
sulfur is activating the same embrittlement process in both materials. For this to be true, it is necessary that equal quantities
of sulfur in the grain boundaries of iron and nickel either reduces
the critical stress for intergranular fracture of iron and nickel
to values less than the homogeneous flow strength or cause the same
hydrogen permeation rate in iron and nickel. Briant and Messmer[7]
found from theoretical bonding calculations that sulfur in a
cluster of nickel atoms caused strong planar bonding with weakened

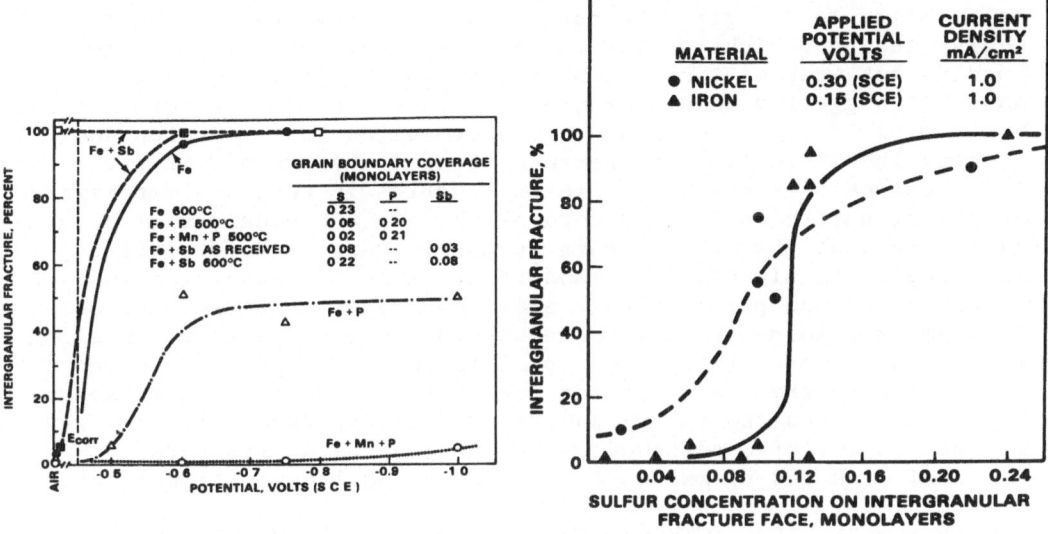

Fig. 1a. Comparison of the effect of Sb, S and P on inter-
          granular fracture of iron as a function of cathodic
          potential.
      b. Comparison of S in iron and nickel.

Table 1.   Comparison of Impurity Effects in Iron

Relative Ranking	Approximate Ranking	Material	Test	Reference
Sb > S >> P	5:1:0	Iron	Fracture at cathodic potentials	1, 1
Sb > S > P	---	Iron	Low temperature fracture	1, 2
Sb > S > P	20:1:0.5	Iron	Fracture appearance transition temperature	3
Sb > P	2:1	Ni-Cr Steel	Critical local tensile stress	4
S > P > Sb	---	Iron	Wire extension from hydrogen absorption	5
S >> Sb	---	Iron	Hydrogen permeation at $>10^{-4}$ moles/ℓ	6
Sb > S	4:1	Iron	Hydrogen permeation at 10^{-5} moles/ℓ	6

bonding normal to this plane. They extrapolated this theoretical calculation to the grain boundary case and proposed that sulfur segregated to a nickel grain boundary strengthens the in-plane bonding but weakens the bonding normal to the interface. Briant[8] has also suggested that sulfur should have the same effect in iron. For sulfur to cause equal hydrogen permeation rates in iron and nickel, the large (10^5) difference in their hydrogen diffusivities must be circumvented. One explanation is that 0.1 monolayers of sulfur sufficiently restructures the grain boundaries of both materials such that their grain boundary hydrogen diffusivities are equal. An alternate explanation is that of Latanision and Opperhauser[9] who proposed that grain boundary impurities act as hydrogen recombination poisons enhancing hydrogen absorption along grain boundaries. It is also possible that hydrogen absorption occurs behind the crack tip and for an intergranular crack, absorption occurs into the intergranular fracture surface. In this way, 0.1 monolayers of S adsorbed on the intergranular fracture surfaces of iron and nickel could result in equal permeation rates.

COMPARISON BETWEEN ANTIMONY AND PHOSPHORUS IN IRON AND NICKEL

The similarity between the effect of S in iron and nickel results in C_s^{eq} (Fe) = C_s^{eq} (Ni). This equality makes it possible to compare γ_{sb} and γ_p in iron and nickel and from equations 1 and 2 it can be seen that Sb is five times more effective in iron than in nickel while P was ineffective in both. If the major contribution of Sb to the fracture of iron and nickel at cathodic potentials was to enhance hydrogen permeation, it should have an equal effect in both materials. Therefore, it is concluded that the major contribution of Sb in iron is to reduce the intergranular fracture stress. The extent to which Sb reduces the intergranular fracture stress of nickel is not known; however, other studies[9] have related Sb to intergranular fracture of nickel tested at cathodic potentials. Recalling the hydrogen absorption data in Table 1, the small P effect in both iron and nickel suggests that grain boundary P does not affect intergranular fracture by an enhanced hydrogen permeation process.

SUMMARY

Evidence that S, Sb and P participate in the embrittlement of iron and nickel tested at cathodic potentials by decreasing the critical intergranular fracture stress or by enhancing hydrogen permeation was examined and it was concluded that some elements may affect one or the other or possibly both processes. The following conclusions were made about S, Sb and P in iron and nickel:

Sulfur: (1) Affects iron and nickel very similarly, (2) probably reduces both the critical intergranular fracture stress and enhances hydrogen permeation in iron and nickel.

Antimony: (1) Primary effect is to reduce the critical intergranular fracture stress of iron, (2) effect in nickel is equal
with sulfur but there isn't enough data to assign a mechanism.

Phosphorus: (1) Negligible affect in both iron and nickel, (2)
does not enhance hydrogen permeation when present in the grain
boundaries of iron and nickel as when added as a compound to an
electrolyte.

REFERENCES

1. R. H. Jones, S. M. Bruemmer, M. T. Thomas and D. R. Baer,
    Effect of Sulfur and Antimony on the Intergranular
    Fracture of Iron at Cathodic Potentials, Met. Trans., in
    press.
2. S. M. Bruemmer, R. H. Jones, D. R. Baer and M. T. Thomas,
    Influence of Sulfur, Phosphorus and Antimony Segregation
    on the Intergranular Hydrogen Embrittlement of Nickel,
    Submitted to Met. Trans.
3. P. Ramasubramanian, Grain Boundary Embrittlement in Iron
    and Steels, Ph.D. Thesis, University of Minnesota 1971,
    University Microfilms, 72-14,362.
4. M. L. Jokl, J. Kameda, C. J. McMahon, Jr. and V. Vitek,
    Solute Segregation and Intergranular Brittle Fracture in
    Steels, Metal Science, 375 (1980).
5. M. Smialowski, "Hydrogen in Steels," Oxford and Addison-
    Wesley Publishing Co., Reading, MA.
6. T. Zakroczymski, Z. Szklarska-Smialowska and M. Smialowski,
    Effect of Promoters on the Permeation of Electrolytic
    Hydrogen Through Steel, Werkstoffe und Korrosion,
    27:625(1976).
7. C. L. Briant and R. Messmer, Phil. Mag. in press.
8. C. L. Briant, private communication.
9. R. M. Latanision and H. Opperhauser Jr., The Intergranular
    Embrittlement of Nickel by Hydrogen: The Effect of
    Grain Boundary Segregation, Met. Trans., 5:483(1974).

DISCUSSION

Comment by M. Pourbaix:

    It seems that the action of S, Sb, P (and As?) on hydrogen
evolution occurs under conditions where a transitory formation of
hydrides (H_2S, SbH_3, PH_3, AsH_3?) is possible. I think that extensive research in this subject would be useful.

Reply:

    I agree with Dr. Pourbaix that the hydrides of S, Sb, P, etc.,

may be stable at the more cathodic potentials of our tests and that
they could enhance hydrogen absorption; however, our tests were also
conducted at potentials where hydrides are not stable and the embrittlement process does not show a discontinuity at the hydride
stability boundaries. For instance, for a 1 N H_2SO_4 solution at
25°C (pH = 0.3), antimony hydride becomes stable at cathodic potentials of -800 mV (SCE) or less. Our results for iron and nickel
do not show a discontinuity between tests conducted at potentials
less than and greater than -800 mV (SCE). Of course, the conditions
at the crack tip are not the same as the bulk solutions; however, in
iron-antimony alloys fracture of the straining electrode samples
appears to be crack nucleation and not crack growth controlled.
Therefore, the bulk electrolyte conditions should be controlling the
embrittlement in the iron-antimony alloys.

A COMPARATIVE STUDY OF THE INFLUENCE OF RARE EARTH AND MOLYBDENUM ADDITIONS ON THE TEMPER EMBRITTLEMENT CHARACTERISTICS OF A LOW ALLOY STEEL

W. Barrett, J.V. Bee and G.G. Garrett

Department of Metallurgy
University of the Witwatersrand
Johannesburg   South Africa

INTRODUCTION

Temper embrittlement is one of the most serious problems associated with the fracture of medium-strength quenched and tempered steels. Considerable progress has been made in recent years towards an understanding of the mechanisms of this embrittlement, which arises due to the segregation of impurity elements (eg P, S, Sb, Sn and As) to prior austenite grain boundaries,[1,2] leading to intergranular failure.

Although many elements can be used to improve the hardenability, and to increase the high temperature strength and creep resistance of low and medium alloy steels, molybdenum initially appears to be unique in also preventing temper embrittlement. Considerable controversy still exists as to the precise mechanism by which Mo inhibits this phenomenon. It has been suggested[3] that its principal role is to scavenge phosphorus by the formation of Mo-P clusters. This will reduce the bulk chemical potential of the active impurity, thereby reducing segregation to grain boundaries and therefore also increasing the resistance of the steel to embrittlement. However, there is also evidence to indicate that the beneficial role of Mo is not only due to its ability to tie up impurities in the grain interior, but that Mo segregated at the grain boundaries must also be responsible in part for the de-embrittlement effect, possibly through influencing the intrinsic cohesion of the boundaries.[4]

The aim of the present study is to investigate alternative methods for the elimination of the embrittling effects of impurities. An obvious means of lowering the activities of the species in a given steel is to reduce significantly the allowable residual impurity content at the primary production stage. Unfortunately this can give rise to considerable cost penalties.

An alternative approach is to attempt to eliminate active impurities by chemical fixation. For example, it has already been predicted,[5,6] and confirmed experimentally from ductile-brittle transition temperature (DBTT) data, that P - induced embrittlement of a 2,25 Cr-1 Mo steel can be reduced to a large extent by the addition of lanthanum. This reportedly acts by the formation of LaP precipates. Other harmful species are also removed from solid solution. The effectiveness of La is governed by the stoichiometry of the precipitates according to the equation:[5]

$$La = 8,7S + 2,3 \, Sn + 4,5P$$

Other rare earth elements (eg Ce, Nd, Pr and Sm) are thought to behave in a similar way and scavenge other prevalent residuals (eg S, Sn, As and Sb).

In an attempt to confirm the beneficial influence of rare earth elements on resistance to temper embrittlement, the present paper reports on a comparison made between the influence of Mischmetall and Mo additions on the DBTT of a medium carbon steel doped with phosphorus to the maximum acceptable industrial level.

EXPERIMENTAL PROCEDURE

The alloys were vacuum melted using an En 18D base steel which had been doped with phosphorus. The compositions are listed in Table I. Ingots were forged to 13 mm thick plate, and from these notched 3-point bend test samples were machined. The specimens were austenitised at 900°C for 20 minutes and quenched in a salt bath. The resultant martensitic structure was tempered for 2 hours at 650°C; then isothermally aged at 510°C for 4 days (to induce embrittlement). The doped steel with no Mo or Mischmetall additions was used as the control alloy.

Table I. Composition of Alloys used.

	C	Mn	P	S	Si	Ni	Cu	Cr	Mo	V	Al	MM
En 18D	0,42	0,76	0,021	0,030	0,26	0,14	0,16	0,99	0,06	0,04	0,01	
W4	0,46	0,75	0,050	0,030	0,28	0,15	0,16	0,97	0,88	0,040	0,015	
W6	0,41	0,72	0,040	0,029	0,26	0,14	0,16	0,97	0,35	0,038	0,007	
W7	0,44	0,73	0,041	0,029	0,27	0,14	0,16	0,97	0,58	0,038	0,016	
W8	0,43	0,73	0,041	0,031	0,28	0,15	0,16	0,97	1,09	0,040	0,019	
W9	0,38	0,72	0,04	0,028	0,27	0,14	0,16	0,97	0,054	0,037	0,010	0,2
W10	0,40	0,71	0,038	0,004	0,26	0,14	0,15	0,93	0,053	0,038	0,050	0,5
W11	0,37	0,71	0,030	0,003	0,26	0,14	0,15	0,94	0,051	0,036	0,03	0,87
Control W12	0,41	0,73	0,04	0,014	0,27	0,14	0,16	0,97	0,051	0,036	0,03	

*Mischmetall (52% Ce, 24% La, 15% Nd, 7% Pr, 1% Sm, plus 1% other rare earth elements)

RESULTS AND DISCUSSION

Transition Temperatures

The ductile-brittle transition temperatures of all the alloys studied are presented in Fig 1.   These were determined using the accurate slow-bend test method described in detail elsewhere.[1]   The DBTT of the control alloy (W_{12}) was in excess of 85°C.   The exact value of this transition temperature was not determined since all the other alloys exhibited substantially lower DBTT values.

In the Mo-containing alloys (W_4, W_6, W_7 and W_8) the transition temperatures for 50% ductility were approximately 65°, -15°, 7° and 30°C respectively.   These correspond with Mo additions of 0,88; 0,35; 0,58 and 1,09%.   The variation in DBTT with Mo-content is plotted in Fig 2, which shows a minimum value at 0,35 Mo.   This is an unexpected result since the accepted optimum molybdenum concentration is approximately 0,7% [4].   This difference may be associated with variations in carbon content. Thus, at the higher carbon levels used in this investigation, where the activity of carbon is greater, it may be expected that Mo-carbides will form at lower Mo concentrations.   The formation of these carbides will remove Mo from solid solution and so release phosphorus which is then free to segregate, promoting embrittlement.

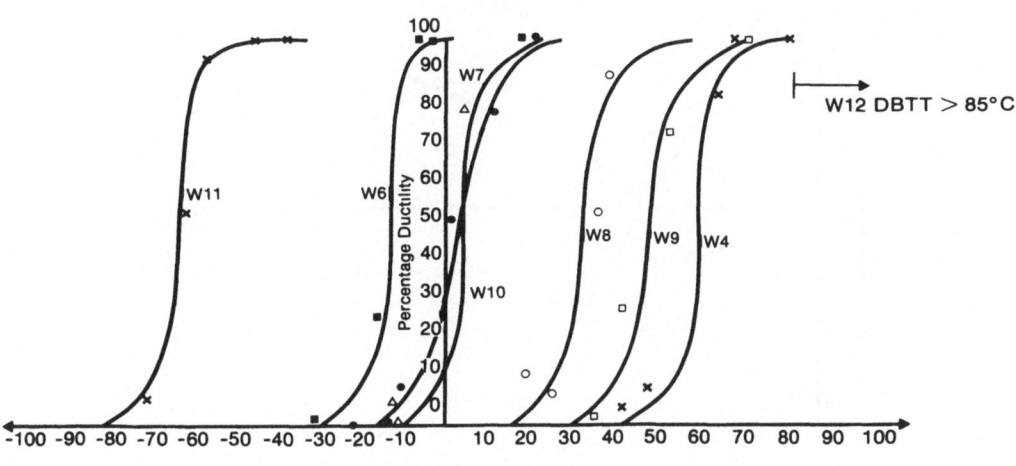

Figure 1.   Ductile-Brittle Transition Temperatures of the alloys.

In the alloys containing Mischmetall (W_9, W_{10} and W_{11}), there is a linear decrease in DBTT with increasing concentration. The observed values at 50% ductility decrease in the order 45°, 7° and -65° for the addition of 0,2; 0,5 and 0,87% Mischmetall. This suggests that progressively more impurities are removed from solid solution by the formation of inclusions. The variation of DBTT with Mischmetall-content is also shown in Fig 2. By comparison with the Mo behaviour it would appear that Mischmetall is more effective in reducing temper embrittlement in this steel. However, it should be appreciated that the results of tensile and hardness tests indicate that the Mo-containing alloys generally exhibited higher strength levels and might therefore be expected to have slightly higher DBTT values. (Subsequent tests will clarify this effect for alloys tempered to the same strength levels).

A recent, elegant analysis by Seah[7] predicts the influence of any segregant atom on the grain boundary cohesion of iron. This work is also developed elsewhere in these Proceedings, where it is shown that the rare earth elements, if free to segregate to grain boundaries, will be expected to increase embrittlement. Thus, at higher Mischmetall levels, the curve shown in Fig 2 would be expected to show a minimum, and work is in progress to confirm this effect.

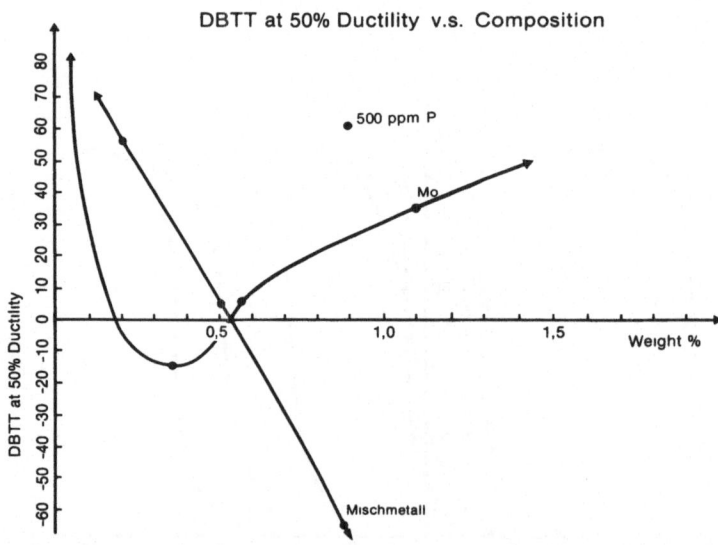

Figure 2.   Variation of DBTT with alloying additions.

Electron Microscopy

Observation of the fracture surfaces of each alloy by SEM confirmed that temper embrittlement did occur in the susceptible alloys.   This is evident from the fractograph in Fig 3 which shows a typical intergranular failure.   This is characteristic of the effect of impurity segregation to grain boundaries.   The Mo-and Mischmetall-containing alloys which had the lowest DBTTs (W_6 and W_{11}) are clearly resistant to temper embrittlement and their fracture surfaces (eg Fig 4) exhibited transgranular fracture.

Using the energy-dispersive X-ray analytical facility on the SEM, it was not possible to identify the mechanism by which Mo produced this effect.   However, in the Mischmetall-containing alloys, analyses of the inclusions found in the bulk material showed them to be rich in rare-earth elements, arsenic, phosphorus and magnesium.   It appears therefore that the rare earths have two major effects.   Impurities such as sulphur are taken out of the steel into the slag during melting (see Table I).   In addition, other impurities are rendered harmless by the formation of chemical compounds.   The net result is a drastic reduction in the activities of the residual impurities in the bulk steel, and hence the elimination of their potential to induce temper embrittlement.

At the present time, using microanalysis in STEM, no grain boundary segregation of rare earths has been detected in the Mischmetall containing alloys.   Thus no corresponding reduction in grain boundary cohesion would be expected at the Mischmetall concentrations used in the present study.

Preliminary results from 1ZOD impact tests show no significant drop in the upper shelf energies for the alloys containing rare earth element additions.   This is contrary to the observations of Seah et al,[5] who found a definite fall in upper shelf energy with increasing rare earth content.   However, it is not thought that this would have any significant effect in service materials processed by conventional routes.

In addition to the influence of rare earth additions in terms of the prevention of temper embrittlement, the formation of compounds can also provide beneficial effects on inclusion shape and distribution.[6] The brittle nature of the inclusion phases prevents the development of deleterious stringers at hot rolling temperatures.

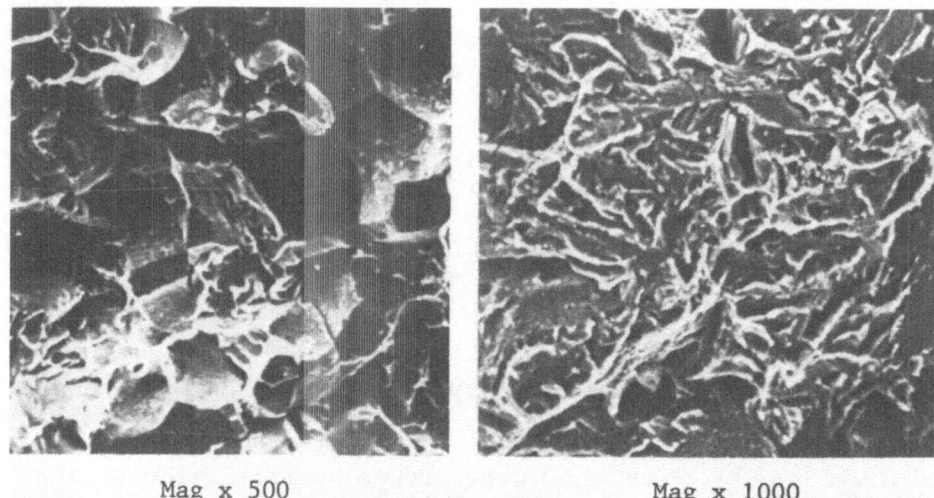

Mag x 500   Mag x 1000

Fig. 3.   Fractograph of Specimen   Fig. 4.   Fractograph of Specimen
          W$_{12}$ Intergranular failure.   W$_{11}$ Transgranular failure.

SUMMARY

The steel studied (EN18D) doped with 400 ppm phosphorus is
susceptible to temper embrittlement. This segregation-induced
embrittlement may, to a large extent, be removed by the addition
of rare earths in the form of Mischmetall. The rare earths form
sulphide, phosphide and other impurity compounds which are
present in the form of inclusions. This leads to a lower
activity of impurities in solid solution and a corresponding
decrease in segregation and therefore in embrittlement.

Molybdenum was also found to inhibit embrittlement although
the optimum concentration was different to that reported in the
literature, which may be attributed to variations in carbon
content.

ACKNOWLEDGEMENTS

The authors gratefully acknowledge stimulating discussions
with Dr Martin Seah and his most useful suggestions. The
financial support of the Council for Scientific and Industrial
Research (Co-operative Scientific Programmes) and the University
of the Witwatersrand for one of the authors (W B) is also acknowledged.

REFERENCES

1.   H Ohtani, H C Feng, C J McMahon JR, and R A Mulford
        "Temper embrittlement of Ni-Cr steel by antimony I.
        Embrittlement at a low carbon concentration"
        Met. Trans. 7A (1976) pp 87-101.

2.   C J McMahon Jnr. "Temper brittleness - An interpretive
        review" Temper embrittlement in steel ASTM STP 407
        (1968) pp 127-167.

3.   Jin Yu and C J McMahon Jnr. "The effects of
        composition and carbide precipitation on temper
        embrittlement of 2,25 Cr, 1Mo steel Part I" Met.
        Trans. 11A (1980) pp 277-289.

4.   P H Dumelin, M Guttmann, M Foucault, M Palmier, M
        Wayman and M Biscondi "Role of molybdenum in phosphorus
        - induced embrittlement." Metal Sci. 14 (Jan 1980) pp
        1-15.

5.   M P Seah, P J Spencer and E D Hondros "Additive remedy
        for temper brittleness" Metal Sci. 13 (May 1977)
        pp 307-314.

6.   Yu V Kryakovskii, Yu I Rybenchik, E I Tyurin and V I
        Yavoiskii "Mechanical properties and the nature of
        non-metallic inclusions in structural steel containing
        rare earth elements" Metallov. 8 (1963) pp 11-18.

7.   M P Seah "Adsorption induced interface decohesion" Acta
        Met. 28 (1980) pp 955-962.

THE EFFECT OF SULPHUR ON HYDROGEN RECOMBINATION ON IRON

P. Marcus, S. Montes and J. Oudar

Laboratoire de Physico-Chimie des Surfaces
Université Pierre et Marie Curie
ENSCP - 11, rue Pierre et Marie Curie - 75005 PARIS
FRANCE

INTRODUCTION

The hydrogen evolution reaction on metals, and particularly on iron, has been widely studied in the past. In addition to fundamental aspects (1), this reaction is most important on the applied level, since it is involved as the cathodic reaction in the corrosion of metals. The hydrogen evolution reaction must also be considered in relation with the penetration of hydrogen into metals (2). It is well known that the presence of hydrogen in metals can drastically change the mechanical properties of the material (hydrogen embrittlement). Non-metallic impurities, such as S, As, are known to favor the penetration of hydrogen in metals. The role of these impurities is still poorly understood, while a number of hypothesis have been suggested. The aim of this work was to clarify the effect of an adsorbed impurity on the hydrogen evolution reaction on various metals. The results reported here concern the effect of sulphur on the hydrogen evolution reaction on iron. Our intention was to take into account the dependence of the reaction on the surface structure, by using single crystal electrodes, and to monitor the surface concentration of sulphur at the various stages of the reaction, by means of a radiochemical technique (^{35}S radiotracer). This last parameter was unknown in most of the previously reported studies.

EXPERIMENTAL PROCEDURE

The single crystal electrodes were prepared from 99.995 % iron (Johnson Matthey) by the strain-anneal technique. The main residual impurity was nitrogen. The samples were oriented by the X-ray Laue backscattering method, to within 1° of the desired orientation. They were mechanically and electrochemically polished, and subsequently

treated at 800°C under purified hydrogen. Some samples were further
treated at 550°C under a gaseous mixture H_2S-H_2 ($PH_2S/PH_2=10^{-4}$,
$PH_2=200$ Torr). Under these conditions a quasi-complete layer of adsorbed sulphur is obtained (3). The samples could then be transfered
under protective gas, by means of a lock device, to the electrochemical cell, which was located in a glove box filled up with purified
Argon. The electrochemical measurements (cathodic i-E curves) were
performed by means of a conventional set up. The electrolyte was
0.1N H_2SO_4, the potential sweep rate was $3Vh^{-1}$. The measurements of
the radioactivity, which allowed us to determine the surface coverage
with sulphur, were performed by means of a calibrated Geiger-Müller
counter.

EXPERIMENTAL RESULTS

     Figure 1 shows the cathodic polarization curves obtained in
0.1N H_2SO_4 on electrodes with crystallographic orientation (111),
(110) and (100), for the sulphur-free and the sulphur-covered surfaces. A partial desorption of the adsorbed sulphur was observed,
at open circuit, after immersion of the electrode. The extent of
this desorption was 10-20 % of a monolayer, depending on the crystallographic orientation. The remaining coverage was found to be
constant over the whole range of cathodic potentials investigated
This coverage was 36 ± 2ng cm^{-2}, irrespective of the crystal plane.
It corresponds to a ratio S/M (ratio of sulphur atoms to metal atoms
in the outermost layer) equal to 1/2 for the (110) face. It is quite
clear that the presence of sulphur i) lowers the anisotropy observed
between the various crystal planes without sulphur and ii) promotes
in all cases the hydrogen evolution reaction. Similar conclusions
were derived from the i-E curves recorded from the (113) face and
from a polycrystalline electrode. The curves are not reported in
fig. 1 in order to keep it clear. It is worth pointing out that the
effect of sulphur is identical for the two studied coverages (36
and 25ng cm^{-2}). From the i-E curves, the Tafel parameter and the
exchange current densities were calculated. They are reported in
Table 1.

Table 1

Plane	Without sulphur		With adsorbed sulphur	
	bmV	$-\log io$ $(A\ cm^{-2})$	bmV	$-\log io$ $(A\ cm^{-2})$
(110)	105	4,8	90	4,6
(100)	105	4,8	90	4,8
(111)	105	5,0	90	4,8
(113)	105	5,0	90	4,6
Polycristal	105	4,6	100	4,6

   The general trends observed when the surface is covered with
sulphur are a decrease of the Tafel slope and an increase of the
exchange current density.

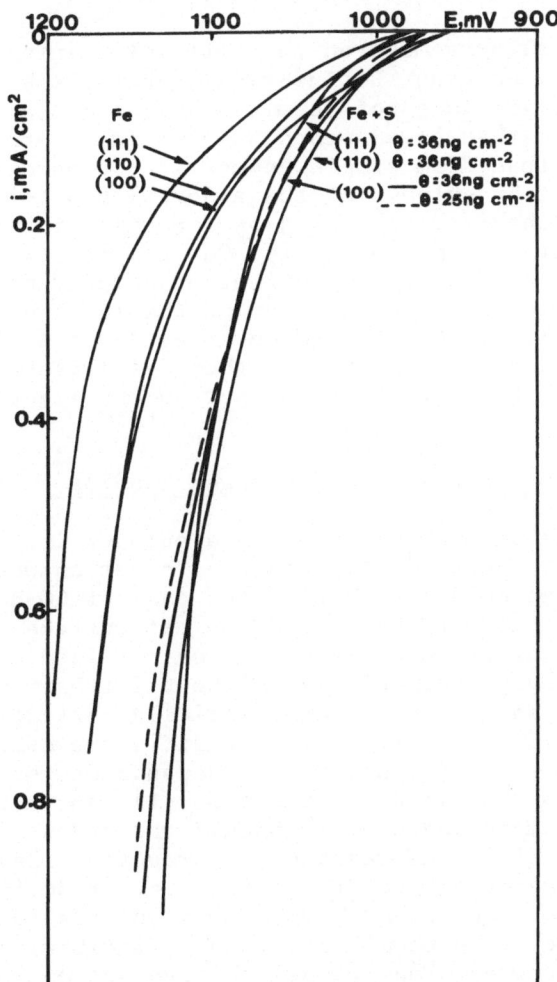

Fig. 1 - Influence of the adsorbed sulphur on the cathodic pola-
         rization curves of iron (0.1N H_2SO_4, 3Vh^{-1}).

DISCUSSION

   The overall hydrogen evolution reaction (h.e.r.) $2H^+ + 2\bar{e} \rightarrow H_2$
can be split into the well known steps that are listed below (4,5).

Proton discharge and adsorption $H^+ + \bar{e} \rightarrow Hads$ (1)

Chemical recombination $2Hads \rightarrow H_2$ (2a)

Electrochemical recombination $Hads + H^+ + \bar{e} \rightarrow H_2$ (2b)

In addition to these elementary reactions, the penetration of hydrogen into the metal can take place (Hads → Habs).

There is no general agreement on the rate determining step (r.d.s.) of the h.e.r. on iron. Let us recall that the simple recording of the i-E curves does not permit a unique mechanism to be derived. We have shown that an adsorbed sulphur layer promotes the h.e.r. on iron. Note that from simple geometrical considerations on the blocking effect by sulphur of the adsorption sites for hydrogen an inhibition of the reaction could be expected, instead of the promoting effect observed experimentally. Our aim is now to discuss some factors to which this promoting effect of sulphur can be attributed, namely i) the influence of sulphur on the metal-hydrogen binding energy (M-H b.e.), ii) the existence of an S-H transient species that favors the approach and the further desolvatation of the solvated proton, iii) the effect of sulphur on the electric field in the double layer.

i) <u>Influence of sulphur on the metal-hydrogen binding energy</u>.

The first question that arises is : does sulphur increase or decrease the M-H b.e.? There is at present no clear answer to this question. There is some evidence that in the case of CO-H_2 coadsorption on nickel (6) the presence of CO on the surface increases the adsorption energy of hydrogen, and that on oxidized platinum hydrogen is more strongly adsorbed (7). Without further evidence, it can only be suggested that S can have a similar effect on iron. Then the second question that arises is : what is the effect of an increase of the M-H b.e. on the h.e.r.. It depends on the rate determining step. Should the r.d.s. be reaction (1), an increase of the M-H b.e. could promote the h.e.r.. Should the r.d.s. be reaction (2a), it would inhibit the reaction. If reaction (2b) is the r.d.s., it further depends on wether the M-H b.e. predominates over the energy required for the approach and the desolvatation of the proton. If the M-H b.e. is predominant, then an inhibition of the reaction should be observed. What we did observe was an increase of the reaction rate. Assuming an increase of the M-H b.e. due to adsorbed sulphur, reaction (2a) as the r.d.s. can be excluded. Reaction (1) is likely to be the r.d.s.. However reaction (2b) can also be the r.d.s. if the activation energy for the approach and desolvatation of the proton predominates over the M-H b.e..

ii) <u>Existence of an S-H transient species</u>.

It is suggested that the presence of sulphur can facilitate

the approach of a solvated proton to the surface and the further
desolvatation of the proton. Such a mechanism can be described in
term of an S-H transient species. It should not affect reaction
(2a), but it could lower the activation energy for reaction (1) and
the activation energy for reaction (2b) provided that it predominates over the M-H b.e..

iii) <u>Effect of sulphur on the electric field at the interface.</u>

In a recent study we have shown that, when sulphur is adsorbed
on single crystal silver electrodes, the double layer capacity is
substantially lowered and the potential of zero charge (p.z.c.)
shifted to more positive potentials (8). An acceleration of the
h.e.r. attributed to sulphur, was also observed. Below the p.z.c.
the surface is charged negatively, thus favoring the approach of a
proton (in an acidic medium) or the orientation of the water molecules with H toward the surface. These conditions are required for
the h.e.r. to take place. Therefore, the increase of the p.z.c.
caused by the adsorbed sulphur, can be responsible for the enhancement of the h.e.r. if we consider that a given potential corresponds
to a higher rational potential E-Ep.z.c. in the presence of sulphur.
This interpretation is supported by the previous results on the
Ag-S/NaF interface, assuming that the adsorbed sulphur also increases
the p.z.c. of iron.

It is well known that sulphur favors the penetration of hydrogen
in the metal. It should be pointed out that the promoting effect of
sulphur on the h.e.r. on iron is not inconsistent with an enhanced
hydrogen permeation. Indeed the considerations developped above in
iii) show that the presence of adsorbed sulphur can increase the
equivalent hydrogen partial pressure on the surface. This will increase the hydrogen equilibrium concentration in the bulk and the
permeation flux resulting from a concentration gradient.

The results presented here may provide a basis for further experiments. Such data as the changes of hydrogen coverage in the presence of sulphur, the influence of sulphur on the metal-hydrogen
binding energy, simultaneous measurements of the hydrogen evolution and permeation rates under carefully controled surface conditions should permit a better understanding of the mechanisms.

REFERENCES

(1) S. Trasatti - J. Electroanal. Chem. 39 (1972) 163 and referen-
    ces therein.

(2) Proc. 2nd Int. Cong. Hydrogen in Metals, Paris France (1977).

(3) Y. Berthier and J. Oudar - C. R. Acad. Sci. Paris 269 (1969)
    1149.

(4) K. J. Vetter - Electrochemical Kinetics, Academic Press, London (1967).

(5) J. O'M Bockris, A.K.N. Reddy - Modern Electrochemistry, Plenum Press, New-York (1970).

(6) G. Wedler and F.J. Brücker - Z. Physik. Chem. Neue Folge, 75 (1971) 299.

(7) R.W. Mc Cabe and L.D. Schmidt - Surface Sci., 65 (1977) 189.

(8) C. Nguyen van Huong, R. Parsons, P. Marcus, S. Montès and J. Oudar - J. Electroanal. Chem., 119 (1981) 137.

DISCUSSION

Comment by R.M. Latanision:

If sulfur inhibits the chemical recombination step even though increasing the electrochemical hydrogen evolution reaction rate, one would expect the permeation flux of hydrogen through your materials to increase. Have you observed anything like this?

Reply:

I agree entirely with your comment. As we have mentioned in our paper, adsorbed sulfur can change the chemical potential of the adsorbed hydrogen and consequently its solubility. We expect in the future to reach a better understanding of this process by permeation measurements.

WORKSHOP SUMMARY: INTERGRANULAR EMBRITTLEMENT

Prepared by D.J. Duquette

Discussion
     Leaders: D.J. Duquette, J.O'M. Bockris, P.L Pratt, and
     G.J. Dienes

Recorders: H.P. Bonzel, H. Gleiter, D.M. Esterling, and R. Messmer

The discussions in this workshop centered on at least three key issues (A) solute segregation (B) grain boundary cohesion vs. decohesion and (C) potential experimental techniques.

A.   Solute Segregation

The specific role of segregation to grain boundaries is still not well understood. However, there is general agreement that the driving force for segregation is not just the lowering of the boundary energy but of the entire system.

Seah has conveniently divided the problem into at least two principal aspects; that of segregation per se, where solute may either weaken or strengthen the boundary and the actual process of grain boundary fracture. According to this view, the metal-solute bond must be stronger than the metal-metal bonds. Additionally, species which form stronger metal-solute bonds than metal-metal bonds will increase cohesion, and vice versa.

This model yields a classification of elements into either grain boundary strengthening or weakening ones but does not provide the parameters that describe segregation induced intergranular fracture (SIIGF) such as, for example, the ductile/brittle transition temperature.

The question was asked whether the relationship (Rice & Thompson) between the reversible work of fracture and free energies of segregation could be obtained experimentally. The general consensus was that it would be a different process.

Birnbaum made the point that 100% S segregation in pure Ni does not cause intergranular failure, although it should according to Seah's calculation (Fig. 10 of Seah's paper). Seah noted that his model only describes a shift in the balance towards SIIGF, but does not give the exact parameters for which it occurs.

The theoretical physicists are not happy with the concept of surface or grain boundary energies, since these are difficult to

915

understand on rigorous grounds. Gadzuk suggests consulting a recent
paper in Phys. Rev. Letters, 42, 989 (1979) dealing with segregation
in AB alloys.

B. Grain Boundary Cohesion

Most of the proposed mechanisms of failure focus on decohesion.
Two stages are important, first segregation (which may or may not be
equilibrium segregation) and then embrittlement depending upon the
materials involved. Do dislocations held up at twins or grain boundaries play an important role? This aspect is still unknown.

While decohesion models are attractive - and indeed the work of
Seah and Hondros supports such a model - dislocation processes cannot
be ignored out of hand. The boundaries may be obstacles for slip
leading to dislocation pile-up and fracture. The segregation of impurities may influence plastic flow near the grain boundary.

Certainly the previous stress state of the boundary is important;
low temperature charging of iron with H_2 to only 3-5 ppm produces
brittle fracture, whereas at 400°C, 40 ppm causes no damage - stress
relief can occur. The importance of grain size in Al was referred
to and the question was raised as to how general this was. Clearly
austenitic grain size is important in temper embrittlement of steels.

It was also felt that calculations should be encouraged for
individual grain boundaries. Cluster models are attractive as descriptions of the polyhedra in a grain boundary. The SW-Xα-SCF cluster
calculations represent one state-of-the-art way to obtain the electronic state of local regions of a boundary. Certain limitations of
these cluster calculations should be noted. The conclusions based
on these calculations come from looking at the high lying orbitals
out of a large group of orbitals (whether their lobes indicate
bonding or antibonding and how the wave functions are influenced by
an impurity atom). The sorting out of the correct high-lying orbitals
requires energetic calculations correct to a few tenths of an electron
volt. The order of the orbitals is a sensitive but important issue
when investigating strengths and types of bonds. It is not clear
that various bonds become weaker just from the evidence in orbital
pictures. Detailed energetics are important. The orbital picture
may also be modified when (1) the cluster geometry is relaxed or
(2) the cluster is embedded in a bulk description.

C. Experiments

Further experiments are needed to characterize grain boundaries
more adequately, both in pure and in commercial materials using
electron microscope techniques. These may not reflect the structure
in the bulk. The analogy between metallic glasses and intrinsic

grain boundary structure (and chemistry) was emphasized.

Experiments on well-defined systems may be helpful in checking the validity of various cluster descriptions. Ideally the systems would have a well-defined geometry for the fracture region - such as layered structures (intercalated compounds) or alkali halides with substitional impurities. The correct prediction of trends in these systems by cluster techniques would be encouraging.

Transition metals are difficult systems for theorists. Lead is a simple metal that can be treated with reasonable accuracy by existing pseudopotential models, and it has industrial interest. Further, under proper conditions it becomes brittle (though perhaps by dissolution). Hence a combined theoretical/experimental assault on lead embrittlement may be a helpful first step towards understanding more complex materials.

The following electro chemical experiment also may be useful: Two crystals of the same metal in contact (i.e., separated by a grain boundary) are placed in a solution containing ions of the metal - for example Ag in $AgNO_3$. Then deposition and dissolution are carried out at very low overpotentials. From the crystals formed, it might be possible to find out something about γ grain boundary/γsolid-solution, provided that the electrochemical conditions are those of fairly high frequency (1000-10000 c.p.s.), low amplitude (5-10 mV) and low current density (1-10 μ amp cm^{-2}).

Other suggested experimental approaches include:

a.   Use of tritium to investigate H accumulation at grain boundaries with and without segregated solute.

b.   Study the relationship between orientation dependence of solute segregation and grain boundary fracture path.

c.   Investigate temper embrittlement and SIIGF under well-defined conditions (orientation of grain boundary, concentration of solute in bulk and grain boundary, etc.), similar to the experimental approach presented by Gleiter for the Cu (Bi) system. The use of single crystal (bicrystal) specimens is strongly recommended.

D.   Other Issues

A secondary issue was addressed concerning the relationship between hydrogen and solute-induced temper embrittlement. The question was put forward whether there is a synergism when hydrogen is added to a system where solute-induced temper embrittlement is observed. Birnbaum stated that SIIGF is in general enhanced by hyrogen. The mechanism, however, seems to be unclear.

On the other hand, an interpretation of the additional effect of hydrogen on temper embrittlement was also offered by Seah. He notes essentially two (inter-related) effects: H entry into the grain boundary is facilitated by the presence of recombination poisons at the grain boundary; and secondly, the element which acts as a recombination poison is also an adsorption promoter, i.e., it makes a stronger bond with H than iron does. Hence the brittleness of the grain boundary is further affected through the presence of Me-solute-H complexes compared to just segregated Me-solute species. It is further pointed out by Seah that elements such as S, As, Sb, Se, Ti and Bi, bond strongly to hydrogen and are, of course, also embrittling elements.

At this point in the discussion it was suggested to measure the H segregation with varying amounts of segregated solute at the grain boundary (S, P, C), and possibly connect these results to the ductile/ brittle transition temperature.

Birnbaum mentioned that experiments of this kind have been done by using SIMS as a monitor for hydrogen species; a peak in H at the boundary was detected where the peak height depended on the amount of segregated sulfur. It would be interesting to continue these experiments, particularly at lower temperatures, in order to increase the H coverage at the surface and the grain boundary.

WORKSHOP SESSION 3

LIQUID METAL EMBRITTLEMENT

Session Chairman: R.A. Oriani

LIQUID AND SOLID METAL EMBRITTLEMENT

Norman S. Stoloff

Materials Engineering Department
Rensselaer Polytechnic Institute
Troy, New York 12181

INTRODUCTION

Low melting metals can interact with metallic substrates in several distinct ways that lead to premature fracture or surface degradation. The most spectacular, and widely studied, manifestation of metal-induced embrittlement (MIE) is the sudden fracture of stressed metals with little or no apparent inter-diffusion or chemical interaction. In this classical form of embrittlement, ductility is minimized at or near the melting point of the metallic surface film, and is then restored at some higher temperature, as shown in Fig. 1 for zinc embrittled by indium (1). Similar embrittlement may be noted when the surface film is solid, but the degree of embrittlement is reduced and then disappears as the temperature falls well below the melting point of the embrittler – 100°C or so in the case of lead on (or in) steel (2). Other manifestations of liquid metal embrittlement are connected with rapid penetration of the environment along grain boundaries (e.g. gallium on aluminum), by preferential chemical reactions (e.g. lithium on iron containing carbon or carbides), and by corrosion, perhaps aided by cavitation, in ferritic steel tubing containing mercury in a thermal gradient. In the latter case, failure of the component may occur either by thinning or ultimate penetration of the wall, or by mass transfer-caused plugging of the tube. In several instances, two or more of these phenomena may appear either simultaneously or under different exposure conditions with the same combination of environment and substrate, for example, aluminum single crystals and polycrystals also may suffer "instantaneous" embrittlement due to gallium.

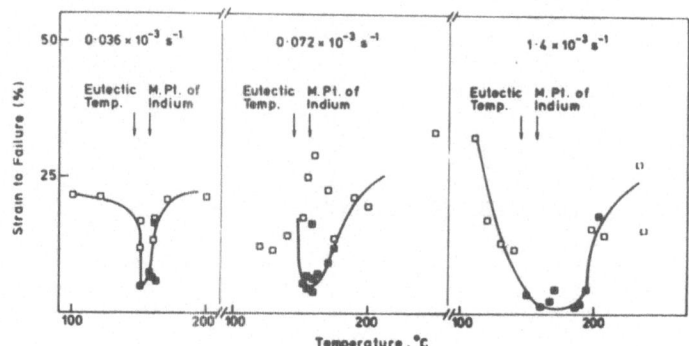

Fig. 1.   Effect of temperature on tensile ductility of zinc tested
          in indium at various strain rates (1).

        Shchukin et al (3) have reported that molten metals may also
reduce fracture strength of covalently bonded materials, e.g. germanium is embrittled by gold and gallium and graphite by alkali metals.
Low melting metals also reportedly reduce the strength of Al_2O_3.

        Practical implications of these various embrittlement or degra-
dation phenomena may be noted in widely diverse industrial situations, e.g.:

    a) embrittlement of alloys containing low melting metallic
       inclusions, as in steels and aluminum alloys to which lead
       is added to improve machinability.
    b) deterioration of alkali metal piping system in nuclear fis-
       sion or fusion devices relying upon liquid metal coolants,
       as in the liquid metal fast breeder reactor or in lithium-
       moderated fusion devices.
    c) embrittlement stemming from plating of structural parts, as
       in cadmium on steel or titanium alloys.
    d) welding, brazing, or soldering operations, as in steels
       where copper contamination (from welding electrodes) may
       occur, or solder contacting stressed iron-base alloys.
    e) various industrial situations where molten metals are han-
       dled or where metal coatings are heated above their melting
       points.

        Although metal-induced embrittlement is usually to be avoided,
workers in the Soviet Union have utilized liquid metals, such as
Pb-Sn eutectics, to facilitate drilling steels, Ni-Cr heat resistant
alloys, titanium alloys, and others (3,4). Improvement achieved with
liquid metals included increased drill life and improvement of the
quality of machined surfaces.

        From a scientific point of view, catastrophic failure without

apparent diffusion, i.e. classical liquid metal embrittlement, has
received the most attention. The first portion of this paper is
devoted to review of the principal features of this type of embrittlement, followed by an examination of the usefulness of various
predictive criteria for identifying embrittlement couples and a
critical evaluation of several proposed mechanisms of embrittlement.
Emphasis will be on monotonic or static loading situations, since
the literature of metal-induced embrittlement under cyclic loading
has been reviewed recently (5). The balance of the paper will summarize the status of our understanding of alkali-metal induced embrittlement processes.

PREREQUISITES FOR EMBRITTLEMENT

It is generally accepted that for metal-induced embrittlement
to occur, a stressed metal must be wetted by the liquid, the stress
state should be one in which either primary or secondary <u>tensile</u>
stresses are generated, and the temperature should be near that of
the melting point of the embrittler. The precise relation between
test temperature and the melting point cannot be predicted, since
for many systems embrittlement occurs well below T_m: for example,
embrittlement of AISI 4140 steel begins at $T/T_m = 0.75$ for cadmium,
and 0.85 for lead and tin environments (2). In a few cases, e.g.
zinc embrittled by tin, the lower limiting temperature for embrittlement is T_E, the melting point of the eutectic between substrate and
embrittler (1). In other cases, e.g. Zn-Hg and Zn-Ga (1), there is
no embrittlement reported at any temperature below T_m, and in fact
embrittlement may only be observed at $T > T_m$, as in the embrittlement of austenitic stainless steel by zinc (6) or pure iron by
indium (7); the width of the temperature range over which embrittlement occurs depends upon strain rate and can be over 300°C (brass-
Hg) (8) or as little as 5°C (Zn-In), Fig. 1 (1). In addition to the
basic conditions for embrittlement set forward above (wetting of the
substrate, tensile stresses, correct temperature range), the degree
of embrittlement depends upon other conditions, such as grain size
of the substrate or imposed strain rate. Therefore, the number of
combinations of environment and substrate that are found to be
embrittlement couples is constantly expanding, as wider variations
in test conditions are imposed. Among the "new" embrittlement couples recently reported have been Zn-In (1), Al-Bi (9) and Al-Pb (9).
Data for the latter systems appear in Fig. 2. This work, in which
aluminum alloys containing inclusions of bismuth, cadmium or lead
were tested in impact, showed that maximum embrittlement required
at least 0.2% of either solute, suggesting that sufficient embrit-
lter must be present to facilitate propagation as well as nucleation
of a crack. The original demonstration of this phenomenon involved
blunting of gallium-initiated surface cracks in cadmium with inten-
tionally small amounts of liquid placed on the surface (10). However, notch-sensitive substrates such as Fe-3.5%Si can be fractured
with very small amounts of embrittler (11).

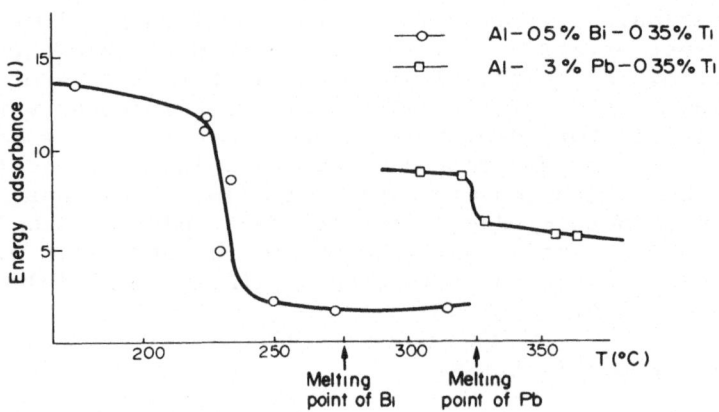

Fig. 2.   Temperature dependence of energy absorbed during impact
          testing of aluminum containing bismuth or lead inclusions
          (9).

Metallurgical and Test Variables

The most widely utilized model for explaining the role of metallurgical variables in brittle fracture processes is the Cottrell (12)-
Petch (13) relation, which was originally stated in the form:

$$\sigma_y k_s d \geq \beta \mu \gamma \tag{1}$$

where σ_y is yield stress, k_s is the slope of a (shear) yield stress
vs. $d^{-\frac{1}{2}}$ plot, d is one-half the grain diameter, β is a geometrical
constant, μ is the shear modulus and γ is the effective surface
energy for fracture. This equation may also be rewritten as:

$$\sigma_i k_s d^{\frac{1}{2}} + k_y k_s \geq \beta \mu \gamma \tag{2}$$

If yielding and fracture are viewed as competitive factors,
ductile behavior is promoted by low yield stress, small grain size,
and low resistance of grain boundaries to propagation of slip, while
fracture is promoted by reduced shear modulus or surface energy. A
ductile to brittle transition is predicted to occur in those materials, principally bcc metals, where σ_i is temperature dependent,
since neither μ nor γ is expected to vary appreciably with temperature.

It has been proposed that liquid metals reduce the energy associated with fracture, and perhaps the shear modulus at a crack tip
as well, thereby inducing brittle fracture in ordinarily ductile
metals and alloys (10). The precise nature of the surface energy
lowering will be described later; for now it is only necessary to
consider the consequences of a reduction in that energy, including

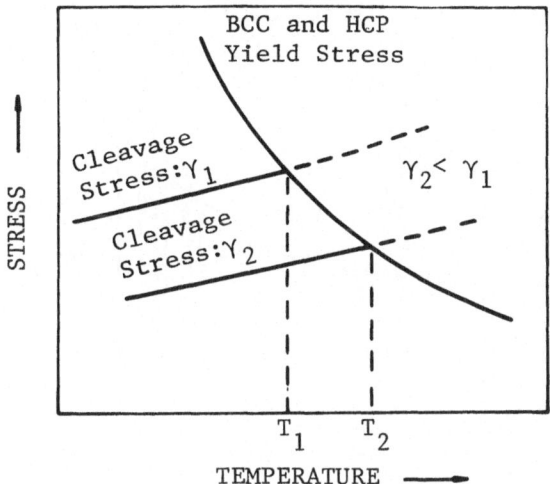

Fig. 3.   Schematic diagram showing influence of surface energy on
          fracture stress and transition temperature (1).

any changes in the plastic work associated with crack propagation.
Figure 3 shows schematically how a ductile to brittle transition can
be obtained in bcc or hcp metals with temperature-dependent yield
stresses, and nearly temperature independent fracture stresses (1,
14). Note that lowering the surface energy from γ_1 to γ_2, as by
testing in a liquid or solid metal environment, causes a significant
increase in the transition temperature; no conclusion need be reached as to the precise mechanism of lowering of γ. In all cases,
coarse grain size increases the likelihood of brittle fracture.

    Equation (2) shows clearly that the transition to brittle frac-
ture should be aided by any factor that raises σ_i. These would generally include decreasing temperature and increasing strain rate.
Recently, there has been considerable attention directed to the
effect of strain rate, as outlined below.

    The ductile to brittle transition temperature of polycrystalline
zinc tested in air ranges between -30°C to +11°C, dependent upon
strain rate. Testing zinc in tin or indium environments at high
strain rates causes the transition temperature to exceed 350°C, as
indicated by the data of Fig. 1 (1). The onset of embrittlement,
however, is near 150°C, independent of strain rate. Similarly, the
onset of embrittlement in aluminum tested in mercury remains constant, at the melting point of mercury, while the transition temperature increased by some 70°C as strain rate increased from 4×10^{-5}
sec^{-1} to 0.503 sec^{-1}, see Fig. 4 (15). Aluminum displayed an extremely sharp transition temperation region ($\sim$ 7°C) and the fracture
mode was intergranular, independent of strain rate. Other systems
exhibiting an (linear) increase in recovery temperature with

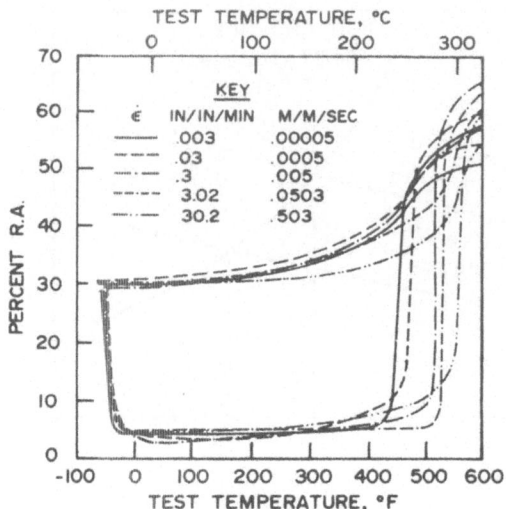

Fig. 4.   Embrittlement of 2024-T4 aluminum by mercury at several
          strain rates (15).

increasing strain rate include Zn-Hg (1), Zn-Ga (1), Ti55A/Cd (16),
2024Al/Hg-3%Zn (17), and 4145 steel/Pb (15).

     The pattern of behavior of the recovery from MIE suggests that
the substrate properties control ductility, rather than a desorption
reaction (17) or a transport problem (18). This approach is supported by work on zinc coated with cadmium (1). At temperatures
above the cadmium melting point, the yield stress of zinc at ordinary strain rates lies below the fracture strength in cadmium. Upon
increasing the strain rate by a factor of ten, the yield stress was
increased and embrittlement was observed.

Specificity

     The strain-rate data illustrate two important points: that
apparently nonsusceptible metals at any temperature may in fact be
severely embrittled by a liquid (or solid) metal environment if the
strain rate is increased sufficiently, and that the melting point of
the embrittler at best can only suggest the temperature of maximum
susceptibility. In the case of Armco iron tested in tin, cadmium,
bismuth or Pb-Bi, the maximum embrittlement occurred near 350°C,
even though the embrittler melting points differ by several hundred
degrees (20). These findings cast serious doubt upon the significance of "specificity", since so many embrittlement and non-embrittlement couples reported and summarized in the literature have been
identified on the basis of experiments at one or a few temperatures,
usually at a single constant strain rate. Similar caution must be
exercised in the choice of substrate grain size, and perhaps the

nature of the environment surrounding the liquid metal (20).

Transport Mechanisms

There has been some controversy in the literature concerning the mechanism of embrittler transport to a crack tip. When present as a liquid, the embrittler is generally believed to approach the crack tip by bulk liquid flow; however, access to the crack tip itself has been thought to occur by some rapid diffusional process, such as second monolayer surface diffusion (21). Gordon (22) has recently proposed, however, that liquid zinc can penetrate to very near the tip of a sharp crack in 4140 steel, based upon both direct observation and calculations. Gordon further calculated that liquid flow transport is very rapid for all liquid metal embrittlers, even in cracks of very small tip radii. However, for high vapor pressure liquids such as zinc, cadmium and mercury, vapor transport also may contribute to embrittlement.

Below the embrittler melting point, transport by bulk liquid flow is, of course, impossible, but transport through the vapor phase also seems to be unlikely for most embrittlers (other than zinc, cadmium and mercury). This is largely a consequence of very low vapor pressures at their respective melting points for lead, bismuth, lithium, indium, tin and gallium (2). Solid state diffusion is left as the only viable transport mechanism. Grain boundary diffusion and first monolayer diffusion of the embrittler over the substrate are likely to be too slow. Second monolayer or multiple monolayer self-diffusion would then be necessary to avoid embrittler-substrate diffusion interaction. Surface self-diffusion rates of 10^{-3} to 10^{-6} cm^2/sec have been measured (23), in good agreement with apparent diffusivities inferred from crack propagation rates in steel.

EMPIRICAL PREDICTIONS OF SUSCEPTIBILITY

Several approaches have been used with varying success in recent years to predict or rationalize susceptibility of specific metals to embrittling environments. Kelley and Stoloff (24) proposed that cohesion of the solid is lowered or shear is enhanced in the presence of a liquid metal to a degree determined by the amount of the solid-liquid bond energy. The Engel (25) - Brewer (26) thermodynamic cycle was used to estimate the interaction energy (IE) between the atoms of the two metals forming an embrittlement couple. [A similar approach, developed independently, has been applied to hydrogen embrittlement of austenitic stainless steels (27).] IE represents the energy released by one-to-one bond formation between atoms of the liquid and the solid, neglecting any effects relating to less than full surface coverage. The fracture surface energy (FSE) was assumed to be proportional to the bond energy, so that the ratio of the solid-liquid bond energy to the solid-solid bond energy (assumed equal to the atomization energy) is a measure of the reduction of

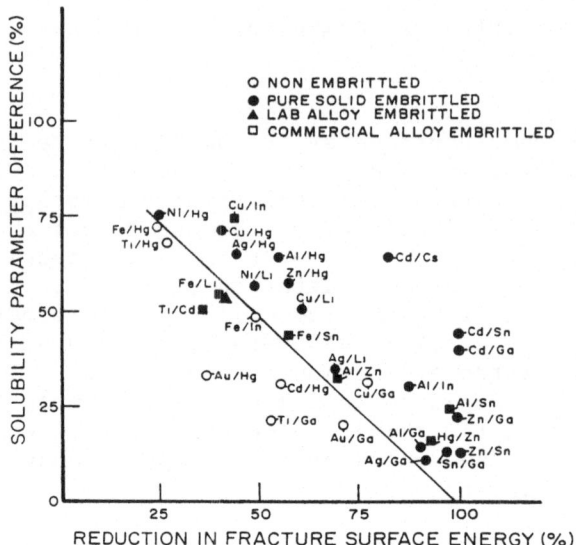

Fig. 5.   Calculated reduction in fracture surface energy relative
         to solubility parameter for various substrate/environment
         combinations (24).

FSE in the presence of the liquid. The likelihood of embrittlement
was related directly to a second factor, the solubility parameter,
defined as the square root of the sublimation energy per unit volume; a small difference in solubility parameters (28) defined as
$\delta = (\Delta E_1/V_1)^{\frac{1}{2}}$ where ΔE_1 = heat of sublimation and V_1 = molar volume
of solid and liquid is indicative of a large mutual solubility. It
was proposed that a plot of solubility parameter difference vs.
%reduction in FSE could identify the boundary between embrittlement
and no embrittlement, Fig. 5 (24). Embrittlement is favored by
increased reduction in FSE and by reduced mutual solubility.
Reduced FSE was considered to arise from decohesion, Fig. 6.

     This model was recently tested by Roth et al (9) who calculated
reduction in FSE and solubility parameter differences for three couples: Al-Bi, Al-Cd and Al-Pb; embrittlement data for Al-Bi and Al-
Pb appear in Fig. 2. The reductions in fracture surface energy for
these systems were calculated to be 82, 67 and 80%, respectively.
Relative solubility parameter differences between aluminum and bismuth, cadmium and lead were computed to be 44, 48 and 40%, respectively, consistent with small solubilities of these elements in
aluminum. The data for these alloys all fell within the embrittling
range of Fig. 5, so that embrittlement was successfully predicted.
However, the calculations did not predict the correct order of the
relative severity of embrittlement of aluminum with the three environments. The predicted loss in cohesion caused by lead and bismuth
is nearly identical, and larger than that predicted for cadmium
(and that for mercury, which is known to cause much more severe

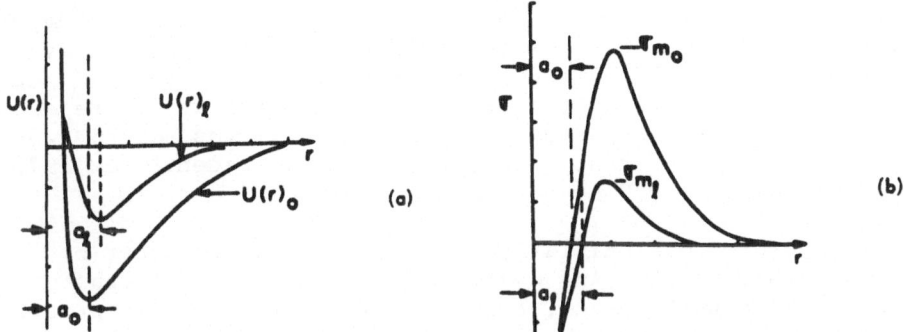

Fig. 6.   Schematic diagram of decohesion model, showing proposed
          influence of liquid metals on bond strength, σ_m, and energy
          to rupture bonds at crack tip, U (r). (10)

embrittlement in aluminum); however, bismuth was much more potent
an embrittler than lead. Thus the model was concluded to be inadequate in predicting susceptibility to embrittlement. It should be
emphasized, however, that the demarcation between embrittlement and
non-embrittlement conditions shown in Fig. 5 is sensitive to metallurgical variations such as grain size and environmental considerations such as temperature and strain rate, consistent with the requirements of Eq. (2) (5,24). It was concluded in an earlier paper
that "specificity is a matter of degree: strong interactions will
lead to embrittlement under a variety of test conditions, while weak
interactions require a very precise combination of conditions" (3).
This caveat must be applied to any model of embrittlement that relies
solely upon bulk thermodynamic or physical properties.

     Several alternative approaches to predicting susceptibility of
LME consider solely the reduction in surface energy, γ, due to the
presence of a liquid at the crack tip. According to Griffith (29),
for a body stressed elastically, and fracturing with no dissipative
processes other than forming a new surface of energy per unit area,
γ_s,

$$\sigma_F = \sqrt{\frac{2E\gamma_s}{\pi a}} \quad \text{(plane stress)} \tag{3}$$

where E is Young's modulus and a is half-crack length. Later,
Cottrell (12) and Petch (13) independently showed that crack nuclea-
at the yield stress could be described in terms of dislocation pile-
ups coalescing to form a crack nucleus by Eqs. (1) and (2), except
that the appropriate γ to use in Eq. (2) is the effective surface
energy, γ_{eff}, which includes the effects of plastic work.

     When attempts are made to explain embrittlement phenomena
through a reduction in surface energy, objections arise, based upon

the seeming difficulty in explaining how the plastic work accompanying fracture, which far exceeds the true elastic surface energy in
metals, can be decreased by the environment. Orowan (30) had suggested that plastic work and elastic surface energy terms are additive quantities. This apparent problem, in fact, was solved twenty
years ago, although apparently not recognized as recently as 1977 by
Birnbaum (31) and in 1980 by Jokl et al (32).

It was pointed out initially by Gilman (33), and later by
Stoloff and Johnston (10) and Westwood and Kamdar (21) in the context
of liquid metal embrittlement, that when a plastic zone is present at
the crack tip, the radius, ρ, of the zone will influence the effective energy to fracture the material, γ_{eff}. The two surface energies
are proportional:

$$\gamma_{eff} \propto \frac{\rho \gamma_s}{a_o} \tag{4}$$

where a_o is the substrate lattice parameter. On this model, the
metal embrittler atoms lower γ_s, as shown in Fig. 6, and by causing
the crack tip to remain sharp during propagation, γ_{eff} is reduced
proportionally. This approach is much more straightforward and
tractable than the more involved path taken by Jokl et al (32) to
come to the same conclusion.

The application of thermodynamic surface energy measurements to
embrittlement will now be considered. When a perfectly brittle single crystal is fractured in an inert atmosphere, γ_{eff} is approximated
by the solid-vapor surface energy γ_{SV}^o. For intergranular fracture of
a polycrystalline alloy:

$$2\gamma = (2\gamma_{SV}^o - \gamma_{GB}^o) \tag{5}$$

where γ_{GB}^o is the energy of the grain boundary. When liquid metal is
present, the equations are $\gamma = \gamma_{SL}$ for single crystals (or transcrystalline failure in polycrystals) and $\gamma = 0.5(2\gamma_{SL} - \gamma_{GB})$ for grain
boundary fractures. For wetting systems $\gamma_{SL} < \gamma_{SV}$ and a coefficient
of embrittlement, η_1:

$$\eta_1 = \frac{\gamma_{SL}}{\gamma_{SV}} \quad \text{or} \quad \frac{2\gamma_{SL} - \gamma_{GB}}{2\gamma_{SV} - \gamma_{GB}^o} \tag{6}$$

may be defined (35,36). The effective surface energy, γ_{eff}, is simply proportional to the appropriate value of γ defined above. As
η_1 decreases, the stresses necessary to initiate or propagate cracks
decrease, and the ductile to brittle transition temperature increases.

Roth et al (9) have applied this macroscopic model to the grain
boundary fracture of aluminum induced by liquid inclusions. The

severity of embrittlement was related to the energies defined above
in terms of the fractional reduction in work of fracture, $\Delta W/W$, due
to the presence of the liquid:

$$\frac{\Delta W}{W} = \frac{2\gamma_{SV}^{o} - \gamma_{GB}^{o} - (2\gamma_{SL} - \gamma_{GB})}{2\gamma_{SV}^{o} - \gamma_{GB}^{o}} \qquad (7)$$

The grain boundary energy of most pure metals and alloys is approximately one-third of γ_{SV}, and γ_{GB} can be expressed in terms of the
dihedral angle, θ, reducing $\Delta W/W$ to the following:

$$\frac{\Delta W}{W} = 1 - 0.2(\sec\theta - 1)(\frac{\gamma_{GB}}{\gamma_{GB}^{o}}) \qquad (8)$$

Roth et al (9) concluded from the large dihedral angles observed with
lead, cadmium, and bismuth on aluminum that these elements do not
strongly adsorb at grain boundaries, so that $\gamma_{GB}/\gamma_{GB}^{o}$ is about 1.
Then $\Delta W/W$ is a function of θ only, resulting in $\Delta W/W = 0.88$ for Al-
Bi, 0.86 for Al-Cd and 0.81 for Al-Pb. These values of $\Delta W/W$ are in
the same order as the severity of embrittlement noted in these materials, as measured by the impact energy absorbed when the inclusions
are liquid (1.5J for Al-Bi, 1.8J for Al-Cd and 5.4J for Al-Pb); on
this basis, the surface energy approach was considered applicable to
LME. In extreme cases, the dihedral angle, θ, approaches zero and
the liquid metal will penetrate grain boundaries even if the solid
is unstressed. In this case $2\gamma_{SL} \leq \gamma_{GB}$. Al-Ga is a system of this
type.

Nicholas and Old (36) also have recently advocated use of γ, determined from wetting studies, to correlate with fracture data. For
example, ratios of γ_{SL}/γ_{SV} measured in moderate vacua are in reasonable agreement with reductions in the effective fracture energy.
These observations suggested that ratios of γ_{SL}/γ_{SV} of less than 0.5
are associated with LME of zinc (e.g. $\gamma_{SL}/\gamma_{SV} = 0.41$ for embrittling
mercury and 0.27 for embrittling gallium, but $\gamma_{SL}/\gamma_{SV} = 1.16$ for non-
embrittling sodium and 0.61 for non-embrittling bismuth). However,
γ_{SL}/γ_{SV} is 0.5 for non-embrittling cadmium, which is very close to
the ratio of 0.48 noted for embrittling lead. It is doubtful that
any model could so sharply define an embrittlement condition, especially in view of the role of metallurgical variables. Also, no
explanation of the significance of a ratio of 0.5 was offered, suggesting that the dividing line is totally arbitrary.

Another problem with the surface energy model is that it does
not predict changes in bond strengths with crystallographic orientation. Studies on several metals suggest that surface energies differ
by only a few % with orientation, while a difference in bond strength
of some 30% between [001] and [111] was inferred for fracture of

aluminum in gallium (37).

Several authors (1,19,37,38,39) have suggested that the surface
energies of pure elements may be connected with high heats of vaporization or high heats of mixing, ΔH_m. Couples which display a high
positive ΔH_m would not be expected to form bonds; the interfacial
energy is then expected to be high and embrittlement is not observed.
Negative values of ΔH_m imply preferred bonding between environment
and substrate, and either significant material solubility or compound
formation will occur. Only when ΔH_m is small and positive is embrittlement to be expected. Limited success in rationalizing embrittlement of zinc by six low melting metals has been claimed, see Table 1
(1). However, the correct order of embrittling tendency was not
predicted. Moreover, the heat of mixing approach predicts non-
embrittlement of aluminum by lead and bismuth (19), contrary to recent experimental results (9). Worse yet, Tetelman and Kunz (19)
identified hydrogen "conclusively" as a non-embrittler of aluminum,
contrary to many recent published results (40). A significant problem in evaluating the model for other alloy systems is the relative
sparseness of ΔH_m data. Nevertheless, the results obtained to date
with this model do not appear to be promising; in fact, Gilman (41)
has pointed out that heat of mixing data provide no better than 50%
success in predicting embrittlement; i.e. there is no predictive
capability. Perhaps the situation could be improved by combining
the heat of mixing data with solubility parameter differences to
establish a boundary between embrittling and non-embrittling conditions.

Toropovskaya (42) has related the heat of fusion, H_F, to interatomic bonding, suggesting that:

Table 1.   Susceptibility and ΔH_m

Couples	ΔH_m	Embrittlement
Zn–Bi	19	Yes
Cd	4	Yes
Ga	0	Yes
Hg	4	Yes
In	12	Yes
Sn	4	Yes
Cu–Na	37	No
Cu–Li	−39	Yes-Compds
mild steel – Li	95*	Yes-decarb, ΔH_m typical
mild steel – Na	236*	No
Al–Cd	13	Yes
Al–Zn	2	Yes

* embrittlers of iron tend to have fairly high $+\Delta H_m$.

$$\frac{H_{F_{L-S}}}{H_{F_S}} = \frac{\sigma_{F_{L-S}}}{\sigma_{F_S}} = \eta_2 \tag{9}$$

where $H_{F_{L-S}}$ is the heat of fusion of a liquid-solid solution. Good agreement for several embrittling elements on copper was claimed between the ratios of the thermodynamic and fracture stress quantities. No data for other than copper substrates have been reported.

Proponents of each of the empirical systems outlined above can claim limited success, sometimes with one substrate, others with a number of base metals. However, there is still no completely satisfactory system for predicting embrittlement couples. Moreover, none offer any insight into the nature of the precise environment-substrate interaction that leads to embrittlement.

MECHANISMS OF CLASSICAL METAL-INDUCED EMBRITTLEMENT

The precise interaction between atoms of an embrittler and a stressed substrate that leads to embrittlement remains unknown. The calculational schemes outlined previously, through which likelihood and degree of embrittlement may be predicted with varying degrees of success do not shed much light on this matter, nor do the transport mechanisms also outlined earlier. We shall now briefly review several proposed mechanisms of embrittlement that focus upon events at the crack tip, as follows:

1) decohesion
2) enhanced shear
3) stress-assisted dissolution
4) embrittlement of grain boundaries by diffusion

Decohesion

The most widely accepted theory of both metal-induced and hydrogen-induced embrittlement postulates that chemisorbed embrittler atoms lower bond strength at the crack tip, thereby facilitating both nucleation and propagation of a crack (10,21). The proposed reduction in cohesion due to an embrittler atom is shown schematically in Fig. 6 (10). For a perfectly elastic fracture the lowering of bond strength has two equivalent effects: the stress to rupture the bonds is decreased and the energy obtained by integrating under the force-distance curve (i.e. the bond energy or true elastic surface energy, γ) is reduced. Note that the diagram implies also reduced elastic modulus at the crack tip. Oriani (43) has pointed out that in non-ideally elastic situations (i.e. most structural metals in contact with liquid metals or hydrogen), a reduction in cohesive stress need not be equivalent to reducing surface (cohesive) energy. For this reason, he has suggested that attention be directed towards the bond strength as being the critical parameter.

The decohesion model, in spite of its wide acceptance, is open
to a number of criticisms, stemming from the inability of researchers to define the interaction that ultimately leads to enhanced bond
breaking.

If there is indeed a finite number of specific embrittler-
substrate combinations, the decohesion model cannot predict such
specificity. However, as pointed out by others (1,44) and demonstrated earlier in this review, the presence or absence of embrittlement depends sensitively upon the choice of test variables such as
temperature and strain rate, and metallurgical variables such as
grain size and stacking fault energy. This factor impacts equally
on other proposed embrittlement mechanisms, to be outlined below,
and therefore cannot be the definitive factor in rejecting any
embrittlement model.

A more specific criticism of decohesion may be found in the
relatively wide occurrence of delayed failure among embrittlement
couples, as recently compiled by Gordon.(45) In the majority of cases,
e.g. 4340 steel in cadmium (46) and Cu-2%Be in mercury (47), readily
recognizable delayed failure (type A behavior) was observed, but in
several instances, all involving mercury or mercury-indium environments on zinc (48), cadmium (48), silver (49), or aluminum (49),
delayed failure did not occur (type B behavior). In type B situations, there appears to be a critical stress above which failure is
virtually "instantaneous" (less than 1-2 sec) but below which no
failure is observed. However, Gordon suggests that even in the
latter category very rapid delayed failure may be occurring, i.e.
that crack nucleation occupies most of the very short failure time.
An incubation period, during which no crack greater than about 0.01
mm long could be detected, was observed in delayed failure experiments on unnotched 4140 steel, in the quenched and tempered condition, embrittled by either liquid or solid indium. Other evidence
on this point is conflicting, an incubation period having been
reported in some, but not all, systems studied. It is difficult to
reconcile delayed failure, especially with an incubation time, with
a surface-induced decohesion phenomenon which should occur instantaneously upon contact of the embrittler atoms with a crack nucleus.

An added criticism of the decohesion model as applied to hydrogen effects on fracture has been made by Clark et al (50) who showed
that the surface free energy of a Ni-300 ppm hydrogen solid solution
is nearly equal to that of pure nickel. However, a decrease in grain
boundary energy of about 20% due to hydrogen in solution was measured . It was concluded that hydrogen can decrease the work for
intergranular or transgranular fracture due solely to adsorption
effects and not by significantly altering the lattice potential
between the solvent atoms. The implications of these findings for
metal-induced embrittlement are not yet clear. Similar measurements
of lattice energies of metal in the presence of liquid metals do not

seem to have been made. Yet the doubt cast upon decohesion of nickel
by hydrogen, a known embrittler, does not seem to bode well for the
decohesion model of LME.

Enhanced Shear

     The most sustained criticism of the decohesion model as applied
to both LME and hydrogen embrittlement is based upon metallographic
and fractographic observations of embrittled samples, most notably
by Lynch (51-55). Decohesion implies a <u>cleavage</u> mechanism of fracture in which atomic bonds defining a crack tip are ruptured by
forces acting normal to the crack plane, leading either to crystallographic cracking or to grain boundary separation. Careful fractographic examination of many LME couples has, however, revealed dimples in the fracture surfaces and there is metallographic evidence
of considerable plastic strains accompanying fracture, as will be
described in the following section. For example, fracture surfaces
in polycrystalline cadmium embrittled by gallium reveal what appear
to be cleavage facets at low magnification, Fig. 7 (a), but at higher
magnification numerous small dimples may be observed, as shown in
Fig. 7 (b) (52). Fractographic examination of embrittled aluminum
single crystals tested under cyclic loading further showed that
cracks propagated along cube planes in the close-packed <110> directions, see Fig. 8 (55); and that extensive slip seemed to occur on
{111} planes intersecting the crack tip. On the basis of these observations, Lynch suggested that chemisorption facilitated nucleation of dislocations from crack tips, Fig. 9 (51,53,55).

     Several means by which weak chemisorption can facilitate dislo-
cation nucleation have been suggested. The transfer of electrons

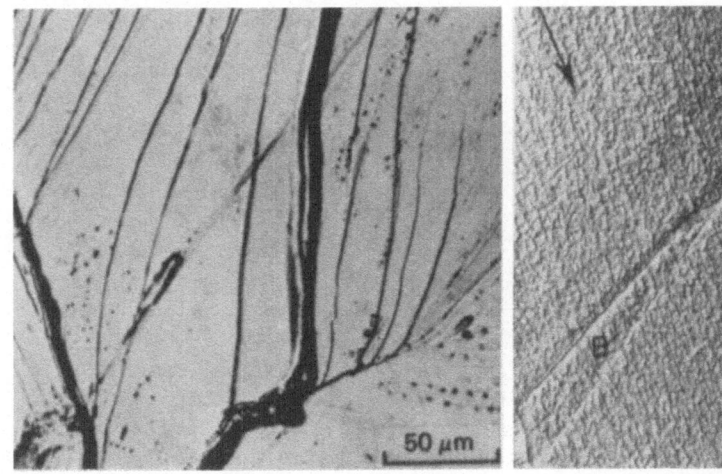

Fig. 7. Fractograph of cadmium after fracture in liquid gallium.
         (a) apparent cleavage facets; (b) shallow elongated dim-
         ples (52).

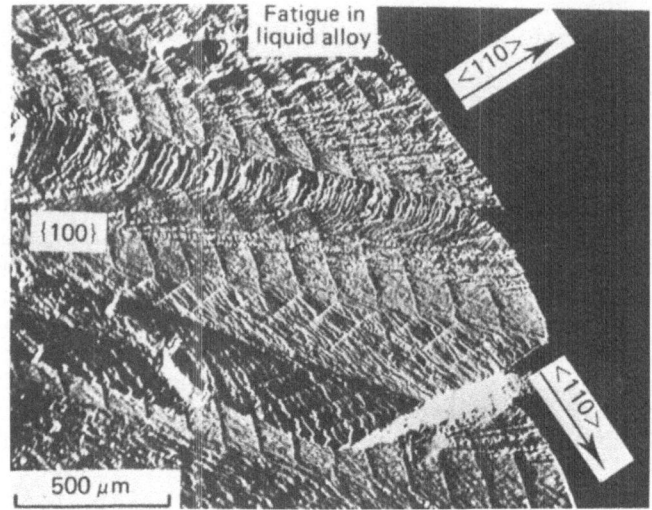

Fig. 8.   Scanning electron micrograph of solution treated aluminum
          alloy, showing {100} facets produced by fatigue.   Zigzag
          crack fronts due to growth in different <110> (55).

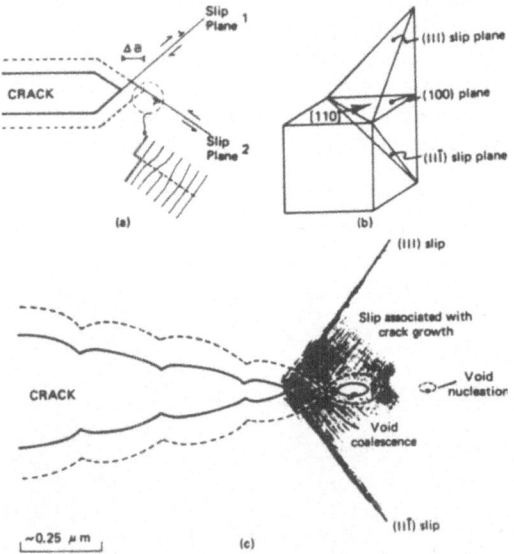

Fig. 9.   Schematic diagram showing:
          (a) crack growth, Δa, by alternating shear on two slip
          planes; (b) orientations of planes and direction in fcc,
          showing that alternate slip on {111} produces crack
          growth along <110>; (c) general slip, in addition to alter-
          nate slip, producing nucleation and growth of voids ahead
          of crack.   Dotted line shows crack position after growth
          Δa (53).

from adsorbed species to atoms of the substrate at the surface may locally increase the electron density and thereby increase the repulsive force between adjacent surface atoms. Reduced cohesion between surface atoms should thereby enable shear between surface atoms to occur more readily. Stoloff and Johnston (10) proposed that the localized Young's modulus (and therefore the shear modulus) at a crack tip may be reduced by an adsorbing species, see Fig. 6. Krishtal made the same point in 1970 (56), as did Matsui et al in 1977 (57) and 1979 (58), the latter in reference to appreciable softening of high purity iron in the presence of hydrogen. Birnbaum has, in fact, reported a substantial drop in the shear modulus of niobium as measured in a torsion pendulum (31). Oriani and Josephic (59), faced with this evidence, conclude that softening by hydrogen can sometimes arise from a reduction in interatomic bond strength at the dislocation cores, leading to a lowering of the Peierls barrier to dislocation motion. Nevertheless, these authors continue to adopt the view that decohesion is responsible for embrittlement, citing other examples of softening attributable to an increase in density of microvoids or internal pressurization effects. Thus, although it is now widely agreed that dislocation nucleation or motion can be facilitated by hydrogen and perhaps by liquid metals, there is still disagreement about the significance of such observations.

A critically important feature of suggestions as to the role of the environment is the concept that moduli are reduced only in the immediate vicinity of the environment. There is no measurable effect of liquid metals on bulk elastic or plastic properties, except that the fracture stress and/or strain are reduced. It seems more likely, based upon the widely divergent fracture observations outlined above, that both tensile and shear strength of interatomic bonds are decreased by films causing LME or SMIE. The observed fracture morphology will then depend upon such additional factors as the availability of a low energy cleavage plane, segregation of impurities to grain or subgrain boundaries, and the orientation of the maximum principal stress axis relative to individual crystals. Similarly, Clum (60) has suggested that enhanced dislocation nucleation in iron at surfaces in the presence of adsorbed hydrogen can lead to barrier formation by interaction of these dislocations with slip dislocations, leading to embrittlement. The distinguishing criteria were suggested to be the number of dislocations and their Burgers vector in any region of the solid. The results of Wanhill (61) on fracturing of a high strength titanium alloy in mercury also are consistent with a model based on liquid metal facilitating slip on the $\{10\bar{1}0\}$ which intersect (0001) cleavage surfaces.

Additional evidence that enhanced shear may contribute to metal-induced embrittlement may be found in the work of Goggin and Moberly (62) on the tensile behavior of single crystal aluminum foils exposed to gallium. Foils generally failed in a ductile manner, except for short (few micron) segments which rapidly propagated before blunting occurred. Subsequent failure was by ductile shear. Cracks did not initiate at dislocation pileups, at grain boundaries (in polycrystals) or at slip band intersections. In the case of cyclic loading of mercury-coated copper, Tiner (63) has reported that the crack apparently propagates at high velocity by transcrystalline shear. This observation also is in accord with an environment-assisted shear model of LME.

In an effort to approach the mechanism of embrittlement from an entirely different point of view, efforts have been made to embrittle amorphous metallic alloys either with liquid metals or with hydrogen introduced cathodically (64). Since amorphous alloys deform and fracture in inert environments by a shear process, it was anticipated that embrittlement by any of the selected environments could lead to improved understanding of the relevance of shear processes to environmentally-induced fracture.

Table 2 lists the compositions of several amorphous alloys tested in mercury or Hg-In environments. When tested in air in tension at room temperature, all of the alloys are macroscopically brittle; but as temperature is increased appreciable plastic deformation occurs. Plastic deformation occurs by intense shear on planes or surfaces inclined at about 45° to the stress axis, see Fig. 10 (a). Deformation is quasi-viscous in nature, and therefore no work hardening occurs. Initiation of shear in one of the bands leads to fracture at the yield stress, with microscopic strains of the order of 1000% or more. Fracture surfaces in air reveal a veining pattern, see Fig. 11 (a). When these alloys are tested in the presence of a liquid metal, fracture stresses are reduced, Table 2, the shear pattern on the specimen surfaces changes, Fig. 10 (b); the fractographic features, at least at low magnification, appear to be very brittle, Fig. 11 (b). However, at high magnification the veining pattern reappears, Fig. 11 (c), showing that only the scale of the pattern has changed. Similar effects are noted in the same amorphous metals after cathodic charging. Consequently, it appears that not only is enhanced shear a viable mechanism for LME of amorphous metals, but that it may account for hydrogen embrittlement of these alloys as well.

Table 2. Embrittlement of Amorphous Alloys at 25°C (64).

Alloy	σ_F (GPa)		
		Hg	Hg-In
$Fe_{81.5}B_{14.5}Si_4$	1.90 ± 0.26	1.36 ± 0.5	1.26 ± 0.33
$Fe_{81.5}B_{13.5}Si_{2.5}C_{2.5}$	1.50 ± 0.37		1.03 ± 0.48
$Fe_{40}Ni_{40}P_{14}B_6$ (2826)	1.30 ± 0.21		1.18 ± 0.16
$Fe_{40}Ni_{38}B_{18}Mo_4$ (2826 MB)	1.60 ± 0.19		1.48 ± 0.17

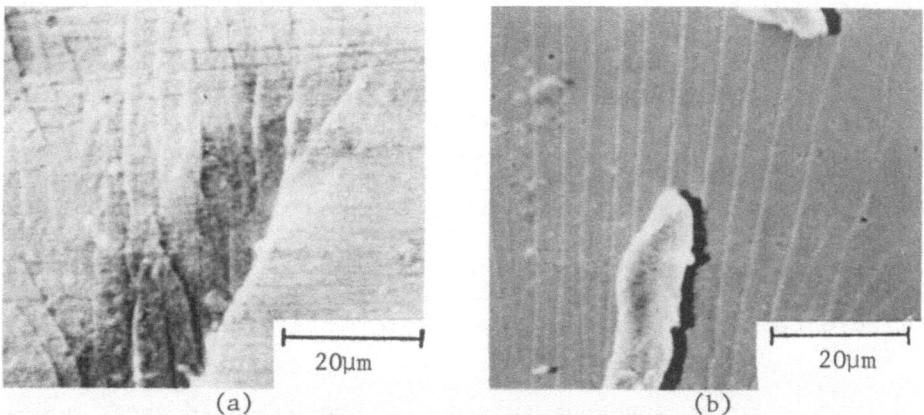

Fig. 10. Shear bands near crack tips in bend specimens of amorphous
          FeBCSi, tested at 25°C (64).
          (a) air; (b) Hg-In.

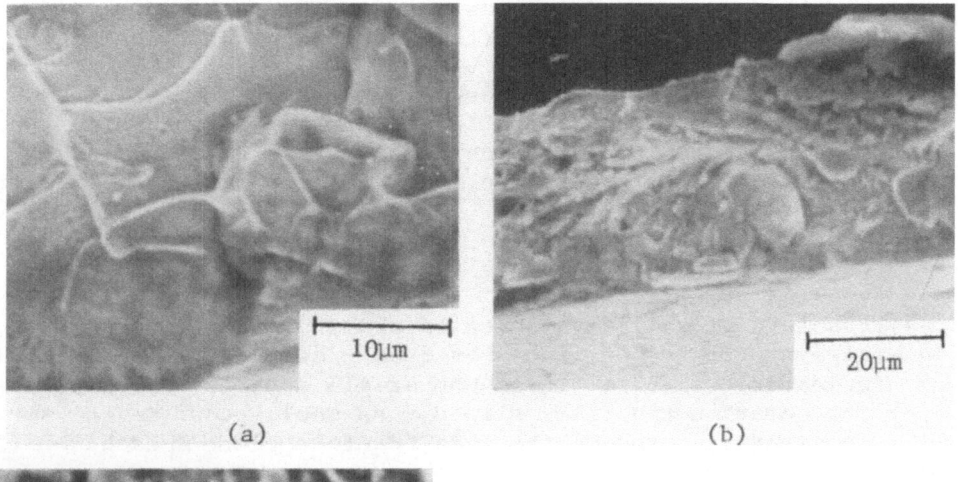

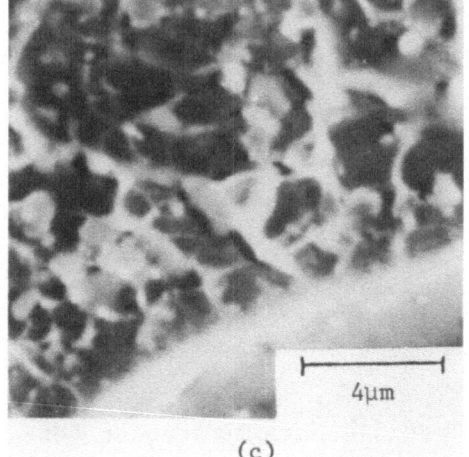

Fig. 11. Fractographs of amorphous
          FeCBSi tested in tension
          at 25°C (64).
          (a) air; (b) Hg-In (low
          magnification); (c) Hg-
          In (high magnification).

A further critical test of the enhanced shear mechanism involved bending of one of the amorphous alloys (Metglas 2826) with liquid lithium on the compressive or tensile side of the bend sample. Severe embrittlement was noted in either case, providing clear evidence that a state of tensile stress is not necessary for embrittlement to occur. This observation is perfectly consistent with a fracture process depending upon shear only, but cannot be explained by a decohesion model. However, this result does not explain the apparent lack of reports of embrittlement of crystalline metals in bending or compression, except for a few cases in which secondary tensile stresses developed as a result of bending. Certainly more attention should be directed to state of stress in embrittlement testing to clarify this point.

In spite of the apparent success of the enhanced shear model with regard to amorphous metals, there remain serious criticisms of the applicability of the model to crystalline metals:

a) The formation of microvoids ahead of a crack tip should not be influenced by liquid metal atoms at or just behind the crack tip.
b) No reasonable mechanism for delayed crack initiation is offered, since the alteration of surface structure of bonding should be virtually instantaneous if it is caused by embrittler atom adsorption, particularly if the embrittler is in liquid form (45).
c) Slip on alternating planes is highly unlikely in zinc and cadmium, which deform solely on the basal plane for most crystal orientations.
d) Fracture of gallium-embrittled aluminum single crystals always occurred on {100} for four widely differing initial orientations (37). It is not at all apparent how alternating shear on two {111} slip planes could occur in the amounts necessary to produce the same fracture plane in each case (31).

The questions surrounding both the decohesion and enhanced shear models are sufficient to require much more experimental and theoretical work before either mechanism can be accepted for crystalline metals.

Dissolution

Robertson (16), and later Glikman et al (65,66), proposed that rapid localized dissolution of highly stressed atoms at the leading edge of a crack is the cause of crack propagation in LME. The dissolved material would need to be carried away by diffusion through the liquid. The problems with this approach have been well known for some time:

(a) dissolution would have to be rapid enough to account for

       crack propagation rates measured in cms/sec; this is highly
       unlikely.
(b)   increasing temperature should lead to increased embrittle-
       ment; in fact, the reverse is observed with most embrittle-
       ment couples, see Figs. 1, 2, 4.
(c)   crack initiation and propagation are not separable events,
       yet the embrittlement of 4140 steel by indium suggests that
       there are distinctly different initiation and propagation
       stages of both SMIE and LME (45).
(d)   the mechanism can hardly account for SMIE at all, but in any
       case would predict an abrupt change in temperature depen-
       dence of crack initiation at the embrittler melting point;
       this also is contrary to results of indium embrittlement of
       4140 (45).
(e)   the severity of embrittlement should increase with increas-
       ing solubility of the substrate in the liquid; this, of
       course, violates one of the principal empirical prerequi-
       sites for LME, namely limited mutual solubility is required
       for embrittlement to occur.

Grain Boundary Embrittlement

    Krishtal (56) has proposed that LME is composed of a two step
process: embrittler atoms must initially diffuse along grain boundaries in the substrate to a critical depth and concentrate on the
boundaries, which are then embrittled. Dislocations produced by
strains resulting from the embrittler atoms' presence as well as dislocations produced by lowering of the base metal surface energy are
responsible for the embrittlement. Ignoring the rather fuzzy references to dislocation generation, Gordon (45) has adopted and extended
the Krishtal mechanism to explain SMIE. Embrittler atoms first
change from the adsorbed to the dissolved (in the surface) state,
and subsequent penetration occurs along grain boundaries or other
preferred paths. Both steps are accelerated by the applied stress.
Crack nucleation eventually occurs in the embrittled zone when the
proper concentration of embrittler atom has been reached. Once the
crack forms, it grows rapidly until a visible crack is produced, or
until the length of the crack is such that the transport process
becomes rate controlling.

    The principal advantage of such a mechanism is the natural pre-
diction of delayed failure phenomena: the model also provides reasonable estimates for the apparent activation energy for crack initiation with indium on 4140 steel. The model does not address specificity, but can explain the onset of embrittlement and the restoration of ductility at some higher temperature on the following basis:
as temperature increases the rates of solution and grain boundary
diffusion first become sufficiently high to cause crack initiation
near the embrittler melting point, T_m. At temperatures higher than
T_m, volume diffusion from the boundaries into adjacent grains reduces

the grain boundary concentration, eventually producing the characteristic brittle-ductile transition.

Gordon (45) also suggests that the grain boundary embrittlement
model can account for strain rate, solute, grain size and cold work
effects, all due to the influence of these variables on volume diffusion, e.g. at higher strain rates, higher temperatures are needed
to dissipate grain boundary penetration zones. Decreasing grain size
reduces pileup-induced stress concentrations at grain boundaries,
requiring greater embrittler penetration to accomplish embrittlement,
while increasing cold work increases dislocation density in base-
metal grains, providing more short-circuit volume diffusion paths to
dissipate embrittler concentrations at grain boundaries.

This mechanism, unfortunately, provides no insight into the many
recorded instances of transgranular (or single crystal) embrittlement
noted in the literature, nor can it explain the embrittlement of
amorphous alloys. Furthermore, and perhaps most importantly, pre-
exposure to the environment, followed by testing in an inert medium,
should cause embrittlement; in fact, this is rarely the case, except
for the well known Al-Ga system in which diffusion along grain boundaries is very rapid (67) and for single crystals of zinc in contact
with mercury or gallium which display embrittlement due to the formation of a brittle alloy layer (68).

SIMILARITIES AMONG EMBRITTLEMENT PHENOMENA

In recent years there has been renewed interest in possible
links between embrittlement phenomena: e.g. impurities that promote
temper embrittlement also have been shown to increase the susceptibility of steels to hydrogen embrittlement. More importantly, there
have been several cases where hydrogen embrittlement and MIE have
been closely linked (19,51,54,69,70), both in crystalline and amorphous substrates, as pointed out earlier. This linkage has been
based on the following conditions which appear to be necessary for
embrittlement of crystalline metals with both species:

a) tensile stress must exist in the base metal (this has now
   been cast in doubt for amorphous metals)
b) pre-existing cracks or stable obstacles to dislocation motion
   must be present
c) embrittling species must be present at the obstacle and subsequently at the tip of the propagating crack
d) low mutual solubilities, little tendency to form compounds
   and strong binding energies are characteristic of embrittlement couples.

In the case of steels, certain other similarities also have
been noted (54,71):

1) embrittlement tends to increase with increasing strength
   level.
2) elements that form stable compounds with the solid, when
   added to the liquid, inhibit MIE, just as oxygen inhibits
   embrittlement by hydrogen.
3) the temperature sensitivities of HE and MIE are similar,
   with maximum embrittlement noted at some intermediate temperature (room temperature for HE, generally near the melting point of the liquid for MIE, see Figs. 1, 4.
4) factors that favor planar glide increase susceptibility.
5) similar crack growth rates have been observed for D6AC steel
   in mercury and in dissociated hydrogen (54).

Temper embrittlement (TE), HE and MIE also may be linked in the
common effects of certain impurities, most notably antimony and tin,
in promoting fracture. Breyer and Johnson (72) have noted the apparent connection between TE and MIE, while Yoshino and McMahon (73)
have noted a relation between conditions favoring TE and HE, thereby
suggesting that there may be a common predisposition of certain
steels to all three types of embrittlement. It should be pointed
out, however, that even if such a common predisposition exists, there
need not be a single mechanism by which bond strength (or fracture
energy) is reduced.

EFFECTS OF ALKALI METALS ON MECHANICAL PROPERTIES

Sodium and lithium are to be used in contact with structural
materials in the liquid metal fast breeder reactor and in controlled
fusion devices, respectively. Much of the research on suitable con-
tainment materials for high temperature lithium has focussed on ferrous alloys, including carbon, low alloy and stainless steels. In
addition, there are some limited data on embrittlement of nickel and
palladium-based alloys.

Two characteristics of lithium-induced embrittlement differ from
those associated with most embrittling metals. Firstly, the degree
of embrittlement produced by lithium is very sensitive to the atmosphere in which the lithium is applied, see Fig. 12 (74), due probably to differences in moisture content as well as due to the variation in wetting behavior of the lithium. Secondly, lithium can cause
embrittlement at stresses much below the apparent or macroscopic
yield stress of the substrate. In fact, the ratio of wetted fracture
stress to unwetted yield stress for Ni-Cu and Pd-Ag alloys can
be as low as 0.2 (74). This contrasts with the situation for most
crystalline substrates, for which the wetted fracture stress is equal
to or greater than the unwetted yield stress. Embrittlement noted
with lithium often is concentrated at grain boundaries (75), leading
to the conclusion that decarburization at grain boundaries can facilitate crack growth. However, recent work on amorphous alloys has
demonstrated severe lithium embrittlement of $Fe_{40}Ni_{40}P_{14}B_6$ in the

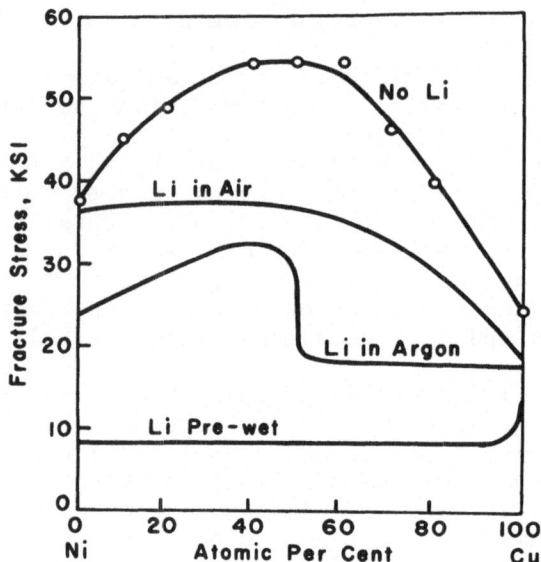

Fig. 12.   Influence of test conditions on fracture stress of Ni-Cu
           alloys in molten lithium (74).

absence of <u>both</u> grain boundaries and carbon (76).   Therefore lithium
must have an intrinsic effect on bond strength as well.

     A limited amount of fatigue data on steels in alkali metal en-
vironments is available.   A 2-1/4Cr-1Mo steel tested in lithium containing 1.5%N shows a much higher crack growth rate at low ΔK relative to the same steel tested in argon at 400°C, 40 cpm (75).   Later
work showed that the apparent embrittling effect of lithium is increased at higher frequency, see Fig. 13 (77).   At low frequencies,
it was suggested, stress-assisted grain boundary penetration, leading
to intergranular cracking, is enhanced, while at high frequencies a
form of strain-rate induced transgranular LME occurs.   Lithium could
increase the crack growth rate relative to argon at ΔK < 40 ksi$\sqrt{\text{in}}$,
by weakening grain boundaries, while the improvement at high ΔK
could arise from screening of oxygen from crack tips, as observed
for 316 SS in sodium, Fig. 14 (78).   Sodium with a low impurity content (0.3 ppm C, 1 ppm O) has been shown to <u>increase</u> LCF life of
316 SS by three to four times, relative to air, while no effect of
environment is noted in 304 SS (79).   However, pre-exposure of 304 SS
to sodium causes LCF life to be reduced at strain ranges above $\Delta \varepsilon_T =$
0.8%, with longer lives at strain ranges below this value.   There is
little effect, however, of pre-exposure of 316 SS to sodium.   The
difference in response of the two steels is attributed to greater
depth of carbon penetration in 304 than in 316 under the pre-exposure
conditions.

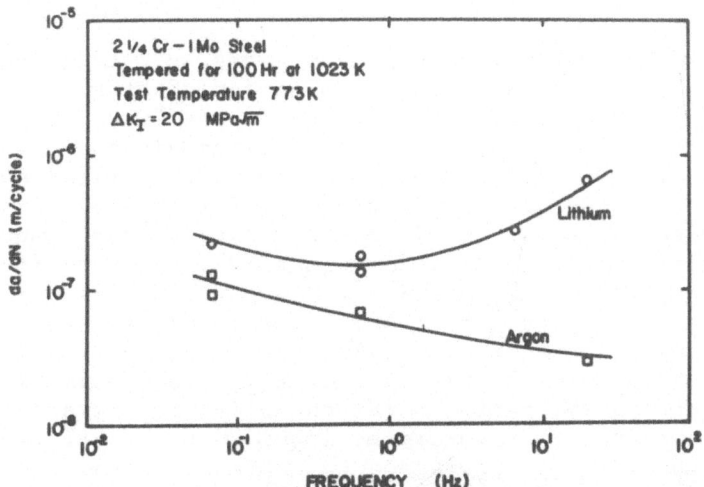

Fig. 13. Effect of frequency on crack growth rate, da/dN, for argon
and lithium environments at 773°K, ΔK = 20 MPa $m^{\frac{1}{2}}$ (77).

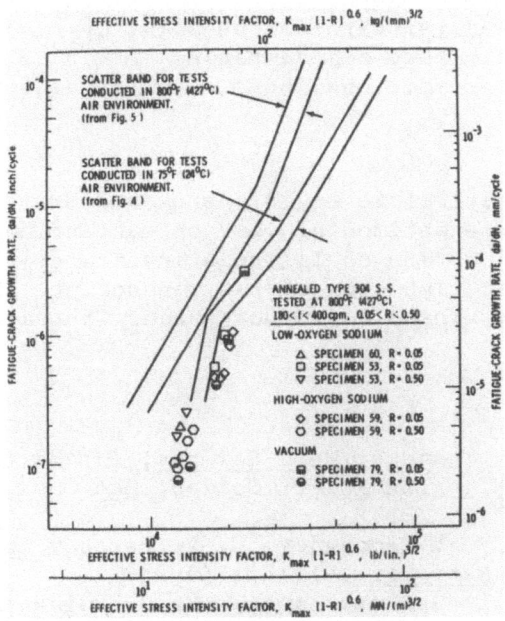

Fig. 14. Fatigue crack growth of 316 SS in sodium and in air at
773°K (78).

SUMMARY

     Metal-induced embrittlement encompasses an intriguing, complex
set of phenomena which have not yet been satisfactorily explained.
The metallurgical and environmental conditions favoring embrittlement
are now well known, and several empirical techniques to rationalize
or predict observed embrittlement have met with limited success.
Our knowledge of possible transport mechanisms has been improved.
The least success has been achieved in identifying an embrittlement
mechanism which simultaneously can describe the interaction between
embrittler and base metal, predict which combinations of environment
and substrate will constitute embrittlement couples, and finally can
explain the varied effects of metallurgical and test variables. As
in the case of hydrogen embrittlement, there is reason to suspect
that more than one mechanism is responsible for "classical" metal-
induced embrittlement; for example, liquid metal embrittlement of
amorphous metals is best explained by enhanced shear, liquid metal
embrittlement of many crystalline metals can be explained by either
enhanced shear or decohesion models, while solid-metal induced
embrittlement and all forms of delayed failure phenomena may best be
explained by a grain boundary penetration model.

     Finally, degradation of structural alloys by alkali metals under
monotonic and cyclic loading often, but not always, is controlled by
the presence or absence of impurities. Experiments with nickel and
palladium alloys as well as with amorphous alloys indicate that lithium, in particular, can have a devastating effect on mechanical
properties even under conditions where impurity effects are minimal.

ACKNOWLEDGEMENTS

     The author is grateful to Dr. S. Ashok and Mr. T. Slavin who have
performed most of the experimental work on embrittlement of amorphous
alloys; to Prof. Paul Gordon of Illinois Institute of Technology for
providing a pre-publication copy of his manuscript, and to the Office
of Naval Research for financial support under Contract No. N00014-
79-C-0583.

REFERENCES

1. C. F. Old and P. Travena, Met. Science, 13:487 (1979).
2. J. C. Lynn, W. R. Warke and P. Gordon, Mat. Sci. Eng., 18:51
       (1975).
3. E. D. Shchukin, Yu V. Goryunov, N. V. Pertsov and L. S. Bryuk-
       hanova, Sov. Mat. Sci., 14:341 (1978).
4. E. D. Shchukin, L. S. Bryukhanova, Z. M. Polukarova and
       N. V. Pertsov, Fiz. Khim. Mekh. Mater., 12, 40 (1976).
5. N. S. Stoloff in Environment-Sensitive Fracture of Engineering
       Materials, Z. A. Foroulis, Ed., AIME, New York, NY, 486
       (1979).

6. G. Herbsleb and W. Schwenk, Werk. und Korrosion, 28:145 (1977).
7. F. A. Shunk, Ph.D. Thesis, Illinois Inst. of Technology (1976).
8. H. Nichols and W. Rostoker, Acta Met., 8:848 (1960).
9. M. C. Roth, G. C. Weatherly and W. A. Miller, Acta Met., 28: 841 (1980).
10. N. S. Stoloff and T. L. Johnston, Acta Met., 11:251 (1963).
11. N. S. Stoloff, R. G. Davies and T. L. Johnston in Environment-Sensitive Mechanical Behavior, Gordon and Breach, New York, 613 (1966).
12. A. H. Cottrell, Trans. Met. Soc. AIME, 212:192 (1958)
13. N. J. Petch in Fracture, Wiley, New York, 20 (1959).
14. A. S. Tetelman and A. J. McEvily, Jr., Fracture of Structural Materials, Wiley, New York, 268 (1967).
15. K. L. Johnson, N. N. Breyer and J. W. Dally, Proc. Conf. on Environmental Degradation of Engineering Materials, Blacksburg, VA, 91 (1977).
16. W. M. Robertson, Met. Trans., 1: 2607 (1970).
17. W. Rostoker, J. M. McCaughey and H. Markus, in Embrittlement by Liquid Metals, Reinhold, New York (1960).
18. C. M. Preece and A. R. C. Westwood, Trans. ASM, 62:418 (1969).
19. A. S. Tetelman and S. Kunz, in Proc. Int. Conf. on Stress Corrosion Cracking and Hydrogen Embrittlement of Iron-Base Alloys, Unieux-Firminy, 359 (1977).
20. V. V. Popovich and I. G. Dmukhouskaya, Sov. Mat. Sci., 14:365 (1978).
21. A. R. C. Westwood and M. Kamdar, Phil. Mag., 8:787 (1963).
22. P. Gordon, Met. Trans. A, 9A:267 (1978).
23. N. A. Gjostein, in Diffusion, American Soc. for Metals, Metals Park, OH, 241 (1973).
24. M. J. Kelley and N. S. Stoloff, Met. Trans. A, 6A: 159 (1975).
25. N. Engel, Powder Met. Bull., 7:8 (1954).
26. L. Brewer, in High Strength Materials, V. F. Zackay, Ed., Wiley, New York, 12 (1975).
27. R. M. German, in The Metal Science of Stainless Steels, E. W. Collings and H. W. King, Eds., 41, AIME, New York, (1978).
28. J. H. Hildebrand and R. L. Scott, Solubility of Non-Electrolytes, 3rd Ed., Prentice Hall, Englewood Cliffs, NJ (1950).
29. A. A. Griffith, Trans. Roy. Soc. Lond., 221:163 (1920).
30. E. Orowan, Welding J., 34:1575 (1955).
31. H. K. Birnbaum, in Environment-Sensitive Fracture of Engineering Materials, Z. A. Zoroulis, Ed., AIME, New York, NY, 326 (1979).
32. M. L. Jokl, V. Vitek and C. J. McMahon, Jr., Acta Met., 28: 1479 (1980).
33. J. J. Gilman, Plasticity - Proc. Sec. Symp. on Naval Structural Mechanics, 43, Pergamon Press, New York (1960).
34. C. F. Old, in Fracture 1977, Proc. ICF4, D. M. R. Taplin, Ed., Waterloo, Canada, 2:331 (1977).
35. M. H. Kamdar, Prog. Mat. Sci., 15:289 (1973).

36. M. G. Nicholas and C. F. Old, J. Mat. Sci., 14:1 (1979).
37. C. F. Old and P. Travena, Met. Sci., 13:591 (1979).
38. M. I. Chaevskii, Sov. Mat. Sci., 1:433 (1965).
39. C. F. Old and P. Travena, Proc. Third Int. Conf. on Mechanical Behavior of Materials 2, Pergamon Press, Oxford, 397 (1980).
40. G. M. Scamans and C. D. S. Tuck, Sec. Int. Cong. on Hydrogen in Metals, Paris, 1977, Pergamon Press, paper 4A-11.
41. J. J. Gilman, Discussion to ref. 19, p. 375.
42. N. Toropovskaya, Sov. Mat. Sci., 6:324 (1970).
43. R. A. Oriani, in Proc. Int. Conf. on Stress Corrosion Cracking and Hydrogen Embrittlement of Iron-Base Alloys, Unieux-Firminy, 351 (1977).
44. F. A. Shunk and W. R. Warke, Scripta Met., 519 (1974).
45. P. Gordon, Illinois Inst. of Technology, unpublished (1981).
46. Y. Iwata, Y. Asayama and A. Sakamoto, J. Jap. Inst. Met., 21: 77 (1967).
47. J. V. Rinnovatore, J. D. Corrie and H. Markus, Trans. ASM, 59: 665 (1966).
48. L. S. Bryakhanova, I. A. Andreeva and V. I. Likhtman, Sov. Phys. Solid State, 3:2025 (1962).
49. C. M. Preece and A. R. C. Westwood, in Fracture, Proc. Sec. Int. Conf. on Fracture, Chapman and Hall, London, 439 (1969).
50. E. A. Clark, R. Yeske and H. K. Birnbaum, Met. Trans. A, 11A: 1903 (1980).
51. S. P. Lynch, Scripta Met., 13:1051 (1977).
52. S. P. Lynch, "The Mechanism of Liquid Metal Embrittlement - Crack Growth in Aluminum Single Crystals, A.R.L. Mat. Report 102, Aeronautical Res. Labs, Melbourne, Australia (1977).
53. S. P. Lynch, Acta Met., 29:325 (1981).
54. S. P. Lynch and N. E. Ryan, Proc. Sec. Int. Cong. on Hydrogen in Metals, paper 3D12, Pergamon, Oxford (1977).
55. S. P. Lynch, in Fatigue Mechanisms, ASTM STP675, Amer. Soc. for Test. and Materials, Philadelphia, PA, 174 (1979).
56. M. A. Krishtal, Sov. Phys. Dokl., 15:614 (1970).
57. H. Matsui, H. Kimura and S. Moriya, Proc. Fourth Int. Conf. on Strength of Metals and Alloys, 291 (1976).
58. H. Matsui, H. Kimura and S. Moriya, Mat. Sci. Eng., 40:207 (1979).
59. R. A. Oriani and P. H. Josephic, Met. Trans. A, 11A, 1809 (1980).
60. J. A. Clum, Scripta Met., 9:51 (1975).
61. R. J. H. Wanhill, Corrosion, 29:435 (1973).
62. W. R. Goggin and J. W. Moberly, Trans. ASM, 59:315 (1966).
63. N. A. Tiner, Trans. Met. Soc. AIME, 221:261 (1961).
64. S. Ashok, N. S. Stoloff, M. E. Glicksman and T. P. Slavin, Scripta Met., 15:331 (1981).
65. E. E. Glikman and Yu. V. Goryunov, Fiz. Khim Mekh. Mater., 14:20 (1978).

66. E. E. Glikman, Yu. V. Goryunov, V. M. Demin and K. Yu. Sarychev, Izv. Vyssh. Uchebn. Zaved. Fiz., 15 (1976).
67. S. K. Marya and G. Wyon, Scripta Met., 9:1009 (1975).
68. L. S. Soldachenkova, Sov. Mat. Sci., 11:251 (1975).
69. H. Nelson and D. P. Williams, Met. Trans., 1:63 (1970).
70. I. M. Bernstein, Mat. Sci. Eng., 6:1 (1970).
71. S. P. Lynch, Proc. Fourth Int. Conf. on Fracture, D. M. R. Taplin, Ed., 2, Waterloo, Canada, 859 (1977).
72. N. N. Breyer and K. L. Johnson, J. Test. and Eval. 2: 471 (1974).
73. K. Yoshino and C. J. McMahon, Jr., Met. Trans. A, 5A : 363 (1974).
74. N. M. Parikh, in Environment-Sensitive Mechanical Behavior, A. R. C. Westwood and N. S. Stoloff, Eds., Gordon and Breach, New York, 563 (1966).
75. D. L. Hammon, S. K. DeWeese, D. K. Matlock and D. L. Olson, Closed Loop, 9:3 (1979).
76. S. Ashok and N. S. Stoloff, unpublished.
77. R. E. Spencey, S. K. DeWeese, D. K. Matlok and D. L. Olson, Met. Trans. A, 11A:1758 (1980).
78. L. A. James and R. L. Knecht, Met. Trans. A, 6A:109 (1975).
79. D. L. Smith and G. J. Zeman, Closed Loop, 9:12 (1979).

DISCUSSION

Comment by S. Lynch:

You have mentioned four possible objections to the "enhanced shear" mechanism for LME of crystalline metals. These objections are considered in the following in the order listed in your paper.

(a) Growth of voids ahead of cracks, which occurs simply because crack growth is accompanied by quite high plastic strains, obviously cannot be directly affected by adsorption at external crack tips. However, if adsorption at external crack tips facilitates crack growth (by whatever means), then general plastic strains ahead of cracks should be reduced and, hence, any nucleation and growth of voids should be indirectly affected.

(b) Delayed crack initiation probably occurs because a notch of a certain depth and accuity is necessary to nucleate rapid crack growth. Such a notch probably develops gradually as a result of slip processes at the surface (cf. fatigue crack initation); an effect of adsorption on the nucleation of dislocations at surfaces is consistent with these observations. In other words, some delay prior to rapid fracture would be expected if a crack nucleus is formed by a slip process but fracture would probably be instantaneous if crack initiation occurred by a tensile decohesion process.

(c)   For CPH metals, alternate–slip on $\{11\bar{2}2\}$ planes (second-
      order pyramidal planes) combined with void growth ahead
      of cracks could produce fracture surfaces macroscopically
      parallel to basal planes. There is some evidence that
      slip can occur on these planes planes in some circumstances
      and, since stresses at crack tips are higher than most
      other circumstances, their operation during crack growth
      is quite possible.

(d)   Crack growth generally occurs approximately parallel to
      $\{100\}$ planes (on a macroscopic scale) in aluminum single
      crystals because roughly equal amounts of slip always
      occur on either side of the crack despite the slip planes
      on either of the crack being unequally stressed. Dislo-
      cations are emitted on slip planes on either side of the
      crack in roughly equal amounts since, if more dislocations
      were emitted on one side of the crack than the other, then
      a larger back stress opposed to subsequent slip would
      counteract further dislocation nucleation on the more
      active side.

Thus, it appears that the objections to the enhanced shear
mechanism are not justified. However, I agree with your conclusion
that LME may occur by more than one mechanism: easier injection of
dislocations from crack tips best explains LME of ductile materials
while easier tensile decohesion probably explains LME of inherently
brittle materials.

Reply:

I generally agree with your comment (a), but must point out
that the sequence of steps by which the voids form in your model
is not clear.

Comment (b) about delayed failure is speculative. I suggest
that work be done on delayed failure of single crystals at a series
of temperatures to identify pre-crack phenomena.

Comment (c) concerning non–basal slip in hcp metals again is
speculative in the absence of direct evidence for such systems ahead
of crack tips in embrittled zinc or cadmium. Zinc is highly unlikely
to exhibit non–basal slip at room temperature, while pyramidal slip
occurs in cadmium in certain orientations only. Therefore, I would
expect that your model would be more likely to explain LME in cadmium
than in zinc. Again, experiments on oriented single crystals clearly
are necessary to resolve this question.

Comment (d) on back stresses relies upon a precise sequence of
events such that alternating slip continuously occurs. Since crack
tips extend in two dimensions, and the plasticity problem is, in

fact, a three-dimensional one, the likelihood of alternating slip occurring all along a crack front seems low. Since this point is crucial to the enhanced shear mechanism I believe that more convincing evidence of continual alternating slip along a crack front is required.

As a general comment, I do not believe that the objections enumerated in the paper to the enhanced shear mechanism should lead the reader to suppose that I do not support this mechanism in certain situations. Therefore, we are basically in agreement that either decohesion or enhanced shear can occur, depending upon the circumstances.

Comment by R. Messmer:

Can you explain the difference between enhanced shear and decohesion from an atomic point of view as a fracture mechanism?

Reply:

I believe, although in both cases bonds must be stretched and broken, in the case of enhanced shear the bond breaking occurs by shear stresses, while in the decohesion model the bond breaking occurs by the action of normal stresses.

Comment by M.W. Finnis:

In a thermodynamic sense the weakening of interatomic bonds and the reduction in surface energy are the same mechanism: the latter is the macroscopic way of saying the former.

I know of no data at present on liquid/metal interfacial energies on clean interfaces against which to test the hypotheses that they correlate with LME. Is this the case?

Reply:

I agree with your first comment in respect to a reversible process. Oriani has made a distinction between the two for metals. You are correct with regard to the lack of data for clean interfaces. When correlations are attempted between interfacial energies, measured on less than clean interfaces, and degree of embrittlement, the results have been unsuccessful.

Comment by H. Vehoff:

If you have decohesion at the tip of a growing crack you will always move dislocations in the metal. Therefore, evidence of slip on a fracture surface will be always present and is neither evidence for decohesion nor evidence of enhanced slip. Are there any direct measurements of enhanced dislocation motion due to the environment?

Reply:

I do not know of any such direct evidence for enhanced dislocation motion due to liquid metals. However, I believe, that such
evidence has been provided by Clum for hydrogen environments (see
my paper for reference).

Editor's Note:

There is considerable evidence that, for example, surface
dissolution during deformation of single crystals affects such
properties as the yield stress, work hardening rate, extent of
easy glide, etc. [See Proc. Advanced Study Institute Surface Effects
in Crystal Plasticity, Noordhoff, Leyden (1977)]. It is often observed
that the flow stress and work hardening rate are decreased and extent
of easy glide increased in the case of fcc metals, for example.

Comment by P. Neumann and P. Haasen:

Indication of slip (dimples, slip lines) by no means prove the
absence of decohesion but prove only the presence of some plasticity
as fracture goes on.

Reply:

I agree. Fractographic information alone cannot distinguish
between proposed embrittlement mechanisms.

Comment by M.P. Puls:

You indicated that fcc metals embrittle intergranularly but
also pointed out that intergranular embrittlement can occur in bcc
and hcp crystals. Can you suggest a reason why apparently intragranular embrittlement does not occur or is not observed in fcc?

Reply:

Intragranular embrittlement is in fact observed in aged polycrystals of fcc alloys. Also, aluminum single crystals undergo
failure on {100} in mercury. I cannot explain why intergranular
fracture is more likely in fcc than in bcc or hcp alloys.

Comment by R.M. Latanision:

It seems to me that of those embrittlement phenomena that we
have been discussing here, liquid metal embrittlement is probably
the most tractable experimentally (since it does not necessarily
involve penetration in depth in advance of fracture). It would,
therefore, seem that if one could perform theoretical (molecular

orbital or other) calculations on appropriate solid metal-liquid metal systems, comparison between theory and experiment might be most informative.

NUCLEATION AND EGRESS OF DISLOCATIONS AT CRACK TIPS

S.P. Lynch

Aeronautical Research Laboratories
Defence Science and Technology Organisation
Department of Defence, Melbourne, Australia

INTRODUCTION

On the atomic scale, crack growth must occur by tensile separation of atoms (tensile decohesion), shear movement of atoms (dislocation nucleation or egress), removal of atoms (dissolution) or combinations of these processes at crack tips. Metallographic and fractographic studies in a wide range of materials and environments suggest that fracture in metals commonly occurs by dislocation nucleation or egress at crack tips and involves nucleation of voids ahead of cracks. In such cases, it is suggested that materials are ductile or brittle depending on the plastic work required for coalescence of cracks and voids.

DUCTILE VS BRITTLE FRACTURE

Fracture should occur by nucleation or egress of dislocations at tips of cracks rather than by tensile decohesion when strains around cracks are sufficiently large that slip exactly intersects crack tips and nucleation of voids occurs just ahead of cracks. Only dislocations which exactly intersect crack tips on planes which are obtusely inclined to the crack can produce crack extension (and opening) (Fig. 1(a)), while other dislocations produce either crack blunting (Fig 1(b)) or contribute to the general strain around cracks (Fig. 1(c)). The fracture resistance should depend mainly on the plastic work required for linking up cracks and voids and, hence, increasing the proportion of mobile dislocations which produces crack extension should decrease the resistance to crack growth.

When dislocation nucleation at crack tips is very difficult relative to general matrix slip, cracks and voids will link up when sufficient egress of dislocations on suitably orientated slip planes

955

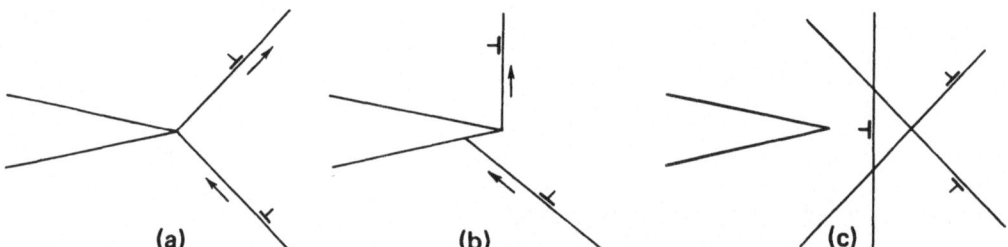

Fig. 1. Schematic diagram showing dislocations which produce
(a) crack extension, (b) blunting, and (c) general strains.

occurs exactly at crack tips. The probability of there being
near-surface dislocation sources, which can be activated at low
stresses, on slip planes <u>exactly</u> intersecting crack tips will initially
be low.   Thus, extensive general dislocation activity also occurs
before there is sufficient dislocation egress at crack/void tips to
produce coalescence. The development of large strains is accompanied by
work-hardening near crack tips and, hence, large plastic zones develop
before strains are sufficiently high in the crack-tip region for
coalescence. Voids form some distance from the crack tip as well as
close to it and small dimples within large, deep dimples are often
observed on fracture surfaces (Fig. 2(a)), although the small dimples
may not be readily apparent in some cases. Most ductile materials
cracked in inert environments exhibit this kind of behaviour.

Now if the crack tip acts as an efficient source of dislocations,
i.e. nucleation and subsequent movement of dislocations from crack tips
on planes obtusely inclined to cracks occur readily compared with
extensive activity of other dislocations, then cracks and voids will
link up with only a small amount of plastic work. In some cases, roughly
equal amounts of dislocation nucleation on planes either side of the
crack could occur since, if more dislocations were emitted on one side
of the crack than the other, then a larger back stress opposed to
subsequent nucleation would counteract slip on the more active side.
Fracture planes would, therefore, approximately bisect the angle
between the active slip planes on a macroscopic scale and directions of
crack growth would be normal to the line of intersection of the slip
planes[1,2].   In   other   cases,   dislocation   nucleation   may   occur
preferentially at grain-boundary/crack-tip intersections and produce
intercrystalline fracture. 'Cleavage' of aluminium in liquid-metal
environments[2] on {100} planes in <110> directions (Fig. 2(b,c)) and
brittle intercrystalline fracture in some steels[3] (Fig. 2(d,e)) fall
into this category.

Fracture probably occurs by tensile decohesion at atomically sharp
crack tips when strains around cracks are sufficiently small that
dislocations do not generally intersect crack tips and voids do not form
ahead   of   cracks.   Crack   growth   should   occur   on   low-index

crystallographic planes and in directions for which the associated
plastic work of fracture (due to slip on planes not intersecting crack
tips) is minimised[4]. Cleavage of zinc at -196°C on {0001} planes
probably occurs by tensile decohesion with associated plasticity[5] and,
as would be expected, fracture surfaces do appear to be flat
(Fig. 2(f)). For very brittle solids, fracture can probably occur by
tensile separation of atoms at atomically sharp crack tips without any
accompanying dislocation activity. Recent transmission-electron-
microscopy studies[6] indicate that this mode of fracture can occur in
ionic and covalent bonded solids such as silicon and alumina.

TENSILE DECOHESION VS SLIP MECHANISMS FOR BRITTLE FRACTURE

It is obviously important to distinguish between brittle fracture
by tensile decohesion <u>accompanied</u> by some dislocation activity and
brittle fracture <u>accomplished</u> by dislocation movement. However, this
distinction may not be easy since fractures in both cases may be highly
reflective, macroscopically parallel to low-index crystallographic
planes and exhibit only very localised plasticity. For fractures
accomplished by dislocation motion, dimples on fracture surfaces may be
as small as ~0.05 m diameter and ~0.02 m depth and may not be resolved
by scanning-electron microscopy even at high magnifications. Thus,
there have been some cases which have been assumed[7] in the past to be

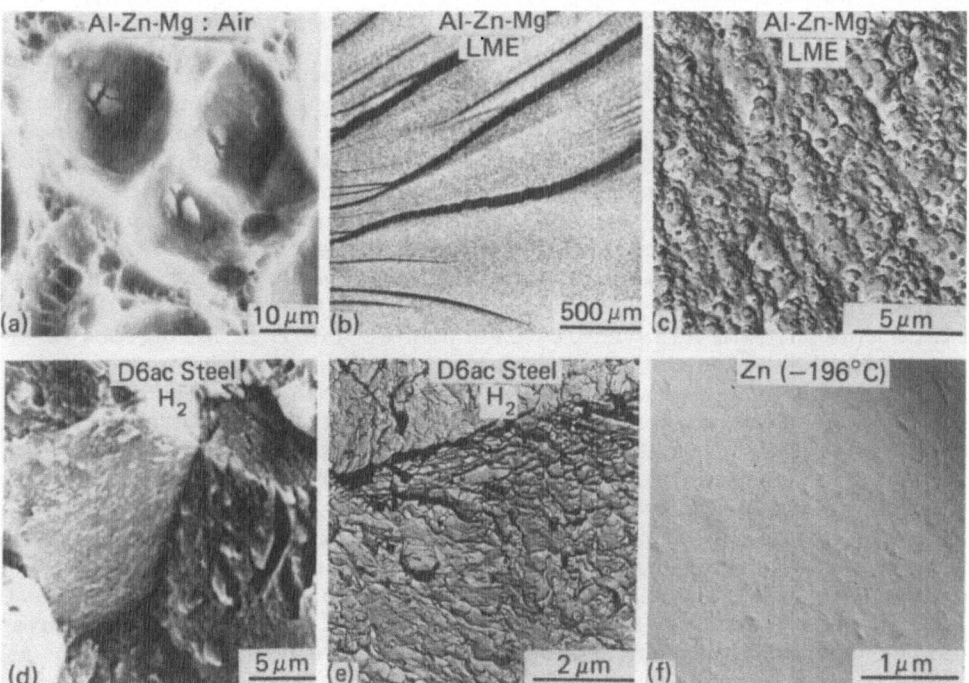

Fig. 2.   Some examples of fracture-surface appearance.

atomically brittle but which probably occur by nucleation or egress of dislocations at crack tips, e.g. 'cleavage' of aluminium in liquid-metal environments. Extremely small, shallow dimples should generally be detected by transmission-electron microscopy of secondary-carbon replicas of fracture surfaces and sensitive techniques should reveal evidence of quite high localised strains for fractures accomplished by dislocation movement. Directions of crack growth may also be a good guide to fracture mechanisms; thus, for fracture involving alternate-slip (dislocation nucleation) at crack tips, directions of crack growth should be normal to the line of intersection of active slip planes.

ENVIRONMENTALLY ASSISTED CRACK GROWTH

In normally ductile materials, fractures produced by liquid-metal embrittlement (LME), hydrogen-assisted cracking (HAC), and stress-corrosion cracking (SCC) often exhibit macroscopically brittle characteristics. However, close examination of such fractures shows that they are generally not atomically brittle but probably result from localised slip at crack tips, e.g. small, shallow dimples are often observed on fracture surfaces and localised strains are quite high. Furthermore, characteristics of LME, HAC and SCC are so similar in some cases that a similar mechanism is probably involved. Only adsorption generally occurs during LME and, hence, it has been proposed[1,3,8] that some cases of HAC and SCC can also be explained by adsorption.

In inert environments, the surface lattice is sometimes contracted to a significant extent[9] and interatomic bonds between the first and second atomic layers at the surface are probably stronger than those in the bulk. In air/oxygen environments, the surface lattice is often disrupted and strong bonds are formed between surface metal atoms and oxygen atoms. Thus, the strong interatomic bonds at surfaces in inert or oxygen environments probably restrict nucleation of dislocations at crack tips so that crack growth occurs predominantly by egress of dislocations and, hence, ductile behaviour is observed (Fig. 3(a)). Surface non-uniformities could also hinder dislocation egress to some extent but the momentum of dislocations and an image force attracting dislocations to the surface should generally allow egress to occur. In environments where weak chemisorption, e.g. of liquid metal or hydrogen atoms, occurs at crack tips, the 'clean-surface' non-uniformity may be reduced or cancelled[9] and the (shear) strength of surface interatomic bonds may be reduced. Hydrogen atoms 'dissolved' just beneath the first atomic layer could also have a similar effect. The reduced shear strength of interatomic bonds should facilitate nucleation of dislocations and, hence, result in macroscopically brittle behaviour (Fig. 3(b)). However, if fracture in inert environments occurs by tensile decohesion, environmentally assisted cracking in these materials could result from a reduction in the tensile strength of interatomic bonds due to adsorption.

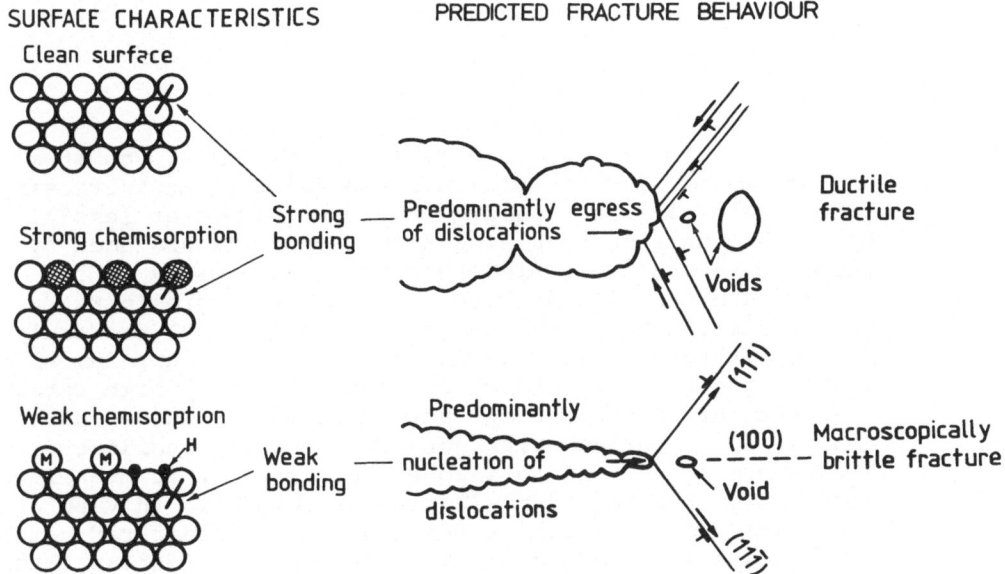

Fig. 3. Schematic diagrams showing possible effects of chemisorption on surface characteristics and fracture behaviour.

REFERENCES

1. S.P. Lynch, Hydrogen embrittlement and liquid-metal embrittlement in nickel single crystals, Scripta Metall. 13:1051 (1979).
2. S.P. Lynch, Liquid-metal embrittlement in an Al-Zn-Mg alloy, Acta Metall. 29:325 (1981).
3. S.P. Lynch, Mechanisms of hydrogen-assisted cracking, Metals Forum 2:189 (1979).
4. W.R. Tyson, R.A. Ayers and D.F. Stein, Anisotropy of cleavage in B.C.C. transition metals, Acta Metall. 21:621 (1973).
5. S.J. Burns, Dislocation motion accompanying cleavage in zinc, Acta Metall. 18:969 (1970).
6. B.R. Lawn, B.J. Hocking and S.M. Weiderhorn, Atomically sharp cracks in brittle solids: an electron microscopy study, J. Mat. Sci. 15:1207 (1980).
7. A.R.C. Westwood, C.M. Preece and M.H. Kamdar, Application of a crack propagation criterion to liquid-metal embrittlement; Cleavage of Al monocrystals in liquid gallium, Trans. ASM 60:723 (1967).
8. S.P. Lynch, A comparative study of SCC, HAC and LME in Al, Ni, Ti and Fe-based alloys, 3rd Int. Conf. on Effects of Hydrogen on Behavior of Materials, Jackson Lake Lodge, Wyoming (1980).
9. M.A. van Hove, Progress in surface crystallography, Surface Sci. 80:1 (1979).

DISCUSSION

Comment by G.G. Garrett:

1.   In the work you report on pure Al-Zn-Mg alloys, you show
      an increasing size of voids corresponding to an increasing
      precipitate size and spacing (minimum ~0.05μm or less?).
      Presumably this varying size distribution is produced by
      aging? Have you identified these precipitates by TEM/
      diffraction techniques? Even assuming these precipitates
      are incoherent (and indeed can this be proven?) the early
      fractographic observations in age-hardening aluminium alloys,
      for example by David Broek, indicate that very high strains
      are necessary to debond age-hardening precipitates, parti-
      cularly in comparison with non-metallic dispersoids or
      inclusions which tend to decohere or fracture at much lower
      strains (in the region ~3-7% tensile strain). In view of
      the fractographic evidence you show in air-fractured samples
      where large voids are clearly in evidence indicating that
      your alloys - even though nominally pure - do contain
      second-phase intermetallics, how do you explain that these
      particles are not also involved in your enhanced-shear
      mechanism for liquid-metal embrittlement? Also, have you
      done any parallel studies, for example, on conventional
      tensile or notched tensile specimens, to establish the
      magnitude of strain necessary to initiate voids from age-
      hardening precipitates? Is this level of plastic strain
      consistent with your model and an 'embrittled' mode of
      fracture?

2.   The experimental observation of voids produced in the
      absence of void-nucleating second phase particles, and
      dislocation models for their occurrence, have yet to achieve
      widespread acceptance in the literature. Is it necessary
      for you to invoke such models to interpret your observations
      of dimples in LME in alloys where there is no obvious second
      phase present?

3.   Both you and Dr. Stoloff talked of 'regimes' under which
      either a decohesion model or your enhanced-shear mechanism
      can operate. Under what conditions do you consider the
      decohesion model to be appropriate?

4.   Could you clarify what tests you have carried out, or have
      in progress, to ensure that the dimples you observe cannot
      be associated with a post-fracture etching/surface prepar-
      ation artifact.

Reply:

1. Details of the work on Al-Zn-Mg alloys (mentioned only
    briefly in the present paper) have been published recently
    (reference 2 of paper). Briefly, the variation in precipi-
    tate size and spacing was achieved by aging but the precipi-
    tates were not identified. However, work by others (J.D.
    Embury and R.B. Nicholson, Acta Met., 13, 403 (1965)) for
    a similar alloy indicates that specimens overaged to various
    extents would contain incoherent precipitates and that
    specimens aged to peak-hardness would contain predominantly
    semi-coherent precipitates. For the example shown in the
    present paper, specimens were grossly overaged and incoherent
    precipitates were as large as the (relatively few) inclusions.
    Metallographic and X-ray techniques revealed the presence
    of quite large localized strains (10-50% depending on the
    heat-treatment) associated with LME fractures and, hence,
    the nucleation of voids at precipitates is not surprising.

2. To produce macroscopically brittle modes of fracture by an
    alternate-slip mechanism, it is necessary to have voids
    ahead of cracks, and there may well be sufficient particles
    to nucleate such voids even in quite high purity materials.
    However, a void-nucleation mechanism not involving particles
    would have to be invoked for ultra-high purity materials
    and further studies are clearly necessary to establish the
    validity and generality of proposed mechanisms for void
    nucleation at, for example, dislocation-cell boundaries.

3. Tensile-decohesion models are probably applicable to ionic-
    and covalent-bonded solids fractured in inert and embrittling
    environments, while the enhanced-shear (nucleation) model
    is probably applicable to the embrittlement of normally
    ductile metals. Fracture of intrinsically brittle metals
    possibly falls into one or the other category depending on
    the particular metal and circumstances.

4. Electropolished surfaces, dimpled fracture surfaces (pro-
    duced by fracture in air), and abraded surfaces have been
    wetted with liquid metals, followed by removal of the liquid-
    metal by techniques used to remove the liquid metals from
    fracture surfaces produced by LME. (Such techniques have
    included evaporation of mercury from LME fractures of steel
    and titanium, and dissolution of a liquid alloy in concen-
    trated nitric acid for LME fractures of aluminum alloys.)
    These tests showed that wetting the solid metals with the
    liquid metals and then removing the liquid metal did not
    produce significant etching or pitting effects which could
    be confused with dimples which were observed on the LME
    fracture surfaces, e.g., see Fig. 4 of a paper in _Fracture_

<u>1977</u>, ed., D.M.R. Taplin, Univ. of Waterloo Press, Vol. 2,
p. 859, 1977. However, further tests will be carried out
in the near future since this aspect of the work is of such
concern (see following question).

Comment by E.N. Pugh:

Some of the fine features which you referred to as dimples looked
reminiscent of corrosive attack either by the liquid metal or by the
etching process used to remove the liquid metal after fracture. I
recognize that you have subjected electropolished surfaces to the
liquid metal and to removal. Is there any evidence of attack at
those surfaces; are comparable micrographs available showing these
surfaces? Further, do you think that electropolished surfaces simulate the actual fracture surface?

Reply:

See above.

Comment by J.R. Pickens:

The dimple sizes on fracture surfaces of the Al-Zn-Mg alloy at
various stages of aging roughly correlate with the precipitate size.
However, did you find a correlation of dimple size and/or spacing
with macroscopic fracture stress or fracture strain?

Reply:

Macroscopic fracture stresses and strains for LME of Al-Zn-Mg
alloys have not been measured but would be worth doing. There is
probably some relationship between these parameters and the degree
of embrittlement (estimated from crack tip angles) which was found
to decrease with increasing amounts of overaging (decreasing
strength).

Comment by P. Neumann:

In our experiments, ductile crack growth in inert environments
produces a crack tip angle which is predicted by a process of
alternate-slip alone. In embrittling environments, just facilitating
the nucleation of dislocations at crack tips should not produce the
observed decrease in crack tip angle.

Reply:

The specially oriented, pre-hardened single crystals (e.g., of
Ni) which you use in your experiments may well fracture in inert
environments by alternate slip alone and I agree that facilitating

the nucleation of dislocation (in embrittling environments) would not change the crack tip angle. However, other explanations for embrittlement, e.g., based on tensile decohesion, also have their problems and further work is clearly necessary to understand this phenomenon. On the other hand, ductile fracture in most circumstances involves blunting (rounding) at crack tips and nucleation and growth of voids ahead of crack tips. In these cases, facilitating the nucleation of dislocations (in embrittling environments) could facilitate the link-up of cracks and voids and produce fractures which are more brittle, i.e., have smaller crack tip angles.

Comment by T.E. Fischer:

H_2S surface measurements support your contention that hydrogen embrittlement is a surface effect. We know (T.E. Fischer, these proceedings) that after some time of exposure to H_2S a surface is inert even to H_2S dissociation. Thus, the increase in embrittlement by H_2S over H_2 indicates that the relevant surface reaction occurs at the very sites where fresh, clean surface is exposed by deformation, namely, the crack tip. It is conceivable that the adsorbed molecule can facilitate the creation of that new surface. Adsorption further from the crack tip and bulk transport of H atoms to the plastic zone is excluded by the gaseous H_2S experiments.

WORKSHOP SUMMARY: LIQUID METAL EMBRITTLEMENT

Prepared by: R.A. Oriani

Disussion
  Leaders: R.A. Oriani, E.N. Pugh, A.R.C. Westwood,
             P. Haasen

Recorders: J.W. Gadzuk, B.J. Berkowitz, J.F. Knott,
             N.H. Macmillan

The embrittlement of solids by liquid metals (as well as by solid metals in some cases) is a subject with only a very few practitioners, and the relatively small number of established facts are not known sufficiently well by the materials community. This limited the scope, but not the enthusiasm, of the workshop discussions on this topic

Interest in liquid metal embrittlement (LME) has been intensified recently by S. Lynch's provocative challenge to the decohesion models of both LME and hydrogen embrittlement (HE). Reminiscent of Beachem's proposal that HE of steels is really a fine-scale hydrogen-induced enhancement of slip, Lynch asserts that in both LME and HE the embrittling species adsorbs upon the crack tip surface and thereby facilitates the nucleation and glide of dislocations along slip planes intersecting the plane of the crack. The crack is kept sharp by intercepting extremely small microvoids that are nucleated ahead of the advancing crack along the projection of the plane of the crack. Thus, embrittlement is the phenomenological result, despite the hypothesized enhanced plasticity, because of the cleavage appearance of the fracture and because of the reduced overall work needed to cause fracture due to the localization of dislocation activity to a narrow zone.

It is not surprising that much of the workshop discussions revolved about questions provoked by the enhanced plasticity concept. Some people doubted that the very small microvoids (<0.1µm) seen by Lynch, and upon whose existence his hypothesized mechanism critically depends, are anything but artifacts of the replica preparation and handling, technique, and strong recommendations were made for attempts at independent confirmation. This will be attempted at the Max-Planck Institu fürEisenforschung, Düsseldorf. Other workshop participants expressed puzzlement at the mechanism of formation of the putative microvoids, given the absence of any relationship to inclusions, the hypothesized restricted orientation of slip, and the fact that metals of very high purity also suffer LME. The often observed very fast rates of cracking and the LME of amorphous metal alloys aggravated these concerns.

965

It was repeatedly pointed out that indisputable fractographic evidence of the occurrence of microvoids and of plasticity cannot alone establish a mechanistic, causal relationship between these phenomena and either LME or HE. Some specific suggestions for experiments were made. Since the choice between the decohesion and the enhanced plasticity points of view, both in LME and in HE, depends on how the embrittling species affects the σ/τ ratio, it was suggested that suitably oriented single crystals of Zn, Cd, and Ge as well as metallic whiskers, be used for experimental materials in order to control the slip response. Since microvoids are mechanistically necessary for Lynch's mechanism, ultra-pure single crystals should be employed, in which microvoid formation is highly unlikely. In situ transmission electron experiments in which dislocation activity, resulting from contacting a liquid metal to a stressed specimen, would be looked for would be very welcome. TEM examination of the dislocation distribution ahead of a crack tip, both without and with the agency of an embrittling liquid metal, should also be done. Other suggested experimental work included the determination of the J-intetral for an advancing crack in the absence and in the presence of LME, the determination of the role of the liquid metal chemistry on the manifestations of LME, more detailed investigations of the LME of metallic glasses particularly by applying the liquid metal to the compressive side of an elastically bent strip, and the investigation of the effects on mechanical properties of well characterized surface phases.

On the theoretical side, it was pointed out that although chemisorption might decrease bond strength across one atomic plane, that across another, differently oriented atomic plane might be increased simultaneously, and slip along some other plane might also be facilitated. Some optimism was expressed that cluster calculations upon clusters of host atoms containing a large foreign atom might be easier to perform than the corresponding calculation for hydrogen as the foreign atom.

In summary, one can say that the same considerations arise, and hence the same sorts of experiments are relevant, both in LME as in HE. It was agreed that Lynch's observations and ideas, whether correct or not, are having the valuable consequences of forcing a critical examination of older ideas and the formulation of new experiments.

WORKSHOP SESSION 4

STRESS CORROSION CRACKING

Session Chairman: E. N. Pugh

STRESS CORROSION CRACKING

R.N. Parkins

Department of Metallurgy and Engineering Materials
University of Newcastle upon Tyne
England

SYNOPSIS

Environment sensitive crack growth by dissolution related mechanisms is capable of prediction on the basis of appropriate electrochemical measurements that define the potential ranges for cracking and the upper bound crack velocities for potent environments. Where the cracking environment is locally generated from an innocuous bulk environment predictability is more difficult, but there are the beginnings of indications of correlations between pitting propensity and the incidence of cracking in such circumstances. There is now sufficient data for a variety of systems to show that the relative time dependencies of film growth and plastic straining of the underlying metal determine whether or not cracking will occur, but the quantification of such a model has not yet resulted in accurate prediction of the limiting plastic strain rate. This is probably because little is yet known about the details of the conditions at crack tips. For intergranular cracking it is likely that chemical heterogeneity at the grain boundaries will play a critical role, but little is known about why segregating species facilitate cracking in such regions. Similarly, in relation to transgranular cracking, the initiation of the latter at emergent slip steps suggests a critical role for plastic strain, but the reasons for the localized embrittlement of the metal in the crack tip region remain to be elucidated.

INTRODUCTION

Mechanisms of slow crack growth in metals, at stresses markedly below their normal fracture stresses, due to the influence of environmental factors are usually related to the ingress of hydrogen

into the metal or to the localized dissolution of the latter. Since
hydrogen related fracture is dealt with in other papers, the present
one will concentrate upon dissolution related crack growth in which
it is important to recognise at the outset the distinction to be
drawn between mechanism and rate controlling process. Thus, while
the mechanism of crack extension may be by localized dissolution,
subsequent film growth over the advancing crack front and the disruption of the film to allow further dissolution may be rate controlling even though they do not directly contribute to crack growth.
It is convenient to divide dissolution related mechanisms of crack
growth into those that promote predominantly intergranular or transgranular crack paths. The former are likely to be related to
chemical heterogeneity at grain boundaries, since in many systems
that display intergranular stress corrosion cracking there is an
inherent propensity towards intergranular corrosion even in the
absence of stress, although in such conditions the grain boundary
attack may not penetrate to any marked extent or the rate of penetration may be appreciably less than when stresses of appropriate
sign and magnitude are present. With transgranular cracking the
frequently observed association of sites of crack initiation with
emergent slip steps suggests a critical role for plastic strain,
although this may be in rupturing otherwise protective surface films
in view of the crystallographic aspects of crack path and slip planes
being different[1].

There are obvious metallurgical overtones associated with these
two simple dissolution related models. Thus, alloying or structural
changes may influence the segregation to grain boundaries of solutes
or precipitated phases, and hence affect local cell action there-
abouts, or they may alter the thickness of surface films and/or the
heights of surface slip steps to modify the propensity towards
transgranular cracking in a given environment. Similarly, the
literature contains many instances correlating environment sensitive
fracture to mechanical properties, the latter having been varied by
changes in composition or structure of the materials concerned.
These general points have been discussed in some detail elsewhere[2]
and need not be repeated, but their import at this stage is essentially phenomenological rather than mechanistic.

The latter situation is probably related to the fact that there
is still little real understanding of the details of the reactions
in the crack tip region whereby metals normally displaying marked
ductility can be caused to behave in an essentially brittle manner,
but with some attendant small amount of plastic deformation, simply
because they are brought into contact with certain environments.
For transgranular cracking, it seems likely that the detailed
mechanisms will involve some element of localized embrittlement of
the metal, for which the ingress of hydrogen is an obvious contender.
However, the conditions for hydrogen discharge from environments are
not invariably met in all instances of environment sensitive

fracture and other possible mechanisms of local embrittlement, involving de-alloying[3] for example, are beginning to receive detailed consideration. For intergranular cracking the lack of understanding relates more to the precise mechanisms whereby chemical heterogeneity promotes cracking. Although there have been many demonstrations of relationships between intergranular cracking sus- ceptibility and the segregation of specific species to grain boun- daries, these frequently amount to no more than empirical correlations, and the mechanisms whereby such chemical heterogeneity promotes cracking have rarely received appropriate attention, especially from the electrochemical standpoint. However, whilst there is still much to be understood about stress corrosion cracking in terms of the detailed interactions between the controlling parameters at the crack tip, a picture is emerging at the macroscopic level, especially in relation to the roles of environment and stress, that should lead to better service predictability of the likelihood of stress cor- rosion cracking and provide a more realistic framework for mechanis- tic developments.

ENVIRONMENTAL INFLUENCES

The concept that stress corrosion cracking only occurs in environments highly specific to a particular alloy persists in the sense that not all environments corrosive to a particular material will promote stress corrosion. However, from the initial recognition that hot hydroxide or nitrate solutions could promote such failure in ferritic steels, or ammoniacal solutions in brasses, the list of potent environments for a given material has grown and will continue to do so with the passage of time. Moreover, for those materials, such as the high strength steels, that fail as the result of hydrogen ingress into the metal, the idea that solutions of specific chemical composition are necessary to promote stress corrosion can be mis- leading since such materials will fail in any aqueous solution, organic medium or gaseous environment from which hydrogen can be released to enter the metal. Nevertheless, for dissolution related crack growth, the concept of solution specificity exists and this is broadly predictable where the bulk environment is the direct cause of crack growth, but where the cracking environment is locally generated from an otherwise innocuous bulk environment predictability remains a difficulty.

To initiate and propagate a crack under environmental influence in the presence of a nominally static stress requires, for the retention of crack geometry, that the rate of advancement of the crack tip shall be markedly greater than any lateral spreading from the crack sides, since otherwise a pit or more general form of cor- rosion will result if the conditions are conducive to metal dis- solution. But what constitute the crack sides at any time were at an earlier stage of the growth part of the crack tip, so that having been active when part of the tip those surfaces must become

relatively inactive, or passive, at a later time. This transition
from active → passive behaviour will only occur for a given metal in
certain environments over particular ranges of potential, hence the
specificity of solutions for stress corrosion cracking by a dissolution related mechanism. There exist a variety of experimental
techniques[4] for identifying active → passive transitions, all of
which involve measuring the current or impedence response with time
at an electrode surface that is essentially bare at the outset and
which condition can be achieved by a variety of methods involving
scratching the surface, rapid plastic deformation or cathodic
reduction to disrupt or remove initial films. Since such measurements need to be performed over a range of potentials to identify
the extent of the latter in which cracking is likely, it is sometimes convenient to use a potentiodynamic method which, in those
cases where the initial metal surface is not already passive, can
define the potential range for cracking as well as giving an indication of the upper bound for the crack velocity. The method, which
has mostly been applied to mild or low alloy steels, involves firstly
sweeping an appropriate range of potentials at a high rate, of about
1V/min, to minimize the time for film formation and to define those
potentials in which the metal-environment combination shows active
behaviour. In a second experiment, the same range of potentials is
scanned but at an appreciably slower rate, of about 10 mV/min, to
allow time for filming, so that comparison of the two curves will
indicate any ranges of potential within which relatively high anodic
activity in the 'film free' condition reduces to insignificant
activity when the time requirements for film formation are met and
this will indicate the range of potentials within which one of the
requirements for stress corrosion cracking by dissolution is fulfilled. Figure 1 shows such curves[5] for a C-Mn steel immersed in

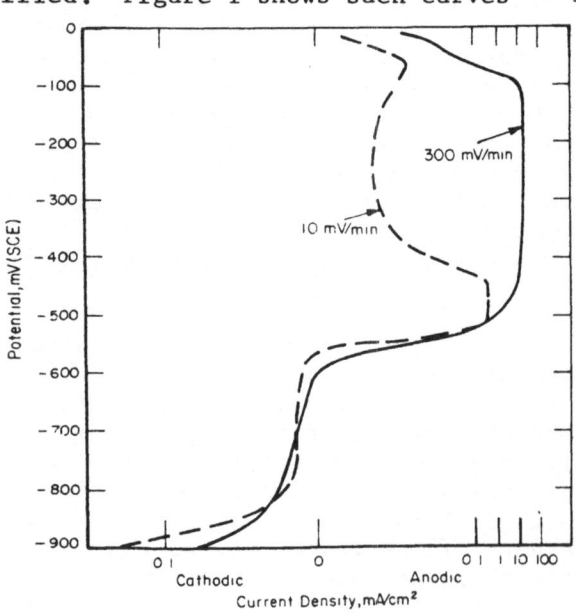

Fig. 1. Potentiodynamic
polarization curves for
C-Mn steel in 1N Na_3PO_4 +
H_3PO_4 at pH 4[5].

1N Na_3PO_4 + H_3PO_4 at pH 4 and indicating the possibility of cracking in the range from about -100 to -470 mV. Slow strain rate stress corrosion tests produced transgranular stress corrosion cracking over a range of potentials reasonably close to those predicted, whilst at higher potentials only ductile failure was observed due to the electrochemical activity of the system at such potentials, as reflected in the curves of Figure 1. At potentials below those that promoted transgranular cracking, failure related to the ingress of hydrogen occurred in slow strain rate tests, reflected in the cathodic response of the steel in phosphate solutions at sufficiently low potentials, and where the phosphate probably poisons the hydrogen combination reaction to facilitate its entry to the steel. Essentially similar results to those for C-Mn steel in phosphate solutions have been observed in other environments, with ferritic steels, involving nitrates, hydroxides, carbonates and acetates[4].

Where the initial air-formed oxide film cannot be reduced prior to initiating the potential sweeps then it is necessary to mechanically disrupt the film measuring the current response at constant potential during and after the exposure of bare metal. Scratching the electrode surface has been applied in studying the cracking of aluminium alloys[6] and rapid plastic straining has been used in studying the cracking of brasses[7]. With the latter potentiodynamic polarization curves give a very approximate guide to the potential range for cracking in acetates, formates, tartrates, nitrates and hydroxide solutions, but straining electrode measurements give a much better indication of the overall potential ranges for cracking, which can be appreciably wider than indicated by the potentiodynamic method. The film disruption methods involve assessing the likelihood of cracking from the maximum current density achieved when the bare metal is initially created and from its rate of subsequent decay. If the latter is very rapid the system is probably too passive, at the particular potential of such an observation, for cracking, whilst if the current density persists at a high value it is probably too active for the localized dissolution that is a necessary requirement for cracking. The latter is likely to be associated with intermediate rates of current decay, but there is as yet no quantifiable definition of what constitute these critical rates. Potentiodynamic methods similarly rely upon empirical approaches to defining cracking conditions in terms of the relative displacement of fast and slow sweep rate polarization curves.

Despite this empiricism, it is to be expected that the boundaries of the potential ranges in which stress corrosion cracking is observed will relate to the stabilities of different species or corrosion products. As the compositions of cracking environments are changed, frequently manifest in pH changes, so also do the potential ranges in which the various modes of behaviour are observed, so that potential-pH diagrams are a convenient means of representing such behaviour. Figure 2 shows the results[5] from slow strain rate tests

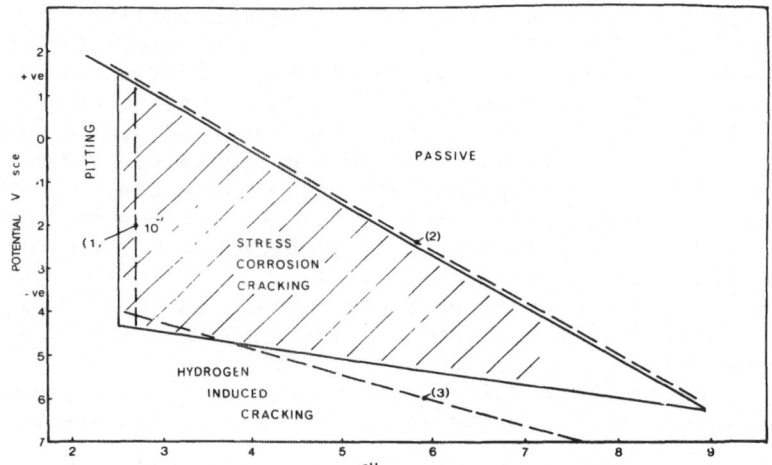

Fig. 2. Observed stress corrosion cracking range (cross hatched) for C-Mn steel in phosphate solutions and domain boundaries calculated from equations (1) to (3).

on a C-Mn steel in phosphate solutions at different pH's, with the various modes of failure indicated. Transgranular stress corrosion cracking in this system is associated with the formation of $Fe_3(PO_4)_2$ + hydrated Fe_3O_4 films, with the cracking domain defined by boundaries where other substances become the stable phases. At low pH's Fe^{2+} fulfills the latter role to promote pitting or more general corrosion, whilst at higher potentials $\gamma-Fe_2O_3$ may be expected to be the stable phase and is associated with ductile fracture. The relevant reactions representing the domain boundaries are

$$3Fe^{2+} + 2H_2PO_4^- = Fe_3(PO_4)_2 + 4H^+$$

for which the equilibrium constant

$$K = \frac{(H^+)^4}{(Fe^{2+})^3 (H_2PO_4^-)^2} = 10^{-7.8} \qquad (1)$$

and

$$2Fe_3(PO_4)_2 + 9H_2O = 3\gamma-Fe_2O_3 + 2H_2PO_4^- + 2HPO_4^{2-} + 12H^+ + 6e^-$$

for which

$$E_{sce} = 0.46 - 0.12 \, pH + 0.02 \log (H_2PO_4^-) + 0.02 \log (HPO_4^{2-}) \qquad (2)$$

At lower potentials cracking is associated with the ingress of hydrogen into the steel, so that the reaction

$$H_2 = 2H^+ + 2e^-$$

and for which

$$E_{sce} = -0.24 - 0.06 \text{ pH} \tag{3}$$

should define the boundary separating dissolution-controlled cracking
from that due to the embrittlement of the steel by hydrogen. Figure
2 shows the positions of the lines calculated from Equations (1-3)
and the agreement with the observed domains is reasonable for such
calculations. Within the region of the lower domain boundary for
dissolution related cracking there should be an additional line
representing the limit of stability of Fe. The obvious reaction,
in view of the substances formed at higher potentials is

$$6Fe + H_2PO_4^- + HPO_4^{2-} = Fe_3(PO_4)_2 + Fe_3O_4 + 11H^+ + 14e^-$$

and, in the absence of a value for the free energy of formation of
hydrated Fe_3O_4, using a value for the anhydrous form gives

$$E_{sce} = -0.45 - 0.05 \text{ pH} - 0.02 \log (M_2PO_4^-) - 0.02 \log(H_2PO_4^{2-}) \tag{4}$$

The line calculated from this equation is too low by about 200 mV.
It is likely however that the free energy of formation of hydrated
Fe_3O_4 will be somewhat smaller than that of the anhydrous form
(-242 Kcal/mol) and to make Equation (4) approximate to the observed
boundary would require the free energy of formation of the hydrated
form to be about 200 Kcal/mol. If such a value is used then the
calculated line has a similar slope to that shown as (3) in Figure 2
and is displaced below it by some 50 to 25 mV.

Somewhat similar results to those shown in Figure 2 have been
achieved in relation to the cracking of ferritic steels in an acetate
solution[8] and in carbonate-bicarbonate solutions, with the intergranular cracking domain in the latter environment being associated
with films of $FeCO_3$ + Fe_3O_4 and ductile failure with Fe_2O_3 films.
Indeed the latter appear invariably to be associated with ductile
failure, whether the films are formed by the addition of inhibitive
substances to otherwise potent cracking environments or more directly
by potentiostatic control. This is presumably no more than an
indication of stress corrosion cracking not being possible unless the
system shows an appropriate level of activity. On the other hand
cracking in ferritic steels appears to be associated with the formation of Fe_3O_4, as in nitrate or hydroxide solutions, sometimes
mixed with another filming substance, as in phosphate or carbonate-
bicarbonate environments, and the reasons for these associations
between the nature of the film and the cracking propensity are not
immediately clear. Similarly, in the stress corrosion cracking of
brasses in a variety of environments, intergranular cracking appears
to be associated with the formation of Cu_2O and transgranular cracking with the domain of stability of CuO, as indicated in Figure 3,
and again for reasons that are not understood.

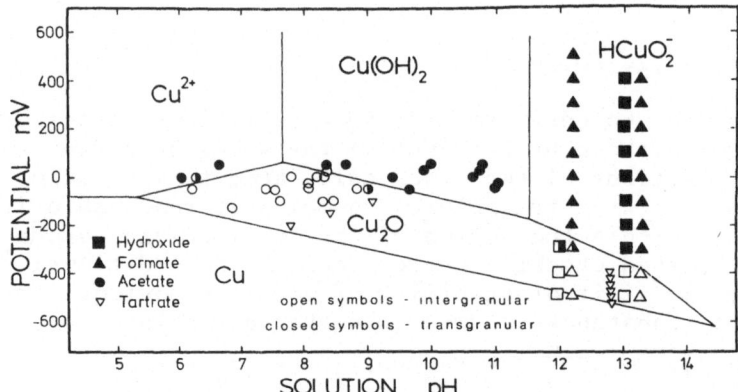

Fig. 3. Potential-pH diagram showing the domains of failure
mode for 70/30 brass in various solutions, together
with the calculated positions of various boundaries
relating to the domains of stability of different
chemical species.

Whilst filming reactions appear to determine the potential
ranges for dissolution controlled cracking, there remains the usual
kinetic related problem of the crack growth rate within the predicted
potential range. For crack growth by dissolution the order of the
growth rate at a crack tip maintained essentially bare is approximated by the Faradaic equivalent, i.e. there is a simple relationship between crack velocity and dissolution rate, as is apparent
from Figure 4 for a variety of systems. In practice this will
constitute an upper bound for crack velocity because it only applies
when the dissolution rate is maintained by the avoidance of any significant time during which the crack tip is filmed and relatively
inactive. The rate controlling process, as opposed to the mechanism
of crack extension, may therefore be the rate of film growth over
the crack tip or the frequency with which that film is disrupted to
reinitiate crack growth, and this constitutes an area in which
quantitative prediction is still poor but one of considerable practical significance if the crack velocity for a particular system is
to be precisely placed between the upper bound, as previously
indicated, and the lower bound of zero corresponding to maintained
crack tip inactivity. Moreover, where intergranular cracking is
involved and arises because of chemical heterogeneity at the grain
boundaries, the relevance of macroscopic current measurements to
preferential grain boundary attack is questionable and constitutes
another area where detailed work would be valuable.

Notwithstanding these points of detail in relation to the
application of electrochemical measurements to the prediction of

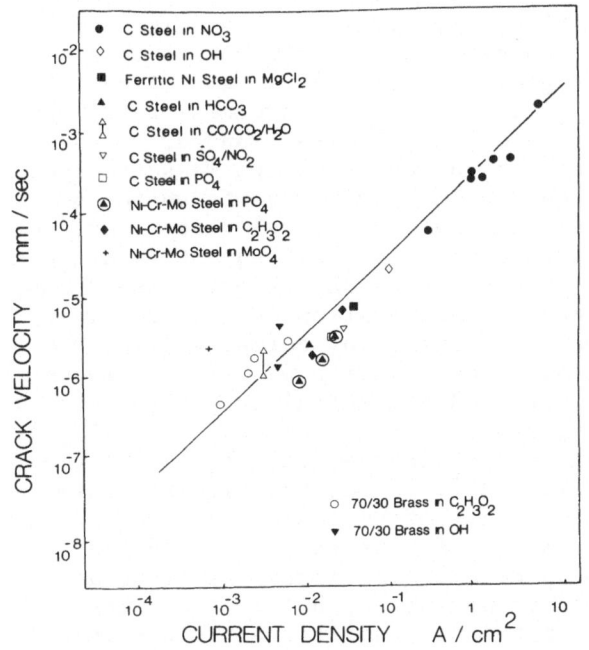

Fig. 4. Measured stress corrosion crack velocities and anodic current densities passed at a relatively bare surface in a variety of systems. The line is for the crack velocities calculated from the Faradaic equivalent.

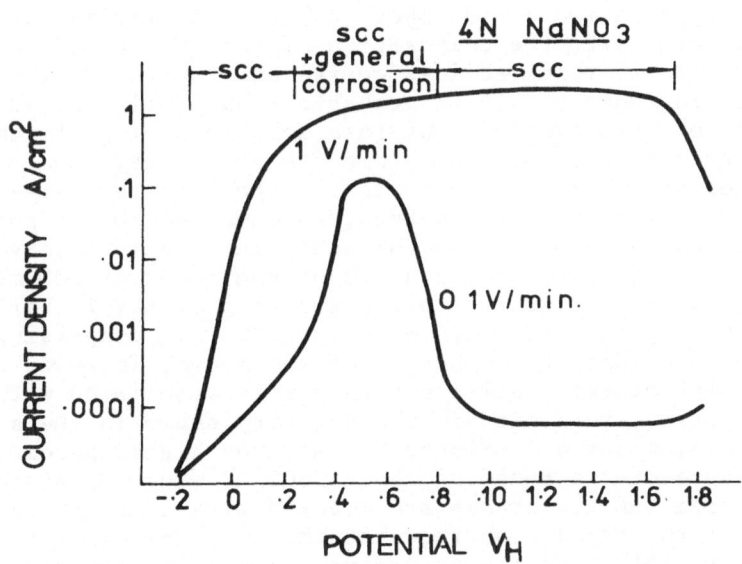

Fig. 5. Potentiodynamic polarization curves at two sweep rates for C-Mn steel in boiling 4 N NaNO₃ showing forms of corrosion in different potential ranges.

potential ranges and crack growth rates for stress corrosion cracking,
they do offer reasonable guidance in assessing the potency of bulk
environments. However, the problems are appreciably greater when
the bulk environment itself is innocuous but may give rise to a
potent environment by localized generation. This is most likely to
involve some species at low level in the bulk environment being concentrated under the influence of heat and/or mass transfer to the
point of potency. Changes in concentration involving mass transport
by diffusion or ion migration can occur in environments where they
have access to metal surfaces containing geometrical discontinuities
and the observation that stress corrosion cracks frequently initiate
within corrosion pits probably arises from such concentration changes,
rather than for reasons of stress intensification in such regions,
and suggests that the incidence of cracking may correlate with the
pitting potential in such regions. The addition of a few grams per
litre of sodium nitrite to concentrated boiling nitrate solutions is
well known to inhibit the intergranular stress corrosion cracking of
mild steels that occurs in nitrate solutions at the free corrosion
potential. Stress corrosion cracking is associated with the formation of black films of Fe_3O_4 over all of the exposed surfaces, but
the addition of the inhibitive amount of nitrite results in the
formation of an invisible film of Fe_2O_3. The latter persists at
potentials above the free corrosion potential, but if the pitting
potential is exceeded then one or two very small black spots are
observed to develop on the otherwise bright surfaces of the test
specimens and from these develop intergranular cracks covered with
Fe_3O_4. The implication would appear to be that, within the confines
of the initiating pit, the composition of the environment changes so
that the nitrite is rendered ineffective and conventional nitrate
related cracking ensues. It is probable that pitting is involved in
the cracking of mild steels in nitrate solutions at sufficiently high
potentials even in the absence of nitrite additions. One of the
peculiarities of cracking by nitrates is the very extensive range of
potentials over which it is observed, as compared to the potential
ranges observed, as compared to the potential ranges observed with
other environments, which are usually of the order of 200 mV.
Figure 5 shows that stress corrosion cracking occurs in boiling
4N $NaNO_3$ over a potential range of about 2000 mV, comprising a band
some 400 mV in extent at the bottom of the range, followed by a
regime in which cracking still occurs but is associated with more
general corrosion, reflected in the activity shown in the slow sweep
rate polarization curve in Figure 5. At even higher potentials,
cracking occurs over a range of about 1000 mV where no visible surface films form but the cracks are covered with a magnetite film,
as are all of the exposed surfaces at the lower cracking potentials.
It appears probable that the extensive range of cracking potentials
above about +0.8V_H is related to the pitting potential being
exceeded.

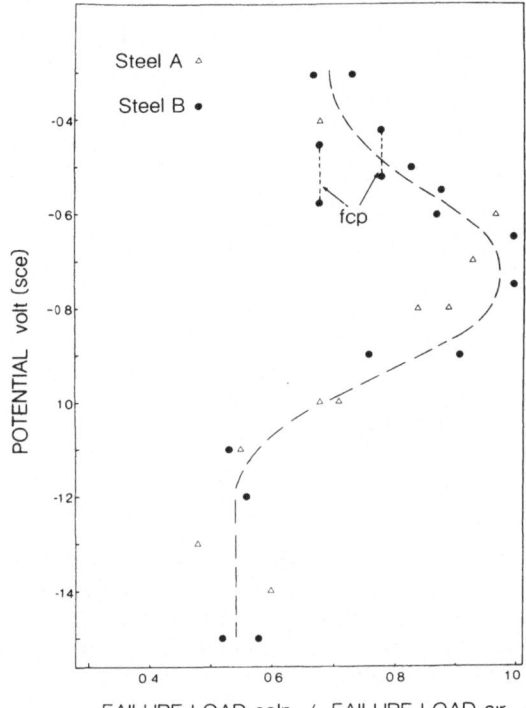

Fig. 6. Effects of applied
potential upon the failure
load ratio of notched specimens of two prestressing
(eutectoid) steels in 1 g/1
$Ca(OH)_2$, pH 6-8.5, in slow
strain rate tests.

A clearer demonstration of the association of cracking with
pitting potentials is provided by the results of tests on prestressing steels of about eutectoid composition immersed in calcium
hydroxide-chloride solutions having bulk pH's in the range 6 to 8.5
and subjected to slow strain rate tests[9]. Figure 6 shows the
effects of applied potential in this system, indicating that at
potentials of about -900 mV (sce) and below there is a failure regime
separated from a second failure regime above about -700 mV by a region
in which ductile fracture occurs. In the lower potential regime
failure results from the ingress of hydrogen into the steel, producing fracture surfaces exhibiting quasi-cleavage and different
from those surfaces associated with the second regime. In the latter
preferential dissolution of the ferrite of the pearlite on the fracture surface suggests an involvement of dissolution, and exposure of
unstressed, polished samples of the same steel to the same environment at potentials such as those associated with the higher regime
of cracking potentials produced pits within which the steel became
etched in a few minutes of exposure. Although much of the exposed
surface of such polished specimens remained unattacked when exposed
to solutions in the pH 6-8.5 range, when the pH was in the range
from about 2-4 the whole of the polished surface was quickly etched
at relevant potentials. The obvious implication is that the pH within

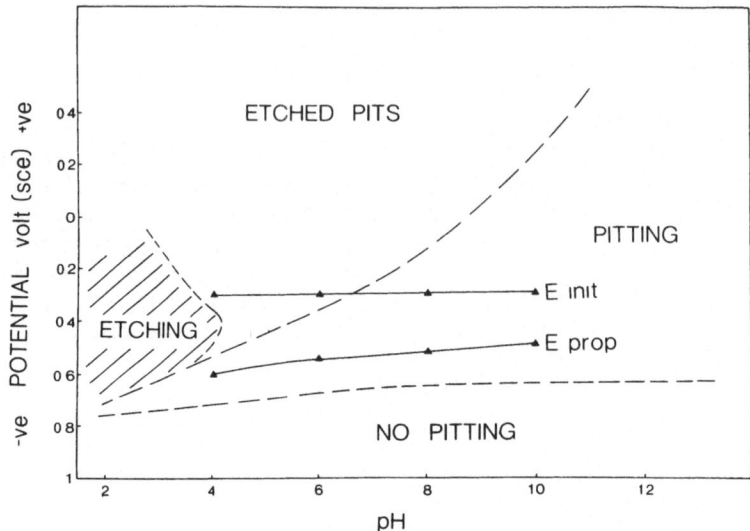

Fig. 7. Modes of attack observed on polished surfaces of
prestressing steel after exposure to 1 g/l Ca(OH)$_2$
solutions of various pH's and at different potentials, together with the pitting initiation and
propagation potentials obtained from reverse scan
polarization curves.

the pits had reduced to about 4 or below from the bulk pH of 6-8.5
that did not produce preferential attack upon the ferrite of the
pearlite. This in turn suggests that the higher potential cracking
regime of Figure 6 is related to the pitting potential being exceeded
and this was supported by measurements of the pitting potentials by
the usual reversed potentiodynamic technique. The results are shown
in Figure 7 for solutions of various bulk pH's, together with the
observations upon the incidence of pitting on the initially polished
surfaces. The potentials measured potentiodynamically for the
initiation of pitting were somewhat higher than those observed by
potentiostatic exposure and subsequent microscopical examination of
the surfaces, as indicated by Figure 7, but the latter measurements
gave good agreement with those lowest potentials at which began the
higher cracking regime.

There have been many experimental and analytical investigations
of solution composition changes within pits or crevices which show
that metal ions produced therein hydrolyze so acidification occurs,
of which those by Pourbaix[10] are deserving of special mention.
Figure 8, based upon work by Galvele[11], shows the concentration of
H$^+$ as a function of the product of the depth and current density in
a unidirectional pit and shows that, for a typical current density
of 1 A/cm^2, a surface discontinuity only 10^{-2} μm deep can allow a

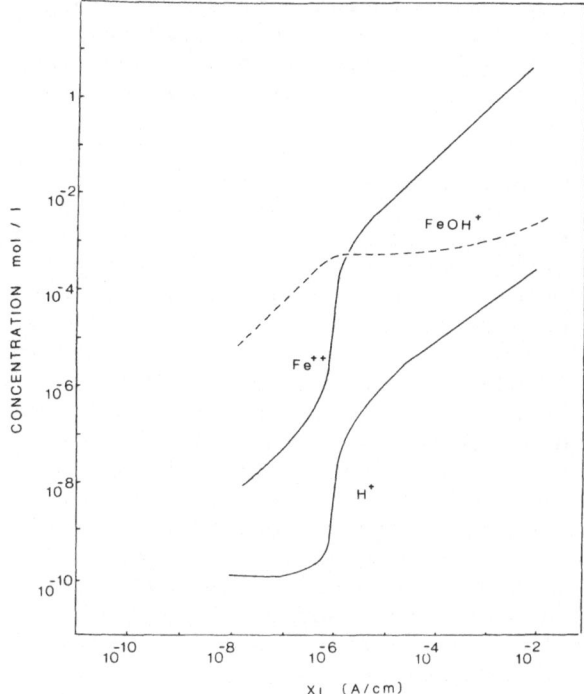

Fig. 8. Concentration of H^+ as a function of the product of the depth (X) and current density (i) in a unidirectional pit[11].

reduction of pH to about 8 and at a depth of 1 μm the pH can fall to 5, from an initial value of 10 where an oxide film should be thermodynamically stable. For different bulk pH's the pH within a pit will change with depth at different rates, but for depths of about 1 mm the pH in the pit is predicted as likely to be in the region of 4 irrespective of the bulk pH, although the changes may also be influenced by the presence of different anions, especially those with a buffering action that will restrict pH changes. It is clear from such studies that surface discontinuities, no matter what their origins, can result in considerable changes in solution composition from those represented by the bulk solution and that to assume that surface discontinuities are only of significance in the context of stress concentration can be misleading. This is particularly so where pre-cracked specimens are employed in laboratory tests or where crevices exist in service situations, i.e. where some pre-existing geometrical feature exists that avoids the necessity for a corrosion pit to form on an initially plain surface to facilitate solution composition changes. Such features can facilitate changes in environment composition so that cracking can occur when the bulk environment is innocuous, but to assume that such composition changes will be invariably present may be as misleading as to ignore the possibility of their existence in some circumstances. Thus, in buffered solutions or where the solubility of the solvated species is

extremely small, it seems likely that solution composition changes
in cracks or crevices will be appreciably more restricted than in
other circumstances, but there is need for detailed considerations
of the crack enclave chemistry in a wider variety of systems than
those studied hitherto.

The sensitivity of environment related fracture to potential
raises questions as to how dangerous potentials are achieved and
whether the crack tip potential is the same as that at the surface
where the crack enclave emerges and where the potential is most
likely to be measured. In many dissolution controlled stress corrosion systems the free corrosion potential will frequently lie at
the more negative end of the potential range for cracking, but since
the free corrosion potential is sensitive to surface condition it is
not invariably the case that it will lie within the cracking range,
which is why it is as important to measure the potential in stress
corrosion experiments as it is to be able to quantify the stress or
metal conditions. The variability of the free corrosion potential
frequently results in tests being conducted at controlled potentials,
but in view of the fact that the current must flow along the thin
film of liquid contained within the crack enclave, the extent to
which the crack tip potential departs from that of the control
potential, usually applied at the outer surface, should be known.
The solution of the differential equation, based upon Ohm's Law, for
conditions in which an IR drop occurs because of the resistivity of
the liquid film within a crack can lead to a variety of conclusions,
depending upon the boundary conditions and other assumptions made to
solve the equation, ranging from there being very small potential
drops (~10 mV) to very large ones (~1 V). It is probable that in
different circumstances potential drops can vary from being negligible to very large and there is circumstantial evidence for such
variability. Thus, the results in Figure 6 suggest that below about
−1100 mV (sce) there is no further increase in susceptibility to
cracking, possibly because the effective potential cannot be further
decreased because of IR drops. On the other hand, the potential
ranges for the intergranular stress corrosion cracking of a C–Mn
steel in a carbonate-bicarbonate solution are essentially the same
for initially plain and pre-cracked specimens[12], yet if the latter
contained appreciable potential drops along the pre-cracks the
potential ranges for cracking would be expected to be displaced by
the amount of the potential drop.

The work of Ateya and Pickering[13] shows that, for unfilmed
metal surfaces, appreciable potential drops, of the order of 100 mV/cm,
can occur if attempts are made to pass relatively large currents
through the liquid contained within a crevice. The results shown in
Figure 9 confirm these results in that for potentials (−0.7 and −0.95V
sce) not far removed from the corrosion potential there are no significant changes in potential along a crevice containing a ferritic
steel and natural seawater. However, when attempts were made to

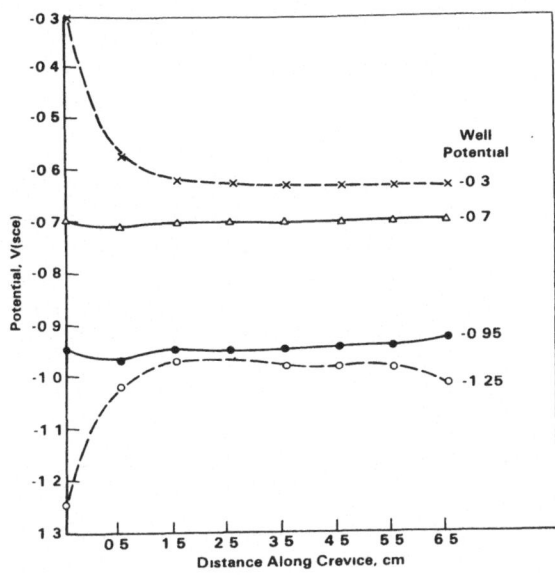

Fig. 9. Potentials of segmented ferritic steel surfaces along a 0.025 cm thick
crevice containing seawater
with the entrance to the
crevice controlled at different potentials.

control the crevice potential at −0.3 or −1.25 V, requiring the
passage of relatively large currents (of the order of mA, as opposed
to the µA required for the potentials nearer to the free corrosion
potential) then large potential drops occurred in the first cm or
so of the crevice. These potential gradients were not markedly time
dependent, in contrast to some measurements made on a C-Mn steel in
a carbonate-bicarbonate environment, a system known to promote
stress corrosion by a dissolution controlled mechanism, unlike the
ferritic steel in seawater, to which Figure 9 refers.

The experiments involving the C-Mn steel in a carbonate-
bicarbonate solution were made with segmented specimens, 1 cm long
and 2 cm wide, separated by a layer of electrically insulating
varnish. There were 7 segments in all to produce a crevice about
7 cm long, 2 cm wide and with the liquid film 0.05 cm thick contained by an acrylic cover overlaying the creviced specimens. A
zero resistance ammeter allowed the current flowing between adjacent
segments to be measured, the potentials of which were determined at
their centres by the introduction of a reference electrode through
holes in the crevice cover. The potential at the entrance to the
crevice could be controlled potentiostatically, with a counter
electrode immersed in a well external to the crevice and filled with
liquid continuous to the crevice film. The surfaces of the segments
were rended film-free by cathodic polarization to produce a free
corrosion potential in the region of −850 mV (sce) after which the
potential at the entrance to the crevice was controlled at −500 mV.
Immediately a potential drop of about 300 mV was observed along the
crevice but this reduced with time, as the results in Figure 10
indicate. At the remote end of the crevice the potential changed

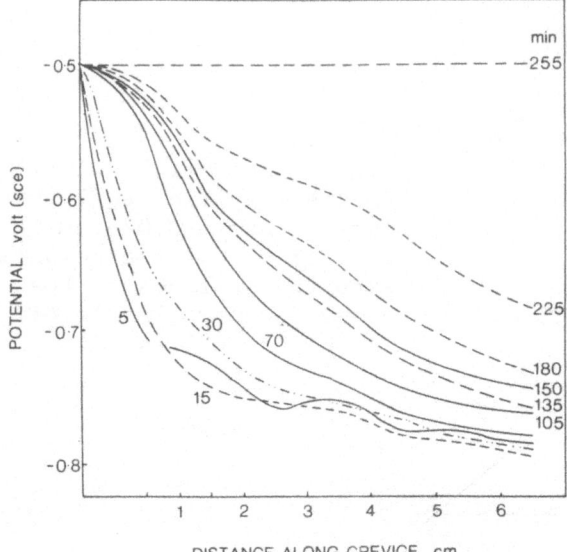

Fig. 10. Potentials of segmented C–Mn steel surfaces along a 0.015 cm thick crevice containing 1N Na_2CO_3 + 1N $NaHCO_3$ solution at room temperature after various times.

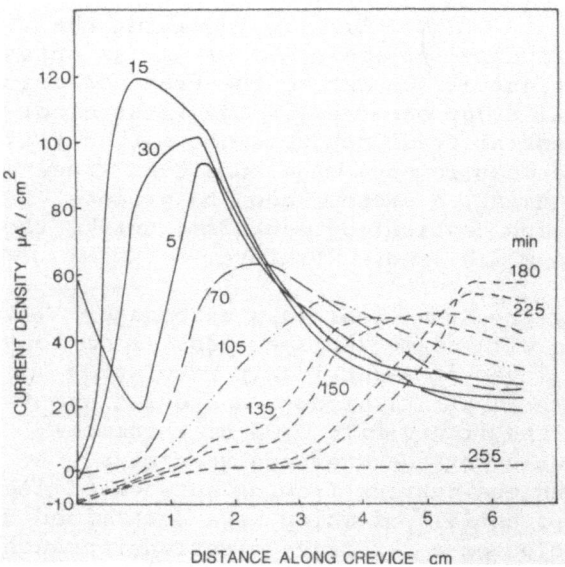

Fig. 11. Current densities flowing between adjacent segments at various times corresponding to the potential-distance curves of Figure 10.

relatively slowly with time to reduce the potential drop to about 200 mV in some 4h and then, quite suddenly, the potential gradient along the crevice disappeared for all practical purposes. The reason for this latter behaviour is apparent in Figure 11, which shows the average current density on each segment as a function of time. From the movement of the maximum current density along the crevice with time it is clear that the metal near the entrance to the crevice passivates initially and then the passivated zone gradually spreads along the crevice. The passivated material is

associated with little current transfer across its surface in contact with the liquid, so that the current density at the remote end
of the crevice increases with time until it is adequate for passivation in those regions, when the potential gradient suddenly
disappears. Although the rates at which these changes occur vary
with the crevice thickness it is likely that similar events will
occur in the context of crack growth, the relatively inactive crack
sides being associated with little current drainage so that the
potential gradients along typical cracks will be small where crack
growth involves dissolution and subsequent filming of the crack
sides. By contrast, cracking due to hydrogen entry does not involve
filming, and potential gradients will not display the time dependence
shown in Figure 10.

THE FUNCTION OF STRESS

It is well established that for a given stress corrosion system
there is a threshold stress below which failure does not occur in
an extended period of time or, in the context of fracture mechanics,
that there is a stress intensity factor below which the crack
velocity reduces from the value sensibly independent of stress
intensity to some very much lower value or to zero. These threshold
stresses or stress intensity factors are well known to be sensitive
to the environment involved in their determination, as well as to
changes in alloy composition or structure. They are not, therefore,
material properties in the sense of, say, the yield strength or
fracture toughness, although the mechanical properties of the
material involved are likely to be important in determining the
threshold stresses for fracture in a given environment. The importance of the film growth rate in dissolution-related stress corrosion
cracking has already been mentioned and it is likely that the role
of stress is linked to the filming rate, which varies with the
environmental conditions to produce variations in the threshold
stress with changes in the latter. Below the threshold the surface
film remains protective because its rate of repair does not fall
below the rate at which it is mechanically disrupted. Above the
threshold stress the film repair rate falls behind the rate at which
it is mechanically disrupted and the limiting condition relates to
the rate at which the bared metal can dissolve to promote crack
growth. At even high stresses, ductile or other essentially
mechanical modes of fracture become kinetically more favourable
than dissolution controlled crack growth and there is a transition
to the former. The net effect is the well known crack velocity-
stress intensity factor relationship.

The essential nature of the role of stress lies in its ability
to promote a deformation rate that, above the threshold, exceeds
the rate at which the bare metal is refilmed at emergent slip steps,
i.e. it is the effective strain rate that is mechanistically important rather than stress itself, although clearly the strain rate

derives from the stress. There are two obvious implications in this
model for constant load situations, namely that whether or not cracking occurs will depend upon the time at which the environmental
conditions are established in relation to the time of loading and,
secondly, that below the threshold stress cracks may initiate but
cease to propagate after some time. The creep rate that follows
initial loading decays with time, so that if creep has become
exhausted before the environmental requirements for cracking are
established, then cracking may not occur, even though the stress
may be appreciably above the threshold stress determined by loading
test pieces after establishing the environmental conditions.
Exceptions to the threshold stress being dependent upon the amount
of prior creep will occur when some corrosive process, such as pitting or intergranular corrosion, reinitiates creep by locally
raising the stress. At stresses below the threshold cracks can
still initiate but later cease to propagate, so the threshold stress
is not that stress below which cracking does not occur but the
stress above which cracks propagate to produce total failure. Cracks
cease to propagate below the threshold when the decay in the creep
rate following initial loading exceeds the increase in creep rate
attendent upon crack growth to the point where the effective strain
rate is no longer sufficient to maintain the crack tip in an active,
unfilmed, condition. Data relevant to these points have been summarized elsewhere[14], including consideration of estimating thresholds from creep data[2,14].

Stress corrosion crack growth may be sustained below the
thresholds observed in constant load situations if the crack tip is
maintained active by either monotonic or cyclic straining.
Consequently, while threshold stresses determined with constant
loads are frequently in the region of the yield or 0.2% proof
stresses they may be reduced to about half these levels with monotonic or cyclic slow straining. Variation of the monotonic strain
rate from test to test indicates that there is a lower rate below
which cracking will not occur because the crack tip remains in a
passive condition. Figure 12 shows some typical results from tests
in which the specimens were loaded above the threshold, allowed to
creep until the rate of the latter reduced below the strain rate to
be applied, and then the environmental requirements for cracking
were established. Clearly, there is a sharply defined strain rate
below which crack growth does not take place and the general form
of Figure 12 is the same as plots of stress intensity factor against
crack velocity. In principle it should be possible to calculate this
limiting strain rate from a knowledge of the lateral growth rate of
the protective film, but if this is attempted, using the two dimensional nucleation and growth model for a surface film due to
Armstrong et al[15], the calculated limiting strain rate is some
two or three orders of magnitude higher than is observed for a C-Mn
steel in a carbonate-bicarbonate solution at a particular potential.
The reasons for this large discrepancy are essentially twofold.

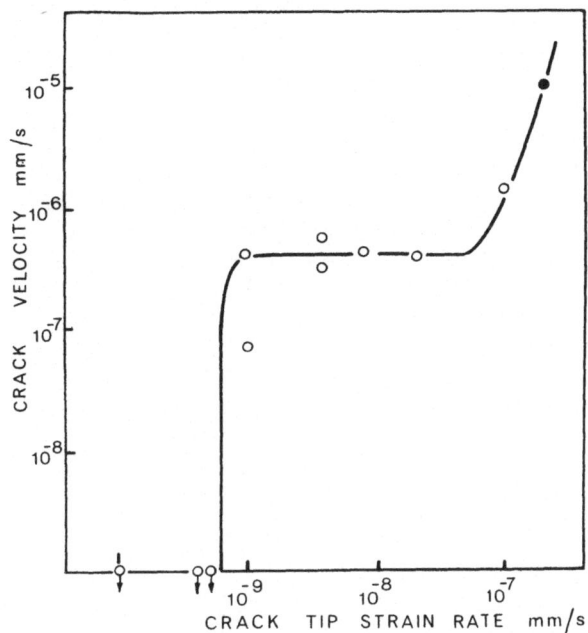

Fig. 12. Intergranular crack velocities for various applied crack tip strain rates in C-Mn steel immersed in 1N Na_2CO_3 + 1N $NaHCO_3$ at -650 mV (sce) and 75°C.

The engineering strain rate is probably much less important than the slip step emergence rate and whilst there are still problems in precisely quantifying the engineering crack tip strain rate, even less seems to be known about slip step emergence rates. The latter are likely to be appreciably higher locally than the averaged engineering strain rate, which is in the appropriate direction for reducing the discrepancy between calculated and observed limiting strain rates. The second difficulty with the model relates to the lack of detailed knowledge of the crack tip conditions in the electrochemical sense. The two dimensional nucleation and growth model for a surface film relates to a metal surface on which there is no initial film, yet the conditions in the vicinity of the critical strain rate will be such that the crack tip region already has a relatively thick precipitated film present. The repair of this film where it is locally disturbed seems unlikely to follow the same kinetics as obtain in the Armstrong model if only because nucleation need not be involved and a better route probably would only involve an appropriate growth law.

Perhaps the most practically significant aspect of these observations of the importance of dynamic straining upon the incidence of environment sensitive fracture is in relation to cyclic loading. The latter can sustain plastic straining[16], even at very low cycle frequencies and with small load changes, and since it is doubtful if any engineering structure ever operates under truly static conditions it is probable that load cycling plays an important

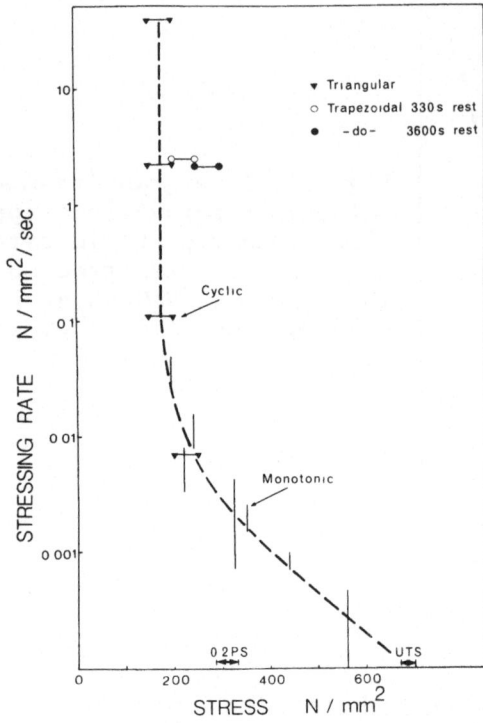

Fig. 13. Threshold stress as a
function of stressing rate for
monotonic and cyclic loading
tests on cast Ni-Al bronze in
seawater at -150 mV (sce).

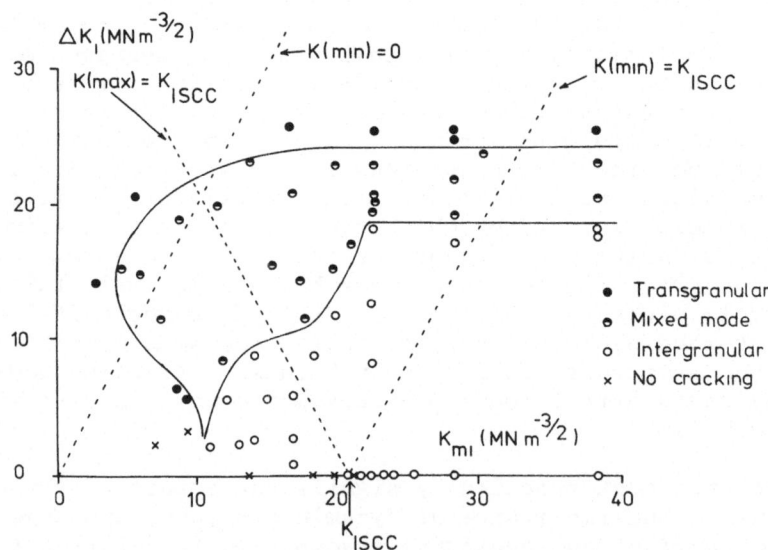

Fig. 14. Modified Goodman diagram indicating loading parameters for
which intergranular, mixed-mode and transgranular cracking were
observed in C-Mn steel immersed in 1N Na_2CO_3 + 1N $NaHCO_3$ solution
at -650 mV (sce), 75°C, 0.19 Hz, R>0[17].

role in service failures. That the effects of monotonic and cyclic
loading, at comparable slow strain rates, in environment sensitive
fracture are the same as indicated by the results shown in Figure 13,
which shows the threshold stresses from monotonic and cyclic load
tests on a cast Ni-Al bronze in seawater at a controlled potential.
There is, of course, a problem in comparing results from monotonic
tests, in which the strain rate is controlled, with cyclic tests in
which the load is controlled, since in the latter hysteresis will
cause the strain rate to vary during any particular load cycle and
the width of the hysteresis loop will change with the number of load
cycles. However, it is possible to provide a rate of stress change
from the monotonic straining tests, as for the cyclic loading tests,
and this affords a means of comparison of the results from the tests
involving the two loading modes. The fact that the same threshold
stresses are observed for the two types of loading at comparable
stressing rates, together with the velocity and fractographic
characteristics of the cracks produced being indistinguishable for
the two loading modes, suggests that the same phenomena are involved.

Essentially similar effects have been observed in a different
system involving a C-Mn steel in a carbonate-bicarbonate solution[17].
Figure 14 shows results from tests upon pre-cracked specimens in
which the intergranular cracking produced was indistinguishable in
constant load or cyclic tests. In constant load tests cracking only
occurred when the stress intensity factor exceeded 21 $MNm^{-3/2}$, but
with cyclic loading involving ΔK values of only about 2 $MNm^{-3/2}$ the
threshold was lowered by a factor of about 2. The crack growth
rates were essentially the same with the two loading modes, but if
ΔK was raised sufficinetly there was a transition through a mixed
mode to a transgranular mode of fracture with markedly higher
velocities than those associated with the intergranular cracking.
These various indications of the importance of time dependent
phenomena in the crack tip region have implications for fracture
mechanics approaches to environment sensitive fracture that, so far,
have received scant attention.

REFERENCES

1.   E.N. Pugh, This conference.
2.   R.N. Parkins, 3rd International Conference on Mechanical
     Behaviour of Materials, Edited by K.J. Miller and R.F. Smith,
     Pergamon Press, Oxford 1980, Vol. 1, 139.
3.   R.C. Newman and T. Burnstein, Corr.Sci, 1981, 21, 119.
4.   R.N. Parkins, loc.cit, 1980, 20, 147.
5.   R.N. Parkins, N.J.H. Holroyd and R.R. Fessler, Corrosion,
     1978, 34, 253.
6.   F.P. Ford, Metal Sci.Jnl, 1978, 11, 326.
7.   R.N. Parkins and N.J.H. Holroyd, to be published.
8.   N.J.H. Holroyd and R.N. Parkins, Corr.Sci, 1980, 20, 707.
9.   R.N. Parkins, M. Elices, V. Sanchez-Galvez and L. Cabellero,
     to be published.

10.   M. Pourbaix, This conference.
11.   J.R. Galvele, Jnl.Electrochem.Soc, 1976, 123, 464.
12.   R.N. Parkins and I.H. Craig, "Mechanisms of Environment
      Sensitive Cracking of Materials", Edited by P.R. Swann, F.P.
      Ford and A.R.C. Westwood, Metals Soc, London, 1977, 32.
13.   B.G. Ateya and H.W. Pickering, Jnl.Electrochem.Soc, 1975, 122,
      1018.
14.   R.N. Parkins, ASTM STP 665, "Stress Corrosion Cracking – The
      Slow Strain Rate Technique", 1979, 5.
15.   R.D. Armstrong, G.M. Bulman and H.R. Thirsk, Jnl.Electrochem.
      Chem, 1969, 22, 55.
16.   J.T. Evans and R.N. Parkins, Acta Met, 1976, 24, 511.
17.   R.N. Parkins and B.S. Greenwell, Metal Sci.Jnl, 1977, 11, 405.

DISCUSSION

Comment by M. Pourbaix:

I wish to congratulate you for this excellent paper, the
text of which needs to be read carefully. I have three questions.

(1)   Did you observe any influence of the composition
         of the bulk solution on the "protection potential"
         (which seems to be about -0.7 to -0.8 volts, SCE)?

(2)   Are you, in this respect, in agreement with Floyd
         Brown?

(3)   Could you look to the influence of the composition
         of the bulk solution on the composition of the
         solution under the cracks?

Reply:

(1)   There was a small effect of solution composition on
         the protection potential, as is apparent from Figure
         7 of my paper.

(2)   The results shown in Figure 6 are for an intermediate
         pH, but they were not markedly different for other
         pH's, except when the latter was reduced to 2-3 when
         the ductile fracture associated with the intermediate
         pH's were not observed. At all other pH's up to 12.6
         the potential at which the hydrogen induced failure
         began (-900mV) was the same, which was surprising in
         view of Brown's results for heat treated alloy steels
         in chloride solutions. It appears that, with the pre-
         stressing steel in $Ca(OH)_2$+HCl solutions, hydrogen
         induced failure only occurs below the potential at

which iron ceases to go into solution.

(3) In our work on crevices containing $CO_3^=$ - HCO_3^- solutions, we have not been able to detect any pH changes in the solution during the tests, although these have lasted for only a few days at the most.

Comment by D.J. Duquette:

The use of differential potential scan rates to determine stress corrosion cracking susceptibility (by measuring $\Delta\Phi$ vs. ΔI "windows") must be approached with some caution. For example, if such experi- ments are performed in carbonate/bicarbonate solutions or pure Fe, 0.05% C-Fe and 0.8% C-Fe the polarization curves, as a function of scan rate, are quite similar. Yet, none of use would suggest that each material would exhibit similar stress corrosion cracking pro- perties in this environment. Thus, it is important to note that your discussion, related to chemical heterogeneity is highly pertinent and again demonstrates that there are few, if any, single tests or observations which will always allow good prediction of stress corrosion cracking susceptibility.

Reply:

I agree that the potentiodynamic, or any other method of pre- dicting cracking potency is not sufficient in itself to provide a definitive answer to whether or not stress corrosion cracking will occur. This is hardly surprising since stress corrosion cracking requires the simultaneous fulfillment of a number of conditions. The original paper on the cracking of ferritic steels in $CO_3^=/HCO_3^-$ solutions (Corrosion, 1972, 28, 313) makes this clear.

Comment by J.R. Pickens:

Could you please comment on the role of pitting in stress corrosion crack initiation, and has evidence of this mechanism been demonstrated in aluminum alloys?

Reply:

I see the primary role of pitting as allowing a localized change in environment that will sometimes lead to cracking, whilst the bulk environment is not conducive to such, as indicated in my paper. I am not aware of any detailed work on pitting leading to stress corrosion cracking in Al alloys but I see no reason why it should not occur in some circumstances. Pitting is more frequently associ- ated with crack initiation by corrosion fatigue, where there is work on Al alloys (L.V. Corsetti and D.J. Duquette, Met. Trans, 1974, 5, 1087), with similar indications of the significance of non-metallic inclusions as sites for pitting and subsequent cracking in steels

(J. Congleton, R.A. Olieh and R.N. Parkins, Metals Technology, to
be published).

Comment by J.R. Pickens:

Do you find any problems with the torsion/tension method to
discriminate between stress corrosion cracking and hydrogen embrittlement?

Reply:

.   The most appropriate way to discriminate between stress corrosion
cracking and hydrogen embrittlement is through electrochemistry, since
if the potential at the crack tip is not below that which will discharge hydrogen from the solution in the crack enclave I cannot see
how hydrogen can enter into the cracking process. I am doubtful of
the ability of torsion versus tension to discriminate unequivocally
between the two modes of failure, because at the atomic level involved
in crack growth the continuum approach on which the torsion/tension
argument is based is without consequence.

Comment by A.R.C. Westwood:

What would you say are the two or three most critical scientific
challenges in the field of stress corrosion cracking at this time?
Do you have any advise on how best they may be approached?

Reply:

Whilst the last two decades have seen a marked improvement in
the definition of the phenomenology of SCC, and that itself should
lead to more realistic mechanistic modelling, we still know very
little about the reactions that occur in the crack tip region whereby
crack growth occurs. The point is touched upon at various points in
my paper but may be underlined by a few of the many questions to
which we have no convincing answers, as yet. In relation to ferritic
steels, why is cracking associated with the formation of Fe_3O_4 films,
with or without other substances present, and immunity with the
formation of Fe_2O_3 films, especially when films similar to the latter
are associated with cracking in the austenitic steels? An how important is crystallography in relation to the latter, when the
addition of a few percent of Ni to a ferritic steel will promote
cracking, that fractographically appears indistinguishable from
cracking in austenitic steels, when both are caused to fail in $MgCl_2$
solutions? If dealloying is the essential embrittling step in transgranular cracking by dissolution, how does this relate to solution
specificity, especially when, for example, environments that cause
dezincification of brasses do not invariably cause SCC? If grain
boundary segregation is responsible for intergranular cracking, what
is the role of the particular species involved in the cracking re-

actions? Maximum crack velocities appear to correlate reasonably with
current densities passed at relatively bare surfaces, but what is the
surface state at the crack tip? These, and many other questions all
relate to our ignorance of the crack tip conditions, but I doubt if
the answers will lead to some all embracing model that explains all
instances of environment sensitive cracking. Our best hope of overcoming this ignorance would appear to me to lie in the application
of those relatively new techniques of surface analysis to guide
electrochemical measurements relevant to crack tip conditions.

Comment by J.O'M. Bockris:

My absence during this session leads me to make three points
which I hope may be included in the symposium.

1. Distinction between stress corrosion cracking and hydrogen
    embrittlement. Although it may be possible to have ac-
    celerated anodic dissolution of the crack tip, as shown
    by Despic, Raicev and Bockris, and the cathodic component
    entirely by O_2 reduction, there seems much evidence that
    many sytems of stress corrosion cracking involve not only
    a partial cathodic contribution by means of hydrogen
    evolution but that this evolution occurs near the crack
    tip (e.g., the photos shown in my lecture) and embrittles
    the tip. I believe, therefore, that one should reduce the
    importance of a distinction between SCC and HE. Often,
    they are coupled and one of the reasons for the increase
    of the anodic current is HE at the tip.

2. Crack initiation. I certainly think there is much to do
    here. However, as suggested by MacBreen in his 1965
    thesis, there will be very considerable stress augmen-
    tation at the emergence of screw dislocations on the
    surface and H present on the metal surface as a result of
    cathodic corrosion reaction will permeate into the metal
    at these stress points. Inside, it will find microvoids
    and open them. Some will reach the surface. Of course,
    this would only explain pit formation: it is conceivable
    that join-up of these embrittled areas occurs near the
    surface to initiate a crack.

3. Need for double layer data. Practically all our double
    layer data is on Hg and some on Pt. Virtually none is
    available on metals of metallurgical interest. Surface
    H concentration, double layer electric field strength,
    nature of the passive layer, adsorption of anions on metals
    or oxides are all necessary if our attitude towards stress
    corrosion cracking mechanisms is to become much less
    speculative. A general point must be added. The electro-
    chemical kinetics relevant is that for the semiconductor

solution interface in the universally present oxide–solution contact. Among a number of implications is that
the rate of corrosion may be light sensitive.

Reply:

1. I think that there are circumstances where hydrogen embrittlement and dissolution can occur together to promote
   cracking (references 5 and 12 of my paper), but I also
   think that they can occur quite separately. And for that
   reason I cannot agree that one should reduce the importance
   of a distinction between SCC and HE, by which you presumably mean that hydrogen is responsible for all cracking.
   I find it difficult to accept that the same mechanism of
   crack growth is operative in a C–Mn steel in a CO_3–HCO_3
   solution when two separate potential regimes are observed,
   one at relatively high potentials, with intergranular
   cracking, and another at low potentials where transgranular
   quasicleavage is produced, with a 300 mV range between,
   in which ductile fracture is produced. The upper cracking
   regime is predictable from anodic polarization curves and
   coincides with the potentials at which intergranular attack
   is observed on unstressed specimens; the lower cracking
   regime occurs at potentials below those for hydrogen discharge. There are many other systems that show essentially
   similar characteristics.

2. Perhaps the emphasis on fracture mechanics in the last
   decade or so, with the incorrect view that it avoids the
   initiation process, has reduced the activity that ought
   to have been spend on SCC initiation and I agree that there
   is room for additional work here. There are many ad hoc
   observations recorded in the literature of preferred sites
   for crack initiation, such as grain boundaries, as
   mentioned above, the interfaces between the matrix and a
   separated phase, as with non-metallic inclusions, or where
   slip steps emerge at a surface, all of which will promote
   initiation in appropriate environments in the absence of
   applied stress, although to sustain the dissolution a
   stress of appropriate magnitude is required.

3. I agree with you that relevant electrochemical measurements
   are one of the greatest needs if we are to understand SCC.
   Apart from a desire to see the electrochemist work on
   surfaces of engineerng interest, it would be equally
   valuable if those involved with the metallurgy or mechanics
   of SCC would give as much attention to defining the electrochemical conditions of their work as they currently do to
   defining the condition of the metal or the stress state.

Not to define the electrochemical conditions, at least in terms of potential and pH, is the equivalent of testing without knowing the state of stress or the composition or structure of the metal.

ON THE PROPAGATION OF TRANSGRANULAR STRESS-CORROSION CRACKS

E. N. Pugh

Chemical Stability and Corrosion Division
National Bureau of Standards
Washington, DC 20234

INTRODUCTION

Both inter- and transgranular stress-corrosion cracking (SCC) occur in aqueous environments, and it has been argued[1] that the mechanisms are basically different. There is broad agreement that the majority of intergranular cases occur by the film-rupture mechanism[2,3], that is, by preferential anodic dissolution at the crack tip where protective films are continually disarrayed by plastic deformation. Transgranular SCC, on the other hand, is poorly understood. It will be shown that the propagation of transgranular cracks involves discontinuous cleavage, but that the manner in which the environments induce this brittle mechanical behavior has not been established for most systems. The role of hydrogen embrittlement (HE) in these failures is critically examined.

FRACTOGRAPHY AND ORIENTATION OF FRACTURE SURFACES

Virtually all transgranular stress-corrosion fracture surfaces so far studied are closely similar in that they are cleavage-like in appearance, being characterized by flat, parallel facets which are separated by steps. The steps are generally crystallographic, exhibiting a serrated appearance, and range in height from several μm to dimensions barely resolvable with the SEM. These fracture surfaces have been fully illustrated elsewhere[1,4], and examples are shown in Fig. 1(a) and in later figures; the fracture surfaces are shown schematically in Fig. 1(b). Extensive SEM and TEM studies have established that the opposite fracture surfaces are matching and interlocking. The orientations of the primary fracture facets have been determined for several FCC and HCP alloys, and the orientations of the faces of the steps are also known in certain cases, Table 1.

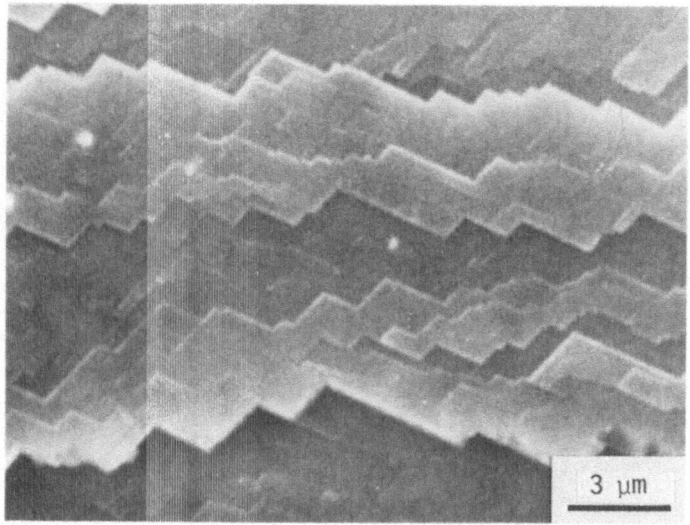

(a)

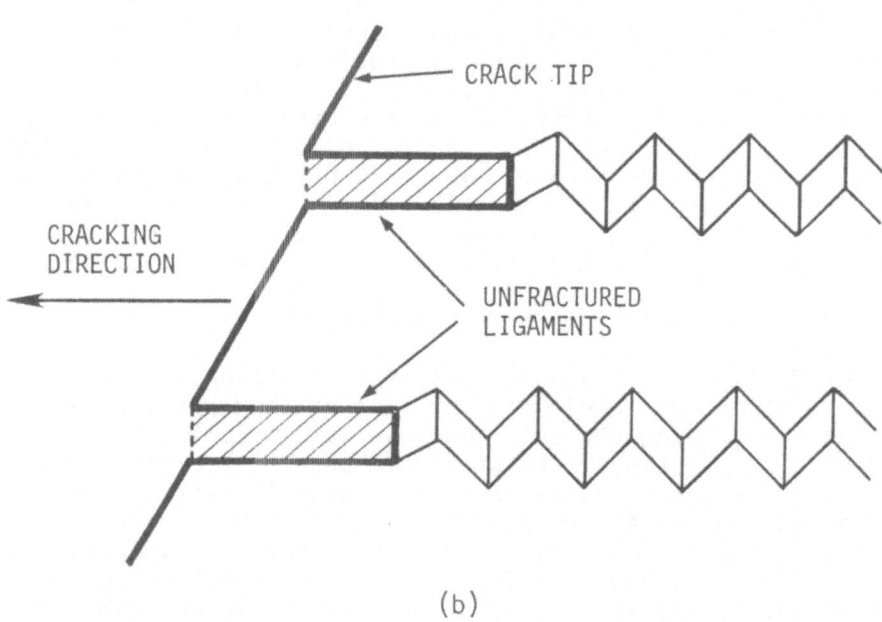

(b)

Fig. 1.   (a) Scanning electron micrograph of transgranular stress-
corrosion fracture surface in an Admiralty-Metal monocrystal tested in aq NH$_3$. (b) Schematic illustration of
transgranular fracture surfaces.

Table 1. Orientations of Transgranular Stress-Corrosion Fractures

System	Primary Facets	Faces of Surface Steps
Admiralty Metal/aq NH_3[5]	110	111
Stable[a] austenitic stainless steels/ aq $MgCl_2$(155 °C)[6,7]	100	Not known
Al-5.5 Zn-2.5 Mg/aq NaCl[5,8]	110	110
Mg-7.5Al/aq NaCl-K_2CrO_4[9]	$31\bar{4}0$	$31\bar{4}0$ and 0001
Ti-5Al-1Mo-1V/aq NaCl[10,11]	$10\bar{1}7$ (also $10\bar{1}0$)	--

[a]Stable with respect to the formation of martensitic phases. The formation of α'martensite in Type 304 steel complicates the fractography and the fracture surface is not (100)[6,7].

The surface steps are closely similar to those observed on conventional cleavage surfaces in brittle solids[1,4]. They are approximately parallel to the cracking direction, and form "fan" patterns when they emanate from a single source and "river" patterns when the crack front crosses a grain boundary. Undercutting is commonly observed at the steps. The steps on the stress-corrosion fracture surfaces differ from conventional cleavage steps in that the latter are not generally crystallographic in appearance, but this has been attributed simply to the difference in stress level operating in the two processes[4]. SCC generally occurs at low stresses, so that a low-energy process is necessary to produce the step. For the steps described in Table 1, those in Admiralty Metal are thought to be produced by shear[12], and those in the Al-Zn-Mg alloy[8] and Mg-7.5 Al[9] by SCC on secondary planes. Recent observations[13] suggest that step formation occurs at some distance behind the crack front as shown in Fig. 1(b). As the front advances, the stresses on the ligaments between adjacent primary surfaces increase until rupture occurs.

It will be seen below that the mechanism of cracking is best understood in the case of the alpha-titanium alloys. However, it is relevant to note that the fracture surfaces in this case differ from other cases of transgranular SCC in that the step heights are smaller and the steps do not display a crystallographic appearance[1].

EVIDENCE FOR THE DISCONTINUOUS NATURE OF CRACK PROPAGATION

The initial view that the propagation of transgranular stress-corrosion cracks is discontinuous was based largely on the observation of crack-arrest markings on the fracture surfaces in certain cases[4]. These markings, which delineate the successive positions

of the crack front, Fig. 2, are not always detected by SEM examination, but are observed under conditions where a relatively large
strain rate exists at the tip of the arrested crack, producing a
large discontinuity. Thus crack-arrest markings are commonly
observed in austenitic stainless steels tested under slow-strain-
rate conditions, and have been reported in alpha-brass tested
under constant load in the region near the transition to ductile
overload[1,4]. In the latter case, the markings became progressively
more difficult to detect with the SEM at areas corresponding to
lower stresses and hence smaller crack-tip strains. That the
presence or absence of crack-arrest markings is a problem of detection is underscored by the finding that they were not visible on
the fracture surfaces of Mg-7.5 Al bend specimens tested under
constant deflection using the SEM but were revealed by the use of
TEM replica techniques[9].

These considerations led to the use of a load-pulsing technique to facilitate detection of the crack-arrest positions on
transgranular fracture surfaces[12-16]. In this method, small load
pulses are superimposed onto an otherwise constant load during
crack propagation, accentuating blunting of the crack and permitting the corresponding positions of the crack front to be detected
with the SEM. The resulting fracture surface is represented schematically in Fig. 3, illustrating the relationship between crack-
front and crack-arrest markings, and fractographs showing examples
of crack-front markings are presented in Figs. 4 and 5. In

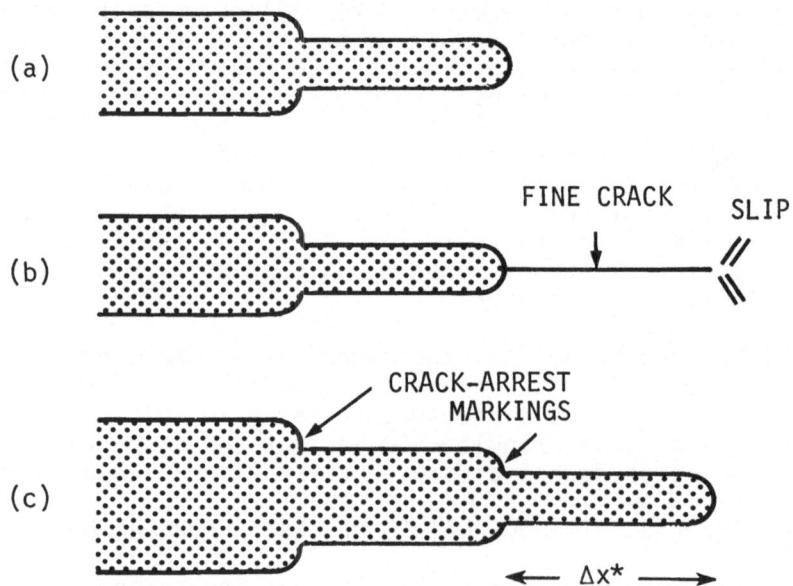

Fig. 2.   Schematic illustration of propagation of transgranular
           stress-corrosion crack.

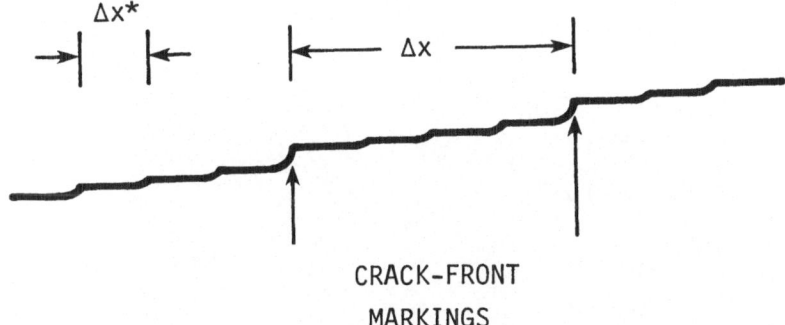

CRACK-FRONT

MARKINGS

Fig. 3.   Schematic of a section of a transgranular stress-
corrosion crack, showing the relationship between
crack-front and crack-arrest markings.

Fig. 5(a), the time interval between pulses was varied systematically
to establish that a one-to-one correspondence exists between pulses
and marking. Such a correspondence is essential when the markings
are used for crack-velocity measurements.

In Fig. 3, the time interval between pulses, Δt, is greater
than the time between crack-advance events, Δt^*. In this case,
the spacing between the imposed crack-front markings, Δx, is
greater than the crack-advance distance, Δx^*, so that several
crack-arrest makings exist between the crack-front markings. As
pointed out above, the crack-arrest markings are not generally
detectable with the SEM, corresponding to Figs. 4 and 5(a). However, at large crack-tip stresses both crack-arrest and crack-
front markings are visible, Fig. 5(b). An interesting case arises
when Δt becomes smaller than Δt^*, since, on the basis of the discontinuous fracture model, Δx would then be expected to attain a
limiting value, viz Δx^*. This has been confirmed for Type 310
steel in hot aqueous $MgCl_2$ (155 °C) in which cracking was in the
stress-independent region of the V-K curve[14,15]. Δt^* was found
to be $\sim$ 15 s and $\Delta x^* \sim 0.5$ µm; as would be expected, the number of
markings became less than the number of pulses for $\Delta t < 15$ s.

In the case of the brass-ammonia system there is direct evidence for the discontinuous nature of crack propagation. Edeleanu
and Forty[17], studying the movement of crack traces in monocrystal
bend specimens by means of optical microscopy, observed that the
crack advanced by the sudden appearance of a fine crack which then
slowly widened with time. This observation has been confirmed in
more recent work[13,16], and a typical sequence of events is shown
in Fig. 6. The study of crack traces by Edeleanu and Forty was
criticized on the grounds that surface-trace observations are not

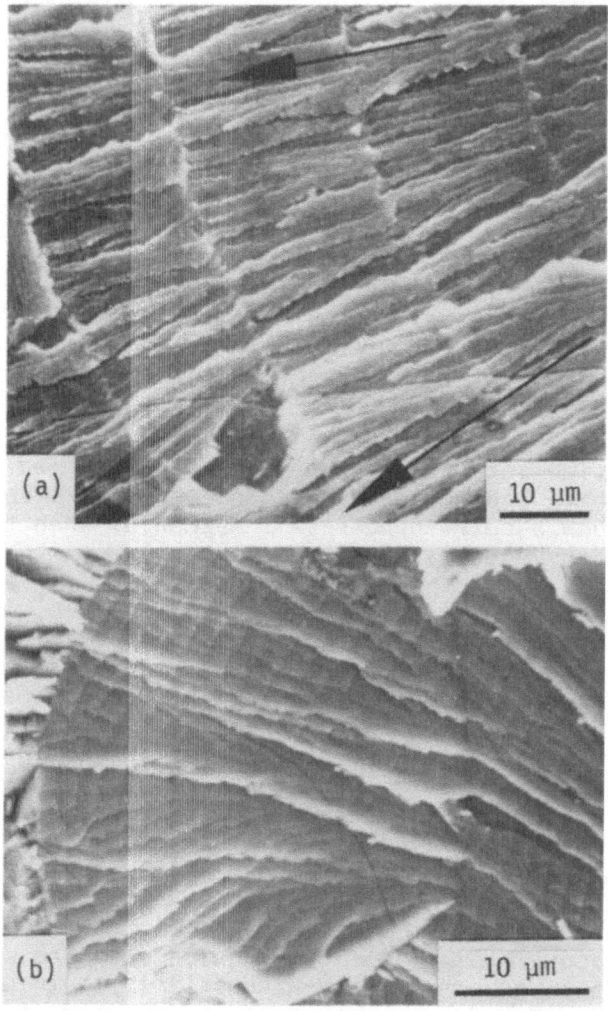

Fig. 4.   Scanning electron micrographs of transgranular stress-
          corrosion fracture surfaces showing crack-front markings
          introduced by load pulsing. (a) Admiralty Metal in
          aq NH_3; $\Delta t = 100$ s[1]. (b) Type 310 steel in aq $MgCl_2$;
          $\Delta t = 50$ s[14,15].

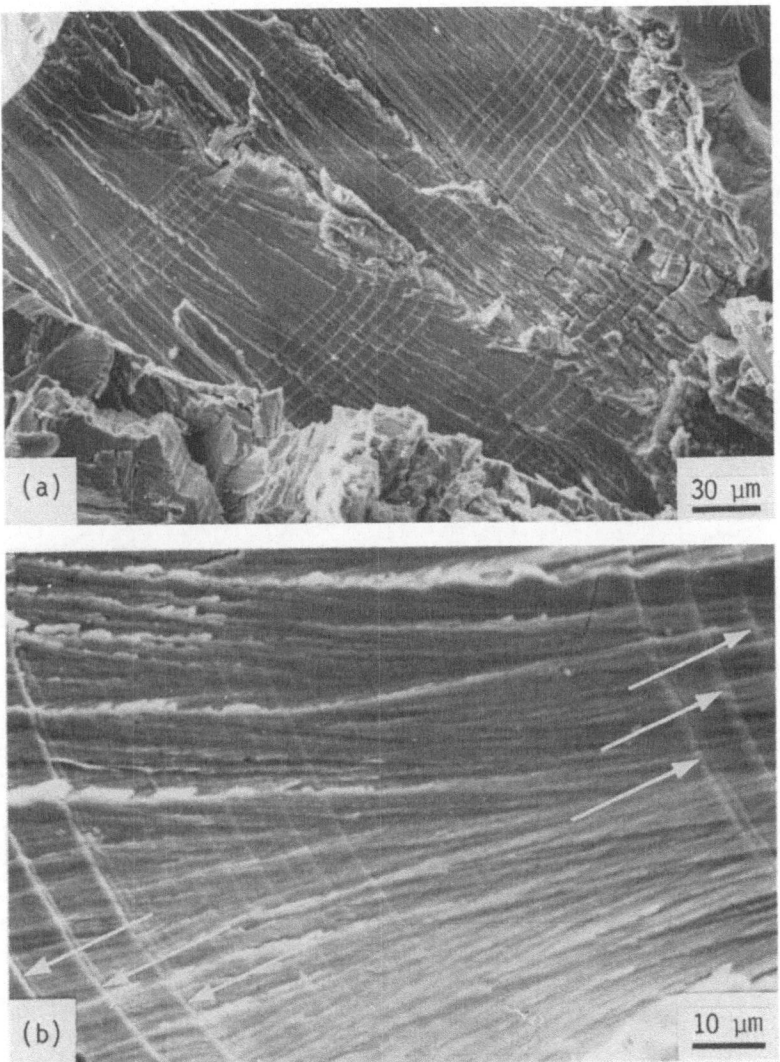

Fig. 5.   Scanning electron micrographs of transgranular stress-
corrosion fracture surfaces in Cu-30Zn monocrystals in
aq NH$_3$ showing crack-front markings.   Two values of Δt
were used in each case, <u>viz</u>. 8 s and approximately 80 s.
The crack-front markings are arrowed in (b); the numerous intervening markings were produced by crack arrest[16].

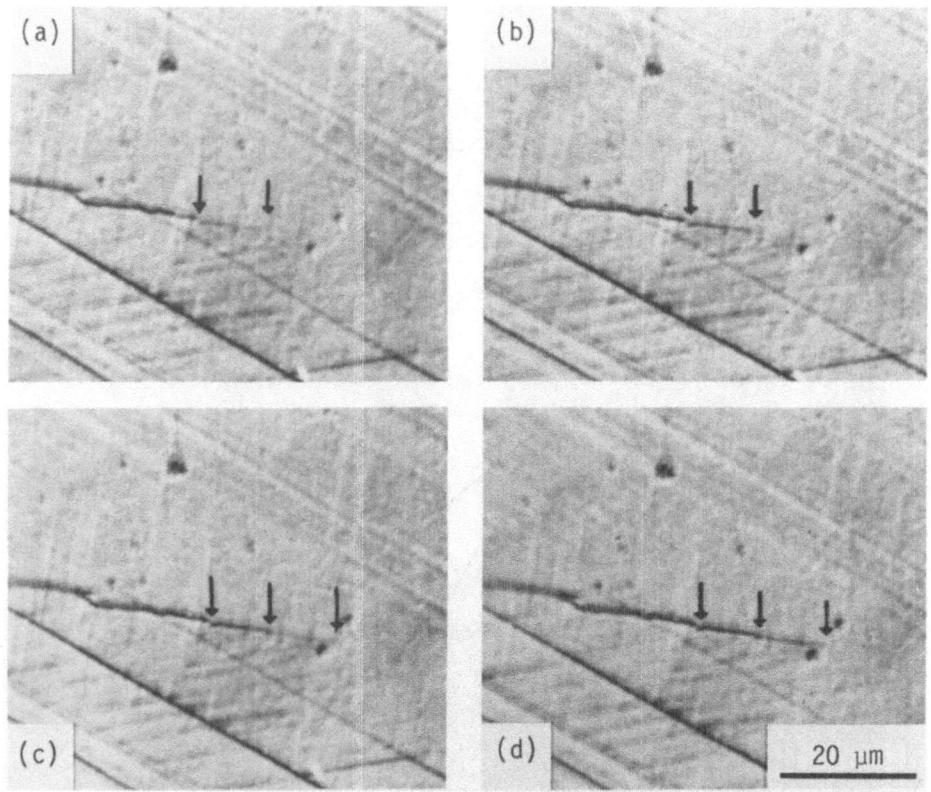

Fig. 6.   Optical micrographs of a crack trace in a Cu-30Zn mono-
           crystal bend specimen tested in aq NH_3, showing succes-
           sive positions of the crack tip (arrowed). Fine cracks
           are visible in (a) and (c), and they have widened in
           (b) and (d). Numerous slip bands are also evident[16].

representative of the behavior of sub-surface cracks. However,
the use of the load-pulsing technique in the recent studies established that the crack front shows little bowing, and extends continuously to the observation surface, so that the behavior of the
surface trace is in fact representative of sub-surface cracking in
this system.

     The crack-advance distance, Δx^*, observed in the crack-trace
studies depended on the stress, varying from 2 µm at large stresses
to 10 µm at low values; the corresponding values of the time interval between advances were 2 to 240 s[13,16]. These values of Δx^* are
consistent with the spacings between crack-arrest markings such as
those illustrated in Fig. 5(b). Anodic dissolution at the crack
tip can be expected to contribute to the observed crack blunting,
but slip bands near the crack tip were observed to become more
pronounced during the blunting process, indicating that creep also
plays some role. Another important observation, made by Edeleanu

and Forty[17] and confirmed in the recent studies, was that the
application of a load pulse immediately after a crack-advance
event did not cause immediate advance, whereas pulsing after a
short time delay (< Δt^*) caused advance. This observation
suggests that solid-state diffusion is necessary ahead of the
crack or that time is required to generate the necessary chemical
conditions in the solution at the crack tip.

The sudden appearance of the fine crack, fully formed, indicates that crack advance is extremely rapid. This is supported
by acoustic-envission studies which demonstrated that each advance
event is associated with a discrete signal whose duration, including ringdown, is $\sim$ 1 ms[13,16].

It is concluded that transgranular cracks propagate by discontinuous cleavage as shown in Fig. 2. Each crack-advance event
is assumed to be initiated at a point on the crack front and the
crack to spread laterally along the crack front[9,12]. Studies of
crack-arrest and crack-front markings indicate that the crack-
advance distance, Δx^*, remains uniform along the crack front, see
Fig. 5(b), and this would seem best rationalized in terms of a
bulk-embrittlement model[1,4], that is, by embrittlement of an uniform zone ahead of the crack front.

MECHANISM(S) OF FAILURE

In the case of the alpha-titanium alloy, Table 1, there is
strong circumstantial evidence that cracking results from hydrogen
absorption, specifically, by formation and rupture of titanium
hydride, TiH_2. Thus slow-strain-rate HE occurs in this material,
producing fracture surfaces indistinguishable from those for SCC[11],
and having the same orientation[18]. Further, both (10$\bar{1}$7) and (10$\bar{1}$0)
fracture planes correspond to known habit planes of TiH_2[19], and
the hydride has been detected at the fracture surfaces in both
SSRHE and SCC[11]. Thus it appears that SCC in alpha-titanium
occurs by the entry of cathodically-generated hydrogen into the
lattice, the formation of the hydride, and fracture through the
hydride or along the metal-hydride interface. The details of this
process have not been established. For example, it is not known
whether fracture follows immediately after nuleation of the first
hydride, or whether a stage of slow hydride growth exists. Further, the crack-advance distance is not known for this system.

The success of the hydrogen model for alpha-titanium makes it
attractive to extend it to the other cases of transgranular SCC.
With the exception of the alpha-brass, the other alloys of Table 1
undergo conventional HE. Hydride formation is well establbished in
the magnesium alloy, and hydrides have been reported for the
aluminum alloy[20] and for austenitic stainless steels[21]. Moreover,
hydrogen is commonly generated during exposure to the aqueous

environments which cause SCC in each case. The hydrogen model is particularly attractive in the cases of the alpha-brass and the austenitic stainless steel, since the bulk diffusivities of hydrogen in both systems can readily account for the hydrogenation of layers of depths corresponding to the observed values of Δx^* in the observed times Δt^*[16].

Despite the attractiveness of the hydrogen model, it faces a number of difficulties. In the Mg-7.5Al alloy, the transgranular fracture surfaces produced by both internal and gaseous hydrogen do not resemble those produced by SCC, and their orientation is not $(31\bar{4}0)$[22]. In austenitic stainless steels, the hydrogen hypothesis is opposed by the well-known experiments of Wilde and Kim for Type 304 steel in aqueous LiCl solution at 146 °C[23]. These showed that no cracking occurred at cathodic potentials where hydrogen absorption took place, whereas transgranular SCC was observed at anodic potentials where no hydrogen absorption was detected. No satisfactory rationale has been put forward to reconcile these observations with the hydrogen model.

The alpha-brasses provide a greater challenge to the HE model, since there is no evidence for the occurrence of conventional HE in these alloys, nor does it appear possible to generate hydrogen in the ammoniacal solutions. The possibility of hydrogen formation by some specific reaction involving NH_3 or NH_4^+, such as dissociative adsorption, has been made irrelevant by the observation that transgranular SCC is observed in a number of non-ammoniacal aqueous solutions[24-26], including those containing the oxyanions SO_4^{2-}, NO_2^-, NO_3^-, ClO_3^-, and MoO_4^{2-}. Anodic polarization was found to be necessary to produce SCC in many of these cases[26], further weakening the hydrogen hypothesis. It should be noted that anodic polarization accelerates cracking in the ammoniacal solutions[27], so that anodic dissolution at the crack tip appears to play a role in the failure.

The serious difficulties faced by the HE model lead to the question of alternative mechanisms by which cleavage-like mechanical fractures can be induced. In terms of the bulk-embrittlement approach, no other species can be identified which have a sufficiently large diffusivity. For example, dezincification has been suggested as the cause of embrittlement in the case of the brasses[28], but the depth of such a layer cannot approach values such as 10 μm. Thus, by such a mechanism, dealloying might permit crack initiation, but the bulk of the propagation would have to occur in the non-embrittled lattice, suggesting that (110) is the "normal" cleavage plane for alpha-brass. The way in which dezincification might induce cleavage is not understood, although Forty[28] suggested that it might involve associated vacancy injection and subsequent vacancy-dislocation interaction. In this regard, it should be noted that restricted cross-slip appears to be necessary for the cleavage-like failures, since they are not observed in alloys with zinc contents less than ~ 15 percent where the stacking fault energy is large and where cross-slip occurs; similarly, transgranular cracking occurs in alpha-phase Cu-Al alloys (low SFE) but not in Cu-Ni alloys (high SFE)[29]. The austenitic steels which exhibit transgranular cracking also display coplanar slip[30].

Another possible mechanism which has been proposed as a generalized model[31] and specifically for the brass-ammonia system[32] is that based on adsorption. This model, also advocated for liquid-metal embrittlement[33], proposes that specific aqueous species adsorb and interact with strained bonds at the crack tip, reducing bond strength and permitting brittle fracture at low stresses. This model cannot be ruled out, but faces a number of difficulties[1]. Rather than conventional brittle fracture, it would appear to predict relatively slow propagation at a rate determined by the transport of the critical species to the crack tip.

It is apparent that our basic understanding of this technologically important set of environmentally-induced fractures is poor, and that this area requires attention from a group such as this.

REFERENCES

1. A. J. Bursle and E. N. Pugh: Environment-Sensitive Fracture of Engineering Materials, p. 18, TMS-AIME, Warrendale, PA, 1979.
2. F. A. Champion: Symposium on Internal Stresses in Metals and Alloys, p. 468, Inst. of Metals, London, 1948.
3. H. L. Logan: J. Res. Natl. Bur. Stand., 1952, vol. 48, p. 99.
4. A. J. Bursle and E. N. Pugh: Mechanisms of Environment Sensitive Cracking of Materials, p. 471, The Metals Society, London, 1977.
5. J. L. Nelson and J. A. Beavers: Met. Trans. A, 1979, vol. 10A, p. 658.
6. R. E. Reed and H. W. Paxton: First International Conference on Metallic Corrosion, p. 301, Butterworths, London, 1962.
7. R. Liu, N. Narita, C. J. Altstetter, H. K. Birnbaum, and E. N. Pugh: Met. Trans. A, 1980, vol. 11A, p. 1563.
8. J. L. Nelson: Ph.D. Thesis, University of Illinois at Urbana-Champaign, 1976.
9. D. G. Chakrapani and E. N. Pugh: Met. Trans. A, 1975, vol. 6A, p. 1155.
10. D. A. Meyn and G. Sandoz: Trans. TMS-AIME, 1969, vol. 245, p. 1253.
11. G. H. Koch, A. J. Bursle, R. Liu, and E. N. Pugh: Met. Trans. A, in press.
12. J. A. Beavers and E. N. Pugh: Met. Trans. A, 1980, vol. 11A, p. 809.

13. D. V. Beggs: Ph.D. Thesis, University of Illinois at Urbana-
    Champaign, 1981.
14. M. T. Hahn and E. N. Pugh: Corrosion, 1980, vol. 36, p. 380.
15. M. T. Hahn and E. N. Pugh: Fractography and Materials
    Science, p. 413, ASTM STP 773, 1981.
16. D. V. Beggs, M. T. Hahn, and E. N. Pugh: Proc. Symposium on
    Hydrogen Embrittlement and Stress-Corrosion Cracking, Case
    Western Reserve University, June 2-3, 1980, in press.
17. C. Edeleanu and A. J. Forty: Phil. Mag., 1960, vol. 5,
    p. 1029.
18. D. A. Meyn: Rept. NRL Prog., September, 1969, p. 34.
19. N. E. Paton and R. A. Spurling: Met. Trans. A, 1976,
    vol. 7A, p. 1769.
20. S. W. Ciaraldi, J. L. Nelson, R. A. Yeske, and E. N. Pugh:
    Proc. Third Internat. Conf. on Effect of Hydrogen on Behavior
    of Materials, Jackson Lake, WY, August 26-29, 1980, in press.
21. Z. S. Smialowska: Private communication, May, 1981.
22. D. G. Chakrapani, G. H. Koch, and E. N. Pugh: Proc. Sixth
    Internat. Cong. on Metallic Corrosion, Sydney, Australia,
    December 3-9, 1975, in press.
23. B. E. Wilde and C. D. Kim: Corrosion, 1972, vol. 28, p. 350.
24. T. R. Pinchback, S. P. Clough, and L. A. Heldt: Corrosion,
    1976, vol. 32, p. 469.
25. A. Kawashima, A. K. Agrawal, and R. W. Staehle: J. Electro-
    chem. Soc., 1977, vol. 124, p. 1822.
26. A. Kawashima, A. K. Agrawal, and R. W. Staehle: Stress
    Corrosion Cracking--The Slow Strain-Rate Technique, p. 266,
    ASTM STP 665, 1979.
27. E. N. Pugh and J. A. S. Green: Met. Trans., 1971, vol. 2,
    p. 3129.
28. A. J. Forty: Physical Metallurgy of Stress Corrosion Frac-
    ture, p. 99, Interscience, New York, 1959.
29. E. N. Pugh, J. V. Craig, and W. G. Montague: Trans. ASM,
    1968, vol. 61, p. 468.
30. P. R. Swann: Corrosion, 1963, vol. 19, p. 102 t.
31. H. H. Uhlig: Physical Metallurgy of Stress Corrosion Frac-
    ture, p. 1, Interscience, New York, 1959.
32. H. H. Uhlig, K. Gupta, and W. Liang: J. Electrochem. Soc.,
    1975, vol. 122, p. 343.
33. N. S. Stoloff: This volume.

DISCUSSION

Comment by R.H. Jones:

Your proposed mechanism of brittle-cleavage like fractures in
Cu-Zu alloys during anodic dissolution by monovacancy and di-vacancy
generation could be substantiated by doing some transmission electron
microscopy to look for vacancy clusters. From room temperature

(20°-30°C) neutron irradiation of copper and nickel, it is known that significant yield strength increases occur only when the interstitial and vacancy concentrations agglomerate to form clusters. Cluster densities on the order of $10^{16} cm^{-3}$ result in a factor of 2 increase in the yield strength of copper and nickel irradiated and tested at room temperature. Therefore, if the dezincification-vacancy gener-ation mechanism can produce a sufficient vacancy and di-vacancy concentration, at distances of 0.1 to 1μm ahead of the crack, to produce clusters, these clusters would harden the material and should be measurable by TEM. Of course, such an observation is not sufficient to demonstrate that a factor of 2 or so increase in the yield strength is sufficient to produce brittle-cleavage like fracture in Cu-Zu, but it would be supportive of your hypothesis.

Reply:

    No reply is necessary, except to thank Dr. Jones for this helpful suggestion.

Comment by R.A. Oriani:

    Externally applied stress pulses give striations that resemble naturally occurring striations. But is that enough to conclude that the SCC crack advances sporadically? For this to be concluded one would have to believe that the mechanism by which both kinds of striation are produced is the same. What is known about that? If the mechanism is indeed the same, it would seem that pulse produced striations should interfere with the frequency of occurrence of the natural striations. If no such interference occurs, I would guess that the two kinds of striations have different mechanistic causes. (Perhaps the pulses produce pumping of electrolyte and through this the markings are produced.)

Reply:

    We do consider that the crack-front markings revealed by load pulsing and crack-arrest markings are produced in the same way, namely, by crack blunting. One might expect that load pulsing would influence the normal crack-advance process, as you suggest. However, our work on the transgranular SCC of austenitic stainless steels and alpha-brass indicates that pulsing does not significantly influence either the crack-advance distance, Δx^*, or the time interval between crack-advance events, Δt^*, for cracking in Stage II of the V-K curve, i.e., when the crack velocity is independent of the stress intensity. On the other hand, Δx^* and Δt^* are observed to be influenced by load pulsing under conditions where V is stress-dependent. The precise relationship between Δx^* and K has not been established, but it appears to resemble that reported by Nelson (Met. Trans. A, 1976, vol. 7, p. 621) for the gaseous hydrogen embrittlement of Ti-6Al-4V. Nelson showed that the crack-advance distance decreased with increasing K in Stage I, was essentially constant in Stage II, and again, decreased with increasing K in Stage III. The significance of this similarity between transgranular SCC and GHE has not been established.

Concerning your final comment, well-defined crack-front markings are produced on the fracture surfaces of high-purity Al-5.5Zn-2.5Mg undergoind slow crack growth in laboratory air. Pumping of electrolyte is clearly not relevant in this case.

THE EMBEDDED DISK PRESSURE TEST (DPT) : A SENSITIVE

TECHNIQUE TO INVESTIGATE MATERIALS EMBRITTLEMENT

J.P. Fidelle, R. Arnould-Laurent

BP. 511, 75752 Paris Cédex 15, France

INTRODUCTION

We describe a mechanical test to characterize embrittlements of materials submitted to inhomogeneous stresses and strains. After a theoretical mechanical analysis supported by experimental data, its scope is envisionned. Finally subsidiaries of the basic standard test are considered as well as their potentialities.

DESCRIPTION OF THE TECHNIQUE (Fig. 1).

In general, disks clamped at their edge are ruptured under a generally increasing gas pressure ; one compares rupture pressures obtained either with reference (e.g. neutral gas) or embrittling conditions. The ratio of the rupture pressures supplies quantitative data on resulting damage. The rupture appearances also inform on embrittlement.

Besides of the standard prequalification test performed at a constant pressure increase rate ($\Delta p/\Delta t$) for a given category of materials, the studies generally carried out involve the influence of temperature (-196 to + 1000°C) and that of $\Delta p/\Delta t$ (0.5 bar.h^{-1} to 50,000 bars.mn^{-1}).

MECHANICAL ANALYSIS

Means of finite elements one has evidenced a number of practical results, which are compared to experimental determinations : deflection or local elongation as a function of pressure. Altogether this provides with a better understanding of the technique sensitivity. Main features are :

1) The disk top is plane-stressed (similarity with a thin sphere),

2) The anchorage region is plane-strained,

3) The onset of plastification appears very early as compared with the rupture pressure,

4) The hydrostatic component is very high at the anchorage,

5) When the general plastification regime is obtained, the strain rate keeps constant until the end of the test,

6) Moreover a triaxial stress stade builds up, right at the beginning of the test, at odds with other tests on smooth specimens.

These reasons explain why the test is especially sensitive to phenomena involving plastification and to embrittlement of materials by interstitials such as hydrogen (H) or oxygen (O).

INVESTIGATION SCOPE

At odds with other techniques, any material ranging from most ductile to brittle is a priori able to be investigated simply with this method. It is also applied easily to non-metallics such as plastics, glasses, etc.,

In a like manner, damaging conditions can be a variety :

• Internal embrittlements or intrinsic brittlenesses : low-temperatures brittleness : ductile to brittle transition curves can be traced. Embrittlements due to cold work, heat treatments, welding ... internal O or H absorbed during a wealth of circumstances. Some are well-known, other more insidious ; corrosion, electrochemical forming, machining ... are especially well evidenced thanks to the high surface to volume ratio for disks.

• Environment embrittlements or degradations : solvents, machining fluids, corrosive media, wet pressurized gases, H_2, liquefied gases e.g. NH_3. Instances of Stress Corrosion Cracking. HGE of disks can be fruitfully compared to the blistering of pressurized cavities.

When materials and/or environments are first hand little known, the influence of strain rate is systematically tested (SRT). For, the transport kinetics of an embrittling species and earlier - if relevant - that of its absorption, are critical factors ; similarly, the kinetics of surface depassivation by plastic strain and those of repassivation by adsorption of impurities present in the embrittling fluid.

Fig. 2 on Hydrogen Gas Embrittlement (HGE) shows the influence of :

- H diffusivity D_H in case of 2 HS martensitic steels :

        D (medium alloy steel) ① > D (maraging steel)

- temperature   for HS steel ① : D (-60°C) < D (RT)

- strength level and correlatively that of H embrittlement (HE)

<u>sensitivity</u> for steel ① heat treated to 2 strength levels :
          HS steel ① / Medium Strength (MS) steel ①
- <u>crystallographic structure</u>, acting on D_H and HE sensitivity (σ)
          FCC Co alloy (D_H 250°C, σ) < (D_H RT, σ) martensitic steel

The curves display a maximum sensitivity area, which corresponds to an optimum for the H-dislocations interaction. The rupture pressure minimum is not far from - but never below the D.F.L. (Fig. 3). Fig. 4 shows that the mitigation of HGE at low strain rates can become very important in case of materials very sensitive to surface recontamination by impurities of the gas.

As the area of maximum sensitivity is identified, one uses this rate range for further investigation such as screen tests on new batches of same material or quality control for products to be exposed to H_2 under stress, during service.

The comparison of Disk Pressure Tests (DPT) with the materials behavior during service life has afforded the specification of a quality criterion either for a material or a specific environment :

---

pHe/pH must be <u>unambiguously</u> lower than 2

---

pHe is the rupture pressure under helium of a standard size disk of uncharged material.

pH, determined in the $\Delta p/\Delta t$ range for maximum sensitivity, is
- either the rupture pressure under helium of an H-charged material
- or the rupture pressure under gaseous H_2, or an H generating fluid.

By comparing laboratory data with the satisfactory behavior or failure of materials in service, it has been shown that the ratio 2 is a landmark between the areas where materials susceptibility is acceptable and where it is too high.

Of course the position with respect to this landmark not only depends on materials, but also on embrittlement conditions such as pressure, purity and temperature of a gas.

Therefore in a test program, at first it is desirable, if possible, to run standard rupture tests (under an increasing gas pressure at RT) to compare materials with the some 130 ones already inventoried the same way : Al,Ag,Au,Be,Co,Cr,Cu,Fe,Mo,Mn,Ni,Pd,Pt,Ta, Ti,U,Y,Zn,Zr base materials.

Next, if necessary, tests will be carried out under conditions corresponding to envisionned service conditions (either less or more severe).

If the ratio is then >2, it is mandatory not to use under an appreciable stress :
- either a given H-charged material, if H content cannot be lowered
- or a given material in a contemplated environment : environment is rejected ...
- or a proposed material in a fixed environment : material is rejected ...

.... if one cannot improve the environment (f.i. with inhibitors),
or the metal (f.i. by heat or surface treatments)

If the ratio is somewhat larger than 2, but still lower than 2.3,
utilization can be tentitavely tried, if specific requirements moti-
vate it, but must be submitted to suitable conditions of various kinds
(e.g. modification of service conditions : pressure, length of cycles,
outgassings ...).

Finally it is easy to study the synergistic effect or the competition between various kinds of embrittlements or brittlenesses.

SUBSIDIARY OR PARENT TESTS

- Permeation-diffusion tests on same specimens
- Static Fatigue (delayed failure) tests : They are also well suited
  to the instance where H is generated by a corrosion reaction and to
  the assay of protective coatings.
- Low-cycle Fatigue : Disks are submitted alternatively to a differential pressure, at an adjustable frequency. Thus one compares the
  t.t.f. of materials submitted to embrittling and reference conditions.
- Center-notched disk (CND) : Then the aggressive medium is located
  downstream. Specimens are thicker. They can simulate a wall submitted to pressure or plain stress and an aggressive environment (hull
  of a ship, tank, etc.,). Of course, this CND can be used for non-
  symetrical low cycle fatigue tests.
  Tests on high cycle-fatigue center precracked disks simulate
  the case of walls including defects. Tests can be easily run with
  radioactive materials or gases and can be supplemented by autoradiography.

CONCLUSIONS

The existence of a plane strained zone at the anchorage and the
very small ratio of this pressure Π at the onset of plastic flow to
the rupture pressure, make the DPT especially sensitive to the effect
of aggressive media on materials and to evidence brittleness.
Among most important applications, let us remind :
  - materials selection
  - choice of their making
  - heat treatments
  - joining
  - surface treatments
  - qualification of various media to use in presence of a given
    material.

The test is also well adapted to materials recipe.

Finally the possibility of using minidisks helps to make the
DPT a tool for failure analysis, and the study of welded zones.

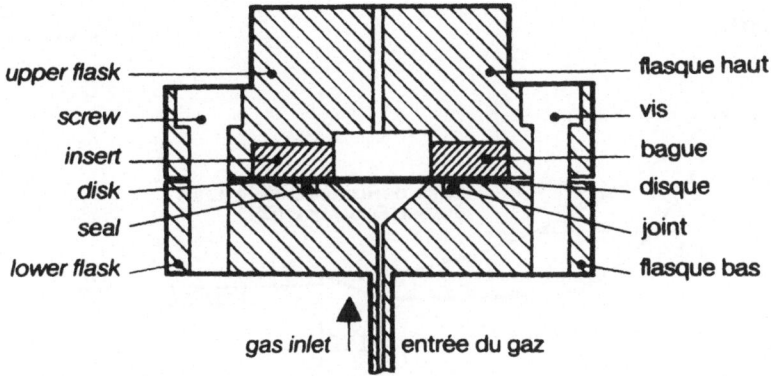

Fig. 1:　Basic Rupture Cell BF 58

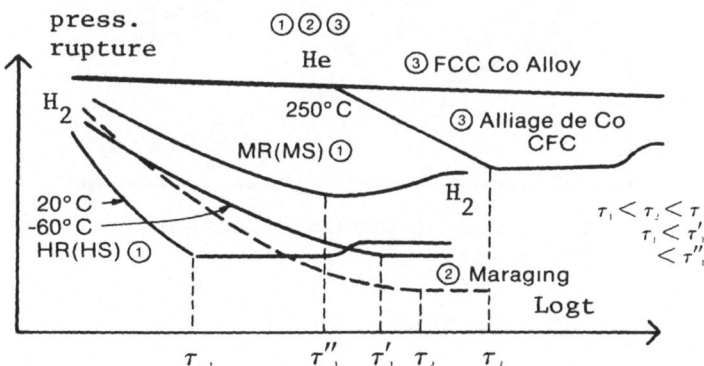

Fig. 2:　Influence of the test duration or $\frac{\Delta P}{\Delta t}$ (pressure increase rate) on rupture pressures.

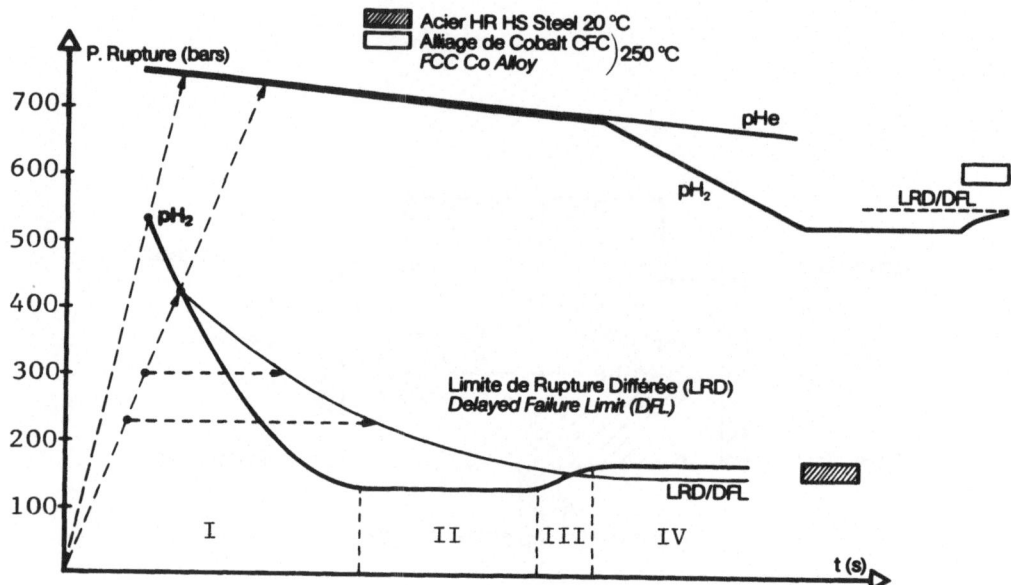

Fig. 3:   Influence of test length and relationship
          with DFL.

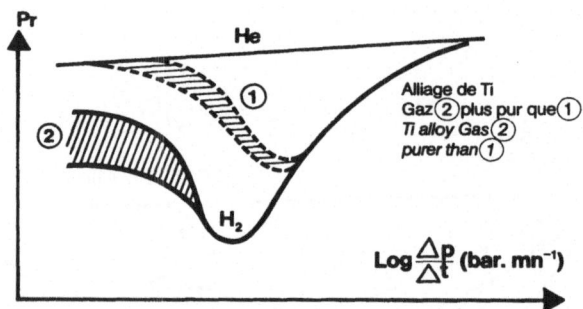

Fig. 4:   Influence of pressure increase rate and
          gas purity.

REFERENCES of work by J.P. Fidelle et al.

1. "H in metals" CEA,91680, Bruyères-le-Châtel,Fr.(1969)pp.131-172.
2. "HE testing" ASTM STP 543 (1974)pp.31-47, 221-253, 267-272.
3. "The Disk Pressure Tests" CEA, B-1e-Ch. (1975) 112 pages.
4. Paper 3D1 in Proc. Internat.Cong.on "H in metals". Paris (1977)
   Pergamon.
5. "H in metals" JIMIS-2 (1980) pp.233-236, 505-508.JIM,Sendai,Japan.
6. "Effect of H on behavior of mtls". A.W.Thompson and I.M.Bernstein,
   ed., AIME (1976) pp.91-101, 507-515, (1981) to appear.

DISCUSSION

Comment by J.O'M. Bockris:

Much has been said at this meeting about the crack tip and I have persistently advocated concentrating experimental methods to determine there the H concentration. However, I should like to point out that, for pure Fe, single crystal Fe, and a few steels, there exist thermodynamic data from the work of my coworkers (Beck, MacBreen, Subramanium) at the University of Pennsylvania in the late 1960's. We showed that the H permeation increased with elastic tensile stress and decreased in compressive stress. The dependence of H solubility on stress was shown to correspond to a $\bar{V}_H$ (for pure Fe) of around 2.67 cc mole^{-1}. No variation of the diffusion coefficient with stress could be detected.

The point of this is that it gives an indication of what hydrostatic stress does to the solubility of H at the crack tip. Neglecting the effects on the local stress due to the presence near the tip of dislocations, but taking into account the effect of crack length and radius of curvature, it is easy to show that the concentration of H very near the crack tip surface can be 10^4-10^5 times the solubility under standard conditions and at zero stress. This is conceptually very important. We should not think of all the marked effects of crack spreading, and embrittlement, as being due to 1 atom of H among 10^7 atoms of iron. It is more like 1 in 100 - or maybe 10 in 100 (taking into account stress due to dislocations). The consequences are considerable. For example, it is not impossible, utilizing these concepts, to understand the effect of 10^{-8} Ats of H on the spread of cracks. Metal-metal bond weakening and subsequent decohesion becomes a reasonable model.

Reply:

Certainly, and the non-equilibrium situation we have when dislcoations drag H isotopes in the proper strain rate range, can only make this segregation even more significant.

H$_2$S ADSORPTION ON PLASTICALLY STRAINED IRON

M. T. Thomas, D. R. Baer and R. H. Jones

Pacific Northwest Laboratory*
P.O. Box 999
Richland, Washington   99352

The purpose of this paper is to present data showing an increased adsorption rate for H$_2$S on plastically strained iron.
There has been considerable speculation in the literature concerning
the effect of strain on enhanced corrosion and surface reactivity,
but there has been no direct experimental evidence that strain does
increase surface reactivity. To understand the possible effects of
strain enhanced adsorption on crack initiation and growth is the main
motivation for these studies. It is also possible that strain enhanced adsorption may be important for catalytic, corrosion and wear
properties of materials.

It is important to consider the influence of strain on adsorption in evaluating any theory of embrittlement as has been pointed
out by Sieradzki and Ficalora[1]. Models of environmentally enhanced
crack growth usually involve either atom transport to the surface or
a reaction at the surface as the limiting step[2]. Therefore, the influence of strain on surface reactivity must be considered for the
models to be realistic. Enhanced adsorption in the regions of local
strain could also be important in promoting crack initiation.

There have been indications in the surface science literature
for many years that surface defects substantially alter surface
reactivity[3,4]. However, only recently have systematic studies of
this phenomena been initiated[5]. Most of these studies consist of
examining carefully prepared single crystal surfaces that have a
known step orientation and density[6]. In addition, there have been

*Operated by Battelle Memorial Institute for the U.S. Department of
Energy under contract DE-AC06-76RLO 1830

experiments reported that showed changes in gas adsorption on surfaces with defects produced by ion sputtering[4,5]. Although these
experiments are helpful in formulating models relating surface chemical activity to defects, they do not involve all of the possible
effects associated with plastic strain. This present study was
undertaken to determine directly the influence of strain on adsorption.

EXPERIMENTAL

    Both the specimen straining and H_2S adsorption were done in situ
in the work chamber of a Model 545 Scanning Auger Microprobe, Perkin
Elmer, Physical Electronics Division. The specimens were prepared
from zone refined iron foil that had been vacuum annealed at 850°C
and given a mechanical polish. After mounting on a specially designed straining stage and insertion into the Auger spectrometer, the
specimen surfaces were cleaned by ion sputtering. Sputtering must be
kept to a minimum to prevent sputter damage from masking the effect
due to straining. The straining stage was attached to a side mount
specimen manipulator, allowing the straining stage to be positioned
in front of the electron energy analyzer (Fig. 1) Research grade
H_2S was introduced on the specimen by means of a small capillary
tube. The gas dosing and straining were performed with the specimens
in the Auger analysis position. Intensities of the Auger signals
from S_{150}, Fe_{45} and Fe_{703} were monitored as a function of time, producing adsorption curves as shown in Fig 2. A large portion of the

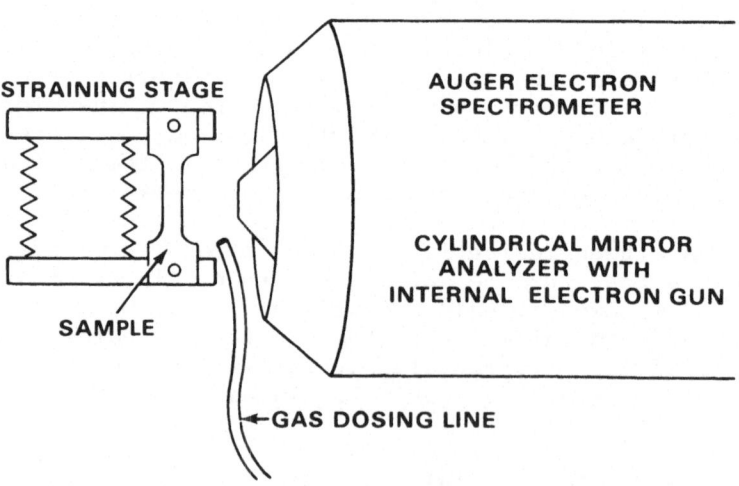

Fig. 1.   Schematic drawing showing relative positions of sample,
           straining stage, analyzer, and gas doser.

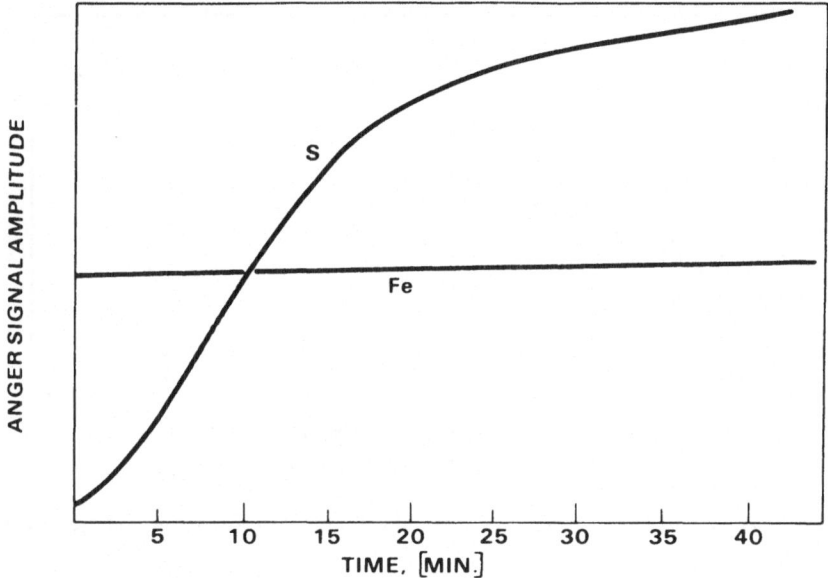

Fig. 2.   Amplitude of sulfur and iron Auger peaks as a function of
          time during H_2S dose.

adsorption curve was linear with time, so the slopes of this linear
region were used to determine the adsorption rates.

RESULTS AND DISCUSSION

     The adsorption rate of H_2S on polycrystalline iron was found to
increase as a function of strain (Fig. 3). Specimen fracture
occurred at roughly 12% full load or just off the scale of the
graph.

     These results are believed to be the first direct measurements
of increased adsorption rate due to plastic strain. However, similar
increases in adsorption rate have been reported for other specimen
treatments. It is well known, for example, that specimen reactivity
can be greatly enhanced at points where defect lines emerge from the
surface[7]. Sulfur, in fact, is observed to react preferentially at
some defect sites[8] and nucleation of two-dimensional phases often
occurs at these defects. Increased adsorption rates are reported for
H_2S on uniformly stepped surfaces generated on single crystal
materials[6]. Similar results are reported for oxygen adsorption on
sputtered surfaces[5]. Finally, Ciftan and Ruck[9] indicate that mechanical stress can significantly alter the chemical potential of
adsorbed atoms, thereby influencing the adsorption rate. Thus,

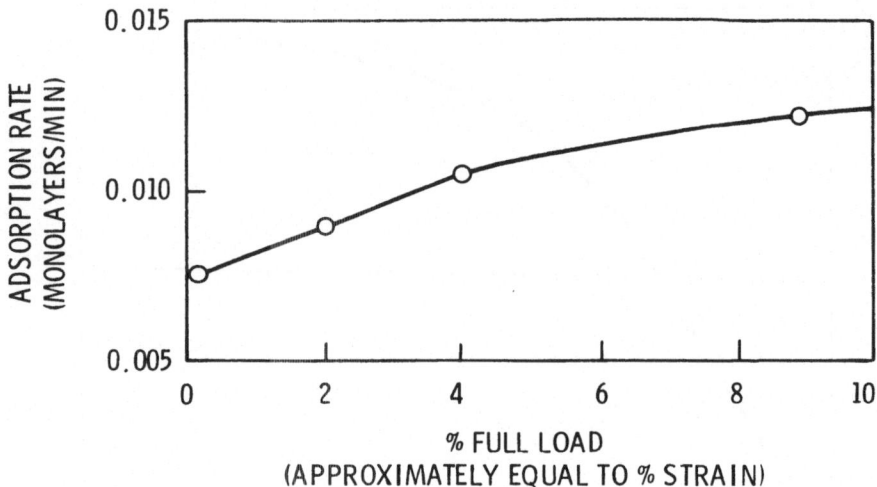

Fig. 3.   Adsorption rate of H_2S on iron as a function of strain.

emerging defect lines, surface steps, elastic strain and displaced
atoms can all increase the adsorption rate. Many of these phenomena
are present on our strained specimens and probably occur also at
crack tips.

Although one might intuitively accept the fact of enhanced
adsorption on defected surfaces, especially in the light of the
above results, a detailed understanding of the phenomena is difficult. For example, there are H_2S adsorption models that postulate
a two-step process with a precursor state[10]. It is not known if
strain enhances adsorption by altering adsorption kinetics, by
changing the number or density of a particular adsorption site, by
producing new adsorption sites, or by encouraging other reactions
to occur. The details of the reaction will determine the final
state of hydrogen and sulfur and thus how the properties of the
material are influenced.

CONCLUSIONS

We have shown directly that plastic deformation increases H_2S
adsorption on iron. These results suggest two important points:
1) adsorption will be enhanced in the deformed region of a crack tip
and, 2) enhanced adsorption at the site of a localized surface defect
may promote crack initiation. Before complete understanding of
strain enhanced adsorption is obtained, additional experimental
studies must be done. Such questions as what is the effect of surface orientation on the increased adsorption rate and by what

phenomena does strain enhance the adsorption must be answered.

ACKNOWLEDGMENTS

This work was supported by the Division of Materials Sciences, Office of Basic Energy Sciences, U.S. Department of Energy.

REFERENCES

1. K. Sieradzki and P. Ficalora, The dependence of the fracture toughness of 4340 steel on an external chlorine gas environment, Scripta Metallurgica. 13:535 (1979).
2. T.W. Weir, G.W. Simmons, R.G. Hart and R.O. Wei, A model for surface reaction and transport controlled fatigue crack growth, Scripta Metallurgica. 14:357 (1980)
3. R.K. Sherburne and H.E. Farnsworth, Activation of a solid nickel catalyst for the hydrogenation of ethylene by heat treatment in a high vacuum, J. Chem. Phys. 19:387 (1951)
4. M.T. Thomas, G. Shimaoka and J.A. Dillon, Jr., Lattic defects and surface properties of clean germanium, Surface Science. 6:261 (1967)
5. R. Miranda, J. Ibañez, J.M. Rojo and M. Salmeron, Influence of argon bombardment on the reactivity of (110) platinum with oxygen, J. Chem. Phys. 72:6614 (1980); R. Miranda, J.H. Rojo and M. Salmeron, Enhancement of oxidation in nickel (001) surface bombarded with argon ions, Solid State Comm. 35:83 (1980)
6. H. Wagner, Physical and chemical properties of stepped surfaces, in: "Solid Surface Physics," G. Höhler, ed., Springer-Verlag Berlin Heidelberg (1979).
7. G.G.T. Guarini and R. Spinicci, Dislocations and their role in the thermal dehydration of $Ba(ClO_3)_2 \cdot H_2O$, in: "Reactivity of Solids: Proceedings of 8th International Symposium on Reactivity of Solids," J. Wood, O. Lindquist, C. Helgesson and N. Vannerberg, ed., Plenum Press, New York (1977).
8. J. Oudar, Metal surfaces: structure and initial stages of reactivity, International Metals Reviews. 228 (1978).
9. M. Ciftan and V. Ruck, The adsorptive chemostress effect, Phys. Stat. Sol. 6. 95:237 (1979).
10. M.R. Shanabarger, Kinetics of H_2S adsorption onto Fe films, in: "Proceedings of the 4th International Conference on Solid Surfaces," Cannes, France (1980).

WORKSHOP SUMMARY: STRESS CORROSION CRACKING

Pepared by: E.N. Pugh

Discussion
    Leaders: E.N. Pugh, P. Neumann, J.C. Scully, B.E. Conway

    Recorders: D.J. Duquette, M.T. Thomas, P. Azou, T.E. Fischer

Two main topics were discussed:

The Mechanism of Transgranular SCC

This was identified as an area requiring particular attention.
It was felt to be poorly understood in comparison to the intergranular case, which was generally considered to occur by a film-rupture
and anodic-dissolution process. The predominant view was that
transgranular SCC proceeds by discontinuous cleavage, and attention
was focused on possible mechanisms by which interaction with the
aqueous environments could induce brittle mechanical fractures.
There has been continuing controversy in recent years concerning
the role of hydrogen in stress-corrosion failures. The consensus
here was that certain cases of transgranular SCC may be due to
hydrogen embrittlement (HE), but that hydrogen cannot be responsible
for all instances, e.g., alpha-brass in aqueous ammonia and
austenitic stainless steels in high-temperature water. Thus alternative mechanisms of cracking were discussed.

The possibility that dealloying or selective dissolution plays
an important role was discussed at length, particularly in the case
of alpha-brass. Vacancy injection associated with dezincification
has long been suggested as a key process, possibly due to vacancy-
dislocation interactions. The brasses which undergo T-SCC have low
stacking-fault energy (zinc contents greater than ∿15 percent), so
that the dislocations are extended and cross-slip restricted. Some
attention was given to the possibility that vacancy-induced climb
of the partials might increase the critical resolved shear stress.
There is evidence for hardening during recovery of cold worked
brasses, but it was generally felt that the magnitude of this effect
is too small to permit the initiation of cleavage. The proposed
hardening effect also is difficult to reconcile with the observation
that considerable blunting occurs during periods of crack arrest.
It was pointed out that the latter observation is consistent with
the report that anodic dissolution of alpha-brass accelerates creep
at room temperature, an effect also attributed to dezincification
and vacancy injection. The assumption that either selective
dissolution or restricted cross-slip are necessary conditions was

1025

also challenged, since it has been reported that unalloyed copper
can undergo transgranular SCC. However, it is not clear whether
the pure metal fails by the cleavage-like process observed in the
alloys. It was also noted that elements added to brasses to combat
dezincification, e.g., Sn, As, and Sb in Admiralty Metal, do not
provide protection against T-SCC.

The behavior of vacancies in the crack tip region was also
discussed. Little information exists on the effect of the stress
field on vacancy motion at room temperature, and this was seen as
an area requiring attention. The question of whether vacancy motion
occurs by single defects or as divancies was considered and the
latter was felt to have little relevance to the situation at the
crack tip. The possibility that vacancy motion might be the rate
determining step led to the subject of activation energies, and the
recognition that there have been no detailed studies of the temperature dependence of transgranular crack propagation.

Analogies with other embrittlement phenomena were brought up.
The use of LME to model T-SCC was considered, but the consensus
was that it probably would not be useful and that each process needs
to be examined separately. Similarities with the complex-ion
embrittlement of silver chloride were noted. This failure can be
transgranular, and the fracture surfaces also exhibit crack-arrest
markings. This effect has been attributed to hardening of the
lattice by large concentrations of Frenkel defects induced by the
adsorption of highly charged complexes, and thus bears some
similarity with the vacancy-injection model discussed for metallic
alloys. The suggestion was made that ionic transport in corrosion
product films may be relevant to SCC, and that adsorption of charged
species from solution might play an important role.

Some attention was given to the behavior of dislocations in the
vicinity of the crack tip, and to the question of whether crack
blunting can occur by homogeneous nucleation of dislocations. This
subject, covered at length by Thomson in this volume, emphasized
the need for further experimental data. In particular, TEM studies
of the surface layers at the fracture faces and of the crack tips
were recognized as being urgently required.

One group queried the view that the propagation of transgranular
cracks is discontinuous. The main evidence for discontinuous
propagation is fractographic and metallographic, and it was felt
that independent experimental evidence is required. Monitoring of
acoustic emission or sensitive temperature measurement were two
methods suggested for obtaining independent evidence. The significance of the serrated steps on transgranular fracture surfaces was
also identified as requiring further experimental study.

No conclusions were reached concerning the orientation of the
fracture surface in the alpha-brass/ammonia system, viz. (110). It
was suggested that studies of transgranular SCC in beta-brass might
be fruitful. The fracture surfaces in these BCC alloys (in water)
exhibit the characteristic cleavage-like appearance, but their
orientation has not been determined. Glide is also planar in these
alloys because of ordering. Copper-gold was identified as a useful
system for mechanistic studies, since it undergoes both selective
dissolution (of copper) and transgranular SCC. The dislocation
behavior in this case can be varied by changing the degree of order.

Chemistry and Electrochemistry Within Cracks

Several groups devoted time to this general topic. It is
generally accepted that the potential, pH and overall solution
composition within cracks can be quite different from bulk values,
and discussion here centered on the ways in which these local changes
are developed. Taking into account the small volume of solution
near the tip of an advancing crack, it was recognized that local
anodic dissolution must give rise to large mass-transport effects.
Moreover, the localization of cathodic reactions to regions external
to the cracks in many cases will set up potential differences, the
magnitude of these IR drops depending on crack geometry, conductivity
of the solutions, and other factors such as the formation of bubbles
within the cracks. Other processes discussed included film formation, adsorption of species on newly formed surfaces, and the
relaxation of the double layer charge distribution. The influence
of stirring effects due to the opening or advancing of the cracks
was also discussed.

While few specific suggestions were made, it was strongly
recommended that further work be directed to defining potential and
compositional gradients within cracks. Both experimental and
theoretical studies were considered necessary. The detailed
calculations of Doig and Flewitt were specifically cited, and it
was recommended that those be extended to more systems, and should
include pH variation. Comparison of crack propagation near the
specimen surface and in the interior was suggested as a means of
studying solution changes.

Another area of discussion was the ratio of atoms which are
mechanically perturbed and which are actively involved in the
crack-advance process to those involved in other Faradaic corrosion
processes within the crack. This question was seen to be relevant
to the important practical matter of the distribution of the Faradaic
anodic current through the crack, specifically to the relation
between active dissolution at the crack tip and current flow
through passive films on the walls.

Miscellaneous Topics

1.   Attention was given to atomistic modeling of dissolution and passivation processes, i.e., involving kinks, ledges, etc., of the type carried out in the 1960's by Conway and Bockris, Despic and Bockris, or involving dislocations as done by Despic and by Budewski. More work is necessary to investigate the role of hydration of the emerging metal ion and its coordination by complexing ligands, nucleating hole defects, etc., at the corroding metal. Little fundamental work appears to exist on the effects of stress on the dissolution process.

2.   Model experiments with well-defined electrochemical systems, the use of single-crystal surfaces, for example, were advocated to study the relationship between mechanical perturbation and both metal oxidation and hydrogen evolution.

3.   The possible significance of low-valence unstable inter-mediates in corrosion of electro-passive metals and related cracking phenomena, e.g., species such as Mg^+, Zn^+, Al^{+3}, or solvated electrons transiently produced.

4.   The need to study the initiation stage of SCC was emphasized. Propagation is more tractable experimentally and this has led to the neglect of the initiation process(es). The identification of sites of crack initiation was seen as important in specific industrial applications. The use of precracked specimens in slow strain rate experiments was recommended.

5.   Further work was thought to be necessary to define the stress states and the local structure (precipitates, etc.,) at grain boundaries. The role of grain boundary misorientation in intergranular SCC has received surprisingly little attention and was seen as a fruitful area of research.

SUMMARY SESSION

Chairman: A.R.C. Westwood

ATOMISTICS OF FRACTURE,

A CONFERENCE SUMMARY

A. S. Argon

Massachusetts Institute of Technology
Cambridge, MA   02139

INTRODUCTION

Without doubt, some of the most challenging and still unsolved
problems in fracture today are of the environmentally affected variety. Thus, when the opportunity presented itself to me to give
the summary lecture of this conference probing the atomistic aspects
of environmentally affected fracture and attended by a handpicked
collection of luminaries of impressive interdisciplinary dimensions,
I accepted the task without hesitation – on the hope that I would
be forced to listen carefully to all the presentations and educate
myself in this field.

The conference fulfilled its promise in every respect and became a stimulating learning experience to me, and I think, to most
others. The short summary that I shall furnish here will make no
attempt to repeat any part of the material covered very ably by the
principal lecturers and the many contributors. For this, the reader
is advised to consult the individual contributions. I shall try to
give instead a synthesis of the topics of this conference from my
own perspective, referring to the specific authors only occasionally to give a direct connection. Thus, I apologize in advance to
the many authors who I shall not directly refer to in my summary.
This is no reflection on the quality or importance of their work,
which in all cases was uniformly and remarkably high, but merely
a necessity for keeping the summary short. As consolation to these
heartbroken friends, I can assure them that summaries of conferences
are practically read by no one, hence this omission is of no consequence at all.

THE FRACTURE INSTABILITY

Fracture is in general a macro phenomenon in which energy balances on a global scale are of concern. Thus, the engineer has found it quite effective to characterize fracture largely in terms of its "driving forces" that can be related directly to readily measurable quantities such as applied stresses and crack lengths, and characterized by stress intensity factors, special contour integrals, etc. In this approach, the crack tip free energy production rates or the dissipation rates are not usually demanded but are fixed by fracture instability experiments where they are just equal to the "driving forces". This non-atomistic approach is very often adequate provided the performance of the material, its fracture toughness to be specific, shows only gentle variations with temperature, rate of loading, and is insensitive to the environment.   Unfortunately, there exist many situations where this is not the case, where dramatic reductions in toughness are brought about by changes in these factors. Under such circumstances, a more mechanistic and searching approach to the crack tip processes becomes essential that often, it is felt, must go down to the scale of atoms. Needless to say, this increases the complexity of the problem manyfold, requiring not only a continued appreciation of the macroscopic nature of the fracture problem but also considering in detail the collective nature of the atomic processes at the crack tip - often in parallel with the plastic dissipation processes.

In view of the above it is instructive to be reminded of the scale of the fracture problem. Thus, consider a simple case of a semi-brittle solid in which crack advance is accompanied by only very limited plasticity. The change in free energy ΔG with and without a penny-shaped crack of radius c in the solid under a tensile stress σ is

$$\Delta G = 2\pi c^2 \chi - \frac{4\pi c^3}{3} \left(\frac{\sigma^2}{2E} \right) \quad , \tag{1}$$

where χ and E are the surface free energy and Young's modulus respectively. For limited plasticity, the additional plastic dissipation associated with the relief of high crack tip stresses can be included by generalizing the surface free energy into the fracture separation work, as done first by Orowan [1]. At the fracture instability, the critical crack length c^* maximizes the free energy and results in

$$c^* = \frac{2E\chi}{\sigma^2} \quad , \qquad \Delta G^* = \frac{8\pi}{3} \frac{\chi^3}{E^2} \left(\frac{E}{\sigma} \right)^4 \quad . \tag{2a,b}$$

The fracture separation work χ can be given in terms of the modulus E and the atomic dimension a, as $\alpha E a$, where α ranges from about 2.5×10^{-2} for a purely elastic solid to 10 or larger with plastic dissipation present at the crack tip. Thus, at the point of fracture instability, the change in the free energy ΔG^* of the overall system, for a representative stress, $\sigma = 10^{-3} E$ ranges from 7.5×10^8 eV for a purely elastic solid to a value several orders of magnitude larger for a solid with some plastic dissipation. Hence, it is easy to conclude that: a) at the fracture instability, very large amounts of energy are being converted from one form into another, and that in this process thermal assist is inconsequential in the case of "dry" fracture, without the effect of a strength degradation process. We note further, that even in the case of a strength degradation process, where the fracture separation work χ is reduced by equilibrium segregation of an embrittling species along an internal interface, along the path of the crack, prior to the fracture, this conclusion still holds. In this case, for the same stress level, the critical crack length for fracture instability is considerably shorter (in the ratio of the contaminated to the uncontaminated interface energy), and the overall free energy change ΔG^* will have been very markedly decreased (by the cube of the same ratio). The overall magnitude, however, will still be many orders of magnitude above the range in which thermal assist is possible. Thus, we conclude that in either case of uncontaminated or fully contaminated and embrittled interfaces, time dependent crack growth is not possible at any measureable rate – even though the fracture strength for the same crack length may have very markedly decreased in the case of the contaminated interface. This leads us to a profound conclusion before even considering any mechanism, atomistic or otherwise, that the study of equilibrium properties of clean or contaminated interfaces can not give any useful information on the rates of crack growth. Such studies are, however, of great value in connection with another problem: the understanding of ductile vs. brittle behavior of cracks, changing the behavior of a material from inherently ductile to inherently brittle. Thus, if time dependent crack growth is observed in a solid, it must be concluded that: a) a strength degradation process operates at the crack tip allowing the crack to advance at the rate at which the degradation front advances; and b) that either the degrading agency, deposited on the crack faces, prevents the subcritical crack from healing, or that an accompanying time dependent plastic separation process at the crack tip separates the crack flanks and prevents them from interacting. In these cases it becomes essential to concentrate attention at the crack tip degradation processes and their rates of advance in the crack tip field. This conclusion, albeit from a chemists viewpoint and with somewhat different emphasis, is advanced by Whitesides [2] in his contribution.

COHESION, DE-COHESION

    In the simplest case of a brittle solid, or an embrittled pre-
viously ductile solid where little or no plastic deformation accompanies the extension of cracks, the cohesive strength is the most
important property. Although considerable strides have been made
in the computation of equilibrium cohesive energies for lattices
of different type, heats of formation, magnetic properties, etc.
of many metallic solids by quantum mechanical approaches [3], with
very few exceptions [4], these developments have not been carried
out far enough to calculate cohesive resistances of solids leading
to cohesive strengths. To remedy this situation, a widespread
practice has been to calculate cohesive strengths from pair potentials, tailored to fit elastic constants, heats of formation,
stacking fault energies, phonon spectra, etc. - in spite of the
fact that the quantum mechanical computations have established
that the cohesive properties of many metals, and particularly transition metals, can not be represented by pair potentials [3]. In
view of this state of affairs, even more elementary approaches relating the cohesive strength to bulk properties of elastic constants, lattice parameters, and surface free energy still find
widespread use [5] and often give remarkably good results [4].

    The problem of loss of cohesion due to the presence of segre-
gated foreign species at interfaces is now receiving some attention at the same level as the fundamental quantum mechanical computations of equilibrium properties of solids referred to above.
Although these computations, discussed by Johnson [6] in his contribution, have been instructive and have shown some distortions
of interactions of certain impurity atoms in their immediate surroundings of host atoms, actual computations of loss of cohesion by
these techniques are no closer at hand than computation of cohesive
strengths in pure substances. Clearly, this area is fertile and
in need of further development.

    The study of equilibrium segregation to surfaces and inter-
faces, and the measurement of the effect of such segregation on
their free energies, is of great value here, where the elementary
concepts based on bulk properties, referred to above, can furnish
very useful answers on loss of cohesion. The very attractive advances in the area of segregation of certain impurity species to
surfaces and interfaces are discussed by Guttman [7] and their
consequences on reduced cohesion by Seah and Hondros [8] in their
contributions respectively.

    It is in this connection of severe embrittlement due to loss
of cohesion, where a usually ductile metal is turned into one of
quite brittle behavior, that the principal atomistic problem in
fracture of inherent ductility or inherent brittleness at a crack
tip arises. As formulated first by Kelly Tyson, and Cottrell [9],

and in a more refined way by Rice and Thomson [10], the state of stress at the tip of an atomically sharp crack is considered to conclude whether the cohesive strength is reached ahead of the crack to give brittle propagation before ideal plasticity can be initiated by punching out dislocation loops to introduce crack blunting and ductile behavior. In its simple form, treating the crack tip region as a continuum and resorting to line formalism of dislocation properties, it has been possible to decide unambiguously that pure f.c.c. metals are inherently ductile while many covalent solids are inherently brittle. To decide on the behavior of transition metals that can cleave and to assess the effect of segregated impurities at interfaces, atomistic models become necessary to represent both the non-linear elasticity near the crack tip and the competition between cleavage and crack blunting. Computer models using pair potentials have been developed for some time to deal with this problem with mixed results [11]. Their development is reviewed by Dienes and Paskin [12] in their contribution. Further developments in this area will be very fruitful and decisive if they use <u>ab-initio</u> computations of cohesive resistances in pure substances, and the computational developments for reduced cohesion due to impurity atoms treated as small clusters in the host atom surroundings under large stresses at the crack tip. Many problems of intergranular embrittlement due to prior segregation of cohesion impairing species such as sulfur, phosphorus, antimony, arsenic, etc., that are known to produce temper embrittlement in steel, are likely to be quantitatively understood through such developments.

It must be emphasized that the Rice-Thomson problem of crack tip de-cohesion vs. crack tip blunting is basically of an athermal nature – even with segregation present – provided that the segregation is previously accomplished. Under these conditions, the advance of the crack by brittle cleavage requires the satisfaction of the Griffith condition for a structureless but contaminated reference continuum [1,13], while the nucleation of a dislocation loop requires reaching locally the ideal shear strength where again thermal motion plays only a very small role in reducing the elastic constants of the solid. Although lattice trapping has been advocated by Thomson [14] as a mechanism for thermally activated advance of a crack, it is quite likely that this effect is fully randomized in a real solid where the crack tip samples many lattice orientations along its length [1,13]. Furthermore, some additional reflection based on arguments of the type presented above under the fracture instability can show that any thermally activated growth of a brittle crack over the lattice traps can only occur <u>at the Griffith stress</u> if the second law of thermodynamics is not to be violated [1,13]. Naturally, when a cohesion reducing species diffuses into the crack tip material from the crack flanks, slow advance of a crack is possible, governed by the rate of advance of such a diffusion front. This brings us to the subject of rate mechanisms of crack advance.

RATE MECHANISMS OF CRACK ADVANCE

In many instances, crack advance is known to occur in a time dependent manner at stress levels substantially less than the Griffith level for the pure solid. In all these cases, as remarked already, the time dependent sub-critical crack growth is associated with one or more fundamental processes of strength reduction such as: a) a cohesion reducing species is diffusing into the crack tip in the crack tip field of concentrated stresses; b) the strength of the crack tip material is being degraded by diffusive cavitational processes of vacancy aggregation, formation of gas bubbles, etc.; c) crack flank contamination, crack flank dissolution or crack tip plastic blunting is preventing crack healing and is making progress of further degradation possible. In the sub-category of processes where cohesion impairment results from segregation of an impurity species on an interface in the extension of the crack tip, well defined conditions of upper and lower critical stress intensity exist for the clean and fully contaminated interface. This had been well appreciated for decades in connection with the static fatigue problem of glass [15], but has been re-emphasized more recently by Rice [16,17] and by Hirth and Rice [18] who have also suggested methods of analysis of time dependent crack growth by time dependent contamination of the path of the crack. In all such instances, a more definitive understanding of the crack advance and its rate depends on the understanding of the activated process at the crack tip. The importance of this point is being made by Whitesides [2] in his contribution by drawing many parallels to chemical reactions at surfaces. Although this point is clear and its importance is comprehended by most workers in the field, the complexity of crack tip processes has not permitted the development of very many definitive mechanisms for phenomena such as hydrogen embrittlement, liquid metal embrittlement, stress corrosion cracking. In spite of this, much impressive progress has been made in these areas and was discussed during the conference.

Hydrogen Embrittlement

It has been suggested earlier by Orioni [19] that embrittlement due to hydrogen results from a reduction of cohesion as in the cases of temper embrittlement. Although this remains a possibility, there has been no new and definitive measurement that is in support of it. In an intriguing experiment, Vehoff and Neumann [20] however, have observed very substantial reductions in crack tip angle in fully plastic ductile crack opening experiments in Fe 3% Si alloy single crystals when the hydrogen partial pressure is increased. Although the effect is dramatic and qualitatively in support of a reduction of cohesion, the explanation is not clear.

On the other hand, there have been several important developments that were discussed at length during the conference. Thus, considerable evidence now exists that hydrogen is transported by dislocations during their motion from the surface inward which affects the hydrogen penetration rate into the crack tip region. Hirth and Johnson [21], however, show that previous estimates of the development of large supersaturations of hydrogen are erronious and that only small supersaturations develop.

Birnbaum [22] reports that in many cases of hydrogen embrittlement, hydride formation has been observed and that such hydrides are associated with premature cracking and thereby advancing the crack. Whether the formation of the hydrides ahead of the crack is aided by hydrogen transport, by moving dislocations arriving from the crack tip, is not clear. Even though this mechanism of crack advance through the formation of a brittle and possibly also fragile phase is a good candidate for modeling of hydrogen embrittlement, it is not widespread enough to be general. In other cases, some evidence exists for transient formation of hydrides at the crack tip stress field, possibly being aided by some crack tip shears. These hydrides disappear upon the passage of the crack, but can nevertheless play an important role as "activated states" promoting crack advance as an extension of the mechanism of crack growth by the formation of stable brittle hydrides. There are, however, other cases of hydrogen embrittlement reported by Birnbaum where no hydrides have been found, requiring a different explanation altogether.

Liquid Metal Embrittlement

The rapid and dramatic fracturing of some metals in the presence of certain liquid metals is reviewed in depth by Stoloff [23] in his contribution. Many factors have been identified that help characterize the phenomenon, but the mechanism of this type of fracture remains unclear.

A controversial but intriguing detail is furnished by Lynch [24] who has found dimples on a very fine scale on the surfaces of liquid metal induced fractures. Although no certainty exists for the dimples not to be after-the-fact artefacts, they suggest a cavitation phenomenon at the crack tip. Whether this is due to complex slip intersection processes as suggested by Lynch or by ideal cavitation due to high short range stresses resulting from a misfit wedge of the penetrating liquid metal remains in the realm of speculation.

Stress Corrosion Cracking

A prevalent mode of rapid crack advance with little observable plastic flow is stress corrosion cracking, where crack tip dissolution is often a prominent factor. There is much in solution

chemistry that is relevant to some forms of this problem which is
reviewed by Convay [25] and Bockris [26] in their contributions,
and is transferred to the crack tip region by Pourbaix [27] and
Scully [28].

The rates of crack advance by stress corrosion have been
measured for many alloys in a number of relevant aqueous environments potentiostatically, and in the context of fracture mechanics,
relating the crack growth rates to the stress intensity factor and
establishing threshold stress intensity factors K_{ISCC} below which
no crack growth is found. Yet much of the description of the crack
tip separation processes, including dissolution under stress or
during deformation, remains qualitative and even speculative.

The specific phenomena of stress corrosion cracking are reviewed in detail by Parkins [30] who discusses the electro-chemical
dissolution aspects of stress corrosion cracking, its relationship
to pitting, the factors that make cracks propagate intergranularly,
transgranularly, or along heavy shear bands. All this gives a rich
picture and useful information for engineering application, but
falls short of providing definitive understanding. Nevertheless,
a picture emerges that indicates that the phenomenon is broad and
encompasses several mechanisms.

CONCLUSIONS

The most positive accomplishment of the conference was the
exchange of divergent views between conferees of different disciplines. These exchanges have established that while in the simple
processes of fracture involving cleavage, and in some cases of intergranular embrittlement, there is a direct connection to atomic
processes of crack tip separation vs. crack tip blunting, in most
other instances of hydrogen embrittlement, stress corrosion cracking,
and liquid metal embrittlement, the fracture process involves large
scale deformation, diffusion, dissolution, and transport processes
during the advance of the crack. Thus, few of these latter problems, if any, have a simple atomistic mechanism that can be solved
some fine day by the friendly physicist, once the proper boundary
value problem can be stated. On the other hand, many of the current approaches have been quite valuable in outlining the nature
of the problem. Thus, approaches based on mechanics have characterized the driving forces; those based on measurement of the
segregating species and their kinetics at grain-boundaries and at
free surfaces have given the worst conditions; and finally, those
based on electro-chemistry have stated the ranges where dissolution
plays a role and where other processes must govern. These have all
served to characterize the phenomena, have indicated the ranges
where certain mechanisms are likely to govern, and have outlined
the nature of the needed studies. It is very likely that when
these milestones are reached, this conference will be identified

as having been of crucial importance in the definition of the
problems.

ACKNOWLEDGEMENTS

The author's research on fracture in creeping alloys is
supported by the U.S. Department of Energy under Contract No.
DE-AC02-77ER04461.

REFERENCES

1. E. Orowan, in "Reports of Progress in Physics", Physical
    Society: London, vol. 12, p. 185 (1949).
2. G. M. Whitesides, in this volume.
3. D. G. Pettifor, in this volume.
4. E. Esposito, A. E. Carlsson, D. D. Ling, H. Ehrenreich, and
    C. D. Gelatt, Jr., Phil. Mag., $\underline{41}$, 251 (1980).
5. N. H. Macmillan, in this volume.
6. K. H. Johnson, in this volume.
7. M. Guttmann, in this volume.
8. M. P. Seah, and E. D. Hondros, in this volume.
9. A. Kelly, W. R. Tyson, and A. H. Cottrell, Phil. Mag., $\underline{15}$,
    567 (1967).
10. J. R. Rice and R. Thomson, Phil. Mag., $\underline{29}$, 73 (1974).
11. P. C. Gehlen, and M. F. Kanninen, in "Inelastic Behavior of
    Solids", edited by M. F. Kanninen et al, McGraw-Hill: New York,
    p. 587 (1970).
12. G. J. Dienes, and A. Paskin, in this volume.
13. A. S. Argon, submitted to Scripta Met.
14. R. Thomson, in this volume.
15. E. Orowan, Nature, $\underline{154}$, 341 (1944).
16. J. R. Rice, in "Effect of Hydrogen on Behavior of Materials",
    edited by A. W. Thompson and I. M. Bernstein, AIME: New York,
    p. 455 (1976).
17. J. R. Rice, J. Mech. Phys. Solids, 26, 61 (1978).
18. J. P. Hirth, and J. R. Rice, Met. Trans., $\underline{11A}$, 1501 (1980).
19. R. A. Oriani, and P. H. Josephic, Acta Met., $\underline{25}$, 979, (1977).
20. H. Vehoff and P. Neumann, in this volume.
21. J. P. Hirth and H. H. Johnson, in this volume.
22. H. K. Birnbaum, in this volume.
23. N. S. Stoloff, in this volume.
24. S. P. Lynch, in this volume.
25. B. E. Convay, in this volume.
26. H. J. Flitt and J. O'M. Bockris, in this volume.
27. M. Pourbaix, in this volume.
28. J. C. Scully, in this volume.
29. R. N. Parkins, in this volume.

THE ADSORBATE-SUBSTRATE BOND AND STRESS-RELIEF NEAR A CRACK TIP

A. W. Sleeswyk

Department of Applied Physics, University of Groningen
Nijenborgh 18, 9747 AG Groningen
The Netherlands

The object of this presentation is to provide a dislocation model of plastic deformation near the tip of a growing crack which shows how the bonding character of the atoms adsorbed at the crack faces may critically affect the geometry of the dislocation configurations, hence of the stresses near the crack tip. Thus, the model may provide a link between bonding calculations of the type presented at this conference by Keith Johnson and Dick Messmer, and the observed dependence of the geometry of the progressing crack on the character of the surrounding medium, presented by Horst Vehoff, Peter Neumann and Stan Lynch.

Obviously, adsorbates must be divided in two categories: 'metallic' adsorbates-hydrogen being considered the simplest metal – and 'non-metallic' ones, which in their effect much resemble that of an uncontaminated surface.

In the case of crack propagation pipe diffusion in dislocation cores is obviously irrelevant; the character of atoms adsorbed at the free surface can then only affect dislocations being nucleated there. In this particular instance, especially perfect edge dislocations. The creation of one of these will result in the formation of a step of unit height at the free surface, which we assume to be covered with a mono-layer of adsorbate. To a first approximation, the creation of this step will not result in fresh metal surface being exposed, but evidently the bonding of the absorbed atoms to the substrate of the rows of atoms at either side of the surface step is significantly changed. In particular, the row in the re-entrant corner of the step will be able to form new bonds with adjacent substrate atoms. As far as known, no ab initio

calculations for bonding of atoms of adsorbed atoms in this position have as yet been performed, so we must for the time make do
with notions such as 'metallic' and 'non-metallic' bonding of adsorbates. The basic assumption is that the two types of bonding
affect the nucleation of subsequent dislocations at the re-entrant
corner of the surface step differently.

Let us first consider 'non-metallic' adsorbates first: these
we are familiar with, as air generally provides an abundance of
them. Then the atoms adsorbed at the re-entrant corner of the step
render that site a favoured one for nucleation of new dislocations
(Fig. 1a). This is corroborated by the well-known coarsening of
the slip system operating near the free surface in air; it is due
to this effect that slip lines bunch together and may be observed
optically. It is enhanced by the stress concentration near the re-
entrant step when it grows in height by emitting dislocations to
the interior of the crystal.

These piled-up dislocations end up far away from the crack,
which consequently is blunted: the stress required for propagating
the crack is high, the angle α is large. Alternating cracking and
blunting is responsible for the coarsely stepped crack faces
(Fig. 1b).

Now consider the 'metallic' adsorbate atoms located in the
re-entrant angle of the step: the assumption made here - and it
must be emphasized that it still has to be verified - is that in
contrast to the previous case, these adsorbed atoms block the subsequent formation of dislocations at the step created by nucleating a dislocation.

Where then, will the next dislocation be nucleated? Consideration of the stress field around the dislocation as it is still
close to the surface provides the answer: there is an octant -
indicated in Fig. 2a - where the stress field will pull a dislocation of the same type inward. The portion of the free surface lying in that octant is therefore favored for nucleation of a new
dislocation. Once this second dislocation has been nucleated, it
will facilitate the creation of a third one, and so on.

The result will be the formation of a tilt wall near the
crack tip (Fig. 2b), or, if the whole arrangement is symmetrical,
two tilt walls. The dislocations in them move away from the crack
faces over a limited distance only; actually, as the stress falls
off behind the crack tip, it is even possible that the tilt wall
is attracted to the crack surface and annihilated, if the friction
stress is low enough. At any rate, because the dislocations remain
close to the crack, the latter does not open widely, the crack
angle α remains relatively small, and the sharp crack may propagate at a low stress.

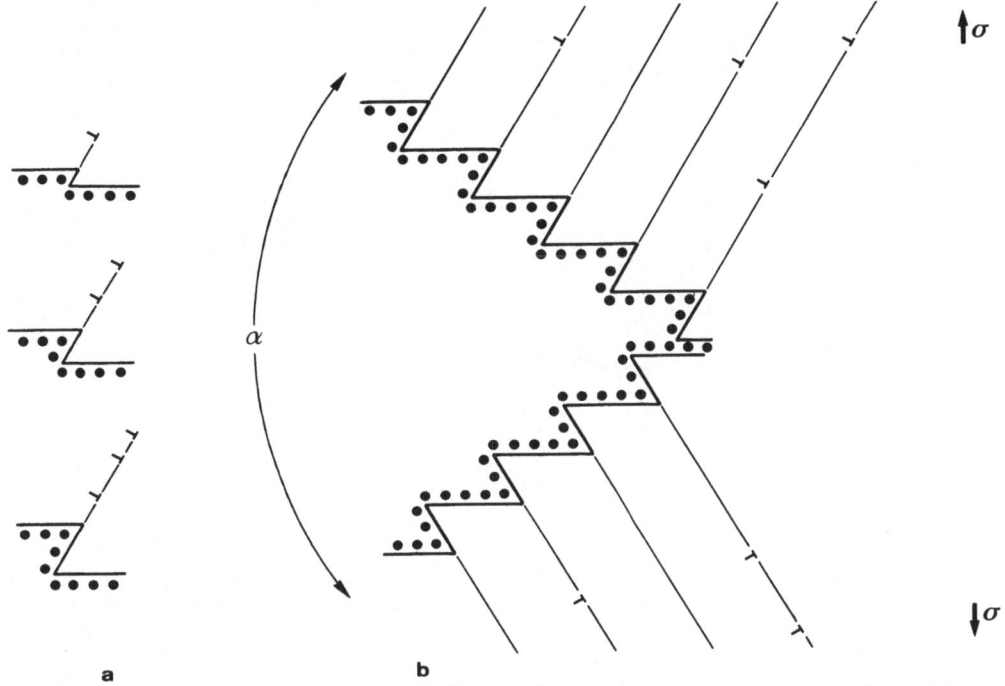

Fig. 1.   Preferred nucleation of dislocations near the re-entrant
          corner of a surface step. a: basic mechanism; b: growing
          blunted crack in 'non-metallic' atmosphere.

A word must be said about the combination: tilt wall-cleavage
crack in a crystal under stress. This assembly is one of lower
energy than either the tilt wall alone ending in the crystal, or
the crack under stress by itself. As a consequence, stress near
the end of a tilt wall may be relieved by formation of a cleavage
crack, or a tilt wall may relieve the stress around a crack;
Orowan (1954).

The important point of the combination is the strong bond
between the individual components: in our case, if the tilt walls
are made to move under stress, the crack will propagate through
the lattice with them. The limiting factor would seem to be the
rate at which the surrounding medium can supply adsorbed atoms to
the re-entrant corners of the ledges left at the free surface by
dislocations nucleated there.

The tilt walls were not (as yet) observed by the workers at
Düsseldorf: this may be due to the Burgers vectors of the dislocations being parallel to the specimen surfaces, which would make
the tilt walls virtually impossible to detect, or to annihilation
at the crack faces when the load is removed.

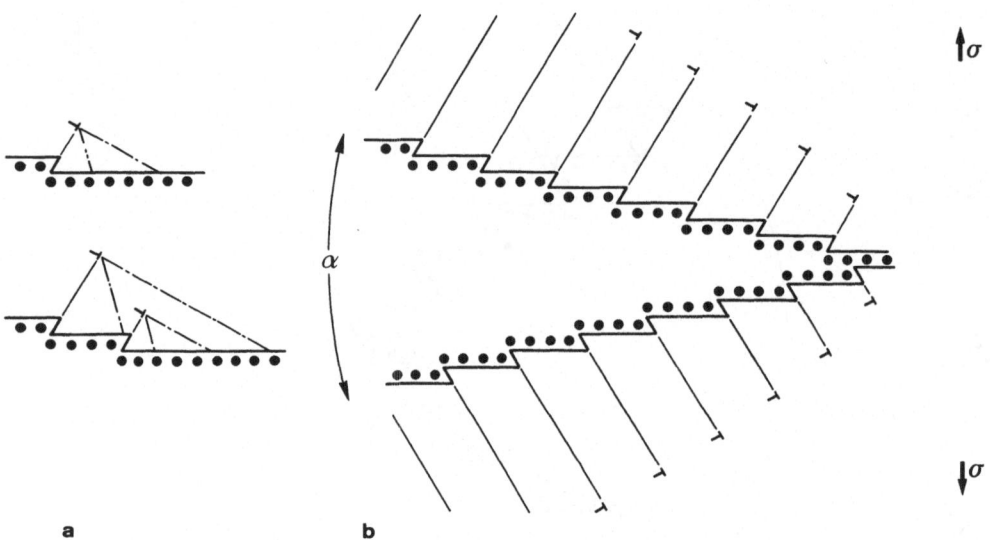

Fig. 2.   Nucleation of only one dislocation per step. a: basic
          mechanism; b: growth of sharp crack in 'metallic' atmos-
          phere.

The contrast between the effect of the two types of adsorbates on the propagation of a crack may be usefully summarized as
follows:

'non-metallic' adsorbates   'metallic' adsorbates

dislocation nucleation takes   adsorbates at surface step block
place preferentially at sur-   nucleation of dislocations at that
face step;   site;

coarse slip lines;   tilt walls;

dislocations pile up and move   dislocations remain in vicinity of
away from fracture faces;   crack;

blunting: large crack angle α;   little blunting: sharp angle α;

small stress concentration;   large stress concentration;

large load at fracture;   small load at fracture;

jagged fracture faces;   smooth fracture faces;

no large effect of rate of ad-   crack growth rate limited by rate
sorption;   of adsorption.

Of course, much of the fore-going is speculative, and a number of points need to be checked, especially the supposed blocking
of dislocation nucleation at surface steps by 'metallic' adsorba-

tes, and the presence of tilt walls or similar configurations near
the tip of a crack growing in a 'metallic' adsorbate atmosphere.
However, it is demonstrated here that the nature of the adsorbate-
substrate bond may influence the fracture process profoundly,
though indirectly, by altering the nucleation of dislocations,
hence the mode of stress relief near the tip of a growing cleavage
crack.

Reference

Orowan, E., 1974, "Mechanical Properties" in: "Dislocations in
    Metals", J. S. Koehler et al. eds., Inst. Metals,New York.

CRACK-TIP BLUNTING VERSUS CLEAVAGE EXTENSION

J. E. Sinclair and M. W. Finnis

Theoretical Physics Division
AERE Harwell
Didcot, Oxon, OX11 0RA

Suppose that a crystal containing a cleavage crack is subjected to a stress. The crystal may fail in a ductile manner, with blunting of the crack, or it may fail in a more or less brittle manner by cleavage extension of the crack. Under what conditions will one or the other behaviour occur? In a classic paper Rice and Thomson[1] addressed this question by considering whether spontaneous emission of a dislocation was possible from the tip of a sharp crack loaded to the critical condition for cleavage, according to the Griffith criterion for supply of the intrinsic surface energy. In practice however, brittle cleavage, especially in metals, is often accompanied by a considerable amount of dislocation activity. Emission of a single dislocation may therefore not be a sufficient condition for continued blunting to occur. We must ask whether limited blunting combined with cleavage extension is possible. To address this question, we attempt in this note to extend the analysis of Rice & Thomson, to see whether the presence of one or more emitted dislocations affects the competition between further emission and cleavage. Note that we do not consider the alternative possibility of cleavage accompanied by dislocation generation and motion in the bulk which does not blunt the tip.

Following Rice and Thomson[1] we consider a crack tip intersected by a single inclined slip plane (figure 1). We consider only pure edge dislocations, and assume that remote loading produces a mode-I opening with stress intensity factor K. For a hypothetical emitted dislocation at distance r from the tip, Rice & Thomson show that the two principal forces are an

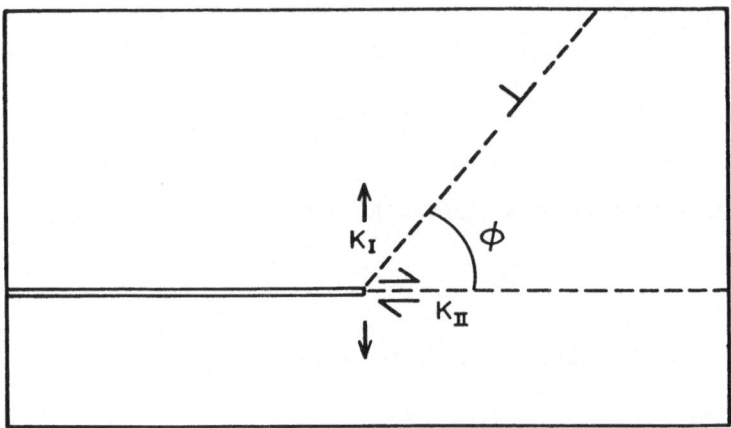

Figure 1

outward one proportional to $Kr^{-\frac{1}{2}}$, and an inward image force
proportional to r^{-1}. If the critical distance at which these
balance is less than a dislocation core radius (r_c), then this
continuum theory result cannot be expected to hold, and
spontaneous emission of the dislocation is taken to be possible.

    This analysis for a sharp, remotely loaded crack is affected
in three ways if one or more blunting dislocations are supposed
to have already been emitted. Firstly, these dislocations modify
the applied stresses, assuming that pinning or other friction has
prevented their removal to infinity. Secondly, the blunting
change to the crack tip geometry changes the local stress
response to the applied and dislocation loadings from $r^{-\frac{1}{2}}$ to a
weaker form. Finally, the image force on a further hypothetical
dislocation being emitted may be somewhat modified by the
geometrical changes. In the following simple analysis, we
neglect the second and third effects. Since the blunting will
reduce both the tendency to cleave and to emit dislocations, this
approximation may not have too severe an effect on the
predictions of the relative likelihood of these two processes.
The effect we consider is that the emitted dislocations combine
with the applied load to produce a mixed mode-I and mode-II field
near the tip. As the applied load is increased, it is found that
the condition for cleavage extension can sometimes be met before
that for further blunting, reversing the initial situation for a
sharp crack.

    Suppose, then, that a number of dislocations have already
been emitted, and lie at distances r_i, $i=1,\ldots n$, along the slip
plane. From results in the appendix to reference 1, we find that
the total mode-I and -II stress intensity factors are given by

$$K_I = K + L \qquad\qquad (1a)$$

$$K_{II} = \eta L/3 \qquad\qquad (1b)$$

where L, the induced component, is given by

$$L = -\frac{Eb(3\sin\phi\,\cos\tfrac{1}{2}\phi)}{4(1-\nu^2)} \sum_{i=1}^{n} \frac{1}{\sqrt{2\pi r_i}} \qquad\qquad (2)$$

and

$$\eta = (1-3\sin^2\tfrac{1}{2}\phi)/(\sin\tfrac{1}{2}\phi\,\cos\tfrac{1}{2}\phi). \qquad\qquad (3)$$

ν is Poisson's ratio, and ϕ the slip plane inclination.

The simple Rice-Thomson criterion for dislocation emission mentioned above can now be applied, and we find that it implies the following condition on K_I and K_{II}:

$$K_I + \eta K_{II} > B, \qquad\qquad (4)$$

where

$$B = \frac{Eb}{2(1-\nu^2)\sqrt{2\pi r_c}\,\sin\phi\,\cos\tfrac{1}{2}\phi}.$$

On the other hand the Griffith condition for cleavage extension takes the form:

$$K_I^2 + K_{II}^2 > c^2 , \qquad\qquad (5)$$

where

$$c^2 = 2\gamma E/(1-\nu^2) \qquad\qquad (6)$$

and γ is the surface energy. The conditions (4) and (5) for blunting and cleavage are illustrated graphically in figure 2.

Figure 2 illustrates a case of $C > B$, in which as K is increased from zero the first blunting dislocation is emitted at K=B (point B_1 in the figure) before the cleavage condition is met. When this occurs the dislocation will move off to some distance r_1 and contribute some value $L(>o)$ to K_I and $\eta L/3$ to K_{II}. This brings us to point D_1 in the figure. As K is increased the dislocation will move further, reducing $|L|$, and at some stage (point B_2) the condition for a second blunting reaction may be met. If dislocation emission on the other side of the crack remains inhibited, a sequence of blunting reactions B_1, B_2, ... will eventually lead to cleavage when point A is

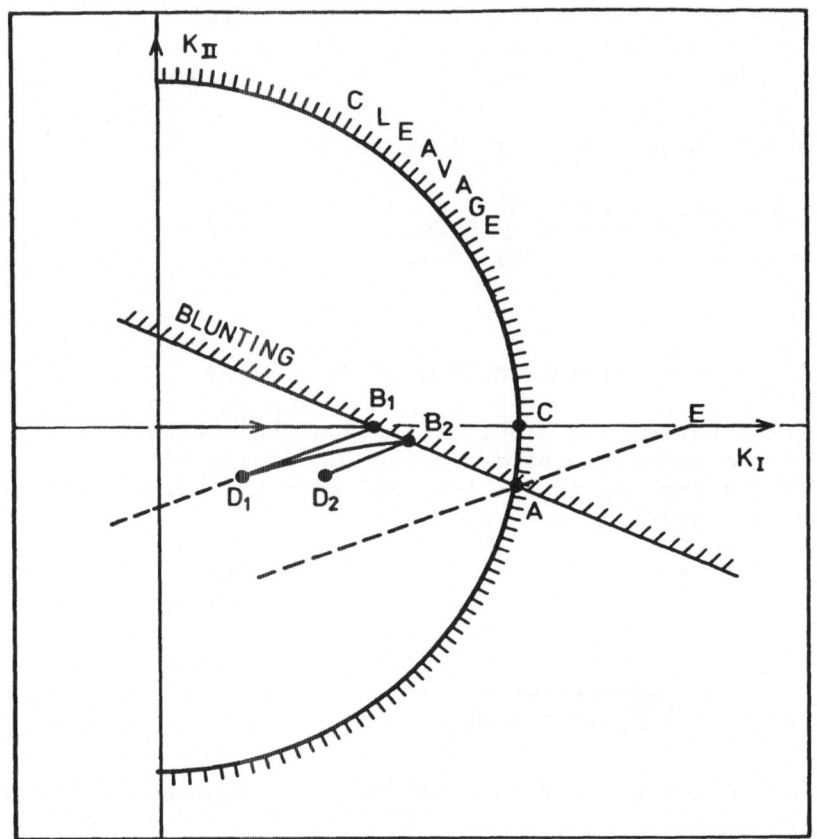

Figure 2

reached. The number n of blunting dislocations which have to
emerge before cleavage depends on the pinning force on the
emitted dislocations, for this will determine how close they
remain to the crack tip and hence the initial magnitude of ⌐L
compared to K.

     We can estimate the number n in the following way. Let us
model the emitted dislocations as a "super-dislocation" of
Burgers vector nb, at a distance r, pinned by a friction stress
τ_o. Neglecting image forces, r is obtained by equating the
applied force on the dislocation to the frictional force, which
gives

$$\tau_o = \frac{K}{\sqrt{2\pi r}} \sin{\tfrac{1}{2}} \phi \cos^2{\tfrac{1}{2}} \phi \ . \tag{7}$$

With r determined by equation (7) the contribution L to the
stress intensity near the tip is given by (2) as

$$L = -\frac{3E \tau_0 nb}{2(1-v^2)K} .$$ (8)

The problem is now reduced to solving equations (1), (4), (5) and
(8) simultaneously for the intersection point A. The solution
determines the smallest value of $|L|$, or n, at which cleavage
will occur. Omitting algebraic details, the result is

$$\frac{\tau_0 nb}{4\gamma} = \frac{(3+\eta^2)(1+\eta^2)-(3-\eta^2)\alpha^2-(5+\eta^2)\eta\alpha(1+\eta^2-\alpha^2)^{\frac{1}{2}}}{\eta^2(1+\eta^2)^2} ,$$ (9)

where
$$\alpha = B/C.$$ (10)

α measures the initial ratio of the K values for blunting and
cleavage. The larger the value of α, e.g. due to a small value of
surface energy, the smaller the value of n from equation (9).
Taking typical values $\gamma = 2$ Jm^{-2}, b = 0.3nm, $\phi = 45°$,
$\tau_0 = 500$ MPa, then for example:

if $\alpha = 0.8$, n ≈ 5;

if $\alpha = 0.5$, n ≈ 15.

It should be pointed out that the above analysis can
predict a reversal if the blunting-cleavage competition only
because a single slip system was considered. If slip on a
symmetrically disposed plane at $-\phi$ were allowed, then blunting on
this system would always be predicted before cleavage.
Consequently, symmetric slip would occur, keeping $K_{II} \approx 0$.
However, the above typical predictions of n=5 to 15 are
sufficiently small that it is quite plausible that random
asymmetries would allow the sequence we have considered to take
place. For instance, especially in inter-granular fracture, the
two (or more) slip systems may not lie at equal angles to the
crack plane. Furthermore, the microscopic effects of the
detailed geometry of a blunted tip, and atomic-scale details of
the blunting and cleavage processes may play a significant role.
Thus, the above analysis presents a plausible basis for the
occurrence of mixed cleavage and dislocation emission.

1. J.R. Rice and R. Thomson, Ductile versus brittle behaviour of
   crystals, Phil.Mag. 29:73(1974).

A COMMENT ON THE INTERPRETATION OF THE GRIFFITH'S SURFACE ENERGY IN TERMS OF LATTICE VIBRATIONS

Giuseppe Caglioti

Istituto di Ingegneria Nucleare
CESNEF-POLITECNICO

20133 MILANO, Italy

Information concerning the surface energy for brittle fracture, γ, in terms of the dynamics of the atoms in the crystal structure is given.

A phenomenological interpretation of γ has been reported few years ago in a number of papers (Refs 1- 5 deal with brittle fracture in metals and Ref 6 deals with alkali halides or ceramics in general).

The basic approach can be simply presented as follows.

Let us refer to a pure metal. The main features of the dynamics of the atoms inside it are implicit in the dispersion relations, $\omega(q,j)$, correlating the frequency ω to the wavevector q and branch index j of the (q,j) vibrational modes of the structure. These frequencies $\omega(q,j)$ can be measured by inelastic scattering of neutrons. In directions q/q of crystallographic symmetry, the knowledge of the $\omega(q,j)$'s allows to obtain the interplanar force constants, i.e. the force K acting on an atom as a consequence say of the unitary displacement along q/q of crystal planes orthogonal to q. Starting from equilibrium, the potential energy corresponding to an increment 2u of separaion of the two-be-split-half-crystals across the fracture plane is

$$\frac{1}{2} K(2u)^2$$

where

$$K = K_1 + 2K_2 + 3K_3 + \ldots.$$

and K_1, K_2... refer to first, second... neighbor planes. The assumption is made that when, as a consequence of pulling apart the two

half-crystals, u reaches the value u_m corresponding to the Linde-
mann mean square atomic displacement at the melting temperature T_m,
a critical configuration of local melting across the fracture plane is reached:

$$u_m = (\frac{9K_B T_m}{m\omega_D^2})^{1/2} ; \tag{1}$$

here K_B is the Boltzmann constant, m the atomic mass and ω_D the
Debye frequency of the crystal. An external work

$$\gamma_m = n \frac{9K_B T_m}{m\omega_D^2} \sum_p pK_p \{erg \ cm^{-2}\} \tag{2}$$

is then required to obtain local melting along the two sides of the
fracture plane. Here n is the number of atoms per unit area belonging to each of the two surfaces to be created as a result of the
fracture.

Once local melting is reached, the additional amount of energy

$$\gamma_b = K_B(T_b - T_m) \ n \tag{3}$$

should be provided to obtain local boiling of the two surfaces
across the fracture plane; here T_b is the boiling temperature.

Finally, the Griffith surface energy γ is obtained summing the
contributions (2) and (3):

$$\gamma = \gamma_m + \gamma_b \tag{4}$$

Detailed numerical computations of γ for several metals are given in references quoted above using the interplanar force constants
K obtained from neutron scattering. By and large both the order of
magnitude and the characteristic anisotropy predicted for γ agree
with the experimental data available for γ at the time.

REFERENCES

1. G.Caglioti,G.Rizzi,J.C.Bilello, The Surface Energy for Brittle
    Fracture in Metals from Phonon Frequencies, J.Appl.Phys 42,
    4271 (1971)
2. G.Caglioti,G.Rizzi,J.C.Bilello, The Surface Energy for Brittle
    Fracture in Metals from Phonon Frequencies, J.Appl.Phys.43,
    3600 (1972)
3. J.C.Bilello,G.Caglioti, On the Fracture Energy of Molybdenum,
    Scripta Met. 6, 1041 (1972)

4. G.Caglioti,G.Rizzi,J.C.Bilello, Reply to "On the Calculation
    of Surface Energies from Phonon Frequencies, J.Appl.Phys 45,
    1468 (1974)
5. G.Caglioti,Caratteristiche Meccaniche dei Metalli e Dinamica
    reticolare, in "Trends in the Physics and Engineering of Tech-
    nological Materials", International Meeting, Roma 17-19/10/73
    Contributi del Centro Linceo Internazionale No.30, p.15
    Accademia Nazionale dei Lincei, Roma 1976,
6. G.Benedek, S.Boffi, J.C.Bilello, G.Caglioti, Surface Energy
    for Brittle Fracture of Alkali Halides from Lattice Dynamics,
    Surface Science, 48, 561 (1975).

PARTICIPANTS

Argon, A. S.   Department of Mechanical Engineering
                            Massachusetts Institute of Technology
                            Cambridge, Massachusetts 02139, U.S.A

Azou, P.   Institut Superieur des Materiaux et de la
                            Construction Mechanique, 3, Rue Fernand-
                            Hainaut, 93407 St. Ouen Cedex, France

Berkowitz, B. J.   Materials Technology Division, Exxon Research
                            and Engineering Company, P.O. Box 101
                            Florham Park, New Jersey 07932, U.S.A

Bernstein, I. M.   Department of Metallurgical Engineering and
                            Materials Science, Carnegie-Mellon University
                            Pittsburgh, Pennsylvania 15213, U.S.A

Bianchi, G.   Universita Di Milano, Istituto di
                            Electrochimica e Metallurgia, Via Venezian, 21
                            20133, Milano, Italy

Birnbaum, H. K.   Department of Metallurgy and Mining
                            Engineering, University of Illinois
                            Urbana, Illinois 61801, U.S.A.

Bockris, J. O'M.   Texas A&M University, College of Science
                            Department of Chemistry, College Station
                            Texas 77843, U.S.A.

Bonzel, H. P.   Institut fur Grenzflächen-forschung und
                            Vakuumphysik, Kernforschungsanlage Jülich
                            KFA-IGV Postfach 1913, D-5170 Jülich 1, F.R.G.

Bottani, C. E.   Istituto di Ingegneria Nucleare, Centro Studi
                            Nucleari Enrico Fermi, Via Ponzio, 34/3
                            20133 Milano, Italy

Bullough, R.	Theoretical Physics Division, AERE Harwell Oxfordshire OX11 ORA, England
Caglioti, G.	Istituto di Ingegneria Nucleare, Centro Studi Nucleari Enrico Fermi, Via Ponzio, 34/3 20133 Milano, Italy
Climent, J.	Centro de Estudios Y Experimentacion De Obras Publicas - Laboratorio Central De Estructuras Y Materiales, 4 Alfonso XII 3, Madrid, 7 Spain
Christmann, K.	Physikalisch-Chemisches Institut, Universitat Munchen, 8 Munchen 2, Sophienstr 11, F.R.G.
Conway, B. E.	University of Ottawa, Department of Chemistry Ottawa, Ontario K1N 6N5, Canada
Cyrot-Lackmann, F.	Centre National de la Recherche Scientifique Groupe des Transitions de Phases, Avenue des Martyrs, Grenoble, 38042 Grenoble Cedex France
Dienes, G. J.	Brookhaven National Laboratory, Department of Physics, 510 B, Upton, New York 11973, U.S.A.
Duquette, D. J.	Materials Engineering Department, Rensselaer Polytechnic Institute, Troy, New York 12181, U.S.A.
Erturk, T.	Department of Metallurgical Engineering Middle East Technical University, Ankara, Turkey
Esterling, D. M.	George Washington University, NASA Langley Research Center, Joint Institute for Advancement of Flight Sciences, Hampton, Virginia 23665, U.S.A.
Fidelle, J. P.	Service Metallurgie, Commissariat a L'Energie Atomique, 91680 Bruyeres-le-Chatel, France
Finnis, M. W.	Theoretical Physics Division, AERE Harwell, Oxfordshire, OX11 ORA, England
Fischer, T. E.	Exxon Research and Engineering Company P.O. Box 45, Linden, New Jersey 07036, U.S.A.

Frankel, G.

Department of Materials Science and
Engineering, Massachusetts Institute of
Technology, Cambridge, Massachusetts 02139,
U.S.A.

Gadzuk, J. W.

Surface Science Division, Center for
Thermodynamics and Molecular Science
United States Department of Commerce
National Bureau of Standards, Washington, D.C.
20234, U.S.A.

Garrett, G. G.

University of Witwatersrand, Johannesburg
Department of Metallurgy, Johannesburg, 2001
South Africa

Gleiter, H.

Universitat des Saarlandes, Fachbereich 12.1
Werkstofftechnologie, Bau 2, D-6600
Saarbrucken, F.R.G.

Guttmann, M.

Ecole Nationale Superieure des Mines de Paris
Centre des Materiaux, B.P. 87, 91003 Evry
Cedex, France

Haasen, P.

Institut für Metallphysik der Universitat
Gottingen, 3400 Gottingen, Hospitalstrasse 12
F.R.G.

Hirth, J. P.

Department of Metallurgical Engineering
Ohio State University, Columbus, Ohio 43210,
U.S.A.

Howard, I. C.

The University of Sheffield, Department of
the Theory of Metals, Mappin Street
Sheffield Sl 3JD, England

Johnson, K. H.

Department of Materials Science and
Engineering, Massachusetts Institute of
Technology, Cambridge, Massachusetts 02139,
U.S.A.

Jones, R. H.

Battelle Pacific Northwest Laboratories,
P.O. Box 999, Richland, Washington 99352,
U.S.A.

Kitagawa, H.

Institute of Industrial Science, University
of Tokyo, 22-1, Roppongi 7 Chome, Minato-ku,
Tokyo 106, Japan

Knott, J. F.

University of Cambridge, Department of
Metallurgy and Materials Science, Pembroke
Street, Cambridge, CB2, 3QZ, England

Latanision, R. M.	Department of Materials Science and Engineering, Massachusetts Institute of Technology, Cambridge, Massachusetts 02139, U.S.A.
Lynch, S.	Materials Division, Department of Defense, Aeronautical Research Labs, 506 Lorimer Street, Fishermen's Bend, Box 4331, Melbourne, Victoria 3001, Australia
Macmillan, N. H.	Materials Research Laboratory, Pennsylvania State University, University Park, Pennsylvania 16802, U.S.A.
Mazza, F.	Universita di Milano, Istituto di Electrochimica e Metallurgia, Via Venezian, 21 20133 - Milano, Italy
Messmer, R.	Chemical and Structural Analysis, Inorganic Materials and Structures Laboratory General Electric Company, Research and Development Center, P.O. Box 8, Schenectady, New York 12301, U.S.A.
Nabarro, F.R.N.	University of Witwatersrand, Johannesburg Department of Physics, Johannesburg, 2001, South Africa
Neumann, P.	Max-Planck-Institut für Eisenforschung GMBH Max-Planck-Strasse 1, 4 Düsseldorf, F.R.G.
Oriani, R. A.	University of Minnesota, Department of Chemical Engineering and Materials Science Minneapolis, Minnesota 55455, U.S.A.
Oudar, J.	Ecole Nationale Superieure de Chemie de Paris Laboratoire de Metallurgie et Physico-Chemie des Surface, 11, Rue Pierre et Marie Curie 75231 Paris Cedex 05, France
Parkins, R. N.	University of Newcastle-upon-Tyne, Department of Metallurgy and Engineering Materials Haymarket Lane, Newcastle-upon-Tyne NE1 7RU, England
Pelloux, R. M.	Department of Materials Science and Engineering, Massachusetts Institute of Technology, Cambridge, Massachusetts 02139, U.S.A.

Pettifor, D. G.   Imperial College of Science and Technology
                     Department of Mathematics, Huxley Building,
                     Queen's Gate, London SW7 2BZ, England

Philibert, J.   Laboratoire de Physique des Materiaux
                     Centre National de la Recherche Scientifique
                     1, Place Aristide Briand, 92190 - Meudon,
                     France

Pickens, J. R.   Martin Marietta Laboratories, 1450 South
                     Rolling Road, Baltimore, Maryland 21227, U.S.A.

Pourbaix, M.   Centre Belge D'Etude de la Corrosion, Avenue
                     Paul Heger, Grille 2, 1050 Brussels, Belgium

Pratt, P. L.   Imperial College of Science and Technology
                     Department of Metallurgy, Royal School of
                     Mines, Prince Consort Road, London SW7 2BP,
                     England

Pugh, E. N.   United States Department of Commerce
                     National Bureau of Standards, Room B254,
                     Building 223, Washington, D.C. 20234, U.S.A.

Puls, M. P.   Materials Science Branch, Atomic Energy of
                     Canada Ltd, Whiteshell Nuclear Research
                     Pinawa, Manitoba ROE ILO, Canada

Ritchie, R. O.   Department of Materials Science and Mineral
                     Engineering, First Mining Building
                     University of California, Berkeley,
                     California 91720, U.S.A.

Scully, J. C.   The University of Leeds, Department of
                     Metallurgy, Leeds LS2 0JT, England

Seah, M. P.   Division of Materials Applications
                     National Physical Laboratory, Teddington
                     Middlesex TW11 OLW, England

Sleeswyk, A.   Department of Applied Physics
                     Rijksuniversiteit, Universiteitscomplex
                     Paddepoel, Nijenborgh 18, 9747 AG Grongingen,
                     The Netherlands

Stoloff, N.   Rensselaer Polytechnic Institute, School of
                     Engineering, Materials Engineering
                     Department, Troy, New York 12181, U.S.A.

Talbot-Besnard, S.   Ecole Centrale des Arts et Manufactures
                        Grande Voie des Vignes, 92290 Chatenay-
                        Malabry, France

Thomas, M. T.   Battelle Pacific Northwest Laboratories
                        Surface Science Group, Materials Department
                        Battelle Boulevard, P.O. Box 999, Richland,
                        Washington 99352, U.S.A.

Thompson, A. W.   Carnegie-Mellon University, Department of
                        Metallurgy and Materials Science, Schenley
                        Park, Pittsburgh, Pennsylvania 15213, U.S.A.

Thomson, R.   Center for Materials Science, National
                        Bureau of Standards, Washington, D.C. 20234,
                        U.S.A.

Vehoff, H.   Max-Planck-Institut für Eisenforschung GMBH
                        Max-Planck-Strasse 1, 4 Düsseldorf, F.R.G.

Westwood, A. R. C.   Martin Marietta Laboratories, 1450 South
                        Rolling Road, Baltimore, Maryland 21227,
                        U.S.A.

Whitesides, G. M.   Department of Chemistry, Massachusetts
                        Institute of Technology, Cambridge,
                        Massachusetts 02139, U.S.A.

1. Westwood
2. Duquette
3. Berkowitz
4. Azou
5. Climent
6. Haasen
7. Bullough
8. Pelloux
9. Scully
10. Pourbaix
11. Argon
12. Pratt
13. Caglioti
14. Furstenberg
15. Bottani
16. Birnbaum
17. Bianchi
18. Martin
19. Pugh
20. Jones
21. Sleeswyk
22. Seah
23. Cyrot-Lackmann
24. Conway
25. Dienes
26. Fidelle
27. Mazza
28. Thomas
29. Bockris
30. Frankel
31. Vehoff
32. Neumann
33. Kitagawa
34. Gleiter
35. Latanision
36. Erturk
37. Oudar
38. Esterling
39. Knott
40. Macmillan
41. Christmann
42. Garrett
43. Bernstein
44. Stoloff
45. Talbot-Bernard
46. Thompson
47. Whitesides
48. Lynch
49. Oriani
50. Ritchie
51. Hirth
52. Nabarro
53. Howard
54. Pettifor
55. Finnis
56. Thomson
57. Parkins
58. Fischer
59. Philibert
60. Bonzel
61. Puls
62. Pickens

PHOTOGRAPHS OF CONFERENCE